28th Center for Chemical Process Safety International Conference 2013

(CCPS)

Topical Conference at the 2013 AIChE Spring Meeting and 9th Global Congress on Process Safety

San Antonio, Texas, USA
28 April - 2 May 2013

Volume 1 of 2

ISBN: 978-1-62748-446-6

Printed from e-media with permission by:

Curran Associates, Inc.
57 Morehouse Lane
Red Hook, NY 12571

Printed by Curran Associates, Inc. (2013)

AIChE
3 Park Avenue
New York, NY 10016-5991

Phone: (203) 702-7660
Fax: (203) 775-5177

www.aiche.org

Additional copies of this publication are available from:

Curran Associates, Inc.
57 Morehouse Lane
Red Hook, NY 12571 USA
Phone: 845-758-0400
Fax: 845-758-2634
Email: curran@proceedings.com
Web: www.proceedings.com

TABLE OF CONTENTS

VOLUME 1

Applicability of the Hazards and Effects Management Process (HEMP) in Research and Development 1
Ahmed J. Allawi

Scale up of Stirred Batch and Semi-Batch Reactors - Gaps and Limitations of the Current Methodologies 12
Olga Juliana Reyes Valdes, Alba Pineda-Solano, Victor H. Carreto, Sam Mannan

Safety, ASAP and ALARP in Fast Track Process Development 28
Joseph B. Powell

An Update On the Use of Process Safety Leading Indicators 33
Dan Sliva, Shakeel Kadri

Role of Leading Indicators in Prevention of Major Accidents: CSB Learnings From BP Texas City and Deepwater Horizon 53
Kelly Wilson, Donald Holmstrom, Cheryl Mackenzie, Marybeth Mulcahy, Kara Kane, Jerad Denton, Chris Lyon, Bill Hoyle, Steve Cutchen

Process Safety Leading Indicators – A Perspective From Europe 54
Piet Knijff, Lee Allford, Peter Schmelzer

So We All Have Been Implementing Process Safety Metrics – What Next? 67
Shakeel Kadri, James Vanommeren, Glen Peters, Martin A. Dennehy, Alvin Mateo, Kenneth Fegley

Process Safety Leading Indicators: A 'Silver Bullet' or 'Bean Counting' for Enhanced Risk Management 80
David E. Cummings

Using Metrics to Drive Process Safety Performance At Maersk FPSOs 96
William Bradshaw, Jeppe Guld

Learning From High Potential Near Misses in the Refining Sector 111
V. M. Ramesh, R. Sakthivel

Felt Leadership: An Essential Ingredient in Successful PSM Implementations 112
Brian D. Rains

Localization of Control Rooms Using a MINLP Approach, Based On Accidental Explosion Scenarios 124
Lina M. Sierra, Sergio E. Rodríguez, Jaime E. Cadena, Manuel A. Ramos, Jorge Mario Gómez, Felipe Munoz

Risk Analysis for Overpressure From Pressure Vessel Burst 138
Seungho Jung, Elizabeth Lutostansky, Joan M. Schork

Comparison of the Peak Relief Rate Estimated by Dynamic Simulators Compared to Traditional Methods for Several Distillation Columns 147
John Wilkins, Dustin J. Smith

A Hybrid Model for Characterizing the Source of Hazardous Material Release 157
Shahryar Khajehnajafi, Anjana Meel

Guidelines for Developing Site-Specific Human Error Rates and Human IPLs for LOPA Using a Safety Climate Approach 175
Farheen Khan, Sunil D. Lakhiani, Delmar Morrison

Benchmarking – an Important Milestone in the Journey towards PSM Excellence 176
Prerna Jain, Ravi Ramaswamy, Cindie Pridy, Robert Cravens

Process Safety Management - A Practical View 190
German Luna-Mejias Sr.

Auditing the Process Hazard Analysis - Key Characteristics to Scrutinize in a Good-Quality HAZOP Study and Supporting Effective HAZOP Study Updates 202
Steven T. Maher, Michael A. Nour, Kristin D. Norton

Vapor Cloud Explosion Test of Air-Inflated Fabric Structure 203
Matthew Edel, Brad Horn

Revisiting the Conditions and Consequences of CO2 Tank Explosions 218
Yu Zhang, Joan M. Schork, Keith Ludwig

New Ignition Source "Exploding Wire" for the Determination of Explosion Characteristics of Combustible Dusts in the 20-L-Sphere 230
Marc Scheid, Volkmar Schröder, Christian Kusche, Alexander Klippel, Rico Tschirschwitz, Stefan Zirker

Operational Discipline and Beyond: An Interdisciplinary Approach to Improving Process Safety Management Implementation Through Behavior and Cultural Change 242
Charles A. Soczek

Risk Analysis for Pressure Relief Devices – Managing Plant Safety and Reliability 249
Jose R. Aguilar-Otero Sr.

Enterprise Risk Management--Adopting Risk-Based Decision Making in a Chemical Enterprise From the Board Room On Down Through All Functions and All Levels 251
Jerry L. Jones
The Chemical Safety Board and the Prevention of Combustible Dust Hazards 253
Manuel R. Gomez, Christina Morgan
Performing HAZOP of Startup/Shutdown Procedures 254
Fareed Ebrahim, German A. Luna Sr.
Chasing the Protection of Intermediate Bulk Containers, Year 10 263
John A. Leblanc
Process Safety Challenges in the Hazardous Waste Industry 264
Hope Luebeck
A Novel Methodology for Maintaining Mechanical Integrity of High Pressure Relief Systems 274
Sandipan Laskar
A Framework for Managing near-Miss Incidents 284
Ghanshyam Shah
HAZOP Analysis and Debottleneck for Laboratory Operation in the Semiconductor Industry 292
Tianxing Cai, Qiang Xu
Best Practice Management of Change & Pssr Process 307
Mike Bearrow
Unifying Consequence Models for Natural Gas Pipelines in Petrobras Transpetro 321
Ana Bahiense, Maria Caputo
Automation Strategies: Avoiding Accidents During the Process of Draining Water From LPG Spheres 329
Carlos André Vaz Jr., Maurício Bezerra Souza Jr., Diego Da Silva Lione, João Márcio Almeida Villaça, Leonardo Salim Damian Aguiar
Complex BMS Made Simple 344
Dave Sheppard, Nikki Bishop
Using ESD Valves As Safeguards, Myth or Reality? 345
German A. Luna-Mejias Sr.
Contours When Pipelines Meets Facilities 356
Karthick Deivasagayam
Ozone Flammability and Its Affect On the Lower Flammable Limit of Simple Hydrocarbons 357
Paul Osterberg, Thomas Johnson, Thomas Frawley
Evaluate the Plant-Wide Safety of Your Interlock Systems 358
Mohammad Naved Khan, James Brigman, Swapnil Pathak
Virtual Reality Applied to Training and Process Safety: A Semi-Industrial Boiler Case Study 372
Carlos André Vaz Jr., Maurício Bezerra Souza Jr., Thiago Souza De Oliveira
Investigations On the Influence of Different Inert Gases On the Laminar Flame Propagation in Lycopodium/Air Mixtures 382
Hannes Kern, Bruno Wimmer, Harald Raupenstrauch
Fire Protection and Life Safety Issues in Semiconductor and Related Facilities 391
Jonathan Eisenberg
Lower Flammability Limits of Mixtures of Hydrogen and Methane At Atmospheric and Sub-Atmospheric Pressures 392
Hai Le, Yi Liu, M Sam Mannan
Informatics Rich Integrated Chemical Safety Risk Assessment (ICSRA) Platform 401
Muthukumarasamy Karthikeyan, Renu Vyas, Hv Prasad
IT Enablement for Effective Implementation and Monitoring of Process Safety Management 405
Pawan Mittal, Ajay Kumar
Sulphur Dust Explosions an Overview 409
Karthick Deivasagayam
Simplified Uncertainty Analysis of LOPA Results 410
Raymond "randy" Freeman
The Impact of Ozone On the Lower Flammable Limit of Hydrogen in Vessels Containing Savannah River Site High Level Waste 437
Paul Osterberg, Carol Sherburne, Thomas Johnson, Thomas Frawley
Process Safety Considerations for New DOE Depleted Uranium Hexafluoride Conversion Plants 452
Peter M. Burban
Minimum Ignition Energy Study of Flowing Heat Transfer Fluid Aerosols: Experimental and Theorectical Approach 459
Szu-Ying Huang, Xinrui Li, M. Sam Mannan

Domino Effect Model Applied to the Hertfordshire Oil Storage Terminal (Buncefield Oil Depot) 460
Esteban Bernechea, Xavier Seguí, Josep Arnaldos
Proven Approach to Investigating Near Misses 470
William Bridges, Gregory Smith
Key Strategies for Implementing a Corporate Fatigue Risk Management System 484
David E. Cummings
Consequences Calculation Tool for Storage Tank Fire Scenarios 495
Juan C. Díaz, Jaime E. Cadena, Felipe Munoz
A Practical View of Risk Reduction Management of Hardware and Software Design for Safety Critical Applications 509
Luis Duran, Mike Krywonos
Metodologías Para Reducir Las Fallas De Causa Común .... N/A
Luis Manuel Duran

VOLUME 2

Establishing a 3 Credit Course in PSM for Undergraduate Engineers 522
Deborah L Grubbe
Investigations On the Self Heating Potential of Garage Waste As Basis for the Hazard Evaluation of Bulk Storage Facilities 523
Hannes Kern, Christoph Spijker, Harald Raupenstrauch
Estimate Vibration Risk for Relief and Process Piping 533
Georges Melhem
Performance Metrics Development Management and Reporting Reaching Sustainability by Design 560
Robert C. Smith
Alarm Management - the Human Factors Challenge 561
Robert C. Smith
Dust Flame Propagation in Industrial Scale Piping - Part 2: CFD Study of a Conveying Vessel-Pipeline System 562
Jim Vingerhoets, Thomas Farrell, Jef Snoeys
The Use of CFD to Evaluate the Interactions Between Multiple-Leak Sources and to Assess Effectiveness of Integral Modeling Techniques in a Multiple-Leak Scenario 584
Joshua Richardson, Ha Nguyen, M. Sam Mannan
Understanding, Managing, and Communicating High Consequence Risk 586
Charles A. Soczek
Frequencies of Accidents As Functions of Design Variables 587
Esteban Bernechea, Josep Arnaldos, Cristian Massaguer
Routes Optimization for Hazardous Materials Transportation in Bogotá, Colombia 599
Jorge Victoria, Nubia Velasco, Eliecer Gutierrez, Felipe Munoz
Mejores Prácticas En Los Procesos De Gestión Del Cambio Para Seguridad De Procesos y La Evaluación De Seguridad Previa Al Inicio (PSSR) 600
J. Luis Aguirre
Lesha – Multi-Layer Progressive Stress & Impact Assessment On Health & Behavior 601
Salvador Ávila Filho
Building Design for Resilience Beyond Design Basis Blast Loads 603
Michael A. Polcyn, Raymond H. Bennett
Reliability Analysis of Man-Machine Response to an Emergency Situation Based On a Detailed-Action Plan 612
Najmeh Vaez, Farshad Nourai, Yaneira Saud, Suresh Yelisetty
Hazards of Gasoline Contamination – Case Study of a Tanker Truck Explosion Resulting From Switch Loading 613
Scott G. Davis, Kees Van Wingerden
Is Your MOC Process Doing All That It Should Be? 614
William Fink, Carolina Herrera
Beyond DSC, Improved Thermal Hazards Screening Methodology 615
Peter J. Ralbovsky Iv
Using Explicit Finite Element Analysis to Simulate the Effects of External Chemical Explosions On Single and Double-Walled Storage Tanks 617
Phillip E. Prueter, David J. Dewees
The Hidden Risks of Deposits, by-Products, and Residues On Exhaust and Waste Streams 634
Carlos A. Barrera, Abid Kemal

Explosion Characteristics of Carbon Nano-Fibers 635
Jiaqi Zhang, Yi Liu, M Sam Mannan
High-Fidelity Combined Dynamic Modeling of Depressurizing Vessels and Flare Networks: Accurately Assess the Effects of Process Changes On Mechanical Integrity and System Capacity 636
James Marriott
An Engineering Method to Mitigate the Impact of Regulatory Focus On Relief System Installations by Prioritizing Risk 637
Dustin J. Smith, John Burgess
Making Relief Load Estimates Match Reality 645
Nicholas N. Cristea, Dustin J. Smith
Explosion Accident Simulation for Petrochemical Plant 656
Xiangdi Zhao, Jiwu Yuan, Chunming Jiang, Zheng Wang
Using Reliability Centered Maintenance (RCM) to Reduce Fire Risk in High Temperature Oil Handling 664
Marc W. Yarlott, Ryan Johnson, Steven Waters
Characterization of the Emissive Power of Hybrid Methane-Air-Dust Flames 679
Brian Elias, Ali S. Rangwala
Are You in Control of Process Safety? 680
Ellis Graeme
Process Hazard Analysis, Layers of Protection Analysis and the Human Factor 688
Paul Baybutt
Benchmarking Industry Practices for the Use of Alarms As Safeguards and Layers of Protection 722
Todd Stauffer, Peter Clarke
The Impact of Human Factors On LOPA (And Other Risk Assessment Methods) 752
William Bridges
Take the Guess Work Out of Testing Your Safety Instrumented System 792
Charles Fialkowski
Tips for the Creation and Application of Effective Operating Procedures 802
Colin Scholtz, Steven T. Maher
Management Discipline: Defining a Process Safety Strategy 815
Jerry Forest
Using Assessments to Improve Process Safety Culture 824
Russell A. Ogle, Delmar Morrison, Sean J. Dee
Using New Technology to Ensure Corporate Standards Compliance and Improved Frontline Operational Risk Management 833
Mike Neill, Scott Lehmann
Conduct of Operations Best Practice Networks 865
Marco Antonio Vela, Jerry J. Forest
Management of Change Indicators 875
Clarissa Sossmeier Da Silva, Americo Diniz
Organizational Change Management 876
John Wincek, Molly Myers, Henry Ozog
Increasing the Agility of Process Safety Management Systems 887
Brian D. Rains
Technical Risk – It's Not a One-Size-Fits-All Proposition 900
Kumar Israni, Joshtin Cooper
Preventing Process Safety Problems of Tomorrow, Today! 913
Kevin Watson, Robert Dayton
Critical P & ID Reviews Using HAZOP/LOPA Methodology – Overkill or Underrated? 921
John T. Perez, Andrew Madewell
Using near Misses to Improve Risk Management Decisions 937
Sean J. Dee, Brenton L. Cox, Russell A. Ogle
Keeping the Memory Alive, Preventing Memory Loss That Contributes to Process Safety Events 949
Barry Throness
Blacks Swans, White Swans and Fifty Shades of Grey: Remembering the Lessons Learned from Catastrophic Process Safety Incidents 980
John F. Murphy, Jim Conner
Lessons Learned From Fire in Air Supply Ducts 994
Alvin Waller
Hydrogen Sulfide Exposure From Molten Sulfur- A Forgotten Hazard? 1002
Mark J. Viz, Russell A. Ogle, Sean J. Dee, Brent L. Cox

Calcium Stearate Deflagrations – A Series of Unfortunate Events 1013
Konanur Manjunath, Albert I. Ness
Effectiveness of Improvised Gas Absorption Techniques for Emergency Responders At Releases of Toxic Gases 1022
Hannes Kern, Matthias Taferner, Harald Raupenstrauch
Case Studies in Process Safety: Lessons Learned From Software-Related Accidents 1029
Terry Hardy
CSB Investigation of Chevron Richmond Refinery Fire 1044
Mark Wingard, Lauren Grim, Steve Cutchen, Kelly Wilson, Roger Evans, Amanda Johnson, Donald Holmstrom, Dan Tillema
Author Index

Applicability of the Hazards and Effects Management Process (HEMP) in Research and Development

Ahmed J. Allawi
Shell International Exploration and Production
Shell Technology Center Houston
3333 Highway 6 South
Houston, TX 77082
ahmed.allawi@shell.com

Prepared for Presentation at
American Institute of Chemical Engineers
2013 Spring Meeting
9th Global Congress on Process Safety
San Antonio, Texas
April 28 – May 1, 2013

UNPUBLISHED

Applicability of the Hazards and Effects Management Process (HEMP) in Research and Development

Ahmed J. Allawi
Shell International Exploration and Production
Shell Technology Center Houston
3333 Highway 6 South
Houston, TX 77082
ahmed.allawi@shell.com

Keywords: Hazards and Effects, HEMP, Bowtie, ALARP, Process Safety in R&D, Pilot Plant Safety.

Abstract

The Hazards and Effects Management Process (HEMP) evolved in the chemical process industry as a methodology for ensuring consistency in risk mitigation. In a nutshell, HEMP defines the manner in which the risk of injury to personnel, or damage to assets, environment, or reputation, is managed to As Low As Reasonably Practicable (ALARP). Although the HEMP concepts are universally valid, the pathways by which one can demonstrate ALARP were developed primarily for industrial production facilities such as refineries and chemical plants. This paper addresses the applicability of the HEMP to the type of pilot plants and laboratory-scale processes typically found in industrial research and development (R&D) facilities. In addition, this paper presents a case study involving the customization of HEMP to mitigate the risk of personnel exposure to hydrogen sulfide (H2S) from a pilot plant.

1. Introduction

The Hazards and Effects Management Process (HEMP) provides a framework for managing HSSE-related hazards. An overview of how the process is implemented at Shell is presented by Salter and Siddoway [1]. The focus of this paper is a specific subset of HEMP, namely, the pathways for mitigating risks to As Low As Reasonably Practicable (ALARP) – in both industrial and R&D settings.

Reaching ALARP can be viewed as a process of selecting the right pathway from a menu of increasingly more complex pathways. These pathways are well-defined for industrial settings, but are somewhat challenging to follow in R&D settings. Despite these challenges, HEMP and its tools represent an invaluable component of a comprehensive process safety program in R&D. The case study presented in the paper demonstrates how the methodical application of HEMP in R&D is likely to highlight safety gaps that would otherwise go unnoticed.

Central to the discussion of HEMP and ALARP is the bowtie concept. This is the best starting point to delve into the details of HEMP and ALARP.

2. What is a Bowtie?

In the HEMP world, a bowtie is a graphical representation of how a hazard can be released, what the consequences of the release might be, and how the consequences might be prevented [2]. As a risk management tool, the bowtie is one of the most concise ways of depicting the complex relationship between hazards, threats, consequences, control barriers, and recovery measures. Figure 1 shows the major components of a bowtie. What follows is a brief definition of each term.

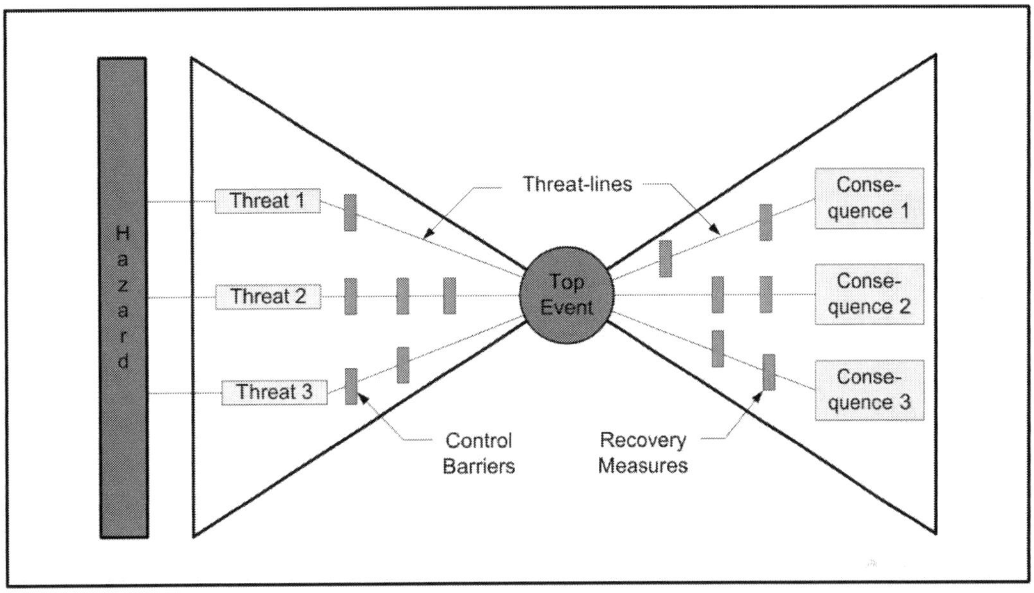

Figure 1: A Typical Bowtie

Barrier: A common name for Control Barriers and/or Recovery Measures. Barriers can be hardware, human intervention, or a combination of both.

Consequence: The final results of the Hazard being released, but not controlled.

Control Barrier: Barrier that reduces the likelihood of a Threat leading to a Top Event.

Hazard: The potential to cause harm, including illness & injury, and damage to the environment, property, products, and/or reputation.

Recovery Measure: Barrier that reduces the likelihood of the Top Event leading to a Consequence.

Threat: The way in which a Hazard can be released to cause an incident.

Top Event: The way in which a Hazard can be released to do harm. Note that the Top Event is not the ultimate consequence of releasing the Hazard.

3. HEMP and ALARP

Figure 2 is a simplified flow chart showing how ALARP may be demonstrated in an industrial setting. The risk assessment block at the top of the figure consists of two steps. First, the hazard is identified and the consequence of releasing the hazard is established. Then, the risk (likelihood and severity) of an injury to personnel, and/or damage to assets, environment, or reputation is assessed. It is the magnitude of the perceived risk that dictates the mitigation strategy specified by HEMP. Low risks hazards are managed using a hierarchy of controls within an existing HSE Management System (HSE-MS). Medium and high risks are managed to ALARP using one of the four other pathways shown in Figure 2. What follows is a brief description of each of these pathways. Note that these pathways had been developed primarily for industrial settings.

Clearly, the outright elimination of a hazard is the ideal method of risk mitigation. Examples include replacing a toxic material with a more benign substitute, operating under ambient conditions as opposed elevated temperatures or pressures, operating at grade instead of at heights, etc. As obvious as it seems, this risk mitigation pathway is often outside the scope of detailed process safety reviews. By design, these types of safety reviews are focused on the details of the design as opposed the rationale for the existence of the hazard in the first place.

Conforming to applicable safety standards is necessary; however, **it is not always sufficient to mitigate a specific risk to ALARP**. For the chemical process industry, applicable safety standards could come from the government (OSHA, NIOSH, CSB, etc.), industry (API, ASME, ANSI, etc.), or the corporation itself. These standards could be generic (e.g., API 520/521 for overpressure protection), or specific (e.g., NFPA-2: Hydrogen Technologies Code). In general, generic standards are not sufficiently granular to mitigate a specific risk to ALARP. Conversely, specific standards might only apply to a narrow spectrum of threat-lines. This is the reason why conforming to applicable standards is shown as an "add-on" requirement rather than an independent pathway to ALARP.

Risk reduction to ALARP can be viewed as a process of selecting the right pathway from a menu of increasingly more complex pathways. If the hazard cannot be eliminated, then the next logical question to pose is whether a corporate "best practice" exists to mitigate the specific risk. One way of capturing a best practice for safeguarding common systems is through the use of a "model bowtie". In general, these are applicable to *common* hazards present in *common* systems. An example might be the mitigation of hydrocarbon overflow from a storage tank located in a tank farm. Deviation from a common practice could render the model bowtie potentially irrelevant. For instance, if the hydrocarbon is highly toxic, or if the storage tank is located inside an operating unit, then the model bowtie described above would not be sufficiently conservative to demonstrate ALARP.

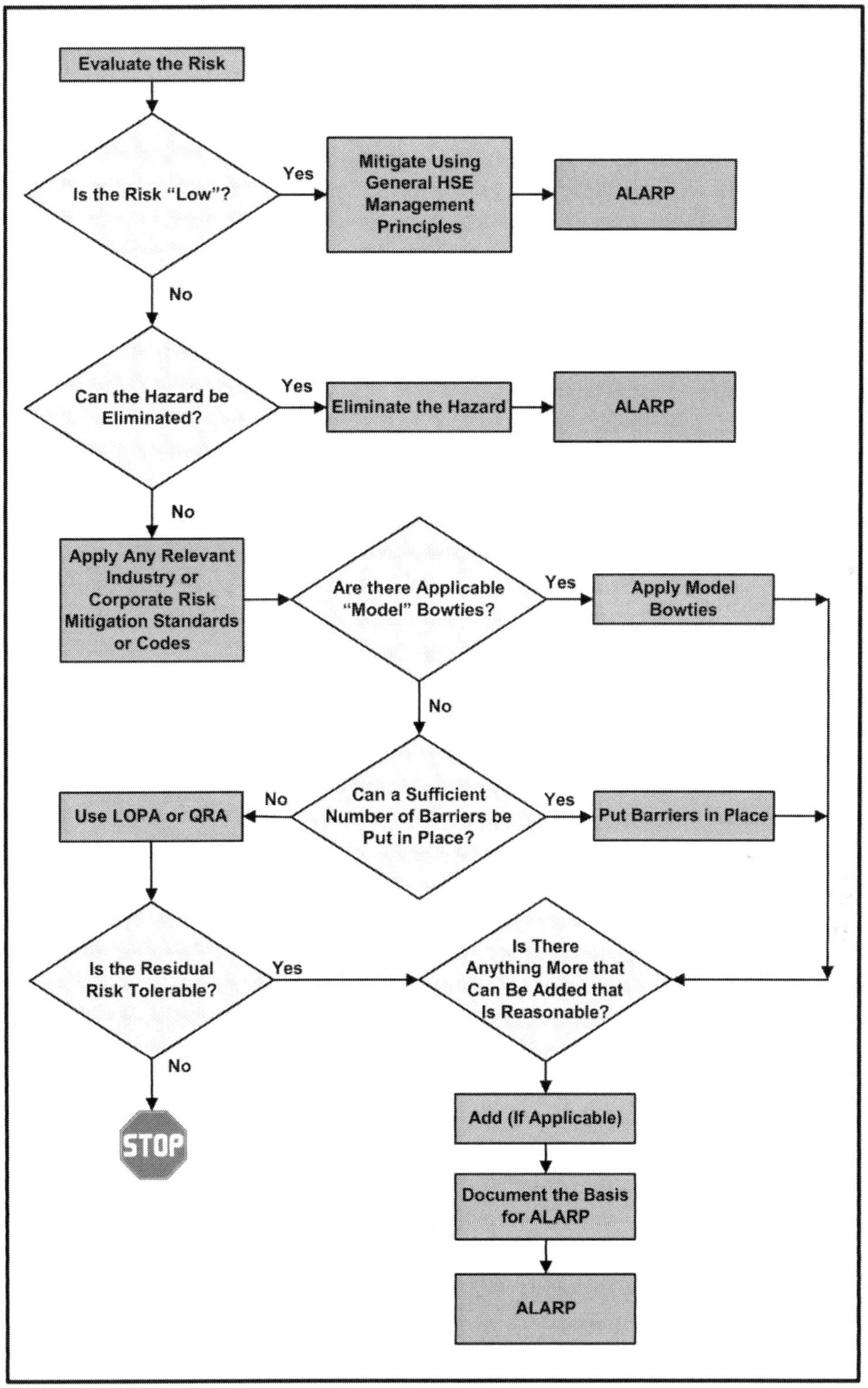

Figure 2: ALARP in HEMP

If the risk cannot be mitigated using a model bowtie, then the development of a unique bowtie would be the next possible pathway to demonstrating ALARP. As discussed in the previous section, a bowtie will show what barriers (control barriers and recovery measures) are relevant in a particular threat-line. However, the bowtie itself does not indicate how many barriers are required to demonstrate ALARP. The number of required barriers depends on the difference between the unmitigated likelihood of reaching an undesirable consequence and the risk tolerability limit: The larger the difference, the more barriers will be required to close the gap. A rather simplistic (yet conservative) approach involves setting a *minimum number of required barriers* for each risk level. By assuming that each barrier reduces the residual risk level by a factor of 10, i.e. each barrier is at least 90% effective, the minimum number of required barriers (N) can be determined as follows:

$$10^N = \frac{Unmitigated\ Liklihood\ of\ Reaching\ an\ Undesirable\ Consequence}{Risk\ Tolerability\ Limit} \quad \text{[Eq. 1]}$$

For example, if the unmitigated likelihood is once every ten years (10^{-1} $year^{-1}$) and the risk tolerability limit is once every million years (10^{-6} $year^{-1}$), then the minimum number of required barriers is 5. This suggests that ALARP can be achieved if *at least* 5 barriers can be placed in that particular threat-line on the bowtie. Corporate guidelines can then be consulted to determine how the minimum number of barriers needs to be split between the control and the recovery sides of the bowtie.

If ALARP cannot be demonstrated using the barrier counting technique described above, then the only remaining pathway to ALARP would be through a full blown Layer Of Protection Analysis (LOPA) or Quantitative Risk Assessment (QRA).

Going through the bowtie or LOPA/QRA pathways may still leave one question unanswered. The question has to do with the "R" component of ALARP. As a final check, we need to ask the question of whether there is anything more that can be done by way of risk mitigation that is "reasonable". By following this train of thought, we ensure that the ALARP methodology strikes a balance between rigor (as defined by the various pathways of Figure 2) and common sense (as encapsulated by posing a completely open-ended question).

4. HEMP Challenges in R&D

What makes HEMP a challenging process to apply in R&D settings? Below are the salient points with reference to the discussion in the previous section:

- Assessing the likelihood of an event requires historical data, e.g., how often do we have a process fluid leak from pump seal failure, or a fire from a pipe rupture, or overpressure from control valve failure, etc? Equipment failure and safety incident data from industrial plants are more readily available than those from R&D facilities. Therefore, the main driver for HEMP, namely, the risk assessment for a potential incident, is likely to be less accurate in R&D settings than in industrial settings.
- The outright elimination of a hazard may be the only pathway that is easier to follow in R&D settings than industrial settings. A change in material, operating conditions, or location is

frequently less cumbersome to pursue in a simple lab-scale process or a small pilot plant than a large industrial plant.

- Safety standards and codes governing process equipment and piping are, typically, more relevant for industrial plants than small pilot plants or lab-scale processes. In pilot plants, for example, vessels are typically constructed of pipes instead of ASME-coded vessels, and tubing is used throughout the process instead of ANSI-rated pipes.
- As pointed out in the previous section, model bowties could be a useful risk mitigation tool for common hazards used in common systems. The very existence of an R&D program, however, implies that things that are not commonly known are being discovered or invented – typically in fit-for-purpose systems built specifically for that purpose. This effectively limits the use of model bowties to only peripheral areas such as analytical equipment, emission control devices, and/or utility systems.
- Operating personnel interface with equipment and tubing in a much more "hands-on" manner in pilot plants/lab-scale processes than in industrial plants. As a result, operating personnel in R&D are in much closer proximity to hazards than their counter-parts in industrial plants. From a bowtie point of view, this compression in the length of the threat-line makes it difficult to accommodate multiple barriers in R&D. Stated differently, the minimum number of required barriers to mitigate a specific risk to ALARP is easier to achieve in an industrial plant than a pilot plant/lab-scale process.
- For LOPA and QRA to provide useful results, representative equipment failure rates and relevant safety incident data are needed. These data are generally difficult or impossible to find for pilot plants and lab-scale processes.

5. Case Study

In 2011/2012, a HEMP study was conducted on a small scale pilot unit at a Shell R&D facility. The purpose of the study was to assess, whether the risk of personnel exposure to hydrogen sulfide (H_2S) from a small pilot plant was being mitigated to ALARP. This was purely a proactive study aimed at finding and eliminating safety gaps before they resulted in safety incidents.

The design aspects of the *process area* of this particular pilot plant had been thoroughly reviewed and vetted as part of a well-established Management Of Change (MOC) work process. Therefore, it did not come as a surprise to the HEMP study team members that the *process area* included a sufficient number of barriers to meet the definition of ALARP. One of the main features of HEMP, however, is that it is not limited to *process areas*. In essence, any location where a hazard could be released to do harm would be within the scope of a HEMP study. This is somewhat different than a traditional process safety review (such as a Process Hazards Analysis - PHA), where the focus of the safety review is what can go wrong *in the process*. For this particular HEMP study, the team focused on two areas that are peripheral to the process. The first was the process vent emission control system and the second was the Gas Chromatograph (GC) calibration system.

Figure 3 shows the process vent emission control system employed at this particular pilot plant. The purpose of this system is to remove trace levels of H_2S from process vents before routing them to a common flare. Scrubbing with caustic is used in a packed bed scrubber to achieve near complete removal of H_2S from the process vents. Long before the scrubbing liquor is "spent", the scrubber is drained and recharged with a batch of fresh caustic.

From a *process* point of view, there is nothing exceptionally hazardous about the scrubbing system shown in Figure 3. In fact, it is a very common system used in many, if not most, R&D facilities around the world. From a HEMP point of view, however, examining how the hazard (H_2S) can be released provides a powerful insight into what can go wrong *outside* the process.

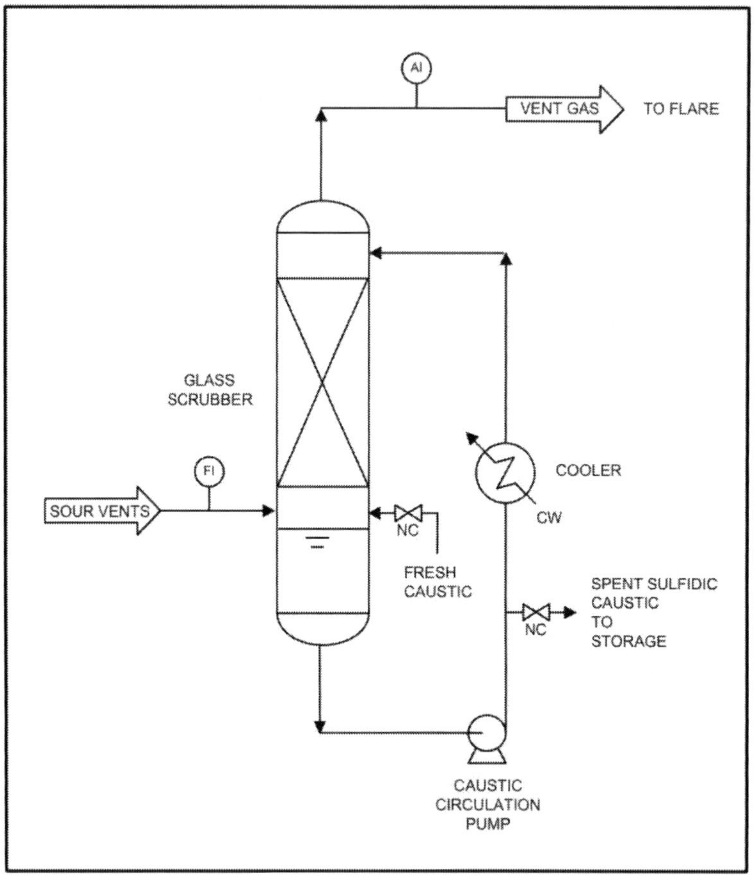

Figure 3: Process Vent Emission Control System

The chemistry behind the removal of H_2S from a process vent stream is well known: H_2S reacts with caustic (NaOH) to form sodium hydrosulfide (NaHS) and water:

$$H_2S + NaOH \rightarrow NaHS + H_2O \quad \text{[Eq. 2]}$$

As the sodium hydroxide gets depleted, the ability of the scrubbing liquor to remove H_2S from the vent gas is reduced. Eventually, the scrubbing liquor must be purged from the system as "spent sulfidic caustic". When a sufficient number of spent sulfidic caustic batches are generated, the material is sent for offsite disposal as a single large batch.

The HEMP study team determined that the handling of the spent sulfidic caustic described above was, in fact, the most hazardous part of the entire process. A loss of containment of this material followed by a drop in pH would cause the sodium hydrosulfide to revert to H_2S. While the reversion of sodium hydrosulfide to H_2S is well known, the ease with which it can happen was under-appreciated prior to this HEMP study. By merely entering a sewer system, for instance, the spent sulfidic caustic could generate significant quantities of H_2S as it gets diluted with water or mixed with low pH wastes. Another way in which H_2S can be released in significant quantities is if the emergency response personnel treat this material as a typical "spent caustic" instead of a *special* form of spent caustic. This is because the standard procedure for dealing with spilled spent caustic is to neutralize the material first, which is *exactly* the wrong thing to do when dealing with spent *sulfidic* caustic.

The HEMP team concluded that the then existing spent sulfidic caustic management system was not at ALARP: An insufficient number of *effective* barriers were in place to adequately mitigate the risk of personnel exposure to H_2S. An interim solution was developed to address the gaps while a more permanent solution was being pursued. The permanent solution involved the outright elimination of the hazard (spent sulfidic caustic) in line with the HEMP requirements as depicted in Figure 2.

The second focus area of the HEMP team was the GC calibration gas system. This is an area that doesn't receive sufficient attention in the typical PHAs conducted on *process areas*. Within the *process area*, extreme care is usually taken in minimizing the quantity and/or concentration of the hazardous material present. In addition, robust recovery measures are put in place such that if the process were to experience catastrophic failure, personnel exposure to the hazard would still be minimized. In many cases, however, the largest quantity of the hazard found in a lab is not in the process itself, but rather, in cylinders containing the standard gases used for GC calibration. As with the process area, these calibration gas systems warrant special safety considerations.

Figure 4 shows an example of how H_2S-containing calibration gas cylinders are handled in a lab at a Shell facility. This figure is intended to show the type of lab-scale barriers used to mitigate personnel exposure to H_2S. In no way should this design be viewed as a recommended (or best) practice for every application involving a hazardous calibration gas.

Unlike the scrubbing system where nothing exceptionally hazardous was previously noted, it was well known that the calibration gas system described in Figure 4 involved a significant quantity of the main hazard (H_2S). From a *process* point of view, the set up shown in Figure 4 would appear to meet all the safety requirements typically identified in a process safety study such as a PHA. In these types of safety reviews, the questions is posed as "*What can go wrong with this process*"? This is in stark contrast to HEMP, which poses the question as "*How can the hazard be released to cause harm*?" As was the case with the scrubbing system, HEMP pin-pointed a risk that was somewhat under-appreciated: The criticality of making a perfect seal between the cylinder and the sampling manifold. By far, this was the most credible threat-line for releasing an uncontrolled quantity of H_2S. The design of the system presented in Figure 4 may have been perfectly adequate to mitigate the risks associated with *most* threat-lines. However, this design is not adequate to mitigate the risk from *the one* threat-line that could result in the largest uncontrolled release of H_2S. Additional barriers were needed to supplement the procedure for connecting the cylinder to the sampling manifold, and to protect the operator in the event of a mechanical malfunction.

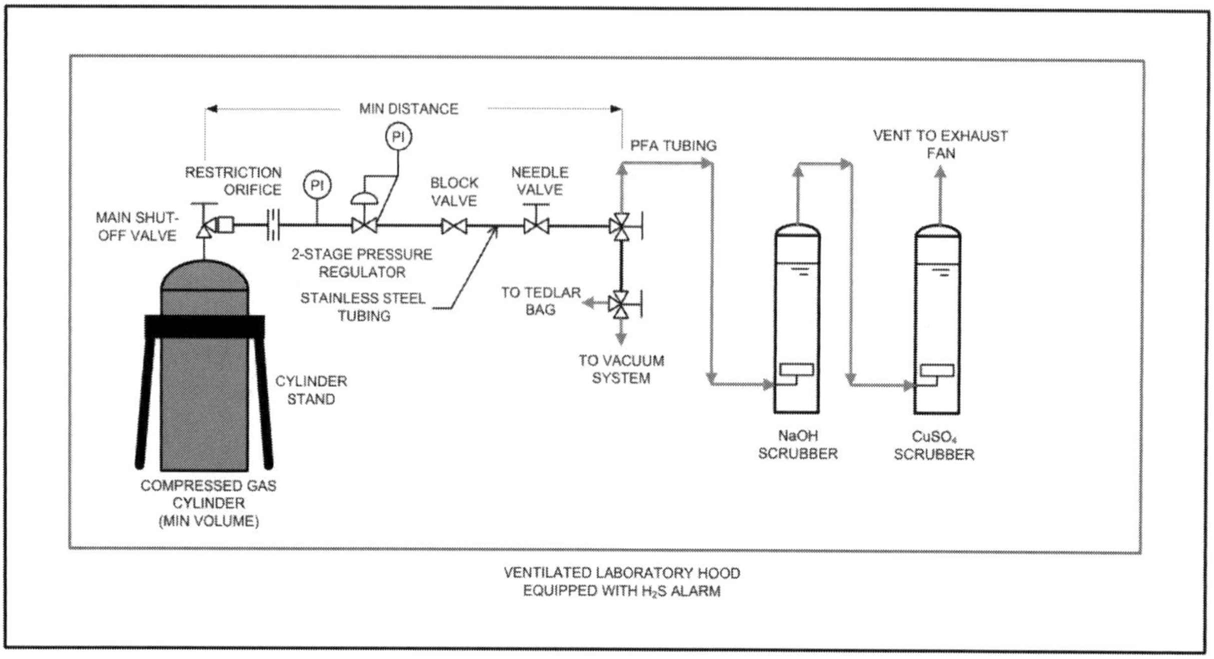

Figure 4: H_2S Calibration Gas System

6. Conclusions & Recommendations

Despite the challenges of applying HEMP in R&D settings, the process and its tools represent an invaluable component of a comprehensive process safety program. On the one hand, it is difficult to imagine how HEMP could identify a *new* hazard that had been overlooked in traditional process safety reviews. On the other hand, a methodical application of the process is likely to highlight safety gaps that would otherwise go unnoticed.

The author recommends that users of caustic scrubbers in H_2S service conduct safety gap reviews on their spent sulfidic caustic handling systems. These systems have quite a few credible scenarios for releasing large quantities of H_2S upon loss of containment and neutralization of the spent sulfidic caustic.

The author also recommends that calibration gas cylinders are included as part of the standard process safety reviews. Frequently, these cylinders contain the largest quantity (or concentration) of the hazard present at an R&D lab.

7. References

[1] Salter, N., and Siddoway, M., "Implementation of the Hazards and Effects Management Process (HEMP)", Presented at the 16th Annual Ethylene Producers' Conference, AIChE Annual Meeting, New Orleans, April 2004.

[2] Crerand, A., "Risk Management – The Bow-Tie Way", *The Chemical Engineer (Suppl),* September, 2005.

[3] "Safety Bulletin: Warning of Dangers of Sodium Hydrosulfide (NaHS) in the Workplace", Chemical Safety Board, Retrieved on January 30, 2013 from http://www.csb.gov/newsroom/detail.aspx?nid=72.

Scale up of Stirred Batch and Semi-Batch Reactors- Gaps and Limitations of Current Methodologies

Olga Juliana Reyes Valdes
Texas A&M University – Mary Kay O'Connor Process Safety Center
Texas A&M University, 3122 TAMU
College Station, TX 77843-3122
july2040760@neo.tamu.edu

Alba Pineda-Solano, Victor H. Carreto and Sam Mannan
Texas A&M University – Mary Kay O'Connor Process Safety Center
albis420@tamu.edu and victor.carreto@neo.tamu.edu

Prepared for Presentation at
American Institute of Chemical Engineers
2013 Spring Meeting
9th Global Congress on Process Safety
San Antonio, Texas
April 28 – May 1, 2013

UNPUBLISHED

Scale up of Stirred Batch and Semi-Batch Reactors – Gaps and Limitations of Current Methodologies

Olga Juliana Reyes Valdes
Texas A&M University – Mary Kay O'Connor Process Safety Center
Texas A&M University, 3122 TAMU
College Station, TX 77843-3122
july2040760@neo.tamu.edu

Alba Pineda-Solano, Victor H. Carreto and Sam Mannan
Texas A&M University – Mary Kay O'Connor Process Safety Center

Keywords: Scale-up, Boundary diagrams, Scale down, Mixing effects.

Abstract

In this paper, a comprehensive review of the current methodologies for the scale-up of batch and semi-batch reactors will be presented, along with the identified gaps, limitations and challenges of these methodologies. The paper will also discuss the importance of considering factors such as heat transfer dynamics and the relationship between mixing and mass transfer kinetics during the scale-up process. Finally, new developments and research needs will be discussed.

1. Introduction

Stirred batch and semi-batch reactors are widely used in many industrial chemical processes, particularly, in the fine chemical and pharmaceutical industries, to perform exothermic reactions. These often multi-purpose reactors present a higher risk for a runaway reaction due to the complexity of the processes, difficult implementation of sophisticated control and multiple variables involved. Therefore, it is important to follow adequate scale-up practices to ensure the safe and appropriate operation of the process.

In 2002, a report published by the U.S Chemical Safety and Hazard Investigation Board (CSB) [1] analyzed 167 accidents occurred in U.S. related to chemical reactivity. According to the report, the most common reactive chemical incidents occurred in the unsteady manufacture of fine chemicals, intermediate organics, plastics, rubbers and resins carried out in batch reactors. The principal causes of batch reactor incidents, have also been analyzed by Verjiws [2]. The author compared the results of three different studies and concluded that proper reactor design and adequate operating conditions could have prevented more than 60% of the reported incidents.

Scale-up is one of the most important and challenging aspects in the development of batch and semi-batch chemical processes [3]. During scale-up, parameters necessary for the safe and

efficient operation of processes at increasingly larger scales are determined. In general, reactors of different sizes present three main differences that must be forecast during the scale-up process, namely, operational times, mixing conditions and heating/cooling capacity.
However, process scale-up is not a straightforward task. The inadequate forecast of scale-up parameters can lead to the loss of process control, runaway reactions, over pressurization, inadequate mixing of solvents and reagents, vapor explosion, fires, etc. For example, the recent incidents at MFG in 2004, Syntron LCC in 2006, and T2 Laboratories in 2007, occurred as a result of improper scale-up. The lack of understanding of the thermochemistry of both, intended and unintended reactions, led to undersized cooling systems, and ultimately, to catastrophic consequences. These accidents clearly highlight the importance of the development and utilization of proper and reliable scale-up methodologies for batch and semi-batch stirred tank reactors.

Currently, many different approaches for the safe scale-up of batch and semi-batch stirred reactors can be encountered in the literature. Unfortunately, many of these approaches are developed based on a particular reaction or specific assumptions and, therefore, miss discussed critical elements of the scale-up process. This fact makes the scale-up process challenging and costly and the applicability of a single approach to different systems almost impossible.

The objective of this manuscript is to review different scale-up approaches and discuss the main gaps that may restrict their applicability. The next sections will describe the traditional scale-up process, followed by two alternative approaches for the safe scale-up of batch and semi-batch reactors, which are available in the literature. Later, the main gaps and limitations of these approaches will be discussed. Finally some conclusions and recommendations will be drawn based on the more important findings.

2. Scale-up Methodologies

The increasing developments of appropriate equipment and methodologies at lab-scale have allowed reducing uncertainties in the early stage of the process development. An enormous amount of information can be extracted from experiments performed at lab-scale, where the experiments are less costly and more flexible and thus reduce the number of experiments required at larger – and more expensive – scale.
Reaction calorimetry is a very useful tool, used in all scale-up methodologies that will be describe below, and which makes a significant impact in the scale-up and assessment of safety aspects of batch and semi-batch reactors [4]. The different applications of calorimetry have been comprehensively described by several authors [4-9]. Calorimetric studies are extensively used for the evaluation of thermal and kinetic parameters of chemical reactions. Data obtained from calorimetric measurements allows finding the rate of reaction, specific heat capacity of the reaction mass and the enthalpy of the reaction, which are scale-independent. However, the appropriate use of calorimetric data, obtained at laboratory scale, for the scale-up of batch and semi-batch reactors to industrial scale, still represents a challenging task [10].

2.1 Traditional Scale-Up

The scale-up of batch and semi-batch reactors have been performed for many years as an iterative methodology. This methodology involves experimental testing, development of heat flow models, simulation of large scale vessels, and also many rules of thumb. The process is repeated for each scale of the process. Usually the scale-up from laboratory scale through pilot plant to industrial scale increases by a factor 10^4 [3].

Traditional reaction scale-up commonly involves the construction of mathematical models based on calorimetric data of the instantaneous heat of release. These heat flow models allow to determine the macrokinetic parameters of a reaction of interest [11]. Generally, the cooling power of the industrial scale reactor is estimated, using reaction calorimetry, by scaling-up the overall heat transfer coefficient (U-value) [12]. The overall heat transfer coefficient of the system is given by the sum in series of three different resistances, namely, the resistance of the liquid film that is in contact with the reactor wall, the resistance of the reactor wall itself and the resistance of the fluid in the jacket.

The internal heat transfer coefficient due to reaction mixture, is estimated at laboratory scale by performing calorimetric calibrations at different stirring rates and using a Wilson plot [13]. The Wilson method uses a plot of $\frac{1}{U}$ vs $\left(\frac{N}{N_0}\right)^{-2/3}$, where N corresponds to the agitation rate and N_0 to a reference agitation rate where the internal heat transfer becomes zero. Then, the heat transfer coefficient of the reaction mixture inside an industrial scale vessel, $h_{r,\ scale}$, can be related to the data obtained at lab-scale, at the same temperature, by:

$$h_{r,scale} = h_{r,lab}\left(\frac{d_{scale}}{d_{lab}}\right)^{1/3}\left(\frac{N_{scale}}{N_{lab}}\right)^{2/3}\left(\frac{\mu_{r,scale}}{\mu_{w,scale}}\right)^{0.14} \qquad \text{[Eq. 1]}$$

When using Wilson's method, it is important to keep in mind that it is based on semi-empirical relations. Consequently, it is necessary to consider the differences in mixing effects when measuring the maximum cooling capacity in pilot experiments and in the industrial scale reactor.

2.2 Combined Boundary Diagrams (BDs) and temperature diagrams (TDs)

Safe operating conditions in exothermic semi-batch reactors are characterized by a low accumulation of reactant within the system [14, 15]. These conditions can be found at lab-scale through the use of boundary and temperature diagrams. Safety boundary diagrams basically consist in a plot of the Reactivity, *Ry*, versus Exothermicity, *Ex* numbers, where the region of runaways is surrounded by a boundary line. *Ry* and *Ex* are defined as follows:

$$Ry = \frac{v_A RE k_{T_c} Da}{(R_H + Co)\varepsilon} \qquad \text{[Eq. 2]}$$

$$Ex = \frac{\gamma \quad \Delta T_{ad,o}}{\tau_c{}^2 (R_H + Co)\varepsilon} \qquad \text{[Eq. 3]}$$

Figure 1 shows a schematic representation of a boundary diagram, where the safe operating condition's dependence on the following two factors can be observed:

1. The cooling number – as the cooling number increases, the boundary line moves to lower exothermicities. Lower exothermicities involve operation at safer conditions.
2. The phase in the reactor where the reaction takes place.

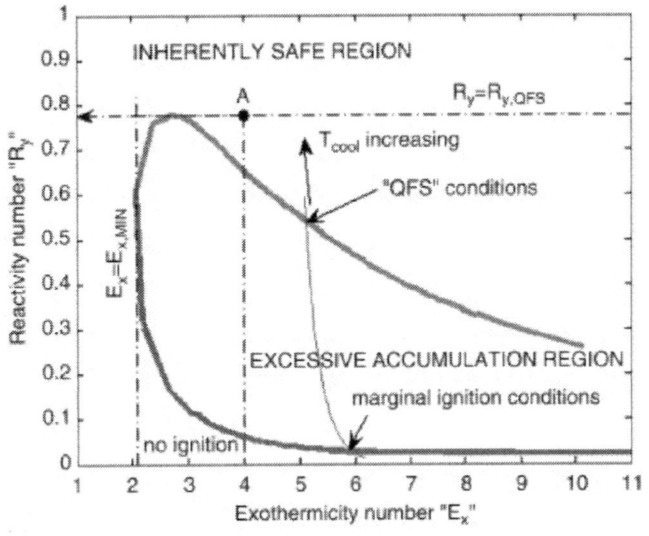

Figure 1: Schematic representation of a boundary diagram. Figure taken from[16]

Inside the runaways region, a target temperature is exceeded. This temperature is given by:

$$\tau_{target} = \tau_c^{eff} + 1.05\frac{\Delta T_{ad,o}}{\varepsilon[Co(1+\varepsilon\vartheta)+R_H]} \qquad \text{[Eq. 4]}$$

Hugo and Steinbach [14] firstly studied the thermally safe operation of homogeneous semi-batch reactors, in which an exothermic reaction of (1,1) order happens. Boundary diagrams for this type of reactors are characterized by a line that separates the runaway region from the region with small harmless accumulation of the dosed reagent. These authors defined this line for different modified Stanton numbers with the following equation:

$$S_t = (\frac{UAt_D}{V\rho C_p}) \quad \text{[Eq.6]}$$

Furthermore, Hugo and Steinbach also empirically determined that in the boundary of the two regions the following relation must hold:

$$D_a \leq S_t \qquad \text{[Eq. 7]}$$

Where D_a refers to the Damköhler number, defined as:

$$D_a = kC_{B,0}t_D \qquad \text{[Eq. 8]}$$

In a later publication [15], the same authors derived empirical correlations to find safe operating conditions in homogenous semi-batch reactors. They concluded that:

1. The thermal relation number $B = (E\Delta T_{ad}/RT_R{}^2)$ have to be minor than the limit value $B = (\sqrt{S_t + 2G} + 1.2)^2$, where G represents the increment in volume due to the addition of the second reagent, and is defined as the ratio of the mass of reagent A to the final mass of the mixture within the reactor.
2. The temperature of the coolant has to be higher than the limiting value T_δ. This limiting value can be found using the following equation:

$$D_a(T_\delta) = S_t \qquad \text{[Eq. 9]}$$

Stneensma and Westerterp [17-19] studied liquid-liquid Semi-batch reactor (SBRs) heterogeneous systems with (1,1) reaction kinetics order, characterized by power law type microkinetic rate expression:

$$r = k(T)C^n{}_A C^m{}_B \qquad \text{[Eq. 10]}$$

This work was later extended to a universal (*n,m*) rate of reaction expression by Maestri and Rota[20, 21], who provided a full range of industrial practice cooling numbers. The authors also showed the sensitivity (mainly with respect to the reaction order of the dosed reagent, n) on the shape and extension of BDs in this kind of systems. The authors highlighted the importance of the reaction kinetics' knowledge, for the properly use of Boundary Diagrams method in heterogeneous SBRs, by demonstrating that the use of BDs calculated for a higher dosed correactant reaction order, than the actual one, can lead to dangerous scenarios. Boundary Safety diagrams for multiple heterogeneous reactions were lately developed by Van Woezik and Westerterp[22]. The discussion of the mathematical model on which boundary diagrams are based can be found in different publications [17-20, 22-25].

With the use of safety boundary diagrams it is possible to differentiate the inherently safe operating conditions (where no accumulation phenomena occurs),without the necessity to solve the mathematical model of the reactor [26]. However, identifying and operating in regions where the accumulation is avoided, may not be enough to classify a set of operating conditions as thermally safe, in cases where the system can achieve a temperature higher than the maximum allowable temperature MAT, and therefore decomposition or side reactions can occur[16]. Based on this fact, Maestri and Rota [16] proposed to couple safety boundary diagrams with temperature diagrams (TDs), which involve the same dimensionless parameters. The combination of BDs and TDs allow determining, for a specific set of operating conditions, the probable maximum increment in temperature during normal operation. This approach was experimentally validated through the analysis of an industrial nitration process of N-(2-phenoxyphenyl) methane sulfonamide to N-(4-nitro, 2-phenoxyphenyl) methane sulfonamide [24].

2.2.1 *Applicability of BDs and TDs to systems of unknown kinetics*

Westerterp and Molga [27] demonstrated that the findings on boundary safety diagrams also apply to batch reactors, and extended the previous work on safety diagrams to a full range of practical cooling number values, for heterogeneous systems with slow reaction rate, as they occur on large industrial scale reactors.
As a result of their study, Westerterp and Molga presented two different diagrams, one for reactions taking place in the dispersed phase and the other one for reactions taking place in the continuous phase. Both diagrams are based on the cooling number and allow finding the minimum values of exothermicity ($E_{x,MIN}$) and reactivity ($R_{y,QFS}$), below and above which, respectively, no runaways will occur. The obtained data can also be used for multiple reactions, but it is not applicable in case of autocatalytic reactions. The safety criterion for autocatalytic behavior was developed later by Maestri and Molga [25], and must consider the amount of catalyst as an additional parameter with respect to the non-catalytic reactions.

The necessity to have detailed understanding of the kinetics is eliminated, because the necessary information can be obtained by lab-experiments in standard calorimeter (*e.g.* DSC and RC1), reactor tests, in the plant and literature. Therefore, even when the detailed kinetics of the reaction is not known, safety assessments can be performed in a short period of time (*e.g.* few days) with the use of a reaction calorimeter [28], as will be described in the following section.

2.2.2 *Scale-up Procedure based on Boundary Diagrams*

Maestri *et al.* [26] developed a simple procedure, based on the method of boundary and temperature diagrams, for the safe and optimum scale-up. The authors represented $R_{y,QFS}$ values for an entire set of cooling numbers of relevance in the industrial practice, and denominated these curves as "inherently safe condition curves" (ISC). This representation can be found in figures 2 and 3 of the cited paper.

Their approach is based on the use of quick onset fair conversion, and smooth temperature profile (QFS) operating condition and generalized forms of temperature diagrams. Under QFS conditions, the curve of the dimensionless maximum temperature rise is defined as:

$$\psi = \left({}^{T_{max}}/_{T_0} \right)_{max} \qquad \text{[Eq. 11]}$$

When plotted against exothermicity (Ex), the dimensionless maximum temperature rise tends to overlap in a small region as can be observed in **Error! Reference source not found.** The upper boundary of this region delineates a curve known as the temperature rise curve (TRC). Under QFS conditions the kinetics loses importance, so the TRC can also be used in the estimation of the maximum peak temperature of autocatalytic reaction systems [26].

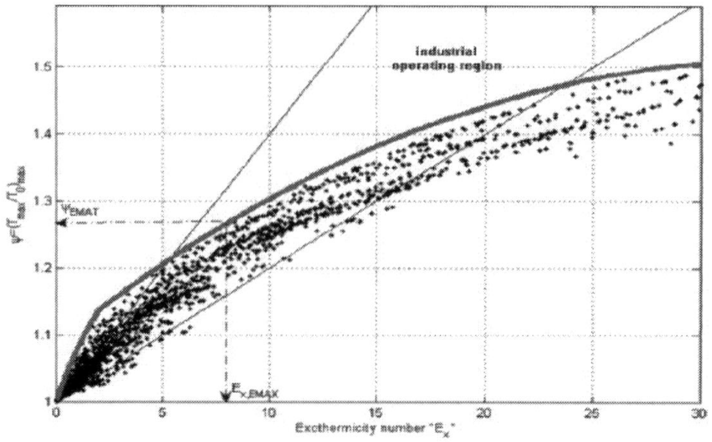

Figure 2: Inherently safe operating conditions for SBRs in which a single exothermic reaction is carried out. Figure taken from [26]

The scale-up procedure proposed by Maestri *et al.,* consists of 6 steps, that are summarized in Figure 3. For the detailed explanation of each of the steps, the reader can go to the cited paper.

2.3 Scale-Down Approach

The first approach to a scale-down methodology was published by Kupr & Hub [29]. They recreated the operating conditions of a large scale reactor in a 1L reaction calorimeter (scale-down). That is, they limited the heat supply to the 1L reaction calorimeter. Their aim was to modify the heating capacity of the laboratory equipment by limiting the heat transfer area of the reactor mantle so that the heat transfer area per volume of the reaction mass of the laboratory and large scale reactor were the same.
Although it is likely that other scale-down approaches have been developed, they are not readily available in the literature, probably due to confidentiality issues. Some recent contributions that can be found in the literature include those of Benoit Zuffrey [3], and Davis & Viswanath [30].

The methodology developed by Zuffrey [3] can be summarized in the following five steps:

1. Using the RC1 calorimeter, make an online observation of the instantaneous heat production rate during a small time interval.
2. Multiply the heat value obtained by the scale-up factor given by equation 12 to simulate numerically the plant reactor dynamics

$$Scale - up\ factor(S_f) = \frac{m_r^{indus}(k)}{m_r^{RC1}(k)} \quad \text{[Eq. 12]}$$

Where m_r^{RC1} is the reaction mass in the RC1, and m_r^{indus} is the hypothetical (defined by the user) reaction mass in the industrial reactor.

3. With the numerically model, infer the evolution of the jacket and reaction mixture temperatures of the plant reactor model.

4. Force the calorimeter (in this methodology, the RC1 calorimeter) to follow this temperature evolution by changing the batch temperature set point of the calorimeter to the new calculated large scale reaction temperature.
5. Repeat the steps 1-4.

The procedure described above comprises the repetition of numerous calculations of the reaction temperature and heat of reaction over a small unit of time (10 seconds).

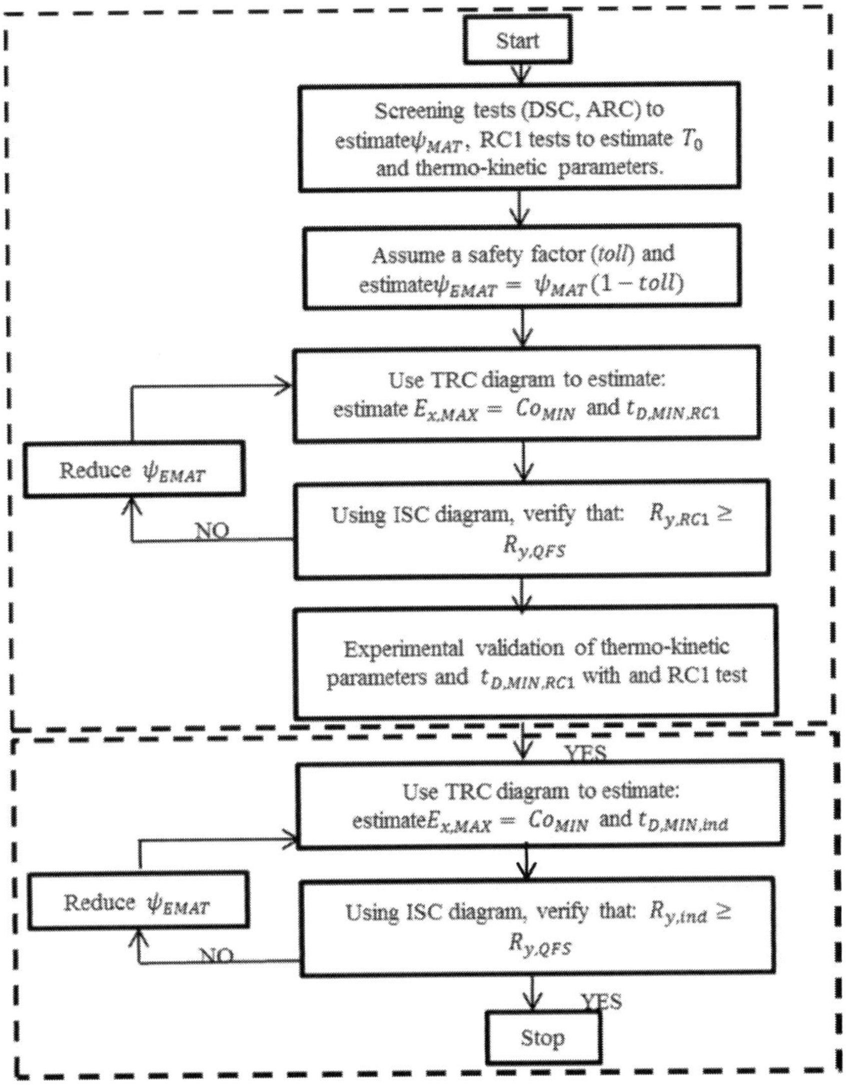

Figure 3: Flow diagram for the selection of safe operating conditions at lab-scale and their optimum scale-up to plant reactor scale [26].

Davis & Viswanath [30] proposed a scale-down approach which does not need the heat flow data, and therefore can be applied to any computer-controlled jacketed reactor, with known vessel constants. Their scale down approach is based on restraining the heat transfer of a laboratory reactor to the mole proportional equivalent for the large scale vessel. Keeping equivalent mixing will allow direct observation of the large scale temperature profile and reaction behavior in the laboratory vessel. The authors successfully applied their approach to four different case studies and demonstrated that their methodology is able to predict reaction temperature profiles for dose as well as non-dose controlled reactions. The full mathematical derivation of this approach can be found in reference [30].

3. Advantages and disadvantages of the proposed approaches

Although all of the methodologies described above try to accurately and safely predict the heat profile of a large scale reactor, all of them fail in considering the relationship of the intrinsic chemistry with each one of the physical parameters listed by Atherton [31], *i.e.* heat transfer, mixing times, processing times, etc. For example, one of the assumptions of the scale-down approach proposed by Davis & Viswanath is that the industrial and lab-scale runs have the same time values (addition times, etc.). Special attention must be paid to that assumption for certain processes in particular, for example, in polymerization reactions where the dispersion process time changes with the increment in circulation paths in larger reactor sizes. In general, as the size of the reactor increase, the circulation times and paths become larger [32].

Table 1 summarizes the main advantages and disadvantages of each of the methodologies that have been described throughout this manuscript.

Table 1. Principal advantages and disadvantages of scale-up methodologies

Methodology	Advantages	Disadvantages
Traditional Scale-up	• Model can be accurate and predict well in the case of simple dose control reactions. • Can be time adjusted to accurately predict the effects of the addition rates on temperature profile [30].	• Only remain valid as long as the originally measured heat of reaction profile is valid. • A detailed knowledge of the reaction kinetics is indispensable. • Expensive and time consuming.
Boundary Safety Diagrams	• Even if detailed kinetics of the reaction is not known, safety assessments can be performed in a short period of time.	• The use of BDs calculated for a higher dosed correactant reaction order, in heterogeneous SBRs; and for a lower dosed correactant reaction order in homogeneous SBRs, than the actual one, can lead to dangerous scenarios • Not applicable for autocatalytic reactions
Scale Down	• The kinetic model of the reaction is not necessary	• Data of the industrial reactor were it will be applied is needed. • Difficult to apply and improve in academic research. • Does not consider mixing effects.

4. New developments

With the aim to include the mixing effects into the scale-up process of batch and semi-batch reactors, recent research on scale-up has focused mainly on the combination of experimental technology (to acquire trustable data) and CFD simulations (which allow simulating reactors at different scale at lower costs). However, these CFD models usually require the input of pilot plant data, which is not always available. Furthermore, it is necessary to be able to analyze the output of these models, because even when they do not match exactly with reality, they can be a powerful tool for screening analysis and decision making.

Regarding process development, a technique that has gained a lot of attention, in the last decade is micro-reaction calorimetry. Work with micro-scale continuous reactors gives researches advantages, over conventional batch and semi-batch reactors, such as rapid mixing and more efficient thermal transfer. Therefore it is possible to obtain high purity products (by suppressing side reaction because of poor mixing or aging in batch reactor), also, highly exothermic reactions can be performed in a safer manner[33]. These facts facilitate researches to investigate different synthetic routes for a product of interest and even the extraction of reaction of kinetics information using micro-calorimetry has been reported[34]. Furthermore, continuous reactions scale-up easier and in more economically advantageous way than batch and semi-batch[35].

However, because of the time and money constraints, the main focus of industry research is to find quick and inexpensive scale-up methodologies, (as the scale down approach described above) that do not need the kinetics to achieve safe and high yield at bigger scale productions.

5. Research needs

Caygill *et al.*,[36] reviewed some of the most common factors that obstruct scalability in the fine chemicals and pharmaceutical industries. Their findings indicated that difficulties associated with mixing and heat transfer are the most common obstacles encountered in scale-up. Nevertheless, most of the literature addressing scale-up try to overcome changes in heat transfer when scaling up, but very few address the problems relating with mixing.

Although a lot of research has been done in the field of industrial mixing, many aspects are still not well understood. Due to this lack of knowledge, the assumption of "perfect macro-mixing" is a common denominator in the scale-up/down approaches mentioned in this review. Improper scale-up of mixing can lead to mass transfer issues, *e.g.*, when working with semi-batch stirred reactors at lab-scale, the dosing of a solution of reagent to the surface of the liquid that is being agitated, usually performs well. But the same procedure at plant scale can lead to poor mixing [37]. In mixing sensitive reactions, poor mixing gives lower yield and selectivity of the reaction [38] and in competitive reactions, it may promote side reactions, and unintended by-products.

Furthermore, in order to develop a robust methodology for the scale-up of batch and semi-batch reactor, it is imperative to build stronger relationships and collaboration between industry and academia.

6. Conclusions

Even though the process of scaling-up batch and semi-batch stirred tank reactors has been widely studied, it is still not fully understood. Traditional scale-up methods require extensive number of experimental testing at different scales, the development of heat flow models, use of rules of thumb and a detailed knowledge of the reaction kinetics, which make the scale-up process costly and time consuming.

Different approaches, such as the use of boundary diagrams and scale-down procedures have been proposed recently in order to overcome these disadvantages. However, due to the complex nature of some of the physical interactions that affect the scale-up process, such as mixing, these approaches are based on assumptions that do not hold for all kind of reactions.

The use of CFD modeling, in combination with experimental technologies, hold a great promise to allow the development of procedures and methodologies that are able to consider and predict the effects of different physical parameters on the reaction kinetics. Ideally, these methodologies will reduce the necessity to perform costly experiments at pilot scale. Ultimately, this will allow the quick, assertive, cheap and safe development of new processes to emerging and small companies.

Symbols

$C_{B,0}$: initial concentration of reagent B

Co: cooling number, given by $Co = \frac{U*(Da)}{\varepsilon}$

$c_{P,0}$: Heat capacity at initial stage

c_P: Heat capacity

D_a Damköhler number

d_{scale}: Diameter of scale vessel

d_{lab}: diameter of lab vessel

Ex: Exothermicity number

G: increment in volume due to the addition of the second reagent, defined as the ratio of the mass of reagent A to the final mass of the mixture within the reactor

$h_{r,\,scale}$: Heat transfer coefficient of the reaction mixture inside an industrial scale vessel

$h_{r,lab}$: Heat transfer coefficient of the reaction mixture inside lab scale vessel

k_{τ_c}: Kinetic coefficient at the coolant temperature τ_c

k : Kinetic coefficient

m: reaction order with respect to the reagent initially charged at the reactor, B

$m_r{}^{RC1}$: Reaction mass in the RC1

$m_r{}^{indus}$: Hypothetical (defined by the user) reaction mass in the industrial reactor.

n: reaction order with respect to the dosed reagent A

$n_{B,0}$: initial mols of the reagent B.

N: Agitation rate

N_0 : agitation rate where the internal heat transfer becomes zero

N_{scale}: Agitation rate at industrial scale

R: Ideal gas constant

RE: Reactivity enhancement factor

R_H: Ratio of the specific heat of the liquid being added to the one already present in the reactor.

Ry: Reactivity number

S_t: Stanton number
t_D: Dosing period
U: Overall heat transfer coefficient
ΔH_r: Heat of reaction
$\Delta T_{ad,o}$: Adiabatic temperature rise at the initial state of the reactor, given by: $\Delta T_{ad,o} = \left(-\Delta H_r n_{B,0}\right) / \left(v_B \rho_0 c_{P,0} V_0\right)$

Greek Symbols
$\mu_{r,scale}$: Dynamic viscosity industrial scale reaction mixture
$\mu_{w,scale}$: Dynamic viscosity industrial scale reactor wall
v_A: Stoichiometric coefficient reagent A
v_B : Stoichiometric coefficient reagent A
γ: Dimensionless activation energy
ε: The relative volume increase given by: $\varepsilon = (V_{end} - V_0)/V_0$
ρ: Densit
ρ_0: Density at initial stage
ψ: Dimensionless maximum temperature

References

1. CSB, C.S.B., *Improving Reactive Hazard Management*, Report No. 2001-01-H, 2002.
2. Verwijs, J.W., *Reactor start-up and safeguarding in industrial chemical processes*, 1994, Twente University: Enshede, The Netherlands.
3. Benoit, Z., *Scale-Down Approach: Chemical Process Optimisation Using Reaction Calorimetry For the Experimental Simulation of Industrial Reactors Dynamics*, in *Chemical and Chemical Engieering*2006, Ecole Polytechnique Federale De Lausanne: Lausanne, France.
4. Landau, R.N., *Expanding the role of reaction calorimetry.* Thermochimica Acta, 1996. **289**(2): p. 101-126.
5. Stoessel, F. and O. Ubrich, *Safety Assessment and Optimization of Semi-batch Reactions by Calorimetry.* Journal of Thermal Analysis and Calorimetry, 2001. **64**(1): p. 61-74.
6. Singh, J., *Reaction calorimetry for process development: Recent advances.* Process Safety Progress, 1997. **16**(1): p. 43-49.
7. LeBlond, C., et al., *Reaction calorimetry as an in-situ kinetic tool for characterizing complex reactions.* Thermochimica Acta, 1996. **289**(2): p. 189-207.
8. Duh, Y.-S., et al., *Applications of reaction calorimetry in reaction kinetics and thermal hazard evaluation.* Thermochimica Acta, 1996. **285**(1): p. 67-79.
9. O'Neill, M.A.A. and S. Gaisford, *Application and use of isothermal calorimetry in pharmaceutical development.* International Journal of Pharmaceutics, 2011. **417**(1–2): p. 83-93.
10. Bonvin, D., *Optimal operation of batch reactors—a personal view.* Journal of Process Control, 1998. **8**(5–6): p. 355-368.
11. Bernhard, G. and R. Rudolf, *Reaction calorimetry for the development of chemical reactions.* Thermochimica Acta, 1987. **114**(1): p. 83-90.

12. Choudhyry, S., Utiger, L,, and Riesen, R, Heat transport in agitated vessels: scale-up methods. Mettler-Toledo publication, (00724218).
13. Wilson, E.E., *A basis for rational design of heat transfer apparatus.* Trans. Am. Soc. Mech. Engrs, 1915. **37**: p. 47-82.
14. Hugo, P.a.S., J., *Practically oriented representation of thermal safe limits for an indirectrly cooled semi-batch reactor.* Chem Ing Tech, 1985. **57**: p. 780-782.
15. Hugo, P.a.S., J., *A comparison of the limits of safe operation of SBR and CSTR.* Chem Eng Sci, 1986. **41**: p. 1081-1087.
16. Maestri, F. and R. Rota, *Temperature diagrams for preventing decomposition or side reactions in liquid–liquid semibatch reactors.* Chemical Engineering Science, 2006. **61**(10): p. 3068-3078.
17. Steensma, M. and K.R. Westerterp, *Thermally safe operation of a semibatch reactor for liquid-liquid reactions–fast reactions.* Chemical Engineering & Technology, 1991. **14**(6): p. 367-375.
18. Steensma, M. and K.R. Westerterp, *Thermally safe operation of a semibatch reactor for liquid-liquid reactions. Slow reactions.* Industrial & Engineering Chemistry Research, 1990. **29**(7): p. 1259-1270.
19. Steensma, M. and K.R. Westerterp, *Thermally safe operation of a cooled semi-batch reactor. Slow liquid-liquid reactions.* Chemical Engineering Science, 1988. **43**(8): p. 2125-2132.
20. Maestri, F. and R. Rota, *Thermally safe operation of liquid–liquid semibatch reactors. Part I: Single kinetically controlled reactions with arbitrary reaction order.* Chemical Engineering Science, 2005. **60**(12): p. 3309-3322.
21. Maestri, F. and R. Rota, *Thermally safe operation of liquid–liquid semibatch reactors Part II: Single diffusion controlled reactions with arbitrary reaction order.* Chemical Engineering Science, 2005. **60**(20): p. 5590-5602.
22. van Woezik, B.A.A. and K.R. Westerterp, *Runaway behavior and thermally safe operation of multiple liquid–liquid reactions in the semi-batch reactor: The nitric acid oxidation of 2-octanol.* Chemical Engineering and Processing: Process Intensification, 2002. **41**(1): p. 59-77.
23. Hugo, P., J. Steinbach, and F. Stoessel, *Calculation of the maximum temperature in stirred tank reactors in case of a breakdown of cooling.* Chemical Engineering Science, 1988. **43**(8): p. 2147-2152.
24. Maestri, F., et al., *Safe and Productive Operation of Homogeneous Semibatch Reactors. II. The Nitration of N-(2-Phenoxyphenyl) Methane Sulfonamide.* Industrial & Engineering Chemistry Research, 2006. **45**(24): p. 8014-8023.
25. Maestri, F. and R. Rota, *Safe and Productive Operation of Homogeneous Semibatch Reactors Involving Autocatalytic Reactions with Arbitrary Reaction Order.* Industrial & Engineering Chemistry Research, 2007. **46**(16): p. 5333-5339.
26. Maestri, F., et al., *Simple Procedure for Optimally Scaling-up Fine Chemical Processes. I. Practical Tools.* Industrial & Engineering Chemistry Research, 2009. **48**(3): p. 1307-1315.
27. Westerterp, K.R. and E.J. Molga, *No More Runaways in Fine Chemical Reactors.* Industrial & Engineering Chemistry Research, 2004. **43**(16): p. 4585-4594.
28. Westerterp, K.R. and E.J. Molga, *Safety and Runaway Prevention in Batch and Semibatch Reactors—A Review.* Chemical Engineering Research and Design, 2006. **84**(7): p. 543-552.
29. Kupr, T.a.H., L., *Industrial reaction calorimeter.* Angew. Chem. Thermodynamic. Thermoanal, 1979(37): p. 334-342.
30. Davis, E.M. and S.K. Viswanath, *Heat Transfer Based Scale-Down of Chemical Reactions.* Organic Process Research & Development, 2012. **16**(8): p. 1360-1370.

31. Atherton, J.H., *Chemical aspects of scale-up*. Pilot plants and Scale-up of Chemical Processes II, ed. W. Hoyle. 1999, Cambridge: Royal Society of Chemistry.
32. Koh, P.T.L., and Xandtidis, F., *CFD Modelling in the scale-up of a stirred reactor for the prodcution of resin beads*, in *Second International Conference on CFD in the Minerals and Process Industries*1999: Melburne, Australia.
33. Wiles, C.a.W., Paul, *Recent advances in micro reaction technology.* The Royal Society of Chemistry, 2011(47): p. 6512-6535.
34. Valera, F.E., et al., *The Flow's the Thing...Or Is It? Assessing the Merits of Homogeneous Reactions in Flask and Flow.* Angewandte Chemie International Edition, 2010. **49**(14): p. 2478-2485.
35. Schaber, S.D., et al., *Economic Analysis of Integrated Continuous and Batch Pharmaceutical Manufacturing: A Case Study.* Industrial & Engineering Chemistry Research, 2011. **50**(17): p. 10083-10092.
36. Caygill, G., M. Zanfir, and A. Gavriilidis, *Scalable Reactor Design for Pharmaceuticals and Fine Chemicals Production. 1: Potential Scale-up Obstacles.* Organic Process Research & Development, 2006. **10**(3): p. 539-552.
37. Laird, T., *How to Minimise Scale Up Difficulties.* Chemical Industry Digest., 2010.
38. Bourne, J.R., *Mixing and the Selectivity of Chemical Reactions.* Organic Process Research & Development, 2003. **7**(4): p. 471-508.

Safety, ASAP and ALARP in Fast Track Process Development

Joseph B. Powell, Ph.D.
Shell Chief Scientist – Chemical Engineering, Houston, TX

Introduction

Microscale methods have become commonplace in high school as well as college or university chemistry curriculum [1-4], as a means to reduce costs and environmental footprint, while also increasing safety by the principle of "as low as reasonable practicable" or ALARP [5,6]. Similar benefits hold for industry. Advances in analytical chemistry and process modeling have reduced the need to weigh and measure yields during process development. Screening of process options or concept validity can often be performed at the microscale at significantly reduced time and cost, given the increasing burden of management-of-change procedures and documentation associated with laboratory, pilot plant, and process safety. Speed is essential in the competitive industry environment, where business customers seek to either commercialize and deploy, or to fail, "as fast as possible" (ASAP), to reduce overall R&D costs.

Safety Management of Change in Industry (MOC)

Industry management of change procedures favor "one size fits all" methodologies, where the same formal procedure is used to either stream a new refinery process, or manage a laboratory distillation. This gives the technology management team confidence in containment of risk, but can lead to business sponsors losing confidence in costs and speed of the R&D organization. Effective risk management is thus a critical issue for research scientists and process development engineers, effectively balancing safety and assurance vs. speed to delivery.

One way to manage the apparent conflict is conduct an umbrella or generalized laboratory MOC review to entail operating limits that maximize allowed temperature and pressure limits to work within the maximum ratings of process equipment, and to generalize flammability and chemical exposure hazards to a range of allowed substances. This one-time comprehensive review follows the full MOC process, but then enables new experiments to be conducted after only a brief review to assess whether the conditions fall under the envelope of existing safety reviews for the laboratory equipment and procedures to be deployed. If additional hazards are identified in future proposed experiments, one can focus on mitigation of the specific features of the new hazard and the safeguarding changes needed, without having to re-approve all other aspects of the experiment.

By using scaling down to smaller-footprint experiments (ALARP), a broader array of new studies can be conducted without tripping risk hazards that would otherwise required a new full safety review.

Example: Mitigation of New Chemistry Exposure Hazard

A process development bench scale laboratory had a 10-year history of handling ethylene oxide, carbon monoxide, and hydrogen using 300-ml or larger stirred reactors in a fume hood, fully instrumented for shutdown in the vent of release to the hood during reactor sampling, or upon failure of reactor gasket. Monitoring of personnel exposure during sampling and upset (gasket failure) had been periodically conducted over the 10-year period, and continuously monitored via a carbon monoxide monitor situated outside the hood.

A new process development entailed generation of hydrogen sulfide for reactions also conducted in the presence of hydrogen. However the quantities of sulfur employed resulted in less of an approach to 8-hour time-weighted average (TWA) exposure limits (relative to ethylene oxide), or to short term exposure limits (STEL) and immediately dangerous to life and health (IDLH) limits relative to the amounts of carbon monoxide deployed in the previous development. Overall exposure risk was assessed as less than those previously managed, with implementation of personnel H_2S monitors. Management of change focused on potential new corrosion issues, and mitigation of nuisance odors given odor thresholds that were far lower than encountered in previous process chemistries.

A second new use for the bench scale unit entailed hydrotreating associated with biofuels process development. In this case, the array of oxygenated intermediates encounted presented no new hazards relative to solvents employed and safeguarded in the previous process development.

Example: Use of Multireactor with Scale-Down

The concept of ALARP was further implemented along with an increase in R&D speed, by adaptation of process screening experiments to an even smaller scale, batch multireactor (Parr-5000). The 75-ml batch reactors were charged with only 0.3 grams of catalyst and 20-ml of solvent, representing a scale down by a factor of 5 or more from the larger, single stirred reactor of the previous example. Use of stir-bar for mechanical agitation provided poorer gas-liquid transport relative to the draft-tube impeller of the previous unit. However, side-by-side comparison of conversion relative to a know catalyst, enabled rapid assessment of 3 different methods for catalyst preparation (Figure-1).

Example: Use of Pulse Micro-reactor for Rapid Screening and ALARP

The pulse microreactor entails injection of sub-microliter quantities of reactant for liquid phase reactants, or less than a milliliter STP of gas phase sample. In this case, the reactor is scaled to the size of the analysis, which represents full ALARP. Safeguarding procedures become those associated with analytical sample handling.

For one example in this category, novel H_2S conversion chemistry was examined using a 1000 ppm lecture bottle standard, and an externally heated catalyst bed fabricated from 6-inches of 1/8-inch diameter tubing. The screening program demonstrated success for the targeted conversion, with less than a month of staff time for equipment setup.

For a second example, liquid samples from a Parr-5000 reactor screening study were injected onto 0.1 milligrams of catalyst retained in a GC injector glass insert, to assess conversion and product formation potential. Ability to screen the second step reaction at microscale represented an enabling technology for the practice of ALARP using the multireactor for the initial step of the conversion.

Detailed kinetics are difficult to assess in the absence of first-order reaction with linear adsorption isotherm [7]. However, upon establishment of a tie-point with known kinetics and conversion in a larger-scale reactor, the method can be used to screen for alternate conditions or catalysts, taking full advantage of automation in sample injection.

Example: Scaled-Down Integrated Process Performance

A final example entails support to a commercial process startup, where poor startup performance required a rapid fix that ultimately led to selection of a new catalyst and reactor flow configuration [8,9. A pilot plant used for process development was available for restart. This would require heating a 55-gal drum of phenol, use of a 9-ft by 1-inch diameter flow reactor, a one-gallon crystallizer, construction of a new finishing stripper for evaluation of trace chemistry during product finishing, construction of a scaled-up unit for assessment of pressure drop using recirculation of hot phenol. Management of change and safeguarding for construction of new pilot plant operations would create delays from 3 – 9 months, and the resulting unit could only study one new process condition or chemistry at a time. Pilot plant operability was challenged via use of solvent, reactants and products which are solids at room temperature, such that extensive heat tracing and monitoring of relief headers was required to maintain and verify process flows and measurement accuracy.

The problem was resolved alternately at the bench scale via use of multi-throughput experimentation, whereby 100-gram samples of phenol were reacted via shaker flask, using an alternate low-volatility promoter. Crystallizations were conducted in parallel at the 10-grams scale using a multi-block heater. Product finishing chemistry was examined via heating in inert Teflon vials to assess reaction chemistry as a function of trace impurities transmitted from previous process steps. Flow scaleup was examined via use of modeling, use of mechanical loads, and substitution of water for phenol as working fluid.

These approaches enabled both ALARP, and fast resolution of commercial problems via use of multi-throughput or enhanced experimentation, including parallel processing of known chemistries to serve as tie-points in assessment.

Summary: ALARP and ASAP Tradeoff

The above examples show that safety ALARP and business ASAP can both be jointly optimized via responsible design and maintenance of laboratory capabilities. To make optimal use of previous construction and safeguarding, it is important to retain bench-scale laboratory equipment intact, for use in future analogous process studies. A known, debugged laboratory system represents ALARP relative to design, construction, debugging and streaming of a new laboratory unit, where idiosyncrasies of operation which can impact safety, are often discovered during new unit startup.

Scale-down enables multiple experiments to be conducted in parallel, faster, at smaller footprint and with reduced safety risk (ALARP). Critical scale-up issued may require a larger pilot unit for resolution if reliable engineering correlations are not available, but multiple options especially in chemistry, are best screened at the bench scale prior to demonstration at pilot scale, both to optimize ALARP and to execute ASAP.

References

1. Aylen Badilla Wargniez,Rita C. Oleas,Kenneth S. Yamaguchi, Improving laboratory safety through mini-scale experiments: A case study of New Jersey City University, *Journal of Chemical Health & Safety*, November/December 2012, pp. 12-23.
2. Maureen Rouhi, Minimalist Chemistry Through small-scale experiments, Asian educators increase access to, and safety of, hands-on lab experience, *Chemical and Engineering News, Volume 89 Issue 40,* pp. 41-42 Issue Date: October 3, 2011.
3. *Safety Considerations in Microscale Chemistry Laboratories, ed. by Ruth Hathaway, ACS Symposium Book, Div. of Chem Health and Safety, ACS.*
4. *The National Microscale Chemistry Center (NMCC, NMC2) currently housed at Merrimack College,* http://www.microscale.org
5. ALARP: http://en.wikipedia.org/wiki/ALARP#Origin_in_UK_law
6. Health and Safety at Work etc. Act 1974 (Commencement No.1) Order 1974, 1974/1439, art.2(a)/ Sch.1
7. W. A. Blanton, Jr., C. H. Byers, and R. P. Merrill, Quantitative Rate Coefficients from Pulsed Microcatalytic Reactor, *I&EC Fundamentals, 7,* 611-617 (1968).
8. J. B. Powell, *Use of Shortcut Methods in Process Development*, **AIChE Fall Annual Meeting,** Cincinnati, OH, Oct. 30 – Nov. 4 (2005).
9. J. B. Powell, *PROCESS DEVELOPMENT: IS SCALE-UP REALLY ABOUT SCALE?, 8th World Congress of Chemical Engineering, Montreal, Canada, August 23rd-27th, 2009.*

Figure 1: Parr-5000 reactor side-by-side comparison of catalyst activity vs. sulfiding method: 0.3 grams catalyst, 20 grams liquid, 750 bar H_2 in each of 4 reactors.

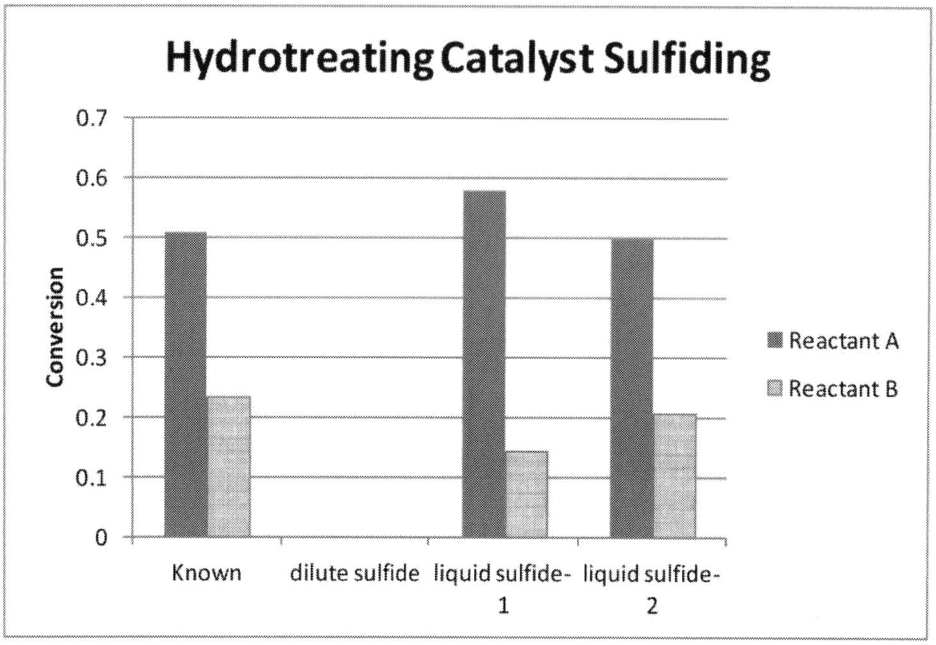

Process Safety Leading Indicators Industry Survey
February 2013
Center for Chemical Process Safety

Presenter
Shakeel Kadri
Air Products & Chemicals, Inc.

Author(s)

Sanjay	Choudhary	Tata Chemicals Ltd.	sgchoudhary@tatachemicals.com
Elroy	Christie	Honeywell	Elroy.Christie@Honeywell.com
Jeffrey	Fox	Dow Corning Corporation	jeff.fox@dowcorning.com
Eric	Freiburger	Praxair, Inc.	Eric_Freiburger@Praxair.com
R. Wayne	Garland	Eastman Chemical Co.	wgarland@eastman.com
Andrew	Goddard	Arkema	andrew.goddard@arkema.com
Shakeel H.	Kadri	Air Products & Chemicals, Inc.	KADRISH@airproducts.com
Darrell	Loewe	Formosa Plastics Corp., USA	DarrellL@ftpc.fpcusa.com
Steve	Marwitz	Formosa Plastics Corp., USA	SMarwitz@ftpc.fpcusa.com
Louisa A.	Nara	CCPS	louna@aiche.org
Tim	Overton	TOPS Consulting LLC	topsconsulting@hotmail.com
Jeff	Philiph	Monsanto	jeffrey.e.philiph@monsanto.com
Philip	Rasch	BASF	Philip.rasch@basf.com
Daniel E.	Sliva	CCPS	sliva@capital.net
Kenan	Stevick	Dow Chemical	KPStevick@dow.com

Prepared for Presentation at
American Institute of Chemical Engineers
2013 Spring Meeting
9th Global Congress on Process Safety
San Antonio, Texas
April 28 – May 1, 2013
UNPUBLISHED

Preface

The Center for Chemical Process Safety (CCPS®) was established in 1985 by the American Institute of Chemical Engineers (AIChE) for the express purpose of assisting industry in avoiding or mitigating catastrophic chemical accidents. In 2012, CCPS updated our mission to eliminate process safety incidents, in all industries. To accomplish this, more than 150 corporate members around the world have created and sustain a community committed to process safety. Together, we drive the activities of CCPS.

In 2006, the CCPS Technical Steering Committee authorized a project committee to develop a Guideline book for the development and use of Leading and Lagging Process Safety Metrics. That committee identified that a key breakthrough opportunity for industry was the establishment of an industry lagging metric that would become the benchmark across the chemical and petroleum industry for measuring process safety performance. To achieve this objective, representatives and members from each of the major chemical and petroleum trade associations as well as other key stakeholders were engaged.

The outcome of that effort was published in December 2007. Many companies and organizations have used those metric definitions since publication. Those definitions established in 2007 were a key input to the creation of a new ANSI/API standard (ANSI/API RP 754), which was finalized and released in April 2010. CCPS and several members of the original CCPS Metric committee were involved in the API standard committee.

In 2011, CCPS updated of the December 2007 publication with the intent to align CCPS and API documents. A copy of this stand-alone document, "*Process Safety Leading and Lagging Metrics.....You Don't Know What You Don't Measure*"[2] is available at the CCPS Website at http://www.aiche.org/sites/default/files/docs/embedded-pdf/CCPS_ProcessSafety2011_2-24-web.pdf. These documents, when used together create a solid foundation for the establishment and use of both Leading and Lagging indicators.

In 2013, CCPS continues its efforts for the development and use of Leading and Lagging Process Safety Metrics with this paper focused on the use of leading process safety metrics.

This paper provides an update on the chemical, petroleum and other process industries use, direction and effectiveness of leading indicators and to provide recommended leading indicators to help drive performance improvement and lead to the reduction in the number and severity of process safety incidents. As the use of leading indicators is in its early stages, it is anticipated that this topic will continue to evolve. Additional surveys will be conducted and updates published periodically.

Keywords: process safety metrics, leading indicators, lagging indicators, survey

Abstract
This paper provides an update on the Chemical and petroleum industries use, direction and effectiveness of leading indicators and provides recommended leading indicators to help drive performance in a common direction. As the use of leading indicators is in its early stages, it is anticipated that additional surveys and updates will be published on a periodic basis.

1. Executive Summary

The purpose of this white paper is to provide an update on the use, direction, and effectiveness of Process Safety leading indicators in the chemical, petroleum and other industries. Leading Indicators are intended to help drive performance improvement and lead to reduction in the number and severity of process safety incidents. The information presented in this white paper was collected through a survey of CCPS member companies. As the use of leading indicators is in its early stages, it is anticipated that this topic will continue to evolve. Additional surveys will be conducted and updates will be published periodically.

A high percentage of companies recognize the value in utilizing leading indicators to support management in focusing engagement and efforts of the engineering and operations personnel. Although the survey would indicate that the industry is still 'experimenting and discovering' as to which indicators provide the greatest value, three different focus areas or approaches were identified as most effective in improving performance. These are as follows:

- **Ensuring follow-up on actions across the spectrum of Process Safety Management Systems**
    - Audit Corrective Actions
    - PHA Actions
    - Completion of Safety Critical Equipment Inspections or Calibrations
    - MOC Actions
    - Unplanned Event Corrective or Preventive Actions
- **Leveraging Learning Experiences and Management of Deviations**
    - Process Safety Near Miss Reporting including fires
    - Challenges to Safety Systems in general and specifically calling out: Safety Instrumented Systems and Relief Device Activations
- **Ensuring Management Engagement**
    - Picking the most pertinent measures to your operation and getting them in front of leadership; including them in agendas of various operational reviews and ensuring action.

The survey indicates there is still work needed to assist companies to reach a better understanding of the definitions intended for certain leading indicators to harmonize the understanding and usage. Most of the companies responding indicated they roll their measures up into meaningful scorecards, prompting management action, and review with senior leadership, in some cases all the way up to board members. Most companies also publish the data on internal review reports, websites and in newsletters for communication and action throughout their company.

Needless to say, it is essential to have management involvement, mentoring (conversation with employees) and support in implementation. Leading indicators by their nature may tend to convey a negative connotation as a weakness in Management Systems but if considered as opportunities for improvement, they will start to drive improvements. As is the case with any company program, senior management support and commitment are essential for the implementation and sustainability of a successful metrics program.
Long term success in making process safety robust and reliable will involve management commitment to widen the scope of leading indicators and to actively share and learn with others in the industry.

2. Introduction: Project Purpose and Scope

As outlined in the CCPS *Process Safety Leading and Lagging Metrics....You Don't Improve What You Don't Measure © 2011*[2] *publication*, an essential element of any improvement program is the measure of existing and future performance. Therefore, to continuously improve upon process safety performance, it is essential that companies in the chemical and petroleum industries implement effective leading and lagging process safety metrics.

This whitepaper summarizes the respondents' survey input regarding:

- Which leading indicators companies are currently utilizing and are finding most effective in driving performance objectives
- Identification of barriers during implementation
- Strategies used to overcome barriers in the implementation

This white paper began with a project to perform a broad survey of chemical companies' use of leading metrics in order to determine:

- Commonalities of approach,
- Areas of potential difference
- Good practices worthy of consideration and
- Areas needing improvement

The anticipated result of this project is an ongoing update of member companies' journey on the effective use of Leading Indicators to drive process safety performance improvement.

2.1 Background

Chemical Process Safety Indicators are generally broken down into the following categories:

"Lagging" Metrics – a retrospective set of metrics that are based on incidents that meet the threshold of severity that should be reported as part of the industry-wide process safety metric.

"Challenges to Safety Systems, Near Miss" and other internal Lagging Metrics – the description of less severe incidents (i.e., below the threshold for inclusion in the industry lagging metric), or unsafe conditions which activated one or more layers of protection. Although these events are actual events (i.e., a "lagging" metric), they are generally considered to be a good indicator of conditions which could ultimately lead to a more severe incident.

"Leading" Metrics – a forward looking set of metrics which indicate the performance of the key work processes, operating discipline, or layers of protection that prevent incidents.

These three types of metrics can be considered as measurements at different levels of the "safety pyramid[1]" illustrated in Figure 1. Although Figure 1 is divided into four separate layers (Process safety incidents, other incidents, near miss, and Unsafe behaviors/Insufficient operating discipline), it is easier to describe metrics in terms of the three categories defined in Figure 1. It is strongly recommended that all companies incorporate each of these three types of metrics into their internal process safety management system.

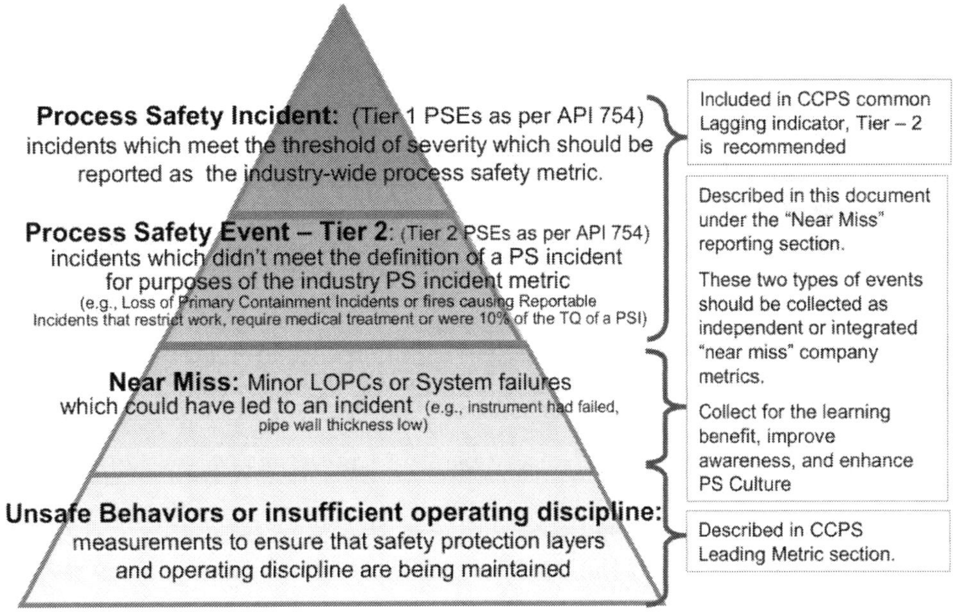

Figure 1

[1] From CCPS, "Process Safety Leading and Lagging Metrics …..You Don't Know What you Don't Measure" AICHE, New York, 2011

Lagging indicators for the industry are fairly well defined and efforts are underway for formally gaining a standardized global definition[1]. CCPS has received input from the industry that publications on Process Safety Metrics are incomplete regarding leading indicator recommendations in its Tier 3 (challenges to safety systems) and Tier 4 (Operating Discipline), which will present a challenge to implementation. Also, in some regions where formal Process Safety Metrics do not apply, there is a growing desire for recommended leading indicators to help drive performance in a common direction.

Tier 3 indicators are an actual event or discovery of a potentially unsafe situation; therefore these metrics fall within the continuum of "leading" and "lagging" metrics. A large number or increasing trend in such events could be viewed as an indicator of a higher potential for a more significant event; therefore, many companies use Near Miss metrics as a surrogate for a "Leading" metric. Tier 3 incidents by definition are a failure of our process safety management systems and give an excellent road map to where management systems need to be strengthened.

Tier 4 Leading indicators monitor the health of important aspects of the process safety management system. If measured and monitored, data collected for leading metrics can give early indication of deterioration in the effectiveness of key safety systems, and enable remedial action to be undertaken to restore the effectiveness of these key barriers, before any loss of containment event takes place.

3. Data Summary

The received survey results were tabulated and analyzed to determine the number of leading indicators used by companies. Of the 43 responding companies (95% of the responding companies), 41 used leading indicators, thus indicating broad use of leading indicators.

The use of leading indicators varies by company, from a low of three leading indicators to as many as 28 leading indicators for a company.

The chart shown in Figure 2 plots the number of leading indicators used by companies; with tier 3 leading indicators shown in red and tier 4 leading indicators shown in blue.

Each of the 25 leading indicators was used by one or more of the responding 43 companies. However 12 or more leading indicators were used by 20 or more of the 41 companies, or over 45% of the companies that responded to the survey.

The red box on the chart (Figure 2) highlights the 12 leading indicators used by the 20 or more companies.

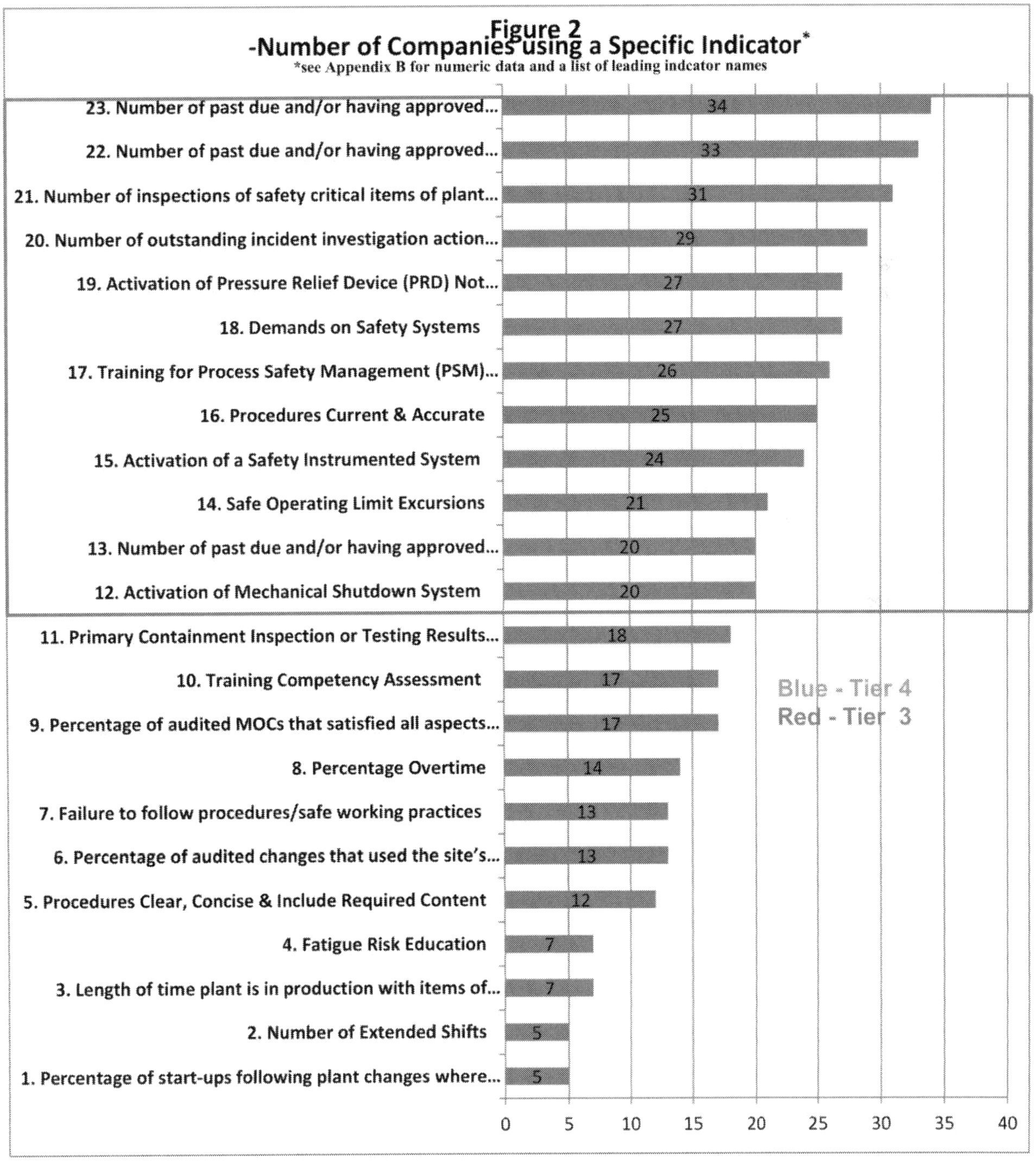
Figure 2
-Number of Companies using a Specific Indicator*
*see Appendix B for numeric data and a list of leading indcator names
23. Number of past due and/or having approved...
34
22. Number of past due and/or having approved...
33
21. Number of inspections of safety critical items of plant...
31
20. Number of outstanding incident investigation action...
29
19. Activation of Pressure Relief Device (PRD) Not...
27
18. Demands on Safety Systems
27
17. Training for Process Safety Management (PSM)...
26
16. Procedures Current & Accurate
25
15. Activation of a Safety Instrumented System
24
14. Safe Operating Limit Excursions
21
13. Number of past due and/or having approved...
20
12. Activation of Mechanical Shutdown System
20
11. Primary Containment Inspection or Testing Results...
18
10. Training Competency Assessment
17
9. Percentage of audited MOCs that satisfied all aspects...
17
8. Percentage Overtime
14
7. Failure to follow procedures/safe working practices
13
6. Percentage of audited changes that used the site's...
13
5. Procedures Clear, Concise & Include Required Content
12
4. Fatigue Risk Education
7
3. Length of time plant is in production with items of...
7
2. Number of Extended Shifts
5
1. Percentage of start-ups following plant changes where...
5
Blue - Tier 4
Red - Tier 3
0 5 10 15 20 25 30 35 40

3.1 Metrics Found Most Effective / how do You Make Them Visible?

We received input from 31 of the 43 companies in this area. Although the survey would indicate that the industry is still 'experimenting and discovering' as to which indicators make the most sense, three different focus areas or approaches were identified as most effective in improving performance. These are as follows:

3.1.1 Follow-up on Actions across the Spectrum of Process Safety Management Systems.

- Audit Corrective Actions
- PHA Actions
- Completion of Safety Critical Equipment Inspections or Calibrations
- MOC Actions
- Unplanned Event Corrective or Preventive Actions

3.1.2 Learning Experiences and Management of Deviations

- Process Safety Near Miss Reporting including fires
- Challenges to Safety Systems in general and specifically calling out Safety Instrumented Systems and Relief Device Activations

3.1.3 Management Engagement

- Picking the most pertinent measures to your operation and getting them in front of leadership, including them in agendas of various operational reviews and ensuring action.

4. Metric Selection

Two companies indicated that they had not started on Tier 3 and Tier 4 measures. However they had just implemented Tier 1 & 2 metrics and were seeing an immediate impact; i.e., improved performance. One company indicated they have seen strong improvements from monitoring "work permit auditing statistics". These points demonstrate the adage: "You don't improve what you don't measure."

Most of the companies responding indicated they roll their measures up into meaningful scorecards, prompting management action, and review with senior leadership, in some cases all the way up to board members. Most also publish the data on internal review reports, websites and in newsletters for communication and action throughout their company.

5. Where More Guidance is Needed

The specific survey question asked if further definition and guidance is needed on the definition for "challenges to safety systems."

As companies start to collect data on metrics, specific situations arise which demonstrate that the original definitions are not sufficient. Through dialogue between Member Companies, agreement can be reached on common definitions and conventions. Consistency is needed between companies and within companies to ensure that there is a common understanding of the meaning of "challenge to a safety system" so that tracking of progress on this metric will be on a consistent basis.

Twenty-five companies responded to this question and the responses were mixed on this topic.

- Fifteen of the responding companies felt that the existing guidance was sufficient. Eight of the companies indicated they had been using these metrics and were comfortable in their understanding. Three companies indicated that they were not yet using these metrics.
- Seven companies felt additional guidance would be useful to ensure that "challenges to safety systems" is used consistently within and between companies. The need for internal understanding and consistency between operators and engineers was also raised by one of the companies that indicated no additional industry guidance is required.
- Two companies included examples of what they would consider as challenges to safety systems in their responses.

The issue of definitions is also addressed more broadly in the section on Barriers to Implementation.

6. Barriers to Implementation

6.1 Introduction

Of the 43 responders to the CCPS PS KPI Survey, 35 either provided feedback on what types of issues/barriers they encountered in their PS KPI implementation activities or shared learning on what approaches worked best for them. As one would expect, there were a number of common issues, so what follows is a summarization of the survey commentaries for those common elements.

6.2 Commitment / Support

As is the case with any company program, senior management support and commitment is essential for the implementation and sustainability of a successful metrics program. Furthermore, it also makes sense to undertake an effort to align the metrics with the company business plan and culture. Too many metrics can result in information overload making it more difficult for executives to understand the information and how to apply it in selecting improvement opportunities.

6.3 Definitions

Another key barrier was the differences in understanding the metrics definitions. Comments would indicate there is just enough vague wording in metrics definitions to cause some companies inconsistencies in their application. Companies with a large number of facilities scattered around

North America had the added issue of reconciling the apparent differences in which individual facilities viewed the definitions. Global companies add yet another layer of complexity to definition application. The challenge of integrating acquisition facilities should not be underestimated.

6.4 Data Collection

Data collection systems often did not readily produce the information companies wanted to track; i.e., designed for injury tracking but not for Process Safety incidents. Developing good metrics often involves changing and standardizing systems so that the scope of what is and isn't included is consistent. Significant time and money are needed in the development of data collection systems to enable easier collection of information relating to leading indicators. Automated data output is also essential if company management is to utilize the information in a timely manner. It is essential to be clear as to why the data is being collected and how it will be used. If the need for measuring performance is not seen, it will not be done.

Communication to the workforce about the criteria for inclusion can take a great deal of time and effort. Training, open communication channels and recognizing that data consistency will not be perfect from the start but will improve over time are important messages that employees at all levels need to hear.

The result from the collected information needs to be presented in a format that those who are expected to act on it can readily understand the messages. Simplified charts and graphic representations as opposed to heavily detailed charts worked best. Likewise, accompanying interpretation statements which are concise and not overly detailed are recommended.

6.5 Resources

In general, the view was that the collection of data to track a metric takes a large amount of resources in order to report them in a timely manner. Those companies that already had some computerized data collection systems in place mentioned resource challenges but did so to a much lesser degree. Maintaining trained resources who understood the definitions and how to extract the data from the computer tracking system presented another challenge due to transfers, turnovers and retirements.

6.6 Reluctance to Implement

Starting a new program or modifying an existing safety program almost always has some resistance to change. Because of the potential performance aspects of tracking KPIs, a few companies indicated the presence of a lingering cultural heritage at some sites that may have discouraged reporting because of the connection to discipline. Progress has been made and will continue in this regard as management demonstrates a clear focus on system weaknesses rather than simply blaming human error. It takes a while to change culture. Take it slow, implement simple metrics and try not to implement too many metrics at once.

7. Metrics Discarded and New Metrics being Considered

The choice of which leading metrics are the most effective for an organization is expected to change over time for any number of good reasons, including diminishing value for a particular metric due to implementation of effective change.

7.1 Revising Metrics which become "Habit Strength"

Thirty-three of the companies responded to the question, "Do you switch metrics after performance improves or the Management System reaches "habit strength"".

Twelve of these companies responded yes and cited the following reasons for doing so:

- Due to substantial improvements, the earlier leading metrics no longer represent areas where focus or improvement efforts were needed. This was decided after periodic reviews of performance and prioritization efforts to identify the areas of higher need.
- Some companies have evolved their leading metrics to better focus on the issues. While staying close to the original issue, the emphasis may be changed to better align with an improved understanding of the underlying issue that was intended to be addressed.
- As performance improved on key metrics, these metrics are often moved to the background and others brought forward for additional emphasis. Even as this is done, the old metrics often are continued to ensure that progress is maintained.
- Some leading metrics are used to help drive the management system improvements – for instance early on there is often an emphasis on data collection in a given topic area. Over time data collection improves and the metric can be redirected at a related issue which is more indicative of the PSM performance.

Those companies who have not yet changed their leading metrics (as those 21 companies) typically felt they were too early in the process to make changes, rather were still gaining experience and assessing what the leading metrics were indicating.

7.2 Discarding Metrics

Twenty-eight companies responded to the question regarding discarded metrics. Twenty-two stated that they had not discarded any metrics at this time, but some metrics had been de-emphasized since tracking had started.

Six of the responding companies stated that they had changed their leading metrics and cited the following reasons for doing so:

- The initial metric was intended to be only useful over the short-term to drive certain initiatives and when sufficient momentum was attained, the metric was intended to be changed.

- The metric was modified to make it more useful, but the underlying intent remained much the same.
- Some metrics were found to have limited usefulness and were of much less value than the effort to collect the data.
- As metric performance shows full implementation or compliance, these offer the opportunity to sunset these metrics and implement others.
- A few companies stated they dropped a few leading metrics because they did not see value in continuing to monitor that particular issue/metric.

The following are metrics which the responding companies have chosen to discard as they were not felt to lead to improved performance:

- Fatigue Risk Management
- Number of MOCs completed
- Maintenance Costs
- Number of contractor audits performed
- Number of audit findings
- Ratio of safety critical emergency work orders to total work orders
- Average time to complete incident investigations
- Presence in PSM related committees

7.3. New Leading Indicators

Thirty-three companies responded to the question regarding which leading indicators are being considered for the future. Twenty-nine stated that they were considering new metrics.
Of the companies who are considering new leading indicators, some had very specific areas that they wanted to focus on and others were vague.

Specific examples included:

- Preventive Maintenance activities which reveal a deficiency
- Alarm management, including nuisance alarms and disabled alarms
- Senior leadership time in the field
- Measure the performance of Independent Protection Layers
- Reporting of Process Safety near misses
- PHA recommendation progress (number open, number closed, time to closure)
- Fatigue risk management

Examples of less specific leading indicators which are being considered include:

- Process Safety Culture
- Tier 3 and Tier 4 metrics
- Improve the quality of the measures currently used
- Monitor procedures [quality]
- Incident investigation quality

In summary, key metrics tend to be more dynamic for companies who have been collecting data longer, as they have a better understanding of what metrics are important to drive the changes they want to effect. Many of these companies continue to collect data on their earlier metrics, but often will move these to the background as new metrics are added to their portfolio.

Companies with less experience are still learning how to capture the data and assess if the data is providing meaningful information and, as such, they tend to keep their initial set of leading metrics.

8. Conclusions

A high percentage of companies recognize the value in utilizing Leading Indicators to assist management in focusing engagement and efforts of the engineering and operations personnel. Although the survey would indicate that industry is still 'experimenting and discovering' as to which indicators make the most sense, three different focus areas or approaches were identified as most effective in improving performance. These are as follows:

8.1. Follow-up on Actions across the Spectrum of Process Safety Management Systems

- Audit Corrective Actions
- PHA Actions
- Completion of Safety Critical Equipment Inspections or Calibrations
- MOC Actions
- Unplanned Event Corrective or Preventive Actions

8.2. Learning Experiences and Management of Deviations

- Process Safety Near Miss Reporting including fires
- Challenges to Safety Systems in general and specifically calling out Safety Instrumented Systems and Relief Device Activations

8.3 Management Engagement

- Picking the most pertinent measures to your operation and getting them in front of leadership, including them in agendas of various operational reviews and ensuring action.

The survey indicates there is still work needed to assist companies to reach a better understanding of the definitions intended for certain Leading Indicators to harmonize the understanding and usage. Most of the companies responding indicated they roll their measures up into meaningful scorecards, prompting management action, and review with senior leadership, in some cases all the way up to board members. Most also publish the data on internal review reports, websites and in newsletters for communication and action throughout their company.

Needless to say, it is essential to have management involvement, mentoring and conversation with employees and support in implementation. Leading indicators by their nature will tend to convey a negative connotation as weakness in Management Systems but if considered as opportunities for improvement they start to drive improvements. As is the case with any company program, senior management support and commitment are essential for the implementation and sustainability of a successful metrics program.

Long term success in making process safety robust and reliable will involve management commitment to widen the scope of leading indicators and to actively share and learn with others in the industry.

9. Future Steps

The following activities are scheduled to continue the focus on the development of leading indicators in the chemical industry:

- Paper Presentation in the 9th Global Congress on Process Safety in April 2013Publication of the White Paper in CCPS website for comments
- Communication to CCPS Member companies to adopt leading indicators and inform CCPS of the list of indicators adopted along with the definitions and examples.
- The members of this Project Team to share their own company experiences in use of leading indicators.
- The review to be included in the 2014 Global Congress on Process Safety.
- Biannual Survey from the CCPS members

10. References

[1] American Petroleum Institute, ANSI/API Recommended Practice 754, Process Safety Performance Indicators for the Refining and Petrochemical Industries, First Edition, Washington D.C., 2010.

[2] CCPS, "Process Safety Leading and Lagging Metrics," AICHE, New York, 2011, http://www.aiche.org/ccps/resources/overview/process-safety-metrics/recommended-process-safety-metrics

11. Appendices

APPENDIX A

Process Safety Metrics Leading Indicator Survey Data

Process Safety Metrics Leading Indicator Survey Data can be provided by contacting **ccps@aiche.org.**

Appendix B-Numeric Data/ Leading Indicators used in Fig.2

Leading indicators*	Number of Companies using Leading Indicators
1. Percentage of start-ups following plant changes where no safety problems related to the changes were encountered during re-commissioning or start-up	5
2. Number of Extended Shifts	5
3. Length of time plant is in production with items of safety critical plant or equipment in a failed state, as identified by inspection or as a result of breakdown	7
4. Fatigue Risk Education	7
5. Procedures Clear, Concise & Include Required Content	12
6. Percentage of audited changes that used the site's MOC procedure prior to making the change	13
7. Failure to follow procedures/safe working practices	13
8. Percentage Overtime	14
9. Percentage of audited MOCs that satisfied all aspects of the site's MOC procedure	17
10. Training Competency Assessment	17
11. Primary Containment Inspection or Testing Results Outside Acceptable Limits	18
11. Activation of Mechanical Shutdown System	20
13. Number of past due and/or having approved extension of regulatory issue	20
14. Safe Operating Limit Excursions	21
15. Activation of a Safety Instrumented System	24
16. Procedures Current & Accurate	25
17. Training for Process Safety Management (PSM) Critical Positions	26
18. Demands on Safety Systems	27
19. Activation of Pressure Relief Device (PRD) Not Counted as a Process Safety Incident (PSI) or Loss of Primary Containment (LOPC)	27
20. Number of outstanding incident investigation action items closed	29
21. Number of inspections of safety critical items of plant and equipment due during the measurement period and completed on time	31
22. Number of past due and/or having approved extension of PHA action items	33
23. Number of past due and/or having approved extension of audit action items	34

*** NOTE: Tier 3 indicators shown in red and italics; Tier 4 indicators shown in blue**

Appendix C– Process Safety Metrics Survey Form

Other Loss of Primary Containment (LOPC) Events and other comments

Process Metrics Survey https://chenected.wufoo.com/forms/process-metrics-survey/

Process Metrics Survey

A survey of Leading indicator Process Metrics

Company Name

Name

Title First Last Suffix

Phone Number

- -

####

Phone Number - If International

Company Email Address

Company Type - Select a Choice

Commodity and Special chemicals

Tier 3 Metrics - Challenges to Safety Systems / Process Safety Near Miss (PSNM)
Do you follow CCPS or API RP 754 definitions for Tier 3?

if so, check the "yes" box for the leading indicator and the check the boxes for the region or regions

Safe Operating Limit Excursions

☐ yes ☐ Company-wide ☐ Business-level ☐ Facility -level ☐ Regional

Primary Containment Inspection or Testing Results Outside Acceptable Limits

☐ yes ☐ Company-wide ☐ Business-level ☐ Facility -level ☐ Regional

Demands on Safety Systems

☐ yes ☐ Company-wide ☐ Business-level ☐ Facility -level ☐ Regional

Activation of a Safety Instrumented System

☐ yes ☐ Company-wide ☐ Business-level ☐ Facility -level ☐ Regional

Activation of Mechanical Shutdown System

☐ yes ☐ Company-wide ☐ Business-level ☐ Facility -level ☐ Regional

Activation of Pressure Relief Device (PRD) Not Counted as a Process Safety Incident (PSI) or Loss of Primary Containment (LOPC)

☐ yes ☐ Company-wide ☐ Business-level ☐ Facility -level ☐ Regional

2:39 PM

Process Metrics Survey https://chenected.wufoo.com/forms/process-metrics-survey/

Percentage of start-ups following plant changes where no safety problems related to the changes were encountered during re-commissioning or start-up

☐ yes ☐ Company-wide ☐ Business-level ☐ Facility -level ☐ Regional

Other Management of Change Leading Indicators and other comments

· Process safety training:

Training for Process Safety Management (PSM) Critical Positions

☐ yes ☐ Company-wide ☐ Business-level ☐ Facility -level ☐ Regional

Training Competency Assessment

☐ yes ☐ Company-wide ☐ Business-level ☐ Facility -level ☐ Regional

Failure to follow procedures/safe working practices

☐ yes ☐ Company-wide ☐ Business-level ☐ Facility -level ☐ Regional

Other Process Safety Training Leading Indicators and other comments

· Fatigue Risk Management

Fatigue Risk Education

☐ yes ☐ Company-wide ☐ Business-level ☐ Facility -level ☐ Regional

Percentage Overtime

☐ yes ☐ Company-wide ☐ Business-level ☐ Facility -level ☐ Regional

Number of Extended Shifts

☐ yes ☐ Company-wide ☐ Business-level ☐ Facility -level ☐ Regional

Other Fatigue Risk Management Leading Indicators and other comments

Section Break

· Operating & Maintenance Procedures

Procedures Current & Accurate

☐ yes ☐ Company-wide ☐ Business-level ☐ Facility -level ☐ Regional

Procedures Clear, Concise & Include Required Content

☐ yes ☐ Company-wide ☐ Business-level ☐ Facility -level ☐ Regional

Other Operating & Maintenance Procedures Leading Indicators and other comments

› PM

Do you switch metrics after performance improves or the Management System reaches "habit strength"
If so, please describe

How do you make these visible and how do you communicate?
Please describe

How do you characterize the level of risk at the site and how the level of maturity might affect the metrics?
Please describe

•Which metrics have not correlated to improved performance and have been discarded?
Please list and discuss

•Did you specifically consider other Leading Indicators from the recommended list of metrics for the 20 elements of Risk Based Process Safety (RBPS), If so please list

APPENIX D

Location of document, *"Process Safety Leading and Lagging Metrics.....You Don't Know What You Don't Measure"*

The document, CCPS, "*Process Safety Leading and Lagging Metrics .....You Don't Know What you Don't Measure*" AICHE, New York, 2011, can be found at this website location:http://www.aiche.org/sites/default/files/docs/embedded-pdf/CCPS_ProcessSafety2011_2-24-web.pdf

(19b) Role of Leading Indicators in Prevention of Major Accidents: CSB Learnings From BP Texas City and Deepwater Horizon

Kelly Wilson[1], Donald Holmstrom[2], Cheryl MacKenzie[2], Marybeth Mulcahy[1], Kara Kane[2], Jerad Denton[1], Chris Lyon[1], Bill Hoyle[2] and Steve Cutchen[3], (1)U.S. Chemical Safety and Hazard Investigation Board, Denver, CO, (2)U.S. Chemical Safety and Hazard Investigation Board, Washington, DC, (3)U.S. Chemical Safety and Hazard Investigation Board, Houston, TX

In July 2012 the Chemical Safety Board (CSB) held a two-day forum to discuss the use of process safety indicators both on and offshore. The CSB examined progress on previous recommendations related to process safety indicators as well as conclusions in our Deepwater Horizon investigation. Representatives from various high-hazard industries, the workforce, and national and international regulators spoke on how indicators shape safety performance in their company, industry or nation. From its investigative work and the forum discussions, the CSB has identified some common pitfalls of indicators programs as well as some "indicators" of a program's success. This paper will discuss those key aspects and explore the role of leading indicators for prevention of major accidents.

Process Safety Leading Indicators – A Perspective from Europe

Piet Knijff
DSM
Heerlen, Netherlands
piet.knijff@dsm.com

Lee Allford
European Process Safety Centre
Rugby, United Kingdom
lallford-epsc@icheme.org

Dr Peter Schmelzer
Bayer Healthcare AG
Leverkusen, Germany
peter.schmelzer@bayer.com

Prepared for Presentation at
American Institute of Chemical Engineers
2013 Spring Meeting
9th Global Congress on Process Safety
San Antonio, Texas
April 28 – May 1, 2013

UNPUBLISHED

Process Safety Leading Indicators – A perspective from Europe

Piet Knijff
DSM
Heerlen, Netherlands
piet.knijff@dsm.com

Lee Allford
European Process Safety Centre
Rugby, United Kingdom
lallford-epsc@icheme.org

Dr Peter Schmelzer
Bayer Healthcare AG
Leverkusen, Germany
peter.schmelzer@bayer.com

Keywords: process safety indicators

Abstract

In 2012 the European Process Safety Centre (EPSC) published *Making the case for leading indicators in process safety* which is a pamphlet on the selection, development and implementation of leading indicators for process safety. The document contains details on the type and spread of indicators which EPSC members have established within their own companies. In addition EPSC provided several practical tips for the do's and don'ts of successful implementation including presentation of data and engagement of staff. This paper is to develop these aspects further whilst exploring the relationship between leading indicators on a plant, business unit and corporate level. Additionally the interaction between leading and lagging indicators is to be highlighted alongside the management space which process safety indicators now inhabit with respect to monitoring and audit/review activities.

1. Introduction

For several years organisations involved in high hazards operations have reported their process safety performance in terms of lagging indicators or "after the fact" measures such as loss of containment. These measures are naturally important to the company, the industry and the wider public. Nevertheless major hazards businesses are increasingly dissatisfied with basing their future strategy entirely on lagging indicators. This has been likened to driving a car using the rear view mirror and can result in equally disastrous consequences. From a control perspective, feed forward as well as feedback control is desirable. In other words whilst learning lessons from past incidents, of both actual and potential impact, is vital, the major hazards sector would be unwise to wait for such events in order to initiate improvement actions. What are needed are indicators that proactively drive improvement. Leading indicators for process safety offer a way of examining the contributing factors to accidents and putting in place a system for monitoring these factors in a tangible process. In this manner early warning signals can be acted upon and the progress towards zero accidents can be assured in a meaningful way.

2. European Context

Following the environmental accident in Seveso in 1976, the European Union adopted the first Seveso Directive in 1982. Following several other major chemical accidents, including Bhopal and the Sandoz Warehouse fire, the original directive was amended and, finally, at the end of 1996, the so called Seveso II directive was adopted. Several further notable accidents then occurred (Toulouse, Baia Mare and Enschede) and the directive was further extended in 2003. Whereas the original Seveso directive focussed on technical measures for controlling hazards, the modified Seveso II Directive focuses more on the management systems needed to ensure the technical measures are in place.

In July 2012 the European Union adopted the Seveso III directive which includes explicit reference to safety performance indicators under those issues to be addressed by the safety management system of a Seveso operator. Member States have to transpose the Directive by 31 May 2015.

The developments just described show that over time Europe has developed a legislative environment for high hazard industries in which technical as well as management system aspects are being addressed. Collecting data on performance is part of this. The Major Accident Reporting System (MARS) was established to handle the information on 'major accidents' submitted by Member States of the European Union in the framework of the Seveso Directives. The criteria for a 'major accident' include damage to the environment, disruption of public life as well as impact on persons (a fatality or a higher number of injured persons).

The context of publicly being held accountable for a range of potential effects from the release of hazardous substances explains why European-based companies see process safety as bit broader than the potential for acute hazards such as fire, explosion or acute intoxication. The prevention of sudden releases with chronic health and environmental impacts is an integral part of process safety management in Europe. The reason for this stems primarily from the high density of companies working with markedly different chemicals and processes and their use of process safety indicators both leading and lagging reflects this broader focus on all types of releases aside from the classical acute process safety hazards. Ultimately the aim of process safety is to keep the materials in the pipe or vessel.

Since the inception of Seveso II EPSC has been closely involved in a number of pan European working groups involving Competent Authorities in an effort to understand how the Seveso directive is implemented and enforced within each Member State across the European Union. The ultimate aim of this work is to harmonise the effects of the Seveso directive whilst recognising that an individual Member State has the discretion to implement more rigorous obligations than that contained in the directive.

3. Process v Personal Safety

Process safety circumscribes those activities within a major hazard environment whose aim is to keep substances within the vessel or pipe. In other words good process safety prevents loss of containment. The consequences of getting process safety wrong include fires, explosions and toxic releases which can result in harm to workers, damage to the environment and interruption to business.

A common pitfall is to see process safety as identical to that of personal safety and use the latter as the sole indicator of workplace safety. Personal safety across industry has enjoyed remarkable improvement in the last two decades and its success is testament to the belief that "what gets measured gets managed". However bitter experience shows that many organisations who have suffered devastating major accidents have in fact reported excellent personal safety.

To the casual observer the benign neglect of process safety appears counter intuitive. This is because process safety incidents have the potential to impact many if not all workers whereas the effects of personal safety incidents are limited often to the injured party. One reason for the pre-occupation with personal safety is that it is readily measured, tracked and published within organisations and is capable of being managed and improved on the frontline as evidenced by safety observation programmes. Remedies to improve personal safety are relatively low cost and good performance in personal safety is often rewarded through bonus payments to senior staff and much of the workforce.

On the other hand a full understanding of how process safety is assured for a complex operation can be confounded by a series of independent safety barriers such that the path between hazard and consequence is obscured to all but a few experts and those with significant operational experience. The effectiveness of good process safety is therefore not readily visible. This hampers communication, on plant level as well as up to corporate level and may makes it more difficult to dedicate sufficient and effective means to ensure continued process safety. One of the aims of introducing metrics for process safety is therefore to enhance its visibility such that a process safety risk is as every bit apparent as that for a personal safety risk.

4. Process Safety: Lagging and Leading Indicators

The most obvious measure of process safety is to record losses of containment such as leaks and spills of hazardous substances which can lead to more catastrophic effects. These are described as lagging indicators. Although there is no doubt that lagging indicators are an authentic reflection of process safety they are in essence a reactive measure in that control has already been lost.

The major problem with simply relying on lagging indicator events to manage process safety is that they occur rarely and therefore by themselves possess little statistical relevance. Absence of incidents is not by itself an indicator of sufficient weight to confirm good process safety management. It therefore makes sense to examine those leading indicators which are predictors of loss of containment events. These precursor events are more numerous and lend themselves more readily to analysis and improvement. Instead of waiting for a loss of control to happen, major hazard organisations can in a pro-active sense manage those activities which have direct influence on loss of control events.

How can we be more proactive? It makes sense to move down the "Indicators pyramid" as depicted (Figure 1) and seek indicators that reflect activities which positively impact on process safety. These are described as leading indicators. Leading indicators give confidence and assurance to site and company leaders that process safety is not only under control but also subject to continuous improvement.

Figure 1[1] – Indicators Pyramid

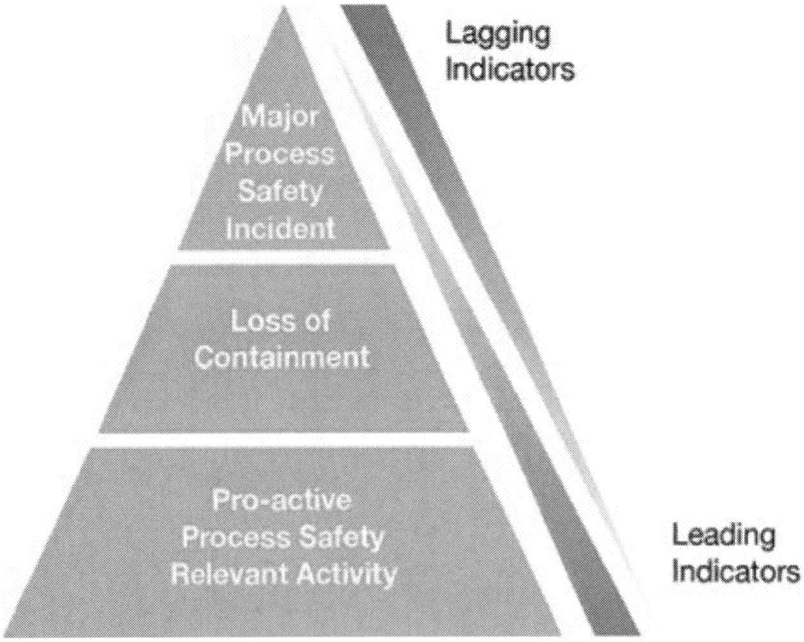

Ultimately the bottom line is satisfactory performance in process safety as measured by lagging indicators. Furthermore good performance in well chosen leading indicators feeds forward to good performance in lagging indicators. It is tempting to enter into rather dry debates in an effort to distinguish between leading and lagging indicators. However it is less important to argue whether indicators are leading or lagging; it is important to be proactive and the focus should be on what they impact, and how meaningful they are in steering improvement actions.

5. Leading indicators: EPSC member survey

In 2011 EPSC invited its members to share details on leading indicators that each has introduced for process safety. A general finding is that no company uses more than six indicators, and one company uses a single leading indicator as (Figure 2) indicates

5.1 Specific findings

The most commonly found indicators are

- Mechanical Integrity

For example the percentage of inspections executed according to schedule, or the percentage of inspections without non-conformities.

- Action Item Follow up

For example the percentage of actions completed by due-date. Differences are apparent in the sources of the action items included in the indicator. Some companies track actions from incident investigation, others include actions from additional sources (e.g. audits and inspections).

- Training/Competence

Often this indicator measures process safety training delivered. As an indicator this might be valid provided the training has process safety specific relevance and individual competence is assessed and documented (by test, demonstration, interview etc). Others use it as an indicator for how complete organizational roles in process safety are defined and assigned. Alternatively you could look at the number of incidents where (lack of) process safety training or competency played a role.

The table illustrates other indicators used that proved to be of value for specific companies in their specific situation. Remember that quality of indicators, not quantity, is more likely to deliver success.

Figure 2 – Type and Spread of Leading Indicators

	Company							
	A	B	C	D	E	F	G	Total
Leading Indicator:								
Mechanical Integrity	●	●				●	●	4
Action Item Follow Up	●		●	●		●		4
Management of Change	●							1
Training/Competence	●	●		●			●	4
Risk Assessment		●	●					2
Overriding/Bypassing		●				●		2
Operating Window Excursions		●				●		2
Activation or Failure of Protective Device		●			●			2
Number of Leak Boxes or Clamps						●		1
Operating Procedures (SOPs) / Critical Task Execution							●	1
Number of Leading Indicators	4	6	2	2	1	5	3	

6. Hazard Specific Indicators

Despite the fact that we saw that different organisations arrive at similar leading indicators, it is vital to review the risk profile of the process safety activities and pinpoint the specific vulnerabilities.

Selected indicators need to be meaningful and reflect the true risk. Only then will efforts to improve be successful. A starting point is to identify chemical hazards throughout a plant and then to map how the hazard can give rise to a process safety incident. A bowtie model such as the one illustrated (Figure 3) is then applied to identify the barriers or key process safety control and mitigation measures. It is important to select indicators that directly show how well the systems are working in practice.

Figure 3 – The Bow Tie (courtesy Shell International Exploration and Production)

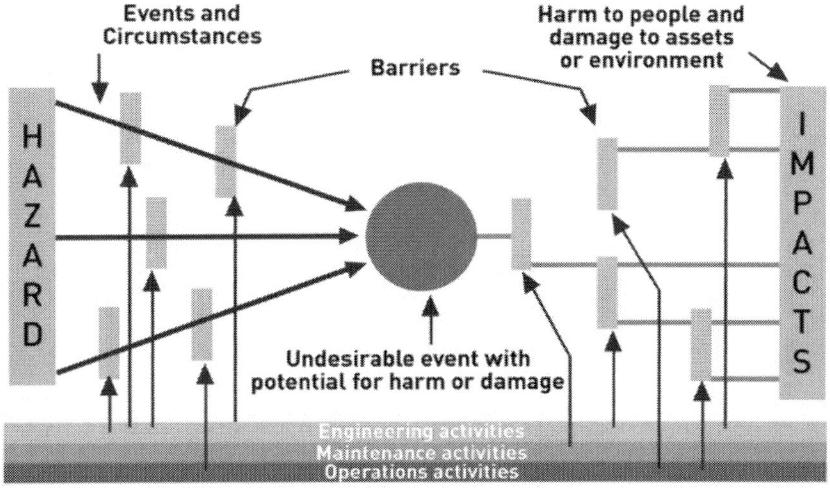

A typical example is the number of operating window excursions or the number of challenges to a specific safety provision in a given time period. This type of indicator is particularly useful when the process safety risks are concentrated in one specific activity or plant operation. When process safety risks are found throughout the operations, leading indicators typically are linked to measuring the functioning of critical safety systems or procedures, such as Management of Change, Risk Assessments, Permit to Work, Mechanical Integrity, etc.

For these generic safety barriers it is conceivable to consider indicators that measure completion of a programme (e.g. the % of inspections of safety critical plant equipment due for inspection and completed on time in a given period) or the quality of the activity result (e.g. the % of safety critical plant equipment that performs to specification when inspected or tested).

Data for certain indicators may be obtained by reviewing on a sample basis the execution of critical tasks. For example the % of work permits issued in which the hazards, risks and control measures were adequately specified. This offers the possibility to engage workers in peer reviews of such safety critical tasks.

An additional reality check can be done by examining actual process safety incidents. We should ask ourselves what caused the incident, and then which leading indicator would have signaled the need for timely intervention and hence could have prevented the incident. In this manner several of EPSC member companies try to turn results on the lagging indicator in to leading indicators. Investigating (smaller) releases or events may give a distribution of causes that then may be used to select leading indicators that steer improvements in these fields.

7. Embedding Indicators

Post introduction it is wise to allow the process for collection of indicators to establish and then observe emerging trends.

- Does the analysis offer any surprises or suggest intervention not only for a specific indicator but also when all indicators are examined as a whole?
- Are the indicators going in an undesirable direction?
- Does the analysis invite the use of additional or realignment of resources?

Be careful not to be too comforted by desirable movements in the indicators unless they can be substantiated by other sources (actual incidents, auditing, etc). Potential improvement opportunities can be identified by comparing and indeed calibrating leading indicators across different sites within the same business.

A necessary condition for introduction of indicators is that the business has a reasonable level of procedures and practices in place to help manage process safety. It is important to note that a system for leading indicators requires a proper level of resourcing in terms of data collection, recording, reporting and review. Data from existing systems should be used as much as possible as opposed to the implementation of new and costly systems.

Crucial for the acceptance may be the choice between a prescriptive approach and one that allows sites to choose their own relevant leading indicators. This choice need not mean a vast array of indicators that are used across an organisation. This remains a balancing act between selecting site specific indicators on the one hand, and more generic indicators used across sites and enabling comparison within the company. Members experience is that, when sites are given a choice in selecting indicators, each independently arrives at similar leading indicators. Still we need to keep in mind that comparability of leading indicators between plants is to be used with caution. Comparability is influenced by culture, a different background in the way of operations, heterogeneity of activities. This poses challenges to the comparability of leading indicators and the preference is to compare only where it really makes sense, otherwise the results should stay on local level.

It is worth noting that even mature multinational organisations operating with well established process safety practices and reasonable hazard awareness throughout its workforce will need to adopt an ethos of continuous improvement in order to keep leading indicators as fit for purpose. Expectations of success should be that of demonstrable progress rather than overnight success.

8. Leading Indicators Do's and Don'ts

EPSC has consulted with its members on the experience in introducing and using leading process safety indicators. They have shared the following insights.

1. Don't try to measure everything, start with a pilot. Gather experience in collecting the data, educating and involving the end-user and in demonstrating the added value in gaining improvement on the selected topics. Benefit from quick wins.

2. Whenever you use several leading indicators it makes sense to aim for a blend of leading indicators comprising both specific operational parameters and functioning of generic safety barriers.

3. As with any reporting, it will appear that performance is becoming worse before it improves. Prior to reporting you had an impression of the tip of the ice-berg. On reporting the submerged parts become visible. Allow time for sustainable improvement actions.

4. Leading indicators originate at plant level where the hazards are. They have a greater relevance for operating staff and lend themselves to a greater degree of involvement from the workforce.

5. Leading indicators are often expressed as a percentage or ratio and not an absolute value. They should be expressed positively (100% is desired instead of 0%, this is in contrast to lagging indicators). They should promote an informed discussion on where to invest resources (money, effort).

6. Only when there are good reasons to compare plants (within a site), or plants across sites, and the indicator is not expressed as a percentage, there is a need for shared indicators and normalization (have common denominator). For normalization often the total number of employee and contractor hours in the reporting period is used. It is not directly related to process safety hazards, but is a measure for the scale of operation.

7. 'Compliance Indicators': The use of leading indicators that demonstrate legal compliance is not recommended. Any outcome other than 100% legal compliance is unacceptable and therefore the indicator is not helpful for steering efforts.

8. Some companies may link performance as measured by indicators to a bonus. However, this method can be found to become ineffective and irrelevant over time due to an undue focus on the numbers themselves Under-reporting can be a negative outcome, and employees can begin to focus on the reward rather than a risk reduced workplace. It is better to try to engage staff rather than try to "buy" commitment.

9. Leading Indicators: How many?

In early 2012, EPSC supported Cefic, the European Chemical Council, in holding an international conference on process safety performance indicators. It is evident that discussion has moved on from questioning the value of leading indicators to discussion on best practice implementation. One of the issues raised is how many leading indicators is considered sufficient.

It would seem that the majority of high hazard businesses take the view that their sites need select and report no more than six leading indicators. (UK HSE speaks of the "rule of six"). Typical implementation consists of getting both the business and its sites to a place where they are capable of selecting meaningful leading indicators relevant to their operation and then reporting these on an ongoing basis. The collector of this regular data then needs to act on the content of such reports as well as initially tolerate differences across sites in their individual definition of indicators.

In contrast and at the same conference, a major energy provider, presented on a UK HSE endorsed approach which uses in the order of 100 leading indicators using standard definitions across each of its sites. The entire business has clear line of sight of process safety performance and from top to bottom of the organisation. This has necessarily involved considerable senior management commitment and workforce engagement as well as introduction of reporting software. Such standardisation is claimed to have facilitated the sharing of best practice across the business and there is no doubt that degree of standardisation of leading indicators is likely to prove a hot topic for future discussion.

10. Leading Indicators – the Management Context

One of the reasons for the interest in leading indicators for process safety is that they place an additional level of scrutiny to those conventional checks in high hazard operations contained in the typical safety management system. These would be monitoring, auditing and review activities as illustrated in (Figure 4). Monitoring activities in high hazards operations include continual checking on operating parameters such as temperatures, pressures, flow rates and fill levels to prescribed safe operating limits. Auditing activities are a less regular check on compliance and effectiveness. Review activities are a periodic check on whether the entire safety management system delivers on major accident prevention and reduces risk to as low as reasonably possible.

Leading indicators can be viewed as occupying a space between that of monitoring and auditing in terms of prescription and frequency and in themselves are informed and subject to these same activities. So for example the number of excursions from safe operating limits would be a credible leading indicator as would be the number of outstanding audit action items as would be the number of engineering changes stemming from a design review. Leading indicators need to evolve to reflect the major accident risks profile and should be subject to periodic audit and review.

Figure 4[2] – Hierarchy of Review, Audit and Monitoring Activities

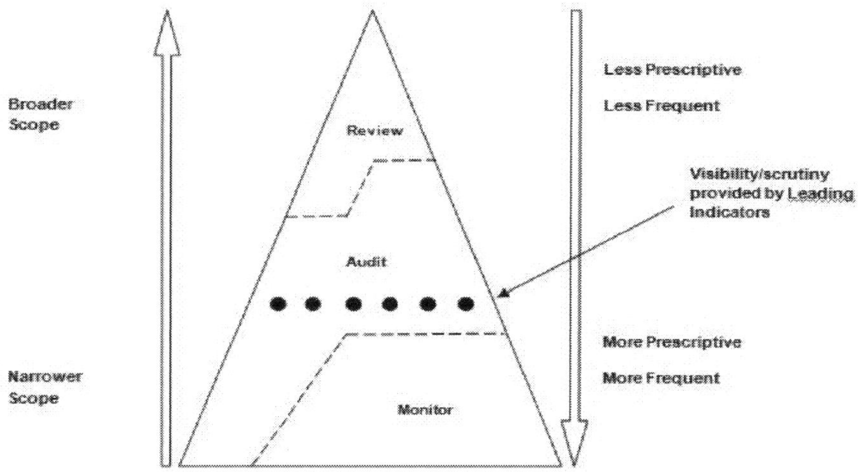

11. Leadership

One of the avowed benefits of leading indicators is to provide information on process safety performance at the right level of detail to corporate and site level management. At a senior level the ultimate ambition is that executives have access to process safety reporting in the same way that they have access to financial reporting. Both are necessary to operate a high hazards business in a risk informed fashion and can operate together to provide a balanced score card. Post implementation the attitude of senior managers towards leading indicators can be considered a critical success factor – some common pitfalls to be avoided at this level include

- Viewing leading indicators as offering the complete picture as far as process safety performance without cross referencing other aspects (incident investigation, audits, safety tours)
- Shooting the messenger and blaming those who collect and present the report for its contents
- Pouncing on "red risks" and calling for their virtual eradication overnight without putting in place the necessary planning and resources

12. Leading Indicators: Final thoughts

Process safety performance finds itself behind that of personal safety. Leading indicators for process safety offer a way of enhancing the visibility of major accident risks and fostering risk reduction. They are now generally considered a necessary tool for effective process safety management but they should not be seen as sufficient. The risk picture provided by leading indicators needs to be supplemented by other safety activities in order to offer a complete view on the state of process safety performance.

EPSC member companies value working together and sharing experiences in the domain of leading indicators. The pro-active steering of improvement actions that the use of leading indicators facilitates has its value mainly within companies. This leaves unchanged the need for reporting on lagging indicators to make our process safety performance transparent in the public domain. A global, harmonized approach is now aimed for by many organisations under the flag of ICCA, including EPSC and CCPS, will hopefully bring that transparency. In that way, we can focus on working with leading indicators, aimed at preventing incidents, which appeals more to our responsible corporate citizenship.

In Europe each member state will now be in the act of transposing the Seveso III directive into national legislation and will be considering ways as to how the clause contained in the directive on safety performance indicators is best enacted into their own law. These will necessarily be minimum requirements and will therefore create a threshold of acceptable practice for the use of leading indicators across business. Those Seveso industries which are pioneering the use of leading indicators will undoubtedly seek inspiration from high reliability domains such as nuclear and aviation sectors and import learning into their operations.

13. References

[1] Process safety performance indicators for the refining and petrochemical industries, API RP 754, 1st ed. April 2010

[2] Safety Assurance: Major Industrial Hazards Advisory Paper No. 1 Department of Infrastructure, Planning and Natural Resources, New South Wales

References not cited

[3] Developing process safety indicators, UK Health and Safety Executive HSG254, 2006

[4] Guidance on developing Safety Performance Indicators, OECD, 2nd ed. 2008

[5] Process Safety Leading and Lagging Metrics, CCPS, revised ed. January 2011

[6] Guidance on Process Safety Performance Indicators, Cefic, 2nd ed. May 2011

So We All have been Implementing Process Safety Metrics – What Next?

Abstract ID# 295006

Co-Authors:
Shakeel Kadri, Glen Peters, James VanOmmeren,
Kenneth Fegley, Martin Dennehy, Alvin Mateo
Air Products & Chemicals Inc.

Abstract

Industry over the last two to three years has put a lot of effort into implementing Process Safety Metrics. This is a good practice but, now that we have metrics data, how are we utilizing them to improve process safety performance? Air Products has been monitoring process safety performance for the last five years using combined Center for Chemical Process Safety (CCPS) and American Petroleum Institute (API) RP 754 metrics. This article will highlight the approach we have applied, both through our lagging and leading indicators, to drive our process safety performance improvement programs. It will review the programs we initiated, leading indicators we identified, and how it also helped to improve our process safety culture. This article will also look at the importance of measuring severity and its impact in driving leadership behaviors. We will also discuss future opportunities to improve process safety performance via metrics analysis and follow-up initiatives.

Key Words: Near miss, accidents, incidents, accident investigations, near miss reporting, leading indicator, continuous improvement, metrics.

I. Introduction

Air Products has been collecting and reviewing process safety events (PSE) for many years. In the last 5 years, we have adopted Process Safety Metrics and severity measures modeled after API [2] and CCPS [3] guidelines to measure our performance. In a previous AIChE paper we presented details of how we have adapted and modified those guidelines for our business needs and how we analyze the data and communicate the analysis results within the company [1]. We are now in the process of developing systems to implement corrective actions based on these data collection, measurement, and sorting processes to obtain meaningful process safety performance improvements. This article describes our journey to drive improvement in our process safety performance through the use of both lagging and leading process safety indicators to minimize significant process safety incidents and keep our colleagues and the communities we operate in safe. We will describe the programs we have initiated, the deep-drill analyses we have made, and explore future opportunities to improve our process safety performance by looking at new leading indicators and PSE system improvements. The ultimate goal is to have a robust program that is able to predict where potential major accidents may occur so they can be prevented rather than reacted to. There is still much work to be done in this area, but we have had some successes in reducing some of our reported process safety events.

There are many ways the process safety event (PSE) data that is being collected can be analyzed and sorted. This article will present as a series of Cases illustrating a few of the data sorts currently used by Air Products to analyze the PSE data we are collecting, a discussion of the methods we used to identify improvement options, and managements reaction to that data presentation.

II. Business Area Sorts

Case 1

In the beginning of Air Products' newly aligned process safety metric reporting program, we analyzed PSEs on global bases that were a compilation of all business areas. About six months into the effort, we started to categorize the data with a relative comparison among business areas. This additional data breakout method was established for two reasons:

- The wide variety of business areas that Air Products manages could "dilute" potential trends that might be occurring in one business across all areas.
- Including business area specific results in the quarterly and yearly global communication package would bring high visibility to business area leaders to emphasize trends where they needed to own performance improvement.

Several reporting periods after this new data characterization started, a certain business area trend (BA-1) was clearly highlighted as having the worst process safety performance across all businesses at the time as measured by PSE lagging indicators (see Figure 1). The chart represents all Tier 1, Tier 2, and Tier 3 events reported for one year. The performance results were also well above industry standard loss of primary containment event rates.

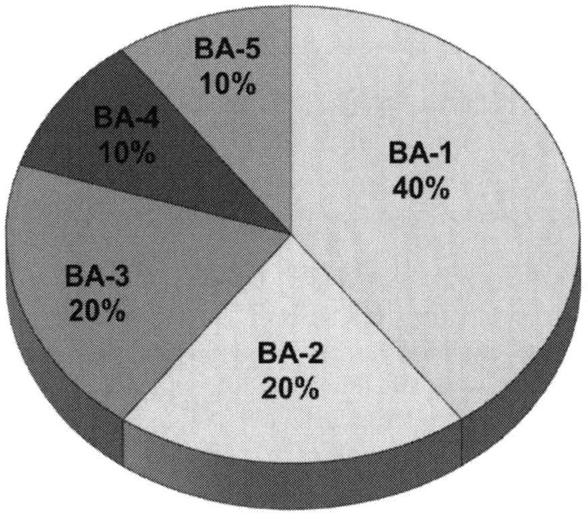

Figure 1 – All Process Safety Events Categorized by Business Area (BA) for One year

Further scrutiny of that business area's events indicated significant numbers of PSEs were related to several reoccurring themes. This enabled business area management to develop a dedicated action plan that included conducting employee specialized hazard awareness training to drive a culture change within the business. One specific process safety event theme was related to the several million pressurized gas cylinders that Air Products manages and that must be periodically retested. Most retesting is contracted to external companies and part of the retest procedure involves removing and reinstalling the cylinder's valve. A significant number of cylinders with loose valves were being returned from our contracted retest companies that were manifest either during incoming inspections or by valve ejections during cylinder filling. The data convinced business area management to develop and implement specific actions in the cultural improvement plan to address contract company performance. Contract companies were re-assessed to qualify as approved suppliers, new auditing programs were established, and contractors who did not qualify were terminated.

This effort was successful for two reasons: First by greatly improving contract company performance regarding returned cylinders and second by increasing the entire business area team's (i.e., managers down to technicians) process safety culture. Near miss reporting went up and PSEs involving cylinder valves decreased significantly. In fact, this top-down driven process safety cultural improvement effort resulted in the business area not experiencing any Tier 1 or Tier 2 process safety events for over one full year, indicating a tremendous turn-around in performance. The business area is now consistently operating at a lagging indicator performance level below industry benchmark and the business area's success was subsequently honored with an Air Products' global EH&S Award.

<u>*Case 2*</u>
We realized early on in the communication packages to management with our various sorting and reporting methods that implementation of performance improvement options was better initiated and accomplished if it came from a top down approach from management. Since we were doing a relative comparison, once the first business area no longer appeared in the Key Performance Indicator (KPI) data, another business area showed a significant contribution to our reported KPI events. The following pie chart (Figure 2) illustrates one yearly KPI report that got considerable attention from management in one of those business areas requiring ideas for process safety performance improvements.

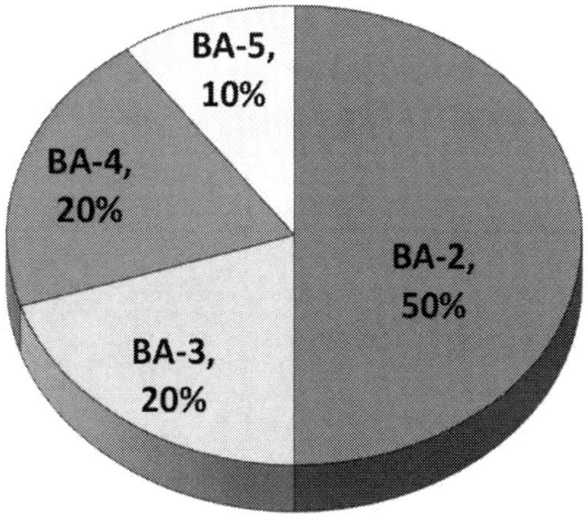

Figure 2 – Yearly KPI Data (Tier 1 and 2) by Business Area (BA)

The management team in BA-2 was very interested in determining whether there were any common themes within their plants causing this large contribution to the "piece of the pie". This particular business area has batch processes and further analysis of the data being reported from the plants indicated that the main contributor to the number of KPIs reported from all the plants was lack of confirmation of system readiness for restart after system cleaning between product changes and after maintenance, such as filter replacements. A continuous improvement (CI) event concentrating on "mistake proofing" and mapping the product and filter change processes was held at one of the plants to determine what could be implemented or changed in the plant work processes to reduce the occurrence of manual valves being left open or flanges not being fully tightened resulting in releases when the system was restarted. The review discovered that many of the events were occurring in one area of the facility where two different operating groups were held responsible for making the system ready for restart. The communication of this readiness was less than optimal. Changes were made to assign one of the operating groups the system readiness responsibility and avoid the "handoff" issues. Other procedural changes were also identified within the plant to confirm system readiness prior to restart to reduce the possibility of valves being left open or filter flanges not being fully tightened.

Implementation of the procedural items has resulted in a 52% reduction in the number of Tier 1 and Tier 2 loss of primary containment events at this particular plant in the past year. There is an understanding that procedural changes may only result in short-term improvements to the performance so engineering controls such as spring-loaded bleed valves, system pressure testing at low pressures before restart, and elimination of some unnecessary valves from the system are also being considered. We are also in the process of taking this "mistake proofing" review process to other plants in this business area for additional system-wide improvements in process safety performance.

III. Sorts by Process Safety Management (PSM) Elements

Another method we use to sort our PSE data is by the elements of Process Safety Management. We chose to use this method of sorting the data because we could easily relate each element of PSM to specific existing work processes and identify any area of weaknesses in those programs where corrective actions could improve our process safety performance. Figure 3 below illustrates the typical breakdown among the PSM elements we are experiencing in our PSE data. It should be noted that a single PSE could relate to multiple PSM elements in the analysis process.

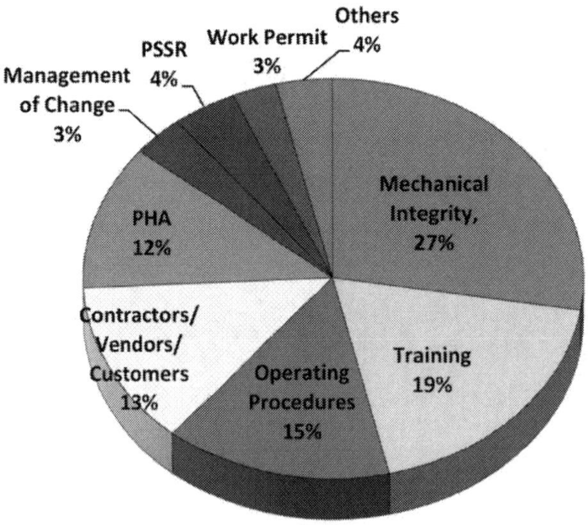

Figure 3 – PSE Data Sorted by PSM Elements

Case 3

This method of classification and sorting of each PSE has consistently identified in quarter-to-quarter and year-to-year comparisons the top three program elements influencing the occurrence of our PSEs to be Mechanical Integrity (MI), Training, and Operating Procedures. Another CI event with a cross functional team was held to identify and track implementation of improvement options for these three areas. To date, this CI event has concentrated on the MI piece of this data with a request to collect more information. There is a belief among the attendees that the MI issues could be related to plant age so a program was implemented to determine age of the equipment involved in the events. The team thought that the age data may indicate a "bath tub" type curve for MI related equipment failures where there may be a high number of events in the early, run in stages of the equipment and in the late, aged stages of the equipment so our improvement programs could concentrate on specific equipment in those age groups. This request was a difficult task to obtain meaningful information considering the lack of availability of age data on each piece of equipment in a plant. The results indicated a rather random failure distribution based on plant age.

Further review using different sorting criteria indicated a trend of failures associated with systems experiencing excessive vibration. The systems included MI issues in our transportation fleet of liquid nitrogen and oxygen containers while transferring the materials as well as in our fixed facilities. Over the road vibrations could be contributing to the failures we are experiencing in the tanker fleet. The data did not provide any information on whether the vibration was resulting from deficiencies in design or increased vibration from years of operation.

One other request that has come out of the study is to determine whether the PSEs in the MI area are associated with acquisitions or plants we have designed ourselves. To date, we are still in the data analysis phases of this study and have not identified any specific implementation tasks to reduce the number of PSEs we are experiencing in the areas of MI, operating procedures, and training.

IV. Severity Data Analysis

Another area of data presentation used to identify potential areas of concern in the company for corrective actions is the severity data reporting. Figures 4a and 4b below show the rolling average severity data experienced over different time periods. (NOTE: In the following two figures PSIR is the Process Safety Incident Rate and PSISR is the Process Safety Incident Severity Rate)

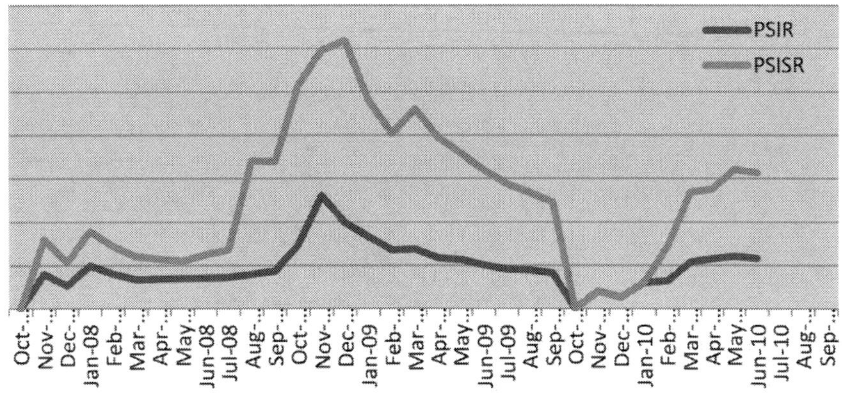

Figure 4a – Rolling Average Incident and Severity Rates for 3 Year Time Period

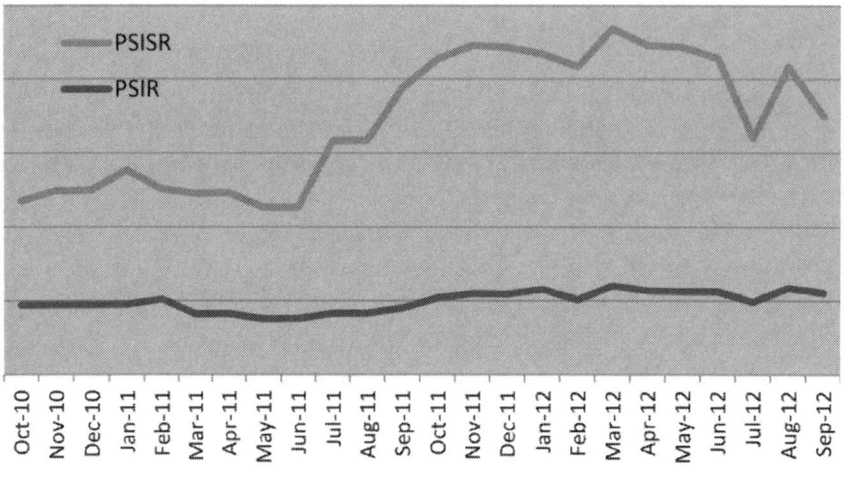

Figure 4b - Rolling Average Incident and Severity Rates for Remaining Years

Case 4

On these 2 charts, there was an obvious increase in the rolling average severity rate (PSISR) in two time periods, the August to December 2008 and August to November 2011 periods. Analysis of the PSE data indicated some significant plant fires contributing to those increases in our severity rates with some significant dollar losses resulting. A program we are calling a "Zero Fire Initiative" has been started with the intent of reducing our fire events in our facilities. We are presently in the data collection portion of that initiative to understand where and why fires are occurring in our facilities. Our first quarter of data collection indicates that a majority of our fires (66%) are non-process related such as electrical, transportation, and office related fires. If "zero" fires is really the objective, we will need to address these since the majority of our reported facility fires are not process safety related, however some of these incidents may have the potential to cascade into a process safety event. The chart below (Figure 5) shows the breakdown of the process safety related fires we are experiencing.

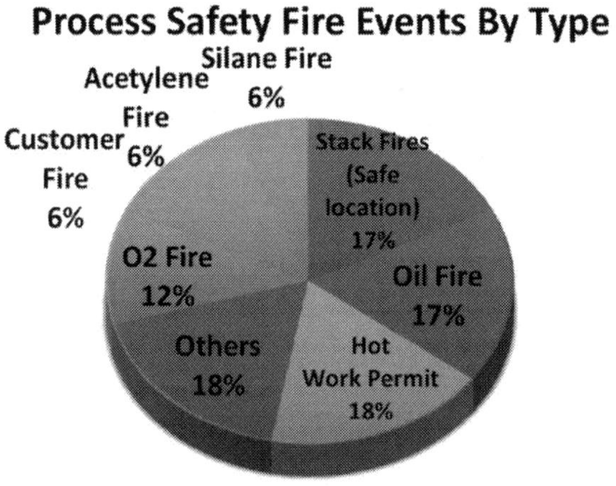

Figure 5 – First Quarter FY13 Process Safety Fires by Type

For our first three months of data collection on fires, a large percentage of process safety related fires are occurring in our machinery lube oil systems and are resulting from less than optimal application of our hot work permit procedures. Many of these fire events are very minor in consequences, such as insulation smoldering from hot work in the area, however, attention to these minor, leading indicator events will help prevent the occurrence of major fires in our plants. Since this data includes only 17 total fire events, additional data collection for this "Zero Fire" initiative is proceeding before any definitive improvement plans can be identified and implemented to reduce the number and severity of process safety related fires occurring in our facilities.

V. Reoccurring Unit Operations KPIs/PSEs in Monthly Reports

During development of our monthly management reports for the Process Safety KPIs, our review team typically notices any repeat KPIs or PSEs from the same unit operations. This occurred in three unit operations areas during the recent years: cryogenic pump fires, cryogenic tanker overfills, and loss of primary containment due to various root causes like overfill and corrosion in our vendor supplied sulfuric acid water treatment systems.

Case 5

During the period 2007 to early 2012, we experienced 5 fires in our cryogenic pump systems that were reported as Tier 1 or Tier 2 KPIs. Although not a significant number of events, the fact that we experienced any considering our protective system designs installed on these units from our previous years of operating experience and operating core competency raised concern. Why were we having any fires in systems using design standards for safety based on our numerous years of operating experience? The data indicated we were not consistently maintaining and operating the safety systems we have installed on these pumps for the prevention of such events. We implemented a retrofit program to clearly define the safety systems installed on the pumps for the prevention of fires to the operating teams so they would be aware of the importance of maintaining the safety equipment and repaired any of these safety systems that were no longer operational. This effort has resulted in zero reported cryogenic pump fires for almost two years.

Case 6

Improvement in one business area's process safety lagging indicators highlighted the next business area (BA-2) on which to focus attention (see Figure 6).

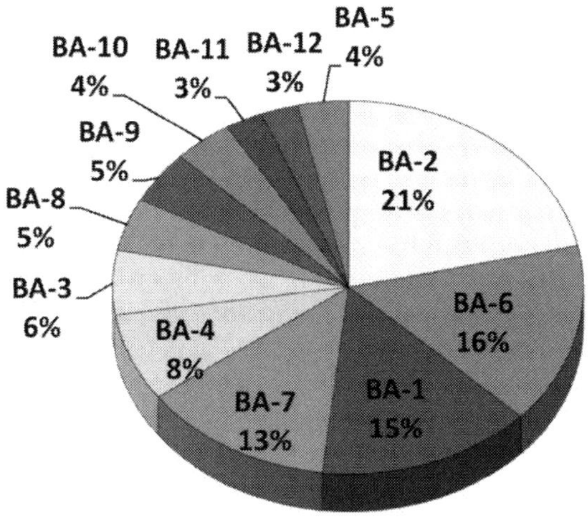

Figure 6 – Process Safety Events Categorization by Business Area for One Quarter

Further analysis of BA-2's specific process safety event types (i.e., hazardous conditions) indicated that a significant proportion of the loss of containment incidents involved overfilling of cryogenic trailers before shipment. This further evaluation prompted two observations for Air Products:

- The API RP 754 list of chemicals did not include the cryogenic fluids like liquid nitrogen, liquid oxygen, or liquid argon and, in our previous years of reporting, loss of containment of these materials did not appear as KPI events in our metrics.
- The automated cryogenic trailer filling systems required re-evaluation of design and improvement to reduce the frequency of future overfills.

To address the first issue, Air Products incorporated atmospheric cryogenic liquefied gases into our internal incident reporting system to customize the program to include these materials that we manufacture and transport in large quantities. Even though API did not include this class of chemicals, the materials may still pose hazards if released in sufficient quantity and represent an Air Products core business requiring proper metrics to achieve our global process safety continuous improvement goal. CCPS has incorporated atmospheric cryogenic liquefied gases into the reporting system and it is expected that API will incorporate them in the next revision. We also benchmarked other companies within the industrial gas industry and developed the threshold quantities on a consistent basis with our US counterparts.

Air Products fills our fleet of over 1,000 cryogenic trailers multiple times per year worldwide. Overfilling of cryogenic trailers can result in a cryogenic liquid release through the trailer relief device(s). We were already in the process of investigating the possibility of trailer overfills creating an overpressure demand on the trailers and identifying the required safety instrumented system (SIS) design. The reported number of overfill KPIs expedited our decision making process to quantitatively evaluate the current automated trailer filling system design, identify actionable items to improve performance meeting our global risk tolerance criteria, update engineering design standards accordingly, and retrofit improvements at existing high-risk facilities as deemed necessary.

Confirmation of success for these business area initiatives and actions is partly determined by evaluating future leading and lagging metrics. After Air Products implemented cryogenic trailer filling improvements throughout our global systems the lagging indicators for that business area confirmed that reported overfill events reduced by an average of 77% for the last year's data, thus meeting management's original improvement goal.

Case 7

Vendor supplied packages to our plants have been a design practice issue for some time now. There has been an effort to perform less and less review on these systems and get the "off the shelf", low cost designs installed in our plants. This is especially true where the vendor is supplying systems associated with unit operations that are outside the main part of the process, such as utility systems. Cooling water treatment systems are one such unit operation where this approach has been applied. In the last several years, we have experienced several KPIs and PSEs associated with overfills and corrosion induced equipment failures on vendor supplied sulfuric acid storage systems in our facilities. In the past, as with our recent incidents, releases from these systems have always remained within secondary containment and did not receive much attention from management. Since we are now reporting these as loss of primary containment events and KPIs in the new metrics system, management awareness of and attention to these incidents has increased. We performed a hazard review on a "generic" vendor supplied sulfuric acid storage and water treatment package to identify required safety systems that must be installed to reduce the possibility of overfills (e.g., high level alarms and/or trips), reduce the amount released in the event of an overfill (e.g., local emergency push buttons to stop the fill process), and define the accepted materials of construction. A design standard has been developed to ensure all our new facilities have these safety systems and proper materials installed. We are in the process of reviewing the need to retrofit these systems into our existing plants. Plans are in place to improve our MI inspections on these systems also.

VI. Specific High Risk Plants Experiencing a Large Number of KPIs

During the development of our monthly management, quarterly trend, and yearly summary management reports, we have identified some of our larger, high risk processing plants as having a significant number of PSEs. This resulted in an effort to deep drill the PSE data in three of our plants for root cause themes.

Case 8
One of the facilities that underwent a rebuild as a result of a previous major fire was one of those plants experiencing a large number of minor releases (PSEs). Prior to the major fire, a significant number of small releases from flanges and other plant equipment were occurring and not reported into our incident tracking database. Following the fire, there was an apparent heightened awareness for incident reporting. The incident investigation team for the major fire noted lack of reporting as a contributing factor to the incident so it was very important to review the reports that were now being entered into the system to "reward" the new culture of improved reporting. A review of the PSE data indicated many of the issues were occurring during the startup phases of the rebuild project. An increased level of equipment inspection to confirm system integrity before introduction of process fluids and increased heat tracing to prevent condensation was implemented resulting in a significant reduction in the number of leaks and instrumented trips from the equipment.

Case 9
Another plant that experienced a major fire was also experiencing a noticeable increase in the amount of PSEs in our incident reporting data. This also may have been attributed to an increased awareness to report based on the learning from the fire incident. A review of the data indicated a significant amount of releases and fires from pumps handling flammable materials were occurring. The vendor supplied package in the one area of the plant experiencing a number of these events had single packing gland (single seal) pumps. During the hazard review, it was recommended to provide this system design with double mechanical seal pumps and a project scope change was approved to accept the single sealed pumps because the loss prevention risk was not significant for this facility. Further review of the current plant design indicated a greater use of this system within the plant. The pumps are being replaced with the double mechanical seals as originally recommended.

Case 10
Based on the reviews from the two plants above, we decided to "deep drill" the incident data from one of our largest plants. That data review discovered a significant number of rupture disk openings in our electrolytic cell reactors. Having another set of eyes separate from the plant reviewing this data provided some value to the analysis. The plant was beginning to "accept" these rupture disk openings as a normal part plant operation (i.e., "normalization of deviation"), although reporting them as PSEs. Previous work on the cause of these incidents did not result in a reduction in the number of cases. An effort to collect additional operating conditions of the reactors is being implemented to determine if the cause(s) of these rupture disk openings can be identified. Previous historical operating condition data collection was not rapid enough to provide any meaningful data. Advances in monitoring instrument design has allowed a shorter

time period between recording of operating conditions so rapid condition changes may be able to be recorded and studied.

VII. Design Near Miss (DNM) Leading Indicator

Air Products discussed the origin and benefits of a design near miss leading indicator metric in a prior AIChE paper [1]. This initiative has proven worthwhile for various reasons, but none more important than the troubling theme we identified from three years of DNM leading indicator data collection. The data indicated a large portion of DNMs were related to pressure relief device engineering/design issues (see Figure 6).

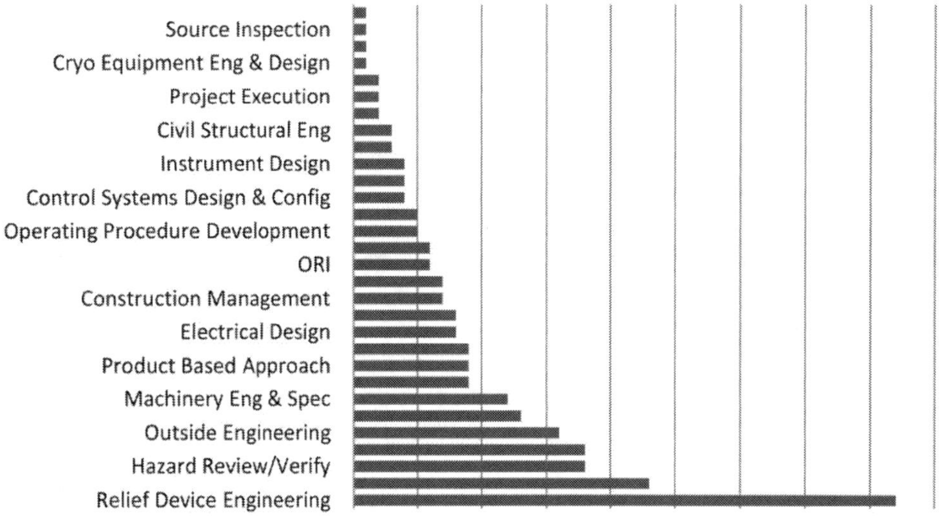

Figure 6 – Design Near Miss Categories for Three Years of Data

Case 11

The importance of proper pressure relief system design and functionality before turning over newly built plants to an operating group cannot be overemphasized. Therefore, this DNM theme gained upper management visibility and several initiatives were developed to scrutinize and improve engineering work processes that were experiencing systemic problems. Air Products is still working on this DNM theme to enact all improvements but there has been focused improvement effort on relief device calculation tools, documentation hand-offs between engineering functions, and improved execution with external engineering contractors. We expect that the improvements being made due to the DNM themes analysis will significantly reduce pressure relief device engineering errors as the new engineering work processes and training are implemented worldwide.

VIII. PSE Metrics Challenges and future opportunities toward predictive techniques

Air Products is still in the early stages of developing work practices and processes to use our KPI and PSE data to identify process safety performance improvement opportunities. There is still much work to be done in this area. Challenges within the Air Products' PSE data analysis programs are presented below so other companies can assess their internal program's effectiveness or understand if similar challenges exist in their organizational structure.

Setting Process Safety Metrics Targets or Goals for Both Lagging and Leading Indicators

You cannot improve what you do not measure. In some cases you will not improve in the right areas if you do not set targets or goals to achieve so you can measure for success or identify areas for improvements if you are not looking at measurable goals or targets. One method currently being used by Air Products to identify where we should concentrate our improvement efforts is by doing a relative comparison among business area performance. This method of reporting our data definitely received meaningful management attention, but such relative comparisons among business areas may not be the best method to identify where to concentrate limited resources on improvement programs. It may be necessary to customize the process safety metrics so that each business area has a unique target metric. Sometimes business areas can be too general or too broad to have meaningful actionable items and the metric may have to be evaluated at the site level. Currently Air Products' leading indicators are globally communicated which cannot provide insight to the underlying factors or weaknesses within the system at the business area level, regional level, or site level.

Process safety performance objectives provide personal accountability from the executive level to technician level. They are a reminder that each employee has a stake in process safety. It raises awareness throughout the company to be safe. However, having a global reduction goal does not always relate to each employee. The company goal should be relevant to each individual and employees should have a plan to achieve that goal, whether it is completing process safety training, identifying process hazards in the work place, or tied to a specific job of conducting critical maintenance inspections on time. This holds each employee responsible for the process safety continuous improvement goals and targets of the company.

Addressing Major vs. Minor Events

One of the challenges we face with near miss reporting and focusing on events at the Tier 3 and Tier 4 levels is that these events can be viewed as minor by the reporter and the facility. Even some of the KPI events can be regarded by the plants as minor risks especially if the materials were contained within secondary containment designed for the releases and flammable materials were released well below their flash point temperatures. Tier 3 and Tier 4 events do not receive the same investigative focus during the incident reporting process as a Tier 1 event at the plant level. Most times, investigations do not properly uncover the root cause of the Tier 3 or 4 events. This poses a challenge to the PSE analyst to provide meaningful improvement options. Our adopted KPI definitions and severity ratings may need to be revised to more closely match our business requirements and needs. Having the plants address identifiable trends at the site level within the reporting system has also been a challenge.

Improving the Incident Reporting Tool and Information from the Field

At Air Products, we have been reporting incidents into our global electronic system since the mid-1990s before the API and CCPS Process Safety Metrics guidelines were issued. Our challenge was to adapt that existing reporting system to the changing Process Safety Metrics from the API and CCPS. Another challenge is the collection of data from our reporting tool. When new information is required to establish a baseline, such as plant age, the means of collecting the data can be challenging. From a global initiative, the baseline can take months or even years to implement a meaningful data collection program. The reporting tool may not have the ability to track new information. It may be necessary to adopt other systems and gather the data. The lack of sufficient information or information entered incorrectly still means that each entry needs to be quickly screened to validate the classification automatically assigned by the tool.

Define Optimal Data Sorts (What information do we want to Use to "Slice and Dice" the Data)

There are many ways to analyze, sort and review the PSE information being developed. Since we are in the early stages of analyzing our Process Safety Metrics information, many levels of management have been requesting different sorts on information that may not be readily available from the incident reporting tool currently in use. This requires the analyst to find applicable data in other databases and then create the information. Our challenge is to define how we would like to analyze and present the data so a single incident reporting system can provide that information is an easily input, useable, and meaningful format for the analyses we would like to perform.

Identification of Meaningful Leading Indicators

Currently one of our major challenges is the identification of meaningful leading indicators that can predict when a major accident may be imminent to prevent it from occurring. To date, our Tier 3 PSEs and engineering design near misses have been our best source of this information, but it would be valuable if we could identify other metrics to support this effort.

IX. Conclusions

Although it may not be possible at this time to directly correlate that our process safety metrics program is responsible for avoiding major incidents, we still believe that such a program has significant benefits in reducing the number of incidents and improved performance through early recognition of incident trends and systemic issues. Several of the examples we have presented here have resulted in measurable reductions in the number of KPIs and PSEs we are experiencing. In the long term, we anticipate that the process safety performance improvement efforts we are implementing today will improve our process safety performance in the future. It should also be recognized that, as companies become more tune to reporting and measuring, there will be a spike in the number of KPIs, especially Tier 2 events that may not be getting recognized.

In order to drive continuous improvement in the process safety measures, a company must have very strong process safety culture, top management support, and adequate resources (both personnel and data collection systems) to efficiently identify and capture data when looking for new leading indicators that are not very obvious in the process. Identification of some new process safety related leading indicators that can provide a predictive rather than reactive measure to these process safety metrics will also be required to drive a significant improvement in our process safety performance. As we have learned, a more focused scope on an area or individual plant may need to be identified prior to implementing process safety improvement programs.

Our industry now has some defined process safety metrics to measure where we are. We now need to take this to the next step and use those measures to quantifiably improve our performance. Although much still needs to be done, in the long term, we anticipate that the process safety performance improvement efforts we are implementing today will improve Air Products' process safety performance in the future.

CCPS has recently completed their review of the Process Safety Leading Indicators Survey [4]. Such industry efforts will significantly enhance our abilities to identify best practices and help standardize key process safety leading indicators that will ultimately help improve the industry's process safety performance.

X. References

[1] Kadri, S., Peters, G., VanOmmeren, J., "One Company's Near Miss Program – Successes, Learning, and Improvements," Global Congress on Process Safety, Paper 78c, April 2012.

[2] Process Safety Performance Indicators for the Refining and Petrochemical Industries, American Petroleum Institute (API) Recommended Practice 754, 1st Ed., April 2010.

[3] Guidelines for Process Safety Metrics, Center for Chemical Process Safety (CCPS), November 2009.

[4] White Paper on Process Safety Leading Indicators Survey, Center for Chemical Process Safety (CCPS), February 2013.

Process Safety Leading Indicators: A "Silver Bullet" or "Bean Counting" for Enhanced Risk Management

David E. Cummings
E.I. DuPont de Nemours & Co., Inc.
1007 Market Street, Wilmington DE 19898
david.e.cummings-1@dupont.com

Prepared for Presentation at
American Institute of Chemical Engineers
2013 Spring Meeting
9th Global Congress on Process Safety
San Antonio, Texas
April 28 – May 1, 2013

UNPUBLISHED

Process Safety Leading Indicators: A "Silver Bullet" or "Bean Counting" for Enhanced Risk Management

David E. Cummings
E.I. DuPont de Nemours & Co., Inc.
1007 Market Street, Wilmington DE 19898
david.e.cummings-1@dupont.com

Keywords: leading indicators, PSM metrics, lagging indicators, ANSI/API RP 754

Abstract

DuPont adopted a broad set of PSM metrics initially in 2007 to measure and track site and unit level status and performance in a consistent and standardized manner across all global operating locations. These performance measurement processes and leading indicators have been refined and expanded over time to include additional criteria and focus areas both within each site and at the corporate level with the explicit goal fulfilling internal policies and commitments, reducing incident frequencies and strengthening process safety management systems. This presentation will review the current site and corporate metrics, specific challenges and limitations with the use of metrics, and suggestions in order to apply and enhance the broader use of leading indicators for improved and effective risk management outcomes.

1. Introduction

The subject of PSM lagging and leading indicators has expanded greatly over the past several years with the publication of multiple industry standards and guidelines, including ANSI/API Recommended Practice RP 754 (2010)[1], CCPS Process Safety Leading and Lagging Metrics (Rev 2011)[2], and the UK Health and Safety Executive HSG 254 Developing Process Safety Indicators (2006)[3]. The term "lagging indicators" in these publications is generally aligned to include a focus on process safety incidents which meet specific severity criteria or "loss of primary containment" hazardous material thresholds. While the definitions of leading indicators may vary more between each of the these standards, the common theme for these types of performance metrics focuses on 1) systemic data reporting and analysis, 2) measuring the degree of implementation of or conformance with policies and procedures that support a PSM management system, and 3) applying a proactive approach to utilize the relevant metric before an incident or negative event occurs. However, in all cases, the specific selection of leading indicators is left up to the discretion of each organization based on internal systems and requirements, and unique needs within the business.

These guidelines do not provide detailed content on how to use or apply the lagging and leading indicators in a manner that achieves maximum benefits for each user at the shop floor and site level, within the business, and across the enterprise. In other words, they define the basic objectives of performance indicators, how and what to consider and select in terms of data (criteria), and provide numerous examples as related to process safety management program variables, however, information on the application of the metrics in order to achieve the desired effects and results is left up to each operating organization.

One could inquire, however, that if an organization embraces these standards and adopts a robust set of Tier 1, 2, 3 and 4 indicators as described in ANSI/API RP 754, will that result in improved process safety risk management outcomes? In other words, is the following simple relationship a valid expectation and correlation for these measures of process safety performance?

The "Perfect" Set of PSM Metrics = Stellar PSM Results and Incident Prevention

This paper will explore this general theme as well as additional issues and suggested practices associated with process safety metrics and leading indicators.

2. DuPont's Process Safety Management Program

E.I. du Pont de Nemours & Co. is a global science-based products and services company with over 200 manufacturing units across 70 countries involving nine business segments. It is engaged in technology solutions and innovation across a range of disciplines, including agriculture and industrial biotechnology, chemistry, biology, materials science and manufacturing, offering a wide range of products and services for markets including agriculture and food, building and construction, electronics and communications, general industrial, and transportation.

Manufacturing technologies and site operations involves a myriad of process hazards, including flammable liquids and gases, highly toxic materials, reactives, combustible dusts, corrosives, asphyxiants, and thermal/mechanical energies. DuPont's philosophy and current practice is to apply process safety management systems and requirements to all manufacturing operations and locations using one global set of internal elements and technical requirements as depicted in the model in Figure 1 below.

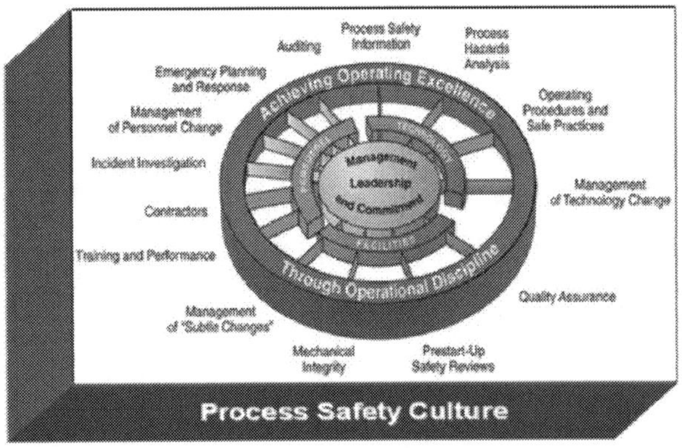

- **14 Technical elements**
- **Applicable to all global operations**
- **Basis for internal auditing requirements**
- **Leadership responsibilities**
- **Operational Discipline and safety culture**
- **Sister plant technology leveraging**
- **Leading and Lagging indicators**

Figure 1: DuPont Process Safety Management Model

The diversity and complexity of DuPont's global operations and technologies requires a consistent and detailed set of process safety metrics in order to measure, report, and ensure conformance with internal policies both at the local site level and across the corporation. DuPont has adopted a systematic set of site, regional and corporate process safety metrics and internal "rituals" which includes lagging indicators (incidents) and a tiered set of leading indicators. The system defines what data elements are required for measurement and reporting, but also routine generic practices to assess and take action on the results. The primary focus of this paper will be on the application of leading indicators as described in Table 1.

Table 1: Summary of DuPont Current Lagging and Leading Indicators

Type of Performance Metric	Elements
Lagging Indicators – reporting and collection of all process safety incidents	• Site level summary and trend analysis (annual) • Corporate level trend analysis (annual analysis at macro level and long term trends)
Leading Indicators – performance against internal standards, systems, and objectives	• Site/Unit Monthly PSM metrics (16) Site level near miss indicators Site level customized KPI's Annual PSM Program Evaluation • Corporate Quarterly PSM metrics (9) and dashboard 2nd Party PSM audit results (scores and trends) Regional 3rd party PSM audit program review (annual) PSM annual goals/critical operating tasks (annual)

3. Site Based PSM Metrics

DuPont developed and implemented an internal policy commencing in 2007 involving the tracking and reporting of a multi-tiered set of PSM metrics at the site, region, and corporate level. These metrics have been modified several times in order to provide perspectives on meeting our global PSM commitments and goals, both at a site and enterprise level. The following sections will briefly describe existing metrics as they are applied at the site level.

3.1 Site PSM Leading Indicators (API Tier 4 Data Elements)

Each global manufacturing site and unit locally tracks and reports the set of sixteen (16) metrics listed in Table 2 on a monthly basis. Additionally, all proposed extensions to any of these metrics also require site management approval, reporting, and tracking. In order to ensure standardization and consistency in site metrics tracking and reporting process, the following items have been developed and implemented as part of the corporate standard:

1) One standardized set of leading indicator definitions for each variable were developed to help ensure there is a clear understanding and standard means to gather and report metrics for accuracy and completeness at each site. For example, specific definitions are important to define what PSM related training is to be tracked for operating technicians and mechanics, when is a scheduled cyclic process hazards analysis considered "overdue", and what are the "PSM critical equipment" components to be included in mechanical integrity tests and inspections.

2) Site specific requirements were adopted to document procedures on how process safety metrics are tracked and locally managed to ensure completeness over time, particularly as changes occur to site personnel, operations, and structures. The following is an excerpt extracted from the corporate PSM standard:

"Sites and process units shall develop and maintain data management and tracking systems that periodically review and analyze the PSM program status through the use of leading and lagging metrics. Each site shall document or describe these data management systems as part of the PSM manual, including specifics on who is responsible to collect, enter, and update data, the review frequency, and how data are collated, validated, and reviewed."

3) Specific internal requirements for leadership review and approval of extensions is also included in the corporate PSM standard as described below. Extensions may be relevant and appropriate

involving specific circumstances such as delays in governmental permitting, procurement of equipment due to supplier issues, extended shutdowns, or unexpected personnel changes. The following is an excerpt extracted from the corporate PSM standard:

"When PSM recommendations and action item target dates are modified or extended beyond their initial target date by the site or operating area, the extensions shall be reviewed and approved by line management. Documentation shall be maintained and shall list the initial assigned target date, the extended assigned target date, the technical basis or rationale for the extension, the name of the person assigned action, and the name and title of the leadership approver. Site manager approval is required for any extension request two or more times, regardless of duration."

4) The PSM Management of Change- Personnel (MOC-P) element includes a specific step to analyze and formally reassign open action items that were the responsibility of personnel who are transferring, retiring, or otherwise leaving the site prior to final checkout as noted below:

"Sites shall also develop and implement procedures to document, analyze, and manage open or incomplete PSM-related individual work activities that must be reassigned as a result of changes to incumbent personnel prior to their departure and to help ensure timely completions of open actions, including assigned corrective action items"

Table 2: Site and Unit PSM Metrics

PSM metric	*HHP*	*LHO*
Number of PHAs scheduled, open, and overdue	*Required*	*Required where applicable*
Number of open and/or overdue incident investigation recommendations	*Required*	*Required*
Number of open and/or overdue PHA recommendations	*Required*	*Required*
Number of open and/or overdue first-party PSM and HTM audit recommendations	*Required*	*Required (NA for HTM audits)*
Number of open and/or overdue second-party PSM, HTM, and sister plant technology (where applicable) audit recommendations	*Required*	*Required (NA for HTM audits)*
Number of open/overdue emergency response drill critique recommendations	*Required*	*Required*
Number of overdue MI tests and inspections for PSM-critical equipment	*Required*	*Required*
Number of tests and inspections temporarily extended or delayed for PSM-critical equipment	*Required*	*Required*
Number of open corrective actions or work orders resulting from completed MI tests or inspections for PSM-critical equipment	*Required*	*Required*
Personnel (i.e., operations and mechanical) skill-based refresher training and re-qualifications completed versus scheduled	*Required*	*Required*
Number of open and/or overdue MOC-T recommendations	*Required*	*Required*
Number of temporary changes or trials open past duration limit	*Required*	*Required*
Number of operating procedures and safe work practices overdue for review and/or revision (three-year review)	*Required*	*Required*
Number of open and/or overdue PSSR recommendations	*Required*	*Required*
Status of MOC-Personnel 90-day competency demonstrations	*Required*	*Required*
Number of Category A, B, and C PSM incidents	*Required*	*Required*

HHP: Higher Hazard Process
LHO: Lower Hazard Operations
NA = not applicable

3.2 Local Site Practices and Rituals Involving PSM Metrics

Local practices to routinely review the status of PSM metrics at each site will vary depending on the size of the site and local operations, however the steps described in this section are typical for each operating location in order to maintain the visibility and focus on strong metrics performance as part of a standardized management process and rituals as described below. This process also provides a practical way to communicate the status of PSM metrics to all operators, mechanics and support staff in a routine, transparent, and efficient manner without any filters or organizational "bureaucracy". Communication of PSM metrics out on the shop floor also helps create personal engagement with each employee in terms of their specific role, involvement, and understanding. The following steps are typical of a DuPont manufacturing site as part of routine review of PSM metrics:

1) Each week during the site Leadership team meeting led by the site manager, the team will review a set of "open" PSM metrics that are due by the end of the month

2) These data are posted on visual management display boards which are used to communicate the status and ensure active engagement with all key functions at the site and unit level. Visual boards are typically provided in the administration building, central control rooms, and maintenance shops. Electronic boards may also be posted on the site server and local website. See figure 2 below for a typical site and unit visual management board for PSM metrics.

3) Visual boards provide an easy method to communicate the status and results with everyone in the operating area. They typically include the following data for each open action item based on the data elements listed in Table 2:
   a) The targeted due date
   b) The name of the responsible person
   c) What is the status and are there any potential roadblocks to completion
   d) Who may have overdue action items
   e) Verifier status before final closure

4) Each operating manager will review open action items with the responsible person to verify it is on track and will be completed (this may include manufacturing technical, first line supervisor, maintenance, and other support personnel). The visual boards are typically updated weekly and always evolving as new action items appear and completed actions are removed. Actions which have been completed but are still open and have reached the verifier stage are also posted to help ensure the verification step is completed by line management in a timely manner. Verification typically involves an independent assessment of the completed action by the applicable line manager with supporting evidence to help ensure the action was satisfied completely prior to final closure.

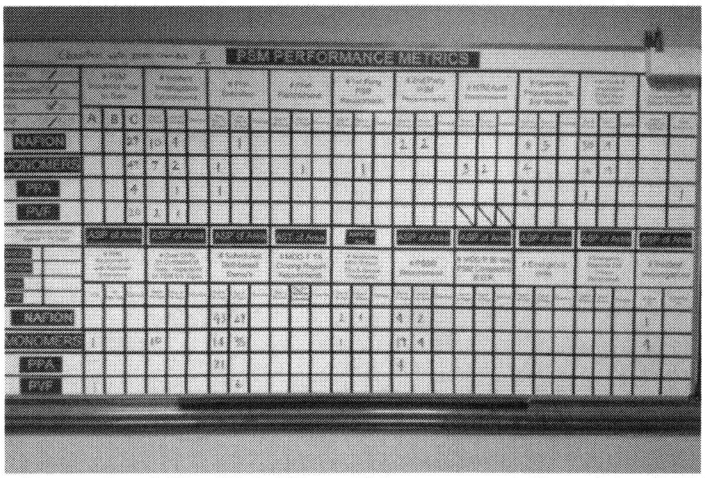

Figure 2: Example of Site/Unit Visual Management Board for PSM Metrics

Senior leadership site visits typically involve interactions with site employees and include a review and discussion on current PSM metrics, site level PSM objectives, and accomplishments.

3.3 Site level Annual PSM Systems Reviews and Auditing

In 2011, DuPont adopted a new internal policy whereas each site conducts an annual PSM performance review using standardized criteria to identify strengths, celebrate successes, and target key improvement opportunities. This review includes a required set of data elements which are listed below, which includes PSM incidents (lagging) as well as performance metrics (leading):

- Analysis and review of results and outcomes versus site and unit annual PSM goals and objectives, including positive recognition of key milestones and accomplishments
- Detailed PSM incident results and trend analysis (location, equipment, types of events)
- PSM first- and second- party audit results
- Review of all quarterly PSM metrics results, including the individual status of all open PSM recommendations or corrective actions that were extended or have been open for an extended period
- Overall PSM performance, including PSM committee effectiveness, employee participation and inputs, key personnel changes, results of regulatory inspections, where applicable, and future objectives to drive additional improvements.
- Other site- or business-related key PSM program attributes and initiatives, as applicable (e.g., site participation in sister plant technology networks and major expansion projects)

DuPont's internal first and second party process safety audit program was updated in 2008 to include a new focus and analysis of site based PSM metrics, including specific verification steps that all relevant PSM action items are included in the local data collection process and are accurately reported at both the site and corporate level. Audits will also evaluate local site procedures which describe the metrics collection and reporting process, communication of metrics to site personnel, and how metrics are used to establish future local PSM goals and objectives. The CCPS reference guidance on "Auditing Process Safety Management Systems"(4) includes a short discussion on metrics in Chapter 21. Wherever PSM metrics are required, internal auditing protocols and practices should include steps to ensure they are complete and accurate.

3.4 API Tier 3 Indicators – Challenges to Safety Systems

DuPont adopted internal guidance on ANSI/API 754 Tier 3 indicators commencing in January 2011 which are tracked and managed at the local site and process unit level. These metrics related to "near misses" or challenges to safety systems include the follow four elements:

1) Number of true demands on safety systems (e.g., activation of a SIS, relief valve, rupture disk, or other active mitigation device; activation of an emergency shutdown system)
2) Number of near misses related to process conditions (e.g., excursions above safe operating conditions)
3) Number of functional failures of PSM-critical equipment during tests and inspections (e.g., relief valves, SIS's, or primary containment inspection results outside of acceptable limits such as vessel measurements below T-minimum thickness calculations)

4) Number of significant non-conformances involving safe work practices (e.g., field audits and program reviews conducted for confined space entry, opening process equipment, hot work permits lockout and/or tag out, and other life- saving procedures)

These metrics are tracked and reported only at the local unit or site level since they tend to be unique and relevant to local process design, operating conditions, management systems, and personnel.

3.5 HSG 254 Customized Site or Unit PSM Metrics

The UK HSE HSG 254 publication provides a step by step guide for developing process safety leading and lagging indicators for the chemical and major hazard industries. One of the key premises in this guideline is that key process indicators (KPI's) should be derived and implemented based on specific site hazards or risks in order to achieve optimum operating results. It suggests decisions on the scope and setting indicators for the risk management system should be based on four key questions:

1) What can go wrong (anticipated incident scenarios and causes)
2) What controls are in place to prevent major accidents (engineering and administrative)
3) What does each control deliver in terms of a safety outcome (and what are the tolerances)
4) How do we know they continue to operate as intended (what variables should be measured)

This approach can yield substantially different and more customized leading indicators across different plant operations and hazardous manufacturing technologies, however, the outcome can be additive and incorporated in the overarching generic leading indicators that are described in the API and CCPS metrics documents. Although not currently required as part of process safety policies, many DuPont sites have adopted local practices that fit the HSG 254 methodology in addition to the prescriptive corporate metrics previously outlined in Table 2. These metrics are unique to the site or each sister plant technology network and are not currently summarized or tracked beyond the site. Table 3 provides an example of customized key process indicators that have been applied at specific sites.

Table 3: Examples of Local site PSM Key Process Indicators

Leading Indicator (KPI)	Rationale	Hazard/Risk
Weekly field audits - flexible hose storage, use, and field identification	High intensity operations associated with loading, unloading and transfer operations	LOPC during transfer of flammable, toxic and corrosive materials
Housekeeping indexes for combustible dust processing	Sustained focus on combustible dust prevention efforts	Secondary dust explosion potential
Vibration analysis trends and summaries (monthly)	Large scale high energy rotating equipment (PSM critical components)	Mechanical energy hazards in close proximity to operating personnel
Zero leak philosophy repair data (steam, air, water leaks)	Ensure effective reporting, barricading, and repair (work orders) for high pressure systems	Thermal and stored energy
Leadership time (weekly) spent in the field	Leadership by example; process safety culture, improvement suggestions; empowerment on the shop floor	Inadequate employee participation and engagement in PSM
Contractor safety field audits – work permits and maintenance procedures	All maintenance and E&I work on PSM critical components conducted by resident contractor	Lack of training and adequate adherence with maintenance procedures/field work

4. Corporate PSM Performance Indicators

4.1 Lagging Indicators

DuPont's existing process safety management policies require that all PSM incidents are reported internally based on a corporate classification system. Each site line management organization determines which incidents require a full investigation and root cause failure analysis based on a standard set of internal criteria. Incidents are reviewed and summarized through trend analysis by a variety of factors, including by PSM element, types of equipment components (physical failures), type of event (fire, explosion, loss of containment, stored/mechanical energy, etc.), status of process operations (normal, start-ups, standby, shutdown), and classes of hazardous materials (highly toxic materials, corrosives, combustible dusts, etc.). Figure 3 provides an example of incident data summaries typically generated each year at the site and regional level, as well as across the corporation.

Multi-year incident trends are also used as inputs to short and long term global improvement strategies, including additional internal procedures, training, resources, and systems. For example, DuPont initiated a global improvement process on mechanical integrity/quality assurance (MIQA) in 2009, in part based on the historical incident trend analysis which illustrated specific improvement opportunities in this area. This global MIQA initiative included staffing additional positions, enhanced technical training, use of site "heat maps" to track local improvements, and enhanced procedures in specific areas such as piping systems, welding, and fitness for service inspections.

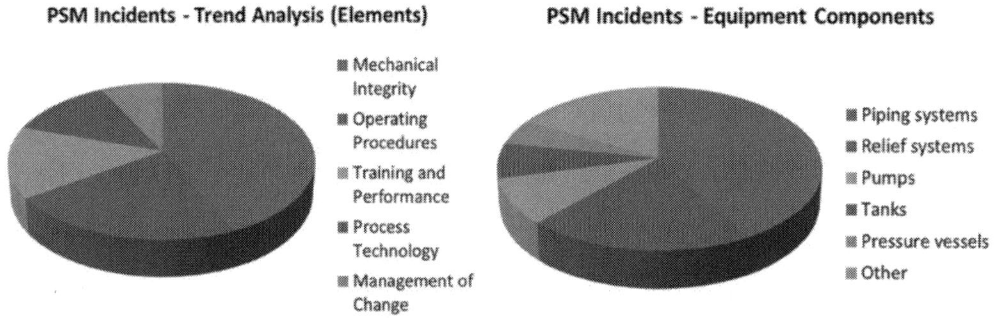

Figure 3: Example PSM Incident Trend Analysis

4.2 Leading Indicators

Corporate reporting from each site and process unit is required on a quarterly basis for the data elements listed in Table 4 which is a subset of the full data set from Table 2. DuPont does not currently utilize one integrated data collection software system at the enterprise so these data are collected in a central master spreadsheet as depicted in Figure 4. These data may be sorted by site, business, region, as well as by "line of sight" for regional operations leaders in order analyze trends. Site based data management systems will vary and can involve web based tracking, Lotus Notes, and SAP for maintenance related data.

Table 4: Quarterly Corporate PSM Leading Indicators

PSM metric	*HHP*	*LHO*
Number of PHAs scheduled, open, and overdue	*Required*	*Required where applicable*
Number of open and/or overdue incident investigation recommendations	*Required*	*Required*
Number of open and/or overdue PHA recommendations	*Required*	*Required*
Number of open and/or overdue first-party PSM and HTM audit recommendations	*Required*	*Required (NA for HTM audits)*
Number of overdue MI tests and inspections for PSM-critical equipment	*Required*	*Required*
Number of operating procedures and safe work practices overdue for review and/or revision (three-year review)	*Required*	*Required*
Number of Category A, B, and C PSM incidents	*Required*	*Required*
Number of PSM Action Item Extensions (approved by leadership)	*Required*	*Required*

NA = not applicable

Once the quarterly data is submitted (typically within the first 15 calendar days after the quarter ends), the results can be collated and reviewed by regional, business, and operations leadership. Any concerns or negative trends at a site or within a region will typically generate follow-up discussions with senior operations leadership to ensure understanding and develop interim improvement plans.

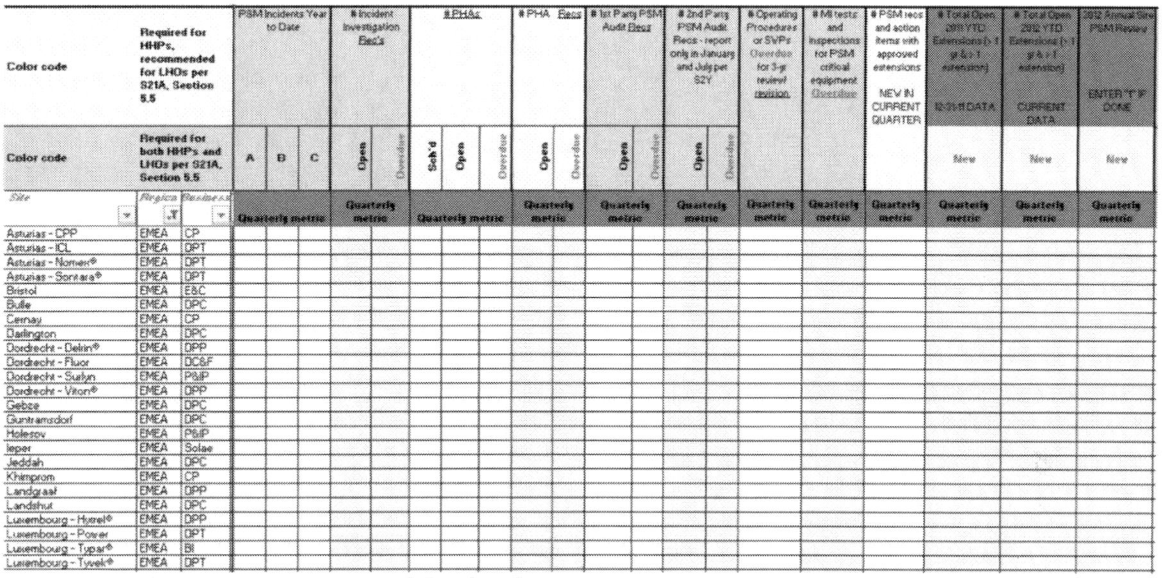

Color code	Required for HHPs, recommended for LHOs per S21A, Section 5.5		PSM Incidents Year to Date			# Incident Investigation Recs		#PHAs			# PHA Recs		# 1st Party PSM Audit Recs		# 2nd Party PSM Audit Recs - report only in January and July per S2Y		# Operating Procedures or SVPs Overdue for 3-yr review/ revision	# MI tests and inspections for PSM critical equipment Overdue	# PSM recs and action items with approved extensions NEW IN CURRENT QUARTER	# Total Open 2011 YTD Extensions (> 1 yr & > 1 extension) 12-31-11 DATA	# Total Open 2012 YTD Extensions (> 1 yr & > 1 extension) CURRENT DATA	2012 Annual Site PSM Review ENTER "1" IF DONE
Color code	Required for both HHPs and LHOs per S21A, Section 5.5		A	B	C	Open	Overdue	Sch'd	Open	Overdue	Open	Overdue	Open	Overdue	Open	Overdue				New	New	New
Site	Region	Business	Quarterly metric			Quarterly metric		Quarterly metric			Quarterly metric		Quarterly metric		Quarterly metric		Quarterly metric	Quarterly metric	Quarterly metric	Quarterly metric	Quarterly metric	Quarterly metric
Asturias - CPP	EMEA	CP																				
Asturias - ICL	EMEA	DPT																				
Asturias - Nomex®	EMEA	DPT																				
Asturias - Sontara®	EMEA	DPT																				
Bristol	EMEA	E&C																				
Bulle	EMEA	DPC																				
Cernay	EMEA	CP																				
Darlington	EMEA	DPC																				
Dordrecht - Delrin®	EMEA	DPP																				
Dordrecht - Fluor	EMEA	DC&F																				
Dordrecht - Surlyn	EMEA	P&IP																				
Dordrecht - Viton®	EMEA	DPP																				
Gebze	EMEA	DPC																				
Guntramsdorf	EMEA	DPC																				
Holesov	EMEA	P&IP																				
Ieper	EMEA	Solae																				
Jeddah	EMEA	DPC																				
Khimprom	EMEA	CP																				
Landgraaf	EMEA	DPP																				
Landshut	EMEA	DPC																				
Luxembourg - Hytrel®	EMEA	DPP																				
Luxembourg - Power	EMEA	DPT																				
Luxembourg - Typar®	EMEA	BI																				
Luxembourg - Tyvek®	EMEA	DPT																				

Figure 4 Corporate Metrics Quarterly Data Sheet (Data Collection)

The example in Figure 5 illustrates recent upgrades which have been developed to provide a web-based management dashboard for the global PSM metrics results. This also allows the spreadsheet data to be extracted, filtered (by quarter, region, business, and corporate), and displayed as part of a broader EHS dashboard to review current performance and focus on year over year trends. DuPont has not adopted specific policies on "tolerance rates' for these leading indicators, however, it is generally accepted that a 95% performance rate or better for open and overdue actions is a typical threshold for acceptable quarter over quarter results.

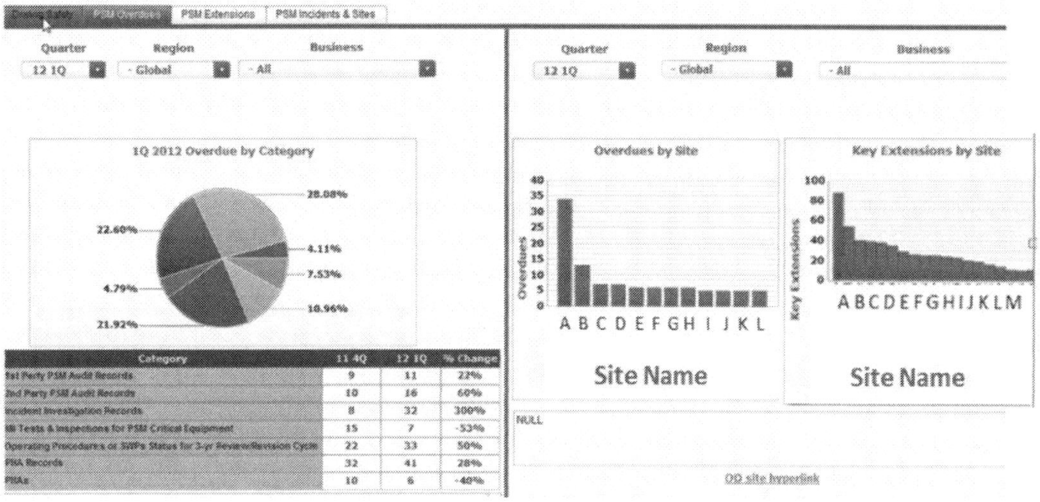

Category	11 4Q	12 1Q	% Change
1st Party PSM Audit Records	9	11	22%
2nd Party PSM Audit Records	10	16	60%
Incident Investigation Records	8	32	300%
MI Tests & Inspections for PSM Critical Equipment	15	7	-53%
Operating Procedures or SWPs Status for 3-yr Review/Revision Cycle	22	33	50%
PHA Records	32	41	28%
PHAs	10	6	-40%

Figure 5: Web based Corporate and Regional PSM Dashboard

5. Annual Inputs to Supplement Long Term Metrics

5.1 Annual Corporate and Site PSM goals/objectives

In addition to a standard suite of year over year leading indicators which relate to ANSI/API RP 754 Tier 4, DuPont typically adopts specific broad based and global PSM related goals and objectives each calendar year throughout operations in order to strengthen the existing risk management program. In specific cases, the objective (or COT's – critical operating tasks) may involve a multi-year focus. For example, a three year initiative was completed in 2011 to strengthen the design and operability of safety instrumented systems across all sites, and in 2012, a global COT on alarm management was adopted and completed in order to meet the new ANSI/ISA 18.02 (2009) standard "Management of Alarm Systems for the Process Industries" (5). Completion of these annual COT's are tracked using standardized data at each site and at the corporate level, and are used as a short term non- traditional leading indicator. Figure 6 provides for an example of the 2012 alarm management COT results which are based on the completion of six key steps and milestones at each site within the region.

	4Q % Complete				
	AP	EMEA	LA	NA	Global
Step 1 - Leadership	100.0%	100.0%	100.0%	100.0%	100.0%
Step 2 - Monitoring	100.0%	87.5%	100.0%	100.0%	97.2%
Step 3 - Benchmarking	100.0%	100.0%	100.0%	100.0%	100.0%
Step 4 - Quick Hits	100.0%	96.9%	100.0%	100.0%	99.3%
Step 5 - Training	100.0%	93.8%	100.0%	100.0%	98.6%
Step 6 - Philosophy	100.0%	100.0%	100.0%	100.0%	100.0%

Figure 6: Annual PSM Objectives - Regional Reporting for Alarm Management Initiative

5.2 Second and Third Party Reviews of PSM Auditing Processes and Results

DuPont conducts second party PSM audits at all higher hazard global manufacturing locations every three years (approximately 60 audits per year). Each audit is scored and the recommendations are tracked in a global action tracking system. This tracking generates a broad set of PSM audit metrics and enables focused data mining to review the types of findings, trends, and potential improvement opportunities across each region, PSM element, and at the corporate level. Any negative trends collated from PSM audit results typically guide an enhanced focus on specific technical areas, which can include additional training, technical guidelines, implementation tools, or annual critical operating tasks.

DuPont also conducts an annual review of the internal PSM audit program to verify its results and overall effectiveness by way of analysis using an independent third party EHS organization. This review also serves as a year over year "leading indicator" as to audit program performance opposite specific internal criteria and policies, as well as accepted industry practices (e.g., an audit of the audit program). The annual review includes a detailed evaluation of ten program elements, including a random review of finalized PSM audit reports to assess completeness and accuracy, a review of audit recommendations and follow-up within each operating region, and a field evaluation of selected internal PSM audit teams during an actual audit. The results and any improvement recommendations are communicated to senior leadership, tracked to completion, re-assessed the following year, and posted on the DuPont sustainability (external) internet site for public sharing. Figure 7 displays a list of the 3rd party audit elements which is provided in addition to a detailed independent program analysis summary.

PROGRAM EVALUATION CRITERIA	Partially Meets Expectations	Fully Meets Expectations	Exceeds Expectations
1. Explicit Top Management Support		✓	
2. Program and Auditor Independence		✓	
3. Appropriate Audit Team Staffing		✓	
4. Comprehensive Written Audit Procedures			✓
5. Defined Program Objectives and Scope		✓	
6. Adequate Program Resources		✓	
7. Audit Frequency Based on Risk		✓	
8. Appropriate Audit Process		✓	
9. Documentation of Audit Findings		✓	
10. Resolution of Corrective Actions		✓	
11. Appropriate Quality Assurance Measures			✓

Figure 7: Summary Results -Annual 3rd Party PSM/EHS Audit Program Evaluation

6. Risk Management Benefits

Use of a range of site and company level PSM metrics (leading indicators) creates tremendous value and focus on effective risk management when applied with a consistent approach by site line management, senior leadership, and PSM professionals. DuPont's experience over the past five years includes the following benefits:

1) Standardized metrics process which is sustainable when operational and personnel changes occur
2) High visibility of process safety performance and direct connections to improvement initiatives
3) Improved predictability and accountability for results on an individual, business and corporate level, including integration of metrics results for individual performance reviews
4) Proper allocation and adjustment of resources when necessary
5) Common focus and perspectives in operations and with the business in order to apply meaning to the numbers and use them to drive the right decisions
6) Help ensure compliance with regulatory requirements and external certifications (RCMS, ISO 14000, etc.)
7) Enhanced mechanisms to measure and drive PSM integration efforts involving acquisitions
8) Routine engagement of leadership at all levels to monitor performance and drive PSM improvements

The application of these metrics should not be applied in a static environment since new initiatives, internal requirements, and changes in performance trends can occur over time. A periodic review of existing metrics and their overall relevance and value to effective risk management should be considered at a minimum of every two years.

7. Limitations and Precautions with Leading Indicator Metrics

As the CCPS "Guideline on PSM Leading and Lagging Metrics"(6) clearly notes, "you don't improve what you don't measure". DuPont's experience with leading metrics (Tier 4) over the past five years has shown a clear improvement in the rigor and organizational focus on PSM related actions, including timely completion and closure of actions and use of the data to further strengthen existing systems and reduce incident severity. However, these and other metrics have specific limitations and precautions which must also be recognized and understood at the site and business level including:

1) Strong performance on metrics provides organizational feedback that things are being done, but that doesn't always equate to the fact that the right things are being done. This question is best answered by a critical analysis of internal policies and procedures vs. industry best practices and operational needs. An organization that maintains only a "compliance mindset" only which does not include all processes and process hazards will not achieve maximum risk management benefits.

2) Metrics alone cannot measure the specific quality and thoroughness of risk management work activities and decision making. For example, metrics which track incident or PHA recommendation closures have little connection by themselves to the completeness and overall quality of the root cause failure analysis or HAZOP review. Metrics involving mechanical integrity data will not correlate to whether the actual test and inspection was completed per the procedure and in accordance with applicable RAGAGEP's, These issues related to the quality of PSM work products are best addressed through training and auditing processes.

3) Metrics do not measure or guarantee the quality and completeness of actions taken to address process safety recommendations or related actions to achieve closure. Actions completed and implemented by site personnel should be evaluated for completeness prior to final closure, typically by another individual as a line manager or functional expert as part of a formal verification step. Management expectations and pressures (perceived or real) to minimize the number of overdue action items can also result in undesirable behaviors (e.g., premature or incomplete closure, low willingness to include actions in the tracking system) and must be avoided.

4) Each organization needs to develop the most appropriate metrics to support internal needs. The CCPS guideline for Process Safety Metrics includes a description of over 350 sample metrics (30 pages) in Appendix 1. Choosing too many metrics can certainly result in non-value adding work, organizational resistance, and metrics "fatigue". Adoption of too many metrics can also overwhelm or confuse the organization and create an environment where the focus shifts from meaningful risk control and reduction activities to data tracking and documentation only (bean counting, paper pushing or a "check the box" mindset). Performance goals tied in with leading indicators are important, however unreasonable goals should not be adopted with could drive reporting underground or sub-optimize the system.

5) Strong results with metrics alone cannot provide an absolute assurance of preventing high consequence, low probability catastrophic events. In fact, month over month excellent results on leading performance indicators may actually create over-confidence or a false sense of security. A balanced organizational mentality of celebrating positive metrics performance coupled with the need to maintain an ongoing a sense of vulnerability and avoiding complacency should be reinforced by site and corporate leadership.

8. Conclusions

Most of the applicable current industry guidelines on metrics identify the "what's" but don't currently describe specific methods or approaches to apply leading and lagging indicators (the "how's") in the most effective manner in order to pursue and achieve performance improvements. While many may believe that this is or should be intuitive, there are specific policies and mechanisms that should be considered as established rituals and standard practices involving metrics, several of which have been described in this paper. Each company should analyze and develop their own internal practices relative to this area based on internal policies, structures, unique hazards, and leadership expectations. Industry is currently working on broad based harmonization of the API 754 Tier 1 and 2 lagging indicators, and it is reasonable that a certain small base load of Tier 4 variables should be standardized and consistently applied across all industrial segments and companies (for example, tracking recommendations associated with incident investigations, audits, and process hazards analyses, and the status of mechanical integrity tests and inspections).

DuPont's practices on metrics, as with those of many other companies, have evolved over time and currently include a multi-faceted approach to help foster continuous improvements in process safety systems and results. As so clearly stated in the CCPS guideline on metrics, "metrics alone will not improve process safety; data must be collected, analyzed, communicated, understood, and *acted upon*". One can expand this concept much more broadly in light of a comprehensive risk management approach – even if the data is acted on as noted above, this alone does not provide a complete picture on the quality and completeness of the data or the content and effectiveness of the overarching risk management system. An integrated and evergreen approach relative to metrics, PSM management systems, personnel, and verification methods should be considered in order to achieve maximum benefits. In this case, the process to apply, manage, and continuously improve the technical elements of process safety has equal importance to the actual selection of the performance variables. Figure 8 provides a macro-summary of the current system for site, regions, and corporate.

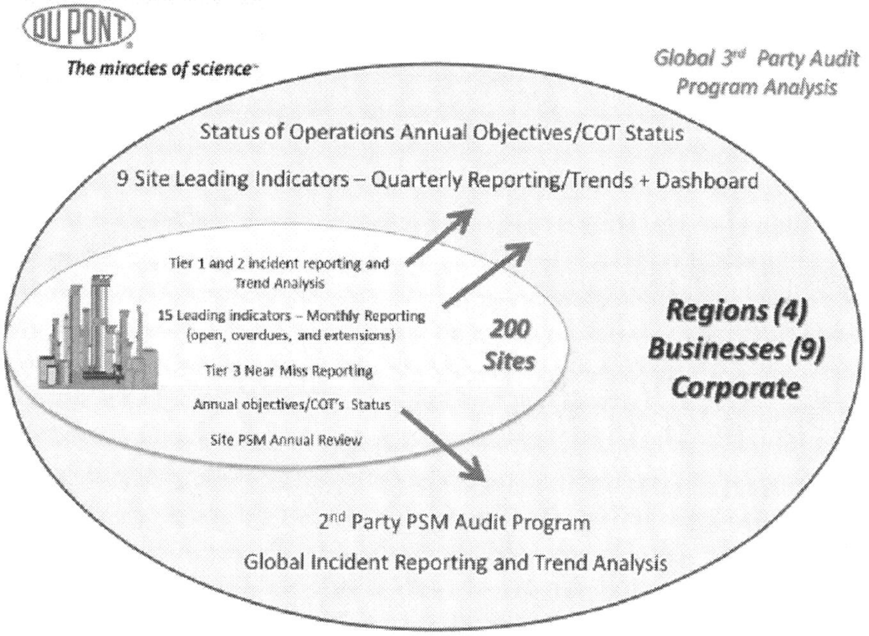

Figure 8: Enterprise Level View of PSM Performance Metrics

Management's visible involvement with metrics and their focused response to both successes and weaknesses in performance are vitally important to the overall metrics approach and the focus to maintain and strengthen process safety. Without actively engaged and vocal leaders at the corporate, regional, and site level who use the metrics to drive organizational performance, the metrics process will only serve as a paper program for "bean counting" and can be significantly compromised by creating a false sense of confidence based on the numbers or actually driving the wrong behaviors.

Heavy reliance on failure data to monitor performance can provide only limited prevention benefits since improvements or changes are only applied after something has gone wrong. Effective risk management requires a dual approach involving a combination of both lagging and leading indicators. Formal processes to adopt, collect, track, report, and apply PSM leading indicators, and the use of year over year annual PSM objectives or critical operating tasks should be considered as part of an overall metrics approach. Properly selected leading indicators that are reported and reviewed at both the site and corporate level with a high degree of transparency can be used to fully engage employees, site line management, and senior leadership as well as identify and address potential needs and issues across operations. There is no "silver bullet" that can guarantee stellar PSM results, however, to achieve and maintain maximum benefits with metrics, they have to be directly coupled with a fully integrated and progressive risk management program similar to the elements noted in Figure 9.

Figure 9: Metrics with a Comprehensive Approach for Integrated Risk Management

9. References

1) ANSI/API RP 754 "Process Safety Performance Indicators for the Refining and Petrochemical Industries" (2010), American Petroleum Institute

2) Center for Chemical Process Safety: "Process Safety Leading and Lagging Indicators", American Institute of Chemical Engineers (2011)

3) UK Health and Safety Executive: "Step by Step Guide to Developing process safety Performance indicators HSG 254 (2006)

4) Center for Process Safety "Guidelines for Auditing Process Safety Management Systems" (2nd edition), American Chemical Engineering Association (2012); ISBN 978-0-470-28235-9

5) ANSI/ISA-18.2-2009 "Management of Alarms in the Process Industries", International Society of Automation

6) Center for Process Safety "Guideline for Process Safety Metrics"; American Chemical Engineering Association (2010); ISBN 978-0-470-57212-2

9TH GLOBAL CONGRESS ON PROCESS SAFETY

Using Metrics to Drive Process Safety Performance at Maersk FPSOs

Jeppe Guld, Senior General Manager, Operations, A.P. Moller – Maersk A/S, Esplanaden 50, Copenhagen K, Denmark, +45 3363 8716, jeppe.guld@maersk.com

William M. Bradshaw, P.E., Senior Engineer, ABSG Consulting Inc., 10301 Technology Drive, Knoxville, Tennessee, 37932, (865) 671-5829, WMBradshaw@ABSConsulting.com

Prepared for Presentation at
American Institute of Chemical Engineers
2013 Spring Meeting
9th Global Congress on Process Safety
San Antonio, Texas
April 28 – May 1, 2013

UNPUBLISHED

Using Metrics to Drive Process Safety Performance at Maersk FPSOs

Jeppe Guld, Senior General Manager, Operations, A.P. Moller – Maersk A/S, Esplanaden 50, Copenhagen K, Denmark, +45 3363 8716, jeppe.guld@maersk.com

William M. Bradshaw, P.E., Senior Engineer, ABSG Consulting Inc., 10301 Technology Drive, Knoxville, Tennessee, 37932, (865) 671-5829, WMBradshaw@ABSConsulting.com

Abstract

Maersk FPSOs operates five offshore assets, including floating oil production and storage ships, in locations ranging from the North Sea to the South Atlantic Ocean. Like many operators in the oil and gas industry, Maersk FPSOs has long understood the importance of identifying, understanding, and managing risks associated with major accidents. But it's difficult to manage something without feedback; many would say it's impossible. Like other operators, Maersk FPSOs struggled to measure process safety performance, and more importantly to help everyone in the organization understand the linkage between the activities they do on a day-in, day-out basis and major accident risk. In addition, senior management struggled to know if process safety performance is "good enough." Other companies have clearly proven that the lack of major accidents is not a reliable predictor of when the next one might occur. Thus, the lack of major accidents at Maersk FPSOs could not be taken as proof that process safety performance was acceptable; it might have been improving, stagnating, or possibly degrading. Therefore, about 2 years ago, Maersk FPSOs embarked on a path to create an effective tool to monitor process safety; one that focused on measuring "what we do," rather than standards or policies that focused on "what we say we should do" or "what we know we should do."

The core of the effort is a dashboard with a series of layered process safety metrics using displays on video monitors located in prominent locations in offices, production areas, and other areas where the indicators provide maximum visibility for Maersk FPSO employees. The aim of this effort is to ensure that process safety indicators not only are reported to senior management, but, more importantly, are also seen and used on the producing units by front line employees whose actions directly influence/control the process safety performance. The metrics chosen by Maersk FPSOs "roll up" so the top of the metrics pyramid is linked to measurable activities, thus providing a clear picture linking what Maersk FPSOs employees and contractors do (or don't do) and what happens in the workplace to Maersk FPSOs' risk profile. Since launching this initiative, Maersk FPSOs has seen measurable improvement in many of the leading indicators. More importantly, the rate and severity of precursor events have fallen as leading indicators have improved.

1. Introduction

Maersk FPSOs (floating production storage and offloading) is part of the A. P. Moller-Maersk Group (Maersk), a global conglomerate established in 1904. In addition to being one of the world's largest shipping companies, Maersk is involved in a wide range of activities in the energy, logistics, retail, and manufacturing industries. Maersk FPSOs was established in 1995 as part of Maersk Contractors, and in 2008 it became a separate business unit. Throughout its history, Maersk FPSOs has worked closely with customers to build and operate floating production units customized to the specific field and meteorological/oceanographic conditions where they will be deployed. As shown in Figure 1, Maersk FPSOs has grown from one asset (known within Maersk FPSOs as a unit) in 1996 to five assets

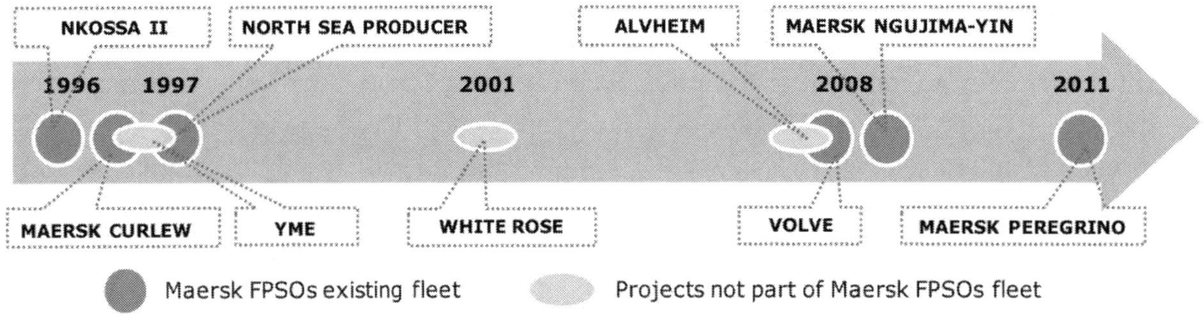

Fig. 1. Maersk FPSOs Fleet Development

As the fleet grew, management became more aware of the risks associated with major accidents. Armed with the knowledge that one of the conclusions in the *Report of the BP U.S. Refineries Independent Safety Review Panel* (aka, the Baker Panel report, Ref. 1) was that industrial safety statistics such as lost-time injury rates may not be a good predictor of the effectiveness of an organization's process safety management program, the management team began taking a closer look at process safety within Maersk FPSOs. The consensus from a 2010 workshop with management representatives from all Maersk FPSOs units was that while the fleet was doing well in occupational health and safety performance, there was a need to increase understanding and awareness of process safety. (As shown in Table 1, injury rates continued to improve between 2010 and 2012, even with the significant increase in emphasis on process safety.)

In fact, there is no proof that good health and safety performance also translate into good process safety performance. The Baker Panel found that "BP's reliance on occupational illness and injury rates such as days away from work case frequency and recordable injury frequency significantly limited BP's perception of process risk within the U.S. refineries." (Ref. 1) Elliott and his colleagues found only weak evidence that good everyday safety performance is a foundation for preventing or mitigating relatively rare major accidents. (Ref. 2)

Table 1. Occupational Injury Rates for Maersk FPSOs

Year	LTI Rate[1]	TCRF Rate[2]	Manhours
2010	2.83	8.50	1,764,498
2011	0	3.72	2,152,135
2012	0	1.59	1,884,111

1 Rate for accidents involving days away from work; data cannot be directly compared to rates in the US because the European method uses different criteria/definitions and a different denominator

2 Rate for accidents involving worker injuries; data cannot be directly compared to rates in the US because the European method uses different criteria/definitions and a different denominator

Based on a request made by the six operation managers who attended the 2010 workshop, senior management within Maersk FPSOs initiated a project to increase the awareness and understanding of process safety within the company, and designated a senior manager within the operations group to oversee the initiative. Unlike many process safety implementation projects where companies start with management system definition and implementation and eventually develop metrics to measure the effectiveness of these systems, Maersk FPSOs embedded metrics and a commitment to present the data in a useful manner in the foundation of the work that ensued. The project charter involved:

1. Developing and implementing a Process Safety Dashboard (PSD)
2. Arranging a rollout campaign for the PSD
3. Performing process safety audits

This paper describes steps one and two (i.e., how Maersk FPSOs developed and implemented the PSD).

2. Getting started

While many organizations are good at collecting data, they often fall short when it comes to analyzing and appropriately reacting to the data. To help avoid this pitfall, Maersk FPSOs decided to use a dashboard to present the current process safety status based on the view that graphical presentation of data and trends is the most efficient way to communicate real-time information to employees at all levels of the organization. There was a general sentiment within the offshore teams that they were reporting a lot of data with minimal benefit; they did not really see the results or understand how the data were used. This view could have undermined the project; dealing with it effectively was key to successful implementation.

The fleet in Maersk FPSOs spans three continents. Age-wise there is a span of 15 years from the oldest to the newest unit. There were diverse unit cultures and history that had to be addressed. These facts influenced the initial approach to selecting indicators and gathering data. At the outset Maersk FPSOs set four governing principles for the project:

We will involve the users
-> initially by use of a reference group
-> subsequently by inviting feedback for improvement

Involvement of the end-users raised the quality of the end product and further ensured a sense of ownership that was crucial in the implementation and operation phases.

We will keep it simple
-> relatively few indicators; use existing data sources

Due to the existing infrastructure, culture, and environment, every attempt was made to maximize use of data and data sources that were currently available. Given a clean slate, it might have been possible to incorporate other data sources and on-line data in the specification, but this was viewed as a minor obstacle rather than a roadblock.

We see a dashboard as being dynamic and the development of a dashboard as being an iterative process
-> implement a system that easily can be adjusted
-> start simple and only expand the complexity if it makes sense
-> use internal resources for coding

Maersk FPSOs chose to use standard software and internal resources to train personnel to use the dashboard. This (1) increased the speed, (2) reduced the cost of development, and (3) facilitated subsequent adjustments to the dashboard. It also provides greater flexibility for future adaptations and improvements.

A Process Safety Dashboard can not stand alone
-> it requires a dedicated rollout effort and active follow-up from management
-> it shall be a tool to assist in driving behavior
-> it will be followed up by process safety audits
-> process safety status will be integrated into unit reporting

This last principle is, in fact, more an alignment of expectations: pointing out that a process safety dashboard is not a magic wand that on its own can ensure good process safety performance. A dashboard is a tool that can (1) assist in visualizing data, (2) provide a good overview of large volumes of data, and (3) reveal trends based on the accumulated effects of subtle changes that otherwise might escape notice. However, on its own, it does not change behaviors or prioritize work activities. Useful information warns key personnel of failures or loss of safeguards that can lead to incidents; it's not just numbers on spreadsheets or lines on charts. Key indicators that measure the health of the equipment and capabilities of the crew, such as a gradual increase in maintenance backlog, or a gradual slide in crew competency, were of great interest.

It was clear that the dashboard would contribute to improvements in process safety performance only through consistent monitoring and timely implementation of actions to address the issues highlighted by the dashboard. Therefore, it was decided that once the project phase was complete, the dashboard would be handed over to operations as a tool to assist in ensuring effective process safety practices over the long run.

Backed with these governing principles, Maersk FPSOs chartered a reference group that included representatives from all units, engineering, and the technical support function to (1) identify useful metrics, (2) design the dashboard, and (3) roll out the dashboard within the company.

3. What is process safety within Maersk FPSOs?

Before Maersk FPSOs could identify useful process safety metrics and design the dashboard, the organization (and the reference group) had to reach a common understanding of process safety. This proved to be a worthwhile exercise, as amongst the 12 members of the reference group there were essentially 12 different perceptions about what process safety meant. Several believed that process safety was:

- Effective safety barrier management
- Only related to what happens at the process plants, and not directly applicable to offshore operations
- Synonymous with asset integrity

The reference group also took the opportunity to align its understanding of process safety with Maersk Oil and Gas, another branch in the Maersk Group that in 2007 adopted the 4-pillar, 20-element process safety model first proposed in the Center for Chemical Process Safety's (CCPS's) *Guidelines for Risk Based Process Safety*. (Ref. 3) This led to some good, and occasionally lengthy, discussions that significantly contributed to achieving one of the overall objectives of the project:

Raising the understanding and knowledge about process safety in the organization.

Eventually, the team aligned on a definition based on what had been presented by the CCPS, tweaked to suit the offshore oil and gas industry:

Process Safety is a blend of engineering and management skills focused on preventing catastrophic accidents, particularly explosions and fires, associated with the handling of hydrocarbons.

4. Choosing among possible process safety indicators

Choosing indicators that are relevant to Maersk FPSO operations was an essential step. There are many good recommendations and guidelines on the topic. Essentially all suggest organizations adopt a blend of leading and lagging indicators. However, the difference can be subtle. For example, a leaking pipe carrying hydrocarbons could rightfully be called both a leading and a lagging indicator:

- It is a leading indicator for a fire/explosion
- It is a lagging indicator for the integrity management program

The reference group adopted the somewhat simplified view that it would not get bogged down in semantics on leading versus lagging indicators. Rather, it focused on identifying important indicators and implementing metrics to see how well they worked. The view was that the initial set of indicators would be adjusted and refined after an initial trial phase.

Some of the operating units held the view that they were already reporting a sufficient amount of data, and others believed they were actually overreporting, with little or no perceived benefit. The reference group members helped drive decisions regarding what should be reported and tracked.

There is a balance between too much and too little information. The optimum level of reporting will always be a topic for discussion, especially from the viewpoint of operating groups who see themselves as being perfectly well informed about the current status of their unit. However, beyond the minimum level of reporting required to comply with regulatory requirements, there are data that are critical to cross learning between units and even between companies. Getting the right balance can be difficult. Certainly one hazard is placing such a burden on an operating unit that it actually becomes a hindrance to exercising good and adequate management and supervision. If the most senior and skilled people on board get tied to their desks reporting data instead of walking the decks, guiding newcomers, and supervising the work being performed, the opportunity loss can far exceed any benefit the metrics may provide.

With this in mind, Maersk FPSOs adopted a governing principle that:

> We will keep it simple
> -> relatively fewindicators derived from existing data sources

This led to a mapping exercise where the operating units identified (1) indicators they believed would be good indicators for the process safety status and development of their unit and (2) where the data for this indicator was being recorded, by whom, and the frequency at which they are updated. The result of the mapping exercise was a list of 70 indicators, which was ultimately pared down to 29 indicators that were already being recorded on all units. The process of going from 70 to 29 indicators forced the team to toss out ideas for some very good indicators, but the desire for a common set of metrics and dashboard dictated that the metrics be optimized for the entire organization. These 29 indicators formed the first set of common process safety indicators used for the dashboard.

The 29 indicators currently being monitored (split among the six "layers" shown in Figure 2) are:

Layer 6 - Process safety accidents
Layer 5 - Process safety incidents
- 5.1 Ignition
- 5.2 Hydrocarbon release

Layer 4 - Precursors
- 4.1 Emergency shutdown system activations
- 4.2 Pressure safety device activation
- 4.3 Safety critical element failed
- 4.4 Level 1 and 2 remedial action recommendations (RARs) overdue
- 4.5 Inhibits
- 4.6 Total number of open RARs

Layer 3 - Management system failures
- 3.1 Overdue action items from incident investigations
- 3.2 Overdue temporary repairs
- 3.3 Temporary operating procedures
- 3.4 Integrity management anomalies – Pressure
- 3.5 Integrity management anomalies – Structures
- 3.6 Overdue integrity management anomalies – Pressure containment
- 3.7 Overdue integrity management anomalies – Structures
- 3.8 Enforcement actions

Layer 2 - Unsafe behaviours and attitudes
- 2.1 Overdue safety critical maintenance jobs
- 2.2 Overdue high criticality maintenance jobs
- 2.3 Overdue management of change action items
- 2.4 Overdue audit findings
- 2.5 Temporary equipment on board

Layer 1 - Culture: Individual and organizational tendencies
- 1.1 Active cards
- 1.2 Active cards addressing breach of procedures
- 1.3 Near-miss reports
- 1.4 Noncompliance with permit to work requirements
- 1.5 Training compliance
- 1.6 Competency
- 1.7 Supervision strain

5. Designing the dashboard

The authors of CCPS's *Guidelines for Risk Based Process Safety* met with Maersk FPSOs personnel and presented the concept of a 6-tier risk pyramid (shown in Figure 2), with accidents (layer 6) at the top of the pyramid and cultural issues (layer 1) at the bottom, that is the foundational level of the risk pyramid. Their suggestion is that Maersk FPSOs should adopt metrics at each level.

Maersk FPSOs adopted this suggestion and split the 29 indicators among the six layers of the pyramid. It also decided to use a pyramid form for the dashboard since the concept of risk pyramids is commonly used and understood in the FPSO industry. The pyramid icon also ties in well with the principles of OGP 456: *Process Safety – Recommended Practice on Key Performance Indicators.* (Ref. 4)

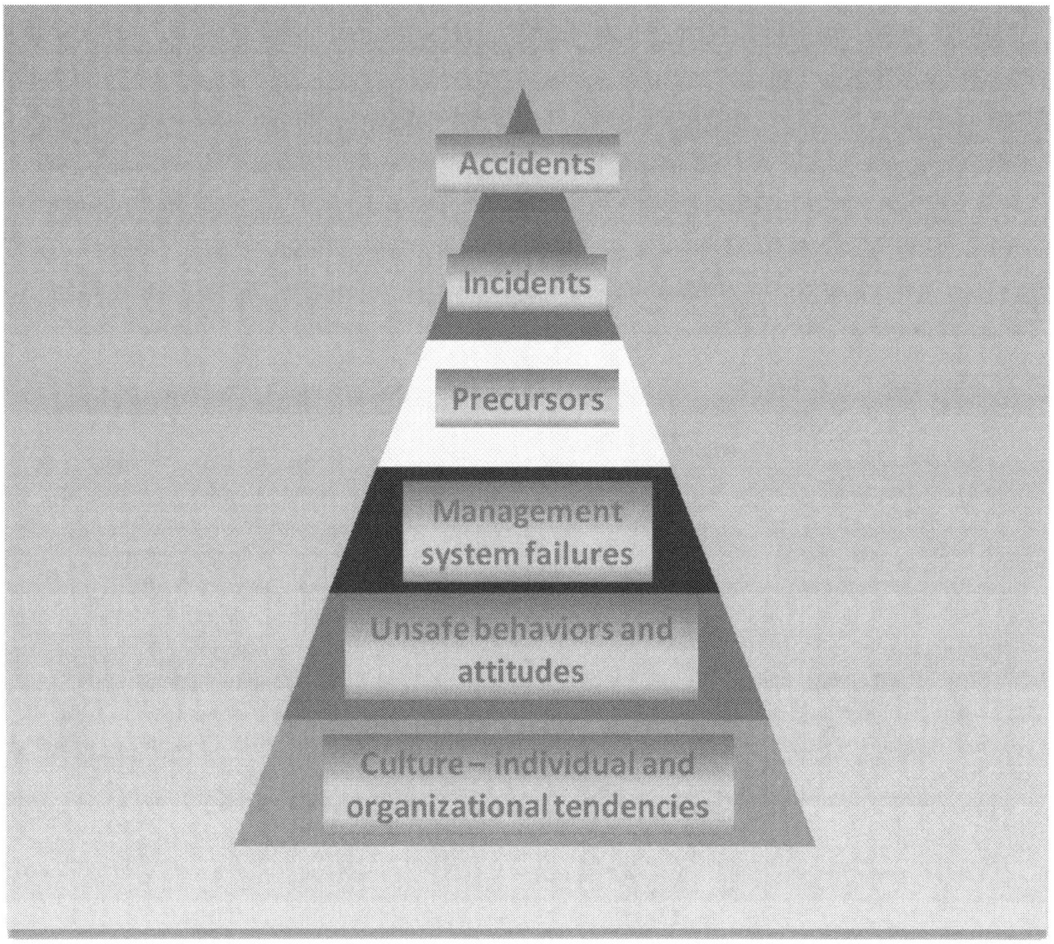

Fig. 2. Six-tier Risk Pyramid

Maersk FPSOs wanted the dashboard to be interactive. The full pyramid became the front layer with the idea that users could drill down into different layers of information that provided additional detail on each layer.

The concept for transforming data into a useable dashboard was done relatively quickly, but the dashboard development phase was more complex and time consuming, requiring a number of iterations. While the software used to build the dashboard was quite advanced, Maersk FPSOs focused on maintaining a simple, clean layout to avoid information overload. The reference group sought and adopted guidance from communication experts regarding how much information can be digested and how to improve assimilation via proper layout, color coding, and so forth.

A picture of the tool is shown in Figure 3. Each level of the pyramid consists of a number of indicators.

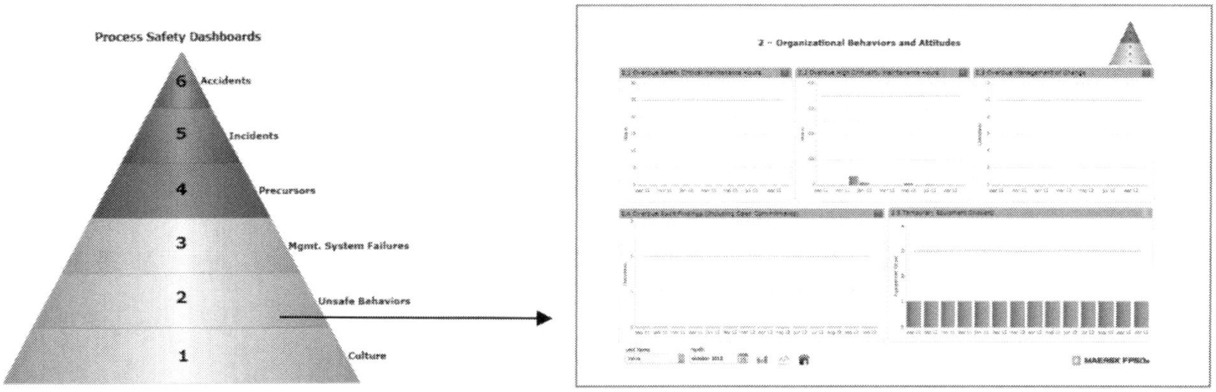

Fig. 3. Screen Shot of the Dashboard

Each of the six layers is color coded to display the status in real time. The color code adopted was based on:

- Red means a process safety risk requires action.
- Yellow means a process safety risk requires attention.
- Green means the process safety risk is tolerable.

When a layer contains multiple indicators, which is currently the case for all layers except the top layer (accidents), it displays the color of the "worst-performing" individual indicator(s) included in that layer. This approach, rather than one based on a simple or weighted average, reflects the view that "the chain is only as strong as its weakest link."

Displaying data in this manner invites users to "dive deeper" and interrogate the level to see which indicator(s) is causing the layer to be yellow or red. For example, Level 2 has five indicators. If the layer is shown as green, then all five indicators are green. However, if it is showing yellow, users can quickly determine which indicator(s) is yellow; if it is red, users can quickly determine which indicator(s) is red (and possibly yellow). The logic is shown graphically in Figure 4.

2.1	2.2	2.3	2.4	2.5
G	G	G	G	G
G	G	Y	G	G
R	G	G	G	Y

Layer color
G
Y
R

Fig. 4. Example of How Indicator Status Rolls Up to the Color for Each Layer

Once the work of the reference group progressed to what it believed was an acceptable level of usefulness and effectiveness, it released the first version of the dashboard and engaged in an "open rollout." The approach was shared with a wider audience to get their feedback on how the dashboard could be improved. The rollout explained how the initial dashboard had been developed, acknowledged that this was not a finished product, and invited reviewers to provide suggestions for improvement.

The open rollout was done on one of Maersk FPSOs North Sea units, which ran the dashboard in a test mode for 3 months. This allowed each shift to interact with the concept, test it, and provide feedback. (The FPSOs where the dashboard was tested is operated with a 2 weeks on, 4 weeks off staff rotation.) This test run provided some further refinement and good suggestions from offshore personnel for improvement. The dashboard was then amended and prepared for a wider rollout.

6. Data sources

As mentioned previously, gathering useful data from pre-existing sources is a challenge. Ideally, one would design the dashboard within the enterprise resource planning (ERP) system and automatically extract the data from the database maintained by the ERP system. In this case, the software had to extract data from a number of sources and place them in a data warehouse that would feed the dashboard. To speed up the implementation, Maersk FPSOs introduced a Web-enabled form to facilitate manual data input to the data warehouse, while working to automate the data transfer in parallel with the rollout plan. These concurrent efforts are depicted graphically in Figure 5.

7. Roll out

When the discussion turned to rollout, one of the operation managers very bluntly stated: "I am sick of crappy implementations; this needs to be done properly!" Indeed, implementations to an offshore workforce are always a challenge because the crew rotation dictates a period of several weeks before everyone sees the new tool. Another challenge was dealing with a group that already had a high safety focus and was constantly exposed to safety messages and campaigns. Considerable effort was applied to getting the message across in a well-orchestrated manner.

Aside from the use of flyers, intranet, merchandise, etc., the rollout also used:

- Films on real process safety events from the units
- Touch screens for the dashboard

Films were shot of real process safety incidents - incidents that under different circumstances could have had a much more severe outcome. These films were made on location using personnel actually involved in the incident, making them much more credible and relevant.

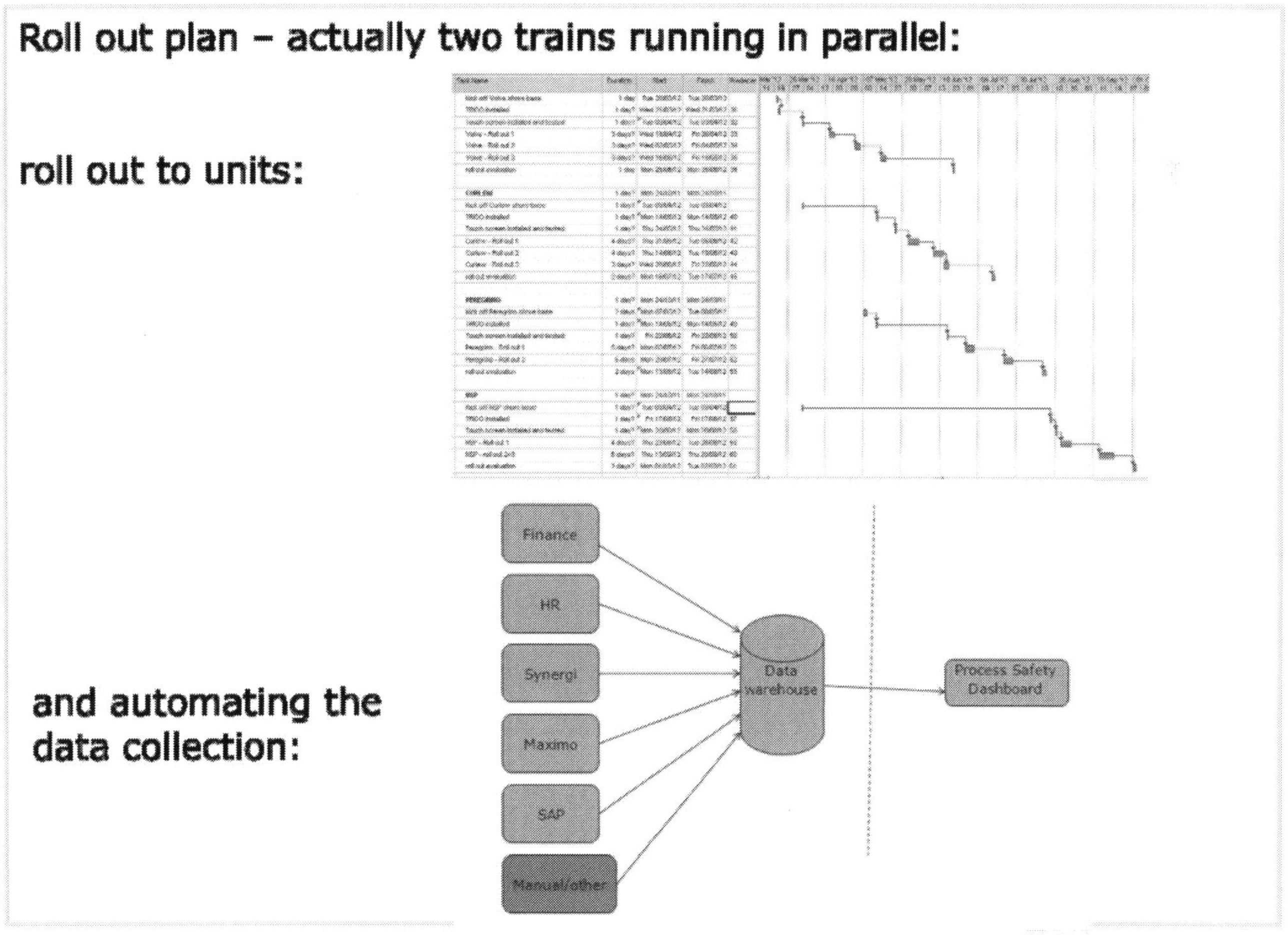

Fig. 5. Rollout Plan and Data Collection Architecture

Touch screens were used in an effort to make the dashboard inviting and interactive. Most offshore units do not have access to high-capacity communication lines, so accessing data directly from Maersk FPSOs' intranet was not a good option. As shown in Figure 6, dashboard displays were placed on board units in locations where most people pass on a daily basis and were powered by a local computer to ensure a quick response time. Having the dashboard so visible on board made it easier to engage everyone. They could also see the safety status of their unit and follow how indicators changed based on what they did, what they failed to do, and events that occurred on their unit.

Fig. 6. Operational Dashboard On Board an Operating Unit

Similar touch screens were placed in onshore bases and the main office to ensure that the entire organization was aware of the process safety status for each unit and the message was consistent.

The dashboard helps ensure that process safety indicators not only are reported to senior management, but, that they are seen and used at the producing units by frontline employees who directly influence/control the process safety performance. More importantly, the dashboard allows everyone to (1) drill down to determine what is causing a particular layer to be yellow (or red) and (2) quickly access the trends for each of the 29 indicators.

8. Results

By selecting 29 important indicators; agreeing to levels for green, yellow, and red performance levels; and displaying this information, Maersk FPSOs simultaneously (1) highlighted the importance of process safety, (2) modified behaviors and reprioritized key work activities, and (3) achieved positive results.

This effort has provided measurable benefits. Figure 7 shows the improvement in a number of indicators that are part of "Layer 2" (organizational behaviors and attitudes). All five are trending in a positive direction.

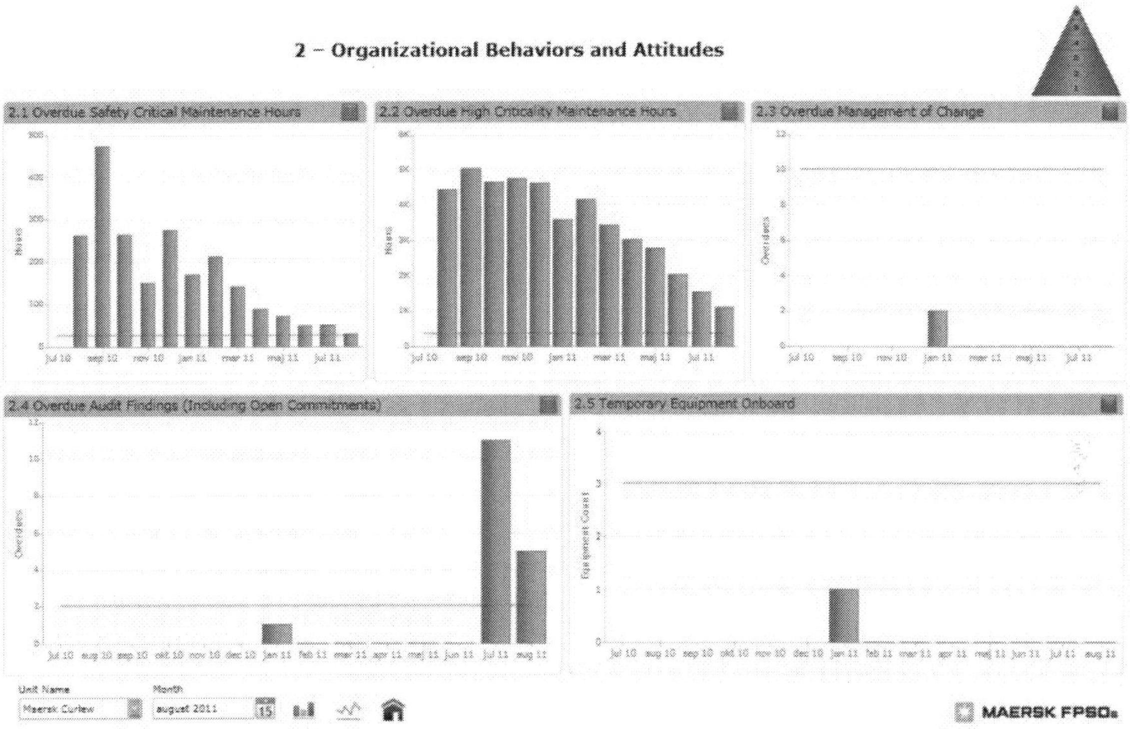

Fig. 7. Trends for Layer 2 Indicators

Figures 8 and 9 provide larger views for metrics 2.1 and 2.2 shown in Figure 7.

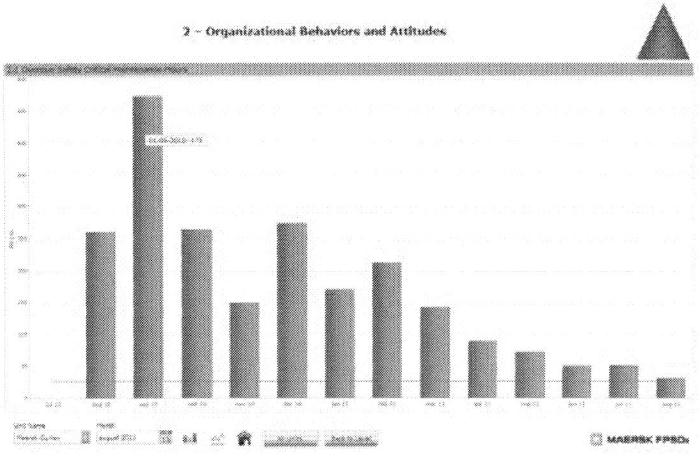

Fig. 8. Overdue Safety Critical Maintenance Backlog Reduced from 473 Hours in September 2010 to 30 hours in August 2011

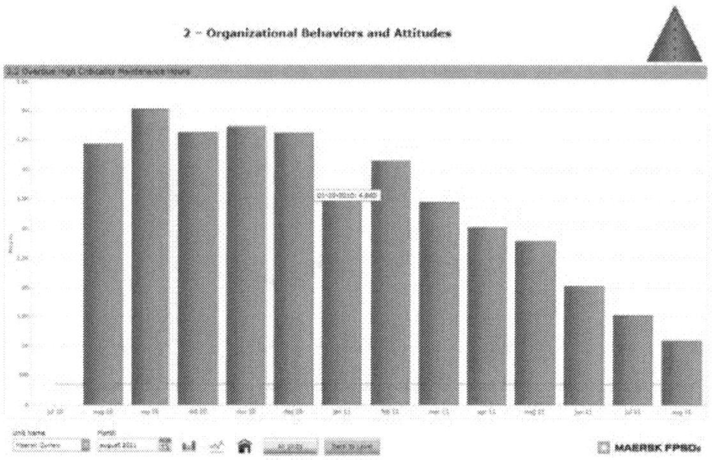

Fig. 9. Overdue High Criticality Maintenance Hours Reduced from 5,037 Hours in September 2010 to 1,097 Hours in August 2011

Since the initiation of this project, metrics have played a critical role in improving process safety performance at Maersk FPSOs. In addition to providing the measurable benefits shown in Figures 7 through 9, the dashboard described in this paper has greatly increased awareness of, and focus on, process safety performance at all levels of the organization.

References:

1. The Report of the BP U.S. Refineries Independent Safety Review Panel, January 2007.
2. Elliott, M.R., Kleindorfer, P.K., Dubois, J., Wang. Y., and Rosenthal, I., "Linking OII and RMP*Info Data: Does Everyday Safety Prevent Catastrophic Loss," *International Journal of Risk Assessment and Management*, 10, 130-146, 2008.
3. Center for Chemical Process Safety, *Guidelines for Risk Based Process Safety*, American Institute of Chemical Engineers, 2007.
4. OGP Report No. 456, *Process Safety – Recommended Practice on Key Performance Indicators*, International Association of Oil and Gas Producers, November 2011.

(43a) Learning From High Potential Near Misses in the Refining Sector

Ramesh VM, Research Scholar, Bharathiar University, 641046, India and Dr. Sakthivel R, Management Studies, Govt Arts College, 641018

Abstract: Safety is the first priority for any industry. Though steps are taken to ensure an accident-free environment, "Near Miss" is one phenomenon whose possibility of occurrence cannot be ruled out completely. What matters here is the way in which such instances are tackled preventing accidents. Moreover, these incidents can be used as an effective tool in education and training. Each near miss thus offers a free learning tool and a whole new experience with a scope to preach and practice efficient safety mechanisms. Near misses can be classified into two types- Low potential near miss and High potential near miss. All near misses whether it is of Low or high potential require in-depth analysis leading to identification of the root cause so as to avoid the recurrence of such incidents in future. The outcome of every near miss is to reach down the line so as to spread the useful information to all working in the industry. Many workers do not report "near misses" due to fear of negative actions from the management that can deprive them of their future progressions in the Industry. Steps need to be taken to overcome this fear as capturing of all near misses will definitely help the industry in bringing out innovative ideas to ensure safety. The exercise of accident prevention will bear fruits only if the effort to spread the message of usefulness of near miss reporting is spread down the line. Some of the near misses that occurred in the refinery sector are identified to be of High potential and major mishaps were averted only because of the timely action in the second of need. The ability of the concerned to do the right job at the right hour can be attributed to the training given to every employee on fire fighting which includes the analysis of near miss as well as the analysis of various other accidents happened in the other industries.I wish to share few high potential near misses which can be a learning to all working the refinery sector.

See more of this Session: Electronic Poster Session
See more of this Group/Topical: Global Congress on Process Safety

Felt Leadership:
An Essential Ingredient in Successful PSM Implementation

Brian D. Rains
DuPont Sustainable Solutions
974 Centre Road; CRP-730
Brian.D.Rains@DuPont.com

Prepared for Presentation at
American Institute of Chemical Engineers
2013 Spring Meeting
9th Global Congress on Process Safety
San Antonio, Texas
April 28 – May 1, 2013

UNPUBLISHED

Felt Leadership: An Essential Ingredient in Successful PSM Implementation

Brian D. Rains
DuPont Sustainable Solutions
974 Centre Road; CRP-730
Brian.D.Rains@DuPont.com

Keywords: Felt Leadership, PSM, High Performing Work Organizations

Abstract

"Felt Leadership" is a term that is frequently mentioned in today's PSM conferences and informal conversations as a quality that must be demonstrated by management if PSM implementation is going to be successful. Interestingly, these two words were first introduced by the DuPont Company over 20 years ago and the context was very different then and not even directly related to safety. Felt Leadership was one of the six attributes of a "High Performing Work System." As one of these 6 attributes, it was deemed essential to achieve outstanding operations and business results through people. This paper will not only describe "felt leadership" in meaningful, practical terms but will also deliver insights as to why it is both appropriate and necessary for effective PSM implementation as well as high performing work organizations.

1. Introduction

One of the clear lessons learned from the Baker Panel Report of the BP-Texas City incident was the role of leadership in successfully implementing and sustaining excellent PSM performance. Of the Panel's ten recommendations, five spoke directly to changes that leadership needed to make. One of the panel members, Dr. Paul V. Tebo, former DuPont Vice President of Safety, Health, and Environment, has actually stated that in his opinion, all ten recommendations speak to leadership. He went on to say that the fundamental, underlying issue at BP-Texas City was "leadership, leadership, leadership."

- **RECOMMENDATION # 1 – PROCESS SAFETY LEADERSHIP ****
- **RECOMMENDATION #2 – INTEGRATED AND COMPREHENSIVE PROCESS SAFETY MANAGEMENT SYSTEM**
- **RECOMMENDATION #3 – PROCESS SAFETY KNOWLEDGE AND EXPERTISE**
- **RECOMMENDATION #4 – PROCESS SAFETY CULTURE ****

- **RECOMMENDATION #5 – CLEARLY DEFINED EXPECTATIONS AND ACCOUNTABILITY FOR PROCESS SAFETY ****

- **RECOMMENDATION #6 – SUPPORT FOR LINE MANAGEMENT ****

- **RECOMMENDATION #7 – LEADING AND LAGGING PERFORMANCE**
- **NDICATORS FOR PROCESS SAFETY**

- **RECOMMENDATION #8 – PROCESS SAFETY AUDITING**

- **RECOMMENDATION #9 – BOARD MONITORING ****

- **RECOMMENDATION #10 – INDUSTRY LEADER**

**** Direct Leadership Responsibility**

As helpful and appropriate as these corporate recommendations were at the time and continue to be, the subject of PSM leadership can, at times, need to be brought down to a very individual and even personal level. In recent presentations at PSM Conferences, the term "felt leadership" has been used to attempt to describe a brand or style of leadership that is also necessary in complex chemical processes and operations. It has been used to describe a quality of leadership that can be sensed or "felt" rather than just heard or seen. It suggests leadership with passion and authenticity and that is genuine and even humble.

2. Origin of Felt Leadership in DuPont

The term "felt leadership" was first used in the DuPont Company in a systemic way in 1991. And the context was very different. In fact, it was not used in connection with safety in any direct way.

A group of Plant Managers, of which I was one, had been formed into a body called the "Plant Managers Board." We had a very specific task and that was to find and identify best practices related to "High Performing Work Systems" or HPWS for short. High Performing Work Systems were described then as "high involvement systems of accomplishing work in which all employees have developed the capability to connect with and drive the quality of business results to be world class."

This board was made up of approximately 10 Plant Managers from all regions of the world. Some were quite young whereas others were more experienced. Some managed very large plants whereas others managed small plants. (I was personally leading a site organization of 20 DuPont employees.) Some spoke English as a second (or third) language whereas others only spoke English. Some were engineers whereas others had educational backgrounds spanning the disciplines. We came together face-to-face at least quarterly for roughly two years. And we almost always visited different operating sites to observe and analyze. Sometimes the facilities were DuPont's whereas in other cases, we visited such diverse facilities as the Honda car assembly plant in Marysville, Ohio, USA.

After a visit to a DuPont facility in rural Illinois, this group met together in a hotel conference room and attempted to synthesize our collective observations into something usable and concise; something that we could attempt to package and leverage across DuPont Operations. We filled tear sheet after tear sheet of paper until we literally covered the four walls in the room. We had seen and observed much, even too much as there was so much data.

Then, the key moment arrived. One of our members, who later rose to become DuPont's Sr. Vice President of Operations (and interestingly spoke English as a second language) took the floor and articulated in simple terms what became to be known as the "Six Attributes of HPWS." The first attribute was "Vision, Mission and Strategic Intent Known by All." The second was "Committed and Dedicated Leadership 'Felt' within the Organization", or simply "Felt Leadership."

Six Attributes of High Performing Work Systems in DuPont

1. Vision, mission, strategic intent known by all.
2. Committed and dedicated leadership "felt" within the organization.
3. Every person knows the paying customer and how their work impacts the customer.
4. Teamwork and people working in teams is part of the structure.
5. Everyone learns, and teaches. "Learn-Teach-Learn".
6. Principle driven.

Phrases that were used at the time to further clarify the meaning of Felt Leadership included:

- Creates vision and energy
- Stretches goals
- Invites positive push back
- Results and output oriented
- "Felt" sense of purpose

Further insight into what was the intended meaning of Felt Leadership was captured in a "Six Attributes of HPWS Self-Assessment Tool." The questions asked were:

- Do the leaders have a clear vision for change based on needs of all stakeholders?
- Are the leaders visible, knowledgeable and open with people?
- Do the leaders value upgrades from anywhere?
- Do the leaders continually set objectives that demand that people develop and grow skills needed to move toward the vision?
- Do the leaders engage people in a process that builds accountability, willingness, and confidence to act in an empowered way?

3. Felt Leadership and PSM

With all this as a backdrop, what really does "Felt Leadership" have to do with Process Safety Management?

A more descriptive term for what process safety is all about might be "Operational Risk Management." Operational Risk Management implies that:

- It is the operation or the materials processing activities that must be controlled.
- It is the reduction of risk to an acceptable level with the ultimate elimination of significant process incidents that is the goal.
- It is the management of these risks that is one of the organization's key responsibilities and challenges.

The degree of success that is achieved in the management of operational risks can be equated with the quality of decision making by and within the operations organization. And leadership is all about influencing and improving the quality of decision making. Thus, Felt Leadership can best be described in the context of PSM as the approach that leaders follow to insure the highest quality of decision making by the organization in the management of the existing operational risks. Decision making links Felt Leadership to PSM Performance. Felt Leadership becomes a necessary ingredient in successful PSM implementation and execution.

4. Demonstrating Felt Leadership

Within the organization of one of our global clients, inspired leaders captured the vision of their roles with respect to PSM and embraced the concept of providing Felt Leadership. But they did not know how to act differently or more effectively in the management of their significant operational risks. They understood the concept of Felt Leadership as it related to PSM but with no role models, they struggled to visualize or project themselves effectively into their/environment or culture in a uniquely additive way. They needed help and they asked for help. They wondered what supportive but probing questions they should even ask to strengthen a culture of PSM Excellence through Felt Leadership. We provided the following as initial guidance:

1. While visiting a control room and speaking to panel board operator, the focus of the discussion could be on safe operating limits for the process:
    - Does he/she know the safe operating limits, and the consequences of operation outside those limits?
    - Is he/she currently operating with any interlocks by-passed? If so, was the by-pass procedure appropriately followed?
    - How long has this situation (interlock bypassed) existed and how long is the by-pass condition expected to last?
    - How many panel alarms are there per shift? Is there any program to reduce the number of panel alarms?

- Is he/she aware of any work going on in the plant, so that he can alert people to any process emergency?

2. While visiting a plant, question the operations supervisor to see what work permits have been prepared for the day. Take time and review them, looking for the quality of the risk assessment and the precautions that have been put in place to mitigate the identified risks.

3. While visiting a plant, a review of a broad range of PSM elements can occur, such as asking operations personnel about:
   - Any incidents that have occurred in other similar operations.
   - The frequency and nature of emergency and disaster mock drills.
   - The number and nature of near misses that have occurred recently and the results of investigations into these and any other incidents that have occurred in the plant.
   - The results of recent PSM audits and what weaknesses were identified.
   - Whether there are any tests or process changes active in the plant.

4. In conversation with operators in the field, discuss the procedure governing the activity underway and how well it is written and followed.

The coaching of these leaders was actually very similar to what had occurred previously inside DuPont as we strived to strengthen our PSM Performance and the quality of Felt Leadership. Simple "Felt Leadership Cards" were printed to help Operations Leaders consider how best to demonstrate their commitment and passion for PSM.

These cards are still in use today and one is duplicated on the next two pages.

Personal "Felt Leadership" – PSM

What is my spoken and unspoken message on PSM expectations to the organization – integral to core values and business success or a "necessary evil"?

What have I done this month to review PSM metrics and performance indicators for the site?

What am I doing to reinforce Operational Discipline?

Is my organization providing adequate "rewards and recognition" for PSM accomplishments?

How is my organization performing versus annual PSM COT's and objectives?

Where do we need to enhance short and long term PSM resources and organizational capabilities?

What is the status of PSM integration improvements at recently acquired sites or businesses?

What are my direct reports doing to reinforce the value, expectations and accountabilities for PSM on an ongoing and consistent basis?

Are we providing enough support and resources to maintain site equipment and infrastructure?

What have I done personally to engage new leaders in my organization that plays key PSM roles to ensure effective MOC- Personnel?

Have I established clear measures on overall PSM program performance (not just a focus on incidents)?

The Goal is Zero

Ten Key PSM Questions for the Line Organization

1) What trends are your PSM tracking systems showing – do you have any overdue PSM action items within the work group?

2) What recent improvements or accomplishments in PSM does your organization feel good about?

3) What is your biggest concern about the area's PSM program right now and what are you doing about it?

4) What additional focus or support do you need, if any, to improve and maintain PSM?

5) What are your first party PSM auditing and incident trend analysis showing across operations?

6) What are you doing to improve your people's knowledge and capability in PSM to eliminate hazards and incidents?

7) How are you driving a consistent "no leak" and "no alarm" operating philosophy in the operating area?

8) What are you personally doing to demonstrate "felt leadership" in PSM on a routine basis and is it effective?

9) How do you ensure all your people training and written operating procedures up to date and complete?

10) Can you show the status of mechanical integrity tests and PSM work orders in the tracking system and are they up to date?

Other words have been used to describe how Felt Leadership applies to PSM Performance. The following ten basic characteristics have been drafted to summarize this thinking.

Ten Characteristics of PSM Felt Leadership

1. Be visible to the organization.
2. Be relentless about time with your people.
3. Recognize your role as a teacher/trainer.
4. Develop your own safety functioning skills and pass them along to the organization.
5. Behave and lead as you desire others to do.
6. Maintain a self-safety focus.
7. Confirm and reconfirm safety as the Number 1 VALUE.
8. Continuous emphasis and clarity around SHE expectations.
9. Show a passion for ZERO injuries, illnesses, incidents.
10. Celebrate and recognize "ZERO" successes.

Some additional guidance for each of these ten characteristics is included in the appendix.

Additional specific actions that management can take to demonstrate Felt Leadership and their commitment to safety include:

- Leading by example in following all safety rules and procedures;
- Showing a passion for safety & a genuine care for individuals;
- Leading safety meetings with direct reports;
- Including safety in all routine meetings and discussions;
- Participating in safety observations regularly;
- Reviewing/critiquing all incident investigations and personally getting involved in high potential and serious investigations;
- Meeting with your organization to express concern and reinforce the learnings from incident investigations; and
- Using company events to recognize/celebrate good safety performance by individuals and groups.

Other attributes of Felt Leadership can be demonstrated by always following this guidance:

- I always practice what I preach with reference to safety.
- I know all potential hazards in my area.
- I audit for hazards.
- I do not permit unsafe shortcuts in order to get a job completed.
- I check on the use of PPE.
- I correct unsafe practices without delay.
- I make sure workers have proper tools.
- I make sure tools are in good condition.

- I encourage reports on unsafe conditions.
- I promptly respond to workers' safety concerns.
- I investigate incidents promptly and fully.
- I encourage safe suggestions.
- I encourage housekeeping for safety.
- I consider my own health and safety.
- I regularly attend safety meetings.
- I set high standards and try to be the example for everyone else.
- I ensure that all workers are properly trained in safety.
- I stop work if necessary to correct unsafe practices.
- I remove unsafe workers from the job.
- I encourage safety communications and follow-up on corrective actions.

5. Conclusion

The successful implementation of robust Process Safety Managing Systems in complex chemical operations requires a high degree of order and high quality decision making. Leadership is fundamentally charged with creating a culture where all members of the organization respond to these needs and perform at a very high level. Although the term Felt Leadership was initially used in DuPont in the context of an essential ingredient of a High Performing Work System, its use as a descriptor of the necessary leadership qualities in Process Safety Management is both appropriate and accurate.

Felt Leadership is essential to the successful implementation and execution of Process Safety Management.

6. Appendix

Ten Characteristics of PSM Felt Leadership

1. **Be visible to the organization:**
    - "Walk the Talk".
    - Provide a positive example to your employees.

2. **Be relentless about time with your people:**
    - Find ways to spend time with employees/contractors.
    - This will encourage you people to do the same.

3. **Recognize your role as a teacher/trainer:**
    - Teach safety through your:
        - Own behavior
        - Interactions with your employees/contractors
    - You are always 'on stage'.

4. **Develop your own safety functioning skills and pass them along to the organization:**
    - Continuously develop and hone skills such as safety observations, incident investigations, improving leadership skills (Felt Leadership), etc.
    - Make sure your people have the time and resources to develop their safety skills.
    - Reinforce and support what they learn when they return from training.

5. **Behave and lead as you desire others to do:**
    - What people see you do is what they believe.
    - When you notice an unsafe situation and do something about it, you are behaving as you want all of your employees to behave.

6. **Maintain a self-safety focus:**
    - Assess yourself in terms of how 'good' or 'bad' a job you are doing.
    - Stop and think about – 'How did that go?'; 'What might I have done differently to get an even better result?'

7. **Confirm and reconfirm safety as the Number 1 VALUE:**
    - Make comments about safety as you walk throughout the work area.
    - Identify and reinforce what is being done right.
    - Identify unsafe conditions and behaviors and correct them before they cause injuries.
    - Learn what safety issues exist on the shop floor.
    - Learn how the issues are being handled.

- Determine how well employees understand and apply the core principles in their work.

8. **Continuous emphasis and clarity around SHE expectations:**
    - Establish and publicize SHE expectations.
    - Include expected safety behavior.
    - Continuously emphasize their importance by:
        - Including them in communications.
        - Weaving them into most, if not all, conversations you have with your organization.

9. **Show a passion for ZERO injuries, illnesses, incidents:**
    - Make clear your expectation that 'zero' is the goal.
    - Put safety upfront on all interactions.
    - Ensure that your safety performance expectations of your contractors is the same as your own employees.
    - Require that everyone work on a steadily improving path in our journey to ZERO.
    - Periodically focus special attention on an aspect of needed safety improvement; be specific.

10. **Celebrate and recognize "ZERO" successes:**
    - Live safety, not just preach it.
    - Recognize and reward improvement and success.
    - Recognize the achievement at all levels and make sure leadership recognizes and shows appreciation.
    - Recognize specific aspects of safety that have improved as a result of targeted efforts.

9TH GLOBAL CONGRESS ON PROCESS SAFETY

Localization of control rooms using a MINLP approach, based on accidental explosion scenarios

Jaime E. Cadena
Chemical Engineering Department, Universidad de los Andes
Cra 1 Este No 19A – 40, Bogotá, Colombia
je.cadena21@uniandes.edu.co

Lina M. Sierra, Sergio E. Rodríguez, Manuel A. Ramos, Jorge M. Gómez, Felipe Muñoz
Chemical Engineering Department, Universidad de los Andes
Cra 1 Este No 19A – 40, Bogotá, Colombia
lm.sierra443@uniandes.edu.co, se.rodriguez68@uniandes.edu.co, ma.ramos43@uniandes.edu.co, jorgomez@uniandes.edu.co, fmunoz@uniandes.edu.co

Prepared for Presentation at
American Institute of Chemical Engineers
2013 Spring Meeting
9th Global Congress on Process Safety
San Antonio, Texas
April 28 – May 1, 2013

UNPUBLISHED

Localization of control rooms using a MINLP approach, based on accidental explosion scenarios

Jaime E. Cadena
Chemical Engineering Department, Universidad de los Andes
Cra 1 Este No 19A – 40, Bogotá, Colombia
je.cadena21@uniandes.edu.co

Lina M. Sierra, Sergio E. Rodríguez, Manuel A. Ramos, Jorge M. Gómez, Felipe Muñoz
Chemical Engineering Department, Universidad de los Andes

Keywords: facility layout, optimization problem, uncertainty reduction, Monte Carlo simulation, UVCE modeling

Abstract

Facility layout for the process industry plants is an important element in the basic engineering stage, which is fundamental in the design of the process, influencing significantly the future safety performance of the operation and its related costs. A control room serves as an operations centre where the facility can be monitored and controlled, therefore, its location and design is a priority for the safety of the process unit operation. Given that the control room is the 'core' of control and monitoring of the process, any significant damage to it resulting from a process incident might jeopardize the whole process and generate subsequent incidents. This work presents a novel methodology intended to serve as an additional tool for aiding the decision-making process regarding the location of control rooms, taking into account the risk of unconfined vapor cloud explosion (UVCE) and several passive barriers against it. The methodology is based on a mathematical model that bonds the design and localization of a control room into a computational tool, which has been developed taking into account key characteristics of a process or plant layout for explosion scenarios. The tool combines stochastic estimation of atmospheric parameters through raw Monte Carlo simulations with a mixed integer nonlinear programing (MINLP) optimization to determine the control room location. This process also helps to choose the best construction strategy for the control room among a portfolio of strategies fed to the model by the user. The final results of the methodology are the optimal location for a control room in an existing facility and the minimum costs related to the risk level to which it is exposed.

1. Introduction

A process incident such as a fire, explosion or dispersion can cause injuries and/or fatalities, but also a disruption of operations and undesired consequences, business-continuity wise. This makes determining an optimal facility layout and the location of the control room, one of the priorities for safe operations within a production plant. Given that facility layout is one of the elements determined in the early plant design stages, it also represents one of the first safety barriers in a hazard-incident representation such the one presented in Figure 1.

Figure 1: Hazard-Incident representation and location of facility layout as a safety barrier (FLB) and additional safety barriers (SB) [1].

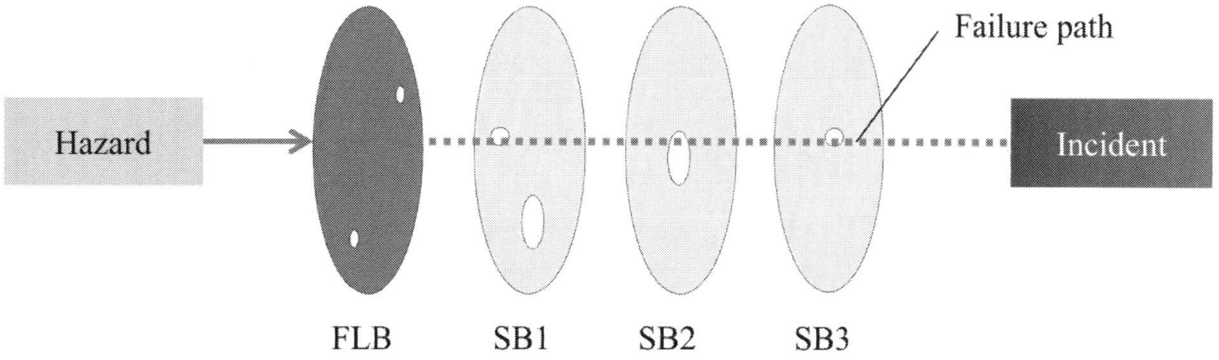

Figure 2 represents the desired function of facility layout as a safety barrier, including the adequate location of the control room. However, this desired representation is not always taken into account by the design philosophy implemented, which leads to process incidents with low probability of occurrence and high consequences, such as Flixborough (1974), Texas City (2005), Buncefield (2005), T2 Labs (2007), Puerto Rico (2009) and Amuay (2012) [2]. Although these incidents involve facility layout issues, Flixborough and Texas City involve serious damage to control rooms and its consequences has made them examples of scenarios and failures that should always be taken into account when determining the final facility layout.

Traditionally the facility layout problem has been approached with heuristic and technical recommendation published by several organizations such as the Center for Chemical Process Safety (CCPS) [3]. However, the integration of such approaches with current edge-cutting engineering tools such as non-linear optimization algorithms and solving platforms allow presenting novel approaches according to characteristics of the problem. In general, three types of problem can be identified: formulation, space treatment, unknown parameters estimation. For each type of problem a different approach is needed and this is summarized in Table 1.

Table 1. Approaches used depending on type of problem

Type of problem	Solution approach	Advantages	Disadvantages
Formulation	Linear [4]	The solution of the problem is usually simple and assures a global optimum under linearity assumptions	Does not allow the modeling of non-linear physic-chemical behavior and/or requires simplifications and strong assumptions
	Non-linear [3][5-7]	Applicable approach to complex systems with a high degree of non-linearity	Obtaining a solution requires more iterations than with a linear approach
Space treatment	Continuous [3][5-7]	Allows determining the exact location of buildings and equipment. Use of different shapes to represent footprints.	Complicated mathematical formulation. Obtaining a solution requires high computational times.
	Discrete [4]	Simple formulation and the solution	Exact locations are not possible. Shapes for

Type of problem	Solution approach	Advantages	Disadvantages
		can be obtained after a few iterations.	footprints are limited to squares o rectangles.
Unknown parameters estimation	Stochastic [3][5-7]	Better representation of real behavior.	Complex procedure that demands large amounts of input data.
	Deterministic [6]	Simple procedure based on worst-case scenario approach.	Derived distribution show few applications.

This work defines describes a methodology which identifies the sets of parameters, variables, functions and restrictions that represent the control room's location problem. Furthermore, the methodology implements a strategy that allows reducing the atmospheric conditions uncertainty when modeling UVCE and dispersion scenarios. The output of the modeling (given by overpressure) allows determining risk values, with consequences on people and infrastructure. Finally a case study is presented and the results analyzed.

2. Methodology

Although no exact and unique solution exists for a highly complex engineering problem such as facility layout, the proposed methodology (Figure 2) intends to support the decision making process in which final layouts are determined and the protection strategies are selected. The methodology is configured as a non-linear optimization of discrete control room's location, having an underlying process of stochastic parameters (atmospheric, risk) estimation. This configuration of the problem requires the following assumptions:

- A square process plant area is selected
- The process plant has only one hazardous process unit, hence only one hazards source
- The process plant area is divided into a uniform computational grid
- A control room can occupy one and only one node of the computational grid
- Historic atmospheric data must be available

The methodology begins by characterizing the hazardous process unit in terms of the possible Loss of Primary Containment (LOPC) scenarios, allowing an estimation of the liberation quantities of hazardous materials (HazMat). From the LOPC, failure trees can be depicted, allowing a classical calculation of the UVCE occurrence probability. Here the liberation of HazMat stops driving the analysis and focus is shifted towards the next driven force: atmospheric conditions. In order to take into account real atmospheric data of the determined location of the plant, it is necessary to acquire the historical atmospheric data, process it, organize it and finally fit it to probability distributions. These distributions are the input for the next step, the Monte Carlo simulation. Since the final aim is to obtain the resulting consequences of the UVCE, both dispersion and an explosion models are implemented within the Monte Carlo simulation that has a fixed number of iterations, *N*. The output of the Monte Carlo simulation is a probability distribution of overpressure-impulse measurements for each node of the grid. This constitutes a risk map, which combined with the calculation of the damages that a determined level of overpressure can cause to humans or infrastructure (i.e. control room) at a specific node, leads to the consequences calculation. The previous steps and costs information on materials, construction and installation of the control room, feed the non-linear optimization algorithm with the required data to run it and obtain an optimum control room's location.

Figure 2: Detailed steps of the proposed methodology

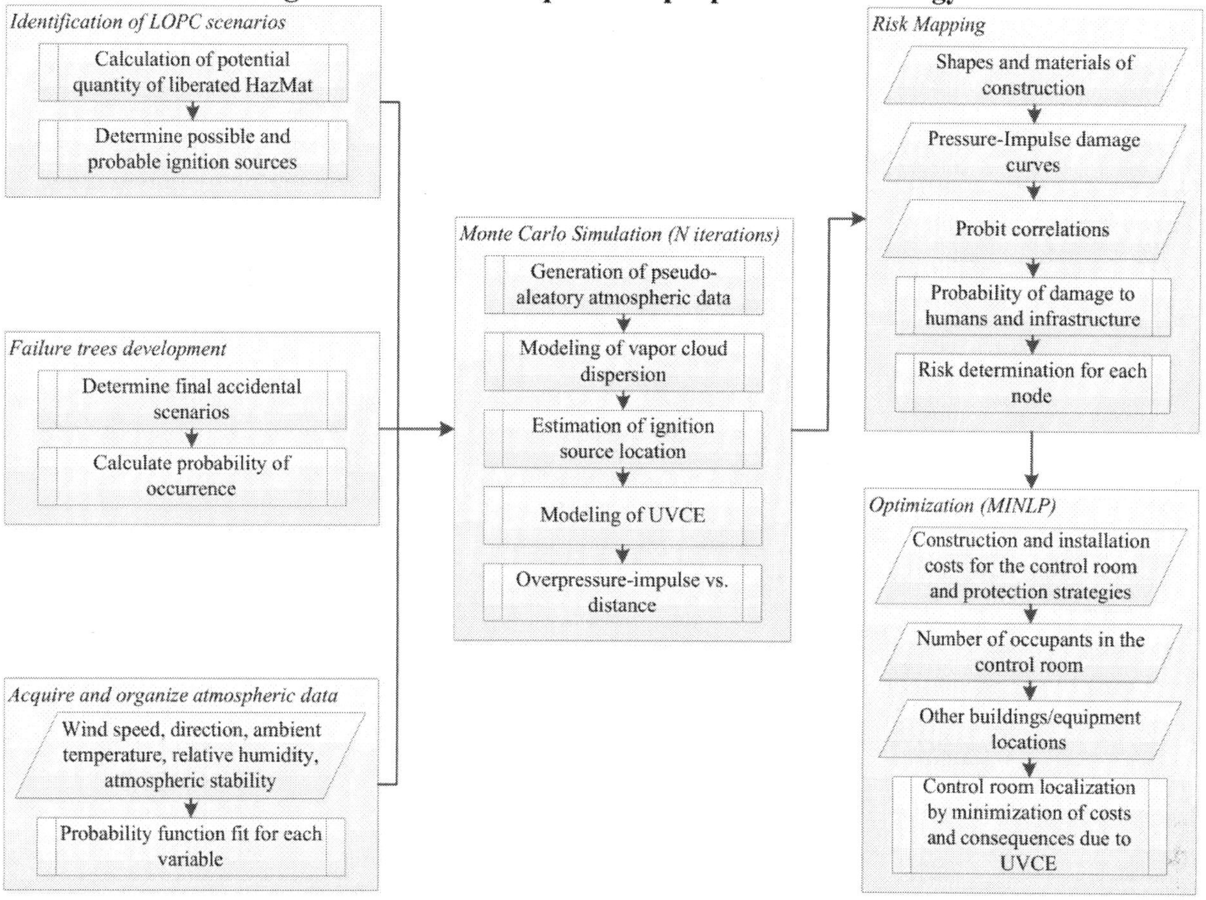

Following the general description of the methodology, it is implemented on a case study and the results are analyzed to establish advantages, disadvantages and improvement opportunities to be implemented in future work.

3. Case study: Hexane-Heptane separation unit

In order to implement the proposed methodology, a facility at Corpus Christi, TX, USA is selected as case study [13]. The hazardous process unit consists of a distillation unit and the following assumptions apply to the subsequent analysis of the system:

- The HazMat consists of 28 tons of flammable material
- The operating temperature of the distillation tower is 130°C and 160°C for top and bottoms, respectively
- The operating pressure is 4 barg
- The vapor transport pipeline has a 0.5 internal diameter and a 25 m length
- The liquid transport pipeline has a 0.15 internal diameter and a 25 m length
- The process plant area is 100 m in length and 100 m wide

- The grid consists of uniform nodes of 10 m by 10 m (see Figure 3)

The previous information and assumptions allow implementing the methodology in 7 sequential steps. The following is a summary of the most important developments in each step.

Figure 3: Representation of the grid and the nodes (labeled) in the process plant area

k01	k02	k03	k04	k05	k06	k07	k08	k09	k10
k11	k12	k13	k14	k15	k16	k17	k18	k19	k20
k21	k22	k23	k24	k25	k26	k27	k28	k29	k30
k31	k32	k33	k34	k35	k36	k37	k38	k39	k40
k41	k42	k43	k44	Hazardous process unit		k47	k48	k49	k50
k51	k52	k53	k54			k57	k58	k59	k60
k61	k62	k63	k64	k65	k66	k67	k68	k69	k70
k71	k72	k73	k74	k75	k76	k77	k78	k79	k80
k81	k82	k83	k84	k85	k86	k87	k88	k89	k90
k91	k92	k93	k94	k95	k96	k97	k98	k99	k100

3.1 Hazards identification

This step aims at identifying a representative group of scenarios that can result in a UVCE accidental event due to the LOPC of the involved HazMat. The found applicable scenarios are catastrophic failure of a vessel within the unit (column, condenser, reboiler), catastrophic failure of the vapor transport pipeline and catastrophic failure of the liquid transport pipeline.

The use of proper source models allow calculating that for both pipeline failures, the total HazMat available would be released in less than 120 seconds in vapor state at averaged atmospheric conditions. Therefore, it is assumed that any of the three mentioned scenarios has a leak/liberation time of 120 seconds. After the characterization of the LOPC, a event tree is developed to obtain the frequency of occurrence of the possible accidental events, using annual occurrence frequencies and event probabilities within the tree, obtaining the results presented in Table 2 [13].

Table 2. Annual occurrence frequencies for the identified accidental events [11]

Scenario	Probability	Frequency (year^{-1})
UVCE	0.34	6.5 x 10^{-6}
Flash Fire	0.34	5.1 x 10^{-6}
BLEVE	0.25	5.1 x 10^{-6}

3.2 Atmospheric data

A 10 year period (2002-2012) of atmospheric data with entries by the hour for the Corpus Christi area was obtained from the National Climatic Data Center (NCDC-NOAA) [19]. Wind speed & direction, ambient temperature, cloud coverage and relative humidity were extracted from the whole data set. This filtered data set allowed determining the Pasquill stability class each entry [8], using the relationship between solar elevation angle (determined an NCDC tool [17]) and cloud coverage as shown in Table 3. The final results of this step allow obtaining 109000 data registries, input for the next step.

Table 3. Relationship between solar elevation angle and cloud coverage

Cloud coverage	Solar elevation angle α		
	60°< α	35°< α<60°	15°< α<35°
Less than 4/8	Strong	Moderate	Minor
Medium altitude clouds	Moderate	Minor	Minor
Low altitude clouds	Minor	Minor	Minor

3.3 Dispersion and explosion models

The SLAB model negative floatability gases, is selected to model the dispersions due to its fit to the case study conditions. Such conditions involve a HazMat (hexane-heptane mixture) with a density higher than that of the air by a 2.5 factor. This model calculates the atmospheric dispersion solving a variant for mass, momentum, energy and species spatial and temporal averages [22]. Additionally, the SLAB model is considered a 'top hat' model, presenting the great advantage of its repetitive use within the Monte Carlo simulation. Once the model is applied, a concentration-distance profile is generated for each data set being evaluated and the ignition point is located as shown in Figure 4. An analysis of the model and the results obtained show that the accuracy of the SLAB model in the near-field is poor [22].

The TNT model is selected for the UVCE intensities calculation (overpressure and impulse) for each node of the computational grid imposed on the process plant area. Even though this model can be considered as the simplest one, its easy implementation in line with the Monte Carlo simulation and its well-known limitations [23] make it an attractive option. This model relies on an explosion efficiency factor (η) that is usually assumed to be between 1% and 10%, however a fixed 1% value is assumed since higher values presented unrealistic results in the near field when compared to typical UVCE overpressure values [4][8][25] .

Figure 4: Assumptions regarding ignition source location

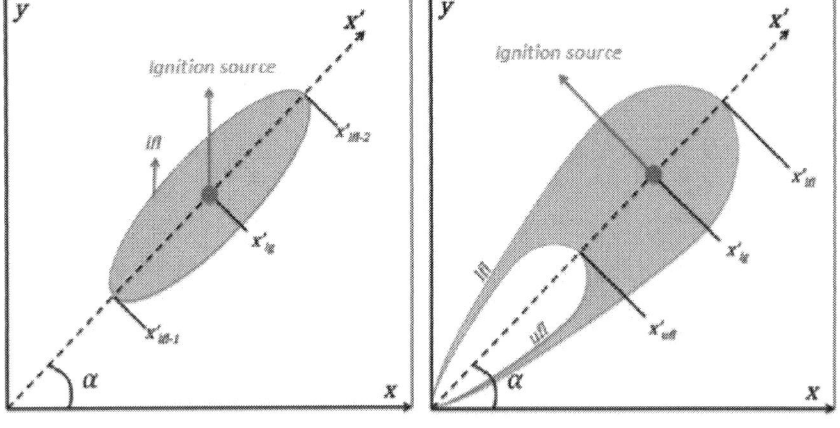

Figure 5: Example of a concentration-distance profile with its ignition source located

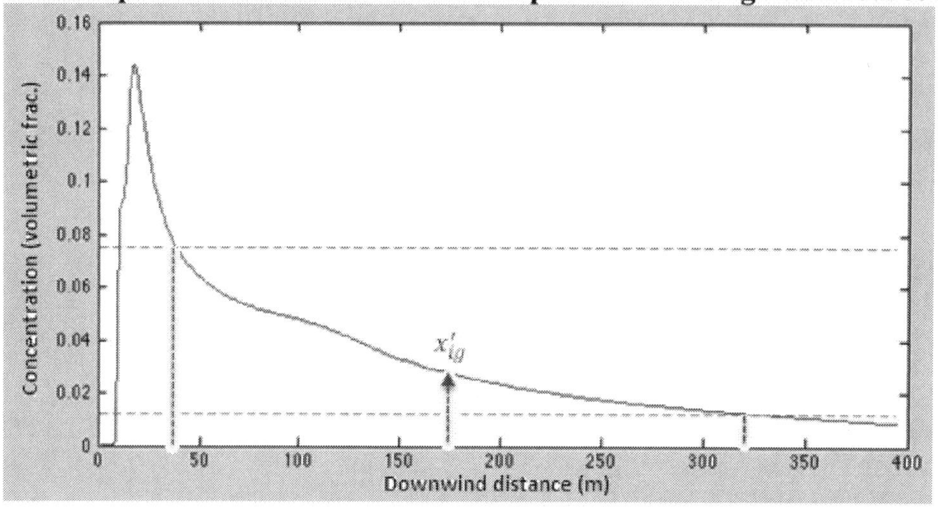

3.4 Monte Carlo simulation and risk mapping

Steps 3.1 and 3.2 are needed inputs for the Monte Carlo simulation and its output is given by the probability distributions of the fitted results of the *N* iterations performed over the dispersion and explosion models. For this case study the number of iterations was taken as 1000 as an initial guess, later proving to be a reasonable number of iterations when comparing computational time against the results' mean behavior. The results (profiles of concentration and overpressure against distance) are organized in form of histograms. From these results the expected values of overpressure (Figure 5) and impulse can be calculated for each node, necessary input to develop the risk maps. It must be noted that the expected value of the wind direction coincides with the non-symmetrical trend in the intensities maps such as Figure 5.

Risk mapping consist on assigning risk values or scores to each node of the grid taking into account structural damage to infrastructures and human affectation. The risk values for infrastructure damage depend on the type of protection strategy selected for the control room, having four different strategies as follows:

1. Steel reinforcement, walls (panels), roof and frame
2. No reinforcement, concrete reinforced walls, concrete roof and steel frame
3. Prefabricated concrete coating, concrete reinforced roof and walls, and steel frame
4. Reinforced masonry walls, light steel roof and no frame

For each construction configuration for the control room, there is an associated risk map calculated from Pressure-Impulse curves (PI) and Probit approach for human affectations. Once this risk maps are established assuming a Risk = Probability x Consequence relationship, the only step left previous to the optimization is the determination of additional parameters such as related costs of installation, construction and the location of buildings and equipment in the process plant area[4][5][12][18].

Figure 5: Overpressure map on the process plant area

35.01	47.41	29.74	29.24	28.53	26.52	24.48	22.61	21.11	20.09
30.49	30.84	30.17	31.36	32.44	31.44	28.44	25.06	22.39	20.70
28.73	29.89	33.48	40.49	48.05	47.33	38.31	29.73	24.43	21.57
28.45	31.39	40.55	64.92	109.75	109.16	63.14	37.52	26.97	22.47
28.60	32.77	48.20	109.78	Hazardous process unit		108.27	45.59	28.86	22.93
28.77	32.17	47.52	109.14			107.84	45.21	28.51	22.59
28.02	29.42	38.43	62.97	108.03	107.67	61.84	36.37	25.94	21.48
27.98	26.05	29.62	37.07	45.07	44.77	36.12	27.84	22.77	20.04
25.63	22.86	23.86	26.08	27.93	27.70	25.36	22.47	20.20	18.80
21.74	20.23	20.29	20.92	21.42	21.29	20.50	19.47	18.66	18.41

3.5 Optimization of the control room's location

The output of the 3.4 step along with some additional parameters, represent the needed input for to carry out the control room's location optimization. This step is performed through a MINLP formulation [24] of the optimization problem and a Simple Branch and Bound (SBB) solving algorithm, implemented on the GAMS software platform [25]. The optimization problem is structured as follows:

$$\begin{aligned} \text{min}: \quad & z = f(x,y) \\ \text{s.t.} \quad & gj(x,y) \leq 0 \qquad j \in J \\ & x \in X \subseteq R^n \\ & y \in \{0,1\}^m \end{aligned}$$

The formulation can be divided into: parameters, sets, variables, constraints and objective function. A brief description of each one is presented as follows:

1. Parameters: the required parameters to solve the optimization problem are comprised of the protection strategies, PI curves, Probit parameters for human affectation, human occupancy, risk maps, construction/interconnection/installation costs, minimum separation distances, existing buildings/equipment & related location and the coordinates of each node within the computational grid (x_k, y_k).

2. Sets: the problem is divided in three sets. The first one (*F*) represents the existing buildings/equipment, the second one (*K*) the node set and the third one (*R*) the protection strategies.
3. Variables: the optimization problem solves two different kind of variables, integer and continuous. The former are given by binary variables that define the location of the control room at a specific node (*k*) and the active protection strategy (*r*). The continuous variables are given by the coordinates of the control room's location (x_c, y_c), but this is merely a formulation requirement since the possible value of these variables is given by the values of the node coordinates (x_k, y_k) specified as parameters.
4. Constraints: the following constraints assure that the obtained solution in the optimization is within physical and logical bounds:
    a. Control room's coordinates: the sum of the node-coordinate by binary variable of location product, of each node allows to determine the location of the control room.
    b. Unique location: the sum of the binary variable of location for each node must equal one.
    c. Protection strategy: the sum of all the binary variables of active protection strategy must equal one.
    d. Separation distances: the control room must be located taking into account the minimum separation distances established. This is a Pythagoras distances' formulation and finds its origins in [26].
5. Objective function: this function comprises the total cost related to the control room's obtained location and the protection strategy selected. Being a cost function, the optimization problem seeks to minimize it.

Once the elements of the optimization problem are defined, it is executed. The results show that the optimum control room's location for the case study is the *k87* node (using the 3rd protection strategy). It is important to note that this location does not correspond to the node with the lowest risk score in the grid, on the contrary, its risk score is 'medium'. The selection of this node balances the cost components involved in the control room's construction, installation and other key parameters evaluated in the objective function.

Regarding the optimum protection strategy, it is found to be the one that resist the highest levels of overpressure even when it also represents the most expensive one. This variable (protection strategy) and its influence demonstrates that it directly affects the decision making process and has a direct relationship with the plant's safety and business continuity.

4. Conclusions

This work proposes a methodology that integrates economic and safety analysis in order to determine the optimum location of a control room through the use of risk scores (risk mapping) and MINLP formulation of an optimization problem. This methodology seeks to aid the decision making process of facility layout.

The proposed methodology was applied to a previously documented case study and the results of it showed that it can effectively locate the control room within a discretized process plant area, while taking into account different protection strategies related to the control room's construction.

Nonetheless, the proposed methodology has diverse and important improvement opportunities in the treatment of the process plant space and the selection of dispersion and explosion models. These models do not allow the modeling of chemical species mixtures and do not take into account the interaction with objects, opposed to a Computational Fluid Dynamics (CFD) approach. The assumption that a building occupies a whole node tends to overestimate the real size of the facilities, inducing uncertainty to the results.

5. Perspectives

Future work should be focused in incorporating risk acceptability within the optimization formulation, as well as the interaction of the dispersion and the blast wave with the facility's buildings and equipment, e.g. through a CFD approach.

A spatially continuous approach should be developed in order to capture in the highest detail the consequences to humans due to structural damages, hence increasing the models sensibility to the number of occupants inside the control room.

Finally, it would be a considerable step forward to include dynamic risk scores related to the location of diverse hazardous process units and the risk represented by these (even if considered fixed).

6. References

[1] Sklet, S. 2006. "Safety barriers: Definition, classification, and performance". Journal of Loss Prevention in the Process Industries, 19, 494-506, http://dx.doi.org/10.1016/j.jlp.2005.12.004.

[2] Vyas, K. & Minaya, E. 2012. "Explosion at Amuay refinery in Venezuela kills 41; fire still burning". Hydrocarbon Processing. Punto Fijo, Venezuela.

[3] C. O. Diaz, R. Vázquez-Román, S. Jung, and M. S. Mannan, "A Comparison of Deterministic and Stochastic Approaches to Solve the Facility Layout Problem with Toxic Releases," in 19th European Symposium on Computer Aided Process Engineering, vol. Volume 26, Elsevier, 2009, pp. 93–98.

[4] S. Jung, D. Ng, C. D. Laird, and M. S. Mannan, "A new approach for facility siting using mapping risks on a plant grid area and optimization," Journal of Loss Prevention in the Process Industries, vol. 23, no. 6, pp. 824–830, Nov. 2010.

[5] S. Jung, D. Ng, J.-H. Lee, R. Vazquez-Roman, and M. S. Mannan, "An approach for risk reduction (methodology) based on optimizing the facility layout and siting in toxic gas release scenarios," Journal of Loss Prevention in the Process Industries, vol. 23, no. 1, pp. 139–148, Jan. 2010.

[6] C. Díaz-Ovalle, R. Vázquez-Román, and M. Sam Mannan, "An approach to solve the facility layout problem based on the worst-case scenario," Journal of Loss Prevention in the Process Industries, vol. 23, no. 3, pp. 385–392, May 2010.

[7] R. Vázquez-Román, J.-H. Lee, S. Jung, and M. S. Mannan, "Optimal facility layout under toxic release in process facilities: A stochastic approach," Computers & Chemical Engineering, vol. 34, no. 1, pp. 122–133, Jan. 2010.

[8] D. A. Crowl and J. F. Louvar, Chemical Process Safety: Fundamentals with Applications, 2nd ed. Prentice Hall, 2001.

[9] C. Cannalire, R. Clarke, B. Mazzarotta, and R. Nicoletti, "Optimizing safety of control rooms in case of flammable releases," Journal of Loss Prevention in the Process Industries, vol. 6, no. 5, pp. 313–317, Oct. 1993.

[10] Q. Baker, Explosion Risk ans Structural Damage Assestment Code (ERASDAC). Arlington, VA: 30th DoD Explosive Safety Seminar, Department of Defense Explosive Safety Board, 2002.

[11] DDESB (Department of Defense Explosive Safety Board), "Technical Paper 14." Approved Methods and Algorithms for DoD Risk-Based Explosives Siting, 2009.

[12] API, "Risk Based Inspection." Base Resource Document. API Publication 581, 2000.

[13] Center for Chemical Process Safety (CCPS), Guidelines for Chemical Process Quantitative Risk Analysis, 2nd ed. Wiley-AIChE, 1999.

[14] S. Mannan, Ed., Lees' Loss Prevention in the Process Industries, Third Edition: Hazard Identification, Assessment and Control, 3rd ed. Butterworth-Heinemann, 2005.

[15] F. Díaz Alonso, E. González Ferradás, J. Francisco Sánchez Pérez, A. Miñana Aznar, J. Ruiz Gimeno, and J. Martínez Alonso, "Characteristic overpressure–impulse–distance curves for the detonation of explosives, pyrotechnics or unstable substances," Journal of Loss Prevention in the Process Industries, vol. 19, no. 6, pp. 724–728, Nov. 2006.

[16] M. Swisdak, Simplified Kingery Airblast Calculations. New Hampshire: Naval Surface Warfare Center, 1980.

[17] Q. Baker and C. Oswald, "Vulnerability model for occupants of Blast Damaged Buildings." 34th Annual Loss Prevention Symposium, Session 3, 2000

[18] Center for Chemical Process Safety (CCPS), Guidelines for Facility Siting and Layout, 1st ed. Wiley-AIChE, 2003.

[19] National Oceanic and Atmospheric Administration, "Climate data online: map and application search," National Climatic Data Center, 24-Oct-2012. [Online]. Available: http://www.ncdc.noaa.gov/cdo-web/. [Accessed: 24-Oct-2012].

[20] National Oceanic and Atmospheric Administration, "Solar Calculation Details," National Climatic Data Center, 2009. [Online]. Available: http://www.esrl.noaa.gov/gmd/grad/solcalc/calcdetails.html. [Accessed: 02-Oct-2012].

[21] Center for Chemical Process Safety (CCPS), Guidelines for Use of Vapor Cloud Dispersion Models, 2nd ed. Wiley-AIChE, 1996.

[22] D. Enmak, User's Manual for SLAB: An Atmospheric Dispersion Model for Denser-than-Air Releases. California: University of California, 1990.

[23] Van der Bosch, C. J. H., Weterings, R. A. P. M. & Stoffen, N. C. P. v. R. d. G. 1997. Methods for the Calculation of Physical Effects: Due to Releases of Hazardous Materials (liquids and Gases): Yellow Book, Directorate-General for Social Affairs and Employment, Committee for the Prevention of Disasters.

[24] J.A. Caballero, I.E. Grossmann. "Una revisión del estado del arte en optimización, Revista Iberoamericana de Automática e Informática Industrial". RIAI, 4 (2007) 5-23.

[25] A. Brooke, D. Kendrick, A. Meeraus, R. Raman. "GAMS User's Guide". Washington DC, USA, 1998.

[26] J.E. Cadena, M.A. Ramos, J.M. Gómez, F. Munoz, "Facility layout optimization detailed research through MILP and MINLP", in AIChE - 2012 AIChE Spring Meeting and 8th Global Congress on Process Safety, Conference Proceedings , 2012.

Risk analysis for Overpressure from Pressure Vessel Burst

Seungho Jung
Air Products and Chemicals, Inc.
7201 Hamilton Blvd.
Allentown, PA 18195
Jungs2@airproducts.com

Elizabeth M. Lutostansky (lutostem@airproducts.com)
Joan M. Schork (shorkjm@airproducts.com)
Air Products and Chemicals, Inc

Prepared for Presentation at
American Institute of Chemical Engineers
2013 Spring Meeting
9th Global Congress on Process Safety
San Antonio, Texas
April 28 – May 1, 2013

UNPUBLISHED

Risk analysis on overpressure of Pressure Vessel Burst

Seungho Jung
Air Products and Chemicals, Inc.
Allentown, PA 18195
Jungs2@airproducts.com

Elizabeth M. Lutostansky, Joan M. Schork
Air Products and Chemicals, Inc

Keywords: Pressure Vessel Burst (PVB), explosion, Blast curves, geometry factor

Abstract

A pressure vessel burst (PVB) is a type of explosion that involves the burst of a pressure vessel containing gas at elevated pressure. Any vessel that is pressured or has the capacity to build high pressure before bursting can be a candidate for generating a PVB. At burst, a blast wave propagates outward from the source by the sudden expansion of a compressed gas. Fragment hazards are also present when the shell of the vessel along with attached external accessories is thrown. PVBs are possible with vessels containing any gas or mixture including inerts and flammables.

Historically, there have been incidents related to PVBs. Work that has been done to predict the overpressure impacts is covered in Guidelines for Vapor Cloud Explosion, Pressure Vessel Burst, BLEVE and Flash Fire Hazards, 2nd Edition by CCPS [1]. Moreover, as Facility Siting has received significant attention recently, Air Products has included PVBs in Facility Siting studies to safely locate occupied building.

This paper focuses on how to select an appropriate PVB model and correctly use it in calculating vulnerability inside an occupied building. It provides an overview of a simplified method for estimating the characteristics of pressure vessel bursting for practicing engineers.

1. Introduction

Explosions can be classified into two types, oxidations such as Vapor Cloud Explosions (VCE), dust explosions and condensed phase explosion and sudden expansions such as Pressure Vessel Bursting (PVB), Boiling liquid expanding vapor explosions (BLEVE), and rapid phase transitions (RPT). Pressure vessel burst (PVB) is involves a rupture of a vessel containing gas at elevated pressure. The source of the energy released in the blast is the compressed gas. A BLEVE is caused by the rupture of a vessel containing a pressurized liquid above its normal boiling point.

With a BLEVE, the liquid in the vessel undergoes a phase transition from liquid to gas. This flashing contributes to the energy released. Many engineers become confused by the difference between PVBs and BLEVEs because of their similarities. Both liquid and gas may be present in vessels undergoing PVB or BLEVE and blast effect calculations look similar. It is important to remember that in the case of PVB, liquids present in the vessel do not undergo phase transition while for BLEVEs, the primary source of energy is the phase transition. .

The scope of this paper is limited to PVBs. It presents a simplified method for obtaining PVB overpressure results for a building siting study.

2. Methods for predicting blast effects from pressure vessel bursts

A blast curve method is highly recommended because it is more easily implemented than doing numerical simulations. There is an outstanding set of blast curves based on a systematic one-dimensional numerical study by Tang et al. [2]. The Baker-Tang curves assume all of the energy is employed to make a blast effect rather than to be used into the fragments as kinetic energy. These curves are for spherical, free air burst which do not include ground reflection.

Figure 1. Positive overpressure curves for various vessel pressures (courtesy copied from Guidelines for Vapor Cloud Explosion, Pressure Vessel Burst, BLEVE and Flash Fire Hazards, 2nd Edition by CCPS [1])

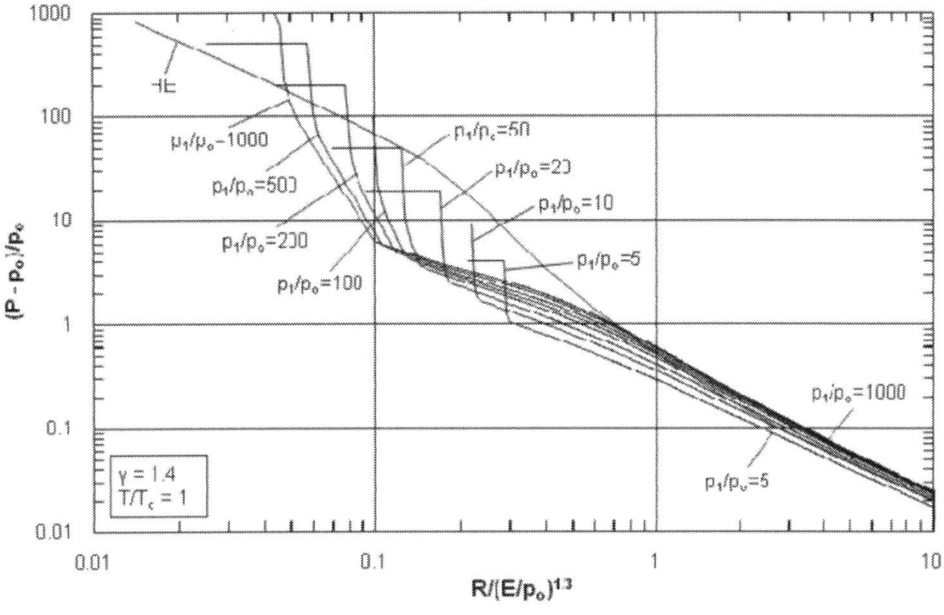

There were not much experimental data for PV burst but Table 1 shows the set of experimental data at relatively lower operating pressure. After early 1980s, there appeared many experimental studies that reported PV burst overpressure measurements but they cover different vessel shapes in different locations such as ground and air [3]. Therefore only data points with spherical free air burst are chosen to compare and validate against the Baker-Tang curves.

Table 1. Summary of literature data

Author	Year	Shape of vessel	Type of burst	Substances	Operating pressure ratio to atmospheric pressure (p1/p0)	Overpressure range
BAKER	1977	Spherical	Free air burst	Air, Argon	6	2.7~50 kPa
BAKER	1977	Spherical	Free air Burst	Freon 12	18 ~ 22	0.73~7.3 kPa

Figure 2 shows good validations against two experimental data sets qualitatively. As the peak overpressure line with the ratio, p1/p0=20 is higher than that with the ratio, p1/p0=5, the experiment data points with Air, Argon are distributed well above the data point with Freon 12.

Figure 2. Validation of Tang's curves on lower pressure cases against experimental data

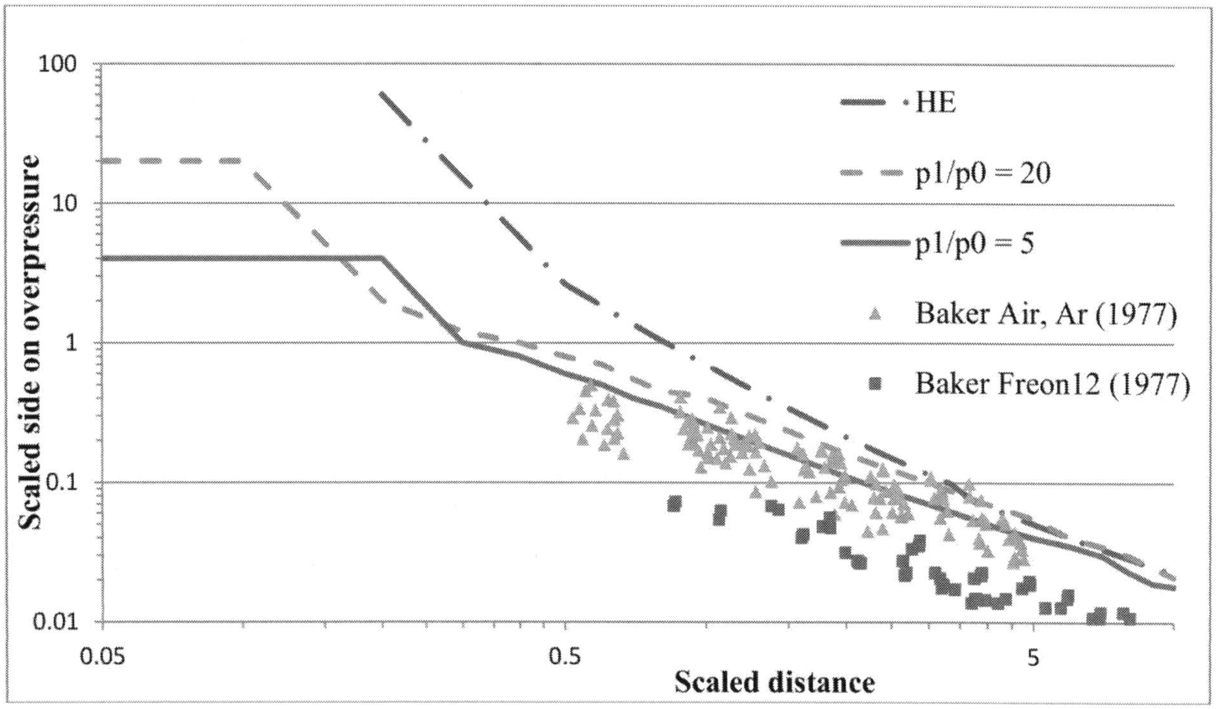

In the far-field, the HE line is close enough to all the other pressure ratio curves as seen in Figure 1. Though a recommendation was made to find the scaled side on overpressure based on p1/p0 ratio, it is quite helpful and faster to use the HE for the far-field further than scaled distance ($\bar{R}$) > 1. 6th order polynomial equation can better fit the HE curve in Figure 1 and suggested as

$$log\bar{P_s} = c_0 + c_1 log\bar{R} + c_2(log\bar{R})^2 + c_3(log\bar{R})^3 + c_4(log\bar{R})^4 + c_5(log\bar{R})^5 + c_6(log\bar{R})^6 \quad \text{[Eq. 1]}$$

where:

$\{c_0, c_1, c_2, c_3, c_4, c_5, c_6\} = \{-0.4102, -1.8282, 0.3661, 0.1122, -0.0830, -0.0042, 0.0046\}$

$\bar{P_s}$ is the dimensionless overpressure, or the overpressure scaled by ambient pressure (Multiply by 14.696 to convert to psig).

3. General calculating procedure for Pressure Vessel Bursts

3.1 Calculate PV energy

The internal energy U of a system is the sum of the kinetic, potential and intermolecular energies of all the molecules in the system. For an ideal gas it is

$$U = \frac{pV}{\gamma - 1} \quad \text{[Eq. 2]}$$

where p is the absolute pressure of the gas, V is the volume of the gas, and γ is the ratio of specific heats of the gas in the system, i.e. C_p/C_v. Therefore, the Brode's PV burst energy is

$$E_{ex,brode} = \frac{(p_1 - p_0)V_1}{\gamma_1 - 1} \quad \text{[Eq. 3]}$$

where the subscript 1 refers to initial state and the subscript 0 refers to the ambient conditions. For ground burst, the effective energy is doubled to account for the ground reflection as

$$E_{eff} = \frac{2(p_1 - p_0)V_1}{\gamma_1 - 1} \quad \text{[Eq. 4]}$$

3.2 Calculate scaled distance, $\bar{R}$

Once the explosion energy of a PV burst is estimated, the scaled distance, $\bar{R}$ can be obtained from

$$\bar{R} = R\left(\frac{P_0}{E_{eff}}\right)^{1/3} \quad \text{[Eq. 5]}$$

where R is the distance from the blast at which blast parameters are to be determined.

3.3 Determine scaled positive side on overpressure, $\overline{P_s}$

To use this graphical method, select the overpressure curve in Figure 1 that corresponds to the vessel burst pressure ratio p1/p0. Read the $\overline{P_s}$ = (p1-p0)/p0 value. For simplicity, if $\bar{R}$ is larger than 1, the equation correlated with the Highly Explosive (HE) line can be used

3.4 Adjust scaled positive side on overpressure, $\overline{P_s}$

There are two adjustment factors; one is for temperature, the other is for geometry. Higher temperature increases the overpressure near the vessel but the effect decreases with distance away from the vessel. If the target is far enough from the vessel, this adjustment effect is negligible and the use of ambient temperature in Figure 1 can be adequate.
Geometry adjustment factors are used with of cylindrical vessels and slightly elevated spherical vessels. The multiplier factors shown in Table 2, for$\overline{P_s}$ initially suggested by Baker et al. [4], are widely used.

Table 2. Factors for geometrical effect

Multiplication factors on $\overline{P_s}$ for slightly elevated spherical vessel	
$\bar{R} < 1$	2
$\bar{R} > 1$	1.1
Multiplication factors on $\overline{P_s}$ for cylindrical vessel	
$\bar{R} < 0.3$	4
$0.3 < \bar{R} < 1.6$	1.6
$1.6 < \bar{R} < 3.5$	1.6
$3.5 < \bar{R}$	1.4

However, these factors were made based upon high explosives experiments, and it was found by Geng et al. and written in the CCPS book [1] as the factors are not directly applicable to PVBs. It was suggested using a group of adjustment factor specifically for PVBs with bursting pressure ranges from 10 to 100 for both the cylindrical and elevated spherical PVBs. Note the factors vary continuously with scaled distance, $\bar{R}$. For simplicity, it is recommended that a factor of 1.8 be used for cylindrical vessels and a factor of 2.0 for slightly elevated spherical vessels as these are the maximum values given by Geng.

3.5 Calculate Vulnerability

Given the adjusted $\overline{P_s}$ the vulnerability of people inside a building is next determined. A vulnerability curve can be generated as Figure 3 using information from Guidance for the Location and Design of Occupied Buildings on Chemical Manufacturing Sites, published by the UK's Chemical Industries Association (CIA). The overpressure-vulnerability relationship for type 3, typical domestic building, is presented in Figure 3.

Figure 3. Vulnerability curve against overpressure (courtesy coped from Marx, et al. [5])

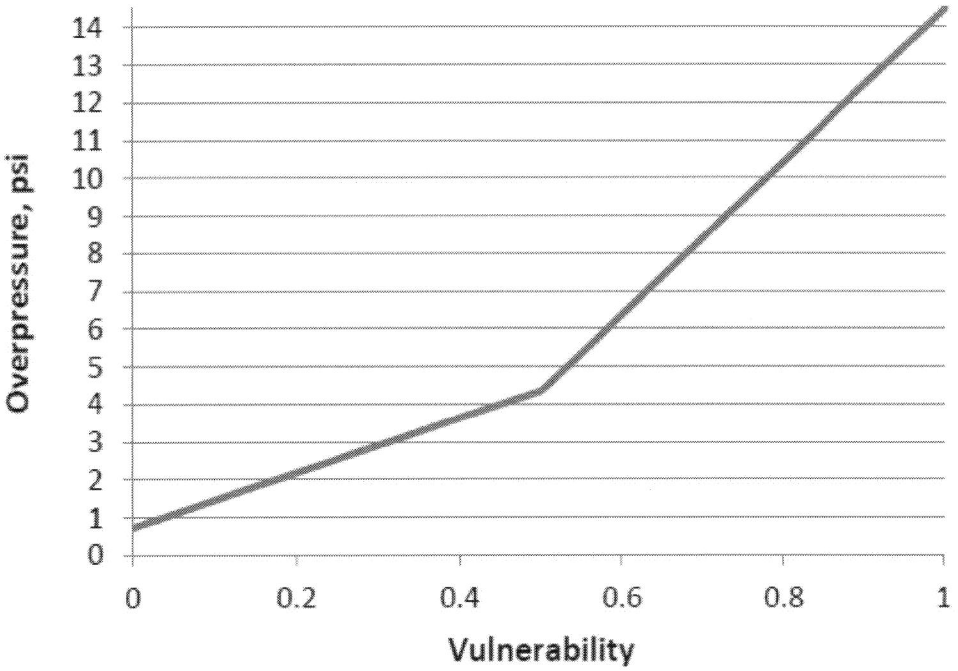

4. Example calculation case study

A $100m^3$-cylindrical pressure vessel at grade was under pressure test with nitrogen gas at 20 barg and it ruptured. There was a control room located 50 me from the vessel. What would be the side on overpressure at the control room? And what would be the vulnerability inside the control room if the building type is CIA type 3.
Table 3 presents the details of PVB calculation and recommendations.

Table 3. PVB calculation flowchart

Procedure	Details of calculation	Note
Collect data	Internal pressure is 21 bara The ratio of specific heats of nitrogen is 1.40 Ground burst, ambient pressure is 1 bara	
Calculate PV burst energy	$E_{ex,brode} = \frac{(p_1 - p_0)V_1}{\gamma_1 - 1} = \frac{(21-1)100}{1.4-1} = 500\ MJ$	
Calculate effective energy	$E_{eff} = \frac{2(p_1 - p_0)V_1}{\gamma_1 - 1} = 1\ GJ$	Multiply 2 if it is ground level explosion
Calculate scaled distance, $\bar{R}$	$\bar{R} = R\left(\frac{P_0}{E_{eff}}\right)^{1/3} = 50\left(\frac{10^5\ Pa}{10^9\ J}\right)^{1/3} = 2.32$	R = 50 m

Read or calculate scaled positive side on overpressure, $\overline{P_s}$	$\overline{P_s} = 0.13$	$\bar{R}$ is larger than 1, equation (1) is proper to use
Adjust scaled positive side on overpressure, $\overline{P_s}$	Multiply 1.8 for a cylindrical vessel geometry factor $\overline{P_s} = 1.8 \times 0.13 = 0.23\ barg$	
Vulnerability inside building	40 % from Figure 3.	CIA type 3 building
Recommendation	If the consequences are judged to be unacceptable, the blast-pressure effects shall be mitigated by the location of the equipment within suitable shields, walls, or barriers. Or the personnel in the control room should be evacuated when the test is ongoing.	

5. Conclusion

Young engineers are often told that BLEVEs are one of the highest impact phenomena in the process industry, even more so than PVBs. PVBs tend to have lower energy releases involved compared to BLEVEs since PVBs do not involve energy from liquid flashing. However, PVBs cannot be ignored since there can be many pressured gas vessels and pressure tests on a gas industry site. This paper provides an overview of method for estimating the characteristics of pressure vessel bursting for practicing engineers. It has been presented to include advanced modeling technology on PVBs and how to properly apply a simplification in adapting blast curves or equation and geometry effects.

6. References

[1] "Guidelines for Vapor Cloud Explosion, Pressure Vessel Burst, BLEVE and Flash Fire Hazards", 2nd ed., CCPS

[2] Tang, M. J., Cao, C. Y., Baker, Q. A., "Calculation of Blast Effects from Bursting Vessels", 1996 AIChE National Meeting

[3] Chen, Fei, Lutostansky, Elizabeth M., Schork, Joan M., "Improvement of BLEVE Overpressure Method Using Historical Data", 2011 CCPS conference

[4] W.E. Baker, J.J. Kulesz, R.E. Richer, R.L. Bessey, P.S. Westine, V.B. Parr, G.A. Oldham, "Workbook for predicting pressure wave and fragment effects of exploding propellant tanks and gas storage vessels", NASA CR-134906, NASA Scientific and Technical Information Office, Washington, 1975 and 1977

[5] Marx, Jeffery D., "API RP 752 Building Siting Using F-N curves", 2013 MKOPSC international symposium

Effects of Process Variables on Peak Relief Rates Estimated by Dynamic Simulation for a Multiple Distillation Column System

John Wilkins
Staff Process Consultant
Smith & Burgess LLC
5700 Northwest Central Drive Ste. 301
Houston, Texas 77092
John.Wilkins@smithburgess.com

Dustin Smith
Vice President of Operations
Smith & Burgess LLC
Dustin.Smith@smithburgess.com

Prepared for Presentation at
American Institute of Chemical Engineers
2013 Spring Meeting
9th Global Congress on Process Safety
San Antonio, Texas
April 28 – May 1, 2013

UNPUBLISHED

Effects of Process Variables on Peak Relief Rates Estimated by Dynamic Simulation for a Multiple Distillation Column System

John Wilkins
Staff Process Consultant
Smith & Burgess LLC
5700 Northwest Central Drive Ste. 301
Houston, Texas 77092
John.Wilkins@smithburgess.com

Dustin Smith
Vice President of Operations
Smith & Burgess LLC
Dustin.Smith@smithburgess.com

Keywords: dynamic simulation, VMG Sim, distillation columns, sensitivity analysis, pressure relief valves

Abstract

Process engineers and designers have recently turned to using dynamic analysis as a more realistic method for sizing relief systems. The primary goal of using dynamic analysis is to ensure that modifications are truly required instead of making unnecessary and costly upgrades of relief systems based on overly conservative calculations. In work previously presented by the authors, it has been shown that, for certain overpressure scenarios, the dynamic relief rates predicted by dynamic simulation are significantly affected by certain operating conditions [1].

This paper shows the effects of process variables on the relief loads estimated by dynamic simulation for multiple columns. This paper will also show how these effects would impact the flare load for a unit with multiple columns. The multiple column system is based on a system the authors have encountered at a refinery. The sensitivity analysis shows that changes in the column liquid levels significantly affect the initial relief times, relief loads, and relief durations.

1. Introduction

To stay in compliance with industry, regulatory, and typically corporate guidelines, it is important for a plant to conduct and document analyses of their relief systems. It also allows for plant engineers and designers to see the effects of their process on existing or proposed relief systems. Relief system analyses traditionally use steady-state equations to determine the relief rate used to size relief devices. For complex systems, these equations are based on conservative simplifying assumptions. This usually results in a larger relief rate that requires a larger relief device. The drawback to using traditional methods, when analyzing an existing system, is that they can overstate the required relief rate. Sometimes this results in the recommendation of costly relief systems upgrades where only minor modifications are needed.

Dynamic analysis of complex systems involves creating a simulation that calculates changes to the system over time. This analysis can be used to better estimate the relief requirements to protect from overpressure for a specific relief scenario. The primary goal of dynamic relief system analysis is to better understand how the system will respond to the upset scenario. An additional benefit is that the detailed analysis can be used to support making limited changes in lieu of more extensive modifications. The fundamental differences between these two methods are important to understand in order to make sure they are both being used to determine relief rates that are conservative.

Equation 1 shows a simplistic steady-state equation that can be used to determine the relief rate through a relief valve installed on top of a distillation column that has lost cooling, feed, and reflux, but has continued heat input (a typical total power failure scenario).

$$\dot{M}_{relief} = \frac{\dot{Q}_{reb}}{\Delta H_{vap}} \quad \text{[Eq. 1]}$$

In order to use the above equation, the right hand side of the equation would be evaluated at relief conditions, which may vary from normal operating conditions [2]. Assumptions such as the composition of the fluid on the process side of the reboiler can have effects such as a lower heat of vaporization or a larger reboiler duty that will increase the predicted relief rate.

Equation 2 is an example of how the above equation would be represented in a dynamic environment. It shows the same equation as functions of time as well as initial process conditions such as pressure, temperature, etc.

$$\dot{M}_{relief}(t, P_i, T_i, etc) = \frac{\dot{Q}_{reb}(t, P_i, T_i, etc)}{\Delta H_{vap}(t, P_i, T_i, etc)} \quad \text{[Eq. 2]}$$

Dynamic simulations calculate the changes in the system based on incremental time steps from an initial starting point. Therefore, the selection of initial conditions is important and will affect the results [1]. These initial conditions are typically based on operating conditions that represent the normal operations of the facility. Operating conditions are controlled in ranges, and depending on the specific item, there may be no such thing as a normal condition. Liquid levels, temperatures, pressures, and reflux rates for distillation columns can vary during operation and can change due to upsets elsewhere in the plant. This is important to consider when global relief scenarios like power failure and cooling failure affect the entire plant, as these variables are all essentially assumptions.

API 521 Section 5.22[3] states that if dynamic simulations are used for column relief system design, the model must be conservative with respect to the maximum relief load. It goes on to state that the assumptions used for the simulation *shall* be checked by sensitivity analyses to assess the impact on the column relief rate. A sensitivity analysis is the study of how the uncertainty of the input variables affects the output of a mathematical model. Therefore, the input variables of a dynamic simulation must be checked in order for the designer to ensure that the simulation is conservative.

2. Methodology

2.1 Setting up the dynamics file

A steady-state model containing 2 debutanizers, a depropanizer, a gasoline splitter, and 12 PSVs was created using VMG Sim. The flow diagram of this system is displayed in Figure 1.

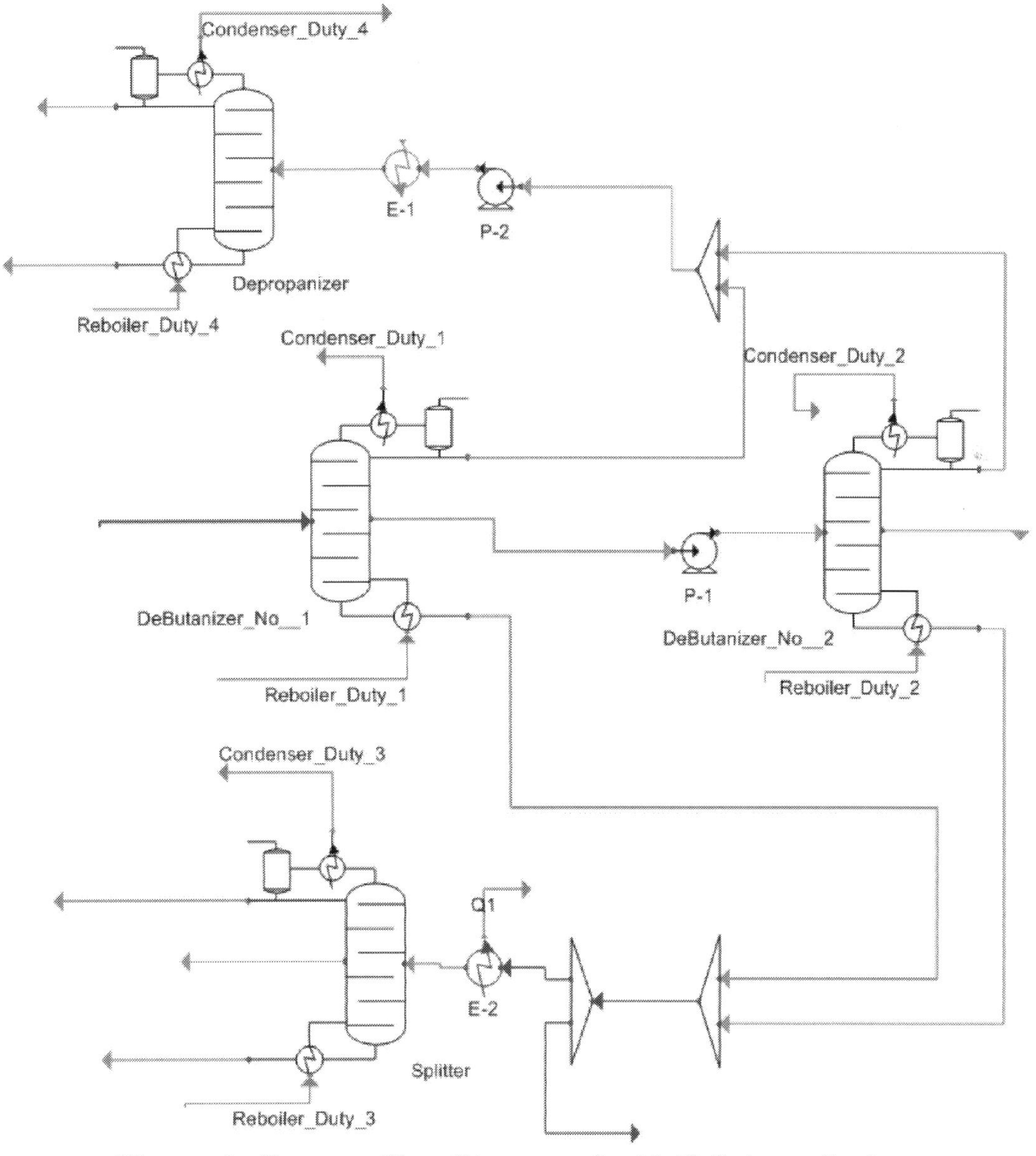

Figure 1 : Process Flow Diagram of a Multi-Column System

The steady-state model was used to create a dynamic model that uses 21 controllers set to match the operating parameters of the equipment. Controllers are used in dynamic simulations to achieve specified flowrates, pressures, temperature, and other physical conditions. The controller attempts to achieve the user's specification by opening or closing a valve located on a material or energy stream. This is different from a steady-state model where flowrates, duties, and physical properties are typically specified or calculated. The action of the controller is based on the tuning parameters provided by the user.

Twelve PSVs were added to the dynamic model to simulate relief. Table 1 displays information on these relief valves.

Table 1: Summary of PSV Information

Name	Location	PRV Size	Set Pressure (psig)
PSV-1	DeButanizer No. 1 vapor overhead	6Q8	184
PSV-2	DeButanizer No. 1 vapor overhead	6Q8	190
PSV-3	DeButanizer No. 1 accumulator	3L4	180
PSV-4	DeButanizer No. 2 vapor overhead	6Q8	190
PSV-5	DeButanizer No. 2 accumulator	1.5H3	190
PSV-6	Splitter vapor overhead	6R8	30
PSV-7	Splitter vapor overhead	6R8	30
PSV-8	Splitter vapor overhead	6R8	30
PSV-9	Splitter vapor overhead	6R8	30
PSV-10	DePropanizer vapor overhead	6Q8	290
PSV-11	DePropanizer accumulator	3K4	274
PSV-12	DePropanizer accumulator	3K4	274

In order to perform the sensitivity analysis, different column liquid levels were used as operating conditions. Setpoints of the respective level controllers for the columns were adjusted to achieve the desired liquid levels. Three simulations were created for a low liquid level of 20%, a medium liquid level of 40%, and a high liquid level of 60%. These percentages represent the fraction of liquid height in the lower section of the column that does not contain trays.

2.2 Initiating relief

Cooling water failure was chosen as the overpressure scenario to review as it affects all of the four columns simultaneously. Total loss of cooling water is expected to result in the loss of overhead condenser duties, which may lead to overpressure in the column due to excess vaporization. This was accomplished by specifying duties of zero for all condensers. It was assumed that all of the condensers lost cooling at the same time.

All controllers in the dynamics model were turned off (set to manual) to avoid taking credit for positive controller action. This was done in accordance with API 521 Section 4.2.4[3]. Changing the controller mode from "automatic" to "off" causes the controlled valve to hold its position during relief.

For each liquid level, the integrator (essentially an electronic strip chart recorder in VMG Sim) was turned on and recorded the pressure in the columns as it slowly increased. The overhead PSVs opened at their respective set pressures and flow data was collected. The sum of the mass flow rates relieved by the PSVs was recorded in order to view the total load into the flare.

3. Results

Graphs of dynamic relief for the three liquid levels are displayed below. PSV-5 was not graphed as it does not contribute a significant relief load. The initial time (t = 0), is the time at which the cooling to the tower was stopped and the overpressure scenario began. Note that the change in liquid levels shifts the magnitude of the relief rate and the peak point in time for some of the towers.

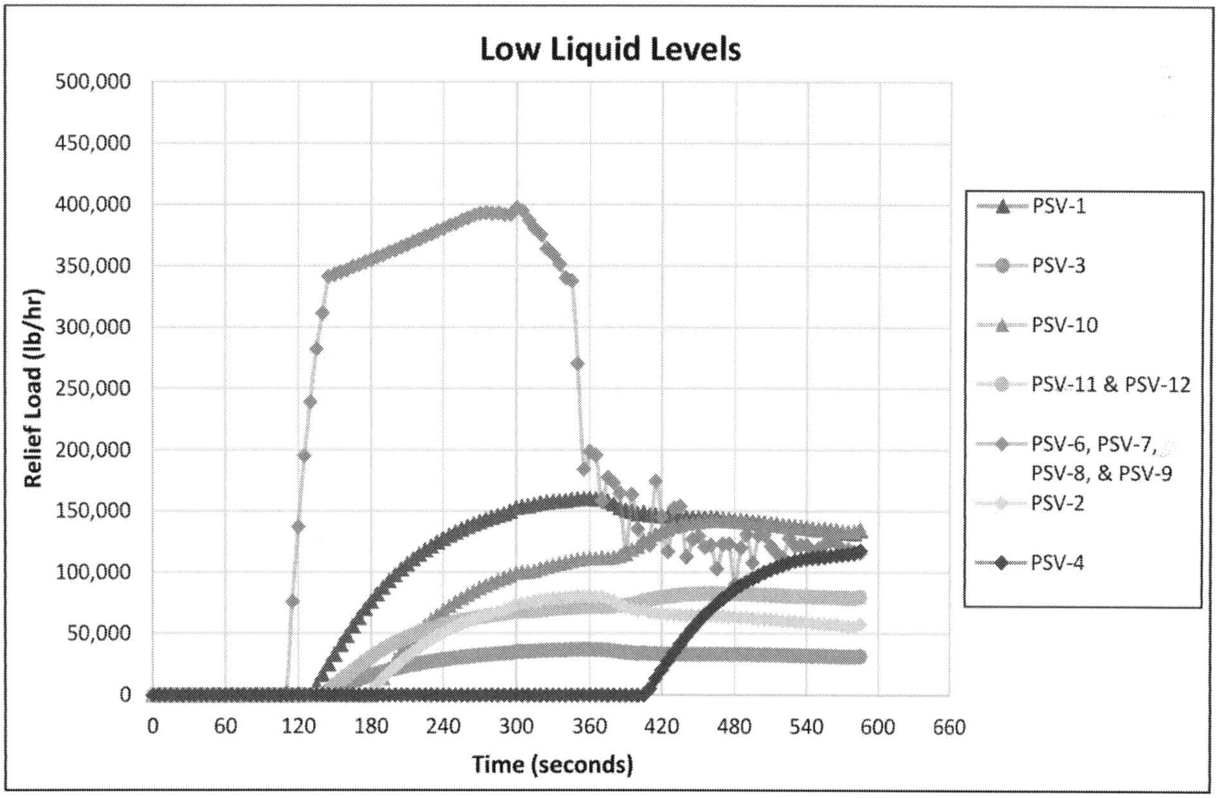

Figure 2: Cooling failure relief rates with low liquid levels

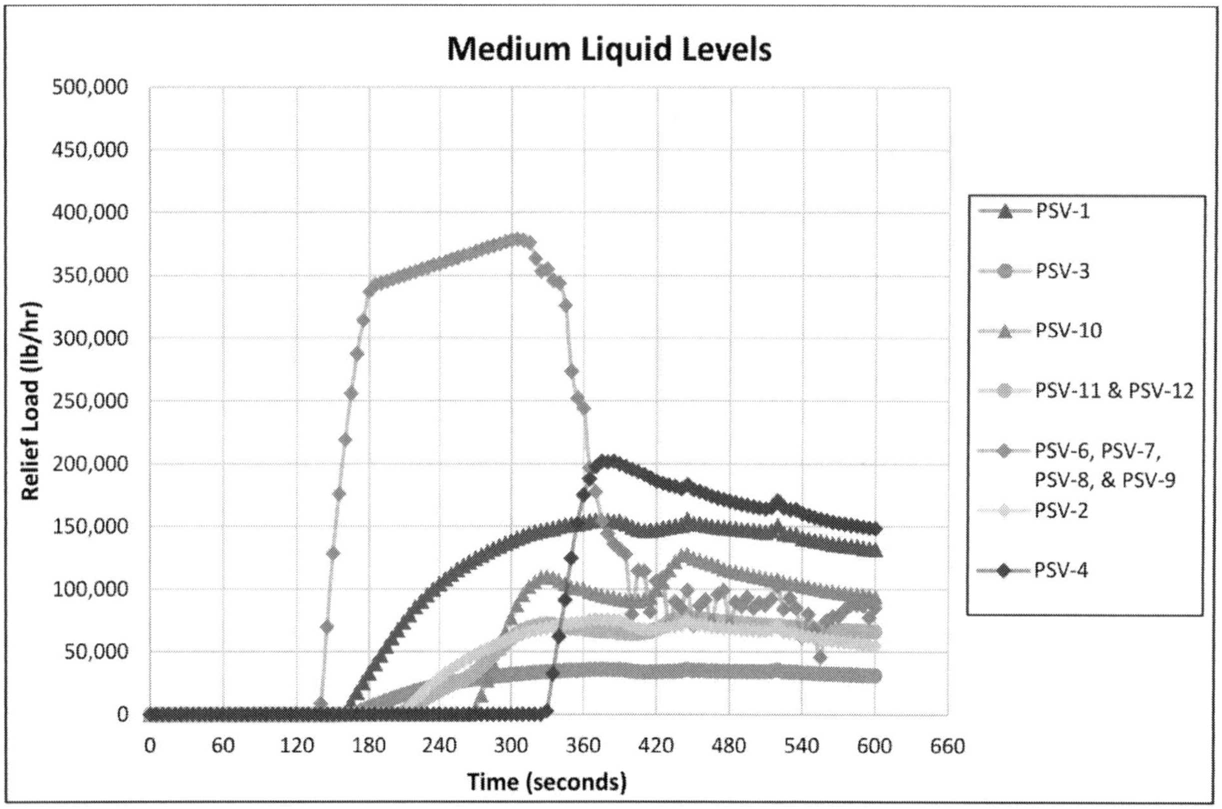

Figure 3: Cooling failure relief rates with medium liquid levels

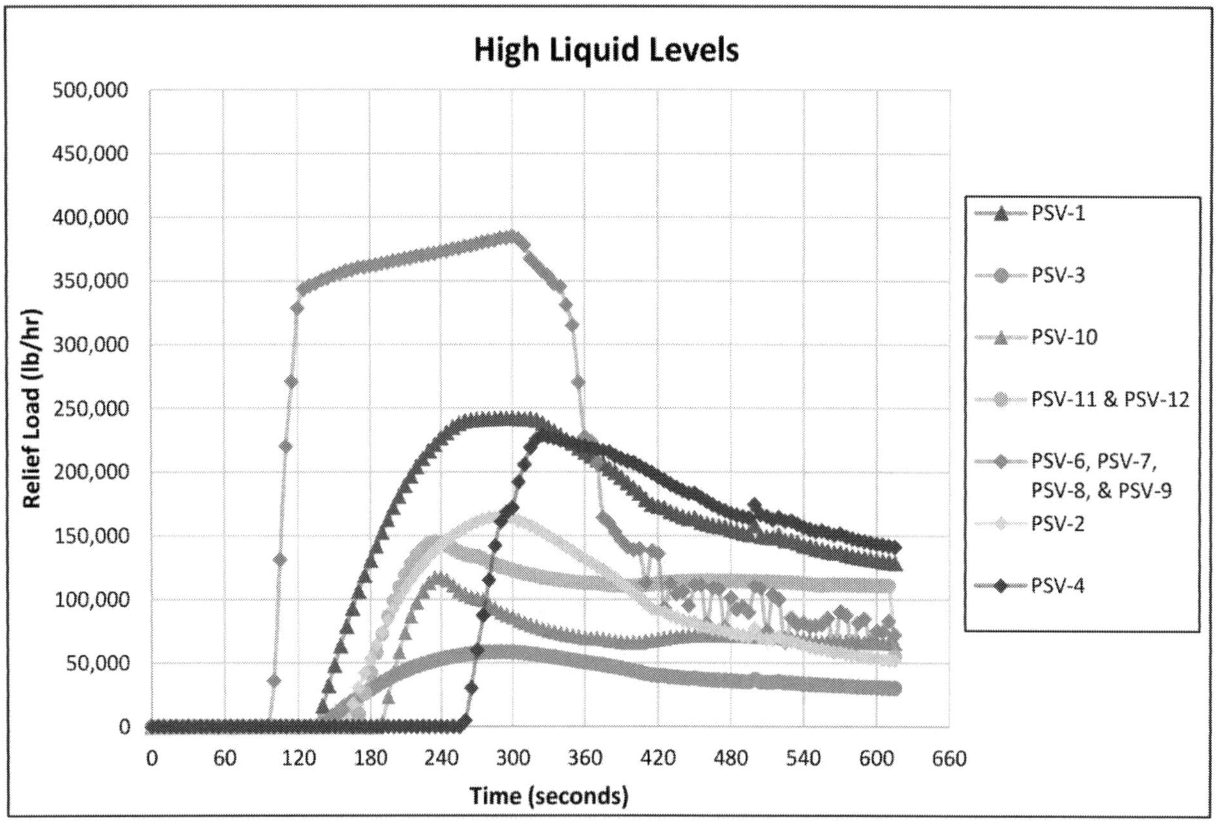

Figure 4: Cooling failure relief rates with high liquid levels

Figure 5 shows the combined flare loads for the three liquid levels. The curves were made by adding the relief loads from the previous graphs over the same time frame. The combined load represents the load that would enter the flare header. The horizontal line represents the steady-state relief load for the four columns estimated by traditional methods.

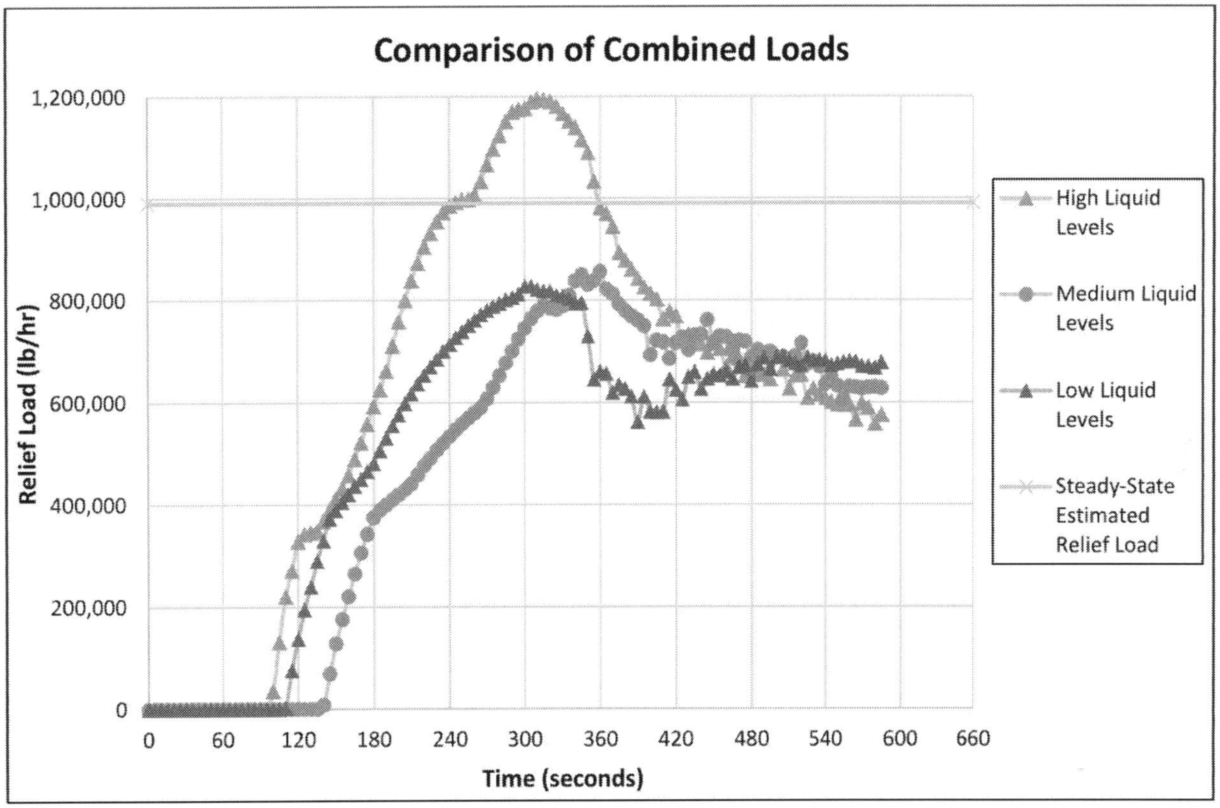

Figure 5: Combined cooling failure loads for different column liquid levels

The steady-state estimation of the combined relief load of the columns assumes that the reboilers have a reduced capacity due to the reduction in the log mean temperature difference (see Reference 1 for an additional explanation). The dynamic model was based on a constant heat input, and thus may over-predict the duties of the reboilers under upset conditions. A further improvement could be made by modeling the reboilers based on their individual UA characteristics.

The three curves in Figure 5 have roughly the same shape as well as time of peak flow. The noticeable difference is the peak relief load for the high liquid level curve. There is a 43% increase in the peak of the combined relief load for all PSVs when the liquid levels are increased by 50%. Lowering the liquid levels by 50% had little effect on the combined peak load.

The increase in the combined peak load appears to be caused by two factors. The peak flows of PSV-1 and PSV-2 located on the 1st debutanizer increased by 57% from normal liquid levels to high liquid levels. The peak time for PSV-4 located on the 2nd debutanizer lowered when the liquid level increased, which resulted in PSV-4 contributing significantly to the combined load peak. Changing the liquid levels in the columns affected the time to initial relief, the time to reach peak load, and the height of the peak for these valves. This behavior is present to a lesser extent in the other relief valves.

4. Conclusions

Dynamics simulation can be a useful tool to show how a system reacts to sudden changes and how the system uses control valves to stabilize itself. Dynamic analysis is very different from hand-calculated methods which are traditionally used to size relief systems. Although dynamic analysis can be used to size relief systems, it must be done knowing how process variations affect the relief rates.

The results show that variations in the liquid levels affect relief times and peaks for a system of multiple columns. Increasing the liquid levels of four columns in a system by 50% increased the combined peak load into the flare by 43%. This was due to increases in the peak flowrate of two PSVs and the shifting of a large peak towards the combined load peak.

This result shows that changes in process variables can have a large impact on the relief rate used to size a flare system. Consideration must be given to the effect process variables can have on the time of initial relief, the time to reach the peak load, and the magnitude of the peak flow.

Because there are multiple variables that can impact the peak relief rate for a flare, a full sensitivity analysis of a system with multiple columns would be time consuming and costly. However, if dynamic analysis is used to size a relief system, a sensitivity analysis must be performed to ensure that the analysis will lead to a safe design.

5. References

[1] Wilkins, John & Dustin Smith, P.E. *Effects of Process Variables on Dynamic Relief Load Estimates for a Depropanizer and Debutanizer*. Oct. 2012. Electronic poster presented at the AIChE Regional Process Technology Conference, Texas City, TX.

[2] Cristea, Nicholas and Dustin Smith, P.E. *Making Relief Load Estimates Match Reality*. 2013

[3] ANSI/API Standard 521 5th Edition, ISO 23251, May 2008

A Hybrid Model for Characterizing the Source of Hazardous Material Release

Shahryar Khajehnajafi, PhD
Anjana Meel, PhD
SAFER Systems LLC
5284 Adolfo Road, Suite 100
Camarillo, CA 93012
shah@safersystem.com

Prepared for Presentation at
American Institute of Chemical Engineers
2013 Spring Meeting
9th Global Congress on Process Safety
San Antonio, Texas
April 28 – May 1, 2013

UNPUBLISHED

A Hybrid Model for Characterizing the Source of Hazardous Material Release

Shahryar Khajehnajafi, PhD
Anjana Meel, PhD
SAFER Systems LLC
5284 Adolfo Road, Suite 100
Camarillo, CA 93012
shah@safersystem.com

Keywords: hazardous chemicals, chemical spill, source of release, source of hazardous material, plume model, Source Area Locator (SAL), source characterization, Gaussian plume model, Bayesian method, Bayesian inference, hybrid model, odor source, source location estimation, burro

Abstract

On Sept 10, 2012, Los Angeles Fire department and 911 operators received many complaints about a strong foul smell. The complaint covered an area of 10,000 square miles. There were many speculations for the source of the release; sewage, chemical spill, pipeline carrying natural gas. However, after many hours of investigation by South Coast Air Quality Management it was discovered that the source of the foul odor was dead fish at the Salton Sea churned up by a thunderstorm that had occurred that evening. The Salton Sea is a 376-square-mile saltwater lake about 150 miles southwest of Los Angeles. The odor remained strong for up to 150 miles downwind of the source, and quite unusual. Fortunately the above case did not pose any health effects.

It is common for industrial sites carrying hazardous chemicals to receive calls from surrounding communities complaining of a chemical presence in the ambient air. Following a complaint, industrial sites must work promptly check and verify if the odor source is related to a chemical release on their property, especially when there are other industrial sites in the vicinity, for mitigating and reducing the extent of the possible hazard. Locating the source and amount of release for predicting the geographical extent of the exposure in a timely manner is, therefore, essential to take an effective response action for evacuation or shelter-in-place.

With only basic knowledge of meteorological information; wind speed, wind direction, and stability along with two or more gas concentration measurements, it is possible to quickly frame the zone of chemical release. This is an inverse problem to the usual forward dispersion calculation in which the source location and release rate is known and the concentrations at sensors positions are predicted. There are many publications in the area of source characterization; location, and amount of chemical being released. The two most cited methods are: Bayesian inference using Markov Chain Monte Carlo (MCMC) sampling methodology, and adjoint advection- diffusion equation. The Bayesian methodology fits well for application to atmospheric dispersion where ambient and gas sensor data are stochastic in nature, however, it takes a long time to converge.

In this paper, we briefly describe a simple least-square based method for source characterization using Gaussian plume model. We apply the Bayesian method to address the same problem and compare the results of the two methodologies for few selected release scenarios as well as a set of field data. Subsequently, we will present a hybrid model using Bayesian inference which utilizes the result of the simple least- square model as a seed for narrowing the search domain and reducing the time for locating source of chemical release. We present the improvement in the run time using the hybrid model which makes it suitable for emergency response and provides a reasonably accurate prediction of source characterization.

1 Introduction

In this paper, the focus is on estimating the source area location of a hazardous release using the sensor measurements obtained from sensors located throughout a facility. If parameters of a system are known, the outcomes of measurements can be predicted. That is known as the 'forward problem'. On the contrary, estimation of the values of the parameters of the system using the results of the measurements is called a 'backward problem'. Identifying source parameters for location, quantity of release using sensor measurements is a backward problem. Tarantola (2005) discussed the inverse problem theory and methods for model parameter estimation in great detail. Usually a forward problem has a unique solution, but the inverse problem can have multiple solutions. Therefore, in an inverse problem, any apriori information should be utilized explicitly and uncertainties in the data should be handled carefully. When nonlinearities are available, it is important to describe uncertainty and apriori information using probability distribution. Bayesian theory based methods can be used to describe uncertainty. When systems are not nonlinear, least-square methods are considered very efficient. Both least-square methods and Bayesian Theory based methods have their own advantages and disadvantages. Least-square method usually minimizes the error between measurements and prediction to calculate the system unknowns while Bayesian theory based methods updates the prior probability distribution of unknown parameters using the sensor measurement data to obtain a posterior distribution of the unknown parameters, which provides the updated information of the system.

Thus the source area location and release rate is indeed a very challenging and nonlinear problem, even for identifying the source area from a single sensor concentration measurement. Even for a single sensor measurement, there can be multiple solutions for release location all of which may lie on a line/curve with varying distance and release rates. For example, a large release rate at a further distance from a sensor may result in the same concentration values observed by a fixed sensor as a small release rate at a location closer to the sensor. When more than one sensor is involved, the problem becomes even more complicated.

The Bayesian theory has been used by several researchers for estimating the source area location. Yee et al. (2007, 2008) developed a Bayesian probabilistic inferential solution for source reconstruction (both location and emission rate) from a limited noise concentrated data and the methods were tested for JU2003 field study in Oklahoma City (urban dispersion) and the ETEX (long-range dispersion on continental scales). Kim et al. (2007) developed a source reconstruction algorithm using dynamic Bayesian inference and applied it on Yonggwang atmospheric tracer experiment in Korea to show that the posterior modes of the release point and the released source rate for this experiment converged to their true values. Keats (2009) developed Bayesian inference technique for source determination in atmospheric environment. Adjoint (backward) dispersion models were used to reduce the computational effort required from calculating one [forward] plume per possible source configuration to calculating one [backward] plume per detector. Markov chain Monte Carlo (MCMC) is used to efficiently sample from the posterior distribution for the source parameters; both the Metropolis- Hastings and hybrid Hamiltonian algorithms are used. These algorithms were applied on a full-scale field experiment (Joint Urban 2003) in Oklahoma City and the results from the algorithms were also compared with the US Environmental Protection Agency's standard CMB model using a test case involving PM data from Fresno, California. Neuman et al. (2005) used an urban puff model [the Defense Science Technology Laboratory's (Dstl) Urban Dispersion Model (UDM)] for prediction plume evolution in complex urban geometries that enabled rapid urban dispersion simulations by combining traditional Gaussian puff modeling with empirically deduced mixing and entrainment approximations. In their work, a preliminary reconstruction of an atmospheric release event was demonstrated using stochastic sampling algorithms and Bayesian inference together with the rapid UDM urban puff model based on point measurements of concentration. This model was applied to a prototype isolated building and for observations and flow conditions taken during the Joint URBAN 2003 field campaign at Oklahoma City. Ideally, the urban puff model is used to approximate the source location and strength. The more accurate CFD model can then be used to refine the solution. Zheng et al. (2011) reviewed inverse calculation approaches for source determination and classified them into probability modeling methods and optimization modeling methods. According to them, probability modeling methods are demonstrated to be insufficient during emergency responses due to their lacking of enough prior information of unknown parameters, while optimization modeling methods are efficient and become a new trend in source determination. Sharan et al. (2011) applied least square data assimilation for identification of point source emissions from limited number of atmospheric concentration measurements. Monache et al. (2008) applied Bayesian inference and Markov Chain Monte Carlo sampling to reconstruct a contaminant source on a continental scale. By running the Bayesian MCMC algorithm on a large parallel cluster the inversion results could be obtained in few hours as required for emergency response to continental-scale releases. Senocak (2008) developed a method for stochastic event reconstruction of atmospheric contaminant dispersion using Bayesian Inference. Cowles and Carlin (1996) wrote a review article describing the markov chain monte carlo convergence diagnostics. Convergence criteria to the right solution and optimum step size for chain movement are important topics in the application of Markov Chain Monte Carlo method. Haario et al. (2001) described the importance of choice of proposal distribution for MCMC methods for convergence of the algorithm. An adaptive metropolis algorithm with Gaussian proposal distribution was introduced.

As shown in the above references, a number of researchers have worked in source location estimation using Bayesian probabilities methods. Since probability based methods alone cannot be sufficient during emergency response due to long convergence time periods, in the approach proposed in this paper, first an initial solution is quickly estimated using a least-square based method based on backward wind corridor. Subsequently, to obtain a more rigorous solution using Bayesian probabilistic method (referred as MCMC), the initial solution based on least square method is provided as a seed solution along with a search area/boundary. Thus, this two-stage method decreases the overall convergence time for probabilistic method by providing an intelligent apriori information. Using this approach, the emergency responders can quickly start responding based on the initial solution and more accurate information using a probabilistic method can be provided to them in due course of time. Thus, a combined least square-Bayesian approach ensures that even a probabilistic method provides a solution in much smaller time frame.

2 Approach

In this section, our hybrid approach for estimation of source area location is described. At first, source area location is estimated using a simple and intelligent technique developed using backward wind corridor and least-square based method (Khajehnajafi and Burla, 2008). The least-square method is an optimization based approach and several key features of this approach are listed in Table. 1. This method is very efficient and gives a good estimate for source location for most of the wind stability conditions, except when the wind is too stable because the slope of wind corridor is very narrow.

Another approach used for source location estimation is a probability based method using Bayesian theory where a probabilistic distribution is assumed for source area location and the sensor measurements and markov chain monte carlo (MCMC) simulations are used to estimate the source area location. Since this is a statistical approach, it takes significantly more time to converge to a value. Key features of this approach are listed in Table 1.

A Hybrid approach is a combination of the two approaches as listed in Table 1. Deterministic approach assumes that the parameters of the system are known completely without any uncertainty while probabilistic approach accounts for uncertainty in the model at the expense of longer time for convergence. Therefore, having a method that can narrow down the initial search zone for the Bayesian approach could be very useful in decreasing the convergence time. Thus, a hybrid approach is suggested in this paper where at first a Source Area Locator (SAL) model is used to narrow down the area and then Bayesian approach (referred as MCMC) is used to zero down on the actual source area location. The idea is that the combination of these two is a great method that accounts for weaknesses of both the models and makes it a very efficient model to narrow down upon a location very close to the actual source for all wind and stability conditions. Each approach on its own can provide reasonable results with some limitation either in terms of a further away solution for certain atmospheric conditions, or really long convergence time. Thus, the hybrid approach is a very valuable approach to estimate the source area location. Next, specific details of each of the methods are provided.

Table 1: Deterministic, Stochastic, and Hybrid Approach to Source Area Location

Deterministic (Least-square based)	Stochastic (Probabilistic)
• Assumes that all the system parameters are known perfectly • Usually faster runtime • Good for emergency response situations • Referred as SAL – Deterministic approach based on least-square based method to narrow down the search area	• Assumes uncertainty in system parameters • Usually longer run-time • Probabilistic approach • Based on Bayesian - Markov Chain Monte Carlo method (MCMC)
Hybrid Approach	
• Initially solved using deterministic approach based on least –square method SAL • Solution of deterministic approach is used to narrow down the search area for Bayesian - Markov Chain Monte Carlo method (MCMC). Thus, reducing the run time and obtaining a more accurate prediction of source parameters.	

2.1 SAL (Least Square Based Model)

The least square based method is referred to as SAL (Source Area Locator) in this paper. In this technique, it was demonstrated that with only the basic knowledge of meteorological information; wind speed, direction, and stability along with two or more gas concentration measurements, it is possible to quickly frame a region for the source of chemical (Khajehnajafi, and Burla 2008). A technique called 'reverse corridor' or 'back trajectory' is used to narrow the search domain for release location as seen in Figure 1 (a). This simple methodology uses the horizontal dispersion obtained from surface data at receptor locations. The construction of the corridor is key to reduce search area, therefore, reducing the computational cost. In Figure 1(a) the yellow region is the search domain if only sensor 1 and 2 were impacted. Introducing sensor 3 value, the search area is further reduced to the meshed zone. Adding additional sensors may narrow the search area; however, it will significantly increase the simulation time and increase the nonlinearity of the problem.

For the simple example shown in Figure 1(a) initially, sensitivity of the model to main parameters for dispersion, e.g. wind meander, sensor concentration fluctuations, and atmospheric stability was studied. A Gaussian model was used for that study. The result of SAL model for different stability classes A through F (1 through 6) showed that the accuracy of the model prediction for source area decreased with the increasing in stability class especially for higher stability classes 5 and 6 and at very low release rates. The results for the source location and the release rates, even when the model was simulated with the noisy sensor data, viz. 40% and 80%, matched very well with the true data when very fine mesh size was use (except for highly stable conditions at low release rates) indicating that the back trajectory method coupled with a simple Gaussian plume model is in fact a very simple yet plausible technique for locating source of pollutant.

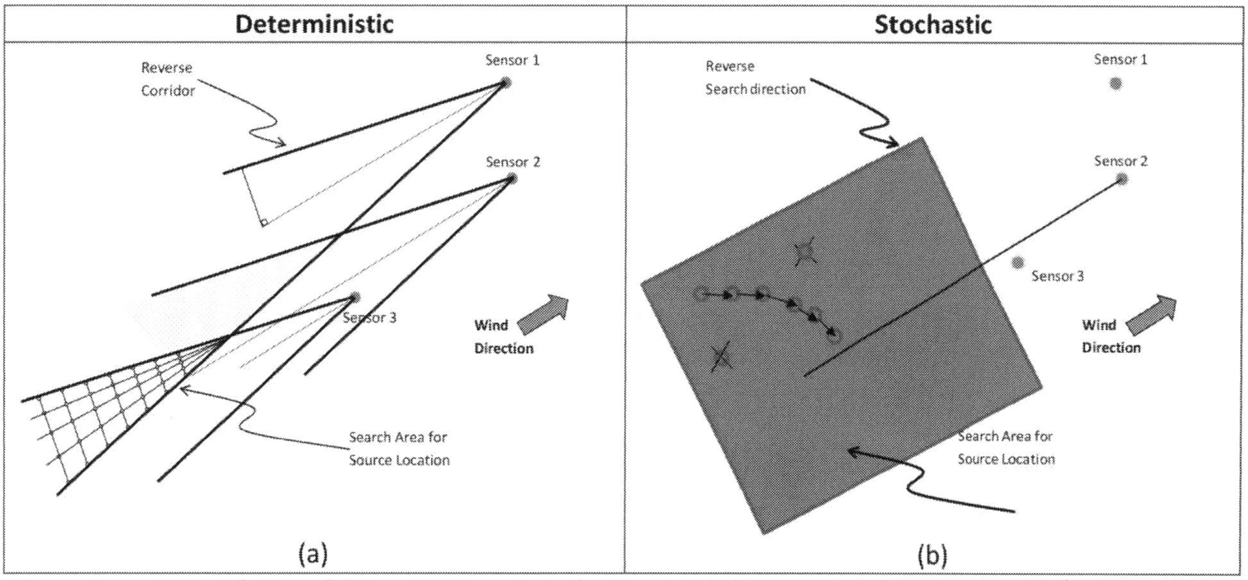

Figure 1: Pictorial view of deterministic and stochastic approach to source location of hazardous material release

2.2 Bayesian Model (Markov Chain Monte Carlo)

As mentioned in the introduction section, MCMC alone takes a lot of time to converge. Herein, based on the solution from the SAL model, an area was outlined as a narrowed search domain for the Bayesian probabilistic method. If the initial guess is chosen at random, the outcome from MCMC may take much longer time to converge and may even fails to converge sometimes because of broader search zone. In Bayesian method, a prior distribution is defined for parameters of interest (For example the location and release rate in this case). The likelihood function is estimated using the predicted concentration and measured sensor concentration. Then both prior distribution and likelihood distribution are used to estimate the posterior distribution. The posterior distribution gives us the latest knowledge of the original parameters of interest. The Bayes's theorem for the source reconstruction can be written as (Monache et al. 2008).

$$p(x,y,q|\,Sensor\ Data) = \frac{p(Sensor\ Data|x,y,q)p(x,y,q)}{p(Sensor\ Data)} \propto p(Sensor\ Data|x,y,q)p(x,y,q) \quad (1)$$

Where p( ) is the probability distribution. The x, y, q are the source coordinates and the release rate. *Sensor data* is the measurements from the different sensors. Bayes theorem relates the posterior distribution, $p(x,y,q|\,Sensor\ Data)$, to the product of the probability of measurements given the source parameters, $p(Sensor\ Data|x,y,q)$, also called the likelihood function, and the probability of the source parameters prior to any knowledge of the measurements, $p(x,y,q)$, also called the prior. For this study, x, y, q are assumed to be unknown parameters. One can include other parameters in the unknown list.

2.2.1 Prior Distribution

Let m be the model parameters i.e. all the parameters for which prior need to be defined. x, y, q are the prior parameters, x coordinate of the release location, y coordinate of the release location and the release rate, respectively. The prior distribution can be written as:

$$p(m) = p(x)p(y)p(q) \tag{2}$$

2.2.2 Likelihood Function and Acceptance Condition

The likelihood distribution is function of CP, predicted concentration for a sensor, and CM, measured concentration by an actual sensor. Here, the likelihood function is assumed to be of the following form (Monache et al. 2008):

$$\ln[(CM|S)] = \ln[L(\sigma, CP, CM)] = -\frac{\sum_i^N [log_{10}(CP_i) - log_{10}(CM_i)]^2}{2\sigma^2} \tag{3}$$

Where L is the likelihood function, CP_i is the predicted concentration values at the sensor location i, CM_i is the sensor measurements for sensor i, σ is an error parameter chosen accordingly to expected errors in the observations and predictions. N is the number of sensors. S is the state vector formed by all the unknown parameters (x, y, q). If the measured concentration values spread over several order of magnitudes, taking the logarithm of both predicted and measured values prevents large concentrations from dominating in the likelihood computation. The above likelihood is relatively simple, where the N prediction errors are assumed independent and expected to be proportional in size to the predicted concentration (as a result of log transforming both the observed and predicted concentration). If large errors are expected in the prediction/or measurements, large values of sigma should be assumed resulting in a broader-flatter posterior distribution. Here a sigma value of 0.8 is assumed. For a more complex scenario, it can also be treated as an unknown parameter and assigned a prior distribution.

The posterior distribution is sampled with an MCMC procedure via a Metropolis-Hasting algorithm.

2.2.3 Concentration Prediction Model

The concentration prediction model is based on a Gaussian plume dispersion model as shown below (Kim et al. 2011).

$$CP_i(x, y, z) = \frac{q}{2\pi U \sigma_y \sigma_z} \exp\left(-\frac{y^2}{2\sigma_y^2}\right) \tag{4}$$

Where, $CP_i(x, y, z)$ is predicted concentration at a particular location (x, y, z). q is the emission rate, U is mean wind speed, H is height of the release, x is distance along the wind, y is distance along the horizontal crosswind direction, z is the distance along the vertical axis, σ_y, σ_z are the standard deviations in the horizontal crosswind and vertical directions that are also called as Gaussian plume dispersion parameters and are defined empirically for different stability conditions. σ_y, σ_z are defined below:

$$\sigma_y = \zeta_1 x\,(1 + 0.0004x)^{-0.5}, \qquad \sigma_z = \zeta_2\, x \tag{5}$$

2.2.4 Sampling Procedure and Algorithm

Various algorithms exist for the sampling with MCMC. We have assumed uniform prior for the prior parameters. For the MCMC method, x is in the range -10,000 to 10,000 and y is the range 10,000 to 10,000 with (0,0) as the release location. q is uniform prior between 0.1 to 10 kg/s. U is the value given for the experiment of interest. H is assumed at the ground level. Ideally, parameter ranges can be assumed for wind direction, wind speed, ζ_1, ζ_2. However, for this study we have kept these constant. Of course, ζ_1 and ζ_2 will vary based on the stability class.
For the hybrid approach, the release location is first determined using the SAL approach. Limits for x, y are assumed to be ($-1000 + x_{sal}$, $1000 + x_{sal}$) and ($-1000 + y_{sal}$, $1000+ y_{sal}$) where x_{sal} and y_{sal} are release location coordinates obtained using the least-square method (SAL). Range for q is 0.001 to 10 Kg/s. The new value of x, y, q is obtained using adaptive sampling based on the current value and a step size.

2.2.5 Algorithm

An initial value for each parameter is sampled from its prior distribution. System parameters are assumed to be uniform prior with range as described above. The prior distributions are not completely uncertain as a good initial range is provided using the SAL method. Using the likelihood function, these sampled values from the prior distribution are selected to be part of the posterior distribution based on some criteria. A number of iterations involving source term sampling and dispersion simulation are executed until convergence to the posterior distribution is reached. Accepted parameter values collected during all the iterations constitute a Markov Chain which is defined as a set of consecutive values of parameter in which the probability of a value depends only upon the previous sampled value._Each iteration of the solution that becomes part of Markov Chain consists of following steps (Monache et al. 2008):

1. New value for source term is proposed (x_{prop}) according to $x_{prop} = x + dx$. Where x is the current value and dx is displacement from the current value modeled as a random walk sampled from a Gaussian distribution with zero mean and a standard deviation specified from the current step size. The Prior distribution $p\,(m)$ is used as a target distribution to estimate a prior likelihood of x_{prop}. If the prior likelihood of x_{prop} exceeds that of x, the proposed value x_{prop} replaces x. If not, a random (Bernoulli) "coin flip" determines whether the new proposal, even with its lower prior likelihood will be accepted. This ensures that sampling doesn't get stuck at a random point.

2. Repeat step 1 for the y and q parameters considered in the prior distribution. The assumption is that independent sampling of each parameter ensures that they reflect their likelihood better than sampling all of them at once and testing their likelihood
3. The predicted concentration for sensors is simulated using the current values of all the prior parameters using equation 4.
4. Likelihood of the current values of all the parameters which forms the proposed state is calculated using the predicted concentration and observed concentration by sensors. If the proposed likelihood is higher than the likelihood of previous state, it is accepted. If not, a random (Bernoulli) coin flip is used to ensure that the search reconstruction procedures continues to search the complete space of proposed source term parameter states, preventing the procedure from remaining indefinitely within the neighborhood of a local extreme.

After the proposed state is calculated, it is accepted if

$$\ln[L_{new}] - \ln[L] \geq \ln[rnd(0,1)] \quad (6)$$

Where L_{new} is the likelihood value of the proposed state and L is previous likelihood value and rnd(0,1) is a random number generated from a uniform distribution in the internal [0,1].

At a particular time instant, likelihood test for proposed state using measured data from all sensors at a particular time instant determines the acceptance or rejection of this state. Steps 1-4 are repeated to create a new proposed state with respect to all the parameters until the Markov Chain converges for a particular time instant. Note that in case of time varying information; step 1 to 4 can be further repeated for each time instant.

3 Case Studies: Comparison with the Field Data

3.1 BURRO EXPERIMENTS

BURRO experiments from the 1980 LNG (Liquefied Natural Gas) spill tests conducted by Lawrence Livermore National Laboratory (LLNL) and Naval Weapons Center (NWC) were used to compare the predictions from the model with the experiments. BURRO series experiments are one of the most comprehensive experimental analyses available in the literature for sensor measurement data of a chemical release (Koopman et al. 1980, Goldwire et al. 1980). In total nine experiments were conducted. Five experiments from the 1980 LNG (Liquefied Natural Gas) spill tests were used for this analysis. It is very difficult to get access to experimental data in this field and BURRO data are indeed very useful to compare modeling predictions with the experimental data for such a complex set of sensors.

This particular experimental setup employs close to 20-30 sensors. Not all sensors will see concentration values all the time so only the sensor with non-zero concentration reading are analyzed in this situation. This in itself is debatable as non-zero sensors may provide some information about the location which otherwise can be missed in case of uncertainty. Ideally, non-zero sensors should be included as well but for simplicity reasons we have not included those here and can be part of a further study. The next level of complication arises from the dynamic nature of the measured sensor values due to plume meander caused by horizontal turbulent fluctuation in the atmosphere. In ideal situation one expects the sensors to measure concentrations reliably based on the release rates. However, the uncertainty in wind direction, wind speed, measurement errors due to hardware and software and variability in atmospheric stability condition can contribute to a lot of uncertainty in the sensor data. For example, see the BURRO experiments report for concentration values measured by sensors for few experiments. Because of these uncertainties in the values and the simplified nature of our modeling approach, careful and intelligent decisions need to be made about the measured sensor readings that are used for this analysis as only a single measured value is used for each sensor in this analysis. Nonetheless, a more complicated model can be created which can include few/all aforementioned uncertainties at the expense of higher simulation time and resources. These may be included in the future studies of the project after carefully analyzing the tradeoffs involved in various approaches.

In our approach, the spill scenarios conditions and measured sensors values were used to obtain solution from SAL and MCMC approaches and then predictions for source area locations from the two models were compared against the actual experiment release location. In particular, five different spill scenarios from BURRO experimental series i.e. BURRO 2, BURRO 3, BURRO 7, BURRO 8, and BURRO 9 covering a range of atmosphere stability conditions are reported in this study (Table 2).

The release location was denoted as (0, 0) position in a new coordinate system, which is about 45 degree from the true north. The wind conditions were measured along the true north conditions. Therefore, wind directions for the new coordinate system are calculated by adding 45 degrees to wind direction values reported in the BURRO experiments. A plot in the new coordinate system containing actual release location and sensor locations is shown in Figure 2. The sensors were mainly distributed in four different arches as shown in Figure 2. Detailed location of these sensors and other towers can be found in the BURRO report part 1. The wind fields for different experiments are reported in the BURRO report part 2. To run SAL, a constant wind field (i.e. constant wind speed and wind direction) is assumed. For these short duration experiments, wind field does not change a lot so this is a reasonable assumption. Undoubtedly including variability in the wind field would result in a better prediction at the expense of more computational time.

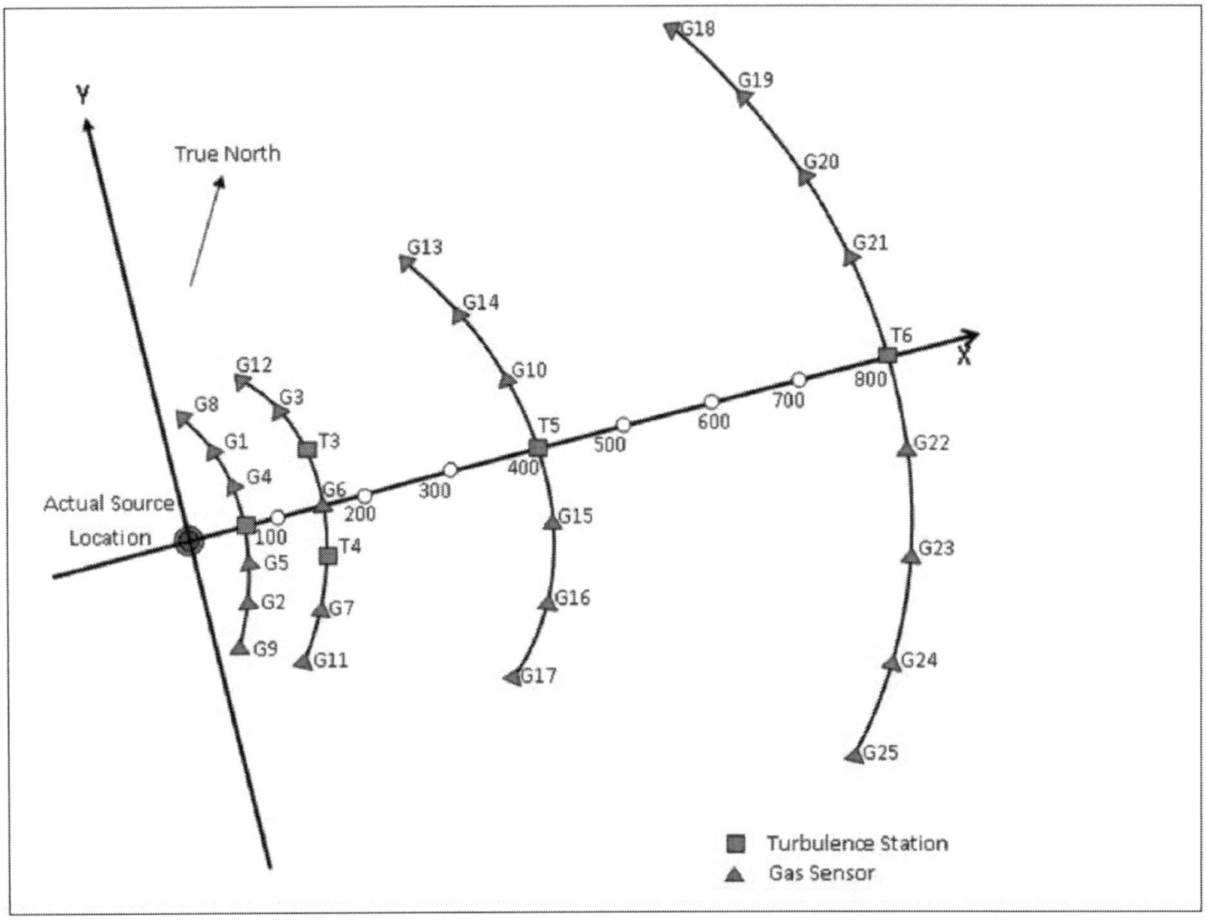

Figure 2: **Burro experiment setup of sensors.**

Table 2: Spill conditions for BURRO 3, BURRO 7, BURRO 8, and BURRO 9

Experiment Name	Chemical	Wind direction	Stability	Wind speed (m/s)	Spill Rate (m3/min)	Spill duration
BURRO 2	LNG	266	2	5.4	11.9	173
BURRO 3	LNG	270	2	5.4	12.2	166
BURRO 7	LNG	253	3	8.4	13.6	174
BURRO 8	LNG	280	5	1.8	16	107
BURRO 9	LNG	277	4	5.7	18.4	79

3.2 SAL Sensitivity to Different Cost Functions

A reasonably good approximation for source area location using SAL is our first target. To accomplish that goal we had to identify the best cost function. Three cost functions were identified; A simple concentration difference objective function **abs**(c_m-c_p), referred as CF 1, Normalized concentration based objective function [**abs**(c_m-c_p)/c_m], referred as CF 2; Distance and concentration normalization based objective function [abs (c_m-c_p)/c_m *$(x_{dist}^2+y_{dist}^2)^{0.5}$], referred as CF 3. Where c_m is the measured concentration and c_p is the predicted concentration.

Only a single measurement value per sensor is used for this analysis. Since it is a dynamic release scenario, various values can be selected from the sensor measurements. A sample of sensors output for Burro 3 is shown in Figure 3. Two set of data was used for SAL study. One was based on the single best sensor value for the entire time zone which reflects the best/average sensor values and the other is based on the maximum value for the entire time. All the sensors with non-zero sensor values were included in the analysis. For SAL simulation, thirteen different values of release rates ranging from 0.01 to 10 are used for source location. A local source point was identified which was average of these thirteen values and was used as an initial point for the next level search. First a coarse grid is used to narrow the search zone. Then a finer mesh grid search was carried out around that zone for various ranges of release rates to identify the source area location.

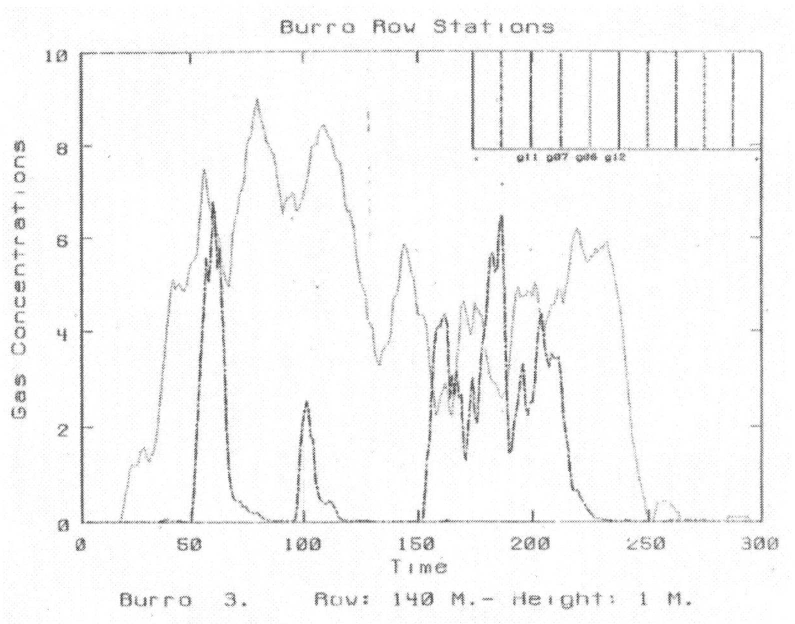

Figure 3: **Gas sensors concentration measurement at 140 m arc.**

The following tables represent the single sensor value obtained from each sensor positioned in the experimental grids for this study along with the source location predicted by SAL using three different cost functions.

3.2.1 Burro 2

Table 3: Sensor's value from BURRO 2 experiment

Sensors	G2	G5	G4	G1	G8	G3	G12	G13
ppm	8	5	1.2	1	2	5	4	0.02

Table 4: Source area location for BURRO 2 using sensor values

WD=266	CF1	CF2	CF3
Stability	2	2	2
X location	-65.1	-57.5	-56.6
Y location	9.2	7.5	8.8

3.2.2 Burro 3

Table 5: Sensor's value from BURRO 3 experiment

Sensors	G2	G5	G4	G1	G7	G6
ppm	9.0	4.0	17	15	4	5

Table 6: Source area location for BURRO 3 using sensor values

WD=270	CF 1	CF 2	CF 3
Stability	2	2	2
X location	-73.4	-117.2	-42.5
Y location	-14.8	40.2	16.3

3.2.3 Burro 7

Table 7: Sensor's value from BURRO 7 experiment

Sensors	G12	G3	T3	G13	G14	G18	G19
ppm	9.0	0.78	0.34	2.8	2.6	0.4	0.05

Table 8: Source area location for BURRO 7 using sensor values

WD = 253	CF1	CF2	CF3
Stability	3	3	3
X location	-111.6	69.5	191.6
Y location	-49.6	59.0	148.6

3.2.4 Burro 8

Table 9: Sensor's value from BURRO 8 experiment

Sensors	G9	G2	T2	G4	G1	G8	G11	G7	T4	G6	T3	G3
ppm	15	6	8	11	5	8	17	12.5	3	6	5	4.8
Sensors	G12	G10	G16	G15	T5	G17	T6					
ppm	2	0.02	0.7	0.7	2	3.8	0.25					

Table 10: Source area location for BURRO 8 using sensor values

WD = 280	CF1	CF2	CF3
Stability	5	5	5
X location	-633.9	-285.4	-355.2
Y location	119.8	70.6	77.9

3.2.5 Burro 9

Table 11: Sensor's value from BURRO 9 experiment

Sensors	G2	T2	G4	G7	T4	G6	T5	G16	T3	T6
ppm	6	1	7	1.2	2.2	6	0.25	0.2	4.5	0.2

Table 12: Source area location for BURRO 9 using sensor values

WD=277	CF1	CF2	CF3
Stability	4	4	4
X location	-70	63.5	66.3
Y location	8.8	-3.9	-5.3

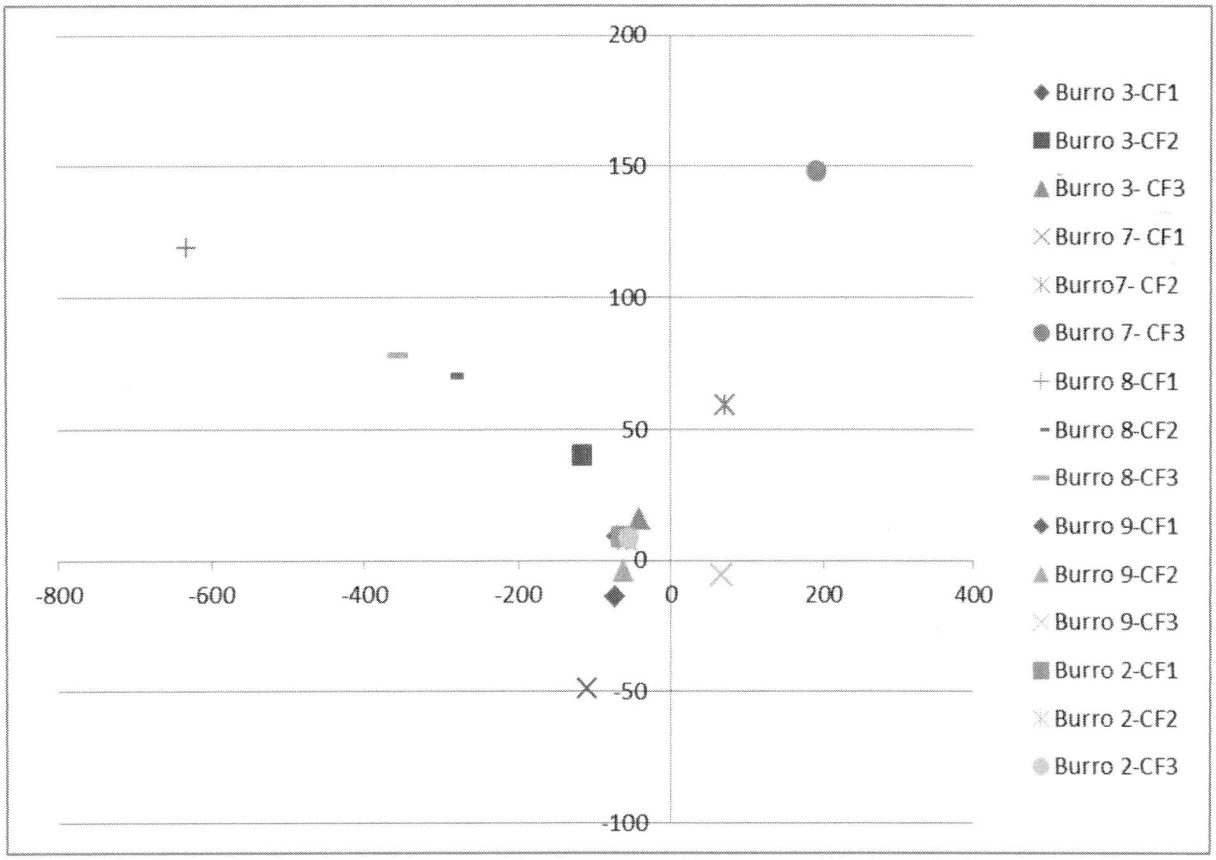

Figure 4: Source location prediction for various Burro experiments using SAL (actual release is at 0,0).

From this analysis we can see that the CF2 give better results out of all three objective functions. It was found that other than stability case 5, in most cases the source area was located within (+/- 70, +/- 70) range of the actual release location. SAL took only few seconds to run for all cases.

3.3 Hybrid Model

Subsequently, the solution from the hybrid approach with an initial guess from SAL to narrow down the search domain using Bayesian theory was obtained. BURRO MCMC refer to the results obtained using the hybrid approach in Figure 5. BURRO SAL refers to results obtained using the least-square based methodology. As evident from Figure 5, BURRO MCMC provides good solution for all stability conditions. Improvements in source location were seen for almost all the cases using the hybrid approach. These simulations were done for a wide variety of stability ranges including unstable to stable atmospheric conditions. Improvements seen in unstable cases using hybrid MCMC approach are more profound than stable cases as shown later in this report. The hybrid model improved the run time by a factor of two to three times.

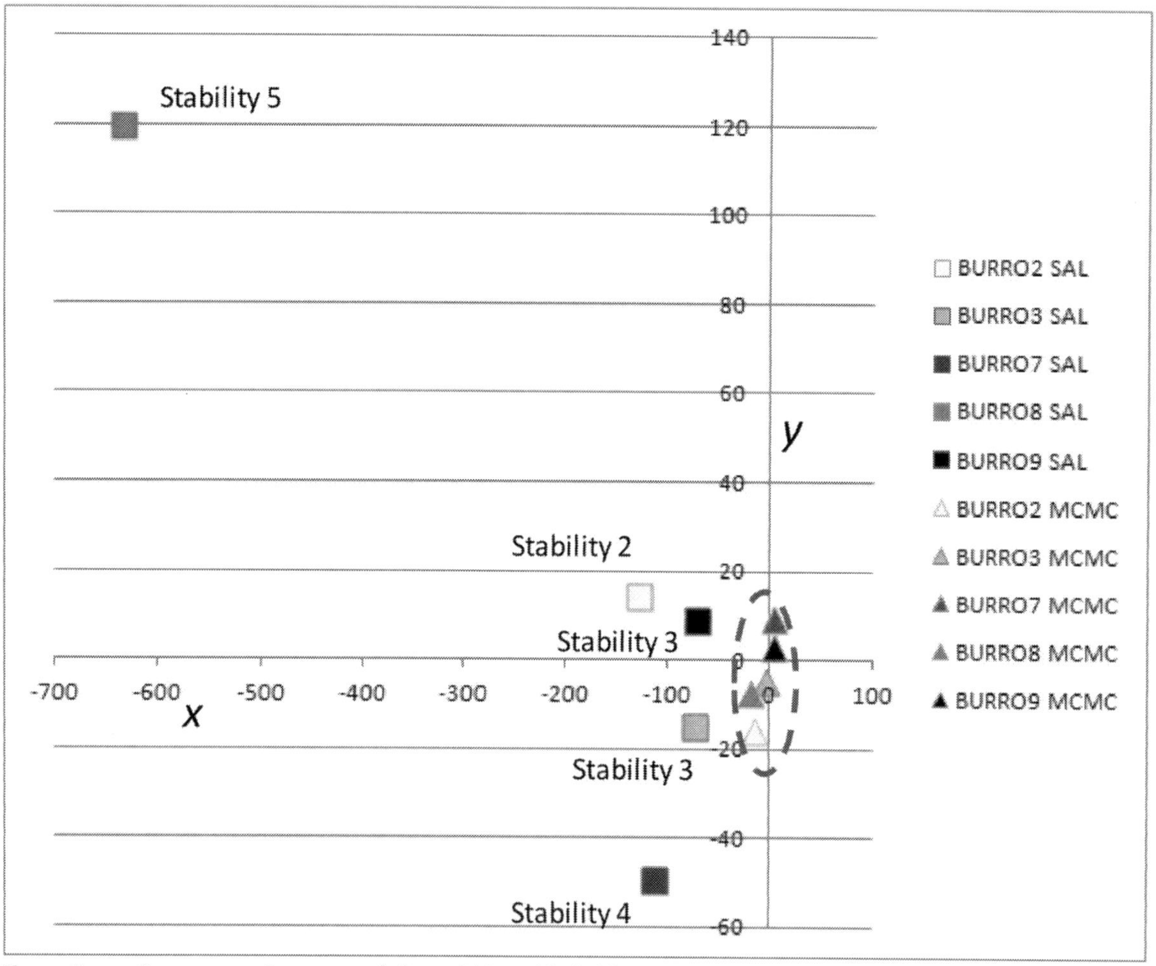

Figure 5: Burro experiment result from SAL and hybrid MCMC approach

Time-varying wind field, uncertainty in wind direction and speed, dynamic sensor readings, release duration, release time may be included in prior parameters for the hybrid approach where the main focus will be on including uncertainties and then identifying the location of the release.

If sensor data is available in dynamic fashion, these hybrid models can be extended to accommodate the dynamic Bayesian model (Johannesson et al. 2004) techniques.

4 Conclusion

The focus of this paper is mainly on estimating release location using the sensor measurements and providing an estimate for the release rate. An important method to estimate source area location using a hybrid approach based on least-square method (SAL) and bayeisan theory (MCMC) was provided. It was shown that the method provides a good solution for BURRO experiment data series for all stability conditions.

5 References

[1] Tarantola A., " Inverse Problem Theory and methods for model parameter estimation", Society for industrial and applied mathematics 2005, Philadelphia.

[2] Johannesson G., Hanley B., Nitao J., Dynamic Bayesian Models via Monte Carlo – An Introduction with Examples -, UCRL-TR-207173, Lawrence Livermore National Laboratory, October 2004.

[3] Cowles M. K. and Carlin B.P. ,Markov Chain Monte Carlo Convergence Diagnostics: A Comparative Review, Journal of American statistical association, vol 91 (434), 1996, pp. 883.

[4] Yee E., Lien F-S, Keats A., D'Amoursc R. , Bayesian inversion of concentration data: Source reconstruction in the adjoint representation of atmospheric diffusion, Journal of Wind Engineering and Industrial Aerodynamics, Vol 96, 2008, pp. 1805–1816.

[5] Yee E., Bayesian Probabilistic Approach for Inverse Source Determination From Limited and Noisy Chemical or Biological Sensor Concentration Measurements, Chemical and Biological Sensing VIII, Proc. of SPIE, Vol. 6554, 65540W, 2007.

[6] Kim J.Y., Jang H.K., and Lee J.K, Source Reconstruction of Unknown Model Parameters in Atmospheric Dispersion Using Dynamic Bayesian Inference, Progress in Nuclear Science and Technology, Vol. 1, 2011, pp. 460-463.

[7] Keats W.A. , Bayesian inference for source determination in the atmospheric environment, PhD thesis, University of Waterloo, Waterloo, Ontario, Canada, 2009.

[8] Neuman S., Glascoe L., Kosovi´c B., Dyer K., Hanley W., Nitao J., Event Reconstruction for atmospheric releases employing urban puff model UDM with stochastic inversion methodology, American Meteorological Society Conference, 2005, Atlanta GA.

[9] Zheng X., Chen Z., Inverse calculation approaches for source determination in hazardous chemical releases, Journal of Loss Prevention in the Process Industries 24 (2011) 293-301.

[10] Sharan M., Singh S. K., and Issartel J.P, Least Square Data Assimilation for Identification of the Point Source Emissions, Pure and Applied Geophysics, August 2011.

[11] Monache L.D. et al., Bayesian Inference and Markov Chain Monte Carlo Sampling to Reconstruct a Contaminant Source on a Continental Scale, Journal of Applied Meteorology and Climatology, pp 2600-2613, vol 47, 2008.

[12] Senocak I., Hengartner N. W., Short M.B., Daniel W.B., Stochastic Event Reconstruction of Atmospheric Contaminant Dispersion using Bayesian Inference, Atmospheric environment, Vol 42 (33), 2008, pp7718-7727.

[13] Koopman R. P., Baker J., Cederwall R. T., Goldwire H. C., Hogan W. J. LLNL/NWC (Lawrence Livermore National Laboratory/Naval Weapons Center) 1980 LNG (Liquefied Natural Gas) Spill Tests. Burro Series Data Report.

[14] Goldiwre H. C., Baker J., Koopman R. P., Cederwall R. T., Hogan W. J. LLNL/NWC (Lawrence Livermore National Laboratory/Naval Weapons Center) 1980 LNG (Liquefied Natural Gas) Spill Tests. Burro Series Data Report: The Appendices.

[15] S. Khajehnajafi, S. Burla, Prediction of Source Location and Rate of a Chemical Release for Emergency Response: Back Trajectory Approach, AIChE Spring Meeting, 2008

[16] Hogan W.R., Cooper G.F., Wallstrom G.L., Wagner M. M. and Depinay J-M, The Bayesian aerosol release detector: An algorithm for detecting and characterizing outbreaks caused by an atmospheric release of *Bacillus anthracis, Statist. Med. 2007;* **26**:5225–5252.

[17] Haario H., Saksman E., Tamminen J., An adaptive metropolis algorithm, Bernoulli, Vol 7 (2), 2011, pp. 223-242.

(43ak) Guidelines for Developing Site-Specific Human Error Rates and Human IPLs for LOPA Using a Safety Climate Approach

Farheen Khan, Human Factors Practice, Exponent, Inc., Chicago, IL, Sunil D. Lakhiani, Human Factors, Exponent, Failure Analysis Associates, Wood Dale, IL and Delmar Morrison, Thermal Sciences, Exponent, Inc., Warrenville, IL

Layer of Protection Analysis (LOPA) is a simple order of magnitude quantitative risk analysis technique commonly applied in the process industries. This powerful tool is especially useful for refining specific hazard scenarios during a process hazard analysis or a risk analysis. In LOPA, a hazard scenario is decomposed into a series of events, which may include the frequency or probability of equipment failures, personnel actions, and physical processes. By one estimate, 80 percent of incidents involve human error; thus, most risk analyses consider the likelihood and effect of human factors on potential scenarios. For LOPA, human error includes events initiated by the human in the work system and events where the human element failed to act as an independent layer of protection against an adverse event.

Although the LOPA method is widely used, there are few guidance sources for refining the human error component. The currently available standard LOPA estimates for the occurrence of human error assume optimal human factors conditions and that work systems are static environments unchanging over time. Work systems are, in fact, dynamic environments, which evolve over time through planned changes, as well as unplanned adaptations. Thus, no work system is completely free of error at any given time. Current LOPA estimates, therefore, fail to account for the dynamic nature of work systems where change from original conditions is not only possible but even necessary from time to time. This paper will examine the deficiencies in the LOPA human error probability estimates and propose a strategy that utilizes a safety climate approach to understanding error causation and incorporating it as a fundamental building block to human error estimation.

9TH GLOBAL CONGRESS ON PROCESS SAFETY

Benchmarking – An Important Milestone in the Journey towards PSM Excellence

Prerna Jain

Reliance Industries Limited

India

prerna.jain@ril.com

[1]CindiePridy, [2]Ravi Ramaswamy, [3]Robert Cravens

[1,3]Phillip Townsend Associates, Texas

[2]Reliance Industries Limited, India

cpridy@ptai.com

ravi.ramaswamy@ril.com

rcravens@ptai.com

Prepared for Presentation at

American Institute of Chemical Engineers

2013 Spring Meeting

9th Global Congress on Process Safety

San Antonio, Texas

April 28 – May 1, 2013

UNPUBLISHED

Benchmarking – An Important Milestone in the Journey towards PSM Excellence

Prerna Jain

Reliance Industries Limited

India

prerna.jain@ril.com

[1]CindiePridy, [2]Ravi Ramaswamy, [3]Robert Cravens

[1,3]Phillip Townsend Associates

[2]Reliance Industries Limited

Keywords: benchmarking, gap analysis, metric, performance gap, process safety management system, risk based process safety element, survey

Abstract

Reliance Industries Limited (RIL) has implemented a Process Safety Management (PSM) system through a structured and systematic approach over the last 6 years. In its PSM journey, RIL realized that there is a continuing need to improve and sustain PSM performance and to develop an understanding of the current status of RIL's existing PSM system, a benchmarking process of PSM practices, with reference to industry peers, is required. This was essential in order to identify and address the gaps to best performers within the peer group, define specific action items to obtain significant performance improvements, and ultimately achieve the best systems in the world in field of Process Safety Management.

To accomplish this goal, RIL conducted a PSM Benchmark Program Study designed by Phillip Townsend Associates (PTAI) and CCPS® at its six largest manufacturing locations in India. By helping to understand the relationship between process safety outcomes and management practices; this benchmarking program provided several key approaches for improving overall PSM performance. The program covered benchmarking of six elements related to a world class process safety management system. They are: process safety culture; compliance with standards, codes, regulations, and laws; hazard identification and risk analysis; asset integrity and reliability; management of change; and measurement and metrics.

The results of the study provided element specific quartile performance of each of the six manufacturing sites and Reliance as a whole in comparison to the industry peers who participated in the PTAI study. The gap analysis report presented by PTAI to the six sites

PSM team and senior management of RIL upon completion of the benchmarking study identified key areas of improvement for implementing and evolving effective Process Safety Management System at RIL.

This paper, presented by RIL and PTAI, discusses the benchmark process, methodology followed for the program, the value proposition of Process Safety Benchmarking, benefits of the program, key applications and lessons learned by RIL and new directions for improving PSM. The paper specifically focuses on the journey to success by realizing the gaps to top performers in the peer group, enhancing PSM knowledge and spreading PSM best practices.

1. Introduction

Chemical production and the associated processing industries dealing with hazardous and toxic materials have a significant commercial and societal responsibility to eliminate and/or mitigate the loss of primary containment or fires related to process materials while also ensuring that adequate safety systems are in place. Process Safety Management systems differ from personnel safety management in that process safety focuses directly on the processes, equipment and technology involved in causing an incident. The broader application of process safety management decision making and leadership behavior needs to address loss of containment events of greater and lesser consequences, safety system challenges and failures that could have led to failure, and operating discipline and management behaviors that ensure adequate layers of protection are being maintained. The focus shifts from the lagging indicators that track incidents to leading metrics that are used in a proactive manner to prevent incidents and continuously improve as shown in Figure 1: Process Safety Indicator Pyramid (API RP 754, 2010).

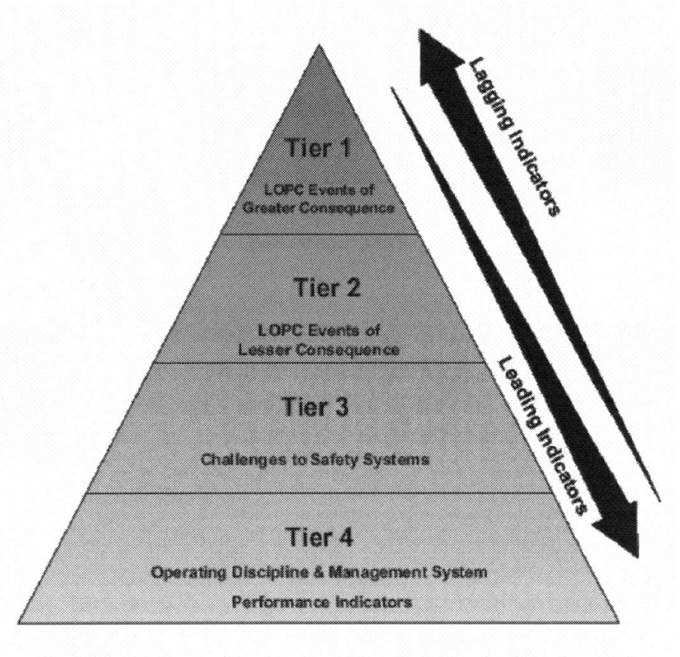

Figure 1: Process Safety Indicator Pyramid

The vision and recommendation of industry-wide process safety originated with the setting up of Centre for Chemical Process Safety (CCPS) by AIChE as a result of the December 3, 1984 Bhopal incident killing more than 3000 people due to toxic Methyl Iso Cyanate gas release. The need for continued focus on process safety got reinforced with the 2005 explosion at the BP refinery in Texas City, Texas. Since then, the Center for Chemical Process Safety (CCPS®) has identified 20 process safety elements to be used as guidelines for a risk based safety process (CCPS®, 2007). In 2009, the CCPS® and Phillip Townsend Associates, Inc. (PTAI) entered into an agreement to develop a benchmarking program for Process Safety Management Systems (PSM).

Viewed from a risk based perspective, the process safety elements can be grouped into four categories: Commitment to Process Safety, Understanding Hazards and Risk, Managing Risk, and Systematic Learning from Experience. PTAI and CCPS® designed the PSM benchmark selecting six of the 20 elements: Process Safety Culture, Compliance with Standards, Hazard Identification and Risk Analysis, Asset Integrity and Reliability, Management of Change, and Measurement and Metrics (CCPS®, 2007). The key goal was to seek an understanding of the relationship between process safety outcomes and management practices.

This paper discusses the details of the PSM benchmark, the value proposition to be gained, and specifically the experience of Reliance Industries Limited (RIL) in developing an understanding of the current status of RIL's existing PSM system, and the practices that will improve and sustain PSM performance. The objective of RIL is to take PSM systems and practices to the top quartile performance through a benchmarking process of industry peers relative to PSM practices. Specifically, the paper focuses on the most important decisions and behavior changes that leadership can make to spread process safety awareness throughout the organization, close the gap to the top performers, develop central and site specific action plans for improvement, and raise the internal level of all organizational units to the that of the top internal performer.

2. Methodology

Benchmarking identifies how well a company is doing compared to its peers in the industry, what the industry top performance is, how large the gap is to the top performers within the industry, and create a stimulus for continuous improvement. Benchmarking programs seek to understand the relationship between outcomes and management practices. The PSM benchmarking program is a systematic study of process safety between companies that will allow sites/companies to compare their management systems with the goal of improving the design and implementation of management practices that impact process safety.

2.1 Value Proposition of Process Safety Benchmarking

PSM benchmarking can elevate a company's process safety performance through four distinct steps:

2.1.1 Identify performance gaps

The identification of a company's performance in relation to Performance Quartiles will highlight key drivers for performance enhancement versus its peers. These quantifiable results replace perceptions with a data-based reality check. In addition, these results identify if the enterprise policies, practices, and procedures are operating as intended and are achieving the desired results.

2.1.2 Create performance enhancement goals

Data driven information provides a measure of the enterprise effectiveness – emphasizing strengths, weaknesses and potential for improvement over time. This helps focus attention on the critical aspects of an enterprise that create risk. It also underscores what improvements should be made – providing quantifiable performance enhancement goals.

2.1.3 Measure the effectiveness of improvement programs

PSM benchmarking can be a starting point for goal setting and performance initiatives aimed at boosting step change improvements in process safety. The clear outcome is that this will add significant shareholder value through reduced risks of catastrophic incidents.

2.1.4 Maintain continuous improvement

The PSM benchmark helps maintain a stimulus for continuous improvement. It provides the necessary motivation to ensure consistent top performance results.

2.2 Benchmarking Process

A peer group is established by PTAI of sites/companies agreeing to compare their process safety programs and performance in an effort to identify best practices and drive improvement. Ensuring consistency and protection of each participant's interests are critical to a successful benchmark. A confidentiality agreement was completed between PTAI and RIL along with execution of a participation agreement. The Data Collection Instrument (DCI), providing clear and consistent definitions for all participating sites/companies was made available. The use of a secure site over the internet ensured security of data and ease of use by participants as they are able to see all the survey questions and select their answers. A kickoff discussion to align key stakeholders at each of the sites is critical. In RIL case, this was done through a series of webinars between PTAI consultants, RIL's Centre for HSE Excellence and stakeholders at each of the participating sites.

RIL chose to benchmark PSM in six of its large Refining and Petrochemical manufacturing sites in India and a team of site focal points, corporate coordinator and PTAI worked in a seamless manner to plan and execute the benchmark program. Following submission of the data, PTAI validates the data on a site and company basis and provides analysis of the results. Validation was enhanced with RIL by individual site visits to each of the six sites by PTAI. This provided the opportunity for discussion and verification of the results reported to the on-line survey. Participation by key RIL personnel during this process added to the effectiveness and accuracy of data reported and provided a seamless voice to the sites between RIL's Center for HSE Excellence and PTAI.

In particular, statistical analysis of the data was performed by PTAI to evaluate various drivers of process safety using a Gap Analysis comparing RIL's performance to the top quartile of the peer group. The quantitative and qualitative results were distributed to participants with numerical scores for each element along with an overall site/company specific score. Strengths, as well as the main opportunities for improvement to close the gap towards the top performers were identified. A discussion of the study results and gap opportunities was held with the executive management in an executive management summary with specific individual gap analysis presentations to the top Line and HSE management of each of the six RIL sites benchmarked.

2.3Benefits of PSM Benchmarking

Management review is an integral part of moving from lessons learned to systematic improvement of PSM. Reviews of the performance gaps can identify potential problems and address them before an incident occurs; as well as, identify opportunities for improvement to move from current levels to the level of top performers. PSM benchmarking helps an organization prioritize the allocation of process safety resources, both financial and human, to focus on the highest identified risk areas.

Among the benefits of PSM benchmarking, there are certain ones that can have a big impact on an organization. There should be a marked increase in staff awareness of process safety issues. This awareness will expand the issues from the Health & Safety group to Engineering to Operations. Based on the organizational level of results discussion, it will potentially increase awareness from site management to impact those that are responsible for setting corporate strategy. A key learning for RIL was the differences between the individual site scores resulting in a changed perception by the Center for HSE Excellence and executive management. PSM benchmarking creates a common understanding and language on process safety matters. Benchmarking can also reduce risks by potentially providing an incident early warning signs.

Sometimes the opportunities for improvement can be very different from the perceived gaps in an organization. Identifying and gaining insight on perceived gaps versus actual gaps is greatly enhanced by an independent benchmark which effectively creates consistency across multiple sites within a company and with the others in the peer group.

3. Key Applications and Lessons Learned

Traditional process safety surveys have identified a list of both quantitative and qualitative issues. Translating the qualitative issues into measurable, actionable steps is a key issue for senior management decision making and allocation of resources. Benchmarking results must be measurable and use a scoring methodology that will not only identify the performance gaps, but also quantify the impact of selected actions. The

PSM benchmarking process designed by CCPS® and PTAI includes a scoring methodology which enables participants to have measurable action plans. Benchmarking is a systematic way to make site-to-site company comparisons and view a company's performance level in relation to that of top performers.

The application of the benchmarking results to identify performance gaps provides meaningful insight for organizations as they develop goals and the associated improvement plans to reach those goals. Repeated benchmarking monitors the rate of improvement. One example of a lesson learned that applies to numerous organizations relates to Process Safety Culture. The findings from the Process Safety Culture section indicate the largest range of results compared to the other sections. The underlying cause of the low scores was a direct outcome of not monitoring the process safety culture through the use of an employee survey. Without a way to assess their work force's values, attitudes and behaviors pertaining to process safety, an organization will have no meaningful basis for making and measuring improvements.

The global process safety incidents have produced tremendous need for the use of PSM benchmark assessment to measure and improve process safety performance. Initially focused on fatality rates, manufacturers quickly learned that having strong PSM system and effective layers of protection were important to prevent any loss of containment scenarios.

At RIL, PSM Benchmark assessment and results helped in identifying following three main interrelated but distinct benefits or observations: (a) Internal Benchmark of various RIL sites, (b) learning through process of understanding the benchmark questionnaire, and (d) need for a detailed analysis of existing systems implementation. We briefly discuss and illustrate examples of each benefit or observation, highlighting how the results could be used.

(a) Internal Benchmark of RIL multiple sites:

PSM Benchmark program conducted for six RIL sites brought to the surface best practices followed at different sites like: Number of indicators tracked and analysis of metrics trend, Quality control reviews of Process Hazard Analysis reports, action plan clearly defined from the Operational Discipline survey results, employee involvement through various PSM campaigns and some other actions related to operational discipline. These practices were beyond the existing standard and system requirements and became good learning for other sites. Also, these best practices will be additional inputs to the group level company documents on PSM. For example, the delta between the site with the highest score and the site with the lowest score in the Compliance with Standards section was 17%; offering a significant opportunity to analyze internal best practices and improve the level of performance to the highest score.

(b) Learning through process of understanding the questionnaire

The benchmarking questionnaire itself generated opportunities for learning new knowledge. Understanding the "why" part of questions called for discussions among various functions at each of the manufacturing sites and reinforced the understanding of PSM. A self-assessment process got started on the compliance to PSM requirement. Following table gives certain examples of learning and new knowledge for the questionnaire itself:

Table 1. Examples of learning and new knowledge from Benchmark questionnaire

Sr. no.	PSM Element	Question	Learning
1	Hazard Identification and Risk Analysis	How are overdue HIRA action items escalated in your organization?	Tracking system to be further strengthened for overdue HIRA action items.
2	Asset integrity and reliability	How is the metal thickness data from the periodic inspections analyzed for trends?	Thickness measurement trends should be analyzed, remaining life projections should be calculated, and adjustments to be made to inspection intervals.
3	Process safety culture	If your organization has utilized a survey tool to assess Process Safety Culture, how many questions were on the survey?	PSM Culture Survey to be established for the organization.

The validation visits from PTAI enhanced the quality of responses and the participation of various functions in the PSM benchmarking program.

(c) Need for a detailed analysis of existing systems implementation

The PSM benchmark results identified areas for further analysis to enhance the systems implementation for improved performance. At RIL, the trend of Process Safety incidents is varying for the last three years and there is no clear trend of reduction in loss of containment events from the metrics, inspite of implementing PSM systems for the last four years. For this, RIL has initiated a group level six sigma project on reduction in loss of containment events with support of an inter-site team. The project follows Define, Measure, Analyze, Improve and Control

(DMAIC) methodology of six sigma. Following is a brief of activities initiated for the above:

- A multisite team has been formed with a six sigma Black Belt guide.
- Literature on Process Safety and Six sigma circulated to team members.
- Project charter was prepared and circulated to team members and other stakeholders for inputs/suggestions and firmed up.
- Cause-Effect Matrix Format developed and Process mapping completed for PSM elements.
- Audit Report Analysis.
- Incident investigation report analysis.

4. New Directions and Next Steps

The PSM benchmark program results gave new insights to the RIL team on the status of PSM within RIL and in relation to the other participants. The gap analysis was converted to actionable recommendations at Group Level and individual site level. Next steps in the PSM excellence journey for RIL have been identified and these are

- Develop a formal system of Process Safety Culture Survey.
- Review of existing PSM metrics and make necessary amendments.
- Review Human Reliability/Human Factors/ Human-Machine Interface requirements, develop best practice requirements and make necessary amendments in Standards.
- Develop program for assessment of compliance with standards/codes/practices for existing equipment.
- Establish system to manage deviations from the industry consensus standards.
- Roll out the Management of Change –Personnel process formally.
- Actual implementation of Standards/Procedures/Practices in field:
  - Meeting target inspection/testing intervals, backlog of overdue inspections
  - Closure of recommendations of PHA, QRA studies
  - HIRA team members to meet qualification requirements
  - Closure of Audit recommendations
  - Implementation of temporary Management of Change
- Review RIL Group PSM Standards and Procedures for gaps and make necessary amendments.

A key benefit of the program for RIL is an Internal Benchmarking among its major manufacturing sites and sharing the best practices within the group. All the sites have developed action plans to close their own gaps in PSM practices.

The collaboration between PTAI and RIL gave new directions for RIL as well as for the Benchmark program itself. A development of two distinct Plan Do Check Act Assess (PDCAA) cycles, one for RIL and the other for PTAI has been the result of this collaboration.

4.1 Overview of PDCAA cycle

PDCAA is a 5-step methodology for process improvement. The stages are shown in Table 2.

Table 2. PDCAA stages

Plan	Establish objectives and how they will be achieved
Do	Put the plan into effect
Check	Verify that the process achieved the desired results
Act	Analyze any differences and their causes. Take action to improve things.
Assess	Analyze and reflect on evidence to make informed and consistent judgments on an ongoing basis for continual improvement.

The PDCAA Cycle for RIL is addressed through the following:

- Planning upgrade of RIL PSM standards to industry best levels and establishing systems to manage deviations, and developing appropriate leading and lagging metrics.
- The do part of the cycle will address the actual implementation of standards and practices in the field at different sites through a process of sharing the best practices by institutionalizing the inter-site network teams for the PSM elements.
- Check part will address strengthen the current PSM metrics tracking and review systems, IT enablement of dashboards for Process Safety at different levels of organization, and reflecting the updated requirements of compliance in the PSM audit protocols.
- A robust system of closure of recommendation and actions arising out of various studies and audits through a system of IT enabled universal work list and linking it to individual performance management system is under development which form the Act step of the cycle.

- Periodic reviews and assessment of PSM will complete the whole cycle of improvement.

4.2 Plan-Do-Check-Act-Assess- RIL Example: (Hazard Identification and Risk Assessment)

Plan

In this step, process needs are analyzed by examining a range of data available from the Benchmark assessment gap analysis relating questions in program to performance results. Data can be analyzed at the individual sites level and the company level. In this process, aspects on HIRA management system, including HIRA methodologies, team composition, criteria for HIRA team leaders/facilitators, level of training, techniques, risk criteria, recommendations and their status were covered.

Do

This step asks what group standards require and how they will be assessed. Forward action plan is made by looking at what is lagging in actual implementation in field ; such as, the timely closure of HIRA recommendations and minimum criteria/qualification for team members. Sites develop individual goals to improve their performance.

Check

In the "check" step, formal and informal assessments take place continually, to strengthen the performance of Metrics on HIRA elements. Sites can make use of comparative data to identify trends to study; for example, overdue HIRA action items.

Act

In this example the "act" step calls for standardization or having a robust system to ensure closure of overdue action items. Results from this cycle become input for the "assess" phase.

Assess

In this step, process of periodic reviews and monitoring of HIRA action items is done and tracked to ensure continual improvement.

4.3 Plan-Do-Check-Act-Assess- PTAI Example: (The benchmarking process and Process Safety Culture)

The PDCAA Cycle has also provided the opportunity for continuous improvement for PTAI and the PMS benchmarking survey.

Plan

Through the validation process the iterative nature of conference calls and site visits proved to be of significant value in providing accurate and consistent understanding of the survey questions. This resulted in comparable and meaningful results among the sites and RIL total scores.

Do

Reviewing the current benchmarking process and seeking feedback from RIL identified that enhanced validation and the creation of a more iterative process could be incorporated into the benchmarking process. A key area of needed improvement supporting the benchmarking process in the Process Safety Culture section is the measurement of Process Safety Culture. Both RIL and PTAI determined the need for a formal system for a Process Safety Culture Survey that provides more consistent measurement and comparable results of the safety culture that can be utilized for RIL internal site to site comparisons and RIL's sites to the external peer group. RIL and PTAI have discussed future collaboration on this project and the possible implementation to all of the RIL sites for internal benchmarking and review.

Check

Informal and formal process assessments are the key to continuous improvement of the PSM Benchmarking Survey and process.

Act

The act step calls for standardization of a formal process assessment tool to measure the benchmarking efforts required by participants and the actual benchmarking survey tool.

Feedback from participants needs to be evaluated and appropriate improvements to the program parameters and the benchmarking process implemented.

Assess

This is the process of continuous improvement and an environment of continuous learning. PTAI will periodically review the process and enhancements with CCPS to ensure an evolving process that grows in value-added elements.

5. Key success factors

The success of the PSM benchmarking survey hinged on four key factors; management support of a systematic approach, the strength of the validation process, the ability to report results with a detailed gap analysis that could be translated into specific action plans at a central and site level, and the seamless voice of RIL's Center for HSE Excellence and PTAI through strong communication and a systematic effort.

The management presentation of results was opened with comments from RIL's President, HSE, emphasizing the strategic importance and focus of PSM as a systematic process and not simply a process safety program. His comments were directed and focused on the areas he felt most important to taking RIL to the next level. This sent a very clear message of the strategic focus placed on PSM at RIL.

The continued dialogue and collaboration between RIL and PTAI has enhanced the double loop, PDCAA cycle. As a critical success factor this creates ongoing and new opportunities for both.

6. Conclusion

Process safety is a mature, but continuously evolving concept with practices and procedures varying widely among the refining, chemical and chemical processing industry. Moving from a compliance model to a risk-based model requires a comprehensive approach to managing process safety. Substantial improvement opportunities exist for even the current top performers. Specifically, no site has made a perfect score on any of the risk-based elements of the PSM benchmark.

The PSM benchmarking program is a unique methodology to allow companies to identify how large the gap is to top performers and to "learn" how to prioritize resource allocations based on the greatest impact opportunity to improve their process safety management system. This shifts the understanding of current practices to the lessons learned that are measurable and actionable. The benchmarking process identifies top quartile management practices while highlighting not only a company's strengths, but also potential development opportunities. The program's scoring system provides a marker to track improvements and measure the gaps within the company sites and against external top performers.

Using a systematic Process Safety Management System benchmark drives lessons learned to strategic and organizational learning.RIL considers periodical assessment of its PSM through repeat benchmarking at regular intervals as a good value proposition to monitor trends of performance and to achieve pacesetter performance compared to industry peers.

7. References

[1] American Petroleum Institute, API RP 754, "Process Safety Performance Indicators for the Refining and Petrochemical Industries, First Edition," 2010.

[2] Center for Chemical Process Safety CCPS®, "Guidelines for Risk Based Process

Safety," John Wiley & Sons, Inc., Hoboken, NJ, ISBN 978-0-470-16569-0: liii-liv, 2007.

"Process Safety Management - A Practical View"

German Luna-Mejias, P. Eng.
ACM Facility Safety
Suite 700 940 6th Ave SW, Calgary, AB, Canada
gluna@acm.ca
guatire2@gmail.com

Prepared for Presentation at
American Institute of Chemical Engineers
2013 Spring Meeting
9th Global Congress on Process Safety
San Antonio, Texas
April 28 – May 1, 2013

UNPUBLISHED

"Process Safety Management - A Practical View"

German Luna-Mejias, P. Eng
ACM Facility Safety
Suite 700 940 6th Ave SW, Calgary, AB, Canada
gluna@acm.ca
guatire2@gmail.com

Keywords: Process, Safety, Management, Practical

Abstract

Process Safety Management (PSM) was developed over 20 years ago to reduce hazards and risk for the Oil Industry. In the last 3 years serious accidents have happened in the Oil Industry that requires us to re-evaluate PSM programs. It has been observed in many HAZOP studies, that the HAZOP (part of a PSM program) does not reflect a real commitment to safety, but the "need" to meet regulations. This paper's initiative is to call organizations/companies to think and be safe instead of just fulfilling project or government rules. Full involvement of the organization, starting with managers, is required for the PSM program to work. In this paper, practical examples will be drawn from the successful implementation of a PSM program from several years ago in a refinery. Additionally, the reasons for this implementation to be considered successful will be discussed at length.

1. Introduction

1.1 History

On June 1st, 1974, a Cyclo-hexane vapor cloud explosion killed 28 workers and injured 53 others; 1800 houses were damaged in the nearby area in Flixborough, England. [1]

On Dec 3, 1984, an uncontrolled release of methyl-isocyanate (MIC) gas from a pesticide plant in Bhopal, India, resulted in more than 3,800 fatalities and approximately 11,000 injuries with different degrees of severity. [2]

In 1985, the American Institute of Chemical Engineers (AIChE) formed the CCPS (Chemical Center for Process Safety) quickly set out to publish its first process safety guideline book: Guidelines for Hazard Evaluation Procedures.

On July 6, 1988, the offshore platform Piper Alpha, located in the North Sea and owned by Occidental Petroleum, suffered a series of explosions and fires that resulted in the deaths of 165 crew members and the total loss of the platform. [3]

On Oct 23, 1989, an explosion in the Phillips refinery in Pasadena, Texas killed 23 people and injured 314 more. Impact of the explosion was registered as a 3.5 on the Richter scale. [4].

In 1990, the American Petroleum Institute issues the recommended practice 750: "Management of Process Hazards" which can be considered the first step onto the Process Safety Management programs. [5].

In 1992, OSHA formally recognized the vital importance for managing highly hazardous chemical processes within the US. A company that is involved in highly hazardous chemical processes or activities is mandated to implement a process safety management system. [9]

In 1994, CCPS modified the elements of the PSM program recommended in API RP-750 [5] to issue a new "standard" for PSM that includes more elements: "Guidelines for implementing Process Safety Management System"

After a period of "relative calm", big accidents in the oil and gas industry occurred again:

- BP Texas city, 2005 [6]
- BP Macondo, Gulf of Mexico 2010 [7]
- San Bruno, California 2011 [8]

Many disasters have impacted the process industry, from Flixborough to Macondo and elsewhere many more. Through the ashes of these tragedies, strong change has been advocated by the public and their government's regulators to hold companies accountable for their negligence.

1.2 Process Safety Management

Process Safety Management (PSM) is a methodology that was developed after the lessons from several major accidents showed that such events can be difficult to prevent using traditional occupational health and safety approach, which tends to focus on the interface between the individual operator and the equipment or process.

The 3 keywords of this philosophy are:

- PROCESS: related to process issues. i.e. fires, explosions, etc
- SAFETY: Safety regulation, safety concern
- MANAGEMENT: anyone is a manager in some degree, requires employment participation.

PSM is the application of management principles and systems to the identification, understanding and control of process hazards to prevent the release of highly hazardous chemicals such as toxic, reactive, flammable and/or explosive substances (process-related events), that may lead to injuries and incidents.

The goal is to minimize process incidents by evaluating the whole process. The PSM standard (OSHA 29- Part 1910.119) [9] is the regulatory framework for process industries in the US and is quickly becoming the industry "best practices" standard for responsible companies worldwide.

The PSM system originally suggested by the CCPS (Guidelines for implementing Process Safety Management System) consists of 12 elements:

1. Accountability: objectives and goals
2. Process knowledge and documentation
3. Capital project review and design procedures
4. Process risk management
5. Management of change
6. Process and equipment integrity
7. Human factors
8. Training and performance
9. Incident investigation
10. Company standards, codes and regulations
11. Audits and corrective actions
12. Enhancement of process safety knowledge.

The intent of this paper is to present some of the elements of PSM that were considered the most important for a specific company; not to debate the number of elements or which one is more important than the other.

Regulations should not be the sole driving force for the process industry to "think safe". Applying the PSM program accurately and effectively requires more than just the requirement to fulfill government regulation but a whole company culture to be involved in making the workplace safe. This paper's initiative is to encourage organizations/companies to think and be safe instead of simply fulfilling a project or complying with government regulations.

The examples given in this paper correspond to a company where PSM was not mandatory by law, but it was part of the company's standard as a way to make sure the company complied with the International O&G industry standards.

2. The practical Approach

In this section, examples of the approach taken by an O&G company to implement PSM are given for what was considered the most relevant part of the program. Not all elements are discussed as the key elements described below will support all other elements. The practical approach also includes unfortunate situations that happened before the implementation of the program to emphasize the need for it.

Elements analyzed in this document are:

- Management commitment (Element 1)

- Process Safety Information (Element 2)
- Process Hazard Analysis (Elements 4 & 5)
- Operating Procedures (Elements 6,7 & 8)
- Hot work Permit (Element 6)
- Training (Element 8)
- Pre-start up Safety Review (Element 9)
- Emergency Planning and Response (Elements 4 & 12)
- Program Audits (Element 11).

2.1 Management Commitment

Management commitment is fundamental to the PSM program. Exercising and eating well is something that people need to do for their own good and not because it is a mandate or a command to obey. Exactly the same happens with the PSM program: if people are not convinced of the importance of safety, the PSM program won't last. The first safety best practice identified, and one which is essential to any good safety program, is a management commitment to safety.

No company can have a successful PSM program if the management of that company is not committed to the concept of safety. Every company has an obligation to ensure, as much as possible, the safety of its employees. Not only does the company have a moral obligation to ensure the safety of their workforce, it is also good business to do so. Keeping the number of incidents down will also help the company to maintain few lost time days, low work injury indexes and increased profitability.

In a particular refinery, the General Manager wanted to develop a PSM program and designated a top manager to be in charge of the PSM program. This designated manager's first action was to convince all top managers of the importance of the PSM program. Convincing the other managers was not based on: "I said so, and I have been appointed by the GM" but with education and conviction.

The training of managers and cascading this knowledge to all personnel with supervisory responsibilities started with the video forum about the Piper-Alpha accident (one of the objectives of the forum was to recognize that this scenario may happen to that facility). The Piper-Alpha video, in conjunction with API 750 video, was presented to all personnel.

"Safety oriented Management behavior" was a key, as the employees know who the managers are, and they observed their commitment to safety. All personnel understood that they were "living by example" instead of "giving lip service to their own safety rules". The exposure of management risk and demonstration of Safety Concept by action prevented the scenario of: "Well, management doesn't need to operate safely, neither do we."

Additionally, financial incentives to personnel were also created during the earlier phases of the program to promote participation within it. After a period of time, people became conscious about safety and they started to behave safely even when "nobody is looking"; not because of a reward but they were convinced of PSM as a way of life.

Part of the management commitment to safety was the implementation of safety related training programs with the same level of importance as technical training. The program was successful because there was support from the entire management team, and management was involved in the formulation, communication and enforcement of its safety program. All levels of management were visibly demonstrating their commitment to safety.

2.2 Process Safety Information.

Process Safety information (PSI) includes design information, mechanical design and working knowledge. These 3 elements are linked to good quality P&IDS, which was the driving reason the company decided to move in the direction to update the P&IDs as the first step under the PSI element.

This was critical to the facility as it was a 60 year old refinery. In several cases the installation of jump-overs, minor changes and similar modifications were done in the field without any analysis or documentation on the P&IDs. Some minor incidents (spills, spray, etc.) were found to be a consequence of inadequate information on P&IDs.

As a result, a team composed with very experienced operators and young engineers was formed to update all the P&IDs of the facility, with two main purposes: pair young engineers with very experienced personnel and teach them how to update P&IDs and to comply with PSM.

A few years later, when most of the 30+ year operators were retired, these updated P&IDs were successfully used to train the new group of operators in the facility.

Another group, supported by Technical Services, began work to update and maintain the MSDS (Material Safety Data Sheets) for the facility. This action proved to be very effective for the firefighters as they were properly trained in the materials they may have to deal with in case of emergency. Before that they were relying on an operator telling them what to do during an emergency scenario.

Nowadays, this situation (outdated PSI) is a concern for many facilities: In a study performed in 2010 by OSHA IN 2010 [10]; 22.2% of the deviations (not compliance with PSM) were associated to non-updated PSI.

2.3 Process Hazard Analysis (PHA)

The purpose of this analysis is to minimize the likelihood of the occurrence and the consequences of a dangerous substance release by identifying, evaluating, and controlling the events that could lead to releases.

The management team and all personnel were asked to answer the following questions:

- Is my facility safe to operate?

- In the case of an event happening, am I protected?
- Is my facility protected?

This is how the HAZOP concept was introduced to the facility. A group of people were trained in HAZOP facilitation and they were in charge of implementing the PHA program.

Implementation of PHA, as with any other type of new procedure, was a challenge. At that time only the HAZOP was considered for implementation; later, additional PHA methods were added to the portfolio. Manual recording of PHA scenarios was a challenge, but the use of PHAs (manual or automated) continues to be vital to gauge whether a facility is protected or not.

In the earlier stages of the implementation, PHA-HAZOP was used as part of Management of Change element, but later was included in the evaluation cycle of all projects.

2.4 Operating Procedures

In the old facility, procedures were transmitted by oral tradition and there were very few written procedures, but even those written were not clear for non-experienced individuals. They were not developed with consideration for new personnel in that corresponding area.

In several opportunities, lack of written procedures led to accidents when one of the "guru's" was not there to execute a particular procedure or when a step was missed in a complicated procedure.

To prevent this from happening repeatedly, a team was built consisting of very experienced operators with experts in writing procedures and process engineers to start writing the procedures for each unit. Managers defined the most critical and most urgent ones to be done first. It was a medium-term effort that produced great results when a few years later the operations crews were modified to include people with technical background but almost zero operational experience.

2.5 Hot Work Permit

In traditional operation, the Operations Supervisor is also in charge of supervising the maintenance activities in his area. About 25 years ago, a request for a work permit was made to the Operations Supervisor; he erroneously determined that the risk was not too high, as he was busy with the day-to-day operation. He did not check the actual situation in the field. Unfortunately, the line where the work was going to be performed was not checked before and was filled with light naphtha (not properly drained) and this led to a fatality when welding.

After this accident and as a part of PSM program, a new work permit procedure was established:

- A dedicated Operations Supervisor for maintenance activities, being the only authority to sign-off work permits.
- A procedure was developed to include documentation and proper sign-off
- Training was developed for all involved parties about how to perform using the new Hot Working Permit, including contractors.

2.6 Training

The training of all personnel is a key element of Process Safety. The training must be on a continuous basis and at all levels of the organization. In this particular company, the kick-off of the PSM was to present the Piper-Alpha accident video to all levels of the organization from the top Management to the bottom of the organization. Also the API-750 Video was presented to all personnel, the objective: understand the risk, study the failures and realize that this situation was very likely to happen to that facility.

The importance of training is that it must to be continuous, refreshed on regular basis. Training must be developed for new employees and all contractors, with a rule of "nobody in" without safety training. The safety training included an exam; and those who do not pass the exam cannot get into the facility. To make sure PSM works, training is a key element and dedicated personnel to fulfill this target is recommended.

2.7 Pre start-up Safety Review

According with API RP 750 [5] the PSSR (Pre Start-up Safety Review) should be performed for new and modified facilities that are subject to this recommended practice to confirm that:

- Construction is according with specifications
- Procedures are in place (including safety procedures)
- PHA recommendations are closed out.
- Training for all operators is provided

It is fairly common to see confusion between PSSR for commissioning and PSSR during normal S/D for maintenance; as API 750 indicates, it is needed under both scenarios.

Before the implementation of PSSR, the responsibility of checking the facility prior to start up was only for operations. Normally this was done by an Operations Supervisor with other duties in the area and sometimes relying on another operator or on a Maintenance Supervisor. In this facility the need for PSSR was identified in an unfortunate incident:

An Operations Supervisor on his last night shift was doing the plant checking for start-up after he was informed by maintenance about completion of work. He was the only person in charge of checking the facility and provided the team with the "go ahead" for start-up 30 minutes before the end of his shift. The heating of the plant started and about 2 hours later a hot naphtha leak spilled over a hot heat exchanger producing a large fire that required 2 months of additional plant shutdown to repair damages. A good example of what "not to do".

A PSSR exercise takes a longer time to perform and needs a multidisciplinary team but will prevent incidents or accidents.

2.8 Emergency Planning & Response

Firefighter training is very important. In 1984, 100 municipal firefighters died when they were called to fight a fire in a heavy diesel tank in another location [11]. They handled it as a regular fire and high quantities of water were directed to the roof of the tank. No effort was made to cool off the walls of the tank and they were not aware of the presence of water in the bottom of the tank. When the diesel in the bottom of the tank reached the appropriated temperature, a boil over situation occurred.

One of the key elements in an Emergency Response plan is the knowledge of the hazards of a system that may create an emergency. As part of a PSM program, an intensive training on Process for non-technical personnel was given to firefighters (permanent and volunteers). Prior to the implementation of the PSM program, in the event of an emergency, the Firefighting Supervisor communicated directly with the Operations Supervisor on site, to obtain information about the potential hazards to people and facility. The Firefighting Supervisor had only a very limited knowledge of the facilities and associated hazards.

Emergency dry-runs were also implemented weekly according with a program developed on the most critical areas. This practice involved regular firefighters and volunteer firefighters.

Another step that was taken as a part of keeping the firefighters engaged was the annual recognition to all firefighters (rewards, celebrations, etc.).

2.9 Program Auditing

All the PSM elements should be audited on a regular basis to ensure the program effectiveness and performance. Programs like PSM require a lot of involvement from the Auditing team, the PSM leading team will define which parts of the program are most critical and it will be the start-up point for the auditing.

Audit process will always be needed, even when every person in the organization is "in the safety mindset".

3. Conclusions

- Implementation of the PSM program is a time consuming effort.
- Dedicated personnel are required for leading the change.
- Management actively involved in the PSM program is a must.

3.1 Tangible benefits from the implementation of PSM

- In a facility with a record of 4 fires per year, they were able to reduce the number of fires down to 1/year.
- Training of operators improved the operations of the facility reaching production records and reliability records.
- The Safety concept was built-in (be safe when nobody is looking)
- Ongoing, non-prescriptive, performance-based, participation at all levels.
- Facility received a Safety Award as "Safest in class" facility.

4. Recommendations

- PSM is a MUST for all O&G facilities
- It is a way of living! I encourage all O&G companies to follow this "way of living".

5. References

[1] Health and Safety Executive, 'The Flixborough Disaster: Report of the Court of Inquiry', HMSO, ISBN 0113610750, (1975)

[2] Chouhan TR "The Unfolding of Bhopal Disaster". Journal of Loss Prevention in the process industry **18** (4–6): 205–208. (2005).

[3] PatC-Cornell, M. E. "Learning from the Piper Alpha Accident: A Postmortem Analysis of Technical and Organizational Factors". Risk Analysis, Vol. 13, No. 2, 1993

[4] Bethea, R.M. "Explosion and Fire at the Phillips Company – Houston Chemical Complex, Pasadena, TX" Chemical Engineering Department, Texas Tech University, Lubbock, TX 79409-3121 (2007)

[5] API Recommended Practice API RP-750. "Management of Process Hazards". American Petroleum Institute, 1990.

[6] CSB. Investigation Report: Refinery Explosion and Fire (15 killed, 180 injured). Report # 2005-04-I-TX; Chemical Safety Board, Washington, DC. 2007

[7] British Petroleum Oil Company (BP). "Deepwater Horizon Accident Investigation Report" (2010)

[8] NTSB "Pacific Gas And Electric Company Natural Gas Transmission Pipeline Rupture and Fire" San Bruno, California. NTSB Report # PAR-11-01 (2011).

[9] OSHA. "Process safety management of highly hazardous chemicals". OSHA 29- Part 1910.119 (1992)

[10] Jordan Barab., "Learning from Industry Mistakes" presented at NPRA National Safety Conference, May 2010

[11] MARAVEN S.A. "Investigacion del Accidente de Tacoa" (Tacoa's Accident Investigation, In Spanish). Caracas, 1985.

(43an) Auditing the Process Hazard Analysis - Key Characteristics to Scrutinize in a Good-Quality HAZOP Study and Supporting Effective HAZOP Study Updates

Steven T. Maher[1], Michael A. Nour[1] and Kristin D. Norton[2], (1)Process Safety, Risk Management Professionals, Irvine, CA, (2)Process Safety, Risk Management Professionals, Inc., Irvine, CA

Although Process Safety Management (PSM), requiring good-quality Process Hazard Analyses, was promulgated in the United States in 1992, and HAZOP Study techniques date back to the mid-1970's, the HAZOP Study technique is still not uniformly applied in a high quality fashion. The objective of this paper is to provide experience-based tips on Process Hazard Analysis auditing, and some key characteristics that the auditor can scrutinize to assure quality implementation. Analogously, a similar set of characteristics that can facilitate future HAZOP Study updates can also contribute to a quality HAZOP Study that supports the spirit of Safety Management System objectives for the identification of hazards potentially impacting plant personnel, the community, and the environment. This paper will also review tips for structuring the HAZOP Study to facilitate later updates and revalidation.

9TH GLOBAL CONGRESS ON PROCESS SAFETY

Vapor Cloud Explosion Test of Air-Inflated Fabric Structure

Brad Horn
Baker Engineering and Risk Consultants, Inc.
3330 Oakwell Court, Suite 100
San Antonio, TX 78218-3024
BHorn@BakerRisk.com

Matthew T. Edel, P.E.
Baker Engineering and Risk Consultants, Inc.
3330 Oakwell Court, Suite 100
San Antonio, TX 78218-3024
MEdel@BakerRisk.com

Prepared for Presentation at
American Institute of Chemical Engineers
2013 Spring Meeting
9th Global Congress on Process Safety
San Antonio, Texas
April 28 – May 1, 2013

UNPUBLISHED

Vapor Cloud Explosion Test of Air-Inflated Fabric Structure

Brad Horn
Baker Engineering and Risk Consultants, Inc.
3330 Oakwell Court, Suite 100
San Antonio, TX 78218-3024
BHorn@BakerRisk.com

Matthew T. Edel, P.E.
Baker Engineering and Risk Consultants, Inc.
3330 Oakwell Court, Suite 100
San Antonio, TX 78218-3024
MEdel@BakerRisk.com

Keywords: vapor cloud explosion, explosion test, air-inflated structure, fabric structure, blast load, thermal load, structural response

Abstract

Portable buildings are gaining widespread use at refining and chemical processing facilities, particularly for facility siting requirements. Although trailers and blast resistant modules (BRMs) can be used for many modular applications, in certain situations fabric structures may provide a viable alternative. However, the blast response of fabric structures is neither clearly defined nor well understood throughout the blast design community.

One such modular fabric structure is an air-inflated structure. A manufacturer of this type of structure recently funded the testing of one of their specimens for the effects of a vapor cloud explosion (VCE). The VCE for the test was initiated using propane and an internal matrix of congestion inside a three-walled cubicle, referred to as the Deflagration Load Generator (DLG) test rig. The test involved a deflagration event that imparted a thermal flux and a maximum free-field pressure of 2.9 psi with an average duration of 31 ms at the front wall of the air-inflated structure.

The reduced blast loads and thermal effects inside the structure were compared to those outside. In addition, the attenuation of blast loads behind the structure was also examined, as well as shielding effects. The structural response of the test specimen was captured with high-speed video, and observations on the motion of the structure are provided.

1. Introduction

Baker Engineering and Risk Consultants, Inc. (BakerRisk) has developed a Deflagration Load Generator (DLG) test rig that can supply VCE blast loads to a variety of structures in order to provide structural response test data to manufacturers (i.e., product proof testing). The DLG rig produces a VCE blast load with a blast wave shape, pressure, and duration similar to the blast loads that are generated by accidental VCEs at chemical processing facilities. Dynamic Air Shelters, Ltd. (DAS) contracted a test of its manufactured air-inflated shelter.

The test involved imparting a VCE blast load on a DAS air-inflated shelter. The shelter was erected to the manufacturer's specifications and then subjected to a VCE blast load using the DLG test rig. The target blast loading for this test was a centrally peaked blast wave with a peak free-field blast pressure of 3.5 psi and 50 ms duration at the shelter wall closest to the DLG rig. Data recorded include pressure histories, thermocouple and heat flux measurements, and video.

2. Test Setup

This section describes BakerRisk's DLG test rig, the DAS air beam shelter, and the instrumentation used to measure and record data during the test.

2.1 DLG Test Rig

The DLG test rig is shown in Figure 1. The test rig is a steel box with one open side that is initially covered with a thin plastic sheet to allow a fuel-air mixture to be developed within the test rig. During testing, the test rig is filled with a propane-air mixture at a desired concentration, which is then ignited at the middle of the back wall. An obstacle array within the test rig causes the flame to accelerate and generate a blast wave. The resulting blast wave propagates out toward the target structure.

2.2 Air Beam Shelter

DAS constructed the air-inflated shelter shown in Figure 1, which was located directly in front of the DLG test rig during the test. The shelter system consisted of a ballistic-resistant wall and an air-inflated shelter with overall dimensions of approximately 46 ft. (14 m) wide by 56 ft. (17 m) long and 16 ft. (4.9 m) tall. The blast wall was made of PVC trapezoidal bags that were filled with material, in this case sand, and then hung between supporting wall frames spaced at approximately 6.7 ft. (2.0 m).

Figure 1. Air-Inflated Shelter (on left), Blast Wall, and DLG Test Rig (on right)

The air-inflated members were continuously pressurized and created the structural support for the shelter. Doors were located on two walls parallel to the direction of the blast propagation. The interior air beams are shown in Figure 2. Except for the instrumentation, the interior of the shelter was vacant during the test.

Figure 2. Air-Beam Columns

A schematic showing the location of the test rig, shelter and instrumentation is provided in Figure 3. The standoff distance between the test rig and the shelter was 21 ft. (6.4 m).

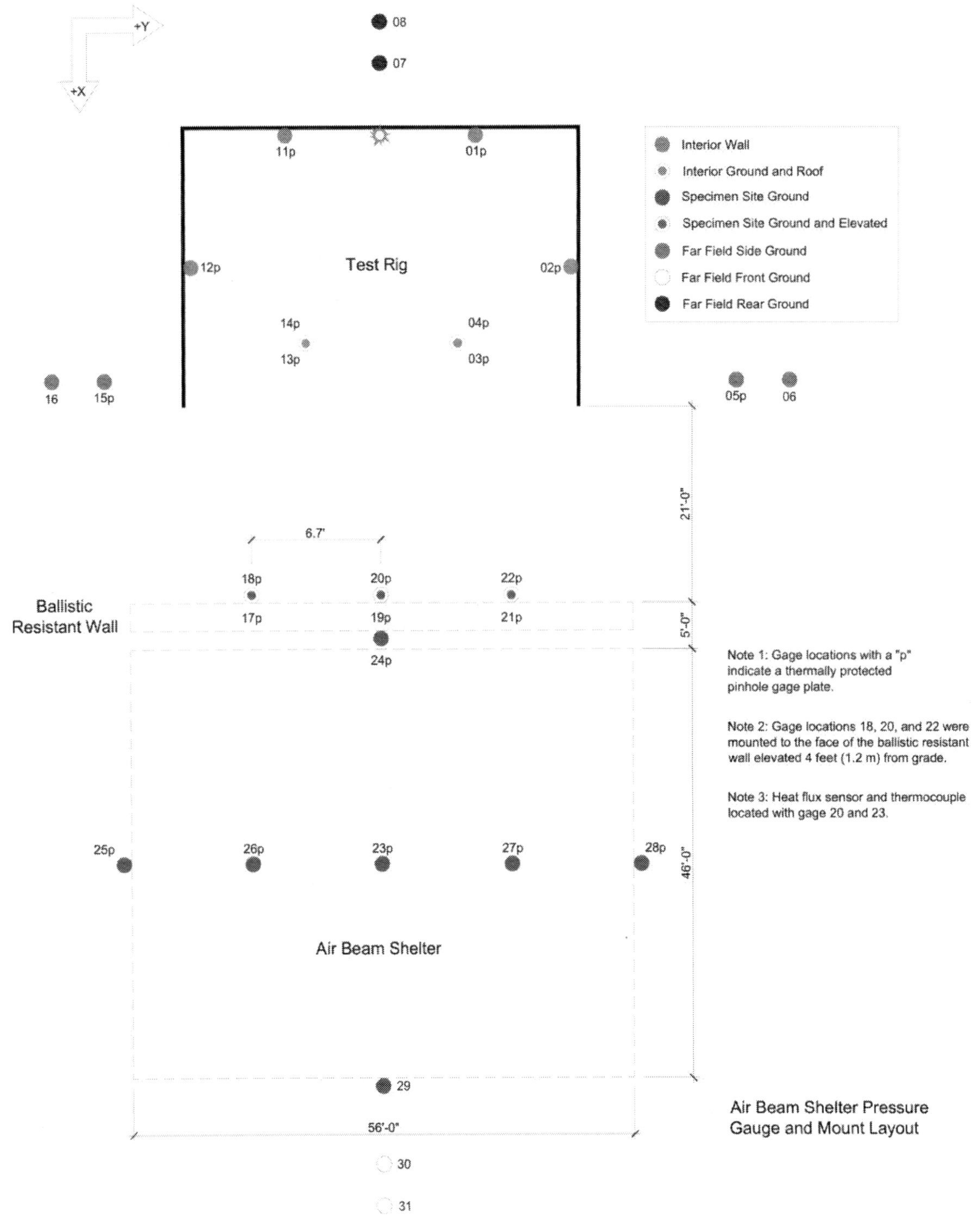

Figure 3. Instrumentation Layout Map

2.3 Instrumentation

During the test, the pressure, heat flux, and temperature histories were measured at locations of interest. Both regular and high speed video of the shelter structural response were captured.

A total of 29 dynamic pressure gauges sampled at 10,000 samples per second were used to measure the blast loads. Pressure gauge locations are shown in Figure 3; the gauge locations relative to the front of the test rig are provided in Figure 5. A total of 13 pressure gauges were positioned to measure the blast load at points of interest on the shelter and its components. Six gauges measured blast loads at the ballistic resistant wall, with three on the ground at the base of the wall and three mounted on the wall 4 ft. (1.2 m) above the ground gauges in order to measure the reflected load. One pressure gauge was located on the ground between the ballistic resistant wall and the shelter wall, and one pressure gauge was located at each shelter doorway. Three pressure gauges were located within the shelter interior, and one pressure gauge was located just outside the rear wall of the shelter.

Thermocouple and heat flux sensors were placed at the ballistic resistant wall and inside the shelter, coincident with pressure gauge locations 20 and 23. Each sensor was elevated 4 ft. (1.2 m) above grade and oriented toward the DLG test rig. Each thermocouple was sampled at 20,000 samples per second.

Video recording devices were used to document the test and record the shelter structural response to the blast loading. The high speed video (HSV) camera was positioned in the same plane as the leading edge of the ballistic resistant wall, while two high definition (HD) video cameras, two exterior spy cameras, and one interior spy camera recorded other angles.

3. Test Results

Figure 4 shows the structural response recorded by the high definition video camera, capturing at 30 frames per second.

Figure 4. VCE Loading Applied to Shelter

3.1 Blast Loads

A peak pressure measurement map with pressure values listed in green font is shown in Figure 5. Table 1 provides the blast load summary which includes the pressure gauge location, standoff distance from the test rig, measured peak pressure, positive impulse, and both the measured and calculated durations. The calculated duration assumes a perfectly triangular blast pressure history (i.e., duration is twice the impulse divided by the peak pressure). Table 2 provides a summary of the peak reflected blast pressures measured at the ballistic resistant wall and the corresponding calculated peak free-field (i.e., side-on, incident) blast pressures.

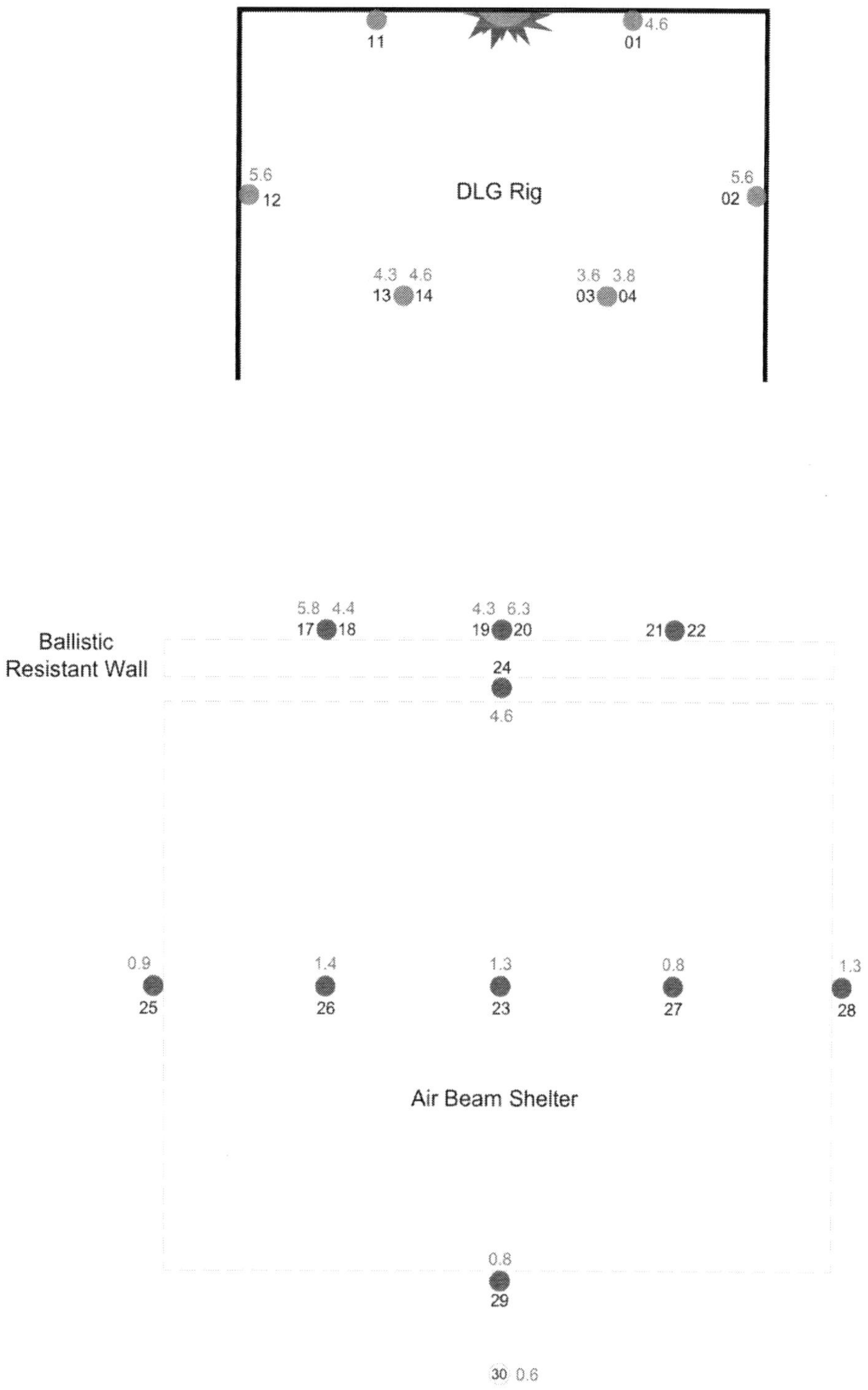

Note: Pressure values directly in front and behind the ballistic-resistant wall are reported as the adjusted peak values.

Figure 5. Peak Pressure Measurement Map (psig)

Table 1. Blast Load Summary

Location	Gauge	Distance From Test Rig (ft.)	Measured Peak Pressure (psig)	Measured Positive Impulse[1] (psi-ms)	Measured Duration (ms)	Calculated Duration (ms)
Rig Interior	2	0	5.6	206	120	74
Ballistic Resistant Wall[1]	17	21	5.4	90	48	31
	18	21	3.9	67	42	31
	19	21	3.9	70	51	32
	20	21	5.8	93	48	30
Front Shelter	24	24	4.6	95	57	41
Shelter Doorway	25	48	0.9	19	57	41
	28	48	1.3	28	57	41
Shelter Interior	23	48	1.3	29	53	45
	26	48	1.4	29	53	41
	27	48	0.8	16	53	41
Rear Shelter	29	71	0.8	16	45	41
Far Field	30	76	0.6	10	30	33

Table 2. Ballistic Resistant Wall Blast Pressure Summary

Location	Gauge	Distance From Test Rig (ft.)	Measured Peak Pressure (psig)	Adjusted Peak Pressure[1] (psig)	Calculated Free Field Pressure[2] (psig)
Ballistic Resistant Wall[1]	17	21	5.4	5.8	2.7
	18	21	3.9	4.4	2.0
	19	21	3.9	4.3	2.0
	20	21	5.8	6.3	2.9

[1] *Pressure values adjusted for pre-event signal ringing*
[2] *Calculated from reflection factor*

The maximum peak pressure in the DLG test rig was 5.6 psig (0.39 bar). Gauges at the ballistic resistant wall were positioned to measure the reflected load on the face of the wall or shelter.

The peak pressures and impulse at the wall were adjusted to account for the signal ringing prior to the arrival of the blast wave and are shown in Figure 7. The maximum adjusted peak pressure was 6.3 psig (0.43 bar). The pressure signals along the centerline of the shelter in the direction of the blast are shown in Figure 6. Two of the six gauges were invalid and were therefore discarded.

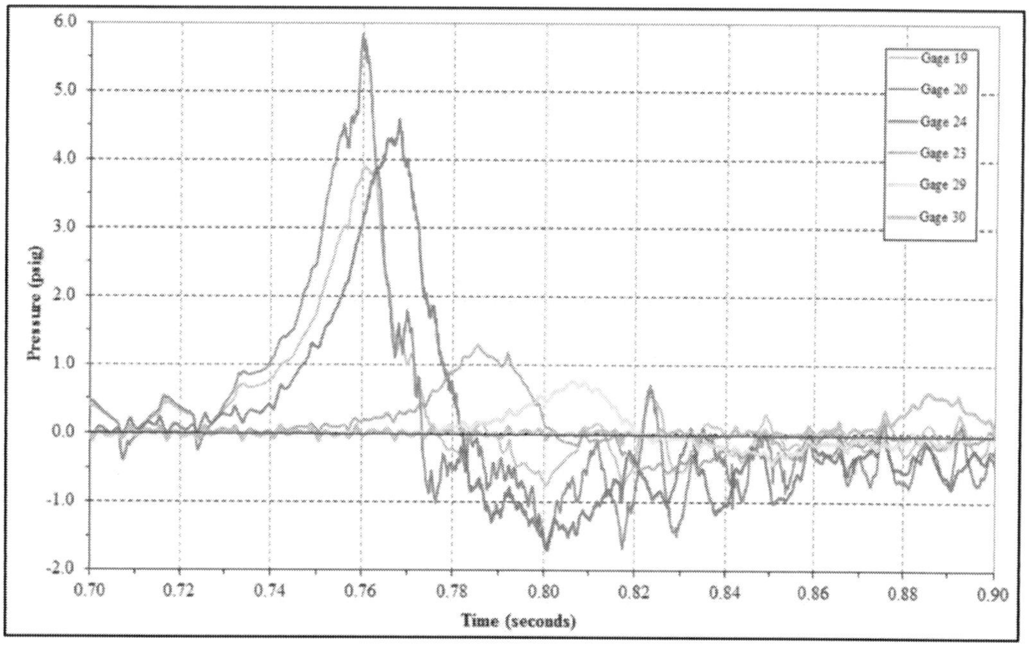

Figure 6. Pressure Histories along the Front Center Gauge Lane

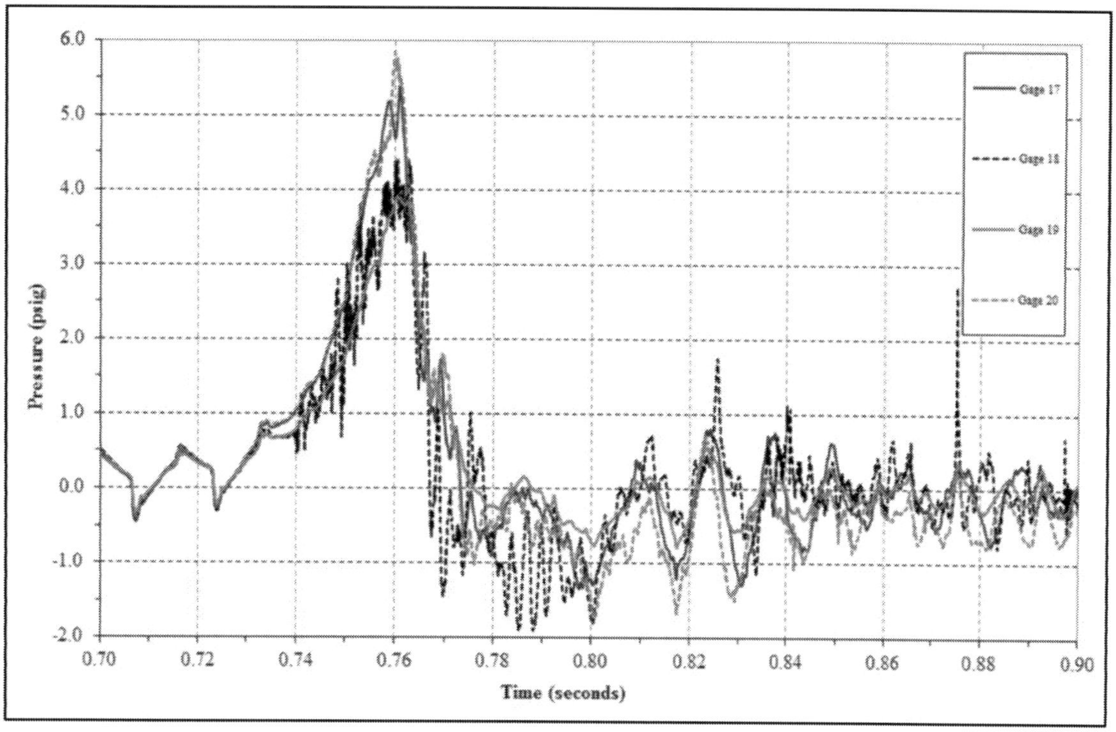

Figure 7. Pressure Histories at the Ballistic-Resistant Wall

The measured peak pressure and impulse between the ballistic resistant wall and air shelter were 4.6 psig (1.4 bar) and 95 psi-ms (6.6 bar-ms), yielding a calculated duration of 41 ms. Pressure histories were measured along the mid-span of shelter. The pressure histories at these mid-span locations are shown in Figure 8. For the doorway exterior locations, the largest measured peak pressure was 1.3 psig (0.09 bar) with an impulse of 28 psi-ms (1.9 bar-ms) and a calculated duration of 41 ms. The largest measured peak pressure among the interior gauges was 1.4 psig (0.1 bar) with an impulse of 29 psi-ms (2.0 bar-ms) and a calculated duration of 41 ms. Behind the shelter, the measured peak pressure was 0.8 psig (0.06 bar) with an impulse of 16 psi-ms (1.1 bar-ms) and a calculated duration of 41 ms.

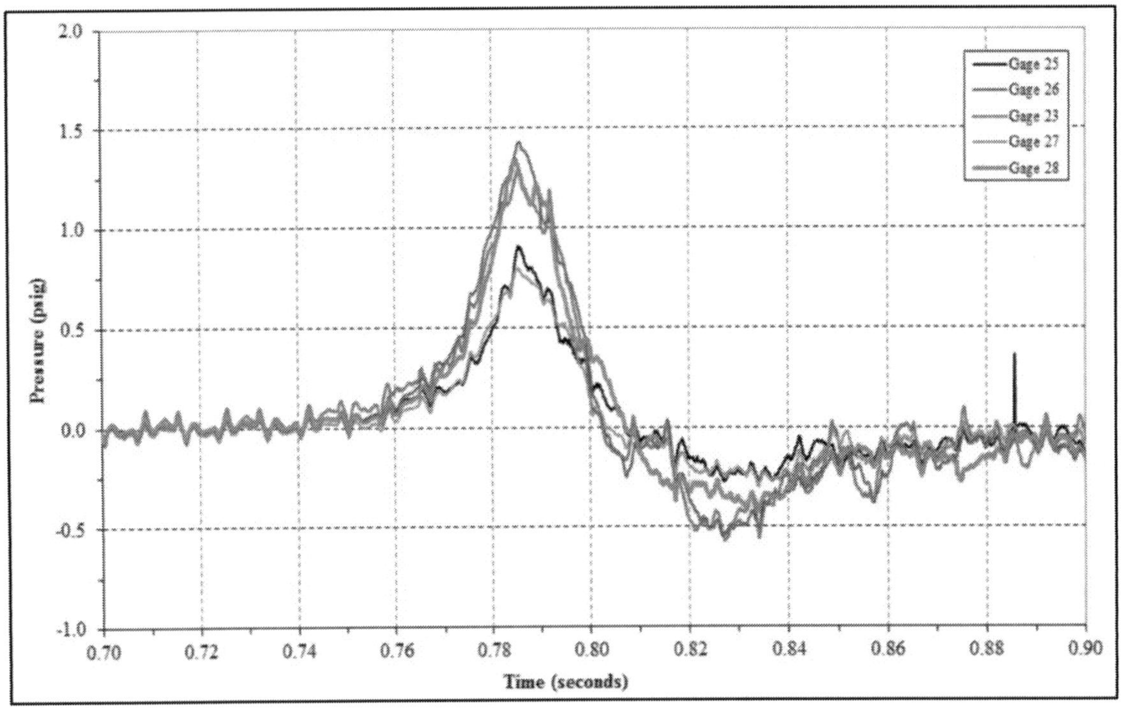

Figure 8. Pressure Histories at Shelter Mid-Span Locations

3.2 Thermal Loads

Figure 9 shows the measured temperature just past the exterior face of the ballistic resistant wall. This peak signal exceeded the range of the signal conditioner and was "clipped" at approximately 1100 °F (590 °C). The estimated actual maximum temperature at this location was just under 1500 °F (815 °C) based on an extrapolation of the signal. The measured signal inside the shelter indicates that the temperature had increased less than 5 °F (3 °C), as shown in Figure 10. That is, while the temperature just outside the shelter increased dramatically due to exposure to the fireball, the temperature within the shelter barely rose.

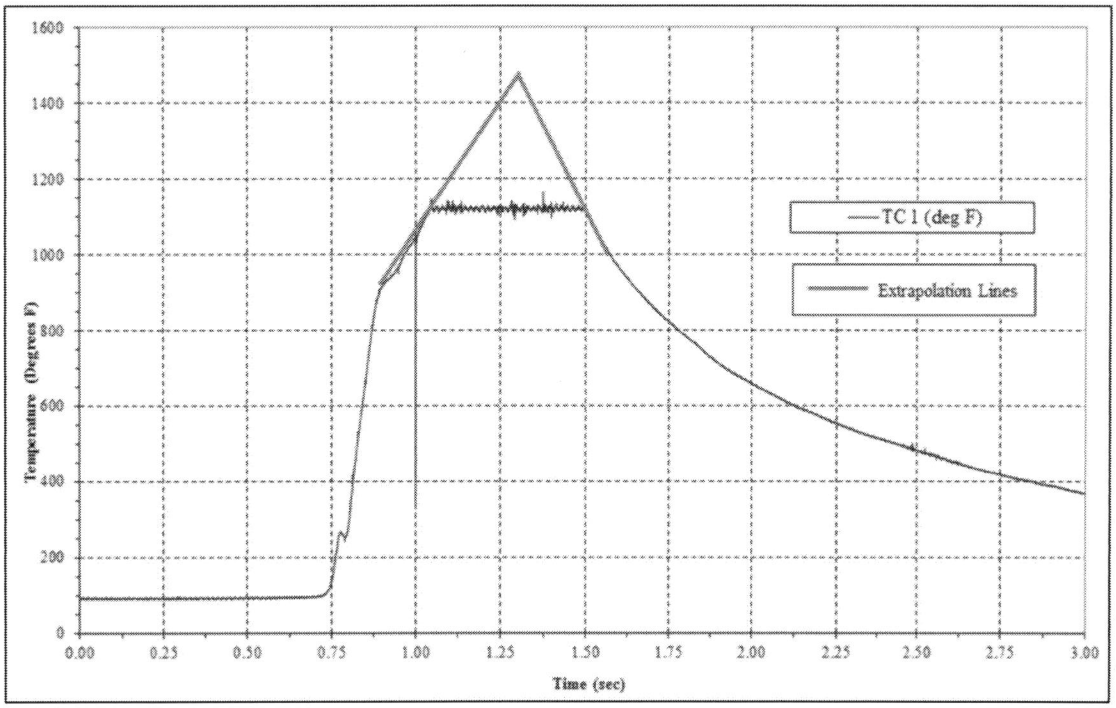

Figure 9. Exterior Temperature History

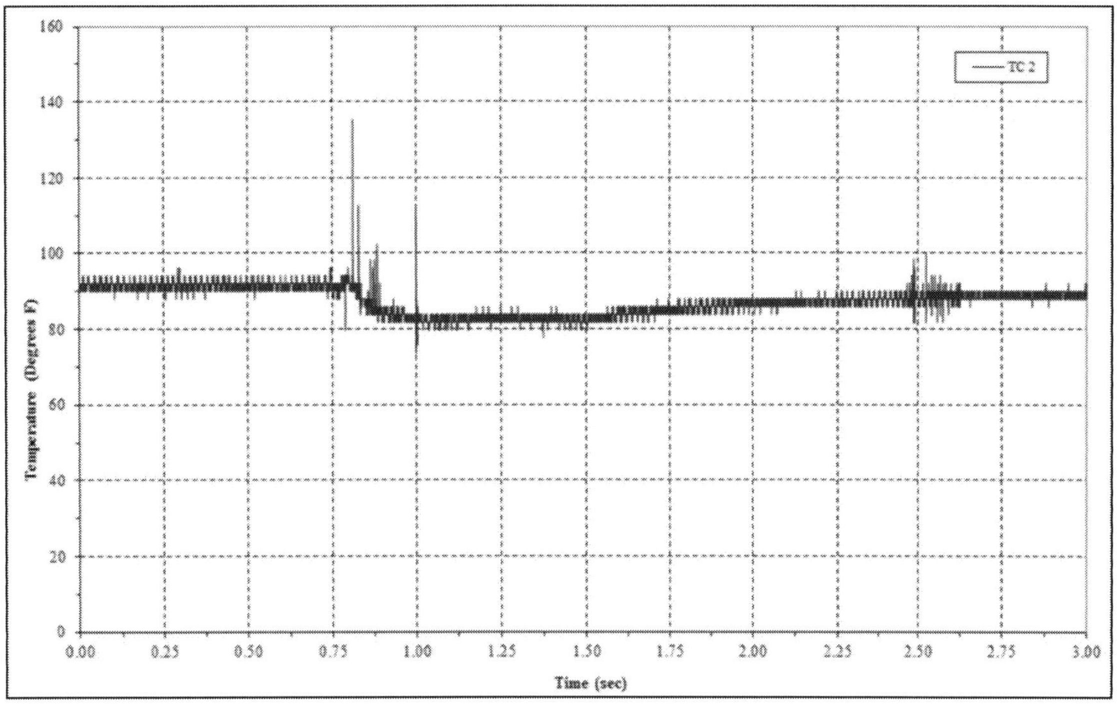

Figure 10. Interior Temperature History

The exterior heat flux signal is shown in Figure 11. This signal was filtered to provide a smooth curve with a maximum heat flux of 6 kW/m^2. The interior heat flux signal was invalid due to the low signal to noise ratio. The virtually zero interior heat flux is consistent with the measured temperature at this location.

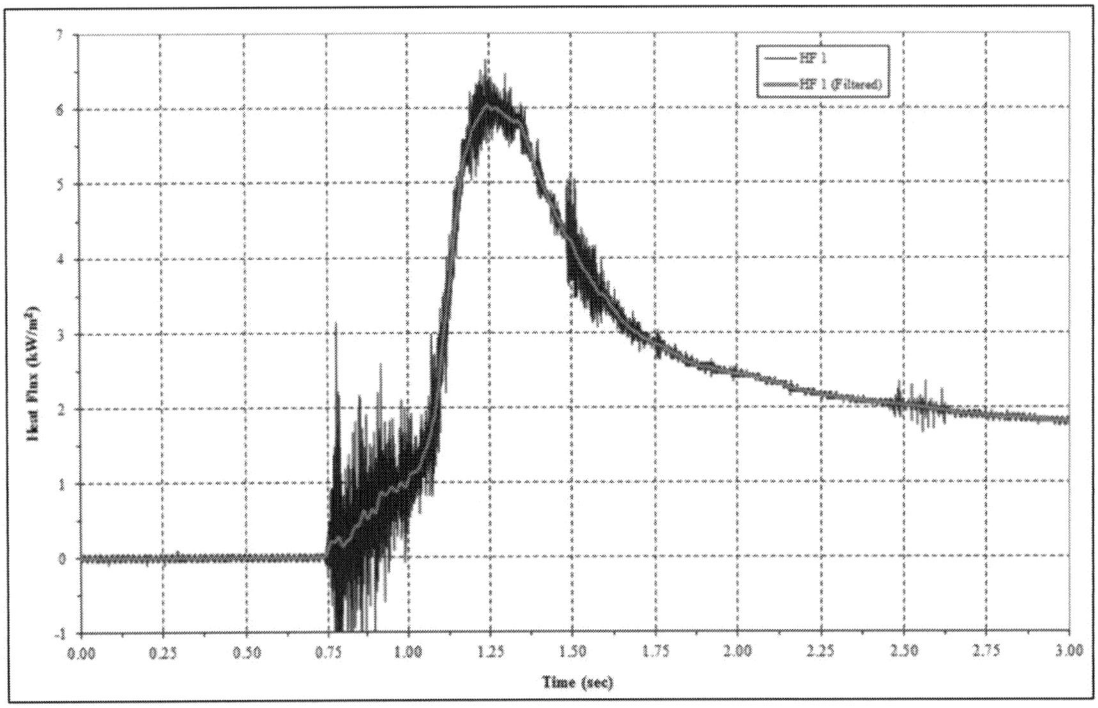

Figure 11. Exterior Heat Flux History

3.3 Structural Response

The air-inflated shelter sustained only minor damage from the VCE blast loading. The air-beams that create the arch section of the shelter experienced rapid temporary deflection as the blast wave propagated across the shelter. The ballistic resistant wall remained functional following the test, allowing safe egress from the shelter after the VCE. Figure 12 shows the wall deflection of about 1 ft. (0.3m) at the top. The yellow grid lines are approximately 1 ft. (0.3 m) apart. The yellow and blue markers indicate the roof deflected approximately 1.5 ft. (0.5 m) downward during the test and returned to the initial position within 2 seconds.

Figure 12. Shelter Deflection Comparison

4. Conclusions

A Dynamic Air Shelters, Ltd. air-inflated structure was subjected to a VCE having a blast load with a peak free-field pressure of 2.9 psi and a 31 ms duration at the wall nearest the DLG rig. The test resulted in a deflagration-type load with a centrally-peaked blast shape. The blast loads inside the structure were reduced from the levels recorded at the exterior ballistic resistant wall by approximately 75% but were comparable to the exterior pressures measured outside of the two doors. Shielding effects of the blast loads were anticipated directly behind the structure, but none were observed at this location relative to the loads recorded closer and further from the rig.

Significant thermal loads were observed on the front face of the structure due to its close proximity to the fireball emanating from the test rig. Yet even at nearly 1500 °F (815 °C) and 6 kW/m^2 at the exterior front face, the interior temperature increased less than 5 °F (3 °C) with virtually no heat flux. The air-inflated structure provided almost complete shielding from the thermal loads with negligible observable thermal damage.

The high speed and high definition video show a maximum of 1 ft. deflection of the ballistic resistant wall and less than 2 ft. deflection of the air-inflated shelter. The structure returned to its initial position within 2 seconds and had no permanent deflection. Internally, the structure had no debris hazards and remained inflated, allowing safe egress from the air beam structure.

9TH GLOBAL CONGRESS ON PROCESS SAFETY

Revisiting the Conditions for a CO2 Tank Explosion

Yu Zhang
Air Products and Chemicals Inc.
7201 Hamilton Blvd, Allentown, PA 18195-1501
zhangy15@airproducts.com

Joan Schork
Keith Ludwig
Air Products and Chemicals Inc.
7201 Hamilton Blvd, Allentown, PA 18195-1501
schorkjm@airproducts.com
ludwigka@airproducts.com

Prepared for Presentation at
American Institute of Chemical Engineers
2013 Spring Meeting
9th Global Congress on Process Safety
San Antonio, Texas
April 28-May 1, 2013

UNPUBLISHED

Revisiting the Conditions for a CO2 Tank Explosion

Yu Zhang
Air Products and Chemicals, Inc.
7201 Hamilton Blvd, Allentown, PA 18195-1501
zhangy15@airproducts.com

Joan Schork
Keith Ludwig
Air Products and Chemicals, Inc.
7201 Hamilton Blvd, Allentown, PA 18195-1501
schorkjm@airproducts.com
ludwigka@airproducts.com

Keywords: Boiling Liquid Expanding Vapor Explosion (BLEVE), Carbon dioxide, Overpressure

Abstract

Carbon dioxide is broadly used in a number of applications including food freezing, beverage production, enhanced oil recovery, supercritical extraction and welding. It is commonly stored as a pressurized liquid (e.g.: -18 °C/0 °F and 2 MPa/300 psia) in carbon steel tanks.

A number of CO2 tank explosions have been reported in the literature over the past 50 years. In order to design appropriate safety systems for liquid CO2 storage tanks, it is important to understand the worst case scenario. Should the energy release be modeled as a rapid vapor phase expansion (PV) or a Boiling Liquid Expanding Vapor Explosion (BLEVE)? The question arises because of a common definition of a BLEVE as an event resulting from the sudden loss of containment of a liquid that is at or above its atmospheric superheat limit temperature. Under the conditions listed above, CO2 is above its atmospheric boiling point but below its superheat limit. In this temperature range, will the vaporization rate be rapid enough to contribute to a blast overpressure in the event of a tank failure?

To answer this question, the authors made a critical review of information available in the literature including laboratory data and incident reports. This paper will describe the analysis conducted and present the results indicating that the potential for rapid vaporization should be included in evaluating the consequences of a CO2 tank rupture below the superheat limit.

1. Introduction

Carbon dioxide (CO2) is broadly used in a number of applications including food freezing, beverage production, enhanced oil recovery, supercritical extraction and welding. Since CO2 at its bubble point becomes a solid at a pressure below 5 bars, CO2 is commonly stored as a pressurized liquid at an elevated pressure of 20 bars absolute in carbon steel tanks.

When a pressurized liquid in an equilibrium state is depressurized instantaneously, it enters a superheated state (a meta-stable liquid above its boiling point) from which the liquid could vaporize in an explosive manner resulting in a blast wave and risk of flying fragments of the vessel. This is called a Boiling Liquid Expanding Vapor Explosion (BLEVE) [1,2]. When the degree of superheat of the liquid is small, the vaporization rate is not large enough to develop a significant blast effect. According to the Superheat Limit Theory developed by Kim and Reid [3], an explosive evaporation occurs when the depressurized liquid enters the spinodal decomposition temperature-pressure range. Based on thermodynamics principles and equation of state of fluids, Kim and Reid [3] predict a so-called BLEVE envelope (Figure. 1) which indicates the range of initial temperatures and pressures which will lead to a BLEVE following instantaneous depressurization. The superheat limit theory is cited in some industrial hazard evaluations as a criterion to determine whether or not a BLEVE should be considered [4,5].

The mechanism of BLEVE based on Kim and Reid's theory is briefly described here. When the boiling point of a liquid is reached, vapor bubbles start to form at active sites such as vessel walls. Nucleation at these sites is called *heterogeneous nucleation*. If bubble nuclei are large (approaching millimeter sizes), a small degree of superheat (defined as heating beyond the boiling point) is needed to make the bubble grow. If the bubble nuclei are small, for example when the vessel is very clean and smooth, they need a large superheat to grow. When there are not many active sites, the bubble nucleation must take place in the bulk of the liquid. Boiling in this case generally takes place at submicron nucleation sites at suspended particles, impurities, or ions. This requires a large degree of superheat to make the bubble grow. This means the liquid normal boiling point can be exceeded without boiling, and the liquid can be significantly superheated. There is, however, a temperature limit at a given pressure above which a phase transition from liquid to vapor has to happen, even though the bubble nulei are small. This superheat limit temperature (SLT) and pressure correspond to the conditions at which a spinodal decomposition of liquid happens. This can produce a very violent phase transition with boiling caused by *homogeneous nucleation* in the bulk of liquid. The superheated liquid state could also be reached by sudden pressure drop. For CO2, the experimental value of SLT at atmospheric pressure is 267 K (-6.2°C) [6].

Although the theory is commonly accepted, some authors question its applicability to a real case. For example, Birk and co-works [7] report that BLEVEs have been observed with propane when the pressure-liquefied gas was at ambient temperature (20 °C), well below its atmospheric superheat limit temperature of 53 °C. In the 2010 edition of the Center for Chemical Process Safety (CCPS) Guidelines to evaluate BLEVE [2], it says "There is no experimental evidence with a real pressure vessel where the phase change was by homogeneous nucleation. Homogeneous nucleation is extremely unlikely in industrial use due to temperature gradients, impurities, surface roughness and other factors that cause localized nucleation."

2. CO2 BLEVE

The end state of liquid CO_2 depressurized to 1 atmosphere is a solid and vapor mixture, which makes its depressurization process unique. CO_2 may evaporate explosively even when its temperature is below the STL, because CO_2 crystallizes at ambient pressure, thus presenting enough nucleation sites for liquid to flash [1].

Several fundamental questions are raised when considering CO_2 system. How does the formation of dry ice affect the blast energy? During the decompression process, does the vapor expand isothermally or isentropically? This initial report will only touch on two questions based on various incident reports on large scale CO_2 tank energy releases:

1. *Could a CO_2 BLEVE happen when the storage conditions (temperature and pressure) are outside the so-called BLEVE envelope predicted by the superheat limit theory?*
2. *If it goes BLEVE, could we use the algorithm presented in the CCPS guide to perform the consequence analysis?*

3. Historical experiences

Table 1 summarizes major CO_2 tank incidents we found with documentation in the open literature. The Worms, Haltern, and Repcelak incidents are summarized in CCPS's guide to BLEVE [1-2]. Information about the incident at Yuhang is collected from news reports in Chinese [8-9], and information about Fukushima's incident is translated from reports in Japanese [10-12].

Table 1. Some major CO_2 incidents

Date	Location	Cause	Initial Pressure [bar]	Initial Temperature [◦C]
13 Nov. 2008	**Yuhang, China***	Brittle fracture	23	-15
21 Nov. 1988	Worms, Germany	Overpressure	20	-20
2 Sept. 1976	Haltern, Germany	Brittle fracture	23	-15
1 Mar. 1969	Fukushima, Japan	Overpressure	20	-20
2 Jan. 1969	**Repcelak, Hungary***	Overfilling	15	-30

* Overpressure tags are extracted from these incidents

Out of the five incidents, only two overpressure damage data points (tags) are obtained, which are used to validate the algorithm employed in the CCPS guidelines. There were many tags on projectiles, but we find these less reliable than the overpressure tags. One data point is from the Repcelak incident: a warehouse about 15-30 meters away from the explosion center was torn down due to heavy damage. The other data point is from the Yuhang incident: Window glass shattered within a 400 meter radius of the explosion center. Criteria on overpressure damage to buildings are obtained from CCPS's guide to BLEVE [1].

3.1 Large scale CO_2 tank energy release

Yuhang, Hangzhou, China, 13 November 2008

A transport ship carrying 130 cu. meter (95% full) of CO_2 exploded on November 13, 2008 in Yuhang, Zhejiang, China [8-9]. The tank's content was stored under a pressure of 23 bar at a temperature of -15 °C. At the moment of explosion, the ship was in the dock belonging to Rongsheng Chemicals. It was not in operation and there were few workers around as it was before regular business hours. The explosion completely destroyed the CO_2 ship and sank H_2SO_4 and H_2O_2 carrying ships nearby. Two workers on the CO_2 ship lost their lives instantly and 3 were injured due to projectiles.

Pieces of the ship were scattered over a 500 meter area around the explosion. The weight of these fragments varied from a few grams to several hundred kilograms. The shrapnel caused damage in several adjacent residential buildings. All the windows of a metal machining factory located around 500 meters away shattered. In the Yujiadouzhen, a village 500 meters away from the explosion, windows shattered in many residential buildings.

The cause of this accident is believed to be overloading and brittle failure of the CO_2 tank. The tank was designed for on-shore use, but was modified by Rongsheng Chemicals for ship transportation. To lower the transportation cost, the company modified the level indicator, locked the relief valve, and overloaded the tank (95% filling level). The tank material, 16 MnR steel, has a working temperature from -20 °C to 475 °C. It was not suited for use under low temperature conditions.

Worms, Germany, 21 November 1988

This is one of the most catastrophic incidents involving a CO_2 tank explosion. The detailed description of Clayton and Griffin is summarized here [13]. The liquid CO_2 was stored in a 30 tons capacity tank at 20bar and -20°C. The vessel was shattered into a number of pieces in the explosion. Most of the tank was propelled into the Rhein River about 300 meters away from the epicenter. Only 20% of the tank was recovered originally. The Worms's investigation concluded that the vessel failed catastrophically, most likely due to overpressure from overheating with a frozen relief valve. The failure pressure was estimated to be between 31 and 51 bara. It was speculated that the failure caused a CO_2 BLEVE [2].

Haltern, Germany, 2 September 1976

A rail car carrying 231,000 kg of CO_2 exploded in Haltern, Germany [1]. The tank's contents were at 7 bar and -15°C. At the moment of the explosion, the car was passing through a railroad shunting yard. As it passed checkpoints, the car was observed to be releasing plumes of CO_2 from the safety valve. It then exploded 15 m in front of a group of other rail cars. Parts of the tank were thrown up to 360 m. One person was killed in the explosion. Analysis of a recovered piece of the tank car showed that failure was due to brittle fracture. Brittle metal fracture is consistent with the near instantaneous overpressure created by a BLEVE event.

Fukushima, Japan, 1 March 1969

A CO_2 storage tank in a steel mill exploded [10-12]. A slate-roofed factory within a 50 m radius was completely destroyed, and only its pillars remained. Windows of houses located within a 500 m radius were damaged. Before the explosion, the main safety valves were closed during repairs. Liquid remained in the tank (50% filling level) and the heat source of the external evaporator was left on. The vapor phase temperature and pressure in the tank kept rising, and eventually the tank exploded when the pressure exceeded the burst pressure of the tank. Based on heating element power, CO_2 usage, and heating time, it is estimated that almost all liquid CO_2 was turned into vapor before the explosion. Thus, this incident is determined to be a PV release.

Repcelak, Hungary, 2 January 1969
Two vessels containing liquid CO2 (15 bar, -30°C) in a CO2 production and filling plant exploded in rapid succession [1, 14]. The explosion completely destroyed the tank yard consisting of four liquid CO2 storage vessels and the equipment of the technological laboratory. Large fragments were scattered in a circle of approximately 400 m radius. A 2800 kg shell landed at a distance of 150 m and a fragment weighing 1000 kg landed 250 m away. During filling, the first vessel exploded. Some minutes later, another close-by vessel exploded. The second vessel probably failed as a result of impact from a fragment from the first vessel. The explosions tore the two vessels into pieces, and tore the third vessel off its foundation bolts. The third vessel was shot into the process laboratory like a rocket due to liquid CO2 rapid expansion through a hole on the bottom. The probable cause of the accident was overfilling the first tank due to a level indicator failure. The explosion of the second vessel was speculated to be caused by a CO2 BLEVE.

3.2 Small scale experimental investigation

Most experimental work on BLEVE has focused on flammable substances like LPG and propane [1,2]. Recently, there were several small scale lab tests performed to study CO2 BLEVE [15-18].

Kim and Reid, 1978
An important reason for considering the possible occurrence of BLEVE with CO2 is this paper by Kim and Reid which presents experimental and modeling work to investigate the phenomena of rapid depressurization of hot, high pressure liquids or supercritical fluids (CO2) [3].

The authors applied a thermodynamic model based on the spinodal fluid state, which represents the limit-of-stability for the liquid phase during expansion. For a BLEVE to occur, the substance has to find itself on the spinodal curve between 1 bar and the critical point where the curve ends. When a catastrophic failure occurs there is not time for heat to pass into the system and so the path during failure is adiabatic. In thermodynamic terms the entropy of the system is constant during failure and the conditions for BLEVE to occur means that the initial entropy has to be higher that the entropy of the intersection of the spinodal curve and the 1 bar isobar. As a result, a region of initial conditions that could give a spinodal state can be identified (Figure 1). Kim and Reid performed some simple experiments on depressurization of saturated liquid and supercritical CO2. None led to explosions regardless of the initial conditions relative to the envelope in Figure 1. The authors suggested the reasons for disagreement between theory and experiments were associated with factors like vapor bubbles on the wall of the vessel (sites for heterogeneous nucleation) and shock waves or disturbances from the mechanism that ruptured the burst disk of the test vessel.

Pettersen, 2002
Pettersen performed an experimental study aimed at clarifying the possible occurrence of BLEVE in CO2 systems [17]. A test vessel was built and instrumented in order to measure transient pressures during initial depressurization, and to record possible subsequent overpressure peak(s). The study concluded that none of the experiments, which were conducted over a wide range of initial conditions (3.5, 4.5, 6.0 and 10.0 MPa initial pressures with varying

liquid fill level and discharge opening), showed signs of BLEVE, which would be characterized by a sudden repressurisation to a level significantly above the initial pressure.

Wei, 2009

This master thesis presents experimental demonstration of CO2 BLEVE which happened below the superheat limit temperature (outside the BLEVE envelop in Figure 1) [18].

The list below highlights (taken from Wei's thesis) the most significant findings:

- The 'Superheat Limit Temperature' theory for predicting occurrence of a BLEVE was not supported with experiments in this work. A CO2 BLEVE can also occur when the initial temperature is below the superheat limit temperature of CO2 (-13.8 °C). Nevertheless, considering influencing factors including CO2 filling level, potential gas leaking and CO2 mixture with different phase compositions, the theory may still be acceptable.
- Two possible approaches for a safer CO2 storage include using an initial storage pressure of around 2 MPa and developing a safety valve that can further slow down pressure drop when an unexpected vessel opening and depressurization occurs.

3.3 Consequence analysis

An established procedure for calculation of the consequence of a BLEVE event can be found in CCPS's guide to BLEVEs [1,2].

1. Calculate the explosion energy. Ongoing debates on the calculation of the explosion energy from a BLEVE event involve discussions about the BLEVE mechanism and are outside the scope of this study. In this work, we consider three options where the explosion energy is
   a. Internal energy difference between the initial and final states of an isentropic expansion process [1],
   b. Internal energy difference between the initial and final states of an irreversible adiabatic expansion [19],
   c. The PV energy (energy from the vapor phase, excluding that from flashing liquid).

   Please note that the explosion energy will need to be multiplied by a factor of 2 if the burst is a ground level, to obtain the effective working explosion energy.
2. Once the explosion energy of a BLEVE is estimated, overpressure can be determined by employing the overpressure- scaled distance curve by Baker.
3. Once overpressure is calculated, a multiplier has to be applied to account for vessel shape effect.

According to CCPS, the method presented above is based on the similarity of the blast waves of pressure vessel bursts and high explosives. This similarity holds only at some distance from the explosion. In the near field (scaled distance < 2), the peak overpressure and impulse from a pressure vessel burst differ greatly from those of a detonation of a high explosive, except when the pressure vessel is filled with a very hot high-pressure gas. Baker et al. developed a method which is used to solely predict blast pressures in the near field. However, it is based on results of numerical simulations and involves solving an implicit equation. The details of this refinement can be found in CCPS's Guidelines [1]. Note that this refinement for the near field cannot be made for non-ideal gases.

4. Results

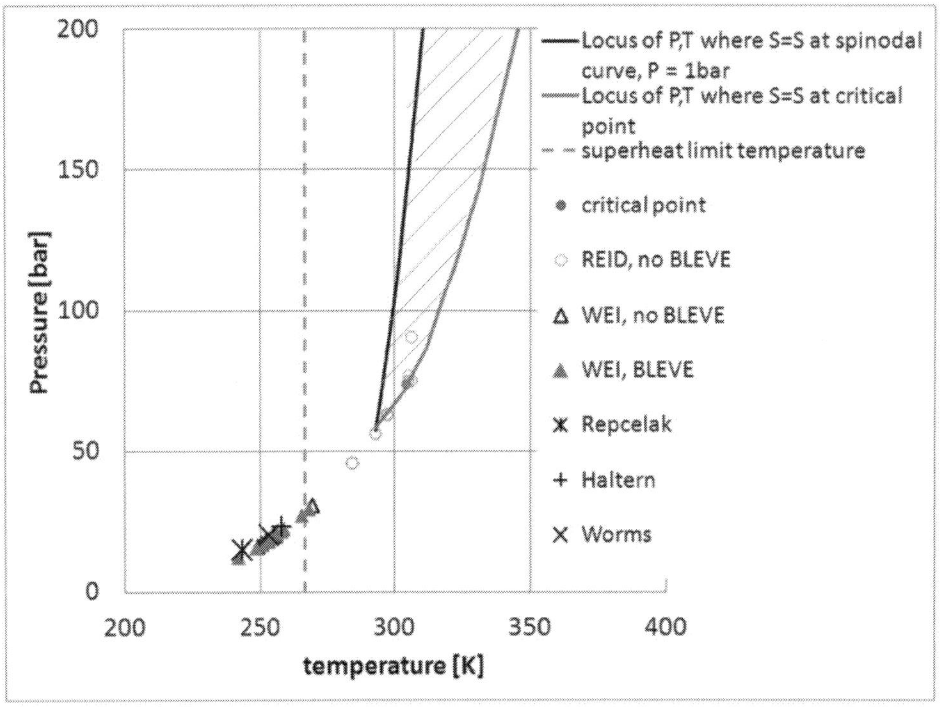

Figure 1. Summary of conditions involving CO2 BLEVE. Empty symbols are experiments with no BLEVE, and solid ones and crosses are experiments/events with BLEVE.

We summarize the literature data discussed in Section 3 in Figure 1. From Figure 1, we see that there is no clear dividing line between a BLEVE and a non-BLEVE event based on the superheat limit temperature, as indicated by experimental results and historical experiences. This might be due to the various processes involved in such an event, including nucleation of bubbles at the vessel wall. It is conclusive that the superheat limit temperature has no special significance to the BLEVE event observed in either past accidents or controlled experiments. Furthermore, Kim and Reid's experimental data along with others failed to produce a BLEVE in the laboratory when initial conditions were within the envelope predicted by the superheat limit theory.

We obtained two overpressure tags, one from Repcelak and the other fromYuhang:

1. The Repcelak: "light structure warehouse was so seriously damaged that it had to be pulled down." [14] The warehouse was 15 meters from the epicenter, and the overpressure was about 0.15-0.25 bar[1];
2. Yuhang: The windows of a metal machining factory located around 500 meters away, all shattered. In the Yujiadouzhen, a village 500 meters away from the explosion, windows shattered in many residential buildings [9]. The overpressure was about 0.035-0.07 bar [1].

Three different explosion energies, as discussed in Section 3.3, for each event are summarized in Table 2. The actual overpressure data *vs.* calculated overpressures using the CCPS's recommended method is shown in Figure 2. If the correct energy is chosen, the points should fall on or near the $y = x$ curve. Clearly, these results suggest that the explosion energy released from the incidents is more consistent with BLEVE type of explosion-energy where energy from the liquid phase contributes significantly.

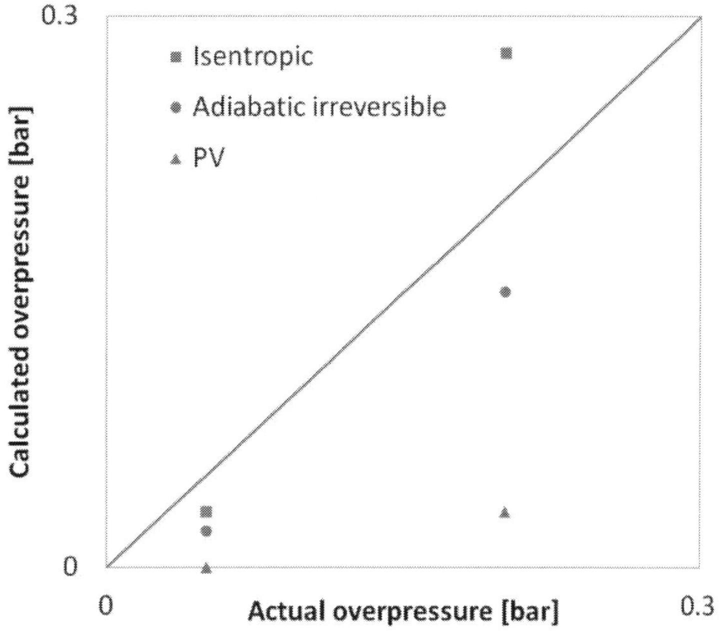

Figure 2. Actual overpressure versus calculated overpressure using different explosion energies for two events.

Table 2. Explosion energy, calculated overpressure (COP), and actual overpressure (AOP) tag for Repcelak and Yuhang incidents

Location	Initial filling level	Isentropic expansion [J]/COP[bar]	Adiabatic irreversible [J]/COP[bar]	PV [J]/COP[bar]	OP tag distance to epicenter R_tag [m]	AOP at R_tag [bar]
Repcelak	80%	3.5E9/0.28	9.9E8/0.15	3.0E7/0.03	15±5	0.20±0.05
Yuhang	95%	1.3E10/0.03	4.9E9/0.02	3.6E7/0.003	400±100	0.05±0.02

5. Conclusion

Through a critical review and analysis of historical and experimental reports of CO2 tank explosion, we conclude that a CO_2 tank rupture could lead to a BLEVE event even when the initial storage temperature is below the superheat temperature limit of CO_2. The consequence of a BLEVE event should be considered when assessing the risk associated with CO2 storage.

6. References

[1] "Guidelines for Evaluating the Characteristics of Vapor Cloud Explosions, Flash Fires, and BLEVEs, Center for Chemical Process Safety", *American Institute of Chemical Engineers*, New York, 1994

[2] "Guidelines for Vapor Cloud Explosions, Pressure Vessel Burst, BLEVE and Flash Fire Hazards, Center for Chemical Process Safety", *American Institute of Chemical Engineers*, New York, 2010

[3] Kim, M.E., Reid, R.C., "The rapid depressurization of hot, high pressure liquids or supercritical fluids", *Chemical Engineering at Supercritical Fluid Conditions*, Ann Arbor Science, 1983

[4] Calculation of BLEVE conditions for carbon dioxide, www.criticalprocesses.com/BLEVE.htm (accessed Jan. 10, 2013)

[5] "Final public report safety study for liquid logistics shipping concept (DNV)", http://cdn.globalccsinstitute.com/sites/default/files/publications/19006/final-public-report-safety-study-liquid-logistics-shipping-concept.pdf (accessed Jan. 10, 2013)

[6] Abbasi, T., Abbasi, S.A., "Accidental risk of superheated liquids and a framework for predicting the superheat limit," *Journal of Loss Prevention in the Process Industries,* 20 (2007) 165-181

[7] Birk, A.M., Cunningham, M.H., "The boiling liquid expanding vapor explosion", *J. Loss Prev. Process Ind.,* 7 (1994) 474–480

[8] Li, W., Di, G., Wang, R., "The cause analysis of a liquid CO2 tank explosion on a ship", *Materials Science Forum,* 689(2011)461-467

[9] Yuhang CO2 cargo ships explosion, http://zjnews.zjol.com.cn/05zjnews/system/2008/11/14/014973390.shtml (in Chinese, accessed Jan. 10, 2013)

[10] "High Pressure Gas Safety Inst. of Japan. Rupture of liquefied carbon dioxide in a low-temperature storage tank", *Accident investigation report (outline),* 1969 (in Japanese)

[11] "High Pressure Gas Safety Inst. of Japan", *Examples of accidents, Collection of examples of high-pressure gas accidents,* 1982 (in Japanese)

[12] Kitagawa, T., "Vapor explosion in liquid carbon dioxide tank", *Analysis of explosion hazard.* 1980 (in Japanese)

[13] Clayton, W.E., Griffin, M.L., "Catastrophic failure of a liquid carbon dioxide storage vessel", *Process Safety Progress* 13 (1994) 202-209.

[14] Voros, M., Honti, Gy, "Explosion of a liquid CO2 storage vessel in a carbon dioxide plant", *Loss Prevention and Safety Promotion in the Process Industries*, 1974

[15] van der Voort, M.M, van den Berg, A.C., Roekaerts, D.J.E.M., Xie, M., de Bruijin, P.C.J., "Blast from explosive evaporation of carbon dioxide: experiment, modeling and physics", *Shock Waves* 22 (2012) 129-140

[16] Bjerketvedt, D., Egeberg, K., Ke, W., Gaathaug, A., Vaagsaether, K., Nilsen, S.H., "Boiling liquid expanding vapour explosion in CO2 small scale experiments", *Energy Procedia* 4 (2011) 2285-2292

[17] Pettersen, J, "Experimental study on boiling liquid expansion in a CO2 vessel", 5th IIR Gustav Lorentzen Conference on Natural Working Fluids, Guangzhou, China, September 2002

[18] Ke, W., "CO2 BLEVE (Boiling Liquid Expanding Vapor Explosion)", Master Thesis, Telemark University College, Porsgrunn, Norway, 2009

[19] Chen, F, Schork, J.M., "Improvement of BLEVE Overpressure Method Using Historical Data. Conference Proceedings", 7th Global Congress on Process Safety. Chicago, IL, 2011.

9TH GLOBAL CONGRESS ON PROCESS SAFETY

New Ignition Source "Exploding Wire" for the Determination of Explosion Characteristics of Combustible Dusts in the 20-L-Sphere

Marc Scheid
Federal Institute for Materials Research and Testing (BAM),
Department 2 "Chemical Safety Engineering"
D-12200 Berlin, Germany
marc.scheid@bam.de

Alexander Klippel, Rico Tschirschwitz, Volkmar Schröder, Stefan Zirker,
Christian Kusche
Federal Institute for Materials Research and Testing (BAM), Department 2 "Chemical Safety Engineering"
D-12200 Berlin, Germany

Prepared for Presentation at
American Institute of Chemical Engineers
2013 Spring Meeting
9th Global Congress on Process Safety
San Antonio, Texas
April 28 – May 1, 2013

UNPUBLISHED

New Ignition Source "Exploding Wire" for the Determination of Explosion Characteristics of Combustible Dusts in the 20-L-Sphere

Marc Scheid
Federal Institute for Materials Research and Testing (BAM),
Department 2 "Chemical Safety Engineering"
D-12200 Berlin, Germany
marc.scheid@bam.de

Alexander Klippel, Rico Tschirschwitz, Volkmar Schröder, Stefan Zirker,
Christian Kusche
Federal Institute for Materials Research and Testing (BAM),
Department 2 "Chemical Safety Engineering"
D-12200 Berlin, Germany

Keywords: dust explosions, dust explosion testing, exploding wire, explosion characteristics, fuse wire, ignition source

Abstract

According to international standards several safety characteristics of dusts are determined in the 20-L-sphere (also known as SIWEK-Chamber). Dust cloud ignition is carried out using pyrotechnical igniters. Due to various disadvantages of pyrotechnical igniters the need for alternative ignition sources arises again and again. An alternative ignition source could be the so-called "exploding wire" or "fuse wire". The paper presents test results of a comparative study between both ignition sources for the determination of the safety characteristics Maximum Explosion Pressure and Maximum Rate of Explosion Pressure Rise of selected dusts in the 20-L-sphere. In addition to that the flame/arc propagation of both ignition sources was analysed by high speed and IR camera recordings. Tests in a windowed autoclave were performed to get information on the influence of turbulence generated due to dispersion of the dust on the shape of the generated flame and its propagation. Turbulence measurements in the windowed autoclave as well as in the 20-L-sphere with a LDA system allowed investigation if the igniter led to turbulence increase at the moment of ignition in comparison to tests without ignition.

1. Introduction

Safety characteristics are essential for the determination of explosion hazards during handling of combustible dusts and for the design of safety measures. The safety characteristics Maximum Explosion Pressure p_{max}, Maximum Rate of Explosion Pressure Rise $(dp/dt)_{max}$ and Lower Explosion Limit LEL are determined in closed vessels such as the 20-L-sphere (also known as SIWEK-Chamber). Tests are preformed according to international standards, for example EN 14034 part 1 and part 2 [2,3] or ASTM E1226 [1]. In order to determine the safety characteristics the dust samples are dispersed in air and ignited using two pyrotechnical igniters with energy contents of 1000 J (LEL) or 5000 J (p_{max}, $(dp/dt)_{max}$). Since the igniters could have an influence on the test results they are defined in the standards.

Due to various disadvantages of pyrotechnical igniters such as high costs, legal requirements concerning its storage and use as well as high energy input in comparison to most ignition sources in practice, the need for alternative ignition sources arises again and again; see also Krietsch [6]. Such an ignition source should be less expensive, readily available and the operator should be able to use it without a certificate of competence. In addition to that, the ignition energy should be adjustable over a wide range and if possible allow energy amounts less than 1000 J. Compared to the ignition energies described in the standards such energy amounts are more realistic for ignition sources found in practice, such as electrostatic discharges, mechanical sparks or hot surfaces.

An alternative ignition source which fulfils these requirements is the so-called "exploding wire" or "fuse wire". This type of ignition source is used for the determination of explosion limits of gases and is described in the standard EN 1839 [4].

This paper presents test results of a comparative study in order to find out, if the exploding wire is suitable as an alternative ignition source for the determination of explosion characteristics of dusts. For the study Maximum Explosion Pressure and Maximum Rate of Explosion Pressure Rise values from different dusts were determined with both ignition sources in the range of 100 J to 2000 J. The dusts were selected such that different combustion mechanisms were considered. High speed and IR camera recordings allowed comparison of flame front and electrical arc generated from the igniters. Calorimetric and electric measurements obtained information on the ignition energy of the igniters delivered to the mixture. With additional tests in a windowed autoclave it was possible to get information on the influence of pre-ignition turbulence on the shape of the generated flame/arc and its propagation as well as information on the turbulence generated from the ignition source. Turbulence measurements in the windowed autoclave were realised with a LDA system. The same system was used for tests in the 20-L-sphere to get information on the turbulence at the moment of ignition.

2. Test assembly and test settings

2.1 20-L-sphere

Main component of the test assembly was the 20-L-sphere (Cesana and Siwek [5]; EN 14034 1-2 [2,3]). In order to reach almost homogeneous dust dispersion into the test chamber the so called "rebound nozzle" was used.

Tests with pyrotechnical igniters as an ignition source were performed using the standard control unit of the 20-L-sphere. For tests with exploding wires an external ignition unit was necessary. Depending on the ignition energy different systems with isolating transformer, or capacitors were used. An isolation amplifier allowed the determination of the ignition energy by measuring current and voltage.

2.2 Windowed autoclave

A windowed autoclave with cylindrical shape was used for testing in order to get information on flame/arc propagation in a closed vessel as well as on the shape of flame/arc generated from both ignition sources. In this autoclave it was also possible to determine the influence of turbulence on the generated flames. The volume of the autoclave was approximately 11 L. Its diameter was 220 mm. Optical accessibility via the window allowed high speed recordings at different positions and turbulence measurements. A fan which was placed on the bottom of the autoclave enabled average turbulent velocity fluctuations (RMS-value) up to 1 m/s. In addition to that the system allowed measurements of turbulence generated from the ignition source.

2.3 Ignition sources

One of the fundamental differences of the two ignition sources pyrotechnical igniter and exploding wire is the ignition mechanism. While for the exploding wire an electrical arc and plasma is generated the pyrotechnical igniter emits flames and burning solids.

The ignition sources were placed in the middle of the vessel and connected to the lid of the vessel, such as it is described in the test standards. According to the European standards for the determination of the explosion characteristics Maximum Explosion Pressure and Maximum Rate of Explosion Pressure Rise two pyrotechnical igniters with energy of 5000 J shall be used, firing horizontally in opposite directions (EN 14034 1-2).

Instead of the ignition energy described in the standards tests were performed with ignition energies of 100 J, 500 J, 1000 J and 2000 J. In the first tests only a single ignition source with ignition energies of 100 J, 500 J and 1000 J was used. Further tests with a modified lid allowed tests with two igniters with ignition energies of 500 J and 1000 J each. For all tests with one ignition source the igniter was mounted such that it produced a flame or an electrical arc downwards. Electrodes and connected igniter for tests with a single ignition source are shown in Figure 1. In tests with two igniters the ignition source was installed such that the flame propagated in nearly horizontal direction such as it is described in the standard.

Pyrotechnical igniter

All pyrotechnical igniters used for this work have the same composition as the standard igniters used for determination of explosion characteristics in the 20-L-sphere. They contained 40 % by weight zirconium metal, 30 % by weight barium nitrate and 30 % by weight barium peroxide.

Exploding wire

The principle of the exploding wire is the evaporation and ionization of metal particles from a wire due to a sufficient current flow within milliseconds. During that process a metal vapour is created in which an electrical arc is generated between the electrodes because of the electrical conductivity of the plasma.

For all tests a nickeline wire with a diameter of 0.12 mm was used. In tests with a single ignition source the wire was tensed between the ends of two auxiliary electrodes which were connected two the main electrodes of the 20-L-sphere, see Figure 1. For tests with two exploding wires the lid of the 20-L-sphere had to be modified. It was designed such that it contained four electrodes.

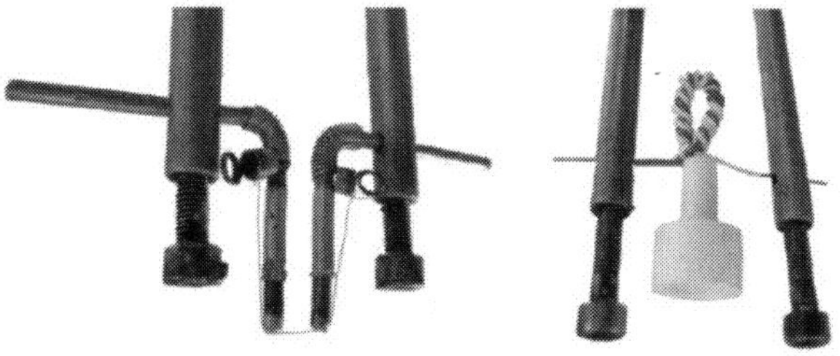

Figure 1. Ignition source for tests with a single igniter: Exploding wire with electrodes (left side) and pyrotechnical igniter (right side)

Ignition unit with isolating transformer

First tests with the exploding wire were performed with ignition energy of 100 J. In these tests the electrical power which is needed for melting the wire and generating the arc was supplied by an isolating transformer similar to that used for the determination of explosion characteristics of gases and vapors according to EN 1839 [4]. This system and its disadvantages such as not reproducible ignition delay are described by Scheid [7].

Ignition unit with capacitors

In tests with ignition energies of 500 J, 1000 J and 2000 J an ignition unit with capacitors was used instead of the isolating transformer. The main advantages of this new type of ignition unit are the higher ignition energies which the unit can produce and that it is able to ensure an exact ignition delay. A choke enabled elongation of the electrical arc during each test.

2.4 Recordings of flame front and electrical arc

In order to record the propagation of flame/arc of both igniters a high speed camera system was used. It allowed recordings of 5000 frames per second. So it was possible to determine maximum flame volume and length as well as the time delay for activating the ignition source and its burning time. In addition to that a fast acting IR camera system was used to estimate the maximum temperature level generated from the igniters. The IR camera allowed recordings with 850 frames per second.

2.5 Measurements of ignition energy (delivered to the mixture)

A combustion calorimeter working with the dry method allowed measurements of the ignition energy of both ignition sources. Dry method means, that the brass bomb of the calorimeter is covered with an insulating styrofoam and not with a water bath. This allows a more sensitive direct measurement of the temperature changes in the metal. Additionally an isolation amplifier was used to measure the burning current in the exploding wire as well as the voltage and enabled the determination of ignition energy generated in the electric arc during each test.

2.6 Turbulence measurements

A one-dimensional laser-Doppler-system (LDA) was used for turbulence measurements in the 20-L-sphere as well as in the windowed autoclave. Due to the big window of the autoclave the turbulence was measured at three different positions around the ignition source. Since the optical accessibility in the 20-L-sphere could only be realized via a windowed flange with a diameter of 40 mm it was only possible to measure the turbulence at one position next to the ignition source.

2.7 Tested dusts

Dust explosions generally show complex reaction mechanism, which depend e.g. on the chemical composition of the dust, the dust concentrations and the flow conditions. Van der Wel [8] differs between different reaction mechanisms depending on whether the reaction takes place in the gas phase due to evaporation or at the solid surface in form of gaseous products, solid or liquid material. The dusts used in this work were chosen such that they represent the different chemical classes and mechanisms as well as the dust explosion classes found in practice. As a result dusts from lignite, maize starch, niacin and anthraquinone were chosen. In addition to that a steel dust which was generated during a shot blasting process was tested.

3. Test results

3.1. Ignition energy measurements

Measurements with the isolation amplifier lead to different test results depending on the boundary conditions of the measurements. Table 1 shows the test results for measurements of ignition energy outside the test vessel without dust and flow, inside the test vessel with flow but without dust and inside the test vessel with flow and with a dust concentration of approximately 1000 g/m³ - 1250 g/m³ maize starch. Ignition unit was the isolating transformer.

Table 1. Ignition energy of igniter determined with the isolation amplifier; ignition source isolating transformer; ignition energy 100 J

Experiment-No.	Determined ignition energies		
	Outside the apparatus [J]	Inside the apparatus without dust [J]	Inside the apparatus with dust [J]
1	93.61	98.42	98.88
2	86.84	94.19	84.14
3	90.92	89.27	88.24
4	89.80	101.59	48.85
5	91.62	82.96	94.13
6	84.46	98.58	83.89
7	90.40	94.74	100.63
8	94.41	91.37	74.36
Ø	**90.26**	**93.89**	**84.14**
SD	**3.30**	**5.97**	**16.67**

Table 1 shows that the ignition energy values for tests without dust (inside and outside the test vessel) were almost similar. As a trend the values with flow (tests inside the vessel) and their standard deviation seem to be slightly higher. Tests with dust led to slightly lower ignition energies and a significant higher standard deviation of the values. Most of the tests led to ignition energies of more than 90 % of the adjusted value (set point adjustment). However, some tests with dust led to lower values. In one test less than 50 % of the adjusted value was measured. As a result the dust particles seem to influence the test results in some tests.

Figure 2 shows the arithmetic mean of the determined ignition energies as a function of dust cloud concentration for maize starch and lignite for adjusted ignition energy of 100 J. From the course of the mean values follows, that there is only a minor influence of the dust concentration on the ignition energy up to concentrations of approximately 1250 g/m³. For higher concentrations the influence is significantly higher.

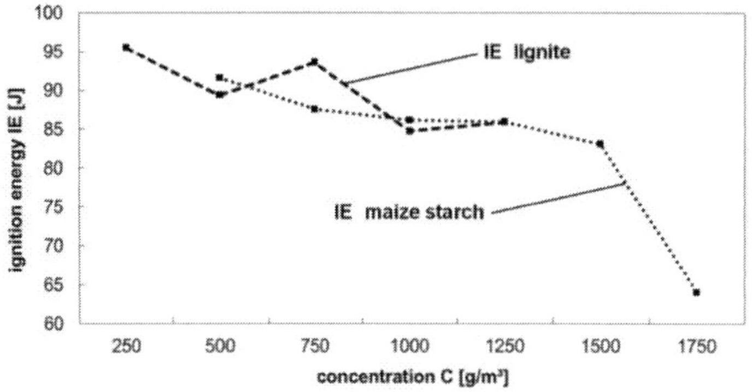

Figure 2. Arithmetic mean of ignition energy of igniter as function of dust cloud concentration for maize starch and lignite; adjusted ignition energy 100 J

Comparable test results were determined for ignition energies of 500 J, 1000 J and 2000 J using an ignition unit with capacitors. But as a trend the influence of the dust cloud on the ignition energy seems to be lower. For all tests the determined ignition energy was at least 80 % of the set point adjustment. Energy measurements with the combustion calorimeter confirmed the test results.

3.2 Propagation of flame/arc of ignition source

Tests in the windowed autoclave with maximum fan speed were performed in order to get information if pre-ignition turbulence has an influence on the flame/arc propagation of a 1000 J pyrotechnical igniter or an exploding wire with the same energy. However, for both ignition sources no visible influence on flame/arc propagation and maximum flame/arc volume was found.
Figure 3 shows the flame propagation for a 1000 J pyrotechnical igniter in comparison to an exploding wire with the same energy for tests in the windowed autoclave (no reaction of dust). The flame recordings show, that the flame of the pyrotechnical igniter propagates 3-4 times faster than the flame of the exploding wire and reaches a much higher maximum volume.

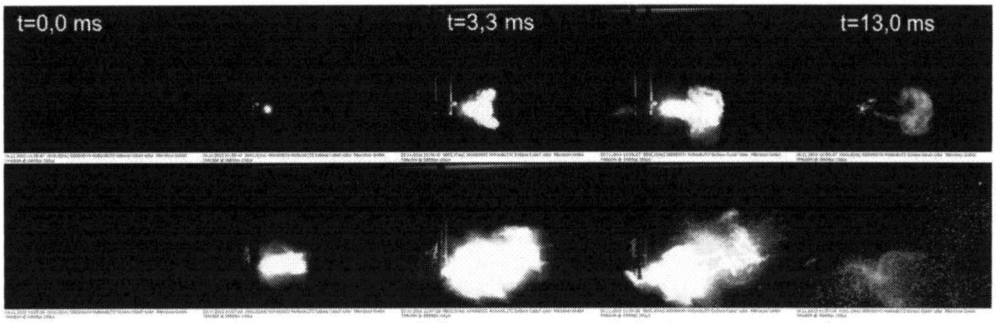

Figure 3. Flame propagation of 1000 J exploding wire (above) and pyrotechnical igniter (below) in a windowed autoclave with a volume of 11 L recorded with high speed camera

Recordings with a fast IR camera allowed an estimation of the maximum temperature level in the generated flame/arc and the maximum volume of flame or arc. Figure 4 shows four sequences of the flame/arc propagation of exploding wire (first and third sequence) and pyrotechnical igniter (second and fourth sequence) recorded with the IR camera. The sequences show a period of approx. 5 ms within the first 15 ms after triggering the igniter. Temperature range in the first two sequences was adjusted such that temperatures between 200 °C and 650 °C were visualized.

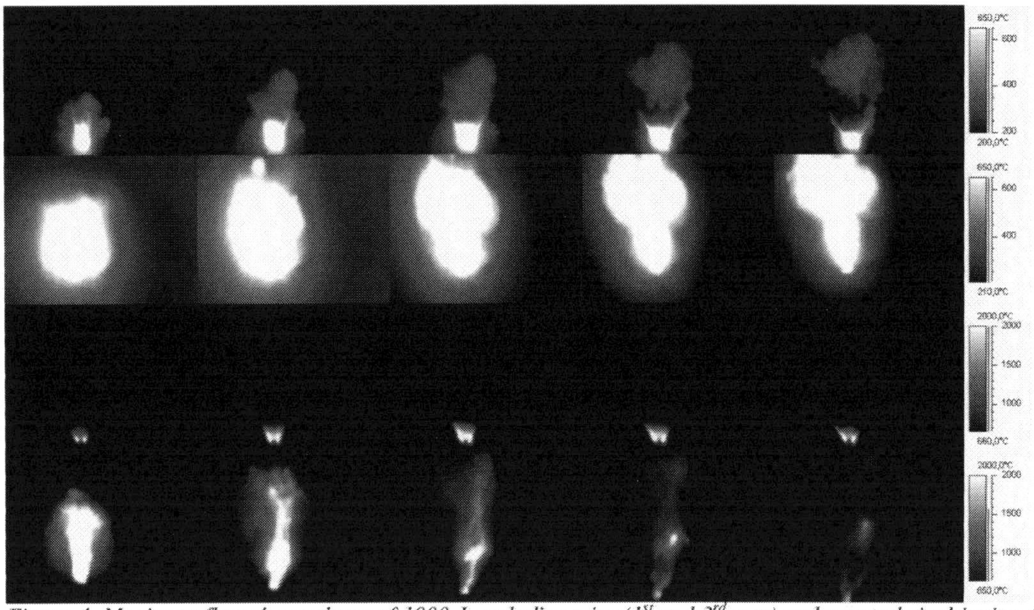

Figure 4. Maximum flame/arc volume of 1000 J exploding wire (1st and 3rd row) and pyrotechnical igniter (2nd and 4th row) recorded with a fast IR camera system

A comparison between sequence one and sequence two shows that the volume with the highest temperatures generated from the pyrotechnical igniter is much larger than the volume with such temperatures generated from the exploding wire. While for the exploding wire the volume with the highest temperatures does not increase with the propagation of the flame, it increases for the pyrotechnical igniter.
Sequence three and four visualize the temperature range between 650 °C and 2000 °C. For the exploding wire the highest temperatures were determined close to the wire. The volume in which such temperatures were determined stays almost constant in the whole sequence. In contrast to that the pyrotechnical igniter seems to generate a large volume with such high temperatures in from of a jet immediately after it was triggered. However, the volume with the highest temperatures seems to decrease significantly within 2 ms.

3.3 Turbulence measurements in 20-L-sphere and windowed autoclave

In tests in the windowed autoclave with operating fan (generated RMS-value of approx. 1 m/s) no differences in the measured RMS-value could be determined between tests without ignition and with a 1000 J exploding wire. Tests with the pyrotechnical igniter resulted in RMS-values of 2.5 m/s - 4.5 m/s depending on the measuring position. In tests in which the fan was turned off the exploding wire led to RMS-values of 0.5 m/s while the pyrotechnical igniter led to values of 2 m/s - 3 m/s. As a result turbulence measurements in the windowed autoclave showed that the turbulence generated from a 1000 J pyrotechnical igniter is stronger than the value generated from an exploding wire with the same ignition energy.
LDA-measurements in the 20-L-sphere showed, that the dispersion of dust with the rebound nozzle generated a flow with a maximum velocity of more than 40 m/s. This flow velocity decreased rapidly with decreasing pressure in the dust storage chamber and with closing the inlet valve. At the moment of ignition the flow velocity reached a value of approximately 10 m/s. The measured RMS-value at the moment of ignition was approximately 15 m/s.
In comparison to the turbulence generated in the windowed autoclave the value determined in the 20-L-sphere is much higher. Hence the test results from the windowed autoclave could not be transferred to the 20-L-sphere. As a result LDA-measurements in the 20-L-sphere were necessary. These measurements showed for tests with 1000 J ignition sources that at the moment of ignition for none of the igniters an increase of the RMS-value could be measured. However, the pyrotechnical igniter showed a not negligible velocity increase in one direction in comparison to the velocity generated due to the dispersion of dust.

3.4 Explosion behavior

Figure 5 presents the explosion pressure of maize starch as a function of the dust cloud concentration determined with both ignition sources with ignition energy of two times 1000 J. Both curves show a comparable maximum of approximately 8.2 bar for the exploding wire and 8.4 bar for the pyrotechnical igniter. For both igniters the maximum was determined for a concentration of 1000 g/m³. Deviations in explosion pressure between the different series were very low for dust explosion testing.
The determined rate of explosion pressure rise values as a function of the dust concentration are found in Figure 6. All test results show a good conformity concerning maximum rate of explosion pressure rise values which were almost similar. However, the reproducibility was not as good as for the explosion pressure. As a result the test values of the different test series differed more for both ignition sources. But this is very often found in dust explosion testing. Comparable test results were found for all tested dusts and ignition energies.

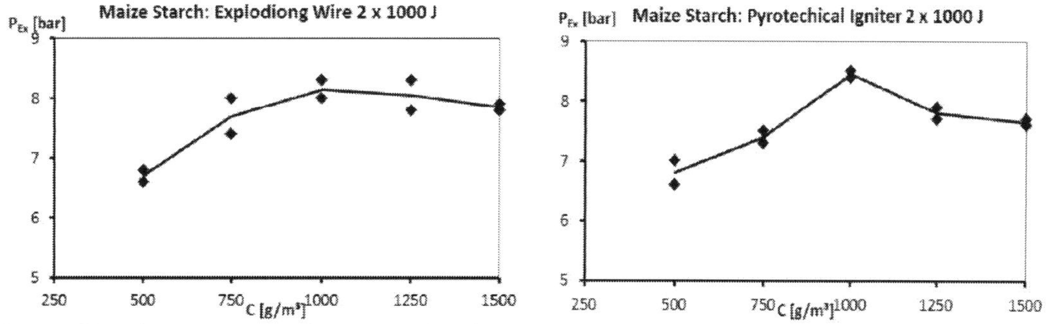

Figure 5. Explosion pressure of maize starch as function of dust cloud concentration determined with pyrotechnical igniters and exploding wires with ignition energies of 2 x 1000 J

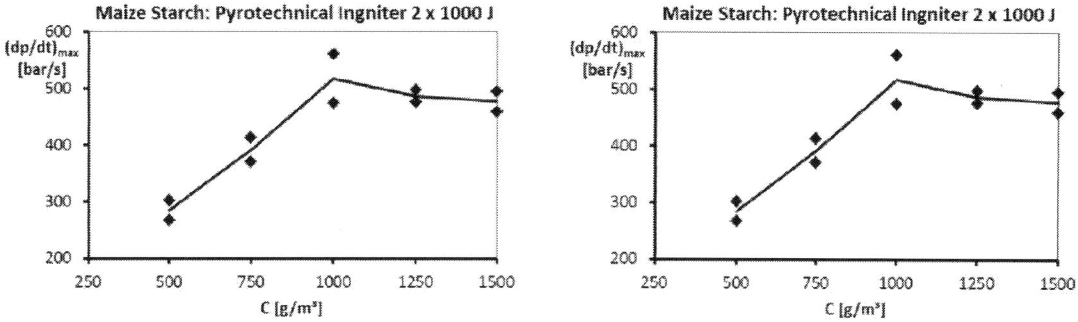

Figure 6. Rate of Explosion pressure rise of maize starch as function of dust cloud concentration determined with pyrotechnical igniters and exploding wires with ignition energies of 2 x 1000 J

Table 2 gives an overview of all determined p_{max} and $(dp/dt)_{max}$ values as well as the K_{St}-values which were calculated from the $(dp/dt)_{max}$ values.

A comparison of the results determined with two 500 J ignition sources with one 1000 J ignition source in order to find out if the use of two ignition sources leads to different test results did not come to a clear result. While for lignite tests with two ignition sources led to slightly higher values for the Maximum Explosion Pressure, for maize starch one ignition source led to slightly higher values. The determined rate of explosion pressure rise values were for both dusts slightly higher in tests with two ignition sources. However, for both tested dusts the deviations in explosion pressure and explosion pressure rise between the different test series seem to be slightly lower for tests with two ignition sources.

Table 2 also includes test results which have been determined with two 5000 J igniters, which is the ignition energy described in the test standard for determining p_{max} and K_{St}. The tests showed that the Maximum Explosion Pressures values determined with ignition energies of 100 J, 500 J, 1000 J and 2000 J are less than 10 % lower than those with 10000 J. For steel dust and niacin the values determined with 10000 J were slightly lower than the values determined with the other ignition energies. However, the determined $(dp/dt)_{max}$ values and therefore the K_{St} values differed a lot from the values determined according to the standard procedure, especially for tests with comparatively low ignition energies. Tests with 10000 J led to values which were up to 30 % higher than the values determined with low ignition energies. Only tests with 2000 J ignition sources led to reasonable test results.

Table 2. p_{max}, $(dp/dt)_{max}$ and K_{St} values for all tested dusts determined with exploding wires and pyrotechnical igniters with ignition energies of 1 x100 J, 1 x 500 J, 2 x 500 J, 1x 1000 J 2 x 1000 J and 2 x 5000 J.

Dust		1 x 100 J		1 x 500 J		2 x 500 J		1 x 1000 J		2 x 1000 J		2 x 5 kJ
		Ex. Wire	Pyro. Igniter	Ex. Wire	Pyro. Igniter	Ex. Wire	Pyro. Igniter	Ex. Wire	Pyro. Igniter	Ex. Wire	Pyro. Igniter	Pyro. Igniter
Lignite median 37 µm	p_{max}	8.4	8.1	8.1	7.6	8.1	8.5	7.9	7.7	8.3	7.9	8.4
	$(dp/dt)_{max}$	601	545	681	630	711	619	671	608	723	705	784
	K_{St}	163	148	185	171	193	168	182	165	196	191	213
Maize starch median 14 µm	p_{max}	8.4	8.5	8.4	8.6	8.0	8.2	8.2	8.4	8.2	8.4	8.7
	$(dp/dt)_{max}$	408	470	528	480	500	527	487	468	504	530	614
	K_{St}	111	127	143	130	136	143	132	127	137	144	167
Niacin median 14 µm	p_{max}	-	-	8.3	8.2	-	-	8.2	8.0	-	-	7.9
	$(dp/dt)_{max}$	-	-	953	942	-	-	920	874	-	-	1051
	K_{St}	-	-	259	256	-	-	250	237	-	-	285
Anthraquinone median 3 µm	p_{max}	-	-	-	-	-	-	8.1	8.0	-	-	8.4
	$(dp/dt)_{max}$	-	-	-	-	-	-	1113	999	-	-	1379
	K_{St}	-	-	-	-	-	-	307	271	-	-	374
Steel dust median 24 µm	p_{max}	-	-	-	-	-	-	4.1	4.1	-	-	3.8
	$(dp/dt)_{max}$	-	-	-	-	-	-	398	394	-	-	432
	K_{St}	-	-	-	-	-	-	108	107	-	-	117

4. Conclusion

Measurements of ignition energy delivered to the mixture of the ignition source exploding wire using two different methods showed that the ignition energy generated with the exploding wire was reproducible. For ignition energies of 500 J and higher, the dust cloud did not influence the ignition energy significantly. However, some tests with ignition energy of 100 J and dust concentrations with more than 1000 g/m³ showed that the dust cloud was able to influence the ignition process such that the electric arc extinguished earlier. This resulted in lower ignition energy.

High speed recordings showed that the flame generated from the pyrotechnical igniter propagated faster than the flame/arc generated from the exploding wire and reached a larger volume. Recordings with a fast IR camera system confirmed these test results. In addition to that the recordings showed that as well the volume with a temperature level comparable to the Minimum Ignition Energy of most dusts as the volume in which very high temperatures were measured was much larger for the pyrotechnical igniter. In contrast to that in tests with the pyrotechnical igniter the temperature decreased earlier as in tests with the exploding wire. As a result of high speed and IR recordings as well as energy measurements for the 100 J exploding wire the igniter should have a significant influence on the determined p_{max} and $(dp/dt)_{max}$ values. However, such an effect was not determined during the measurements of p_{max} and $(dp/dt)_{max}$ of five different dusts. All tests resulted in comparable values of p_{max} and $(dp/dt)_{max}$ for both ignition sources.

A comparison of p_{max} values determined with two 500 J ignition sources with one 1000 J ignition source in order to find out, whether the use of two ignition sources will lead to different test results, did not come to a clear result. The determined pressure rise values were for both dusts slightly higher for tests with two ignition sources. However, for both tested dusts deviations in explosion pressure and rate of explosion pressure rise seem to be slightly lower for tests with two ignitions sources.

The p_{max} values determined with ignition energies of 100 J, 500 J, 1000 J and 2000 J were less than 10 % lower than values determined according to the test standard with two igniters of 5000 J each. As a result the influence of the ignition energy of the igniter on the Maximum Explosion Pressure seems to be almost negligible.

In contrast to that $(dp/dt)_{max}$ values determined according to the test standard led to 30 % higher values, especially in tests with ignition energies of 100 J and 500 J. Only tests with 2000 J ignition sources led to reasonable test results. The reason for that could not be solved totally. It is assumed that a combination of different effects has to be considered such as turbulence generated from the ignition source and the flame/arc volume of the ignition source. However, the effect of the flame/arc volume should result in different test results for exploding wire and pyrotechnical igniter. However, this could not be determined. Probably the turbulence generated from the igniters has a strong influence especially on the $(dp/dt)_{max}$ values. Tests in the 20-L-sphere showed for all ignition sources with energies up to 2000 J that the turbulence generated from the igniters seems not to have a significant influence on the turbulence field generated due to dispersion of the dust (pre-ignition turbulence). This is probably different for higher ignition energies.

In a next step tests with ignition energies up to 10000 J with two pyrotechnical igniters and two exploding wires will be performed. In addition to that a validation of the exploding wire with a greater number of dusts is planned.

5. References

[1] ASTM E1226-12, "Standard Test Method for Explosibility of Dust Clouds," ASTM International, US, 2012

[2] EN 14034–1, "Determination of explosion characteristics of dust clouds - Part 1: Determination of the maximum explosion pressure p_{max} of dust clouds," Beuth-Verlag, Berlin, Germany, 2011.

[3] EN 14034–2, "Determination of explosion characteristics of dust clouds - Part 2: Determination of the maximum rate of explosion pressure rise $(dp/dt)_{max}$ of dust clouds," Beuth-Verlag, Berlin, Germany, 2011.

[4] EN 1839, "Determination of explosion limits of gases and vapours," Beuth-Verlag, Berlin, Germany, 2003.

[5] Cesana C., Siwek R., "Manual KSEP 7.0," Kühner AG, <kuhnersafety.com/tl_files/kuhner/product/safety/PDF/B000_071.pdf>, Accessed 2013-01-04

[6] Krietsch A., Scheid M., Lasry M., Suter G., Teske H.-J., Stahmer K. W., Arnold A., Nemcak O., Cesana C., Siwek R., "Validation of alternative igniters for the determination of explosion characteristics of dust clouds in 20-l-sphere and 1m^3-vessel," Proceedings 13th Int. Symp. on Loss Prevention and Safety Promotion in the Process Industries 6th – 9th June 2010, Brugge, 2010

[7] Scheid M., Kusche C., Schröder V., Barth U., "Tests on Suitability of the Ignition Source "Exploding Wire" for the Determination of Explosion Characteristics of Combustible Dusts in the 20-L-Sphere," Chemical Engineering Transactions, vol. 31, 2013 (in print)

[8] Van der Wel P. G. J., "Ignition and Propagation of Dust Explosions," Dissertation, Delft University Press, the Netherlands, 1993

Operational Discipline

Improving Process Safety Management Implementation through Behavior and Cultural Change

Charles A. Soczek
Principal Consultant
DuPont Sustainable Solutions
E.I. DuPont De Nemours and Company

Abstract

In 2007, US OSHA initiated a Petroleum Refinery Process Safety Management National Emphasis Program. This was followed in 2009 with a similar program focused on Chemical operations. These programs developed a series of findings, many of which were results of failure to implement OSHA 1910.119 PSM regulatory requirements. Many facilities had well documented programs, but there was poor implementation. In DuPont it was recognized that implementation is an Operational Discipline issue.

Operational Discipline (OD) as defined by DuPont is "the deeply rooted dedication and commitment by every member of an organization to carry out each task, the right way, and each time". OD is recognized as the key to successful implementation of a Process Safety Management program, turning a paper program into reality. Typically, companies have addressed OD improvements through behavioral modifications, but while the gains from programs to modify behavior can be substantial, they are difficult to sustain. Cultural change, while far more difficult to achieve, can produce more sustainable results. When done together, substantial improvements in PSM performance can be both rapid and sustained. This presentation will discuss the key attributes of operational discipline as well as provide an outline of the process to address behavior and culture change is provided.

Introduction

Operational Discipline (OD) as defined by DuPont is "the deeply rooted dedication and commitment by every member of an organization to carry out each task, the right way, and each time". Per the Business Dictionary, OD is defined as assurance that specific activities necessary for successful daily operation of a company's operations are running as they should be. This can include specific goals, missions, and guidelines. Whichever definition is chosen, OD is important to success of all aspects of an operating facility.

Process Safety Management has been in place for over twenty years. Many companies have good PSM systems, but based on the CSM Special Emphasis studies, implementation is poor. It is being increasingly recognized that Operating Discipline is a key factor that drives PSM implementation. Operating Discipline helps turn a paper PSM program into a reality

Improving Operating Discipline

Advances in Social/Psychological theory are showing that operating discipline improvements are influences by two factors, Behavior Change and Culture change. The Dejoy model helps explain this.

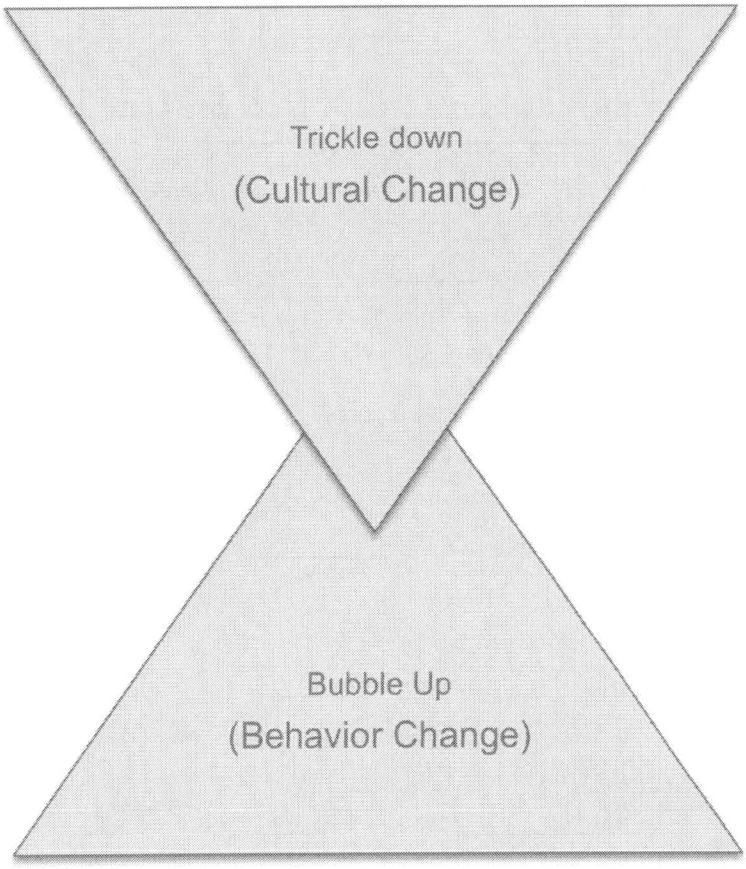

The basis for the Dejoy[1] model is that behavior change can be made quickly, but improvements are not sustainable without a large continuing effort, and that even with this, results will plateau (Bradley Curve). Safety culture change takes much longer, but once in place, superior performance can be both achieved and sustained when coupled with behavior improvement. In the Dejoy model, cultural change is driven by leadership whereas behavior change is driven by supervisors. This is an important distinction and helps focus the role of leadership and the role of supervision.

In a study of some DuPont sites, it has been found that multi-faceted Operating Discipline programs are in place. These programs address both behavior change and cultural change. Programs vary, and can consist of consists of a large number of small efforts or fewer larger programs involving all members of the organization. Either can be successful.

Dejoy[1] did a comparison of a cultural based program and a behavior based program.

Characteristic	Behavior -Based changes approach to safety	Cultural Based changes approach to safety
Background/origin	Operant psychology/ behavior modification	Organizational behavior/ anthropology
Key aspects	Bottom up approach Analytic/data driven Setting Specific continuous process	Top-down Intuitive/ethnographic Setting Specific Self-sustaining
Typical Implementation	Identify and define critical behaviors	Assess aspects of culture (values, beliefs, assumptions, etc.)
	Set performance goals	Devise alternative vision
	Observe/sample behavior	Work with leadership and employees to implement change
	Provide contingent feedback and reinforcement	
Principal Strengths	Specific technology	Emphasizes organizational change
	Objective/empirical Shop floor focus Participatory (usually) Positive	Focuses on basic causes participatory (often) Comprehensive
Principal Weaknesses	Victim-blaming Minimizes environment Focuses on immediate cause	Diffuse technology Subjective/intuitive Indirect

The above chart has highlighted a few of the key differences between the behavioral and cultural approach. In the bottom up, the emphasis is on the operators, maintenance workers, and contractors to follow procedures and rules. This requires good processes and standards and includes items as good communications, housekeeping, documentation, following procedures, and adequate resources. The top-down cultural improvement process is driven by leadership and is reflected in efforts such as leadership by example, building pride in the organization, expression of values, risk sensitivity, and being a learning organization that is innovative, agile, and continuously improves..

Some Definitions

- Operant conditioning psychology (sometimes referred to as instrumental conditioning) is a method of learning that occurs through rewards and punishments for behavior
- Psychological anthropology is an interdisciplinary subfield of anthropology that studies the interaction of cultural and mental processes. The subfield tends to focus on ways in which humans' development and enculturation within a particular cultural group—with its own history, language, practices, and conceptual categories—shape processes of human cognition, emotion, perception, motivation, and mental health. It also examines how the understanding of cognition, emotion, motivation, and similar psychological processes inform or constrain our models of cultural and social processes. Each school within psychological anthropology has its own approach
- Ethnography is a qualitative method aimed to learn and understand cultural phenomena which reflect the knowledge and system of meanings guiding the life of a cultural group

In behavior based approach, performance goals are set, whereas in a cultural based approach leadership works to build a vision of a better place (zero incidents).

Behavior programs are focused on the shop floor, whereas cultural change is focused on everyone in the organization.

Operating Discipline Characteristics

Organizations that have a high level of operating discipline exhibit an identifiable set of characteristics. DuPont has defined these characteristics as follows:

OD Characteristics

» Leadership by example
» Sufficient & capable resources
» Employee involvement
» Active lines of communication
» Strong teamwork
» Shared values
» Up-to-date documentation
» Practice consistent with procedures
» Absence of shortcuts
» Excellent housekeeping
» Pride in the organization

» Risk sensitivity
» Innovation, agility, and continuous improvement
» Justice and fairness

It is recognized that behavior change of people and cultural change are very complex. It is not proposed that the set of characteristics above is comprehensive, and it is expected that over time additional characteristics will be recognized and added to this list.

Implementation of an Operating Discipline Improvement Program

The process developed to implement an operating discipline program starts with an assessment based on 6 Sigma methodology. In the assessment, each characteristic is defined by a set of CTQs (Critical to Quality) and CTXs that define the CTQ.

These steps are

1.Pre-work

- Define scope of the assessment
- Form client OD team
- Develop assessment schedule
    - Capture input from 50% or more of the population
    - Capture current relevant data (see appendix A)
- Assessment team orientation
- Develop current state hypothesis

2. Field data collection (reach 50% of the workforce)

- Focus Groups – operators, mechanics, first line (2 hours)
- One-on-One- managers and leaders (60 minutes)
- Field Tours- employees actively performing work

3. Analysis

- Review and summarize key findings
- Apply a rating to each CTQ per rating guidance

4. Recommendations

- Each low rate CTQ has a recommendation(s) based on best practices. These best practices come from DuPont sites.
- The completed Assessment Model is used as the report to management

5. Action Plan

- Owned by line management based on findings

6. Continuous Improvement

- Conduct reviews of the assessment and periodically update the action plan with additional "best practices"
- Conduct re-assessments no more frequently than once per year

Examples of the Assessment Questions

Below is an example of the assessment tool developed that is used to understand each of the fourteen OD characteristics.

OD Characteristic: Leadership by Example

Leaders are demonstrating unrelenting passion for SHE and continuously model the behavior they expect of others

CTQ-1 Visible leadership presence and involvement across the organization

CTX – Build value in leaders that this is a critical part of their role and hold them accountable with specific expectations

- Is visible leadership evaluated as part of the overall personnel review process for line leaders, and if so, how ? Do Leaders at all levels have COTs relating to Operational Discipline?
- Do line leaders consistently take specific corrective action steps for non-conformance to SHE requirements?
- Line management is actively involved with planning, leading safety meetings & communications, and emphasizing broad participation
- Is visible leadership included in new supervisory/manager training and if so, is it effective?
- Is there consistency in approach and focus over time as leaders change?
- Does leadership review action plans to assure alignment with goals and objectives?
- Does leadership allocate adequate resources to support the safety program?
- Does leadership provide a personal example of following standards and procedures? (Wearing required personal protective equipment when in the field; using handrails, etc.)

CTX – Define and communicate what is "Visible Leadership" throughout the organization

- Does the site have formal/informal guidelines for leadership time spent in the field? How does the site measure this and is it effective
- Has the site clearly defined opportunities for visible leadership and personal leadership participation at all levels?
- Is leadership at all levels routinely in work areas conducting Safety Observations audits or other field evaluations?
- Has site leadership clearly articulated a vision and goals for Operational Discipline?
- Does leadership conduct periodic reviews of activities with all safety related teams?
- Does leadership critique audit and reports and follow-up to assure implementation of recommendations?

This characteristic has three CTQs and six CTXs. This is similar to the other characteristics.

The assessment process generates a large volume of information. The assessment team takes this information and develops action items. These items must be prioritized. Below are some examples of action items for leaders taken from an operating facility that could be generated from the process described above.

1. Increase field time and double safety audits
2. Devote 50% of PM meeting to safety
3. Off hours audits and safety coaching
4. Continue small things matter theme
5. Establish designated time in the field for leadership on safety
6. Revamp weekly safety audit meeting
7. Focus team on elimination of slips trips and falls
8. Review all letters of commendations

Summary

Sustainable improvements in Operating Discipline take time, and require facilities to continuously work on improving the safety culture and the behavior of people. This continuing focus on Operating Discipline is a common characteristic of high performing facilities.

References

(1) "Behavior change versus culture change: Diverging approached to managing workplace safety" David M. Dejoy Safety Science 43 2005.

(43g) Risk Analysis for Pressure Relief Devices – Managing Plant Safety and Reliability

Jose R. Aguilar-Otero Sr., Technology Application, COMIMSA, Saltillo, Mexico

Risk Analysis for Pressure relief devices – Managing Plant Safety and Reliability

Pressure relief devices are one of the most critical assets for plant risk and safety, therefore it is essential to be able to ensure PSV's operation, reliability and availability through the implementation of risk mitigation or prevention actions, either, corrective, preventive or predictive; these actions will allow plant owners to manage the risk associated with the device operation (or failure). This paper is intended to show a risk assessment using Weibull analysis focused on the identification for risk reduction opportunities and the selection for the best chance and dates to implement risk mitigation actions. Risk assessment methodology is based on API RP 581, Risk Based Technology.

Historically, maintenance planning for a PSV was carried out, based on pressure tests and engineering judgment; a pressure (POP) test is usually conducted to evaluate device's availability and functional condition and its frequency is linked to the protected equipment's maintenance program, regardless whatever the failure mode. Nevertheless, now a days, more quantitative and rigorous methodologies are required not just for maintenance planning but for developing a risk management strategy.

The methodology is a risk assessment. We evaluate the probability of failure (POF) times the consequence of failure (COF), for both, a valve failure and a protected equipment failure. Two failure modes are evaluated: fail to open on demand (FOD) and leak (LEAK). POF is calculated analyzing three main issues:

1. Historical inspection records and time in service, using a two parameters Weibull distribution, as follows:

$$F(t) = P_{fod}^{PSV} = 1 - \exp\left[-\left(\frac{t}{\eta}\right)^{\beta}\right] (1)$$

2. Probability of failure from the protected assets,

$$P_f(t) = ffg * D_f(t) * F_{SA} \quad (2)$$

3. Probability of failure from process over pressure scenarios.

$$P_{f,j} = P_f(t) + \left(\frac{1-ffg}{3}\right)\left(\frac{P_{o,j}}{MAWP} - 1\right) (3)$$

Consequences of failure (FOD, LEAK) are evaluated as a function of consequence area and as a function of its economic impact, for a valve failure and for a protected equipment failure.

FOD consequences are evaluated for four leak scenarios (1/4", 1", 4" and catastrophic rupture) for each protected equipment and using an overpressure as the discharge pressure. This overpressure is a higher pressure than the normal operating conditions. In this approach impact zones are used to evaluate fire, explosion and toxicity effects, in the same way we do it for a traditional quantitative consequence analysis. Meanwhile, for the leak failure mode, consequences are evaluated using a financial approach, no impact area is considered. Consequences categories for the leak mode are as follows:

- Loss of production costs
- Maintenance cost (repair costs)

- Product costs

The leak mode includes both, a small leak and a moderate leak. Moderate leak means valve stuck open, while small leak means leakage past valve (No seal).

Therefore, the global risk is calculated using the two risks approaches,

$$Riesgo^{PSV} = Riesgo^{PSV}_{FOD} + Riesgo^{PSV}_{LEAK} (4)$$

Risk calculation involves a straightforward review of the process operating conditions using a layer of protection analysis (LOPA), it also requires analyzing the process conditions and initiating events than can challenge or demand the PSV and the layers affecting the initiating event frequency. The methodology also involves a process for evaluate as many overpressure scenarios as possible; this scenarios will also challenge the pressure relief device.

This risk analysis methodology has the following benefits:

a) Perform a What if...? Scenarios analysis, and evaluate risk mitigation actions

b) Update analysis modeling, using new data

c) Understanding risk drivers

d) Understanding event sequence for risk contributors

e) Set up a risk management strategy

The risk assessment was performed for 100 PSV, located on several offshore facilities. Results show risk contribution for every failure mode, risk ranking for the whole set of PSVs, new maintenance dates as well as reliability and availability estimations. The use of this approach is very useful, because we are able to deal with large data samples and also fits well with small data samples. Results can be expressed using failure mode information or time to failure information.

(43h) Enterprise Risk Management--Adopting Risk-Based Decision Making in a Chemical Enterprise From the Board Room On Down Through All Functions and All Levels

Jerry L. Jones, PE, Consultant, Menlo Park, CA

Enterprise risk management (ERM) practices have been in place for decades in many businesses. However, following a number of failures of large financial, energy and telecommunications ventures, the interest in broader adoption of effective ERM systems has increased dramatically. During the last decade in response to recommendations and guidance by leading professional societies concerned with financial accounting and corporate governance practices, many more publicly owned non-manufacturing and manufacturing businesses have begun implementing very structured and rigorous ERM systems.

While risk has historically been defined in relation to hazards (including natural perils) as a function of the probability and consequences (harm and financial loss) of an undesired exposure event, under various recent ERM guidelines risk has been more broadly defined in relation to business objectives:

o **Risk isthe effect of uncertainty on meeting objectives** (ISO 31000-2009: Risk Management Principles and Guidelines adopted in the US as ANSI/ASSE Z690.2-2011)

o **Risk isthe possibility that an event will occur and adversely affect the achievement of objectives** (COSO Framework for Integrated Risk Management, (Committee of Sponsoring Organizations for the Treadway Commission, 2004)

A key argument made for adopting ERM principles by both public and private companies and using risk-based decision making processes throughout the enterprise is the opportunity the approach offers to more effectively and efficiently protect current value (preventing losses, the downside of risk) and to create value (the upside of some risk exposures) for owners and other stakeholders. A further incentive is the potential to establish long-term strategic advantage by being able to more effectively and efficiently manage risks than competitors.

Both of the ERM system guidelines cited above recommend rigorous and comprehensive oversight of risk management practices by the Board of Directors and active engagement of the executive leadership team in managing the key risks. Such practices are exactly what have been recommended in relation to chemical process safety by a number of investigation panels or boards in the US and abroad following major chemical process industry accidents during the last decade.

In this presentation the wide range of risk exposures that must typically be managed by a chemical enterprise (showing that process safety is one of many requiring leadership attention) will be briefly reviewed with some specific examples provided for relevant consensus standards or guidelines for managing those risks. The following list of objectives that an ERM should achieve will be discussed along with some the challenges they present.

The ERM system should assure that:

o The company's significant existing risk exposures are identified, assessed and managed in a consistent manner by the various functional organizations and the results of such risk management activities regularly reviewed by the executive leadership team and the Board of Directors,

o The "right" risks are being taken based on consideration of risk/reward tradeoffs and the "risk appetite" of the organization, and resources for risk management are allocated commensurate with the risk exposures and risk control needs,

o Individual at all levels are actively engaged in risk-based decision making and risk management so as to leverage off the knowledge and experience of the entire organization,

o Individual responsibilities and accountabilities for ongoing risk control and monitoring are clearly defined in all functional organizations at all levels, including risk management related reporting requirements with metrics (both leading and lagging indicators) that go to the executive leadership team and to the Board of Directors,

o Internal auditing and self-assessment processes are in place to regularly evaluate the effectiveness of risk management processes and controls so as to keep risk exposures below the agreed upon risk tolerance limits.

A four-step business process for risk identification, analysis, risk-based decision making, and risk controls monitoring/reporting will be described that is consistent with ERM consensus guidelines and others including those for risk-based process safety. This generic business process can be readily adopted by various functional organizations within companies of all sizes to manage a wide range of types of risk exposures and used to support other business processes requiring risk-based decision making (e.g. process hazard analysis, product safety analysis, product quality assurance, management of change, incident investigations, supply chain security, mechanical reliability analysis, and business continuity analysis).

(43i) The Chemical Safety Board and the Prevention of Combustible Dust Hazards

Manuel R. Gomez, Chemical Safety Board, Washington , DC and Christina Morgan, Office of Recommendations, US Chemical Safety Board, Washington, DC

THE CHEMICAL SAFETY BOARD AND THE PREVENTION OF COMBUSTIBLE DUST HAZARDS

This presentation will:

- Briefly summarize the CSB's rationale for recommending an OSHA general industry combustible dust standard;
- Review OSHA actions up to date;
- Comment on the difficulties OSHA may be facing in the conduct of combustible dust rulemaking; and,
- Describe the CSB's ongoing and planned efforts to advocate successful rulemaking.

Performing a HAZOP on Start-up/Shutdown Procedures

Fareed Ebrahim
ACM Automation Inc.
#700, 940 – 6th Avenue SW
Calgary, Alberta
febrahim@acm.ca

German Luna-Mejias, P. Eng
ACM Automation Inc.
gluna@acm.ca

Prepared for Presentation at
American Institute of Chemical Engineers
2013 Spring Meeting
9th Global Congress on Process Safety
San Antonio, Texas
April 28 – May 1, 2013

UNPUBLISHED

Performing a HAZOP on Start-up/Shutdown Procedures

Fareed Ebrahim
ACM Automation Inc.
#700, 940 – 6th Avenue SW
Calgary, Alberta
febrahim@acm.ca

German Luna-Mejias, P. Eng
ACM Automation Inc.

Keywords: Procedure HAZOP, Start-up/Shutdown

Abstract

Statistics show that approximately 65% of process safety related incidents occur during the start-up/shutdown operations of the plant. Yet, the bulk of the risk assessments performed today focus on the normal operation of the plant. A major contributor to the lack of start-up/shutdown risk assessments is the lack of a strongly documented methodology. One method of performing this assessment is through the HAZOP of start-up/shutdown procedures. This HAZOP involves a step by step review of the procedure with consideration given to the sequencing, delay, or missing of steps involved in the procedure. By performing this review, required permissives are identified, contingencies are developed and essential safeguards are identified. This will help ensure the operations team is confident that they are performing the start-up/shutdown in a safe manner. This paper focuses on the framework for guidelines, procedures and templates that are necessary to perform this HAZOP. A practical approach with examples is presented.

1. Introduction

Currently, it is common practice to reserve a deviation in a continuous process HAZOP for start-up and shutdown issues. Unfortunately, during analysis, this deviation if often not explored to its full potential. This abrupt analysis is often due to the lack of available information and expertise, lack specificity within the deviation, and time constraints. Yet, statistics provide a compelling argument to improve the rigour associated with start-up/shutdown analysis. Approximately 65% of process safety related incidents occur during these modes of operations1. To improve the rigour associated with the assessment of start-up/shutdown, the HAZOP technique must be further extended to the procedures for these modes of operation. Many sources are available which explore the fundamentals of performing such a HAZOP1,2, and therefore, this paper will focus on developing a methodology by expanding upon the commonly used HAZOP technique for normal operating modes.

2. Preparing for the HAZOP

In order to ensure that a complete and accurate assessment is performed, compiling the correct documents and team members for the HAZOP is of paramount importance. Just as there are key documents required for a process HAZOP, specific documents are also required for the start-up/shutdown procedure HAZOP.

2.1 Required Documentation

Firstly, the procedure for start-up/shutdown must be available. Just as the P&IDs are the central document during a process HAZOP, the procedure will be the main focus in the start-up/shutdown analysis and therefore, should be the most up-to-date revisions available. This procedure can then be split into "nodes" or subsystems for analysis. Nodes or subsystems may be defined as each individual step in the procedure.

Depending on the complexity of each step, it may be required to split the step into simplified actions, which in turn become the node for analysis. Conversely, procedure steps may be combined to develop a node based on timing requirements and the requirement for actions needing to be performed simultaneously. Timing concerns associated with procedures will be further discussed with the HAZOP deviations. As the implications of each procedure step may not be immediately clear to the HAZOP facilitator, it is of utmost importance that nodes are developed in conjunction with the operations personnel.

Once the nodes have been identified, and the complexity of the procedure understood, a decision must be made on the degree of analysis for the HAZOP to be performed. Two methods of analysis can be employed for the HAZOP – a "What-If?" approach, or a guideword method. The "What-If?" approach follows the same method as that used for a continuous process. If only an extremely high level analysis is required, the "What-If?" methodology may be applied to the entire procedure instead of applying it to nodes. Conversely, the application of the guideword method provides a more structured approach. Generally, as the complexity of the procedure increases, more guidewords should be employed in order to ensure a thorough analysis is completed. PHA software3 contains pre-packaged guidewords to be used for analysis. Explanation and application of the guidewords will be explored in Section 3.

Supporting documents for the procedure HAZOP include the following:

- As-Built P&IDs
- Isolation markup drawings - showing spec blind placement and position as per start-up/shutdown modes
- Shutdown keys - including permissives
- Alarm lists
- Alarm/Trip bypass procedures - including contingency plans

It is important to note that the shutdown keys and alarm lists may not outline which systems may be bypassed during a start-up and shutdown. Therefore, it is critical to obtain the bypass information, as well as the operating window for all instrumentation that will be referenced during the start-up/shutdown operation.

2.2 Required Personnel

In addition to compiling the correct documents for the session, it is of equal importance to ensure the correct personnel are in the room during the HAZOP. Similar to a process HAZOP, representation from the process team, instrumentation, and operation is required. It is suggested to have extra representation from the operations group to provide a more diverse perspective for the brainstorming process. Furthermore, if the HAZOP is being performed at the commissioning phase, it is recommended for the construction manager or equivalent representative to be present.

3. Performing the HAZOP

3.1 Introducing Guidewords

Risk assessment software[3] often includes preloaded libraries of guidewords for the HAZOP of procedures. Table 1 below lists seven guidewords that can be used for the HAZOP. These guidewords are listed in a suggested order for use in the HAZOP. Guidewords may be changed, added or deleted based on the HAZOP team experience, process being analyzed, and corporate guidelines. Note that the performance of a start-up/shutdown procedure HAZOP does not negate the requirement to perform a pre-start-up safety review.

Table 1. Suggested Guidewords and Description for Procedure HAZOP

Guideword[3]	Description
Less	Not enough is done to adequately execute procedure action.
More	Action is performed to a further extent than specified.
No	Action is fully omitted.
Reverse	Actions in procedure are performed out of specified sequence.
Other than[3]	Action is performed that does not align with what is specified.
As well as[3]	The operator does another action, in addition to the specified step being done correctly.[2]
Sooner/Later[3]	Timing of action is incorrect.

All seven guidewords may not be used for each start-up/shutdown procedure being analyzed. Guidewords should be applied based on the complexity of the procedure. An example of a simplified method is the two-guideword approach[2] is contained in Table 2 below.

Table 2. Two-Guideword approach for Procedure HAZOP[2]

Guideword	Description
Omit	Step is not done or part of the step is not done.
Incorrect	Step is not performed as intended.

3.1 Application of Guidewords

In order to further illustrate the application of each guideword, a subsection of a mock, simplified procedure (listed below) for a Reformer will be used. Guidewords from the two word approach will not be further explored as they can be considered a subset of the initial seven. Note that nodes will be assumed to be the individual procedure steps for the purpose of this illustration. Consequences are not to be considered complete for use in industry HAZOPs but for the purpose of example only.

Simplified Excerpt of Reformer Procedure:

1. Once pilots are lit, start lighting main burners (two per cell).
2. After completing the first ten pilots, light another ten but evenly distributed per cell.
3. Wait for 30 minutes or until the heater outlet temperature is stable
4. Continue lighting burners to increase the heater temperature at a rate of 50°C/hour until a temperature of 380°C is reached.
5. Once temperature is stable and maintains at 380°C, start injecting HP Steam and continue lighting burners as required to prevent the heater temperature from dropping below 350°C.

3.2.1 Less

Example Deviation from Step 3: Adequate time is not allotted by operator for heater outlet temperature to stabilize in an attempt to speed up start-up.

Consequence: Potential for a sudden temperature increase in heater leading to refractory damage or flame impingement.

3.2.2 More

Example Deviation from Step 1: More than two burners are lit per cell in order to speed up start-up of unit.

Consequence: Potential for a sudden temperature increase in heater leading to refractory damage or flame impingement.

3.2.3 No

Example Deviation from Step 4: Additional burners are not lit by operator.

Consequence: If steam is injected into the heater when temperature is below 350°C there will be deactivation of catalyst within heater tubes.

3.2.4 Reverse

Example: Deviation from Step 3: Additional burners are lit prior to stabilization of heater outlet temperature.

Consequence: Potential for a sudden temperature increase in heater leading to refractory damage or flame impingement.

3.2.5 As well as

Example: Deviation from Step 2: Steam is injected by operator as well as lighting the additional burners.

Consequence: If steam is injected into the heater when temperature is below 350°C there will be deactivation of catalyst within heater tubes.

3.2.6 Other than

Example: Deviation from Step 5: Natural Gas is injected into the heater instead of Steam.

Consequence: Overheating of tubes in reactor leading to tube damage and potential explosion.

3.2.7 Sooner/Later

Example: Deviation from Step 5: Steam injected to heater prior to temperature reaching 350°C.

Consequence: If steam is injected into the heater when temperature is below 350°C there will be deactivation of catalyst within heater tubes.

3.3 Applying a Risk Ranking

As HAZOPs have developed over time, there has been a continued emphasis to add more semi-quantitative aspects to the study. A major element of this semi-quantitative aspect is the application of a risk matrix. Unfortunately, risk matrices are designed using a per year time frame. This per year concept is of course not applicable to a procedure. Therefore, when performing the procedure HAZOP, two options are available. One, is to avoid the risk ranking and perform a fully qualitative analysis. Second, is to apply the risk matrix by avoiding the yearly references contained in the likelihood, and setting a default likelihood of making an error in the procedure to correspond with highest likelihood available (ie. frequent) on the matrix. This is to quantify the fact that there is a possibility to make a mistake in executing the procedure at any given opportunity for execution. For example, if a procedure step is incorrectly worded, there is an opportunity for error every time that step is executed. The consequence levels outlined in the risk matrix are not time constrained and therefore are easily applied to a start-up/shutdown procedure HAZOP. Safeguards can then be applied in a fashion similar to process HAZOPs by dropping the likelihood by one level.

3.4 Adding Safeguards

When applying safeguards to a procedure, it is of course of utmost importance to adhere to the guidelines set out in IEC-61511 if they will be used as independent protection layers. In addition to this, it is extremely important to ensure that the safeguard will not be bypassed during the start-up or shutdown. In some cases, safeguards through a safety instrumented system may be disabled during start-up or shutdown. It is extremely important to identify these bypasses and ensure safeguards are not credited that will not be active. If a safeguard is bypassed, refer to contingency planning documents in order to identify an alternative. Furthermore, it is of utmost importance to ensure that the safeguards will in fact be within their operating range during start-up and shutdown. One control related safeguard that is far more predominant in start-up/shutdown procedure HAZOPs when compared to process HAZOPs are permissives. It is therefore important to have these permissives identified, optimally on the shutdown key. In addition to instrumentation safeguards, a double check to the operator performing the specified action also acts as a valid safeguard.

4. Conclusion

The application of the HAZOP technique to a start-up/shutdown procedure is an effective way to improve the quality of analysis in these modes of operation. Just as in a normal operating mode HAZOP, specific documentation and team members are required. Furthermore, the level of rigour of the study may be adjusted by using a "What-If?" approach or by varying the number of guidewords to be used. A risk ranking may be applied during a procedure HAZOP, but caution must be applied in order to avoid diluting the risk of a situation by ensuring the procedure is analyzed on a per opportunity (instead of a per year) basis. Safeguards must also be applied with caution due to the potential for bypasses or conditions being out of the operating window of certain instruments during start-up/shutdown. It is hoped that the application of the HAZOP methodology to start-up and shutdown procedures will greatly reduce the number of incidents that occur during these modes of operation.

5. References

[1] Bridges, Bill. "Paper: How to Efficiently Perform the Hazard Evaluation Required for Non-Routine Modes of Operation (Start-up, Shutdown, Online Maintenance)." Proc. of 2011 Global Congress on Process Safety., 2011

[2] CCPS. "9. Extensions and Special Applications." *Guidelines for Hazard Evaluation Procedures.* New York: Center for Chemical Process Safety of the American Institute of Chemical Engineers, 1992. 257-67.

[3] Dyadem International Ltd. *PHA-Pro.* Computer software. Vers. 8.2.2.0.

(43p) Chasing the Protection of Intermediate Bulk Containers, Year 10

John A. Leblanc, Engineering Standards, FM Global, Norwood, MA

The challenges of protecting composite intermediate bulk containers (IBC) storing liquids that burn have been under investigation by a number of groups for more than 10 years. The path to resolving this challenge chosen by FM Global looked at defining the fire hazard created by the liquids and then understanding how the containers reacted to fire. This effort has resulted in a strong understanding of the fire hazards and protection options for several liquid-IBC combinations:

Rack storage of Oil and alcohol in IBCs

Silicone fluids in palletized IBCs

Very high flash point liquids in palletized IBCs

Propylene glycol in palletized IBCs

Storage of Empty IBCs

Defining an FM Approved IBC

This paper will provide an overview of the fire hazards associated with storing liquids that can burn in IBCs and present the new protection options and their associated test programs.

Process Safety Challenges in the Hazardous Waste Industry

Hope A. Luebeck
Veolia Environmental Services, L.L.C.
700 East Butterfield Road, Suite 201
Lombard, IL 60148
Hope.Luebeck@Veoliaes.com

Prepared for Presentation at
American Institute of Chemical Engineers
2013 Spring Meeting
9th Global Congress on Process Safety
San Antonio, Texas
April 28 – May 1, 2013

UNPUBLISHED

Process Safety Challenges in the Hazardous Waste Industry

Hope A. Luebeck
Veolia Environmental Services
700 East Butterfield Road, Suite 201
Lombard, IL 60148
Hope.Luebeck@Veoliaes.com

Keywords: Process Safety, Hazardous Waste Industry, Recyclers' Safety Council

Abstract

The OSHA Process Safety Management (PSM) Standard, 29 CFR 1910.119, was promulgated in 1992 to prevent the release of highly hazardous chemicals (HHCs) following a series of tragic industrial accidents. During the last 20 years, the emphasis of this standard has primarily been focused on refineries, chemical plants, and similar processes. However, companies in the Hazardous Waste Industry (HWI) also handle HHCs and face a special set of challenges as they strive to be PSM-compliant in their operations.

Hazardous Waste Industry facilities are often smaller than typical petrochemical, chemical and pharmaceutical plants. They have fewer resources dedicated to their operations and, as a result, generally make slower progress in advancing their PSM programs. Reasons for these tendencies will be discussed, as will initiatives which are being taken to strengthen their programs.

1. Introduction

History has taught us that the most influential changes result from tragedies. The Process Safety Management (PSM) Standard 29 CFR 1910.119 [1] was established by the Occupational Safety and Health Administration (OSHA) in 1992 after a series of deadly industrial accidents including the methyl isocyanate release at a Union Carbide pesticide plant in Bhopal, India, in 1984 [2] and the vapor cloud explosion at the Phillips 66 chemical complex in Pasadena, TX in 1989 [3]. The 14-element safety standard includes specific requirements for managing risks associated with processes which utilize Highly Hazardous Chemicals (HHCs) such as were involved in the noted catastrophes. More recent events such as the BP Texas City, TX disaster in 2005 [4] have motivated OSHA to renew its scrutiny of process safety practices of PSM-covered organizations through a series of National Emphasis Programs [5].

There have been serious accidents in the Hazardous Waste Industry (HWI) as well, though these have not had the significant impact or notoriety as those in the chemical or petrochemical industries. Such recent HWI events include The Environmental Quality Company fire in 2006 [6] and the Veolia Environmental Services explosion and fire in 2009 [7] (See Table 1). As these sorts of incidents continue in the HWI as well as in other industries, companies must continue to further develop their PSM programs and prepare for potential future NEP inspections.

Table 1. Some Noteworthy Industrial Process Safety-Related Incidents

Industry	Company	Location	Year	Deaths	Injuries
Petrochemical	Union Carbide	Bhopal, India	1984	2000+	Unknown
	Phillips 66	Pasadena, Texas	1989	24	132
	BP	Texas City, TX	2005	15	180
Hazardous Waste	The Environmental Quality Company	Apex, NC	2006	0	30
	Veolia Environmental Services	West Carrollton, OH	2009	0	4

Many of the same issues that challenge the petrochemical industry in their quest to be PSM-compliant are shared by the HWI. However, the size and nature of their operations are often quite different; thus, their approach to PSM compliance can vary significantly.

This paper explains some unique aspects of PSM which apply to companies in the Hazardous Waste Industry (HWI). When the acronym HWI is mentioned, this is intended to apply broadly to all companies having business operations involving hazardous waste. A subset of HWI companies is Hazardous Waste Treatment (HMT) facilities. The HMT acronym is used to emphasize operations related to the treatment (e.g. purification or neutralization), recycling, or destruction of hazardous wastes.

2. Operational Differences by Industry

Companies in the refinery, chemical or pharmaceutical industries typically utilize the same types and quantities of raw materials on a daily basis. They tend to manufacture or store a limited number of products or families of products. Their manufacturing operations occur according to cycles or routine production runs. A given plant has one or more primary technologies which are key to their particular business. These facilities often have recognized that PSM was applicable to their operations long ago and thus have reasonably mature PSM programs.

Companies in the HWI have more variable operations. They utilize an assortment of unit operations to handle, process, recycle or destroy the chemical and biological wastes they receive on a regular or periodic basis. They combine, decant, recycle, fuel blend, distill, incinerate or otherwise dispose of the materials they receive from their customers. These are shipped to their facilities in a variety of container types and material phases. PSM challenges shift on a daily basis since the nature of what they need to process changes frequently and can fluctuate widely. PSM programs in the HWI must be adept at effectively managing these inconsistent hazards.

3. General PSM Challenges for the HWI

Companies in the HWI have to be prepared for a wider variety of process hazards due to the nature and kind of substances they accept from their customers. They refer to waste shipments of a given type they receive as a 'stream'. Even trace amounts of certain chemicals can have a tragic effect if these contaminate incoming streams without the HWI recipients' knowledge. HMT facilities need close ties with their customers to truly understand the nature of the wastes they receive and know how repeat streams may vary in chemical concentrations of major components and may contain different kinds of constituents.

The number and volume of these streams are always changing, even from the same customer. Customers of HWI companies don't intentionally generate waste regularly and the amounts of each are often related to either their production volumes or the number of process upsets they have experienced. So companies in the HWI can't always anticipate how much of what kinds of streams they will process in a business cycle. The generation of customer waste is extremely unpredictable. HWI companies have to be prepared for almost anything and have equipment which is designed for just that. Since waste can be generated at any time, HMT facilities need to be nimble and be able to respond quickly to new business or business for a newly-generated stream of waste from an existing customer. Compliance concerns which may result from new business need to be promptly addressed.

HWI companies are usually comprised of small facilities (with less than 100 employees per site) which have a multitude of regulations which they need to meet, including RCRA, RMP, and numerous others in addition to PSM. Personnel in these more-compact organizations tend to wear many hats and are distracted by the assorted duties cast upon them during the normal workday. They many even include the owner/operator of such a facility. Expediency in addressing PSM compliance issues may be affected by the amount of resources available in such organizations.

HMT facilities tend to have smaller budgets than comparable size facilities in larger companies or those in other industries. As such, monies to upgrade facilities, build infrastructure, or address hazard reduction projects can be more limited. Business cycles have a huge impact on what funding is available to address all compliance needs. And with the HWI downstream of the manufacturing industry, the HWI often goes as manufacturing does.

Companies and operating sites in the HWI have had multiple owners over the years, perhaps more than those in the chemical or petrochemical industries. This has led to mixed safety cultures both at individual sites and across organizations. Dealing with PSM issues may be controversial, since senior personnel or management might have experienced one or more safety cultures and younger personnel others. Companies must apply extra effort to address identified deficiencies since these cultural viewpoints may lead to disagreement regarding the extent of work necessary to effectively resolve compliance issues.

Organizations in the HWI industry previously treated their activities as simply waste-related operations, even if they were recycling or creating renewed end-products through processes such as distilling or fuel blending. But they really are another kind of chemical plant, albeit a more unique sort. These facilities need to be just as focused on maintaining their waste processing equipment as though they are generating a virgin product. Operations can have deadly consequences from improperly processing or having uncontrolled releases of highly hazardous chemicals, which is the reason for the PSM regulation in the first place. HWI companies have improved their performance in this arena in recent years, especially as OSHA continues its Chemical NEP activities [5].

4. Specific Challenges by PSM Element

Employee Participation

Effective Employee Participation can be limited by the number of personnel available at HMT facilities to participate in, lead and drive improvements and PSM-related activities. Companies must allot time and resources for their employees to participate in training, hazard analyses, safety committees, operating procedure reviews, incident investigations, emergency drills, etc. required for their PSM program to be successful. Site management may need to be resourceful in giving personnel time to meet PSM program obligations.

Process Safety Information (PSI)

Efforts must be ongoing to keep PSI current. Personnel must be encouraged to review and update P&IDs, equipment files, chemical and process hazard information, consequences of deviation in procedures and similar documents so they remain accurate and useful. One of the hardest challenges seems to be keeping good records of PSM-covered equipment in easy to access files. Older facilities have limited equipment design or code information available, which can prompt re-inspection or re-certification of equipment to verify these critical details. Sometimes design literature may need to be requested from a manufacturer if it becomes damaged or misplaced. If changes of service (say of storage tanks) have occurred but were poorly documented, records may need to be re-created and verified before inclusion in PSI databases.

Some HMT facilities have difficulty describing their 'process chemistry' per the requirements of the PSM standard. This is a dilemma unique to the HWI. True process chemistry is often not intended in typical HWI operations. If this applies to a given location, the site's process safety documentation must clearly explain this circumstance.

Process Hazard Analysis

A comprehensive Process Hazard Analysis (PHA) program is critical to understanding any site's hazards and ensuring that appropriate controls and acceptable levels of risk are assumed. Many PSM-covered sites in the HWI are older facilities which suffer from process designs which did not adequately cover facility siting concerns at the time of construction. If rebuilding such sites is not possible, locations instead have to focus on administrative and engineering controls to reduce or mitigate risk to their personnel and property from their operations. And facilities with multiple sites operated independently can face inconsistencies in risk tolerances when coming under a corporate PSM umbrella.

Some HWI organizations heavily rely on consultants to lead their PHAs. This can create a double-edge sword: a site gets support for a needed activity, but if they don't define the requirements for the PHA adequately nor hire a suitable consultant, the PHA results yielded may be substandard. The PHAs may not cover enough hazardous aspects of a process if their consultant is not properly informed and knowledgeable. PSI details and reports for these PHAs may be limited in their scope and may omit critical information (background or otherwise) relevant to the study. Sites must be vigilant to ensure that their PHA activities be thoroughly completed as required.

Operating Procedures
HMT facilities typically have very good basic operating procedures due to the nature of their operations. There would either be more accidents or more hazardous spills if these facilities were not being run safely. But, as is sometimes the case in any industry, the written procedures which exist may not be adequately comprehensive in their descriptions of consequences of deviation or emergency actions to take, including explaining what conditions warrant the shutdown of a process. Some procedures may direct the operator to consult with supervision during atypical or emergency operations. While this is not necessarily incorrect, what does an operator do when the process goes haywire on a weekend or off-shift when less supervision or technical support is available? Authors of such procedures need to clearly spell out common out-of-control responses and/or define when emergency shutdown should take place to provide the operators with the best ready guidance at all times.

On the maintenance side of operations, many more procedures often need to be composed to describe inspection, testing and preventive maintenance (ITPM) activities and how repairs and replacement of equipment ought to take place. A significant number of procedures should exist to define ITPM for PSM-critical equipment systems and controls. When predictive and reliability-centered maintenance (RCM) programs become utilized by more advanced organizations, the number of maintenance procedures should increase proportionally to define and address these specialized activities.

Training
Training can be executed in a variety of ways. Some companies depend upon computer-based training for much routine or refresher training, since employees can determine when they have time available to meet the required obligations and can fit in training modules while they are present during normal work hours. When facilities do not have this option available, they may need to be resourceful with training solutions. A familiar pattern is to schedule hourly workers for day long training sessions when they have a day off. Even this becomes an issue for companies who do not have excess staff to backfill work slots for other employees undergoing training. Companies can also organize special 'training day or days' when operations are intentionally curtailed so that particular training or orientation programs may be held for the entire facility at one time. Possibly other creative solutions may be sought to balance staffing with training needs and requirements as situations dictate.

Contractors
Managing contractors can be tricky, depending on the number and type of crafts or specialty groups on which a facility may rely. Overseeing the screening, performance and insurance

requirements, among other factors, can require constant review. Staff dedicated to birddogging records and updates of same can help to keep these appropriately current. Ensuring that contractors have proper desired skills, certifications, and current site orientation may be another story. Some sites find it cumbersome to track down qualifications of contractors and have regular orientation sessions with them. That is why many smaller sites often partner with local/regional safety and health groups which specialize in OSHA and other common safety program training. In the greater Chicago area, one such group is the Three Rivers Manufacturing Association, which has a specialty Safety Center [8]. Similar such organizations exist in other parts of the country. For an annual membership fee, the safety groups can ensure their customers that a variety of contractors serving a region have certain basic safety and health skills, which then make it easier for member companies to quickly orient contractor personnel to their PSM-covered sites when jobs are required.

Sites in the HWI may also have their own employees serve as contractors for customer facilities when they work on special projects or are embedded at customer sites. These HWI personnel may need additional PSM-related training to be appropriately compliant both for their employer as well as for the customer. This is another unique aspect of the HWI business.

Pre-Start Up Safety Reviews (PSSR)

It is less common in the HWI for companies to debottleneck or re-design their processes. They don't generate as many cases where PSSRs are required. As a result, the PSSR process tends to be less well-known and less understood by site personnel. More time needs to be taken whenever PSSR opportunities exist to refresh personnel's memory of the purpose of the review and to ensure that a thorough and proper study is made of the new equipment or system before it is commissioned and HHCs are introduced to the process.

Mechanical Integrity

Maintaining the mechanical integrity of PSM-critical equipment is complicated for any organization. What makes this particularly difficult for smaller companies is establishing and staying on top of inspection, testing and preventive maintenance (ITPM) for what is often older equipment that had historically been run until it was broken. Emphasis is needed on implementing aspects of preventive and predictive maintenance to enhance the reliability and service life of their critical equipment, whether this is done in-house or via contract resources. Maintenance employees need appropriate training and/or certification according to the ITPM skills expected of them. Improving ITPM practices may even involve a type of 'culture change' which shifts focus from reactive to preventive/predictive maintenance practices. Management influence is essential to support this enhanced execution of maintenance activities.

Management of Change (MOC)

The MOC process is not only to be used for new equipment, capital projects, or changes in personnel with PSM responsibilities, but is also for new streams to be processed in an HMT facility. As waste compositions change from allowed norms, reviews need to occur to ensure that new streams or significant variations in existing ones will not result in any new hazards in the anticipated handling of the materials or the associated process operations. Changes in streams should likewise result in re-evaluating the materials of construction of the process equipment which may handle the new or different stream so that unanticipated failures of equipment do not occur. Furthermore, since newer equipment is less common in the HWI, employees need to be diligent in understanding the difference between 'replacement in kind' (RIK) and non-RIK situations so they know when to apply MOC appropriately.

Hot Work

Hot Work operations in an HMT facility can be confusing to personnel. Some sites have difficulty defining hot work areas due to the fact that handling of waste streams may occur in many more locations around the site than at a standard manufacturing plant. Employees may need additional orientation and training to recognize Hot Work situations and zones where Hot Work operations may apply. As employees become more knowledgeable of their site's Hot Work requirements, proper preparation for and permitting of all Hot Work activities should occur as expected.

Incident Investigation

HMT facilities tend to have well-established incident investigation and emergency response planning. Many of their environmental permits already require these to be fairly robust. HWI organizations, like many other organizations, need to place more emphasis on leading indicators rather than lagging indicators. The new focus on standard API RP 754 *Process Safety Performance Indicators for the Refining and Petrochemical Industries* [9] is further emphasizing this trend. One leading indicator that is easy to track is near misses. Good PSM programs already have discipline in identifying incidents or spills to report. Now these need to identify events which may not have had an adverse result, but may have had the potential for the release of a highly hazardous chemical. Understanding the dangerous potential of these near miss situations can help sites to avoid tragic consequences from ignoring such warning signs.

Emergency Planning and Response

There needs to be diligence in conducting enough annual emergency response drills and in holding debriefings to help identify weaknesses in response programs and make necessary changes to improve. HWI organizations focus heavily on spills due to the nature of their environmental permits, but should equally consider drills on equipment failures or releases from pressure relief valves, for example. This 'balanced approach' to emergency drills should result in more comprehensive emergency response procedures and staffs which are better prepared to handle a greater range of adverse events.

Compliance Audits

Sites in the HWI are used to compliance audits from regulators on EPA and similar programs and tend to be well-versed in preparing for such audits. PSM auditing activities likewise need to be ongoing, including self-audits, to support PSM gap analyses which ought to be routinely occurring. Management should regularly review the results of these audits and put in place systems to successfully address identified deficiencies.

Trade Secrets

HMT sites do not tend to have many trade secrets, but, as with other organizations, they need to ensure that these are made available to personnel who need them to conduct PSM compliance activities. These personnel may need to enter into confidentiality agreements regarding access to this critical information to perform their normal duties.

5. Advancing PSM Programs at any PSM-Covered Facility

PSM programs are only as good as their corrective actions and improvement efforts. If you find problems in your systems but don't address them, it is as though you had no knowledge of them. Shortcomings identified in audits but which were not addressed have been contributing factors to major incidents, such as that which occurred in Texas City, TX in 2005 [4]. Management teams in every company need dedication to follow through on the recommendations from hazard analyses, incident investigations and compliance audits, among others, to address deficiencies and to correct issues. They must ensure continual improvement activities and projects exist and are led by appropriate champions. A variety of solutions may be needed to engage the work force to achieve better success with these aspects of PSM.

Sister plants can work together or company divisions can pool resources when there is an inadequate number of technical staff at a given location to lead PSM efforts. Today's internet-based applications and array of electronic gadgetry make these interactive communications easier than ever without excessive travel burdens.

Companies may consider utilizing process safety consultants to supplement their programs. This is a great way to get new ideas and access to industry best practices from trusted sources. Third-party auditors may also be employed to make unbiased suggestions for improvement. However, be sure your company actively partners with these groups for more successful PSM program advancement.

Organizations should get involved with trade organizations, professional societies and other groups that conduct seminars and training on process safety topics. Technical conferences are a great way to network with others in your own industry and learn about current PSM topics and newly identified best practices. The Global Congress on Process Safety (GCPS), which is a part of the Spring Meeting of the American Institute of Chemical Engineers (AIChE), is a fine example of an ongoing conference series dedicated to Process Safety education [10].

Companies in the HWI also have a new organization to supplement their PSM compliance efforts. The Recyclers' Safety Council (RSC) was created in 2011 by a consortium of industrial companies who recycle waste solvents and purify them for reuse. The intent of the RSC is to share best practices related to safety programs and compliance with the PSM standard. No similar group previously existed specifically for companies who recycle waste solvents. The RSC promotes collaboration among organizations who understand the unique set of challenges they face in achieving safety and process safety goals. My company, Veolia Environmental Services, is one of the founding members of this Council. Please contact me, the author of this article, for more information on participating in RSC activities, including an annual safety-themed conference.

6. Conclusions

HWI facilities, and HMTs in particular, require special diligence to effectively manage and progress their PSM programs. The size and nature of their operations require them to seek creative and possibly atypical measures to meet PSM compliance challenges. Though they are sometimes less advanced than their counterparts in other industries, they remain dedicated to addressing process safety gaps and continually improving their PSM systems. The recent formation of the RSC is an example of this commitment to process safety excellence.

PSM program success for any company is a function of the availability and dedication of the employees and the leadership of management. Limited resources, while a challenge, should not be an impediment to compliance. Be proactive and focused on your program and it will be possible to have a thriving process safety culture in your organization, regardless of size or the nature of your company's business.

7. References

[1] Process Safety Management of Highly Hazardous Chemicals, 29 CFR 1910.119, Occupational Health and Safety Administration , Occupational Safety and Health Standards.

[2] Jackson B. Browning, Jackson Browning Report, "Union Carbide Disaster at Bhopal (1993)" Available at www.bhopal.com/reports-and-studies/union-carbide-disaster-at-bhopal.

[3] "OSHA: Phillips 66 Company Houston Chemical Complex Explosion and Fire: A Report to the President, U.S. Department of Labor, Washington, D.C. (1990)".

[4] Baker, James et al 2007 "The Report of the BP U.S. Refineries Independent Safety Review Panel".

[5] OSHA National Emphasis Programs: Available at: http://www.osha.gov/SLTC/processsafetymanagement/index.html

[6] U.S. Chemical Safety Board Case Study 'Fire and Community Evacuation in Apex, N.C.', April 16, 2008. Available at: www.csb.gov/investigations .

[7] U.S. Chemical Safety Board Case Study 'Veolia Environmental Services Flammable Vapor Explosion and Fire', July 21, 2010. Available at: www.csb.gov/investigations .

[8] Three Rivers Manufacturing Association; Available at www.trma.org.

[9] API RP-754 Process Safety Performance Indicators for the Refining and Petrochemical Industries, 2010. Available at: www.api.org .

[10] Global Congress on Process Safety; Available at: www.aiche.org/resources/conferences.

9TH GLOBAL CONGRESS ON PROCESS SAFETY

A Novel Methodology for Maintaining Mechanical Integrity of High Pressure Relief Systems

Sandipan Laskar
Sr. Process Safety Manager
Clean Harbors Inc.
2027 Independence PKWY S
La Porte, TX 77571
laskar.sandipan@cleanharbors.com

Prepared for Presentation at
American Institute of Chemical Engineers
2013 Spring Meeting
9th Global Congress on Process Safety
San Antonio, Texas
April 28 – May 1, 2013

A Novel Methodology for Maintaining Mechanical Integrity of High Pressure Relief Systems

Sandipan Laskar
Sr. Process Safety Manager
Clean Harbors Inc.
2027 Independence PKWY S
La Porte, TX 77571
laskar.sandipan@cleanharbors.com

Keywords: Relief System, Mechanical Integrity (MI)

Abstract

Safe operation and maintenance of emergency systems are of primary importance to the process industries, including upstream oil and gas. The upstream production facilities are usually protected from overpressurization by high integrity pressure protection systems (HIPPS) which shuts in the well when local pressure levels exceed a certain threshold. The pressure relief system is often installed as a last safety barrier, when HIPPS fails. One of the critical elements of emergency system is proper design, documentation and to maintain high reliability of this system.

In particular, the pressure relief system is an integral part of upstream design and operation; prevention of an accidental release of toxic, flammable or noxious materials from a production facility requires correct design and construction, safe operation, and proper documentation of this system.

This paper proposes a basic change in the way relief systems are documented - a requirement in order to comply with OSHA 1910.119 and thereby describes a new methodology and a change in documentation to meet OSHA requirements and other industry standards (such as API 520/521, API-RP 75, 14J). This paper also describes the steps involved in a high pressure relief system evaluation and introduces concept of ever-greening to maintain a high reliability of this system. The paper illustrates how an up-to-date relief system leverages management of change (MOC) and helps in setting up the foundation for a strong overall process safety management (PSM) program. The paper will identify the drivers and the benefits of the proposed methodology for pressure relief documentation as well as explore the interdependency of the proposed program elements.

The paper will assist upstream facility personnel to utilize good engineering practice in design and documentation of safety relief systems, to meet both onshore and offshore process safety requirements, and to develop consistent company best practices for process safety management. In addition the paper will describe a comprehensive equipment based analysis protocol, preparation of documentation sufficient for design, operational, and audit purposes, and discuss maintenance both of the physical system and design data.

1. Introduction

Most oil and gas production, refining and petrochemical organizations have some kind safety management program that stems from the requirements outlined under OSHA's CFR 1910.119. Some have a very extensive program that covers most of the 14 elements of Process Safety Management (PSM) and is managed via an integrated platform while others follow an ad hoc approach that does not span all of the elements or is disjointed.

Maintaining accurate and current Process Safety Information (PSI) is a cumbersome and difficult task to accomplish due to the numerous sub-elements of Safety Management Program (SEMP). One of the key sub-elements of PSI maintenance is the management of the pressure relief documentation. Historically this is an element that has not been given the attention it deserves. However, recent incidents in the industry have elevated the importance of maintaining accurate and current pressure system documentation.

The ultimate goal is to allow all plant personnel to understand the basis of the current plant relief and flare header design and ensure that this current plant relief and flare header design provides adequate overpressure protection for all equipment. Thus, a new methodology for relief system documentation is created that is customized to fit the needs and resources of the operating company and had the ability to relate to the other Process Safety Information elements.

2. Pressure Relief Analysis

Pressure relief systems are applied in various installations in the oil and gas industry. The function of a pressure relieving system is to provide a positive means of protection for plant equipment from overpressure above limits permitted by the relevant codes and to dispose safely of any resultant relief of the process fluids.

The pressure relief analysis provides an efficient way to document the basis for relief system design basis. The major reasons for undertaking the pressure relief analysis by an operating company are listed below.

- Safety Initiatives
- Design/Capital Projects
- Audit/Compliance Regulations
- Corporate Standardization
- Acquisitions

Identification of the causes of overpressure, quantification of relieving rates required complying with codes, sizing of relief devices and the design of the relief disposal system is a complex task. Use of a systematic method for identification of contingencies, together with a set of project specific standards, provides the basis for sound relief system design. Quantification of contingencies using established calculation methods and process simulation is then used to size pressure relief devices and the associated relief header piping network. This approach provides an auditable safe design for a new plant and existing facilities. The different steps involved in a pressure relief analysis are illustrated below.

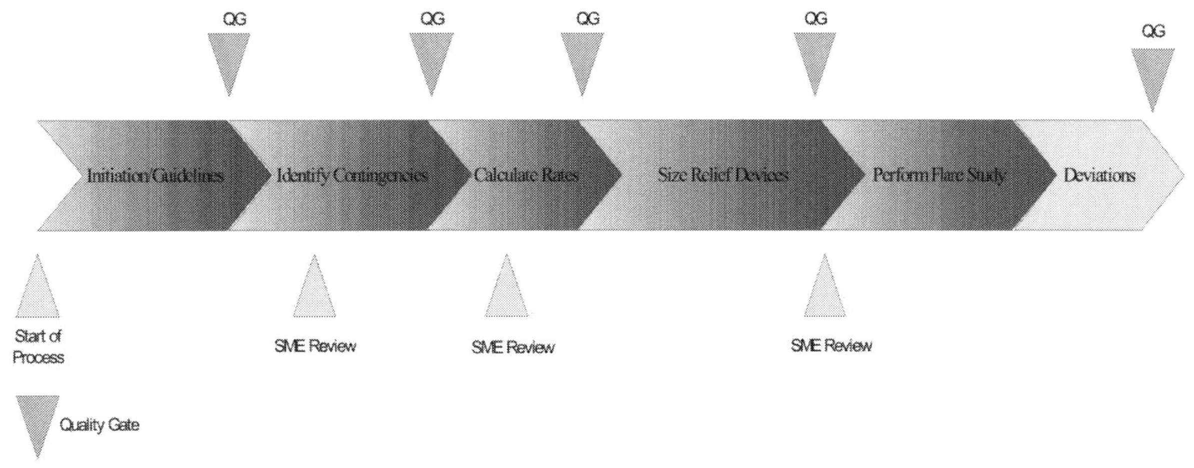

Figure 1 –Typical Phases of Pressure Relief Analysis

Mostly the pressure relief analysis is done as a one-time comprehensive project during the design phase or if there are any changes in the process. The major issues with one time comprehensive studies are listed below.

- Reactive instead of proactive approach
- Based on snap shot
- Underutilized valuable dormant data
- Missed training opportunities
- Out-of-sync with typical MOC programs
- Spend millions in 1-2 years for complete overhaul
- Increase risk of missing safety issues

3. Mechanical Integrity

Mechanical integrity is defined as the program or activities which ensure that equipment in a specified process is designed, constructed, installed and maintained to minimize the risk of accident release of high hazardous chemicals. Mechanical integrity can be achieved through the use of followings.

- Proper design to assure that equipment is suitable for its intended service
- Procedures to assure that we know how to properly operate, inspect and maintain equipment
- Training to assure everyone understands the procedures and the importance of doing things the right way
- Inspection to find flaws or deficiencies that may occur over a period of time or to confirm the status of equipment if nothing has changed
- Repair to maintain integrity
- Quality assurance to make sure that whenever something is done it is the done the right way.

Though there are many codes, standards and recommended practices that are available as a guide to ensure compliance with this element of PSM but still Mechanical Integrity (MI) presents the greatest implementation challenge. Mechanical integrity can be defined as an element of process safety management used to encompass activities that are designed to assure that mechanical equipment in an operating facility has the integrity to perform in the manner and service for which it was designed, locate, characterize, and rectify areas of deficiency that might lead to a condition that would be unsafe and endanger human life, health and/or property, assure that equipment is built and maintained in such a way as to perform it's desired function and ensure the ongoing durability and functionality of equipment achieved through a systematic approach. Although maintenance departments have successfully installed, repaired, and replaced plant equipment for decades, many of these same maintenance departments have struggled with OSHA's PSM requirements. One major challenge is prioritizing resources. Opportunities to improve will always exist; however, it is often difficult to effectively allocate money and manpower. Another challenge is simply getting organized. To meet the requirements of maintaining mechanical integrity, it is necessary to develop a program of testing safety devices and inspecting for both internal and external corrosion and erosion.

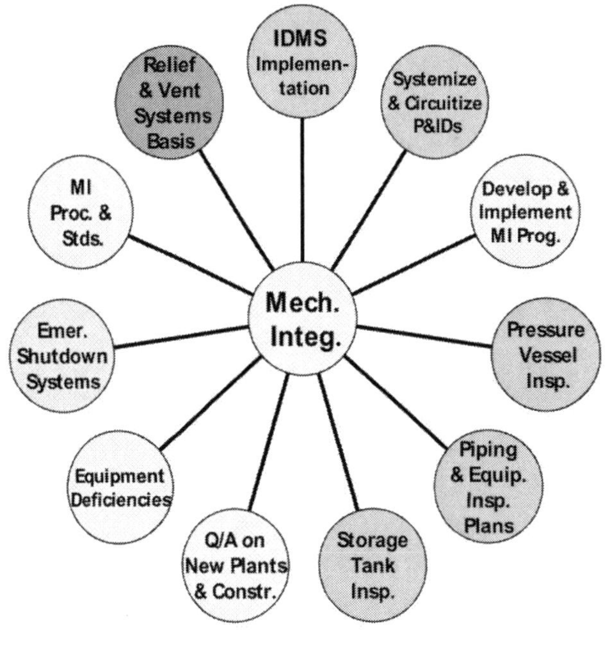

Figure 2 -Elements of Mechanical Integrity (OSHA's 29 CFR 1910.119, Para. j)

Mechanical integrity assures integrity and reliability of pressure equipment which in turn results in fewer safety incidents; reduction of mechanical failure (minimizes downtime and enables better planning), less process upsets and improves maintenance consistency and efficiency.

OSHA 1910.119 was written so that it would cover highly complex and hazardous processes handling a wide variety of hazardous chemicals. Production facilities are some of the simplest processes covered under this regulation. On the other hand, API RP 75 is written specifically for oil and gas production. It addresses offshore operations but much of the information is applicable to onshore operations as well. API helps in interpreting the application of OSHA's requirements to oil and gas facilities.

Many of the mechanical integrity activities are performed as part of a plant's maintenance program. Generally the OSHA guidelines rely on recognized and generally accepted good engineering practice (RAGAGEP), but allows companies to utilize alternate methods as long as they can prove that performance resulting from that method meets the standard. The specifications and techniques that form RAGAGEP are generally understood to mean recommended practices from the American Petroleum Institute (API), codes from the American Society of Mechanical Engineers (ASME), standards from the International Standards Organization (ISO), certification agency standards (DNV, Lloyd's, ABS), and other sources in regular use. The different elements of the PSM and SEMP program, is as shown below.

OSHA's PSM	SEMP
1. Employee Participation 2. Process Safety Information 3. Process Hazard Analysis 4. Operating Procedures 5. Training 6. Contractors 7. Pre-startup Safety Review 8. Mechanical Integrity 9. Hot Work 10. Management of Change 11. Incident Investigation 12. Emergency Planning and Response 13. Compliance Audits 14. Trade Secrets	1. Safety and Environmental Information 2. Hazard Analysis 3. Operating Procedures 4. Training 5. Pre-Startup Review 6. Assurance of Quality and Mechanical Integrity of Equipment 7. Safe Work Practices 8. Management of Change 9. Investigation of Incidents 10. Emergency Response and Control 11. Audit of Safety and Environmental Management Program Elements 12. Records and Documentation

The OSHA PSM guidelines are enforceable throughout the United States and specific US holdings for facilities having a threshold volume of hazardous chemicals, but are not applicable where another federal agency has occupational safety and health jurisdiction, including in federal waters controlled by Bureau of Ocean Management Regulation and Enforcement (BOEMRE, formerly Minerals Management Service (MMS)). Oil and gas drilling and servicing operations and normally unmanned remote facilities are specifically excluded from the standard, but typical facilities in the upstream industry such as production facilities, gas oil separation plants, and gas plants are covered if the threshold chemical volumes, typically hydrocarbons, are present.

In today's economic climate, we are operating with fewer people and demanding more from them than ever before. Therefore, it is imperative that the amount of documentation required be reduced to the minimum necessary to improve communication and increase safety. Excessive documentation places a time requirement that distracts people from their true and safe work practices. Documentation must be kept simple and concise so that it will be easy to use, with essential information readily at hand. The goal of the methodology presented in this paper is to improve safety without creating a large paperwork burden.

4. Methodology

The main goal of the proposed methodology is to develop a program that impacts the culture of the organization, thereby, drives and sustains accurate pressure relief information on a continual basis. The new program must leverage the overall, existing Safety Management Program with focus on the specific task at hand; keep perspective of risk in the current design. The new methodology eradicates the short comings of a one-time comprehensive pressure relief analysis and provides a standardized approach to ensure high reliability of the pressure relief system. The proposed methodology is further sub-divided into six different elements. These elements are Workflow, Resources and Processes, Tools, Training, Key Performance Indicators and Auditing. All these elements are interdependent for an overall success of the proposed methodology. The workflow process starts with a relief system review and generates the PRV Impact checklists, this checklist can be handed over to the MOC program. The actual workflow of the process is shown below.

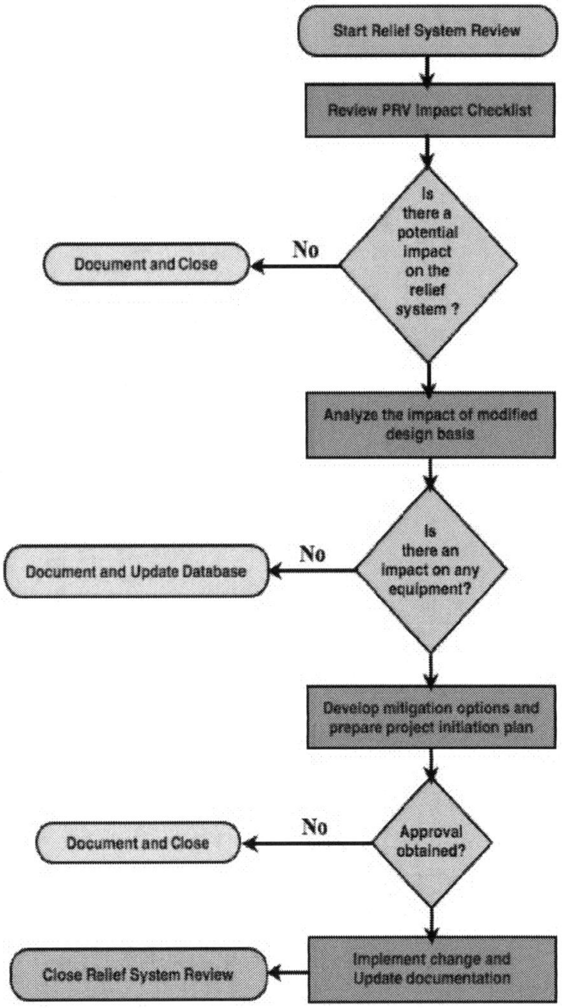

Figure 3 –Pressure Relief Valve Program Workflow

The steps involved in maintaining the data integrity and control of the proposed methodology are mentioned below.

- Develop a Relief System Impact checklist
- Incorporate checklist in the MOC program
- Control access to electronic databases
- Establish within MOC an exception check process for reconciling and validating data changes
- Develop a revalidation process for the plant-wide flare/relief system evaluation
- Complete a revalidation of each study within five years of the previous study
- Perform a complete redo of the study on a cyclical basis
- Manage the resolution of recommendations
- Establish a system of records for data
- Pool data into a central repository

The other elements of the program are to designate an owner for the program and appoint subject matter experts for engineering and information technology. An advisory group both internally at the site and externally at corporate level should be formed to support the program. There are readily available software in the market today to facilitate the actual calculations and sizing of pressure relief valves. Care should be taken to ensure that the tools used meet corporate standard, could be tailor-fitted to compliment internal company practices and also adapt to industry technology and technical practices.

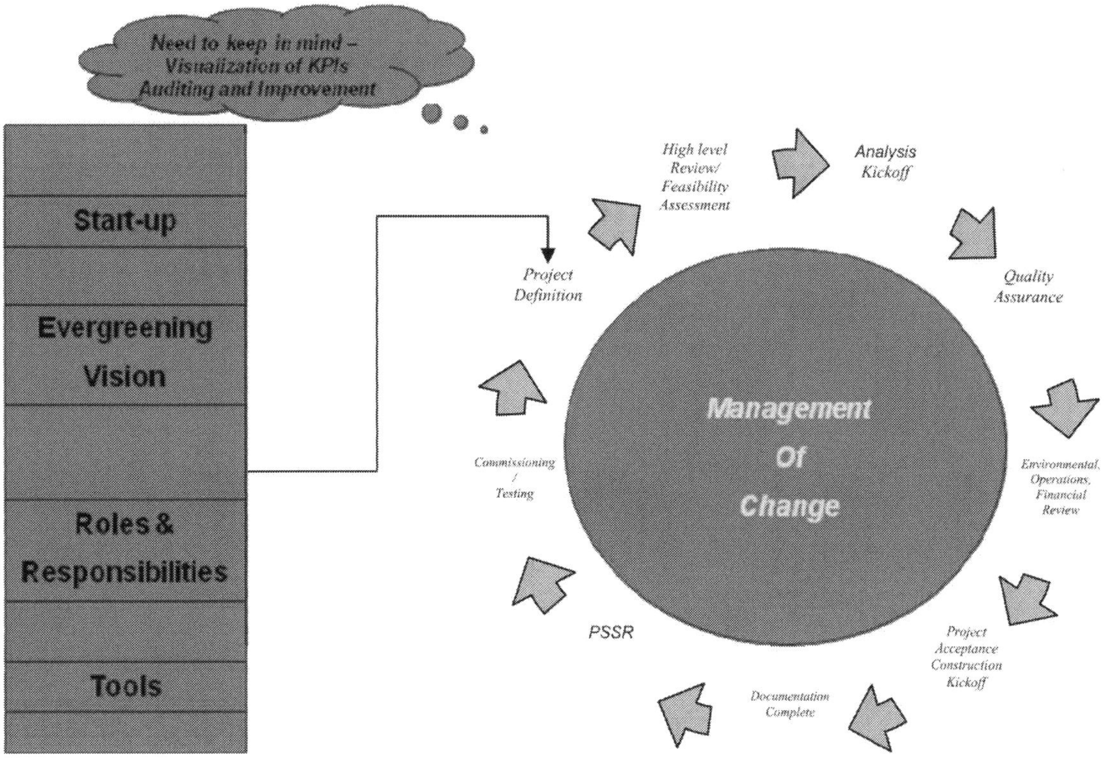

Figure 4 –Pressure Relief Valve Program Leverage Management of Change (MOC)

The above figure enunciates the importance of the proposed methodology in leveraging the existing management of change (MOC) program. The key performance indicators of the program are the data accuracy between different repositories, leverage existing information to measure performance and the program pull dormant data from relief system analysis to measure other PSM metrics. The major benefits of the proposed program are shown below.

- Standard approach
- Continual training
- Defined success factors
- Repeatable process
- Known roles and responsibilities
- Updated and accurate documentation

5. Conclusion

A good program work process starts from the projects to operations handover step on new construction or debottlenecks and runs all the way through the lifecycle of the plant encompassing modifications as a result of small and large projects. It provides a platform to communicate the PSI requirements through all disciplines at the operating facility in a concise manner and holds everyone at a high standard to meet these set requirements by continuously auditioning its effectiveness and providing training and improvement measures.

The cause of overpressure and the amount of fluid that is required to be removed from the system (vessel and piping) is unique for each installation. The proposed methodology ensures process safety without creating a large paperwork burden. The proposed methodology provides operating and maintenance staff, supervisors and foreman with the information, training and motivation they need to conduct their business in a safer manner. It is also intended that the information presented will help in proper documentation and thereby enable a faster response in case of an emergency.

The above mentioned methodology ensures most cost effective process for maintaining the mechanical integrity of high pressure relief systems while contributing up-to-date documentation of the emergency systems and thereby leveraging the existing MOC program. Particularly for high profile vertically integrated companies operating both onshore and offshore facilities, coordination of programs and practices provides an opportunity to streamline internal processes, comply with the new offshore regulatory environment, and operate more safely than ever before.

6. References

[1] "API RP 75, Recommended Practice for Development of a Safety and Environmental Management Program for Offshore Operations and Facilities", third edition. 2004. Washington, DC: API

[2] "Oil and Gas and Sulphur Operations in the Outer Continental Shelf--Safety and Environmental Management Systems", Bureau of Ocean Energy Management, Regulation and Enforcement Implementation (BOEMRE), *Final Rule*, 30 CFR Part 250, Volume 74, Number 199, (October 15, 2010)

[3] "Petroleum Refinery Process Safety Management National Emphasis Program", U.S. Occupational Safety and Health Administration (US OSHA), Directive CPL 03-00-004, (June 7, 2007)

[4] "PSM Covered Chemical Facilities National Emphasis Program", U.S. Occupational Safety and Health Administration (US OSHA), Directive CPL 09-06-002, (July 27, 2009)

[5] "Survey for API RP 75", Bureau of Ocean Energy Management, Regulation and Enforcement Implementation (BOEMRE), (September. 17, 2010)

[6] "API 520 Part 1, Standard for Sizing, Selection, and Installation of Pressure-Relieving Devices in Refineries, eighth edition. 2008. Washington, DC: API

9TH GLOBAL CONGRESS ON PROCESS SAFETY

A Framework for Managing Near-Miss Incidents

Presenter: GC Shah/Principal Author: GC Shah
Affiliation: Wood Group Mustang inc.
Address: 16401 Park Ten Place Blvd.; Houston, TX. 77084
email address: ghanshyam.shah@mustangeng.com

[

Prepared for Presentation at
American Institute of Chemical Engineers
2013 Spring Meeting
9th Global Congress on Process Safety
San Antonio, Texas
April 28 – May 1, 2013

UNPUBLISHED

A Framework for Managing Near-Miss Incidents

Presenter: GC Shah/Principal Author: GC Shah
Affiliation: Wood Group Mustang Inc.
Address: 16401 Park Ten Place Blvd.; Houston, TX. 77084
email address: ghanshyam.shah@mustangeng.com

Keywords:

ANSI/API RP-754-2010: Standard that creates "performance indicators" for process safety.
HAZID: Hazard Identification.
HAZOP: Hazard and Operability Study.
Lagging Indicators: Failures which have resulted in loss.
Leading Indicators: Indicators or evaluations that help predict the risk of unsafe events.
Near-miss incidents: Incidents that did not result in any loss, but have a potential for loss.
PHA: Process Hazard Analysis.

Abstract

Today, there is considerable emphasis being placed on learning from near-miss incidents and applying additional safeguards which would minimize possibility of escalating adverse impact of these events in the future. Not all near-miss incidents have potential for major adverse impact. It is important to focus on near-miss incidents which could result into incidents with major adverse impact. This paper applies a modified "Risk Assessment Matrix," that is commonly used for conducting PHA assessments such as HAZID and HAZOP, to estimate potential future risk indication conveyed by near-misses. In a strategic sense, near-miss management spans all work systems of an organization and its culture: IT system, operations, legal, and many others.

1. INTRODUCTION

In the hydrocarbon industry, as well as in many other industries for that matter, the "minor" events occur fairly often. While most of these events probably would end there and not result into major adverse impact, some of the events, if they recur perhaps under different circumstances, could have major impact on safety, environment, asset or a company's reputation. For instance, the event where a worker misses a rung on the ladder and falls and gets bruised may not develop into a major event. On the other hand, an event where there is a pipe leak which creates a cloud of flammable vapor that gets dissipated because of high wind, could under slightly different circumstances, result into a major event (potential fire and explosion). Even small leaks of flammable gases, if allowed to accumulate, could cause a fire and/ or an explosion.

OSHA defines near miss as "*incidents where no property was damaged and no personal injury sustained, but where, given a slight shift in time or position, damage and/or injury could have occurred.*" The definition seems to suggest that, the existing safeguards failed to work; but luckily the incident turned out to be a minor event. However, sooner or later luck may run out and a near miss can turn into a major adverse event. In terms of probabilities, one may say that statistically, there is a greater probability of occurrence of 'small impact" events and much smaller probability of high consequence or high impact events. From cost considerations, high impact events with high frequency simply can't be sustained financially.

Once near-miss events which have a potential to turn into "major" events have been identified, resources can be allocated to either strengthen the existing safeguards or add new safeguards in order to minimize the likelihood of occurrence of a major event.

2. IDENTIFYING POTENTIAL "TIME BOMBS."

Some experts suggest that near-miss is an accident waiting to get you into big trouble. The goal is to let it keep waiting!

How do we identify with a good degree of certainty that a near-miss will grow into a large-impact event? Experience with operations and maintenance could be a big help. You can develop checklists of incidents likely to develop into major events and incidents which are not likely to develop into large-scale events. Before we look at these lists, it is worth considering the meaning of the term "large-scale?" Stated broadly, the events that could cause one or more of the following may be regarded as the events that are likely to develop into "large-scale events."

OSHA recordables
Lost-time
Environmental RQ (Reportable Quantity) events
Environmental or safety violations
Events that could become a legal liability to a company
Public complaints

Some examples of specific events which could potentially turn into events with major adverse impact are:

Pin-hole leaks (e.g., pipes or flanges and fittings handling flammable, explosive or toxic materials; steam or condensate leaks on old boiler systems);
Leaking flanges, leaking mechanical seals;
Leaking tanks with flammables and/or toxic materials;
Spills under Reportable Quantity;
Lack of or poor lubrication of large pumps and/or compressors because pre-startup checks were not made or there was no checklist;
A worker narrowly escapes being asphyxiated in a confined space where nitrogen is being compressed;
Purge cycle on a furnace not working properly;

Welding sparks falling on hydrocarbon containing pipes;
Workers fail to understood a procedure (and therefore don't follow it) as it contained high-level technical jargon;
An isolation valve fails in open position;
Security infraction incidents (including cyber security);
Others.

It may be convenient to categorize the above events into the following groups:

- *Failure to follow procedures (OR poorly written or nonexistent procedures)*
- *Control system (and safety system) failures*
- *Old (older) equipment in challenging service*

Similar other systems of classification may be developed for your company.

Examples of events which will not likely develop into major events are:

Minor bruises or cuts;
Falls with a minor injury;
Potable water leaks;
A worker experiencing headache and taking non-prescription medication;
Other similar examples may be cited.

3. A FRAME WORK FOR NEAR-MISS MANAGEMENT

To be effective, the framework should be efficient. Workers should be encouraged to report ALL near-misses. Near-miss management start NOT after a near-miss, but way before such an event occurs. Guidance on leading indicators (as well as lagging indicators) is developed by API-754-2010. As Figure 1 shows near-miss is not a narrowly defined safety issue of simply reporting events; it spans through the entire organization and systems.

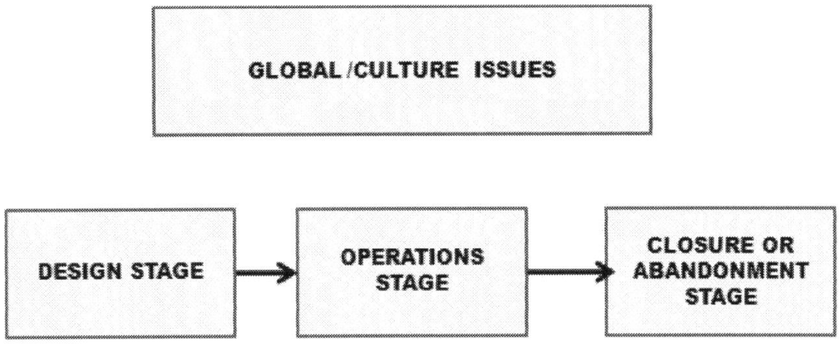

FIGURE 1: NEAR-MISS INCIDENT MANAGEMENT STAGES

3.1 Global and Cultural Issues

- Follow "good neighbors" policy. Establish good rapport with neighbors and regulators.
- Strategically important documents should receive thorough checks so as to avoid mis-interpretation.
- Encourage mutual trust between workers and management.
- Avoid bureaucratic systems in near-miss management.
- Encourage high-tech professionals (e.g., IT networks, advanced controls) to translate technical jargon into lay-person's language. The key is that persons, for whom the procedures (developed by technical professionals) are intended, should understand clearly.
- Ensure there are adequate resources to address near-miss incidents in reasonable time. Do not "dilute," safety.
- Acquisitions and divestitures also should (and typically they do) receive due diligence.

3.2 Design Stage:

- Ensure design is in accord with the applicable industrial standards and regulations.
- As far as practicable, follow the philosophy of fault-tolerant design. Stated differently, design should a mishap occur, there should be appropriate mitigative measures to constrain impact resulting from the mishap.
- Ensure documentation is easy to access and not ambiguous. (Workers referring to the documents in the future should get a clear understanding of the meaning and intent of the design).
- Perform periodic risk assessments; address findings thoroughly, and document.

3.3 Operations Stage:

- Encourage a collaborative rapport between your technical experts or engineers and operations and maintenance workers and contractors.
-
- Introduce New Technology only after a thorough risk assessment.
- Implement frequent training.
- Ensure the procedures and instructions are CLEAR to workers. Seek their input before implementing them.
- Conduct periodic audits of the facility. Some companies term these as "Loss Control Audits."
- Encourage workers to report ALL near-misses.
- Identify near-miss incidents which have potential to develop into major events. You may work out checklists workers can use to identify such incidents or consider the Risk Matrix shown later in this section. Of course checklists should be thorough so that workers won't miss out a near-miss that has a potential to grow into a major event.
- Provide resources to address near-misses within a reasonable time.

3.4 Closure and Abandonment

- Develop closure and abandonment policies. Apply Risk Management criteria to this phase of the organization.

3.5 Risk Matrix for Near Miss Incident Management

Figure 2 below shows a proposed Risk Matrix for near miss incident management. Risk is defined as a product of likelihood and consequence (severity of the outcome of an incident). Risk is grouped into three levels:

- High
- Medium
- Low

All near-miss incidents must be looked into. However, incidents identified as those having a potential for developing into "major" events require further investigation and possibly adding new safeguards or making existing safeguards more reliable or robust. Near miss management should be efficient. A small team of experienced workers can apply Risk Matrix (or checklists).

Figure 2: Risk Matrix for Near Miss Incident Management

		Risk Matrix for "Near-Miss" Incidents		
Consequence	3	3	6	9
	2	2	4	6
	1	1	2	3
		1	2	3
		Likelihood		

		Risk Matrix for "Near-Miss" Incidents		
		Likelihood		
		1	Relatively low likelihood of occurrence once in 10 yrs.	
		2	Relatively moderate likelihood of occurrence once every yr.	
		3	Relatively high likelihood of occurrence several times a year.	
		Consequence		
		1	Relatively "low" or "potentially low" level of impact: This events are NOT likely to develop into recordable or lost time or RQ events or higher.	
		2	Moderate level of impact or potential for medium level of impact such as OSHA recordable or lost time incident or RQ events.	
		3	High impact or potential for "high" impact: For example, large scale environmental or safety impact, or potential for damage to reputation.	
		Suggested Corrective Action		
		BLUE	Document the event. Further resources for corrective action may not be warranted.	
		YELLOW	Investigate and implement corrective action in "reasonable "time.	
		RED	Investigate. Should receive top priority and urgency in implementing corrective action.	

4. CONCLUSION

Near miss incident management starts way before a near miss occurs. In a strategic sense, the near miss management pervades all layers of an organization and its interactions with external systems. Strong safety culture and trusting relationships are important components of an effective near miss management framework.

HAZOP Analysis and Debottleneck for Laboratory Operation in the Semiconductor Industry

Tianxing Cai
Dan F. Smith Department of Chemical Engineering
Lamar University, Beaumont, TX 77710, USA
tcai@lamar.edu

Qiang Xu
Dan F. Smith Department of Chemical Engineering
Lamar University, Beaumont, TX 77710, USA
qiang.xu@lamar.edu

Prepared for Presentation at
American Institute of Chemical Engineers
2013 Spring Meeting
9th Global Congress on Process Safety
San Antonio, Texas
April 28 – May 1, 2013

UNPUBLISHED

HAZOP Analysis and Debottleneck for Laboratory Operation in the Semiconductor Industry †

Tianxing Cai
Dan F. Smith Department of Chemical Engineering
Lamar University, Beaumont, TX 77710, USA
tcai@lamar.edu

Qiang Xu*
Dan F. Smith Department of Chemical Engineering
Lamar University, Beaumont, TX 77710, USA
qiang.xu@lamar.edu

Keywords: HAZOP Analysis, Semiconductor Industry, Laboratory Hazards, Operation Debottleneck

Abstract

Unlike the semiconductor manufacturing process with fully automatic equipment, the laboratory operation in the semiconductor industry will still involve multiple types of manual handling for chemical analysis, reliability test, material characterization and failure analysis. The test operators or lab engineers have different levels of exposure to a lot of chemicals and physical hazards in multiple phases. This paper will describe the straightforward HAZOP (Hazard and Operability Study) process to identify the hazards and assess the risks of laboratory operations. The analysis will be deployed in the ways of manpower, machine, materials, method and environment. The analysis output can be regarded as the aid to safer laboratory design, operator training and real-time fault diagnosis. This will help to debottleneck and achieve operation safety and occupational health in the semiconductor industry.

† For presentation at AICHE 2013 Spring Meeting ,9th Global Congress on Process Safety.

* All correspondence should be addressed to Prof. Qiang Xu (Phone: 409-880-7818; Fax: 409-880-2197; E-mail: Qiang.xu@lamar.edu).

1. Introduction

Semiconductor industry revenue in 2012 is expected to reach $323.2 billion, up a slight 3.3 percent from last year's revenue of $312.8 billion, according to an IHS iSuppli Global Manufacturing Market Tracker report from information and analysis provider IHS.
Based on the professional estimation, the growth in the semiconductor industry from 2013 to 2015 will average between a more encouraging 6.6 to 7.9 percent, as shown in the figure below, with total semiconductor revenue by 2015 rising to some $397.7 billion. In the coming several years, the strong semiconductor growth can be seen.

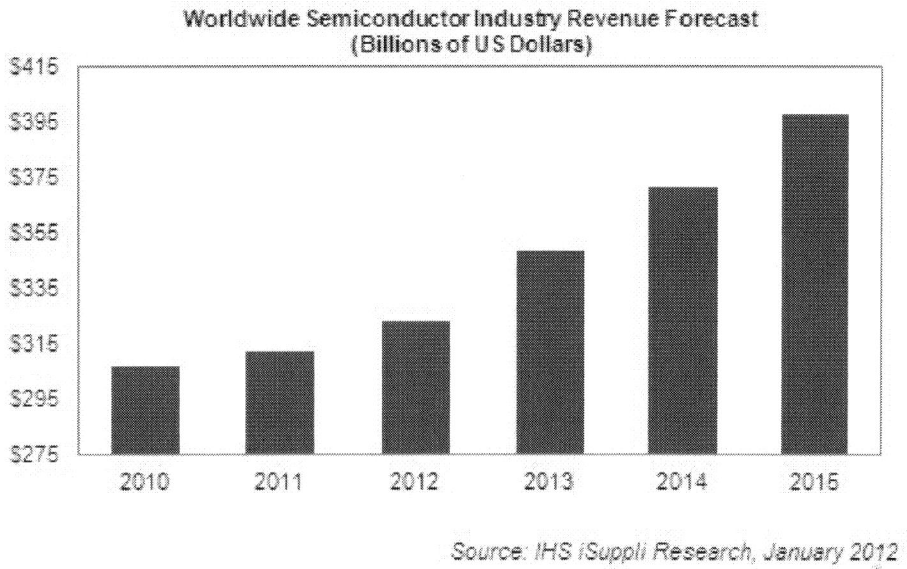

Figure 1: Worldwide Semiconductor Industry Revenue Forecast

Coupled with the fast development, the semiconductor employment has globally stretched. In USA, recently on Dec 11th, 2012, the Semiconductor Industry Association (SIA) announced that U.S. semiconductor industry jobs stretch across almost every region of the country and into the majority of states, according to an analysis of government data. California leads all states with 47,100 semiconductor jobs, followed by Texas (28,800), Oregon (23,400), Arizona (18,800), Massachusetts (10,100) and New York (7,600). According to BLS, the U.S. semiconductor industry's workforce grew by 3.7 percent over the previous year. In comparison, jobs throughout the broader U.S. economy increased by 1.2 percent over the same time period.

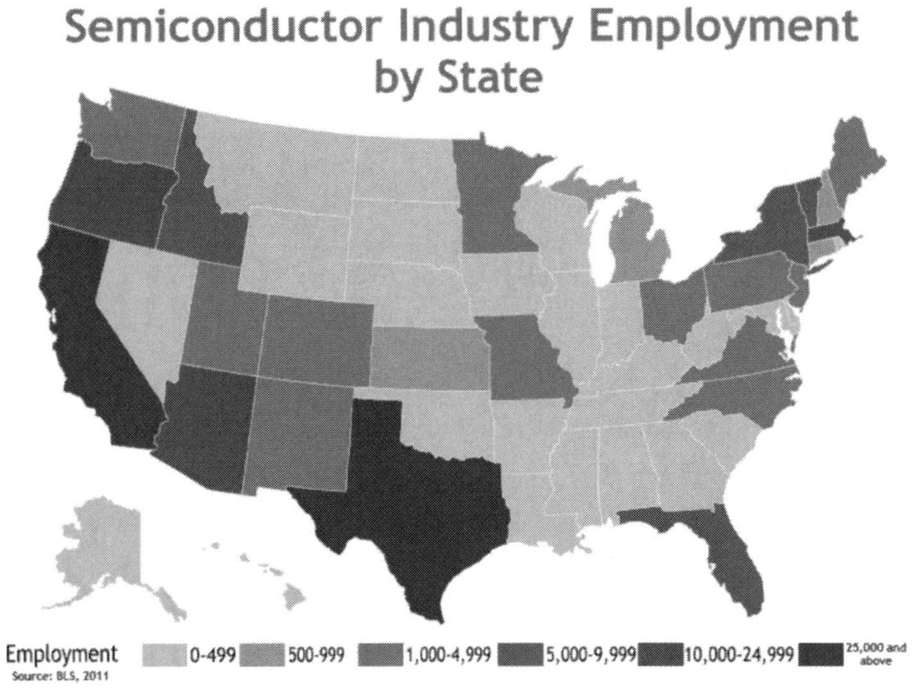

Figure 2: Semiconductor Industry Employment by State

Ever since the invention of transistor by the Bell Telephone Laboratories in 1948, the semiconductors industry has expanded greatly. The continuous fast development and innovations have expedited the increase of the semiconductor industry employment, which has further invoked an unavoidable concern of occupational safety to the semiconductor manufacturing operation.

Unlike the semiconductor manufacturing process with fully automatic equipment, the laboratory operation in the semiconductor industry will still involve multiple types of manual handling for chemical analysis, reliability test, material characterization and failure analysis. These operations will involve the handling of advanced and complex equipment or instruments such as Gas Chromatography (GC), Gas Chromatography-Mass Spectrometry (GC-MS), Fourier Transformed Infra Red Spectroscopy (FT-IR), Ultraviolet Visible Spectroscopy (UV/Visible), Liquid Chromatography (LC), Thermal and Elemental Analysis (TEA). Furthermore, due to rapid changes in this industry, manufacturing processes and their associated hazards may change completely every few years. These changes make hazard assessments more difficult to complete and require that they be conducted more often. Common hazards may include exposure to solvents, acid and caustic solutions, toxic metals, and radiation. Below Table 1 has listed the typical hazardous chemicals with their roles in the manufacturing process and potential health problems linked to the exposure. The test operators or lab engineers have different levels exposure to a lot of chemicals and physical hazards in multiple phases.

Table 1:Hazardous Chemicals in the Semiconductor Industry

Chemical name	Role in manufacturing process	Health problems linked to exposure
Acetone	Chemical-mechanical polishing of silicon wafers	Nose, throat, lung, and eye irritation, damage to the skin, confusion, unconsciousness, possible coma
Arsenic	Increases conductivity of semiconductor material	Nausea, delirium, vomiting, dyspepsia, diarrhea, decrease in erythrocyte and leukocyte production, abnormal heart rhythm, blood vessel damage, extensive tissue damage to nerves, stomach, intestine, and skin, known human carcinogen for lung cancer
Arsine	Chemical vapor deposition	Headache, malaise, weakness, vertigo, dyspnea, nausea, abdominal and back pain, jaundice, peripheral neuropathy, anemia
Benzene	Photoelectrochemical etching	Damage to bone marrow, anemia, excessive bleeding, immune system effects, increased chance of infection, reproductive effects, known human carcinogen for leukemia
Cadmium	Creates "holes" in silicon lattice to create effect of positive charge	Damage to lungs, renal dysfunction, immediate hepatic injury, bone defects, hypertension, reproductive toxicity, teratogenicity, known human carcinogen for lung and prostate cancer
Hydrochloric acid	Photoelectrochemical etching	Highly corrosive, severe eye and skin burns, conjunctivitis, dermatitis, respiratory irritation
Lead	Electroplated soldering	Damage to renal, reproductive, and immune systems, spontaneous abortion, premature birth, low birth weight, learning deficits in children, anemia, memory effects, dementia, decreased reaction time, decreased mental ability
Methyl chloroform	Washing	Headache, central nervous system depression, poor equilibrium, eye, nose, throat, and skin irritation, cardiac arrhythmia
Toluene	Chemical vapor deposition	Weakness, confusion, memory loss, nausea, permanent damage to brain, speech, vision, and hearing problems, loss of muscle control, poor balance, neurological problems and retardation of growth in children, suspected human carcinogen for lung and liver cancer
Trichloroethylene	Washing	Irritation of skin, eyes, and respiratory tract, dizziness, drowsiness, speech and hearing impairment, kidney disease, blood disorders, stroke, diabetes, suspected human carcinogen for renal cancer

In order to mitigate the potential risk during the chemical process operation, HAZOP study is always conducted. A hazard and operability study (HAZOP) is a structured and systematic examination of a planned or existing process or operation in order to identify and evaluate problems that may represent risks to personnel or equipment, or prevent efficient operation. The

HAZOP technique was initially developed to analyze chemical process systems, but has later been extended to other types of systems and also to complex operations and to software systems. A HAZOP is a qualitative technique based on guide-words and is carried out by a multi-disciplinary team (HAZOP team) during a set of meetings.

This paper will describe the straightforward HAZOP process to identify the hazards and assess the risks of laboratory operations. The analysis will be deployed in the ways of manpower, machine, materials, method and environment. The analysis output can be regarded as the aid to safer laboratory design, operator training and real-time fault diagnosis. This will help to debottleneck and achieve operation safety and occupational health in the semiconductor industry.

2. General Methodology

Several tools and methodologies are adopted to deploy the HAZOP Analysis and debottleneck for laboratory operation in the semiconductor industry. This has included Laboratory Operation Flow Diagram (LOFD), HAZOP Meeting, Parameters and Guide Words

2.1 Laboratory Operation Flow Diagram (LOFD)

The method applies to existing or planned laboratory operation for which design information is available. Similar with the commonly used process flow diagram, this commonly includes a laboratory operation flow diagram, which is examined in small sections, such as individual items of laboratory test instruments, glassware or chemicals between them. For each of these, a operation design intention is specified. For example, in a failure analysis laboratory, a standard solution of aqua regia (the combination of HNO_3 and HCl with the volume ratio of 1:3) may have been used. The intention to remove gold layer in the integrated circuit so that the further visual inspection can be conducted to check the failure mode. The HAZOP team then determines what are the possible significant deviations from each intention, feasible causes and likely consequences. It can then be decided whether existing, designed safeguards are sufficient, or whether additional actions are necessary to reduce risk to an acceptable level. Besides the operation safety, HAZOP can also give the detailed instruction on the opinion whether the potential deviation will affect the quality of experiment or test analysis result.

2.2 HAZOP Meeting

When HAZOP meetings were proceeded by the involved laboratory engineers, technicians and operators, they were generally scheduled for one to two hours per day. For a medium-sized laboratory where the total number of test items to be considered is 600 (items of equipment and pipes or other transfers between them), about 20 such meetings would be needed.

2.3 Parameters and Guide Words

The key feature is to select appropriate parameters which will apply to the design intention. The general words used in the laboratory operation are usually solution concentration, solution temperature, sampling volume, and instrument detection limit. In the above example, it can be seen that variations in these parameters could constitute deviations from the design intention of experiments. In order to identify deviations, a set of guide words is applied to each step of the experiment process. The current standard Guide Words are as follows:

Table 2: Guide Words for HAZOP Study of Laboratory Operation in the Semiconductor Industry

Guide Word	Meaning
NO OR NOT	Complete negation of the design intent
MORE	Quantitative increase
LESS	Quantitative decrease
AS WELL AS	Qualitative modification/increase
PART OF	Qualitative modification/decrease
REVERSE	Logical opposite of the design intent
OTHER THAN	Complete substitution
EARLY	Relative to the clock time
LATE	Relative to the clock time
BEFORE	Relating to order or sequence
AFTER	Relating to order or sequence

2.4 HAZOP Software

Various software programs are now available to assist in HAZOP meetings. Even though they are always used in the chemical industry, the principles will be quite applicable to laboratory operation in the semiconductor industry.

2.5 HAZOP Team

HAZOP is normally carried out by a team of people with roles as follows

Table 3: Team Members for HAZOP Study of Laboratory Operation in the Semiconductor Industry

Name	Position	Role
HAZOP Group Leader	Lab Manager	someone experienced in HAZOP and laboratory operation, to ensure that the method is followed carefully
Recorder	Lab Engineer 1	someone to ensure that problems are documented and recommendations passed on
Designer	Lab Engineer 2	someone to explain any design details or provide further information
User	Lab Operator	someone to consider it in use and question its operability, and the effect of deviations
Specialist	Lab Engineer 3	someone with relevant technical knowledge
Maintainer	Lab Technician	someone concerned with maintenance of the process.

3. HAZOP Management System

For the purpose of effective implementation of the HAZOP management policy and solid application of introduced general methodology, the laboratory HAZOP system should be set up. The system has coupled general management policy (operation specification and work instructions), environment control (temperature and humidity), operation safety (chemical storage, hazards assessment, chemical identification and MSDS), equipment management (PM, calibration, MSA), chemical control (data analysis and shelf life time control), training and certification (orientation, OJT, regular qualification). The whole system has covered multiple layers of analysis factors. The first layer will include manpower, machine, materials, method and environment.

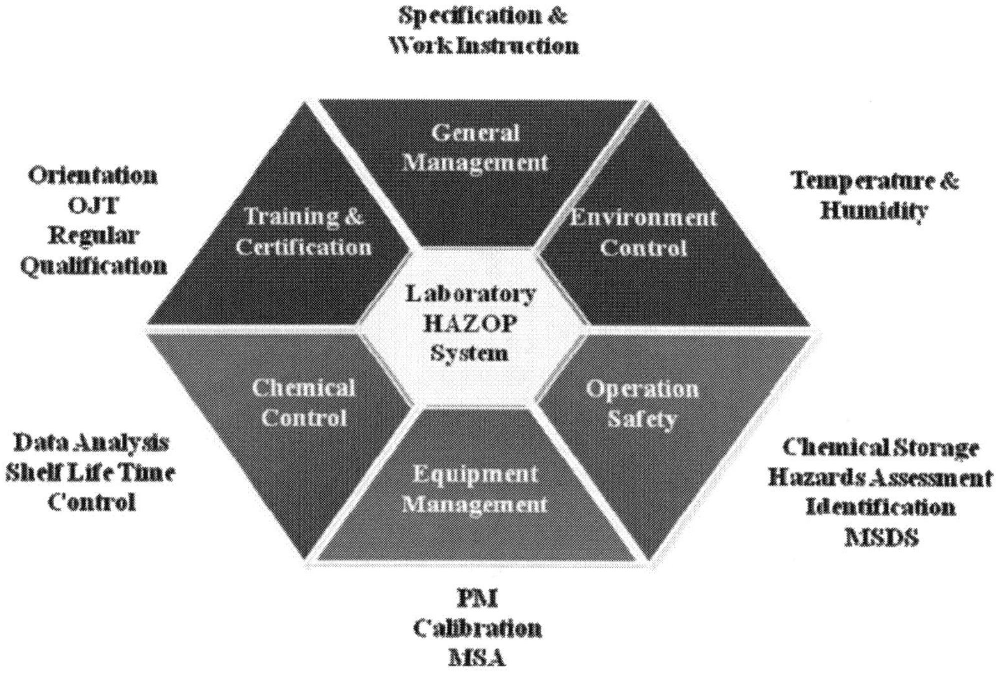

Figure 3: HAZOP Management System for Laboratory Operation in the Semiconductor Industry

These factors can be expanded with below fishbone diagrams so that the second layer of factors will be included into HAZOP matrix system: for example, sample collection and sample preparation will be considered in the material part; training, parallax, reproducibility, practice and ergonomics will be considered in the manpower part; samples, test methods, workmanship and standards will be considered in the method part; discriminations, repeatability, bias, linearity and calibration will be considered in the instrument part; vibration, lighting, temperature and humidity will be considered in the method part.

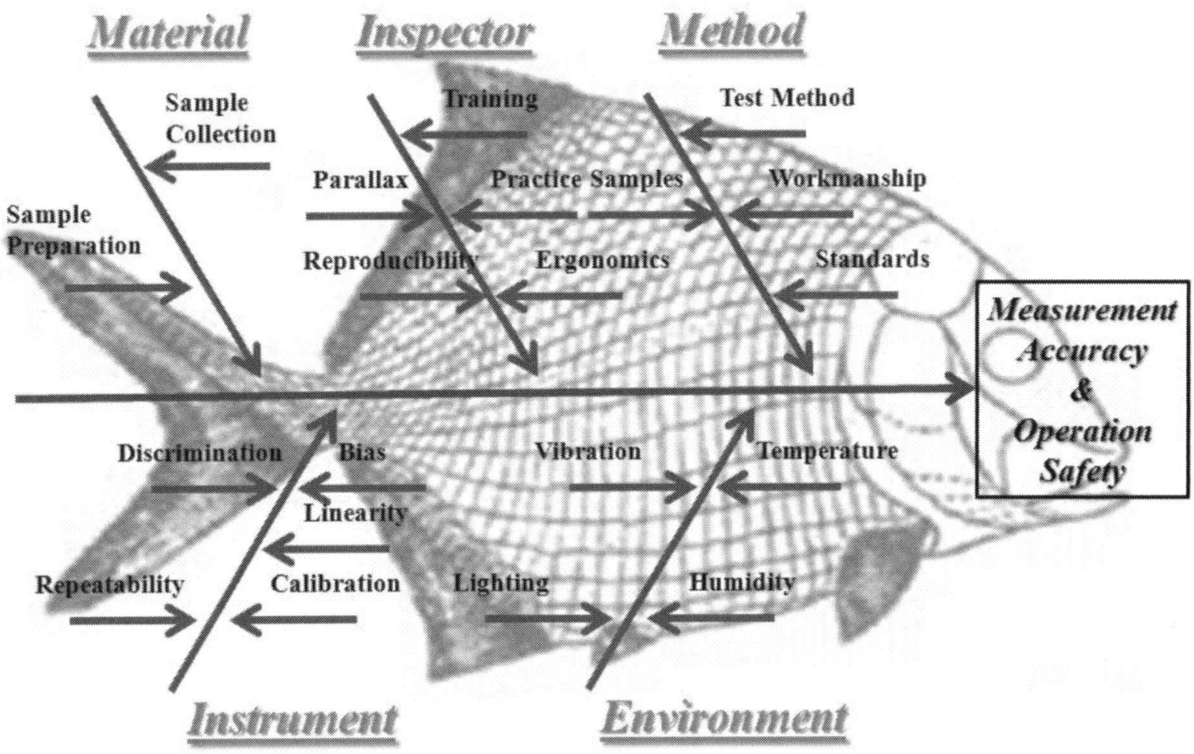

Figure 4: Fish Bone Diagram for HAZAP Study for Laboratory Operation in the Semiconductor Industry

Next, these practices can form a closed PDCA (Plan, Do, Check and Act) loop for continuous improvement of HAZOP activity. The close loop include four phases: plan, do, check and act. In the first phase of planning, the procedures and schedules to conduct the study should be identified and the input data should be collected. Then, the deviation should be identified by the application of key words and the causes and consequences should also be determined in the second phase. Next, the improvement actions need to be followed up. It is better to restudy the necessity of current HAZOP system. After the verification of the above items, the study report needs to be generated. Then the activity of HAZOP study should be evaluated to confirm whether the current scope, objective and responsibility need to be updated. Furthermore, the team should decide whether they need more team members to cover a larger scope.

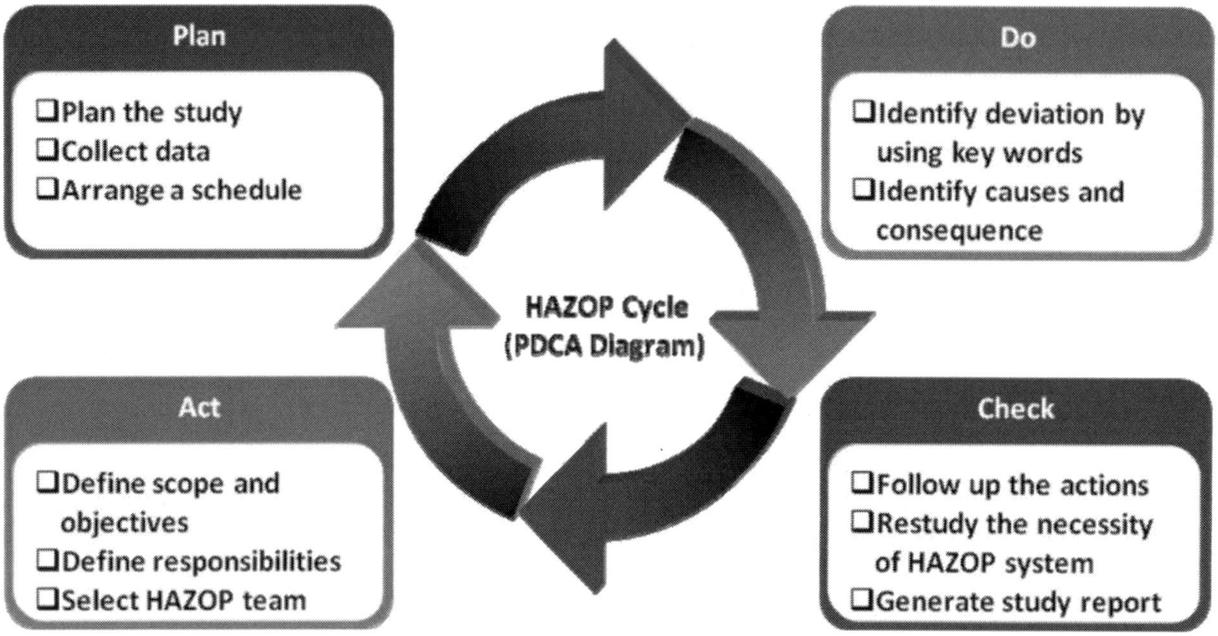

Figure 5: PDCA (Plan, Do, Check and Act) Diagram for HAZAOP Study for Laboratory Operation in the Semiconductor Industry

The analysis output can be regarded as the aid to safer laboratory design, operator training and real-time fault diagnosis. This will help to debottleneck and achieve operation safety and occupational health in the semiconductor industry.

4. Application Illustration

To demonstrate the efficacy of the developed systematic methodology, one case study has been conducted as the illustration of the methodology application in the real practice of plating bath monitoring of wafer bumping in the semiconductor industry. Wafer bumping is an advanced packaging technique that attaches solder spheres or gold bumps to the bond pads on a selected die. Gold bumping is generally used for Tape Automated Bond (TAB) assembly or Anisotropic Conductive Film (ACF) flip-chip assembly. Solder bumping is primarily used for reflow flip-chip assembly which unites the components face down, directly with the substrate or board through conductive bumps on the bond pads. These bumps deliver the electrical, mechanical and thermal interconnection which provides direct contact between the chip package and the device. Another process is called stud-bumping. Unlike gold and solder bumping, this process can be done at the die level. Wafer bumping also known as flip chip has traditionally been used for high-end niche applications. However in the last few years it has gone mainstream to include a wider range of consumer applications geared towards both the performance driven market and the form factor driven market.

Wafer bumping provides the electrical and mechanical connection between a die and substrate, which is one of the most critical elements of any flip chip package structure. Predominantly lead-tin and lead-free solders at the present time, these connections - or bumps - must exhibit superior adhesion to the die, minimal resistance, and result in high assembly yields. Solder bumps are formed by using either thin film metal deposition or ball loading techniques. Currently the advanced semiconductor assembly and test company such as Amkor offers state-of-the-art capability in electroplated solder technologies including the electroplating capability to support fine pitch bumping and copper pillar structures.

Wafer bumping utilizes thin film deposition, photolithography and plating techniques to place solder or gold onto IC (Integrated Circuit) lead pads. The wafer solder bump process is to produce the solder bumping of the wafer pad. Then it is going to melt the bumping by heat in assembly. This technology can substantially reduce IC size and have the advantages such as high density, low sensitivity, low cost and effective heat dissipation. This bumping process is suitable for flip chip assembly for LCD, memories, microprocessors, and microwave RF ICs applications. The process includes sputter UBM (under Bump Metallurgy), Photolithography, Plating and Etching etc., This bumping is suitable to apply on TAB, Flip Chip. The detailed process flow chart has been illustrated in below flow chart.

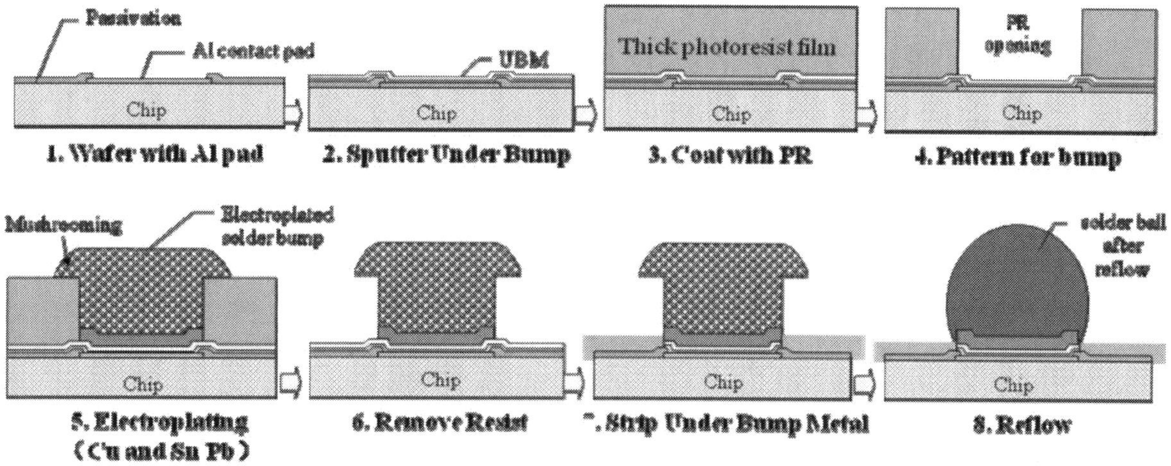

Figure 6: Process Flow Chart for Wafer Bumping Process

One of the key test item for wafer bumping operation is the electroplating bath monitoring and composition control of solder materials and additives. Plating solutions must be routinely analyzed in order to maintain the recommended bath formulation and to preempt the occurrence of problems related to improper levels of bath constituents. Contaminant levels in the solutions must also be monitored. Manufacturers of plating systems establish optimum specifications to ensure maximum solution efficiency and uniformity of deposits. The various factors that cause the concentrations of bath constituents to deviate from their optimum values are as follows: 1. drag-out; 2. solution evaporation; 3. chemical decomposition; and 4. unequal anode and cathode efficiencies. Regular analysis of plating solution becomes the responsibility of process engineering and quality control group.

The techniques employed for the quantitative analysis of plating solutions are classified as volumetric (titrimetric), gravimetric, and instrumental. Volumetric and gravimetric methods are also known as "wet" methods. The analyst must select the method that is best suited and most cost effective for a particular application. The wet methods outlined here are simple, accurate, and rapid enough for practically all plating process control. They require only the common analytical equipment found in the laboratory, and the instructions are sufficiently detailed for an average technician to follow without any difficulty. The determination of small amounts of

impurities and uncommon metals should be referred to a competent laboratory, as a high degree of skill and chemical knowledge are required for the determination of these constituents. Instrumental methods differ from wet methods in that they measure a physical property related to the composition of a substance, whereas wet methods rely on chemical reactions. The selection of an instrument for the analysis of plating solutions is a difficult task. Analysts must decide if the cost is justified and if the analytical instrument is capable of analyzing for the required substances with a high degree of accuracy and precision. Instruments coupled to computers can automatically sample, analyze, and record results. Mathematical errors are minimized and sample measurements are more reproducible than with wet methods. Instrumental methods are also extremely rapid when compared with wet methods. Analytical instruments frequently used in the analysis of plating solutions can be categorized as spectroscopic, photometric, chromatographic, and electro-analytical. Spectroscopic methods (flame photometry, emission spectrometry, X-ray fluorescence, mass spectrometry, and inductively coupled plasma) are based on the emission of light. Photometric methods (spectrophotometry, colorimetry, and atomic absorption) are based on the absorption of light. Chromatographic methods (ion chromatography) involve the separation of substances for subsequent identification. Electro-analytical methods (potentiometry, conductometry, polarography, amperometry, and electrogravimetry) involve an electric current in the course of the analysis. The instrumental methods, comprehensively reviewed below, are most applicable to plating environments.
Here, the manual titration procedures to determine copper concentration in the sample plating bath is used as illustration because the manual titration will have more potential safety concerns compared with the instrument analysis methods. Below is the procedure of the test.

PROCEDURE:

1. Preparation of EDTA (Ethylenediaminetetraacetic Acid) (0.01M). Dry EDTA for 2 hours at 80°C. GMW = 372.2., or check the reagent bottle. Dissolve needed grams in a 500 ml vol. flask. Store in a conditioned polyethylene container. EDTA solutions attack glass and deteriorate after 1 or 2 days.
2. Preparation of known copper solution. Place 0.2 g to 0.25 g of pure Cu. wire (or shot) (weighed on an analytical balance) in a 100 ml beaker. Add 15 ml conc. HNO3 and 1 ml conc. H2SO4. Place a watch glass over the beaker. Boil until dense white fumes of SO3 cease. Transfer carefully to a 200 ml vol. flask. Remember to rinse the watch glass. Dilute to the mark.
3. Preparation of unknown copper solution. Weigh 0.5 to 0.6 grams of unknown into a 200 ml vol. flask. Add 30 ml of 1:5 nitric acid. Heat until all particles are dissolved. Dilute to the mark.
4. Titration with EDTA - murexide indicator. Do this step for 4 samples of known and 4 samples of unknown.

KNOWN: Place 10 ml of Cu known solution (step 2) in 250 ml Erlenmeyer flask. Add pH buffer drop-wise until the solution turns completely blue and the precipitate first formed dissolves. Add about 100 ml. Add more buffer if solution becomes cloudy. Add 1 tablet of murexide indicator and titrate with EDTA until color changes from yellow (green) to deep violet. If the volume of EDTA is less than 10 ml, repeat using 25 ml liquid of Soln. 2.

UNKNOWN: Place 25 ml of Cu unknown solution (step 3) in 250 ml Erlenmeyer flask. Add pH buffer drop-wise until the solution turns completely blue and the precipitate first formed dissolves. Add 1 tablet of murexide indicator and titrate with EDTA until color changes from yellow (green) to deep violet. If the vol. of EDTA solution is less than 10 ml. use a larger liquid of Soln. 3.
Remarks: The copper-murexide complex is pure yellow, but the color at the starting point of the titration is greenish depending on the amount of copper present. The green is the mixture of yellow (indicator complex) and blue (copper-tetramine complex). As the titration proceeds the tetramine complex is transformed to the less intensely colored EDTA complex and the yellow of the indicator complex becomes permanent.
CALCULATION: Calculate the molarity of the EDTA. Calculate percent copper in the sample using the value of the Molarity of the EDTA found. Include the spread sheet for the calculation.

Based on the available procedures to determine the copper concentration in the sample solution of plating bath, HAZOP factors can be expanded so that the second layer of factors will be included into HAZOP matrix system: for example, sample collection and sample preparation will be considered in the material part; training, parallax, reproducibility, practice and ergonomics will be considered in the manpower part; samples, test methods, workmanship and standards will be considered in the method part; discriminations, repeatability, bias, linearity and calibration will be considered in the instrument part; lighting, temperature and humidity will be considered in the method part. These factors are listed as the column. Then, the HAZOP guide words will be listed as the row. The related information of the deviation stored in LOD(Laboratory Operation Database) and HPD(HAZOP Project Database) can be selected into the blank spaces from the drop-down list box in the above expanded table. Next, the data record of regular monitor results will be proceeded into data analysis to validate the impact of the above deviation . The data analysis result will couple with the expanded HAZOP table to quantitatively identify the potential deviation of operation safety and test accuracy.

5. Conclusion

The laboratory operation in the semiconductor industry will have more possibilities of manual handling for chemical analysis, reliability test, material characterization and failure analysis. Therefore higher attention should be paid to the laboratory operations. As illustrated in this paper, the methodology of HAZOP study procedure can be applied to identify the potential deviations in both experiment accuracy and operation hazards. Multiple layers of factor consideration by the application of 4M1E (Man, Machine, Material, Method and Environment) will help the brainstorm to give the aid to safer laboratory design, operator training and real-time fault diagnosis. This will help to debottleneck and achieve operation safety and occupational health in the laboratory operation semiconductor industry.

6. Reference

[1] Chepesiuk, R., "Where the Chips Fall: Environmental Health in the Semiconductor Industry." Environmental Health Perspectives, Volume 107, pp.452-457, 1999

[2] Swann, C. D., & Preston, M. L., "Twenty-five years of HAZOPs", Journal of Loss Prevention in the Process Industries, Volume 8, No 6, pp.349-353, 1995

[3] Amkor Technology Wafer Bumping Service Introduction, http://www.amkor.com /go/turnkey-services/wafer-bumping

(accessed Jan. 5, 2013)

[4] Nolan, D.P., "Application of HAZOP and What-If Safety Reviews to the Petroleum, Petrochemical and Chemical Industries". William Andrew Publishing/Noyes, 1994

[5] Lawley, H. G., "Operability studies and hazard analysis", Chemical Engineering Progress, Volume 70, No 4, pp.45, AIChE,1974

[6] Chemical Industries Association., "A Guide to Hazard and Operability Studies", 1977

[7] IHS iSuppli Research., "Semiconductor Industry Revenue to Ensure Slow Growth in 2012",http://www.isuppli.com/Semiconductor-Value-Chain/News/Pages/Semiconductor-Industry-Revenue-to-Endure-Slow-Growth-in-2012.aspx

(accessed Jan. 5, 2013)

[8] U.S. Bureau of Labor Statistics (BLS) Data Report,2011

[9] Semiconductor Industry Association, http://www.sia-online.org/

(accessed Jan. 5, 2013)

[10] Bolmen, Richard A., "Semiconductor Safety Handbook: Safety and Health in the Semiconductor Industry", Noyes Publications, 1998

Best Practice Management of Change & PSSR Process

Michael Bearrow
Optimized Systems and Solutions Inc.
8303 McHard Road, Suite 200, Houston, TX 77053
Michael.E.Bearrow@o-sys.com

Prepared for Presentation at
American Institute of Chemical Engineers
2013 Spring Meeting
9th Global Congress on Process Safety
San Antonio, Texas
April 28 – May 1, 2013

UNPUBLISHED

Best Practice Management of Change & PSSR Process

Michael Bearrow
Optimized Systems and Solutions Inc.
8303 McHard Road, Suite 200, Houston, TX 77053
Michael.E.Bearrow@o-sys.com

Keywords: MOC, process safety code, recommendation,

Management of Change (MOC) and Pre Start-up Safety Review (PSSR) are still the most challenging elements of OSHA's Process Safety Management (PSM) standard, the EPA's Risk Management Program (RMP) rule, and now, the US Department of Interior's Bureau of Safety and Environmental Enforcement's Safety and Environmental Management Systems (SEMS). Even though the PSM standard has been around since 1992 and the industry has been managing change for several decades, we still can get it wrong, sometimes with disastrous results. The mindful or diligent efforts of many working in concert are necessary to ensure that change is identified, analyzed and executed in a quality way. Just when we get it right on paper and get the workforce up-to-speed, we have employee turnover, neglect and sometimes regulatory change. New actors and a constantly changing script make it hard to manage change efficiently and effectively. This paper discusses how the chemical process industry has defined the MOC and PSSR best management practices and how they should be automated.

Management of Change (MOC) and Pre Start-up Safety Review (PSSR) are still the most challenging elements of OSHA's Process Safety Management (PSM) standard, the EPA's Risk Management Program (RMP) rule, and now, the US Department of Interior's Bureau of Safety and Environmental Enforcement's Safety and Environmental Management Systems (SEMS). Even though the PSM standard has been around since 1992 and the industry has been managing change for several decades, we still can get it wrong, sometimes with disastrous results. The mindful or diligent efforts of many working in concert are necessary to ensure that change is identified, analyzed and executed in a quality way. Just when we get it right on paper and get the workforce up-to-speed, we have employee turnover, neglect and sometimes regulatory change. New actors and a constantly changing script make it hard to manage change efficiently and effectively.

Many high-profile companies have addressed this challenge by creating and automating standard processes across their manufacturing or production facilities. MOC and PSSR are two of these major processes that have been successfully automated. Technology has been used to automate each step of the MOC workflow, enabling the collection of change ideas, formalizing the analysis of the ideas, and documenting the approval and execution of the change. Reports and dashboards can provide visibility of the health or current state of your MOC and PSSR processes. The use of software can enable the standardization of best practices. This paper discusses how the chemical process industry has defined the MOC and PSSR best management practices and how they should be automated.

What is a best practice?

"Best Practice" is an overused buzzword, like so many, and they are hard to define. Similar terms include Industry Standard, Best of Breed, Lesson Learned, Core competency, and Customer-centric, Downsizing, Drinking the Kool-Aid, Eating your own dogfood, Granular, Herding cats, Holistic, Low Hanging Fruit, Mindshare, Mission Critical, Management Visibility, Pain point, Return on Investment, Seamless, Touch point, Value-add and Visibility. Fortunately, there is a working definition of a best practice.

A **best practice** is a method, process or technique that consistently shows results superior to those achieved by other means. It is hard to prove that you have developed a standard process--like MOC--that enables your company, facility or organization to achieve repeatable superior performance, but it has been done.

Some other terms synonymous with best practices are good operating practice, good agricultural practices, good manufacturing practice, good laboratory practice, good clinical practice, good distribution practice, and **R**ecognized **A**nd **G**enerally **A**ccepted **G**ood **E**ngineering **P**ractices (REGAGEP). In the chemical process industry, the PSM standard relies on REGAGEP to ensure facilities adopt best practices. REGAGEP is specified in federal statutes from OSHA and EPA, and is also mentioned in the American Chemistry Council (ACC) Responsible Care® Process Safety Code.

How do you identify a Best Practice?

Everyone would like to be known as the best practice company. Even if you did not create the best practice, identifying them and using them can achieve the same results. If you can tick the box next to these five requirements, then you may have a best practice:

1. Gives a company or facility a tangible competitive advantage.
2. Takes advantage of technology.
3. Improves EHS performance, product quality, and also lowers cost.
4. Gives management more visibility, control and influence over outcomes.
5. Must be a working system; i.e. not a theoretical best practice

The benefits of having such a best practice might separate your company from your competition or keep you from costly business disruption, fines and/or law suits. Many of the best practices have common themes, including: ensuring consistency; standardization; simplifying processes; zero redundancies; leveraging technology and web-based software solutions; and single source for all data (one source of truth). It may be difficult for you to prove you have adopted/created a best practice. However, it is easy to identify a best practice when a confluence of large and small companies adopt the same technique or process and are successful.

In the context of this paper, we are focused on best practice MOC and PSSR automation. Now that we know what a best practice is, let's define change and discuss how to manage it.

What is Change & Why Do We Manage it?

"Change" means to transform or cause to be different. In the chemical process industry, change is omnipresent and required to stay competitive. William Edwards Deming, a famous American statistician, professor, author, lecturer and consultant, once said, "It is not necessary to change. Survival is not mandatory." In order to survive, a company must learn, change and continuously improve to survive. Successful change requires you to learn from your experience and the experience of your peers and make changes that will improve your competitiveness.

Change is a normal part of business, but uncontrolled or poorly managed change has resulted in some of the worst disasters in the chemical process, oil and gas and refining businesses. According to Ian Sutton, a noted PSM expert, *"The root cause of all accidents is uncontrolled change. Leaving aside sabotage and other malicious acts, all industrial facilities are designed and operated to be safe, clean and profitable - yet incidents continue to occur. In every case, the fundamental cause of the incident is that someone, somewhere lost control of the operation, i.e., they allowed operating conditions to deviate beyond their safe range. Hence, the proper management of change is the foundation of all safety and accident prevention programs; an effective Management of Change (MOC) creates an atmosphere of "no surprises".*

Whether it is Bhopal, Piper Alpha or Deepwater Horizon, time after time the difficulty of managing change confounds some of the most proven processes of the most successful companies in the world. As a consequence of these disasters and hundreds more, we have rules, standards, regulations and laws to help ensure that companies manage change better. Guidance documents on managing change include the following sources:

- *US EPA's Risk Management Programs for Chemical Accident Prevention (40 CFR PART 68) or EPA RMP*
- *US OSHA's Process Safety Management of Highly Hazardous Chemicals standard (29 CFR OSHA 1910.119) or PSM,*
- *US Department of Interior's Bureau or Safety and Environmental Enforcement (BSSE) - 30 CFR Part 250 Subpart S - Safety and Environmental Management Systems (SEMS),*
- *OHSAS 18001, 2007 revision - ANSI/AIHA Z10-2005.*
- *European Union Law Seveso II Directive (see Annex III of EU Council Directive 96/82/EC).*
- *API-691 - Risk Based Machinery Management (RBMM) (in development)*

What kinds of change are there?

All ideas for improvement require change. No one ever suggests that we should do exactly what we are doing today, tomorrow. This fact is lost on a segment of the chemical process industry who do not think of recommendations as changes. Recommendations should be treated as "in-kind" changes because they are. They do not have to follow a formal and rigorous MOC and PSSR process, but they are certainly recommendation for change.

All change can be separated into three categories: corrective actions; preventive actions; and continuous improvements. Corrective actions are meant to fix something that is broken, or at least to improve efficiency and functionality. These ideas for change originate from accident investigations, incidents and negative operating experiences. Preventive actions can also come from incidents, but are more commonly created from near misses, safety observations, audits, hazard reviews, inspections, emergency response critiques, and/or general surveillance or observations. These ideas for change help ensure that we address upstream issues or causes before a negative outcome (incident, fine, loss) results. The final kind of change is continuous improvement. These are generally ideas for increasing production, reducing manufacturing costs or increasing product quality.

No matter what kinds of changes you have, they all should be managed. No matter how good an idea is, it must be analyzed for impact (positive and negative) and executed expertly. Ideas can also be rejected or denied if the risk or cost of implementation exceeds the benefits or value of the change. If the change affects the design and/or operation of a facility, it is likely going to require formal change management or MOC. The diagram below illustrates three kinds of change in the idea universe.

Figure 1.0

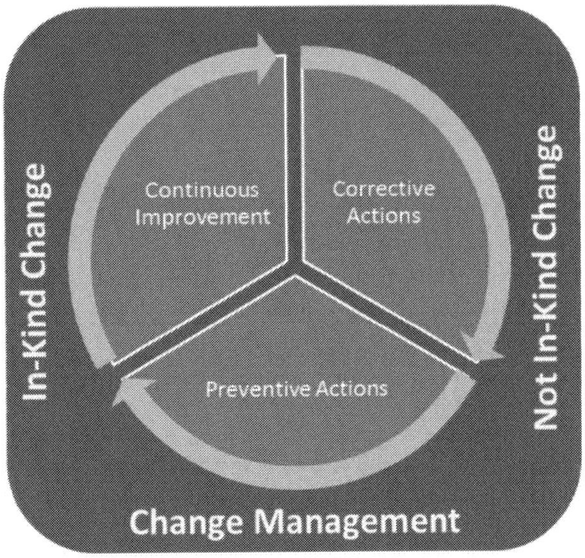

OSHA'S Two Categories of Change

The common way to refer to an idea that requires formal change management as defined by OSHA PSM or EPA RMP is an MOC or Management of Change (Not in Kind). MOCs are managed for changes to equipment, materials, technology, personnel, processes, procedures or anything that deviates from the original design, processes, operating procedures, or maintenance program.

In the chemical process industry, we refer to two categories of change. These are "Replacement In-Kind" and "Replacement Not In-Kind". These odd-sounding categories can have ambiguous meanings, but are explained by Ian Sutton, a noted process management expert, as follows:

"If an equipment item is to be replaced with one that is functionally identical, i.e., if the new item is built to the same specification as the old one, then the change is "in-kind". Otherwise it is "not-in-kind", and the MOC process has to be followed before the change can be implemented."

... The most challenging aspect of managing change is identifying that the proposed modification is in fact a change.

1. *Management of Change is needed if the change is Not-In-Kind.*
2. *A Not-In-Kind change is one where Management of Change is needed.*

In other words, the terms "Management of Change" and "Not-in-Kind" tend to be defined in terms of one another.

The second difficulty to do with the In-Kind/Not-In-Kind decision, noted above, is that all changes are, when analyzed deeply enough, not-in-kind."

What is the role of PSSR in the MOC process?

Pre-Startup Safety Review, or PSSR, is not actually part of the MOC process per se. It is intended to be a redundant check or review to ensure that an MOC has been implemented in a quality and safe manner. The PSM standard states:

The employer shall perform a pre-startup safety review for new facilities and for modified facilities when the modification is significant enough to require a change in the process safety information. Process safety information includes all information about the process. That covers about everything. The pre-startup safety review shall confirm that prior to the introduction of highly hazardous chemicals to a process:

- *Construction and equipment is in accordance with design specifications;*
- *Safety, operating, maintenance, and emergency procedures are in place and are adequate;*

- *For new facilities, a process hazard analysis has been performed and recommendations have been resolved or implemented before startup; and modified facilities meet the requirements contained in management of change, paragraph (l);*
- *Training of each employee involved in operating a process has been completed.*

When is Management of Change (MOC) Required?

OSHA requires the employer to establish and implement written procedures to manage changes (except for "replacements in kind") to process chemicals, technology, equipment, and procedures, as well as changes to facilities that affect a covered process. The MOC procedures shall assure that the following considerations are addressed prior to any change:

- The technical basis for the proposed change;
- Impact of change on safety and health;
- Modifications to operating procedures;
- Necessary time period for the change;
- Authorization requirements for the proposed change;
- Employees involved in operating a process and maintenance and contract employees whose job tasks will be affected by a change in the process shall be informed of, and trained in, the change prior to start-up of the process or affected part of the process;
- Updates to process safety information and operating procedures.

The industry has been practicing process safety for over 20 years. The definition of MOC and PSSR are well understood. Some companies have developed great procedures for addressing both of these elements in an integrated fashion using Internet-based web applications and powerful database engines. These software systems simplify the complex MOC process and enable standardization, ensuring a higher level rigor which is difficult to impose with paper-based execution. While the best practices for MOC and PSSR work practices continue to evolve, the following sections reveal what is currently accepted as best practice.

Best Practice MOC Part 1 – Collecting Change Ideas from the Idea Generators

As established above, continuous improvement is impossible without implementing change ideas that can come from anywhere. The problem is how can organizations empower their workforce to initiate ideas, capture and evaluate them then act on the ones that will improve the business? Can you imagine how successful your business would be if you could leverage all of the grey matter at your disposal to solve production problems?

In order to efficiently and effectively capture these ideas, we need to leverage technology. Using Internet–based software can make it easy for ideas to be collected at the source. The easier we make it for staff members to pass on their ideas, the more of these ideas we will receive with time to act on them. One best practice that has surfaced to address this need is to design an electronic suggestion or idea box. Web-based quick entry screens can be accessed using a browser on any company computer to easily suggest a change or improvement. Once these ideas are entered, they are automatically communicated to idea reviewers (supervisors) for vetting, categorization (in-kind or not in-kind) and prioritization. No matter what happens to these ideas, the reviewer must provide feedback to the idea initiators on what happened to their ideas. If an idea results in tangible savings or competitive advantage, a tangible reward will serve to prime the idea generator for the future.

You can visualize the best practice MOC process using a series of funnels (below). Changes begin as ideas and they can be entered by anyone in the company into a single quick entry screen and collected in the main funnel. The supervisor is automatically notified that a new idea has been entered for their unit or area. Each idea is meticulously reviewed and categorized as a recommendation or a MOC. Using electronic submission, this centralized collection funneling system is infinitely more efficient and effective than word-of-mouth, a paper system and/or multiple systems that collect and manage ideas. One common system for collecting, vetting, prioritizing and categorizing ideas is a best practice.

Figure 2.0

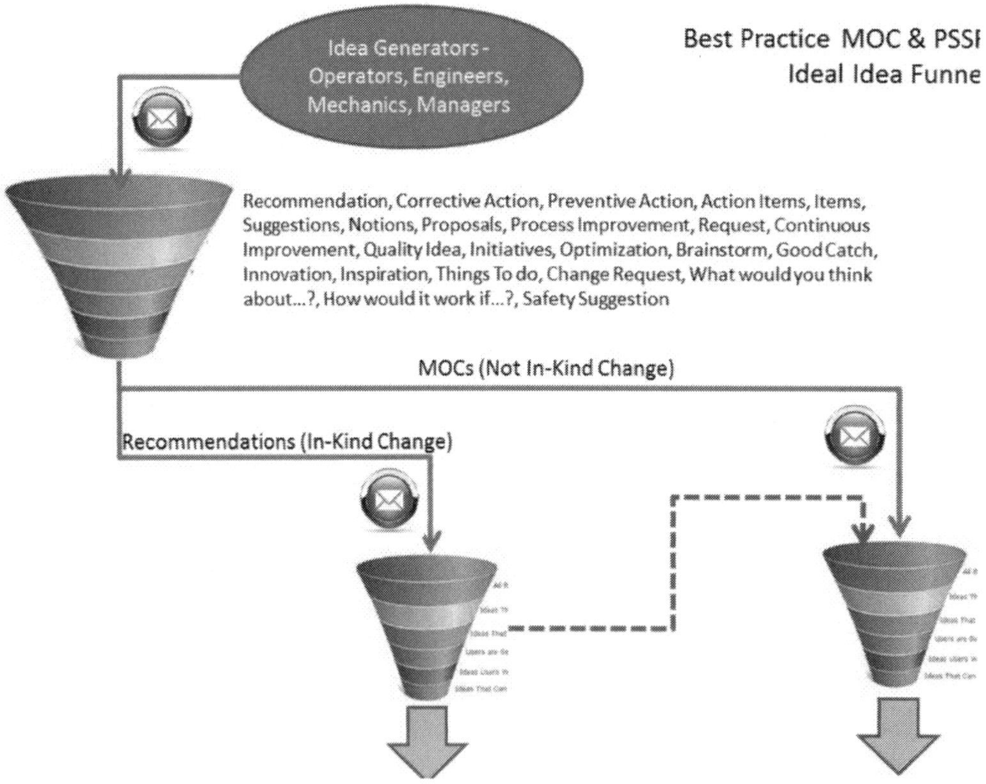

Best Practice MOC Part 2 – Managing the Recommendations (In-Kind Change)

Recommendations can come from audits, incidents, risk assessments, process hazard analysis, emergency response critiques, inspections, reviews, and surveillances. These recommendations are generally ideas for continuous improvement aimed at improving safety, environmental performance, quality or efficiency. Automating and standardizing the recommendation management process makes it repeatable, measurable and more readily understood. Utilizing software that illustrates the correct process and visually progresses users through the workflow is key to the acceptance and the eventual success of the MOC program.

Once the idea has been reviewed, assigned and categorized as a **recommendation**, or in-kind change, a simple workflow can be leveraged to ensure the idea is addressed in an efficient manner. A simple database can be used to document the steps in this process, but a best practice is to use software that illustrates the work process on screen as it is executed. The use of a visual workflow to navigate from step to step, showing progress with check marks or color-coding and enabling steps only when precursors (predecessor steps) are satisfied is a best practice. In addition, email notifications should be used to alert persons assigned to manage, approve and execute actions along the workflow. The more proactive the communication and visual the workflow is, the more intuitive it is to the end-user. The workflow presented below is used for hundreds of thousands of recommendations per year. The entire work process is automated from beginning to end.

Figure 3.0

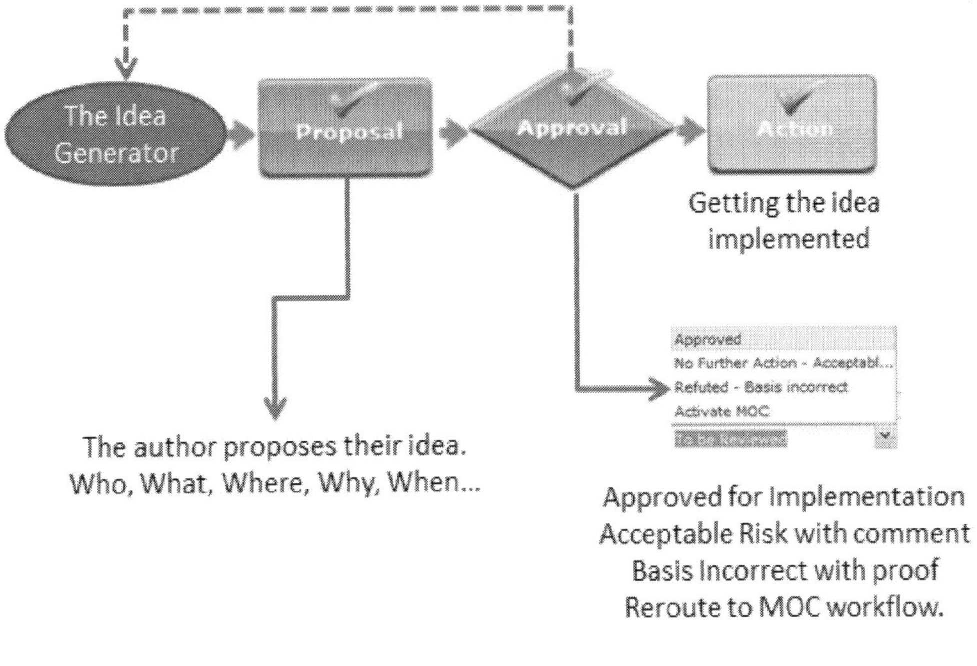

The workflow begins in the proposal step or in the "idea generator's" mind. The recommendation proposal must contain a title, a description (what needs to be done), a justification (why it is important) and a deadline (when it needs to be done) for getting this idea accomplished and a responsible person (RP). The reviewers of the proposal will ensure the accuracy and the resources are available to address the idea. The next step is to accept the idea into the system and automatically notify the RP. The acceptance is illustrated by a check or tick mark in the proposal icon.

After a review of the recommendation, the RP moves to the Approval step. It is in this step that they must decide whether to approve the idea, refute the idea, close the idea as an acceptable risk or decide it needs to be treated as an MOC. Approval of the idea will enable the creation of action items. Refuting a recommendation or addressing it as an acceptable risk should require proof or explanation. These comments must be documented for future reporting and analysis and automatically communicated to the idea generator.

Best Practice MOC Part 3 – Managing MOCs (Not In-Kind Change)

If the idea needs to be managed as a "Not in-Kind" change or MOC, a more complex workflow is required. Similar to the "In-Kind" change, it still requires a proposal, acceptance and assignment to a RP, but additionally will need a formal hazard review and planning. In almost all cases, a MOC also requires the pre-startup safety review (PSSR) step, a Startup Approval step and some post startup steps. Because of its complexity and the need to enforce a rigorous process, a "live" workflow showing what steps are necessary, which ones have been accomplished and which ones are left to do is essential. The workflow below is considered the best practice MOC and PSSR. There are no steps left out and no superfluous steps either.

Figure 4.0

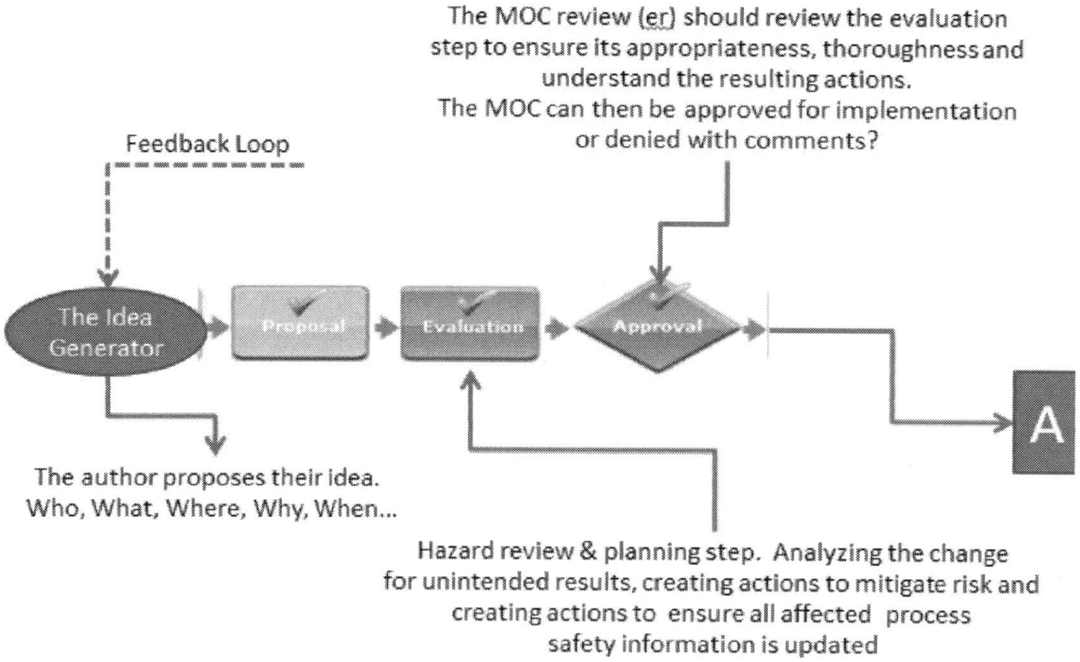

After the acceptance of the idea into the system, the RP receives a notification that they have been assigned a MOC to manage. The next step in the workflow is to "Evaluate" the MOC. This evaluation process is meant to ensure that the hazards of the change are reviewed, understood and mitigated, or at least a plan for mitigation is created. It is also meant to reveal the process safety information that will be affected and the training that will be necessary. Of vital importance is the creation of an action plan to make sure hazards are addressed and that appropriate training is done and all process safety information is completed. Just identifying the need falls very short of hitting the target. In fact, if you identify the need for action and then do not follow through, you have created the proverbial "smoking gun" and a red flag for auditors.

A smart checklist of questions should be presented to the RP in this step which are based on the category of the change. This eliminates the need to answer questions that have nothing to do with the change. Automation gives you the ability to tailor the evaluation process so that it is more effective and efficient to use. Asking evaluation questions unrelated to the change is very inefficient and can also make finding the right questions difficult. Even more automation can be leveraged to pre-assign reviewers to questions that only they should be answering and configuring automatic/predictable actions with associated questions. Here in this step we can make the comments and closure of all questions mandatory for moving to the next step. Until all questions are closed, we cannot move to the next step, the Approval step.

The Approval (to implement) step can be done by a person or groups of people. Their job is to look back at the MOC proposal, review the evaluation that was undertaken and judge whether it was done effectively by the right people. They should also take notice of the mitigating actions that resulted from that review. If more evaluation is required, the reviewer can loop back in the process and create more evaluation questions to satisfy. They should also ensure that actions have been tagged appropriately as pre- or post-startup. If the resulting actions or time have otherwise made this MOC a bad idea to execute, the MOC can be denied at this point with closure comments. Those comments are sent to the idea generator and should be documented in the system. If they are satisfied, then approval can be selected and we can begin implementing the MOC.

The MOC Action Step

Now the real work begins--updating procedures, marking up drawings, making sure all affected personnel are trained, writing work orders, getting an environmental permit, and all of the other actions required for completion to do a quality job. Once all actions (pre-startup) are completed, then you get a check mark over the action icon and we move on to the PSSR step.

Figure 5.0

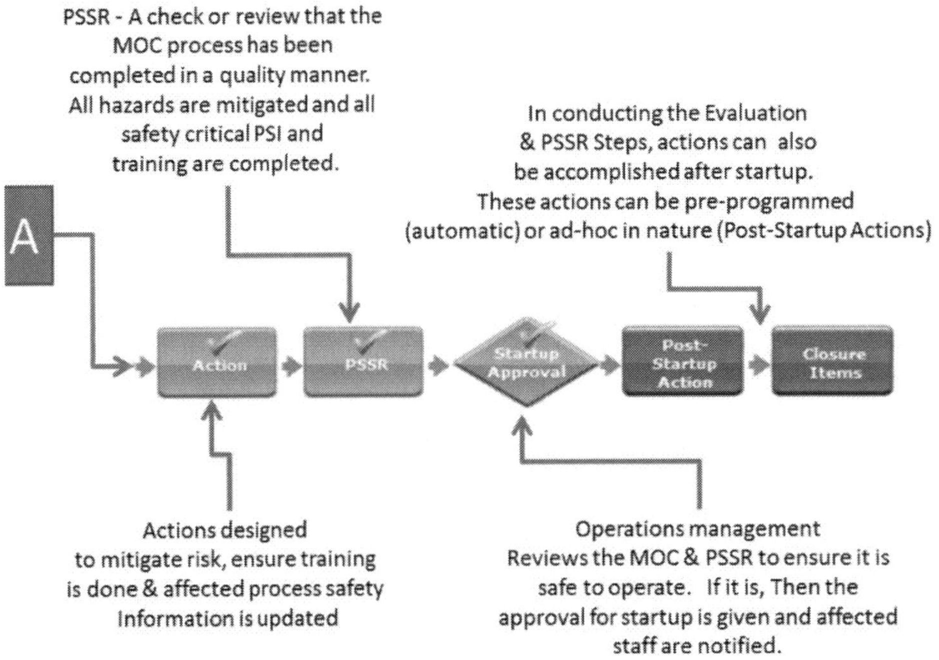

The PSSR Step

As previously stated, the PSSR is a redundant check to ensure the MOC has been properly executed. Like the Evaluation step, once we have chosen a category of change, automation can be used to ensure that the appropriate PSSR checklists/questions are presented to the RP. This increases the efficiency and effectiveness of this step as well. Once the PSSR is completed, we receive our check mark in that workflow step, and an automatic email is sent to the appropriate manager asking for permission to Startup (Startup Approval).

Startup approval is the formal acceptance of the change and the acceptance of the risk associated with it by operations or production. The person who completes this step should review all of the previous work steps in the workflow as well as physically review the work in the field.

The last two steps in the MOC workflow are closure items and post-startup actions. Although these are not referred to in any of the governing regulations, there will be non-safety related activities that are not required to be completed by startup and still need to be managed. This best practice will ensure these actions are accomplished and documented. Past Due email notifications are sent appropriately if the MOC becomes past due.

There are some distinct advantages to adopting the best practice MOC and PSRRs workflows discussed herein. As the best practice evolves, so must the users of the technology. Refresher training, upgrades and constant tuning and tweaking of the workflow are all part of continuous improvement of an MOC and PSSR program. This minimal investment is a small price to pay to avoid poorly managed changes or the opportunity cost of not collecting ideas from your idea generators.

How does Automation help?

Automation of the MOC and PSSR process helps in several ways. First, we can collect ideas at the source of the risk, whether it is the plant or on the platform. This is where the risk is and where continuous improvement is accomplished. Furthermore, it is also where most of the practical/achievable ideas for continuous improvement come from. The best practices must be useful and used where the money is made. Technology makes that possible.

Today's technology is perfectly suited for collecting, analyzing and sharing ideas. Accomplishing these actions without automation is nearly impossible. The days of paper-based MOCs are almost non-existent because automation makes these tasks more efficient and effective as well as less likely to be overlooked.

Reporting metrics and automatic alerts when MOC and PSSR actions are assigned, closed and past due ensure diligence and accountability. Technology can spread the news quickly to those who can create and execute a plan of action and do something with new ideas. Morse code, paper reports and snail mail do not get the job done like email communication and automatic reminders.

A central corrective and preventive action tracking software can help ensure that identified issues are dealt with in a safe and timely manner. Idea or recommendation management and action tracking can be accomplished with one database engine worldwide. That means one system to learn, maintain and continuously improve. It also means you have one global change management system to get things done, communicated and for generating reports. Managers can get the visibility they need to measure performance without bothering the people making the money.

An automated best practice MOC and PSSR system helps ensure that once you get it "right", you keep it "right". It also makes sure that change is appropriately identified, evaluated, approved and executed. It also makes poor change management easy to identify and correct through management surveillance.

Automation is not the goal; best practice is. Like reporting and management visibility and lots of other byproducts of this best practice, the goal is managing change efficiently and effectively.

Words of Caution

A cardinal sin of MOC & PSSR process automation is using a finance or maintenance technology that you already have to automate your best practice, because you already have it. This will sub-optimize a best practice faster than anything you can do, and it will likely defeat the purpose of the automation. Remember that 95% of your users are engineers, operators and mangers--not accountants, IT "geeks" or maintenance planners.

And finally, you can develop your own best practice MOC and PSSR, but to be successful as best practice software, you must have a full-time MOC best practice expert leading the development. Otherwise there will be several months of trial and error. Make no mistake; software development is difficult enough with unlimited resources, expert developers and patient end users. You cannot automate a business process you do not understand. Remember that managing change is a complicated management system for operations or production personnel. You cannot write a requirements document so thoroughly and clearly that a software development team can automate your MOC best practice workflow without constant oversight from the MOC experts.

9TH GLOBAL CONGRESS ON PROCESS SAFETY

Unifying Consequence Models for Natural Gas Pipelines in Petrobras Transpetro

Ana Paula Rodrigues Bahiense
PETROBRAS TRANSPORTE - TRANSPETRO
Av Presidente Vargas 328
Centro CEP 20091060
Natural Gas Division
Rio de Janeiro, Brazil
anapaulabahiense@petrobras.com.br

Maria Beatriz Winter Caputo
PETROBRAS TRANSPORTE - TRANSPETRO
Av Presidente Vargas 328
Centro CEP 20091060
Natural Gas Division
Rio de Janeiro, Brazil
Email: beatriz.caputo@petrobras.com.br

Prepared for Presentation at
American Institute of Chemical Engineers
2013 Spring Meeting
9th Global Congress on Process Safety
San Antonio, Texas
April 28 – May 1, 2013

UNPUBLISHED

Unifying Consequence Models for Natural Gas Pipelines in Petrobras Transpetro

Ana Paula Rodrigues Bahiense
PETROBRAS TRANSPORTE - TRANSPETRO
Av Presidente Vargas 328
Centro CEP 20091060
Natural Gas Division
Rio de Janeiro, Brazil
anapaulabahiense@petrobras.com.br

Maria Beatriz Winter Caputo
PETROBRAS TRANSPORTE - TRANSPETRO
Av Presidente Vargas 328
Centro CEP 20091060
Natural Gas Division
Rio de Janeiro, Brazil
Email: beatriz.caputo@petrobras.com.br

Keywords: pipeline, natural gas, consequence model

Abstract

Petrobras Transpetro has been contracting quantitative risk analysis for natural gas pipelines and their associated installations as a legal and internal requirement. These studies have been carried out by a series of consultants that estimate the magnitude of possible damage using different software available in the market. This fact led to a variety of results based on the premises adopted by consultants (most of them depending on the risk analyst) and differences regarding the peculiarities of mathematical models.

Petrobras Transpetro team, mainly field personnel, must use these data to carry out preventive actions inside vulnerable areas, and also to be aware of these ranges in case of a real emergency, thus providing evacuation and escaping actions, as necessary.

Based on this scenario, Petrobras Transpetro's Natural Gas Division had invested in purchasing a consequence-analysis software in order to model the effects of high-consequence loss events as an effort to unify consequence models for natural gas transportation pipelines at Petrobras Transpetro.

As per the above description, this paper aims to present results and benefits of the practical usage of unified consequence modeling information in risk management at Petrobras Transpetro's Natural Gas Division.

1. Introduction

Transpetro, a Petrobras wholly owned subsidiary, is the Brazilian leading logistics company dealing with transportation and storage of oil and byproducts, biofuels and natural gas. It operates 48 terminals (land & water), a fleet of 60 cargo ships, and a network of 15,000 kilometers of oil and gas pipelines interconnecting these facilities.) In addition, Petrobras Transpetro provides services to several distributing companies and to the petrochemical industry. Although the contribution of natural gas to the country's energy matrix is still comparatively low, it is increasing in accelerated rate. On the other hand, the extent of the gas transmission pipeline network has doubled in the last 4 years. Nowadays, Petrobras Transpetro's Natural Gas Division is responsible for operating and maintaining 7.327km of onshore natural gas pipelines, 12 Compression Stations and 129 City Gates, all of them remotely operated by the National Operation Control Center (CNCO), located in Rio de Janeiro.

In Brazil, State & Federal Environmental Protection Agencies require every installation to carry out a Quantitative Risk Analysis including consequence and vulnerability analysis, frequency estimation and societal and individual risk calculation. These Agencies establish risk tolerability criteria and have, therefore, the responsibility of issuing operational licenses based on the risk results.

Beyond that, ANP (Agência Nacional do Petróleo, Gás Natural e Biocombustíveis), the Federal Agency responsible for regulating the oil & gas industry in Brazil, also requires risk assessments to be implemented for onshore pipelines, which may be either qualitative or quantitative, depending on the installation's complexity.

Based on governmental requirements and on internal procedures, Petrobras Transpetro has carried out Quantitative Risk Analysis (QRA) of natural gas pipelines, conducted by different Consultants, in different periods of time. The first QRA is often carried out when the Company requests an Environmental License, and then it's updated based on a regular basis. The review cycle for these studies depends on internal procedures and the License validity.

The results of consequence and vulnerability analysis, which are part of QRA studies, are the basis for pipelines' Emergency Plans. These results are used basically to assess whether there is population inside the vulnerable area and if it's necessary to evacuate people.

In case of onshore pipelines, that cross several communities and points of interest, such as schools, churches, hospitals and others, the information related to vulnerable areas is vital to adopt more adequate emergency response procedures.

2. Methodology

2.1 Software

Petrobras Transpetro uses PHAST - Process Hazard Analysis Software Tools, by DNV Risk Management Software (DNV RMS), to perform consequence and vulnerability analysis. PHAST contains discharge, dispersion and effects models to assess major hazards, foreseeing complex consequences from possible releases of hazardous materials.

The algorithm for evaluation of jet fire scenarios considers product characteristics (liquid, gas or two-phase), release rate and duration, height of release, orifice diameter and angle of the formed

jet (0° for horizontal jets and 90° for vertical jets). The software estimates flame dimensions and intensity.
For fireball scenarios, the model calculates its radius and duration. The critical parameters to be defined for a fireball model are: product mass involved in the fireball; strength of the thermal radiation source; and the fraction of the total heat that is released as thermal radiation.
The model simulates flash fires due to ignition of a vapor cloud, on an integrated basis, to the algorithm of atmospheric dispersion. The flammable cloud's size is used to model a transient front of flames that travels through the cloud. Thermal radiation in the receptor is calculated as a transient function due to variation of the flames' sizes inside the cloud.
This algorithm considers the speed, width and height of the flame, and concentration of the flammable vapors of the cloud during calculation time. The distance which corresponds to the Lower Flammable Limit (LFL) is assumed as the flame length. Result is calculated in terms of the distance reached by the flash fire.

2.2 *Premises*

2.2.1 Accidental Scenarios

Failure of a high-pressure natural gas pipeline can lead to various outcomes, some of which can pose a significant threat to people and property in the immediate vicinity of the failure location. For a given pipeline, the type of hazard, and the damage or injury potential associated with the hazard, will depend on the mode of line failure (i. e., leak vs. rupture), nature of the gas discharge (i.e., vertical vs. inclined jet, obstructed vs. unobstructed jet) and the time to ignition (i.e., immediate vs. delayed). The various possible outcomes are summarized in Figure 1. In mathematical simulations, methane was used as a representative for natural gas.

Figure 1: Possible outcomes from failure of a high-pressure natural gas pipeline

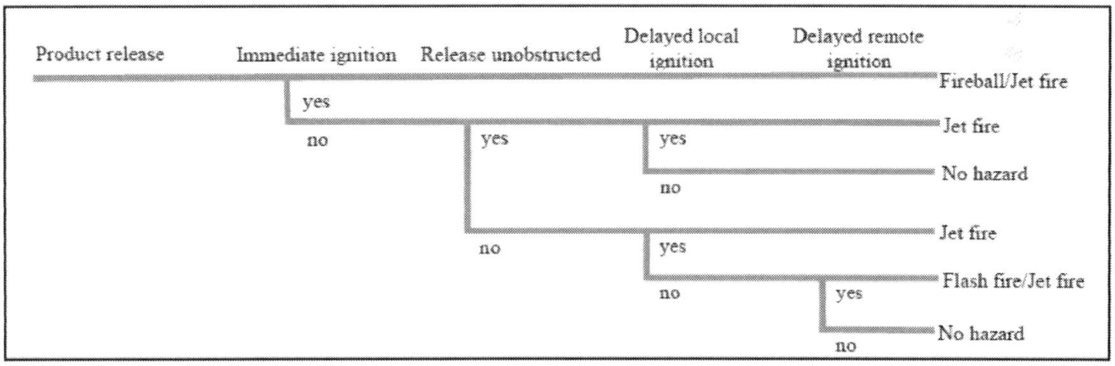

Although there is a great variety of outcomes that can result from failure of a high-pressure natural gas pipeline, the present study aims to analyze thermal effects due to jet fire.

2.2.2 Typical Pipelines

A set of 11 typical natural gas transportation pipelines was established to estimate physical effects due to natural gas possible scenarios. Typical pipelines considered in this study are listed on Table 1.

Table 1. Typical pipelines considered in this study

Pipeline	Pipeline Diameter	Operational Pressure
Pipe 1	8 in	70 bar
Pipe 2	10 in	100 bar
Pipe 3	12 in	100 bar
Pipe 4	14 in	100 bar
Pipe 5	18 in	100 bar
Pipe 6	20 in	120 bar
Pipe 7	22 in	74 bar
Pipe 8	26 in	100 bar
Pipe 9	28 in	100 bar
Pipe 10	38 in	100 bar

2.2.3 Meteorological data

The meteorological data considered in the mathematical simulations are presented in Table 2.

Table 2. Meteorological Data

Parameter	Day Condition	Night Condition
Ambient Temperature	25°C	20°C
Wind Speed	3 m/s	2 m/s
Atmospheric Stability classification*	C	E
Humidity	80%	80%
Soil Temperature	30°C	20°C

* Pasquill-Gifford Stability Categories

2.2.4 Physical Effects Considered

The Eisenberg Vulnerability Model, developed by the United States Coast Guard, was used for calculation of the exposed community vulnerability. This model uses probabilistic functions of the PROBIT type. It is a widely used approach to relate the phenomena intensity (like fire heat radiation) to their consequences.

Probit is a number related to the probability of damages, injuries or fatalities, expressed through the percentage of the population or physical installations affected.

It is relevant to emphasize that, conservatively, the vulnerable area corresponds to the area that could be affected in case of an accident scenario. The area predominantly affected will be influenced by local climatic and geographical conditions at the instant of the leak, such as interference by wind direction and speed, and may vary from person to person and equipment to equipment.

The following levels of interest were considered in this study:

1. Thermal radiation level of $4 kW/m^2$ is frequently suggested in specialized literature as the pain limit after exposure of 30 seconds;
2. Thermal radiation level of $12.5 kW/m^2$, which represents the limit to cause deaths - 1% fatality probability for an exposure time of 30 seconds;

The effects of thermal radiation arising from eventual jet fires were estimated for a height of one meter, understanding that this is a representative height for evaluating physical effects on people.

2.2.5 CETESB Requirements

CETESB is a State Environmental Protection Agency, from São Paulo State, publisher of the Quantitative Risk Analysis Manual: "Accidental risk from technological origin. Method for decision and reference terms", extensively used in Brazil, which establishes most of the premises adopted in this case study.

Although CETESB is a "local" Agency, it's considered as a benchmark all over the country. So, most of State Environmental Protection Agencies base their rules on CETESB requirements, including IBAMA, the National Environmental Protection Agency.

Premises based on CETESB Manual and used in this study are listed below:

- Simulations were carried out considering only pipeline buried stretches, calculating the effects of angular (45°) jet fires;
- Simulations also considered pipe rupture, considering the leak hole as 100% of pipe diameter.

3. Results

Results are presented in Table 3.

Table 3. Reaches (m) for Jet Fires Scenarios

Pipeline	Pipeline Diameter	Operational Pressure	Jet Fire Reaches (m)			
			Daily (kW/m^2)		Nightly (kW/m^2)	
			12,5	4	12,5	4
Pipe 1	8 in	70 bar	154	245	151	252
Pipe 2	10 in	100 bar	228	360	227	369
Pipe 3	12 in	100 bar	271	425	271	436
Pipe 4	14 in	100 bar	312	490	315	503
Pipe 5	18 in	100 bar	395	616	400	634
Pipe 6	20 in	120 bar	433	675	440	695
Pipe 7	22 in	74 bar	477	743	486	766
Pipe 8	26 in	100 bar	554	860	565	887
Pipe 9	28 in	100 bar	592	919	606	949
Pipe 10	38 in	100 bar	782	1209	803	1251

4. Conclusion

The results presented in item 3 show that the higher operational pressure and hole diameter, the higher would be the reaches of an accidental scenario of jet fire. The reaches shown in Table 3 could not be considered as an absolute value, instead it should be considered as a reference for the decision making process.

All models, including consequence models, present uncertainties. Input parameters that are modeled using the consequence analysis software may, and probably will, be different in an eventual emergency real case. Therefore, the results presented should be used as guidance to the emergency response team, so that adequate response is provided considering possible effects of

a jet fire on the pipe neighborhood, such as community evacuation, roads blocking, among others.
This study may be completed by simulating other possible outcomes, such as flash fires and fireballs, and by calculating thermal radiation levels.
As originally intended, the results present unified consequence modeling for jet fire scenarios involving natural gas pipelines, providing a quick reference to response preparedness and field personnel use in decision making process, in case of eventual emergencies.

5. References

[1] Manual de Orientação para a Elaboração de Estudos de Análise de Riscos, P4.261, CETESB, May 2003.

Additional references not cited:

[2] 8th Report of the European Gas pipeline Incident data Group - EGIG, December 2011.

[3] AMERICAN INSTITUTE OF CHEMICAL ENGINEERS (AICHE), 2000, "Guidelines for Chemical Process Quantitative Risk Analysis", New York, USA, 2nd Edition.

[4] "A Model for Sizing High Consequence Areas Associated with Natural Gas Pipelines", Gas Research Institute (GRI), 2000.

9TH GLOBAL CONGRESS ON PROCESS SAFETY

Automation Strategies: Avoiding Accidents During the Process of Draining Water From LPG Spheres

Carlos André Vaz Junior
Departamento de Engenharia Química, Escola de Química, Universidade Federal do Rio de Janeiro
Centro de Tecnologia, Bloco E, Sala E-209, Ilha do Fundão
cavazjunior@eq.ufrj.br

Maurício Bezerra de Souza Junior
Departamento de Engenharia Química, Escola de Química, Universidade Federal do Rio de Janeiro
Centro de Tecnologia, Bloco E, Sala E-209, Ilha do Fundão
mbsj@eq.ufrj.br

Diego da Silva Lione
Escola de Química, Universidade Federal do Rio de Janeiro
Centro de Tecnologia, Bloco E, Sala E-209, Ilha do Fundão
diegolione@ig.com.br

João Marcio Almeida Villaça
Escola de Química, Universidade Federal do Rio de Janeiro
Centro de Tecnologia, Bloco E, Sala E-209, Ilha do Fundão
joaovillaca@hotmail.com

Leonardo Salim Damian Aguiar
Escola de Química, Universidade Federal do Rio de Janeiro
Centro de Tecnologia, Bloco E, Sala E-209, Ilha do Fundão
leonardosdaguiar@gmail.com

Prepared for Presentation at
American Institute of Chemical Engineers
2013 Spring Meeting
9th Global Congress on Process Safety
San Antonio, Texas
April 28 – May 1, 2013

UNPUBLISHED

Automation Strategies: Avoiding Accidents During the Process of Draining Water From LPG Spheres

Carlos André Vaz Junior
Departamento de Engenharia Química, Escola de Química, Universidade Federal do Rio de Janeiro
Centro de Tecnologia, Bloco E, Sala E-209, Ilha do Fundão
cavazjunior@eq.ufrj.br

Maurício Bezerra de Souza Junior
Departamento de Engenharia Química, Escola de Química, UFRJ

Diego da Silva Lione
Escola de Química, UFRJ

João Marcio Almeida Villaça
Escola de Química, UFRJ

Leonardo Salim Damian Aguiar
Escola de Química, UFRJ

Keywords: LPG spheres, automation strategies, process of draining water

Abstract

The present paper has as its main objective to evaluate different strategies of automation, focusing on the reduction of risks on the water drainage procedures in spheres of liquefied petroleum gas (LPG). The condensation phenomenon of the fraction of the water found in gases during the storing process is largely observed in chemical plants. The condensed water will be accumulated in the bottom of the storing sphere and needs to be removed through a drainage procedure. Although being considered an operational routine process, such event was the main cause of at least two severe accidents: in 1966, in France, and in 1972, at the Refinaria Duque de Caxias (Duque de Caxias Refinery), in Rio de Janeiro, Brazil. One of the most critical decisions viewing to avoid that a new accident was to interrupt the drainage before the water could get completely drained. When the drainage continuous after all the water has been removed, the LPG begins to leak. The gas leaking phenomenon promotes the formation of a toxic and inflammable cloud, *par et passu* with a strong reduction of the temperature. In order to analyze the system, mathematical simulations of the drainage were carried out using the UniSim@ Design (Honeywell) simulator. Variations of the different properties (parameters) of the system were studied along the drainage. Understanding the dynamic of the process it is possible to develop automation systems which can preview and anticipate the best moment for the interruption of the water draining process. Traditional control architectures were suggested based on the PID control architecture. The "monitoring processes", that had been proposed by Seborg et al. [2], were tested as well. The monitoring system was considered more efficient, avoiding that natural fluctuations would anticipate and/or retard the action of the controller.

1. Introduction

The present study has as its objective to propose and to evaluate different controlling strategies for water draining systems in LPG (Liquefied Petroleum Gas) spheres. This analysis obtained its stimulus from the study of the accident that occurred in the Refinaria Duque de Caxias/Duque de Caxias Refinery, in Rio de Janeiro, in 1972. Such accident caused the explosion of a LPG sphere, resulting in several fatal victims and a large number of wounded people. The sphere's water draining system consisted, at the time, of a manual opening of the valve localized in the lower part of the reservoir. The employee in charge of such operation should be waiting for the estimated period of time so that practically all the water would have been removed. After this time, the employee should close the valve, interrupting the drainage and avoiding the gas leak. During the whole time of the drainage, the employee should stay in the area, following the process. In the accident of 1972, the employee left the place, not returning for the closing of the valve, causing the LPG leak which formed an inflammable cloud and the freezing of the equipment nearby. The low temperature generated by the liquefied gas leak promoted the freezing of the drainage valve, not allowing its closing. After the accident, the sphere's drainage system received several improvements, such as the inclusion of the blockage double system and safety instruments, even though the drainage procedures continued to be carried out essentially in a manual way.

Proposing to offer a more efficient and safe solutions for the drainage procedures of the water, this study used dynamic simulation. Through the simulator were made several drainage simulations, evaluating the behavior of several parameters of the process. Therefore, the automation strategy was based on the fact that the dynamics of the variables of the process can anticipate under a safe and exact way the ideal moment in which the end of the drainage should happen.

2. The Accident

The description made here is based on statements of the professionals that were working in the location at the time of the accident. The accident occurred in 1972, in the park of the LPG spheres of the Duque de Caxias Refinery, in Rio de Janeiro, Brazil. The storage of the LPG occurs under elevated pressure, allowing the maintenance of the gas in its liquefied state. Nevertheless, the elevated pressure also generates the formation of a condensed water layer in the bottom of the sphere. The valve and pipes localized in the lower part of the sphere are used for the water drainage. To remove the water is fundamental to ensure that the final product will meet the required specifications when in use. At the time of the accident, the space where the drainage pipes sat had a sole manual valve.

The conventional procedure to be carried out in 1972 by the employee on duty was to connect the hose to the drainage pipes. This hose was then positioned in a nearby ditch. The server would return to the place in order to open the valve. The drained water would be routed this way to the ditch. As the amount of water formed during the compression is known, as well as the flow of drained water, it is possible to estimate the necessary time for the water to be removed from the sphere.

According to instructions being followed at the time, it was expected that the employee on duty would stay in the location, following the process. Nevertheless, it is said that on the day of the accident the operator was absent at the exact moment of the end of the drainage, therefore not closing the valve as expected. Due to this fault, not only the water, but the gas as well, was sent to the draining pipes. Besides the formation of an inflammable cloud, the quick decompression of the liquefied gas promoted a strong reduction of the temperature of the line, causing the freezing of the valve. From this moment on, the operator became incapable to stop the leak. The emergency brigade of the company was called, but did not succeed in stopping the gas leak.

It was reported that a dense and large cloud of LPG was formed over the whole park of the spheres of the Refinery. The presence of an ignition source, never identified, detonated the cloud. This detonation, although causing light burning wounds in a few employees, was the main cause of the improvements that followed the accident. The detonation created a jet fire, thus forming a flame that touched the sphere, heating it.

Due to the seriousness of the accident and its duration, the fire was battled not only by the emergency brigade of the refinery, but also with the help of the employees of the companies of the neighborhood. Moreover, the first emergency teams were helped by the next shift of workers. This fact contributed to the increase of the number of victims of the accident.

After almost two hours of fire, the sphere that was leaking exploded in a phenomenon known as BLEVE. The explosion sent fragments over the other spheres, thus propagating the destruction. A sequence of explosions totally destructed the park of spheres of the refinery. Pieces of the reservoirs were found 800m away from their original place.

2.1 Implemented improvements

Similar to the accident reported, a few others occurred in other industrial plants in the same period, as the one in Feyzin, in France, in 1966 [1]. Although the task to remove the water formed in storage spheres of LPG continues to exist until today, several improvements were implemented. The ground of the spheres is now concreted including a slope, avoiding the accumulation of LPG under the sphere in case of a leak (pool fire). Other important changes were implemented in the drainage piping, which now present a larger number of valves, as shown in Figure 1. The opening-and-closing of the valve nearest to the sphere became remote, driven from the control room. The functioning from a distance has as its goal to allow that, even in case of fire, the valve will be controlled. The valve counts on a thermo-sensitive fuse, which promotes its closing in case the external temperature increases. Finally, the length of the drainage piping was also extended. The physical distance between the drainage point and the valves is set so that–even in a leaking scenario– the low temperatures would allow the freezing of the equipment. A larger number of valves in the system also contribute for a safer process. A project of "closed drainage" is being tested in Brazil, nowadays. This system should eliminate the gas release to the atmosphere, in case a leak happened. If a gas leak occurred during the drainage process, the system would send the gas to flare.

Figure 1: Drainage piping with a larger number of valves

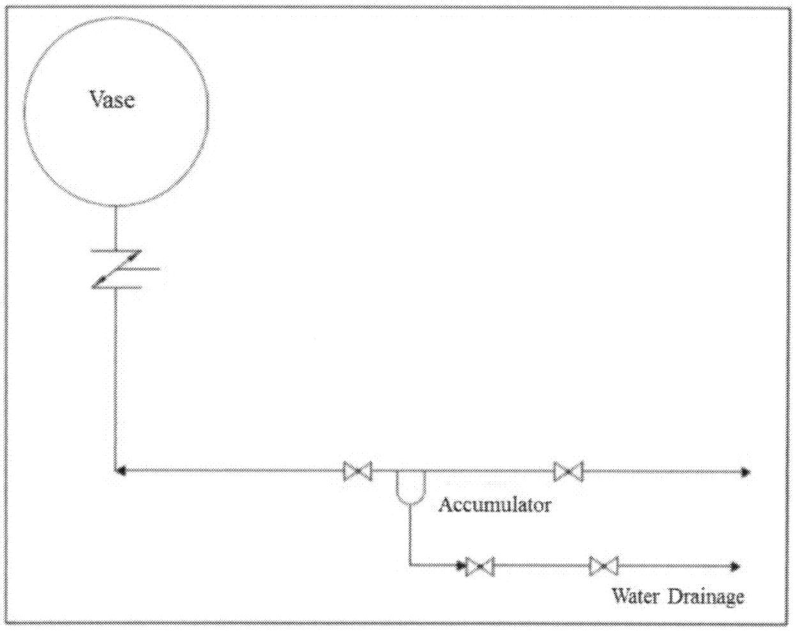

3. Simulator

Through the use of a dynamic simulator, it was possible to carry out several simulations of the drainage system. These results permitted the evaluation of the dynamic behavior of different variables of the process. From the behavior of the process parameters, different control strategies were proposed. The automation would allow the interruption of the drainage process in its ideal moment, thus avoiding the LPG leak.

In this paper we applied the software UniSim® Design. The simulation considers: the storage sphere, the valve directly connected to the bottom of the sphere, and the last blocking valve of the drainage line, as seen in Figure 2. For the simulation it was applied the Peng-Robinson's state equation. The LPG composition was presumed as the mixture formed by 70% of propane and 30% of butane. The amount of the condensed water in the bottom of the sphere, before the beginning of the drainage, was estimated in approximately 5% of the total volume. The diameter of the simulated sphere was 18.25 m. The diameter of the valve nearest to the sphere (VLV-100) was estimated to be 10 in, while the diameter of the most remote valve (VLV-101) was 1 in. The operating pressure of the sphere was presumed to be 8.5 kgf/cm^2 and its initial temperature, 29°C (environmental temperature). In order to simulate the drainage, the outlet flow (flow 5) pressure was adjusted for 1atm (atmospheric pressure). Finally, at the start of the drainage process the initial opening of both valves was presumed to be 100%.

Figure 2: System without PID control

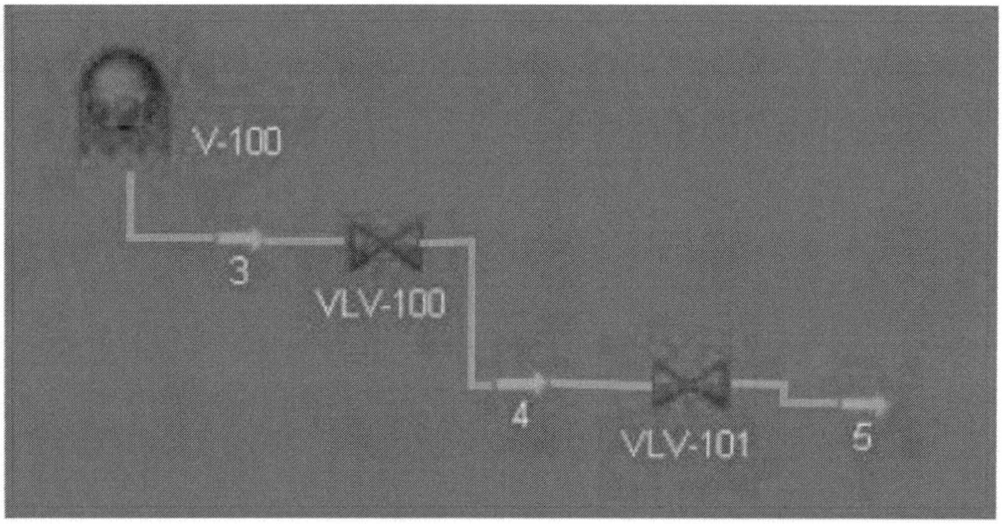

The system represented in Figure 2 does not have the PID controller to interrupt the process. This way, once both valves open, all the condensed water will be removed. After that, the gas release will occur in a situation similar to that observed in the 1972 accident.

After the simulations of the system represented in Figure 2, a PID controller was implemented (Figure 3). The control system sees, as its main goal, to allow that the drainage procedure be carried out safely and efficiently. To achieve this, it is necessary that the controller interrupts automatically the drainage before a leak of the LPG occurs.

Figure 3: System with PID control

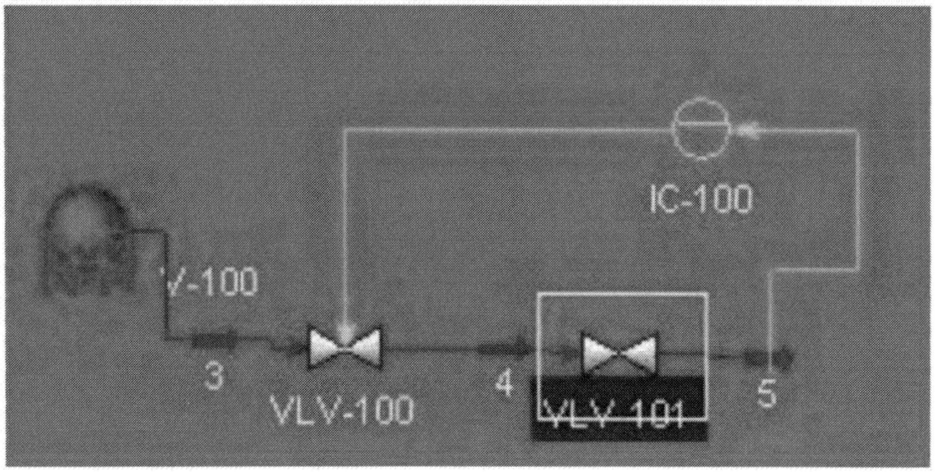

From the simulations study of the flow-diagram without PID controller (Figure 2), the controlled and manipulated variables were defined. The out-flow temperature (flow 5) was defined as being the controlled variable. Such choice was due to the fact that such process parameter presented high sensitivity to the beginning of the LPG leak, varying abruptly. The manipulated variable was defined as being the out-flow drainage of the tank, through the opening control of the valve (VLV-100).

4. Results

4.1 Process simulation without PID controller

First, the simulation of the system was done without the controller (Figure 2). In this simulation, the two valves were maintained opened during all the simulated time (1h40m). This time window was sufficient to simulate not only the water drainage, but the LPG leak as well.

The fractions of the phase water and LPG at flow 3 (closest to the sphere) are illustrated in Figure 4. One may notice that initially there is only a way-out for the water from the spherical tank. This occurs because the water is denser than the LPG, being accumulating so far at the bottom of the sphere. At instant t=1852 seconds, i.e., approximately 30 min. after the beginning of the simulation, the water fraction begins to decrease, initiating the way-out of the liquid LPG from the sphere. From this moment on, the water fraction of flow 3 decreases.

Figure 4: Fraction of the Phase x Time

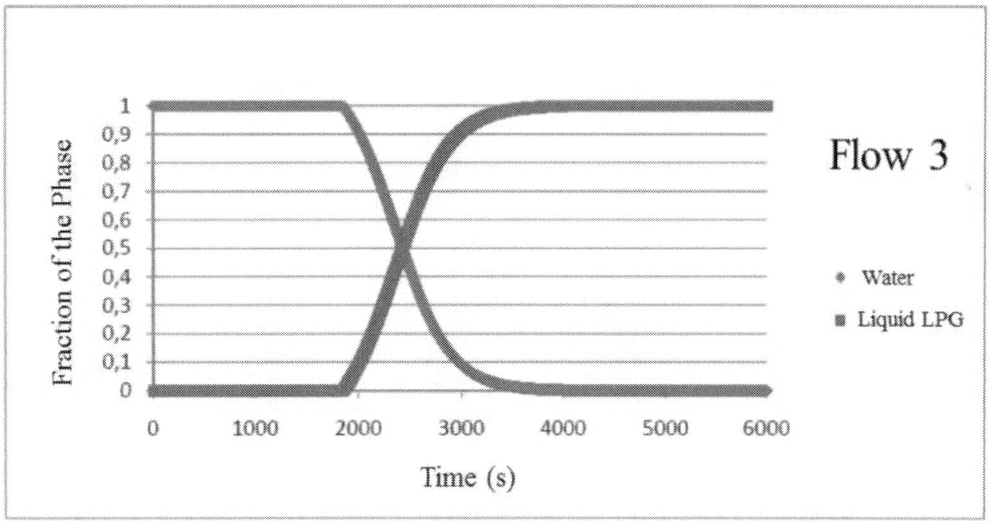

On the other hand, the fractions of flow 5 are represented in Figure 5. In a similar way to flow 3, one may notice that during the first moments of the simulation, the water fraction in the flow equals 1. From instant t=1855 seconds, the water fraction decreases, while the LPG fraction under a vapor state increases. The LPG presence under a vapor state occurs because this flow is being discharged on the environment, with an atmospheric pressure and an initial temperature of 29°C. Under such conditions, the LPG is found in a vapor phase. Therefore, the liquid LPG originated from the tank will vaporize after valve VLV-101.

As observed in the 1972 accident, the vaporization promotes a strong reduction of the line temperature, reaching -30º. C, as we can see in Figure 6. From instant t= 2170 seconds, the flow 5 temperature is already registering -23°C, and the LPG starts to leak in its liquid form. We shall highlight that, in a real scenario, this condition should be sufficient to promote the freezing of the nearest valve.

Figure 5: Fraction of the Phase x Time

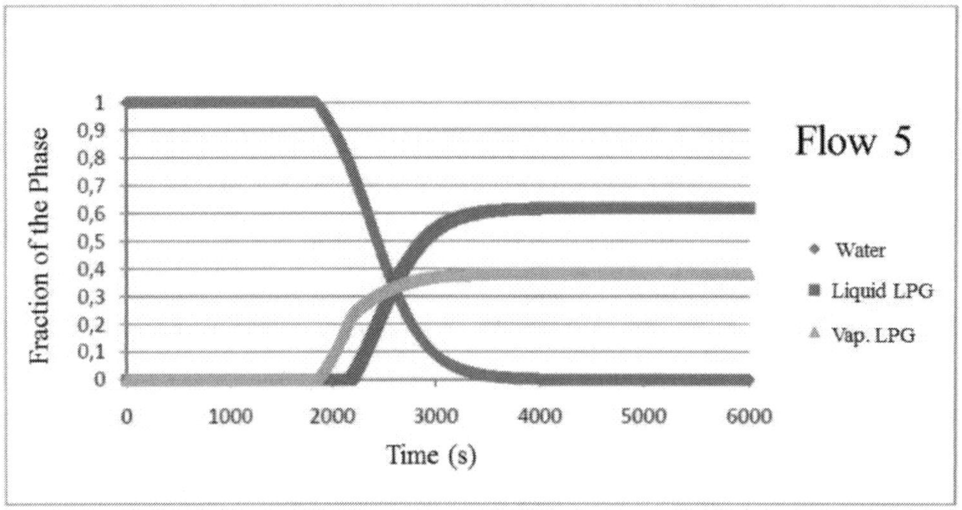

Figure 6: Temperature x Time

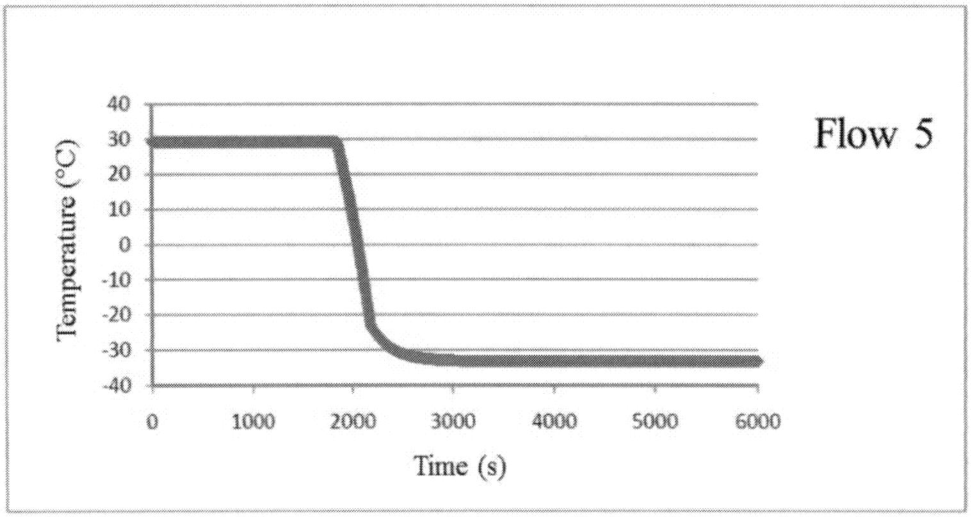

Through Figure 6 we can observe that the out-flow temperature (flow 5) is stable, approximately at the environmental temperature of 29° C, during the first 1855 seconds. In this period of time, only the water of the reservoir leaves it (Figure 5); from then on, the temperature decreases abruptly. At 1915 seconds, i.e., 1 minute after the beginning of the LPG leaking, the temperature was 20°.C, a fall rate of 10 degrees per minute. After 10 minutes, the temperature of the flow reached -30°C. This strong change of the temperature occurs due to the fast change of the phase suffered by the LPG.

From the simulation it was possible to observe that within the variables analyzed, the most sensitive to the beginning of the LPG leaking was the flow 5' temperature. This way, as mentioned, the control system was based on the temperature measures of flow 5.

4.2 Simulation of the process with different tunings of the PID controller

Intending to propose and test different automation strategies for the sphere's water drainage, the use of the PID methodology was initially investigated. Viewing to maintain the controller as simple as possible in a first approach, we opted for the use of a controller known as "P" (proportional). The tuning of the controller had, as its main goal, to maximize the withdrawing of the water from the spheres, thus avoiding the occurrence of a significant LPG leak. The chosen *modus operandi* for the controller was "direct": if the temperature starts to decrease the controller should be activated in order to reduce the flow by closing the valve.

Several simulations were made varying the gain (Kc) of the proportional controller. The set-point was adjusted for 19°.C, 10 degrees lower than the environmental temperature (29°C). In each simulation the following variables were evaluated: opening of the valve, the water fraction at flow 5 and the temperature in flow 5. The results are shown in Figure 7.

Figure 7: Water fraction in the flow 5 x Time

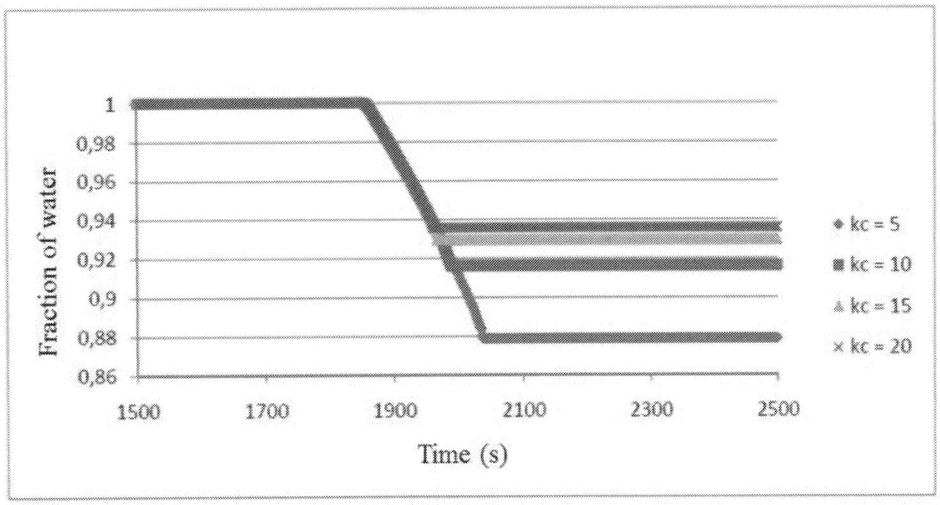

Figure 7 illustrates the behavior of the water fraction in flow 5 as time goes by, for different tunings of the control network. First of all, one may observe that only the water is being drained. From then on, a fall of this fraction occurs, indicating the beginning of the LPG leaking. As the water fraction decreases, the amount of LPG in the flow increases. As a consequence, the temperature decreases severely, according to Figure 8. It is possible to notice that the behavior of the temperature in the outflow follows the same profile of the water fraction in this flow.

Figure 8: Temperature of flow 5 x Time

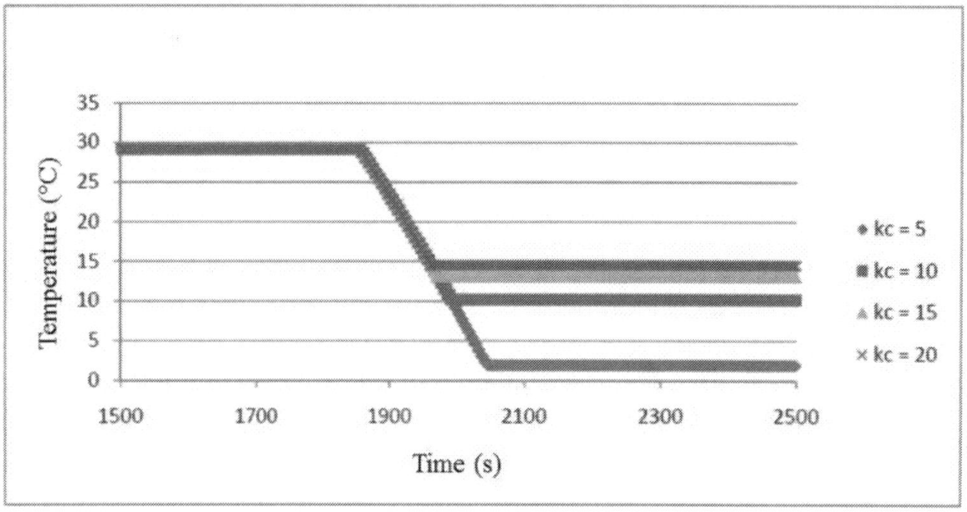

Figure 9 shows the percentage of the valve VLV-100 opening, as a function of time. Initially, while only the water is being removed from the sphere, the valve is kept in a totally open position (100%). As the simulation is carried out, such percentage decreases at the same time that the water content and the temperature decrease. The valve movement proceeds until the valve is totally closed (0% opening). The closing of the valve indicates that the controller is acting expecting the decrease of the leak, since temperature of flow 5 is coming down.

Figure 9: Opening of the valve x Time

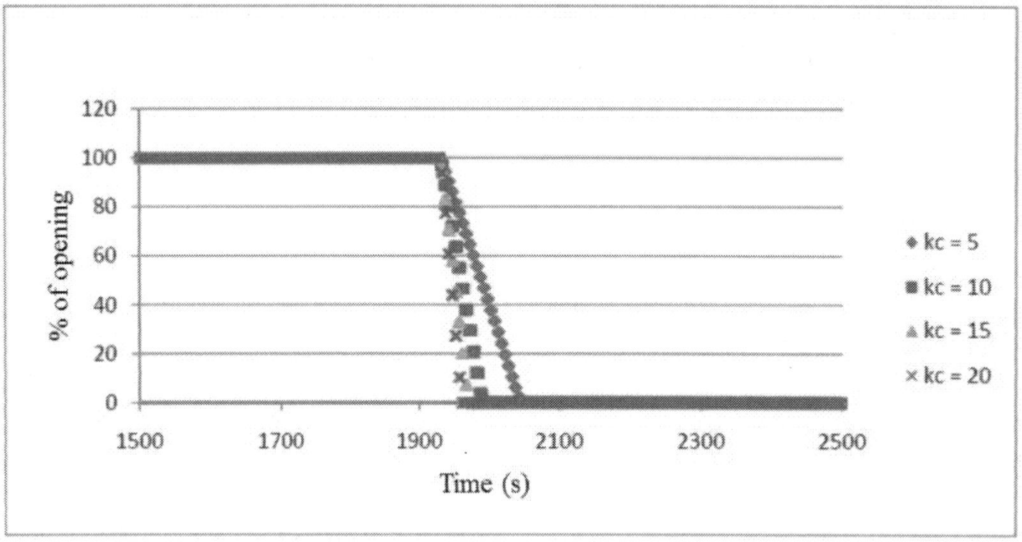

Comparing the behavior of the analyzed variables in relation to the values of the proportional controller gains (Kc), one may observe that the larger the gain, the quicker the valve closes and, therefore, the temperature and the water fraction are higher in flow 5 (Table 1). High gains minimize the amount of the LPG leaked, reducing the risks of frozen valves. Nevertheless, it is worth to highlight the fact that the closing time of the valve is still a function of the valve dynamics, which was not considered in this paper.

Therefore, for a given set-point, the increase of the gain of the controller promotes a quicker closing of the valve and drainage with a smaller LPG leak. It is interesting to notice that for a determined value of gain an increase on the value of the set-point would cause the same results.

Finally, we shall say that very elevated gains of the controller take the valve to act practically as a valve on-off. In this case, there are only two possible positions for the valve: totally closed or totally opened. Other types of control action (integral - I or derivative - D) were not tested, once the results based on the PID methodology took us to choose the on-off action.

Table 1. Dynamic behavior of the parameters of the process

Kc	Final Temperature - Flow 5 (°C)	Final Water Fraction - Flow 5
5	1.95	0.879
10	10.21	0.916
15	13.09	0.929
20	14.49	0.938
100	18.04	0.952
1000	18.74	0.955

4.3 Processes monitoring

In industrial processes, the maintenance of the operational conditions in general uses strategies of feedback or feed forward controller. A different approach was proposed by Seborg et al [2] and denominated "processes monitoring". This approach can be employed with the purpose of following the dynamical behavior of the variables. The monitoring, when associated to tools of supervision and control, guarantees that the variables stay within specific limits and rapidly detect abnormal conditions of operation, implementing corrective actions, etc. This way, the monitoring contributes to the reduction of the variability of the process. In given scenarios, this strategy may present advantages on the use of traditional approaches of control.

In the present study, the variable measured is the temperature of the out flow. The methodology of "process monitoring" is based on the realization of a series of measures of the temperature of the out flow carried out in the beginning of the draining. During the initial phase, the parameter stays practically stable, its average value being used as reference. In a real situation this initial value would correspond approximately to the environmental temperature. As the drainage proceeds and there is a space for a small fraction of LPG, the temperature decreases. The values measured are then compared in real time with the reference value and the differences are then added. When the sum overcomes a specific value, the control action will occur; in this case, the closing of the valve. The strategy of processes monitoring is known as "Cumulative Sum" [2].

It is important to highlight the fact that the temperature of flow 5 – before and at the beginning of the draining process – is subjected to natural fluctuations, such as alterations of the environmental temperature. This normal variation can reduce the efficiency of the traditional approach of control based in a fixed set-point.

In this study, through a simulation the average of the temperature of the out flow was measured during the first 10 minutes of drainage, thus obtaining the average value of 29.17°C. Through previous simulations it was observed that the decrease of the temperature, caused by the beginning of the leak, was of approximately 0.5°.C This was the maximum value of the cumulative sum adopted to determine the closing of the valve.

Using the cumulative sum, two seconds after the beginning of the LPG leak, the cumulative term became larger than the value stipulated as maximum, this way determining the closing of the valve. The quick closing of the valve, immediately after the beginning of the LPG leak, prevents the freezing of the valves. This approach seems to be much more stable and efficient than the proportional controller.

So, it is important to highlight the fact that in real operations the existence of fluctuations noises (interference) and fluctuations of temperature, or the composition of the LPG yet, could well difficult these parameters. The adoption of higher values as the maximum allowed at cumulative sum could appear adequate.

5. Conclusion

The large accidents of the past reveal the catastrophic potential of the events involving large LPG storage spheres. The draining procedure of the water accumulated in the bottom of the spheres was the main objective of this paper. Focusing on the automation of the procedure, we tried to simulate such process, based on dimensions, parameters and typical industrial conditions. We searched for the identification of a variable that could be indicative of gas leaking, being able to be applied as reference to indicate the best time to interrupt the draining. The temperature of the out flow seemed to be the most adequate parameter.

Different tunings of the proportional controller were tested. It was noticed that the larger the gain value (Kc), the smaller was the amount the LPG liberated at the end of the drainage. This way, the results converged to an action of control of the type on-off. This approach seemed to be strongly influenced by noise (interference) on the variable measured and by the environmental temperature observed.

The use of the strategy of process monitoring was also tested, based on the cumulative sum method. The results obtained were higher than those found for the proportional controller. The drainage was immediately interrupted after the occurrence of a small leak of LPG through the out flow. A "cumulative sum" of only 0.5° C, was adopted as the best. A quick interruption of the drainage process would have reduced the risks of freezing in the line and the emission of LPG to the atmosphere. Therefore, it seems to be an interesting option to increase the safety of the procedure.

6. References

[1] Kletz, Trevor. *Learning from Accidents*. 3rd edition, Gulf Professional Publishing, Woburn, MA, 2001.

[2] Seborg, Dale; Edgar, Thomas; Mellichamp, Duncan, *Process Dynamics and Control*, New York, 2nd edition, Wiley, New York, 2003.

(44aa) Complex BMS Made Simple

Dave Sheppard, Emerson Process Management, Hazelwood, MO and Nikki Bishop, Industry Solutions Group, Emerson Process Management, Roswell, GA

Complex BMS systems with multiple fuels and multiple burners have long presented a struggle for engineers and operators alike. This presentation will cover the methods used to simplify the implementation of a complex BMS in DeltaV SIS. The approach to design and use of standardized sequence modules and graphics greatly increases operator awareness of the operating conditions of the BMS. Included in this presentation will be a description of the design philosophy, the configuration standards, and the approach to operator graphics.

A three step approach is implemented. Break the BMS sequences down in to incremental parts that are easy to understand. Develop standard modules that implement these parts individually. Use a graphical interface the represents these individual parts in an easy to understand way for the operator. This approach is highly scalable to meet various complex BMS implementations.

With this approach it is easy to take a complex BMS and break it down in to small and simple to understand pieces. This allows greater alignment and understanding between the safety engineer, configurator, and operator. Ease of use and troubleshooting for the operator is also greatly increased through a standard set of graphics that show actions needed and first outs for trips.

Through this approach it is possible to take the design of how a complex BMS needs to function and actually implement that design cleanly and simply. An accurate system to capture the first out for a trip or loss of a permissive and also indication of the current state and actions needed to progress forward is needed to increase operator knowledge of what is going on in the BMS at all times.

"Using ESD Valves as Safeguards, Myth or Reality?"

German Luna-Mejias, P. Eng.
ACM Facility Safety
Suite 700 940 6th Ave SW, Calgary, AB, Canada
gluna@acm.ca
guatire2@gmail.com

Prepared for Presentation at
American Institute of Chemical Engineers
2013 Spring Meeting
9th Global Congress on Process Safety
San Antonio, Texas
April 28 – May 1, 2013

UNPUBLISHED

"Using ESD Valves as Safeguards, Myth or Reality?"

German Luna-Mejias, P. Eng.
ACM Facility Safety
Suite 700 940 6th Ave SW, Calgary, AB, Canada
gluna@acm.ca
guatire2@gmail.com

Keywords: Process, Safety, Emergency, Shutdown, valve, bypass

Abstract

An Emergency ShutDown (ESD) valve is a safety device to be actuated during an emergency situation to prevent hazardous consequences. The ESD valves are to be designed not to create additional hazards when actuated. Sometimes, due to operational constraints ESD valves are bypassed which contradicts best practices and creates an incremental risk for the facility, people and environment. Lack of regulations sometimes drives ESD valves' users to overlook or ignore best practices with the associated incremental risk. This paper presents experiences during HAZOPs/field observations where installed ESD valves are claimed as safeguards but are not (can be bypassed). The intent of this paper is to call the attention of designers, operators and managers of the potential hazardous consequences they may face if ESD valves are not used as per recommended practices.

1. Introduction

Our police force has to be right every time to prevent worst case scenarios, the "bad guys" only have to be right once for a major impact to occur. Exactly the same happens with safety; safety devices have to work when required; only one failure is enough to cause a major impact.

The use, absence or abuse of safety devices –e.g. emergency isolation valves- have been the cause of several accidents in the industry:

- A.P. Green Refractories Company – Explosion at Brick Kiln during start-up due to a main gas isolation valve bypass line leaking with 2 injured. Georgia, USA 1985. [1]
- Oil & Gas, Sonat Exploration Company - Explosion and fire due to misalignment of manual valves when feeding a low pressure system with HP Gas with 4 fatalities. Louisiana, USA. March 1998.[2]

- Ammonia Plant: Fire and explosion with substantial damage to a reformer-furnace due to bypass of control valve -with shutdown function- inadvertently open. No injuries, India, May 1998.[3]
- Esso-Australia Gas Plant Explosion due to the improper use of bypass valve around safety valve as the safety valve was not operable with 2 fatalities and substantial public welfare impact. Victoria, Australia. Sept 1998.[4]

As per these four examples, problems can happen anywhere in the world. In the last 3 years, when performing Process Hazards Analyses (PHA), the PHA teams indicated no concerns about bypasses around ESD valves, because there are no regulations that prevent this. Common responses were: "this is not regulated, why worry about it" or even worse: "recommended practices are recommendations, they are not mandatory". These practices are not only recommended practices but common sense. Additionally, as AIChE members (including professional denominations), the Code of Ethics indicates that welfare of the public and environment is paramount.

2. Definitions

Emergency isolation valves, emergency block valves, Emergency Shutdown valves or Automated Shutdown valve are different names for the same type of function valve.

A paper presented to GCPS in 2012 [5] indicates that the Emergency Block Valves (EBV) are namely different as per different codes, regulations and companies.

2.1 Additional definitions:

Emergency Block valves are used to control a hazardous incident. These are valves for emergency isolation and are designed to stop the uncontrolled release of flammable or toxic materials [6]

The ASV (Automated Shutdown Valve) is most commonly use in the Pipeline Safety Act 2011 [7]. The concept of the ASV is to mitigate the release of flammable or toxic materials, not the prevention of it. Pipeline Safety concern is H&SE not Process Safety.

2.1 ESD valve:

A valve with the following characteristics:

- Actuated valve
- Actuate upon a detection of a dangerous event defined by designers/operators and confirmed during PHA study.
- Provide protection against possible harm to people, assets or environment.
- Not used for control

An Emergency Shutdown valve –for the scope of this paper- is a valve that will prevent a hazardous situation from occurring (i.e. closing to prevent the overpressure of the vessel downstream of the valve) or mitigating the impact after the situation occurred.

2.2 Bypass valve:

Definition of bypass:

- go around something
- channel carrying a fluid around a part and back to the main stream

In the oil and gas industry is sometimes common to find bypasses around control valves; this situation is commonly used when it is expected that the valve may need maintenance during the operation of the plant. The plant can continue to run for a short period of time with a manual controlled function using a globe valve. This is a common accepted practice.

In this paper, the word "bypass" is a key element of the study, and follows the second definition: "channel carrying a fluid around a part and back to the main stream"; more specific: "around the ESD valve".

2.3 Norms and Standards

Depending on the system under review, there are several norms, recommended practices, regulations, etc. associated with the operation and installation of ESD valves.
Nevertheless, none of these documents indicates or dictates providing a bypass around an ESD valve. A list of these documents is provided in the appendix A.

3. Installation and Use of ESD valves

In Viviano's work [5], the use of the "5W" (What, Who, When, Where and Why) helps to identify the need for emergency isolation valves. In this paper, it is considered that all 5W are being taken in consideration and the evaluation is moving onto the next questions: Is it correctly installed? It is a valid layer of protection? Does it protect against a hazardous event? Does it mitigate the impact?

Is my safeguard device reliable? In the O&G business a spurious trip is also a problem, but does not affect public welfare. Once the ESD valve is included in the detail design, how do we ensure it is going to work in the different scenarios that led to the installation of the ESD valve?

The first phase of the analysis it to perform the risk evaluation or risk assessment.

3.1 Risk Analysis

During a PHA, analysis will be done to ensure the different ESD valves are valid safeguards to prevent the hazardous event from happening. In this phase of the study a qualitative analysis is performed and team members will agree, or disagree, on the protection being used for prevention. A good facilitator should challenge the team about the effectiveness of the safeguard (e.g. is there enough Process Safety Time, Is the actuation of the ESD valve creating another hazardous condition(s)?).

On the other hand, since the PHA is a qualitative analysis, each safeguard is considered to be of the same "weight" or of the "same importance". If required, an additional safeguards evaluation may be carried out: LOPA.

LOPA (Layers of Protection Analysis) is a semi-quantitative analysis to determine the PFD (probability of failure on-demand) of the safeguards to meet the tolerable frequency requirements of each company.

Based on LOPA an Emergency Shutdown Valve may require being a SIL rated device to achieve the tolerable frequency for a specified event. SIL rated means that the associated elements of the safety device has to be designed, constructed and maintain to a minimum level of "probability of failure when demanded to operate" [8]

Previously it was mentioned that ESD valves should not be used to perform a control function as the reliability or SIL rating of the ESD valve may be compromised. The PFD of a typical BPCS (Basic Process Control System) is 0.1 (1 failure every ten times it is demanded) and the ESD valve may require a higher reliability (e.g. SIL 2).

3.2 Use of ESD for Accident/Incident Prevention

When referring to Emergency Block Valves or Emergency Shutdown valves, two functions (or purposes) should be studied: 1. Use of ESD valves for preventing the hazardous situation from happening (preventive barrier). 2. Mitigating the consequence (acting after the hazardous event happens).

3.2.1 Accident/Incident Prevention

Valves in this category are mainly used in industries (e.g. Oil & Gas) to prevent the hazardous event from happening.

In several scenarios briefly described in the introductory section of this paper, ESD valves were installed to prevent a hazardous incident from occurring as per the original plant design concept.

3.2.2 Accident/Incident Mitigation

Valves in this category are mainly used to prevent public exposure to a hazardous event, after the initiating event occurs. These scenarios are largely described and analyzed in documents such as Pipeline Safety Act 2011 [7].

The concerns of the regulator are focused more on spurious trips that may create a public impact due to the effect of cutting the gas supply to communities. In this case, the regulator may recommend a manually activated Isolation Valve. For this scenario, a LOPA is recommended to ensure that the Isolation valve will have a low PFD (probability of failure on demand).

4. ESD Valve Configuration

Sometimes for "ease the operation" the bypassing of ESD valves are permitted as there are no regulations or norms against it but common sense. These "ease operation" scenarios may remove the protection intent of the ESD valve.

4.1 Creating additional hazards.

Installing an ESD valve to reduce the risk is a good idea unless this addition creates additional hazards. If the ESD valve closes, it will isolate two systems: one at high pressure and other at lower pressure. What happen if the high pressure system cannot withstand the pressure? This analysis should be part of the PHA analysis.

Sometimes, a risk analysis is improperly done just to complete a "check mark" in a paper that means the activity was performed. Therefore a proper evaluation is not accomplished and the operator may believe the facility is safe to operate and it is not.

Unfortunately in many projects the PHA is just a "check mark" to achieve the project approval and the PSM is not strong enough to ensure a proper PHA is accomplished. In an evaluation done by OSHA [9] it has been found that improper or inaccurate hazard assessment is one of the main deviations from the safety programs.

In many PHAs studies it has been found that as no regulations "mandate" the owner to avoid bypassing the ESD valves they are frequently bypassed as it makes the start-up much easier and faster.

If the ESD valve is designed to "Provide protection against possible harm to people, assets or environment" or "actuated upon a detection of a dangerous event" it should not be bypassed. A general rule of Safety indicates that: "the important thing to do is to be safe when nobody is watching"

4.2 Bypassing ESD valve for start-up.

After the PHA and LOPA team agrees to install an ESD valve with specific SIL rating and confirms that no added risk will be created if valve is actuated, what is next?

Even if no regulation "mandates it" common sense indicates that if a bypass is required around an ESD valve the bypass has to be equipped with similar protection or equivalent to prevent the inadvertent operation of bypass or minimize the possibility of leaking.

What are the problems associated with improper bypassing of ESD valves?

- Operator error may remove protection
- Valve leakage may remove protection
- Any of these scenarios will invalidate the PHA previously performed and "due diligence" may be difficult to justify if an accident happens.

Filling the lines with products before start-up is a common practice, occasionally done by opening the bypass around the ESD valve. The filling of lines is not a problem but the increased pressure on the downstream side of the ESD valve may be a problem. That is what happened in the Esso accident (Victoria, Australia) as a valve misalignment created an overpressure of the receiving vessel.

Use of CSC (car seal closed) valves in the bypass around ESD valves is a partial protection as the isolation valve may leak overtime without the operator knowing it. Pressure equalization is a bigger issue than flow.

In the incident that occurred in the Brick Kiln the "bypass valve" around the main isolation valve was "somehow inadvertently open"; it means that to start-up the facility the protection was removed.

5. Conclusions

Potential damages to public and environment can be serious, including fatalities, as it has been shown in several accidents around the world.

In the Pipeline Safety Act 2011 [7], the major concern is the amount of gas release and the impact on public after the incident happens. It does not direct the users to provide preventive safeguards.

As there are no specific regulations or codes, the bypassing of safety devices is a concern that is very likely to be overlooked.

6. Recommendations

6.1 General

- Accident prevention requires ACTIVE leadership by management of safety issues.
- Effective implementation of the right controls to manage, mitigate or eliminate hazards and reduce risk.
- Management systems and attitudes towards safety go hand-in-hand in creating robust defenses.

This paper is written with the intention that the reader perform "due diligence" to make sure the area under his/her responsibility is safe.

6.2 Safe Design

- As a general rule, ESD valves should be installed based on PHA/LOPA
- The actuation of ESD valves must not create additional hazards.
- The required reliability of an ESD valve (PFD) must be evaluated by a LOPA or similar analysis.
- Bypass around ESD valve should only be permitted if an additional ESD valve with the same functionality is installed in the bypass.

6.3 Operating practices

In the accidents reviewed at the beginning of this paper, in most of the cases the inadvertent opening or closing of a valve was a key element to the problem. It is highly recommended to implement a PSSR (Pre Start-up Safety Review) before any start-up.

7. Final Remarks

- Bypass of a safety device may save time for operator, but may cost lives!
- Follow PSM or equivalent because you want to be safe, not because a regulation requires it!
- A general rule of Safety indicates that: "the important thing to do is to be safe when nobody is watching"

8. References

[1] OSHA. Accident Report: Explosion in Brick Kiln. Accident # 14542070, report ID 0418100. Washington, DC. 1985

[2] CSB. Investigation Report: Catastrophic Vessel Overpressurization (4 deaths). Report # 1998-002-I-LA; Chemical Safety Board, Washington, DC. 2000

[3] R.M. Chopde & K.M. Patel ., "Ammonia Plants: Past accidents and lessons learnt", Krishak Bharati Cooperative LTD, India. Presented at International Fertilizer Industry Association (IFA) Technical Conference, Louisiana, 2000.

[4] A. Hopkins, PhD ., "Lessons from Esso's Gas Plant Explosion at Longford"; Australian National University, 2000.

[5] J. Viviano ., "Emergency Block Valves"; presented at VIII Global Congress on Process Safety, 2012.

[6] " API Recommended Practice API RP-553:"Refinery valves and accessories for Control and Safety Instrumented Systems", Chapter 7. American Petroleum Institute, 1998.

[7] "Pipeline Safety Act 2011" U.S. Congress, Law 112-90 – Jan 3, 2012.

[8] CCPS., "Layer of Protection Analysis: Simplified Process Risk Assessment", Center for Chemical Process Safety. New York, NY. 2001.

[9] Jordan Barab., "Learning from Industry Mistakes" presented at NPRA National Safety Conference, May 2010

APPENDIX A

Regulations, codes and recommended practices for the installation of emergency block valves.

- API RP-553 "Refinery valves and accessories for Control and Safety Instrumented Systems" American Petroleum Institute, 2007.
- NFPA 30 "Flammable and Combustible liquid Code", National Fire Protection Association, 2012
- NFPA 497 "Recommended Practice for the Classification of Flammable Liquids, Gases, or Vapors and of Hazardous (Classified) Locations for Electrical Installations in Chemical Process Areas" National Fire Protection Association, 2012
- API RP-500 "Electrical Area Classification" American Petroleum Institute, 2000
- NFPA-58 "Liquefied Petroleum Gas Code" National Fire Protection Association, 2011
- API RP-2510 "Design and Construction of LP-Gas Installations at Marine and Pipeline Terminals, Natural Gas Processing Plants, Refineries, Petrochemical Plants, and Tank Farms", 4th Edition, 1978. American Petroleum Institute; recommended practice.
- NFPA 85 "Boiler and Combustion System Hazards Code", National Fire Protection Association, 2011
- API RP-560 "Fired heaters for General Refinery Services" American Petroleum Institute, 2001
- API RP-556 "Instrumentation, Control and Protective Systems for Gas Fired Heaters", American Petroleum Institute, 2000.
- API RP-520 "Part I: "Sizing, selection, and installation of pressure-relieving devices in refineries". Part II – "Installation" American Petroleum Institute, 2000
- API RP-521 "Guide for pressure-relieving and de-pressuring systems", American Petroleum Institute, 1998
- Pipeline Safety Act 2011: U.S. Congress, Law 112-90 – Jan 3, 2012.

(44ac) Contours When Pipelines Meets Facilities

Karthick Deivasagayam, Loss Prevention, SNC Lavalin Inc, Calgary, AB, Canada

Pipelines involved in oil & gas sector carries risk of fires and explosion along with the actual pipelines. This risk of fire and explosion are vulnerable in the places it could affect and become even more vulnerable in places of proximity of facilities where the risk could turn out into initiating of a major catastrophic events.

This should be addressed in the way of doing things.

QRA done for the most of the projects are specific to their scope alone. Boundary limits of the QRA Study is the same as that as the contractual scope of work. Whereas the risk and hazard boundaries are not contractual but as long as it could affect.

This paper will discuss about few pipeline accidents their consequences and the probable effect if they occurred in a different location (close to the facilities) and from discussing about the potential of what a pipeline accident can do and the focus will shift to the way these risks are approached the way what we do.

This paper will also discuss the way the risks of pipeline projects are captured and also drawbacks and grey areas of the existing approach. This paper also recommends the proposed approach that could be followed for the future risk assessments.

(44ae) Ozone Flammability and Its Affect On the Lower Flammable Limit of Simple Hydrocarbons

Paul Osterberg, Thomas Johnson and Thomas Frawley, Fauske & Associates, Burr Ridge, IL

As a result of ozone's strong oxidizing capabilities and ease of generation on an as-needed basis, ozone applications in process operations are gaining momentum. Previous work has demonstrated that ozone is flammable at high concentrations with a lower flammable limit of 8.4 vol.% in oxygen. However, there is limited data regarding the effect of ozone on the lower flammable limit of other gases known to be flammable in air. This work investigated the lower flammable limit of methane, ethane, and propane in various concentrations of ozone at atmospheric pressure and temperature. The tests were conducted in a 2-L vessel using a fuse wire ignition source at 1 percent ozone intervals spanning from no ozone to the LFL of ozone in the vessel. It is anticipated that the addition of ozone will decrease the LFL of each flammable gas as the concentration of ozone increases.

Evaluate Plant-Wide Safety Of Your Interlock System

Principal Author: Mohammed Naved Khan
Ingenero Technologies India Pvt Ltd.
Mumbai, India
nkhan@ingenero.com

Presenter: Swapnil Pathak
Ingenero Technologies India Pvt Ltd.
Mumbai, India
spathak@ingenero.com

Co-author: Jim Brigman
Ingenero Technologies Inc.
Houston, Texas, USA
jbrigman@ingenero.com

Prepared for Presentation at
American Institute of Chemical Engineers
2013 Spring Meeting
9th Global Congress on Process Safety
San Antonio, Texas
April 28 – May 2, 2013

UNPUBLISHED

Evaluate Plant-Wide Safety Of Your Interlock System

Principal Author: Mohammed Naved Khan
Ingenero Technologies India Pvt Ltd.
Mumbai, India
nkhan@ingenero.com

Presenter: Swapnil Pathak
Ingenero Technologies India Pvt Ltd.
Mumbai, India
spathak@ingenero.com

Co-author: Jim Brigman
Ingenero Technologies Inc.
Houston, Texas, USA
jbrigman@ingenero.com

Keywords: IRM, interlock relationship, interlock-process interference, holistic, isolated information, PHA, mitigation

Abstract

Interlocks serve as important safety systems in industrial settings, where they protect equipment from damage and employees from toxic and harmful releases from that equipment arising out of unsafe conditions. The same safety systems can lead to damage of process systems, if not analyzed properly. In some of the cases we encountered, initiation of interlocks in equipment caused severe damage to other equipment (due to change in composition) or initiation of interlock caused damage to the same equipment it was intended to protect (due to incorrect sequencing of interlocks).

The information available with the interlock design documents (like cause and effect diagrams) is limited to the vicinity of that particular interlock. It does not give any information about the relationship of the interlock, under study, with other process equipment or even with other interlocks (except for initiation). The relationship between safety interlocks and process system arises out of the behavior of the process system (e.g. change in composition) to the actions taken by the interlock. This paper proposes how a holistic view for these behavioral responses can be obtained by developing the IRM (Interlock Relationship Matrix). IRM is a single process document (excluding instrumentation and logic details) that provides information on the effects of all interlocks on each element of process system. Developing IRM is crucial in evaluating the interlock-process interference (behavioral responses) and is beneficial in managerial decision making to assess the likely damage to equipment and provide mitigation thereof.

1. Introduction

Interlocks form the most critical and independent layer of safety in industrial settings. They are primarily designed for:

- Equipment - protect the equipment from damage due to unsafe conditions (eg. over speed) and the resulting economic loss arising out of replacement and downtime.
- Personnel - protect the personnel from exposure to harmful and toxic materials that may get released from equipment damage (e.g. Seal leaks from rotating equipment).
- Plant - maintain the integrity of the plant (e.g. flare mitigation methods employed to limit the flaring during global utility failures and avoid exposure of un-burnt hydrocarbons to the general public).

The design intention of the approaches employed for interlock systems are two-fold; first is the isolated view to protect the equipment at hand and second is the global view to maintain integrity of the plant. Although the approach seems to be looking fair, they (even the global interlock system) lack in considering the interlock-process systems relationships in designing interlock systems.

The very definition of an independent protection layer provided by interlock systems (whether a single or global interlock) has an inherent element of relationship between interlock actions and the behavior of process systems i.e. the actions of the interlock system are taken independent of the responses of process system. This relationship may or may not be relevant as far as the functions of that particular interlock is concerned, but in no case it should cause upset in other process units that could jeopardize their integrity and cause irreparable damage. In the following sections, we will look into some of the incidences that lead to irrevocable equipment damage as a direct result of the interlock-process interference and we will suggest an approach as to how a holistic view for these behavioral responses of the process systems to interlock actions can be obtained for effective mitigation. Also the major points of differences between global interlock systems and holistic view of individual interlock systems will be highlighted.

2. Interlock Design Information

The design of interlock systems starts with the development of interlock diagrams or narratives. They provide key information as to the analysis and identification of scenarios that can cause damage to particular equipment (initiating causes) and the actions that are required to be taken for the protection of that equipment. They are of the following three types:

2.1 Narratives

Narratives provide descriptive statements for the various elements of the interlock system. Narratives are a short description stating the primary function of the interlock, events that can lead to unsafe conditions (initiating causes) and finally actions required by the interlock to mitigate these events. They are written in a simple but precise manner without making use of detail identification tags.

2.2 Cause and Effect (C&E) Diagram

C&E diagrams are the next level of detail information about the interlock. C&E diagrams are presented in a tabular format, detailing the initiating causes and the actions taken in the form of individual tag numbers, usually initiating causes are activation of suitable switches due to unsafe conditions and corresponding actions taken (opening or closing a valve). Actions taken are segregated for each initiating cause. They also provide information about time delays. C&E diagrams provide detail accounting of each element of the interlock.

2.3 Logic and Ladder Diagram

The information provided in logic and ladder diagrams is specific to the instrumentation part of interlock system. They provide the circuit diagrams of the relay logic hardware used for the interlock system under consideration.

3. Limitations of Design Information

Interlock design information is limited to the vicinity of the particular interlock under consideration. The actions taken (or required) are bound by the isolated view of the equipment to be protected. They are limited in their ability to evaluate and analyze the effect of actions taken by the interlock systems on other process systems.

The structure of the C&E and ladder-logic diagrams is incompatible with identifying and analyzing such kind of relationships. Interlock narratives can incorporate this information in a piece-meal fashion (similar to the way information is presented in a narrative - one interlock at a time), however the utility of such disintegrated information is questionable due to the following facts –

- Language - The information presented in narratives is written in easy to understand language and hence lacks a structured and objective approach.
- Lack of specific information - They are prepared usually at a stage where system specific detail information is not available and lot of re-work might be required to make them consistent and compatible to receiving detail information.

4. Need for Holistic View

We will discuss two incidences where interlock activation in an equipment led to severe damage in the same or other equipment.

4.1 Incident one – Activation of Interlock in Acetylene Hydrogenation Reactor of Ethylene Unit

Ethylene cracker unit was operating with an acetylene hydrogenation reactor located between fourth and fifth stages of cracked gas compressor after Deethanizer and before the chilling train that produces tail gas (front end configuration). Tail gas, usually, is a methane-rich stream that is used as fuel in cracking furnaces. Fuel gas system for crackers also has backup from imported LPG to handle contingencies.

For this case, interlock was designed for shutting down, boxing up and depressurizing acetylene reactor to purge out reactive hydrocarbons from the catalyst bed and prevent runaway reactions that might cause a meltdown of the reactor. However, shutting down the reactor also meant stopping of tail gas flow to furnaces. Thus LPG from backup system was lined up, which changed the calorific value of fuel to the furnaces. The cracking furnace temperature controller was working without duty correction to account for change in composition and, therefore, was sluggish in taking appropriate corrective actions and closing the valves. This time gap led to increased heat input to the furnaces which resulted in a spike in the tube metal temperatures of the coils and subsequent coil damage. Thus to protect acetylene reactor, interlock actions lead to damage of cracking furnace coils due to inappropriate handling of fuel gas composition change.

In this case, the behavioral response of process system to the actions of interlock was change in fuel gas composition to the cracking furnaces. This information was not apparently or directly visible from interlock design information.

4.2 Incident two - Activation of Total Shutdown interlock of Cracking furnace of Ethylene Unit

Cracking furnaces in Ethylene units usually have two levels of interlock shutdown in case of upset, first is steam standby with isolation of hydrocarbon feed and operating the firebox in hot condition and second is the total shutdown where firebox is cooled.

Usually, first interlock is designed for faster restart after a partial shutdown by sustaining close to operating conditions inside the firebox and maintaining minimum inventory of BFW in the steam drum. The second interlock is designed to maintain integrity of the furnace and cool the firebox, which might be operating close to 1100 deg C, to a safe condition. Both interlocks have different functional requirements and are designed in a way so as to be independent of each other. However, from process point of view, initiation of first interlock also provides the necessary time lag for controlled rate heat release from the firebox, by maintaining minimum firing, before the cooling down of the firebox is initiated by total shutdown interlock.

For the case under consideration, in order to maintain minimum BFW inventory for first shutdown, the trigger was set at significantly low-level point. This resulted in activation of both interlocks with minimum time delay resulting in rapid cooling of firebox. The coke layer deposited inside the coils has very different coefficient of expansion than coil metallurgy leading to unequal contraction and subsequent mechanical failure of coils.

The incident gives the interlock interference or dependence that may be sometimes necessary due to process behavior. Design information of both interlocks, when observed in isolation, was sufficient to perform their designated tasks. However without a holistic view of looking at them, integrity of the equipment was jeopardized.

With the above-discussed instances, it is evident that sole reliance on interlock design information to provide the hindsight necessary to avoid damage to other equipment is insufficient. It requires a holistic view, encompassing all individual interlock systems, to study the interlock-process relationship for each element of the process system to better understand these relationships rather than piece-meal and isolated interlock design information.

One can argue that global interlock systems responsible for maintaining the integrity of the entire plant can take care of these relationships. We think that such a system is practically infeasible. The reasoning starts with the very definition of an interlock which states that it is the actuation of the initiating causes that triggers the interlock responses irrespective of the process. For global interlock systems to take care of interlock-process relationships, all the elements of the process system must be allowed to trigger them (along with the upsets in global utility networks).

Another approach could be to trigger specific interlock systems put in place to protect these relationships. The practicality of such an interlock system is bound to fail because of two reasons. First the investment required in putting up these extra interlock systems or triggers is huge as the permutations and combinations of initiating causes from all process elements is gigantic. Second it can lead to loss of earnings and operational prudence due to the higher cost of production for increased number of instances of unsteady plant operation due to spurious activations.

5. IRM

We propose to develop IRM as a tool to give the necessary holistic view for understanding the interlock-process relationship. IRM is an acronym for Interlock Relationship Matrix. It is the final outcome of the study for evaluation of interlock actions and is a single document that shows the effect of actions of each interlock systems on all elements of the process system.

It is prepared in a tabular format with interlock systems arranged in rows and process system elements arranged in columns. A typical IRM is shown in Table 1 for reference layout.

Table 1. Typical IRM Layout

Elements of Process System	Element 1	Element 2	Element 3	Element 4	...
Interlock Systems					
Interlock 1					
Interlock 2					
Interlock 3					
...					

The effect of each interlock action is documented in the corresponding column for process elements. For example, if the actions of Interlock 2 change the feed composition of Element 3, then "change in feed composition" is typed in cell corresponding to Interlock 2 and Element 3.

IRM should include only those effects of interlock actions that have the potential to cause significant damage to process elements. This will avoid ambiguity and redundancy of information. Populating IRM with all effects of interlock actions reduces the utility of the tool. Existing safeguards should also be considered for evaluating the potential damage to the equipment. If damage is possible after considering existing safeguards, only then IRM should be updated with the relevant effect. For earlier example, it can be analyzed whether change in feed composition can cause damage to the equipment if it is not corrected (by the absence of density or composition meter) for changed composition.

However, reference of existing safeguards, for effects that have potential for damage, can be made in IRM to ensure and document that the analysis of the interlock actions has been conducted.

6. Why IRM

There are generic approaches available for evaluating hazards for process system. They can be studied under two categories: Software Evaluation and PHA.

6.1 Software evaluation techniques

These techniques include Software Systems Hazards Analysis like Fault/Event Tree Analysis, Failure Mode and Effects Analysis, Software Failure Mode, Effects, and Criticality Analysis, etc. They are deemed to analyze and identify whether software system components' operation or failure (functioning or lack of functionality) could result in hazards for process systems. It begins when the components are designed sufficiently, and is updated as their design matures to take care of the potential hazards.

Functional difference between these analyses and IRM is that they are more concerned with the effects of software (interlock) systems on process elements in case the software goes rogue and malfunctions (lack of functionality). Moreover, the analysis of components' operation (functionality) on process systems is unable to understand interlock-process relationship as it is concerned only for segregating and avoiding intermixing interlock (safety actions) and process (control actions) systems.

Whereas, in the case of IRM, it analyzes the effects of interlock actions on process element considering that software (interlock) system has fulfilled its functionality and there is no sharing of functions from either side. It is not intended to find any design fault in software (interlock) systems but to evaluate and understand the probable responses of process system to software (interlock) system actions.

6.2 PHA

PHA techniques include operational and process hazards analysis like HAZOP, LOPA, HAZAN, etc. These techniques use guidewords in combination with process parameters to evaluate deviation scenarios from normal operation and verify the utility of existing safeguards. If the existing safeguards are not sufficient, new safeguards are suggested. Safeguards can include interlocks, process control, alarms, relief valves, operating procedures, etc.

Key difference between PHA and IRM is that PHA considers only the process parameters as the potential causes for deviation and subsequent consequences. To mitigate such consequences, PHA provides the scope for defining actions required by a safety system. However, it does not evaluate the actions of a safety system like an interlock as probable causes of deviation.

IRM fills this gap of analyses of actions of an interlock as potential causes of deviation for the process elements and gives insight to unearth latent harmful events of interlock actions.

7. Developing IRM

IRM can be developed in a structured way such that the aftereffects of interlock actions can be analyzed under the considerations of core parameters critical to the industrial setting. This approach will avoid ambiguity and subjectivity.

Simple IRM layout discussed in Table 1 can be modified to include critical parameters as shown in Table 2.

Table 2. Typical IRM Layout with critical parameters (cp)

Elements of Process System	Element 1	Element 2	Element 3	Element 4	...
Interlock Systems					
Interlock 1	*cp1:*	*cp1:*	*cp1:*	*cp1:*	...
	cp2:	*cp2:*	*cp2:*	*cp2:*	...
	cp3:	*cp3:*	*cp3:*	*cp3:*	...
	...	...	...	...	...
Interlock 2	*cp1:*	*cp1:*	*cp1:*	*cp1:*	...
	cp2:	*cp2:*	*cp2:*	*cp2:*	...
	cp3:	*cp3:*	*cp3:*	*cp3:*	...
	...	...	...	...	...
Interlock 3	*cp1:*	*cp1:*	*cp1:*	*cp1:*	...
	cp2:	*cp2:*	*cp2:*	*cp2:*	...
	cp3:	*cp3:*	*cp3:*	*cp3:*	...
	...	...	...	...	...
...	...	...	...	...	...

For the refining and petrochemical industry, critical parameters are pressure, temperature, flow and composition. Other parameters like chemical reaction, liquid level, etc. can be evaluated from these four parameters. Each action of an interlock system should be evaluated with respect to above four parameters for understanding the relationship with each process element.

We will prepare IRM for the two incidents discussed in Section 4 as case study (Table 3 and 4). These tables will form a subset of the overall IRM for the plant.

Table 3. IRM for interlocks and equipment involved in Incident 1

Elements of Process System	Cold box	Cracking Furnace
Interlock Systems		
Front End Acetylene Reactor shutdown interlock	***Pressure:*** **no change, cold box is boxed up**	***Pressure:*** **no change, OSBL pressure controller is designed for backup**
	Temperature: **no effect, temperature will rise gradually due to ambient heating**	***Temperature:*** **no effect**
	Flow: **no flow to cold box resulting in loss of tail gas and H2**	***Flow:*** **flow of tail gas will be replaced by LPG**

Elements of Process System	Cold box	Cracking Furnace
Interlock Systems		
	Composition: no effect on cold box streams composition	*Composition:* methane-rich tail gas will be replaced by propane-butane rich LPG increasingly the volumetric calorific value, no heat value correction available. Likely to increase firing duty due to sluggish temperature control of furnace coils.

Table 4. IRM for interlocks and equipment involved in Incident 2

Elements of Process System	Cracking Furnace
Interlock Systems	
First shutdown interlock of Cracking Furnace	*Pressure:* no change in firebox and furnace coil pressure
	Temperature: firebox temperature is maintained at lower level by minimum firing.
	Flow: less flue gas flow due to low firing
	Composition: NA
Second shutdown interlock of Cracking Furnace	*Pressure:* firebox pressure will rise to ambient due to unavailability of ID fan
	Temperature: firebox temperature will reduce rapidly to ambient due to fuel gas shutdown. Controlled rate heat release is required to avoid rapid cooling by maintaining minimum firing conditions (first shutdown interlock) for some duration or stopping flue gas instantly by closing the damper. However, neither the duration of firebox operation under first shutdown is sufficient, nor the damper will close to allow controlled rate heat release and avoid thermal shock. Likely damage to coils.
	Flow: no flue gas flow
	Composition: NA

8. Integration of IRM with PHA

A structured IRM described in Section 7 provides compatibility to integration with PHA techniques such as HAZOP, LOPA, etc. PHA technique like HAZOP analyzes each and every parametric deviation on process system that is applicable for an industrial setting. For refining and petrochemical industry, the parameters include pressure, temperature, level, flow, composition, chemical reaction, mixing, contaminants, special procedures, etc.

Development of IRM along with PHA study starts by identification of a subset of critical or primary parameters out of the PHA parametric set on which other parameters depend (like pressure, temperature, flow and composition). It then incorporates and analyzes actions of a safety system (interlock) as potential sources of deviation (in addition to the parametric deviations) for each process element (equipment or node) where the deviations of interlock actions are evaluated under the guidance of critical parameter subset.

9. Conclusion

Interlock design information, in the form of narratives, cause and effect diagrams or ladder/logic diagrams, is limited to the vicinity of the interlock system. It is unable to evaluate and analyze the effect of actions taken by the interlock systems on other process systems. This inability might result in irrevocable damage of equipment or potential personnel exposure to toxic materials as evident from the two case studies presented in Section 4. These damages can be avoided by understanding the interlock-process relationship to gauge probable process responses to interlock actions.

The interlock-process relationship is not visible explicitly in evaluation techniques like Software Systems Hazards Analysis (Fault/Event Tree Analysis, Failure Mode and Effects Analysis) and PHA (HAZOP, LOPA, HAZAN). These techniques intend to eliminate operational hazards caused by either malfunctioning of the software (interlock) system or deviations in process parameters. However, they do not take into consideration the probable causes that can occur in process elements after the actions of a software (interlock) system have successfully executed.

IRM (Interlock Relationship Matrix) can overcome the isolated nature of interlock design information by providing a holistic view of all interlock systems and process elements in the plant (in the form of a matrix) and evaluate critical interlock-process relationship under the guidance of critical parameters applicable for the industrial setting. It can also fill the gap in other evaluation techniques by analyzing actions of a safety system (interlock) as potential causes of deviation for the process elements. It provides efficient incorporation of new interlock systems or functions for existing process systems and vice-a-versa by giving a ready platform for analyzing the interlock actions on the process system. IRM is a managerial decision making instrument that provides valuable information to assess the likely damage to equipment and provide mitigation thereof.

Understanding interlock-process relationship is crucial to avoid jeopardizing mechanical integrity of the process system by the very interlock system installed to protect them and IRM makes it happen.

10. References

1. "Guidelines for Hazard Evaluation Procedures", 3rd edition, Center for Chemical Process Safety (CCPS), John Wiley & Sons, 2008.
2. "Lees' Loss Prevention in the Process Industries, Hazard Identification, Assessment and Control", Volume 1, 3rd edition, Sam Mannan, Elsevier Butterworth-Heinemann, 2005.

Additional references not cited:

1. "Instrumentation and Control systems", Princeton Plasma Physics Laboratory, ES&HD 5008 Section 2, Chapter 10, Revision 6, 2005.
2. "Code Walk-Through", 1984, Dunn, Robert, Software Defect Removal, McGraw-Hill, Inc.
3. "Event Trees and their Treatment on PC Computers", Limnious, N. and J.P. Jeannette, Reliability Engineering, Vol. 18, No. 3, 1987.
4. "A Guide to Hazard and Operability Studies", Chemical Industry Safety and Health Council of the Chemical Industries Association, Alembic House, London, UK.
5. "HAZOP and HAZAN", Klutz, T.V., Institution of Chemical Engineers, UK, 1986.
6. "Nuclear Surety Design Certification for Nuclear Weapon System Software and Firmware", Air Force Regulation 122-4, Department of the Air Force, 24 August 1987.
7. "Petri Net Theory and Modeling of Systems", Peterson, J.L., Prentice Hall, 1981.
8. "Procedures for Performing A Failure Mode and Effect Analysis", MIL-STD-1629A, Department of Defense, 24 Nov 1980.
9. "Generic Techniques in Reliability Assessment", Fussel, J., Noordhoff Publishing Co., Leyden, Holland, 1976.

9TH GLOBAL CONGRESS ON PROCESS SAFETY

Virtual Reality Applied to Training and Process Safety: A Semi-Industrial Boiler Case Study

Carlos André Vaz Junior
Departamento de Engenharia Química, Escola de Química, Universidade Federal do Rio de Janeiro
Centro de Tecnologia, Bloco E, Sala E-209, Ilha do Fundão
cavazjunior@eq.ufrj.br

Thiago Souza de Oliveira
Escola de Química, Universidade Federal do Rio de Janeiro
Centro de Tecnologia, Bloco E, Sala E-209, Ilha do Fundão
thiagowsousa@gmail.com

Maurício Bezerra de Souza Junior
Departamento de Engenharia Química, Escola de Química, Universidade Federal do Rio de Janeiro
Centro de Tecnologia, Bloco E, Sala E-209, Ilha do Fundão
mbsj@eq.ufrj.br

Prepared for Presentation at
American Institute of Chemical Engineers
2013 Spring Meeting
9th Global Congress on Process Safety
San Antonio, Texas
April 28 – May 1, 2013

UNPUBLISHED

Virtual Reality Applied to Training and Process Safety: A Semi-Industrial Boiler Case Study

Carlos André Vaz Junior
Departamento de Engenharia Química, Escola de Química, Universidade Federal do Rio de Janeiro
Centro de Tecnologia, Bloco E, Sala E-209, Ilha do Fundão
cavazjunior@eq.ufrj.br

Thiago Souza de Oliveira
Escola de Química, Universidade Federal do Rio de Janeiro

Maurício Bezerra de Souza Junior
Departamento de Engenharia Química, Escola de Química, Universidade Federal do Rio de Janeiro

Keywords: virtual reality, process simulation, training, 3D plant model

Abstract

The short-term time given to learn a particular task associated with the demands for higher levels of production to increase profits are potential reasons for the generation of accidents. Seeking to overcome this situation in training a large number of operators in a similar environment in which they will work, virtual applications were developed. A 3D reality, integrating phenomenological simulation developed in MATLAB, and Virtual Reality (VR) using Plant Design Management System© (PDMS), were applied in this work. Process simulators using virtual reality are applied in several areas, like flight simulators and firefighters' training. Additionally, the developed 3D plant model based tool allows different safety tasks, such as the optimum location of fire extinguishers, fire hoses and escape routes, among others. As a case-study, a semi-industrial boiler, located in Universidade Federal do Rio de Janeiro, Rio de Janeiro, Brazil, was chosen. The interaction between the user and 3D model was addressed by developing a process data interpreter, so that the model may been modified on the basis of actions taken by the operator and by the phenomenological simulation. As the developed tool improves the operator's understanding of the process, it may minimize rework, accidents and damage. Additionally, this methodology may be replicated in other industrial plants.

1. Introduction

Combine safety process, higher levels of production and environment cares is a permanent challenge to be overcome. Face this fact becomes much harder in the current days due to the increasing complexity of plant process, much ambitious quality range and the short-term time given to learn a particular task. In particular, the complexity of the chemical industry also affects in process safety.

Modern chemical processes use several electronic controls and layers of protection. However, even with all the complex components and systems, the human error still are the responsible by an expressive number of incidents in the industry [1]. It is understood as essential to the operator the full knowledge of the system in which he works. But with the growing demand for increased productivity, professionals have with this feature becomes increasingly rare.

For this scenario, industries are increasingly opting for the use of Virtual Reality (VR) training, where the VR can be defined as a computer-generated three-dimensional environment, updated in real time, allowing human interaction across multiple input-output devices [2]. Virtual reality is used in several studies that allow a greater understanding of safety and process control in many areas of engineering, such as analyzes of alarms allocation, analysis of electrical area classification, determination of routes of leakage, among others.

This work has as objective the development of a simulation tool of operational dynamics of industrial equipment, engaging in 3D graphical representation through an electronic model. To improve the study and the application, this tool was implemented to a plant containing a semi-industrial boiler, as a case study.

2. Material and Methods

In this work, the integration between process simulation and 3D graphical representation was done to provide the interaction with the user. Two specific tools were used: MATLAB and AVEVA Plant Design Management System (PDMS). The MATLAB was used to develop the phenomenological model of the boiler studied in the plant. The PDMS was used to develop the 3D model of the plant. PDMS also work as a bridge between data generated from process simulator (MATLAB) and the 3D model itself in PDMS. The system was used to train operators and safety projects.

2.1 virtual reality

The PDMS is a software engineering that has the capability of modeling three-dimensional CAD (Computer-Aided Design), allowing multiple concurrent users on a single model. The software can be used in several areas. He has a complete interactivity with the environment detailing dimensional, possessing style similar to that adopted by the software of the Windows environment.

The software has a number of benefits that among the most important, is the ability to extract a list of quantitative component containing its data of materials, weights, and other information. Importantly, such a list, depending on the size of the unit may have more than a thousand pages, which greatly increases the chance of inconsistencies, considered the possibility of creating this document manually. Thus, the software allows an important gain in terms of time, eliminating delays, the automated generation of various documents and reports, plans and two-dimensional views and isometric piping, and other documents. The user can check the information contained in the 3D model and compare them to equivalent 2D designs.

2.2 phenomenological simulation

The boiler operation was simulated in MATLAB software. To provide this, was used the model created by ÅSTRÖM & BELL [3], who presented a dynamic model for a boiler, based on basic principles of conservation of mass and energy. Justifies the adoption of this model here for its simplicity and because it has been validated with real data by the authors [3].

2.3 case study: boiler studied

The usability of the tool developed in this work has been tested and applied in a small boiler, model CV-500-VDM. This boiler is located in a pilot plant, which in turn is in the Utilities Central Laboratory of Chemical Engineering, Escola de Química, Universidade Federal do Rio de Janeiro, Rio de Janeiro, Brazil. In the Figure 1 will be shown a picture taken from the plant. All the related equipment was identified in it.

Figure 1: Studied pilot plant containing the boiler simulated

3. Results and discussion

3.1 process simulation

Using the MATLAB and the developed program was possible integrate "process simulation" and "virtual reality". To verify if the data is trustful was produced some graphics and analyze their behavior through steps given in some input variables. The Figure 2 represents the temperature behavior after some different steps given in heat flow, Q.

Figure 2: Temperature behavior after steps in heat, Q

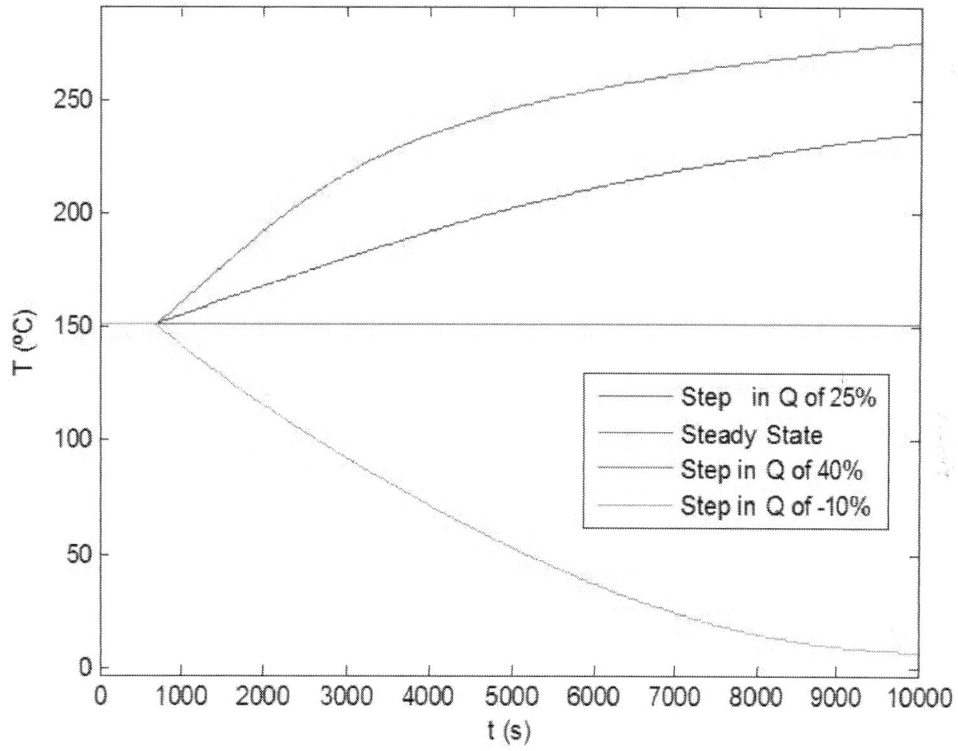

It is possible to see that if the heat is increased, the temperature consequently increase as well. By other side, if the heat is decreased, the temperature also decreases, following the dynamics of the boiler from energy and global mass balance. In the Figure 3, a similar result is obtained, where a step is given in Q and the pressure is observed after it.

Figure 3: Pressure behavior after steps in heat, Q

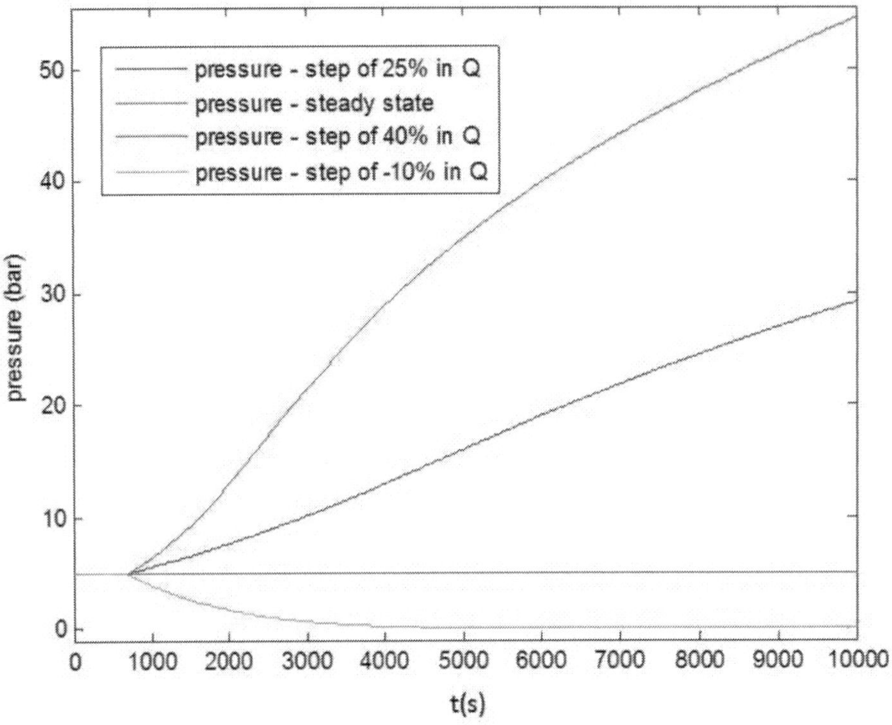

3.2 virtual reality tool

The developed tool has in summary three main functions as results. The first one is provide the integration between the simulations taken in MATLAB. The 3D model shows to the operator if the plant is operating in normal situation or if there is some equipment in alarm status. The model can shows data tables too. In the Figure 4, is possible to see some equipment analyzed by the integration system and showing if they are in normal operation (green), in maintenance (gray), in stand by (yellow) or in alarm (red). Beyond that, it is possible to see relevant data process, like pressure, temperature and level, directly in the model.

The second function is providing the training of the operator. In this work, the training was applied to “boiler starting”. During the training the operator can see instructions and complementary texts. The 3D model change through the actions taken, interacting with the operator. The user has the free movement possibility during the training. In the Figure 5 is possible to see the virtual operator during the training. The water level and pump status are shown in the Figure 5 too.

Figure 4: Equipment analyzed by the integration tool

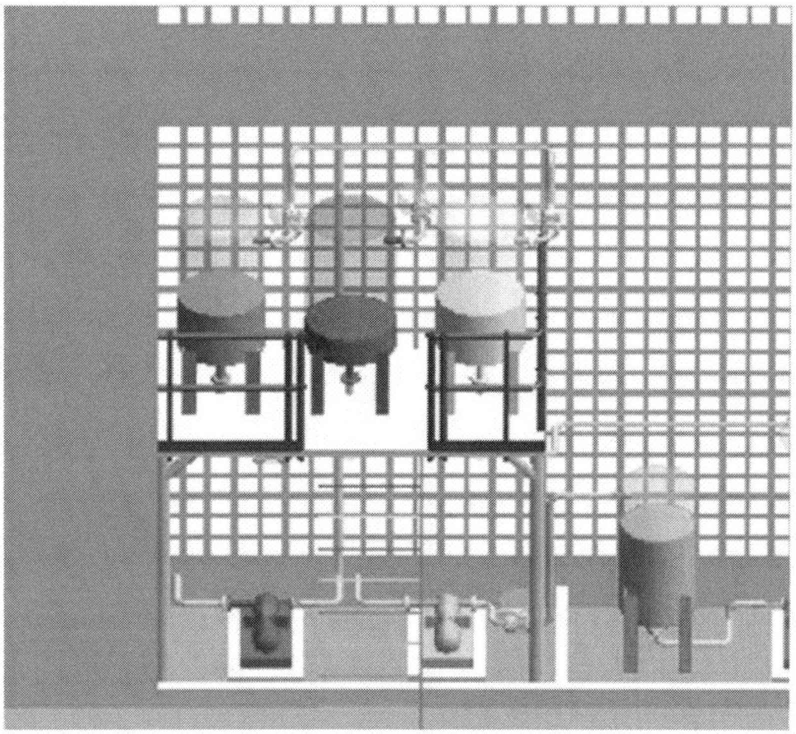

Figure 5: Model during the training

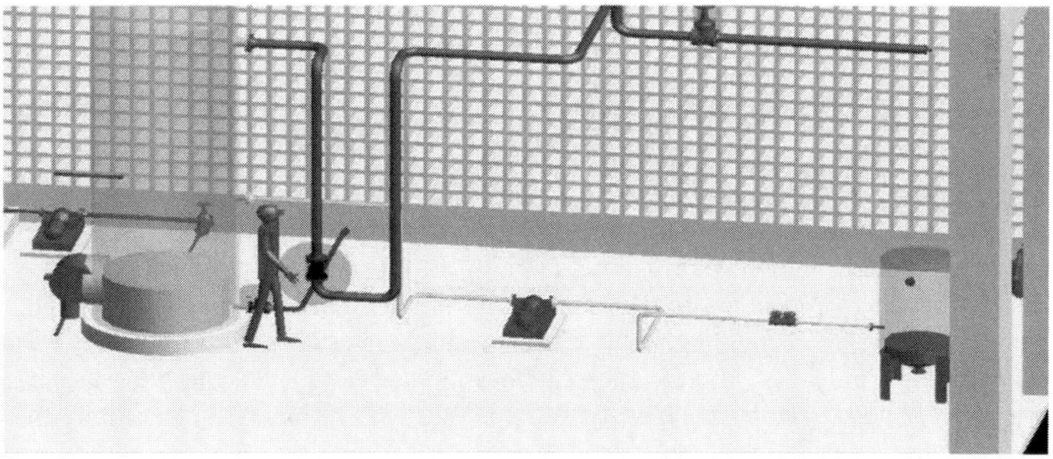

The third function is work as a "support tool", helping the safety analyzes in the plant. The safety analyzes can be applied during project of new process. The virtual reality can be used to design escape routes and emergency signaling. An example of the escape route chooses is showed in the Figure 6.

Figure 6: Virtual tool used to know the best exit to be taken in a case of emergency

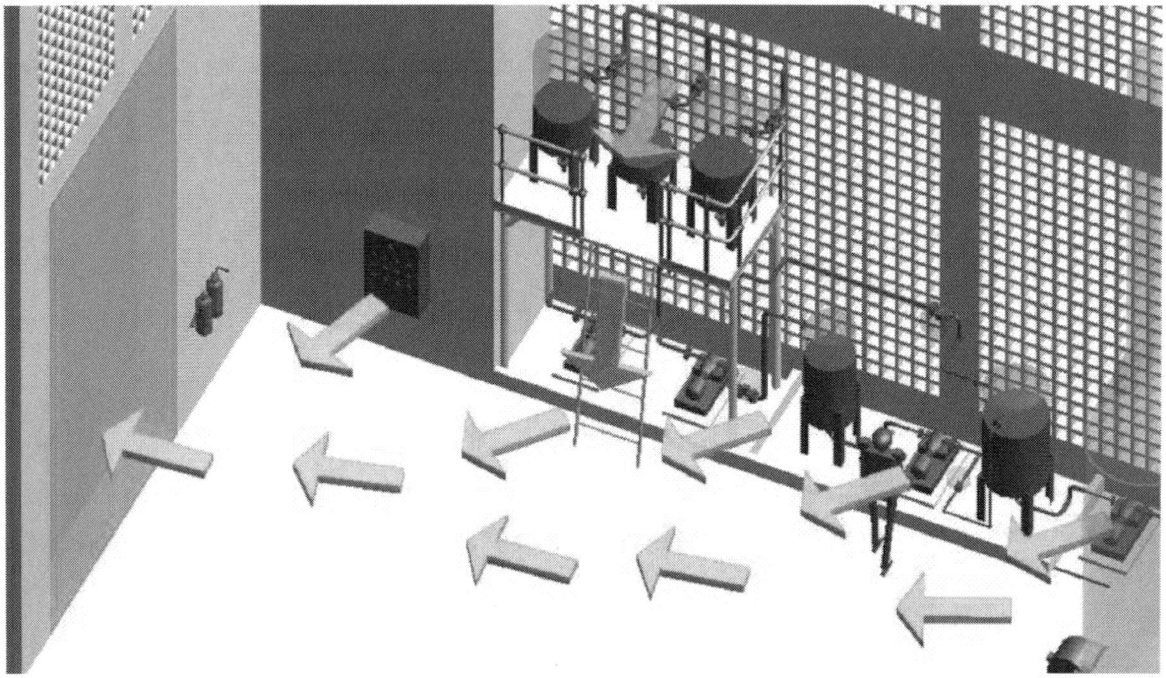

4. Conclusion

The integration of virtual reality and process simulation allow plant operators acquire knowledge about real process. This solution offers low cost and it's safer than use real equipment during operation training. In other words, performing training in the virtual reality tool developed, the user can get information about the plant without actually going to it.

About the development of a three-dimensional supervisory system prototype, it is concluded that it is valid and interesting for training purposes. With the training tool, it is possible to obtain many advantages, due to the operator interaction with the virtual three-dimensional model. So, it is concluded the development of a similar system for all industrial plants can facilitate the training of new operators, facilitating the determination of operations and/or maintenance actions in the plan. Through this, it is possible study the best way to access the site without the need to perform a complete analyzes of the activity at the site to be developed. Furthermore, the safety analyzes can be very useful for the operation team.

About study in general, it can be concluded the use of virtual reality, primarily for training purposes and in addition as a support tool to the operation unit, has much contributing to healthy plant operability. The implementation of virtual reality is becoming an increasingly viable option in view of the development of systems software and hardware capable of supporting the tools developed so that the cost of such equipment is not high as it would at the last decade of previous millennium.

5. References

[1] Lasala, K.P; Filgueiras, L.V.L; Gigley, H.; Ullman, R., *Designing Systems for Reliable Human Performance*, IEEE Tutorial Video, 1997.

[2] Pamplona Jr, A. R. S., *Hidrogeradora Virtual: Utilização de Técnicas de Realidade Virtual Desktop para o Estudo de uma Unidade Hidrelétrica de Energia*, Dissertação de Mestrado, Programa de Pós Graduação em Engenharia Elétrica, Centro Tecnológico, Universidade Federal do Pará, 2006.

[3] Åström, K. J.; Bell, R.D., *Drum Boiler Dynamics*. Automatica, v.36, n.3, p. 363-378, 2000.

Investigations on the influence of different inert gases on the laminar flame propagation in lycopodium/air mixtures

Hannes Kern
Montanuniversität Leoben
Franz-Josef Strasse 18, 8700 Leoben
hannes.kern@unileoben.ac.at

Bruno Wimmer, Harald Raupenstrauch
Montanuniversität Leoben
Franz-Josef Strasse 18, 8700 Leoben
bruno.wimmer@stud.unileoben.ac.at, harald.raupenstrauch@unileoben.ac.at

Prepared for Presentation at
American Institute of Chemical Engineers
2013 Spring Meeting
9th Global Congress on Process Safety
San Antonio, Texas
April 28 – May 1, 2013

UNPUBLISHED

Investigations on the influence of different inert gases on the laminar flame propagation in lycopodium/air mixtures

Hannes Kern
Montanuniversität Leoben
Franz-Josef Strasse 18, 8700 Leoben
hannes.kern@unileoben.ac.at

Bruno Wimmer, Harald Raupenstrauch
Montanuniversität Leoben
Franz-Josef Strasse 18, 8700 Leoben
bruno.wimmer@stud.unileoben.ac.at, harald.raupenstrauch@unileoben.ac.at

Keywords: dust explosion, flame velocity, limiting oxygen concentration

Abstract

The laminar flame velocity is a key parameter in the description of explosions. The determination of the flame velocity (laminar or turbulent) of dust/air flames is relatively complex due to the characteristics of dusts in general. In context of investigations on the limiting oxygen concentration of different powders, the influence of the oxygen concentration and different inert gases on the laminar flame propagation in lycopodium/air mixtures was investigated. In the course of this study, a new apparatus for the determination of the flame velocity was developed. The apparatus mainly consists of a vertical duct with a diameter of 140mm that is split in three main sections. The dust input system lies in the head section of the apparatus and is realized by a small conveyor screw. The bottom section of the apparatus houses the electrical spark ignition system and a gas inlet system. The gas inlet creates a steady flow of a chosen gas. The flame propagation is detected by IR photodiodes along the 2m long main section of the apparatus. The dust concentration is measured by light extinction. The determination of the flame velocity was carried out at dust concentrations of 100g/m³, 200g/m³ and 300g/m³ as well as around the lower explosion level of lycopodium. The oxygen concentration was varied in steps of 1Vol.-% from 21Vol.-% down to the limit where no ignition could be achieved. To investigate the influence of the inert gas on the flame propagation carbon dioxide and nitrogen were used as inert gases. The results of the recent studies are used for further studies on the limiting oxygen concentration of different dusts and the validation of a new CFD model for the simulation of dust explosions in large vessels.

1. Introduction

Dust explosion research mostly relies on sum parameters like the minimun ignition energy (MIE), the maximum explosion pressure or the limiting oxygen concentration (LOC) for the description of explosion processes and material characterization. For mathematical modeling these sum parameters show a lot of uncertainties due to the influence of experimental design. Therefore parameters like the laminar flame velocity are commonly used for the description of the combustion process itself. For combustible gases the determination of the flame velocity is relatively simple compared to combustible dusts. The biggest challenge concerning dusts is the creation of a homogeneous dust cloud with low turbulence.

In the course of this experimental study about the influence of the oxygen concentration on the flame velocity of lycopodium/air flames a new testing device was designed. With this apparatus it is possible to measure the flame speed within a 2m combustion tube, a new method for the calculation of the laminar flame speed from the results is yet to be developed. For current research we relied on present methods. Even if the laminar flame velocity can't yet be measured directly with this device, the influence of changing atmospheric conditions can be observed very well.

2. Experimental investigations

2.1 Method

Various apparatus for the measurement of the flame speed of dust/air mixtures are found in literature [1]. The testing apparatus used for current research consists of three major parts. As shown in Figure 1 the main sections of the apparatus are the top section, the main explosion tube and the bottom section. All elements are built out of a Plexiglas (PMMA) tube with a diameter of 140mm and a wall thickness of 5mm.

The head element houses the dust input system and is equipped with a flame arrestor (Figure 2). The dust input is carried out by a screw conveyor that can be regulated by the operating voltage. The input system allows the generation of dust concentrations from $50 g/m^3$ to $500 g/m^3$ (Lycopodium).

The bottom section contains a gas inlet system and the ignition source. The ignition source is an electric spark igniter (~200mJ), consisting of two 3mm steel electrodes and a high voltage generator. The gas inlet is designed as a circular tube with multiple nozzles to generate a homogeneous gas flow along the main section of the apparatus. The 2000mm long main section of the apparatus is equipped with 5 photodiodes to measure the flame propagation along the tube. The first diode is mounted 10cm above the ignition point, further diodes are placed every 40cm.

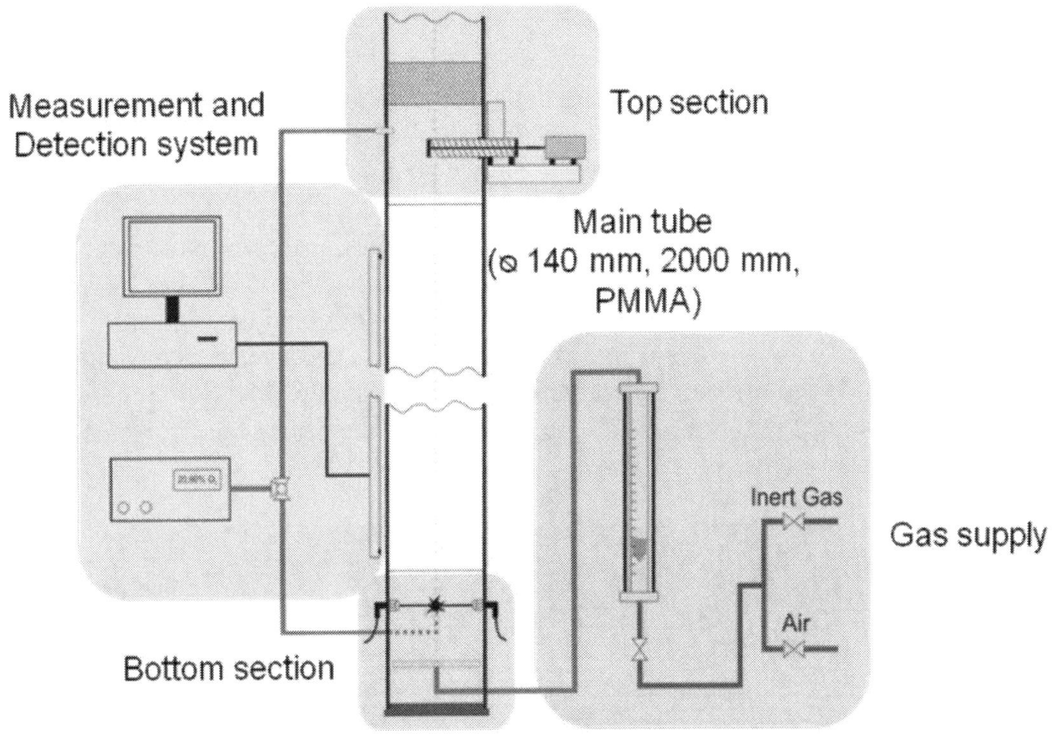

Figure 1: Experimental setup for the determination of the flame velocity

Two of the photodiodes are used to monitor the dust concentration in the tube before ignition by the extinction of IR light emitted by IR diodes. The emitter diodes are turned off shortly before the ignition by the electric spark.

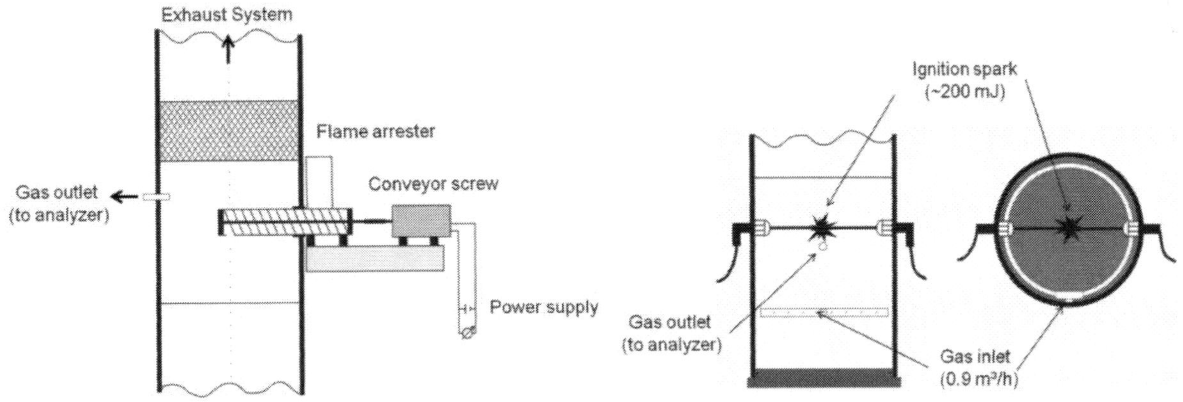

Figure 2: Top section (left) and bottom section (right) of the experimental setup

In the measuring mode a gas flow of 0,015m^3/min is provided through the bottom to realize different atmospheric compositions. The dust concentration is observed throughout IR extinction. The oxygen concentration inside the tube is measured by an oxygen monitor at the head section of the apparatus. When the extinction reaches a constant level along the tube, the emitter diodes are switched off and the ignition source is activated.

2.2 Analysis of flame propagation

For dust explosion research in general, laminar flame velocity is a relevant parameter. Different methods for the determination of the laminar flame velocity are known, like using burners to create laminar dust flames or tube methods like described above. The data sets produced from all of these methods are more or less deviating from laminar conditions due to the experimental conditions. Figure 3 shows a sequence of pictures taken during the measurement of the flame velocity at a dust concentration of 300g/m³ and an oxygen content of about 15Vol.-% using nitrogen as inert gas. The picture shows a more or less turbulent combustion close to the ignition point and a relatively closed flame shape at around 800mm. At lower concentrations the flame tends to be more closed throughout the whole measuring distance. The flame therefore shows a more laminar character at lower concentrations. The turbulence effects may have occurred due to the falling particles themselves.

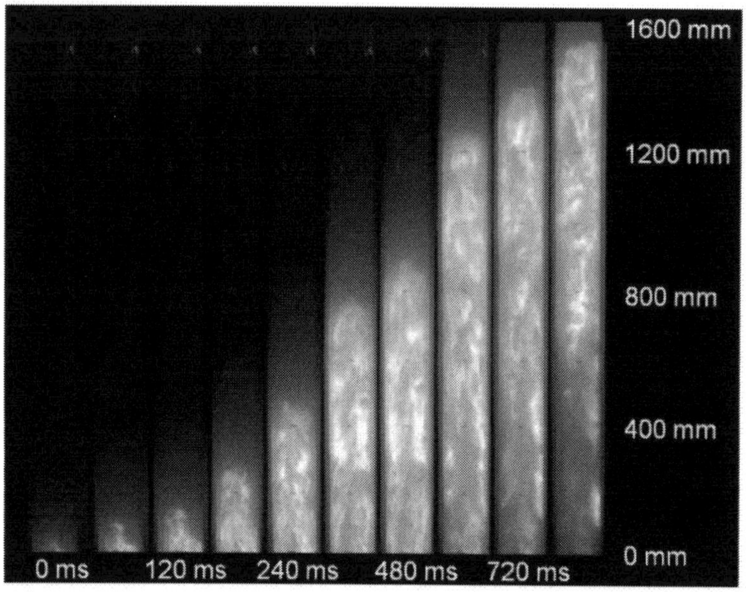

Figure 3: Flame propagation within the testing device

Andrews and Bradley [2] described a method for the calculation of the laminar flame velocity out of relations regarding shape and surface of the flame. Andrews and Bradley's approach (1) was taken into account for further investigations. A video analysis system was installed to determine the flame geometry.

$$S_L = \frac{A'}{A_{fl}} \cdot \left(S_{fl} - u\right) \qquad (1)$$

For a lycopodium concentration of 50g/m³, the flame surface (A_{fl}) and the area cross section (A') were interpolated and calculated by a Matlab code. The measured flame velocity (S_{fl}) was 0.62m/s. Calculations with Andrews and Brandly's relation resulted in a laminar flame velocity (S_L) of around 0.20m/s.

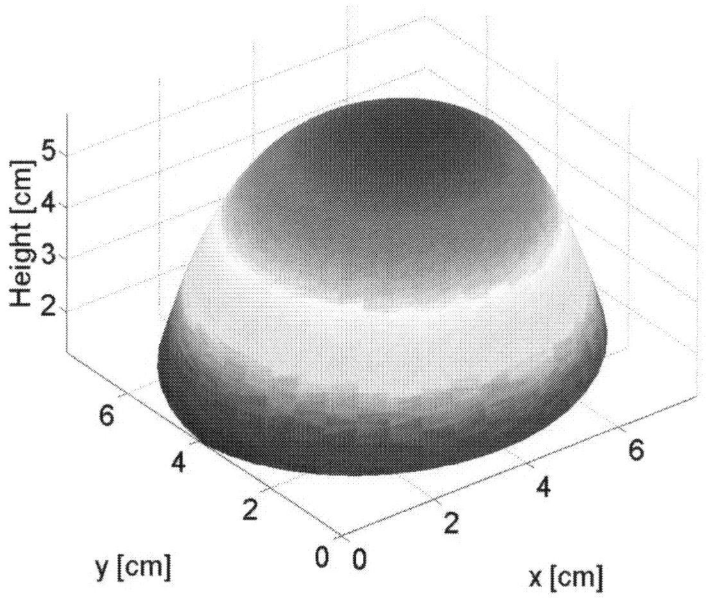

Figure 4: Flame shape at 50g/m³, surface area around 150cm²

Figure 4 shows an example for the flame shape of a 50g/m³ lycopodium/air flame. Compared to the results described by Eckhoff [1], the measured flame velocity is well in the range of velocities measured in other apparatus. For concentrations of around 50g/m³ the flame became very unstable in some measurements, so that higher concentrations were chosen for further research on the influence of the oxygen concentration. Due to the clear connection between actual and laminar burning velocity, only the actual flame speed was measured further on, to avoid deviations from the flame shape analysis.

2.3 Results

Measurements of the flame velocity (S_{fl}) have been carried out at three different dust concentrations. In order to keep the flame stable, concentrations of 100g/m³, 200g/m² and 300g/m³ were investigated. The oxygen concentration was lowered by nitrogen and carbon dioxide as inert gases. Measuring sequences started at ambient conditions (~21Vol.-% oxygen) and were lowered in steps of 1Vol.-% down to the point where no ignition could be reached.

Table 1: Flammability limits

	N_2	CO_2
Dust conc. [g/m³]	cO_2 [Vol.-%]	
100	12.0	15.0
200	12.0	15.0
300	13.0	15.5

As described in Table 1, no ignition could be achieved below 12Vol.-% during nitrogen inerting and 15Vol.-% when carbon dioxide was used. Bartknecht [3] described a relation between the oxygen concentration and the required ignition energy for lycopodium/air mixtures. Regarding the observed ignition limits during the experiments, the results show very good accordance with Bartknecht's values. For an oxygen content of around 12Vol.-% Bartknecht estimated an ignition energy between 150 and 250mJ which is well in the range of the ignition source used for the experiments. Figure 5 shows the development of the flame velocity for nitrogen inerting.

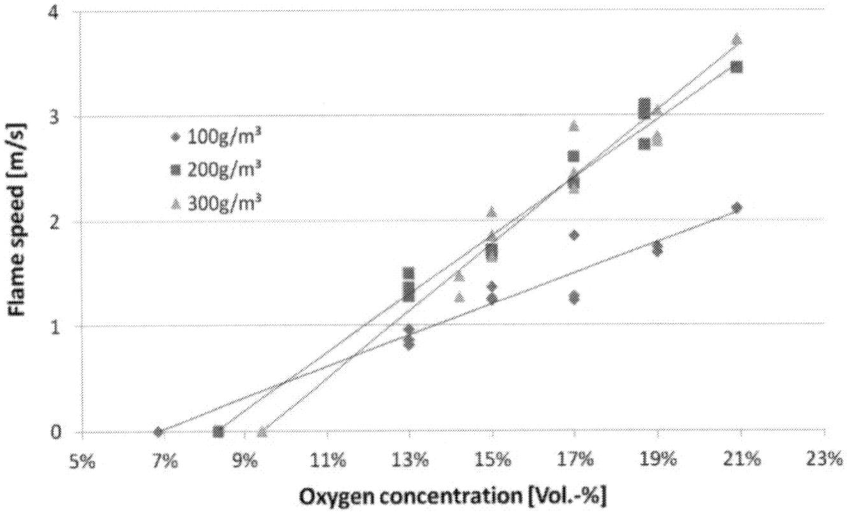

Figure 5: Development of flame speed during nitrogen inerting

At ambient conditions the flame velocity increases with the dust concentration. The increase of the flame velocity in the range from $100g/m^3$ to $200g/m^3$ is significantly bigger than from $200g/m^3$ to $300g/m^3$. This effect can be explained by stoichiometry [4].

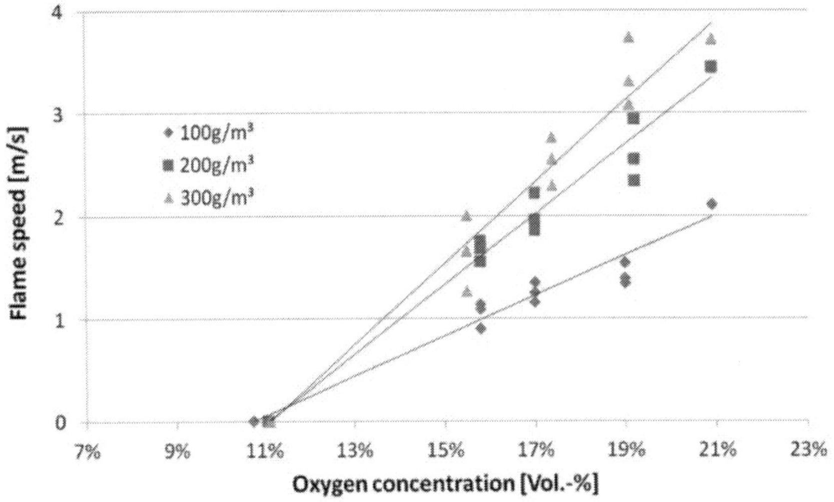

Figure 6: Development of flame speed during carbon dioxide inerting

At lower dust concentrations the combustible material is the limiting component for combustion while at higher dust concentrations oxygen becomes the limiting factor, which is why the flame velocity nearly stagnates. For all dust concentrations the flame velocity decreases with lower oxygen concentrations. The flame velocity drops faster at higher concentrations than at 100g/m³. That is also true for carbon dioxide inerting, as it can be seen in Figure 6. At Figure 5 the lines for 300g/m³ and 200g/m³ intersect at around 17Vol.-%. This is presumably due to problems with the gas measuring system and could not be observed at the carbon dioxide experiments. The values for zero flame speed are estimated.

As mentioned above values for zero flame speed were estimated to be able to compare possible results with the limiting oxygen concentrations of lycopodium. The extrapolated values are given in Table 2.

Table 2: Estimates limiting oxygen concentrations

	N_2	CO_2
Dust conc. [g/m³]	cO_2 [Vol.-%]	
100	6.8	10.8
200	8.4	11.1
300	9.4	11.1
average	8.2	11.0

Siwek [5] found a relation between the limiting oxygen concentrations (LOC) measured during nitrogen inerting and the LOC values for carbon dioxide inerting (2).

$$LOC_{CO2} = 1.36 \cdot LOC_{N2} \quad (2)$$

Due to the described problems with measurement attempts were made to recalculate the LOC for nitrogen inerting out of the measured values for carbon dioxide inerting. The estimation resulted in a value of 8Vol.-% for nitrogen inerting which is highly consistent with the 7.5Vol.-% published in BIA Report 12/97 [6] and the average value that was measured.

3. Conclusions

During the investigations on the flame speed of lycopodium/air mixtures a new apparatus for the determination of the flame velocity was developed. The results from the method regarding laminar flame speed showed good accordance with published values at ambient conditions. With the present experimental setup it was also possible to determine the influence of varying atmospheric conditions on the flame propagation in a vertical duct. Carbon dioxide and nitrogen were used as inert gases. Due to the relatively low ignition energy provided by the spark igniter, the flame propagation around the known limiting oxygen concentrations for lycopodium could not be reached. Extrapolations on the limiting oxygen concentrations based on the measured data have shown good accordance with values measured in the ISO 20L sphere. In addition to the the influence of inert gases on the flame propagation other deviations from atmospheric conditions will be investigated in future research. The results from the current experimental work are used for the validation of a new CFD model for the calculation of dust explosions in complex geometries.

4. References

[1] Eckhoff R. K. (3rd edition) (2003) Dust Explosions in the Process Industries. Gulf Professional Publishing

[2] G.E. Andrews, D. Bradley (1972), Determination of burning velocities: A critical review, Combustion and Flame, Vol. 18/1, Elsevier 1972

[3] Bartknecht W. (1993), Explosionsschutz – Grundlagen und Anwendung. Springer – Verlag

[4] Crowl D. A. (2003), Understanding Explosions, Center for Chemical Process Safety (CCPS) of the American Institute of Chemical Engineers, 1st edition, Wiley-AIChE, New York 2003

[5] Siwek R. (1989), Zuverlässige Bestimmung explosionsschutztechnischer Kenngrößen in der 20l-Laborapparatur, VDI Berichte 701, S215-262

[6] BIA – Report 12/97, Brenn- und Explosionskenngrößen von Stäuben, Hauptverband der gewerblichen Berufsgenossenschaften, November 1997

(44al) Fire Protection and Life Safety Issues in Semiconductor and Related Facilities

Jonathan Eisenberg, Rolf Jensen & Associates, Framingham, MA

Semiconductor and related process facilities contain numerous fire and explosion hazards that impact life safety and operations. This presentation discusses the major fire protection and life safety design aspects for such facilities, including specific requirements contained in the International Building & Fire Codes, as well as applicable NFPA codes and standards. Project examples will be used to illustrate how these aspects are addressed in facility and process design.

The presentation will include the following:

Location and design of Use Group H-5 occupancies – Major IBC Chapter 4, IFC Chapter 18, and NFPA 318 fire protection and life safety requirements, including vertical and horizontal location, occupancy separation, code approach for associated mechanical spaces including interstitial levels, electrical classification, ventilation, fire suppression (overhead and local), fire and gas/vapor detection, and emergency alarms.

Location and design of Use Group H-2, H-3 and H-4 storage areas – Building and fire code requirements for vertical and horizontal location, occupancy separation, electrical classification, ventilation, fire suppression, fire and gas/vapor detection, and spill control/secondary containment. Approaches and ideas on the number, location and function of flammable liquid, flammable and pyrophoric gas, and highly toxic gas storage areas, including piping and delivery philosophy.

Special suppression and detection – Fire protection system approaches for hazards found in a clean room/fab area, eg. wet bench chemistry involving flammable liquids, and processing that uses hydrogen and/or silane mixtures. Discussion of local suppression options, as well as specialty gas and flame detection. We will give our thoughts and experiences on how these systems interact with the rest of the building and the responding fire department.

Industrial Issues – Hydrogen detection; combustible dust hazards; NEC Article 500 electrical classification; explosion protection; and several examples from our industrial/chemical project experience

Licensing and permitting – Discussion of issues faced by owners of fab and clean room facilities that store and use hazardous materials, where the hazards are not clearly understood outside of the semi-conductor industry.

9TH GLOBAL CONGRESS ON PROCESS SAFETY

Lower Flammability Limits of Mixtures of Hydrogen and Methane at Atmospheric and Sub-atmospheric Pressures

Hai Le
Mary Kay O'Connor Process Safety Center, Artie McFerrin Department of Chemical Engineering, Texas A&M University
Room 200, Jack E. Brown Building, Texas A&M University, 3122 TAMU College Station, TX 77843-3122
haile@tamu.edu

Yi Liu, M. Sam Mannan
Mary Kay O'Connor Process Safety Center, Texas A&M University
Room 200, Jack E. Brown Building, Texas A&M University, 3122 TAMU
College Station, TX 77843-3122
mannan@tamu.edu, liuyi.chemical@mail.che.tamu.edu

Prepared for Presentation at
American Institute of Chemical Engineers
2013 Spring Meeting
9th Global Congress on Process Safety
San Antonio, Texas
April 28 – May 1, 2013

UNPUBLISHED

Lower Flammability Limits of Mixtures of Hydrogen and Methane at Atmospheric and Sub-atmospheric Pressures

Hai Le
Mary Kay O'Connor Process Safety Center, Artie McFerrin Department of Chemical Engineering, Texas A&M University
Room 200, Jack E. Brown Building, Texas A&M University, 3122 TAMU College Station, TX 77843-3122
haile@tamu.edu

Yi Liu, M. Sam Mannan
Mary Kay O'Connor Process Safety Center, Texas A&M University

Keywords: lower flammability limit, hydrogen, methane, pressure

Abstract

The knowledge of the LFLs of hydrogen mixtures with hydrocarbons is essential since hydrogen is present in various mixtures with hydrocarbons in many processes and applications. The lower flammability limits of various mixtures of hydrogen and methane in air were determined experimentally in a closed cylindrical stainless steel vessel (inner diameter 10.22 cm, length 100 cm) with upward flame propagation at room temperature (20 ^oC) and initial pressures of 1.0 atm, 0.5 atm, and 0.1 atm. It was found that the LFLs of mixtures of the same hydrogen fuel fractions did not change much when the initial pressure decreased from 1.0 atm to 0.1 atm. And pressure has more impact on the LFLs when there is more hydrogen in the fuel mixture. The application of Le Chatelier formula to predict the LFLs of the mixtures was also evaluated. Overall, Le Chatelier formula can predict the LFLs of mixtures of hydrogen and methane very well at all initial pressures.

1. Introduction

Lower flammability limit (LFL) is an important parameter used in the prevention of fire/explosion when handling combustible gas mixtures. By definition, LFL is the minimum concentration of gas in air capable of propagating flame upon ignition;[1] in other words, if the concentration of a gas is smaller than its LFL, it is no longer flammable. Therefore, LFL is widely used in the calculation of the level of inerting needed to prevent the formation of flammable mixtures.[2] In these calculations and other fire risk assessments, it is advised to select and use the proper value of LFL for the operating condition since LFL is affected by pressure, temperature, direction of flame propagation and other conditions.[1]

The literature has a vast number of LFLs of various gases determined with different apparatuses at both atmospheric and non-atmospheric conditions, e.g. high temperatures, pressures, etc. LFLs of pure gases are more readily available than LFLs of mixtures of gases. Whenever LFLs of mixtures of gases are not available, methods have been developed to predict them. One of the widely used methods is Le Chatelier formula. It is an empirical mixing rule to predict the LFL of a mixture of gases as below: [3]

$$LFL_{mix} = \frac{1}{\sum_i^n \frac{y_i}{LFL_i}}$$

where LFL_{mix} is the LFL of the mixture, y_i is the mole fraction and LFL_i is the LFL of the ith component. For mixtures at non-atmospheric condition (e.g. high pressure, temperature) the LFLs of the components at the same condition should be used to accurately estimate the LFLs of the mixtures.

The knowledge of the LFLs of hydrogen mixtures with hydrocarbons is important since hydrogen is present in various mixtures with hydrocarbons in many processes and applications. Examples of such mixtures are refinery gas streams which contain various percentages of hydrogen (up to 90%) and hydrocarbons (methane, ethane, C_3, C_{4+}.).[4] Hydrogen is also added to natural gas, consisting mostly methane, to reduce CO, CO_2 and NO_x emissions[5-7], improve flame stability,[6, 7] and extend the lean operating limit of spark ignited engines.[8] It is critical to understanding the LFLs of these hydrogen mixtures in order to handle them safely.

Reference to the literature shows that studies on the LFLs of mixtures of hydrogen and hydrocarbons at non-atmospheric conditions, particularly sub-atmospheric pressure conditions, are very limited. Therefore, to enhance our understanding about the LFLs of hydrogen mixtures, this study focuses on investigating the LFLs of various mixtures of hydrogen and methane at atmospheric and sub-atmospheric pressures. Specifically, the LFLs of mixtures of hydrogen and methane were determined experimentally at room temperature (20 °C) and initial pressures of 1.0 atm, 0.5 atm and 0.1 atm. All experiments were carried out in a closed cylindrical vessel with upward flame propagation. The application of Le Chatelier formula to predict the LFLs of mixtures of hydrogen and methane at atmospheric and sub-atmospheric pressures was also verified using the LFLs of pure hydrogen and methane generated by the same experimental method and conditions.

2. Experimental setup

The experimental apparatus and procedure is illustrated schematically in Figure 1 and briefly described below.

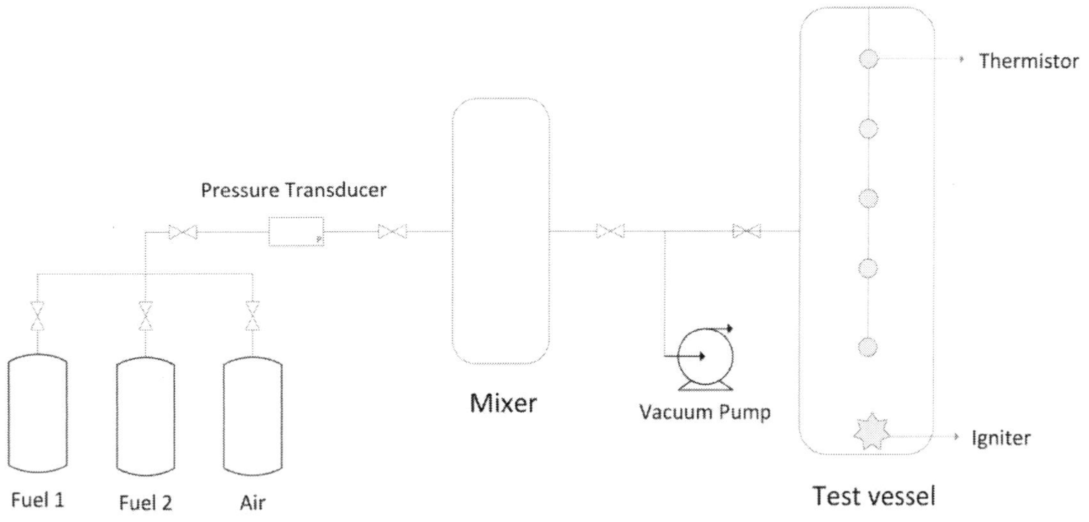

Figure 1: Experimental Setup

The LFLs of mixtures of hydrogen and methane were determined in a test vessel which was a closed stainless steel cylindrical vessel whose design was similar to that developed and used by the U.S. Bureau of Mines.[1] The vessel was 100 cm long with the inner diameter of 10.12 cm and outer diameter of 11.43cm. The igniter was installed at the bottom of the vessel and was capable of providing 10 J of energy consistently. The pressure inside the test vessel was measured by a pressure transducer placed at the top of the vessel. A set of 5 thermistors was positioned along the center axis of the test vessel to detect flame propagation. The distance between the top thermistor and the igniter was 75 cm. Further details about the configuration of the test vessel and its components can be found in our previous published work.[9]

High purity fuels and air from pressurized cylinders were loaded into the mixing vessel. The amount of the fuel and air were determined based on the partial pressure basis. The fuels and air were thoroughly mixed in a mixer which was a stainless steel vessel containing a Teflon block which slid along the vessel when it rotated lengthwise to enhance mixing. After the mixing, the test mixture was permitted to flow to the test vessel at the desired initial pressure measured by the pressure transducer. Then the mixture was ignited from the bottom, and the flame propagation was detected by the 5 thermistors inside the test vessel. Signals from the thermistors were obtained by a Keithley data acquisition card (Keithley® KPCI-3102, 225 signals per second at 0.05% accuracy) installed in a desktop computer, and the experimental process was controlled by a LabView® program. All experiments were carried out at room temperature (20 ^{o}C) and initial pressure of 1.0 atm, 0.5 atm and 0.1 atm. After each experiment, combustion products were purged from the test vessel which was then flushed with nitrogen and vacuumed to be ready for the next experiment.

A mixture is considered flammable if it can propagate flame to the top thermistor upon ignition;[9, 10] this behavior was defined as continuous flame propagation.[9, 10] For each mixture composition, 10 experiments were repeated and the probability of continuous flame propagation was recorded. Between two concentration values different by one experimental step size, one has the probability of continuous flame propagation equal or less than 50%, the other has the probability more than 50%. The former was chosen as the LFL, unless the former's probability is 0%, the later was chosen as the LFL. [9, 10]

3. Experimental Results and Discussion

3.1 Lower flammability limits (LFLs) of mixtures of hydrogen and methane at atmospheric and sub-atmospheric pressures

The experimentally obtained LFLs of various mixtures of hydrogen with methane at room temperature (20 °C) and initial pressure of 1.0 atm, 0.5 atm and 0.1 atm are shown in Figure 2. Overall, the LFLs of the mixtures of the same hydrogen fuel fractions did not change much when the initial pressure decreased from 1.0 atm to 0.1 atm. This is understandable since the LFLs of pure hydrogen and methane also changed very little with the initial pressure. When the fraction of hydrogen in the fuel mixture increased from 0% to 100%, the LFLs of the mixtures showed larger deviation from each other under the change of pressure from 1.0 atm to 0.1 atm. This suggests that pressure has more impact on mixtures of hydrogen and methane when there is more hydrogen in the mixture. However the impact is not significant since the degree of changes of the LFLs is small. It is worth to note that the LFLs of the mixtures was the lowest at 0.5 atm which means that there is an increased risk of fire/explosion when the initial pressure is reduced from 1.0 atm to 0.5 atm. Thus, more attention and precautions are recommended when handling mixtures of hydrogen and methane at pressures lower than 1.0 atm.

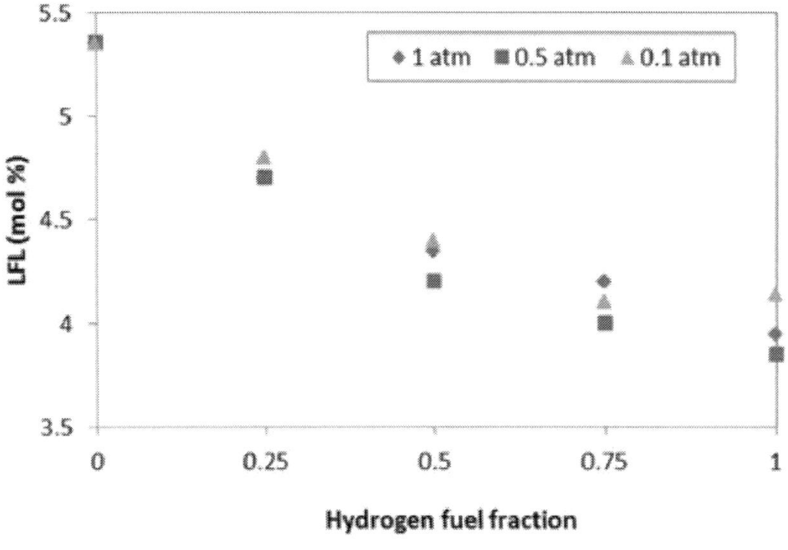

Figure 2: LFLs of mixtures of hydrogen and methane at atmospheric and sub-atmospheric pressures and room temperature (20 °C)

3.2 The application of Le Chatelier formula to the LFLs of mixtures of hydrogen and methane at atmospheric and sub-atmospheric pressures

The LFLs of mixtures of hydrogen and methane were calculated based on Le Chatelier formula using the LFLs of pure hydrogen and methane determined experimentally with the same experimental setup and conditions. Comparisons between the calculated LFLs and experimental LFLs are presented in Figure 3, Figure 4, and Figure 5 for 1.0 atm, 0.5 atm and 0.1 atm respectively. In general, Le Chatelier formula can predict the LFLs of the mixtures very well at all initial pressure conditions; in other words, the prediction of Le Chatelier formula for hydrogen and methane mixtures is not affected by changes in the pressure from 1.0 atm to 0.1 atm. The maximum deviation between experimental and calculated LFLs is 7.0 %. And the deviation is the largest when the fraction of hydrogen in the fuel mixture is 0.5.

The application of Le Chatelier formula to the upper flammability limits (UFLs) of the mixtures of hydrogen and methane will be evaluated in the future when we complete the determination of the UFLs of the mixtures experimentally.

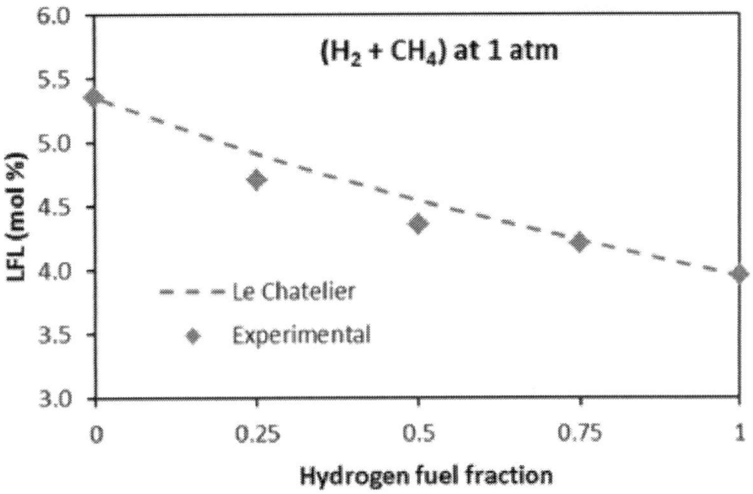

Figure 3: LFLs of mixtures of hydrogen and methane at room temperature and 1.0 atm (experimental and calculated by Le Chatelier formula)

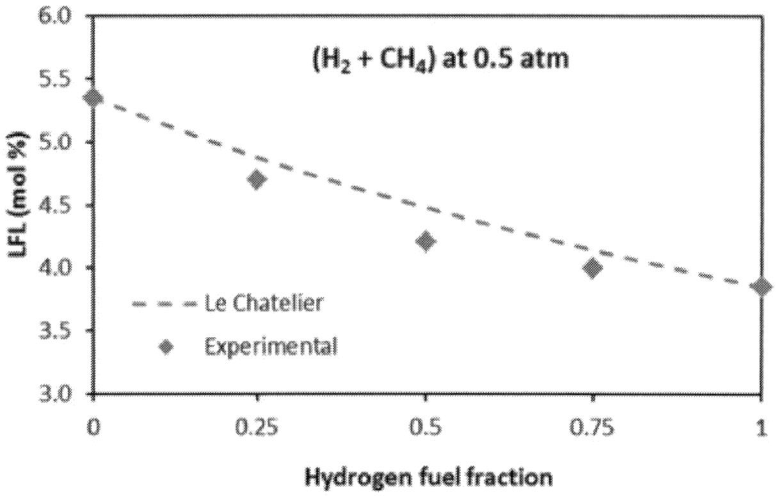

Figure 4: LFLs of mixtures of hydrogen and methane at room temperature and 0.5 atm (experimental and calculated by Le Chatelier formula)

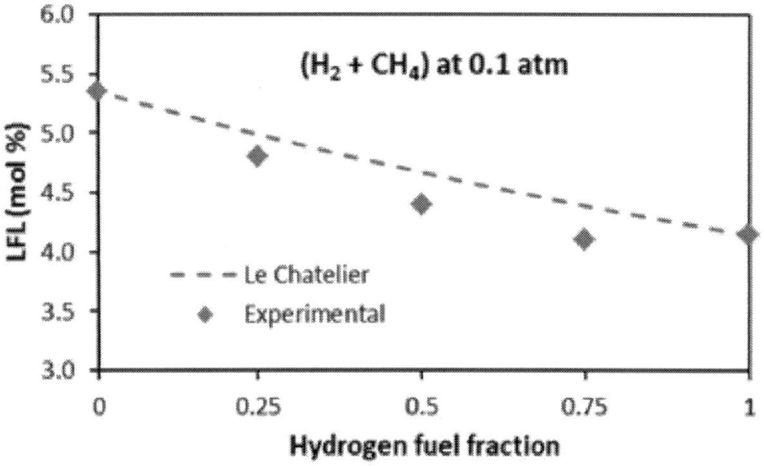

Figure 5: LFLs of mixtures of hydrogen and methane at room temperature and 0.1 atm (experimental and calculated by Le Chatelier formula)

4. Conclusion

The lower flammability limits (LFLs) of mixtures of hydrogen and methane in air were determined experimentally at room temperature and initial pressure of 1.0 atm, 0.5 atm and 0.1 atm. The application of Le Chatelier formula to predict the LFLs of the mixtures was also evaluated. It was found that:

- The LFLs of mixtures of the same hydrogen fuel fractions did not change much when the initial pressure decreased from 1.0 atm to 0.1 atm. Pressure has more impact on the LFLs of the mixtures when there is more hydrogen in the fuel mixture. The LFLs of the mixtures were the lowest at 0.5 atm which implies an increased risk of fire/explosion when the initial pressure decreases from 1.0 atm to 0.5 atm.

- Le Chatelier formula can predict the LFLs of mixtures of hydrogen and methane very well, and the prediction of Le Chatelier formula is not affected by changes in the pressure from 1.0 atm to 0.1 atm.

5. Acknowledgement

This research was sponsored by the Mary Kay O'Connor Process Safety Center, Artie McFerrin Department of Chemical Engineering, Texas A&M University.

6. References

[1] Coward, H.F. and G.W. Jones, *Limits of flammability of gases and vapors*, in *Bureau of Mines Bulletin,* 1952.

[2] Crowl, D.A. and J.F. Louvar, *Chemical Process Safety: Fundamentals with Applications, Second Edition*. 2001: Prentice Hall.

[3] Mashuga, C.V., *Determination of the combustion behavior for pure components and mixtures using a 20L sphere*, in *Chemical Engineering,* 1999: Michigan.

[4] Kaldis, S.P., G.C. Kapantaidakis, and G.P. Sakellaropoulos, *Simulation of multi-component gas separation in a hollow fiber membrane by orthogonal collocation — hydrogen recovery from refinery gases.* Journal of Membrane Science, 2000. **173**: p. 61-71.

[5] TerMaath, C.Y., et al., *Emissions reduction benefits from hydrogen addition to midsize gas turbine feedstocks.* Int J Hydrogen Energy, 2006. **31**(9): p. 1147-1158.

[6] Guo, H., et al., *The effect of hydrogen addition on flammability limit and NOx emission in ultra-lean counterflow CH4/air premixed flames.* Proc Combust Inst, 2005. **30**(1): p. 303-311.

[7] Hawkes, E.R. and J.H. Chen, *Direct numerical simulation of hydrogen-enriched lean premixed methane–air flames.* Combust Flame, 2004. **138**(3): p. 242-258.

[8] Bell, S.R. and M. Gupta, *Extension of the lean operating limit for natural gas fueling of a spark ignited engine using hydrogen blending.* Combust Sci Technol, 1997. **123**(1-6): p. 23-48.

[9] Le, H., S. Nayak, and M.S. Mannan, *Upper Flammability Limits of Hydrogen and Light Hydrocarbons In Air at Sub-atmospheric Pressures.* Ind. Eng. Chem. Res, 2012. **51**: p. 9396-9402.

[10] Wong, W.K., *Measurement of flammability in a closed cylindrical vessel with thermal criterion*, in *Artie McFerrin Department of Chemical Engineering, Texas A&M University*, 2006.

Informatics Rich Integrated Chemical Safety Risk Assessment (ICSRA) Platform

Muthukumarasamy Karthikeyan1, Renu Vyas2, HV Prasad3

1 Centre of Excellence in Scientific Computing and Digital Information Resource Centre, CSIR-National Chemical Laboratory, Pune, India, CSIR-National Chemical Laboratory, Pune, India
2 Chemical Engineering and Process Development, CSIR-National Chemical Laboratory, Pune, India, CSIR-National Chemical Laboratory, Pune, India
3 Chemical Engineering and Process Development, CSIR-Indian Institute of Chemical Technology, Hyderabad, Habsiguda, Hyderabad-500 607, India India

Abstract

Inherently Safer Practices (ISP) is a systematic innovative scientific methodology integrated with the process, design and operations for facilitating the improvement of process safety at all stages of process life cycle. ISP constitutes intensification (minimization), substitution, attenuation, limitation of effects and simplification as procedural options for implementation in process plants.

For this study we conducted literature data search focused on recent domestic and international incidents where the primary cause was related to chemical reactivity and runaway reactions; Several parameters were taken into consideration including chemical manufacturing (i.e., raw material storage, chemical processing, and product storage) and other industrial activities involving bulk chemicals, such as storage/distribution, waste processing, and petroleum refining.

For purposes of the incident search, only reactive incidents that caused serious consequences would be examined. Source of data: Four decades of data collection involve searching over 40 data sources, focusing on incidents where the primary cause was related to chemical reactivity. For the purposes of the investigation and analysis, an "incident" was defined as a sudden event involving an uncontrolled chemical reaction—with significant increases in temperature, pressure, and/or gas evolution etc.,

The proposed experimental and simulation platform will serve as a repository for plant-specific information, and facilitates the classification and ranking of Indian chemical industry - for the first time - through a Risk index.

Introduction

Computational and Simulation of Industrial chemical reactions to predict runaway reactions and conditions as safety measures. To our knowledge, no systematic studies on inherent safety are yet reported in India. The studies comprise important research areas e.g., process evaluation, estimation of reactivity, hazard identification and evaluation, assessment and prevention of thermal runaways, early fault diagnosis etc. The computational risk assessment platform intends to collect new chemical process safety and risk data, and plant-specific incident data from participating industries across the world for model building and simulation.

To integrate and include a dynamic interactive information platform, viz., chemical safety risk ranking and incident reporting system, by the application of High performance Computing (Cloud, Grid and Distributed systems) for simulation, modeling and response for use by Govt., Industry, Emergency Responders, Research and academic communities, and the public. Upgradation of Thermo chemical Laboratories to assess the proposed/existing industry from the context of implementing ISP, i.e., towards safer process chemistry including selection of solvents, safer process routes, safer process operating conditions for preventing Runaways, early detection of runaways, evaluation/selection of process and designs to handle lesser inventories etc.

Methodology

The methodology devised comprises recognition of the hazards and risks posed by the process, and efforts to reduce or control them to the lowest practical levels, keeping in view the principle objectives of the business. Here we studied data based on literature to identify and collect chemically significant industrially important data pertaining to physico-chemical, biological and toxicological profiles of chemicals (reactants, reagents, catalysts, solvents, products, side products) for their suitability and reusability in simulation and modelling. This includes development of suitable algorithms, chemical reaction component descriptors and methods to simulate experimental conditions and validation through experiments to improve the predictions of risk assessment and prevention of disaster in industrial conditions (Lab-scale, pilot plant and large scale manufacturing).

This process also includes mapping known literatures (Publications, patents and reports) to the existing knowledge for interactive risk assessment of chemicals and incidents in industrial environment. The computational approaches include classical methods, semi-empirical methods, and quantum methods to determine properties at the molecular level. We are developing a computational problem-solving platform and these tools, data and methods to build ICSRA a cloud-computing infrastructure for integrated chemical safety risk assessment of chemicals manufactured, imported and exported from world chemical sources. Proper documentation and accurate predictive models on chemicals of Indian origin would facilitate compliance with other global compliance on chemicals being exported especially in REACH, EPA, FDA and other global federal regulation context.

Development of QSPR (Quantitative Structure Property Relationships) models to relate macroscopic behaviour to molecular properties with detailed molecular computations to obtain thermo chemical data for reactive materials and to predict calorimetric data based on molecular properties. These open source based tools are integrated to this platform (ChemInfoCloud) includes identification and assessment of hazards due to chemical reactivity and risk assessment.

For this systematic study we considered major potential chemical group (CG) which are frequently encountered in industrial reagents like Organic peroxides, Organic Nitrites, Unsaturated Hydrocarbons, Epoxides, Hydrides and Hydrogen, Metal acetalides etc. Efforts to study and model major Chemical Reaction Types (CRT) include Decomposition, Polymerization, Pyrophoric, Peroxide Former, Water sensitive and Oxidation reactions. Molecular computing would help to predict heat of formation, electronic energy, core-core repulsion, dipole, no. of filled levels, ionization potential, molecular weight, topological, electronic and three dimensional descriptors for critical components of reactants, products, solvents.

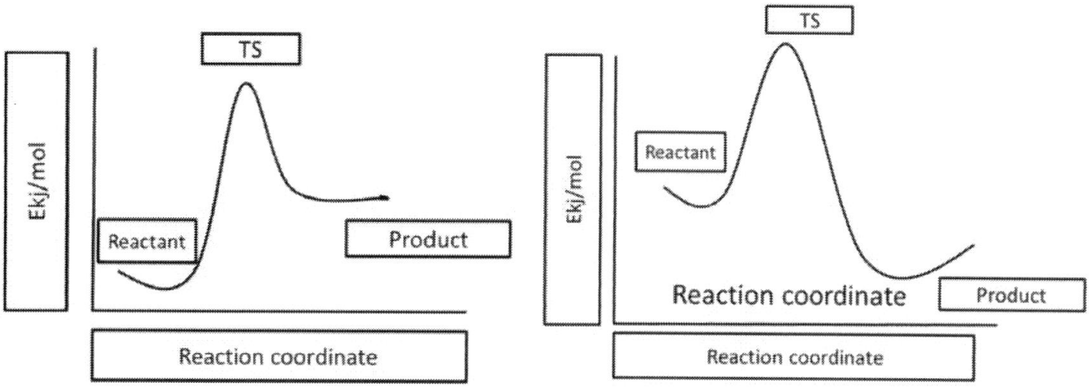

Figure-1: Computational Modeling generic types of chemical reactions along the reaction co-ordinates

Theoretical study of transition states of chemical reactions, computation of activation energies and experimental validation to predict that one reaction pathway is favoured over another reaction pathway. In order to predict the type of polymerization and runaway reactions enthalpy of reactions and kinetics was calculated using the semi-empirical quantum methods, molecular modelling and dynamics simulations of reactants under simulated diverse experimental conditions of Industrial Chemical Processes (ICP), including homogeneous and heterogeneous catalysis.

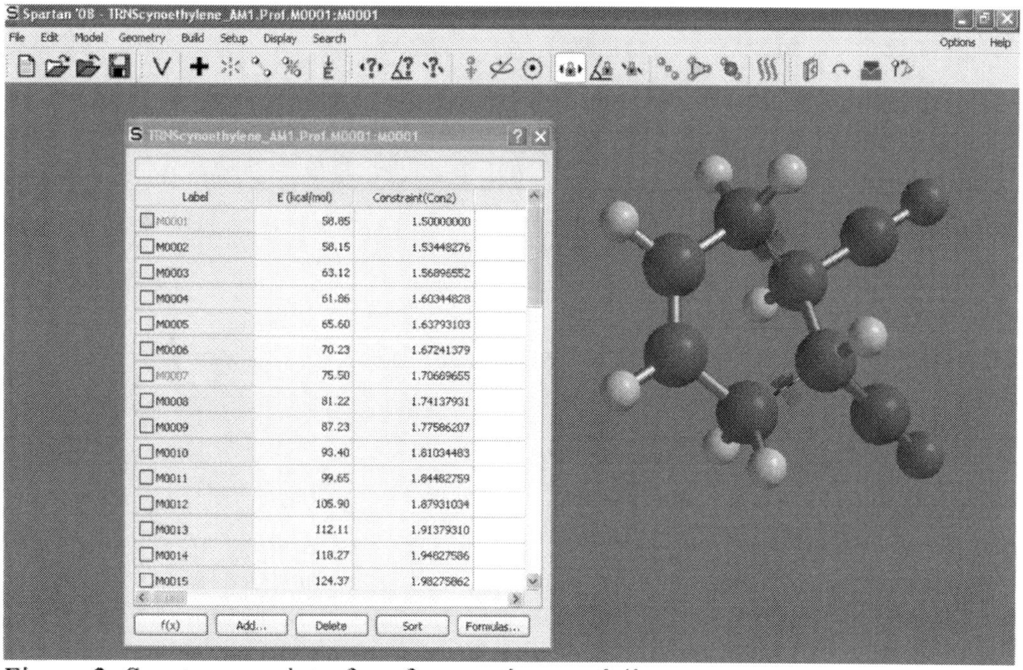

Figure 2: Spartan user interface for reaction modeling

Reactants **Transition State** **Product**

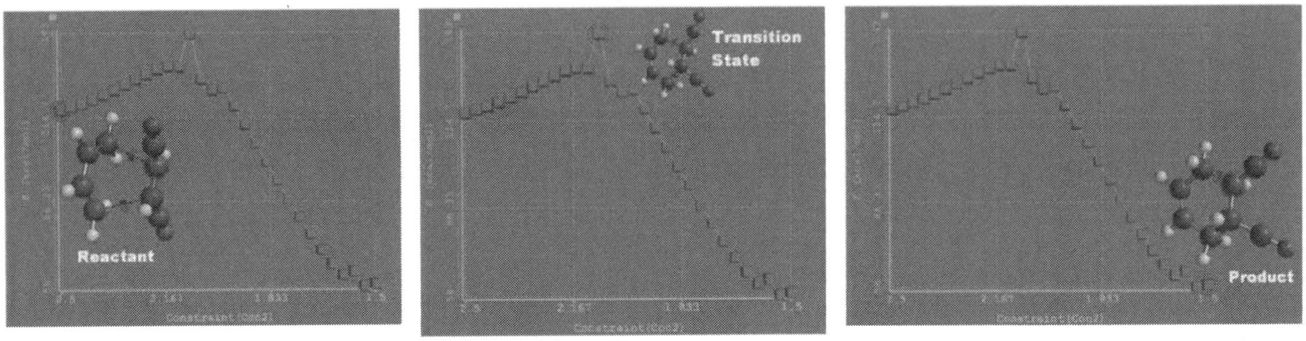

Figure: 3 Computational modelling of organic reactions (Exothermic)

Reactants	Transition State	Product
Heat of Formation: **489.651** kJ/mol Solvation Energy: SM5.4/A -9.308	Heat of Formation: **595.583** kJ/mol Solvation Energy : SM5.4/A -21.665	Heat of Formation: **246.229** kJ/mol

Table:1 Sample data of predicted Heat of formation for reactants, transition state and product

E (TS-Reactant)	E (Reactant-Product) ** heat generated!
E1=595.583-489.65 =105.933 kJ/mol =25.32 kcal/mol	E2=489.65-246.229 =243.421 kJ/mol =58.19 kcal/mol

Table:2 Computed energy difference

Digital mapping of incidental data concerning process safety and their availability to decision makers would help effective risk assessment and rapid response during unexpected events. During the presentation the detailed roadmap, current scenario and future direction of the chemical safety risk assessment platform will be discussed and the input from delegates for this meeting would be considered for integration into the tool if required.

References

1) Organic Reactions Series. John Wiley & Sons
2) http://downloads.wavefun.com/Spartan10Manual.pdf

IT Enablement For Effective Implementation And Monitoring of Process Safety Management

Ajay Kumar/ Pawan Mittal

Reliance Industries Limited

Jamnagar

Ajay.kumar@ril.com/ Pawan.mittal@ril.com

IT Enablement for various PSM elements

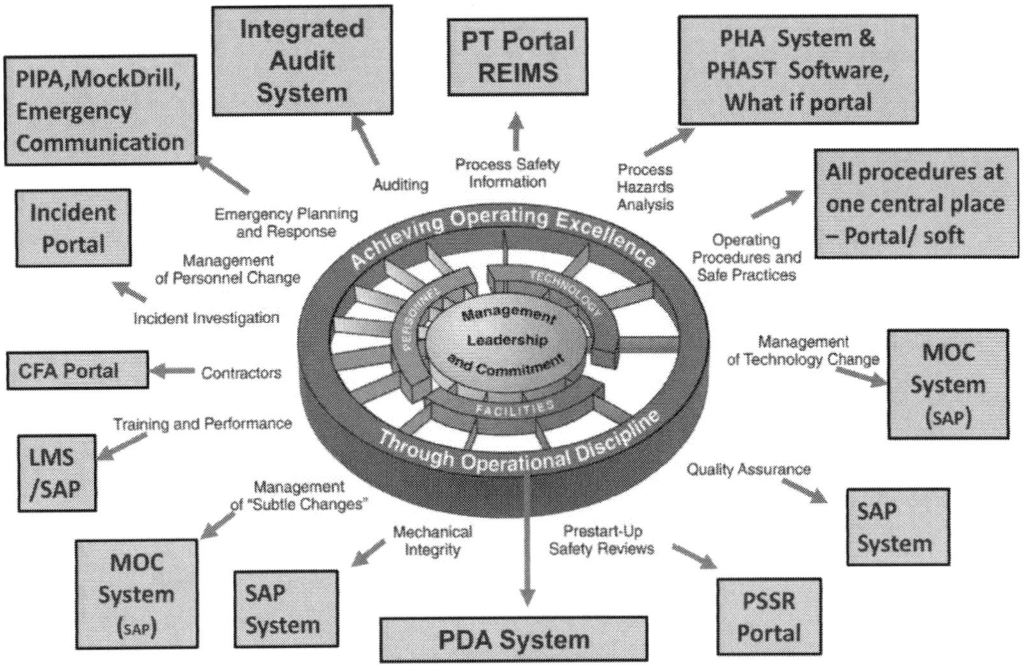

Governance and Control : Corrective Action Tracking System

Advantages of IT Enablement of PSM Elements

- **Process Technology (PT)**
  - ✓ Single repository for storage of all PT related documents
  - ✓ Easy access to PT documents to all concerned.
  - ✓ Easy updation of PT documents

- **Process Hazard Analysis (PHA)**
  - ✓ Automation of all 11 steps of PHA
  - ✓ Linkage of all relevant information required for PHA
  - ✓ Easy Tracking of PHA study progress
  - ✓ Consequence Modelling using PHAST software
  - ✓ Easy recordkeeping of PHA study reports & effective networking
  - ✓ Recommendation management in CAT system

Advantages of IT Enablement of PSM Elements

- **Management of Change Technology & Subtle (MOC – T & S)**
  - ✓ Single repository for recordkeeping of all changes
  - ✓ Controls provided for Hazard Analysis (Pre Screening PHA Checklist)
  - ✓ Tracking & recordkeeping of approval process
  - ✓ Tracking of PT document updating while MOC close out

- **Pre Start up Safety Review (PSSR)**
  - ✓ Automation of PSSR Checklist
  - ✓ Control provided for steps of PSSR e.g. Self Assessment , Level II & Level I
  - ✓ Pending checklist point tracking in CAT system
  - ✓ Permanent records of PSSR points & its close out status

Advantages of IT Enablement of PSM Elements

- **Learning Management System (LMS)**
  - Online nomination for the training program is available
  - Online training modules & validation sheets are available
  - Easy retrieval & tracking of employees' training records
  - Tracking of the pending training modules for planning training programs
- **Incident Reporting Investigation & Learning**
  - Reporting of incident in iRIL system.
  - Online team formation.
  - RCA tool interface
  - Uploading of investigation report in system.
  - Interface with CAT system for Recommendation till closure
  - LFI sharing to all concerned
  - Incident Investigation Quality
  - SAP based incident management system is under development.

Advantages of IT Enablement of PSM Elements

- **PIPA (Pre Incident Planning & Assessment)**
  - PIPA software used for management of emergency scenarios
  - Each plant selected credible scenarios & developed plan with the help of PIPA software
  - Rehearsal of PIPA scenarios & plans is ensured through emergency mock drills
- **Contractor Field Audits (CFA)**
  - Online scheduling & recordkeeping of field audits for various contractor agencies
  - Effective records of audit observations e.g. % Good Citizens observed during audit round, Score achieved by contractor agencies
  - Overall Tracking of Severity Index (Plant/Site wise)
  - Tracking of Contractor Field Audits e.g. Scope Vs. Completed

Advantages of IT Enablement of PSM Elements

- **Progressive Disciplinary Action System (PDA)**
  - ✓ Developed for recordkeeping of violation / negligence of safety rules & procedures
  - ✓ Easy retrieval of records
  - ✓ Uniformity in record maintenance of PDA at all RIL sites.
- **Corrective Action Tracking (CAT)**
  - ✓ Interface with iRIL, Audit Management, PSSR, Mock Drill
  - ✓ Tracking of Recommendation at one location
  - ✓ Mail Alert for Pending and Overdue Recommendation

Thanks

(44at) Sulphur Dust Explosions an Overview

Karthick Deivasagayam, Loss Prevention, SNC Lavalin Inc, Calgary, AB, Canada

This paper will focus on the sulphur dust, its explosion energy, available models and capabilities of modeling its explosion energy, the equipment susceptible for sulphur dust explosion, various prevention& mitigation strategies of sulphur dust explosion, and industrial applied technology to sulphur dust explosion mitigation. This paper will begin with increase of solid sulphur production from the refineries, its present forms of products and case studies.

The technologies and methodologies for sulphur dust explosion, frequency data, consequence modeling, prevention and mitigation of sulphur dust as applied in a present day sulphur plant.

Discussion for future developments will be discussed in this paper as well.

Simplified Uncertainty Analysis of LOPA Results

Raymond "Randy" Freeman
S&PP Consulting
12303 Lake Shore Ridge
Houston, TX 77041

rafree@yahoo.com

ABSTRACT

Layer of Protection Analysis (LOPA) is a semi-quantitative risk evaluation tool. The methodology is widely used in setting Safety Integrity Level (SIL) targets for Safety Instrumented Systems (SIS). The methodology is often used to evaluate the adequacy of existing protective systems against corporate risk targets. The LOPA method is based on the use of conservative estimates of the failure probability of protective systems.

Freeman previously presented a method for the quantification of uncertainty in a LOPA study based on the application of variance contribution analysis (VCA) techniques. The use of VCA to evaluate the uncertainty in a LOPA study requires the evaluation of the sensitivity of the LOPA equation to changes in each of the variables. Freeman used direct analytical computation of the partial derivatives of the LOPA equation for each of the uncertain variables. While mathematically correct, this use of a partial derivative analysis during an actual LOPA review is unlikely. Faster and easier methods are needed if LOPA uncertainty analysis is to be applied by practicing engineers.

This paper explores the use of approximations to compute the required uncertainty information without the use of the partial derivative computations. Numerical approximations (perturbation methods) are found to allow for the estimate of the partial derivatives with a maximum error of 16%. When evaluating simple safety systems designed per IEC 61511, the error is found to be no more than 10% for SIL-1 or SIL-2 capable systems. Compared to direct Monte Carlo simulation methods, the numerical approximation method is much more efficient. Monte Carlo simulation takes approximately 10,000 function evaluations to arrive at a result. For the base line problem studied, the numerical approximation method will take only 5 function evaluations.

BACKGROUND

Layer of Protection Analysis (LOPA) [Ref. 1] is a widely used method for the evaluation of the frequency of undesired events. LOPA is one of the methods proposed in the ANSI/ISA 84.00.01 standard [Ref. 2] for the determination of Safety Integrity Level (SIL) targets. The SIL target [Ref. 3] is used by process and instrument engineers for specification of equipment to reduce the frequency of undesired events such as overpressures, spills from tanks, releases to flare systems, fires, etc.

Freeman [Ref. 4, 5] developed a method for evaluating the uncertainty in a LOPA study using a mathematical method called Variance Contribution Analysis. The VCA method outlined in Freeman's paper require the evaluation of the partial derivative of the basic LOPA equation for each of the uncertain variables. There are several problems with the application of this method in an actual LOPA study, including:

1. Most engineers would not be comfortable evaluating a series of partial derivative functions while leading a LOPA study.
2. The LOPA team leader must develop the uncertainty parameters for each initiating event and each IPL during the analysis. A set of pre-calculated tabulations of the uncertainty parameters would be needed to speed the analysis.
3. The application of Monte Carlo simulation as used in Freeman's paper is not a topic familiar to most process engineers - nor is Monte Carlo simulation a simple methodology to implement for a real LOPA scenario.

The fundamental issue with the use of VCA in an actual LOPA study is the evaluation of the variance contribution from each uncertain parameter. This paper is directed at rectifying these deficiencies in the application of VCA to the evaluation of uncertainty in a LOPA study.

REVIEW OF LOPA METHODOLOGY

A LOPA [Ref. 1] review is a semi-quantitative risk evaluation of undesired events in a process. The LOPA begins with the definition of the set of undesired events (fires, explosions, toxic chemical releases, etc.). Next, potential safeguards (such as pressure relief valves, high/low level or pressure safety shutdown systems, emergency shutdown systems, etc) are reviewed to determine if there are independent protection layers (IPLs) that will prevent the undesired event from occurring. The frequency of the undesired event is then computed by combining the frequency of initiating event with the combined probability of failure on demand (PFD) of the IPLs as:

$$\text{Frequency of undesired event} = (\text{Initiating event freq}) \times \prod_{i=1}^{n} \text{PFD of the IPLs} \qquad \text{(Eqn. 1)}$$

Or in more compact form:

$$\text{Freq} = (\text{IE}) \times \prod_{i=1}^{n} P_i \qquad \text{(Eqn. 2)}$$

Where:

Freq = Frequency of the undesired event being studied by the LOPA method
IE = Initiating event frequency or the starting event frequency for the scenario under study

P_i = Probability of Failure on Demand (PFD) of an Independent Protection Layer (IPL) counted as reducing the frequency of the undesired event.

i = Subscript indicating the ith IPL

n = Total number of IPLs counted in the LOPA study as satisfying the criteria for an IPL

Equation 1 is referred to as the basic LOPA equation. The basic LOPA equation of Eqn. 2 can be restated as:

$$\text{Freq} = (\text{IE})\ P_1\ P_2\ P_3\ \ldots\ P_n \qquad \text{(Eqn. 3)}$$

All of the variables in Eqn. 3 are random variables with inherent uncertainty. The resulting calculated frequency of the LOPA scenario will therefore be uncertain.

APPLICATION OF VARIANCE CONTRIBUTION ANALYSIS TO THE BASIC LOPA EQUATION

Using variance contribution analysis, the mean and variance for the LOPA equation are found directly per the work of Freeman [Ref. 4, 5]. For the expression:

$$Y = F(x_i) \qquad \text{(Eqn. 4)}$$

Where:

Y = dependent random variable computed from x_i
$F(x_i)$ = function that converts (maps) random variable x_i into Y

The mean of Y can be estimated using the following approximation:

$$E(Y) = F[E(x_i)] \qquad \text{(Eqn. 5)}$$

Where:

E(Y) = expected value of random variable Y = mean of Y
$E(x_i)$ = expected value of random variable x_i = mean of x_i

The variance of Y can likewise be estimated as:

$$V(Y) = \sum_{i=1}^{n} \left[\frac{\partial Y}{\partial x_i} \right]^2 V(x_i) \qquad \text{(Eqn. 6)}$$

Where:

$V(Y)$ = variance of random variable Y
$V(x_i)$ = variance of random variable x_i

The application of the VCA method to the evaluation of the uncertainty in a LOPA calculation requires the evaluation of the partial derivative term in Eqn. 6. Previously, Freeman computed the partial derivative directly using the methods of calculus and partial differential equations. However, the direct analytical evaluation of each of the terms in a LOPA expression is time-consuming and prone to error.

The basic LOPA equation of Eqn. 3 is restated as:

$$Freq = (IE)\ P_1\ P_2\ P_3\ \ldots\ P_n \qquad \text{(Eqn. 3)}$$

Note that Freq is first order with respect to each random variable. Therefore, the partial derivative of the Frequency (Freq) with respect to a random variable will become a function of the form:

$$\frac{\partial Freq}{\partial P_1} = (IE)\ P_2\ P_3\ \ldots\ P_n \qquad \text{(Eqn. 7)}$$

The partial derivative may be evaluated numerically rather than analytically. The numerical approximation to the partial derivative may be found by simple evaluation of the Freq near the mean values of each random variable. For all cases to be reviewed in this paper, a 10 percent perturbation about the expected value of the random variable will be used to complete the numerical approximation calculations for the partial derivative term in a VCA study.

For the simple case where all of the IPLs are independent and share no common inspection or repair times, the numerical evaluation of the partial derivative of Freq with respect to the variable P_1 in Eqn. 3 may be estimated as:

$$\frac{\partial Freq}{\partial P_1} \approx \frac{\Delta Freq}{\Delta P_1} = \{[E(IE)\ 1.1E(P_1)\ E\ (P_2)\ldots E(P_n)] - [E(IE)\ E(P_1)\ E(P_2)\ldots E(P_n)]\}/0.1E(P_1)$$
(Eqn. 8)

Where:
$E(P_i)$ = expected value of random variable P_i = mean of P_i

The numerical estimate reduces to:

$$\frac{\partial Freq}{\partial P_1} \approx \frac{\Delta Freq}{\Delta P_1} = (\frac{1.1-1}{0.1})\ [E(IE)\ E(P_2\,)\ E(P_3)\ \ldots\ E(P_n\,)] \qquad \text{(Eqn. 9)}$$

$$\frac{\partial Freq}{\partial P_1} \approx \frac{\Delta Freq}{\Delta P_1} = [E(IE)\,E(P_2)\ \ldots\ E(P_n)] \qquad \text{(Eqn. 10)}$$

Equation 10 is exactly the same as the analytical expression presented in Eqn 7. Thus, for simple LOPA expressions involving multiple independent protection layers, the result obtained by a numerical estimate will be the same as that obtained by the analytic expression obtained from the partial differentiation of the scenario basic LOPA equation. No error will be introduced by using the numerical approximation method to evaluate the partial derivative.

There are occasions in a LOPA review where the resulting basic LOPA equation may become non-linear. For example, if two or more IPLs are implemented in a safety instrumented system, the resulting LOPA equation becomes non-linear with respect to inspection frequency. For example a SIL-2 interlock design might involve two redundant instrumentation systems installed in parallel to reduce the frequency of an undesired event.

Appendix A to this paper evaluates the cases up to, and including, a SIL-4 capable design with 4 safety interlocks all tested on the same test period. The results of the analysis of Appendix A are summarized in Table 1. For most practical LOPA problems, a numerical analysis of the partial derivative term (Eqn. 6) will produce a result within 16% of that obtained using analytical methods.

Table 1.
Error Introduced in LOPA Uncertainty Calculations
by the Use of Numerical Estimates of the Partial Derivative Term

Case	Type of System	Error In LOPA Uncertainty* %
n = 1	SIL-1 Design or BPCS interlock	0
n = 2	SIL-2 Design or two SIL-1 interlocks	5
n = 3	SIL-3 Design or three SIL-1 interlocks	10
n = 4	SIL-4 Design or four SIL-1 interlocks	16

* Error = (Numerical Approximation Estimate – Analytical)/Analytical

The second problem with the application of the VCA method to LOPA uncertainty analysis is the need for pre-calculated tabulations of the mean and variance of the initiating event frequencies and IPL PFDs. Appendix B of this paper presents the method used as well as the results of calculating the needed uncertainty parameters using the data presented in the CCPS LOPA book. These tabulations may be used directly in a LOPA uncertainty study. The application of these simplifications is best illustrated using an example.

EXAMPLE PROBLEM

Previously, Freeman [Ref. 4, 5] presented an example problem to demonstrate the direct application of VCA to the uncertainty in the LOPA calculations. The same example is used for comparison in this work.

Consider the process system shown in Figure 1. The bottom sump level in separation tower (T-200) is controlled by a level controller (LIC-101) that regulates a control valve LCV-102. The tower is normally operating at a pressure of 500 psig and is designed for a maximum pressure of 750 psig. The liquid flow from T-200 goes to an intermediate vessel (Hold Drum D-205) prior to additional processing. D-205 is designed for a maximum pressure of 100 psig and normally operates at 75 psig.

Of concern is the failure of the level control system in T-200 to maintain the liquid seal at the bottom of the tower. If the liquid level in the T-200 sump is lost due to an instrumentation failure, the Hold Drum could be exposed to pressures as great as 500 psig. A pressure of 500 psig is a factor of 5 above the design pressure of the Hold Drum. Catastrophic failure of the Hold Drum is probable.

This problem is generically referred to as a high-low pressure interface problem or as a liquid blow-through case. Two protective systems are in place to prevent the Hold Drum from being over pressured.

1. The Separation Tower T-200 is provided with a SIL-1 low level interlock to prevent the loss of the liquid seal in the tower sump.
2. The Hold Drum D-205 is provided with a relief valve (PSV-206) sized to prevent overpressure if the tower liquid seal is lost.

LOPA EVALUATION OF THE EXAMPLE

For the example problem, the initiating event of concern is a basic process control system (BPCS) failure, resulting in loss of level control in the Separation Tower. From the CCPS LOPA book [Ref. 1], the initiating event frequency of a BPCS failure is 1 event in 10 years.

The low level protective interlock system was designed and installed in accordance with the ANSI/ISA 84.00.01 standard [Ref. 3] as a SIL 1 protective system. By definition, the probability of failure on demand (PFD) of a SIL-1 system is between 0.01 and 0.1. The LOPA book recommends using 0.1 as the PFD of a SIL-1 system.

The relief valve on the Hold Drum D-205 is designed for this hypothetical event. A value of 0.01 is suggested by the LOPA book as an estimate of the PFD for the relief valve of not opening on demand. However, for this analysis, a more conservative PFD value of 0.016 is used based on a review of the work by Bukowskia [Ref 6].

The LOPA formulation of the frequency of the Hold Drum being overpressured is:

$$\text{Freq} = \text{IE x SIS failure x PSV failure} \qquad \text{(Eqn 11)}$$

$$\text{Freq} = (0.1\ \text{event/yr}) \times (0.1) \times (0.016) = 1.6 \times 10^{-4}\ \text{events/yr} \qquad \text{(Eqn 12)}$$

In a LOPA, the frequency of 1.6 x 10^{-4} events per year would typically be compared with applicable risk acceptance criteria to determine if additional safeguards are needed. Many LOPA teams might recommend the addition of a critical high level alarm outside of the basic process control system, with adequate time for operator intervention, to reduce the risk of this liquid blow-through scenario.

Due to conflicts in turnaround schedules, the site management is interested in extending the test interval on the SIS system by 3 months, from 2 years to2.25 years. The management asks the LOPA team:

> "Will extending the SIS test interval by 3 months require the provision of additional IPLs to protect the system?"

At this point in the analysis, the engineer would not be able to explain to management the magnitude of the underlying uncertainty in the computed frequency of the Hold Drum overpressure. The next section of this paper attempts to answer this, and related, questions.

LOPA UNCERTAINTY

The basis for the evaluation of the LOPA uncertainty comes from two sources:

1. The LOPA model presented in Equation 11
2. The uncertainty in the parameters of the LOPA model

Rewriting Equation 11 in shorter terms, we obtain:

$$\text{Freq} = (\text{IE})\,(\text{SIS})\,(\text{PSV}) \qquad \text{(Eqn 13)}$$

Where:

Freq is the frequency of the Hold Drum D-205 being over pressured to the point of failure
IE is the initiating event frequency of the basic process control system failing in a dangerous state
SIS is the PFD of the SIL-1 low level interlock on the Separation Tower T-200
PSV is the PFD of the relief valve (PSV-206) opening and relieving the overpressure event

The variables in Equation 13 are all random variables. The variable SIS is actually the result of a simple calculation involving the equipment failure rate and test interval. For a negative exponentially distributed failure rate for a non-repairable device, the PFD is approximately:

$$\text{SIS} = \lambda\, T/2 \qquad \text{(Eqn 14)}$$

Where:
λ is the failure rate of the SIL-1 low level interlock on the Separation Tower T-200
T is the inspection or test interval for validation that the interlock is working and is functional
Equation 14 is an approximation of the negative exponential distribution. For λ T/2 < 0.2, the approximation is very good (error of 10%) and will be used throughout this paper.

Substituting the SIS failure rate and test interval into Equation 13 yields the basic LOPA equation for this scenario as:

$$\text{Freq} = (\text{IE})\,(\lambda)\,(T/2)\,(\text{PSV}) \qquad \text{(Eqn 15)}$$

The initiating event frequency in the basic LOPA analysis was taken directly from the LOPA book as 0.1 events per year. Discussions with site operations group indicates that the IE frequency is uncertain and might be as low as 0.05 events per year or as high as 0.5 events per year.

Likewise, the failure rate of the SIS interlock (λ) has significant uncertainty. The original plant design targeted this interlock as SIL-1 with a risk reduction factor of 50 based on a two-year testing interval, or a SIS failure rate of 0.02 failures per year. However, the failure rate may be as low as 0.01 events/year or as high as 0.1 events/year. From Appendix B (Table B-6) for a SIL-1 interlock designed with an RRF=50, the mean PFD is 4.3E-2 and the variance is 4.1E-4.

The test interval (T) may be targeted as once every 2 years. However, it might be as short as once every 1.5 years if the site takes advantage of an unexpected shutdown to test the system. It might be as long as 2.5 years, if the production run of the system is extended.

Appendix B has a discussion of relief valve PFDs. The Bukowskia [Ref 6] Process Safety Valve (PSV) PFD of 0.016 is higher than the CCPS LOPA book value and was used in the above LOPA calculation of the frequency of overpressure of the Hold Drum. For this uncertainty analysis, a mean PSV PFD of 0.016 with a standard deviation of 4.7×10^{-3} was used as the best estimate of the probability of failure of the relief valve to open on demand. The PSV variance would be $[4.7 \times 10^{-3}]^2$ or 2.2×10^{-5}.

These data may be used to establish the probability distribution for each of the uncertain variables in Eqn. 15. Assuming a triangular probability distribution (Ref. 7), the mean and variance may be calculated as:

$$\text{Mean} = (A + B + C)/3 \qquad \text{(Eqn 16)}$$
$$\text{Variance} = (A^2 + B^2 + C^2 - AB - AC - BC)/18 \qquad \text{(Eqn 17)}$$

Where:
A = Minimum Value
B = Mode
C = Maximum Value

The mean and variance for the initiating event (IE), the failure rate of the SIS (λ) and the inspection interval are calculated and presented in Table 2.

Table 2
Triangular Probability Distribution Parameters

Model Variable	Units	Minimum A	Mode B	Maximum C	Mean by Eqn 16	Variance by Eqn 17
Initiating Event (IE)	Events per Year	0.05	0.1	0.5	2.167E-01	1.014E-02
λ of SIS (From Table B-6 with RRF=50)	Failure per Year	0.01	0.02	0.1	4.3E-02	4.1E-04
Inspection Interval (T)	Years	1.5	2	2.5	2.000E+00	4.167E-02

The variance contribution for each uncertain (random) variable may now be computed using the numerical approximation for the partial derivative term presented above and derived in detail in Appendix A to this paper. Restating Eqn 15 as the basic LOPA equation for this scenario:

$$\text{Freq} = (\text{IE})\,(\lambda)\,(T/2)\,(\text{PSV}) \qquad \text{(Eqn 15)}$$

Using a 10% perturbation about the means of each random variable, the partial derivative term may be estimated. For example the partial derivative portion of the contribution from the uncertainty in the initiating event may be estimated as:

$$\frac{\partial Freq}{\partial IE} \approx \frac{\Delta Freq}{\Delta IE} = \{[1.1\ E(IE)]\ E(\lambda)\ E(T/2)\ E(PSV) - E(IE)\ E(\lambda)\ E(T/2)\ E(PSV)\}/(0.1\ E(IE)) \qquad \text{(Eqn 16)}$$

$$\frac{\partial Freq}{\partial IE} \approx \frac{\Delta Freq}{\Delta IE} = [1.1]\ E(\lambda)\ E(T/2)\ E(PSV) - E(\lambda)\ E(T/2)\ E(PSV)/(0.1) \qquad \text{(Eqn 17)}$$

$$\frac{\partial Freq}{\partial IE} \approx \frac{\Delta Freq}{\Delta IE} = [(1.1 - 1)/0.1]\ E(\lambda)\ E(T/2)\ E(PSV) \qquad \text{(Eqn 18)}$$

$$\frac{\partial Freq}{\partial IE} \approx \frac{\Delta Freq}{\Delta IE} = E(\lambda)\ E(T/2)\ E(PSV) \qquad \text{(Eqn 19)}$$

Likewise the partial derivative contribution to the variance of the other random variables may be found as:

$$\frac{\partial Freq}{\partial \lambda} \approx \frac{\Delta Freq}{\Delta \lambda} = E(IE)\ E(T/2)\ E(PSV) \qquad \text{(Eqn 20)}$$

$$\frac{\partial Freq}{\partial T} \approx \frac{\Delta Freq}{\Delta T} == E(IE)\ E(\lambda)\ (1/2)\ E(PSV) \qquad \text{(Eqn 21)}$$

$$\frac{\partial Freq}{\partial PSV} \approx \frac{\Delta Freq}{\Delta PSV} = E(IE)\ E(\lambda)\ E(T/2) \qquad \text{(Eqn 22)}$$

It should be noted that these results are identical to those obtained by taking the partial derivative of the basic LOPA equation for this scenario (Eqn 15) with respect to each of the random variables. With these estimates for the partial derivatives, the variance contribution from each of the uncertain random variables may now be computed (see Table 3). Note that the results in Table 3 are identical with those previously obtained by Freeman [Ref. 4, 5] using more formal mathematical methods. The same result could be obtained by simple numerical substitution. Equations 19 through 22 and those in Table 3 are presented to allow the reader to understand why the numerical results of a numerical approximation are the same as obtained using analytical methods.

Table 3.
Contribution of Each Model Variable to Frequency of Drum Overpressure Variance

Model Variable	Units	Equation for Contribution	Variance Contribution	Percent of Total Variance
Initiating Event (IE)	Events per Year	$[E(\lambda)\ E(T/2)\ E(PSV)]^2\ V(IE)$	4.874E-09	41%
λ of SIS	Failure per Year	$[E\ (IE)\ E(T/2)\ E(PSV)]^2\ V(\lambda)$	4.874E-09	41%
Inspection Interval (T)	Years	$[E\ (IE)\ E\ (\lambda)\ (1/2)\ E(PSV)]^2\ V(T)$	2.351E-10	2%
Relief Valve (PSV)	none	$[E\ (IE)\ E(\lambda)\ E(T/2)]^2\ V(PSV)$	1.947E-09	16%
		Totals	**1.193E-08**	**100%**

Since the variance contribution and total variance are the same whether computed via an analytical method or by a numerical approximation method, the accuracy of the resulting will be the same as previously reported by Freeman [Ref 4, 5]. Table 4 presents a comparison of the computed results with those previously obtained by Freeman using both an analytical and a Monte Carlo method for estimating the frequency of overpressure of the Hold Drum in the example problem.

Table 4
Comparison of Results via Numerical Approximation,
Using Theoretical and Direct Monte Carlo Simulation of Hold Drum Overpressure

Method	Mean event/yr	Standard Deviation event/yr	Value of 70% Upper Limit event/yr	Value of 95% Upper Limit event/yr
Theoretical	1.50E-04	1.09E-04	2.07E-4	3.29E-04
Numerical Approximation	1.50E-04	1.09E-04	2.07E-4	3.29E-04
Monte Carlo	1.51E-04	1.19E-04	1.78E-4	3.90E-04
Basic LOPA	1.6E-04	Not Computed	Not Computed	Not Computed

With the results of the uncertainty analysis now available, we can answer the managements question in the example problem on the impact of extending the inspection and test interval by 3 months. As seen in Table 3, the variance contribution of the inspection interval is only 2% of the total. Increasing the inspection interval by 3 months will not appreciably impact the computed 95% upper limit on the frequency of the Hold Drum overpressure.

SIMPLIFIED METHOD FOR QUANTIFYING LOPA UNCERTAINTY

Based on the results obtained using the numerical approximations outlined in this paper, a simplification of the LOPA uncertainty method outlined by Freeman [Ref. 4, 5] is possible. With the exception of the method used to compute the partial derivatives for the variance contribution, the revised procedure will be the same as that which was previously presented. The change to the previously presented method is indicated in *italic font* below.

1. Complete the LOPA study using the methods outlined in the CCPS LOPA book [Ref. 1].

2. Write the basic LOPA equation for the scenario of interest in the form of the product of initiating event frequency and the PFD of the IPLs. An example of the basic LOPA equation is given in Equation 15.

3. Define the uncertainty in the terms of the expanded LOPA equation. The uncertainty can be given as the upper and lower range of the possible values (uniform probability distribution), as the upper, lower and recommended values (triangular distribution), or as a mean and standard deviation (normal distribution). See the example above for guidance in the evaluation of safety instrumented system interlocks.

4. Compute the expected value of each variable in the LOPA equation. The equations for the mean and variance for the uniform, triangular and normal probability distributions are presented for various probability distributions in Vose [Ref. 7]. *For commonly used LOPA initiating events and IPLs, the mean and variance are tabulated in Appendix B to this paper and may be used directly.*

5. Compute the expected value or mean of the LOPA equation using the mean value of each of the variables in the basic LOPA equation.

6. *Compute the sensitivity of the result from the LOPA equation by estimating the partial derivative of the basic LOPA equation with respect to each of the variables using a 10% perturbation about the mean or expected value of each random variable. See the problem worked above as an example of this step. The theory for this simplification is presented in Appendix A to this paper*

7. Compute the variance of the scenario frequency of the basic LOPA equation by use of the variance contribution of Equation 6. This entails multiplying the variance of each of the uncertain variables in the basic LOPA equation by the square of its sensitivity (obtained in step 6), as evaluated at the mean. Sum the resulting terms to obtain the overall variance of the result of the basic LOPA term.

8. Compute the 95% upper confidence limit on the computed frequency by use of the standard normal factor, Z, [Ref. 8] as:

   $$Z = [x_i - E(x)]/\sigma \qquad \text{(Eqn 23)}$$

   Where:

   σ = standard deviation of the frequency of the scenario of interest from the basic LOPA equation obtained from step 7. Note that the variance of a random variable is the square of the standard deviation of the random variable.

   $E(x)$ = the expected value of the frequency of the scenario of interest from the basic LOPA equation obtained from step 5

   For the 95% upper limit, $Z = 1.64$. Rearranging Eqn. 10 allows for the direct calculation of the corresponding value of the 95% upper confidence limit as:

   $$X_{95\%} = 1.64\,\sigma + E(x) \qquad \text{(Eqn 24)}$$

   Where

   $X_{95\%}$ = is the upper 95% limit on the computed frequency of the scenario of interest from the basic LOPA equation.

9. Compare the 95% upper confidence limit on the frequency of the scenario of concern with that obtained using the basic LOPA method (step 1). If the difference is greater than one order of magnitude, revise the LOPA study accordingly to incorporate better information obtained in the uncertainty analysis. Improve the integrity of the LOPA IPLs or identify additional IPLs to drive the risk to a tolerable level.

RESULTS AND CONCLUSIONS

A numerical approximation can be used to estimate the variance contribution of each uncertain variable in a basic LOPA. Use of a 10% perturbation about the mean or expected value of each random variable was found to be adequate to define the variance. The numerical approximation is exact for cases where all of the IPLs are independent. In the case where 4 IPLs share a common test or maintenance frequency, the resulting error in the uncertainty estimate will be less than 16%.

A simplified method has been developed to evaluate the uncertainty in the basic LOPA equation. Using available data, the simplified numerical approximation method estimates the mean and variance of the basic LOPA equation and then computes the uncertainty range in the resulting frequency of the scenario of concern. For the example studied with 4 random variables, a total of 5 evaluations of the basic LOPA equation were all that were required to characterize the uncertainty. The results are found to be in excellent agreement with those obtained using analytical methods. When compared to a Monte Carlo simulation of 10,000 function evaluations, the calculation speed improvement is approximately a factor of 2,000.

A set of tables for the mean and variance of initiating events (IE) and independent protection layers (IPLs) are presented in Appendix B of this paper. These can be used directly by the analyst to estimate the uncertainty in an IE or IPL.

The methodology presented in this paper is simple and direct. It may be programmed into any standard LOPA software package to allow the users the ability to estimate the uncertainty in their calculated result.

ACKNOWLEDGMENTS
This paper is an expanded version of a paper entitled ''Uncertainty in LOPA; Part 2 – Simplified Methods Applied to Safety Systems'' presented at the 2012 Mary Kay O'Connor Process Safety Symposium, October 23-25, 2012, College Station, TX. The author wishes to thank Kathy Kas, Angela Summers, Robert Johnson, Dan Crowl and other individuals who gave comments and suggestions for improving this article.

LITERATURE CITED

1. CCPS, *Layer of Protection Analysis – Simplified Process Risk Assessment*, American Institute of Chemical Engineers, New York, 2001

2. Instrumentation, Systems and Automation Society, Functional Safety: Safety Instrumented Systems for the Process Sector—Parts 1, 2, and 3, Standard ANSI/ISA 84.00.01, 2004. USA implementation of IEC 61511.

3. R.A. Freeman, "Using Layer of Protection Analysis to Define Safety Integrity Level Requirements, pp 185 – 194, *Process Safety Progress*, Vol. 26, No. 3, 2007

4. R.A. Freeman, "Uncertainty in LOPA Studies," pages 623-634, Proceedings of Mary Kay O'Connor Process Safety Center International Symposium held October 25-27, 2011 in College Station, Texas

5. R.A. Freeman, "Quantifying LOPA Uncertainty," pp 240 – 247, *Process Safety Progress*, Vol. 31, No. 3, 2012. Published in Wiley Online Library (wileyonlinelibrary.com), DOI 10.1002/prs.11493

6. Julia V. Bukowskia and William M. Goble, "Analysis of Pressure Relief Valve Proof Test Data," pp 24-29, *Process Safety Progress*, Vol 28, No.1, 2009

7. David Vose, *Risk Analysis – A Quantitative Guide, 3rd Edition*, John Wiley, 2008

8. Paul L. Meyer, *Introductory Probability and Statistical Applications*, John Wiley, New York, 1972

Appendix A – Numerical Approximation to Partial Derivatives

The basic LOPA equation of Eqn. 2 can be restated as:

$$Freq = (IE)\ P_1\ P_2\ P_3\ \ldots\ P_n \qquad \text{(Eqn. A-1)}$$

Where:
Freq = Frequency of the undesired event being studied by the LOPA method
IE = Initiating event frequency or the starting event frequency for the scenario under study
P_i = Probability of Failure on Demand of the Independent Protection Layer (IPL) counted as reducing the frequency of the undesired event.
n = Total number of IPLs counted in the LOPA study as satisfying the criteria for an IPL

The probabilities of failure on demand of the IPLs may be restated in terms of a failure rate and a time between tests as:

$$Freq = (IE)\ (\lambda_1)\ (T_1/2)\ (\lambda_2)\ (T_2/2)\ \ldots\ (\lambda_n)\ (T_n/2) \qquad \text{(Eqn A-2)}$$

Where:

λ_1 is the failure rate of the safety instrumented system, i

T is the inspection or test interval for validation that the interlock is working and is functional

i is a subscript indicating the ith IPL

If all the IPLs are tested or inspected at the same test interval

$$T_1 = T_2 = T_3 = T_n \qquad \text{(Eqn A-3)}$$

Forming a new parameter K as:

$$K = (IE)\ (\lambda_1)\ (\lambda_2)\ \ldots\ (\lambda_n) \qquad \text{(Eqn A-4)}$$

The basic LOPA equation may now be re-written as:

$$Freq = K\ (T/2)^n \qquad \text{(Eqn A-5)}$$

When evaluated at the expected value of the random variables, the parameter K will not change and can be treated as a constant. How good is the numerical approximation for the partial derivative of Freq with respect to time as n increases from 1 to n?

The numerical approximation to a derivative requires the selection of a step size for the change in the variable of interest. In the limit as the step size goes to zero, the resulting answer will become the exact analytical expression for the derivative. However, from a practical point of view the selection of the numerical step size is constrained by two problems:

1. Too large and the function may change too much and the resulting numerical approximation will be very bad.
2. Too small and the numerical approximation will be limited by round-off error of the computer.

The step size of 0.1 T was chosen as the balance point between the two numerical problems. For the following cases (1 through 4), a step size of 0.1T was used as the basis for estimating the partial derivatives required for computations in the variance contribution analysis. Note that the step size for the evaluation of the derivative is arbitrary. A smaller step size could have been selected (say 0.01 T) and the resulting errors would decrease. However, the potential for numerical calculation errors would increase.

Case 1, n=1

For n = 1, restating the LOPA frequency becomes:

$$Freq = K\ (T/2) \qquad \text{(Eqn A-6)}$$

$$\frac{\partial Freq}{\partial T} = 0.5\ K \qquad \text{(Eqn A-7)}$$

The numerical approximation would become:

$$\frac{\Delta Freq}{\Delta T} = K\ (\frac{1.1^1 - 1}{0.1T})\ (T/2)^1 \qquad \text{(Eqn A-8)}$$

$$\frac{\Delta Freq}{\Delta T} = K\ (\frac{0.1}{0.1})\ 1/2 \qquad \text{(Eqn A-9)}$$

$$\frac{\Delta Freq}{\Delta T} = 0.5\ K \qquad \text{(Eqn A-10)}$$

This numerical approximation is exactly the same as the analytical result. For n =1, there will be no error in the computation of the partial derivative.

Case 2, n = 2

For n = 2, restating the LOPA frequency becomes:

$$Freq = K\ (T/2)^2 \qquad \text{(Eqn A-11)}$$

From basic calculus, the analytical expression for the partial derivative of equation A-5 when n=2 is:

$$\frac{\partial Freq}{\partial T} = 0.5\ K\ T \qquad \text{(Eqn A-12)}$$

The numerical approximation would become:

$$\frac{\Delta Freq}{\Delta T} = (\frac{1.1^2 - 1}{0.1T}) \text{ K } (T/2)^2 \qquad \text{(Eqn A-13)}$$

$$\frac{\Delta Freq}{\Delta T} = (\frac{0.21}{0.1}) (1/2)^2 \text{ K T} \qquad \text{(Eqn A-14)}$$

$$\frac{\Delta Freq}{\Delta T} = (\frac{2.1}{4}) \text{ K T} \qquad \text{(Eqn A-15)}$$

$$\frac{\Delta Freq}{\Delta T} = (0.525) \text{ K T} \qquad \text{(Eqn A-16)}$$

The error in the numerical approximation is:

Error = (Estimate - Actual)/Actual x 100 (Eqn A-17)

Error = (0.525 -0.5)/0.5 x 100 = 5% (Eqn A-18)

Thus, for a SIL-2 type of design, we would expect that no more than a 5% error in the LOPA uncertainty estimate would be introduced by the use of the numerical approximation for the partial derivatives.

Case 3, n =3

For n = 3, restating the LOPA frequency becomes:

$Freq = K (T/2)^3$ (Eqn A-19)

From basic calculus, the analytical expression for the partial derivative of equation A-5 when n=3 is:

$$\frac{\partial Freq}{\partial T} = K (3/8) T^2 \qquad \text{(Eqn A-20)}$$

The numerical approximation would become:

$$\frac{\Delta Freq}{\Delta P_1} = (\frac{1.1^3 - 1}{0.1T}) \text{ K } (T/2)^3 \qquad \text{(Eqn A-21)}$$

$$\frac{\Delta Freq}{\Delta P_1} = (\frac{0.331}{0.1}) (1/2)^3 \text{ K } T^2 \qquad \text{(Eqn A-22)}$$

$$\frac{\Delta Freq}{\Delta P_1} = (\frac{3.31}{8}) \text{ K } T^2 \qquad \text{(Eqn A-23)}$$

$$\frac{\Delta Freq}{\Delta P_1} = (0.3875)\ K\ T^2 \qquad \text{(Eqn A-24)}$$

The error in the numerical approximation is:

Error = (Estimate - Actual)/Actual x 100 (Eqn A-25)

Error = (3.31/8-3/8)/(3/8) x 100 (Eqn A-26)

Error = (3.31-3)/(3) x 100 = 10% (Eqn A-27)

Thus, for a SIL-3 type of design, we would expect no more than a 10% error in the LOPA uncertainty estimate to be introduced by the use of the numerical approximation for the partial derivatives.

Case 4, n =4

For n = 4, restating the LOPA frequency becomes:

$$Freq = K\ (T/2)^4 \qquad \text{(Eqn A-28)}$$

From basic calculus, the analytical expression for the partial derivative of equation A-5 when n=4 is:

$$\frac{\partial Freq}{\partial T} = (1/4)\ K\ T^3 \qquad \text{(Eqn A-29)}$$

The numerical approximation would become:

$$\frac{\Delta Freq}{\Delta P_1} = \left(\frac{1.1^4 - 1}{0.1T}\right) K\ (T/2)^4 \qquad \text{(Eqn A-30)}$$

$$\frac{\Delta Freq}{\Delta P_1} = \left(\frac{0.464}{0.1}\right) (1/2)^4\ K\ T^3 \qquad \text{(Eqn A-31)}$$

$$\frac{\Delta Freq}{\Delta P_1} = \left(\frac{4.64}{16}\right) K\ T^3 \qquad \text{(Eqn A-32)}$$

$$\frac{\Delta Freq}{\Delta P_1} = (0.290)\ K\ T^3 \qquad \text{(Eqn A-33)}$$

The error in the numerical approximation is:

Error = (Estimate - Actual)/Actual x 100 (Eqn A-34)

Error = (0.29 – 0.25)/(0.25) x 100 = 16% (Eqn A-35)

Thus, for a SIL-4 type of design, we would expect no more than a 16% error in the LOPA uncertainty estimate to be introduced by the use of the numerical approximation for the partial derivatives.

Table 1 presents a summary of the results of these calculations. In summary, the use of a numerical approximation for estimating uncertainty in a LOPA analysis will introduce no greater than a 16% error in the results. A 16% error in the LOPA uncertainty is well within the accuracy of the Variance Contribution Analysis method for estimating the LOPA uncertainty. This result eliminates the need for time-consuming evaluation of partial derivatives during a LOPA uncertainty analysis. Instead, a simple numerical analysis using a 10% perturbation about the mean of the inspection frequency is adequate to evaluate the partial derivatives needed in a LOPA uncertainty study. The results of this appendix are general and will apply to any basic LOPA equation formulated up to a 4th degree in one or more random variables.

Appendix B – Derivation of Tables for LOPA Uncertainty Analysis

The use of Variance Contribution Analysis to estimate the uncertainty in an estimate of the frequency of an accident requires a source for estimating the statistical parameters of the initiating events (IEs) and independent protection layers (IPLs). This appendix presents a methodology for estimating the required statistical parameters using available published data.

Initiating Events

The initiating event frequencies commonly used in a LOPA study are presented in Table 5.1 CCPS LOPA book [Ref. 1]. Included in Table 5.1 are ranges for the frequencies of each initiating event as determined from a literature search during development of the book. The uncertainty in the frequency data is expressed as a range (minimum to maximum), with a particular value assigned for use during a LOPA review. This type of uncertainty can be described by the triangular distribution [Ref. 7]. The triangular distribution has three parameters:

A = Minimum value
B = Mode
C = Maximum value

For a triangular distribution, the mean and variance are:

$$\text{Mean} = (A + B + C)/3 \qquad \text{(Eqn. B-1)}$$

$$\text{Variance} = A^2 + B^2 + C^2 - AB - BC - AC \qquad \text{(Eqn. B-2)}$$

For the data in Table 5.1, the minimum value becomes A, the maximum value becomes C, and the value used in LOPA becomes the mode, B. The mean and variance of the initiating event frequencies of the CCPS LOPA book Table 5.1 are calculated using Equations B-1 and B-2 and are presented in Table B-1 below. Note that the triangular distribution parameters (A, C and B) are defined in Table B-1 as the Min A, Max C, and Used in LOPA B respectively.

Table B-1. Mean and Variance of LOPA Initiating Events

Initiating Event	Units	Min A	Max C	Used in LOPA B	Mean per unit	Variance per unit^2
Pressure Vessel residual failure	per year	1E-07	1E-05	1E-06	3.7E-06	5.0E-12
Piping residual failure - 100 meter - full breach	per year	1E-06	1E-05	1E-05	7.0E-06	4.5E-12
Piping Leak - 10% section - 100 meter	per year	1E-04	1E-03	1E-03	7.0E-04	4.5E-08
Atmospheric Tank Failure	per year	1E-05	1E-03	1E-03	6.7E-04	5.4E-08
Gasket Packing Blowout	per year	1E-04	1E-02	1E-02	6.7E-03	5.4E-06
Turbine / diesel engine overspeed with casing breach	per year	1E-04	1E-03	1E-04	4.0E-04	4.5E-08
External impact by backhoe, vehicle, etc.	per year	1E-04	1E-02	1E-02	6.7E-03	5.4E-06
Crane Drop	per lift	1E-04	1E-03	1E-04	4.0E-04	4.5E-08
Lightning strike	per year	1E-04	1E-03	1E-03	7.0E-04	4.5E-08
Safety valve operates spuriously	per year	1E-04	1E-02	1E-02	6.7E-03	5.4E-06
Cooling water failure	per year	1.E+00	1E-02	1E-01	3.7E-01	5.0E-02
Pump seal failure	per year	1E-02	1E-01	1E-01	7.0E-02	4.5E-04
Unloading / loading hose failure	per year	1E-02	1E-01	1E-01	7.0E-02	4.5E-04
BPCS failure	per year	1E-02	1E-01	1E-01	7.0E-02	4.5E-04
Regulator Failure	per year	1E-01	1E+00	1E-01	4.0E-01	4.5E-02

The CCPS LOPA book Tables 6.3, 6.4, and 6.5 present the probability of failure on demand of passive, active mechanical and human operator IPLs. Using the same of data as for the initiating events allows for the uncertainty of IPLs to be calculated using Equations B-1 and B-2. The results of these calculations are summarized in Tables B-2, B-3, and B-4 below.

Table B-2. Mean and Variance of LOPA Passive IPLs

IPL Description	Min A	Max C	Used in LOPA B	Mean per unit	Variance per unit^2
Dike	1E-03	1E-02	1E-02	7.0E-03	4.5E-06
Underground Drainage System	1E-03	1E-02	1E-02	7.0E-03	4.5E-06
Open Vent (no valve)	1E-03	1E-02	1E-02	7.0E-03	4.5E-06
Fireproofing	1E-03	1E-02	1E-02	7.0E-03	4.5E-06
Blast Wall / Bunker	1E-03	1E-02	1E-03	4.0E-03	4.5E-06
Flame /Detonation Arrestors	1E-03	1E-01	1E-02	3.7E-02	5.0E-04

Table B-3. Mean and Variance of LOPA Active Mechanical IPLs

IPL Description	Min A	Max C	Used in LOPA B	Mean	Variance
Relief Valve*	1E-05	1E-01	1E-02	3.7E-02	5.1E-04
Rupture Disk	1E-05	1E-01	1E-02	3.7E-02	5.1E-04
BPCS	1E-02	1E-01	1E-01	7.0E-02	4.5E-04

Note the asterisk (*) in Table B-3 for relief valves The issue of relief valves not opening when needed is s topic that has been discussed extensively by Bukowskia [Ref. 6]. Bukowskia found that there is a significant probability that a relief device is not working when installed. A PFD value of 0.016 was suggested as an estimate for the PSV not opening on demand due to initial installation problems. Table 1 in the paper by Bukowski presents an estimate for the uncertainty in the PFD of the relief valve in terms of the standard deviation. For the data sets evaluated, the mean PFD of the PSV may be taken as 0.016 with a standard deviation of 4.7×10^{-3}. The corresponding PSV variance would be $[4.7 \times 10^{-3}]^2$ or 2.2×10^{-5}. The data of Bukowski indicate a PFD of about half that of the mean based on the range of the CCPS book Table 6.4 with a smaller variance. The LOPA value of 0.01 is lower than the value obtained using Bukowski's paper. The analyst completing a LOPA uncertainty study should carefully evaluate which data set to use for representing the uncertainty in the relief valve PFD.

Table B-4. Mean and Variance of LOPA Human Operator IPLs

IPL Description	Min A	Max C	Used in LOPA B	Mean	Variance
Human action with 10 minutes response time	1E-01	1E+00	1E-01	4.0E-01	4.5E-02
Human action with 40 minutes response time	1E-02	1E-01	1E-01	7.0E-02	4.5E-04

Safety Instrumented Systems
Safety instrumented systems are often provided in process design to control undesirable events such as overpressure, runaway reactions, or loss of level interface control. Note for the purposes of this paper the use of the phrase "safety interlock" means:

> "A Safety Instrumented Function (SIF) implemented in a Safety Instrumented System (SIS) to achieve a target Safety Integrity Level (SIL) designed and installed in accordance with the requirements of ANSI/ISA S84 – 2004 [Ref. 2]"

The S84 standard defines a range of acceptable performance operation of the interlocks in terms of a safety integrity level (SIL). The various SIL levels are defined by the probability of failure on demand (PFD) of the interlock, where the RRF is related to the PFD as:

RRF = 1/PFD (Eqn. B-3)

For example, an interlock designed for SIL-1 must have a PFD between 0.1 and 0.01, or by Equation B-3, an RRF between 10 and 100. Table B-5 presents the defined range of PFDs and RRFs for various SIL classes.

Table B-5. Probability of Interlock Failure on Demand of Various SIL Classes

IPL Description	Min PFD	Max PFD	Used in LOPA	Min RRF	Max RRF
SIL-1	1E-02	1E-01	1E-01	10	100
SIL-2	1E-03	1E-02	1E-02	100	1,000
SIL-3	1E-04	1E-03	1E-03	1,000	10,000
SIL-4	1E-05	1E-04	1E-04	10,000	100,000

There are two strategies available for the interlock design engineer to use:

1. Complete the design using rules of thumb, evaluate the RRF, and accept the design if the RRF falls in the desired range (as specified by the standard).
2. Complete the design to a desired risk reduction factor (RRF)

These two different design strategies will result in different probability distributions for the resulting uncertainty in the performance of the interlock. In the first case, any design that falls in the desired RRF range will be acceptable. The resulting uncertainty will be uniformly distributed from the minimum to maximum acceptable RRF. The uniform probability distribution may be characterized by two parameters, A and C:

A = Minimum value
C = Maximum value

The mean and variance of the uniform distribution are:

Mean = (A + C)/2 (Eqn. B-4)

Variance = $((C-A)^2)/12$ (Eqn. B-5)

In the second case, the design is targeted to a particular value of the RRF. This RRF becomes the mode (parameter B) of the triangular probability distribution of outcomes. Equations 1 and 2 may be used to compute the mean and variance of the resulting distributions.

Since a range of RRFs are available for each selected interlock SIL, a set of three possible targets was selected to represent the SIL range. These are:

Target RRF = 25 percent of allowed range (Eqn. B-6)

Target RRF = 50 percent of allowed range (Eqn. B-7)

Target RRF = 75 percent of allowed range (Eqn. B-8)

The uncertainty in the interlock performance was calculated for both design strategies for SIL-1, SIL-2, SIL-3 and SIL-4 at each of the three targeted SIL levels (Eqns. B-6, 7, and 8). The results of these calculations are presented in Tables B-6, B-7, B-8, and B-8.

For design strategy 1, Table B-6 presents the uncertainty parameters for each interlock SIL class when no target RRF is assigned prior to design. For example, such a case could occur if the design engineer were told to simply provide a SIL 2 interlock with no specification of the corresponding RRF.

Table B-6. Uncertainty Parameters for Interlock PFD for Various SIL Classes
(No target RRF set prior to design)

IPL Description	Mean PFD	Variance of PFD
SIL-1	5.5E-02	6.8E-04
SIL-2	5.5E-03	6.8E-06
SIL-3	5.5E-04	6.8E-08
SIL-4	5.5E-05	6.8E-10

For design strategy 2, the interlock RRF is selected prior to the start of design. This selected RRF then determines the design target PFD, which becomes the mode (parameter C, Equations B-1 and 2) for the triangular distribution. Tables B-7, B-8 and B-9 present the results for the computation of the mean and variance of the PFD of interlocks designed for various RRFs.

Table B-7. Uncertainty Parameters for Interlock PFD for Various SIL Classes with Target RRF at 25% of Allowed Range

IPL Description	Min PFD A	Max PFD C	Used in LOPA	Design RRF	Mode B	Mean PFD	Variance of PFD
SIL-1	1E-02	1E-01	1E-01	25	4.0E-02	5.0E-02	3.5E-04
SIL-2	1E-03	1E-02	1E-02	250	4.0E-03	5.0E-03	3.5E-06
SIL-3	1E-04	1E-03	1E-03	2,500	4.0E-04	5.0E-04	3.5E-08
SIL-4	1E-05	1E-04	1E-04	25,000	4.0E-05	5.0E-05	3.5E-10

Table B-8. Uncertainty Parameters for Interlock PFD for Various SIL Classes with Target RRF at 50% of Allowed Range

IPL Description	Min PFD A	Max PFD C	Used in LOPA	Design RRF	Mode B	Mean PFD	Variance of PFD
SIL-1	1E-02	1E-01	1E-01	50	2.0E-02	4.3E-02	4.1E-04
SIL-2	1E-03	1E-02	1E-02	500	2.0E-03	4.3E-03	4.1E-06
SIL-3	1E-04	1E-03	1E-03	5,000	2.0E-04	4.3E-04	4.1E-08
SIL-4	1E-05	1E-04	1E-04	50,000	2.0E-05	4.3E-05	4.1E-10

Table B-9. Uncertainty Parameters for Interlock PFD for Various SIL Classes with Target RRF at 75% of Allowed Range

IPL Description	Min PFD A	Max PFD C	Used in LOPA	Design RRF	Mode B	Mean PFD	Variance of PFD
SIL-1	1E-02	1E-01	1E-01	75	1.3E-02	4.1E-02	4.3E-04
SIL-2	1E-03	1E-02	1E-02	750	1.3E-03	4.1E-03	4.3E-06
SIL-3	1E-04	1E-03	1E-03	7,500	1.3E-04	4.1E-04	4.3E-08
SIL-4	1E-05	1E-04	1E-04	75,000	1.3E-05	4.1E-05	4.3E-10

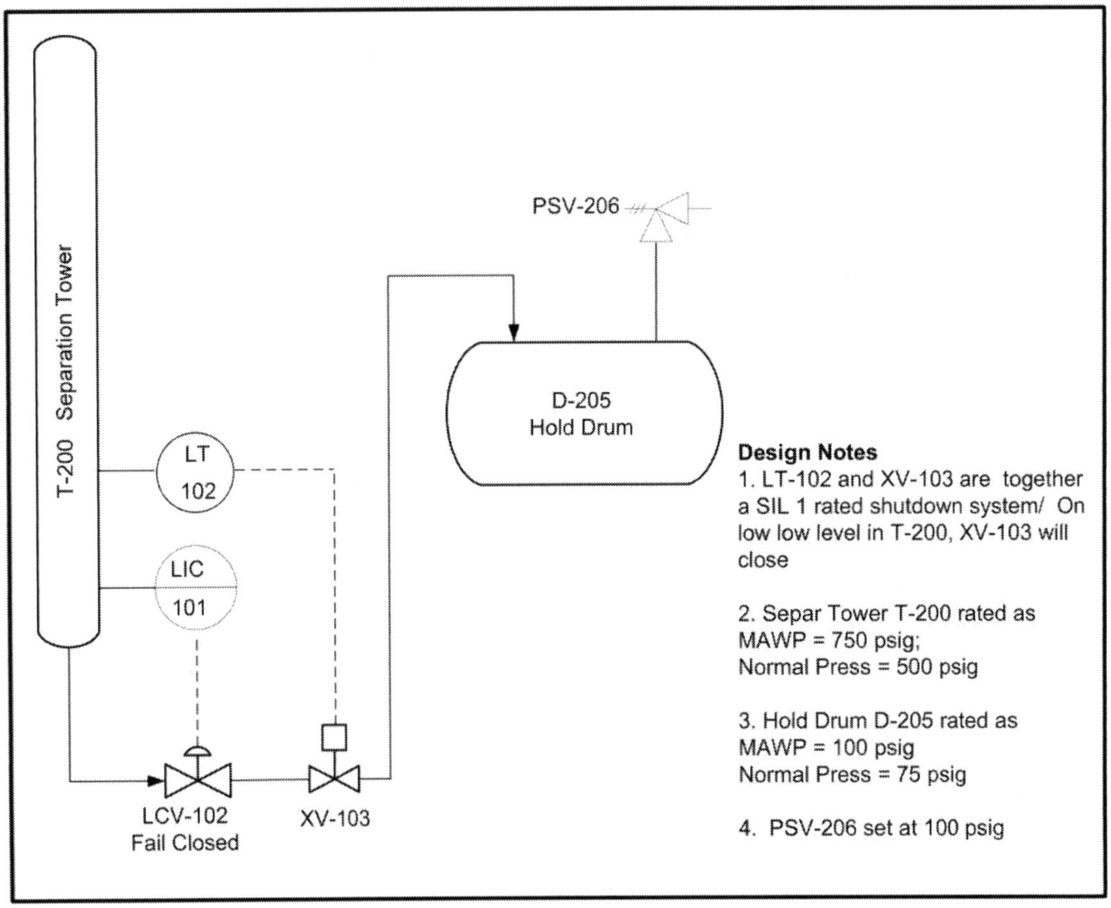

Figure 1. Piping and Instrument Diagram for Example

9TH GLOBAL CONGRESS ON PROCESS SAFETY

The Impact of Ozone on the Lower Flammable Limit of Hydrogen in Vessels Containing Savannah River Site High Level Waste

Paul Osterberg
Fauske & Associates, LLC
16W070 83rd Street
Burr Ridge, IL 60527
Osterberg@fauske.com

Carol Sherburne
Savannah River Remediation, LLC
Savannah River Site, Aiken, SC
Carol.Sherburne@srs.gov

Tom Johnson, Thomas Frawely
Fauske & Associates, LLC
16W070 83rd Street
Burr Ridge, IL 60527
Johnson@fauske.com, Frawely@fauske.com

Prepared for Presentation at
American Institute of Chemical Engineers
2013 Spring Meeting
9th Global Congress on Process Safety
San Antonio, Texas
April 28 – May 1, 2013

UNPUBLISHED

The Impact of Ozone on the Lower Flammable Limit of Hydrogen in Vessels Containing Savannah River Site High Level Waste

Paul Osterberg
Fauske & Associates, LLC
16W070 83rd Street
Burr Ridge, IL 60527
Osterberg@fauske.com

Carol Sherburne
Savannah River Remediation, LLC
Savannah River Site, Aiken, SC
Carol.Sherburne@srs.gov

Tom Johnson
Fauske & Associates, LLC
16W070 83rd Street
Burr Ridge, IL 60527
Johnson@fauske.com

Thomas Frawely
Fauske & Associates, LLC
16W070 83rd Street
Burr Ridge, IL 60527
Frawely@fauske.com

Keywords: hydrogen, ozone, flammability, lower flammability limit

Abstract

The Savannah River Site, in conjunction with AREVA Federal services, has designed a process to treat dissolved radioactive waste solids with ozone. It is known that in this radioactive waste process, radionuclides radiolytically break down water into gaseous hydrogen and oxygen, which presents a well defined flammability hazard. Flammability limits have been established for both ozone and hydrogen separately; however, there is little information on mixtures of hydrogen and ozone. Therefore, testing was designed to provide critical flammability information necessary to support safety related considerations for the development of ozone treatment and potential scale-up to the commercial level. Since information was lacking on flammability issues at low levels of hydrogen and ozone, a testing program was developed to focus on filling this portion of the information gap. A 2-L vessel was used to conduct flammability tests at atmospheric pressure and temperature using a fuse wire ignition source at 1 percent ozone intervals spanning from no ozone to the Lower Flammable Limit (LFL) of ozone in the vessel, determined as 8.4%(v/v) ozone. An ozone generator and ozone detector were used to generate and measure the ozone concentration within the vessel *in situ*, since ozone decomposes rapidly on standing. The lower flammability limit of hydrogen in an ozone-oxygen mixture was found to decrease from the LFL of hydrogen in air, determined as 4.2 % (v/v) in this vessel. From the results of this testing, Savannah River was able to develop safety procedures and operating parameters to effectively minimize the formation of a flammable atmosphere.

1. Introduction

Currently, in support of the Federal Facilities Agreement Act, tanks at the Savannah River Site (SRS) used to collect high level nuclear waste are being emptied, cleaned and grouted; and the waste permanently dispositioned in a series of specialized treatment processes. The majority of this waste removal is accomplished by a Bulk Waste Removal (BWR) campaign which empties the tank of easily accessible liquid supernatant and other liquid, followed by salt dissolution/removal and, finally, heel removal. At the end of the BWR campaign, fewer than 20,000 liters of sludge (less than 2 inches depth) remains in the tank. Currently, this final residual material is cleaned from the tank during a process called Bulk Oxalic Acid Cleaning (BOAC), whereby heated, 6 weight percent oxalic acid is added to the tank to a sufficient level to enable mixing. While BOAC has been effective in sludge dissolution and removal, the metal oxalates formed during oxalic acid dissolution cause downstream complications during evaporation (e.g. foaming) in addition to requiring excessive sludge washing (thereby greatly increasing the overall volume of waste to be dispositioned) to meet feed qualifications of the Defense Waste Processing Facility. This prompted research, testing and acceptance of a treatment process that utilized the proven effectiveness of oxalic acid for sludge dissolution and mobilization, yet removed the offending metal oxalates from the solution without adding liquid volume or contaminants. The Enhanced Chemical Cleaning (ECC) process was thus developed to replace BOAC stage of waste removal.

The Enhanced Chemical Cleaning (ECC) process uses ozone to promote the oxidation of metal oxalates produced during the dissolution of sludge in the Savannah River Site (SRS) waste tanks. The process is held at 70°C to improve decomposition kinetics and 10 psig to improve ozone solubility. A simplified depiction of the ECC process is shown in Figure 1, below.

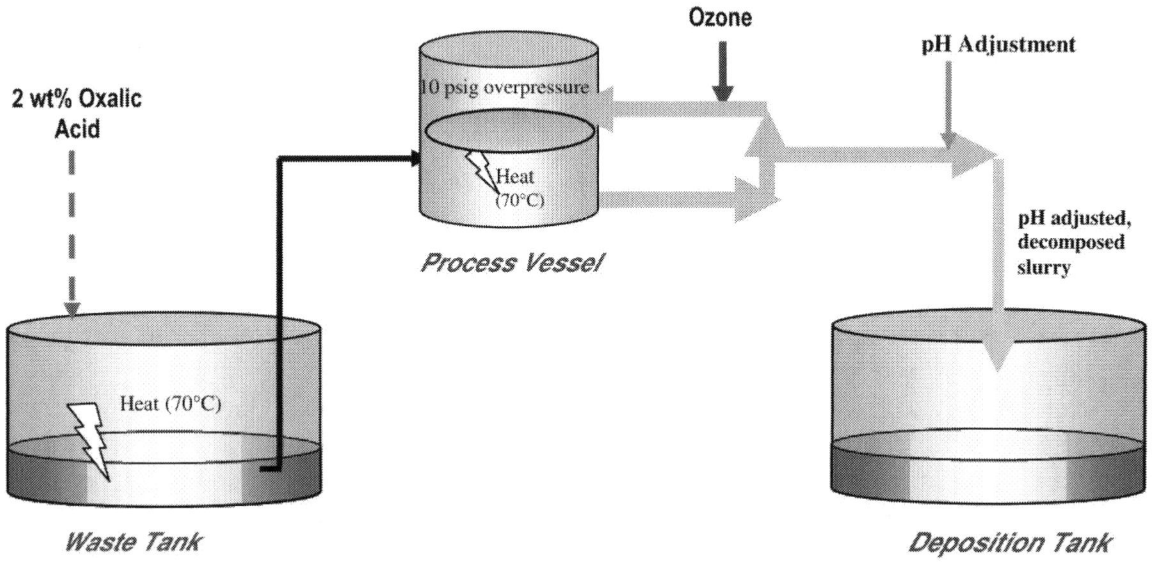

Figure 1: Schematic of the Enhanced Chemical Cleaning Process

In this process, the ozone reacts with the metal oxalates to form metal oxide and hydroxide precipitants, and the CO_2, O_2, H_2O and any unreacted O_3 gases are discharged into the vapor space. The major chemistry employed in the ECC process vessel is the reaction of ozone with the metal oxalates to form metal oxides, metal hydroxides, CO_2 and H_2O by the reactions shown below:

$$O_3 + H_2O \rightarrow O_2 + 2OH\bullet$$
$$Fe_2(C_2O_4)_3 + 6\ OH\bullet \rightarrow 6CO_2 + Fe_2O_3 + 3H_2O$$
$$H_2C_2O_4 + O_3 \rightarrow 2CO_2 + H_2O + O_2$$

While only the Iron oxalate reaction is listed above, formation of the powerfully reactive hydroxyl radical initiates many oxidizing reactions, effectively putting most waste constituents into their highest oxidation state.

In addition to the non-radioactive metals in the waste, however, the SRS radioactive waste also contains a variety of radionuclides, hence, hydrogen gas is also present in the vapor space of the ECC system. In order to prevent a buildup of hydrogen, and, hence, a flammability/explosion concern, waste tanks are purged and the hydrogen concentration monitored to ensure that the vapor space of each waste tank and all related processing equipment remains below the Lower Flammability Limit (LFL) of hydrogen. Furthermore, a comprehensive flammability control program is utilized to ensure that the LFL of the vapor space of a waste tank will not be attained for at least 7 days without forced ventilation. Because the use of ozone/oxygen in the processing equipment may also introduce ozone/oxygen into the vapor space of the waste tanks, the impact of this gas stream on the LFL of hydrogen must be understood for all possible operating scenarios of both normal and off-normal situations, with particular emphasis at the elevated temperatures and pressures of the typical process operating conditions. Oxygen is a known accelerant in combustion reactions, but while there are data associated with the behavior of hydrogen/oxygen environments, relevant studies addressing the effect of ozone on the flammability limit of hydrogen proved scarce. Further, discussions with industry experts verified the absence of data in this area and indicated that laboratory testing, specific to defined operating parameters was needed to comprehensively address the issue. Testing was thus designed and commissioned to provide the data necessary to support safety related considerations for the ozone treatment process.

2. Experimental

2.1 Flammability Apparatus

In order to ensure consistency between this testing and other documented flammability testing, the test methodology utilizes the guidelines established in ASTM E-918-83 (2005), "Standard Practices for Determining Limits of Flammability of Chemicals at Elevated Temperature and Pressure" [1]. Using this standard as the testing basis, a test setup was developed and is given in Figure 2.

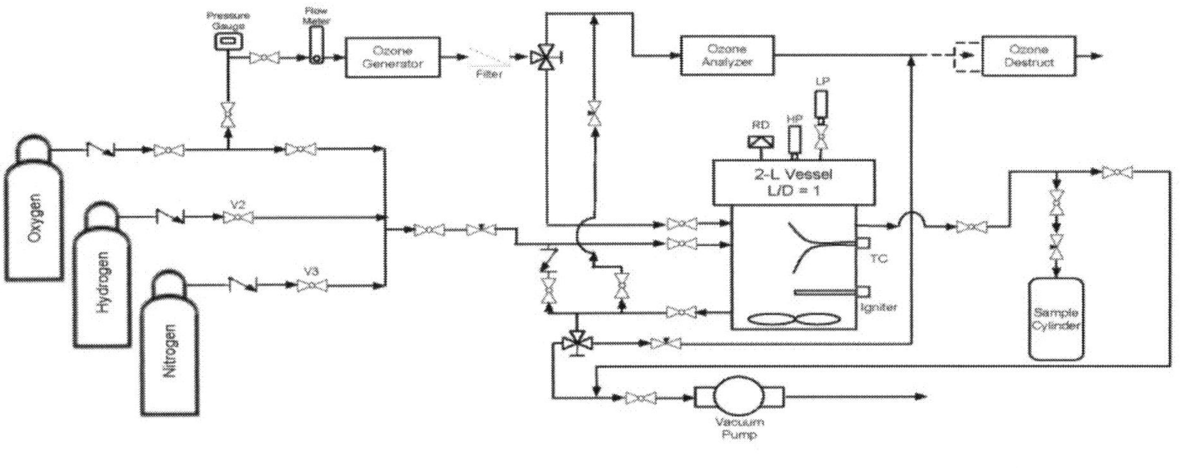

Figure 2: Test Apparatus for Testing the Lower Flammability Limit of Specified Gas Mixtures

The testing was conducted in a 2-L vessel rated to 3,000 psi at 500°C with an L/D ratio of one. There are two pressure transducers mounted on the vessel: a high accuracy (0.05% full scale) low range pressure transducer (0-25 psia) used for preparing the gas mixtures by partial pressure and a high accuracy (0.05% full scale) high range pressure transducer (0-200 psia) used to monitor the pressure after the mixture is ignited. High purity research grade gases were used for the preparation of the gas mixtures: nitrogen (99.9999%), oxygen (99.999%), and hydrogen (99.9999%). Two thermocouples were placed in the vessel to measure the temperature at the center and at the upper area of the vessel. A stir bar was placed at the bottom of the vessel to provide mixing of the gases. The gas mixture was ignited using an exploding fuse wire positioned a third of the way from the bottom. The energy provided by the fuse wire was approximately 10 joules. An ignition energy of 10 J was chosen because it produces very little wire spatter and a cylindrical shock wave is formed upon the vaporization of the wire because it is above the 5 J/cm threshold [2].

Ozone is generated on-site using a high ozone concentration generator. The ozone concentration of the ozone/oxygen stream is analyzed with an in-line UV absorption unit designed to enable accurate measurements of ozone from 0.01 wt. % ozone in oxygen to 20 weight percent.

In order to minimize undesired ozone decomposition, the materials used for this testing were carefully chosen. The test vessel, tubing, and fittings are all made of 316 stainless steel and passivated per ASTM A967 [3]. The packing material and/or seals for the valves were made of PTFE or fluorocarbon FKM.

2.2 Experimental Method

Prior to testing, the thermocouples, pressure transducers, ozone analyzer, and data acquisition instrumentation are checked to ensure they are functioning properly. The 2-L vessel is opened and wiped down with acetone. A 10mm 40 gauge tinned copper wire is wrapped around the igniter electrodes. The vessel is sealed and checked for leaks through pressurizing the vessel. This is followed by a vacuum and pressure purging procedure with a diluent gas (N_2 or O_2) depending on the test conditions. The gas manifold and piping are vacuumed and pressure purged with the respective gas prior to addition into the vessel. The purging helps minimize trace contamination of other gases as well as to remove residual moisture.

By assuming that these gases behave ideally, the gas concentration is determined based on partial pressure addition. Prior to gas filling, the vessel is evacuated down to approximately 0.2 psia. The magnetic stirrer is turned on to provide mixing during the gas filling step. Each gas is added slowly to the desired concentration based on the partial pressure calculation. During addition, there is a slight temperature rise due to the compression of the gases; therefore, after each gas addition, the mixture is allowed to reach equilibrium. The gas manifold and piping are purged with the subsequent gas prior to being added to the mixture.

For the ozone tests, oxygen is fed into a corona discharge ozone generator. A potentiometer on the unit controls the power to the elements. Thus, by adjusting the oxygen flowrate through the generator and tuning the potentiometer, the precise ozone/oxygen ratio specified in the test matrix can be produced. The ozone concentration is measured with an in-line ozone analyzer that is placed on the outlet line of the vessel.

Early scoping tests performed to optimize the test setup indicated that if ozone was added to the test vessel while the vessel was under a vacuum, the ozone decomposed very rapidly, making it difficult to establish the desired test conditions. By feeding the ozone/oxygen mixture through the process vessel at atmospheric pressure, then monitoring the ozone concentration at the discharge, an accurate determination of the ozone concentration in the test vessel, just prior to the addition of hydrogen, may be made. The test vessel can then be evacuated just enough to add the prescribed volume of hydrogen for the test, minimizing loss of ozone.

Once the mixture has been prepared, the magnetic stirrer is turned off. The mixture is allowed to relax prior to ignition so that a quiescent mixture is tested rather than a turbulent mixture. The low pressure transducer valve is closed and the data collection is started. The mixture is manually ignited using the fuse wire providing a nominal 10 joules. The data acquisition records the pressure and temperature time traces in the vessel. A mixture is considered to be flammable if the pressure rise after the ignition event is greater than or equal to 7% in accordance with ASTM E918. The procedure is repeated until there are a minimum of two demonstrations conducted at the lowest fuel concentration to be considered flammable and three at the highest fuel concentration considered non-flammable.

3. Results and Discussion

3.1 Hydrogen Lower Flammable Limit

The lower flammable limit for initially quiescent hydrogen-air mixtures was determined in order to validate the test equipment setup and the procedure. The LFL for hydrogen in air was determined as 4.2 volume percent based on the ASTM 7% pressure rise criteria. A comparison of this result to data obtained by other researchers is summarized in Table 1.

Table 1: Hydrogen LFL in Air, this testing compared to others

Source	Hydrogen Concentration (vol. %)	Comments
LFL using 5% pressure rise criteria		
SAFEKINEX [4]	4.1	2.8-L vessel
	3.9	6-L vessel
	4.1	14-L vessel
	4.7	40-L vessel
	5.8	1250-L vessel
	4.1	2000-L vessel
LFL using 7% pressure rise criteria		
Present Work	4.2 ± 0.1	2-L vessel (L/D = 1)
Crowl and Jo [5]	5.7	20-L spherical vessel
Cashdollar et al. [6]	5.0 ± 0.5	8-L chamber
	6.0 ± 0.5	20-L chamber
	6.5 ± 0.5	120-L spherical vessel
LFL using a visual criterion		
Coward and Jones [7]	4.0	Flammability Tube

The experimentally measured LFL is in close agreement with the accepted lower flammable limit of 4 vol. % which was obtained through visual criterion [7]. Many other researchers have shown that it is difficult to obtain this value using the 7% pressure rise criteria. This has been explained through hydrogen's high diffusivity as compared to oxygen and nitrogen [6].

The majority of tests using the pressure rise criteria are performed with the ignition source in the center of the vessel. Central ignition is used for the determination of the explosion severity of flammable mixtures because it gives the maximum rate of pressure rise and the maximum overpressure. However, central ignition may not be useful when determining the limits of flammability. It has been shown that ignition source location has an impact on the flammability limits of methane [8]. As the location of the ignition source is lowered, a larger percentage of the combustible mixture participates in the reaction with minimal heat loss to the vessel wall. This results in a higher explosion overpressure, thereby widening the flammable region. The present work uses an ignition source placed approximately one-third from the bottom of the vessel.

3.2 Ozone Lower Flammable Limit

Initially, tests were conducted to determine the extent of ozone decomposition over time inside the combustion vessel at ambient conditions. This was accomplished by preparing an ozone-oxygen mixture and holding the mixture in the combustion vessel for approximately 30 minutes while stirring. After the specified hold time the ozone concentration was measured. It was found that there were negligible amounts of ozone decomposition that occurred in the vessel.

The LFL of ozone in oxygen was determined as 8.4 volume percent. Previous studies on the flammability of ozone in oxygen are presented in Table 2.

Table 2: Ozone LFL in Oxygen, this testing compared to others

Source	Ozone Concentration (vol. %)	Comments
Present Work	8.40 ± 0.17	2-L vessel (L/D = 1)
Mizutani et al. [9]	10.5	1-L cylindrical vessel, Electric spark
Koike et al. [10]	10.0	1-L cylindrical vessel, 1 J electric spark
Cook et al. [11]	9.5 ± 0.5	10 cm i.d. glass bulb, 0.47 J electric spark

The present work reveals that the explosion limit of ozone in oxygen is consistent with previously reported values; however, it is lower than other literature data which can be attributed to the differences in the experimental method and procedure. The lower explosion limit obtained in this work is most likely a result of a combination of factors: the type of ignition source, location of the ignition source, and vessel size and geometry. In these tests, the ignition source energy was setup to give approximately 10J which is much greater than that used in previous work. Flammability limits are a function of ignition energy, thus, by increasing the strength of the ignition source the flammable limits tend to widen [12]. There also tends to be some variation in the flammable limits which is influenced by the type of ignition source: a fuse wire as compared to an electric spark [13]. This may be an effect of the energy density provided by each type of ignition source and the amount of energy actually being transferred to the mixture. As discussed previously, the ignition source location may be playing a part in the lower value obtained.

The differences in the size and dimensions of the explosion vessels could have affected the lower flammable limit [13]. As the vessel size is increased, the heat losses to the wall become negligible. Through minimizing heat losses to the vessel wall, more heat is transferred to the combustion reaction, therefore, promoting flame propagation. This will result in a widening of the flammable region and combustion can occur at lower temperatures.

3.3 Lower Flammable Limit of Hydrogen in Ozone/Oxygen Mixtures

The experimental results for the lower flammable limits of hydrogen in the ozone/oxygen are shown in Figure 3. The solid circles represent a flammable mixture and the open circles represent a non-flammable mixture.

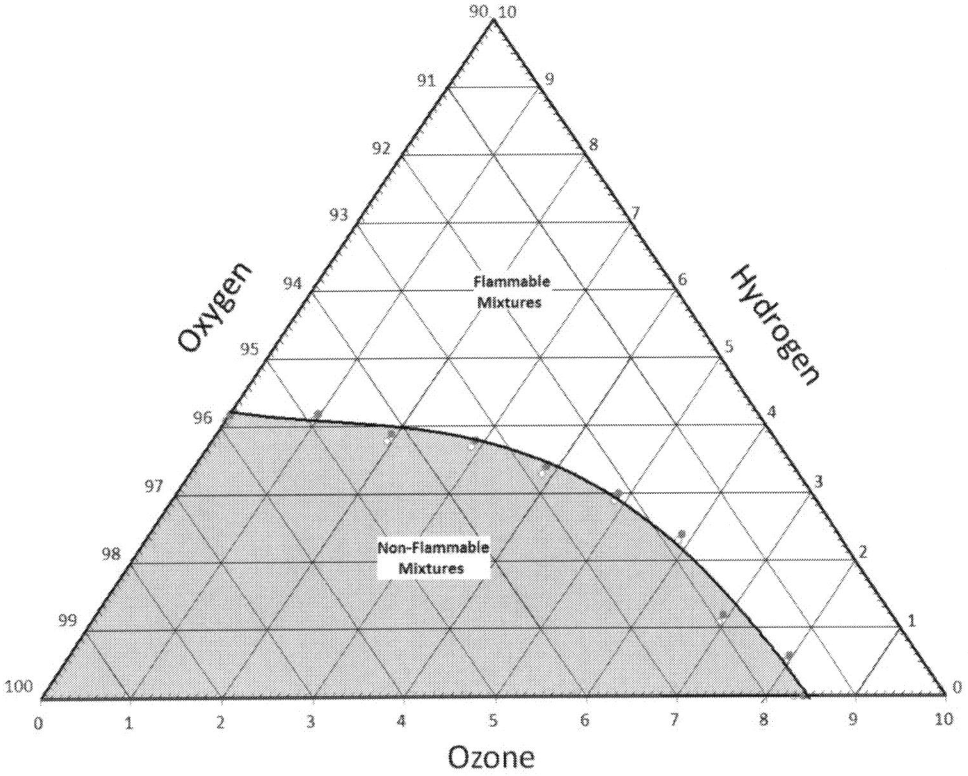

Figure 3: Flammability envelope for hydrogen/ozone/oxygen mixtures (all values are in volume percent).

Typically, it is of interest to estimate the flammability limits of multi-component mixtures. In the absence of experimental data, estimation methods can be useful if the flammable limits of pure components are known. Le Chatelier [14] developed an empirical equation for predicting the flammable limits of fuel mixtures,

$$LFL_{MIX} = \left(\sum_{i=1}^{n} \frac{y_i}{LFL_i} \right)^{-1} \qquad \text{Eq. [1]}$$

where y_i is the mole fraction of the ith component on a combustible basis and LFL_i is the lower flammable limit of the ith component in volume percent.

A comparison of the LFL predictions to the experimentally determined LFL of hydrogen in various ozone-oxygen mixtures is shown in Figure 4. It was found that the LFL of hydrogen decreases as the ozone concentration in the mixture increases. This result is consistent with the values approximated through Le Chatelier's rule; however, the experimental results are higher than the predicted calculations. Previous studies have shown that Le Chatelier's rule provides close approximations for many simple hydrocarbons; however, it is less accurate for mixtures with hydrogen and generally not applicable for substances that undergo decomposition [15]. However, from a safety stand point, Le Chatelier's rule provides a conservative estimate of the LFL.

The experimental data was further compared to flammability predictions from the calculated adiabatic flame temperature (CAFT) method. The flame temperature criteria used to assess whether a mixture is flammable or not was established by first calculating the adiabatic flame temperature at the determined LFL of hydrogen in air and ozone in oxygen. Three flame temperatures were used – the flame temperature of 8.4% ozone in oxygen, the flame temperature of 4.2% hydrogen in air and an average of these two flame temperatures. The LFL of hydrogen in various concentrations of ozone/oxygen was then calculated from these flame temperatures.

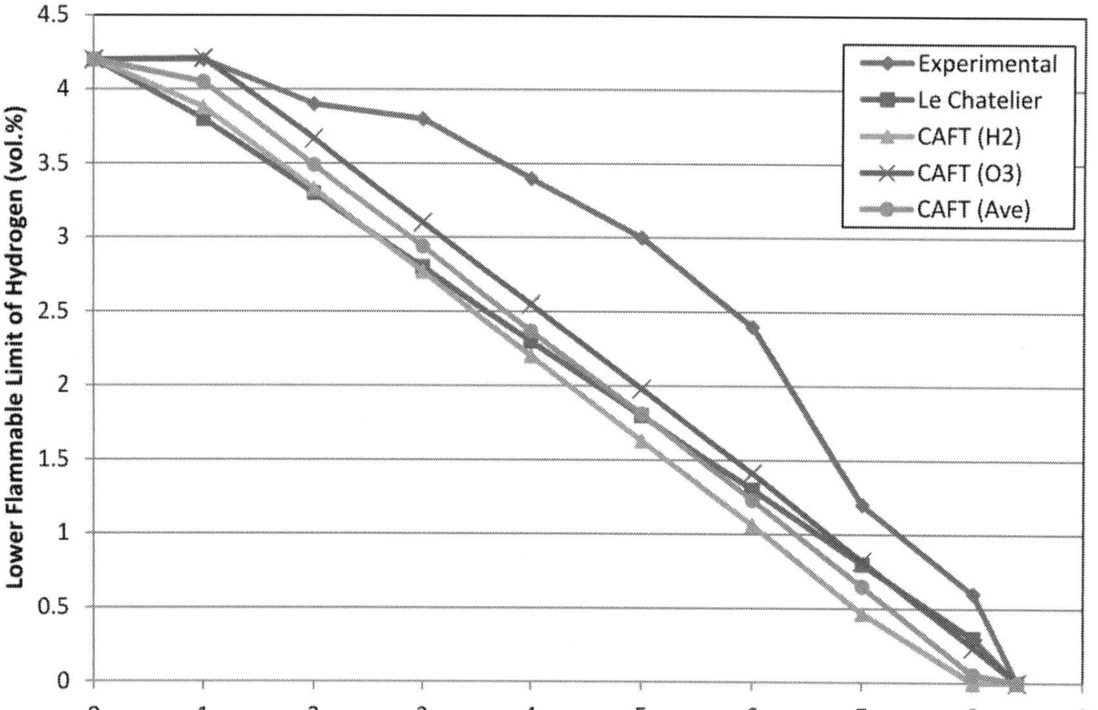

Figure 4: Comparison of lower flammability limit of hydrogen in ozone/oxygen to calculated adiabatic flame temperature predictions.

The calculated adiabatic flame temperature (CAFT) predictions are very similar to the predictions given by Le Chatelier's empirical equation. The deviations of the experimental data from the estimations in the flammability limits of a hydrogen/ozone/oxygen mixture lie within the unpredictability of ozone. This unpredictability comes from the instability of ozone, which tends to readily decompose to form oxygen. Furthermore, there are many different reactions that could be taking part in these experiments: H_2-O_2, O_3 decomposition, O_3-O_2, and H_2-O_3. The data seems to suggest that as the ozone concentration is increased, the reaction mechanism that takes precedence changes.

The deviation between the predicted and the experimental values may be two-fold. First, the reaction mechanism taking preference during the combustion reaction changes as the ozone concentration is increased. Secondly, the concentration of ozone participating in these reactions may be changing as well. Tests conducted on ozone decomposition in oxygen have shown that only 5.3% of the ozone decomposes for a gas mixture containing 6.7 mole % ozone and as high as 90% decomposition when the ozone concentration was 11.2% in the mixture [11]. In any case, additional work is needed to fully understand the nature of these experiments.

3.4 Temperature and Pressure Effect on the Lower Flammable Limit of Hydrogen-Ozone-Oxygen Mixtures

It has been reported previously that the ozone decomposition rate increases at the temperature of its environment increases [16]. Therefore, stability tests were performed again to determine the extent of ozone decomposition at the elevated test conditions. This time it was found that the ozone decomposition rate could not be considered negligible. With initial mixtures containing 5 vol.% ozone in oxygen at approximately 75°C, the ozone decomposition rate was approximately 0.1 vol.% ozone every 10 minutes.

The results from the tests performed at elevated temperatures and pressures are shown in Figure 5. It was found that as the temperature is increased the LFL decreases and the opposite was observed when the pressure was increased.

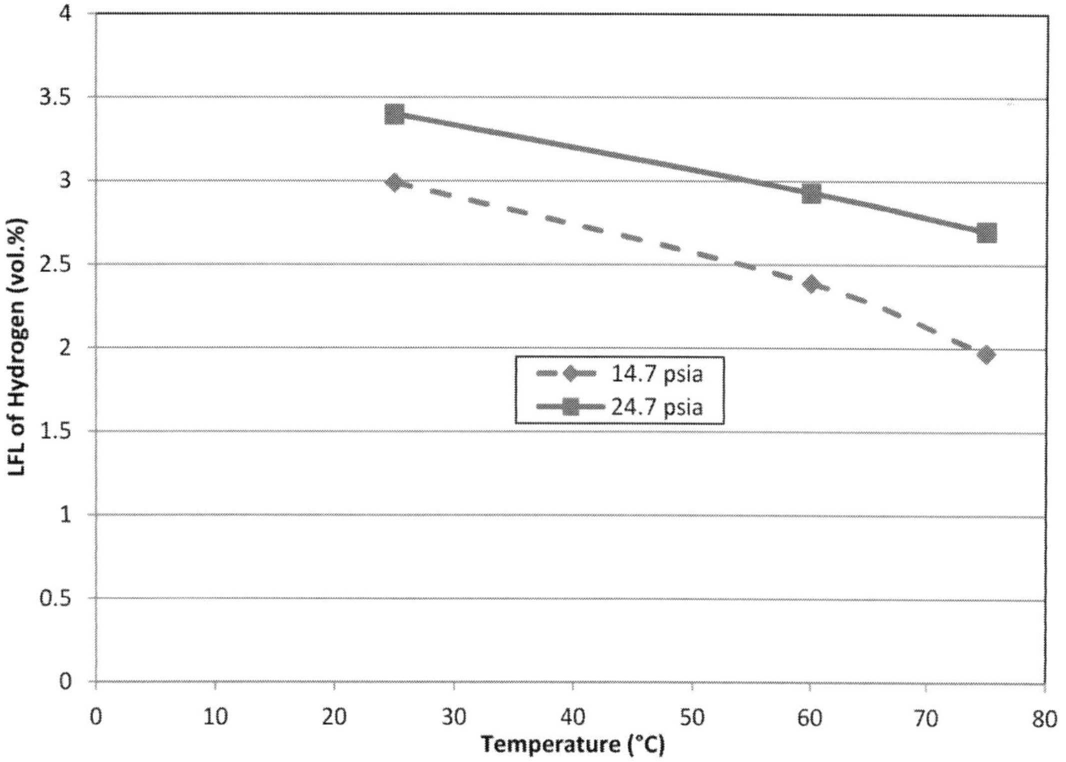

Figure 5: Effect of temperature and pressure on the flammability limits of hydrogen-ozone-oxygen mixtures. Tests were performed with 5 vol.% ozone in oxygen.

The increase in the LFL as a result of the increase in pressure is generally not characteristically observed for flammable gases; however, this is a characteristic of hydrogen flammability [17]. This result leads to the conclusion that the reaction mechanism responsible for the increased LFL as the pressure is increased for hydrogen is still the dominate mechanism when ozone is added. Furthermore, it is noticed that as the temperature is increased the decrease in the LFL is fairly linear throughout the range of temperatures tested. The LFL results from these tests show that the LFL is more susceptible to elevated temperatures when compared to results of hydrogen-air or hydrogen-oxygen. One explanation for this is that at the increase temperatures more of the ozone participates in the reactions that occur in the vessel. Previous research on ozone decomposition has found that only a part of the ozone decomposes when the ozone concentration is below its LFL. At the higher temperatures, the reaction rates increase as well as the instability in the ozone molecule.

3.5 Effect of Relative Humidity on the Lower Flammable Limit of Hydrogen-Ozone-Oxygen Mixtures at Elevated Temperatures and Pressure.

Tests conducted with the addition of water vapor have shown a slight increase in the LFL as shown in Figure 6. Since there is an additional component (water vapor) added to the mixture, the results should be compared on a total combustibles (hydrogen plus ozone) basis rather than just on the concentration of hydrogen in the mixture. This is because as more water vapor is added to the mixture, the concentration of ozone decreases; therefore, even if more hydrogen was needed to be added to the mixture, the total amount of combustibles may have remained constant to produce a flammable mixture.

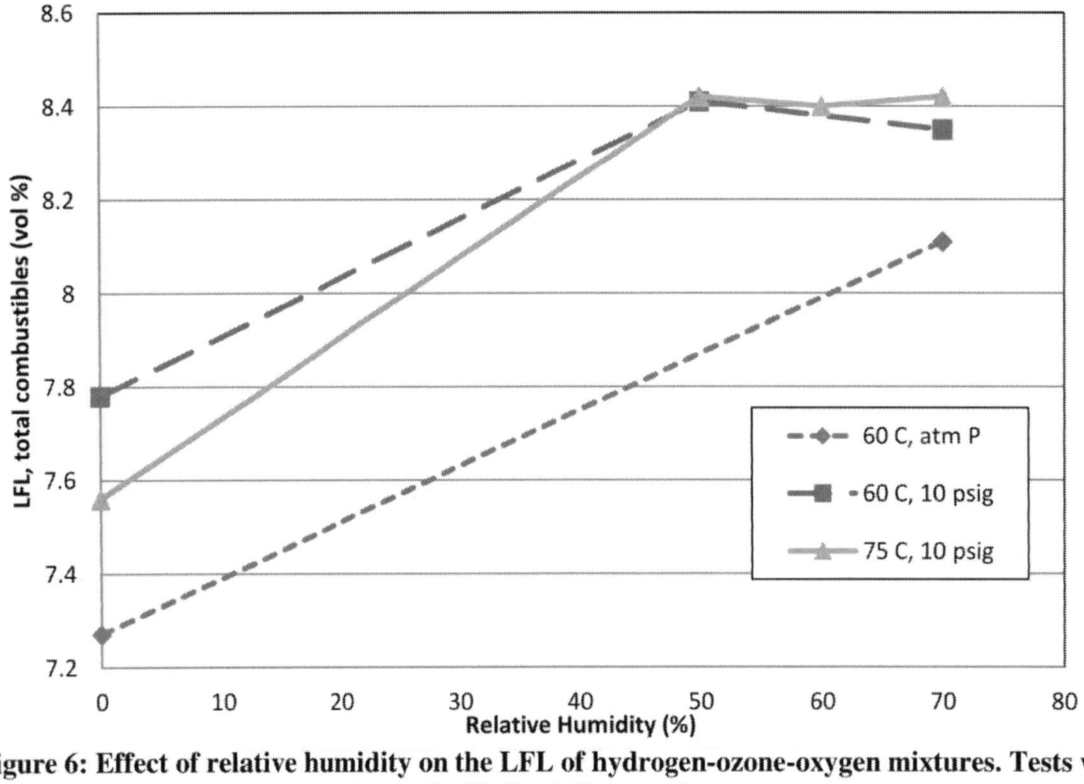

Figure 6: Effect of relative humidity on the LFL of hydrogen-ozone-oxygen mixtures. Tests were performed with 5 vol.% ozone in oxygen.

The slight increase in the LFL could be a result of either the addition of the water vapor or decrease in ozone concentration. The addition of water vapor to the mixture helps extract heat from the reactions occurring in the vessel (combustion or decomposition), thereby, quenching the flame propagation by lowering the reaction rate. Furthermore, the observed increase in the LFL could be a result of a decrease in the ozone concentration. It is known that both hydrogen and water act as a catalyst for ozone decomposition; therefore, the actual concentration inside the vessel prior to igniting the mixture is lower than the reported concentrations.

4. Impact on Operation

This testing was designed to provide data under the conditions considered most optimal to produce a deflagration. The geometry and materials of construction of the testing vessel; the location of the fuse wire; the magnitude of the supplied energy; the careful minimization of diluents and other contaminants; and meticulous procedural detail to maintain integrity of the ozone to the maximum extent practical, result in data that reflect not the expected process conditions, but those that enhance the possibility of flame propagation. For this reason, there is believed to be considerable conservatism in the indicated results. Per the vendor, the maximum possible ozone concentration producible by the planned ECC Ozone generator is 8 volume percent (typical maximum expected concentration is 6.8 vol%), less than the 8.4 minimum volume % concentration shown to be flammable in a pure O_2 environment at the optimally conservative conditions established in this testing. As both water vapor and nitrogen, together constituting approximately 6.5 % of the actual feed stream, are powerful diluents, it is believed not credible that deflagration can occur at this maximum feed concentration condition. In addition, once the ozone stream contacts the waste stream, the many simultaneous oxidizing reactions will rapidly decompose available ozone to well below flammable levels. Further, because the radiolytically generated hydrogen quantity is negligible compared to the supplied ozone/oxygen stream (0.0004 moles per minute H_2 vs 76 moles per minute ozone/oxygen), even the total H_2/O_3 mixture, without crediting decomposition reactions, does not approach flammable concentrations. Finally, even at the "end" of the ECC batch cycle, when most of the metal oxalates have been decomposed, testing has indicated that the ozone concentration in the vapor space of the ECC process vessel reaches a concentration of no more than 3 vol%, remaining well below concentrations of concern.

The major issue for the ECC operation established by this testing is the impact of the data when applied to off normal conditions. While it is possible to discontinue ozone addition to the reaction vessel at any time, the radiolytic hydrogen generation rate continues, varying slightly as ambient pressures and temperatures change. Relative to the data generated and analyzed in this testing, the ECC hazards analysis team will re-evaluate off normal conditions (e.g. those during which process exhaust ventilation is lost) such that issues involving mixtures of hydrogen and ozone in the vapor space can be appropriately controlled.

5. Conclusion

The LFL of hydrogen in air was determined and is in good agreement with the literature data. Ozone-oxygen mixtures were found to be flammable at concentrations above 8.3 vol.% based on the ASTM E918 7% pressure rise criteria for flame propagation. This result is lower than previously reported values which can be explained through the variations in the test setup and procedure. Tests performed with hydrogen in various concentrations of ozone in oxygen have shown that the LFL of hydrogen decreases as the concentration of ozone in the mixture increases.

The estimations of the LFL for various mixtures of hydrogen and ozone in oxygen was performed using Le Chatelier's rule as well as the CAFT method. It was found that both methods underestimate the LFL for mixtures of hydrogen and ozone.

The effects of temperature, pressure, and relative humidity were found for hydrogen-ozone mixtures. The trends observed through these tests were consistent with the general trends in reported results for hydrogen. Overall, the difficulty with some of these tests ultimately lies with the unpredictability of ozone.

6. References

[1] ASTM. (2005). ASTM E 918-83 Standard Practice for Determining Limits of Flammability of Chemicals at Elevated Temperature and Pressure. Philadelphia, PA: American Society for Testing and Materials.

[2] Mashuga, C. V. (1999). *Determination of the Combustion Behavior for Pure Components and Mixtures using a 20-Liter Sphere.* Phd Thesis. Michigan Technological University. Ann Arbor: UMI Company.

[3] ASTM. (2005). ASTM A967-05, Standard Specification for Chemical Passivation Treatments for Stainless Steel Parts. Philadelphia, Pa: American Society for Testing and Materials.

[4] Project SAFEKINEX, Contract No. EVG1-CT-2002-00072, Deliverable No.8, *Report on experimentally determined explosion limits, explosion pressures and rates of explosion pressure rise – Part 1: methane, hydrogen, and propene,* March 2006.

[5] Crowl, D. A., & Jo, Y.-d. (2009). A Method for Determining the Flammable Limits of Gases in a Spherical Vessel. *Process Safety Progress , 28* (3), 227-236.

[6] Cashdollar, K. L., Zlochower, I. A., Green, G. M., Thomas, R. A., & Hertzberg, M. (2000). Flammability of methane, propane, and hydrogen gases. *Journal of Loss Prevention in the Process Industries , 13*, 327-340.

[7] Coward, H. F., & Jones, G. W. (1952). *Limits of Flammability of Gases and Vapors.* BuMines, 503.

[8] Van den Schoor, F., Norman, F., & Verplaetsen, F. (2006). Influence of the ignition source location on the determination of the explosion pressure at elevated initial pressure. *Journal of Loss Prevention in the Process Industries* , 459-462.

[9] Mizutani, T., Matsui, H., Sanui, H., & Yonekura, M. (2001). Decomposing detonation and deflagration properties of ozone/oxygen mixtures. *Journal of Loss Prevention in the Process Industries , 14*, 559-565.

[10] Koike, K., Nifuku, M., Izumi, K., Nakamura, S., Fujiwara, S., & Horiguchi, S. (2005). Explosion properties of highly concentrated ozone gas. *Journal of Loss Prevention in the Process Industries, 18*, 465-468.

[11] Cook, G. A., Spadinger, E., Kiffer, A. D., & Klumpp, C. V. (1956). Explosion Limits of Ozone-Oxygen Mixtures. *Industrial and Engineering Chemistry , 48* (4), 736-741.

[12] Bartknecht, W. (1981). *Explosions, Course Prevention Protection.* New York: Springer-Verlag.

[13] Project SAFEKINEX, Contract No. EVG1-CT-2002-00072, Deliverable No. 2, *Report on the experimental factors influencing explosion indices determination*, September 2003.

[14] Le Chatelier, H. (1891). Note on the Determination of Firedamp by Flammability Limits. *Ann. Mines , 19*, 388-395.

[15] Babrauskas, V. (2003). *Ignition Handbook.* Issaquah: Fire Science Publishers.

[16] Koike, K., Inoue, G., Suzuki, M., Saida, J., & Komatsu, I. (1998). Decomposition Characteristics of Concentrated Ozone. *Journal of Chemical Engineering of Japan , 31* (2), 195-200.

[17] Schroeder, V., & Holtappels, K. (2005). Explosion Characteristics of Hydrogen-Air and Hydrogen-Oxygen Mixtures at Elevated Pressures. *International Conference on Hydrogen Safety.* Pisa.

9TH GLOBAL CONGRESS ON PROCESS SAFETY

Process Safety Considerations for New DOE Depleted Uranium Hexafluoride Conversion Plants

Peter M. Burban, Ph.D.
Department of Energy
Portsmouth Paducah Project Office
Peter.Burban@lex.doe.gov

Prepared for Presentation at
American Institute of Chemical Engineers
2013 Spring Meeting
9th Global Congress on Process Safety
San Antonio, Texas
April 28 – May 1, 2013

UNPUBLISHED

Process Safety Considerations for New DOE Depleted Uranium Hexafluoride Conversion Plants

Peter M. Burban, Ph.D.
Department of Energy
Portsmouth Paducah Project Office
Peter.Burban@lex.doe.gov

Keywords: depleted uranium hexafluoride, dry conversion, fluidized bed, uranium oxide, hydrofluoric acid, layers of protection, independent safety system, defense in depth, basic process control system, HF detectors, passive safety systems, Department of Energy

Abstract

The Department of Energy (DOE) OH & KY depleted uranium hexafluoride (DUF6) facilities are first of a kind operating plants designed to convert DUF6 into a safer form for reuse and/or disposal. Safety played a lead role in process design with HAZOP results guiding implementation of layers of protection to reduce risks associated with several materials including UF_6, H_2 and HF. A stand-alone independent safety control system functions to provide independent monitoring and control capability for those processes that have significant safety considerations.

1. Introduction

The Department of Energy Piketon, OH and Paducah, KY DUF6 facilities, Figure 1, are new fluid bed operating plants designed to convert approximately 700,000 metric tons of depleted uranium hexafluoride into a safer form for reuse and/or disposal. Depleted uranium hexafluoride has been generated in the United States since the mid-1940's as a byproduct of enriching uranium for both civilian and military applications at three gaseous diffusion plants located near Oak Ridge, Tennessee; Paducah, Kentucky; and Piketon, Ohio. This paper will highlight important process safety design considerations implemented that are critical to ongoing operations. Key aspects of the process hazards analysis and some lessons learned will be shared regarding chemical hazards associated with HF.

Figure 1. Portmouth (top) and Paducah DUF6 facilities

2. DUF6 Conversion Process and Safety Considerations

2.1 Process Description

The source DUF6 material, stored at both plants in approximately 63,000 large steel cylinders, is fed direct from the storage cylinders. A DUF6 feed cylinder, heated in a specially designed autoclave, vaporizes DUF6 which is fed with steam and hydrogen, blended with nitrogen, to a high temperature fluidized bed reactor for conversion to uranium oxide powder and hydrofluoric acid vapor. Some elements of the chemical process are described by Urza [1] and a general schematic of the process is given on the internet [2]. The uranium oxide powder transfer system loads powder into modified feed cylinders. The HF recovery system condenses the water and HF vapors in the process offgas, producing aqueous HF acid which is temporarily stored in tanks and later loaded for shipment to HF vendor facilities. The environmental green aspects of this approach include reuse of the feed DUF6 cylinders, regeneration of KOH used in scrubber operation and recovery of aqueous HF for commercial sale.

2.2 Overall Process Safety

A major safety advantage of the conversion process is low pressure reactor operation. The 'dry' conversion process uses superheated steam and nitrogen diluent to ensure safe reactor hydrogen concentrations. Administrative control of the feed cylinders ensures only depleted UF_6 is processed, eliminating any nuclear criticality event.

Hazard evaluations were performed to develop accident scenarios with the potential for exposure to HF, the greatest chemical hazard. Unmitigated consequences and risks were estimated for the credible events and compared to DOE risk guidelines. This paper focuses how safety significant engineered features and administrative controls are integrated in layers of protection to prevent or mitigate accidents associated with the major HF risks [3].

An Integrated Control System (ICS) controls and monitors the facility automated systems and associated components during normal operations. The system also mitigates accidents involving the conversion process and any of the systems or activities that have an interface with the process. The ICS consists of BPCS (Basic Process Control System) and ISS (Independent Safety System) subsystems. The purpose of the BPCS is to provide monitoring and control for normal operations from the Control Room and/or local operator workstations. The purpose of the ISS is to provide independent monitoring and control capability for those processes that have significant safety considerations. The ISS does not rely on the ISS human machine interface or BPCS to perform interlock actions.

3. Safety Features

3.1 Feed System Safety Features

A major component of the feed system is the autoclave, which serves to vaporize DUF6. The autoclave also serves as a pressure containment boundary (at several multiples of the supply pressure) to mitigate the consequence of an accidental release of DUF6 from a breached feed cylinder/piping in the autoclave. Since UF_6 reacts with atmospheric moisture to generate HF vapor, both HF and UF_6 releases have similar hazards classifications.

In the event of an over pressurized autoclave leading to activation of the pressure relief, the vent gases are directed to a controlled plenum. Automatic interlocks shut off DUF6 flow and/or electrical power to the cylinder heating system for an autoclave high pressure condition, as well as the following abnormal operating conditions: cylinder over-temperature, DUF6 header high-pressure, and an open autoclave door. The ISS monitors UF_6 and HF detectors located in the autoclave room which, upon alarm, terminate DUF6 flow to the downstream piping via an autoclave isolation valve. Additional administrative controls lead to shutdown of the reactor system. These layers of protection provide 'defense in depth' to reduce worker exposure from an abnormal UF_6 feed cylinder/system release.

3.2 Reaction System Safety Features

Each reactor serves as a primary containment boundary for reactants and products including DUF6 and HF. The ISS monitors UF_6 and HF detectors located in the reactor system room and adjoining HF condenser room which, upon alarm, terminate DUF6 flow to the downstream piping via the autoclave isolation valve. The BPCS in-line process oxygen monitors alarm on detection of oxygen in the process off gas. Additional administrative controls lead to shutdown of the reactor system.

Upon significant UF6 or HF leak detection, specific administrative controls instruct control room operations to terminate hydrogen flow via an automatic hydrogen feed isolation valve (a redundant manual isolation valve is also closed). Additional administrative controls lead to shutdown of the reactor system. These layers of protection provide 'defense in depth' to reduce worker exposure resulting from breach of the reactor system.

3.3 HF Recovery System Safety Features

Off gases from each reactor pass through individual partial condensers which produce an aqueous HF stream collected in individual primary receivers located in the HF recovery room. Off gases continue flowing to first a water and then to a KOH scrubber. All scrubber outlet streams merge and flow to a final backup KOH scrubber prior to venting out the facility stack. Multiple scrubbers reduce the likelihood of stack HF emissions. A secondary overflow HF receiver for each primary mitigates primary receiver level control problems. The HF recovery system units are skid mounted and reside in a common secondary containment dike mitigating the effect of HF leaks.

The ISS monitors several HF detectors located in the HF recovery room which, upon alarm, terminate DUF6 flow to the specific reactor via the autoclave isolation valve. The ISS also monitors HF detectors located in the HF recovery area for HF leaks in the HF transfer line which, upon alarm, terminates flow from the HF receiver tanks. Upon significant HF leak detection, specific administrative controls instruct control room operations to terminate DUF6 and hydrogen flow via an automatic hydrogen feed isolation valve (a redundant manual isolation valve is also closed). Additional administrative controls lead to shutdown of the reactor system. These layers of protection provide 'defense in depth' to reduce worker exposure resulting from breaches in the HF recovery system.

An important series of lessons learned include proper maintenance of HF process lines to eliminate chronic leaks. In addition, care was taken for stress relief of all piping flange connections, and modifications were made to address piping areas experiencing vibration. The use of HF vapor highly sensitive paint was helpful to identify the onset of leaks. Additional isolation valves were installed to improve the safety and ease of corrective maintenance.

3.4 HF Storage Tank Farm Safety Features

The HF storage area includes several passive, active and administrative safety controls. Each HF storage tank is situated in a secondary liquid dike that can hold the entire capacity of the 10,000 gallon tank in case of a leak. The tanker truck and rail loadout areas have a gravity drain equipped with a pumping system to transfer spilled HF to either a containment dike or a storage tank. This minimizes the spill area to reduce the potential on and off site consequences of an HF leak.

The ISS also monitors HF detectors located in the HF storage area which, upon alarm, terminates flow via the HF storage tank automated isolation valve. In addition, the ISS terminates HF flow from the HF receiver tanks. Specific administrative controls instruct an operator to be in required attendance for all transfers. This operator observes each HF storage tank to mobile tanker transfer and can terminate the transfer via an emergency button which stops HF transfer flow even before HF detection by the ISS. Upon alarm, following administrative control, the operators in the control room isolate the HF recovery system and then shutdown the reactor system.

These layers of protection provide 'defense in depth' to reduce worker exposure resulting from leak accidents at the HF storage tank system.

3.5 Additional Safety Features

Several fire suppression systems minimize the likelihood of a fire causing a cylinder breach and release of HF and/or DUF6. Strategically placed vehicle barriers add "defense in depth" to reduce consequences associated with collision with storage tanks and staged feed cylinders. Safety design elements were also incorporated for natural event accident initiators so that process buildings and equipment meet design criteria for forces such as seismic and high winds. Safety management requires the use of appropriate personal protection equipment for all work activity with potential HF exposure. These layers of protection provide 'defense in depth' to reduce worker exposure resulting from HF leak accidents in various systems described above.

4. Discussion

The process safety approach at the DUF6 facilities blends active, passive and specific administrative controls. Considering the other chemical and radiation hazards at the facility, the ability to maintain a safe operation depends on reliable machines and human performance of required operator actions. A rigorous preventative maintenance program is administratively controlled to maintain the reliability of all active safety systems at the facility in line with DOE safety requirements. The Integrated Safety Management Program at the DUF6 facility includes systematic approach to training for Operations. Training includes preparing each individual with the toolbox needed to address all specific administration controls, responses to alarms and recovery from abnormal operating conditions. Monitoring of human performance is an essential element for maintaining safe operations and continuous improvement.

Major facility start up challenges included accurate procedure development and updates, initial training and continuous training as the Operations knowledge base grew. Unique lessons were learned for operations in the four season climates of Ohio and Kentucky, necessitating adaptation of procedures, and emphasis on minor equipment changes to minimize the need for corrective maintenance. Both sites have had a good start to their multiple decade operations campaigns built on the cornerstone of a defense in depth process safety approach.

5. References

[1] Urza, Inaky J., "Conversion of uranium hexafluoride to uranium dioxide" US Patent 4,830,841

[2] "The DUF6 Process," www.bwconversionservices.com, accessed on Feb. 11, 2013

[3] DOE, Final Environmental Impact Statement for Construction and Operation of a Depleted Uranium Hexafluoride Conversion Facility at the Portsmouth, Ohio, Site, DOE/EIS-0360, prepared by Argonne National Laboratory for U.S. Department of Energy- Oak Ridge Operations, Office of Environmental Management, Oak Ridge, Tenn. June 2004. energy.gov/nepa/downloads/eis-0360-final-environmental-impact-statement

(44b) Minimum Ignition Energy Study of Flowing Heat Transfer Fluid Aerosols: Experimental and Theorectical Approach

Szu-ying Huang[1], Xinrui Li[1] and M. Sam Mannan[2], (1)Mary Kay O'Connor Process Safety Center, Artie McFerrin Department of Chemical Engineering, Texas A&M Univerisity, College Station, TX, (2)Mary Kay O'Connor Process Safety Center, Artie McFerrin Department of Chemical Engineering, Texas A&M University, College Station, TX

Liquid-gas aerosol has been a significant concern in process industries for its complexity and unpredictability of formation, ignition probability, combustion mechanisms, and flame propagation with impact on onshore/offshore facilities. Heat transfer fluid, specifically, has been proved to have the possibility to form aerosol from pressurized leaking, spraying, and potential liquid impingment from surfaces. Having more combustible matter and higher energy density, the high-flash point aerosol may induce more hazardous fires and explosions even operated or stored under the safety limits. Among the indicators describing flammability, minimum ignition energy (MIE) is one of the most widely used and studied ones. It is capable to reflect the internal chemical properties of the material, as well as to make connection to analyze the possible facility ignition sources. This paper demonstrates MIE experimental result of commercial heat transfer fluid aerosols, and develops COMSOL-Multiphysics models as theoretical approach on MIE prediction of similar chemical species. The influence of aerosol droplet sizes, initial spray velocity, fuel-air equivalence ratio, and fluid temperature are analyzed. The characteristic ignition temperature, which is the "flash point" of aerosol system, is also developed to compare with single-phase material. Results of the study may be used for industry liquid handling, working fluid selection, mitigation design, and ignition elimination to prevent fire and explosion of aerosols.

Domino Effect Model Applied to the Hertfordshire Oil Storage Terminal (Buncefield Oil Depot)

Manuscript ID: 295031

Esteban Bernechea
Centre for Technological Risk Studies (CERTEC). Department of Chemical Engineering.
ETSEIB. Universitat Politècnica de Catalunya.
Av. Diagonal, 647. 08028-Barcelona (Spain).
esteban.bernechea@estudiant.upc.edu

Xavier Seguí
Centre for Technological Risk Studies (CERTEC). Department of Chemical Engineering.
ETSEIB. Universitat Politècnica de Catalunya.
Av. Diagonal, 647. 08028-Barcelona (Spain).
xavier.segui@upc.edu

Josep Arnaldos
Centre for Technological Risk Studies (CERTEC). Department of Chemical Engineering.
ETSEIB. Universitat Politècnica de Catalunya.
Av. Diagonal, 647. 08028-Barcelona (Spain).
josep.arnaldos@upc.edu

1. Introduction

On December 11, 2005, a series of explosions occurred at the Buncefield Oil Storage Terminal, located in Hemel Hempstead, Hertfordshire, England. The explosions resulted in large fires which spread throughout the installation, causing more than 40 injuries, the destruction of most of the installation and the emission of large clouds of black smoke to the atmosphere; the accident also caused significant damage to residential and commercial properties in the vicinity of the terminal, the evacuation of population centers near the terminal, and an overall economic loss estimated at close to £ 1 billion [1].

The Buncefield disaster is a perfect example of domino effect in the oil industry; the initial explosion, caused due to the overflow of one of the tanks, caused fires that engulfed more than 20 storage tanks, and resulted in the destruction of a significant part of the terminal. The domino effect is defined as a cascade of events in which the consequences of previous accidents are increased both spatially and temporally by following ones, thus leading to a major accident [2]. This definition describes perfectly the situation that occurred in the Buncefield terminal.

One of the characteristics that made this accident the object of many researches was that the overpressure wave ensuing from the first explosion was much higher than would be expected from a Vapor Cloud Explosion (VCE). However, the aim of this manuscript is not to model the exact explosion, or to determine the reasons that caused the overpressure resulting from it to be higher than expected; much works have been dedicated to this endeavor, producing satisfactory conclusions to the phenomenon. The objective of this work is rather to use the Buncefield Terminal layout to test a domino effect model proposed by the authors [3]; to achieve this, possible accident sequences after an explosion will be simulated, to ascertain if they are similar to the chain of accidents that took place during the real event. Also, Quantitative Risk Assessments (QRAs) will be performed using the layout: one in a traditional fashion [4], without accounting for domino effect, and another taking domino effect into account through the application of the proposed model. This is done to analyze the change in the iso-risk curves obtained for both QRAs.

First, a more detailed description of the accident is presented, followed by an explanation of the proposed domino effect model. After this, the way in which the model has been applied to the layout of the installation is described and the results obtained are presented and discussed. Finally, the conclusions reached by the end of the study are shown.

2. Description of the accident

A brief abstract of the incident as described in Volume 1 of The Final Report of the Major Incident Investigation Board [1] is presented in this section.

On Saturday 10, December 2005, a delivery of unleaded petrol started to arrive at Tank 912 in bund A. At about 05:30 am on 11 December, the safety systems that prevented the overfilling of the tank failed to operate. Petrol cascaded down the side of the tank and was collected in Bund A. As the released continued, a vapor cloud mixture of fuel and vapor was formed and started dispersing westwards. Up to 300 tonnes of petrol were released, and about 10% turned to vapor and mixed with air, eventually reaching dangerous concentrations. At 6:01 am on Sunday 11, December 2005, the first, in a chain of explosions, occurred. The explosions caused a huge fire which engulfed more than 20 tanks and burned for 5 days, destroying great part of the terminal.

3. Domino effect model

The model used in this work [3] was developed to assess the impact that the domino effect could have on the frequencies of the accidents used to perform QRAs on storage installations; however, it can also be used to produce stand-alone accident sequences.

The model is an algorithm that has been programmed in MATLAB; it follows the steps used to perform a QRA in a cyclical manner to produce accident sequences which, in the end, will produce frequencies of accidents in all the tanks that are part of the installation.

The first step to obtain the final frequencies of accidents is to input some initial data that describes the installation; these are: the number of tanks in the installation, the substances stored and their properties, the mass contained in each tank, the sizes of the containment dikes, the types of tanks (atmospheric, pressurized), the initial frequencies of accidents per tank and the distances between tanks.

After the characteristics of the installation have been defined, the initial Loss of Containment events (LOCs) that will occur in each of the units, along with their frequencies have to be specified. Using these LOCs, and knowing the substance stored, event trees can be used to propose sets of possible accidents, together with their probability of occurrence; when these probabilities are multiplied by the frequency of the LOC, the initial frequency of accident is obtained. The initial LOCs and the event tree used in this work are those proposed in "The Purple Book" [4]. The LOCs for atmospheric storage tanks and their frequencies are presented in Table 1.

Table 1: Initial LOCs in atmospheric storage units

Code	Definition	Frequency (y^{-1})
G.1	Instantaneous release of the complete inventory.	$5 x10^{-6}$
G.2	Continuous release of the complete inventory in 10 min at a constant rate of release.	$5 x 10^{-6}$
G.3	Continuous release from a hole with an effective diameter of 10 mm.	$1 x 10^{-4}$

The steps followed up to this point are the same as for a regular QRA. To produce the accident sequences, one of the tanks will be selected to suffer the initial LOC, and the effects and consequences of the accidents in this unit will be calculated for all other tanks. Depending on the consequences, a level of damage, corresponding to a new LOC will be assigned to each of the affected units. This was done using the threshold values presented in Table 2.

Table 2: Threshold values applied in the model

Accident	Effect	Threshold Values		Consequence
		Pressurized Tanks	**Atmospheric Tanks**	
BLEVE	Thermal radiation (kW/m2)	≥ 100	≥ 100 [5]	G.3 type release
		< 100	< 100 [5]	No consequence
Explosion	Overpressure wave (kPa)	≥ 100	≥ 25 [6]	G.1 type release
		$53 < \Delta P < 100$ [7]	$20 < \Delta P < 25$ [7]	G.2 type release
		$30 \leq \Delta P < 53$ [7]	$14 \leq \Delta P < 20$ [7]	G.3 type release
		$\Delta P < 30$ [5]	$\Delta P < 14$ [8]	No consequence
Jet fire	Jet length (m) Heat Radiation (kW/m2)	$dj \geq d$ [5]	$dj \geq d$ [5]	G.1 type release
		$dj < d$ & $I \geq 40$ [5]	$dj < d$ & $I \geq 10$ [5]	G.3 type release
		$dj < d$ & $I < 40$ [5]	$dj < d$ & $I < 10$ [5]	No consequence
Flash fire	-	-	-	No consequence
Pool fire	Thermal radiation (kW/m2) Pool duration (min)	$dp \geq d$ [5]	$dp \geq d$ [5]	G.1 type release
		$60 \leq I$ & $tp \geq 10$ & $dj < d$ [9]	$15 \leq I$ & $tp \geq 10$ & $dj < d$ [9]	G.2 type release
		$I \geq 8$ & $dj < d$	$I \geq 8$ & $dj < d$	G.3 type release
		$I < 8$ & $dj < d$	$I < 8$ & $dj < d$	No consequence
dj = jet length; dp = pool diameter; d = distance between tanks; tj = duration of jet fire; tp = duration of pool fire				

Once new LOCs resulting from an initial accident have been defined in the affected units, the effects and consequences of subsequent events on surrounding remaining units will be estimated; this will lead to other LOCs in other tanks, producing more and more branches of possible events. For each of the levels of propagation, the frequencies of accidents will be calculated, using the frequency of the previous accident as the initiator and the same event tree for the probabilities of possible accidents. Once one of the accidents in a branch reaches a frequency lower than $1x10^{-9}$, the sequence will stop, as it is deemed that this frequency is too low to be taken into account in the QRA.

Figure 1. Domino effect model algorithm

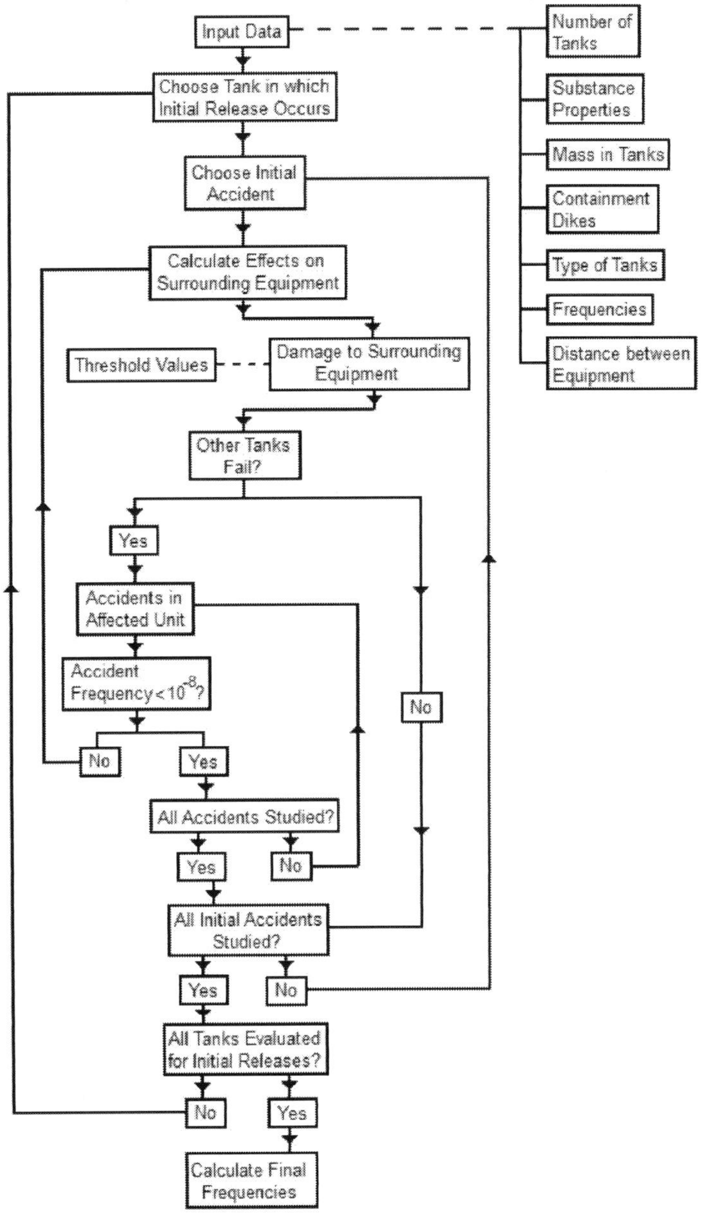

The process has to be repeated in loops, until all sequences have been explored, and all the tanks have been selected as initiators (suffering the initial LOC). In each step, the frequencies with which the different accidents occur in the different tanks are stored. Finally, these frequencies are added, by accident and by tanks, (for example, adding all the times that "accident 1" possibly occurs in "Tank 1"), whether due to an initial LOC event or due to the effect of an accident in another tank; these frequencies are referred to as final, and are the ones used in the QRA. A diagram of the algorithm that is followed in the methodology is presented in Figure 1.

The results obtained using the model can be compared against those obtained from a traditional QRA in order to assess the impact of domino effect on the risk associated to the installation. The model can also be used to produce a singular accident sequence, resulting from a specific accident in one of the units.

Figure 2. State of the terminal and proposed layout

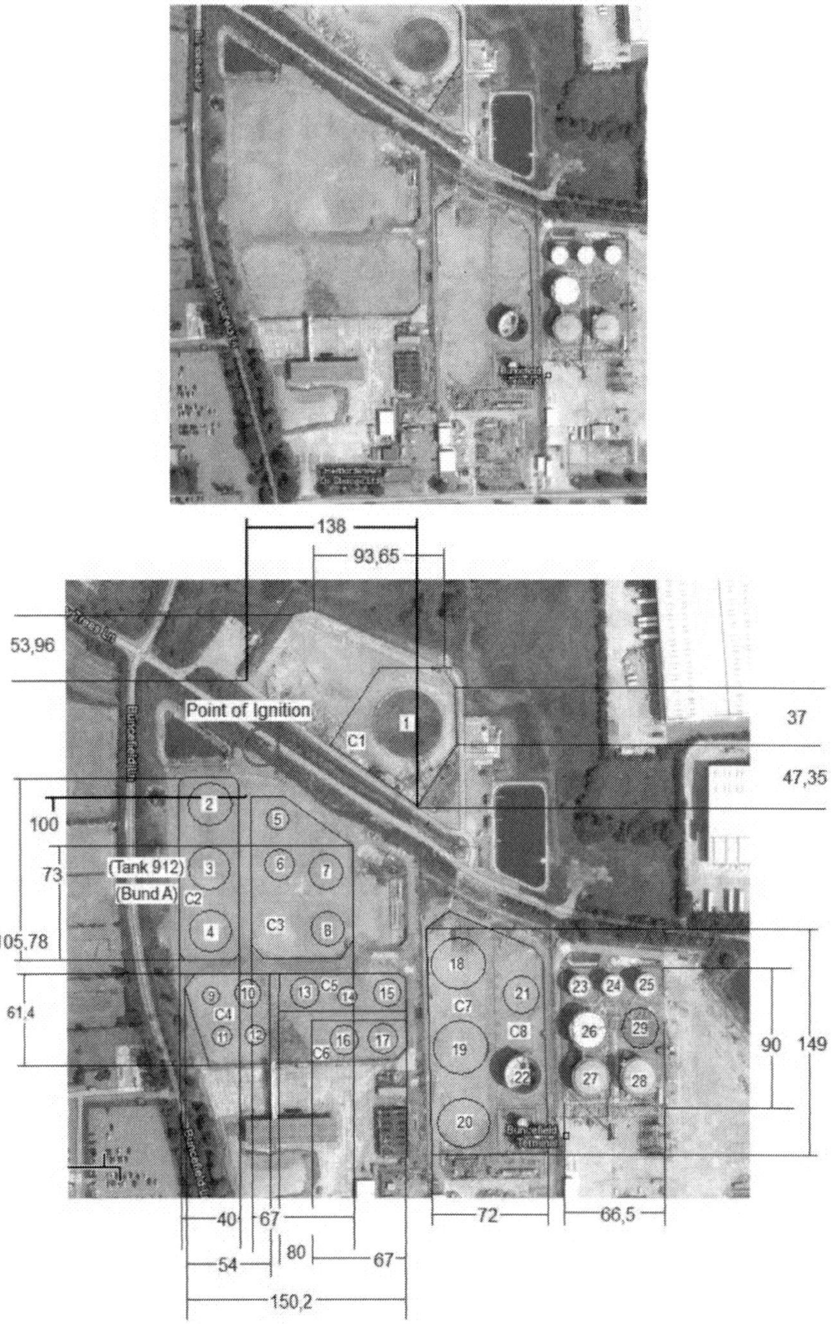

4. Applying the model to the Buncefield terminal

In this section the domino effect model will be applied to the terminal. First, to produce an accident sequence in a situation similar to the real accident and later, to produce iso-risk curves in two different QRAs.

4.1 Modeling the installation

To model the installation, the "Google Maps" tool was used to view the current state of the terminal, and obtain measures similar to those of the tanks that were destroyed and of the tanks that are still standing, as well as the distances between the tanks and the dimensions of the containment bunds. Figure 2 shows, on top, the latest image found of the state of the installation and on the bottom, the distances and dimensions used to model the installation. After estimating the dimensions of the installation, the tanks and bunds were renumbered for the application of the model; for example, Tank 912 becomes Tank 3, and Bund A is called C2.

4.2 Approaches to Accident Modeling

The vapor cloud that resulted in the explosion in the Buncefield Terminal is thought to have ignited at the emergency pump house located close to the containment bund A (in which Tank 912 was located), as shown in Figure 2. Since the domino effect model works by going through a fixed set of possible accidents in each of the tanks, the approach followed to produce the accident sequences was to define a fictional tank "Number 30" in the area in which the ignition of the flammable cloud is thought to have occurred; this tank is modeled containing a mass close to 150,000 kg of pentane.

This approach was followed to produce the accident sequence, but not to develop the QRAs, which were completed using the layout proposed in Figure 2, with all the tanks filled at full capacity.

4.3 Accident Sequence

Figure 3 shows the most probable incidents in the sequence produced when the domino effect method is applied to the case of an explosion derived from a G.2 type release in Tank 30 (the fictional tank created to model the initial explosion). The tanks are represented inside circles, while the possible accidents are shown inside diamonds (E stands for explosion, P for pool fire). The lines generated from the accident diamonds represent the level of damage caused by an event; these lines later divide, reaching different tanks, which means that the tank suffers the associated type of release after the previous incident. All the lines shown in the figure represent possible accidents identified by the methodology. The continuous lines represent the most probable accidents, which derive in the destruction or heavy damage of the affected unit, while the dotted lines represent accidents that do not cause catastrophic damages, or that are less probable; however, they are presented to show other possible scenarios produced when applying the algorithm. Once a tank is reached by a continuous line, it means that it has suffered a level of damage that will result in its destruction.

When an explosion (from a G.2 LOC) occurs in tank 30, the overpressure wave affects tanks 2, 3, 5, 6 and 7 in a catastrophic way; tank 1 suffers damages that lead to a G.2 release and tanks 4 and 8 are affected in a minor way. This is the first sequence of accidents, and it is understood that the failures and later releases occur almost simultaneously. After the mentioned tanks fail due to the explosion, other accidents will generate in each of them, specifically, explosions, flash fires or pool fires; since these last type of accident will be the most probable one (according to the event trees used to produce the sequences), it will be shown for each of the affected tanks.

Figure 3. Accident sequence developed using the domino effect model

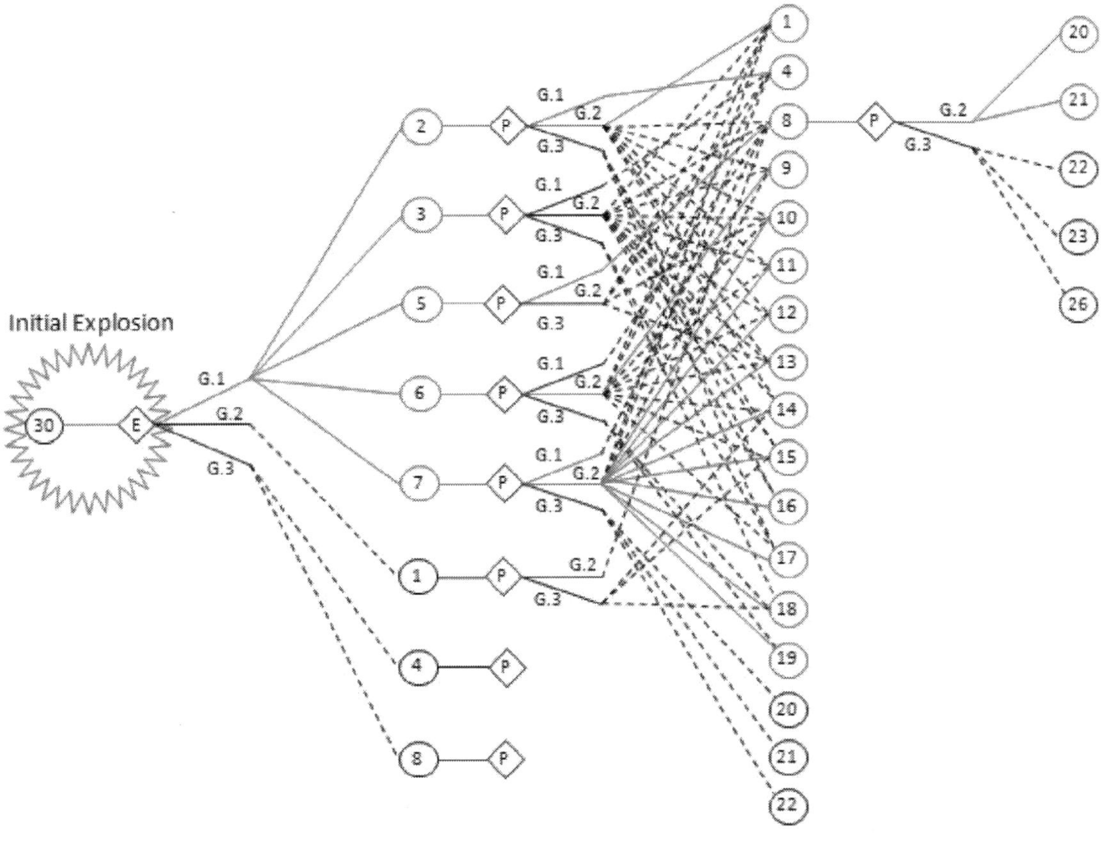

In the case that a pool fire occurs in Tank 2, Tank 4 will fail catastrophically due to the thermal radiation load it receives, while Tank 1 will also be affected in a major way, and also fail. There are other affected tanks, but these possible paths are not explored in detail here. It has to be noted that by the time Tank 2 fails, Tank 3 is also failing. A catastrophic failure in Tank 3 would also result in a pool fire that would further the chain of accidents, but none of these occurrences is explored. Pool fires in tanks 5, 6 and 7 would, according to some paths of the sequence obtained, result in the failure, and grievous damage, of tanks 8 through 19. If the sequence is furthered following a pool fire in Tank 8, tanks 20 and 21 would be damaged beyond repair, while tanks 22, 23 and 26 would suffer minor damages. It has to be noted that in the sequence explored, tanks 24, 25, 27, 28 and 29 do not suffer damages.

The sequence of accidents explored, which is the one that presented higher frequencies in each of its events, results in the destruction of 21 out of the 29 tanks that are part of the installation; this is in accordance to the level of damage in the real Buncefield terminal after the situations resulting from the explosion were controlled. It has to be said that this is one of many possible sequences produced by the algorithm, several of which are different from the real occurrence; however, branches of the possible sequences could be trimmed, in order to choose occurrences that are deemed to be more probable.

4.4 Application to QRA

Two QRAs were performed using the layout presented in Figure 2, the first using the normal procedure and the regular frequencies obtained from the original LOCs and the event trees; the second, using the frequencies of accident obtained from applying the domino effect model. The iso-risk curves obtained for the studies are presented in Figures 4 and 5.

Figure 4. Regular iso-risk curves

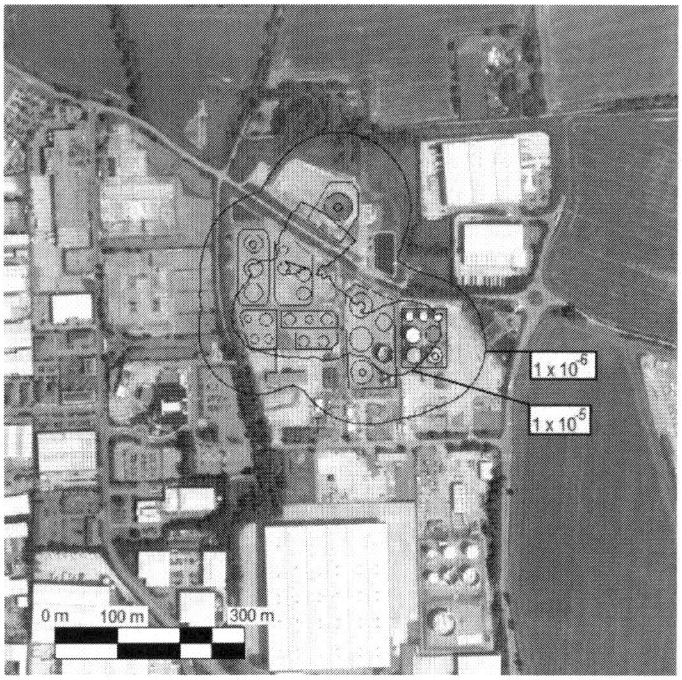

It is clear when observing the figures presented above that the individual risk associated to the installation increases when the domino effect methodology is applied. When regular frequencies are used, only 10^{-5} and 10^{-6} curves appear, while a 10^{-4} curve appears when using the modified frequencies; it encompasses the complete layout of the installation, and in some points, extends some 80 m from the limits of the terminal. The 10^{-4} increases it size so much because many lesser accidents with low frequencies appear in the accident sequences generated by the algorithm. The 10^{-5} curve also increases its area significantly, growing to the point of covering the complete evaluated installation, and reaching close to a 100 m from the limit of the terminal on its furthest point. The 10^{-6} curve is the one that presents a less appreciable change; to the north of the installation the curve suffers almost no alteration, to the south, it increases its length some 30 m and to the east and west, it increases some 20 m in each direction.

Figure 5. Iso-risk curves accounting for domino effect

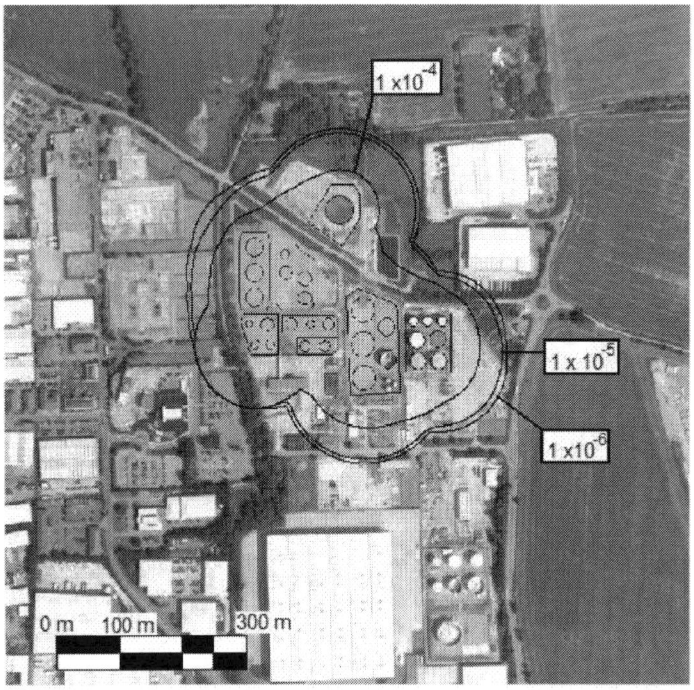

It is interesting to note that when performing a QRA of a storage installation composed of atmospheric tanks, the accidents with highest frequencies of occurrence are the pool fires, followed by flash fires and explosions. This means that the iso-risk curves related to the type of installation studied will not extend over great distances, as occurs when evaluating installations that store pressurized flammable liquids or toxic gases. Therefore, although the increase in the 10^{-6} curve, which, in many European countries is the most representative at the moment of evaluating external affectation, seems marginal, it is important, due to the type of installation that has been evaluated.

5. Conclusions

The domino effect model applied to a layout similar to that of the old Buncefield terminal produced many accident sequences; one of them is thought to be similar to the real accident that occurred in the facility in 2005. This means that the model is indeed suitable to be applied on real layouts, and that it can be used to help assess the risk related to a projected or existing installation, taking into account the possibility of domino effect occurrence.

The model can be applied with a minimum cost of time, and the frequencies obtained can be introduced in commercial software like "Riskcurves" without additional cost; QRAs could be performed taking domino effect into account, without the need to spend more time and money in the development of the study.

An increase in the risk associated to the installation was observed when the domino effect frequencies were used. The 10^{-6} curve does not suffer a significant increase in size due to the nature of the installation and its equipment.

Domino effect should be taken into account when performing QRA of industrial installations, as it can have an appreciable effect on the risk associated to the plant. This could lead to an impact on the Land Use Planning of the zone in which the facility is located.

References

[1] The Buncefield Incident 11 December 2005. The Final Report of the Major Incident Investigation Board. Volume 1, 2008. Buncefield Major Incident Investigation Board.

[2] Delvosalle, C, "Domino Effects Phenomena: Definition, Overview and Classification", First European Seminar on Domino Effects, Leuven, 1996.

[3] Bernechea, E.J., Vílchez, J.A., Arnaldos, J., 2012. A model for estimating the impact of the domino effect on accident frequencies in quantitative risk assessments of storage facilities. Process Safety and Environmental Protection, http://dx.doi.org/10.1016/j.psep.2012.09.004

[4] CPR 18E, 2005. Guidelines for Quantitative Risk Assessment (Purple Book). Directorate General for Social Affairs and Employment, The Hague, Netherlands.

[5] G. Antonioni, G. Spadoni, V. Cozzani, Application of domino effect quantitative risk assessment to an extended industrial area, Journal of Loss Prevention in the Process Industries, 22, 614-624, 2009.

[6] F. I. Khan, S. A. Abassi, Estimation of probabilities and likely consequences of a chain of accidents (domino effect) in Manali Industrial Complex, Journal of Cleaner Production, 9, 493-508, 2001.

[7] V. Cozzani, E. Salzano, The quantitative assessment of domino effects caused by overpressure: Part I. Probit models, Journal of Hazardous Materials, 107, 67-80, 2004.

[8] V. Cozzani, E. Salzano, Threshold Values for domino effects caused by blast wave interaction with process equipment, Journal of Loss Prevention in the Process Industries, 17, 437-447, 2004.

[9] G. Landucci, G. Gubinelli, G. Antonioni, V. Cozzani, The assessment of the damage probability of storage tanks in domino events triggered by fire, Accident Analysis & Prevention, 41, 1206-1215, 2009.

Proven Approach to Investigating Near Misses

Greg Smith, Senior Engineer
Process Improvement Institute, Inc. (PII)
7502 Chad Road
Harrison, TN 37341
Phone: (423) 413-0038
Fax: (865) 671-4944
e-mail: gsmith@piii.com

Gerald D. Burch, Senior Engineer
Process Improvement Institute, Inc. (PII)
e-mail: gburch@piii.com

William G. Bridges, President
Process Improvement Institute, Inc. (PII)
e-mail: wbridges@piii.com

Prepared for Presentation at
9th Global Congress on Process Safety
San Antonio, TX
April 28-May 1, 2013

UNPUBLISHED

9TH GLOBAL CONGRESS ON PROCESS SAFETY

Proven Approach to Investigating Near Misses

Steven (Greg) Smith, Senior Engineer
Process Improvement Institute, Inc. (PII)
7502 Chad Road
Harrison, TN 37341
Phone: (423) 413-0038
Fax: (865) 671-4944
e-mail: gsmith@piii.com

Gerald D. Burch, Senior Engineer
Process Improvement Institute, Inc. (PII)
e-mail: gburch@piii.com

William G. Bridges, President
Process Improvement Institute, Inc. (PII)
e-mail: wbridges@piii.com

Keywords: investigation, near miss, accident, incident, root cause, RCA

Abstract

Getting near misses reported is a major hurdle for most companies, one reason being is that the investigation system is not ready for the large influx of reports. Workers and management respond by being reluctant to potentially overload the investigation system (and overload themselves). This paper explains approaches to manage efficiency and effectiveness in near miss investigations that have been successful for handling near misses reporting within both large and small companies and to ensure high value from the investment of reporting and analysis. This paper builds on the updated paper presented in 2012 at GCPS on "Gains from Getting Near Misses Reported."

1. Introduction

Much attention has been given to the detection and correction of early warning signals as a means of improving safety performance.

Process safety events often have entirely different causal factors (human error, equipment malfunction, natural causes) and root causes (management system weaknesses) from those found in more traditional personal injuries.[1,2] Process safety incidents are more technical and complex in nature, usually involving multiple failures of protective barriers. The human actions and errors which cause them are often several steps removed from the incident itself. The human factors involved are not as obvious to investigators; the causes are more systemic than individual in their nature.

Even though personal injury rates have declined, significant injuries and financial losses have continued in a series of high profile, catastrophic (multiple injuries) process safety events. Because of inconsistencies in reporting standards and definitions, it is difficult to tell if these events are actually increasing in frequency or if their visibility has simply increased due to media attention and heightened public concerns. Either way, improved understanding and enhanced tools are required to improve our process safety performance.

In the process industries one such tool is near miss reporting, the early warning system equivalent to unsafe behaviors for process safety. Companies which have successfully developed the ability to recognize, report, investigate, and learn from near misses have shown dramatic reductions in loss events, losses, and injuries due to process incidents. In some cases, loss reductions of more than 90% have been achieved.[2]

2. Definitions

The following definitions are used throughout this article:

> *An incident is either an accident or a near miss.*
>
> *An accident is a sequence of unplanned events and conditions that result in harm to people, environment, process, product or image.*
>
> *A near miss is an unplanned sequence of events that could have caused harm if conditions were different or is allowed to progress, but did not in this instance.*
>
> *A causal factor is a human error (typically an error by the at-risk employee performing a task/job in the process) or a component fault/failure. Note that these human errors and component failures are probably caused by other humans making mistakes, and all errors are controlled by management systems. An incident typically has multiple causal factors. Natural phenomena can also be a causal factor.*
>
> *A root cause is a management system weakness that results in a causal factor. A casual factor typically has multiple root causes.*

The definitions above are the same as used in *Guidelines for Investigating Chemical Process Incidents, Second Edition* (CCPS, 2003).

3. Importance of reporting and learning from near misses

Investigating near misses is critical to preventing accidents, because near misses share the causes and root causes of accidents; they are one or two barriers away from the loss/accident. We are very likely preventing many apparently unrelated accidents when we prevent the ones that are obviously related. Figure 1 illustrates a hypothetical relationship between causal factors and root causes of accidents and those for near misses. From our experience, this relationship appears to be valid; see the case study section of this paper to see some of the proofs of this concept.

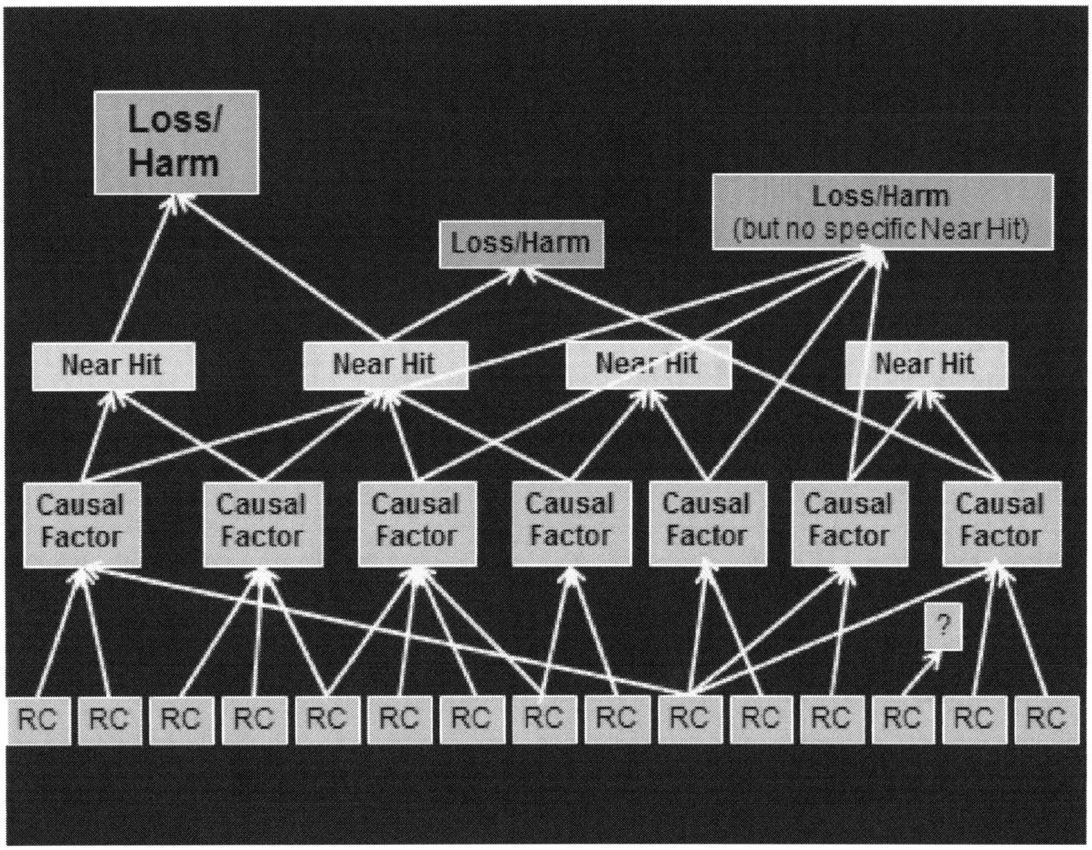

Figure 1 Interrelationship between the causes of Accidents (Losses) and the causes of Near Misses

As a brief explanation of Figure 1, root causes (management system weaknesses) make it more likely for a causal factor to occur. Combinations of causal factors (or in rare cases, perhaps a single causal factor) result in near misses. Additional causal factors result in losses (accidents) in increasing severity. A root cause can increase the likelihood of seemingly unrelated causal factors.

Formal and informal organizational surveys[3] detail the difficulties of implementing an effective near miss reporting system. In this paper, a number of barriers are outlined and potential solutions are proposed. The identified barriers are:

1. *Fear of disciplinary action*

2. *Fear of teasing by peers (embarrassment)*
3. *Lack of understanding of what constitutes a near miss versus a non-incident*
4. ***Lack of management commitment (no training provided on investigation techniques and procedures) and lack of follow-through once a near miss is reported (time is not allocated to investigate near misses, or corrective actions not implemented)***
5. ***An apparently high level of effort is required to report and to investigate near misses compared to low return on this investment***
6. ***There is No Way to investigate the thousands of near misses per month or year***
7. *Disincentives for reporting near misses (e.g., reporting near misses hurts the department's safety performance)*
8. ***Not knowing which accident investigation system to use (or confusing reporting system)***
9. *Company discourages near-miss reporting due to fear of legal liability if these are misused by outsiders*

4. Primary Barriers to Effective Investigation of Near Misses

Four of the barriers (4, 5, 6, and 8) tie directly to the organization's ability to effectively and efficiently investigate and resolve near miss incidents.

Of the four barriers related to getting near misses reported, the two that are most related to effective near miss investigations are **Barriers 6 and 8.**

Overcoming these barriers is described below by looking at the experience of organizations as they implement these concepts.

4.1 Barrier 6. There is No Way to investigate all of the thousands of near misses per month or year!

Normally, when the discussion of having a huge number of near misses reported, such as four per worker per month, the reaction will be shock and then a statement such as Barrier 6. This barrier is closely related to Barrier 5*: An apparently high level of effort is required to investigate near misses compared to the small gain perceived*. If a site has 500 staff as operators and maintenance craftsmen, then likely 25,000 near misses could be reported. At first glance, it can appear impossible to cope with, let alone investigate that number of near misses (incidents).

Maybe the most important solution to Barrier 5 and 6 is to first decide which near misses and losses/accidents need to be investigated. The best solution is normally stated as:

> ***Let front line foremen or supervisors decide if a near miss or accident needs to be investigated to root causes; the decision is made on the apparent Learning Value of the incident. How do you decide it a high or low learning value? The foremen or supervisors will easily decide based on their hands-on experience.*** *For instance, an additive pump failure may directly result in a*

minor and temporary impact to quality or production rates but does not appear to have the potential to cause a large loss. So the foreman of the shift decides to designate this as a low learning value incident and therefore not do an immediate RCA. But, then other similar pumps fail over a six month period. A query shows that after 12 months, there have been 3 failures of the same additive pump and 8 failures of 5 other similar pumps in different services. Based on the number of failures and cumulative losses, the company performs an RCA of the 11 failures, resulting in findings which when applied to all similar pumps and services, greatly reduces the production/quality impacts. The RCA of the grouping of incidents proved very cost effective.

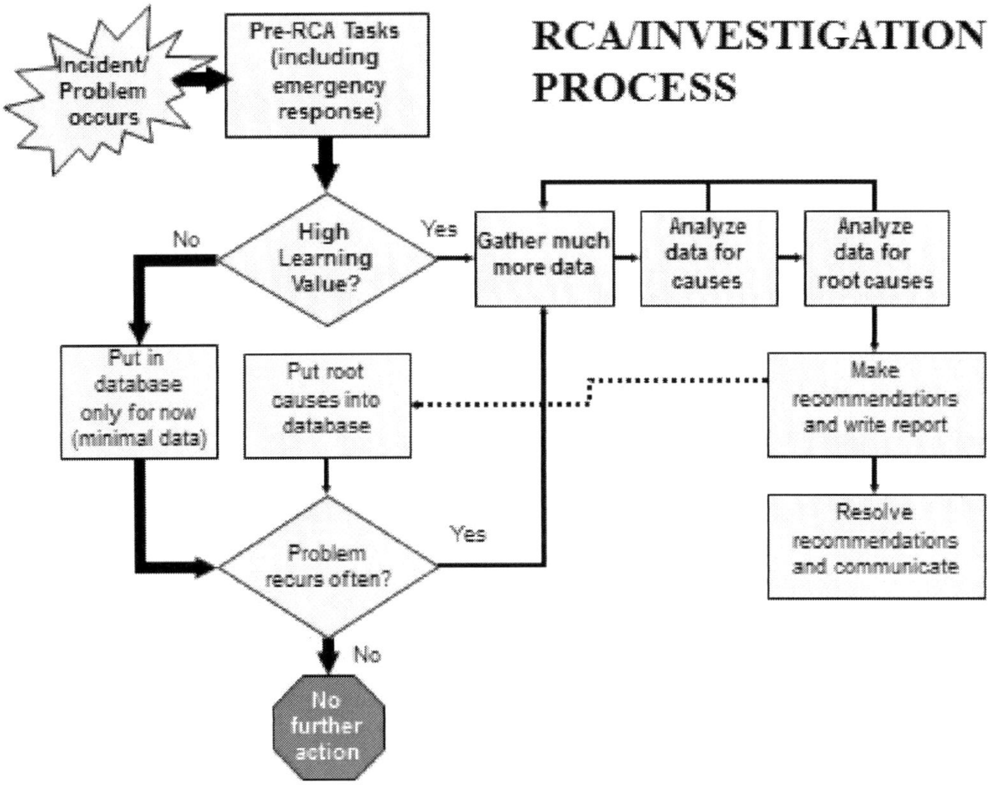

Figure 2: RCA/Investigation Flow Process

Figure 2 illustrates the process flow for an investigation system than can handle a large volume of near misses and losses/accidents. For this process to work:

- Be prepared for investigations by having enough staff trained in root cause analysis methods (or to help in the analysis, such as being able to interview peers).
- When the near miss, etc., is first noticed or reported by staff, let the frontline supervisor or foreman decide if it has high learning value.
- For high learning value incidents, investigate now. For low learning value incidents, put in the database now, along with the little data you have and any obvious causes. Do not investigate yet!

- Query the database every one to six months and perform Pareto or similar analysis to help decide which recurring events need to be analyzed in more detail.
- Take the root causes from investigations/RCAs and put in the database as well.
- Query the database every one to six month

4.2 Barrier 8. Not knowing which accident investigation system to use

One consideration that is not related to any of the barriers mentioned above, except marginally to Barrier 3 (lack of understanding of what a near miss is), is the scope of the investigation program. Some companies have one investigation system for occupational safety incidents, another one for process safety incidents, another for environmental releases, another for reliability issues, and yet another for quality and customer services issues. We have found that the same investigation approach and investigator training works well for incidents in any facet of a business. We believe there is merit in combining the systems and, in particular, in combining the incident databases. Combining the incident systems will require more work on defining near misses and in determining success in reporting near misses.

Consider having ONE incident reporting system with ONE approach for teaching employees the definition of a near miss and with ONE approach for doing incident investigations (including one approach for root cause analysis).

5. Effective and Efficient Incident Investigation and Root Cause Analysis

The incident investigation and root cause analysis (II/RCA) process should enhance organizational learning, reduce accidental losses, invest wisely in protective and mitigative systems, better protect people and equipment from harm, and continuously improve EHSS performance. An efficient and effective process ensures that scarce organizational resources are used appropriately in activities that matter.

The overall II/RCA process consists of the following steps:

- Data collection
- Causal factor analysis
- Root cause analysis
- Develop recommendations
- Reporting and follow through

5.1 Data collection

Every investigation begins by collecting, organizing, and preserving all relevant data as quickly as possible after the incident has been stabilized. Data comes in the form of personnel interviews, physical evidence, and documentation. All data is fragile and susceptible to loss or damage if it is not properly collected and preserved in a timely manner. All front line supervisors should be trained in the II/RCA process with particular emphasis on initiating the investigation, interviewing people, and collecting the physical evidence. Provide front line supervisors with an investigation tool kit containing a camera, barrier tape, collection bags and bottles, and appropriate forms and checklists.

5.2 Causal factor analysis

Once collected, data must be organized and analyzed to determine what happened – the chain of events (causal factors) that led to an undesirable occurrence. The most effective and proven methods for analyzing data are Event and Causal Factor Charting (ECFC) supplemented, if needed, with a Simplified Fault Tree Analysis for complex, chronic, or acute events. These are only effective when the rules are followed for their use – but with training and practice most people can be taught to identify causal factors using these basic tools for most of the incidents they will encounter.

Event and Causal Factor Charting: When an investigation team begins a root cause analysis (investigation), the first step is to prepare an Event and Causal Factors Chart (ECFC).

- Arranges event building blocks (sticky notes) to graphically depict cause-effect relationships between known events
- Allows for testing of "necessity" and "sufficiency"
- Best method when timing of events is important

Events make up the backbone of the ECFC. Event statements describe specific occurrences (e.g., "4-12 shift operator filled Tank 123," or "Control room operator acknowledged level alarm for Tank 123"). Conditions usually provide descriptive information (e.g., "Pressure was 1000 psig" or "The operator did not have hands-on training in this task") as opposed to stating action (e.g., "Operator placed Valve ABC into open position"). Primary events form the basic sequence in the diagram. Secondary events are actions that impact the primary events, but which are not directly involved in the situation. Figure 3[3] provides the steps for developing a casual factor chart.

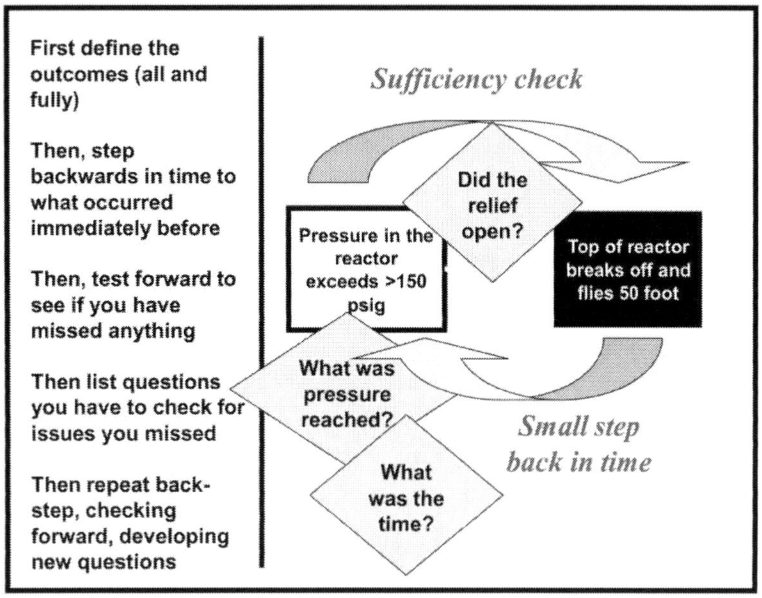

Figure 3: Steps for Casual Factor Charting

If applied correctly, ECFC will aid the investigators in generating a great many new questions. These will lead to more data gathering. Above are a couple of examples of how we show these questions on an ECFC. The answers to the questions sometimes become new events or conditions on the ECFC, as illustrated in Figure 4[3]:

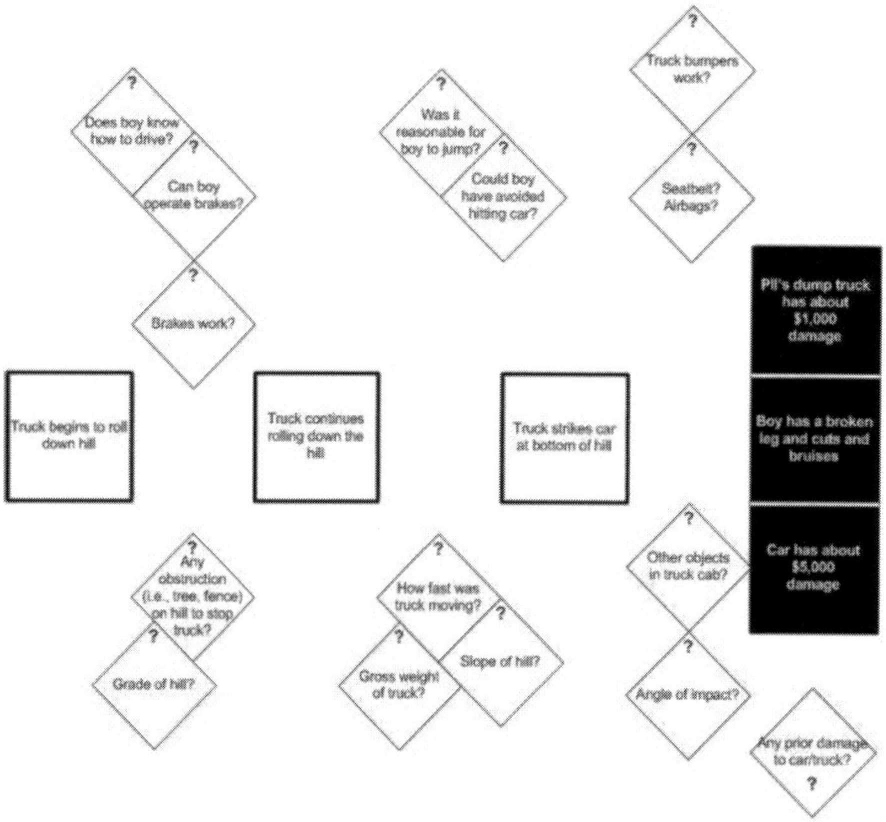

Figure 4: Example of a Partial Casual Factor Chart

The investigators will go back and forth between doing the ECFC, generating new questions, getting new data to answer the questions, and taking more steps backwards in time and sequence. This process continues until all of the data is exhausted and/or the investigators fully understand the sequence of the events and all of the mistakes and failures that led to the final ending condition or loss.

After the ECFC has been completed, the investigators are in a good position to identify the negative factors that influenced the course of events. These components are labeled causal factors. Causal factors, which may be in the form of events or conditions, are those items that are considered to be major contributors to the incident and as defined earlier, are either component failures or human errors or severe environmental problems.

The logic rules for constructing an ECFC are critical to success. These govern how small of step backwards the investigators take backwards in time and how many questions they generate as they test to ensure they understand each pair of building blocks.

5.3 Root cause analysis

When all the causal factors have been identified, the investigation team should begin asking "Why?" each causal factor occurred. Using a combination of the "5 Whys" and a Root Cause Chart™ (derived from the management oversight and risk tree process developed for US DOE's Savannah River Laboratory) shown in Figure 5, step down each of the root cause paths by:

- Select a causal factor from the causal factor chart or fault tree
- Work through the map for each causal factor
- Ask Why to help gathering the right data to get to the lowest level (root causes)
- Step down paths, identifying:
    - causal factor type (equipment, personnel, other)
    - root cause category
    - near root cause
    - root cause
- Record results on forms, at each step (see next page)

The Root Cause Chart™ structures the reasoning process for identifying root causes, identifies detailed root causes for each causal factor, facilitates consistency across all investigations, supports trending of "root causes" and "categories", and gives more thorough coverage of root causes by reminding investigator of all possible root causes.

Figure 5: Root Cause Chart™

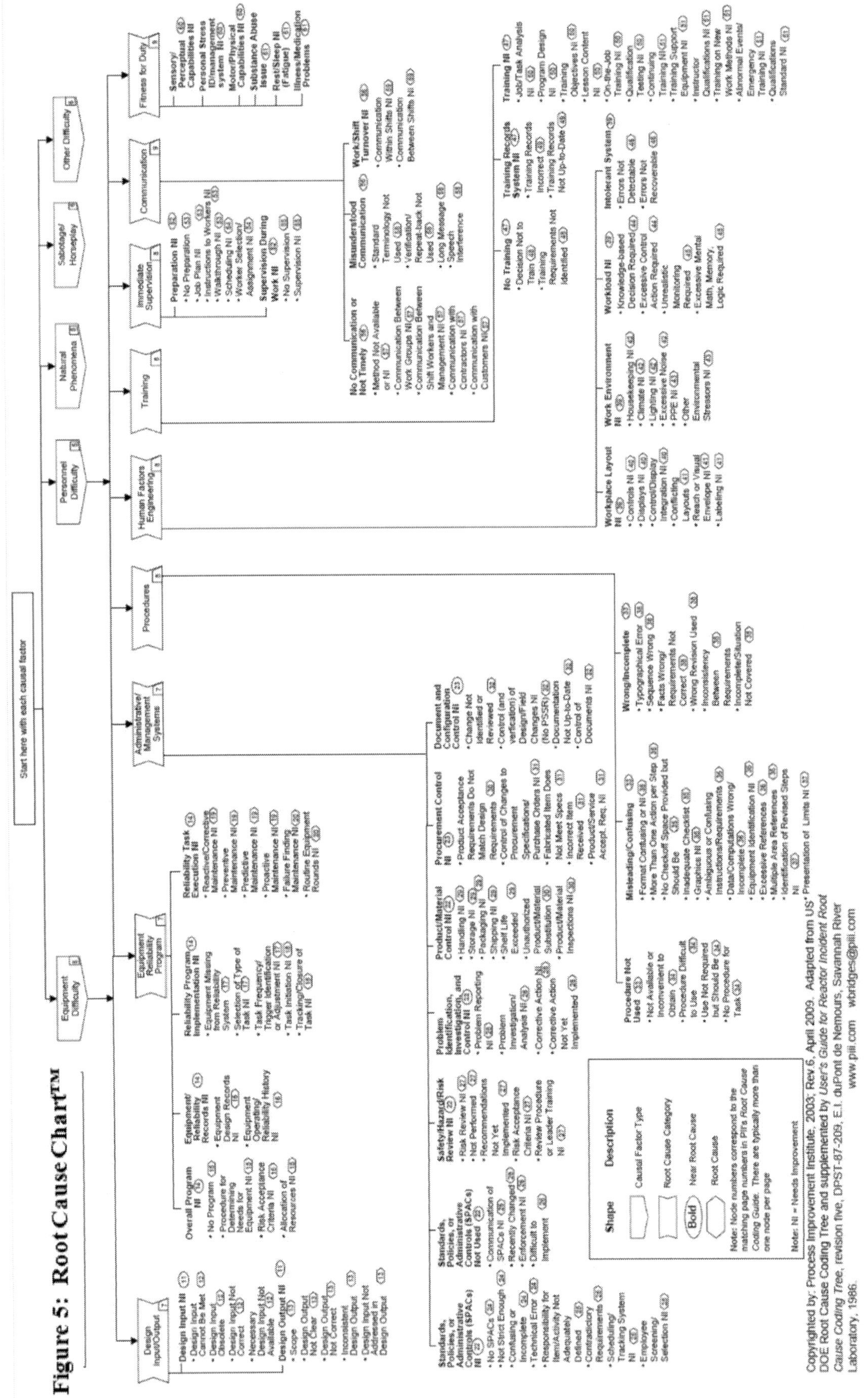

5.4 Develop recommendations

The investigation team should develop recommendations to remedy each root cause of each causal factor. Recommendations should describe the general objective to be accomplished and the consequences if not completed. Recommendations can also include possible solutions considered by the team and an example of how this could be accomplished.

5.5 Reporting and Follow Through

5.5.1 Simplify the reporting of the investigation/RCA results to the bare minimum needed

Part of the reason for the belief that it is impossible to investigate large numbers of near misses stems from the large reports currently required by the company for investigating incidents. Some companies insist on producing what they call "professional" reports of accidents, and these grow to 50 or 100 pages (half of the pages are attachments). Why produce such a large report? What is the use of that large of a report? What makes "size" of a report equivalent to "professionalism" of a report? Think about every aspect of the report and make sure it is needed. Normally, all that is needed is a:

- Cover sheet that includes the date, time, location, one or two sentence description of the near miss or incident, and a title that summarizes the incident at a glance. The cover sheet should also list the team members.
- Forms that have the causal factors filled in and the root causes filled in, with perhaps one or two sentences that explain the root cause. These forms can also contain the recommendations necessary to correct the root causes.
- That's It!

So, most near miss investigation results will be two pages or so and most loss/accident reports will be four pages or so.

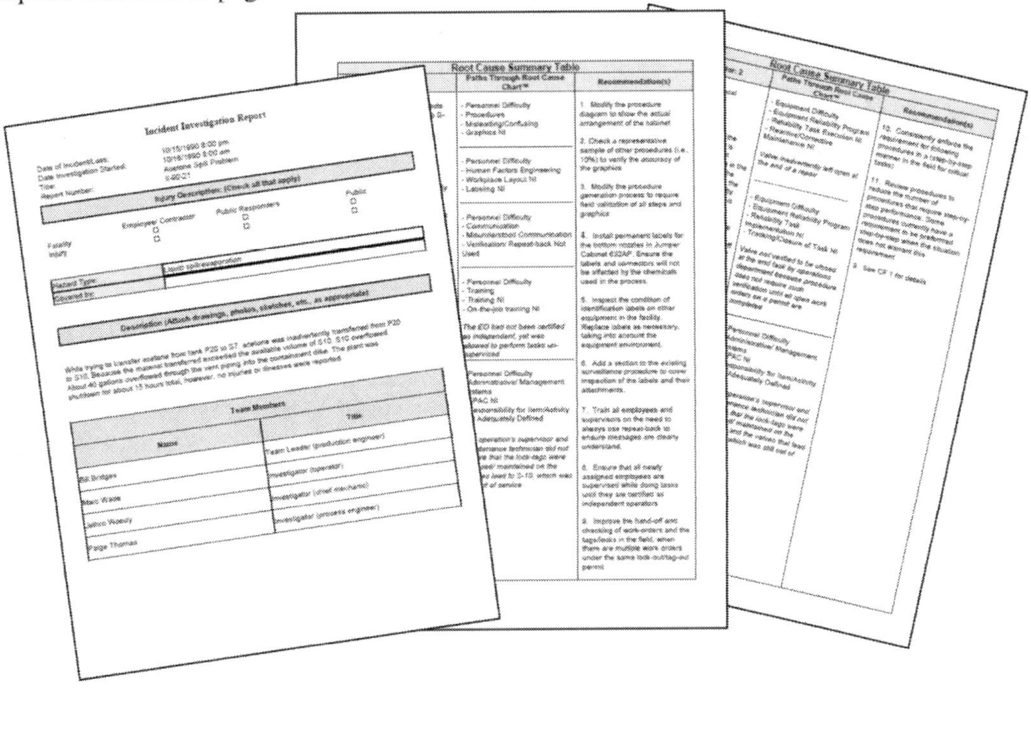

The complexity of results reporting has grown from the legacy of only investigating losses/accidents. When an organization gets a large ratio of near misses report and therefore a large number of investigations going on, the reports must shrink. This is a good thing. However, if you still have a major accident (which you won't have if you get a large number of near misses reported and investigated), then add more documentation to meet the needs related to litigation, regulatory interface, etc.

5.5.2 Deciding which incidents should be investigated

All near miss incidents should be reported and cataloged with basic information and apparent causes, but not all near miss incidents will warrant a full investigation. Only those incidents with either high potential impact or high learning value should be investigated using the II/RCA process. This is an excellent opportunity to engage staff and give them control over the process by granting front line supervisors the authority to decide which incidents have high learning value. For example, establish a shift target of investigating 15-20% of the near miss incidents reported and then let them decide which ones they want to investigate to meet their goal.

6. CASE STUDY: One Company's Experience with II/RCA

A large pharmaceutical company (more than 50,000 employees) initiated a drive to report and analyze large numbers of near misses. Before this initiative, they had about 3 near misses reported per loss event. Their loss categories included quality impacts, reliability issues, safety mishaps, and environmental issues. Also before the initiative, they had five incident reporting systems, investigation protocols, and databases. One of the first steps the company took towards the stated goal was to consolidate the five reporting and investigation systems into one system. They tried to use one of the existing database systems for this consolidation, but decided that all were deficient for this purpose. So, they modified the reliability incident database to make it suitable for all types of incidents. This took several months of effort (not full time though).

In parallel, they adopted a new investigation approach, which is the one described in this paper. This involved changing the database structure to provide pull-down menus for each branch on the Root Cause Chart. The chart was not modified. Next, a training course for lead investigators was customized to meet their new needs and to use their examples in the exercises and case studies. Then, this course was presented seven times in the USA and UK, thereby training an initial group of 200 investigators but more importantly training 7 instructors within the company. The eventual goal was to train 15% of the operating staff how to perform root cause analysis (RCA) of near misses and losses. The new database and consolidation of reporting and analysis systems (developed in parallel with the development of the customized training materials), was rolled out at the same time as the initial training courses. The reporting system was incorporated into their Lotus Notes e-mail system. The reporting forms were very simple and any worker could report a near miss or incident. Initial reports were routed automatically by the e-mail system to the worker's supervisor and viewable by any qualified investigator. A lead investigator on the same shift rotation and from the site (but not from the same group) was assigned to each incident; investigations typically began within one hour and on average were completed within the next hour.

These initial efforts were extremely successful and within about one year, the company achieved a ratio of more than 100 misses being reported per loss event. All near misses and loss events were captured in the database of incidents and about 20% were high value and immediately investigated. Typical reports were 2-3 pages. After the first year, the sites performed queries of the databases and based on Pareto analysis, performed RCA on the highest number of repeated "similar" incidents. In subsequent years, the queries were performed every three months.

This new system was much more productive in reducing losses than the previous collection of five systems. Overall, production and quality losses were reduced about 95% with little or no increase in time spent in RCAs (because the number and magnitude of losses decreased dramatically and so there was much less time spent on fire-fighting and major investigations).

7. Conclusions

Near miss reporting and investigation has been very effective in reducing losses due to process safety incidents. However, implementation has proven difficult for many companies due to several implementation barriers. The proven methods described will improve both the efficiency and the effectiveness of the II/RCA process. Consistently applying these methods will enable near miss reporting, providing early detection and correction of process safety warning signals, thereby reducing losses and improving process safety performance.

8. References

1. Busch, Carsten, *Heinrich's Common Cause Hypothesis: A tool for creating safety*, July 30, 2012, http://www.predictivesolutions.com/safetycary/heinrichs-common-cause-hypothesis-a-tool-for-creating-safety/
2. Wright, Linda and van der Schaaf, Tjerk, *Accident versus near miss causation: a critical review of the literature, an empirical test in the UK railway domain, and their implications for other sectors*, Journal of Hazardous Materials 111 (2004) 105–110.
3. Bridges, William, *Gains from Getting Near Misses Reported,* AIChE 8th Global Congress on Process Safety, 2012.
4. *Incident Investigation and Root Cause Analysis Course Materials*, Process Improvement Institute, Inc., 2003-2013.

Key Strategies for Implementing a Corporate Fatigue Risk Management System

David E. Cummings
E.I. DuPont de Nemours & Co.,Inc.
1007 Market Street, Wilmington DE 19898
david.e.cummings-1@dupont.com

Prepared for Presentation at
American Institute of Chemical Engineers
2013 Spring Meeting
9th Global Congress on Process Safety
San Antonio, Texas
April 28 – May 1, 2013

UNPUBLISHED

Key Strategies for Implementing a Corporate Fatigue Risk Management System

David E. Cummings
E.I. DuPont de Nemours & Co.,Inc.
1007 Market Street, Wilmington DE 19898
david.e.cummings-1@dupont.com

Keywords: fatigue, fatigue management, ANSI/API 755, human factors

Abstract

The First Edition of ANSI/API Recommended Practice 755 "Fatigue Risk Management Systems for Personnel in the Refining and Petrochemical Industries" was published in April 2010 as a result of the BP Texas City catastrophe of March 2005. In the last two years, many companies in the petroleum and chemical processing sectors have been working to implement the basic tenants of this new practice. An effective fatigue management program involves a multi-disciplined team approach to engage various functions of an organization, including PSM/EHS, HR, Medical, and Operations. This paper will explore the process, associated elements, and key issues integral to the design, development, and implementation of a formal fatigue management system across DuPont as a global manufacturing organization with a diversity of high hazard processes. This includes internal policies, training and development, standardized procedures and tools, and internal collaboration strategies.

1. Introduction

ANSI/API Recommended Practice 755 [1] was issued in 2011 based on a recommendation issued to the American Petroleum Institute by the U.S. Chemical Safety Board as an outcome of the BP Texas City 2005 incident investigation report. Prior to that time, there was little industry guidance or standards on the subject of fatigue management (FM). There is and has been substantial literature published and technical resources developed on the general issues and concerns with fatigue management in terms of circadian rhythms and the effects on human performance variables, and many industries such as aviation and commercial transportation have adopted specific standards with accompanying regulatory requirements.

E.I. du Pont de Nemours & Co. is a global science-based products and services company with over 200 manufacturing units across 70 countries involving nine business segments. It is engaged in technology solutions and innovation across a range of disciplines, including agriculture and industrial biotechnology, chemistry, biology, materials science and manufacturing, offering a wide range of products and services for markets including agriculture and food, building and construction, electronics and communications, general industrial, and transportation. Manufacturing technologies and site operations involves a myriad of process hazards, including flammable liquids and gases, highly toxic materials, reactives, combustible dusts, corrosives, asphyxiants, and thermal/mechanical energies. DuPont's philosophy and current practice is to apply process safety management systems and requirements to all manufacturing operations and locations using one global set of internal elements and technical requirements. Based on the company's current risk profile, process safety systems that help maintain a high degree of operational discipline and reduce the potential for human error, including those that could be related to employee fatigue, are key elements of an integrated risk approach.

DuPont's process safety management model and system has included a long standing element and requirements for "Training and Performance" where the performance portion of the element requires a specific focus on physical capabilities (fitness for duty) as well as diminished capacity (mental and psychological capabilities for process safety sensitive jobs, including the effects of fatigue) for operating and maintenance personnel working at all manufacturing locations. DuPont updated the "Training and Performance" element to specifically include an internal fatigue management policy as an integral part of the global process safety management program in January 2011. Initial implementation of the new technical requirements has been progressing since that time. The program embraces the following key elements of fatigue management as described in ANSI/API 755:

- Maximum hours of service
- Training for employees and line management
- Staff workload balance
- Work environment evaluations
- Medical Evaluations (personal risk factors)

Implementing a fatigue management system was integrated into corporate process safety management systems; however, active involvement of other key functions within the organization is important in this area. Within DuPont, the process safety competency serves as the corporate "owner" on FM, while other key functions such as Human Relations (HR), Integrated Health Services, and Occupational Health (EHS) all have played a meaningful and contributing role in developing and implementing portions of the FM process in the initial design stages, during implementation, and in an ongoing role as part of the "run and maintain" phase of this work. This collaboration has contributed greatly to the overall results.

2. Foundational Activities for FM Implementation

Building an internal fatigue management system based on the elements in the API standard requires a number of key strategic decisions and initiative planning prior to formal roll-out, as well as ongoing support, training, and tools during the active implementation phase. The following provides examples of many questions that were posed as part of DuPont's initial design strategy for fatigue management:

- Should the fatigue management program be a global or U.S. based only focus?
- What is the current state at all sites (does anyone already have an effective FM system)?
- What will be the major impacts on Operations at each site (resources, costs, tracking tools)?
- Who are the key stakeholders who should be involved in the design of the system?
- Which sites will be required to implement a local fatigue management system?
- How will fatigue management requirements be "codified" in corporate policies?
- What elements are consider mandatory versus suggested practices/guidance?
- What site roles and positions will be included in FM (and not included)?
- What is the required internal timing for implementation and how will it be tracked?
- When and how will the PSM auditing process incorporate FM elements during site reviews?
- How will interpretations and guidance on FM be captured and distributed?
- Will the program be required for contractors, and if so, which ones?
- Are there leveraged solutions and tools that can be developed to assist sites for effectiveness and efficiency?
- Who has primary responsibility to ensure compliance with Table 1 (employees or line managers)?

2.1 Site Surveys to Define Current State FM Practices

The first step that was taken to develop a fatigue management program across DuPont was to conduct a detailed survey across all manufacturing locations to accurately determine the present state relative to the main elements of ANSI/API 755. This included the following self-assessment parameters:

- Existence of site level fatigue management written programs and policies
- Existence of local policy limits on overtime and maximum consecutive work hours
- Past and present fatigue management training
- Existence of tools or tracking systems to measure work hours and overtime by work group and employees
- Local site systems and capabilities to integrate fatigue management into existing medical programs
- Forecasted requirements and needs with local labor organizations and contracts for fatigue management

Once the site survey was completed, the data was collated and utilized as inputs for the development of corporate policies, timing and implementation plans, management training, and development of internal FM training and guidance documents. The survey was collected and analyzed in a simple excel® tool similar to that depicted in Figure 1.

Site Fatigue Management Survey – Current State

What is your site's current level of implementation of Fatigue Management requirements? List all numbers that apply. 1. Leadership is aware. 2. The site is tracking hours. 3. Employees have received general awareness training. 4. The site is actively managing hours for Shutdowns/TurnArounds, Special Circumstances, or Normal Operation. 5. A Fatigue Management policy has been written. 6. An Exception Management policy has been developed.	If you listed No. 2 above, how is the site tracking hours? Is the tracking historical or real time? What system or tools are being used?			If your answered No. 4 above, when and how is the site managing hours?		Have you considered which employees would be included in a Fatigue Management program? If yes, how many at your site? Does this number include employees in LHO's? If yes, how many of the total are in LHO's?	How are your shifts set up, ie. 8 hour, 10 hour, 12 hour, rotating, 5 days, 7 days, etc.?	If your site has changed any policies, procedures, tracking, people's work times, etc. to move toward compliance, can you describe what you have done?	Site Changes
Level of Implementation	Tracking Hours	Tracking Hrs.?	How Tracked ?	Managing Hours	Mgt. Hrs.?	Which Employ.	Shifts	Changes	Changed Schedule/ Pay?

Figure 1: Sample Site Survey Baseline Questions and Spreadsheet

2.2 Corporate PSM Standard Policies and Requirements

DuPont drafted and included a new fatigue management section in the global process safety management standard as part of the" Training and Development" element as depicted in Figure 2. Specific new content in the corporate standard was intended to incorporate the minimum mandatory requirements involving fatigue management elements described in ANSI/API 755, and not a be a re-statement of the guideline's full technical content. It was also defined with the standard scope that new internal fatigue management requirements would be applied to vital PSM roles within DuPont sites as well as to specific types of contractors, but only at higher hazard site operations (DuPont classifies all global manufacturing locations as higher hazard processes or lower hazard operations). The standard changes went through a broad internal peer review and was published in January 2011.

Figure 2: DuPont's Process Safety Management Model

2.3 Education and technical guidance to drive consistency

Internal webinars were developed and delivered which were targeted to several different key audiences, including site PSM leaders, operations leadership, site HR leaders, and contract administrators in order to develop foundational knowledge of the new FM program elements across the organization. The initial ANSI/API 755 standard as published is a performance based standard with specific prescriptive FM requirements such as Table 1 maximum hours of service limits, however, it became apparent early in the FM implementation process that numerous technical and interpretive questions existed both on the standard, as well as within DuPont's local site implementation efforts. In order to attain as much consistency as possible, a series of Frequently Asked Questions (FAQ's) were collected and formally documented with specific guidance. Once vetted and resolved, these FAQ's were distributed to sites and posted on the internal PSM website for retention and routine access. Additional questions and needs for interpretations have continued to surface from sites and functions and been added to the FAQ "evergreen" document even after almost two years of implementation progress. There is currently no effective platform within industry to collect and address these types of interpretations in a consistent manner.

As another step to implement consistency across all locations, DuPont developed and distributed a standard site based written fatigue management program or template which can be applied by each site as a base document and modified to add in local responsibilities, procedures, and line management practices.

There are also unique situations that may develop for plant operations that do not fit neatly into the existing definition and guidance in ANSI/API 755 involving "outages" as it applied to extended consecutive work shifts in Table 1. The existing definition for outages includes the following:

> *"Planned or unplanned interruption in the normal operations of a unit or plant, including mobilizing or de-mobilizing. Outages include but are not limited to such things as turnarounds, unit shutdowns, operational responses, etc."*

This definition does not cover all situations where an extended work set and shift schedules may be applicable and necessary to minimize fatigue, including commissioning new plants or substantive MOC projects where construction, PSSR's, start-ups, and lining out the process may be necessary involving operations, technical and R&D resources, response to natural disasters (e.g., hurricanes, earthquakes, etc.), mothballing or dismantlement activities, new product trials in pilot plants, etc. Each company should define when and how these types of situations should be managed in terms of unique conditions and the application of extended work hours and shift schedules with potential fatigue factors.

DuPont adopted an internal policy at the site level to develop a formal exception process for situations where the Hours of Service Limits (ANSI/API Table 1) may not be met due to specific circumstances. An exception process is discussed in the standard in Section 4.8.5 with general criteria, however; the specific procedures and details should be developed by each company. For DuPont, the exception approval process has been designed to be executed at the local level for a case-by-case exception by site or unit management. For longer term, permanent, or "standing" exceptions, off-site leadership approvals are required. In both cases, exceptions will document the situation, the rationale, and any relevant alternative controls or methods to be taken to mitigate the potential effects of fatigue. This includes use of a standardized form and line management approvals for each exception situation.

3. Shift Schedules and Maximum Hours of Service

API 755 includes specific requirements for maximum hours of service provided in Table 1 of the standard. The hours of service element in FM has received the most internal discussion, review and implementation work. A detailed analysis of existing shift rotation schedules at all U.S. site operations was conducted by DuPont HR to ascertain if any inherent conflicts existed in the current schedule structures that were not be fully compatible with Table 1. This included a specific emphasis on the number of consecutive shifts and the total hours of recuperative time for both day and night shift rotations. This review revealed that DuPont had over one hundred (100) different shift rotation schedules in use at all U.S sites for 8, 9, 10, and 12-hour operations and maintenance shifts. In a very limited number of cases, an inherent conflict in the site shift structure existed which did not fully meet the maximum "hours of service" in Table 1 (mostly involving the transition from day to night 12 hour shift schedules). Figure 3 describes the shift schedule analysis results at all US sites, and figure 4 provides an example of the detailed shift schedules which were generated and reviewed at a typical multi-unit site.

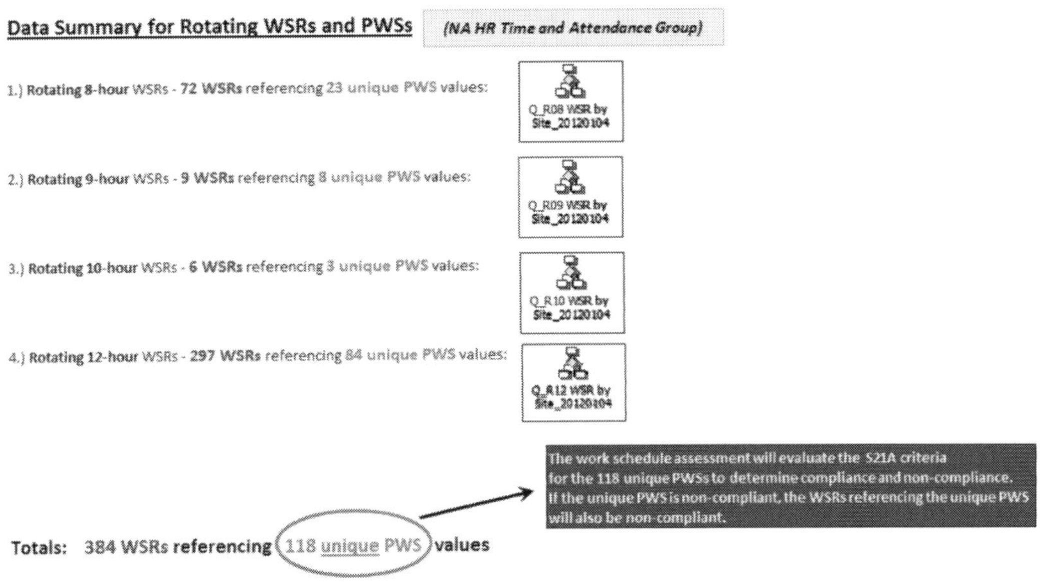

Figure 3: Data Summary for Plant Rotating Work Schedules

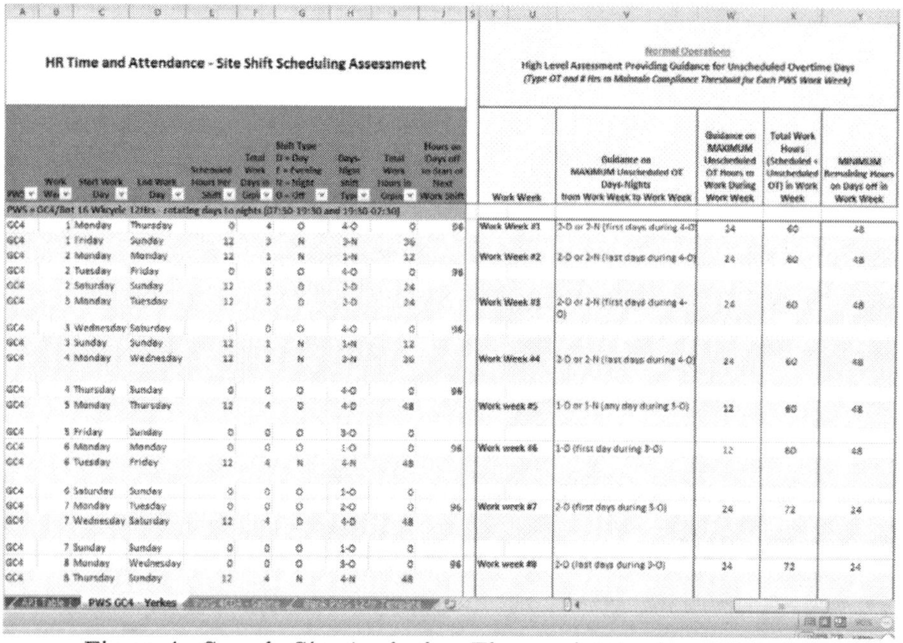

Figure 4: Sample Site Analysis – Time and Attendance Schedules

A significant part of DuPont's initial evaluation of existing shift schedules and site data survey questions focused on the types of tracking and software (scheduling) tools that were either already in use or could require development to track work hours, including scheduled and unscheduled overtime call-ins. In most cases, site systems tracking overtime rates for work groups but not specific employees other than annual salary statistics. Site level tracking systems which were cataloged that could be modified and used to monitor maximum hours of service and off site recuperation times included the use of an HR work hours and pay status, web based tools, security gate electronic logs, and access data bases. None of these existing systems as structured were deemed an acceptable and reliable means to proactively schedule, manage, and track individual employee work hours. A six sigma black belt resource and project was initiated in order to review each of the work hour and overtime tracking system attributes, and develop a standardized solution that could be used at all sites to help line management ensure that the maximum hours of work in ANSI/API 755 Table 1 maximum hours of work. As part of the six sigma project, the main attributes of an overtime tracking system were captured as illustrated in figure 5.

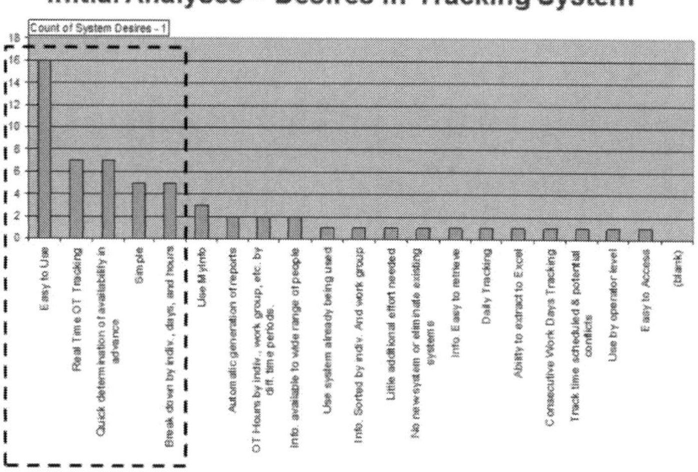

Figure 5 : Site Survey Data – Desirable Attributes for Electronic Scheduling and Tracking

4. Training

DuPont developed a standardized and leveraged training package which could be applied at each site in order to provide initial and refresher training on fatigue management. Two packages were developed ; one to target employees who were subject of FM (operators and mechanics), and one for line managers who supervise these employees. The guidance and criteria in ANSI/API 755 was used to create the technical content of each package. Initial training was designed to be one hour of information and cover the following topics:

- Basics of sleep and fatigue including causes, symptoms, and prevention strategies
- Understanding risk of on the job fatigue and impacts on safety
- Overview of the site fatigue management system including maximum hours of service criteria

DuPont occupational health resources have also previously partnered with external groups such as the Porcupine Health Unit and the Sudbury & District Health Unit in Canada to develop additional sources of fatigue management training such as the Shiftwork like Clockwork (SWLCW) training modules illustrated in Figures 6 and 7 . These materials provide significant details over nineteen (19) modules that can be used for both initial training of new employees and refresher training for existing employees. DuPont has adopted an internal FM refresher training frequency of three years.

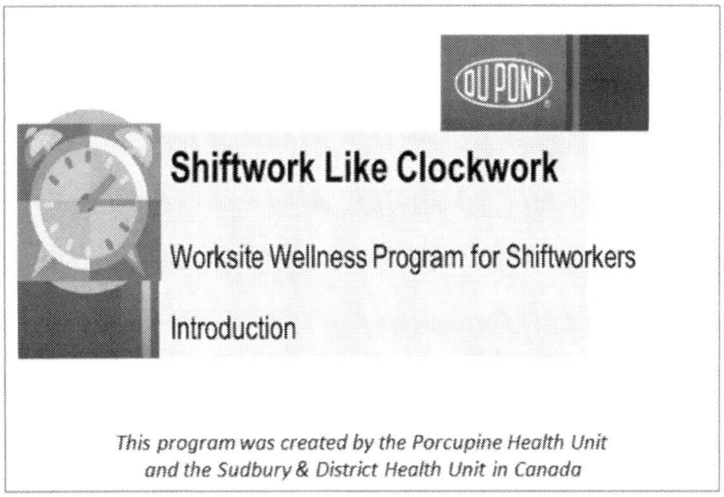

Figure 6: "Shiftwork Like Clockwork" Training Program

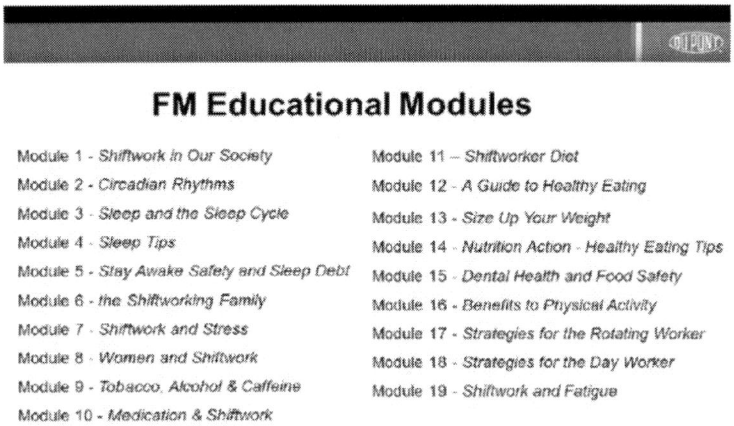

Figure 7: "Shiftwork Like Clockwork" Training Modules

5. Staff Workload Balance

Another key tenant of the ANSI/API 755 standard includes a focus to help ensure a staff workload balance is achieved and maintained within each operating area at each site. This should include an analysis of work activities on each shift, and a focus on considerations and impacts related to employee turnover, long term disability, extended time off due to personal reasons (vacation, leave of absence), staffing vacancies, and other sources of long term absenteeism. DuPont has adopted a policy that each site will conduct and document an annual review of staff workload balance within each higher hazard operating unit to periodically evaluate the status and performance in this area. A standardized set of evaluation criteria for the annual review was developed and distributed to all sites as a leveraged tool to help ensure consistency across all locations.

6. Work Environment

DuPont has utilized a series of PHA related tools and checklists to conduct baseline, cyclical and MOC related evaluations for over the past twenty years. A human factors checklist has been one of these historical tools. DuPont's strategy to address FM work environmental factors involved adding content to these existing tools in order to help ensure that fatigue management workplace factors were considered and evaluated during all cyclic and project related PHA's. Integrating this review into existing PHA processes has proven to be a more sustainable and effective practice rather than creating a "stand alone" fatigue management work environment assessment process.

7. Medical Evaluations (Personal Risk Factors)

DuPont Integrated Health Services, in coordination with the process safety competency, developed a fatigue management medical screening approach and screening form which could be used as a voluntary tool during periodic health exams for shift employees who may be subject to fatigue risks. The health services group also provided targeted FM training to site medical professionals (nurses, doctors, and NP's) in order to consistently apply FM medical aspects, including both the proactive (voluntary) approaches as well as "for cause" evaluations where potential fatigue has been observed by line management in the work place.

DuPont Medical also worked with the corporate HR, labor relations, and legal functions to develop and communicate consistent medical guidance and case management methods where employee fatigue concerns have been identified. This is an area where emerging issues, specific interpretations, and case law should be considered by each entity such as the following (examples only):

1) Where additional FM medical testing is prescribed based on an employee's physical status and personal risk factors (obesity, concerns about sleep apnea) which could relate to fatigue in the workplace, should the costs for further medical evaluations be borne by the employee as part of personal health care or as part of the occupational medical program for a given company?
2) How should an employee be managed if they refuse to participate in additional medical screening or specific diagnostic tests that focus on fatigue management?
3) Are certain medical conditions that can cause fatigue in the work place subject to the 1990 Americans with Disabilities Act (ADA) in the United States and if so, which ones, and what may be the expectations for reasonable accommodations?

8. Contractors

DuPont has also adopted internal process safety requirements for a fatigue management program applied to selective contractors who work in higher hazard process areas which typically is limited to resident maintenance service providers, test and inspection consultants, and specialized services resources who may be on-site only during major turnarounds.

Contractors are required to develop their own internal programs with the basic elements of fatigue management as described in ANSI/API 755. DuPont will typically audit their systems and performance at the site level on a periodic basis to verify the program is adequately implemented. Many major contractors have embraced an enhanced focus to manage fatigue risks. In one example, a resident maintenance contractor developed a modified approach for their employee work scheduling during an annual shutdown with a DuPont site in Protection Technologies in order to improve fatigue management practices. The shutdown typically lasts 15 days involving 12 hour shifts and the contractor reduced the shift duration to 11 hours and inserted a day off for each employee during the process. Feedback from the employees was very positive, and this modifications were managed so that there were no additional costs or delays in the shutdown schedule. This unique approach serves as a good example where the owner/operator (DuPont) worked in collaboration with contactor partners to achieve a common objective which also supported the business and operations.

9. Incident Investigation Processes

DuPont maintains a global internal incident investigation reporting, classification and investigation policy for both EHS and process safety incidents which addresses communications, investigation teams, root cause failure analysis (RCFA), recommendations tracking, and overall documentation. The existing policy does not specifically include or consider fatigue management components as they may relate to contributing factors and human error as part of the overall RCFA process. At some point in the future, the policy will be updated to include a focused consideration of fatigue during all formal investigations, particularly when human error has been identified. This may include a standardized review of employee work shift schedules and overtime for at least the previous two weeks prior to the incident date.

10. Conclusion

Based on the "lessons learned" from the BP Texas City incident in 2005 and the "new" ANSI/API 755 Recommended Practice, internal company initiatives to integrate fatigue management into existing risk management systems is a prudent and necessary step to help minimize the potential for human error and the associated episodic incidents in high hazard process operations. The process safety function should be the logical "process owner", however, the initial systems design and implementation activities involving fatigue management necessitate a collaborative effort between other key company and site level resources and functions, including operations, human relations, health services and occupational health, in order to develop an integrated system and achieve maximum risk management benefits. Since the ANSI/API RP 755 standard includes primarily performance based content, implementation also may involve a substantial number of internal corporate or local policy decisions, and potential issues which require focused discussions and interpretations to promote consistency with internal applications at each site.

The initial FM implementation process may span two to three years in order to plan, educate, develop, implement and monitor site level organizations and accompanying results. Organizations that include fatigue management requirements in existing PSM policies should recognize that implementation progress may vary at each site depending on size, complexity, shift schedules, the existence of labor unions with collective bargaining agreements, and other factors. These potential variations on the rate of progress require a simplified process to track implementation across all applicable operating sites (see figure for an example of tracking based on the six main program elements). Integration of fatigue management evaluation criteria into existing PSM audit protocols and systems is typically the last step in a long term FM implementation and performance measurement process.

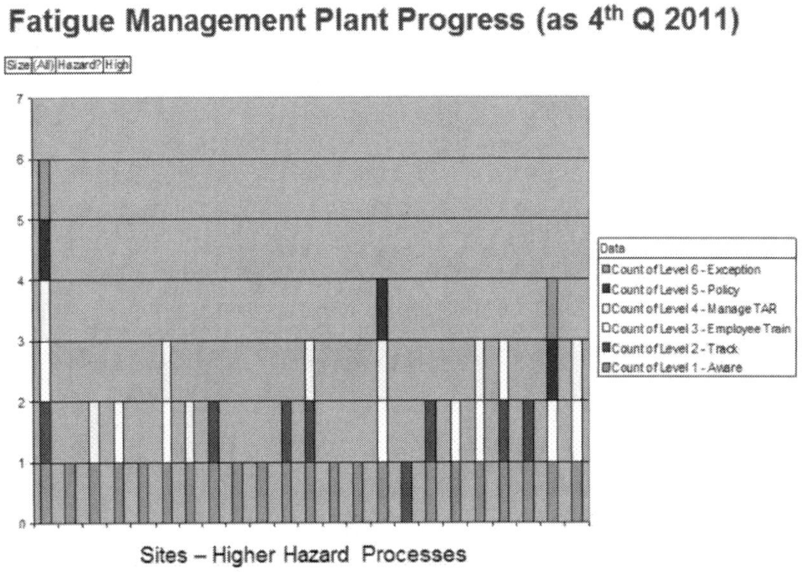

Figure 8: Example of Visual Management Board – Site Progress

11. References

(1) American National Standards Institute (ANSI)/American Petroleum Institute (API) Fatigue Risk Management Systems for Personnel in the Refining and Petrochemical Industries - RP 755 (First Edition, April 2010)

Consequences calculation for boilover and storage tank fires

Jaime E. Cadena
Chemical Engineering Department, Universidad de los Andes
Cra 1 Este No 19A – 40, Bogotá, Colombia
je.cadena21@uniandes.edu.co

Alejandra Cano[1], Juan Carlos Díaz[2], Alexis Vignes[3], Felipe Muñoz Giraldo[4]
[1,2,4]Chemical Engineering Department, Universidad de los Andes
Cra 1 Este No 19A – 40, Bogotá, Colombia
[3]INERIS
Technologique ALATA, 60550, Verneuil-en-Halatte, France
[1]na.cano744@uniandes.edu.co, [2]jc.diaz111@uniandes.edu.co, [3]Alexis.VIGNES@ineris.fr,
[4]fmunoz@uniandes.edu.co

Prepared for Presentation at
American Institute of Chemical Engineers
2013 Spring Meeting
9th Global Congress on Process Safety
San Antonio, Texas
April 28 – May 1, 2013

UNPUBLISHED

Consequences calculation for boilover and storage tank fires

Jaime E. Cadena
Chemical Engineering Department, Universidad de los Andes
Cra 1 Este No 19A – 40, Bogotá, Colombia
je.cadena21@uniandes.edu.co

Alejandra Cano[1], Juan Carlos Díaz[1], Alexis Vignes[2], Felipe Muñoz Giraldo[1]
[1]Chemical Engineering Department, Universidad de los Andes
Cra 1 Este No 19A – 40, Bogotá, Colombia
[2]INERIS
Technologique ALATA, 60550, Verneuil-en-Halatte, France

Keywords: Storage tank fires, boilover modeling, process safety, emergency response

Abstract

During the past decade, high loss incidents have been recorded due to fires in hydrocarbon storage facilities worldwide. An example is the storage tanks fire at a refinery in Amuay, Venezuela (August 25 2012, under investigation), which led to the loss of 650000 barrels of naphtha, around 41 fatalities and 80 injured. These incidents have high magnitude consequences, therefore an a-priori consequences analysis of these events become a key element for the decision-making processes to prevent, control and mitigate fire risks. The risk evaluation and mitigation for fire scenarios usually follows a prescriptive approach, therefore this work presents a tool for obtaining quick and conservative calculation of the thermal risks due to fire scenarios of hydrocarbons storage tanks facilities. These intensities are translated into consequences, emergency response requirements and key information for Contingency Plans development. To complement the fire scenarios, the tool also includes the calculation of effected distances for boilover phenomena. The results obtained by the fire tool were within an acceptable margin of error but it still conservative in comparison with real data. For the boilover tool different models were implemented among them the only one conservative for all the validation cases was the one proposed by Casal et. al. [1] for the feedback heat and the one proposed in the book Risk analysis in industrial facilities [2] for the boilover time.

1. Introduction

Storage tanks are a key element of any Oil & Gas operation since they constitute intermediate steps in the upstream operation e.g. in a refinery or a pipeline operation, and also in the downstream operations in which they are the source of final products to be delivered to the clients and customers. Nevertheless, even when hydrocarbons are stored without physico-chemical changes, risks are still present such as fire and boilover risks [3]. A fire in a storage tank is usually assumed as a pool fire in the surface of its contents, while a boilover is the ejection of the stored HazMat due to the fire and posterior boiling of a liquid with lower boiling point and higher density than the HazMat [2]. Storage tank fires and boilover scenarios represent low probability but high consequences events, making it necessary to apply available knowledge and strategies to analyze, assess, control and mitigate the related risks in order to avoid high capital and human loss as illustrated in Table 1 [4], which highlight that regardless of location or date, such events continue to occur. This makes it a key topic of study to avoid major hazard events and irreparable loss, especially for a region with a high industrial development such as Latin America [5, 6], in which the deployment of new Oil & Gas operations not only present emerging risks but also modify the existing ones [7].

Table 1. Storage tank fires and/or boilovers and related loss

Date	Location	Event	Related loss
13/12/1982	Bogotá, Colombia	Esso oil storage facility fire [8]	• 8000 people evacuated • Loss of 3 storage tanks
19/12/1982	Caracas, Venezuela	Explosion and subsequent oil tank fire [9]	• 153 fatalities • 7 missing • Severe damage to the facility
24/02/1986	Thessaloniki, Greece	Fire at an oil storage tank facility [10]	• 10 out of 12 tanks destroyed • US$330 M
11/12/2005	Buncefield, UK	UVCE and fire at hydrocarbon storage facility [11, 12]	• 43 injured • Over 2000 people evacuated • Low and moderate damage to surrounding buildings • Total destruction of most of the complex • Possible Detonation to Deflagration Transition (DDT) [13]
26/10/2009	Bayamon, PR, USA	Caribbean Petroleum Company fire (currently under investigation) [14]	• 1 injured • Costs around US$6.4 million • Around 5.6 million of product loss • Around 600 people evacuated • Damage to buildings in a radius over 1 mile (1.6 km)
25/08/2012	Amuay, Venezuela	PDVSA Amuay refinery explosion and fire (currently under invetsigation) [15]	• 48 fatalities (latest information) • Over 80 injured • Stop of the refinery with 640000 bpd capacity • Damage to at least two contiguous storage tanks, generating fire • 18 national guard and 15 civilian fatalities • Damage to around 200 residential buildings and 11 businesses • 33 families displaced • Damage to a military base

Although the subject has been addressed through different approaches [16, 17], in the field of process safety it is fundamental to understand the phenomena underlying the events of interest. For tank fires and boilover events this has been carried out for several decades, and a wide variety of empirical, semi-empirical and CFD models have become available to model such events [1, 18-23]. This work aims to develop a methodology to apply a set of the available models to calculate the consequences of the events of interest through its implementation in a user-friendly computational tool that allows obtaining quick and conservative results for either a preliminary risk analysis stage in which critical scenarios are identified or during actual emergencies in which knowledge of the possible outcomes is needed to establish mitigation strategies.

2. Methodology

In order to achieve the objectives stated above, a methodology (Figure 1) was proposed for the development and validation of the tool. First step is defining the scenario to evaluate, the tool use variables that influence tank fire and a boilover like some specific hydrocarbon properties such as viscosity, initial and final boiling point temperatures and combustion heat as input variables, next the tool implements models for the calculation of heat released by the fire the feedback heat and boilover and duration of fire times. Based on the last calculations the tool use models to determine the safe distances for both phenomena (boilover and fire) and in the case of the fire the tool also returns as output variables some mitigation strategies like refrigeration rings.

Figure 1. Methodology

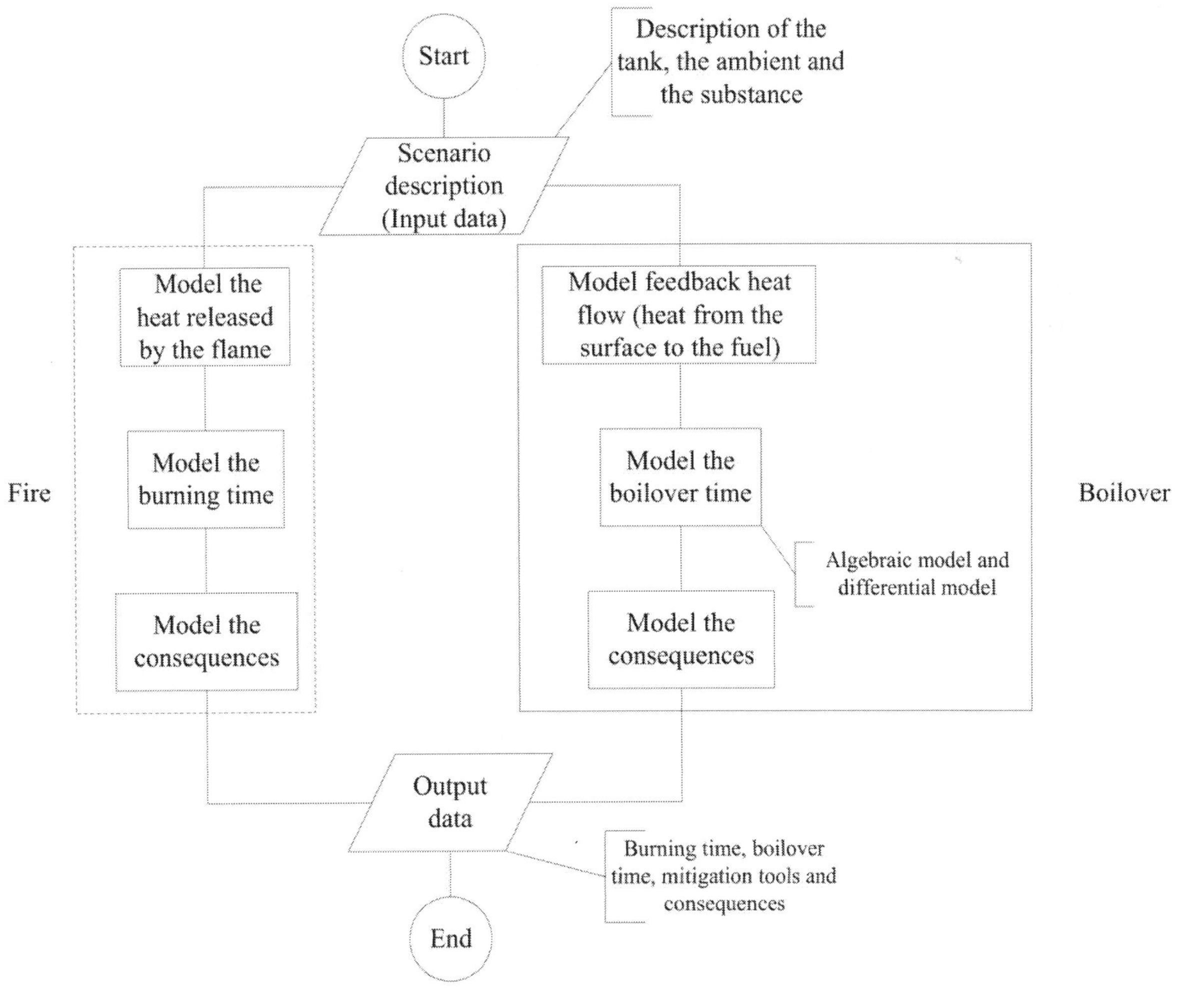

3. Tool implementation

The tank fire tool was developed in excel (visual basic) using macros for both, the heat released by the fire and the burning time. For the boilover there are multiple approximations, for the feedback heat there is an average value reported [2] and models for calculating it using a fraction of the total heat released by the fire, the difference among the last ones is the calculation of the total heat released that can be done using models with combustion heat and experimental corrections depending on the fuel [24] or heat balances between the tank and the surroundings [1], for the boilover time there is an algebraic approximation derived from a heat balance [2] and differential approximations; varying among them the initial and boundary conditions [1] . Some of them were implemented in excel and the ones involving differential approximations were implemented in matlab as this software allows multiples ways for solving differential equations in a short and easy way. Both Excel and Matlab have a user-friendly interface. The models selected for the two tools are presented on Table 2.

Table 2. Models used for the tool implementation.

		Model	Assumptions	Ref
Fire	Time calculation	$t_b = 4V/\pi D^2 v$ $v = m''/\rho$	-	[24]
	Heat calculation	$Q = m''\Delta H_{c,eff}(1-e^{-\kappa\beta D})A_{dike}$ $D = (4\ A_{dike}/\pi)^{1/2}$	- Experimental correction depending on the fuel	[24]
Boilover	Time calculation	1) $t_{boilover} = \rho\, c_p\, h_{HC}\,(T_{hw} - T_a)/(Q_f - m''(H_v + c_p\,(T_{0av} - T_a)))$ $T_{0av} = (T_{0min}\, T_{0max})^{1/2}$		[2]
		2) $\delta^2 T/\delta y^2 = (1/\alpha)\delta T/\delta t$	-Unidimentional	[21]
	Heat calculation	1) $Q_f \approx 60$ kW	-	[2]
		2) $Q_f = (4\chi/\pi)\,\rho_\infty\, c_{p\infty}\,(T_\infty\, g\,(T_f - T_\infty)\, D)^{1/2}$ $\chi = H_v/H_c$	- A heat conduction mechanism is assumed, no conduction in the fuel layer -The radiation is fully absorbed at the surface	[21]

The values for T_f in the heat calculation for the boilover, change depending on the assumptions used for the estimation, some authors use the adiabatic flame temperature as an approximation and others recommend approximated values like 1100 K [1] or values ranging from 1600 to 1900 K depending on the substance.

To solve the differential equation for time of boilover the following boundary conditions were established.

- Initial condition. [1]

$$\mathbf{T(y,0) = T_\infty} \quad \text{[Eq. 1]}$$

- Boundary condition 1.[1]

$$\mathbf{T(y(t),t) = T_s} \quad \text{[Eq. 2]}$$

Where **y(t)**is defined by the following equation.

$$\mathbf{\delta y(t)/\delta t = m'' / \rho_f} \quad \text{[Eq. 3]}$$

- Boundary condition 2.

$$\mathbf{-k_f * \delta T / \delta y |_{fuel} = -k_w * \delta T / \delta y |_{water}} \quad \text{[Eq. 4]}$$

To achieve boundary condition 2, the modeling of the water layer is needed and it can be solved by the same equation for the fuel reported on Table 2. The first boundary condition for the water layer is stated on Equation 4 and the second boundary condition stated in Equation 5. This boundary condition implies assuming adiabatic conditions in the limit of the water layer with the bottom of the tank.

$$\mathbf{\delta T / \delta t |_{y=0} = 0} \quad \text{[Eq. 5]}$$

3.1 Input and output variables

For the fire tank tool is needed from the tank the diameter, the height, the type of roof and the capacity; the tool have all the characteristics for the fuel, it only need to be specified with the volume present in the tank. In addition if the studied facility has more storage tanks near to the one in fire (storage farm tanks) the distances from the one in fire to the others must be known. The output variables are the duration of the fire, the area involved on the fire, the zone classification (hot, warm, cold) and the mitigation and prevention actions [24] like if the refrigeration ring should be activated.
For the boilover the input values are the diameter of the tank the ambient conditions, specifying the fuel involved in the fire, the initial height of the fuel in the tank before the ignition the temperature of the heat wave when the boilover occurs and for one of the models the flame temperature. The output data consists of the boilover time and the distances at which lethally and irreversible consequences occur.

3.2 Consequence estimation

The consequences for both tank fire and boilover were calculated using Excel and both have dependence from the heat irradiated from the fire in the case of the tank fire and for the resulting fireball in the case of the boilover.
The consequence of the fire depends on the heat radiation determining the hot, warm and cold zone. In the hot zone the equipment will fail and the heat release will be lethal for human beings. The warm zone can be accessed with protection equipment, the firemen attack the fire there. Finally the cold zone is the one where the equipment is displayed, there is no need of protection equipment in this zone. The hot zone is determined by values where the heat is over 10 kW/m^2, the warm zone is the one with heat values between 2,5 kW/m^2 and 10 kW/m^2 and the cold zone is the one with values lower than 2,5 kW/m^2 [24]. The consequences of a boilover event are fireballs and in the ejection of ignited fuel around the tank. According to the radiation heat

release of the fire there can be determined lethal and irreversible (serious burns) effect distances corresponding to radiation heats of 1000 kW $m^{-2}$and 600 kW m^{-2} each one, they are estimated using the conservative assumption that the entire fuel in the tank participates to the fireball. The models used for the safe distances and zones are reported on Table 3 [2].

Table 3. Models used for consequence calculations.

	Consequence	Model	Ref
Fire	Irradiated heat	$d_{hotzone}=(QXr/4\pi\, q''_{hotzone})^{1/2}$ $d_{hotzone}=(QXr/4\pi\, q''_{warmzone})^{1/2}$	[24]
Boilover	Fireball	$d_{lethality}=k_{lethality}W^{0.45}$ $d_{irreversible}=k_{irreversible}W^{0.45}$	[2]

4. Results and analysis

To validate the tool it was compared with selected cases for each event (fire and boilover), the case selection was made using two criteria, the first one is searching for cases that are not affected by other events after the fire starts and the second one is searching for cases with complete information about the outputs and inputs for the tool mentioned above. The cases and the comparison with the tool results are described in this section with the comparison among the boilover time and the burning time.

4.1 Tank fire tool

For the tank fire tool the case selected was an accident that occurred at ConocoPhillips Company's Glenpool South tank farm, Glenpool, Oklahoma on April 7 2003. The event involved the tank 11 of the facility during the transport of Diesel from another tank [25] . The results obtained by the tool are reported on Table 4 and Table 5.

Table 4. Results related to the tank and the fire variables

Burning Tank Name:	11
Type of Roof:	CONE
Product:	Diesel
Nominal Volume (m3):	8420,00
Contents (m3):	1338,67
Height (m):	14,63
Diameter (m):	33,22
Area of Fire (m2):	1526,84
Time of Fire (hr):	8,39
Time of Fire (day):	0,35
Refrigeration Range (m):	71,89

Table 5. Results related to the consequences of the fire

	From (m)	To (m)
Hot Zone	0	64,298
Warm Zone	64,298	128,596
Cold Zone	128,596	N/A

The variable to compare among the case and the tool was the burning time because that is the variable reported in the few cases that also have all the input parameters. The percentage of error among the real case and the results reported by the tool is lower than 5%. In the case of the burning time the percentage of error reported may be due to the model, because the variables involved in its calculation (volume of the fuel and dimensions of the tank) don't require any assumptions and are directly specified by the case to evaluate. The possible source of error for this validation case may be directly related to it, as the tank was being filled at the moment of the fire the initial volume of fuel is not an accurate value.

4.2 Boilover tool

For the boilover tool different models were used, a comparison among them is also included in this section.

Case 1:

For the boilover, a large scale experiment was used for the study of boilover in large tanks using crude oil [26]. The results obtained by the tool with different models are presented on Table 6 and the distance effects are reported on Table 7. They are independent from the model used for heat and time of boilover so there is just one value for $d_{lethality}$ and one for $d_{irreversible}$ reported for the case.

Table 6. Case 1 results for the boilover time with the models proposed on Table 2.

	Tflame (K)		**tbo (s)**		**treal (s)**	**%error**	
Heat model 1	NA	NA	4067,67		4728	13,96637056	
Heat model 2	1100	1900	1127,07357	692,031935	4728	76,1617264	85,3631147
Time model 2	NA	NA	60604		4728	NA	

Table 7. Safe distances for the boilover consequences for case 1

dletal (m)	14,7314678
dirrev (m)	20,0281754

Case 2:

A fire on tank 3 containing crude at the Czechowice-Dziedzice Refinery on June 26, 1971 at 19:50 occurred due to a lightning hit [27]. For this case the results obtained by the tool are reported on Table 8.

Table 8. Case 2 results for the boilover time with the models proposed on Table 2.

	Tllama (K)		tbo (hr)		treal (hr)	%error	
Heat model 1	NA	NA	29,77614629		5,3	461,8140809	
Heat model 2	1100	1900	2,4275687	1,64560344	5,3	54,196817	68,9508784
Time model 2	NA	NA	1068,095		5,3	NA	

Table 7. Safe distances for the boilover consequences for case 2

dletal (m)	350,936098
dirrev (m)	477,115369

For case 1 the most accurate value reported by the tool was the one calculated by using an approximated value of 60 kWm^{-2} for the feedback heat, while the most accurate values in case 2 were the ones calculated by using the heat model 2, for this case the value using the heat model 1 is not conservative. This can be explained by the fact that the heat proposed by literature is very small compared with the one calculated by the second heat model, it worked well for a tank with an small amount of fuel but with great amounts of fuel that heat is not enough and will take more time to increase the temperatures in fuel layer and to take the water to its boiling point, making the boilover time bigger. So the second model may be better for modeling large tanks with big amounts of fuel, keeping the time in the same order of magnitude and conservative compared to the real time.

In the case of the differential model mentioned before for both cases the boilover time calculated with the tool is four orders of magnitude or more above the real time, this can be explained by the assumption around the model, first the model is one dimensional assuming that all the heat is transferred from the top to the bottom of the tank when it exist also a radial heat transference, this can also be seen by comparing the results from case 1 with the results from case 2 where the difference among the real time and the calculated time is proportional to the amount of fuel and the tank dimensions. Also the thermal diffusivities along with the rest of the fluid properties were assumed constant during the fire, despite the changes of temperatures in the fuel and water layers. One of the assumptions of the model implies assuming only a heat conduction mechanism, no convection inside the layer occurs, this last phenomena implies the change of density in different heights of the fuel layer and this difference in densities implies that the fuel in the top of the tank will eventually move to the bottom and the fuel in the bottom will eventually move to the top making the heat transference quicker and the boilover time smaller.

4.3 Comparison among boilover time and burning time

The case 2 described for the validation of the boilover tool have the complete information to use also the tank fire tool and can be used to compare the time of burning if the boilover doesn't occur with the time at which boilover occurs. Giving a little notion of what fraction of the total burning time takes the boilover to occur.

The results obtained by tool for the fire are presented on Table 8 and the consequences on Table 9.

Table 8. Results related to the tank and the fire variables for case 2

Burning tank name:	1
Type of roof:	cone
Product:	Crude Oil

Nominal volume (m3):	80000,00
Content (m3):	10142,89
Height (m):	14,7
Diameter (m):	33,00
Area of fire (m2):	1523,99
Time of fire (hr):	93,33
Time of fire (day):	3,89
Refrigeration rate (m):	57,11

Table 9. Results related to the consequences of the fire for case 2

	From (m)	To(m)
Hot zone	0,000	51,083
Warm zone	51,083	102,167
Cold zone	102,167	N/A

The boilover results are reported in the above section, the comparison was made using the real value reported in literature for the case and the only conservative value reported by the tool that was calculated using the heat model 2. The percentages reported on Table 10 are the fraction that takes the boilover to occur with respect to the total time of fire.

Table 10. Percentage boilover time/ burning time

	%
Real Value	5,67877424
Heat model 2	2,59294975

The boilover time is a little fraction of the total burning time, showing how important is to estimate the first one, modeling only the fire will give more time for mitigation actions but if a boilover occurs this time is reduced on a 95 % and being conservative on a 97%, the consequences are also greater for the boilover where the safe distances are superior than 477,11 m while the cold zone for the fire is above 102,167 m, about a fifth times less the boilover distance. Making the boilover a more dangerous event than the fire itself and a matter to be worried about.

4. Conclusions

The fire tool's validation results are within an acceptable error margin but still conservative, which implies that the used model describes the event and can be used for real calculations and being also simple to manipulate for the user because there is no need of complicated interfaces or calculations. In the case of the boilover the recommended model based on the results reported on this work is the one reported by Casal et. al. [1] for the feedback heat and the algebraic approximation for time because with them there were obtained for both cases used for validation a conservative result. This implies that some of the proposed models can actually be implemented on tools that can be used in real life cases helping to generate and design mitigation strategies.

5. Perspectives

For future work the differential solution for time will be studied calculating parameters like the thermal diffusivities depending on the temperature profiles and solving a two-dimensional model. Also validating the tool against more cases including cases that involve alternative fuels.

Nomenclature

Q: Pool fire heat release rate (kW)
m": Mass burning rate of fuel per unit surface area (kg $m^{-2}sec^{-1}$)
$\Delta H_{c,}$: Effective heat of combustion of fuel (kJ/kg)
$A_f = A_{dike}$:Surface area of pool fire (area involved in vaporization)
κβ: Empirical constant (m^{-1})
D: Diameter of the pool fire (m^2)
t_b: Burning duration of pool fire (sec)
V: Volume for liquid (m^3)
D: Pool diameter (m)
v: Regression rate (m sec^{-1})
ρ:Liquid fuel density (kg m^{-3})
c_p: Specific heat of the fuel at T_a (kJ $kg^{-1}K^{-1}$)
h_{HC} : Height of the fuel in the tank after ignition (m)
ΔH_v : Vaporization rate atT_{0max} (kJ kg^{-1})
T_a : Ambient temperature (K)
T_{hw} : Heat wave temperature when the boilover occurs (K)
T_{0av} : Average boiling temperature of the fuel (K)
Q_f: Thermal flow entering the fuel from its surface (kW m^{-2})
T_{0min}:
T_{0max}:
α: Crude thermal diffusivity
ρ_∞ : Air density (kg m^{-3})
$c_{p\infty}$: Specific heat of the air (kJ $kg^{-1}K^{-1}$)
T_∞: Ambient temperature (K)
T_f : Average flame temperature (K)
Xr:
$q''_{hotzone}$: Heat per unit area that humans and equipment can't tolerate (10 kW/m^2)
$q''_{warmzone}$: Heat per unit area that can be tolerated with protection equipment (2,5 kW/m^2)
W: Mass of the fuel at the beginning of the fire.
$k_{lethality}$: Lethality constant depending on the fuel.
$k_{irreversible}$: Irreversible consequences constant, depends on the fuel.

N. References

[1] J. Hristov, E. Planas-Cuchi, J. Arnaldos, and J. Casal, "Accidental burning of a fuel layer on a waterbed: a scale analysis of the models predicting the pre-boilover time and tests to published data," *International Journal of Thermal Sciences,* vol. 43, pp. 221-239, 2004.
[2] J. Casal, *Análisis del riesgo en instalaciones industriales*: Edicions UPC, 2001.
[3] K. Trevor, "Chapter 5 - Storage Tanks," in *What Went Wrong? (Fifth Edition)*, ed Boston: Butterworth-Heinemann, 2009, pp. 97-121.
[4] B. Zheng and G.-h. Chen, "Storage tank fire accidents," *Process Safety Progress,* vol. 30, pp. 291-293, 2011.

[5] E. Corportation, "2012 The Outlook for Energy: A view to 2040," 2012.

[6] CIA. (2011, 11/01/2012). *The World Factbook :: Oil Production*. Available: https://www.cia.gov/library/publications/the-world-factbook/fields/2173.html

[7] G. Gómez, S. Rodríguez, J. Cadena, and F. Muñoz, "Kletz's legacy for developing countries: Simple systems based on inherently safer design," *Journal of Loss Prevention in the Process Industries,* vol. 25, pp. 843-847, 2012.

[8] AP, "8000 flee Colombia gas fire," in *The Telegraph*, ed, 1982, p. 1.

[9] S. Mannan, F. P. Lees, and Knovel (Firm). (2005). *Lee's loss prevention in the process industries hazard identification, assessment, and control (3rd ed.)*. Available: http://www.knovel.com/knovel2/Toc.jsp?BookID=1470

[10] J. I. Chang and C.-C. Lin, "A study of storage tank accidents," *Journal of Loss Prevention in the Process Industries,* vol. 19, pp. 51-59, 2006.

[11] D. Bradley, G. A. Chamberlain, and D. D. Drysdale, "Large vapour cloud explosions, with particular reference to that at Buncefield," *Philosophical Transactions of the Royal Society A: Mathematical, Physical and Engineering Sciences,* vol. 370, pp. 544-566, February 13, 2012 2012.

[12] B. Investigation Board, "The Buncefield Incident," Health and Safety Commision, HSC2005.

[13] B. J. Lowesmith, G. Hankinson, and D. M. Johnson, "Vapour cloud explosions in a long congested region involving methane/hydrogen mixtures," *Process Safety and Environmental Protection,* vol. 89, pp. 234-247, 2011.

[14] U. S. C. S. a. H. I. B. CSB, "CSB Conducting Full Investigation of Massive Tank Fire at Caribbean Petroleum Refining; Investigative Team Plans to Thoroughly Examine Facility Safety Practices," in *News Room* vol. 2012, ed. Washington DC, USA, 2009.

[15] K. Vyas and E. Minaya. (2012) Explosion at Amuay refinery in Venezuela kills 41; fire still burning. *Hydrocarbon Processing*. Available: http://www.hydrocarbonprocessing.com/Article/3080380/Environment-and-Safety/Explosion-at-Amuay-refinery-in-Venezuela-kills-41-fire-still-burning.html

[16] C. Crippa, L. Fiorentini, V. Rossini, R. Stefanelli, S. Tafaro, and M. Marchi, "Fire risk management system for safe operation of large atmospheric storage tanks," *Journal of Loss Prevention in the Process Industries,* vol. 22, pp. 574-581, 2009.

[17] C. D. Argyropoulos, M. N. Christolis, Z. Nivolianitou, and N. C. Markatos, "A hazards assessment methodology for large liquid hydrocarbon fuel tanks," *Journal of Loss Prevention in the Process Industries,* vol. 25, pp. 329-335, 2012.

[18] C.-I. Lin and H.-F. Wang, "Research on Safety and Security Distance of Flammable Liquid Storage Tank," *Procedia Engineering,* vol. 11, pp. 51-60, 2011.

[19] M. Arai, K. Saito, and R. A. Altenkirch, "A Study of Boilover in Liquid Pool Fires Supported on Water Part I: Effects of a Water Sublayer on Pool Fires," *Combustion Science and Technology,* vol. 71, pp. 25-40, 1990/05/01 1990.

[20] T. Inamura, K. Saito, and K. A. Tagavi, "A Study of Boilover in Liquid Pool Fires Supported on Water. Part II: Effects of In-depth Radiation Absorption," *Combustion Science and Technology,* vol. 86, pp. 105-119, 1992/11/01 1992.

[21] I. M. Shaluf and S. A. Abdullah, "Floating roof storage tank boilover," *Journal of Loss Prevention in the Process Industries,* vol. 24, pp. 1-7, 2011.

[22] M. Muñoz, J. Arnaldos, J. Casal, and E. Planas, "Analysis of the geometric and radiative characteristics of hydrocarbon pool fires," *Combustion and Flame,* vol. 139, pp. 263-277, 2004.

[23] M. Muñoz, E. Planas, F. Ferrero, and J. Casal, "Predicting the emissive power of hydrocarbon pool fires," *Journal of Hazardous Materials,* vol. 144, pp. 725-729, 2007.

[24] NFPA, *SFPE handbook of fire protection*, 3rth ed., 2002.

[25] "Storage Tank Explosion and Fire in Glenpool, Oklahoma," National Transportation Safety Board, Washington DC2003.

[26] H. Koseki, Y. Natsume, Y. Iwata, T. Takahashi, and T. Hirano, "Large-scale boilover experiments using crude oil," *Fire Safety Journal,* vol. 41, pp. 529-535, 2006.

[27] H. Persson and A. Lönnermark, "Tank Fires Review of fire incidents 1951–2003," SP Swedish National Testing and Research Institute.

A practical view of risk reduction management of hardware and software for safety critical applications

Luis Duran
Product Marketing Manager Safety Systems
ABB
3700 West Sam Houston Parkway South suite 600
luis.m.duran@us.abb.com

Mike Krywonos
Safety Execution Center Manager
ABB
3700 West Sam Houston Parkway South suite 600
mike.krywonos@us.abb.com

Prepared for Presentation at
American Institute of Chemical Engineers
2013 Spring Meeting
9th Global Congress on Process Safety
San Antonio, Texas
April 28 – May 1, 2013

UNPUBLISHED

A practical view of risk reduction management of hardware and software for safety critical applications

Luis Duran
ABB
3700 West Sam Houston Parkway South suite 600
luis.m.duran@us.abb.com

Mike Krywonos
ABB
3700 West Sam Houston Parkway South suite 600
mike.krywonos@us.abb.com

Keywords: Safety, Risk Reduction, Safety Systems, Independent Protection Layers, Common Cause Failures, Functional Safety Management System, IEC61508, IEC61511, TÜV, third party certification, Human Factors

Abstract

Best in Class companies (both National and Global) continue to invest in safety systems; this in most cases is because safety is at the core of their production process and among their key performance metrics, closely linked to their productivity.

Many unfortunate industrial accidents such as the recent explosion of the Deep Water Horizon with the largest marine oil spill in history are evidence of the negative consequences they have on people, environment, production assets and corporations.

This paper discusses the accepted definition of safety and risk reduction in the industry and will discuss the best practices applied by engineers in the design of products for mission critical functions and the implementation of those products in projects.

Both in product design and implementation, this session will dedicate special attention to methodologies or work processes used to reduce potential common cause failures that might impair the operation of the safety system and ultimately expose the facility to a higher risk; and the human factor considerations recommended to ensure proper operation upon a critical condition or an emergency in the industrial process.

1. Introduction

Best in Class companies worldwide, defined as such for their high percentage of Overall Equipment Effectiveness (OEE) and low injury frequency rate; continue to show strong initiatives intended to expand the safety of their operation. From executive sponsored programs at the corporate level to the definition of proactive risk management strategies, these companies are investing in safety systems and processes; simply because in most cases, safety is identified among the core values of those companies, at the center of their production process and among their key performance metrics, closely linked to their productivity.

There are numerous publications and records of Industrial Accidents, a large number of them with terrible consequences, including not only loss of human life or environmental impact but also affecting the production assets in the site. As an example, in researching for this paper, the authors found record of at least 20 "notable" oil and gas offshore blowouts between 1980 and 2010.

As an example for this paper, the authors compared two serious incidents in the Oil & Gas sector; Piper Alpha operated by Occidental Petroleum and Deep Water Horizon leased to BP. Although both incidents are 22 years apart from each other, occurred in different geographies and have many differences in causes and consequences. They illustrate the risk and the potentially catastrophic dimension they can take in this industry and the impact on people, environment and corporations, which might cease to exist after these incidents.

2. Defining Safety and Risk

Safety is defined in the industry as a reduction of existing risk to a tolerable or manageable level while risk is a combination of the probability of a harmful incident and magnitude of the harm. Traditional design practices allocate the risk reduction across different and independent protection layers, the rationale behind it is simple "any system that can fail will fail" so the engineering best practice is to distribute the risk reduction tasks across multiple independent functions or systems one of these systems is a Safety Instrumented System.

2.1 Safety Instrumented Systems one of many Independent Protection Layers to Reduce Risk

A Safety Instrumented Systems (SIS) is a mission critical system designed following international design practices such as IEC61508 [1] Functional Safety Standard to reduce risk to the people in and around the production environment, the environment, the production asset and the business. In many cases the Safety Instrumented Systems are the last resource to prevent disaster.

SIS performance is measured by Safety Integrity Level (SIL) SIL 1 low, SIL 3 high o Risk Reduction Factor.

As mentioned earlier, the Safety Instrumented System is one of many functionally independent systems each intended to perform a task, as shown on Table 1 typically referred to as Independent Protection layers [2]

Table 1: Expected Functionality of Independent Protection Layers

Layer of Protection	Expected Functionality
BPCS:	Keeping the process under control
Alarms:	Alerting the operator of abnormal conditions and providing guidelines for appropriate operator response
SIS:	Automatically taking the process to a safe condition in case the abnormal condition goes out of control and the operators can't take any corrective action in time
Other layers:	Intended to mitigate the consequence of the hazard

The concept of Independent protection layers (IPL) can also be illustrated in Figure 1

3. Common Cause Failures

The idea of functional independent systems is an attempt to avoid Common Cause Failures (CCF).

Common Cause Failures [3] are failures that might affect more than one of these protection layers at the same time. As engineers, the tendency will be to focus on elements such as Heat, Humidity, Shock, and radio interference among others. However this paper also considers with special interest the human elements related to the design of safety critical systems which can contribute to additional failures.

As with other systems, some SIS problems are related to the Commercial Off the Shelf (COTS) products designed for a specific function (i.e. hardware limitations or poor documentation) while other problems are related to the use of those COTS product such as misapplication, user application programming or poor maintenance practices. Both areas can be addressed by implementing appropriate design best practices to reduce risk.

The industry has conceived these "best practices" as a series of steps that must be performed before putting one of these systems in operation (design, installation and commissioning). This paper will describe the two areas:

- Product Design and Implementation of the Commercial Off-the-Shelf (COTS) Product
- Application Design

Figure 1: Independent Protection Layers

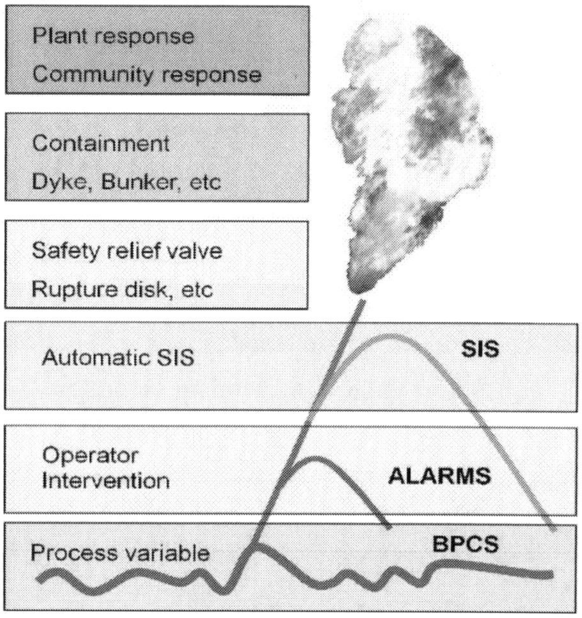

3.1 Reducing Common Cause Failures by Design

Traditionally or historically, Systems designed for SIS functions have relied on physical separation and redundancy [4] to reduce common cause failures. The authors identified technology changes across multiple generations of SIS over a period of 30 years. However over the same time frame the industry applied lessons learned from incidents to develop best practices in design and implementation of systems, the international industry standards used more often are IEC61508 and IEC61511 [5] (Figure 2)

3.2. Hardware Fault Tolerance (1st Generation SIS)

1st Generation systems rely on Hardware Fault Tolerance or redundancy to achieve reliability and availability as required for these applications, this practice is rooted in the late 1970s and early 1980s technology. As the majority of these systems were designed prior to the release of the Functional Safety standards, they do not follow the best practices found with regards to design, documentation and testing found in those standards.

Figure 2: Evolution of National Prescriptive Standard to International Performance Base Standards

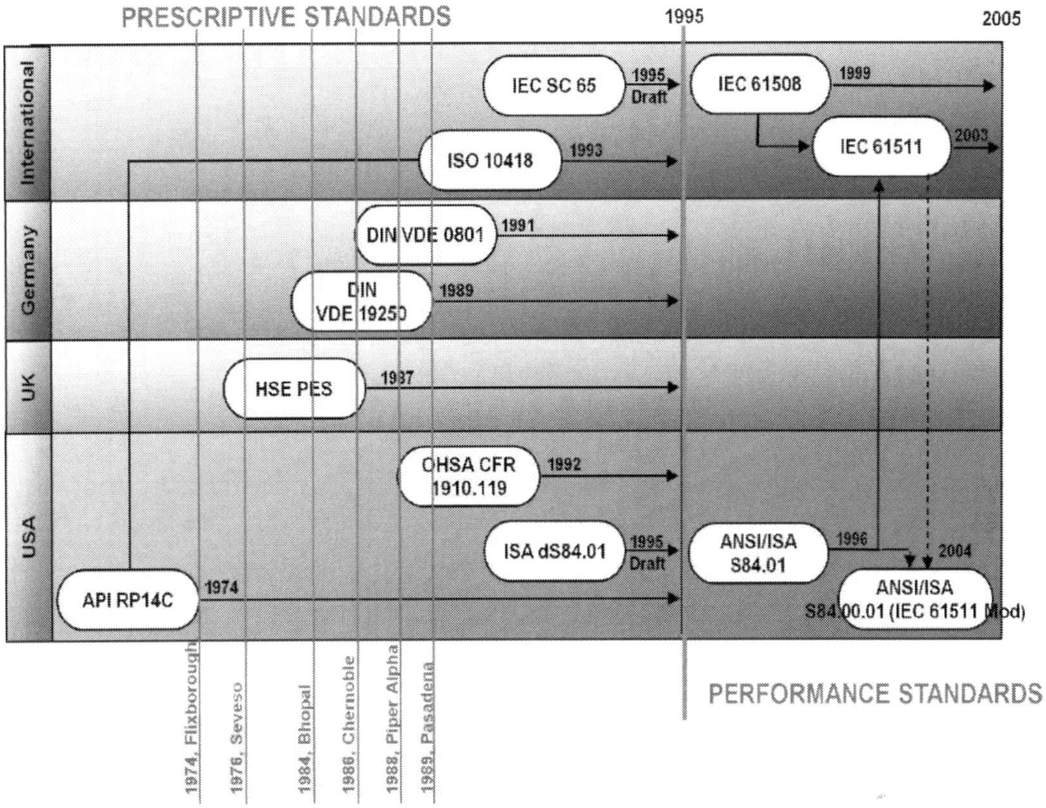

3.3. 2nd Generation Systems

As the standards became available (mid 1990s), vendors started to apply the design best practices and to pursue certification to those standards. This generation of systems was produced having third party assessment and certification but continued to rely on Hardware Fault Tolerance to satisfy the performance requirements. The authors found that a large portion of the systems applied as SIS in the market today, although designed to satisfy Functional safety standards haven't change dramatically in their use of hardware fault tolerance and basic software and hardware diagnostics.

3.4. Diverse Architecture and Implementation a 3rd Generation Systems

A more recent and different type of system, uses diverse redundancy, diverse implementation and active diagnostics [6] and can deliver not only reliability and availability required for the application while minimizing common cause failures found by the use of the same technology but also introduce additional protection to systematic failures normally related to human factors

3.4.1 Hardware Design – Diverse Hardware

This newer generation of systems use diverse processing hardware (multiple technologies) such as diverse operating systems and diverse hardware and diverse redundancy, both enhanced with diverse implementation (different implementation teams).

3.4.2 Application Execution in a SIL Compliant Environment

The use of diverse operating systems (use of different technologies) including the use of third party certified COTS and different execution path with different compiler rules extends the risk reduction found in the hardware design to the firmware/software environment in the system, including the use of limited variability software functions and provision of systematic capabilities as specified in the Functional Safety standards for critical applications.

3.4.3. Systematic Capabilities and Human Error

The concept of systematic capabilities and its associated performance measurement was introduced in the latest revision of IEC61508 standard (didn't exist as such in 1st Generation systems and was emerging in 2nd generation systems).

3.5. Safety and Network Security

1st and 2nd generation Safety Systems were not designed to reside on a networked plant, for the most part these systems allowed some communication interface or gateway but were not intended for integration, therefore in general are not equipped to counter security threats as those the industry has experienced in recent years.

A brief analysis of the definitions will show a clear differences Safety issues are random in nature and statistic analysis is possible security intentional in nature and statistical analysis will not measure performance.

The authors concur with industry standards and experts in the field. Network security may affect system performance and the safety of the installation. A system can't be safe without it being secure.

3.6. Design Best Practices and V-Model

As indicated earlier, industry standards in Functional Safety introduced the concept of Functional Safety Management System (FSMS) which includes a series of steps in the design, documentation and testing of the system, and should include Network Security as well.

Figure 3: Product Development V-Model

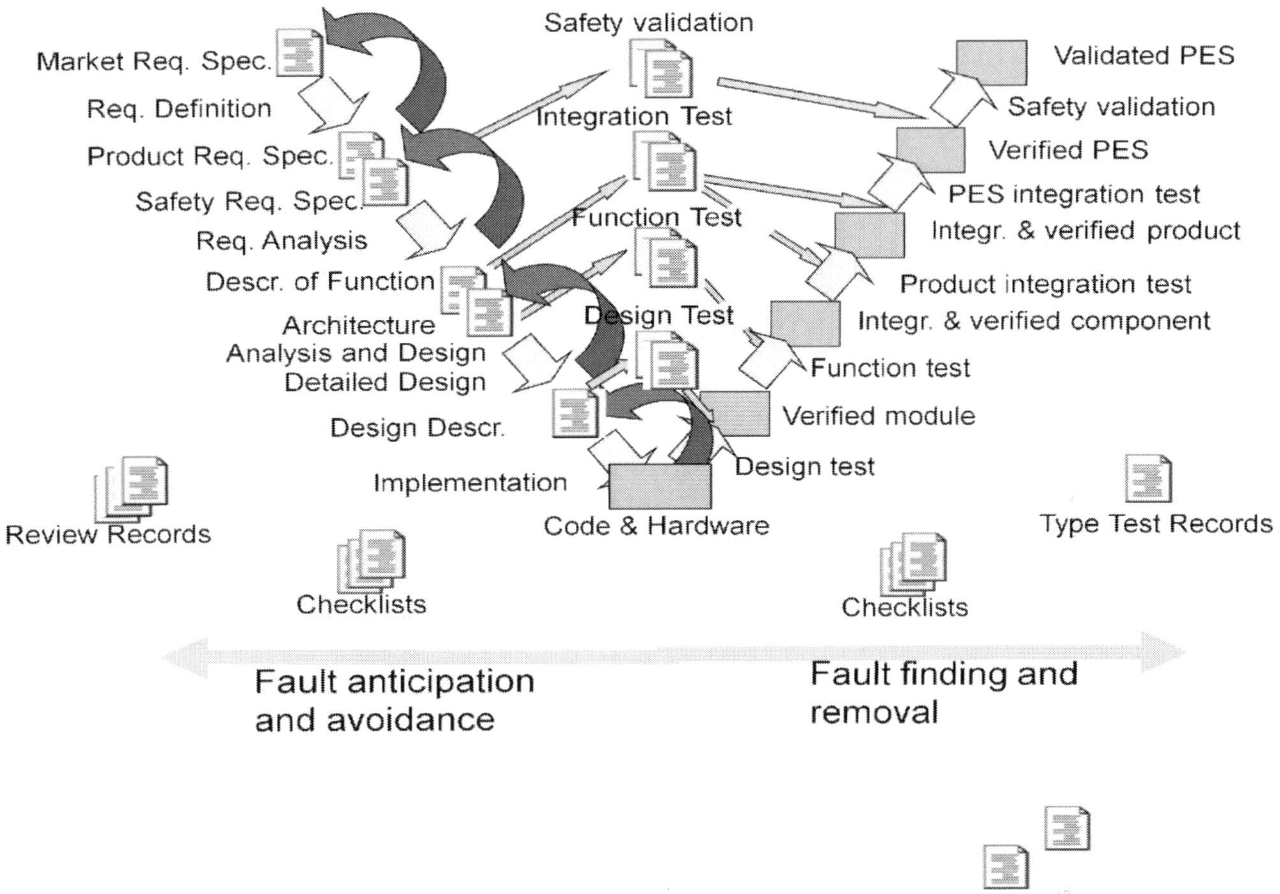

Today product development or design organizations responsible for releasing COTS products (hardware and software) intended for safety applications use design best practices as the V-Model (Figure 3) and are evaluated by a third party organization (i.e. TUV) and their FSMS is approved prior to the assessment of their products.

4. Engineering best practices to reduce risk in application design

As indicated previously, best practices in design apply both in the design of a Commercial Off the Shelf (COTS) Product and the Application Design or the application of a COTS tailored to reduce a particular risk in an industrial installation.

Best Practices in Application Design will apply a model similar to the V-Model used in the development of a product (Figure 3) and reinforced by a structured Quality Management system appropriate for Functional Safety applications.

Application Design starts with Hardware Design, which starts with a design that follows the COTS Product's Safety Manual and considers the particularities of the application as described on the Application Safety Requirements Specification to design a system that is less susceptible to product failures by considering aspects such as

- Power feeds
- Power Supplies
- CPUs
- I/O Modules Hardware
- Application Software
- Networks

4.1 Hardware

It's clear that hardware is not perfect, it's subject to fails, and the application design must consider fail recovery modes or additional risk reduction methodologies or options including diverse methods of performing shutdowns as:

- Hardwired wired pushbutton,
- Hardwired logic systems (with appropriate SIL rating)
- Solid State relays
- Etc.

4.2 Software

One critical aspect that has gained visibility in recent years is the software design of the application. Efforts in this area include the development of a well documented Software Functional Design, appropriate selection of software tools and libraries that meets the Safety Integrity Level (SIL) required for the application, Code Review practices including Independent Validation and Verification depending on the SIL, competence of the programmer and other elements described later in section 5.1. The features available in the COTS product to address the systematic capabilities requirements are a valuable asset for the programmer and contribute to reducing potential errors that will later affect the safety of the installation.

4.3 Functional Testing and Periodic Proof Testing

Another critical aspect is Functional Testing and Periodic Proof Testing, particularly because Not ALL System Faults are Self-Revealing and Covert Faults that may inhibit SIS action on Demand can only be detected by testing the entire system, this requires not only a Full Functional Test prior to commissioning, typically conducted as an integration test in the vendor or System Integrator facility and later in the final instrumentation room in the plant; but Periodical Functional Tests using a documented procedure to Detect Covert Faults and covering the entire SIS.

Functional Testing should Record and Analyze activation of SIS functions, and spurious activation of an Emergency Shutdown Valve due to a Process Shut Down, but this does not test the Entire Function of the same valve during an ESD action.

5. Automation can't check for human intervention

Once the design is complete, all of these systems will have different degrees of interaction with users, and different type of users with different levels of competence depending to their role (i.e. operation, maintenance personnel and engineers)

The most sophisticated automation can't prevent human error, industrial plants are designed with the highest accuracy in mind, with several separate safety loops often checking the integrity of process systems. This apparatus however can't check for human intervention and a small human error could cause an enormous catastrophe. 70% of reported incidents in the oil and gas industry worldwide are attributable to human error and account for in excess of 90% of the financial loss to the industry.

Results from research conducted by the Health and Safety Executive and published in the book Out of control: Why control systems go wrong and how to prevent failure [7] show the impact of the human element in industrial incidents throughout project lifecycle. This is shown on Figure 4.

International Functional Safety Standards (IEC 61508 and IEC61511) and previously ISA 84 introduced the concept of the Safety Lifecycle, describing the phases that should take place from concept to design to implementation and operation of a Safety Instrumented System. The Safety Lifecycle is a step in the direction to reduce the impact of human factors by establishing the proper design best practices, documentation reviews and validation and verification steps in the execution of a safety project. Additionally the standards introduce two important elements: Competence of Personnel mentioned briefly in Section 4.2 and Functional Safety Management System (FSMS).

Recent changes to International Functional Safety Standard IEC61508 have turned the requirements on FSMS and Competence of Personnel into normative clauses of mandatory compliance instead of a recommendation as was presented in the previous version of the same standard.

5.1 Functional Safety Management System (FSMS) and Competence Requirements

A Functional Safety Management System (FSMS) consists of proven and validated procedures, methodologies, templates and report outlines, covering those essential elements of hazard and risk assessment; SIL determination; SIS design and SIL capable hardware, SIS installation, SIS commissioning and validation and SIS operations and maintenance. A FSMS is used in the design of a COTS product and during the Application Design. It's typically an extension of the existing Quality Assurance/Quality Management process and lives from the conception to decommission of the SIS and beyond to become part of the organization best engineering practices and work processes.

Figure 4: Root Cause of Control System Failures

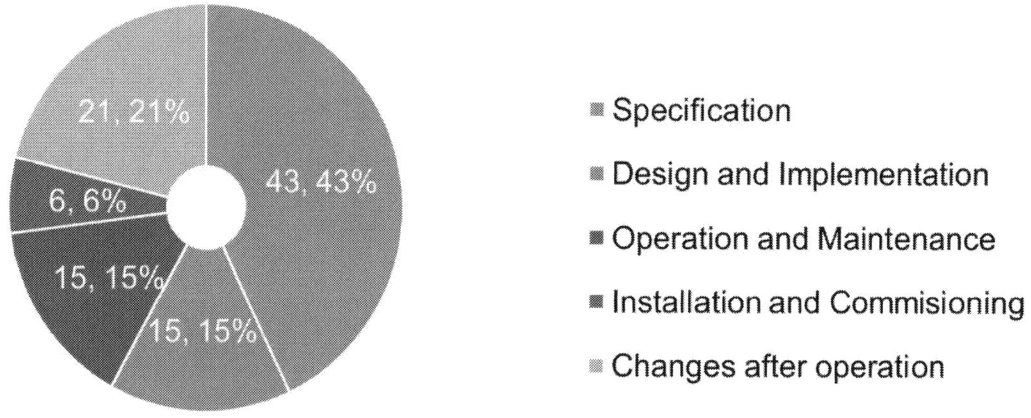

As indicated, the safety standards call for competence of personnel. This requirement shouldn't be interpreted as a need for safety experts in every aspect of the design, commissioning or operation and maintenance of a SIS, but instead as the need for an evaluation of the appropriate expertise and experience required to perform a particular job function in any phase of the safety lifecycle. In practical term those involved with activities related to a safety system should have demonstrated "competence" either by formal training, certification or accreditation or on-the-job relevant experience.

The reader might conclude that the standards today require competent and independent review groups because system designers don't trust product designers or other application designers and therefore we would rather be safe than sorry.

6. Summary

In summary

- Best in Class companies link safety to their success and invest in programs and systems to reduce their risk
- Engineers have attempted to reduce risk by minimizing the potential for common cause failures

- Common cause failures can occur in products (hardware or software) or the implementation of the application
- The industry has conceived best practices (i.e. FSMS) to minimize the impact of human error
- Human factors can't be ignored in the design, particularly in the application design
- Enforcement of these best practices via formal FSMS has proven to be a way to reduce the risk introduced by engineers… it's vital!

References

[1] IEC 61508 Functional safety of electrical/electronic/programmable electronic safety-related systems International Standard 2010

[2]Integrating Control and Safety: Where to draw the line.
Robin McCrea-Steele, TÜV FSExpert, Invensys-Premier Consulting Services

[3] Common Cause and Common Sense, Designing Failure Out of Your SIS
Angela E. Summers, Ph.D. and Glenn Raney, Oct 2000

[4] Integrated but separate, Advances in integrated and safety control,
Roger Prew, ABB , June 2009

[5] IEC 61511 Functional safety - Safety instrumented systems for the process industry sector
International Standard 2003

[6] Providing Independent Layers of Protection with Integrated Safety Systems
Luis M. Duran, ABB and Ron Johnson Dow Chemical, Oct 2009

[7] Out of control: Why control systems go wrong and how to prevent failure

HSE Books ISBN 0-7176-2192-8

AUTHOR INDEX

Aguiar, Leonardo Salim Damian 329
Aguilar-Otero Sr., Jose R. 249
Aguirre, J. Luis 600
Allawi, Ahmed J. 1
Allford, Lee 54
Arnaldos, Josep 460, 587
Bahiense, Ana 321
Barrera, Carlos A. 634
Baybutt, Paul 688
Bearrow, Mike 307
Bennett, Raymond H. 603
Bernechea, Esteban 460, 587
Bishop, Nikki 344
Bradshaw, William 96
Bridges, William 470, 752
Brigman, James 358
Burban, Peter M. 452
Burgess, John 637
Cadena, Jaime E. 124,
Cai, Tianxing 292
Caputo, Maria 321
Carreto, Victor H. 12
Clarke, Peter 722
Conner, Jim 980
Cooper, Joshtin 900
Cox, Brent L. 1002
Cox, Brenton L. 937
Cravens, Robert 176
Cristea, Nicholas N. 645
Cummings, David E. 80, 484
Cutchen, Steve 53, 1044
Da Silva, Clarissa Sossmeier 875
Davis, Scott G. 613
Dayton, Robert 913
De Oliveira, Thiago Souza 372
Dee, Sean J. 824, 937, 1002
Deivasagayam, Karthick 356, 409
Dennehy, Martin A. 67
Denton, Jerad 53
Dewees, David J. 617
Díaz, Juan C. 495
Diniz, Americo 875
Duran, Luis 509
Ebrahim, Fareed 254
Edel, Matthew 203
Eisenberg, Jonathan 391
Elias, Brian 679
Evans, Roger 1044
Farrell, Thomas 562
Fegley, Kenneth 67
Fialkowski, Charles 792
Filho, Salvador Ávila 601
Fink, William 614
Forest, Jerry 815, 865
Frawley, Thomas 357, 437
Freeman, Raymond 410
Gómez, Jorge Mario 124
Gomez, Manuel R. 253
Graeme, Ellis 680
Grim, Lauren 1044
Grubbe, Deborah L 522
Guld, Jeppe 96
Gutierrez, Eliecer 599
Hardy, Terry 1029
Herrera, Carolina 614
Holmstrom, Donald 53, 1044
Horn, Brad 203
Hoyle, Bill 53
Huang, Szu-Ying 459
Israni, Kumar 900
Iv, Peter J. Ralbovsky 615
Jain, Prerna 176
Jiang, Chunming 656
Johnson, Amanda 1044
Johnson, Ryan 664
Johnson, Thomas 357, 437
Jones, Jerry L. 251
Jung, Seungho 138
Kadri, Shakeel 33, 67
Kane, Kara 53
Karthikeyan, Muthukumarasamy 401
Kemal, Abid 634
Kern, Hannes 382, 523, 1022
Khajehnajafi, Shahryar 157
Khan, Farheen 175
Khan, Mohammad Naved 358
Klippel, Alexander 230
Knijff, Piet 54
Krywonos, Mike 509
Kumar, Ajay 405
Kusche, Christian 230
Lakhiani, Sunil D. 175
Laskar, Sandipan 274
Le, Hai 392
Leblanc, John A. 263
Lehmann, Scott 833
Li, Xinrui 459
Lione, Diego Da Silva 329
Liu, Yi 392, 635
Ludwig, Keith 218
Luebeck, Hope 264
Luna Sr., German 190, 254, 345
Lutostansky, Elizabeth 138
Lyon, Chris 53
Mackenzie, Cheryl 53
Madewell, Andrew 921
Maher, Steven T. 202, 802
Manjunath, Konanur 1013
Mannan, M. Sam 12, 392, 459, 584, 635
Marriott, James 636

AUTHOR INDEX

Massaguer, Cristian 587
Mateo, Alvin 67
Meel, Anjana 157
Melhem, Georges 533
Mittal, Pawan 405
Morgan, Christina 253
Morrison, Delmar 175, 824
Mulcahy, Marybeth 53
Munoz, Felipe 124, 495, 599
Murphy, John F. 980
Myers, Molly 876
Neill, Mike 833
Ness, Albert I. 1013
Nguyen, Ha 584
Norton, Kristin D. 202
Nour, Michael A. 202
Nourai, Farshad 612
Ogle, Russell A. 824, 937, 1002
Osterberg, Paul 357, 437
Ozog, Henry 876
Pathak, Swapnil 358
Perez, John T. 921
Peters, Glen 67
Pineda-Solano, Alba 12
Polcyn, Michael A. 603
Powell, Joseph B. 28
Prasad, Hv 401
Pridy, Cindie 176
Prueter, Phillip E. 617
Rains, Brian D. 112, 887
Ramaswamy, Ravi 176
Ramesh, V. M. 111
Ramos, Manuel A. 124
Rangwala, Ali S. 679
Raupenstrauch, Harald 382, 523, 1022
Richardson, Joshua 584
Rodríguez, Sergio E. 124
Sakthivel, R. 111
Saud, Yaneira 612
Scheid, Marc 230
Schmelzer, Peter 54
Scholtz, Colin 802
Schork, Joan M. 138, 218
Schröder, Volkmar 230
Seguí, Xavier 460
Shah, Ghanshyam 284
Sheppard, Dave 344
Sherburne, Carol 437
Sierra, Lina M. 124
Sliva, Dan 33
Smith, Dustin J. 147, 637, 645
Smith, Gregory 470
Smith, Robert C. 560, 561
Snoeys, Jef 562
Soczek, Charles A. 242, 586
Souza Jr., Maurício Bezerra 329, 372
Spijker, Christoph 523
Stauffer, Todd 722
Taferner, Matthias 1022
Throness, Barry 949
Tillema, Dan 1044
Tschirschwitz, Rico 230
Vaez, Najmeh 612
Valdes, Olga Juliana Reyes 12
Van Wingerden, Kees 613
Vanommeren, James 67
Vaz Jr., Carlos André 329, 372
Vela, Marco Antonio 865
Velasco, Nubia 599
Victoria, Jorge 599
Villaça, João Márcio Almeida 329
Vingerhoets, Jim 562
Viz, Mark J. 1002
Vyas, Renu 401
Waller, Alvin 994
Wang, Zheng 656
Waters, Steven 664
Watson, Kevin 913
Wilkins, John 147
Wilson, Kelly 53, 1044
Wimmer, Bruno 382
Wincek, John 876
Wingard, Mark 1044
Xu, Qiang 292
Yarlott, Marc W. 664
Yelisetty, Suresh 612
Yuan, Jiwu 656
Zhang, Jiaqi 635
Zhang, Yu 218
Zhao, Xiangdi 656
Zirker, Stefan 230

28th Center for Chemical Process Safety International Conference 2013

(CCPS)

Topical Conference at the 2013 AIChE Spring Meeting and 9th Global Congress on Process Safety

San Antonio, Texas, USA
28 April - 2 May 2013

Volume 2 of 2

ISBN: 978-1-62748-446-6

Printed from e-media with permission by:

Curran Associates, Inc.
57 Morehouse Lane
Red Hook, NY 12571

Printed by Curran Associates, Inc. (2013)

For permission requests, please contact AIChE
at the address below.

AIChE
3 Park Avenue
New York, NY 10016-5991

Phone: (203) 702-7660
Fax: (203) 775-5177

www.aiche.org

Additional copies of this publication are available from:

Curran Associates, Inc.
57 Morehouse Lane
Red Hook, NY 12571 USA
Phone: 845-758-0400
Fax: 845-758-2634
Email: curran@proceedings.com
Web: www.proceedings.com

TABLE OF CONTENTS

VOLUME 1

Applicability of the Hazards and Effects Management Process (HEMP) in Research and Development 1
Ahmed J. Allawi

Scale up of Stirred Batch and Semi-Batch Reactors - Gaps and Limitations of the Current Methodologies 12
Olga Juliana Reyes Valdes, Alba Pineda-Solano, Victor H. Carreto, Sam Mannan

Safety, ASAP and ALARP in Fast Track Process Development 28
Joseph B. Powell

An Update On the Use of Process Safety Leading Indicators 33
Dan Sliva, Shakeel Kadri

Role of Leading Indicators in Prevention of Major Accidents: CSB Learnings From BP Texas City and Deepwater Horizon 53
Kelly Wilson, Donald Holmstrom, Cheryl Mackenzie, Marybeth Mulcahy, Kara Kane, Jerad Denton, Chris Lyon, Bill Hoyle, Steve Cutchen

Process Safety Leading Indicators – A Perspective From Europe 54
Piet Knijff, Lee Allford, Peter Schmelzer

So We All Have Been Implementing Process Safety Metrics – What Next? 67
Shakeel Kadri, James Vanommeren, Glen Peters, Martin A. Dennehy, Alvin Mateo, Kenneth Fegley

Process Safety Leading Indicators: A 'Silver Bullet' or 'Bean Counting' for Enhanced Risk Management 80
David E. Cummings

Using Metrics to Drive Process Safety Performance At Maersk FPSOs 96
William Bradshaw, Jeppe Guld

Learning From High Potential Near Misses in the Refining Sector 111
V. M. Ramesh, R. Sakthivel

Felt Leadership: An Essential Ingredient in Successful PSM Implementations 112
Brian D. Rains

Localization of Control Rooms Using a MINLP Approach, Based On Accidental Explosion Scenarios 124
Lina M. Sierra, Sergio E. Rodríguez, Jaime E. Cadena, Manuel A. Ramos, Jorge Mario Gómez, Felipe Munoz

Risk Analysis for Overpressure From Pressure Vessel Burst 138
Seungho Jung, Elizabeth Lutostansky, Joan M. Schork

Comparison of the Peak Relief Rate Estimated by Dynamic Simulators Compared to Traditional Methods for Several Distillation Columns 147
John Wilkins, Dustin J. Smith

A Hybrid Model for Characterizing the Source of Hazardous Material Release 157
Shahryar Khajehnajafi, Anjana Meel

Guidelines for Developing Site-Specific Human Error Rates and Human IPLs for LOPA Using a Safety Climate Approach 175
Farheen Khan, Sunil D. Lakhiani, Delmar Morrison

Benchmarking – an Important Milestone in the Journey towards PSM Excellence 176
Prerna Jain, Ravi Ramaswamy, Cindie Pridy, Robert Cravens

Process Safety Management - A Practical View 190
German Luna-Mejias Sr.

Auditing the Process Hazard Analysis - Key Characteristics to Scrutinize in a Good-Quality HAZOP Study and Supporting Effective HAZOP Study Updates 202
Steven T. Maher, Michael A. Nour, Kristin D. Norton

Vapor Cloud Explosion Test of Air-Inflated Fabric Structure 203
Matthew Edel, Brad Horn

Revisiting the Conditions and Consequences of CO2 Tank Explosions 218
Yu Zhang, Joan M. Schork, Keith Ludwig

New Ignition Source "Exploding Wire" for the Determination of Explosion Characteristics of Combustible Dusts in the 20-L-Sphere 230
Marc Scheid, Volkmar Schröder, Christian Kusche, Alexander Klippel, Rico Tschirschwitz, Stefan Zirker

Operational Discipline and Beyond: An Interdisciplinary Approach to Improving Process Safety Management Implementation Through Behavior and Cultural Change 242
Charles A. Soczek

Risk Analysis for Pressure Relief Devices – Managing Plant Safety and Reliability 249
Jose R. Aguilar-Otero Sr.

Enterprise Risk Management--Adopting Risk-Based Decision Making in a Chemical Enterprise From the Board Room On Down Through All Functions and All Levels 251
Jerry L. Jones
The Chemical Safety Board and the Prevention of Combustible Dust Hazards 253
Manuel R. Gomez, Christina Morgan
Performing HAZOP of Startup/Shutdown Procedures 254
Fareed Ebrahim, German A. Luna Sr.
Chasing the Protection of Intermediate Bulk Containers, Year 10 263
John A. Leblanc
Process Safety Challenges in the Hazardous Waste Industry 264
Hope Luebeck
A Novel Methodology for Maintaining Mechanical Integrity of High Pressure Relief Systems 274
Sandipan Laskar
A Framework for Managing near-Miss Incidents 284
Ghanshyam Shah
HAZOP Analysis and Debottleneck for Laboratory Operation in the Semiconductor Industry 292
Tianxing Cai, Qiang Xu
Best Practice Management of Change & Pssr Process 307
Mike Bearrow
Unifying Consequence Models for Natural Gas Pipelines in Petrobras Transpetro 321
Ana Bahiense, Maria Caputo
Automation Strategies: Avoiding Accidents During the Process of Draining Water From LPG Spheres 329
Carlos André Vaz Jr., Maurício Bezerra Souza Jr., Diego Da Silva Lione, João Márcio Almeida Villaça, Leonardo Salim Damian Aguiar
Complex BMS Made Simple 344
Dave Sheppard, Nikki Bishop
Using ESD Valves As Safeguards, Myth or Reality? 345
German A. Luna-Mejias Sr.
Contours When Pipelines Meets Facilities 356
Karthick Deivasagayam
Ozone Flammability and Its Affect On the Lower Flammable Limit of Simple Hydrocarbons 357
Paul Osterberg, Thomas Johnson, Thomas Frawley
Evaluate the Plant-Wide Safety of Your Interlock Systems 358
Mohammad Naved Khan, James Brigman, Swapnil Pathak
Virtual Reality Applied to Training and Process Safety: A Semi-Industrial Boiler Case Study 372
Carlos André Vaz Jr., Maurício Bezerra Souza Jr., Thiago Souza De Oliveira
Investigations On the Influence of Different Inert Gases On the Laminar Flame Propagation in Lycopodium/Air Mixtures 382
Hannes Kern, Bruno Wimmer, Harald Raupenstrauch
Fire Protection and Life Safety Issues in Semiconductor and Related Facilities 391
Jonathan Eisenberg
Lower Flammability Limits of Mixtures of Hydrogen and Methane At Atmospheric and Sub-Atmospheric Pressures 392
Hai Le, Yi Liu, M Sam Mannan
Informatics Rich Integrated Chemical Safety Risk Assessment (ICSRA) Platform 401
Muthukumarasamy Karthikeyan, Renu Vyas, Hv Prasad
IT Enablement for Effective Implementation and Monitoring of Process Safety Management 405
Pawan Mittal, Ajay Kumar
Sulphur Dust Explosions an Overview 409
Karthick Deivasagayam
Simplified Uncertainty Analysis of LOPA Results 410
Raymond "randy" Freeman
The Impact of Ozone On the Lower Flammable Limit of Hydrogen in Vessels Containing Savannah River Site High Level Waste 437
Paul Osterberg, Carol Sherburne, Thomas Johnson, Thomas Frawley
Process Safety Considerations for New DOE Depleted Uranium Hexafluoride Conversion Plants 452
Peter M. Burban
Minimum Ignition Energy Study of Flowing Heat Transfer Fluid Aerosols: Experimental and Theorectical Approach 459
Szu-Ying Huang, Xinrui Li, M. Sam Mannan

Domino Effect Model Applied to the Hertfordshire Oil Storage Terminal (Buncefield Oil Depot) 460
Esteban Bernechea, Xavier Seguí, Josep Arnaldos
Proven Approach to Investigating Near Misses 470
William Bridges, Gregory Smith
Key Strategies for Implementing a Corporate Fatigue Risk Management System 484
David E. Cummings
Consequences Calculation Tool for Storage Tank Fire Scenarios 495
Juan C. Díaz, Jaime E. Cadena, Felipe Munoz
A Practical View of Risk Reduction Management of Hardware and Software Design for Safety Critical Applications 509
Luis Duran, Mike Krywonos
Metodologías Para Reducir Las Fallas De Causa Común .... N/A
Luis Manuel Duran

VOLUME 2

Establishing a 3 Credit Course in PSM for Undergraduate Engineers 522
Deborah L Grubbe
Investigations On the Self Heating Potential of Garage Waste As Basis for the Hazard Evaluation of Bulk Storage Facilities 523
Hannes Kern, Christoph Spijker, Harald Raupenstrauch
Estimate Vibration Risk for Relief and Process Piping 533
Georges Melhem
Performance Metrics Development Management and Reporting Reaching Sustainability by Design 560
Robert C. Smith
Alarm Management - the Human Factors Challenge 561
Robert C. Smith
Dust Flame Propagation in Industrial Scale Piping - Part 2: CFD Study of a Conveying Vessel-Pipeline System 562
Jim Vingerhoets, Thomas Farrell, Jef Snoeys
The Use of CFD to Evaluate the Interactions Between Multiple-Leak Sources and to Assess Effectiveness of Integral Modeling Techniques in a Multiple-Leak Scenario 584
Joshua Richardson, Ha Nguyen, M. Sam Mannan
Understanding, Managing, and Communicating High Consequence Risk 586
Charles A. Soczek
Frequencies of Accidents As Functions of Design Variables 587
Esteban Bernechea, Josep Arnaldos, Cristian Massaguer
Routes Optimization for Hazardous Materials Transportation in Bogotá, Colombia 599
Jorge Victoria, Nubia Velasco, Eliecer Gutierrez, Felipe Munoz
Mejores Prácticas En Los Procesos De Gestión Del Cambio Para Seguridad De Procesos y La Evaluación De Seguridad Previa Al Inicio (PSSR) 600
J. Luis Aguirre
Lesha – Multi-Layer Progressive Stress & Impact Assessment On Health & Behavior 601
Salvador Ávila Filho
Building Design for Resilience Beyond Design Basis Blast Loads 603
Michael A. Polcyn, Raymond H. Bennett
Reliability Analysis of Man-Machine Response to an Emergency Situation Based On a Detailed-Action Plan 612
Najmeh Vaez, Farshad Nourai, Yaneira Saud, Suresh Yelisetty
Hazards of Gasoline Contamination – Case Study of a Tanker Truck Explosion Resulting From Switch Loading 613
Scott G. Davis, Kees Van Wingerden
Is Your MOC Process Doing All That It Should Be? 614
William Fink, Carolina Herrera
Beyond DSC, Improved Thermal Hazards Screening Methodology 615
Peter J. Ralbovsky Iv
Using Explicit Finite Element Analysis to Simulate the Effects of External Chemical Explosions On Single and Double-Walled Storage Tanks 617
Phillip E. Prueter, David J. Dewees
The Hidden Risks of Deposits, by-Products, and Residues On Exhaust and Waste Streams 634
Carlos A. Barrera, Abid Kemal

Explosion Characteristics of Carbon Nano-Fibers ... 635
Jiaqi Zhang, Yi Liu, M Sam Mannan

High-Fidelity Combined Dynamic Modeling of Depressurizing Vessels and Flare Networks: Accurately Assess the Effects of Process Changes On Mechanical Integrity and System Capacity ... 636
James Marriott

An Engineering Method to Mitigate the Impact of Regulatory Focus On Relief System Installations by Prioritizing Risk ... 637
Dustin J. Smith, John Burgess

Making Relief Load Estimates Match Reality ... 645
Nicholas N. Cristea, Dustin J. Smith

Explosion Accident Simulation for Petrochemical Plant ... 656
Xiangdi Zhao, Jiwu Yuan, Chunming Jiang, Zheng Wang

Using Reliability Centered Maintenance (RCM) to Reduce Fire Risk in High Temperature Oil Handling ... 664
Marc W. Yarlott, Ryan Johnson, Steven Waters

Characterization of the Emissive Power of Hybrid Methane-Air-Dust Flames ... 679
Brian Elias, Ali S. Rangwala

Are You in Control of Process Safety? ... 680
Ellis Graeme

Process Hazard Analysis, Layers of Protection Analysis and the Human Factor ... 688
Paul Baybutt

Benchmarking Industry Practices for the Use of Alarms As Safeguards and Layers of Protection ... 722
Todd Stauffer, Peter Clarke

The Impact of Human Factors On LOPA (And Other Risk Assessment Methods) ... 752
William Bridges

Take the Guess Work Out of Testing Your Safety Instrumented System ... 792
Charles Fialkowski

Tips for the Creation and Application of Effective Operating Procedures ... 802
Colin Scholtz, Steven T. Maher

Management Discipline: Defining a Process Safety Strategy ... 815
Jerry Forest

Using Assessments to Improve Process Safety Culture ... 824
Russell A. Ogle, Delmar Morrison, Sean J. Dee

Using New Technology to Ensure Corporate Standards Compliance and Improved Frontline Operational Risk Management ... 833
Mike Neill, Scott Lehmann

Conduct of Operations Best Practice Networks ... 865
Marco Antonio Vela, Jerry J. Forest

Management of Change Indicators ... 875
Clarissa Sossmeier Da Silva, Americo Diniz

Organizational Change Management ... 876
John Wincek, Molly Myers, Henry Ozog

Increasing the Agility of Process Safety Management Systems ... 887
Brian D. Rains

Technical Risk – It's Not a One-Size-Fits-All Proposition ... 900
Kumar Israni, Joshtin Cooper

Preventing Process Safety Problems of Tomorrow, Today! ... 913
Kevin Watson, Robert Dayton

Critical P & ID Reviews Using HAZOP/LOPA Methodology – Overkill or Underrated? ... 921
John T. Perez, Andrew Madewell

Using near Misses to Improve Risk Management Decisions ... 937
Sean J. Dee, Brenton L. Cox, Russell A. Ogle

Keeping the Memory Alive, Preventing Memory Loss That Contributes to Process Safety Events ... 949
Barry Throness

Blacks Swans, White Swans and Fifty Shades of Grey: Remembering the Lessons Learned from Catastrophic Process Safety Incidents ... 980
John F. Murphy, Jim Conner

Lessons Learned From Fire in Air Supply Ducts ... 994
Alvin Waller

Hydrogen Sulfide Exposure From Molten Sulfur- A Forgotten Hazard? ... 1002
Mark J. Viz, Russell A. Ogle, Sean J. Dee, Brent L. Cox

Calcium Stearate Deflagrations – A Series of Unfortunate Events 1013
Konanur Manjunath, Albert I. Ness
Effectiveness of Improvised Gas Absorption Techniques for Emergency Responders At Releases of Toxic Gases 1022
Hannes Kern, Matthias Taferner, Harald Raupenstrauch
Case Studies in Process Safety: Lessons Learned From Software-Related Accidents 1029
Terry Hardy
CSB Investigation of Chevron Richmond Refinery Fire 1044
Mark Wingard, Lauren Grim, Steve Cutchen, Kelly Wilson, Roger Evans, Amanda Johnson, Donald Holmstrom, Dan Tillema
Author Index

(44bj) Establishing a 3 Credit Course in PSM for Undergraduate Engineers

Deborah L Grubbe, Executive Mgmt, Operations and Safety Solutions, LLC, Chadds Ford, PA

Considering academic direction and industrial needs, the Purdue Chemical Engineering faculty and an industrial alumni team jointly sponsored, developed, executed and audited a 3 credit course in Process Safety Management for undergraduates in Engineering. This "team taught" course is now a required in the Purdue ChE curriculum, and has received enthusiastically high marks from the students. Originally conceived in 2010, the course went from concept to required in less than 24 months.

One of the alumni team members, with over 30 years in the process industries, will describe each step of the process: creating the support, drafting the syllabus, designing the course, executing the lectures, planning the ancillary support activities, and auditing the results of the 15 week, 45 lecture series. The final exam was the simulation of an emergency situation, which was modeled after a real life process safety incident. Actual student feedback will be shared. An eBook text has been developed for this course, and the speaker will conduct a short demonstration of this product as part of the overall presentation.

Investigations on the self heating potential of garage waste as basis for the hazard evaluation of bulk storage facilities

Hannes Kern
Montanuniversität Leoben
Franz-Josef Strasse 18, 8700 Leoben
hannes.kern@unileoben.ac.at

Christoph Spijker, Harald Raupenstrauch
Montanuniversität Leoben
Franz-Josef Strasse 18, 8700 Leoben
christoph.spijker@unileoben.ac.at, harald.raupenstrauch@unileoben.ac.at

Prepared for Presentation at
American Institute of Chemical Engineers
2013 Spring Meeting
9th Global Congress on Process Safety
San Antonio, Texas
April 28 – May 1, 2013

UNPUBLISHED

Investigations on the self heating potential of garage waste as basis for the hazard evaluation of bulk storage facilities

Hannes Kern
Montanuniversität Leoben
Franz-Josef Strasse 18, 8700 Leoben
hannes.kern@unileoben.ac.at

Christoph Spijker, Harald Raupenstrauch
Montanuniversität Leoben
Franz-Josef Strasse 18, 8700 Leoben
christoph.spijker@unileoben.ac.at, harald.raupenstrauch@unileoben.ac.at

Keywords: self heating, garage waste, differential reactor

Abstract

The self heating of bulk material is a major point to be considered in hazard analysis especially for waste management and waste treatment companies. The self heating potential for bulk material mainly depends on the composition of the material itself and on systemic influences like the geometry of the storage area, weather conditions or the storage temperature. Besides microbiological activity, chemical oxidation at low temperatures is the main reason for the self heating and auto ignition of bulk material. Data from the measurement of the chemical and biological oxidation kinetics can be used as basis for the simulation of storage areas by computational fluid dynamics. The measurement of the low temperature oxidation kinetics is carried out in a differential reactor, which is characterized by a small amount of sample material, isothermal conditions and very low turnover rates. Due to the fact that chemical oxidations reactions at low temperatures show very low turnover rates themselves, the experimental setup is laid out as a cycle process. The oxidation kinetics are measured by the analysis of the carbon monoxide (CO) and carbon dioxide (CO2) production rate at different temperatures. Usually a temperature range from 40°C up to 120°C is investigated. Waste materials from workshops contain significant amounts of grease and oil as well as residuals of metal dusts or chips dispersed on the surface of paper or cloth material used for cleaning purposes. The large surface combined with the fine dispersion of grease and oil accelerates the chemical oxidation significantly. To determine the influence of the oil and grease content on the oxidation kinetics, measurements of paper and cloth samples with different contents of oil were carried out in a differential cycle reactor. For the investigations biological cutting oil and oil with a petrochemical origin was investigated. Furthermore the influence of metal dusts on the oxidation characteristics of the different materials was investigated. The results of the investigations were compared with the oxidation kinetics of garage waste material with a mean diameter of 0- 40mm.

1. Introduction

The self heating of bulk material is a major point to be considered in hazard analysis especially for waste management and waste treatment companies. Besides safety issues the auto ignition of bulk storage areas is also a serious economical threat and occurs quite frequently. In industrial branches where material with an inhomogeneous composition (like waste treatment facilities) are processed the problem of self heating is even more present. Especially at residual derived fuel storage areas fires are also a serious ecological threat, due to composition of the materials themselves. The measurement of the low temperature oxidation kinetics combined with CFD [1] modeling used for the prediction of the self heating potential of bulk storage areas is an important element of hazard analysis and loss prevention. In the present study the self heating potential of garage waste was investigated. Garage waste shows a high tendency of self heating and therefore auto ignition is a major hazard. In the course of this study different influencing parameters on the self heating potential of garage waste were investigated, to be able to evaluate the risk of auto ignition. Self heating in general is a complex process that mainly depends on the parameters presented below [2]:

Material specific parameters

- composition
- chemical kinetics
- particle size distribution
- bed porosity
- heat conductivity

Apparatus specific boundary conditions

- geometry of the storage area
- size surroundings

Operating conditions

- temperature
- gas velocity

The chemical oxidation at low temperatures is besides microbiological activity the main reason for the self heating and auto ignition of bulk material. Basicly two main oxidation reactions occur at low temperatures:

$$a_{Fuel(s)} + b_{O_2(g)} \rightarrow c_{CO(g)} + d_{\mathrm{Rest}(s)} \tag{1}$$

$$e_{Fuel(s)} + f_{O_2(g)} \rightarrow g_{CO_2(g)} + h_{\mathrm{Rest}(s)} \tag{2}$$

The reaction of the solid fuel with oxygen gives carbon monoxide (1) and carbon dioxide (2). In a simple kinetic approach it is assumed that the solid fuel can be seen as bulk with a constant concentration compared to the oxygen concentration. After some simplifications the relations (3) and (4) can be derived from the Arrhenius relation for chemical reaction kinetics:

$$r_{CO} = k_0 \cdot \exp\left(\frac{-E_A}{R \cdot T}\right) \cdot y_{O_2}^n \quad (3)$$

$$r_{CO} = k_0 \cdot \exp\left(\frac{-E_A}{R \cdot T}\right) \cdot y_{O_2}^n \quad (4)$$

For further research it is assumed that the postulated reactions undergo only chemical limitation, because chemical limitation exceeds transport limitation by several orders of magnitude. The determination of the kinetic parameters is carried out by the observation of the CO and CO_2 production rates.

2. Experimental investigations

2.1 Sample Material

Waste materials from workshops contain significant amounts of grease and oil as well as residuals of metal dusts or chips dispersed on the surface of paper or cloth material. The large surface combined with the fine dispersion of grease and oil accelerates the chemical oxidation significantly [3]. To determine the influence of the oil and grease content on the oxidation kinetics, paper and cotton cloth samples with different contents of oil were chosen as model substances for further investigations. The calorific values as well as the chemical composition of the investigated materials can be seen in Table1. The oil content of the garage waste sample material was determined by Soxhlet extraction with cyclohexane as solvent. The sample had an oil content of around 15% and a particle size of up to 30mm.

Table 1: Composition of the sample material

	Carbon [%]	Hydrogen [%]	Nitrogen [%]	Calorific Value [kJ/kg]
Paper	40.41	6.70	0.61	16277
Cotton cloth	41.81	6.80	0.35	16484
Oil (arecal)	78.7	12.1	0.01	35911
Oil (WD 40)	85.7	13.3	0.04	42258

The product "arecal" is a cutting oil based on rape oil while "WD40" is a cutting oil derived from fossil hydrocarbons. Pre tests in the differential reactor described in 2.2 showed that arecal is far more reactive than WD40. That may be explained by the high content of unsaturated carbon bonds in rape oil, which are more reactive in terms of oxidation processes. Further investigations on the oxidation behavior have been carried out only with the more reactive arecal oil.

2.2 Method

The measurement of the low temperature oxidation kinetics is performed in a differential reactor (Figure 1), which is characterized by a small amount of sample material, nearly isothermal conditions and very low turnover rates. Due to the fact that chemical oxidations reactions at low temperatures show very low turnover rates themselves, the experimental setup is laid out as a cycle process. The oxidation kinetics is measured by the analysis of the carbon monoxide (CO) and carbon dioxide (CO_2) production rate at different temperatures. Usually a temperature range from 40°C up to 120°C is investigated.

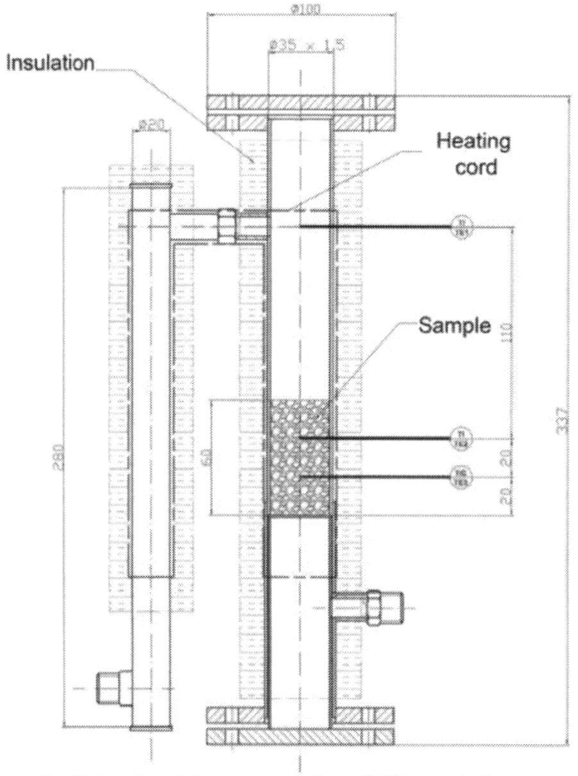

Figure 1: Principal layout of the differential reactor [2]

Figure 2 shows a typical of temperature development and gas production (e.g. carbon dioxide). At the beginning of the measurements the sample material inside the reactor is heated up above the highest planned measuring temperature to desorb carbon dioxide and carbon monoxide that may be adsorbed at the surface of the sample material. After the desorption process the sample material is cooled to the first measuring temperature and constantly purged with synthetic air (80%N_2, 20%O_2). When the sample reaches a constant temperature value the measurement process is started and the measuring cycle is closed. The areas of constant CO_2 increase in the concentration graph of Figure 2 represent these measurement cycles at different temperatures. At higher temperatures an increase of carbon dioxide production can be observed, which is directly linked to oxidation kinetics. The calculation of the kinetic parameters is carried out by the Arrhenius relation after the measuring procedure. A measuring cycle normally takes about 30min for one temperature.

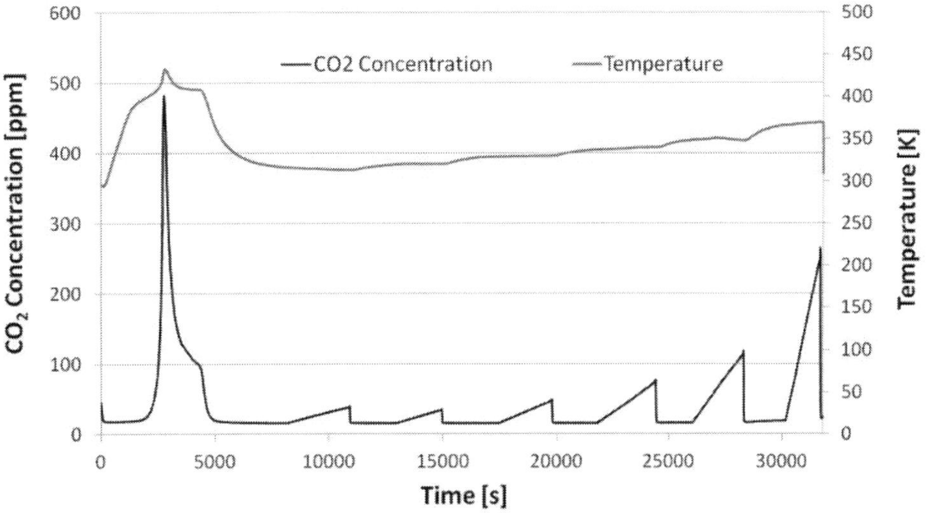

Figure 2: Principal layout of the differential reactor

2.3. *Paper*

Figure 3 below shows the CO production rates of paper with different oil contents. The "0% Oil content" line represents paper material not contaminated with arecal oil. The CO production rate increases with the oil content of the sample. The progress of the CO production rate with an oil content of 6.9% is widely comparable to the results of garage waste (~15% oil content).

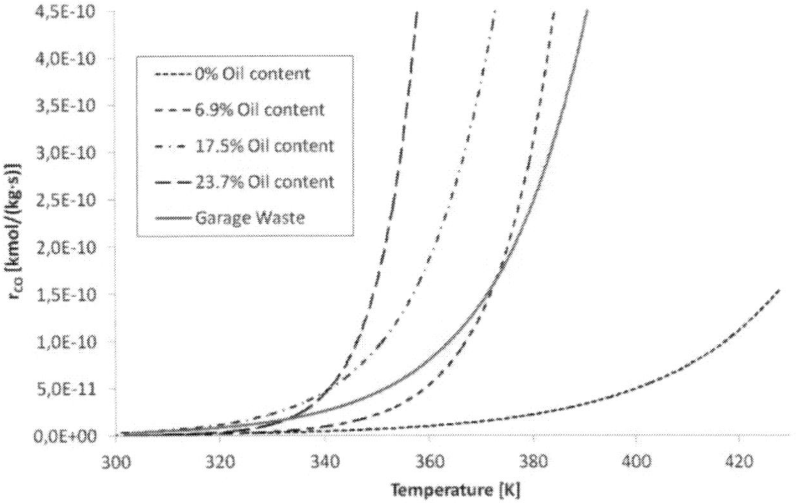

Figure 3: CO production rates of paper with different oil contents

For CO_2 production, as shown in Figure 4, the results were a bit different. The CO_2 production rate of garage waste was higher than all other rates measured. Interestingly the CO_2 production rate for the paper with an oil content of 17.5% was higher than for 23.7%. That may be explained by kinetic effects and is maybe linked to the significant amounts of CO that were produced by this paper sample.

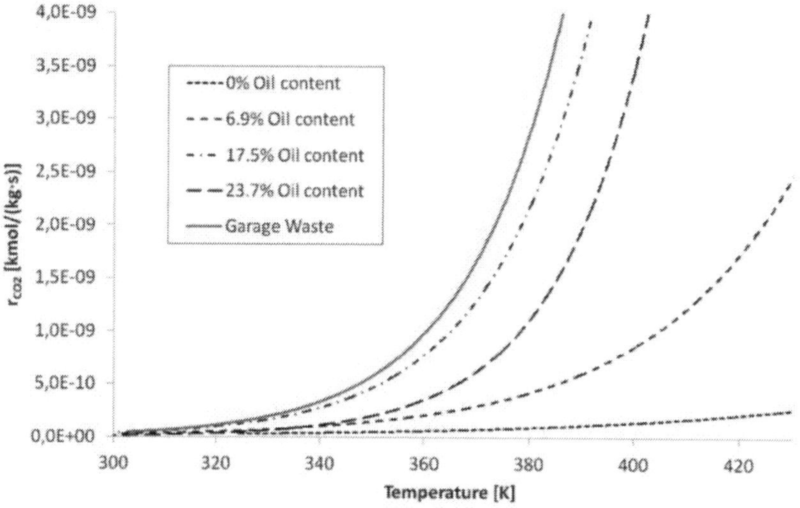

Figure 4: CO_2 production rates of paper with different oil contents

The kinetic data resulting from the calculations based on the Arrhenius relation are presented in Table 2.

Table 2: Kinetic data of the paper sample material

	CO		CO_2	
	E_a [kJ/mol]	K_o Kmol/kg*s	E_a [kJ/mol]	K_o [Kmol/kg*s]
Paper 0.0% Oil cont.	4.89E+04	3.25E-09	2.52E+04	9.74E-12
Paper 6.9% Oil cont.	8.70E+04	7.45E-03	3.45E+04	6.63E-10
Paper 17.5% Oil cont.	6.69E+04	3.17E-05	4.94E+04	3.84E-07
Paper 23.7% Oil cont.	1.02E+05	1.91E-00	5.48E+04	1.04E-06
Garage waste	5.59E+04	3.35E-07	5.22E+04	1.14E-06

2.4 Cotton cloth

During the preparation of the cotton sample material oil contents that were similar to the paper samples have been tried to be reached. Although there are smaller deviations in the oil content, the results are comparable.

Figure 5 shows the results for the investigation of the cotton cloth material. The results do not show good accordance with the garage waste material. But again it can be observed, that the sample materials with a high oil contents also show higher reactivity than the sample with lower contents.

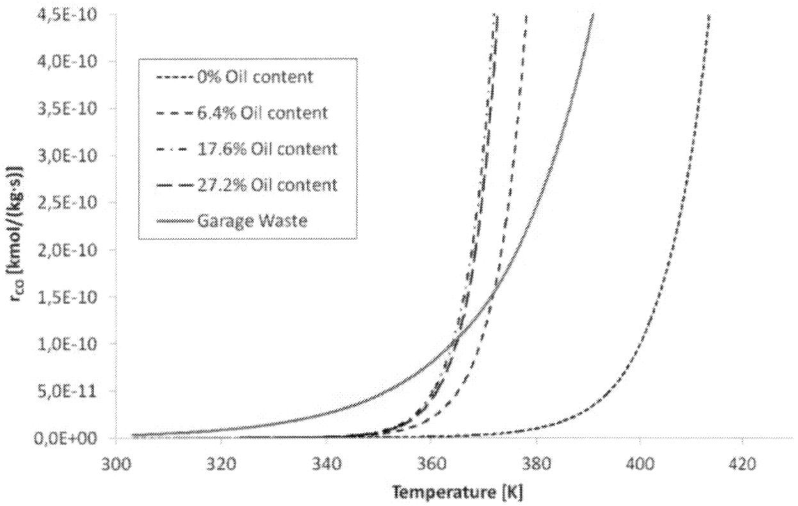

Figure 5: CO production rates of cotton cloth with different oil contents

The production rates from Figure 6 do not show a big deviation regarding the oil content, but the samples contaminated with oil clearly show a higher reactivity than the cotton material itself. In general all CO_2 production rates are comparable with the original garage waste sample material, which again indicates a strong link between oil content and reactivity.

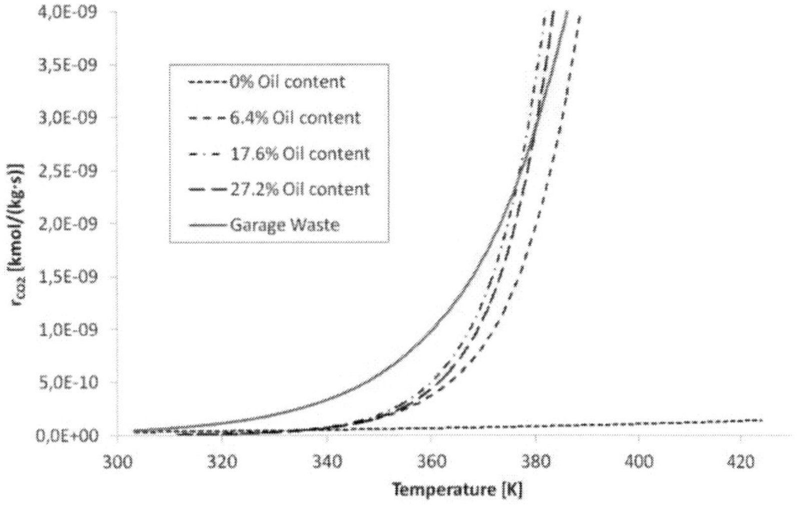

Figure 6: CO_2 production rates of cotton cloth with different oil contents

The kinetic data of the cotton cloth samples is presented in Table 3.

Table 3: Kinetic data of the cotton cloth sample material

	CO		CO_2	
	E_a [kJ/mol]	K_o Kmol/kg*s	E_a [kJ/mol]	K_o [Kmol/kg*s]
Cotton 0.0% Oil cont.	3.68E+04	1.66E-10	1.34E+04	1.99E-13
Cotton 6.4% Oil cont.	1.90E+05	2.90E+12	9.01E+04	1.52E-01
Cotton 17.6% Oil cont.	1.27E+05	3.14E+04	9.96E+04	4.82E+00
Cotton 27.2% Oil cont.	1.02E+05	1.80E+00	5.48E+04	9.78E-07
Garage waste	5.59E+04	3.35E-07	5.22E+04	1.14E-06

2.5 *Influence of iron dusts*

Due to normal workshop operations garage wastes always contain significant amounts of metal residues. To investigate the influence of metal residues on the oxidation kinetics small amounts of iron dust were added to a cotton cloth sample with an oil content of around 11%.

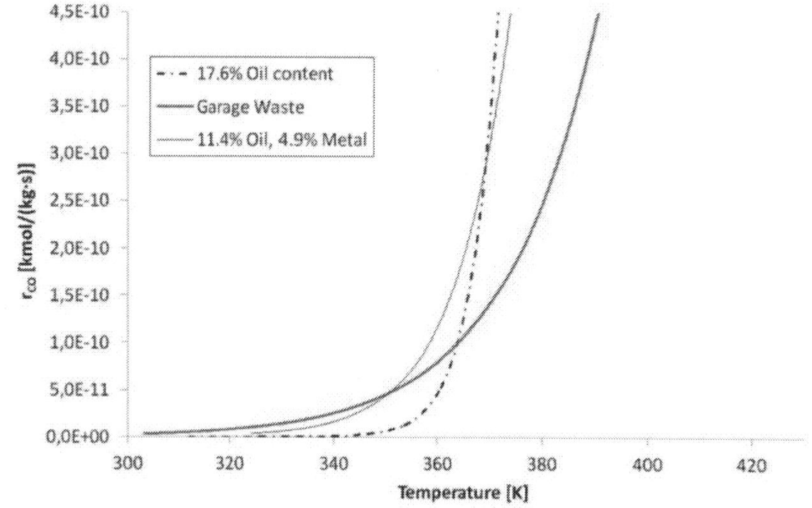

Figure 7: CO production rates of cotton cloth contaminated with oil and iron dust

Figure 7 shows the production rates for the cotton/oil/iron dust sample and also the results for pure cotton cloth with 17.6% oil content. The results show that the iron dust has an influence on the reactivity of the material. Both materials show comparable CO production rates even though the sample containing iron dust contains less oil. The same effect can be observed for the CO_2 production rate in Figure 8. The CO_2 production rate in general shows good accordance to the results of the garage waste sample which indicates the representativeness of the chosen sample materials.

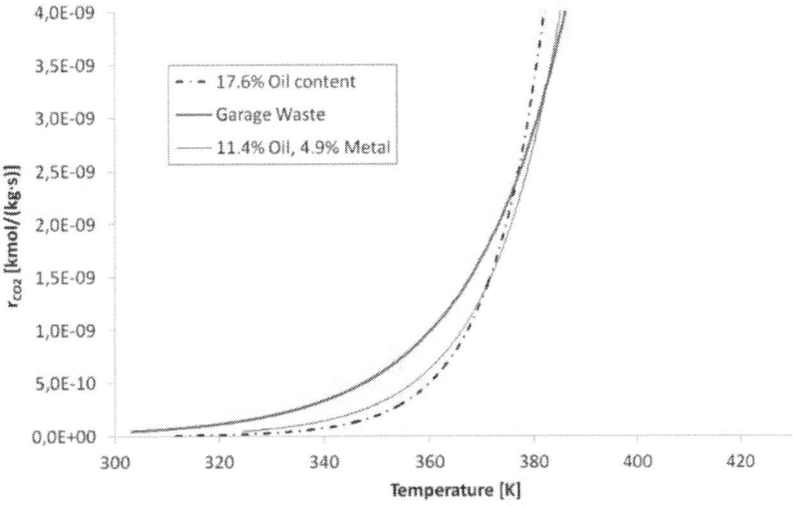

Figure 8: CO_2 production rates of cotton cloth contaminated with oil and iron dust

3. Conclusions

The results of the investigations of garage waste and different reference materials show a clear connection between the oil content of garage waste and its reactivity. Garage waste in general should be considered as highly reactive. Field investigations show, that a significant temperature rise can be observed within days from the beginning of the storage period. High contents of oil and metal residues lead to a significant fire risk. To determine the permissible storage conditions for garage waste CFD models could be applied. The results from measured kinetics can be used as the basis for CFD modeling in combination with the layout of the storage area and the further surrounding conditions [4].

4. References

[1] Wallner, S. (2004). Ein Modell zur Untersuchung des Selbsterwärmungsverhaltens von Schüttgütern. Dissertation, Technische Universität Graz

[2] Walkner, R. (2005) Untersuchung des Selbsterwärmungsverhaltens eines Lagers für aufbereiteten Substitut Brennstoff (ASB). Diplomarbeit, Technische Universität Graz

[3] Wallner, S., Walkner, R., Raupenstrauch, H. (2004). Untersuchung des Selbsterwärmungsverhaltens gelagerter Kunststoffmaterialien mittels numerischer Simulation. Endbericht für das FFF- Projekt „Entwicklung neuer Brennstoffqualitäten aus Abfall". Technische Universität Graz

[4] Raupenstrauch, H. (1997) Gasdurchströmte, chemisch reagierende Schüttschichten. Habilitationsschrift, Technische Universität Graz

9TH GLOBAL CONGRESS ON PROCESS SAFETY

Estimate Vibration Risk for Relief and Process Piping

Georges A. Melhem, Ph.D.
ioMosaic Corporation
93 Stiles Road | Salem, NH 03079
melhem@iomosaic.com

Prepared for Presentation at
American Institute of Chemical Engineers
2013 Spring Meeting
9th Global Congress on Process Safety
San Antonio, Texas
April 28 – May 1, 2013

UNPUBLISHED

Estimate Vibration Risk for Relief and Process Piping

Georges A. Melhem, Ph.D.
ioMosaic Corporation
93 Stiles Road | Salem, NH 03079
melhem@iomosaic.com

Keywords: Acoustic induced vibration (AIV), flow induced vibration (FIV), vibration risk, resonance, singing relief valve.

Abstract

Current API, AIChE/CCPS, and AIChE/DIERS pressure relief and flare systems guidelines and standards do not formally address vibration risk. They do not offer specific guidance on velocity limitations other than backpressure calculations and they do not offer guidance on acoustic induced or flow induced piping vibration fatigue failure. This paper provides a summary of experience based methods for the estimation of vibration risk in relief and process piping. In addition, this paper extends the applicability of the experience based methods to two-phase flow.

1. Introduction

Fatigue failure of relief and/or process piping caused by vibration can develop due to the conversion of flow mechanical energy to noise. Factors that have led to an increasing incidence of noise vibration related fatigue failures in piping systems include but are not limited to (a) increasing flow rates as a result of debottlenecking which contributes to higher flow velocities with a correspondingly greater level of turbulent energy, (b) frequent use of thin-walled piping which results in higher stress concentrations, particularly at small bore and branch connections, (c) design of process piping systems on the basis of a static analysis with little attention paid to vibration induced fatigue, (e) and lack of emphasis of the issue of vibration in piping design codes. Piping vibration is often considered on an ad-hoc or reactive basis. According to the UK Health and Safety Executive (HSE), 21 % of all piping failures offshore are caused by fatigue/vibration. Typical systems at risk include large compressor recycle systems and high capacity pressure relief depressuring systems. For relief and flare piping, flow induced turbulence and high frequency acoustic excitations are key concerns.

2. Flow Induced Vibration (FIV)

Fluid flow in pipes generates turbulent energy (pressure fluctuations). Dominant sources of turbulence are associated with flow discontinuities in the piping systems (e.g., partially closed valves, short radius, mitered bends, tees or expanders). The level of turbulence intensity is a function of pipe size, fluid density, viscosity, velocity, and structural support. High noise levels are generated by high velocity fluid impingement on the pipe wall, turbulent mixing, and if the flow is choked, shock waves discontinuities downstream of flow restriction. This can lead to high frequency excitation / vibration risk.

3. Acoustic Induced Vibration (AIV)

High frequency acoustic energy is often generated by a pressure reducing device such as a relief valve, control valve, or orifice plate. Acoustic induced piping failure is of a particular concern for safety related systems (e.g. relief and blowdown/depressuring). The severity of high frequency acoustic excitation is primarily a function of the pressure upstream and downstream of the pressure reducing device, pipe diameter, and the fluid volumetric flow. Acoustically induced piping failures are known to occur at non-asymmetric discontinuities in the downstream piping such as small bore and branch connections and welded supports.

4. Additional Causes of Vibration

Additional causes of piping vibration include mechanical excitation, pulsation, vortex shedding, surge or momentum changes associated with valves, cavitation, and flashing. Mechanical excitations are often associated with pipes connected with reciprocating compressors, pumps, or rotating machinery. Such connection machines cause vibration of the pipe and its support structure. Thermowells are intrusive fittings and are subject to static and dynamic fluid forces. Vortex shedding is the dominant concern as it is capable of forcing the thermowell into flow-induced resonance and fatigue failure. This is particularly important at high fluid velocities.

5. Relief and Depressuring Systems

Depressuring systems are often subjected to acoustic energy (rapidly fluctuating pressure forces) generated by flow turbulence which is accentuated by flow restricting devices within the flow path. The magnitude of pressure fluctuations depends on the mass flow rate, speed of sound, and density. For choked flow, intense noise due to large shock discontinuities and pressure fluctuations is generated. The generated noise is non-periodic due to the randomness of the pressure fluctuations. Choked flow typically leads to a wide frequency spectrum with peak values than can exceed 1000 Hz. Vibrations associated with small fittings and branch connections are of special concern because they introduce discontinuities and stress concentration points.

In many situations resonance can onset which can lead to magnification of static piping loads up to a factor of 50 times. The presence of discontinuities such as tees and welded pipe supports can further increase these loads.

6. Noise Generation

A pressure reducing device or relief device controls flow by converting internal energy into kinetic energy. Some energy is converted to heat through friction (viscous forces) by intense turbulence and shock formation. Some of the energy is also transferred to the pipe wall as vibration, and a portion of this is radiated as noise. The primary noise generating mechanism is the confined jet of fluid formed between the upstream and downstream locations. Flow noise can be modeled as noise of a confined jet. As a result, the noise-generation mechanisms are turbulent mixing, turbulence boundary interaction, shock, shock/turbulence interaction and flow separation.

Since the noise is generated downstream of a flow restriction, most of the acoustic energy is radiated to the downstream piping, which becomes the transmitting medium. As the noise travels downstream along the inside of the pipe in the fluid it radiates through the pipe wall along its entire length.

7. How to Calculate Sound Power Level

We can calculate the sound power level in a fundamental way by assigning an efficiency factor to provide the fraction of flow mechanical energy that is converted to noise:

$$W = \eta \frac{1}{2} \dot{m} u^2 \tag{1}$$

$$L_W = 10 \log_{10} \left[\frac{W}{10^{-12}} \right] = 10 \log_{10} W + 120 = 10 \log_{10} \left[\eta \frac{1}{2} \dot{m} u^2 \right] + 120 \tag{2}$$

where W is the flow mechanical power or energy, L_W is the sound power level in dB, and η is defined as the acoustical efficiency factor. If the flow is choked, then W becomes:

$$W = \eta \frac{1}{2} \dot{m} u_{sonic}^2 \tag{3}$$

If a value of η can be estimated for single and multi-phase flow, then the sound power level can be easily calculated not only for pressure reducing devices but also for pipe flow. Attenuation due to friction and temperature changes can be calculated from pipe flow equations in a more detailed manner. Computer codes such as SuperChems can then calculate the sound power level at every axial location for flow piping for single and multi-phase flow. For incompressible flow, the value of flow velocity for an ideal nozzle can be calculated from the mechanical energy balance:

$$u = \sqrt{\frac{2}{\rho_l} \Delta P} \tag{4}$$

Substituting the above equation for u in Equation 1 yields the following Equation for W for liquid flow through an ideal nozzle:

$$W = \eta \dot{m} \frac{\Delta P}{\rho_l} \tag{5}$$

For liquid flow, a typical value of η is approximately 10^{-8}.

All gas flow acoustic efficiencies have been reported to approach 1 % of the total flow mechanical energy for rockets[i]. This is illustrated in Figure 1. When the acoustic efficiency is plotted against the flow mechanical power Figure 1 is obtained. The measured curve indicates that the acoustic efficiency falls off as the flow mechanical energy gets larger while the calculated curve[ii] indicates increasing acoustic efficiencies.

Figure 1. Acoustic efficiency trends

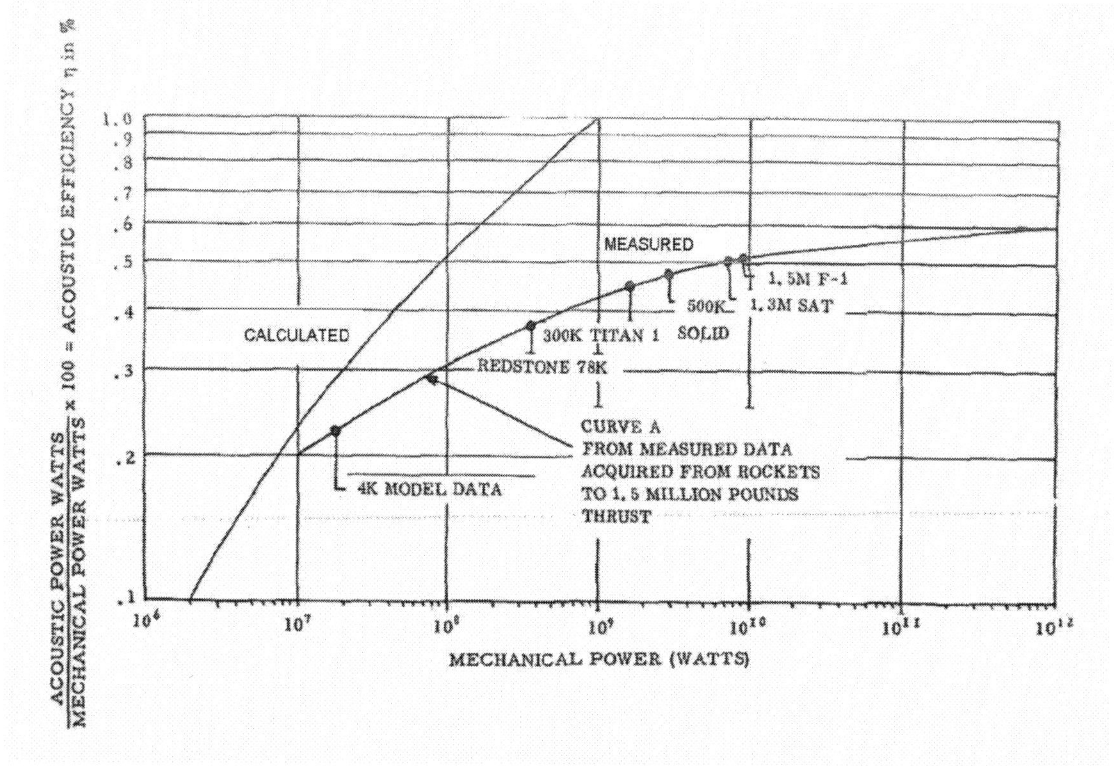

Two-phase flow sound power level can also be calculated using Equation 1. Recent work by Singh et al.[iii, iv] demonstrated that more attenuation of noise is exhibited by two-phase flow. Therefore, using the gas acoustic efficiency values will over predict sound power and noise levels for two-phase flow.

8. Existing Methods and Guidance

Current API, AIChE/CCPS, and AIChE/DIERS pressure relief and flare systems guidelines and standards do not formally address vibration risk. They do not offer specific guidance on velocity limitations other than backpressure calculations and they do not offer guidance on acoustic induced or turbulence induced piping vibration fatigue failure. The Marine Technology Directorate Limited (MTD) has published in 1999 "Guidelines for the Avoidance of Vibration Induced Fatigue in Process Pipework"[v]. A second edition of these guidelines were published in 2008 by the Energy Institute[vi]. The methods outlined in these guides have been incorporated in the SuperChems Expert ioVIPER modules. The MTD/Energy Institute guidelines provide qualitative and quantitative methods for the assessment of piping vibration failure risk and depending on the calculated risk level they provide generic guidance for the mitigation of vibration risk.

Many operating companies have established their own internal guidance for evaluating and minimizing piping vibration risk. Although these criteria vary from company to company, they all in general include a limit of flow velocity in some form.

A common criteria used in relief systems is to limit the value of flow Mach number to a value ranging from 0.3 to 0.9:

$$0.3 \leq M = \frac{u}{u_{sonic}} \leq 0.9 \tag{6}$$

Another common criteria often encountered is to limit the dynamic pressure component or kinetic energy of a flow stream to a value of 100,000 Pascals for gas flow:

$$\rho u^2 \leq 1 \times 10^5 \ \text{kg/m/s}^2 \tag{7}$$

and 50,000 for two-phase flow:

$$\rho u^2 \leq \frac{1}{2} \times 10^5 \ \text{kg/m/s}^2 \tag{8}$$

The 2006 fifth edition of the NORSOK Process Design Standard P-001 limits the flow velocities for all flare lines to $\rho u^2 \leq 200{,}000 \ \text{kg/m/s}^2$ for single and multiphase flow. The piping for flare headers and sub-headers are designed for a maximum Mach number of 0.6 and lines downstream of pressure relief valves to the first sub-header are designed for a maximum Mach number of 0.7. For process lines the maximum design velocity for gas pipes is limited to 60 m/s or $u \leq 175 \frac{1}{\rho^{0.43}}$ whichever is less. The maximum velocity for two-phase lines is limited to $u \leq 183 \frac{1}{\rho^{0.5}}$ or:

$$\rho u^2 \leq 183^2 \leq 33{,}489 \ \text{kg/m/s}^2 \tag{9}$$

9. The Method of Carucci and Mueller

Carucci and Mueller have published guidance for the estimation of sound power levels for control valves and pressure reducing devices[vii, viii].

$$L_W = 10 \times \log_{10}\left[\left(\frac{P_1 - P_2}{P_1}\right)^{3.6} \times \dot{m}^2 \times \left(\frac{T_1}{M_w}\right)^{1.2}\right] + 126.1 \tag{10}$$

$$= 10 \times \log_{10}\left[4 \times \left(\frac{P_1 - P_2}{P_1}\right)^{3.6} \times \dot{m}^2 \times \left(\frac{T_1}{M_w}\right)^{1.2}\right] + 120 \quad (11)$$

$$= 36\log_{10}\left(\frac{P_1 - P_2}{P_1}\right) + 20\log_{10}(\dot{m}) + 12\log_{10}\left(\frac{T_1}{M_w}\right) + 126.1 \quad (12)$$

where P_1 is the upstream pressure or source pressure, P_2 is the downstream pressure, T_1 is the source temperature, $\dot{m}$ is the mass flow rate, and M_w is the gas molecular weight.

Attenuation of noise due to friction and heat conduction losses is estimated from the following equation:

$$L_{W,At} = 0.06\frac{L}{D_i} \quad (13)$$

where L is the pipe length and D_i is the internal pipe diameter. For an L/D of 50 the attenuation loss is 3 dB. Abrupt changes in flow area (expansion) in the piping can also lead to attenuation. The decrease in sound power level can be estimated from the following equation:

$$L_{W,Ex} = 2\left(\frac{D_2}{D_1} - 1\right) \quad (14)$$

where $D_2 > D_1$. A 3 dB reduction is typically applied to the flow leaving a tee in each direction or entering into a header or a large drum or vessel. The Carucci and Mueller Equation 12 is used by the Energy Institute Guidelines for the assessment of failure likelihood for high frequency acoustic excitation [vi]. Note that the Carucci and Mueller equation cannot be easily applied to complex piping systems, multi-phase systems, and relief piping with multiple chokes. Some companies also add 6 dB to the sound power level estimate when sonic flow exists at a branch connection to account for amplified dynamic strain response.

9.1 Analysis of the Carucci and Mueller Data Set

Careful analysis of the Carucci and Mueller data set (see Table 1) indicates that the source pressures were high enough to produce choked (sonic) flow through a flow limiting orifice or valve upstream of the failure point further downstream, often at a branch or line connection. This important point was missed by Eisinger who used the downstream (non-choked) flow velocity to establish his Mach number based failure criterion. We were also able to reproduce Eisinger's estimates of the original Carucci and Mueller data set based on his published paper.

Carucci and Mueller provided enough data in their original paper to allow the calculation of the upstream flow limiting flow area and choke point conditions (this is possible because they reported the actual flow rates) as well as the conditions downstream of the flow limiting device in the discharge piping. Note that the sound power level estimate upstream of the choke points should be based on the pressure difference (or pressure ratio) of the source pressure and the choke point. The sound pressure level downstream of the choke point (the primary cause of acoustic induced fatigue failure in downstream piping of the choked point) should be based on the difference (or pressure ratio) of the choke point and the shock discontinuity pressure downstream of the choke point. The original equation proposed by Carucci and Mueller used the pressure difference (or ratio) between the source pressure and the exit pressure downstream of the choke point. This yields an overestimate of the sound pressure. This overestimate is somewhat tempered by the fact that the pressure difference ratio to the source pressure is raised to the 3.6 power.

The following information was provided by Carucci and Mueller regarding the five failures and two high vibrations cases observed in Table 1.

A. Failure occurred during startup. Sonic velocity is achieved at the 6 inch branch connection of a 24 inch pipe run downstream of the recycle valve. The sound power level estimate considers the combined acoustic energy generated by the letdown valve and the sonic condition at the branch connection. The upstream pressure at the branch connection was estimated to be 98.8 psia.

B1/B2. This system consisted of a control valve letting down pressure to a safety valve / flare header system. The failure occurred after five to ten hours of operation as a crack at a 10 inch branch connection to 28 inch header. Sonic velocity was achieved at both the control valve and the 10 inch branch connection to the 28 inch flare header. As such, two pertinent data points are established:

1. Acoustic energy within the 10 inch line as generated by valve (not the failure point), and
2. Combined acoustic energy within the 28 in header as generated by valve and sonic condition at the 10 in branch connection (the failure point).

C. High vibrations, no failures.

D. Failure during startup. Cracks were observed in a 24 inch line downstream of the compressor recycle valve after twelve hours of operation. The cracks were near the branch connections of a 6 inch line and a 3/4 inch drain valve and at an I-beam support welded to the pipe at the elbow immediately downstream of the control valve.

E. High vibration, no failures.

F. Failure at severely undercut weld on a 300 mm (12 in.) line made to fasten conduit support clip. Points of high stress concentration were later eliminated.

G. This system consisted of six, parallel, high pressure steam letdown valves, each with a downstream, three pass contra-flow attemporator and an in-line silencer. Failure of the 18 inch attemporator shells occurred after four hundred hours of operation.

H. This system included four parallel desuperheaters located downstream of control valves. Several cracks developed in this system after two to three months of operation. Two to three of the four letdown valves were discharging steam through a 10 inch connection to a 20 inch header which swaged up to a 30 inch diameter. Longitudinal cracking occurred at the bottom of the 20 inch header at a transverse guide 0.5 meters downstream of the fourth 10 inch branch connection. In addition, a 1 inch bypass line for a block valve downstream of the third letdown valve cracked and the 20 inch header cracked around a pressure tap downstream of the transverse guide.

Table 1 compares the acoustic efficiencies implied in the Carucci and Mueller method (at the upstream choke point) vs. acoustic efficiencies established using IEC methods described later on in this paper. The acoustic efficiencies implied by the Carucci and Mueller method at the upstream choke point ranged from 0.1 to 5 percent while the IEC acoustic efficiency ranged from 0.67 to 1 percent. The Carucci and Mueller method is based on acoustic energy theories encompassing both jet and choked flow noise as well as test data from work performed by Exxon research and engineering. As discussed in later sections, the experience based failure criteria originally developed by Carucci and Mueller can only be used with sound power levels calculated by Equation 12.

The implied acoustic efficiency by Equation 12 is proportional to the mass flow rate and inversely proportional to u^2:

$$\eta = 8\left(1-\frac{P_2}{P_1}\right)^{3.6}\left(\frac{T_1}{M_w}\right)^{1.2}\left(\frac{\dot{m}}{u^2}\right) \tag{15}$$

If we consider the case of an ideal gas undergoing isentropic choked (sonic) flow through a restriction orifice, we can establish the following expressions for variables of interest at the choke point:

$$T_2 = 2\frac{T_1}{\gamma+1} \tag{16}$$

$$P_2 = P_1\left(\frac{2}{\gamma+1}\right)^{\frac{\gamma}{\gamma-1}} \tag{17}$$

$$\rho_2 = \frac{P_2 M_w}{R_g T_2} \tag{18}$$

$$u_2 = \sqrt{\frac{\gamma R_g T_2}{M_w}} \tag{19}$$

$$\dot{m} = \rho_2 u_2 A_o \tag{20}$$

Substituting the values of mass flow rate $\dot{m}$, choked flow velocity u_2, and choke pressure P_2 in Equation 15, we obtain the following simplified expression for the Carucci and Mueller acoustic efficiency after some algebraic manipulations:

Table 1. Vibration risk data set used by Carucci and Mueller and reproduced by Melhem

CASE	SERVICE	T_1, K	P_1, bara	T_2, K	P_2, bara	D_2, m	u_2, m/s	Exit Mach Number, M_2	Mass Flow Rate, kg/s	Heat Capcity Ratio, C_p/C_v	Restriction Orifice / Valve Flow Speed, m/s	Carucci and Mueller Acoustic Efficiency, %	IEC Acoustic Efficiency, %	PWL_cm, dB	PWL_mel, dB	PWL_cm - PWL_mel, dB	PWL_mel (IEC), dB
A	Propylene Compressor Recycle	319.26	20.55	238.91	1.59	0.610	32.50	0.137	29.74	1.13	224.30	0.55	0.85	168.5	165.0	3.5	161.5
B1	Natural Gas Pressure	318.71	42.75	226.00	6.27	0.254	125.60	0.339	48.26	1.22	322.95	2.54	0.91	170.9	168.1	2.8	163.6
B2	Natural Gas Letdown to Flare	318.71	5.17	270.38	2.07	0.711	56.20	0.159	48.26	1.22	357.00	0.41	0.69	171.4	168.9	2.5	163.3
C	Natural Gas Letdown to Flare	308.15	47.57	195.68	4.14	0.406	64.40	0.218	47.63	1.23	311.25	4.17	0.95	171.8	169.8	2.0	163.4
D	Nitrogen Compressor Recycle	310.93	6.96	280.31	3.72	0.610	93.80	0.297	119.70	1.20	324.75	1.05	0.79	168.3	168.2	0.1	167.0
E	Nitrogen Compressor Recycle	310.93	16.82	267.59	7.03	0.457	60.10	0.154	84.80	1.21	328.35	0.79	0.66	168.8	165.6	3.3	164.8
F	Natural Gas Compressor Recycle	330.37	51.09	220.53	13.86	0.305	28.80	0.093	35.28	1.45	272.52	0.59	0.82	164.9	159.5	5.4	160.3
G	Desuperheater Steam	602.59	99.97	480.32	10.34	0.457	45.80	0.093	36.89	1.29	619.65	2.91	0.94	175.5	173.1	2.4	168.2
H	Desuperheater Steam	658.15	43.09	476.11	10.34	0.508	45.60	0.093	45.99	1.29	586.00	0.36	0.83	163.9	168.3	-4.4	168.9
1	Process Gas Compressor Recycle	308.15	13.17	204.98	1.03	0.406	122.80	0.451	27.09	1.19	256.81	2.22	0.95	165.9	164.2	1.7	160.6
2	Process Gas Compressor Recycle	311.48	38.75	265.88	10.00	0.305	14.20	0.048	13.86	1.13	258.54	0.32	0.83	156.6	151.8	4.8	155.9
3	Propylene Compressor Recycle	316.48	20.55	262.14	3.24	0.406	15.70	0.055	12.60	1.11	209.52	0.31	0.91	156.0	153.5	2.5	154.0
4	Process Gas Compressor Recycle	300.37	2.48	257.91	1.03	0.508	145.60	0.450	34.78	1.21	325.00	0.26	0.77	161.3	156.7	4.6	161.5
5	Process Gas Compressor Recycle	310.93	15.86	215.12	2.21	0.406	68.70	0.226	25.33	1.23	357.01	1.21	0.91	165.2	162.9	2.3	161.7
6	Process Gas Compressor Recycle	300.93	39.44	250.67	15.51	0.254	26.10	0.075	20.66	1.24	330.10	0.20	0.78	158.3	153.6	4.7	159.4
7	Propylene Compressor Recycle	310.93	16.69	233.88	1.38	0.356	131.30	0.575	38.05	1.13	216.35	3.52	0.95	166.6	165.4	1.2	159.3
8	Propylene Compressor Recycle	310.93	16.69	234.72	1.38	0.508	59.20	0.259	35.08	1.13	216.35	3.51	0.95	165.9	164.7	1.2	158.9
9	Waste Steam Dump	403.71	2.76	370.82	1.03	0.406	245.10	0.556	18.90	1.10	436.60	0.15	0.77	160.5	154.4	6.1	161.4
10	Process Gas Compressor Recycle	302.59	10.69	207.90	1.03	0.406	95.80	0.357	20.92	1.19	295.15	1.49	0.94	163.2	161.3	1.9	159.3
11	Process Gas Compressor Recycle	308.15	54.81	263.43	27.51	0.305	7.20	0.016	9.15	1.29	441.77	0.10	0.79	150.5	149.7	0.8	158.5
12	Natural Gas Compressor Recycle	308.15	3.79	282.54	2.48	0.914	28.40	0.075	39.69	1.26	315.95	0.19	0.73	155.4	155.7	-0.3	161.6
13	Natural Gas Compressor Recycle	308.15	6.48	279.42	4.00	0.914	17.30	0.046	39.44	1.25	334.77	0.24	0.75	157.2	157.2	0.0	162.2
14	Natural Gas Compressor Recycle	308.15	11.45	282.82	6.76	0.762	14.40	0.039	38.93	1.19	346.05	0.28	0.76	158.1	158.1	0.0	162.5
15	Natural Gas Compressor Recycle	308.15	21.79	266.72	11.65	0.660	10.10	0.027	36.04	1.30	359.11	0.33	0.78	159.4	158.8	0.6	162.6
16	Natural Gas Compressor Recycle	308.15	40.75	272.56	21.99	0.610	6.30	0.017	35.91	1.25	344.77	0.34	0.78	159.2	158.6	0.6	162.2
17	Natural Gas Compressor Recycle	308.15	80.12	271.44	40.89	0.508	4.80	0.013	35.41	1.23	324.89	0.40	0.78	160.1	158.8	1.3	161.6
18	Natural Gas Compressor Recycle	308.15	172.99	240.53	79.98	0.406	3.80	0.010	35.41	1.47	326.41	0.71	0.82	161.5	161.3	0.2	161.9
19	Natural Gas Compressor Recycle	308.15	3.86	280.10	2.34	0.914	27.60	0.074	37.17	1.24	340.74	0.24	0.75	157.3	157.1	0.2	162.1
20	Natural Gas Compressor Recycle	308.15	6.21	281.91	3.93	0.914	16.50	0.044	36.79	1.24	325.37	0.20	0.74	155.9	155.9	0.0	161.6
21	Natural Gas Compressor Recycle	308.15	10.62	274.09	6.27	0.762	14.30	0.038	36.04	1.29	345.37	0.26	0.76	157.4	157.4	0.0	162.1
22	Natural Gas Compressor Recycle	308.15	19.72	266.35	10.62	0.660	10.10	0.027	33.01	1.31	359.23	0.30	0.78	158.4	158.0	0.4	162.2
23	Natural Gas Compressor Recycle	308.15	36.82	263.14	19.79	0.610	6.40	0.017	32.89	1.34	346.54	0.31	0.78	158.4	157.9	0.5	161.9
24	Natural Gas Compressor Recycle	308.15	64.47	270.22	36.96	0.508	4.90	0.013	32.82	1.31	319.94	0.32	0.77	157.3	157.2	0.1	161.1
25	Natural Gas Compressor Recycle	308.15	139.48	320.30	64.54	0.406	4.20	0.012	32.76	0.95	311.67	0.56	0.80	160.7	159.5	1.2	161.1
26	Propane Compressor Recycle	308.15	12.07	234.24	0.97	0.457	143.50	0.644	50.40	1.12	217.75	4.82	0.96	168.9	167.6	1.3	160.6
27	Steam Desuperheater	658.15	43.09	522.78	10.34	0.762	22.00	0.041	45.99	1.19	564.55	0.91	0.83	161.2	168.3	-7.1	167.9

$$\eta = (4.7950\gamma - 2.5882)\left(\frac{M_w}{T_1}\right)^{0.3} P_1 A_o \tag{21}$$

where A_o is the effective restriction orifice flow area in m^2, P_1 is the upstream pressure in bara, T_1 is the upstream temperature in Kelvin, and η is the calculated Carucci and Mueller acoustic efficiency in percent. This expression shows that the Carucci and Mueller acoustic efficiency is directly proportional to upstream pressure, and the restriction orifice flow area (flow rate). It can be shown that the Carucci and Mueller acoustic efficiency produces unrealistic values for high pressure systems, large flow area, and/or mixtures with high molecular weights.

If we consider the case of methane discharging through a restriction with an effective flow area of 0.1 m^2, upstream temperature of 373.15 K, and an upstream pressure of 100 bara, the calculated acoustic efficiency is:

$$\eta = (4.7950 \times 1.282 - 2.5882)\left(\frac{16}{373.15}\right)^{0.3} \times 100 \times 0.1 \tag{22}$$

$$= 3.559 \times 0.388 \times 100 \times 0.1 = 13.8\% \tag{23}$$

The same value can also be calculated by calculating $\dot{m}$, u_2, P_2, and η directly from the above ideal gas flow equations and Equation 15:

$$T_2 = 2\frac{T_1}{\gamma+1} = 2\frac{373.15}{1.282+1} = 327.05 \text{ K} \tag{24}$$

$$P_2 = P_1\left(\frac{2}{\gamma+1}\right)^{\frac{\gamma}{\gamma-1}} = 100\left(\frac{2}{1.282+1}\right)^{\frac{1.282}{1.282-1}} = 54.9 \text{ bara} \tag{25}$$

$$\rho_2 = \frac{P_2 M_w}{R_g T_2} = \frac{54.9 \times 100000 \times 16}{8314 \times 327.03} = 32.306 \text{ kg/m}^3 \tag{26}$$

$$u_2 = \sqrt{\frac{\gamma R_g T_2}{M_w}} = \sqrt{\frac{1.282 \times 8314 \times 327.03}{16}} = 466.76 \text{ m/s} \tag{27}$$

$$\dot{m} = \rho_2 u_2 A_o = 32.306 \times 466.76 \times 0.1 = 1507.63 \text{ kg/s} \tag{28}$$

$$\eta = 8\left(1 - \frac{P_2}{P_1}\right)^{3.6}\left(\frac{T_1}{M_w}\right)^{1.2}\left(\frac{\dot{m}}{u^2}\right) \tag{29}$$

$$= 8 \times (1 - 0.549)^{3.6} \times \left(\frac{373.15}{16}\right)^{1.2} \times \left(\frac{1507.63}{466.76^2}\right) \tag{30}$$

$$= 8 \times 0.0568 \times 43.78 \times 69.2 \times 10^{-4} \tag{31}$$

$$= 0.1376 \text{ or } 13.76\% \tag{32}$$

A similar equation to 21 can be derived for the acoustic efficiency downstream of the choke point:

$$\eta = (68.63 - 4.836\gamma)\left[1 - (1.0397 + 0.6096\gamma)\frac{P_b}{P_1}\right]^{3.6}\left(\frac{M_w}{T_1}\right)^{0.3} P_1 A_o \tag{33}$$

where P_b is the superimposed backpressure downstream of the choke point. Applying Equation 33 to the same example above we calculate an acoustic efficiency of 227 percent at a backpressure of 1 bara:

$$\begin{aligned}\eta &= (68.63 - 4.836\times 1.282)\left[1 - (1.0397 + 0.6096\times 1.282)\frac{1}{100}\right]^{3.6}\left(\frac{16}{375}\right)^{0.3}\times 100\times 0.1 \\ &= 62.430\times 0.9359\times 0.388\times 100\times 0.1 \\ &= 226.70\%\end{aligned} \tag{34}$$

It is evident from the actual data reported by Carucci and Mueller and from the above theoretical proof that the acoustic efficiency used by Carucci and Mueller in their equation can produce unrealistic values, well in excess of 1 percent.

10. Failure Criteria

Several methods are now available for screening and analyzing the potential failure risk of piping caused by vibration. These methods include:

- Experience Based - D/t method or D Method
- Experience Based - Mach number method (Not recommended)
- MTD methods
- Detailed structural dynamics methods

The experience based methods center around correlating likelihood of failure based on actual experience and/or test data. The most widely cited reference is that of Carucci and Mueller (see vii and viii) for steel pipe.

Figure 2 illustrates the failure criteria developed originally by Carucci and Mueller and later modified by Eisinger. This failure criteria establishes a design sound power level vs. the ratio of pipe diameter to thickness. This criteria is fundamentally better than the other criteria based on pipe diameter only since thicker wall pipes are stronger than thinner wall pipes.

The original work by Carucci and Mueller suggests a limit provided by the following equation (Figure 2):

$$L_{W,limit} = 184.6 - 0.215\frac{2D_i}{D_o - D_i} = 184.6 - 0.215\frac{D_i}{\delta} \tag{35}$$

where δ is the pipe thickness.

The lower allowable limit developed by Eisinger also shown in Figure 2 is given by the following equation:

$$L_{W,limit} = 173.6 - 0.125 \frac{2D_i}{D_o - D_i} = 173.6 - 0.125 \frac{D_i}{\delta} \quad (36)$$

The NORSOK standard published guidance in 2006 using the same equation for $L_{W,limit}$ as proposed by Eisinger above.

Note that the above limits are based on sound power levels calculated using the source pressure and the downstream exit pressure. We have recalculated the sound power levels based on the correct pressure ratios downstream of the choke point. This data is shown in Figures 3, 4, and 5.

Another design limit criterion that was originally proposed by Carucci and Mueller is based on pipe diameter only for steel pipe. This criterion applies to pipe diameters ranging from 10 inch to 36 inch (approximately between 200 mm to 800 mm) and with wall thicknesses ranging from 0.219 in to 0.439 in (approximately 5.5 mm to 10 mm). The design limit is shown in Figure 3 for the corrected data and can be approximated by:

$$L_{W,limit} = 192.8 - 9.8 \ln D \quad (37)$$

where D is the nominal pipe diameter in inches and L_W is the sound power limit in dB.

Recent theoretical and experimental work by Chiyoda Engineering[ix] indicates that AIV risk will increase for smaller pipe diameters at the same diameter to thickness ratio. As a result, the limiting criteria established above should be decreased by 6 dB for an 8 inch pipe, or by 20 $\log_{10}$ (16/D).

The design limit correlations developed in this paper can be used with computer programs such as SuperChems Expert which calculates the sound power level at every axial locations for single and multiphase flow to decide if the steel piping inside diameter to thickness ratio exceed the allowable limits as shown by Figure 6.

11. Other Methods for Calculating η

The simplest method for calculating η is to assume the same efficiency as an expanded jet (see Figure 7) and apply it equally to single and multi-phase flow. The efficiency shown in Figure 7 is the same as what is currently used by API for the estimation of flare noise.

The flow regime types are defined by varying jet shapes in the area upstream and downstream of the vena contracta. These jets change their shape when certain differential pressure ratios are exceeded. In regimes II to IV, higher Mach numbers arise downstream of the vena contracta. Yet, M at the vena contracta itself remains unchanged at 1. The IEC method produces equivalent values of efficiency to the data shown in Figure 7 (as $F_L \rightarrow 1.0$) as shown in Figure 9. Note that Figure 9 uses $x = 1 - P_2 / P_1$ as the X axis while Figure 7 uses $P_1 / P_2 = \frac{1}{1-x}$.

Figure 2. Piping failure limit as originally proposed by Carucci and Mueller and by Eisinger for steel pipe

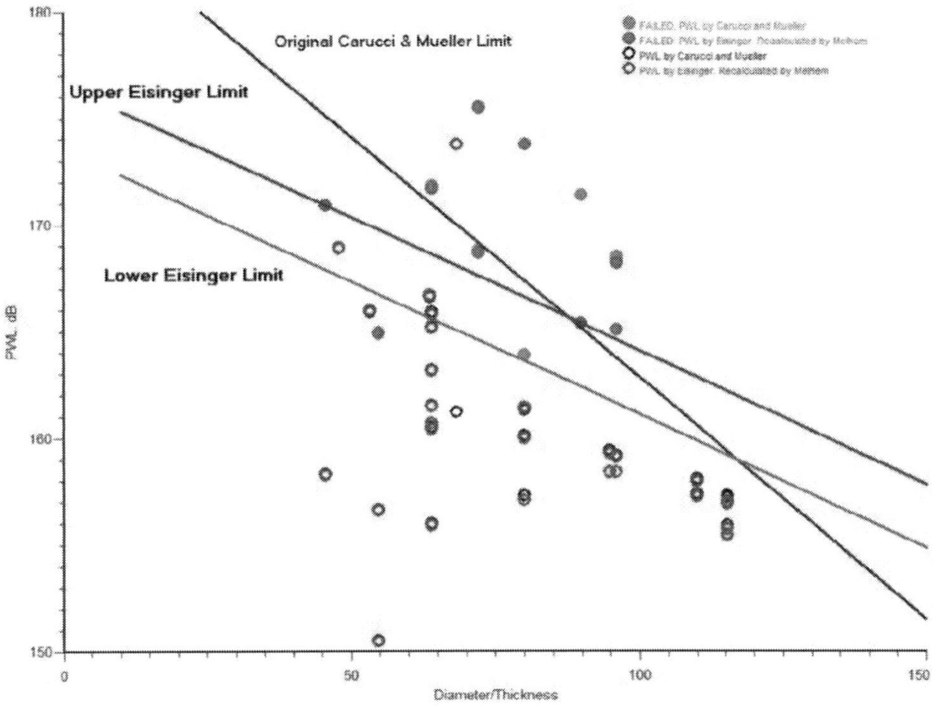

The efficiency is related to the flow pressure ratio and the flow regime as well. IEC calculates the efficiency based on five different flow regimes (see Figure 8) depending on the value of the downstream pressure, P_2:

Regime I The flow is sub-sonic. The sound generation has the character of a dipole jet. The highest Mach number is reached at the vena contracta, not exceeding Mach 1 at the maximum. Downstream of the vena contracta, the jet expands, leading to partial pressure recovery (F_L factor).

Regime II Sonic and supersonic flows exist together, which means that strongly turbulent flow and shock cell structure dominate. Pressure recovery drops until the top limit of regime II is reached.

Regime III The rise in pressure is non-isentropic. The flow is supersonic and shear turbulence predominates.

Regime IV The shock cells disappear and a Mach disk forms. The dominant mechanism is the interaction between shock cells and turbulence.

Regime V The acoustical efficiency is constant.

Figure 3. Piping failure correlation developed by Melhem using Carucci Mueller acoustic efficiency and corrected sound power level estimates for steel pipe. PWL vs. D

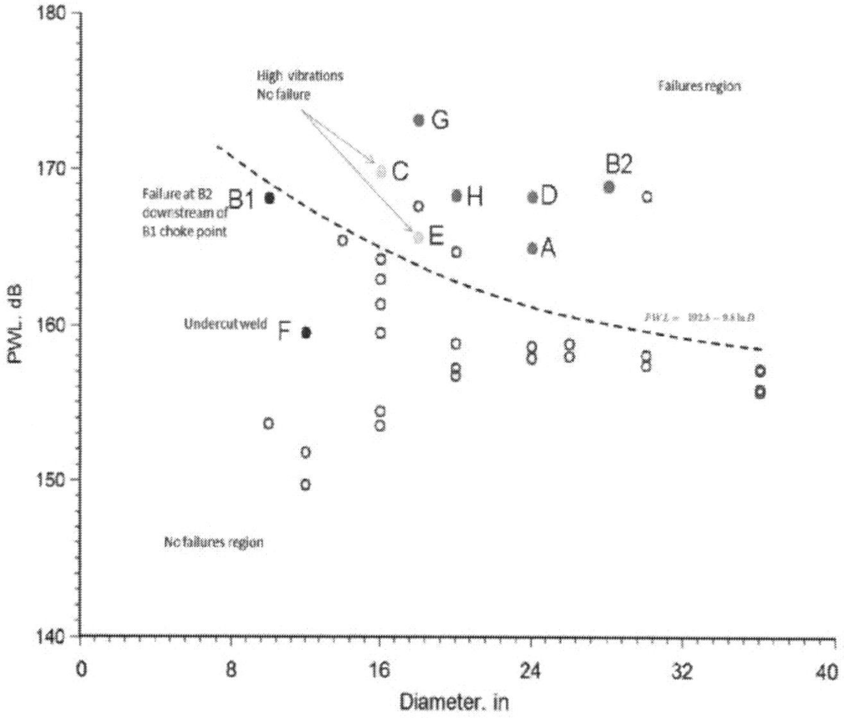

Figure 4. Piping failure correlation developed by Melhem using Carucci Mueller acoustic efficiency and corrected sound power level estimates for steel pipe. PWL vs. D/t

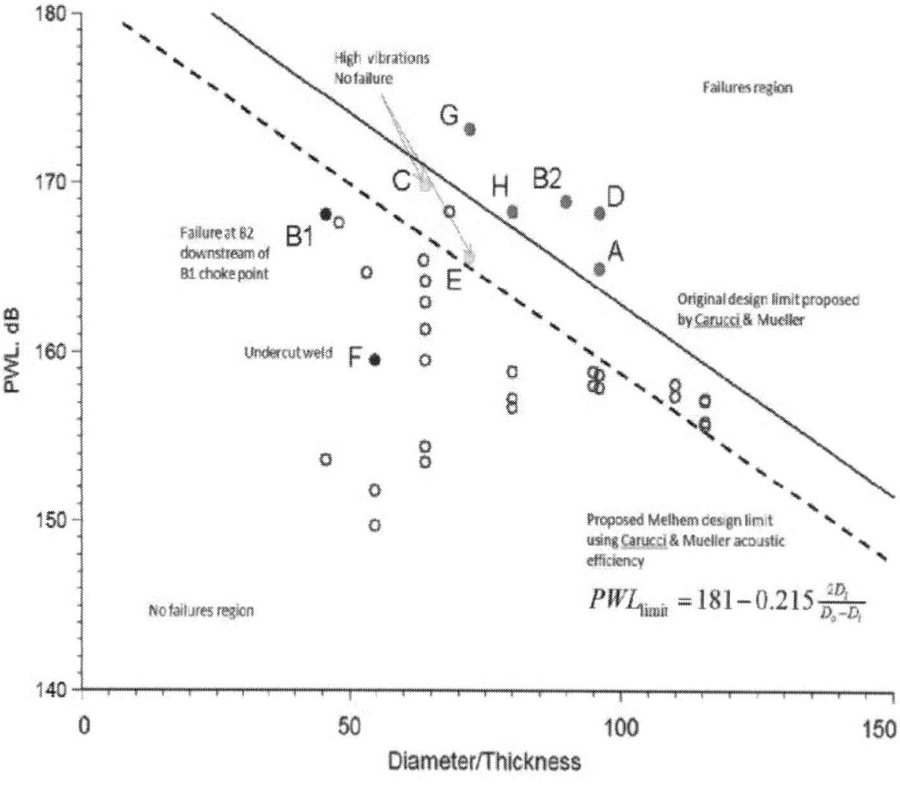

12. Vibration Frequencies

As a fluid moves through piping components there is typically a separation of the fluid from the constraining wall as the fluid changes flow direction. As a result a vortex is formed and then swept into the main stream. This vortex shedding occurs at fairly well defined dimensionless frequencies. The strength of the vortex varies but does not need to be very strong to cause damage especially if the shedding frequency couples with the natural frequency of the piping system. The shedding peak frequency for a vortex for subsonic and sonic flows (Regimes I and II up to a Mach number of 1.4) is given by:

$$f_p = \frac{N_{Str} u}{D} = 0.2 \frac{u}{D} \tag{38}$$

Figure 5. Piping failure correlation developed by Melhem using IEC acoustic efficiency and corrected sound power level estimates for steel pipe. PWL vs. D/t

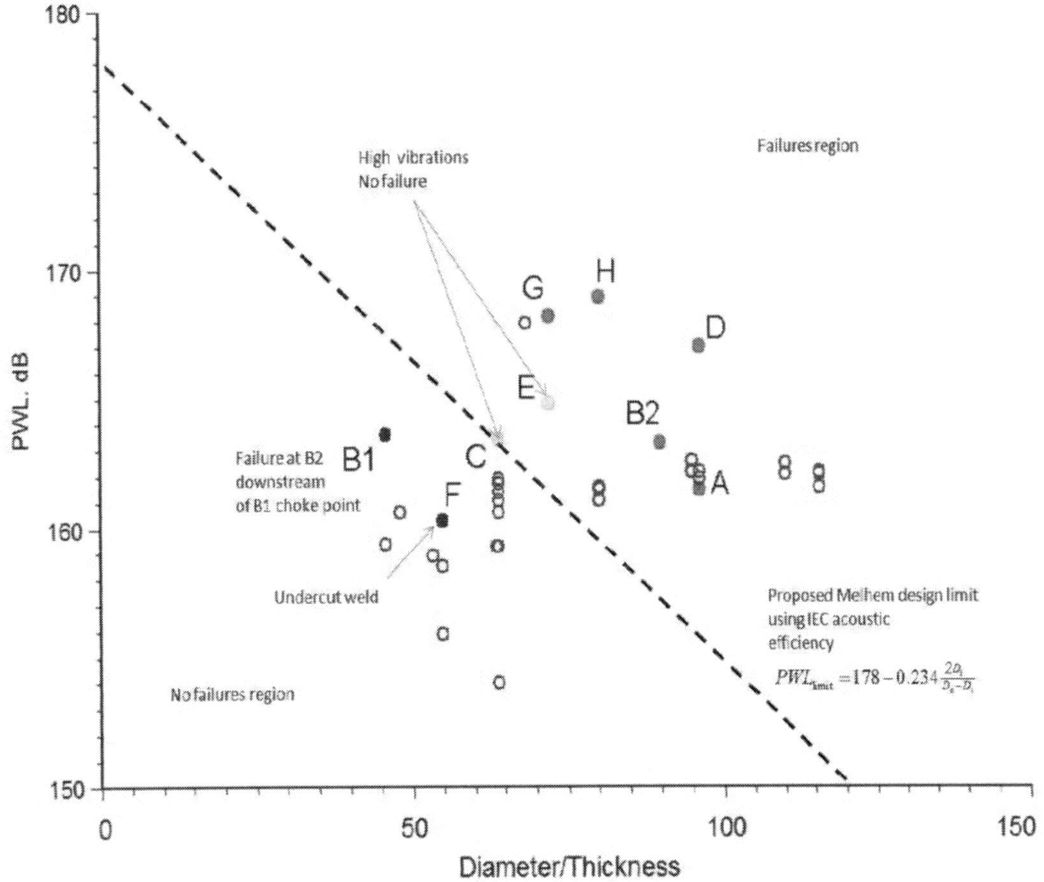

where u is the flow velocity in m/s, D is a characteristic flow dimension (perpendicular to flow), f_p is the peak frequency in Hz and N_{Str} is the Strouhal Number which varies depending upon the geometry causing the separation of the boundary layer. For a circular cylinder its value is 0.2 over a wide range of Reynolds numbers. It is usually between 0.1 and 0.3. Vortex shedding frequencies are generally more than 30 Hz [upper limit for most piping system natural frequencies].

For flows with Mach numbers larger than 1.4 ($M > 1.4$, Regimes III to V), the peak frequency f_p is given by:

$$f_p = \frac{0.4u_{sonic}}{1.25D\sqrt{M^2 - 1}} \tag{39}$$

For pipe flow, D is equal to the inside flow diameter. For a control valve, D is given by:

$$D = 0.0046\sqrt{\frac{C_v F_l}{N_o}} \tag{40}$$

Figure 6. Sample piping sound power level calculated by SuperChems Expert v6.4mp vs. experience based allowable limit for steel pipe

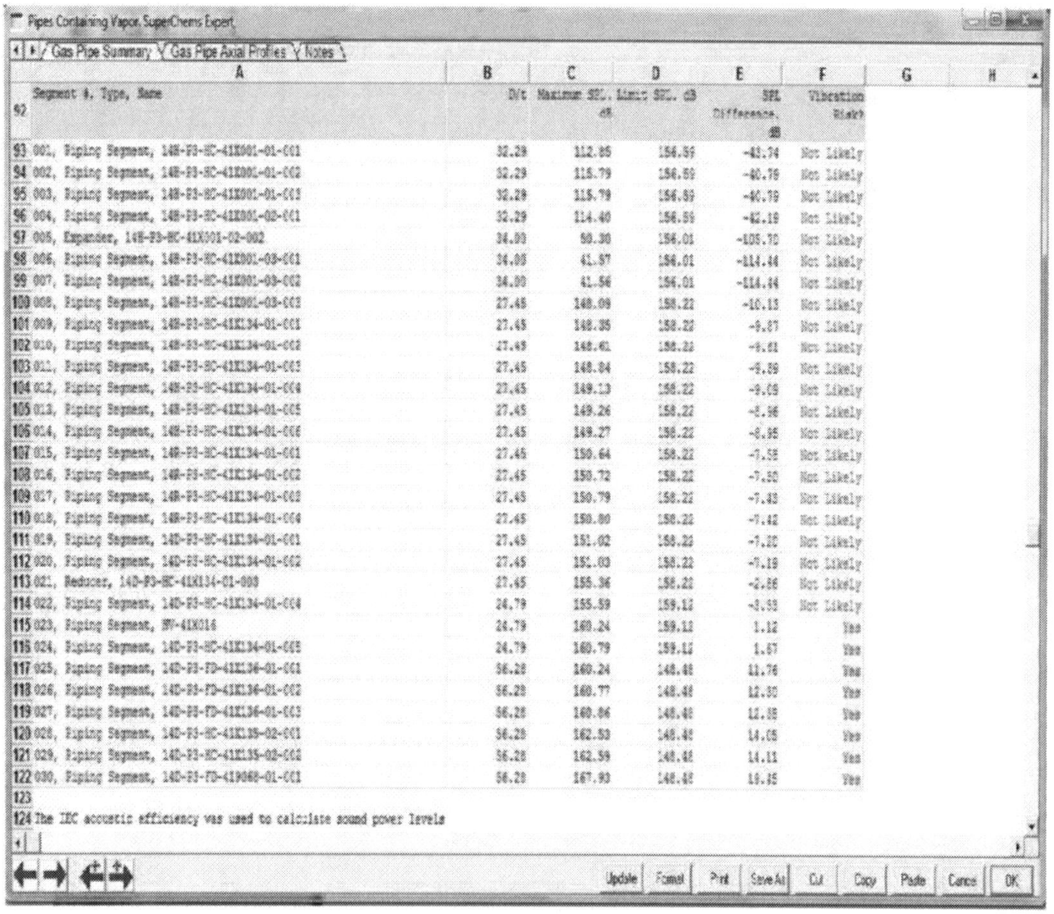

Pipes Containing Vapor, SuperChems Expert

Gas Pipe Summary | Gas Pipe Axial Profiles | Notes

	A	B	C	D	E	F
92	Segment #, Type, Name	D/t	Maximum SPL, dB	Limit SPL, dB	SPL Difference, dB	Vibration Risk?
93	001, Piping Segment, 14B-P3-HC-41E001-01-001	32.29	112.85	156.59	-43.74	Not Likely
94	002, Piping Segment, 14B-P3-HC-41E001-01-002	32.29	115.79	156.59	-40.79	Not Likely
95	003, Piping Segment, 14B-P3-HC-41E001-01-003	32.29	115.79	156.59	-40.79	Not Likely
96	004, Piping Segment, 14B-P3-HC-41E001-02-001	32.29	114.40	156.59	-42.19	Not Likely
97	005, Expander, 14B-P3-HC-41X001-02-002	34.03	50.30	156.01	-105.70	Not Likely
98	006, Piping Segment, 14B-P3-HC-41E001-03-001	34.03	41.57	156.01	-114.44	Not Likely
99	007, Piping Segment, 14B-P3-HC-41E001-03-002	34.03	41.56	156.01	-114.44	Not Likely
100	008, Piping Segment, 14B-P3-HC-41E001-03-003	27.45	148.09	158.22	-10.13	Not Likely
101	009, Piping Segment, 14B-P3-HC-41E134-01-001	27.45	148.35	158.22	-9.87	Not Likely
102	010, Piping Segment, 14B-P3-HC-41E134-01-002	27.45	148.61	158.22	-9.61	Not Likely
103	011, Piping Segment, 14B-P3-HC-41E134-01-003	27.45	148.84	158.22	-9.39	Not Likely
104	012, Piping Segment, 14B-P3-HC-41E134-01-004	27.45	149.13	158.22	-9.09	Not Likely
105	013, Piping Segment, 14B-P3-HC-41E134-01-005	27.45	149.26	158.22	-8.96	Not Likely
106	014, Piping Segment, 14B-P3-HC-41E134-01-006	27.45	149.27	158.22	-8.95	Not Likely
107	015, Piping Segment, 14B-P3-HC-41E134-01-001	27.45	150.64	158.22	-7.58	Not Likely
108	016, Piping Segment, 14B-P3-HC-41E134-01-002	27.45	150.72	158.22	-7.50	Not Likely
109	017, Piping Segment, 14B-P3-HC-41E134-01-003	27.45	150.79	158.22	-7.43	Not Likely
110	018, Piping Segment, 14B-P3-HC-41E134-01-004	27.45	150.80	158.22	-7.42	Not Likely
111	019, Piping Segment, 14D-P3-HC-41E134-01-001	27.45	151.02	158.22	-7.20	Not Likely
112	020, Piping Segment, 14D-P3-HC-41E134-01-002	27.45	151.03	158.22	-7.19	Not Likely
113	021, Reducer, 14D-P3-HC-41X134-01-003	27.45	155.36	158.22	-2.86	Not Likely
114	022, Piping Segment, 14D-P3-HC-41E134-01-004	24.79	155.59	159.12	-3.53	Not Likely
115	023, Piping Segment, BV-41X016	24.79	160.24	159.12	1.12	Yes
116	024, Piping Segment, 14D-P3-HC-41E134-01-005	24.79	160.79	159.12	1.67	Yes
117	025, Piping Segment, 14D-P3-FD-41E136-01-001	56.28	160.24	148.48	11.76	Yes
118	026, Piping Segment, 14D-P3-FD-41E136-01-002	56.28	160.77	148.48	12.30	Yes
119	027, Piping Segment, 14D-P3-FD-41E136-01-003	56.28	160.86	148.48	12.38	Yes
120	028, Piping Segment, 14D-P3-HC-41E135-02-001	56.28	162.53	148.48	14.05	Yes
121	029, Piping Segment, 14D-P3-HC-41E135-02-002	56.28	162.59	148.48	14.11	Yes
122	030, Piping Segment, 14D-P3-FD-41P068-01-001	56.28	167.93	148.48	19.45	Yes
123						
124	The IEC acoustic efficiency was used to calculate sound power levels					

Update | Format | Print | Save As | Cut | Copy | Paste | Cancel | OK where D is in meters, N_o is the number of separate flow passages, C_v is the valve flow coefficient, $\frac{C_v}{N_o}$ is the channel flow coefficient, and F_l is the valve recovery factor. The pressure recovery factor is defined as:

$$F_l = \sqrt{\frac{P_1 - P_2}{P_1 - P_{vc}}} \tag{41}$$

where $P_1 - P_2$ is the pressure differential across the valve and P_{vc} is the pressure at the vena contracta (note that P_2 is larger than P_{vc} since the pressure at the exit of the valve recovers):

$$P_{vc} = P_1 - \frac{P_1 - P_2}{F_l^2} \tag{42}$$

For liquid flow, P_{vc} can reach the vapor pressure of the liquid and cause choked flow.

$$P_{vc} = F_f P_v \approx \left[0.96 - 0.25\sqrt{\frac{P_v}{P_c}}\right] P_v \tag{43}$$

Figure 7. Acoustic efficiency of shock noise generated by choked jets, η vs. jet pressure ratio P_1 / P_2x

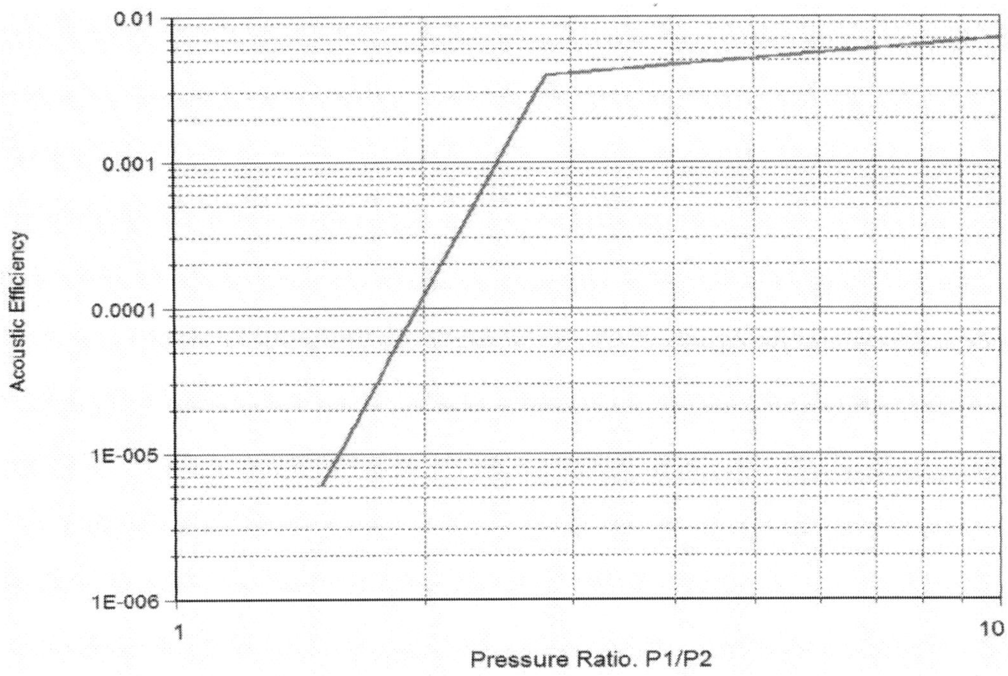

where F_f is the liquid critical pressure factor, P_v is the liquid vapor pressure, and P_c is the liquid critical pressure.

The lowest possible value of P_{vc} in a valve flowing liquid would be vacuum or 0. For this particular limiting case, Equation 41 can be solved for the pressure ratio P_1 / P_2:

$$P_d = \frac{P_1}{P_2} = \frac{1}{1-F_l^2} \tag{44}$$

where P_d is the damaging pressure ratio. If the valve is operated at a pressure ratio exceeding P_d, the flow will be choked, noisy, and subject to excessive vibration. Table 2 shows typical valve values for F_l and P_d.

Within every flowing pipe there will also be a standing wave moving axially back and forth in the pipe. The frequency of this wave depends on the effective acoustic length of the pipe and the sonic velocity of the fluid in the pipe. The effective acoustic length of the pipe is the distance between obstructions or acoustic barriers. Examples of obstructions would be valves, pumps, and orifices. An acoustic barrier would be an opening into a larger pipe, a reservoir, the end of a pipe run such as a 'T' intersection where the branch of interest requires a right angle turn. Piping components such as expanders or reducers could be an obstruction. An assessment of vibration risk should consider both the frequencies with and without the presence of expanders as obstructions.

The frequency of the standing wave can be calculated as shown below and then compared to the natural frequencies of valve components and the piping system to determine if there is a potential for vibration and/or resonance.

Closed End Pipe $f = \frac{i*u_{ac}}{4L}$

Open End Pipe $f = \frac{i*u_{ac}}{2L}$

where $i = 1, 2, ...$ is the wave number, L is the length between acoustic barriers, and u_{ac} is a characteristic acoustic speed throughout the pipe contents defined as follows:

$$u_{ac} = \frac{u_{sonic}}{\sqrt{1+\frac{dK}{\delta E}}} \tag{45}$$

where u_{sonic} is the speed of sound in the fluid, K is the bulk modulus of elasticity in the fluid, d is the pipe diameter, δ is the pipe thickness and E is the pipe material of construction modulus of elasticity.

The above equation is derived from a more general form that depends on the elastic properties of the pipe:

$$u_{ac} = \frac{u_{sonic}}{\sqrt{1+\frac{K}{E}\psi}} \tag{46}$$

where ψ is a function of the elastic properties of the pipe. Typical expressions of ψ are shown in Table 4. Typical material properties for ψ are shown in Table 3. Note that materials properties change with temperature as shown in Figure 10.

The speed of sound will change as a function of pressure, temperature, and composition. The presence of dissolved gas nuclei (such as air or other gases) in compressed liquids can significantly reduce the sonic velocity following a pressure drop which leads to the formation of gas bubbles. Up to 40 % reduction in sonic velocity has been observed[xi].

To control the vibration caused by a standing wave it is necessary to change the magnitude and/or the frequency of the standing wave or to change the natural frequency of the pipe or components being excited by the wave. The best approach is to address the magnitude of the standing wave. The magnitude is related to the fluid turbulent energy that is enforcing the wave. The most dominant source of this turbulence is the kinetic energy generated by the fluid jet exiting the valve trim. Thus a valve change with a trim that reduces this jet energy will eliminate this wave influence. Trying to change the frequencies is usually not beneficial. There is such a wide range of frequencies present in the turbulent flow that excitation can continue to establish a strong wave at the new frequency and continue the piping vibration.

Figure 8. Flow regimes considered for the estimation of acoustical efficiency

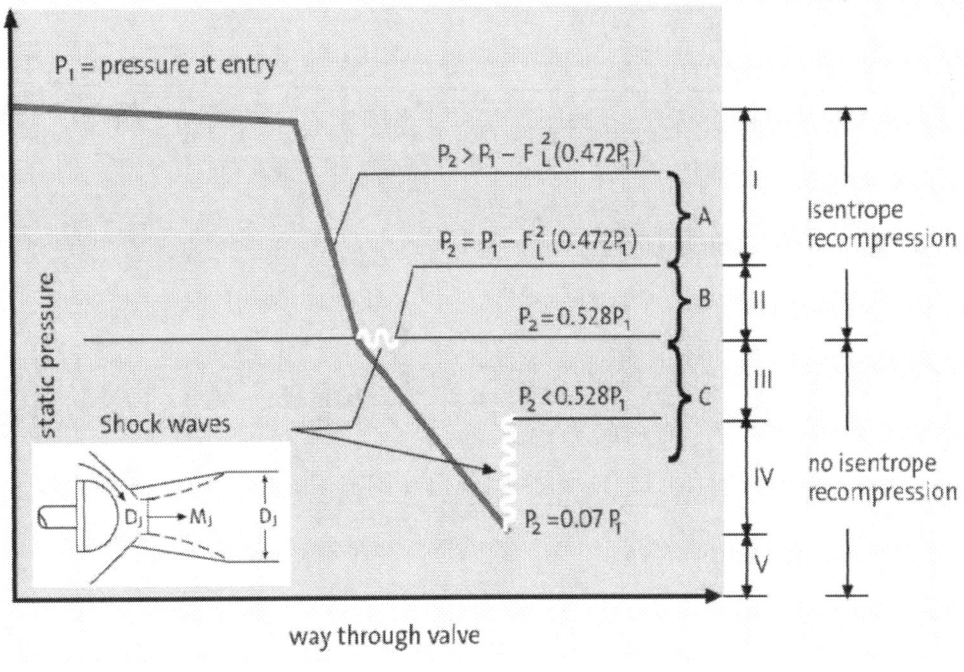

Figure 9. IEC flow acoustical efficiency as a function of F_L and differential pressure ratio

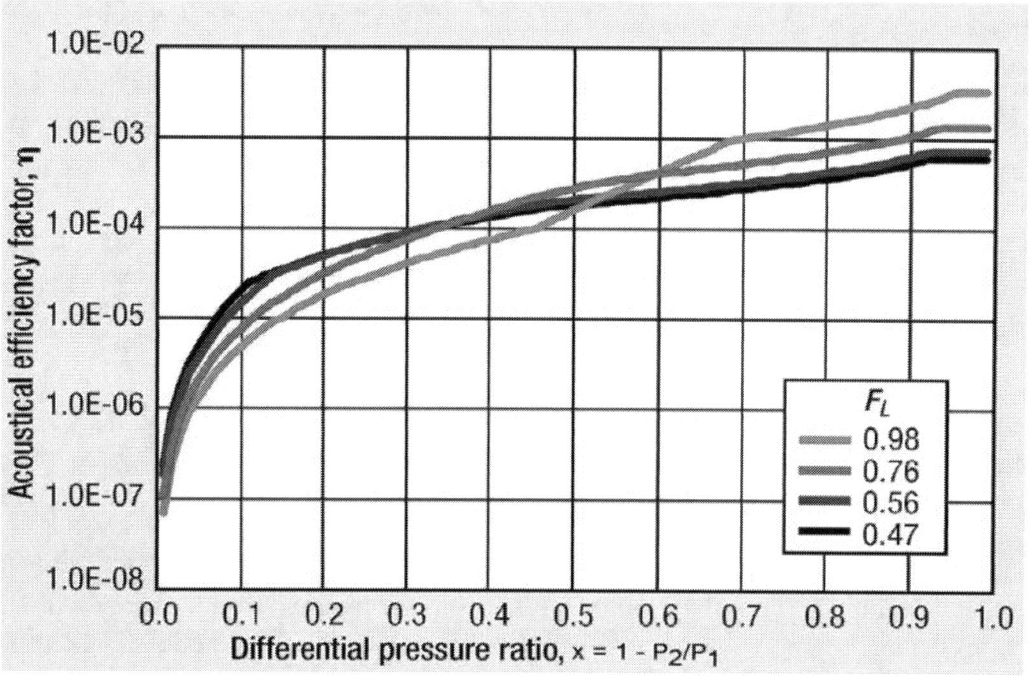

13. The Singing Safety Relief Valve Problem

As a result of increasing steam flow rates, several boiling water reactor (BWR) nuclear power plants have recently experienced the excitation of acoustic standing waves in closed side branches, e.g., safety relief valves (SRVs), due to vortex shedding generated by steam flow in the main stream lines. Flow past a valve entrance cavity excites a standing wave, resulting in noise and vibration[xii]. A similar tone is produced when air is blown across the mouth of a glass bottle.

Table 2. Typical valve values for F_l and P_d

Body	Trim	Flow Direction	F_l	$P_d = \frac{P_1}{P_2}$
Single Seat Globe	Cage	Open	0.90	5.3
	Cage	Closed	0.80	2.8
	V Plug	Open	0.90	5.3
	V Plug	Closed	0.90	5.3
	Contoured	Open	0.90	5.3
	Contoured	Closed	0.80	2.8
Double Seat Globe	V Plug		0.90	5.3
	Contoured		0.85	3.6
	Standard Bore		0.55	1.4
	Characterized		0.57	1.5
Angle	Cage	Open	0.85	3.6
	Cage	Closed	0.80	2.8

Ball	0.8 dia. Orifice		0.55	1.4
Butterfly	60°	Open	0.68	1.8
Butterfly	90°	Open	0.55	1.4

The amplitude of the acoustic pressure waves can be several times higher than the dynamic pressure present in the system (see Figure 12). The acoustic waves propagate in the steam lines, eventually reaching sensitive components such as steam dryers and turbine stop valves. In addition, the acoustic waves generated in the side branches may generate vibration problems locally and may lead to complications such as valve-seat wear. Therefore, the structural components are subjected to high-cycle fatigue loads, which over time may severely impact those components' functionality and safety.

Resonance occurs when the vortex shedding frequency coincides with the acoustic frequency of the standpipe or the valve components. The natural frequency of the standpipe/valve combination for a closed end pipe is given by the following equation:

$$f_a = \left(\frac{2n-1}{4}\right)\frac{u_{ac}}{L+L_e} = \left(\frac{2n-1}{4}\right)\frac{u_{ac}}{L+0.425d} \tag{47}$$

where n is the mode number (1 for 1st mode, 2 for 3rd mode, etc.), u_{ac} is the acoustic speed through the pipe contents as defined earlier, and L_e is an end correction corresponding to Rayleigh's upper limit.

The frequency of pressure oscillations (sound) created by vortex shedding, the energy source for the standing waves, is given by the following equation:

$$f_s = N_{str}\frac{u}{d+r} \approx 0.33(n-0.25)\frac{u}{d+r} \tag{48}$$

Typically peak oscillations occur at a Strouhal Number around 0.4 as shown in Figure 12. Note that the root mean square pressure amplitude shown in Figure 12 is the ratio of pressure oscillations divided by dynamic pressure ($\frac{1}{2}\rho u^2$). RMS begins increasing at a specific onset Strouhal Number and flow velocity depending on acoustic speed, pipe diameter, and pipe length, reaches a peak value and then decreases.

There are many similar installations of pressure relief valves in the process industries where the valves are mounted on large process lines such as overhead lines for distillation columns. In order to avoid fatigue failure from resonance caused by the coupling of normal flow vortex shedding frequency and the acoustic frequency of the standpipe ($f_s = f_a$), the normal flow velocity in the main line has to be limited to less than this critical value:

$$u_{max} < \frac{f_s}{N_{st}}(d+r) < \frac{1}{4}\left(\frac{u_{ac}}{L+L_e}\right)\left(\frac{d+r}{N_{str}}\right) \tag{49}$$

$$< \frac{1}{4}\left(\frac{u_{ac}}{L + 0.425d}\right)\left(\frac{d+r}{0.6}\right)$$

As shown in Figure 12 the pressure fluctuations start to increase at a Strouhal number of 0.6 and then decrease after they reach a peak value around a Strouhal number of 0.4. The same approach can be applied to the flow through the inlet and/or discharge line of a pressure relief valve:

$$u_{max} < \frac{1}{2}\frac{u_{ac}}{L}\frac{D}{0.6} \tag{50}$$

where L is the acoustic length of the inlet or discharge line. Resonance can also be checked by comparing an open pipe/contents frequency with the natural frequency of the pressure relief valve, f_n:

$$f_n = \frac{1}{\tau_n} = \frac{\omega_n}{2\pi} = \frac{1}{2\pi}\sqrt{\frac{K_s}{m_D}} \tag{51}$$

$$\tau_n = \frac{2\pi}{\omega_n} = 2\pi\sqrt{\frac{m_D}{K_s}} \tag{52}$$

where τ_n is the undamped natural period in s, and f_n is the undamped natural frequency in Hz where one Hz equals 1 cycle/second, K_s is the spring constant in $\mathrm{N/m}$, m_D is the mass of the valve disc and moving parts in kg, and ω_n is the undamped circular natural frequency in $\mathrm{radians/s}$.

14. Conclusions

A general method for the estimation of vibration risk for piping systems for single and multiphase flow is developed and implemented. An experience based vibration risk failure criteria is outlined using the work of Carucci and Mueller for gas flow. Because the Carucci and Mueller sound power level equation can produce unrealistic values of acoustic efficiency, well in excess of 1 percent for high pressure systems and/or systems with large mechanical flow energy, the revised experience based failure criteria developed in this paper (see Figure 5) based on the IEC acoustic efficiency is recommended for multiphase flow.

Figure 10. Temperature effects on material of construction

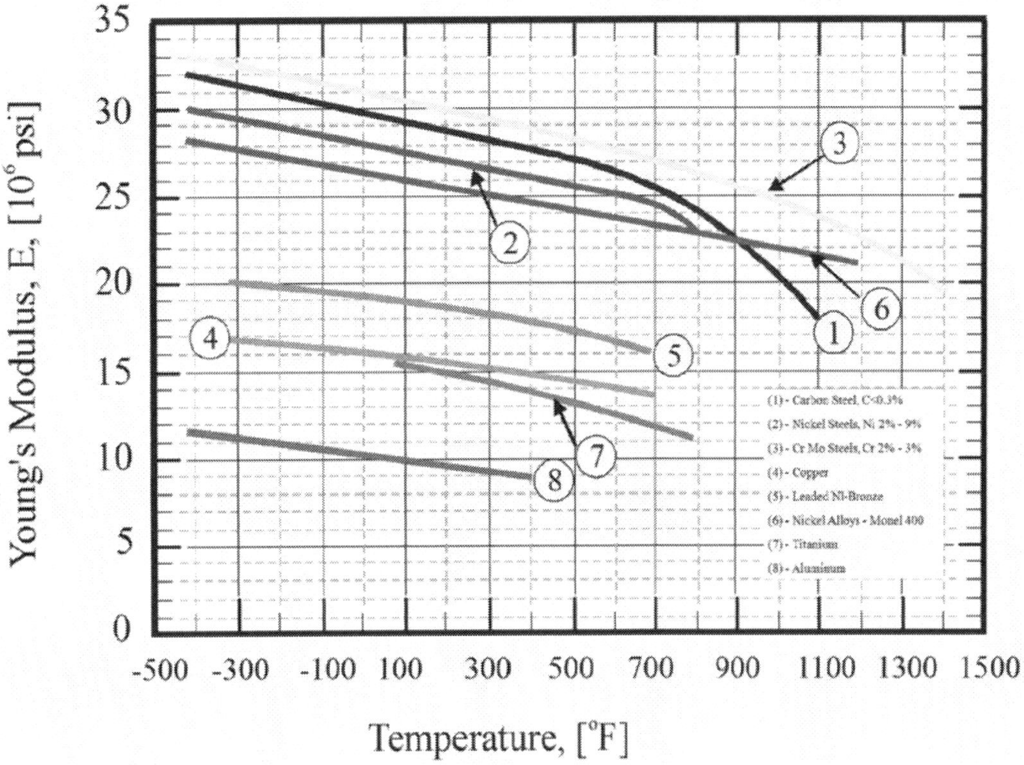

Figure 11. The singing safety relief valve[xiii]

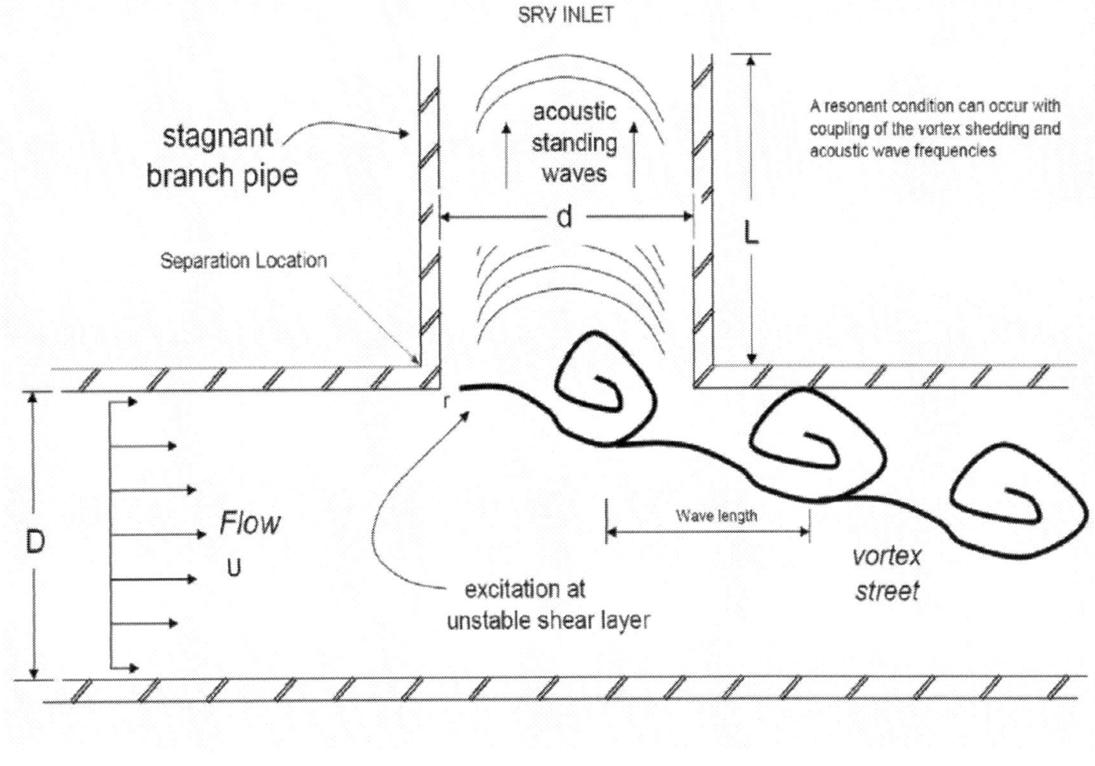

Figure 12. Comparison of calculated and measured pressure fluctuations as a function of Strouhal number[xiv]

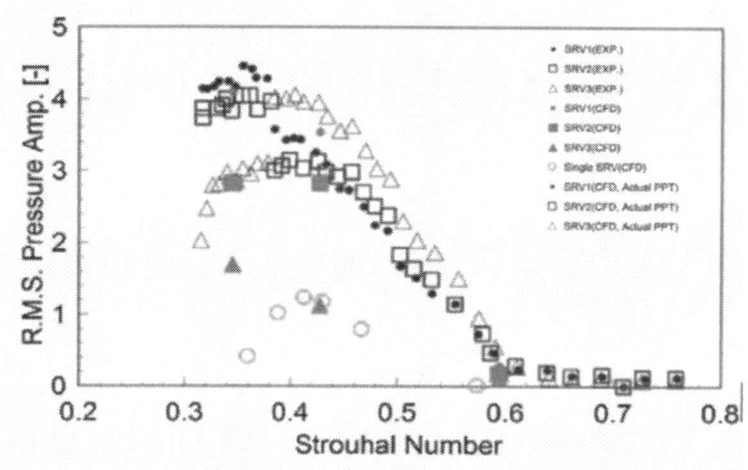

E is typically referred to as Young's modulus of elasticity

G is typically referred to as modulus of torsion, $G = \frac{1}{2}\frac{E}{1+\nu}$

Table 3. Typical data used in the estimation of sonic velocity in pipelines

Material	E (GPa)	Poisson's Ratio ν	$K = \frac{1}{\kappa}$ (GPa)	ρ in kg / m3
Aluminum	69	0.33		
Brass	78-110	0.36		
Carbon steel	202	0.303		
Cast iron	90-160	0.25		
Concrete	20-30	0.15		
Copper	117	0.36		
Ductile iron	172	0.30		
Fibre cement	24	0.17		
High carbon steel	210	0.295		
Inconel	214	0.29		
Mild steel	200-212	0.27		
Nickel steel	213	0.31		
Plastic / Perspex	6.0	0.33		
Plastic / Polyethylene	0.8	0.46		
Plastic / PVC rigid	2.4-2.75			
Stainless steel 18-8	201	0.30		
Water - fresh			2.19	999 at 20 C
Water - sea			2.27	1025 at 15 C

Table 4. Typical expressions for ψ

Pipe condition	ψ
Rigid	0
Anchored against longitudinal movement through its length	$\frac{d}{\delta}\left(1-\nu^2\right)$
Anchored against longitudinal movement at the upper end	$\frac{d}{\delta}(1.25-\nu)$
Frequent expansion joints present	$\frac{d}{\delta}$

References

[i] W.A. Skipwith. Acoustics technology. Survey SP-5093, NASA National Aeronautics and Space Administration, 1970.

[ii] C.M. Harris. *Handbook of Noise Control.* cGraw-Hill Book Co., Inc., 1957.

[iii] N.R. Miller, G.M. Singh, E. Rodarte and P.S. Hrnjak. Modification of standard aeroacoustic valve noise model to account for friction and two-phase flow. In *ACRC TR-162*, pages 1-11. Mechanical and Industrial Engineering Department, University of Illinois, 2000.

[iv] N.R. Miller, G.M, Singh, E. Rodarte and P.S. Hrnjak. Prediction of noise generated by expansion devices throttling refrigerant. In *ACRC TR-163*, pages 1-13. Mechanical and Industrial Engineering Department, University of Illinois, 2000.

[v] MTD. *Guidelines for the Avoidance of Vibration Induced Fatigue in Process Pipework.* Marine Technology Directorate Limited (MTD), 1999.

[vi] Energy Institute. *Guidelines for the Avoidance of Vibration Induced Fatigue in Process Pipework.* The Energy Institute, London, England, 2008.

[vii] V.A. Carucci and R.T. Mueller. Acoustically induced piping vibration in high capacity pressure reducing systems. In *82-/WAPVP-8*, pages 1-13. American Society of Mechanical Engineers, ASME, 1982.

[viii] F.L. Eisinger. Designing piping systems against acoustically-induced structural fatigue. In *PVP-VOL 328, Flow-Induced Vibration*, pages 397-404. American Society of Mechanical Engineers, ASME, 1982.

[ix] Hisao Izuchi, *Investigation of Pipe Size Effect Against Vibration Risk*, Presentation to ioMosaic Corporation, March 2013, Houston.

[x] I. Heitner. How to estimate plant noises. *Hydrocarbon Processing*, 47, 67-74, 1968.

[xi] V.L. Streeter and E.B. Wylie. Water hammer and surge control. *Annual Reviews in Fluid Mechanics*, pages 57-74, 1974.

[xii] T.M. Mulcahy, S.A. Hambric and V.N. Shah. Flow-induced vibration effects on nuclear power plant components due to main stream line valve singing. In *Ninth NRC/ASME Symposium on Valves, Pumps, and Inservice Testing, NUREG/CP-0152,* volume 6, pages 3B:49-3B:69. NCR/ASME, 2006.

[xiii] P. Sekerak. Potential adverse flow effects at nuclear power plants. In *16th International Conference on Nuclear Engineering*. ICONE16-48900, 2008.

[xiv] K. Okuyama, F. Inada, Y. Ogawa, R. Morita, S. Takahashi and K. Yoshikawa. Evaluation of acoustic and flow induced vibration of the BWR main stream lines and dryer. *Journal of Nuclear Science and Technology,* 48 (5): 759-776, 2011.

(44bt) Performance Metrics Development Management and Reporting Reaching Sustainability by Design

Robert C. Smith, ABB, Inc., Houston, TX

Overview

The saying is, "You can't manage, what you can't measure". This presentation will discuss various performance management measurement frame-works; How to use organizational goals and objectives to develop appropriate performance measures; Design effective performance measures to implement your organization's goals and strategies; Link performance measures to a variety of management processes; Develop a performance management framework that drives results. Presentation goals

Create a clear understanding of setting and managing key performance metrics surrounding highly hazardous environments.

- Create a practical base knowledge of leading and lagging metrics and how to practically apply them into your environment.
- Learn about the most common errors people make when selecting metrics and you'll gain new ideas around how metrics can be visualized better.
- This session will give you an in-depth look at the often confusing and misunderstood world of performance metrics. You'll learn the truth about metrics, and how an effective performance measurement system can help you make better decisions.
- Recognize management's role in leading and implementing performance metrics.
- Have a foundation to build metrics and scorecards by using the tips, techniques, and tools acquired in the presentation.

Who should attend?

This presentation is intended for any executive, manager or employee with responsibility for implementing or maintaining score carding performance management systems, and is interested in learning more about how to establish and measure metrics. What you'll learn:

- Detailed knowledge of leading and lagging metrics
- Measuring program success
- Key points continuous improvement and sustainability of measures

- Understand the purpose of process management
- How to analyze an existing metric system
- How to create well designed metrics that support strategy
- How to steer away from some of the classic pitfalls of metric development

(44bu) Alarm Management - the Human Factors Challenge

Robert C. Smith, ABB, Inc., Houston, TX

Refinery alarm systems are supposed to support the operator, yet in several notable accidents they have proven to be ineffective. Part of the problem is that these systems are not being designed or managed with human needs in mind. Do alarm systems help the operator or are they a hindrance? This paper is a short review of how human factors contributed to several incidents including:

- July 1994 explosion at Texaco, Milford Haven, UK
- September 1998 explosion at Esso, Longford, Australia
- March 2005 explosion at BP, Texas City, US.

Human operators are needed to ensure plants and processes continue to operate safely. The control room and control system interfaces must therefore be built with human needs in mind. Includes a review of how implementation of ISA 18.2 and/or EEMUA 191 can improve alarm management practice.

9TH GLOBAL CONGRESS ON PROCESS SAFETY

Dust Flame Propagation in Industrial Scale Piping
Part 2: CFD Study of a Conveying Vessel-Pipeline System

Jim Vingerhoets
Fike Europe
Toekomstlaan 52, 2200 Herentals, Belgium
jim.vingerhoets@fike.com

Thomas M. Farrell
Fike Corporation, Industrial Protection Group
704 SW 10th Street, Blue Springs, MO 64015, USA
tom.farrell@fike.com

Jef Snoeys
Fike Europe
Toekomstlaan 52, 2200 Herentals, Belgium
jef.snoeys@fike.com

Prepared for Presentation at
American Institute of Chemical Engineers
2013 Spring Meeting
9th Global Congress on Process Safety
San Antonio, Texas
April 28 – May 1, 2013

UNPUBLISHED

Dust Flame Propagation in Industrial Scale Piping
Part 2: CFD Study of a Conveying Vessel-Pipeline System

Jim Vingerhoets
Fike Europe
Toekomstlaan 52, 2200 Herentals, Belgium
jim.vingerhoets@fike.com

Thomas M. Farrell
Fike Corporation Industrial Protection Group
704 SW 10th Street, Blue Springs, MO 64015, USA
tom.farrell@fike.com

Jef Snoeys
Fike Europe
Toekomstlaan 52, 2200 Herentals, Belgium
jef.snoeys@fike.com

Keywords: Dust explosion, flame propagation, vented vessel, pipelines, CFD

Abstract

The use of pipes to connect vessels and to transport hazardous materials, such as flammable gases or dusts, is a common practice in the process industry. The potential of fires and explosions has been recognized for decades and is addressed in standards such as NFPA 68 [1] and NFPA 69 [2]. Unlike for gases, the experimental evidence on dust flame propagation is scarce, and this study is particularly aimed at expanding the knowledge in this field. The purposes of this paper are to present the results of a parametric computational fluid dynamics (CFD) study validated by experimental evidence and to further develop a detailed understanding of combustion propagation in industrial scale piping.

The results of an ongoing research program which involves a piping system connected to a 5 m3 vented vessel are presented in the accompanying paper [3]. CFD was used to broaden the field of application beyond the tested configurations, and to study the influence of vessel volume and pressure, fuel characteristics and flow conditions.

Upstream and downstream explosion propagation behavior is studied as a function of fuel reactivity and type, vessel fuel concentration, pipeline fuel concentration, flow velocity, vessel volume, vent area, vent location, ignition location, pipe diameter and length.

1. Introduction

The dust explosion hazard continues to represent a constant threat to process industries that manufacture, use and/or handle powders and dust of combustible materials. In the process industry, the majority of process equipment is connected via pipes or ducts, and accidents have shown that the propagation of the explosion through the connection pipe work worsens the consequences of the explosion. Understanding the mechanisms of explosion flame propagation is a prerequisite for designing effective explosion protection solutions.

The problem of dust explosions in large-scale connected equipment has been much less investigated than isolated equipment, mainly due to cost and complexity to conduct experiments. The information in the public domain on dust explosion propagation is scarce; refer to the bibliography section at the end of the paper for relevant contributions. An alternative to large scale testing is the use of mathematical modeling and simulating the problem using CFD - computational fluid dynamics. CFD uses numerical methods and algorithms to simulate the combustion phenomena defined by boundary conditions. Although the use of CFD in gas and dust explosion applications is relatively new, this technique is irrefutably gaining interest.

The present research investigates, by experiment and CFD, flame propagation from a vented vessel through connected pipes. The aim is to examine the mechanisms that govern the velocity and acceleration of flame through pipelines, thus providing valuable information for the design of explosion protection solutions and systems. Once sufficient correlation between experimental and numerical results was demonstrated, the scope was extended with a numerical sensitivity study to quantify the influence of several application parameters on the flame propagation behavior. These parameters include:

- fuel type and reactivity,
- fuel concentration in vessel and pipelines,
- flow velocity and direction,
- vessel volume, pipe diameter and length,
- vent area, location,
- ignition location.

2. Simulation code

The numerical explosion simulations have been performed using DESC - Dust Explosion Simulation Code. DESC resulted from a research project supported by the European Commission under the Fifth Framework Program and was reviewed by Skjold in 2007 [4]. The main purpose of the project was to develop a simulation tool based on computational fluid dynamics (CFD) that could predict the potential consequences of industrial dust explosions in complex geometries. DESC is presently based on the existing Flame Acceleration Simulator (FLACS) code for gas explosions. The dust particles are modeled as gaseous components that follow the flow of the other gaseous components in the system. The combustion model used in FLACS and adopted by DESC is the so-called β flame model developed by Arntzen [5]. The turbulent burning velocity used by the β model is a function of the root mean square of the turbulent velocity fluctuations and the Karlovitz stretch factor. More specifically, the turbulent burning velocity is formulated as:

$$S_{\mathrm{T}} = 15.1 S_{\mathrm{L}}^{0.784} u'^{0.412}_{\mathrm{rms}} l_{\mathrm{I}}^{0.196}$$

with S_{T} the turbulent burning velocity, S_{L} the laminar burning velocity, u'_{rms} the root mean square of the turbulent velocity fluctuation, also called the turbulence intensity and l_{I} the turbulence integral length scale. The inputs for the combustion model are laminar burning velocities extracted from the measured K_{st} and P_{max} values in standardized 20 L sphere tests. The code has been validated against large scale tests reported in the public domain. It was found that DESC is sensitive to the numerical grid resolution, the presence of bends and interconnected vessels in general. The grid resolution in the current research has been chosen based on recommendations from this validation, and bends were excluded from the study.

3. Experimental setup

A test rig was built at the test site of Fike Corporation consisting of an inlet pipeline, a vessel, and an outlet pipeline, arranged as shown in the following image. The inlet and outlet pipelines were oriented 90° to each other, with the outlet pipeline aligned with the major axis of the vessel. All explosion venting was constrained to a single panel on the vessel, on the wall opposite the outlet pipeline interconnection. The system was equipped with a fan, enabling adjustable pipe flow velocities. By using separate constant mass flow rate dust injectors on the pipelines in addition to traditional dust injectors on the vessel, the vessel dust concentration and pipeline dust concentration could be varied independently of each other.

Figure 1: Photo of experimental setup

In order to cover several industry branches, different fuel types were used to conduct the research. The properties of the fuels used in this program are listed in Table 1.

Table 1: Closed vessel explosion characteristics of the dust fuels used

Dust type	Dust concentration (g/m^3)	20 L sphere	
		K_{st} (bar.m/s)	P_{max} (barg)
Wood flour	250	50	5.7
	500	108	7.0
	750	109	7.2
Phenolic dust	250	203	7.3
	500	217	6.9
	750	200	6.6
Corn Starch	250	93	6.6
	500	199	8.7
	750	205	8.6

Many tests were conducted to determine explosion pressures and flame arrival times as a function of venting area, vessel fuel concentration, pipeline fuel concentration and conveyance velocity. More information can be found in the accompanying paper [3].

4. Simulation setup

Figure 2 shows the basic setup of the explosion simulations. It represents as much as possible the experiment setup. The vessel itself consists of two identical cubes of 2.55 m^3 each, attached to each other without a partitioning wall in between. The cube assembly is equipped with dust jet injectors, 4 vent panels, a 10 kJ igniter and flanges to connect two pipelines. Each of the 4 vent panels can be closed off (to vary the total vent area) or opened at a controlled burst pressure. In some simulation scenarios, the location of the vent panels was moved to another wall of the vessel.

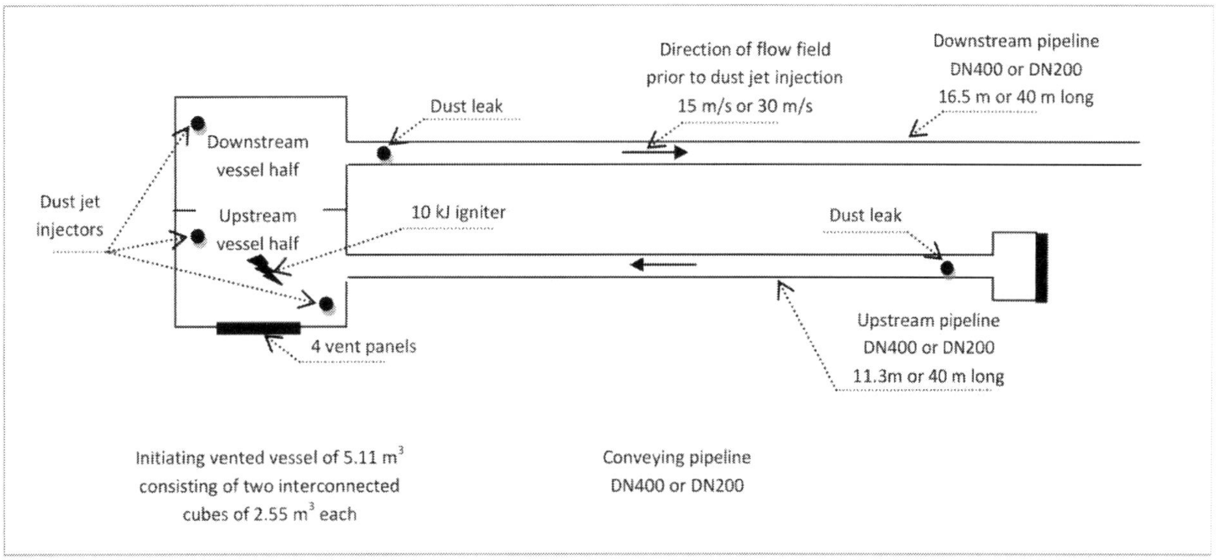

Figure 2: Sketch of basic setup of the explosion simulations.

Figure 3: Three dimensional CAD drawing picture of the simulation setup

The default location of the igniter is in the centre of the upstream vessel half, but in several simulation scenarios the igniter was moved to the centre of the downstream vessel half or to the interconnection with one of the pipelines. The delay time between start of dust jet injection and ignition was identical to the experimental program. The pipelines were either 400 mm or 200 mm in diameter and 11.3 m, 16.5 m or 40 m long. In order to restrict the simulation domain and associated computer processing time, the downstream pipeline was rotated 90° with respect to the experimental setup. Flow of 15 m/s or 30 m/s was established in the pipelines, depending on the simulation scenario. Dust leaks were programmed to fill the pipelines with a dust/air mixture of known concentration. Other simulations were done in a 62 m^3 vessel with proportionally scaled vent panels.

All simulation scenarios have been numerically solved on cubical grid cells with optimized size conforming to DESC guidelines.

5. Comparison between experiments and numerical simulations

Figure 4 to Figure 6 display comparative plots of vessel pressure and flame position as a function of time for three example scenarios including each of the three dusts of the program. Flames propagating towards the blower are represented by negative pipeline distances.

Table 2: Comparison between experiments and simulations giving the main characteristic values of the explosion propagation

Comparison of main characteristic values		Example 1		Example 2		Example 3	
		Exp	Sim	Exp	Sim	Exp	Sim
Vessel peak pressure		3.2 psig	2.0 psig	4.2 psig	6.7 psig	8.6 psig	2.3 psig
Time to reach 1,5 psig in vessel		64 ms	161 ms	46 ms	56 ms	54 ms	110 ms
Average flame speed in first 10 meters of pipe	Downstream	90 m/s	36 m/s	90 m/s	116 m/s	154 m/s	116 m/s
	Upstream	79 m/s	55 m/s	116 m/s	143 m/s	221 m/s	112 m/s
Time to reach 2,7 meters in the pipe	Downstream	110 ms	382 ms	114 ms	117 ms	88 ms	214 ms
	Upstream	127 ms	233 ms	106 ms	95 ms	85 ms	135 ms

The simulations under-predict the explosion peak pressure and pressure rise time in the vessel for two out of the three scenarios. The modeled explosion peak pressure for example 2 is higher (6.7 psig) than its experimental equivalent (4.2 psig), with similar pressure rise time (1.5 psig @ 56 ms and 1.5 psig @ 46 ms). Similar observations are made for the flame front pipeline propagation: the modeled explosions for example 1 and 3 propagate at lower speed (36 m/s downstream, 55 m/s upstream and 116 m/s downstream, 112 m/s upstream respectively), whereas the modeled dust explosion of example 2 propagates faster. Finally, the modeled explosions of example 1 and 3 enter the pipes later whereas the modeled dust explosion pipe entrance time of example 2 is similar to the experiment.

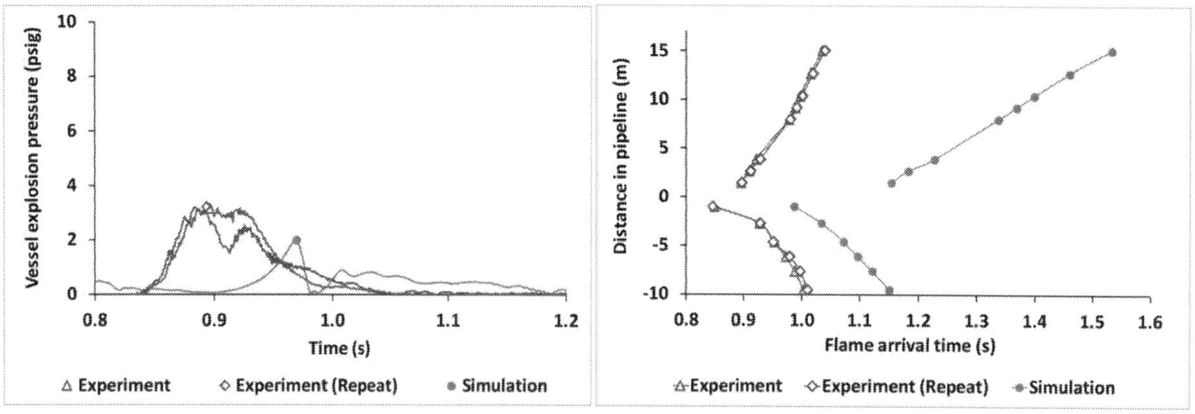

Figure 4: Comparison experiment and simulation: example 1

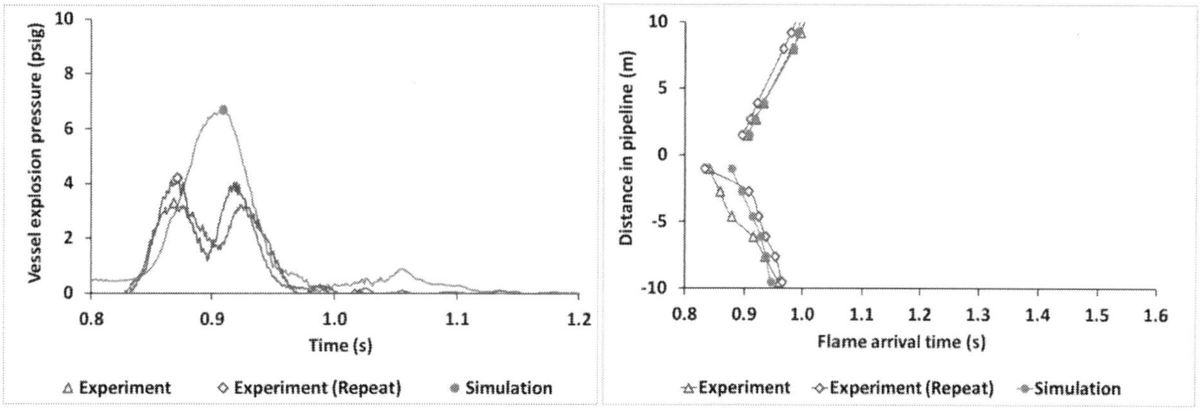

Figure 5: Comparison experimental and simulation: example 2

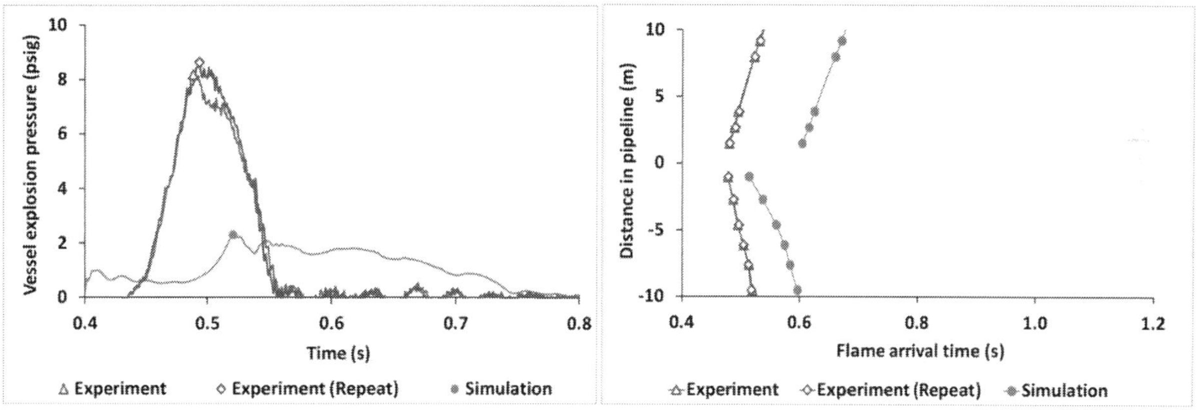

Figure 6: Comparison experimental and simulation: example 3

A likely explanation for the divergence between experimental and simulated results is that the modeled dust jet injectors generate different turbulence conditions than the injectors used in the experiments. This is a plausible explanation since the dust particles are modeled as gaseous phase fuel. Figure 7 shows the turbulence intensity in the centre of the upstream vessel half during simulated dust injection for different vessel dust concentrations and for a pipe flow of 15 m/s. The simulated dust jet injectors create an average turbulence intensity level between 2.0 and 3.4 m/s in the first 200 ms after ignition. 200 ms is the typical duration of an explosion in a 5 m^3 vessel. Although slightly higher than the reported turbulence in 20 L and 1 m^3 tests, the numbers are quite reasonable. Unfortunately, the turbulence could not be measured during the experiments.

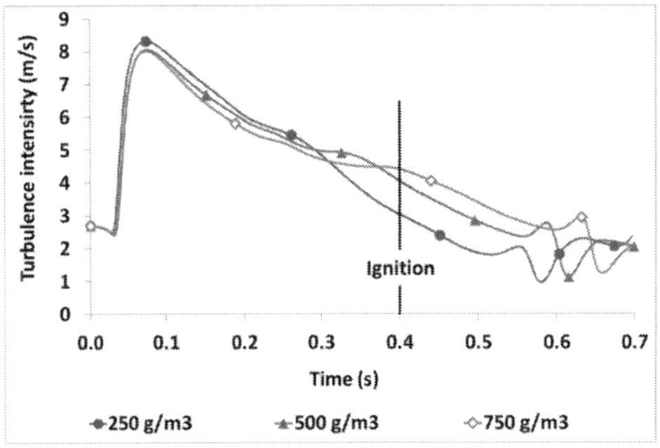

Figure 7: Turbulence intensity in the centre of the upstream vessel half during simulated dust injection for different vessel dust concentrations; initial pipe flow = 15 m/s

Another possible reason is that the 20 L flame data used by DESC do not scale up well to larger volumes. Several researchers postulated scaling problems related to turbulence and the powerful igniter used in the closed vessel 20 L and 1 m^3 standardized tests. This problem was investigated by additional explosion experiments in a closed $1m^3$ vessel with varying ignition delay [3]: the K_{st} of two of the three dusts were more or less independent of volume, although strongly affected by ignition delay. The other dust showed an opposite behavior and is relatively insensitive to decreased ignition delay.

Figure 8 shows the effect of increasing the laminar burning velocities which was used as a surrogate for boosting the turbulence. The agreement becomes much better which indicates that the prediction potential of DESC is improved, if both the fundamental combustion parameters (the laminar burning velocity derived from K_{st} and P_{max}) and the initial process turbulence are estimated correctly.

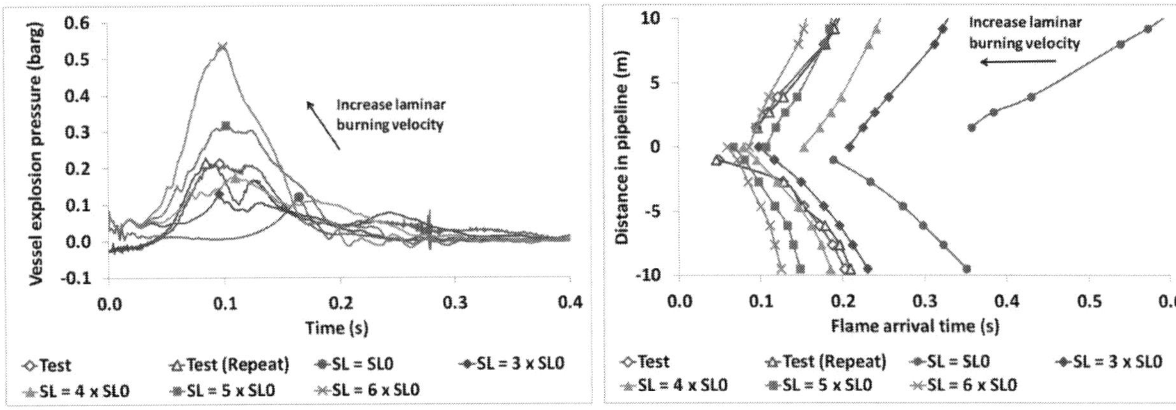

Figure 8: Sensitivity of vessel explosion pressure and flame propagation to the modeled laminar burning velocity and comparison with experiment (example 1)

In order to deal with the uncertainties in turbulence and combustion parameters for different vessel volumes and injector assemblies, simulations were performed in the reference test volume with all vent panels and connections to the pipelines closed off, and the K_{st} and P_{max} values were recorded. The results obtained by DESC should be referenced to K_{st} and P_{max} as obtained by DESC for the closed vessel. By following this approach, together with additional specific considerations as pointed out further, it is believed that simulations can represent industry reality with similar level of confidence as experiments.

6. Mechanisms of flame propagation

A thorough analysis of 94 explosion simulations for different fuel types, vessel and pipeline fuel concentrations, vent areas, flow velocities, significantly contributes to our understanding of the fundamental mechanisms which govern the propagation of dust flames through pipelines which are connected to a vented vessel. For all these scenarios the vessel and pipeline explosion pressures have been recorded as well as the flame arrival times at several distances along the pipelines. Besides pressure and flame, other physical variables, such as flow velocity and dust concentration have also been recorded. In addition to dust fuels listed in Table 1, also other dust fuels were used in the simulation study, so covering a wider range of K_{st} values, from 28 bar.m/s up to 281 bar.m/s.

In all scenarios the flame propagated with high speed up until the end of both pipelines, including the ejection of a large fireball, even against strong initial flows of 30 m/s. The highest recorded flame speed was 434 m/s. This was at the end of the downstream pipeline for the dust and concentration with highest reported K_{st} value and with the smallest venting area programmed. The lowest recorded flame speed was 34 m/s, for the dust and concentration with lowest K_{st} value and the largest vent area programmed.

It was confirmed that flame enters both pipes and propagates through a dust cloud that is pushed from the vessel into the pipes, as shown in Figure 9 below. In this scenario, the pipelines were not filled with dust fuel prior to ignition.

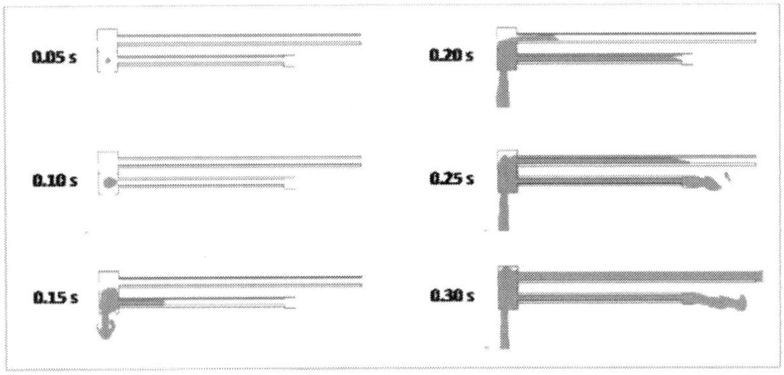

Figure 9: Flame propagation shown at subsequent time steps during an example explosion. No flow through pipe, no dust in pipe. 2 vent panels open. Vessel dust concentration = 500 g/m^3.

Figure 9 also illustrates that flame accelerates as it enters the pipe: It takes 0.1 seconds for the flame to reach the pipe and in only another 0.1 seconds it manages to travel the full pipeline distance.

In the vessel the freely expanding flame front propagates at a moderate speed of about 5 to 10 m/s. When the explosion flame enters the connected pipes, it will accelerate because the flame front will travel along with the combustion gases flowing at high speed out of the vessel:

$$S_{\text{f, pipe}} = S_g + S_T . E$$

The pipe gas flow velocity induced by the explosion pressure in the vessel (S_g) is often a lot larger than the speed of turbulent burning and associated local expansion ($S_T.E$) and can reach up to 300 m/s. For low to moderate explosive reactive materials the vessel pressure induced gas flow will dominate the pipe flame speed.

Figure 10 shows the correlation between gas flow velocity in the pipelines and flame front velocity, for three different fuels and for three different dust concentrations. The more reactive the fuel and the larger the gas flow velocity, the more the flame velocity exceeds the gas velocity. This is because the turbulent burning velocity increases with fuel reactivity and with increased flow turbulence. The flow turbulence is higher at higher gas velocities.

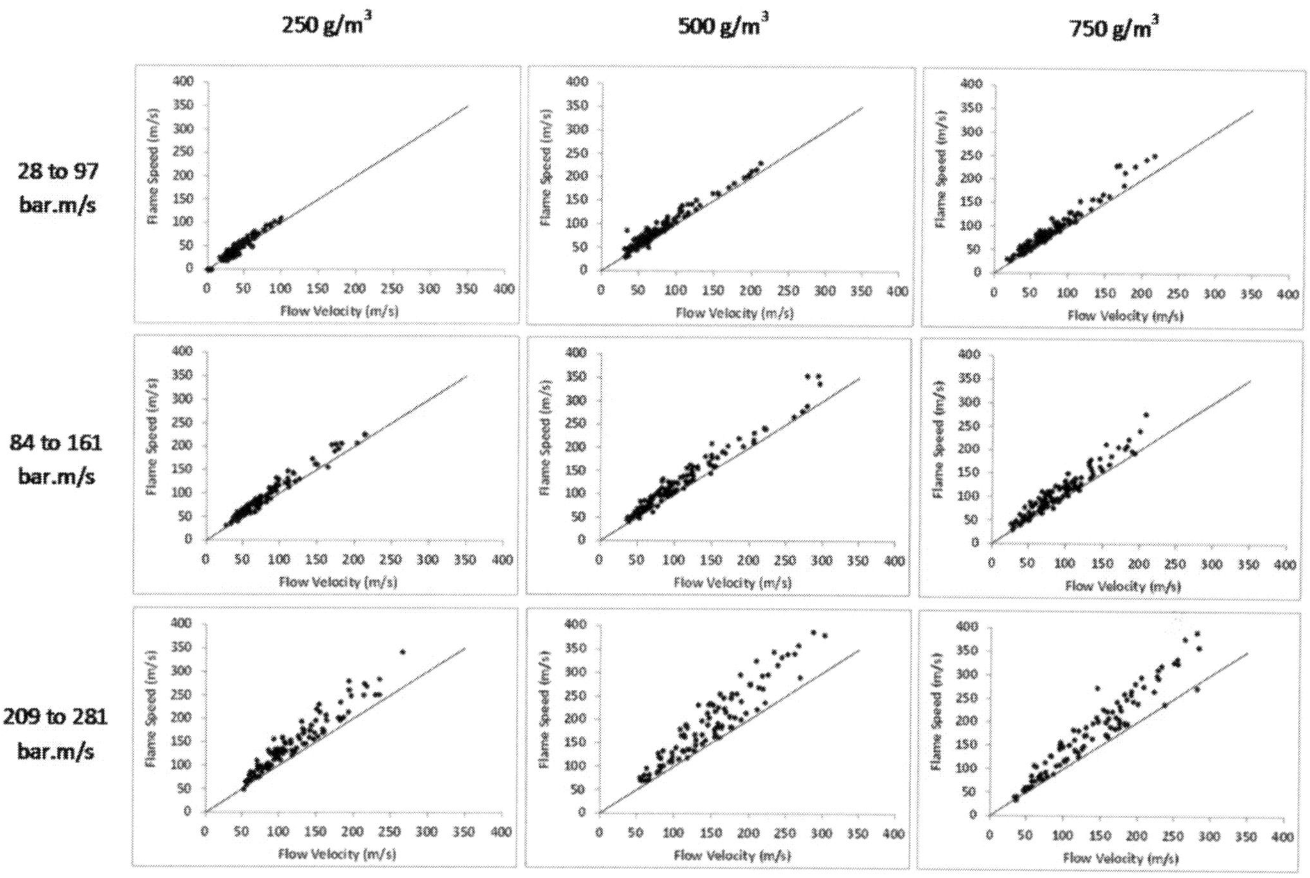

Figure 10: Correlation between flame speed and explosion induced gas velocity in the pipelines for different fuel types and dust concentrations

7. Sensitivity study

To enable extrapolation of the findings to real industry applications, 172 additional scenarios have been simulated in which the flow direction, vessel volume, vent area, vent location, ignition location, pipe diameter and length were varied.

7.1. Effect of vent area

In all cases, the flame speed decreases with larger vent areas. Figure 11 shows example simulation results for two dust types. This finding was consistently confirmed by experiments [3] and is attributed to lower vessel reduced pressures associated with larger venting areas. The decrease in flame speed is less pronounced in the upstream pipe, because the ignition was located closer to the upstream pipe, so that the flame could enter the upstream pipe before the vent panel opened.

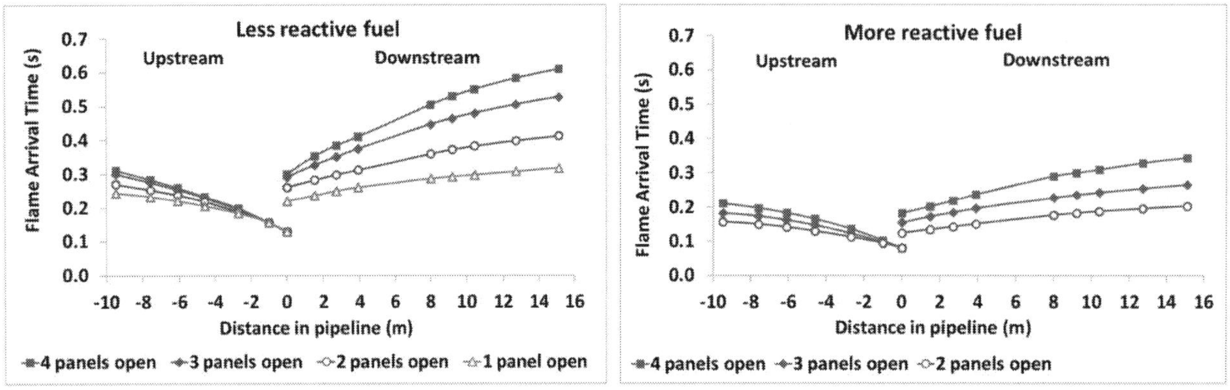

Figure 11: Effect of vent area on flame propagation. Flow = 15 m/s; Vessel and pipeline dust concentration = 750 g/m^3; Ignition location = centre of upstream vessel half.

7.2. Effect of pipeline dust concentration

Many different types of dust transport exist in the industry, including dust aspiration and dust extraction and lean to dense phase transport. Some dusts will stick to the pipe wall and build layers that may become airborne by the explosion propagating through the pipeline, resulting in combustible dust-air cloud generation in transport systems that typically operate at or below the minimum explosible concentration. It is therefore interesting to study the effect of pipeline dust concentration on the explosion propagation. Figure 12 shows examples for two dust types with the pipeline dust concentration changed from 500 g/m^3 to 50 g/m^3. The vessel dust concentration was kept constant at 500 g/m^3. Although the fuels are more reactive at the higher dust concentration, flames are slower at high pipeline dust concentrations. This effect is attributed to the change of the inertia of the fluids in the pipelines. Also this finding was confirmed by the experiments [3].

Figure 13 demonstrates that the combustion in the first meters of the pipeline is fueled by dust ejected from the exploding vessel and not so much by the pipeline dust.

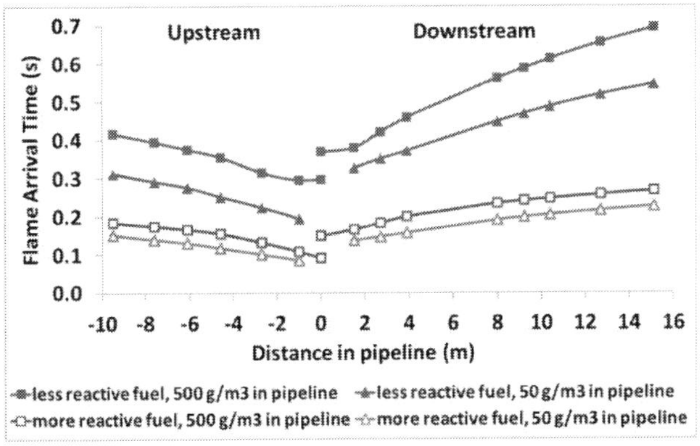

Figure 12: Effect of pipeline dust concentration on flame propagation for two dust fuels; Vessel dust concentration = 500 g/m^3.

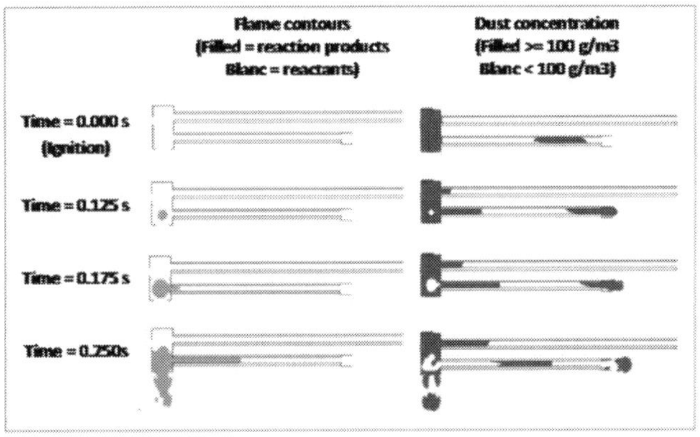

Figure 13: Illustration of how dust is ejected out of the initiating vessel before the flame front arrives for an example explosion; Vessel dust concentration = 500 g/m3; Pipeline dust concentration = 50 g/m^3.

7.3. Effect of ignition location

Scenarios testing the effect of ignition location were run with the igniters located in the three positions indicated in Figure 14 below, including ignition near the vent (1), at the interconnection with the upstream pipeline (2), and in the center of the downstream vessel half, far from the vent (3)

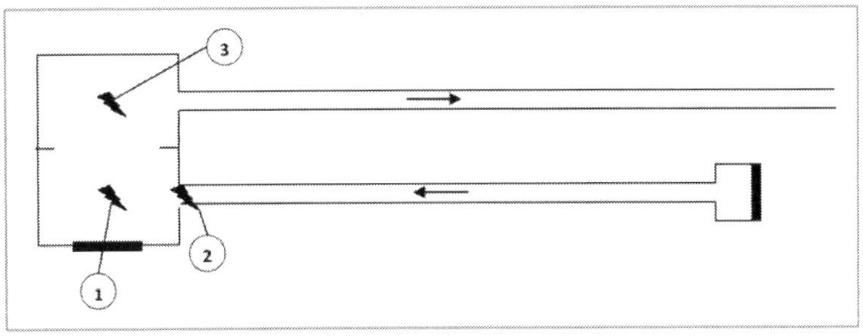

Figure 14: Ignition locations

The effect of the ignition location is twofold: the closer the ignition point is located towards the pipe entry, the sooner the explosion flame will enter the pipeline; and the further the ignition point is located from the pressure venting device, the higher will the explosion pressure in the vessel be, with faster flames as a result. These phenomena are illustrated in Figure 15, where for examples with two dust types, the flame arrival times are compared for three explosion scenarios with varying ignition location. It is clearly shown that flames enter the pipeline earlier for pipeline ignition (2), and that ignition far away from the vent panel (3) generates significantly faster flames

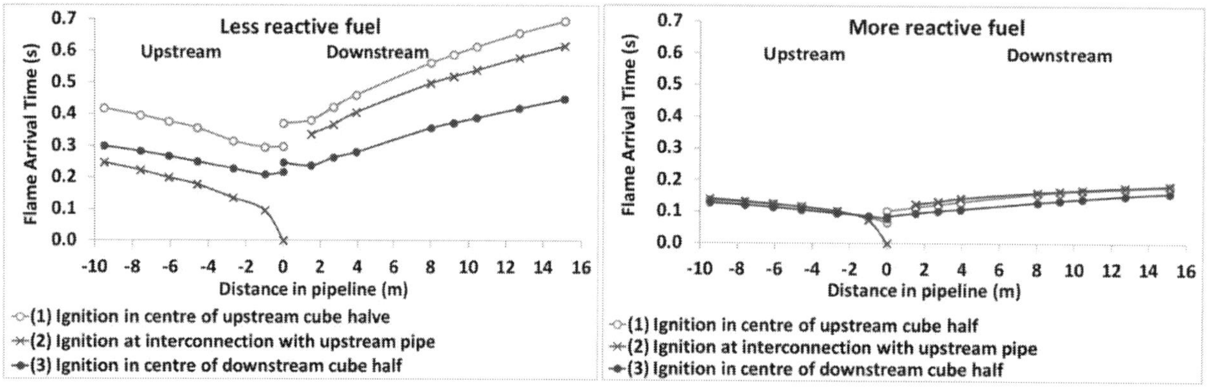

Figure 15: Effect of ignition location on flame propagation. Examples for two dust fuels.

7.4. Effect of vent location

Ideally, the explosion vent panel should be positioned as close as practically possible to the expected ignition location. This will provide an early escape of the combustion gases and will limit the resulting explosion overpressure in the vessel and consequently the flame speed in the pipeline. This is illustrated in Figure 16 and Figure 17: the further the vent panel is moved away from the ignition position (– with vent location alternative 3 further than alternative 2, and alternative 2 further than alternative 1), the faster the explosion flame propagates in both pipelines. The driving factor is the increased explosion intensity in the initiating vessel.

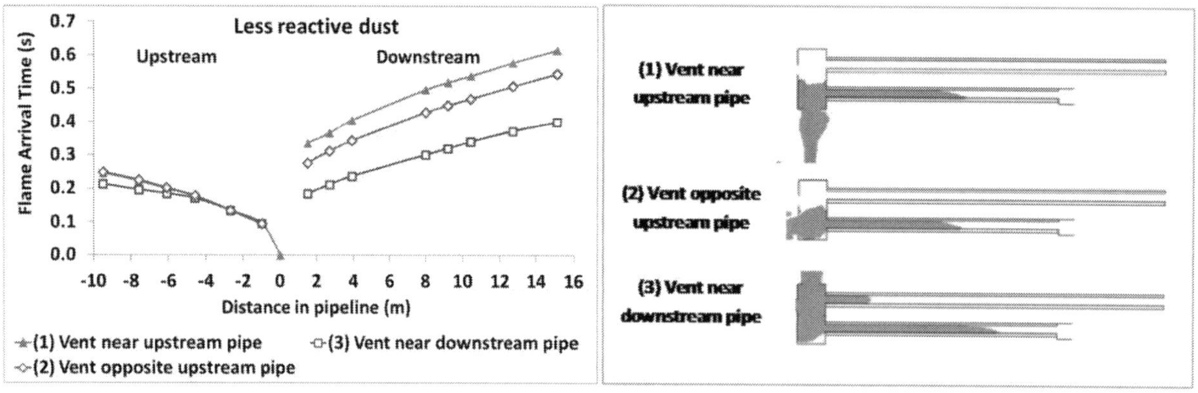

Figure 16: Effect of vent location on flame propagation for an example explosion scenario. Ignition location = interconnection of vessel with upstream pipeline. Left: flame arrival times. Right: flame contours at 0.2 s after ignition.

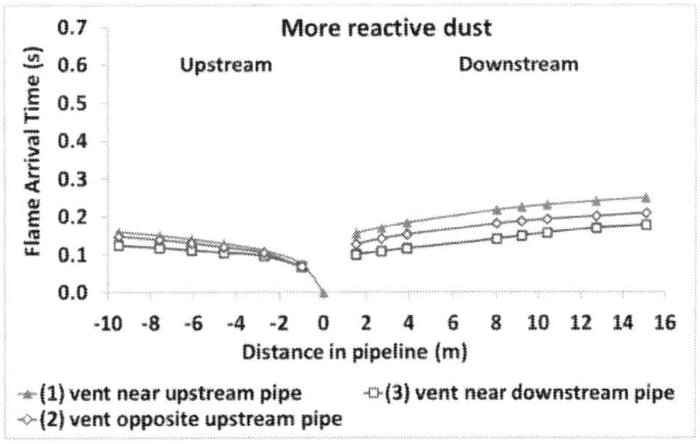

Figure 17: Effect of vent location on flame propagation for another example explosion scenario. Ignition location = interconnection of vessel with upstream pipeline.

7.5. Effect of pipe length

Longer pipes will increase the flow resistance and gas column inertia, resulting in a less aggressively accelerating flame along the pipeline. The effect is most pronounced for cases with low vessel explosion pressure, where flow resistance and gas column inertia play a relatively stronger role, and for cases with low K_{st} value, where the flame velocity is stronger linked with the gas flow velocity. Figure 18 shows the simulated flame arrival times for three example explosions, where the length of the piping was varied from the 11.3/16.5 m of the basic configuration to 40 meters.

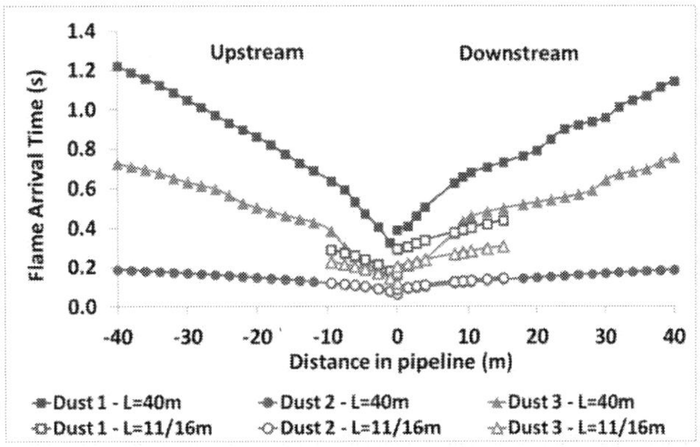

Figure 18: Effect of pipe length on flame propagation for three example explosions.

7.6. Effect of pipe diameter

As with the pipe length, changing the pipe diameter affects the flow resistance. But, in addition to this, the pipe diameter also changes the Reynolds number and thus the gas turbulence in the pipeline. Flames will accelerate faster in smaller pipes as a result of higher turbulence. Note that heat losses of the hot burning gases to the cold pipe wall material have not been modeled. Figure 19 shows the effect on flame propagation when the pipe diameter is decreased from 0.4 m to 0.2

m. For the example with lowest fuel reactivity and generating relatively low vessel explosion pressures, the flame propagates slower in the smaller diameter pipes than in the larger diameter pipes. For the example with highest fuel reactivity and generating relatively high vessel explosion pressures, the flame propagates faster in the smaller diameter pipes than in the larger diameter pipes. For the example with medium fuel reactivity, it is seen that in the initial phase of the explosion, the flame propagates slower in the smaller diameter pipe geometry, but later on, when the explosion pressure develops and the fluid turbulence enhances, the flame will propagate faster.

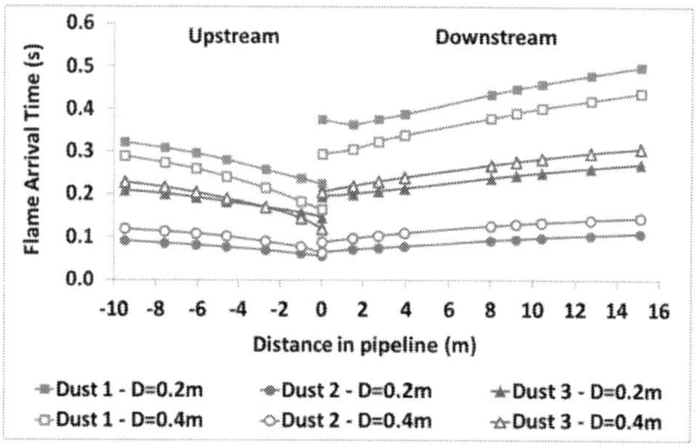

Figure 19: Effect of pipe diameter on flame propagation for three example explosions.

7.7. Effect of vessel volume

When the volume of the initiating vessel is increased, delayed pipeline flame propagation can be expected, because the explosion pressure rise in the vessel will be slower. This is clearly illustrated in Figure 20, where for three example explosions the flame propagation is compared for the cases when initiated in a 5 m^3 vessel and in a 62 m^3 vessel, all other parameters either kept identical or scaled with the volume. The ignition was located in the centre of the 5 m^3 upstream vessel half, which is 0.68 m away from the interconnection with the upstream pipeline, which is then of course out of the centre for the 62 m^3 vessel half. To make the comparison explosion pressure independent, the vent area on the 62 m^3 vessel was increased by factor of 4, which approaches the factor $\left(\frac{62}{5}\right)^{2/3} = 5.36$ which would be required theoretically. Initially the pressure rise is slower which leads to a delayed flame entry into the pipes, but as soon as the vessel pressures in the 62 m^3 exceed the ones recorded in the 5 m3, the flame acceleration rate for the 62 m^3 vessel cases becomes higher.

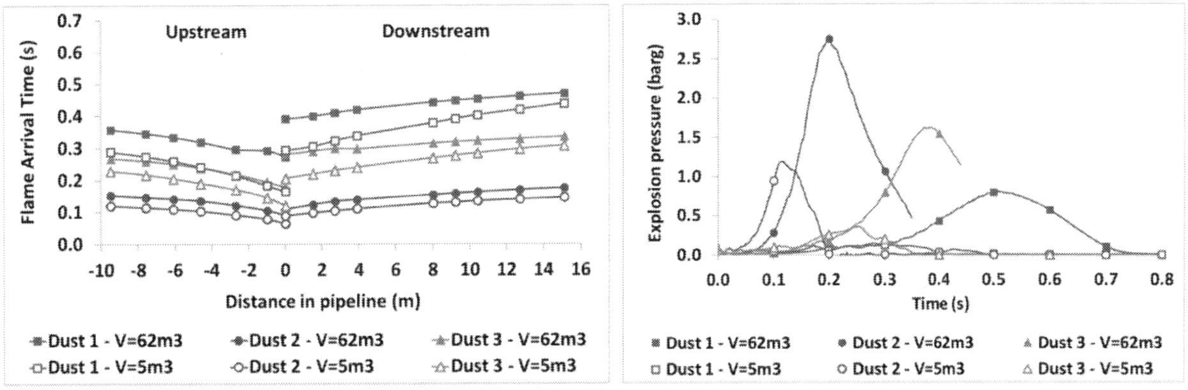

Figure 20: Effect of vessel volume on flame propagation for three example explosions. Left: flame arrival times in pipelines. Right: vessel explosion pressures.

8. Summary

DESC, a CFD code with an integrated combustion model, has been used to study the behavior of propagation of dust explosions through pipelines connected to vented vessels. A 5 m^3 vented vessel in which explosions were initiated, was connected to two DN400 pipelines, respectively 11.3 m and 16.5 meter long. The system was equipped with a fan to establish a flow field prior to dust injection and explosion ignition. Ignition always occurred in the vessel; neither the present study nor the accompanying empirical program [3] considered ignition in the pipeline. Input for the combustion model are laminar burning velocities extracted from K_{st} and P_{max} values determined by standardized 20 L tests.

Comparison between experimental and simulation results revealed that for two out of the three dust fuels, the experimental dust explosions were more intense than the simulated equivalents. A method was devised to improve the confidence level of DESC simulations. By following this approach, and by understanding the underlying physics, it is believed that simulations can represent industry reality with similar level of confidence as experiments.

A total of 94 explosion simulations have been performed in the basic configuration, where the fuel type, vessel and pipeline fuel concentration, vent area and flow velocity were varied. The effects of the parameter variations on flame arrival times in the pipelines were fully in line with the experimental results [3]. In all cases the flame propagated along the full length of the pipes, both upstream and downstream. A high degree of correlation was revealed between explosion induced expansion flow and flame speed.

172 additional simulations were performed in which flow direction, vessel volume, vent location, ignition location, pipe diameter and length were varied. This extended scope enabled to confirm the pronounced effect of the vessel reduced explosion pressure and pipeline geometry.

9. Conclusion

Understanding the mechanisms of explosion flame propagation is paramount for designing effective explosion protection solutions for interconnected process equipment that handle flammable substances. The present research has demonstrated that dust explosions do propagate through pipelines, even against strong flows and that many different application parameters besides K_{st} and P_{max} considerably influence the speed of dust flame propagation. When these parameters are not taken into account, the protection system may fail to protect against the explosion hazard. It has been irrefutably demonstrated that controlling the explosion pressure in the initiating vessel is very effective in controlling the speed of flame propagation. The explosion protection design should therefore always consider the explosion protection problem in its totality: a combination of vessel explosion pressure protection and pipeline explosion propagation protection.

10. References

[1] NFPA 68, Standard on Explosion Protection by Deflagration Venting, 2007

[2] NFPA 69, Standard on Explosion Prevention Systems, 2008

[3] T.M. Farrell, J. Vingerhoets, J. Snoeys, J.E. Going, "Dust Flame Propagation in Industrial Scale Piping - Part 1: Experimental Study in a Conveying Vessel-Pipeline System", Submitted for 9th Global Congress on Process Safety, San Antonio, April 28 - May 1, 2013

[4] T. Skjold, "Review of the DESC project", *Journal of Loss Prevention in the Process Industries*, Volume 20, p291-302, 2007

[5] Arntzen, B.J., "Modeling of turbulence and combustion for simulation of gas explosions in complex geometries", Dr. Ing Thesis, NTNU, Trondheim, Norway, 1998

Additional references not cited:

[6] A.E. Dahoe, J.F. Zevenbergen, S.M. Lemkowitz, B. Scarlett, "Dust explosions in spherical vessels: the role of flame thickness in the validity of the cube-root law", *Journal of Loss Prevention in the Process Industries*, Volume 9, p33-44, 1996

[7] A.E. Dahoe, R.S. Cant. and B. Scarlett, "On the decay of turbulence in the 20-litre explosion sphere", *Turbulence and combustion*, Volume 67, p159-184, 2001

[8] A.E. Dahoe, K. van der Nat, M. Braithwaite and B. Scarlett, "On the sensitivity of the maximum explosion pressure of a dust deflagration to turbulence", *KONA*, Volume 19, p178-195, 2001

[9] G.A. Lunn, "Dust explosions in totally enclosed interconnected vessel systems", *Journal of Loss Prevention in the Process Industries*, Volume 9, p45-58, 1996

[10] P. Holbrow, S. Andrews, G.A. Lunn, "Dust explosions in interconnected vented vessels", *Journal of Loss Prevention in the Process Industries*, Volume 9, p91-103, 1996

[11] M. Roser, A. Vogl, S. Radandt, W. Malalasekara, R. Parkin, "Investigations of flame front propagation between interconnected process vessels. Development of a new flame front propagation time prediction model", *Journal of Loss Prevention in the Process Industries*, Volume 12, p421-436, 1999

[12] A. Vogl, "Explosionsübertragung aus Behältern in Rohrleitungen", *VDI Berichte*, Volume 1272, p215-236, 1996

[13] J. P. Pineau J.P., G. Ronchail, "Propagation of dust explosions in ducts", Symposium on the control and prevention of dust explosions, Bâle, 1982

[14] A. Vogl, S. Radandt, "Explosionsübertragung durch dünne Rohrleitungen", FSA Report F-05-9903, 2005

[15] U. Bielert, M.Sichel, "Numerische Simulation von Staubexplosionen in pneumatischen Saug-Flug-Förderanlagen", *VDI Berichte*, Volume 1601, p449-472, 2001

[16] W. Jian, L. Xinguang, Z. Shengjun, S. Randandt, W. Fuli, R. Chunli, "Flame propagation through potato starch/air mixtures in pipe of interconnected vessels", *Measurement and Control of Granular Materials*, Volume 8, p376-379, 2009

(44c) The Use of CFD to Evaluate the Interactions Between Multiple-Leak Sources and to Assess Effectiveness of Integral Modeling Techniques in a Multiple-Leak Scenario

Joshua Richardson, Ha Nguyen and M. Sam Mannan, Mary Kay O'Connor Process Safety Center - Texas A&M University, College Station, TX

In many practical applications within the chemical industry, it is of vital importance to know the concentration profile that may occur from the dispersion of a material. For example, effect footprints for the release of a toxic material are needed to properly assess risk to both a facility and its surrounding communities and are useful for emergency response planning. Likewise, knowing the concentration gradient of a flammable gas can affect, among many other things, the layout of a facility to prevent or lessen the consequence of explosion and fire scenarios or for the placement of detection systems.

There are two main tools for the modeling of dispersion scenarios. Computational fluid dynamics (CFD) is a useful tool for accurately assessing the dispersion of materials, whether toxic, flammable, pollutant, or otherwise. It has found extensive use in academic as well as industrial settings when scenarios and geometries are well-defined because it takes into account the surroundings in the release environment. However, it is relatively slow and computationally expensive to use.

Also used extensively in the chemical industry are integral consequence modeling tools (such as Phast) which are generally more user-friendly and have the advantage of delivering results in a more time-efficient manner, but cannot take into account complex geometry considerations. A particularly interesting problem to consider in practice is that of multiple simultaneous release sources. This could be caused by a common failure of two units causing concurrent release, a domino effect where an event occurs that causes another event leading to multiple releases, or even an intentional event such as a terrorist attack. While CFD has the capability to resolve the possible interactions between the multiple release sources, integral modeling does not.

A simple current approach to the problem of multiple-source scenarios is to model them independently and superimpose the results upon one another. This may be a good approximation in some cases, but it does not take into account the complex interactions that may take place between the sources. The practical question then becomes whether it is acceptable in certain cases to model a certain set of simultaneous releases as an independent set of sources or whether it could be possible to model them as a single combined source.

This study aims to utilize CFD modeling of a set of varying scenarios to classify multiple-release dispersion scenarios into one of three different categories:

- Releases that are independent and may be modeled as such
- Releases that may be combined and modeled as one source
- Releases that must be considered together due to inherent interactions between them

Based upon criteria such as material, distance between sources, dispersion rate and duration, weather conditions, and geometry considerations, a general methodology for classifying multiple-release scenarios into these three different categories will be proposed so that these releases can ultimately be modeled in the appropriate manner.

(44ca) Understanding, Managing, and Communicating High Consequence Risk

Charles A. Soczek, DuPont Sustainable Solutions, Wilmington, DE

This presentation discusses the relationship between the tools used to identify, understand, mitigate and communicate high consequence risks. The understanding and reporting of high consequence risks, along with providing management, senior leadership, and the Board of Directors with the information necessary to ensure successful risk management, are fundamental outputs to a risk management process. Tools such as the Risk Matrix, Bow Tie Analysis, and Risk Register will be discussed along with key issues such as risk mitigation and risk information flow from the operating level to senior leadership to the Board of Directors. Fundamental to this discussion is the concept that all hazardous events are managed to an acceptable risk through engineering and administrative controls, and a robust Process Safety Management system.

Frequencies of Accidents as Functions of Design Variables

Manuscript ID: 295029

Esteban Bernechea
Centre for Technological Risk Studies (CERTEC). Department of Chemical Engineering. ETSEIB. Universitat Politècnica de Catalunya.
Av. Diagonal, 647. 08028-Barcelona (Spain).
esteban.bernechea@estudiant.upc.edu

Cristian Massaguer
Department of Chemical Engineering. ETSEIB. Universitat Politècnica de Catalunya.
Av. Diagonal, 647. 08028-Barcelona (Spain).
cristian.massaguer@gmail.com

Josep Arnaldos
Centre for Technological Risk Studies (CERTEC). Department of Chemical Engineering. ETSEIB. Universitat Politècnica de Catalunya.
Av. Diagonal, 647. 08028-Barcelona (Spain).
josep.arnaldos@upc.edu

1. Introduction

The mainstream definition of Risk is that it is a measure of the damage caused on humans, the environment or material property, in terms of the probability of occurrence of an accident and the magnitude of the damage it causes. This can be expressed mathematically as follows:

$$Risk = Frequency \cdot Consequences \quad \text{[Eq. 1]}$$

Risk is then, a function of the frequency of occurrence and consequences of a possible accident.

Often, a Quantitative Risk Assessment (QRA) of a specific equipment or installation has to be performed; in Europe the Seveso Directive makes it compulsory for installations that work with a certain amount of dangerous materials. In order to perform a QRA it is necessary to define initial Loss of Containment Events (LOCs) in the studied equipment, to later estimate the frequencies of the accidents that may result from these releases, and the consequences of said events. These values can finally be combined in order to obtain a measure of the risk that the installation poses over individuals or communities.

There is a wide variety of techniques that can be applied to estimate the frequency of an accident, like historical analysis, failure trees or event trees, subjective estimation, Bayesian analysis or Monte Carlo simulations. In Catalonia and Spain it is common to use the values of frequency provided by "The Purple Book" (CPR18E) [1] for the possible LOCs that can occur in different equipment. Later, event trees which depend on the probabilities of occurrence of certain events (like the immediate or delayed ignition of a flammable release) are used to obtain the final frequencies of the set of accidents that will be studied in the QRA.

However, the base frequencies presented in The Purple Book do not explicitly include the possibility that the LOCs will occur due to a domino effect; this phenomenon is defined as a cascade of events in which the consequences of previous accidents are increased both spatially and temporally by following ones, thus leading to a major accident [2]. The complexity of domino effect makes it difficult to model, and more so to include in the calculations performed in a QRA. This leads to the problem of how to take into account a phenomenon that can have a significant impact on the risk associated to the installation, but that is virtually impossible to include (without significantly increasing the number of scenarios to be considered). In some cases in which there are various similar equipment, the common criteria is to select a representative scenario, and multiply its frequency by the number of units and a factor of 2. In this way, domino effect may be included in the calculations, but not in a rigorous way, which could lead to the under or over-estimation of risk.

In this work, a methodology proposed by the authors to produce accident sequences in a domino effect has been exploited to obtain equations that describe frequencies of accidents in a specific installation. The studied plant stores pressurized liquefied propane in horizontal cylinders. The equations will depend on the number of units and the mass that is stored in the installation; they will also depend on the layout, which in turn depends on the size of the tanks, and therefore on the mass and number of equipment. Distances between tanks were defined by using the NFPA58 standard [3].

By finding equations that can be used to estimate the frequency with which a certain accident will occur in the installation, the time required to include domino effect into a QRA could be decreased importantly; only the frequency obtained through the equations, applied to a representative scenario would be necessary to take the phenomenon into account.

The main objective of this work is to use a mathematical model to find equations that relate the frequency of accidents in an installation to some of its design variables. This could be useful not only at the moment of performing QRA, but also at the moment of designing the installation; also, it can help understanding how the frequency of the diverse accidents is affected by the design of the plant.

In the next sections, the model that allows developing the accident sequences is briefly explained, after this, the design of the experiment, which is how the mentioned model was exploited, is presented; later, the results obtained are shown, and finally, some conclusions are reached.

2. Model for the development of accident sequences

The model used to develop the domino effect sequences [4] is based on the use of event trees in a sequential way; it evaluates the damage suffered by one tank in the case that an accident occurs in another, and establishes the possible accidents that may occur afterwards. The LOCs and event trees used to develop the sequences are those presented in the Purple Book [1], and are shown in Table 1 and Figures 1 and 2.

Table 1: LOCs used to define initial accidents

Code	Definition
G.1	Instantaneous release of the complete inventory.
G.2	Continuous release of the complete inventory in 10 min at a constant rate of release.
G.3	Continuous release from a hole with an effective diameter of 10 mm.

The first step to produce the accident sequences is to select one of the units, choose an initial LOC and define the possible accidents that may occur. After this, one of the accidents must be selected, and its effects must be calculated on all surrounding units and compared against the threshold values presented in Table 2.

Depending on the effects of the previous accident each of the affected tanks will suffer a specific LOC, which will in turn, generate new possible accidents; the sequence will continue to unfold until all the tanks have suffered catastrophic events, or the frequencies of the accidents become lower than 10^{-9} y^{-1}. This procedure has to be repeated for each of the possible accidents and for every one of the initial LOCs, starting the sequence in each of the tanks. Finally, the frequencies of all the occurrences of a specific accident in all the sequences, (for example, a flash fire derived from an instantaneous release), has to be summed, in order to know the number of occurrences of the incident.

The frequencies of the accidents down the sequence are calculated using the event trees and probabilities of accidents; each time an accident occurs caused by a previous incident, its initial frequency will be that of the event that caused its failure, multiplied by its probability of occurrence.

Figure 1. Event Tree for the Instantaneous Release of a Pressurized Flammable Gas

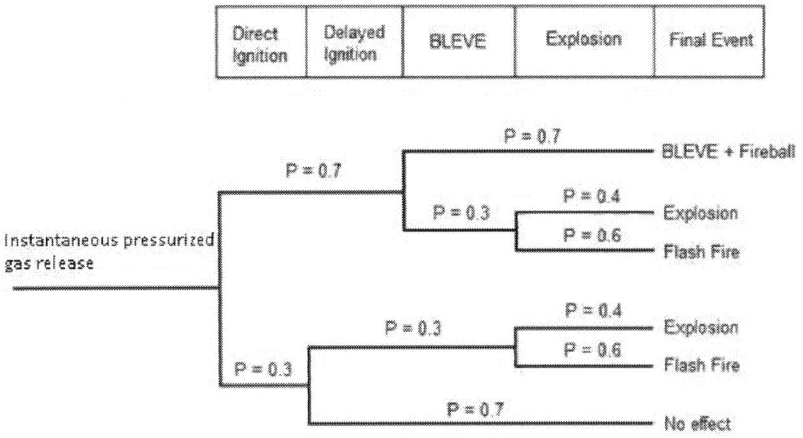

Figure 2. Event Tree for the Continuous Release of a Pressurized Flammable Gas

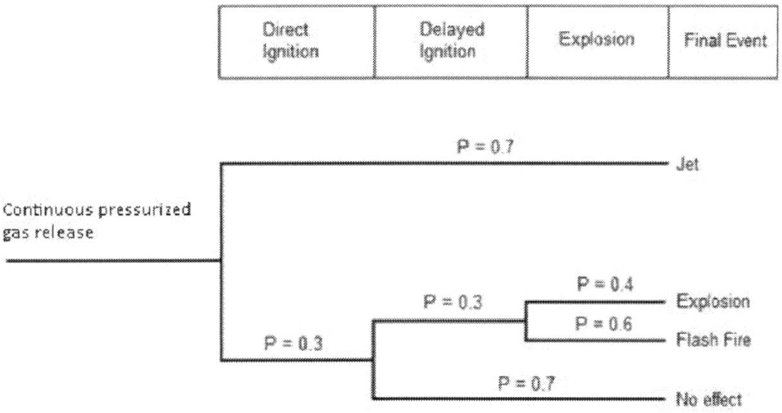

3. Experimental design

The equations that describe the frequencies of the different accidents will be developed for storage terminals of pressurized liquefied propane that use horizontal cylinders. The equations will be functions of the number of tanks used and the total mass of propane stored in the terminal. The number of tanks will be varied from 1 to 4, while 5 values of stored mass will be used: 5,000 kg, 10,000 kg, 20,000 kg, 50,000 kg and 100,000 kg; these possibilities for number of equipment and stored mass will be used to develop different layouts (for example 3 tanks which store 50,000 kg of propane) in which the total mass stored will be equally divided amongst all the tanks (all of which will have the same dimensions). The propane will be said to be stored at the conditions presented in Table 3.

Table 2: Threshold values for tank damage and associated LOCs

Accident	Effect	Threshold Values: Pressurized Tanks	Threshold Values: Atmospheric Tanks	Consequence
BLEVE	Thermal radiation (kW/m2)	≥ 100	≥ 100 [5]	G.3 type release
		< 100	< 100 [5]	No consequence
Explosion	Overpressure wave (kPa)	≥ 100	≥ 25 [6]	G.1 type release
		$53 < \Delta P < 100$ [7]	$20 < \Delta P < 25$ [7]	G.2 type release
		$30 \leq \Delta P < 53$ [7]	$14 \leq \Delta P < 20$ [7]	G.3 type release
		$\Delta P < 30$ [5]	$\Delta P < 14$ [8]	No consequence
Jet fire	Jet length (m) Heat Radiation (kW/m2)	$dj \geq d$ [5]	$dj \geq d$ [5]	G.1 type release
		$dj < d$ & $I \geq 40$ [5]	$dj < d$ & $I \geq 10$ [5]	G.3 type release
		$dj < d$ & $I < 40$ [5]	$dj < d$ & $I < 10$ [5]	No consequence
Flash fire	-	-	-	No consequence
Pool fire	Thermal radiation (kW/m2) Pool duration (min)	$dp \geq d$ [5]	$dp \geq d$ [5]	G.1 type release
		$60 \leq I$ & $tp \geq 10$ & $dj < d$ [9]	$15 \leq I$ & $tp \geq 10$ & $dj < d$ [9]	G.2 type release
		$I \geq 8$ & $dj < d$	$I \geq 8$ & $dj < d$	G.3 type release
		$I < 8$ & $dj < d$	$I < 8$ & $dj < d$	No consequence

dj = jet length; dp = pool diameter; d = distance between tanks; tj = duration of jet fire; tp = duration of pool fire

As the frequency of the accidents will depend, not only on the number of tanks and the mass stored, but also on the distances between the units, this parameter has also been taken into account; this was done by using the distances proposed in the NFPA58 [3], which depend on the dimensions of the tank.

Figure 3 shows an example of the used layouts. In it, L is the length of the cylinder, while ϕ stands for the diameter; B represents the sum of the diameters of the tanks and the distances between them (d). The complete width of the facility will come from multiplying B by a factor of 1.1, and its length, from multiplying L by an equal factor.

Table 3: Atmospheric and Storage Conditions Used for the Application of the Model

Parameter	Value
Ambient Temperature (K)	298.15
Relative Humidity (%)	70
Atmospheric Stability Class	D
Wind velocity (m/s)	5
Ground Roughness Coefficient (cm)	10
Tank Pressure (bar)	9.51
Tank Temperature (K)	298.15

Since the substance stored in the facility is a liquefied pressurized flammable gas, the LOCs presented in Table 1, as defined in The Purple Book are used. The base frequencies of the different accidents for each tank are obtained by using the event trees presented in Figures 1 and 2 for the instantaneous and continuous releases.

Figure 3. Layout example

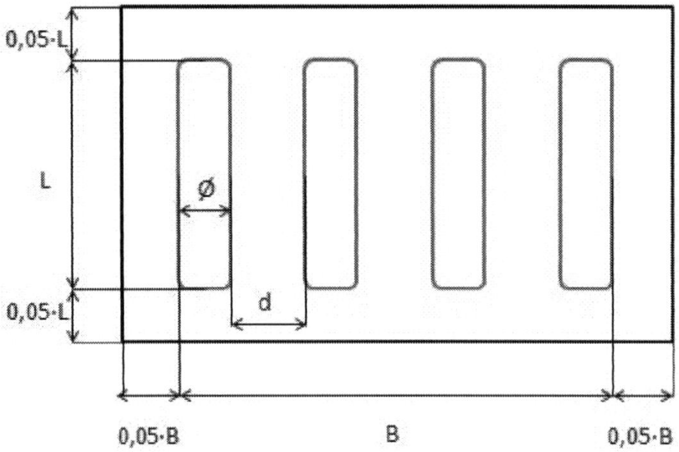

4. Results

It was necessary to use the model a total of 180 times to develop accident sequences for every possible initial accident, combining the number of tanks and values of stored mass. However, since all the proposed layouts are symmetrical, it was only necessary to perform 90 simulations.

Figure 4. G.1 accident sequences for two tanks storing 50,000 kg each

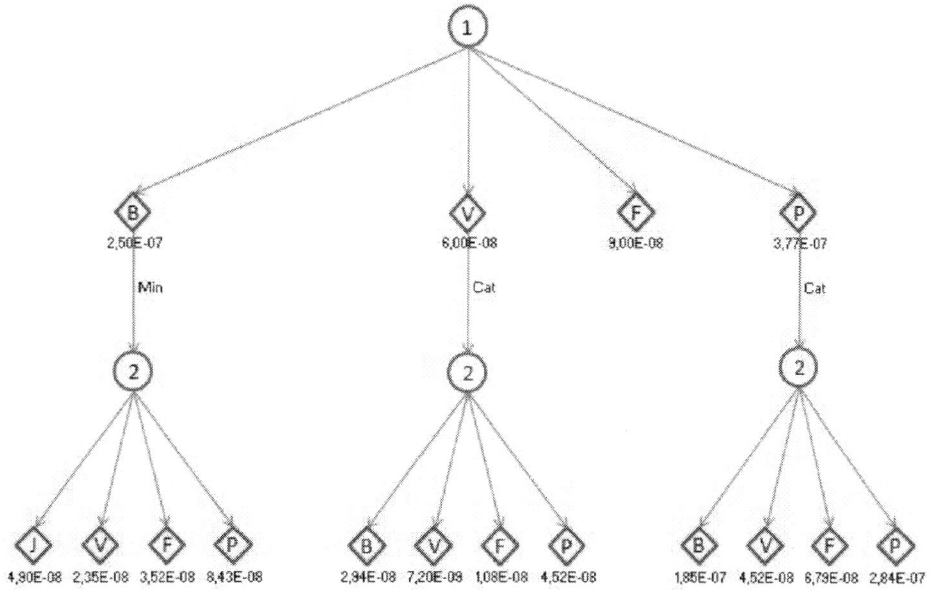

Figure 4 presents the possible accident sequences obtained by applying the model for the case of an instantaneous release from one of two tanks, each storing 50,000 kg of propane. In it, it can be seen that four possible accidents can occur after the G.1 release, a BLEVE (B), a VCE (V), a flash fire (F) or a pool fire (P); the frequencies with which these accidents occur is presented, as well as the level of damage they cause on the second tank, and the possible accidents that may occur in it, due to the initial ones.

The final values of frequency obtained for the case of BLEVE, depending on the mass and number of tanks are presented in Table 4. They serve as examples of the complete results obtained when the model is applied.

After the final frequencies for all possible combinations of design variables have been obtained, they can be plotted in 3D graphs, (in which one of the axis represents the number of tanks, another, the stored mass, and the last one, the frequency of occurrence). The plots obtained for all the studied accidents are presented in Figures 5 through 7.

Table 4: Final frequencies for the BLEVE case

Mass (kg)	Number of Tanks			
	1	2	3	4
5,000	$1.75x10^{-7}$	$1.90x10^{-6}$	$5.77x10^{-6}$	$1.04x10^{-5}$
10,000	$1.75x10^{-7}$	$1.90x10^{-6}$	$4.12x10^{-6}$	$6.89x10^{-6}$
20,000	$2.45x10^{-7}$	$3.44x10^{-6}$	$5.13x10^{-6}$	$6.87x10^{-6}$
50,000	$2.45x10^{-7}$	$3.44x10^{-6}$	$9.95x10^{-6}$	$2.21x10^{-5}$
100,000	$2.45x10^{-7}$	$3.44x10^{-6}$	$1.18x10^{-5}$	$2.10x10^{-5}$

Figure 5. Frequency vs. Number of tanks and Stored Mass for G.1 accidents

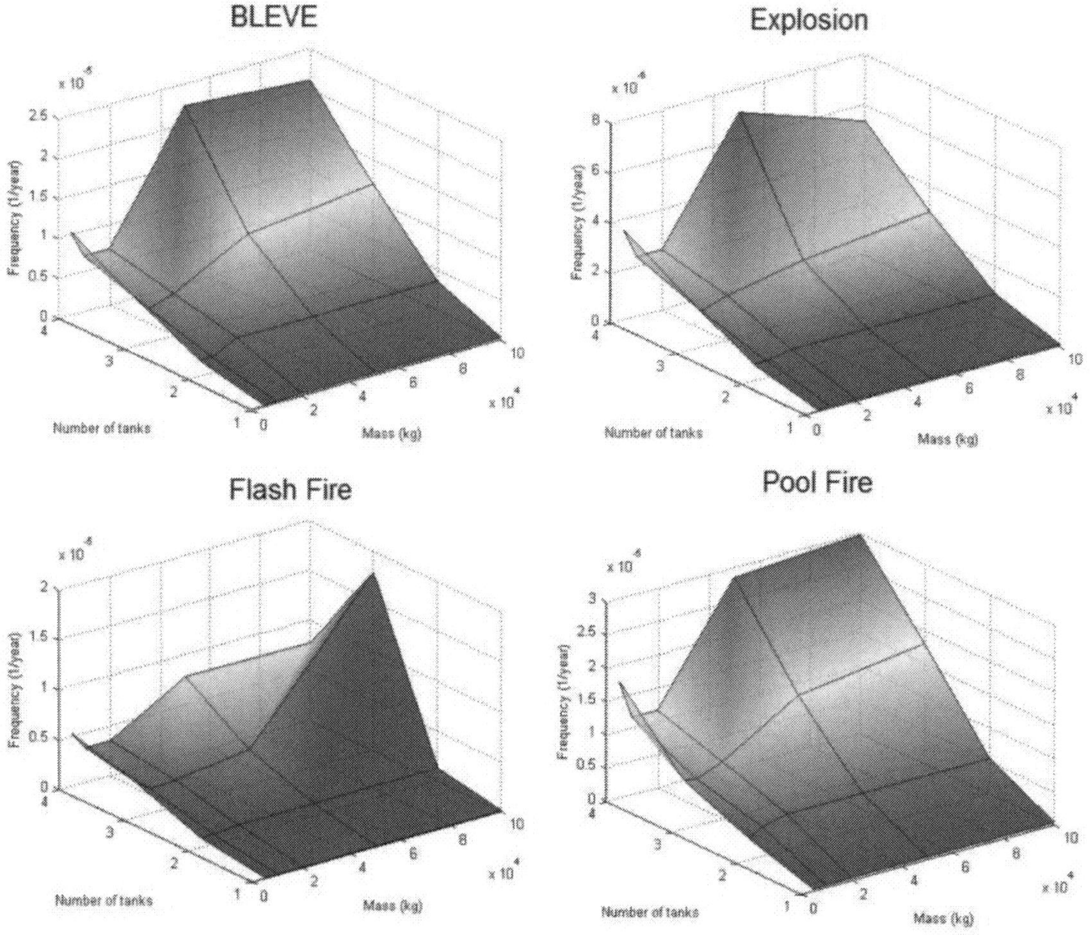

After the plots were obtained, they were fitted using different types of equations, like power, exponential or logarithmic among others. Finally, a Gaussian equation of the form presented in Eq. 2 proved to be the best fitting.

$$f = A \cdot exp\left(\frac{(n-B)^2}{C} + \frac{(m-D)^2}{E}\right) \quad \text{[Eq. 2]}$$

Where n is the number of tanks, m the stored mass in kg and A, B, C, D and E are dimensionless coefficients; these coefficients for all the studied accidents, as well as the values of R^2 obtained for the fitting are presented in Table 5, and the plots obtained when using the Gaussian equation with the presented coefficients are shown in Figures 8 through 10.

Figure 6. Frequency vs. Number of tanks and Stored Mass for G.2 accidents

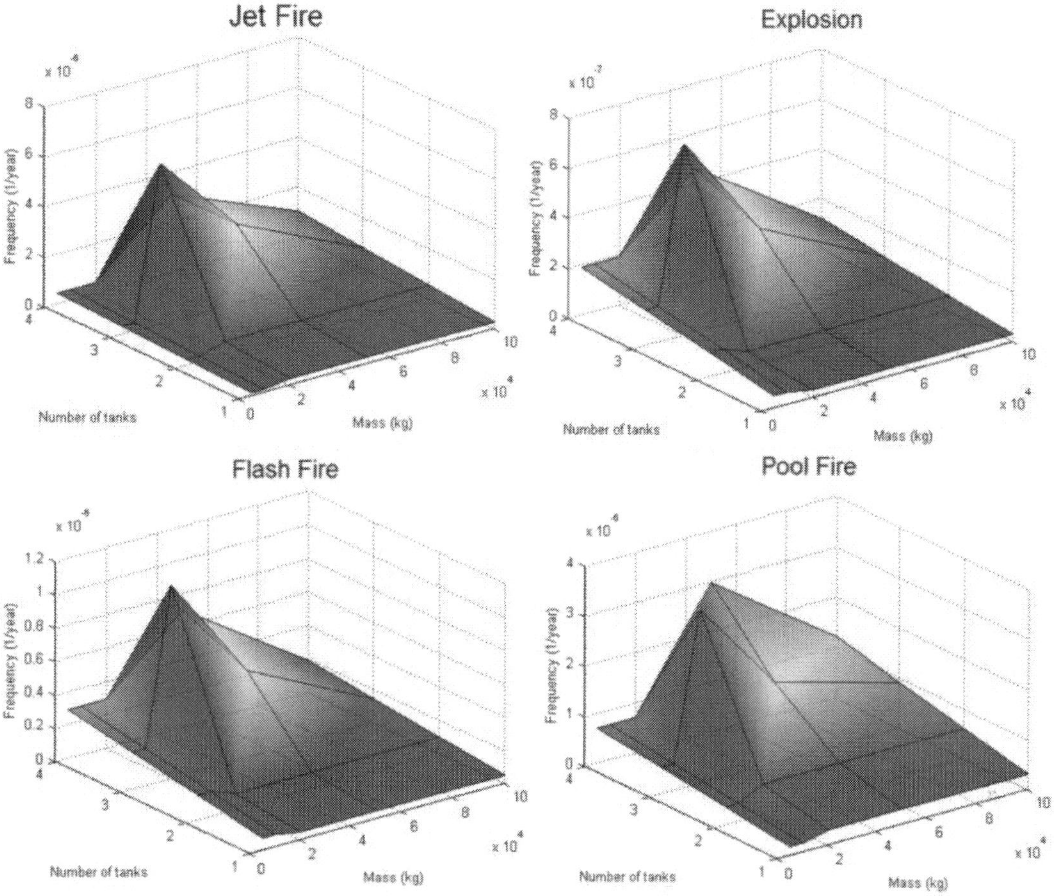

Table 5: Coefficients for the Gaussian equation for all accidents

Accident	Coefficients					R2
	A	B	C	D	E	
BLEVE	2.8x10-5	4.79	-4.15	7.7x104	-4x109	**0.94**
G.1 Explosion	9.9x10-6	5.37	-5.10	7.0x104	-4x109	**0.91**
G.1 Flash Fire	2.8x10-2	3.33	-0.8	1.0x102	-1.2x1011	**0.85**
G.1 Pool Fire	3.4x10-5	4.57	-3.96	8.3x104	-6.3x109	**0.95**
G.2 Jet Fire	2.2x10-5	3.14	-0.56	3.4x104	-1.5x108	**0.84**
G.2 Explosion	1.1x10-6	3.28	-0.82	3.3x104	-3.9x108	**0.77**
G.2 Flash Fire	1.6x10-6	3.27	-0.82	3.3x104	-3.9x108	**0.77**
G.2 Pool Fire	2.7x10-6	3.53	-2.17	5.5x104	-2.3x109	**0.57**
G.3 Jet Fire	1.5x10-4	6.13	-0.15	2.2x106	-1.8x1012	**0.98**
G.3 Explosion	7.8x10-5	6.04	-0.15	2.3x106	-1.9x1012	**0.99**
G.3 Flash Fire	9.2x10-5	6.10	-0.15	2.1x106	-1.7x1012	**0.98**
G.3 Pool Fire	2.2x10-4	6.10	-0.15	2.1x106	-1.7x1012	**0.99**

Figure 7. Frequency vs. Number of tanks and Stored Mass for G.3 accidents

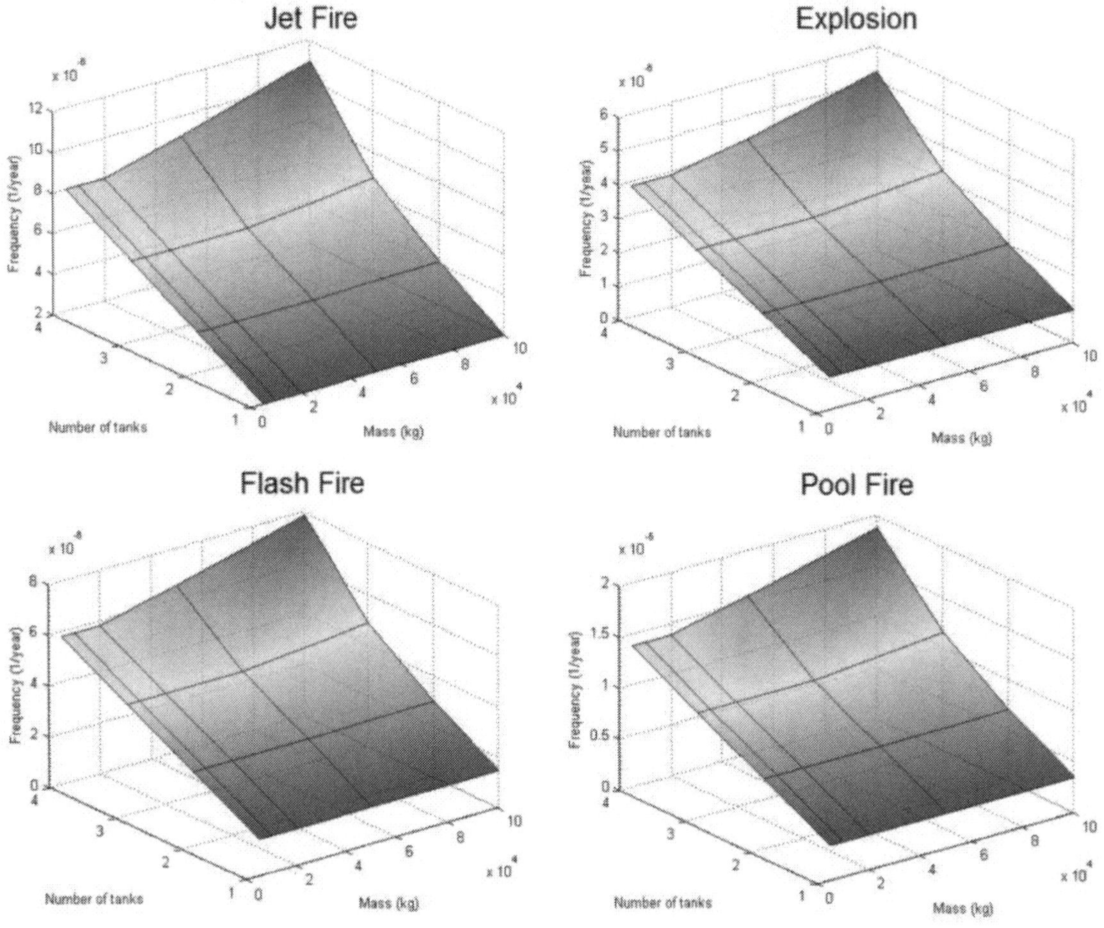

5. Conclusions

Equations that depend on the number of tanks and mass stored in the installation have been found, which, in many of the cases, present good tendencies in replicating the data obtained in the experiment.

The equations obtained could be used to take domino effect into account when performing a QRA on a storage installation of characteristics similar to those presented in the experiment, without increasing significantly the time required to complete the study.

Also, equations of this kind could be used during the design stages of a storage installation to help decide the number of equipment and quantities of substance stored; they could be used to know how risk behaves depending on different designs, and which design is inherently safer.

Figure 8. Frequency vs. Number of tanks and Stored Mass for G.1 accidents using the Gaussian Equation

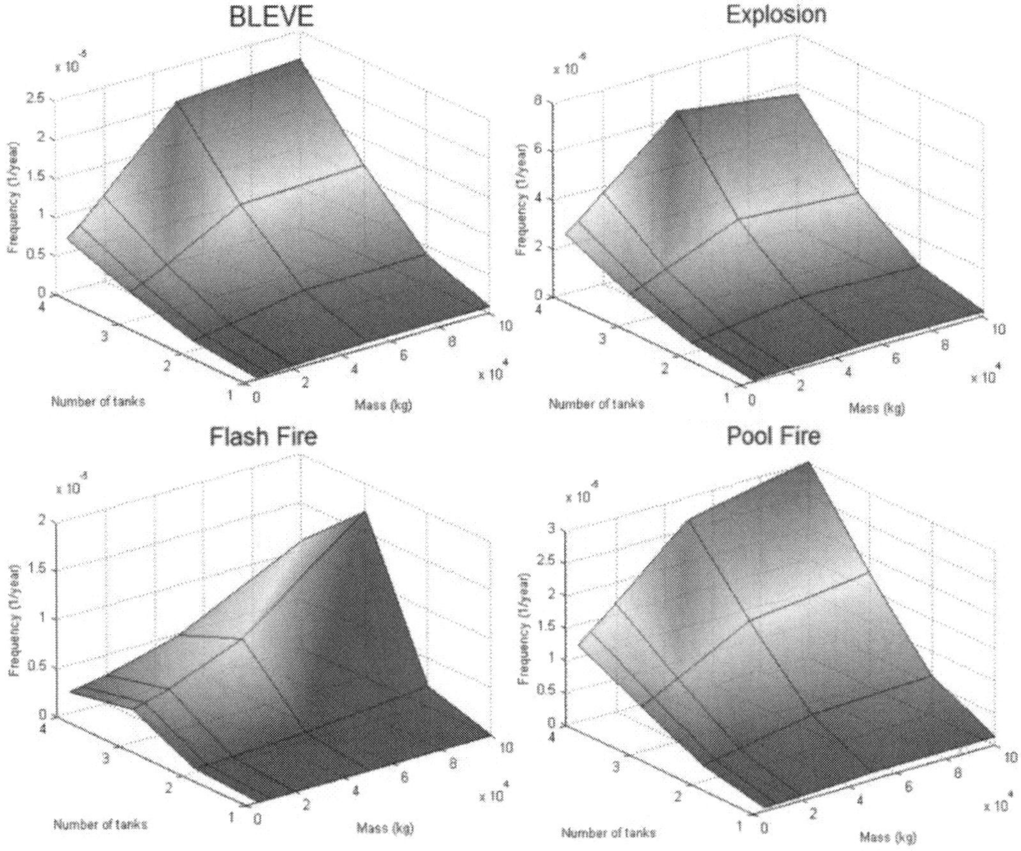

Figure 9. Frequency vs. Number of tanks and Stored Mass for G.2 accidents using the Gaussian Equation

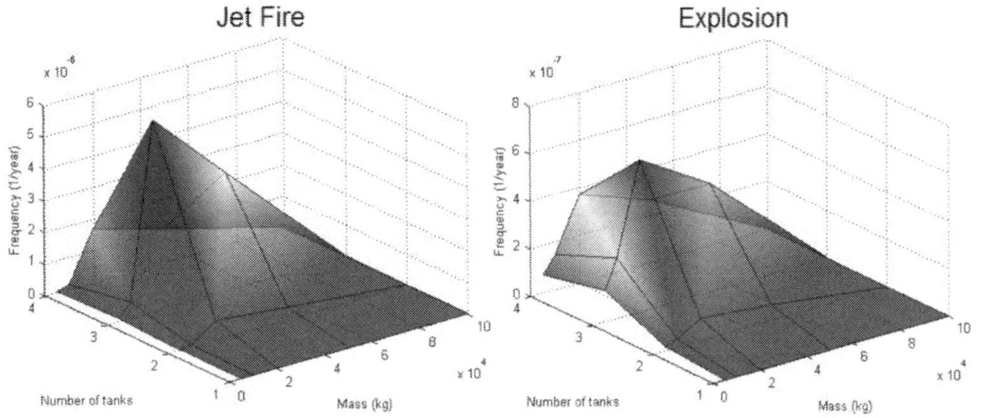

Figure 9. Frequency vs. Number of tanks and Stored Mass for G.2 accidents using the Gaussian Equation (Cont.)

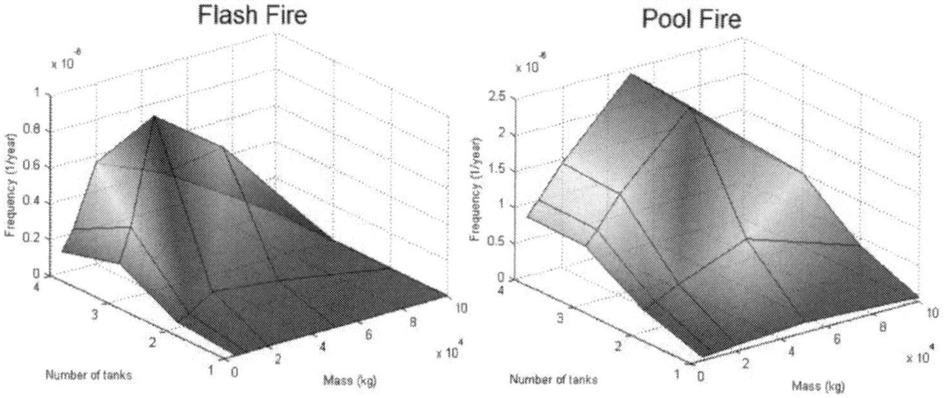

Figure 10. Frequency vs. Number of tanks and Stored Mass for G.3 accidents using the Gaussian Equation

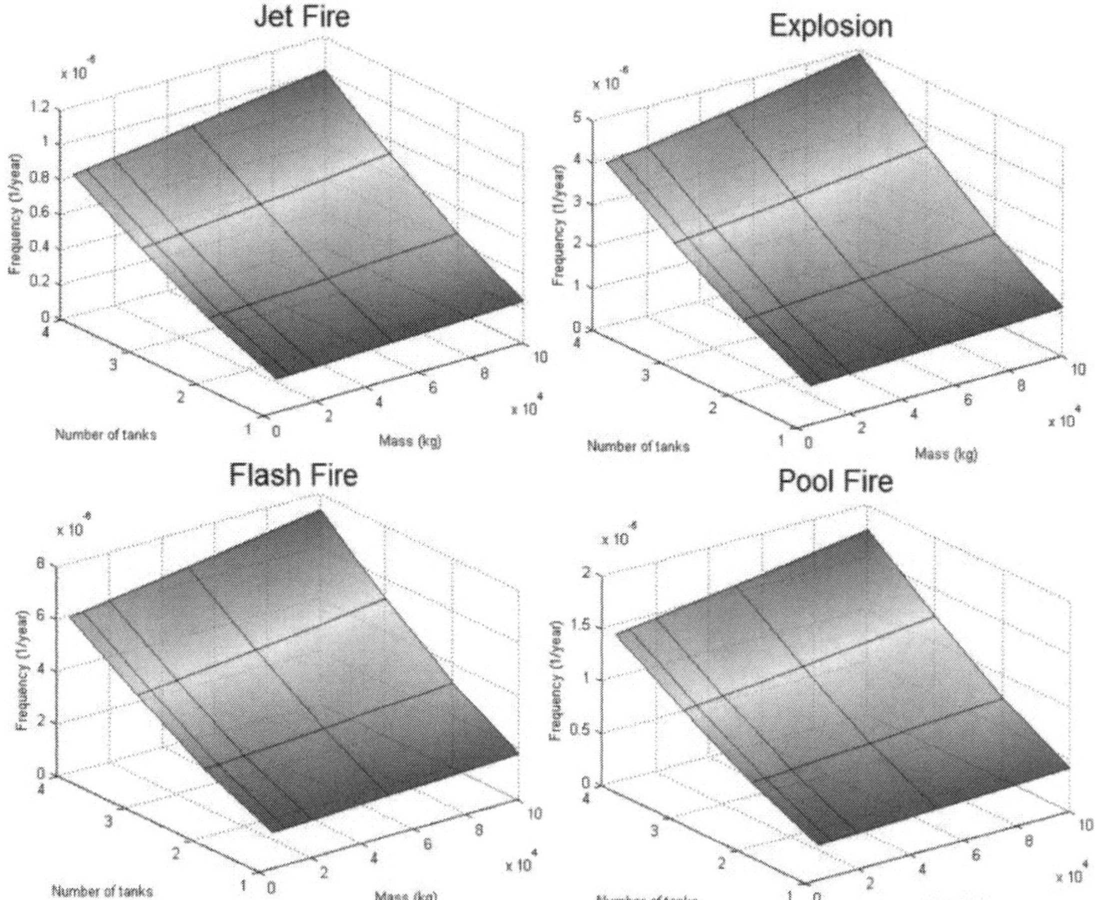

References

[1] CPR 18E, 2005. Guidelines for Quantitative Risk Assessment (Purple Book). Directorate General for Social Affairs and Employment, The Hague, Netherlands.

[2] Delvosalle, C, "Domino Effects Phenomena: Definition, Overview and Classification", First European Seminar on Domino Effects, Leuven, 1996.

[3] "Liquified Petroleum Gas Code". National Fire Protection Association (EEUU), 2001.

[4] Bernechea, E.J., Vílchez, J.A., Arnaldos, J., 2012. A model for estimating the impact of the domino effect on accident frequencies in quantitative risk assessments of storage facilities. Process Safety and Environmental Protection, http://dx.doi.org/10.1016/j.psep.2012.09.004

[5] G. Antonioni, G. Spadoni, V. Cozzani, Application of domino effect quantitative risk assessment to an extended industrial area, Journal of Loss Prevention in the Process Industries, 22, 614-624, 2009.

[6] F. I. Khan, S. A. Abassi, Estimation of probabilities and likely consequences of a chain of accidents (domino effect) in Manali Industrial Complex, Journal of Cleaner Production, 9, 493-508, 2001.

[7] V. Cozzani, E. Salzano, The quantitative assessment of domino effects caused by overpressure: Part I. Probit models, Journal of Hazardous Materials, 107, 67-80, 2004.

[8] V. Cozzani, E. Salzano, Threshold Values for domino effects caused by blast wave interaction with process equipment, Journal of Loss Prevention in the Process Industries, 17, 437-447, 2004.

[9] G. Landucci, G. Gubinelli, G. Antonioni, V. Cozzani, The assessment of the damage probability of storage tanks in domino events triggered by fire, Accident Analysis & Prevention, 41, 1206-1215, 2009.

(44ce) Routes Optimization for Hazardous Materials Transportation in Bogotá, Colombia

Jorge Victoria, Universidad de los Andes, Bogota, Colombia, Nubia Velasco, Universidad de los Andes, Eliecer Gutierrez, Industrial Engineer, Universidad de los Andes, Bogota, Colombia and Felipe Munoz, Universidad de los Andes, Bogota D. C., Colombia

The present work presents an optimization approach to design the set of routes for the transport of hazardous materials (HazMats) at Bogotá. HazMats transportation importance lies in the social, environmental and economic impact that might happen when an accident, during the transportation occurs. This paper is focused in developing a decision support system supported on optimization tools with the purpose of find the routes, -from gas distributor to gas stations-, that minimize: the total travel distance of heterogeneous tank trucks, the social risk and the infrastructure risk around the routes created.

The problem is modeled as a vehicle routing problem and it is solved using the principle of Route first, Cluster second. In a first stage the algorithm generates a big tour where all clients are visited, ignoring the vehicle capacities. Secondly, thanks to a Split procedure the tour is segmented considering the capacities. Then routes are improved with a MultiObjective Metaheuristic called MultiObjective Variable Neighborhood Search (MO-VNS). The algorithm is tested on two set of instances, the first one randomly generated and the second from Bogotá Geographic Information System (GIS). Several approaches to generated solution are proposed.

(44cf) Mejores Prácticas En Los Procesos De Gestión Del Cambio Para Seguridad De Procesos y La Evaluación De Seguridad Previa Al Inicio (PSSR)

J. Luis Aguirre, KMS, Optimized Systems and Solutions, Houston, TX

La Gestión del Cambio continúa siendo el elemento más retador/complicado de la Gestión de Seguridad de Procesos (PSM), del Programa de Gestión de Riesgos (RMP) y de los requerimientos regulatorios de los Sistemas de Gestión de Medio Ambiente y Seguridad (SEMS). A pesar de que el estándar PSM ha existido desde 1992 y que la industria ha gestionando cambios desde hace mucho tiempo, aún es común no obtener el resultado deseado. Los esfuerzos conscientes de muchas personas trabajando en conjunto son necesarios para garantizar que el cambio sea identificado, analizado y ejecutado con calidad. Justo cuando logramos capturarlo en papel y mantener informado al equipo de trabajo involucrado, se presentan cambios, negligencias o las regulaciones cambian. Nuevos involucrado y responsabilidades dificultan la gestión del cambio como sabemos que debe hacerse.

Sin embargo, la tecnología ayuda en todas las etapas del flujo de la Gestión del Cambio (MOC) al permitir la recopilación de ideas, formalizando el análisis de las mismas, su aprobación y ejecución del cambio. También puede ayudar a proporcionar visibilidad del estado actual y análisis de tendencias, así como la identificación de peligros a los que están expuestos los diferentes equipos, sitios, y bienes. Correctamente aplicada, la tecnología podría ayudar a predecir el futuro. Además, la tecnología facilita y ayuda a implementar sus estándares resultando en Mejores Prácticas de Gestión del Cambio, Evaluación de Seguridad Previa al Inicio (PSSR) y el seguimiento al cumplimiento de acciones, notificaciones de alerta y el cierre.

(44cg) Lesha – Multi-Layer Progressive Stress & Impact Assessment On Health & Behavior

Salvador Ávila Filho, PEI TECLIM, UFBA, Salvador, Brazil

The chemical process industry seeks, constantly, to achieve the demands social and environmental (mandatory and voluntary) within the international economic chain. The Integrated Management Systems meet issues in the areas of quality, environment, safety, health, social responsibility, now including risk management (ISO31000). This excess of valuation in integrated management system just prioritizes management functions in place of the functions related to core activity. By integrating management activities into the routine of work, the task will be performed in a more committed form avoiding the domain of management system in place of production skills.

Mandatory Standards for work in the areas of health, hygiene and safety are concerned in dealing the most damage, searching through punishment, correct managerial failure, or correct the ignorance. The preventive tools are not fully developed leading to leaders to be worried about maximum allowable values in the areas of safety, health, hygiene, and also the environment.

The Voluntary activities are not developed due to difficulties initiating deployment of management systems that were born in a TOP-DOWN philosophy. The human and operational reliability was borned in a shape BOTTOM-UP avoiding the occupation of operator, most part of time, doing management systems reports. Operational risks are analyzed including the role of man, the role of group in the planning and execution of tasks. The techniques involve subjectivities as effect of cultures fight in the body of work(social - mass culture, regional culture, organizational, and security). The analysis of dominant cultures involves identifying the various vectors impacting cultural influences in the workplace. Psychological stress causes physical and behavioral changes (affecting the decision) and in the body, both, can cause the loss of quality of life.

Various techniques were developed and presented in GCPS to assess the hazard and the risk of social conflicts and that can cause human error. This error initiates or increases the process of failure in equipment or processes through the intervention performed on the steps of the task. Thus, social conflict in new business (APRSH), cognitive traps and human error in critical task (Social HAZOP), causality analysis of cognition / decision, task and process losses (COGNOSP), analysis of organizational and cultural factors as root cause for accidents, decision analysis operation in progressive stress (LODA), and analysis of cognitive failure of operator in the control of equipment and components (FMEAH).

At the present moment we present a technique that try to study the effect of progressive psychological and physical stress on the body and quality of life, here dubbed as LESHA (Stress Analysis & Progressive Effect on Human Health and Behavior). This study highlights the importance of analyzing the life cycle of faults that impact on health by studying the trend of biological indicators resulting from psychological and physical stress. We also emphasize the search for signs of failure before occupational diseases being installed, showing the most suitable installation of barriers against failure, bringing the barrier closer to the root cause.

The LESHA technique therefore intends to monitor the level of physical and psychological stress of environmental factors, and stress from events related to task, which can turn a pathogen resident in symptom (Holnagel and Reason) and hence in occupational disease.

The events that initiate failures, and the stressed environments can cause stress progressive state on individuals, activating, or causing the trigger of failure through the stages: pathogen resident; uncontrolled organic process; symptom; disease; replication of the disease; and fatality or permanent labor shortage. Each level of stress is analyzed for signs and / or indicators of failure that impact over worker health permitting taking preventive or corrective action (biological safety barriers).

The following levels of analysis are proposed: Level 1 - Identification of Human Type, and the respective pathogen resident or signs that may cause the future disease in the body and in the behavior. Level 2 - initiators of failure in the task, or, bad social relationship that can cause the lack/ fault of organic metabolism, generating growth signals, variables still toward the disease. Signs of lost productivity interlinked with organic symptoms ... Level 3 - Shares of the social environment (society) or technical (technical knowledge), and secondary causal events and their impact on the body by increasing the metabolism of organic lack/ fault of symptom onset with fixed initial stage of the disease. Signs of lost productivity interlinked with organic symptoms...

Level 4 - Emergence of social ills (irritation, and communication difficulties), emergence of occupational diseases (slips related to the task and can cause accidents), and biological diseases like ulcer, high blood pressure and other diseases related to workplace using indirect biological ways (early onset of diabetes or other). Level 5 - Replication of the disease by adopting bad habits in the workplace where colleagues, "introject" behavior and even biological processes of "friends." Level 6 - fatality or permanent labor shortage causing low morale in the team, and developing or activating resident pathogen not active or nonexistent.

Understand the signs of this biological process is challenge of safety, hygiene and occupational medicine in conjunction with studies of human reliability. Different disciplines come together to prevent the disease from the body (regardless of the root cause) had been evident. The diseases are originate from labor, directly, or resulting from the influence of the job, and, from society influence. One should avoid firing the trigger of failure and the development of the pathogen resident.

The result of this analysis allows build the programs to avoid diseases. Prevention and correction programs depend on the intensity of the interference signals and the stress level reached in this process. It is proposed for each level and intensity of the biological and behavior signals following through risk mapping:
* Intensity Level X frequency.

9TH GLOBAL CONGRESS ON PROCESS SAFETY

Building Design for Resilience Beyond Design Basis Blast Loads

Michael A. Polcyn
Baker Engineering and Risk Consultants, Inc.
3330 Oakwell Court, Suite 100
San Antonio, TX 78218
MPolcyn@BakerRisk.com

Raymond H. Bennett, Ph.D., P.E.
Baker Engineering and Risk Consultants, Inc.
3330 Oakwell Court, Suite 100
San Antonio, TX 78218
RBennett@BakerRisk.com

Prepared for Presentation at
American Institute of Chemical Engineers
2013 Spring Meeting
9th Global Congress on Process Safety
San Antonio, Texas
April 28 – May 1, 2013

UNPUBLISHED

Building Design for Resilience Beyond Design Basis Blast Loads

Michael A. Polcyn
Baker Engineering and Risk Consultants, Inc.
3330 Oakwell Court, Suite 100
San Antonio, TX 78218
MPolcyn@BakerRisk.com

Raymond H. Bennett, Ph.D., P.E.
Baker Engineering and Risk Consultants, Inc.

Keywords: blast resistant design, consequence-based design, risk-based design

Abstract

One of the guiding principles of API RP-752 is that buildings intended for occupancy are to be designed, constructed, installed, modified, and maintained to protect occupants against explosion, fire, and toxic material releases. Owners/operators may use either a consequence-based approach or a risk-based approach to evaluate the siting of both new and existing buildings for blast loads. While a consequence-based approach should be based on a maximum credible event (MCE) for each building and type of hazard considered, a risk-based approach may consider numerous scenarios, accounting for both the consequences and the frequencies of the scenarios. However, when designing either new buildings or upgrades to existing buildings, designers are typically provided only a single set of blast loads and a single response criteria (for example, medium response per ASCE "Design of Blast-Resistant Buildings in Petrochemical Facilities"). A lower response level implies lower occupant vulnerability, while higher response levels imply higher occupant vulnerability. A design for higher magnitude loads to low response criteria would result in very conservative design. On the other hand, a design for lower magnitude loads to higher response limits would result in a building without capacity to protect occupants in the event of a more severe explosion.

A better approach would be to consider multiple levels of blast loads when developing a design. Lower magnitude loads (usually from higher probability events) should require designs to meet the lower response limits, while higher magnitude loads (usually from lower probability events) should require designs to meet the higher response limits. This paper demonstrates this approach by evaluating typically designed building systems for a range of blast loads commonly encountered at refining and chemical plants. Recommendations are provided for the design of new buildings or upgrades to existing buildings. By including the concept of resilience in the design process, the designer may provide protection from future yet-to-be-determined hazards or scenarios.

1 Introduction

The structural design of buildings for conventional loads, including wind, snow, and seismic loads, is a well-defined process that is followed by engineers on a routine basis. Loads are specified by various codes or design manuals (e.g., International Building Code [1] or ASCE 7-05 [2]). Similarly, the design approach for determining the capacity of various types of structures is defined in corresponding design manuals. For example, the American Concrete Institute publishes "Building Code Requirements for Structural Concrete" [3], and the American Institute of Steel Construction publishes the "Manual for Steel Construction" [4].

In contrast, the design of buildings subjected to blast loads is not as formalized. Design case blast loads are not specified in any guidance documents and are left for the owner to define appropriate loads based on the processes and materials in place at their facilities. There are several guidance documents that describe approaches for designing and analyzing structures subjected to blast loads, which include dynamic analysis methodology, material properties, and structural response limits. Response or deformation limits, such as those defined by the American Society of Civil Engineers [6], are given ranges of response, where Low Response refers to a component with no more than slight visible permanent damage, Medium Response corresponds to a component with some permanent deflection, but is generally repairable, and High Response indicates that the component has not failed, but has undergone significant permanent deflections and is likely unrepairable following the design case loads. It is up to the owner to select the response level, and thus the level of protection to building occupants, for their building design.

This paper provides an overview of the design process, highlighting the differences between a consequence-based design approach and a risk-based design approach. The discussion continues by presenting features to the design that will enhance the structure's ability to be usable following an explosion event and to provide some degree of protection to occupants for events exceeding the selected design load cases. It should be noted that only explosion hazards are addressed in this paper; building and upgrade designs should also consider hazards from fire and toxics.

2 Dynamic Analysis Method

When designing buildings or building upgrades for blast loads, structural components are usually assessed using nonlinear dynamic Single-Degree-of-Freedom (SDOF) analytical methodology. The SDOF methodology summarized in Biggs [6] is a widely accepted approach for estimating damage to structural components subjected to time-varying blast loading. The SDOF approach is a government and industry accepted methodology that has been described in many guidelines such as UFC 3-340-01 [7], UFC 3-340-02 [8], and ASCE [5].

The dynamic response of structural components from prescribed blast loads is determined by modeling the components as an SDOF system, such as the equivalent spring-mass system shown in Figure 1. Structural components such as walls, windows, beams, doors, and panels will deform and vibrate when loaded with a shock wave. The SDOF model for each component is constructed using the component's physical structural properties so that the model will exhibit the same displacement history as the point of maximum deflection in the component. The displacement history of the SDOF model is obtained with numerical integration techniques using a time-stepping computer program to solve the equation of motion of the equivalent system at discrete time steps.

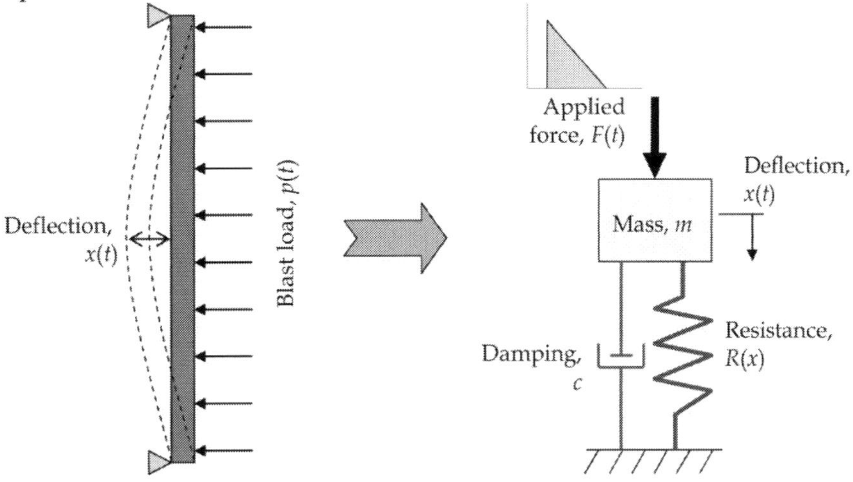

Figure 1. Equivalent Spring-Mass SDOF System

The calculated peak deflection (see Figure 2 for example) is used to obtain the support rotation and ductility ratio, which represent the deformation limit criteria most commonly used in blast design. As shown in Figure 2, the support rotation is the angle between the original un-deflected shape of a component and a straight-line segment between the point of maximum deflection and the support. The ductility ratio expresses the maximum deflection in terms of the minimum deflection of the component when it reaches its ultimate capacity. Therefore, ductility ratios that are greater than unity (1.0) indicate that permanent deformations have been sustained.

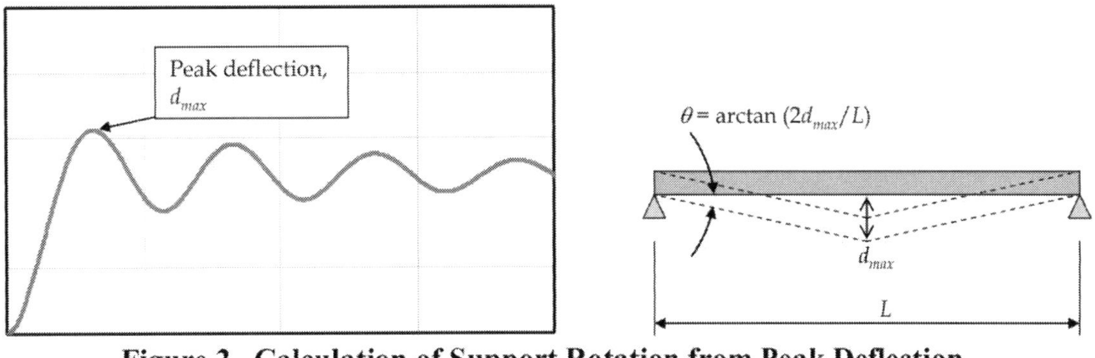

Figure 2. Calculation of Support Rotation from Peak Deflection

The design process for blast loads is an iterative process, where the pertinent design blast loads and response criteria are defined, "trial" structural members are selected based on the building geometry, and the structural components are analyzed using the dynamic analysis methods discussed above. If the response of the component does not meet the required response criteria, a heavier section is evaluated or the building geometry is modified until the analysis shows that the response criteria are met. Connections are typically designed to resist the maximum end reactions predicted from the dynamic analysis.

Pressure-impulse (P-i) diagrams are commonly used to assess building or component response to multiple sets of blast loads. P-i diagrams are plots of constant damage levels for various combinations of blast overpressure and impulse. Each individual curve represents the suite of all pressure and impulse pairs that will result in the same structural damage to a given component. The combination of the three individual curves forms a complete component P-i diagram to define Low, Medium, and High Damage Levels. An example of a P-i diagram is provided in Figure 3. As shown, the region above the upper curve would imply component failure.

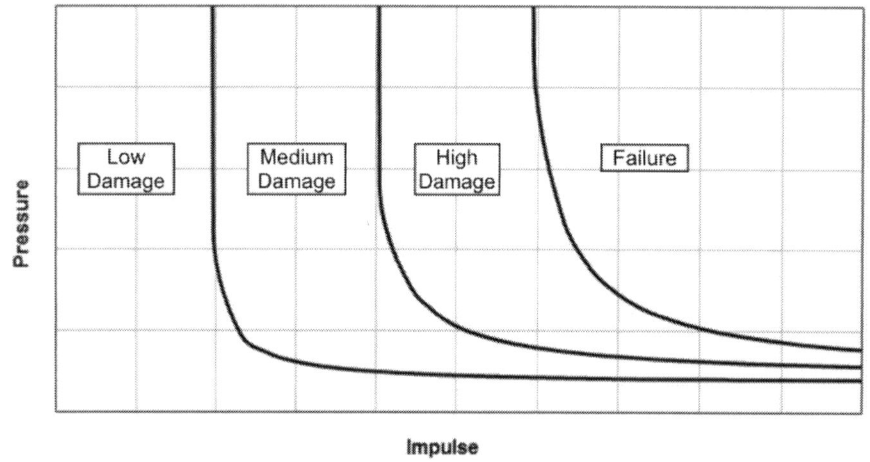

Figure 3. Example of a Pressure-Impulse Diagram

The use of P-i diagrams allows for quick consideration of a variety of blast loading situations. If the blast load of interest falls below or to the left of a particular damage curve, then the given component is predicted to receive damage less than or equal to the level represented by that P-i curve. In addition, component curves can be combined to develop a set of building P-i damage curves, or BDL curves.

3 Consequence-Based vs. Risk-Based Design

Traditionally, design of buildings or upgrade to buildings has been performed using a consequence-based approach where the frequency of events is not considered. According to API RP-752, a consequence-based approach is to be based on maximum credible events (MCEs) for each building. For example, a precast concrete wall for a control room could be designed to meet the ASCE Medium Response criteria when subjected to the applied blast loads from the MCE determined for that building. The resulting P-i curves for the wall would be similar to those shown in Figure 4.

The figure also demonstrates the concept of resiliency in the design. Four pressure-impulse pairs are shown, corresponding to two Design Load Cases and two Potential Overload Cases. The potential overload is 1 psi greater than the design pressure for both cases, while the impulse is held constant. In both Design Cases, the component just meets the Medium Response criteria. However, the curves show that for Design Load 1, the component has the capacity to resist the overload, albeit with larger deformations, but without failing (the component undergoes a high level of response). For low probability events, such behavior might be acceptable. However, the implied resiliency of the component cannot be determined solely by setting the response limits. For example, in Design Case 2, the corresponding overload will result in the failure of the component, even though the Design Case load is farther below the boundary between medium and high response. Thus the true resiliency of a design has to be evaluated in light of the potential for an overload.

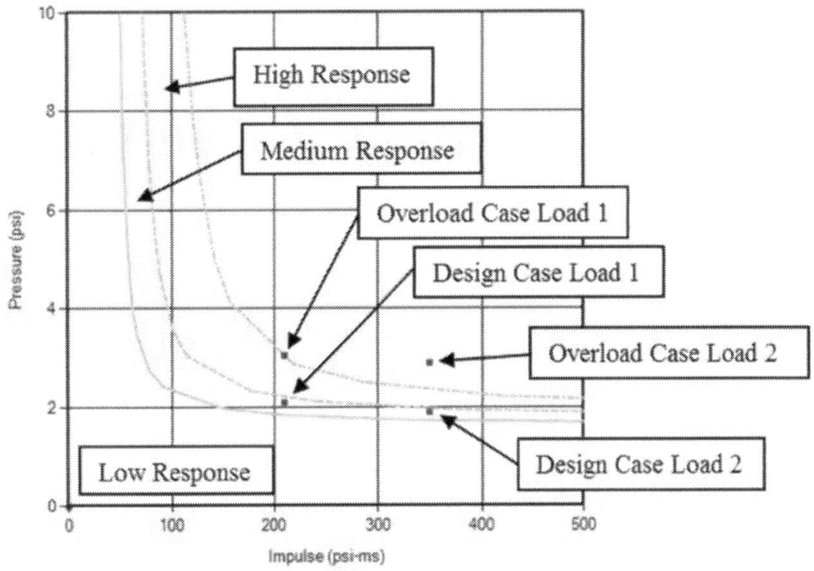

Figure 4. Component P-i Curves

In more recent years, companies have conducted quantitative risk assessments (QRAs) in support of facility siting study efforts. Such studies include estimates of the expected frequency and consequences of potential accidents, typically resulting in an estimate of risk, either as an aggregate risk, an individual risk, or a relationship of the expected frequency to the consequence (e.g., an F/N curve). This approach, which considers a large number of possible explosion scenarios, can be extended to a risk-based design approach for buildings or building upgrades.

When using a risk-based approach, it is necessary to consider the overall building damage rather than the response of the individual components independently. By doing so, the predicted building damage for each scenario can be related to the expected occupant vulnerability [9], which is used to perform the risk calculations. For example, the Building Damage Level (BDL) curves in Figure 5 show that the building is expected to undergo moderate damage, or BDL 2a, when the components are designed to meet the ASCE Medium Response criteria. Note that the inflection points in the BDL curves are caused by a change in the governing building component. For example, the upper left portion of the BDL 3-4 curve may be set by roof purlin response and the lower left portion may be set by a roof girder response. If the design case loads are defined by the MCE for all of the postulated explosion scenarios, the BDL plot in Figure 6 can be generated showing the loads from "all" postulated scenarios with the BDL curves. As shown, none of the scenarios will result in greater than moderate damage (BDL of 2a), and the resulting risk will be very low. It can be concluded that such a design has significant resiliency to withstand loads greater than those postulated.

Alternatively, the building could be designed for higher response limits under the design load cases, possibly allowing partial building failure under the most severe load cases. The BDL curves in Figure 7 demonstrate a case where the components were designed to meet the ASCE High Response limits for the MCE and ASCE Medium Response Limits for the majority of the load cases. As shown, the expected damage for the MCE approaches higher BDLs, but building collapse (BDL of 4) is not expected. In this case, the expected risk to building occupants would still be relatively low, although somewhat higher than the risk demonstrated in Figure 6. However, there is less resiliency in the design should there be an overload case. Note particularly that if the scenario shown in Figure 5 were 0.5 psi greater, the BDL for the building portrayed in Figure 6 would still be 2A, while the BDL for the design represented in Figure 7 would be increased to BDL of 3.

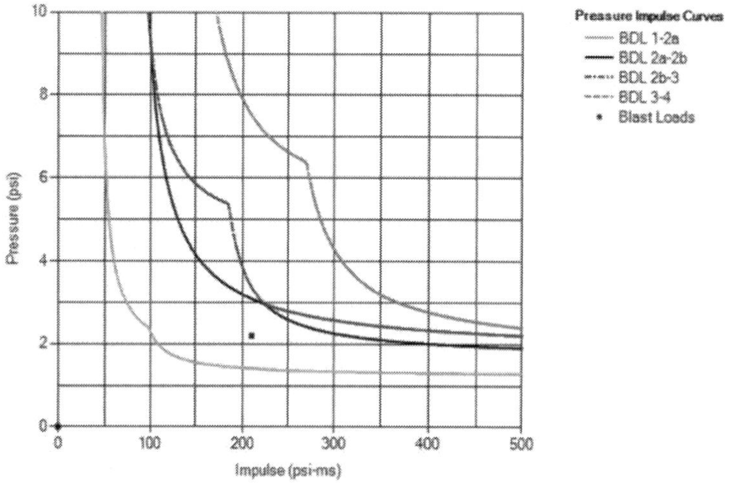

Figure 5. Building Damage Level (BDL) Curves

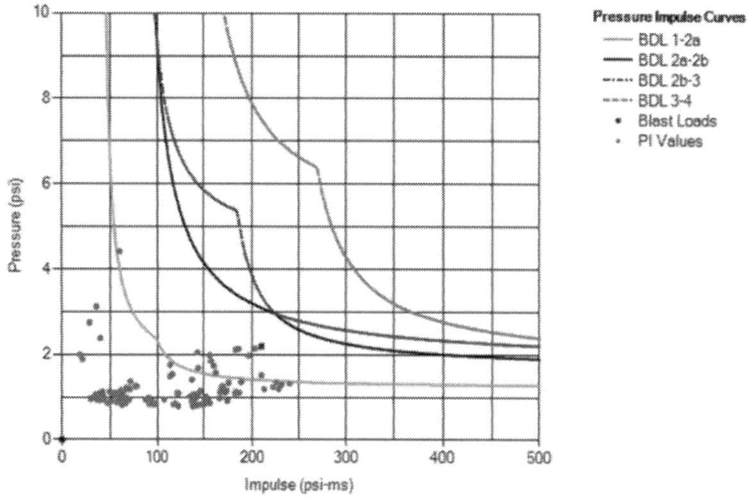

Figure 6. BDL Curves with Loads from "All" Postulated Scenarios

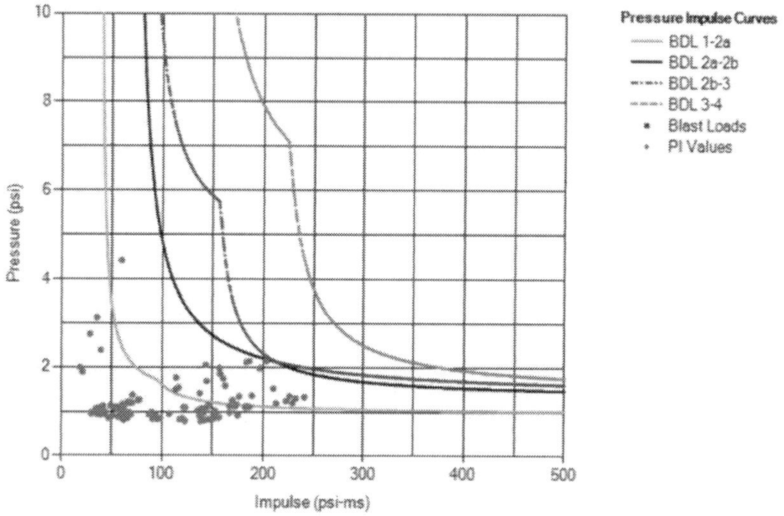

Figure 7. BDL Curves for "Weaker" Building

4 Resiliency in Design of Occupied Buildings

The examples discussed in the previous section demonstrate three approaches to providing a resilient structural design:

1. Structural systems that can develop a high degree of ductility are greatly preferred over structural systems that undergo a more brittle response. Most materials commonly used for blast resistant structures (steel and concrete) can undergo significant plastic deformations, typically well beyond those associated with published response limits, before failing. This plastic deformation allows the building to absorb the energy applied by the blast load. In addition, structural components designed to undergo large plastic deformations will have lower support reactions than those designed to behave elastically or with small plastic deformations, thus resulting in lower demands for connections and the building lateral load-carrying systems.

2. When selecting response limits, the likelihood of the selected design loads should be considered. Typically, high probability scenarios result in lower magnitude loads, while more severe load cases are the result of lower probability scenarios. This would imply that higher response limits could reasonably be used when designing for low probability, high magnitude load cases. Similarly, when designing for higher probability / lower magnitude loads, or loads with considerable uncertainty associated with their prediction, lower response limits should be considered.

3. Since one of the objectives of a blast resistant building is to protect building occupants, the layout of the building and the location of occupants relative to the sources of the potential explosions should be considered. By placing unoccupied areas of the building on sides facing the explosion source, localized failure of non-loadbearing components may be acceptable.

5 Conclusion

The design of blast resistant buildings does not follow an approach that is as prescriptive as the approach used for conventional design. Blast loads must be defined based on the conditions at the facility. Appropriate response limits must be selected for the components comprising the building. In addition, process conditions and building use may change to meet future requirements. Due to the uncertainty that is inherent in this design process, it is important to provide a building design that protects building occupants with a degree of resiliency. This can be developed by selecting the building structural system, building layout, and required response limits in a manner that is consistent with the postulated loads and the expected potential for these loads. In addition, the potential for more severe loads should also be considered in the design process.

6 References

[1] 2012 International Building Code, published by the International Code Council, Inc., May 2011.

[2] ASCE/SEI 7-05, "Minimum Design Loads for Buildings and Other Structures," published by the American Society of Civil Engineers, 2006.

[3] ACI 318-11, "Building Code Requirements for Structural Concrete and Commentary," reported by ACI Committee 318, published by the American Concrete Institute, 2011.

[4] "Steel Construction Manual," 14th Edition, Second Printing, published by the American Institute of Steel Construction.

[5] "Design of Blast Resistant Buildings in Petrochemical Facilities, Second Edition," Task Committee on Blast Resistant Design, American Society of Civil Engineers, 2010.

[6] Biggs, J., Introduction to Structural Dynamics, McGraw-Hill, Inc. New York, NY, 1964.

[7] UFC 3-340-01, "Design and Analysis of Hardened Structures to Conventional Weapon Effects," June 2002. (Formerly Technical Manual for Army TM 5-855-1, Air Force AFPAM 32-1147(I), Navy NAVFAC P-1080, and Defense Special Weapons Agency DAHSCWEMAN-97, August 1998.)

[8] "Structures to Resist the Effects of Accidental Explosions," UFC 3-340-02, 5 December 2008.

[9] Oswald, Charles J. and Baker, Quentin A., "Vulnerability Model for Occupants of Blast Damaged Buildings," presented at the 34th Annual Loss Prevention Symposium, March 6-8, 2000, Session 3.

(44g) Reliability Analysis of Man-Machine Response to an Emergency Situation Based On a Detailed-Action Plan

Najmeh Vaez, North America Risk, Environmental Resources Management (ERM), Houston, TX; Environmental Science and Safety Department, University of Houston Clear Lake, Houston, TX, Farshad Nourai, Aftab Imen Parto Consulting Engineers Co., Tehran, Iran, Yaneira Saud, North America Risk, ERM Southwest, Houston, TX and Suresh Yelisetty, Americas Risk Practice, Environmental Resource Management (ERM), Houston, TX

The primary emergency response actions in oil and gas industries are the functions of the process safety control system and the actions completed by the control room operators. The performance of the safety control system and the control room operating crew has a significant role in proper emergency response. To overcome each critical scenario, an emergency response operating procedure is provided with specific instructions, that in this research paper it is named a detailed action plan (DAP). This term is recommended for this purpose by the Center for Chemical Process Safety (CCPS). The reason of using a DAP is to reduce the probability of error in the emergency response of control room operators by providing scenario-specific written instructions. However, most of the DAP's are developed during design phase when there is no opportunity to test them through emergency drills. Therefore, a framework is proposed to analyze the reliability of the DAP's. This framework should be able to determine reliability of every single human and equipment action as well as the combined human-machine reliability, modeled by means of a Reliability Block Diagram to calculate the reliability of integrated automatic-operator emergency actions. The framework focuses on incorporating the operational and cognitive operator's errors in the process/equipment reliability analysis. For this purpose, a human reliability analysis method named SPAR-H was utilized. The proposed framework was applied for analyzing DAPs of an emergency well control operation in order to check and improve the applicability of this framework. The presented framework facilitates the DAP assessment process by performing task analysis modeling the detailed tasks, and completing a human-machine reliability analysis. Besides, the application of the presented framework, its findings and recommended actions enables a considerable improvement in the reliability of DAPs and therefore proves its efficiency and effectiveness.

(44i) Hazards of Gasoline Contamination – Case Study of a Tanker Truck Explosion Resulting From Switch Loading

Scott G. Davis, GexCon US, Bethesda, MD and Kees Van Wingerden, GexCon AS, Bergen, Norway

When a tanker truck is drained of a Class I liquid (e.g., gasoline) and refilled with a similar liquid, there is no flammable mixture on the surface of the rising fluid because Class I fluids create mixtures at their surface that are too fuel rich to burn. Similarly, if a tanker truck is drained of Class II liquid (e.g., diesel) and refilled with a Class II liquid, there is no flammable mixture created because Class II liquids do not produce ignitable mixtures at ordinary temperatures. An extremely hazardous condition occurs, however, if the liquids are *switch loaded*, which is when a Class II or III liquid is loaded into a tank vehicle that previously contained a Class I liquid. Under these conditions, the atmosphere above the fluid surface can become ignitable. An explosion can occur if a spark occurs from the fluid surface to another object, igniting the flammable mixture in the headspace of the tank.

This paper describes a recent incident involving an explosion during the filling of a tank of a transport truck with diesel fuel. This particular tank was previously filled with gasoline and was not completely emptied of the gasoline prior to the switch loading. After only filling one-quarter of the programmed load, a spark ignited the flammable mixture within the headspace. The resulting explosion and overpressure caused significant damage to the truck and neighboring tanks. The study will address issues and lessons learned regarding spark promotion with the tank, reduction of static generation by the fluid being filled into the tank, and elimination of the flammable mixture within the tank prior to switch loading.

(44j) Is Your MOC Process Doing All That It Should Be?

William Fink and Carolina Herrera, Sage Environmental Consulting, L.P., Austin, TX

OSHA 1910.119 Process Safety Management of Highly Hazardous Chemicals was the driver for refining and chemical industry to implement Management of Change (MOC) programs some 20+ years ago. However, companies have now recognized the benefits of implementing and maintaining an effective MOC program beyond regulatory compliance. Effective MOC programs enhance a company's ability to respond to changing "market" or processing needs with up to date process safety information. Management of Change has even been incorporated into staffing levels and personnel expertise by some Companies.

Management of Change programs typically focus on changes that are "not in kind" changes. Significant effort is placed on proper evaluation of these changes and the potential impacts on the operation. This presentation emphasizes both the typical and non-typical areas of MOC that have been found to be overlooked. MOC programs rigorously evaluate the potential to introduce new hazards of the process such as fire, explosion, toxicity, and environmental impacts. Is that far enough to ensure that you are meeting all of your performance needs?

Focus on environmental impacts such as water contamination, air pollutants, and hazardous waste generation may not be going far enough. Gradual increases in highly hazardous material inventories may cross the lines of "threshold quantities" and introduce new OSHA or EPA requirements. Does your MOC program look at this potential? In addition, management of "in kind" changes may also not be keeping pace with the changes required by the EPA Leak Detection and Repair (LDAR) requirements or the Benzene Waste Operations NESHAP (BWON) requirements. These requirements as applied to MOC are further explained in this presentation. Several "gaps" in MOC programs will also be examined and suggestions for improvements provided.

(44I) Beyond DSC, Improved Thermal Hazards Screening Methodology

Peter J. Ralbovsky IV, Netzsch Instruments North America, LLC, Burlington, MA

Thermal Hazards Screening is a primary part of any chemical process safety program. Differential Scanning Calorimetry (DSC) is used extensively in studying the thermal properties of materials and is often used to screen for thermal hazards. This method is quick and inherently safe as it can be done with a small amount of material (typically milligrams). These two benefits cannot be understated as there may be hundreds of screening test which a lab may have to do in a given year and sometimes little or nothing is known about its thermal behavior, its toxicity, or the toxicity of the decomposition products. DSC also has some short comings as it is unable to measure pressure, which can be a significant hazard, even when the heat release is relatively low. Also it can be difficult to handle multiphase samples or samples which react together at room temperature.

The Multiple Mode Calorimeter (MMC) is an excellent tool to be used in conjunction with the DSC to provide a more comprehensive look at thermal hazards. The MMC uses samples sizes up to about 3 grams so it is possible to do injection and stirring. In the MMC temperature and pressure are recorded as a function of time. Sample energies measurement of both endotherms and exotherms is possible. Heat capacity can be calculated as a function of temperature. This method offers a quick and inexpensive way to perform isothermal tests which can be used to help design larger scale tests done in 2L+ reaction calorimeters or pilot plant operations.

The Multiple Mode Calorimeter consists of an electronic base and a detachable calorimeter assembly. There are multiple types of calorimeters available which can be operated by the electronic base. This allows the user to select the right calorimeter for the type of experiment that is needed. The calorimeters can also be tethered to the electronic base with an electronic cable. This allow for the calorimeter to be placed remotely from the electronics. The calorimeter can then be easily placed in the hood and/or behind explosion barriers or even placed inside a freezer for sub-ambient operations. This instrument has been described in detail elsewhere.

There are a number of calorimeters available; the best two for Thermal Hazards Screening is shown in Figure 1 & 2. The first calorimeter (Figure 3) is capable of operating as a low phi adiabatic calorimeter but more importantly, for thermal hazards screening, can run in constant power or constant temperature rate mode. The MMC external heater module does not run in adiabatic mode but can be operated in scanning mode. Both MMC modules incorporate the VariPhi technology and can run true isothermal calorimetry tests. These MMC modules can measure endotherms and exotherms as well as calculating heat capacity during those periods where no transitions or reactions are occurring. The MMC measures sample pressure as a function of time/temperature which is critical in screening for thermal hazards.

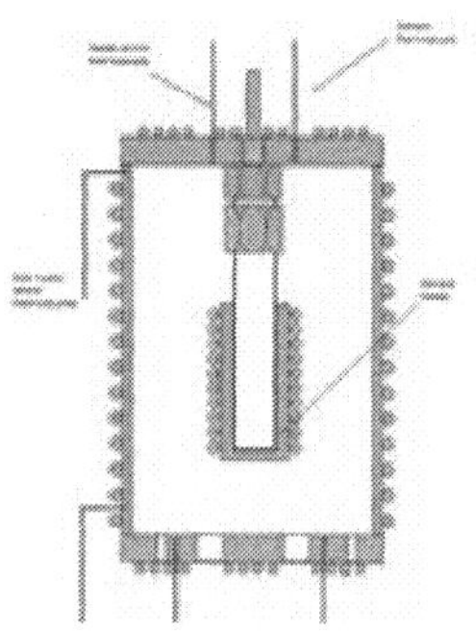

Figure 1

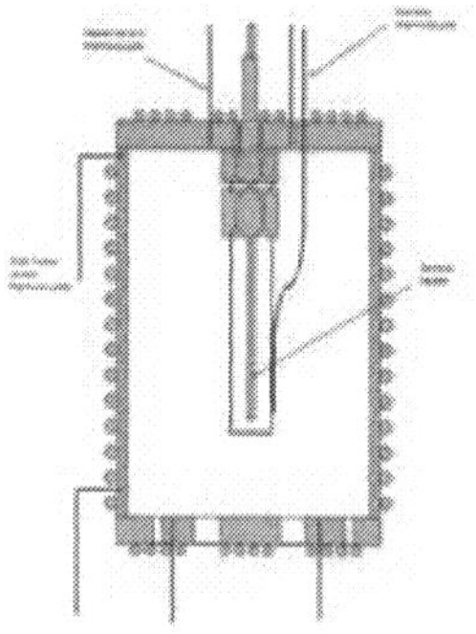

Figure 2

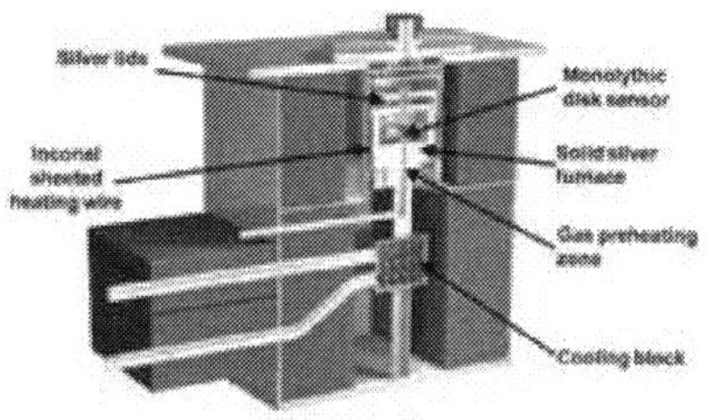

Figure 3

MMC VariPhi Calorimeter MMC External Heater Netzsch DSC 200 F3 *Maia®*

DSCs, like shown in Figures 3, have been used routinely as an initial step in the thermal hazards screening process. They have some unique advantages which make them suitable for this task. Scanning tests can be done very rapidly and with an auto-sampler they can be used to screen many samples efficiently. They sample size is small which makes them intrinsically safe in case of explosion or toxic gas production. These systems are small, inexpensive and have low operating costs. There are numerous DSCs available on the marketplace and they are used in many different applications. Two features are important from the thermal screening application. The first is an auto-sampler as discussed earlier. The second is the material of construction used in the DSC. It is likely that a DSC system used for thermal hazard screening will see explosions and the release of corrosive materials. DSC systems that can run hundreds of polymer melt tests without fail may have a limited life in the thermal hazards screening field. This is often unlooked. Figure 5 shows the cut-away of a robust DSC system.

Their continues to be a need to be able to safely and effectively screen thermal hazards for organizations involved in the manufacture, processing, storage or transportation of chemicals. Numerous screening tools are available in the marketplace, perhaps the most widely used is the Differential Scanning Calorimeter (DSC). This method has many advantages but the largest disadvantage is that the sample pressure is not measured. Thermal hazards are normally associated with pressure increases and these pressure increases are what ultimately provide the hazard to equipment and personnel. A storage drum does not melt due to a thermal runaway it ruptures due to the pressure increase because of the thermal runaway. Screening tools should be sensitive enough to see small exotherms that in near adiabatic conditions can fuel a runaway, they should be easy to use, provide quick results, and be as flexible as possible to handle different scenarios.

[1] Chippett, A New, Low Φ Factor, Scanning, Adiabatic Calorimeter, 18th Annual CCPS International Conference, 2003

[2]Chippett, US Patent Number 7,021,820 B2, April 4, 2006

9TH GLOBAL CONGRESS ON PROCESS SAFETY

Using Explicit Finite Element Analysis to Simulate the Effects of External Chemical Explosions on Single and Double-Walled Storage Tanks

Phillip E. Prueter
The Equity Engineering Group Inc., Shaker Heights, Ohio
peprueter@equityeng.com

David J. Dewees, P.E.
The Equity Engineering Group Inc., Shaker Heights, Ohio
djdewees@equityeng.com

Prepared for Presentation at
American Institute of Chemical Engineers
2013 Spring Meeting
9th Global Congress on Process Safety
San Antonio, Texas
April 28 – May 1, 2013

UNPUBLISHED

Using Explicit Finite Element Analysis to Simulate the Effects of External Chemical Explosions on Single and Double-Walled Storage Tanks

Phillip E. Prueter
David J. Dewees, P.E.
The Equity Engineering Group Inc., Shaker Heights, Ohio
pepprueter@equityeng.com

Keywords: Explicit Finite Element Analysis, Blast Loading, CONWEP, Accidental Explosions, Storage Tanks

Abstract

Accurately simulating the overpressure or shock wave associated with a given far-field chemical explosion is extremely valuable in assessing the structural response and possible failure modes of critical process equipment such as storage tanks, piping, or pressure vessels. Furthermore, assessing the potential damage from an overpressure wave can provide valuable information about protecting structures from external explosions and improving designs that advocate blast damage mitigation. Such analyses become all the more important should the contents of the tank or vessel exposed to such a potentially catastrophic event pose a risk to humans or the surrounding environment. This paper discusses the underlying theory and examines the practical application of multiple finite element based explicit computational techniques for simulating the load acting on a structure due to an external blast. Explicit three-dimensional blast analysis of single and double-walled storage tanks that carry an extremely high consequence of failure is performed. The structural response of the tanks due to postulated accidental explosions is investigated and likely failure modes are discussed.

The two explicit computational blast loading methods discussed in this paper are the incident wave loading model and the Conventional Weapons Effects Blast Loading Model or CONWEP. The CONWEP model is appropriate for detonations of conventional explosions, and in this case automates many crucial features of analysis, such as developing the overpressure time history (positive and negative phases), defining overpressure spatial decay, and accounting for spatially varying reflection effects. Large vapor cloud explosions are known to behave very differently from conventional explosives, and for these cases, the more manual, but also general, incident wave loading approach permits the user to define their own time-varying overpressure amplitude. Additionally, this method also allows for coupled structural-acoustic analysis (should reflection effects need to be considered in detail). While it is not necessary to model the fluid medium using acoustic elements when using the incident wave loading model, it does become essential if reflection effects dominate the response of the given problem. If the fluid medium is modeled (air in this case), the acoustic elements can be coupled to the structural elements and the loading surface becomes the boundary of the acoustic mesh closest to the blast source. While this approach can be accomplished, it is extremely resource and time intensive, and is generally not necessary (which will be discussed later in this paper). Three-dimensional examples of the above blast loading approaches are presented in this paper; incident wave, and CONWEP. Comparisons are made between these methods and the corresponding computational results.

Furthermore, commentary on the structural response and likely failure modes of the storage tanks is provided and a discussion regarding design features that supplement blast damage mitigation and process safety is rendered. The advanced computational methods discussed herein permit realistic evaluation of complex, three-dimensional process equipment subjected to accidental explosions. Simulating and understanding the possible failure modes due to blast loading of critical process equipment in the petrochemical and related industries is very beneficial in not only improving process safety, but also providing plant operators and safety personnel with crucial information regarding the possibility of equipment failure should a catastrophic explosion occur.

1. Introduction

An explosion in air creates a blast wave as a result of the atmosphere surrounding the explosion being pushed back. In general, the pressure of the compressed air at the blast wavefront decays as it moves away from the blast source. As discussed in references [1-6], a typical blast wave, as observed at a location removed from the center of a chemical explosion, reaches a peak value of overpressure and begins to decay exponentially (as shown in Figure 1), eventually decreasing below atmospheric pressure. This peak overpressure decreases as the distance from the explosion source increases. The amount of time it takes the wavefront to reach a given point is known as the arrival time. The overpressure profile can be divided into the positive pressure portion (positive phase) and negative pressure portion (negative phase). Furthermore, the amount of time it takes the peak overpressure to decay below atmospheric pressure is known as the positive phase duration and is closely related to the damage capabilities of a given blast wave. Similarly, the amount of time it takes for the negative pressure to return to atmospheric pressure is the negative phase duration [1]. The area under the curve during the positive phase duration is the positive blast wave impulse.

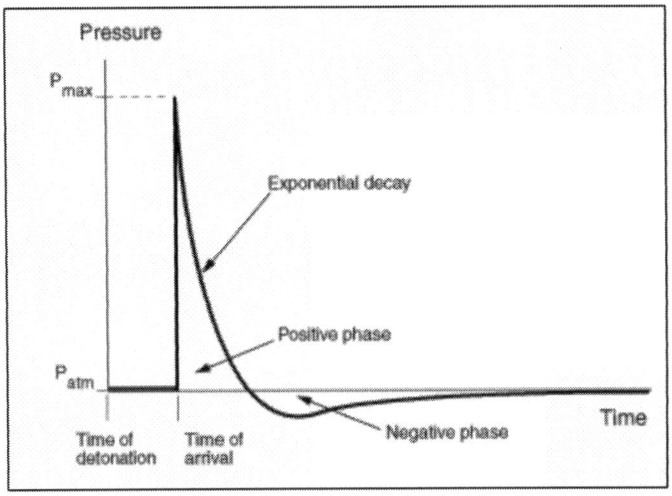

Figure 1: Typical pressure-time amplitude for a chemical explosion [7].

In some cases, the positive phase impulse is calculated and an equivalent triangular impulse with the same peak overpressure is used to approximate the realistic overpressure amplitude shown in Figure 1. In many cases, this approximation is appropriate because the positive phase portion of the incident pressure wave is typically the most damaging. Generally, structural response is more sensitive to capturing the correct incident wave reflection. Accurately simulating a realistic blast wavefront on a structure for a given explosive scenario is challenging because the peak overpressure, positive phase duration, amount of explosives, and distance from the explosion source all affect the overpressure amplitude. Furthermore, accounting for reflection effects on a loading surface adds further complication to the blast loading model.

Two different explicit finite element techniques used to simulate blast loading due to external chemical explosions in air will be discussed herein; incident wave loading and the Conventional Weapons loading model (CONWEP) [8]. CONWEP is well suited for simulating the overpressure wave associated with the detonation of conventional explosives while incident wave loading permits reasonable analysis of vapor cloud explosions. The underlying methodology of each approach will be discussed as well as the advantages and applications of each. Furthermore, both computational methods will be compared for the explicit dynamic simulations of realistic explosive overpressure waves (from a postulated concentrated explosion source and a vapor cloud) impacting single and double-walled storage tanks that carry an extremely high consequence of failure.

2. Blast Curve Generation

In order to develop the blast parameters required to generate realistic overpressure curves for conventional explosives, the equivalent TNT method is most commonly used. TNT equivalency of an explosive is simply the ratio of the mass of TNT to the mass of explosive such that both yield equal pressure and impulse [9]. In order to compare shock wave parameters to available published data, a scaled distance from an explosion source to the loaded structure may be used. As discussed in [10], the most common form of blast scaling is the Hopkinson-Cranz or cube-root scaling method following

$$Z = R/W^{(1/3)}$$

where Z is the scaled distance, R is distance from the explosion source to the loading point, and W is the weight of TNT. This formulation implies that self-similar blast waves are produced at identical scaled distances when two different sized explosive charges (of similar geometry and chemical composition) are detonated in the same atmosphere. Using published empirical data, this scaled distance from the explosion source to a loading surface can be used to determine specific blast parameters. The CONWEP blast loading model discussed herein is based on a TNT explosion (assumed surface blast). Thus, when using CONWEP, it is necessary to convert the mass of explosives in question to an equivalent amount of TNT.

As outlined in [11], a major limitation associated with the equivalent TNT method is that the blast yield is a function of the amount of explosives and not the combustion mode; making this method not preferred when analyzing vapor cloud explosions. As discussed in References [12-

16], vapor cloud explosions differ from conventional explosive detonations in that flame accelerations, which can be exacerbated by confinement due to intervening structures, define the overpressure associated with a given explosion more so than simply the amount of fuel ignited. In order to quantify blast curves associated with vapor cloud explosions several widely accepted methods have been developed; TNO [15] blast curves and Baker-Strehlow [17] blast curves among others. In this study, triangular impulses of varying peak overpressures are used for simplicity when analyzing the structural response of a double-walled storage tank subjected to a postulated vapor cloud explosion (discussed below). This also permits investigation into the sensitivity of structural response to varying blast wave parameters.

3. CONWEP and Incident Wave Loading

A significant challenge associated with computational blast modeling is accounting for reflection effects. The peak blast pressure that a structure is subjected to is greater than the peak overpressure and is known as the reflected pressure. This pressure is a function of the peak overpressure and the angle of incidence (the angle between the loading direction and the surface normal of a point on the loading surface). Neglecting incident wave reflection can yield non-conservative results, particularly for high peak overpressures, while assuming the maximum reflection effect at all points on a (cylindrical) structure can potentially be overly conservative.

The CONWEP blast loading model is based on a collection of conventional weapons effects calculations from the equations and curves of TM 5-855-1 [8] and is implemented into the commercial finite element program, Abaqus. The main advantage of using CONWEP over the incident wave loading approach is that realistic overpressure amplitudes (including positive and negative phase) and other blast wave parameters are inherently calculated based on a defined amount of TNT at a given distance from the explosion source. Furthermore, the CONWEP blast loading model does not require modeling the fluid medium (air) to account for reflection effects.

Incident wave loading is an explicit finite element technique in which an incident wave overpressure at a point can be defined as a function of space and time. This pressure amplitude is defined at a standoff point, located between the detonation point and the loaded structure. When simulating blast loading, realistic overpressure amplitudes need to be defined by the user. An incident wave can be defined as planar or spherical. A schematic showing the incident wave loading model in Abaqus is given in Figure 2 below. For a more rigorous overview of the underlying theory behind both the CONWEP blast loading model and the incident wave loading as well as a benchmark of the overpressure reflection calculations of CONWEP, refer to Reference [18].

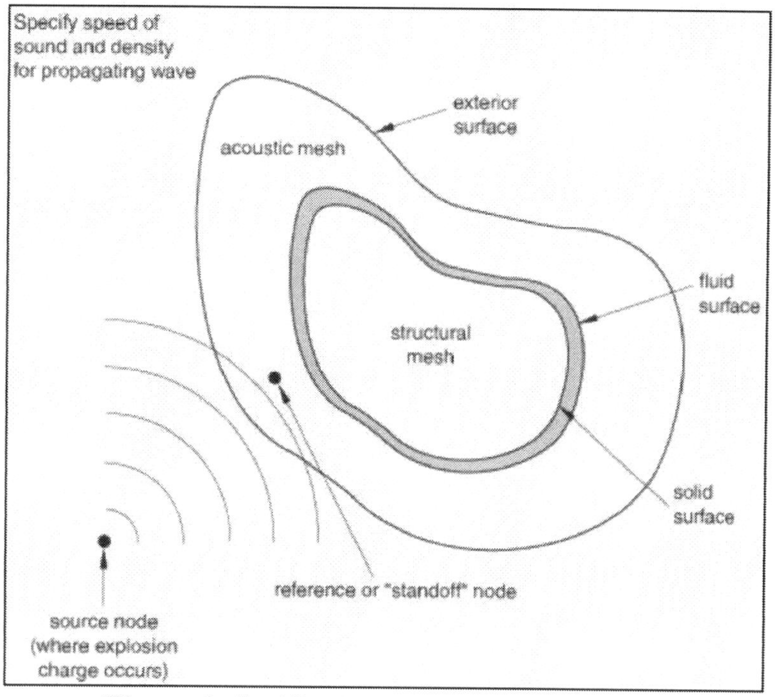

Figure 2: Incident wave loading model [7].

4. Explicit Analysis of a Storage Tank Subjected to a Vapor Cloud Explosion

As discussed in [19, 20] the consequences of storage tank failure in chemical or petrochemical facilities can be dire. Furthermore, it is estimated that 85 percent of storage tank accidents are due to fire and explosion [21]. Analyzing the structural response of storage tanks subjected to potential blast loading from conventional detonations or vapor cloud explosions is of great interest to the process safety community. The explicit finite element analyses discussed below provide real-world examples of the usefulness that computational blast analysis provides in the petrochemical and related industries.

The first computational blast analysis examined herein relates to a double-walled storage tank subjected to a postulated external vapor cloud explosion. The rough size of the outer tank is a 140 foot diameter with a shell height of approximately 125 feet. A complete, elastic-plastic, three dimensional finite element model was constructed using linear shell elements. Figure 3 shows the half-symmetry model with detailed geometric features. The baseline finite element model includes all major structural features, including the suspended deck, bracket supports, and the lower support straps. Quadratic beam elements are used to model the suspended deck hanger rods and the strap supports, and both the inner and outer tanks are modeled for completeness. Contact is defined between the suspended deck and the shell wall.

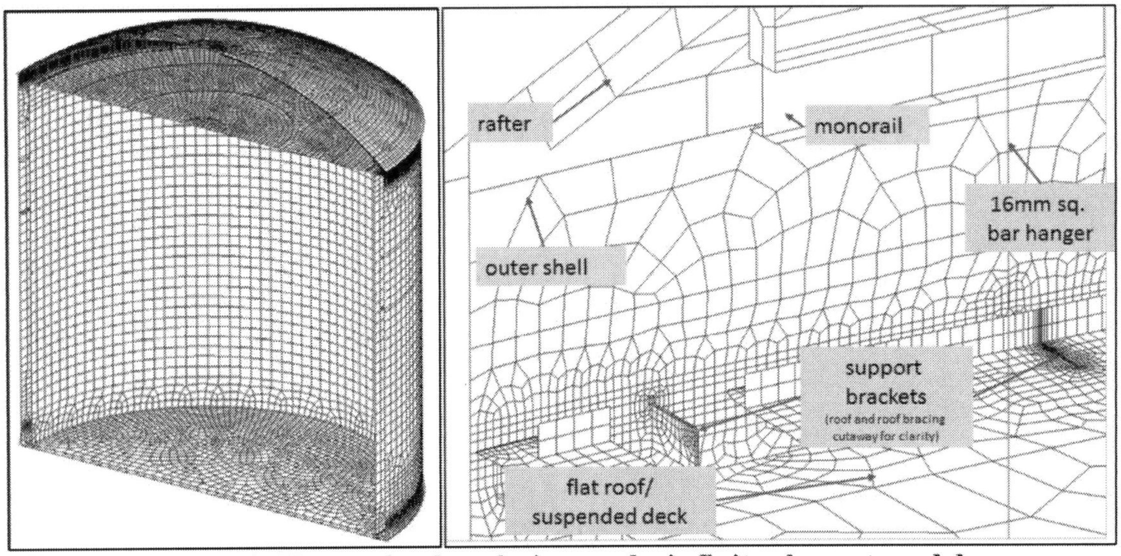

Figure 3: Vapor cloud explosion analysis finite element model.

A simplified, lumped-mass model is also analyzed. In this model, the mass associated with the suspended deck is lumped into the domed roof and the mass of the inner tank shell is lumped into the outer tank shell. The mesh is coarsened in the simplified model to aid in computational efficiency. Incident wave loading is used to analyze a range of peak overpressures (assumed to be peak reflected pressure because acoustic elements are not used) applied as planar triangular impulses. The effects of gravity (self-weight) are included.

In this case, because of the range of potential vapor cloud explosive scenarios, iterations are performed to determine the maximum permissible overpressures and loading durations at the tank to ensure containment and safe operability. That is, the focus is on establishing a sort of blast rating for the structure independent of a specific explosion scenario. This provides an essential characteristic of the structure that can be referred to for any future scenarios or concerns that might arise, and is similar in some sense to the maximum allowed working pressure (MAWP) of a pressure vessel. Peak reflected pressures of 5, 8, and 11 psi are applied for an impulse duration of 50 milliseconds to the lumped model. Figure 4 below compares maximum deflections and stresses in the lumped model from the domed roof (opposite impact side) for peak applied overpressures of 5, 8 and 11 psi. This figure demonstrates the relatively constant nature of the largest displacements. In the 5 psi case, maximum deflections are generally less than 12 inches with relatively little accumulation in plastic strain.

These results indicate that a peak reflected pressure of 5 psi is near the limit to achieve tank containment following blast loading. Furthermore, for the 8 and 11 psi overpressure cases, results indicate that plastic collapse of the domed roof and tank shell would be likely and containment of the vessel contents would be jeopardized. Figure 4 also shows an example transient deformation corresponding to the full model results for 5 psi. In general, there is reasonable agreement between the simplified, lumped parameter model and the complete model with the suspended deck across cases. However, the suspended deck is a critical feature of the tank, and is also an indicator of safe operability. As Figure 4 illustrates, the compete model indicates that the suspended deck is one of the weakest points of the tank internals; results from the complete model series of analyses show that a maximum peak reflected pressure of 2 psi is advisable to ensure safe tank operability (again, impulse duration assumed to be 50 milliseconds). This corresponds to a maximum side-on overpressure of 1 psi (uniform/constant reflection coefficient of 2).

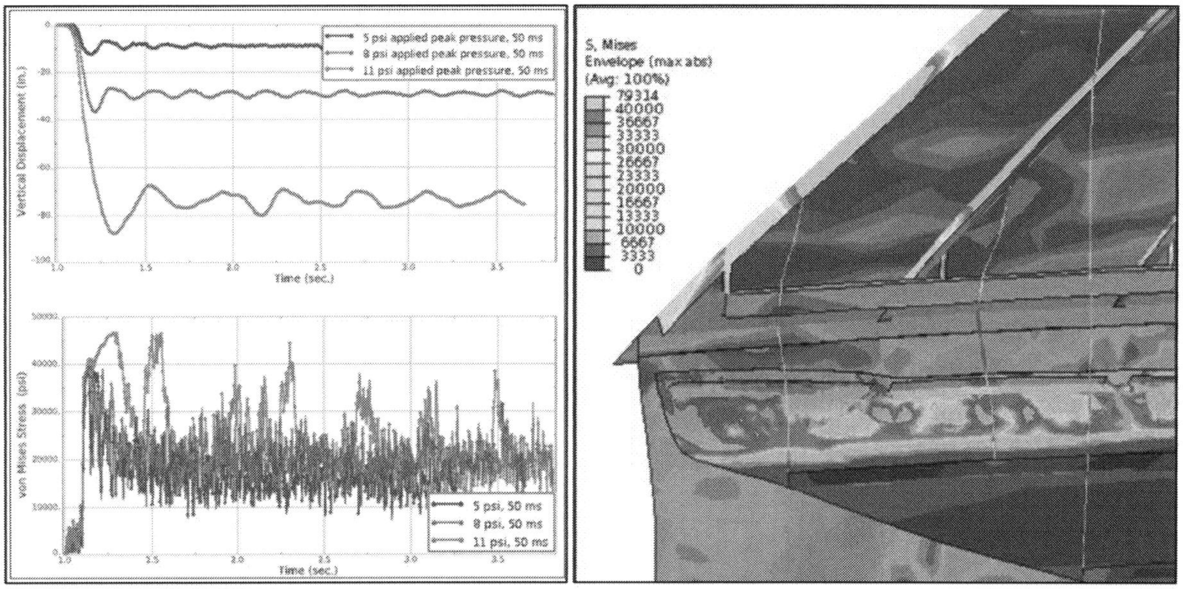

Figure 4: Deflection/stresses in domed roof opposite of impact side for varying peak overpressures and suspended deck deformation.

Additionally, in order to compare the effect of varying blast wave impulse durations, a simulation at 11 psi peak applied overpressure with an impulse duration of 10 milliseconds is carried out. Figure 5 compares maximum domed roof deflections and stresses for the 50 and 10 millisecond loading durations. As expected the duration of the impulse has a significant effect on the overall structural damage. It is noted that much lower peak overpressures acting at longer durations could also be damaging.

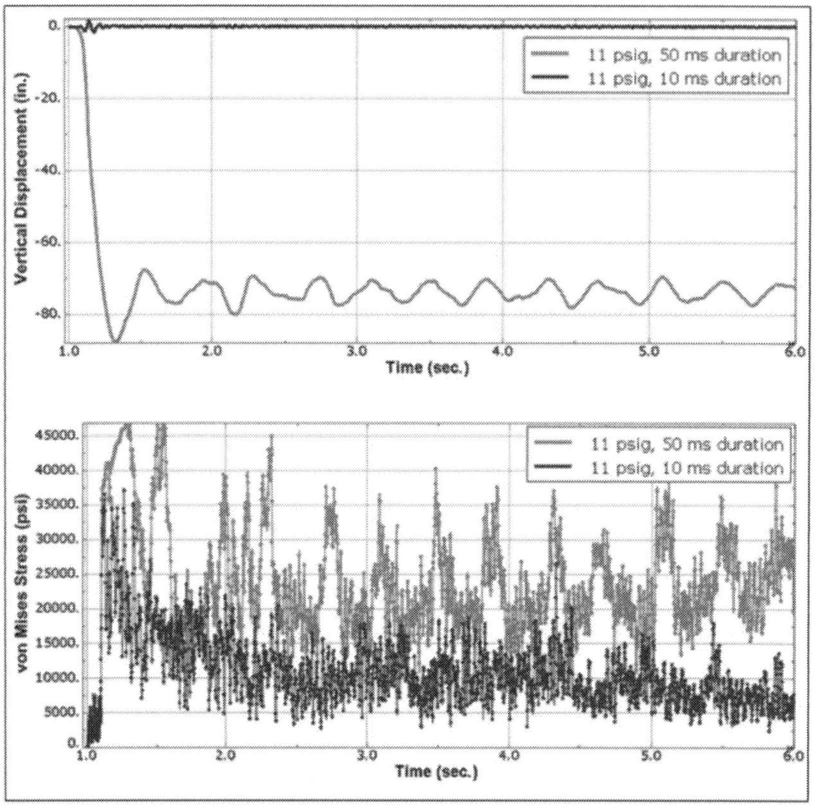

Figure 5: Deflection/stresses in domed roof opposite of impact side for varying impulse duration.

5. Explicit Analysis of a Storage Tank Subjected to Conventional Explosion

A different double-walled storage tank is analyzed; this time the postulated blast loading is due to accidental detonation of a known amount of concentrated, combustible fuel. The approximate outer tank diameter is 100 feet, and the outer tank height is roughly 75 feet. Again a complete, elastic-plastic, three dimensional finite element model is constructed using linear shell elements. Figure 6 shows a cutaway view of the model with contours of shell element thickness. The inner tank has four stiffening rings that protrude into the annular space between the inner and outer tank. The outer tank roof support structure consisting of channels and angles is also modeled in detail and included in the finite element model. Contact is modeled in between the tank structures and the effects of self-weight are included.

An initial analysis is performed on just the outer tank to investigate the plastic strains and wall deflections due to the applied blast wave. Both the CONWEP and incident wave loading models are compared (the equivalent TNT method is used to determine the overpressure profile in accordance with CONWEP). Due to the size of the finite element model and the potentially high computational cost, the fluid medium is not explicitly modeled for the incident wave simulation. An overpressure amplitude that accounts for reflection effects is used (peak reflected pressure applied in the incident wave model). Furthermore, in order to mimic the behavior of the CONWEP simulation as much as possible, the overpressure profile defined for the incident wave simulation as a function of time does, in fact, consist of positive and negative phases and decays exponentially once it reaches its peak value. The side-on reflection coefficient for the overpressure magnitude in this case is 2.

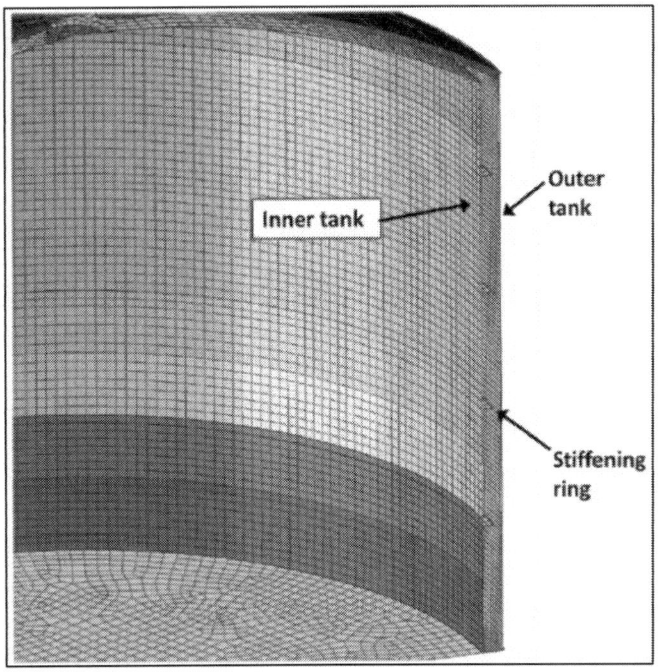

Figure 6: Cutaway view of the complete finite element model with shell thickness contours shown.

Even for the relatively small peak reflected overpressure of roughly 1.9 psi, the outer tank wall deflections prove to be significant, to the point of local collapse. A static post-buckling analysis (using the Riks solver), with surface pressure applied using an analytical field definition to match surface pressure in the dynamic simulations confirms that the outer tank wall buckles before the peak reflected pressure is reached. Riks post-buckling analyses are discussed in more detail below.

For comparison, Figure 7 shows contours of outer tank plastic deformations (identical scale) due to the overpressure wave for both methods. As expected, the incident wave approach produces higher deflections than CONWEP because the applied peak reflected overpressure does not decrease as angle of incidence increases. However, the general agreement between methods is actually quite good. This suggests that the simplification of a single maximum reflection coefficient in this case for the incident wave approach is not excessively conservative, and that reasonable results were obtained without modeling the entire acoustic domain (in which case reflection effects would be rigorously accounted for).

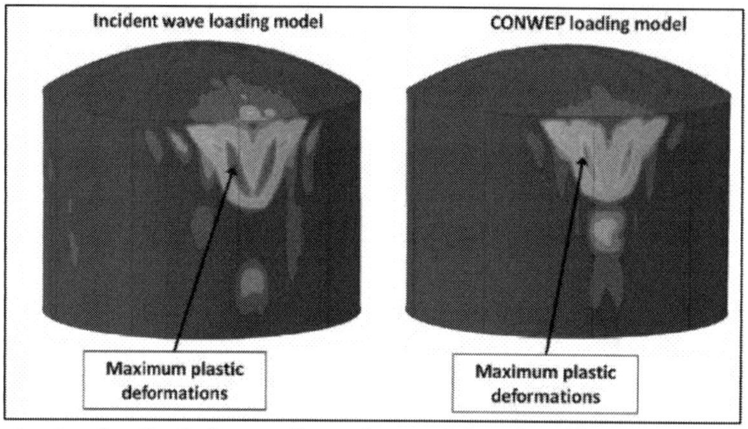

Figure 7: Outer tank plastic deformations due to blast loading using the CONWEP and incident wave loading models (contour scales identical).

For the blast analysis of the complete finite element model that includes both the inner and outer tanks with contact defined between shells, the outer wall deflections are much less significant because the outer tank impacts the inner tank stiffeners. Essentially, the inner tank stiffening rings absorb the impact from the outer tank wall and plastically deform. Furthermore, the deflections in the inner tank wall are minimal. This is demonstrated in Figure 8, which shows the maximum deflection during blast loading from the CONWEP simulation of the inner tank and the outer tank. The contour scale for the inner tank corresponds to 10 percent of the outer tank contour scale. Additionally, the maximum inner tank wall deflection is less than 10 percent of the maximum outer tank wall deflection, and the maximum outer tank deflections occur in the regions in between the inner tank stiffening rings. The outer tank wall never directly impacts the inner tank wall. The minimal inner tank deflections suggest that the sloshing effects of the liquid contained by the inner tank should also be minimal.

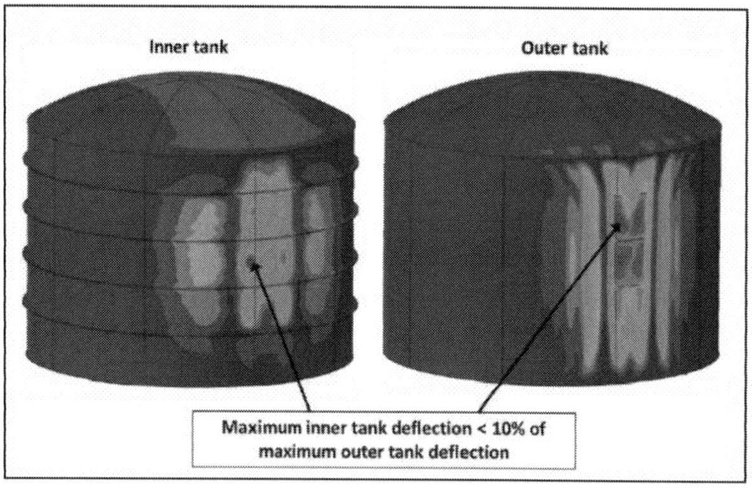

Figure 8: Maximum deflection of the inner and outer tanks due to blast loading from the CONWEP model (inner tank contour scale corresponds to 10 percent of outer tank scale).

These results indicate that from a blast damage mitigation perspective, the four inner tank stiffening rings prove to be very beneficial in reducing the severity of the impact between the inner and outer tanks. While both explicit computational techniques used to analyze this complete, three dimensional model give reasonably similar results, from a convenience standpoint, CONWEP proves to be advantageous for conventional detonations because the appropriate overpressure and spatial decay are automatically calculated for the given equivalent amount of TNT. Additionally, due to more rigorous accounting for reflection effects, some conservatism is removed in the CONWEP case, making it the natural choice for well-defined, conventional explosive detonation scenarios.

6. Explicit Analysis of a Cluster of Single-Walled Storage Tanks

The postulated blast source in this case is the same source as discussed in the previous example (conventional detonation). However, this cluster of single-walled tanks is located much closer to the blast source, meaning peak overpressures are much higher than the above example. All three tanks are roughly 30 feet in diameter and 35 feet high. The tanks also have self-supporting roofs and are anchored to their respective foundations via hold down clips and hold down bolts. As is the case for the aforementioned storage tanks in this study, these tanks carry an extremely high consequence of failure due to their process contents.

Three dimensional, elastic-plastic finite element models of each tank are constructed using linear shell elements and explicitly analyzed using CONWEP. Peak reflected overpressures at each tank as a function of time are provided in Figure 9 below. Again, the effects of gravity are included. A node at the center of each hold-down clip is fixed in all degrees of freedom and then coupled to the surrounding bolt hole to simulate the restraint of the bolts as shown in Figure 10. A rigid surface representing the tank foundation is constructed and included in the model, and contact between this rigid surface and the tank bottom is defined. This ensures that the tank will not unrealistically sag due to gravity alone (essentially penetrate the foundation due to self-weight and induce non-realistic stresses in the tank floor/walls) while allowing the tank bottom to physically lift off its foundation when the blast load is applied. In order to account for the weight of the process fluid, an equivalent pressure is applied to the tank bottom that corresponds to the weight of fluid up to the maximum fill height. This pressure is applied throughout the dynamic simulation (sloshing effects of the fluid are not considered; this assumption is discussed below).

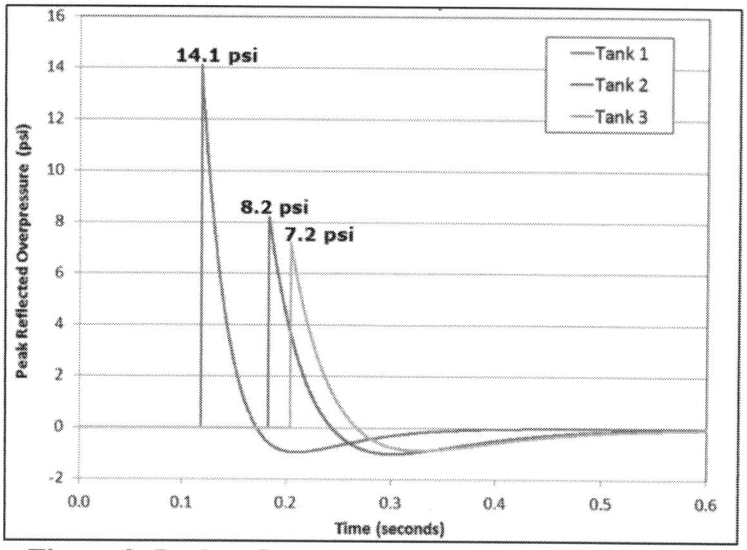

Figure 9: Peak reflected overpressures at each tank.

After the onset of the blast wave, the wall and the roof of each tank significantly deform. In order to better visualize the tank behavior following the onset of the blast wave impact, Figure 11 shows contours of Tank 1 deflection as time elapses (time scale given is seconds after detonation). These large deflections are more indicative of plastic collapse like behavior than a dynamic or vibratory failure. Furthermore, the stresses observed in the tank shortly after blast wave impact are significant.

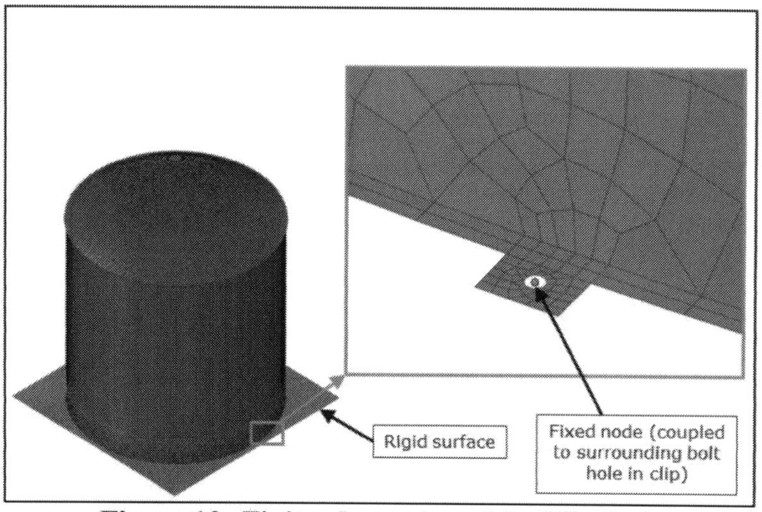

Figure 10: Finite element model of Tank 1.

Additionally, the stresses in the tank at the hold-down clips and reaction forces at the assumed bolt locations predict likely failure of either the bolts themselves or the hold-down clips. Figure 12 shows equivalent plastic strains in Tank 1 after the impact of the blast wave (contour scale capped at 5%). The hold-down clips and portions of the tank wall near the hold-down clips exceed 100% equivalent plastic strain. Results for Tanks 2 and 3 are similar.

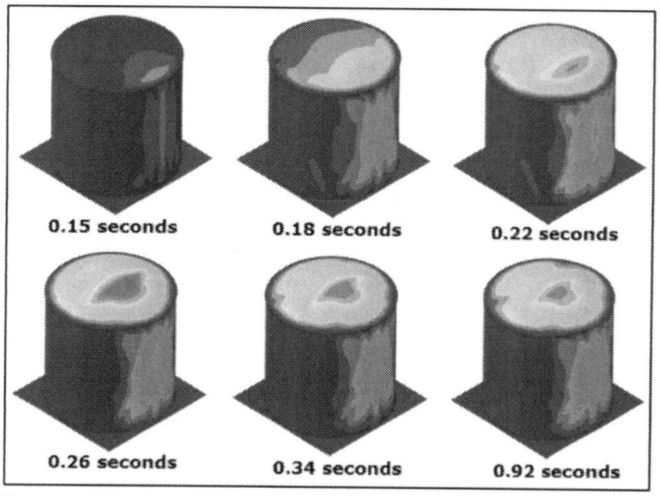

Figure 11: Tank 1 deflections during blast loading.

In order to further investigate the possibility of plastic collapse or buckling, static Riks post-buckling analyses are performed on each tank (a static pressure distribution is applied to mimic peak blast loading pressures). A Riks analysis is useful for simulating non-linear instability problems, and includes the effects of large deformations and elastic-plastic material behavior. It can also be used to look at post-buckling or collapse shapes. The applied load in a Riks analysis is actually a variable and part of the solution for each increment, and is always relative to the current deformed configuration of the structure. If no converged solution is possible for an incremental increase in load, then the equilibrium solutions can be searched for with a decreased load. This is the mechanism that allows post-buckled shapes to be predicted and visualized. In extreme cases, no equilibrium solution can be reached, even with decreased load, and only load reversal is predicted to lead to stability. This is actually the case for the tanks analyzed here, where once collapse has initiated, displacements are predicted to be essentially unbounded for any amount of positive load and thus, surface pressure becomes negative before peak overpressure values are achieved.

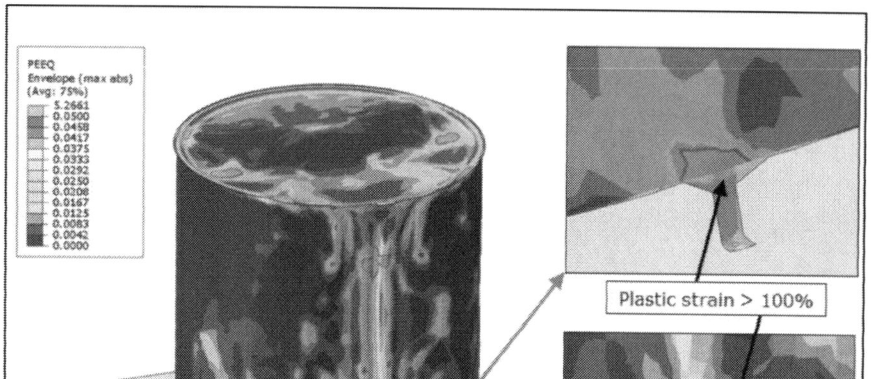

Figure 12: Equivalent plastic strain in Tank 1 due to blast loading (scale capped at 5%).

Based on the results discussed above, the three storage tanks subjected to the assumed postulated blast load are likely to fail via several possible failure modes. The first likely failure mode is plastic collapse of the tank walls and/or roofs. All three tanks see large deflections (the most severe occurring in Tank 1). Riks post-buckling analyses confirm that the magnitude of deflections seen in the tank walls would likely result in unstable collapse behavior before the peak blast overpressure is achieved. A second likely failure mode is extreme damage in the tanks at the hold-down bolts and/or regions of the tanks that are connected to the bolts. In all three tanks, equivalent plastic strains exceed 100% in these regions. Failure of multiple hold-down bolts or hold-down clips could cause a given tank to slide off its foundation and/or overturn. Furthermore, sloshing effects of the fluid inside the tanks are neglected in the simulations discussed herein, but would increase the likelihood of failure.

In short, the close proximity of all three tanks to the postulated blast source causes the corresponding overpressures acting on each tank to be relatively severe. Design changes to the tanks to supplement blast damage mitigation would likely have to be drastic in order to achieve a significant reduction in the probability of failure due to the postulated explosion considered in this study. Finally, in order to achieve practical simulation times, all of the explicit analyses discussed herein are conducted across multiple CPUs (often 16 or more) simultaneously. Utilizing parallel computing permits evaluation of complex three dimensional structures, such as the tanks discussed above, in reasonable amounts of time.

7. Conclusions

This study compares two different explicit computational methods for modeling blast loading on structures; incident wave loading and the CONWEP charge property. The underlying methodology for each method is discussed. CONWEP is well suited for conventional detonations and uses the equivalent TNT method, and it inherently calculates the appropriate blast wave parameters for a given explosive event. Additionally, CONWEP achieves realistic overpressure amplitudes, spatial decay, and reflection effects without having to model the fluid medium, as is the case for incident wave loading. Incident wave loading provides the user with the flexibility to define a time-varying overpressure amplitude. This is useful when performing sensitivity studies to blast loading parameters or when attempting to characterize the blast wave associated with a vapor cloud explosion that does not behave like a conventional detonation. The development and implementation of the CONWEP and incident wave blast loading models into commercial finite element programs makes accurately analyzing the response of complex three dimensional structures subjected to blast loading feasible.

Additionally, explicit, three dimensional finite element analyses of both single and double-walled storage tanks that carry an extremely high consequence of failure are discussed herein. Incident wave loading is used iteratively to determine the maximum permissible overpressure for a double-walled storage tank subject to a postulated vapor cloud explosion. The CONWEP blast loading model is used to analyze single and double-walled storage tanks subjected to a postulated chemical explosion (conventional detonation). The explicit computational methods, such as the ones examined herein, can provide valuable insight into the blast resistance of existing structures and can guide design decisions regarding how to mitigate blast damage for new structures in the petrochemical and related industries. Gaining a better understanding of the possible failure modes due to blast loading of critical process equipment is very beneficial in not only improving process safety, but also in enhancing safety protocol for catastrophic accidental explosions in industrial environments.

8. References

1. DOD, "Unified Facilities Criteria (UFC) 3-340-02: Structures to Resist the Effects of Accidental Explosions," *Department of Defense, United States of America*, 2008
2. Smith, P., and Hetherington, J., "Blast and Ballistic Loading of Structures, 1st ed," *Butterworth-Heinemann*, Oxford, 1994.
3. Kinney, G., and Graham, K., "Explosive Shocks in Air, 2nd ed," *Springer-Verlag*, Berlin, 1985.
4. Dusenberry, D., "Handbook for Blast-Resistant Design of Buildings, 1st ed," *John Wiley and Sons, Inc.*, Hoboken, New Jersey, 2010.
5. ASCE, "Design of Blast-Resistant Buildings in Petrochemical Facilities, 1st ed," Task Committee on Blast-Resistant Design, *American Society of Civil Engineers.*, Reston, Virginia, 1997.
6. Bulson, P., "Explosive Loading of Engineering Structures, 1st ed," *Taylor and Francis Group*, London, 1997.
7. Abaqus 6.10-1, DSS Simulia, "Analysis Users' Manual," 2010.
8. Hyde, D., "Users' Guide for Microcomputer Programs CONWEP and FUNPRO – Applications of TM 5-885-1," *U.S. Army Engineer Waterways Experimental Station*, Vicksburg, VA, 1988.
9. Huntington-Thresher, W., and Cullis, I., "TNT Blast Scaling for Small Charges," *19th International Symposium of Ballistics*, 2001.
10. Baker, W., "Design Considerations for Toxic Chemical and Explosives Facilities – Chapter 1, Blast Pressure Effects: An Overview," *American Chemical Society*, 1987.
11. Tang, M.J., and Baker, Q.A., "A New Set of Blast Curves from Vapor Cloud Explosion," *Process Safety Progress*. Vol. 18, No. 3, 1999.
12. American Institute of Chemical Engineers (AiChE), "Guidelines for Evaluating the Characteristics of Vapor Cloud Explosions, Flash Fires, and BLEVEs," *Center for Chemical Process Safety of the American Institute of Chemical Engineers.*, New York, 1994.
13. Lees, F.P., "Loss Prevention in the Process Industries," *Butterworth-Heinemann*, Oxford, 1996.
14. Mercx, W.P.M., van den Berg, A.C. and van Leeuwen, D., TNO report PML 1998-C53, "Application of correlations to quantify the source strength of vapor cloud explosions in realistic situations, Final report for the project: GAMES," *Netherlands Organization for Applied Scientific Research (TNO)*, 1998.
15. Van den Bosch, C.J.H. and R.A.P.M. Weterings, eds., "Methods for the calculation of physical effects: due to releases of hazardous materials (liquids and gases) ("TNO Yellow Book"), CPR 14E, Parts 1 and 2, 3rd Edition," Director-General for Social Affairs and Employment, *Committee for the Prevention of Disasters*. The Hague, 1997.

16. Woodward, J. L., "Estimating the Flammable Mass of a Vapor Cloud," *Center for Chemical Process Safety of the American Institute of Chemical Engineers*, New York, 1998.
17. Strehlow, R.A., Luckritz, R.T., Adamczyk, A.A., and Shimpi, S.A. "The Blast Wave Generated by Spherical Flames," *Combustion and Flame*, Vol. 35, 297-310, 1979.
18. Prueter, P.E., Dewees, D.J., and Brown, R.G. "Examining Computational Techniques to Simulate the Effects of External Chemical Explosions Using Explicit Finite Element Analysis," *In ASME Pressure Vessels and Piping Conference.* PVP2012-78566, 2012.
19. Maremonti, M., Russo, G., Salzano, E., and Tufano, V., "Post-Accident Analysis of Vapor Cloud Explosions in Fuel Storage Areas," *Trans IChemE*, 77(B), 1999.
20. Buncefield Investigation Board, "The Buncefield Incident 11 December 2005: The Final Report of the Major Incident Investigation Board," Volume 1, 2008.
21. Chang, J., and Lin, C., "A Study of Storage Tank Accidents," *Journal of Loss Prevention in the Process Industries*, 19(1) 51-59, 2006.

(44o) The Hidden Risks of Deposits, by-Products, and Residues On Exhaust and Waste Streams

Carlos A. Barrera and Abid Kemal, Thermal Sciences, Exponent, Menlo Park, CA

The safety risks associated with reactive compounds are usually captured and addressed during the conceptual design and pre-construction process safety reviews as, and for well-established chemistries these risks are well understood for the main processes and reactions. Lest time, if any, is devoted to the analysis of the reactive nature of the compounds present waste streams or exhaust lines. These lines are often overlooked during HAZOPs because of their secondary nature, or because they go to a scrubber or flare, which erroneously is considered a "catch-all" device that will "take care of everything." Waste streams and exhaust lines from certain physicochemical processes contain highly reactive compounds in the form of residues or by-products that in the right conditions can accumulate and form dangerous deposits. Some examples of such deposits are pyrophoric, shock-sensitive, and corrosive compounds. The purpose of this paper is to create awareness in the process safety community about the importance of a complete understanding of the chemistry of the reactive species present in these streams and the possible physical mechanisms than can lead to the formation of dangerous deposits, as well as the importance of including these streams in the process hazard analysis. Examples drawn from multiple processes will be presented.

(44q) Explosion Characteristics of Carbon Nano-Fibers

Jiaqi Zhang, Artie McFerrin Department of Chemical Engineering, Texas A&M Univerisity, Mary Kay O'Connor Process Safety Center, College Station, TX, Yi Liu, Mary Kay O'Connor Process Safety Center, Department of Chemical Engineering,Texas A&M University, College Station, TX and Dr. M Sam Mannan, Artie McFerrin Department of Chemical Engineering, Mary Kay O'Connor Process Safety Center, College Station, TX

Dust explosion is a hazard with serious consequences. While the main principle of dust explosion is still being exploring, the gap of dust explosion between particles of micro-size and those of nano-size also needs tremendous efforts. The explosion characteristics of CNFs are studied in Mary Kay O'Connor Process Safety Center. These experiments are conducted in a 36-L explosion vessel, which has been calibrated with a standard 20-L explosion vessel. Among the experiments, different kinds of CNFs of different grades and different conditions (milled and unmilled) are used. Compared with collected published data of graphic particles of micro-scale, the experimental data shows that the carbon nano-fibers create a higher Kst than the micro-scale ones. The results also show that explosion of the XT-PS grade (Pyrolytically Stripped) is much stronger than that of the XT-LHT grade (Low Temperature Heat-treated) and XT-HHT grade (High Temperature Heat-treated). Also, the explosion of milled CNFs is stronger than the unmilled ones. Further study to seek the causes is undergoing.

(44r) High-Fidelity Combined Dynamic Modeling of Depressurizing Vessels and Flare Networks: Accurately Assess the Effects of Process Changes On Mechanical Integrity and System Capacity

James Marriott, VP PSE Oil & Gas, Process Systems Enterprise Ltd, London, United Kingdom

The use of high-fidelity dynamic modeling for combined depressurization and flare network analysis can result in identification of additional capacity in flare networks without compromising process safety, presenting potentially significant opportunities for reducing or saving capital expenditure (CAPEX). The same analysis can simultaneously provide accurate assessment of temperatures in vessels and pipework, to identify areas where close approach to or violation of minimum metal temperatures may occur, thus compromising vessel and pipework mechanical integrity.

The flare networks for major plants represent a non-negligible part of the overall capital investment. Current industrial practice for their design is primarily based on steady-state analysis, as described in standards such as API 521 and supported by widely-used software tools. However, it is a widely recognized fact that the application of steady-steady considerations to what is fundamentally a dynamic system inevitably requires the use of conservative assumptions which often result in significant oversizing of flare headers and other components of the network. Another major contributor to capital cost is the use of special materials for the parts of the system that may be exposed to low-temperature fluids – typically the tail pipes attached to high-pressure process vessels – and which are therefore at risk of embrittlement. The key to being able to limit the additional capital expenditure is the accurate estimation of the length of piping that is subject to "abnormal" temperatures.

In addition to dynamic flare network analysis, the detailed dynamic modeling and simulation of the rapid depressurization ("blowdown") of high-pressure vessels is a key factor in determining both the load imposed on the pressure relief system (e.g. flare network) and also the temperatures of the vessel and downstream pipework walls. Current methods typically use 'lumped' equilibrium approaches that do not adequately predict these quantities; indeed, in contrast with the current over-conservative practice in flare network design, they may provide non-conservative solutions. By contrast, the high-fidelity approach incorporates a multi-phase non-equilibrium representation of the vessel fluid contents, as well as a 3-dimensional model of the metal walls that takes into account the transfer of heat between regions of the wall in contact with different phases. This allows a much more accurate estimation of the wall temperatures, and the direct computation of thermal stresses.

In the past it has been difficult to assess these important design criteria effectively because of the challenges in modeling such complex systems. This paper describes how the combination of high-fidelity depressurization capabilities and an advanced model-based system for flare system network design, coupled with evolving best practice in using these tools, represents a transformation for process safety management. The new techniques allow accurate assessment of the effects of process changes on mechanical integrity, thus enabling informed safety and CAPEX decision support.

9TH GLOBAL CONGRESS ON PROCESS SAFETY

An Engineering Method to Mitigate the Impact of Regulatory Focus on Relief System Installations by Prioritizing Risk

Dustin Smith
Smith & Burgess
5700 NW Central Dr. Ste 301 Houston, TX 77092
dustin.smith@smithburgess.com

Prepared for Presentation at
American Institute of Chemical Engineers
2013 Spring Meeting
9th Global Congress on Process Safety
San Antonio, Texas
April 28 – May 1, 2013

UNPUBLISHED

An Engineering Method to Mitigate the Impact of Regulatory Focus on Relief System Installations by Prioritizing Risk

Dustin Smith
Smith & Burgess
5700 NW Central Dr. Ste 301 Houston, TX 77092
dustin.smith@smithburgess.com

Keywords: Pressure Relief Valve Installation, API 3% Rule, Chatter, Relief Device Stability

Abstract

Since around 2005, regulators in the United States have put greater emphasis on relief device installations meeting the 3% rule. Spending large amounts of money to "fix" relief device installations that pose no safety risk decreases the plant's overall safety. To date, more incidents have occurred during facility construction / modifications than due to relief device chatter. This paper presents a method to assist engineers in determining if relief devices are susceptible to chatter. The methodology in this paper provides an engineering study to determine if existing installations are safe, which are allowed in the relevant engineering standards (API STD 520), and shows the research data by which it was validated. The model is used as a screening method that places the relief devices into two categories: (1) those installations that may chatter and (2) those installations that need no further review. The goal of any experimental comparison is that the model will error on the side of predicting chatter, but will be reliable enough to screen valves.

In addition to presenting the model, this paper will compare instances of known chatter to research conducted by API and work done by the Electric Power Research Institute in the 1980s. Thus far, based on research and acquired information, the method predicted all instances of chatter known to the authors. By providing a screening methodology that is supported by experimental data, plants can focus their spending on fixing real safety issues by identifying which relief installations are not expected to chatter. The paper will close by giving a brief explanation of the on-going research in relief valve stability.

1. Introduction

In November 2011, *Hydrocarbon Processing* published a paper that documented a method to determine if relief devices were susceptible to chatter. Other methods are being developed to determine the chances of chatter for a specific installation; however, the model discussed in the published paper is the only screening method that places the relief devices into two categories: (1) those installations that may chatter and (2) those installations that need no further review. The goal of any experimental comparison is that it will error on the side of predicting chatter, but will be reliable enough to screen valves. Since the publication of that article, the Oil & Gas industry has continued to struggle with the issue of relief device stability, so much so that API delayed issuance of *API STD 520 Part II Sizing, Selection, and Installation of Pressure-Relieving Devices in Refineries-Part II, Installation.* This paper compares instances of known chatter to research conducted by API, and uses this comparison to evaluate the model. Thus far, based on research and all acquired information, the method predicted all instances of chatter known to the authors.

This paper summarizes the analysis methods and validation for vapor service valves. The November 2011, *Hydrocarbon Processing* paper documents analysis methods for two-phase and liquid services. The focus on vapor / gas service valves is based on the validation data for these services analyzed herein.

2. Summary of the Stability Screening Method

The following is a summary of items used to confirm if chattering is likely to occur. Refer to the published paper for an extensive discussion of the methodology:

1. Confirm that the inlet line length is within the criteria to ensure that the pressure wave decay is not high enough to cause valve instability. The following equation (from Frommann & Friedel, Source 4) is used for this criteria:

$$L_i < 45{,}390\frac{d_i^2}{w_{\%O}}\left(\frac{P_S - P_{RC}}{P_S}\right)(P_S - P_B)\,t_o \qquad \textbf{Equation 1}$$

2. Confirm that the system's frictional and acoustic losses are less than the difference between the operating pressure and the valve's closing pressure. The formula for estimating the acoustical losses (from Singh, Source 5) is presented below:

$$\Delta P_{Acoustic} = \frac{Lw_{PSV}}{12.6d_i^2 t_O} + \frac{1}{10.5\rho}\left(\frac{w_{PSV}L}{cd_i t_O}\right)^2 \qquad \textbf{Equation 2}$$

3. Ensure that stability will not be compromised due to vortex shedding in the inlet line (this phenomenon occurs during normal operation and is generally not associated with relief conditions).

$$L_i < \frac{d_i c}{2.4U}$$ **Equation 3**

4. Ensure that the relief rate or system size is at sufficient capacity to confirm that chattering will not occur from being oversized. For vapor service relief devices, this is assumed to not occur if the open-close cycle time for the system is greater than 1 second. This cycle is broken into two components: the time it takes for the system pressure to increase from the relief devices' blowdown/closing pressure to the set pressure and the time it takes for the system to depressure through the relief valve from the set pressure to the blowdown/closing pressure. This is represented by the following equation.

$$1 > t_{cycle} = t_{P_{RC} \to P_S} + t_{P_S \to P_{RC}}$$ **Equation 4**

5. Ensure that the relief device is installed per the manufacturers' and/or general installation guidelines. These guidelines are shown in the catalogs, ASME B&PVC Section VIII, API STD 520, etc.

6. Review the valve's operational history to ensure that there is no known history of chatter.

For a complete discussion on the methodology of this analysis, the reader is referred to the work published by Smith Et. al, as the discussion is not further expanded in this paper.[1] Note that equation 17 does not follow the verbiage, and the *less than* sign should be replaced with a *greater than* sign.

3. Acceptance Criteria

This model is the only screening method to place the relief devices into two categories: (1) those installations that may chatter and (2) those installations that need no further review. The goal of any experimental comparison is that it will error on the side of predicting chatter, but will be reliable enough to screen valves. The author considers this model to be reliable as an engineering screening tool if the following criteria are met:

- *High Correlation to Experimental Data* – If the method predicts that valves will chatter and valves won't chatter most of the time, then it is a viable screening tool given the low cost of performing the analysis compared to making piping modifications.

- *No False Negatives* – The screening tool will be considered valid so long as it does not predict any installations as safe (no chatter) when in fact chatter was experimentally found.
- *Limited False Positives* – Given that the method is a conservative screening tool, some level of false positives are to be expected. This includes instances when the model would indicate the possibility of chatter when experimental results show valve stability.
- *Screens Valves as Acceptable* – The model will only be of value if the results of the model result in existing installations not requiring further modification. Based on the results presented in the referenced paper, 50% of the installations with inlet pressure losses greater than 3% were deemed acceptable as is.

The model, therefore, will be deemed conservative and reliable if the previous four criteria are met.

4. Comparisons / Validations

To confirm the validity of the procedure, the procedure has been reviewed against all cases of chatter known to the authors. The cases are divided into three different categories:

1. ***Known Installations*** – These installations are those that are known by the authors to have failed with a loss of containment and those installations that have sufficient information to perform the analysis. The information for some of the installations is not available in the public domain, as the confidentiality of these installations has been preserved.
2. ***ASME Studies*** – In the 80's, Zahorsky performed a set of research for the nuclear industry in which they experimentally set the blowdown as the minimum percentage needed for stable operation[2]. In order to meet the acceptance criteria, therefore, this method should determine that these installations are unstable. The results were re-checked with an increased blowdown of 2% above what the author listed to see what would happen.
3. ***API PERF Studies*** – In 2011 the results of the API PERF study were presented at the API 520 Committee meeting. In this work, the API subgroup tested 18 different valves with three different inlet line lengths to determine when chatter would occur. The results of the 54 trials were introduced in their presentation (some combinations were not tested).

5. Results

The following are the results of the comparisons:

Table 1 lists the results of the model comparison with the valves that have had a loss of containment where inlet line chatter was the cause or a contributor to the incident. The model predicted the potential for chatter for each of these installations.

Table 1: Summary of known installations that have chattered

Installation	Service	Chatter Predicted	Data Source
Refinery, North America	Liquid	Yes	Internal company incident investigation
Gas Plant, Middle East	Vapor	Yes	Internal company incident investigation
Refinery, North America	Liquid	Yes	Internal company incident investigation
Refinery, North America	Liquid	Yes	API Published Document [3]

Table 2 lists the blowdown at which the relief devices do not chatter as experimentally determined by Zahorsky. The table also lists the blowdown at which the model would predict stability. The difference between the two values can be taken as an estimate to which the model is conservative. Since the blowdown was experimentally set as the minimum percentage, the model is accurate, as it predicts false positives for these devices.

Table 2: Comparison of minimum predicted blowdown to the results experimentally derived by Zahorsky

Run Case	Experimentally Determined Blowdown	Predicted Acceptable Blowdown	(Δ Blowdown) Predicted – Expermintal)
1	3.9%	4%	0.1%
2	3.9%	5.6%	1.7%
3	5.6%	9.7%	4.1%
4	8.4%	16.7%	8.3%
5	8.3%	12.6%	4.3%
6	4.3	5.3	1.0%

Table 3 lists the comparison of the model results to the experimental PERF results. The model had agreement ~72% of the time if the cases that were not tested were excluded. If one assumes that the results for the untested cases would have chattered, the agreement is ~76%. Of the cases that the model did not agree with the experimental results, the model always predicted chatter on stable valves; thus, the model is accurately screening and is slightly conservative.

Table 3: Comparison of the model to the API PERF Study Results

Model Correlation	**PERF Results**	**Model Prediction**	**No. Of Cases**
Agreement	Chatter	Chatter	9
Agreement	Stable	Stable	25
False Negative	Chatter	Stable	0
False Positive	Stable	Chatter	15
Agreement[1]	Not Tested	Chatter	7

Note 1: There are a number of cases that were not tested, but were assumed to chatter as the reason for not being tested was not included.

6. Conclusion

The model results always predicted when valves may chatter (need modifications). There are no instances of valves that have chattered where the model predicted stability.

- *High Correlation to Experimental Data* – There is a 70% to 75% agreement with the API PERF Studies and all industry installations with known chatter where identified with the model. This model predicted if a valve would chatter accurately 75% of the time (see the next item on false negatives).
- *No False Negatives* – No cases of chatter were found when the model predicted stable operation.
- *Limited False Positives* – 25% of the PERF study valves that were stable indicated that chatter was possible. Also, the work by Zahorsky indicated that the model predicts stable operation with an inlet blowdown generally 2% to 4% greater than needed.

The model, therefore, is a reliable method to screen relief devices based on the comparison to 60+ experimental and industrial data points.

7. Notation

The equations presented in this paper are dimensional and the following listing explains the variables and unit set.

c = speed of sound (ft/s)
d = diameter (in)
L = length (ft)
P = pressure (psig)
t = time (s)
U = process fluid velocity as it passes the PSV nozzle (ft/s)
w = mass flow rate (lb/s)

<u>Greek Letters</u>
ρ = fluid density (lb/ft^3)

Subscripts
B = backpressure on relief device
i = inlet
o = opening
PSV = process safety valve
RC = valve reclosing pressure
S = relief device set pressure
%O = flow rate at the valves percent open

8. References

[1] Smith, Dustin P.E., John Burgess P.E., and Craig Powers PhD. "Relief Device Inlet Piping: Beyond the 3% Rule." *Hydrocarbon Processing.* 2011.

[2] Zahorsky, J.R. "Testing and Analysis of Safety/Relief Valve Performance." The ASME Winter Annual Meeting. November 15-19, 1982. Atlanta, GA

[3] Myers, DF, Conoco. "Denver Refinery Fire." *API.* 1986.

[4] Frommann, O. and Friedel, L. "Analysis of safety relief valve chatter induced by pressure waves in gas flow." *Journal of Loss Prevention in the Process Industries.* 1998.

[5] Singh, A. "On the Stability of a Coupled Safety Valve-Piping System." The ASME Winter Annual Meeting. November 15-19, 1982. Atlanta, GA.

Making Relief Load Estimates Match Reality

Nicholas N. Cristea
Smith & Burgess, LLC
5700 Northwest Central Dr. Ste 301 Houston, Texas 77092
Nicholas.Cristea@smithburgess.com

Dustin Smith
Smith & Burgess, LLC
5700 Northwest Central Dr. Ste 301 Houston, Texas 77092
Dustin.Smith@smithburgess.com

Prepared for Presentation at
American Institute of Chemical Engineers
2013 Spring Meeting
9th Global Congress on Process Safety
San Antonio, Texas
April 28 – May 1, 2013

UNPUBLISHED

Making Relief Load Estimates Match Reality

Nicholas N. Cristea
Smith & Burgess, LLC
5700 Northwest Central Dr. Ste 301
Houston, Texas 77092
Nicholas.Cristea@smithburgess.com

Dustin J. Smith, P.E.
Smith & Burgess, LLC

Keywords:
Relief load estimates, relief system analysis, process simulations, reduced duty, inlet control valve failure, column systems

Abstract

A pressure relief system designer must make some assumptions when looking at a relief systems analysis. Some assumptions may result in costly modifications of the system, so knowing when to apply certain assumptions is paramount with the ability to accurately size a relief valve. When looking at systems, industry standards should be consulted before taking certain credits based on the given scenario. For example, the designer may be able to reduce the relief load by looking at a reduced reboiler duty and a temperature pinch during an overpressure event. When evaluating an inlet control valve failure, some credits may be taken for the minimum turndown for the outflow. However, when looking at how a downstream system may respond to the overpressure event, credit for flow continuing may not be acceptable and flow may be assumed to be lost. The application of assumptions is common practice in relief systems analysis; however, an assumption that does not match the true operational limits may lead to problems with the relief system that do not actually exist. These invalid assumptions could cost millions of dollars to fix the problem.

1. Introduction

When doing a Relief Systems Documentation Analysis, the ability to accurately model a "worst case" relief scenario is paramount to ensuring the system is protected from the overpressure event. The engineer is undoubtedly limited to the availability of equipment data and accuracy of the process simulation developed to predict upset conditions. However, if some conservative yet unrealistic assumptions are used, the relief device can be labeled as providing inadequate protection in the event of the initiating event. [1] Physically modifying the system to fix the problem, as well as the loss of profit from having to shut down the unit to perform the update, can cost the facility large sums of money. However, performing a detailed analysis on the system may better predict the "worst case" scenario, avoiding unnecessary modifications. The ability to make relief load estimates match what may happen out in the field can still be accomplished, but knowing the difference between a realistic conservative assumption and an unrealistic conservative assumption is paramount. The pressure relief system designer should consult industry standards (e.g. API 521or NFPA) to see how to estimate relief loads for given overpressure scenarios and what assumptions for mitigating items/actions are acceptable. This paper will look at two cases where relief load estimates were reduced and one where by matching the physical constraints of the system to reality resulted in a larger relief load estimate.

2. When Realistic Assumptions Lower Relief Estimates

Oftentimes relief estimates are based on normal operating flow rates, heat exchange, and/or flow paths continuing during the upsets. With the amount of optimization and debottlenecking, these assumptions may not be realistic. Equipment capacities that are right on the design limit may not be able to produce the same flow rate or heat transfer under upset conditions. Flow paths that are available under normal operation may be isolated or otherwise not be available during upsets. The following are three examples that show these effects in a typical plant situation. The first two are examples of assumptions that tended to increase the relief rate while the last example illustrates an optimistic assumption that under-predicted the relief requirements.

2.1 Reduced Reboiler Duty

When looking at a column system, the primary relief loads tend to consist of a combination of continued heat input with the loss of cooling. The relief load estimates can be complex due to the very nature of the system. The ability to do a detailed analysis of the column may be able to better predict upset conditions. The authors have seen many cases where a detailed column analysis prevented the relief device modifications from being made based on inaccurate sizing basis. The ability to accurately model the relief scenario is paramount because an error could lead to events such as an unnecessary facility modification or a loss of containment from an undersized relief device. When a pressure relief system designer looks at the loss of overhead cooling on a column system, the scenario is a transient event. The case can start with the loss of overhead condensing with the reflux continuing for a short duration. After some time, the loss of condensing and continued overhead liquid product output results in the loss of level and reflux flow. To estimate this relief condition, the column system is modeled as two steady state cases to bracket the relief load. Under either of these conditions, the bottoms temperature may increase due to the increase in system pressure, but this phenomenon may be offset by the reduction in separation; therefore, a reduced duty calculation should be performed for both calculation sets to ensure that the reboiler is capable of transferring the same heat into the column system as during normal operating conditions. When looking at the loss of overhead cooling with the reflux continuing, the reboiler duty may not be able to be reduced due to increased light component traffic in the tower (as compared to the top tower reflux failure) when compared to the high temperatures in the reboiler and lower duties expected during the top-tower reflux failure scenario due to the loss of the light ends liquid traffic in the column. Taking credit for the reduced duty is still a reasonable assumption as an upper conservative limit when the assumption is based on the maximum duty possible for the installed reboiler and is limited by equipment capacity. The design equation for heat exchangers is given below [2]:

$$Q = UA\Delta T_{LM}$$

Reduced duty calculations, limited to process/process or process/utility exchangers (e.g. shell and tube exchangers), are not typically performed on fired heater reboilers. The exchanger data sheet, if up-to-date, is the best resource for determining the heat transfer characteristics of an exchanger. Most process simulators can simulate the duty of shell and tube exchangers by specifying the clean UA and, if applicable, the utility stream (e.g. 50# steam). The duty for the exchanger may change as reflux is lost to the system, which the upset may result in hotter bottoms temperature to the reboiler. The reduced duty may reduce the relief load estimate, as there is less vapor generation as compared to using the process duty under normal conditions. Figure 1 shows a screenshot from a process simulator that estimates the rate for a column system with continued feed and heat input, but a loss of cooling and reflux.

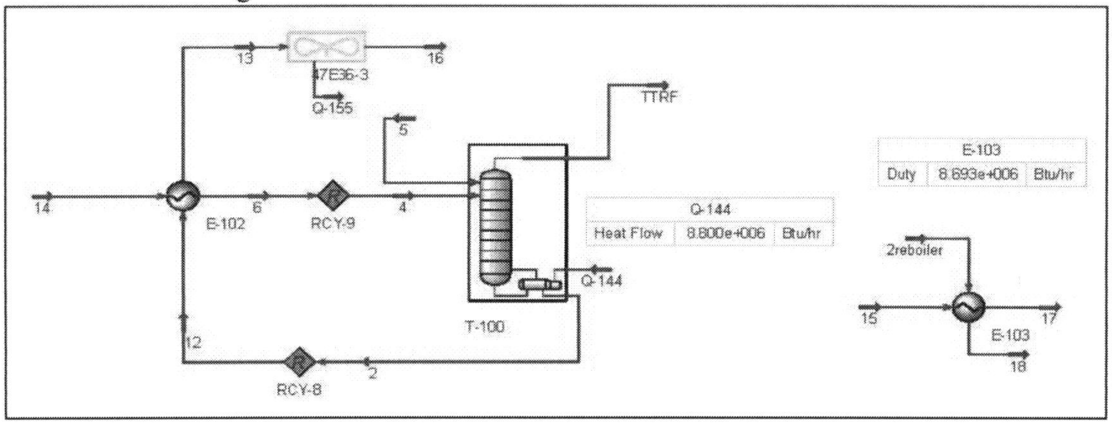

Figure 1: Example of a simulation for a column with loss of cooling and top-tower reflux failure with a reduced duty calculation performed.

2.2 *Using the correct composition and physical properties*

Generally the capacity of the reboiler (e.g. the reduced or LMTD pinched duty) will vary greatly if the calculation is based on the normal bottoms product as compared to the overhead reflux/liquid product. In the case of system upsets, the actual liquid inventory in the bottom of the tower may be lighter than normal. The pressure relief system designer is cautioned to use a composition in the reboiler to estimate the reduced duty that is consistent with the overpressure scenario being reviewed.

2.3 *Inlet Control Valve Failure Credits*

Another common overpressure scenario is the failure of automatic controls. Per API 521, this scenario is the complete failing open of a control valve (irrespective of its fail safe position).[3] If this upset condition results in overpressure, a relief device may be required to protect downstream equipment. When looking at an inlet control valve failure, the first common assumption pressure relief system designers make is to estimate the full capacity of the control valve and then verify if the relief valve can pass that fluid at the appropriate overpressure. The designers verify if the capacity of the relief device is greater than the control valve. If this estimation results in insufficient relief valve capacity, then some credits may need to be taken to make the relief load estimate more realistic.

> Per API 521§ 5.10.2,
> "in evaluating relieving requirements due to any cause, any automatic control valves that are not under consideration as causing a relieving requirement and that would tend to relieve the system should be assumed to remain in the position required for minimum normal processing flow. In other words, no credit should be taken for any favourable instrument response." [3]

For example, if the outlets from the system are expected to remain open, the calculation may assume that some of the fluid will flow through the normal path, see Figure 2. This credit is usually at the minimum turndown rate, as common industry practice is to assume that other control system remains in place and does not respond (unless it exasperates the single failure).[3] If the relief scenario is liquid, credit can be taken for the liquid outlet to reduce the fluid from the control valve that needs to be relieved by the pressure relief device by the minimum turndown rate. If the liquid outlet does not provide enough credit, the vapor outlet can be sized to take some relief as long as the vapor line is not lost by passing the liquid. A common example of this would be that a high-level alarm would trip a compressor; in this case, the relief device would have to pass both the liquid and the vapor feed to the vessel as the increased liquid level caused multiple failures. As long as the pressure relief system designer verified the liquid could flow through the vapor line and accumulate in the downstream system, credit can be taken through the vapor line. If the relief scenario is vapor, credit may be taken for flow through the vapor line at the minimum turndown rate (though the potential for credit for vapor flow through the normal liquid outlet is normally not considered). Consideration should be given for the difference in the relief fluid and the normal fluid and to adjust the credit based on the actual volume rate.

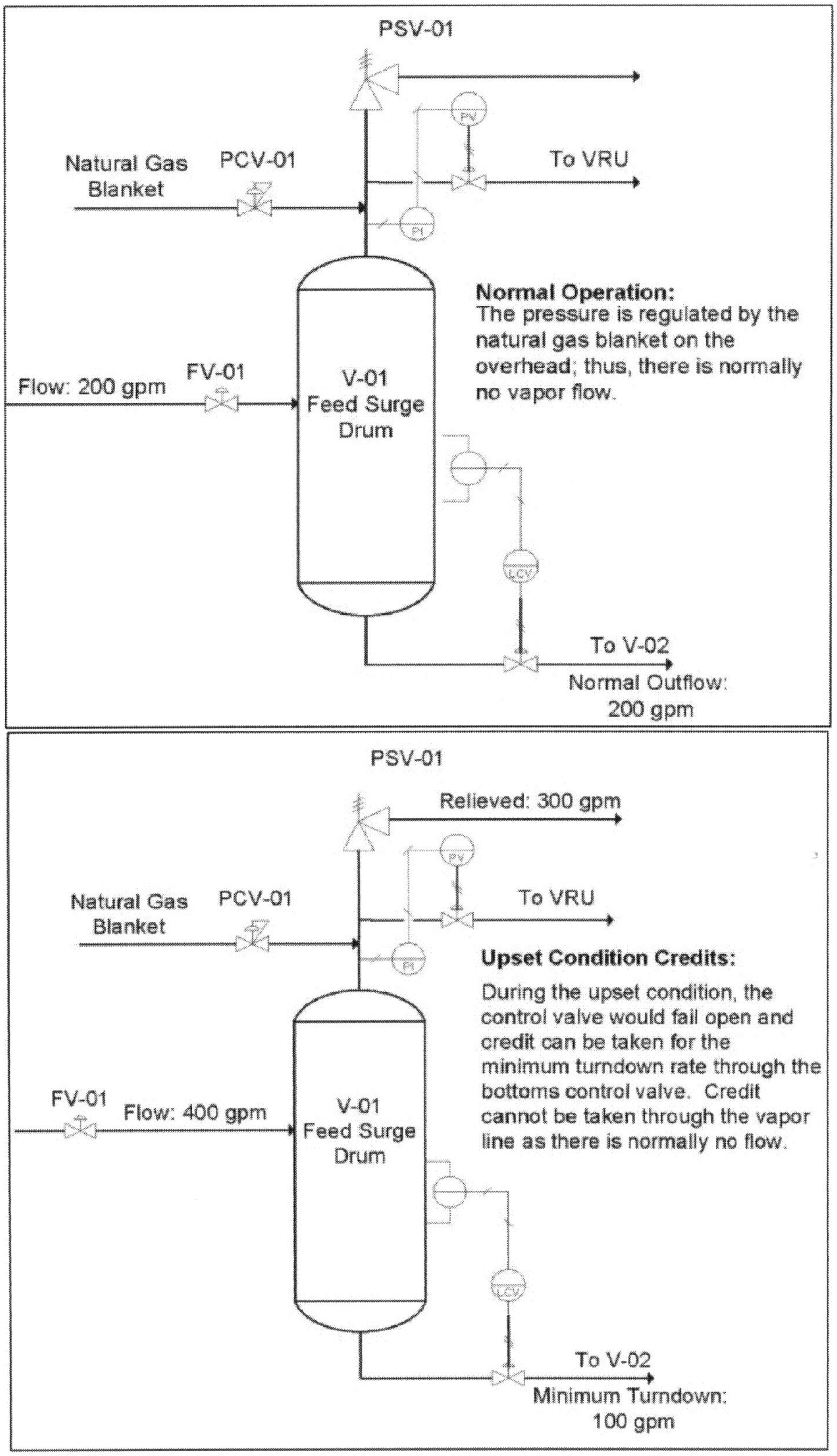

Figure 2: Looking at a feed surge drum during inlet control valve failure can lead to credits to be taken as long as they are reasonable for the system. Credit cannot be taken through the vapor line as there is normally no flow.

There is a special control valve failure case that pressure relief system designers need to be aware of. When a liquid control valve fails open, the vessel may de-inventory and gas may flow through the liquid control valve into the downstream vessel. This scenario, typically called a gas blow-by, should be reviewed for all control valve failures where there is liquid and gas in the upstream vessel. Figure 3 shows a typical system where gas blow-by may occur.

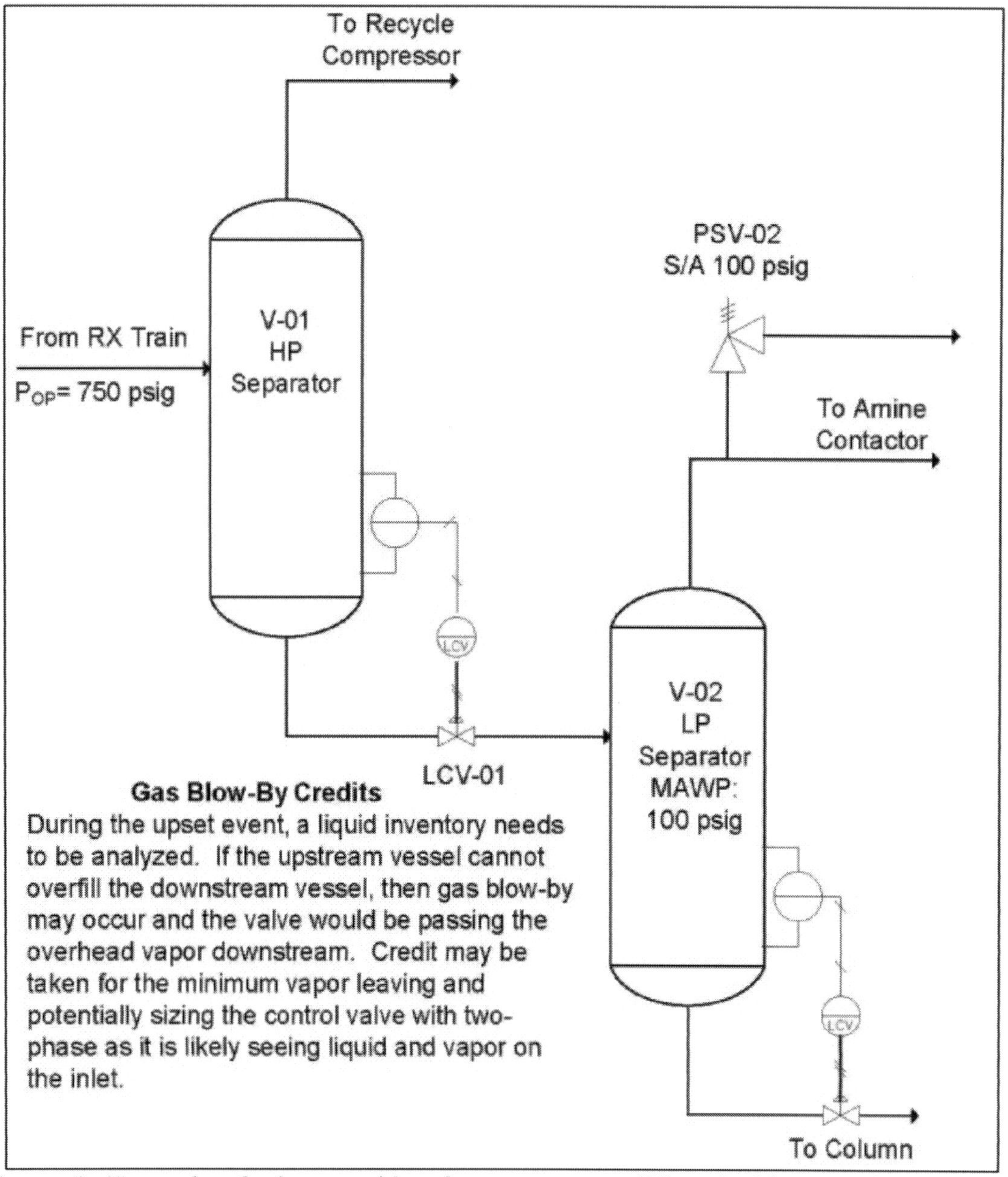

Figure 3: Example of when gas blow-by may occur. When looking at a low pressure separator, the sizing event should be vapor breakthrough the control valve from the upstream high pressure separator.

3. When Realistic Assumptions Make the Estimates Worse

3.1 Compressors Downstream of the System

In some instances, making relief load estimates match closer to reality may lead to a larger relief load estimate. In these cases, the pressure relief system designer must ensure that the relief system is accurately modeled and that the designer uses the larger rates and does not allow designs based on nonsensical assumptions to be used as the basis. There are many different assumptions that could cause this case to be true; the example that we are going to highlight is when the cascading effect of one failure causes the normal plant control system to take an action and increase the relief requirements for a pressure relief device. Oftentimes, a pressure relief system designer may want to take credit for some of the relief effluent to outflow from the system through a compressor. As stated above, this would not be a viable flow path for liquids, as almost all compressors have a high level trip on the inlet drum, but what about vapors? Depending on the scenario, the compressor may pass vapors at the upset conditions or the compressor could trip and result in a blocked outlet. Examples that may cause a compressor to trip under upset conditions include a higher inlet gas temperature trip, a high amp / power requirement trip (due to the increased suction pressure), high temperate trip (due to the potential loss of cooling to the compressor), vibration trips, and etc.[4,5] It is useful for the pressure relief system designer to discuss the scenario with personnel who are familiar with the operation of the compressor (usually either the operators or unit process engineer) to ensure that any assumptions made are consistent with operational history. Figure 4 shows a case where a compressor trip would not result in a larger relief load.

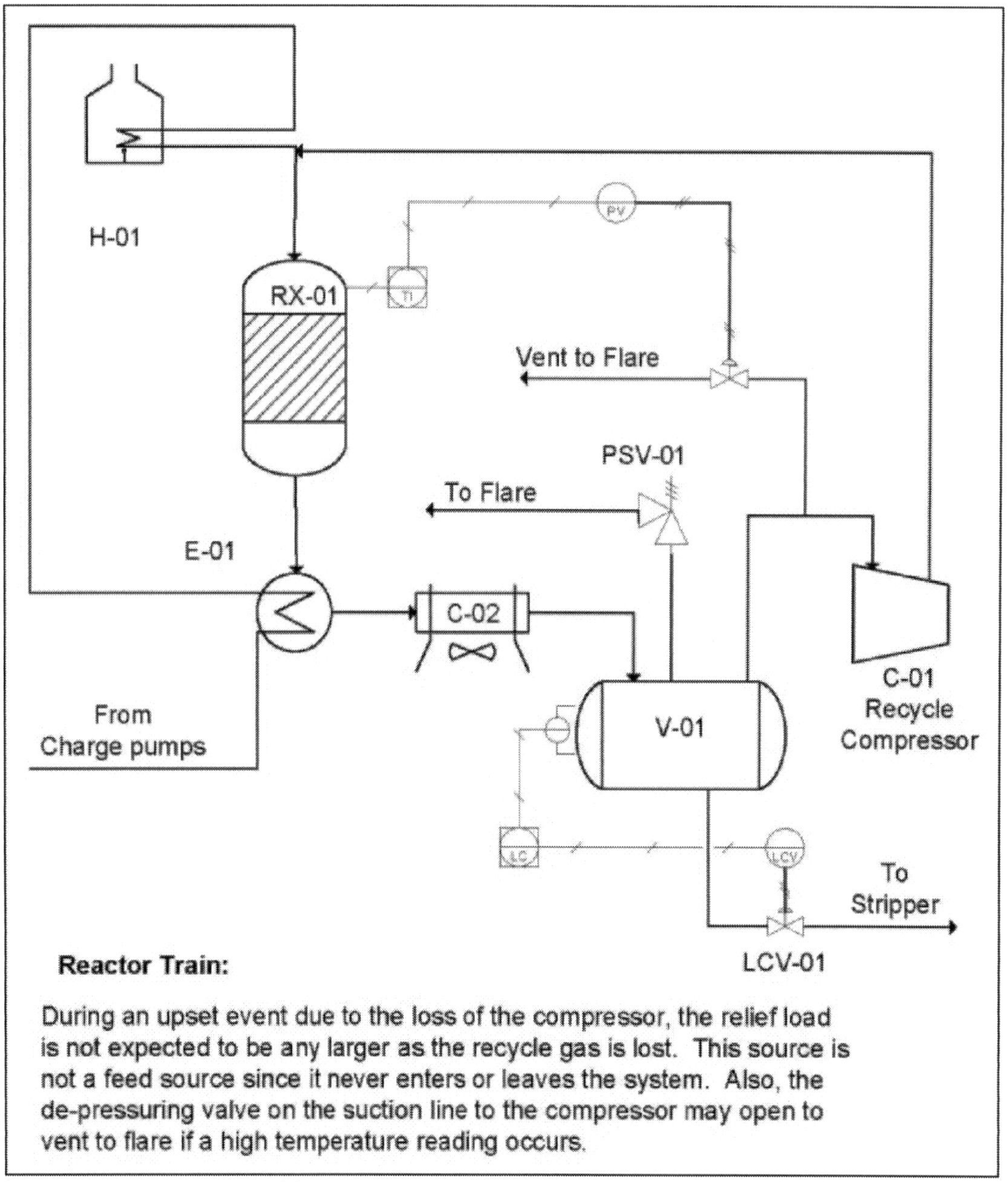

Figure 4: An example of how compressors can affect the relief load. The loss of the compressor would not result in any excess relief as the recycle gas is lost.

3.2 Example of Compressors Shutdown Analysis

A larger estimated rate may occur in a scenario where there is a partial loss of cooling in a column system that uses a compressor to move the off gas from the overhead accumulator to the downstream system (e.g. a FCC Main Fractionator, Coker Bubble Tower, Atmospheric Tower, and etc.). After the simulation is developed to model the column operation during the partial loss of cooling, the modified characteristics of the off gas (e.g. temperature, density) are consistent with conditions that trip the compressor.

Additionally, the pressure relief system designer must ensure that the scenario itself (e.g. a unit wide power or cooling tower failure) could result in a compressor trip and result in this outlet being blocked. The relief load estimate would then need to include the cascading effects of the compressor shutting down and may be much larger than would be otherwise predicted (see Figure 5).

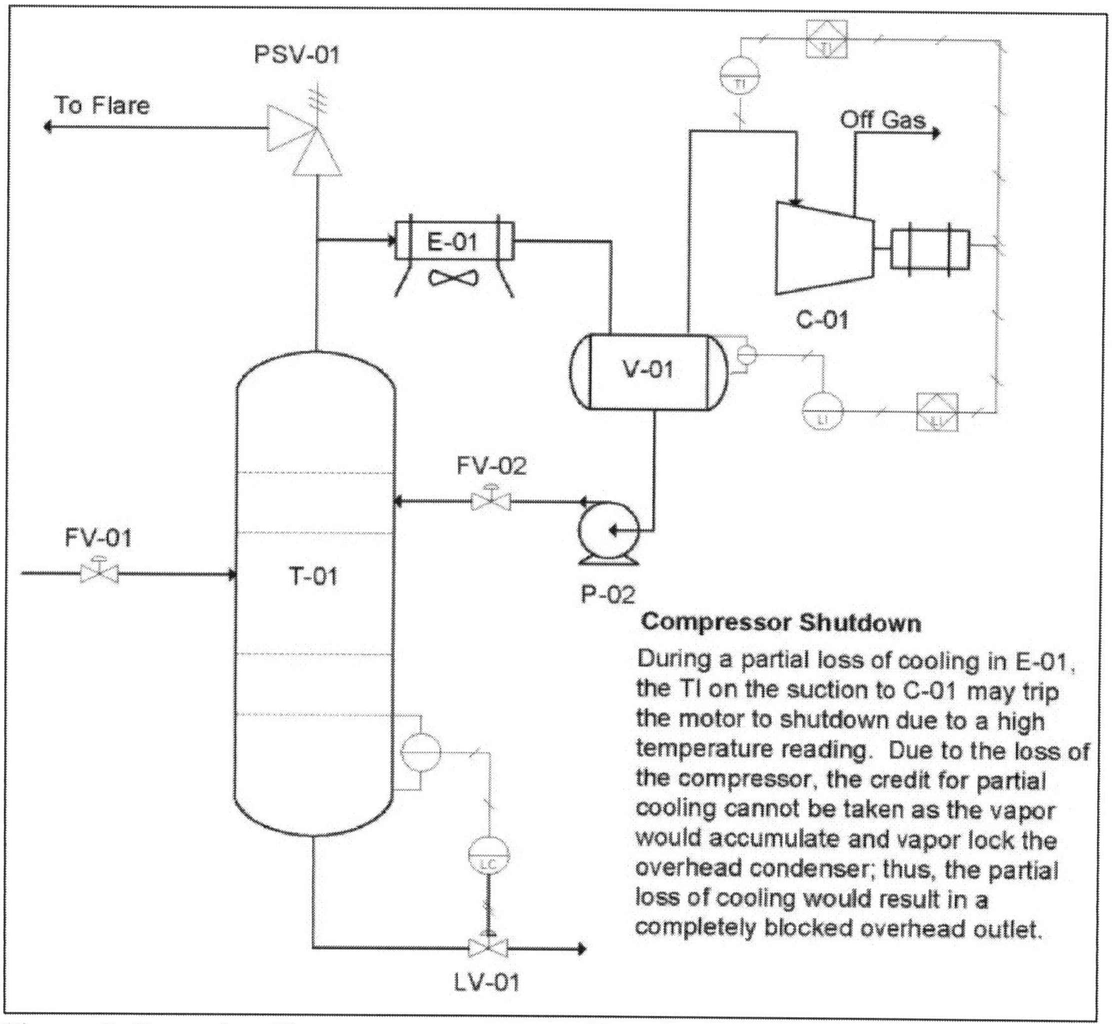

Figure 5: Example of how compressors can affect the relief load. On the right, the loss of the compressor would result in a larger relief as the vapor outlet is completely blocked.

4. Conclusion

To realistically represent the relief load estimate, the pressure relief system designer must understand the system, how it is operated, the limits and shutdowns of the associated equipment, and how these parameters interact during the specific overpressure event being reviewed. The pressure relief system designer must also ensure that all assumptions are consistent with industry, company, and regulatory requirements for the facility. There are many ways to check to see if the relief load estimate is based on realistic assumptions or on conservative/simplifying assumptions. In many cases, conservative/simplifying assumptions are acceptable as long as the assumptions don't result in deviations from standards or practices that result in spending more money. Oftentimes, conservative/simplifying assumptions do result in deviations, and the pressure relief system designer must then use more realistic assumptions to generate the relief load.

The pressure relief system designer must also be vigilant to not use simplifying assumptions that are not conservative. These assumptions tend to result in relief rates that may under-predict the true requirements in an overpressure event. Oftentimes industry, company, and regulatory requirements for the facility are set up to eliminate these scenarios, but the example in Section 3.2 illustrates the pressure relief system designer must check these assumptions (many of which can be implicit).

5. References

[1]. Smith, Dustin and John Burgess, "Relief Valve and flare action items: What plant engineers should know," *Hydrocarbon Processing*, Volume 91, NO.11, 2012

[2]. Middleman, Stanley, *An Introduction to Mass and Heat Transfer: Principles of Analysis and Design*, John Wiley and Sons, New York, NY. 1998

[3] ANSI/API Standard 521, "Pressure-relieving and Depressuring Systems," American Petroleum Institute, 5th ed., 2008.

[4]. GPSA, *Engineering Data Book: Volume I 10th Edition*, Gas Processors Suppliers Association, Tulsa, OK. 1987

[5]. Lieberman, Norman and Elizabeth Lieberman, *A Working Guide to Process Equipment 3rd Edition*, McGraw-Hill, New York, NY. 2008

9TH GLOBAL CONGRESS ON PROCESS SAFETY

Explosion incident simulation for Petrochemical plant

Zhao Xiangdi
State Key Laboratory of Safety and Control for Chemicals, SINOPEC Safety Engineering Institute
No.218, Yan'an San Road, Qingdao, P.R.China 266071
cheeryde@gmail.com

Yuan Jiwu
State Key Laboratory of Safety and Control for Chemicals, SINOPEC Safety Engineering Institute
No.218, Yan'an San Road, Qingdao, P.R.China 266071
Yuanjw.qday@sinopec.com

Jiang Chunming
State Key Laboratory of Safety and Control for Chemicals, SINOPEC Safety Engineering Institute
No.218, Yan'an San Road, Qingdao, P.R.China 266071
Jiangcm.qday@sinopec.com

Wang Zheng
State Key Laboratory of Safety and Control for Chemicals, SINOPEC Safety Engineering Institute
No.218, Yan'an San Road, Qingdao, P.R.China 266071
wz.qday@sinopec.com

Prepared for Presentation at
American Institute of Chemical Engineers
2013 Spring Meeting
9th Global Congress on Process Safety
San Antonio, Texas
April 28 – May 1, 2013

UNPUBLISHED

Explosion incident simulation for Petrochemical plant

Keywords: numerical simulation, explosion incident, TNT equivalent, Multi-Energy

Abstract

A new method of explosion simulation which is based on FLACS software was used for simulating and rebuilding an explosion incident which occurred in a coal gasification company. The other two commonly used methods which are based on TNT equivalent method and Multi-Energy method were considered to calculate the explosion overpressure also. Simulation results indicated that this new method can calculate explosion overpressure in three dimensional. It can make up for the shortfall of traditional assessment techniques. The result is more stable and reasonable when compared with the scene of the incident.

1. Introduction

Petrochemical plants may happen explosion incident easily than other facilities since they have many equipments and devices with lots of hazard material. Once the device exploded, it will cause large casualty and property loss. For example, Alon big spring refinery(2007) and Guangwei(2008) occurred fire and explosion incidents and caused many people dead. So it is important for our to calculate plant's capability against explosion. We could use the results to determine whether those plants should be firmed and changed or not. Researchers around the world found a series special assessment methods based on incident consequences or risk. Someone used those methods to design and renovate process safety of petrochemical plant[1][2]. However, researches on those technology were lags behind in China. We used traditional assessment technology which based on empirical formula. Those traditional methods did not consider the actual situation of factory. It couldn't show the those factor's effect on explosion incident such as the layer equipment, wind, fire protection system and other condition [3][4][5].

Nowadays the Computational Fluid Dynamics analysis technique is becoming more sophisticated. Someone in industry and academia studies explosion simulation used CFD. Herrmann, Hansen studied the consequences of gas cloud explosion using FLACS computational fluid dynamics software, then compared results with experiment, and found the results matched each other well[6][7]. Windhors, used AutoReaGas software to simulate the gas explosion and determine safe distance between petrochemical cracker and control room [8]. P.Hoorelbeke studied explosion risk assessment techniques commonly used [9]. However, there were only a little study on numerical simulation model for explosion incidents. Therefore, assessment on explosion incident by using FLACS were performed in the work.

2. Explosion incident and Simulate model

2.1 Explosion incident

23 December 2008 an explosion incident occurred in Hunan province in China. Gas released from one pipe in the carbon monoxide shift.

Then the gas cloud was ignited and caused explosion. The scenes after this incident with the insulation layer damaged was shown in figure 1.

(a)F2012，F2013, F2014

(b) T2101，R2103

(c) Office building with windows and glass damaged

Figure 1: The scene of the explosion incident with the insulation layer damaged

The gas which flowed in the pipeline was hydrogen. The loss rate of contain was 562kNm3/h in the release phase. The wind speed in-situ was about 3m/s with the direction was NE. The key equipment in this unit for simulation were shown in table 1.

Table 1. Information for the key equipment

Number	Name	Size(m)	Description
F2102	Gas-liquid separator	Φ4.2×13.4	40% of insulation damaged
F2103	Gas-liquid separator	Φ3.7×11.6	25% of insulation damaged
F2104	Gas-liquid separator	Φ3.7×15.6	Little effect
T2101	Regulate tower	Φ3.2×26.0	70% of insulation damaged

R2103	Converter	Φ4.9×13.8	50% of insulation damaged
R2104	Converter	Φ4.9×11.7	30% of insulation damaged
Control room		15.0×46.8×4	Some of the windows damaged; Lots of the windows broken
Office building		15×46.8×19.5	Some of the windows damaged; Lots of the windows broken

2.2 Model for simulation

3-D model which was based on the factory's actual layout was shown in figure 1. This model was built using the actual sizes of the carbon monoxide shift unit. The total size for simulation was 100m×80m×40m. Ignition position was in the center of carbon monoxide shift unit nearby leak point. The distance from ignition position to control room and office building was 65m. Shapes of equipment were modified for simulation. Monitor points were set for key equipments, as shown in figure 2.

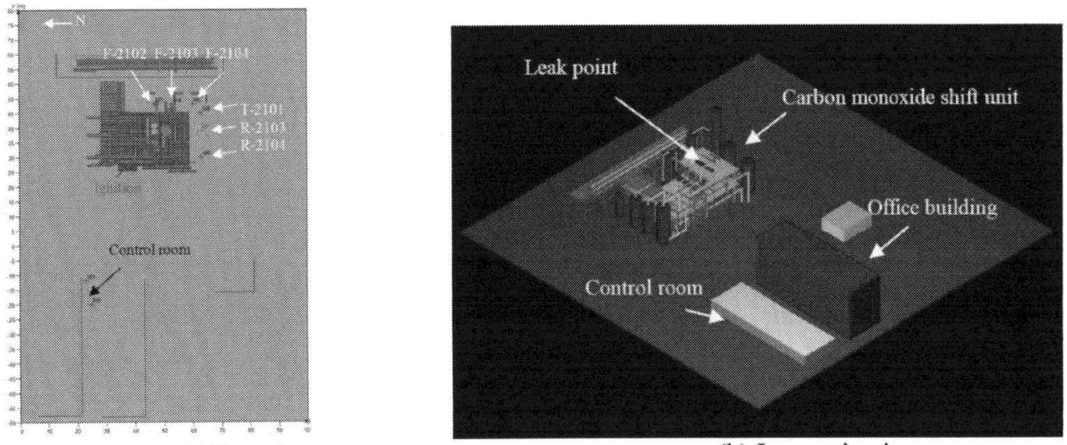

(a)Top view (b) Isometric view

Figure 2: Three dimensional model for simulation

2.3 Results and discussion

2.3.1 Leak and dispersion

The material in pipeline was hydrogen. Loss rate for contain was 562kNm^3/h. The time was 24 second between leak started and gas ignited. Wind speed was about 3m/s with direction was NE when the incident occurred. The new method which was based on FLACS was used to rebuilt dispersion process for hydrogen. Simulation results of fuel size after dispersion were shown in figure 3. The whole volume of the hydrogen was about 10×13×7 m^3 in a mixture concentration with equivalence ratio(ER).

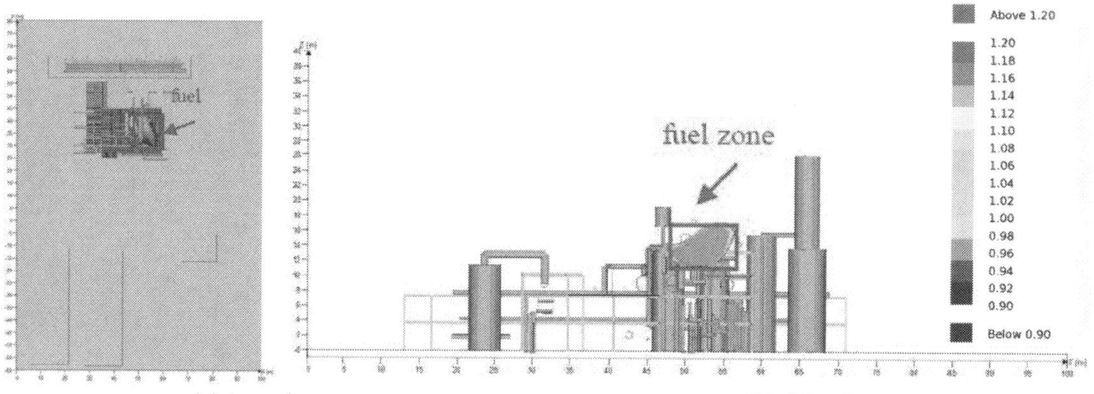

(a) top view (b)side view

Figure 3: The fuel size in this unit after dispersion (0.9~1.2 ER)

2.3.2 Explosion

Simulation assumes that hydrogen and air were in a mixture concentration with equivalence ratio. TNT equivalent method, Multi-Energy method and FLACS method were used to assess the process for gas explosion impact on the equipments and facilities. However, TNT equivalent method was simple and did not consider the effect of actual situation. Although the Multi-Energy method considered the explosion strength as the effect of actual situation, but it also did not consider reflection of complex equipment and did not predict the effect of risk mitigation system such as water spray.

Results of explosion overpressure using those three method were shown in table 2. It can be seen that TNT equivalent method and Multi-Energy method could only take into account explosion overpressure in two-dimensional. For example, overpressure calculated in the position 48m from center of explosion was 5.2 kPa using TNT equivalent method, however, the results increased to 7.9 kPa using Multi-Energy method. But, simulation results using FLACS were different. It could give overpressure in three-dimensional. The overpressure in position 48m from center of explosion was 6.7 kPa with 10m above ground, and 4.7 kPa with 19m above ground , as shows in table 2.

Table 2. Results of explosion overpressure using different methods.

		Overpressure in the different position from the ignition point (kPa)		
		48m (office building)	65m(control room)	74m
TNT equivalent method		5.2	3.48	2.95
Multi-Energy (blast strength 5~10)		7.9 ~ 24.7	5 ~ 16	5.1 ~ 13.4
FLACS method the height above ground(m)	1m	7.8	6.5	3.9
	10m	6.7	-	3.6
	19m	4.7	-	1.9

The overpressure in different monitor points(as shown in figure1) were shown in figure 4. The maximum explosion overpressure in carbon monoxide shift unit was about 62kPa, as well as, the overpressure was about 6.5kPa near the control room and office building.

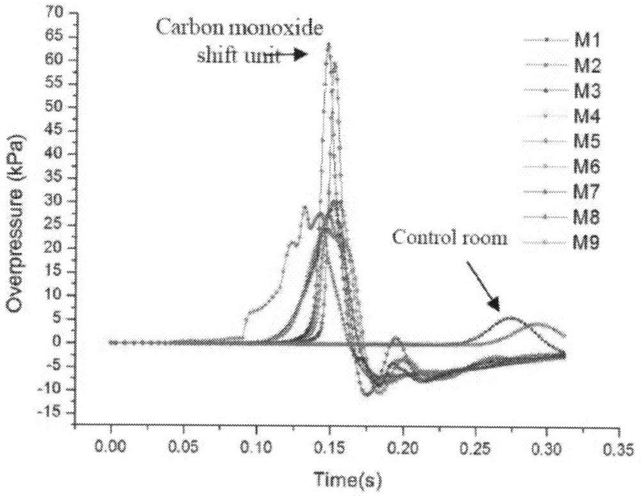

Figure 4: Overpressure of explosion in the different monitor points

The process for explosion overpressure wave affecting on equipments and facilities were shown in figure 5. As soon as explosion starting, a strong shock wave (a), causing major damage to the surrounding equipments, then it would gradually expand and weak (b-c), and subsequently affected office building and control room (d).

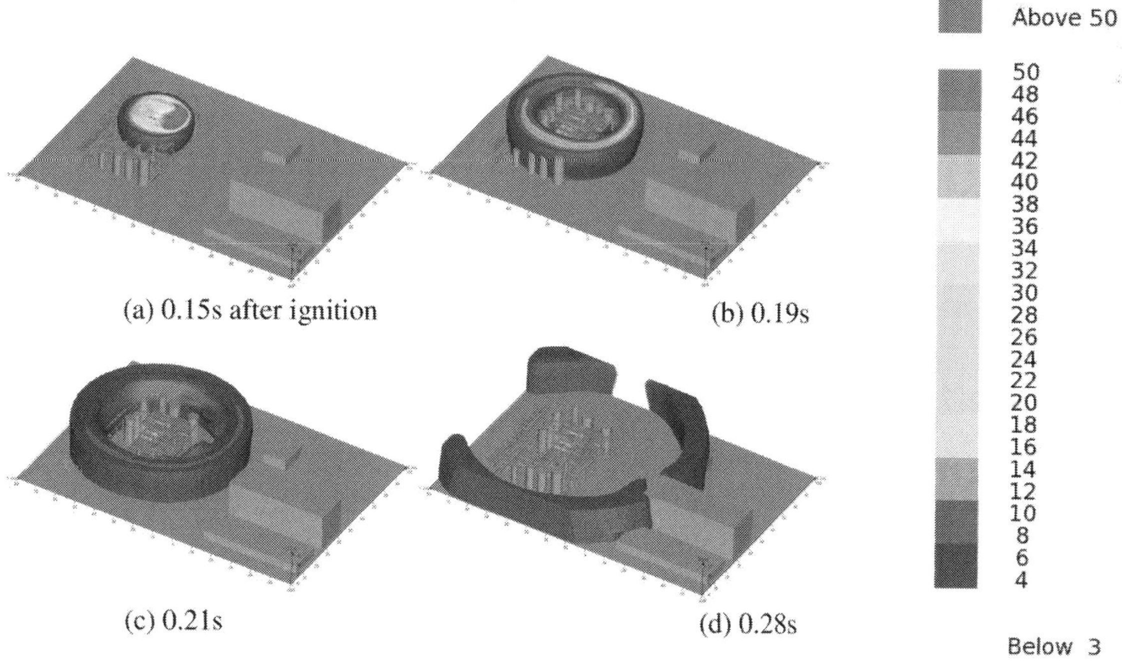

Figure 5: Explosion overpressure wave changed and expanded process (3-50 kPa)

3. Conclusion

The shortfall of traditional assessment techniques is that they can't predict the explosion effects with the layer equipment, wind, fire protection system and other condition's effect on the flame acceleration. A method of explosion simulation which is based on FLACS software for petrochemical plants was introduced in this paper. The results were as follows:

This method can calculate all parts of the space after explosion, and can make up for the shortfall of traditional assessment techniques. The overpressure simulation results using this method are between the results calculated by TNT equivalent method and Multi-Energy method. The result is more stable and reasonable when compared with the scene of the incident.

4. References

[1] D.Bjerketvedt, J.R.Bakke and K. Wingerden(1997), "Gas explosion handbook ,"*Journal of Hazardous Materials*, Volume 52, pp.1-150.

[2] S.B. Dorofeev(2007), "Evaluation of safety distances related to unconfined hydrogen explosions ," *International Journal of Hydrogen Energy*, Volume 32,p.2118-2124.

[3] B.H.Hjertager, I.Moen, J.H.S.Lee, R.k.Eckhoff, k.Fuhre, O. krest,(1981), "The influence of obstacles on Flame propagation and pressure development in a large vented tube ," *CMI Report No. 803403-2,* Chr. Michelsen Institute, Bergen, Norway.

[4] "Guidelines for chemical process quantitative risk analysis,"*American institute of chemical engineers*, NY,2000.

[5] A.C. van den Berg (1985), "The multi-energy method. A framework for vapour cloud explosion blast prediction ,"*Journal of Hazardous Materials*, Vol.12, pp.1-10.

[6] Herrmann, D.D. (1999), " Improved Vapor Cloud Explosion Predictions by Combining CFD Modeling with the Baker-Strehlow Method ," *International Conference and Workshop on Modeling the Consequence of Accidental Releases of Hazardous Materials.* CCPS/AIChE, Sep. 28 ,pp.479-494.

[7] Hansen, O.R., Talberg, O. and Bakke, J.R (1999), "CFD-Based Methodology for Quantitative Gas Explosion Risk Assessment in Congested Process Areas: Examples and Validation Status ," *International Conference and Workshop on Modeling the Consequence of Accidental Releases of Hazardous Materials.* CCPS/AIChE, Sep. 28 ,pp.457-477.

[8] Windhorst, J.C.A (1999), " Deriving Engineering Specifications from Release Dispersion and CFD Explosion," *International Conference and Workshop on Modeling the Consequence of Accidental Releases of Hazardous Materials.* CCPS/AIChE, Sep. 28 ,pp.495-514.

[9] P.Hoorelbeke, C.Izatt, J.R.Bakke, J.Renoult, R.W.Brewerton(2006), "Vapor cloud explosion analysis of onshore petrochemical facilities," *American Society of Safety Engineers Middle East Chapter 7th Professional Development Conference& Exhibition kingdom of Bahrain.* March 18-22,pp.1-15.

9TH GLOBAL CONGRESS ON PROCESS SAFETY

Using Reliability Centered Maintenance (RCM) to Reduce Fire Risk in High Temperature Oil Handling

Marc W. Yarlott, P.E.
Veolia Water North America
200 E Randolph St,
Chicago, IL 60601
marc.yarlott@veoliawaterna.com

Ryan Johnson
Veolia Water North America
200 E Randolph St,
Chicago, IL 60601
ryan.johnson@veoliawaterna.com

Steven Waters
Veolia Water North America
200 E Randolph St,
Chicago, IL 60601
steven.waters@veoliawaterna.com

Prepared for Presentation at
American Institute of Chemical Engineers
2013 Spring Meeting
9th Global Congress on Process Safety
San Antonio, Texas
April 28 – May 1, 2013

UNPUBLISHED

Using Reliability Centered Maintenance (RCM) to Reduce Fire Risk in High Temperature Oil Handling

Marc W. Yarlott, P.E.
Veolia Water North America
200 E Randolph St,
Chicago, IL 60601
marc.yarlott@veoliawaterna.com

Ryan Johnson
Veolia Water North America
Steven Waters
Veolia Water North America

Keywords: reliability, maintenance, RCM, critical spares, uptime

Abstract

Veolia Water North America contracted for the technically difficult Biosolids handling process Design, Build, Operate project with the Metropolitan Water Reclamation District of Greater Chicago at the Stickney Water Reclamation Plant. The plant design uses a high temperature oil to operate the Biosolids drying equipment. Several other plants using a similar high temperature oil system have had fires related to leakage of the oil. Having successfully employed the RCM analysis at many of Veolia's projects world wide, Veolia applied the RCM process to gain improvements in the overall reliability of the plant, and particularly the high temperature oil system, by optimizing maintenance strategies and identifying spare parts, procedures, and plans to maximize the uptime of the system. The RCM process resulted in identification of 425 spare parts at a cost of $525,000 which mitigated over $33 million dollars worth of potential risk. For a very technical process with significant risk of failures, the value of a complete RCM study will have a very high return on investment..

1. Introduction

Veolia Water North America (Veolia) operates a 150 dry ton per day biosolids pelletizer operation within the Stickney Water Reclamation plant in Chicago, Illinois. The facility is comprised of 4 independent pelletizer trains. The facility receives thickened solids (22-28%) and the solids are pumped to one of 4 large Seghers pelletizer units. The facility utilizes a gas fired boiler to heat the thermal oil that is used to transfer the heat needed to evaporate the moisture from the biosolids. The systems are very complex in nature and as with most solids processing facilities these systems are a challenge to operate and maintain.

As with any major start-up and commissioning, the operations staff gained a great deal of knowledge during the commissioning and start-up of the system. The process of converting the solids to a pellet is not just about heat transfer. Veolia's contract is a performance based one with plant uptime being the most significant driver for success. Veolia successfully completed its performance testing in late 2009. The management team at Veolia carefully evaluated each part of the performance test data and determined that it would be prudent to focus technical resources to build an uptime focused maintenance strategy.

In the Spring of 2010 Veolia initiated a comprehensive Reliability Centered Maintenance (RCM) review at the facility. This review was facilitated by internal Veolia employees and included the valuable input of the operations and maintenance staff at the facility. Veolia also brought in staff from other projects to help in analyzing specific functional areas. This RCM review incorporated a critical spares analysis that would be used to make cost/risk based storage decisions.

2. METHODOLOGY

2.1 Reliability Centered Maintenance

RCM is "a process used to determine the maintenance requirements of any physical asset in its operating context" [1]

Reliability Centered Maintenance has its roots in the civil aviation industry. In the late 1960s United Airlines invested a great deal of time and money in studying how and why equipment failed. The United Airlines management team understood that safety and reliability were of paramount importance to the operational mission. United Airlines mapped out a plan to better understand equipment failures, and focused the project on improving overall reliability of its aircraft. This analysis resulted in a startling conclusion that "there was no correlation of equipment age and failure" and that "equipment failures could not be prevented." In 1978, United Airlines published its findings in a document titled Reliability Centered Maintenance [2]. This publication was a significant document that challenged maintenance professionals to take a hard look at how and why maintenance tasks were performed.

The study concluded that equipment failure conformed to one of six failure patterns, see Figure 1 [1]. These curves plot the probability of failure on the y-axis and time on the x-axis. The failure curve analysis yielded very surprising results in that the statistics proved conclusively that equipment failures were not primarily dependent upon equipment age. The equipment failure distribution of these patterns is as follows:

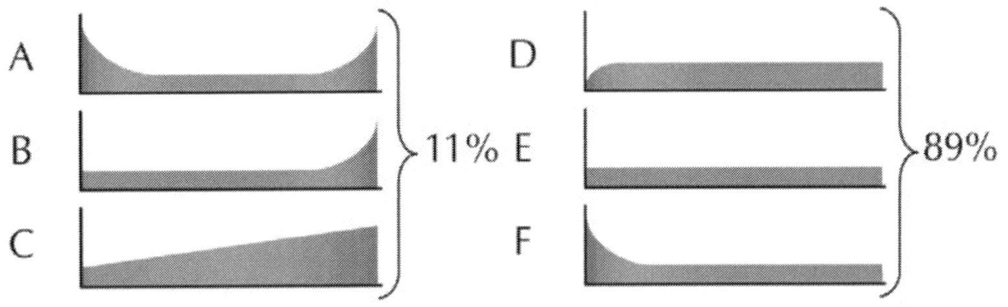

Pattern A	**4%**
Pattern B	**2%**
Pattern C	**5%**
Pattern D	**7%**
Pattern E	**14%**
Pattern F	**68%**

Figure 1, Six Patterns of Failure [1]

Based on this data, government regulators and the airlines decided to revamp the maintenance strategy of the civil aviation industry. In reviewing the failure patterns, one can quickly deduce that patterns A, B, and C have some dependency on equipment age, while patterns D, E, and F are definitely not age related.

It became quite obvious that performing calendar time based overhaul/intrusive maintenance was actually contributing to equipment failures rather than preventing them as planned. This realization resulted in drastic change to the airline maintenance programs and aircraft designs. This was also the beginning of the reliability-centered maintenance movement.

RCM will result in the correct maintenance mix of preventive maintenance (PM), condition assessment, repair (also called reactive maintenance) and proactive maintenance techniques to ensure an increase in the probability that a machine or component will function in the required manner over its design life cycle with a minimum of maintenance. RCM requires that maintenance decisions be based on maintenance requirements supported by sound technical and economic justification. As with any strategies, there are many ways to achieve the end goals.

It is important to note that RCM is not a "silver bullet" that can be applied in a lackadaisical manner. RCM requires discipline and attention to detail in the evaluation process. An RCM analysis is based on a detailed Failure Modes and Effects Analysis (FMEA) and includes probabilities of failure and system reliability calculations. The RCM process ensures that all the critical failure modes are being managed. The analysis is used to determine appropriate maintenance tasks to address each of the identified failure modes and their consequences. RCM is a proven approach that reduces risk and focuses limited maintenance resources on important assets.

In its simplest form the RCM process requires that a group knowledgeable in the operation and maintenance of the facility review systems in a methodical and organized manner. The goal is for the Operations and Maintenance (O&M) staff to learn from one another and freely exchange information related to the process, equipment, reliability, and operation. The RCM process requires that the O&M staff evaluate the following seven questions:

- **Functions:** What does this asset /system do (what do the users want this asset to do)?
- **Functional Failures:** In what ways can the asset/system fail to deliver the functions described above?
- **Failure Modes:** What causes the functional failures to occur?
- **Failure Effects:** What happens when the failure occurs?
- **Consequences:** Does this failure even matter?
- **Proactive Tasks:** Is there anything that can be done to predict or reduce the likelihood of the failure?
- **Default actions:** Otherwise, what other action is available to improve reliability?

Each failure mode is put through a decision tree (Figure 2) to determine its impact on environmental, safety, and financial risk. All of the critical failure modes are subsequently examined and assigned a proactive task (PM, PdM, or failure finding) to manage the risk.

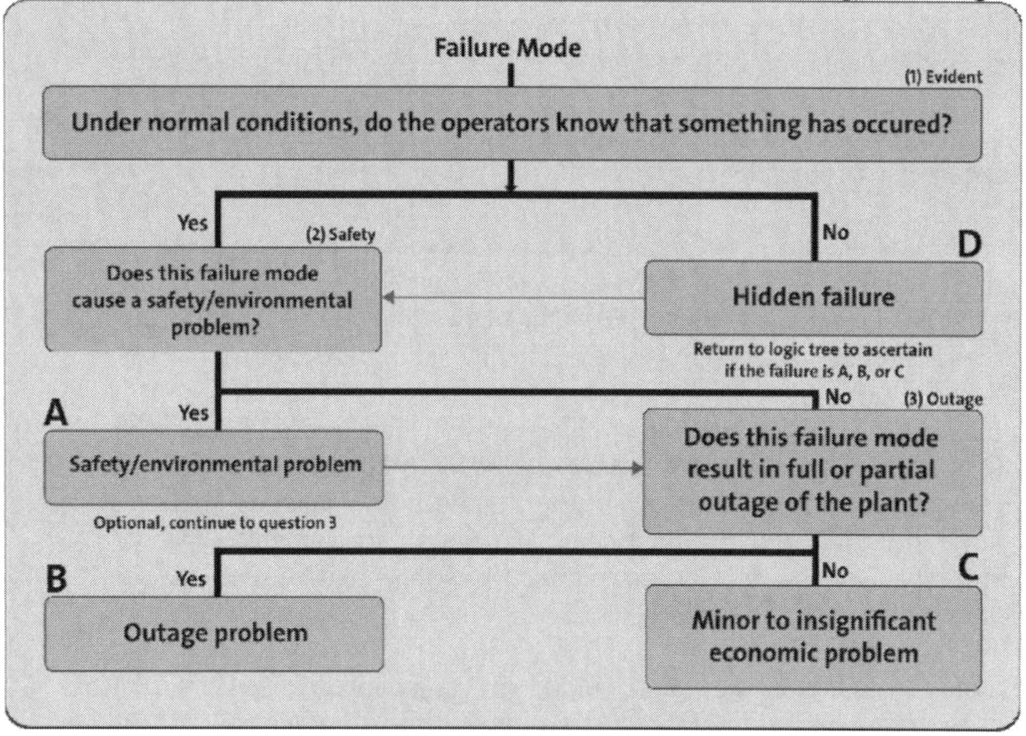

Figure 2, Decision Diagram [3]

3. RESULTS

3.1 Implementation of RCM

Prior to performing the RCM review, the project team was experiencing mechanical failures of system components that were resulting in some periods of downtime. The facility uptime was being negatively impacted by asset failures, and the need for rapid repairs resulted in little to no root cause analysis. The facility uptime had dropped off after performance testing and uptime was running on average of 50% June 2010. Staff worked out a plan that prioritized the RCM reviews based on the principle of analyzing problem systems first.

Initially the RCM project focused on one train of the processing facility. The train was further separated into 13 systems. Over a 16 week period, staff completed 13 classical RCM Reviews. In total over 1784 failure modes were analyzed in total.

One of the benefits of conducting the RCM review was that it also gave the project staff the opportunity to understand in more detail where the system choke points were, and why the failures were occurring. Many useful suggestions on improving uptime and reliability were recorded and noted in the RCM review as "items of interest". In total over 104 items were identified as recommended improvements to the existing installation. For example, one item of interest was related to the biosolids cake pumps. These pumps had some highly unusual failure modes and had become quite problematic from an O&M perspective. During the RCM review there were several failure modes that pointed the review team to questions if the hydraulic unit that was provided with these pumps was adequately sized. The hydraulic unit reservoir was running hot, and there were wide fluctuations and pressure swings in the system. After the RCM review the facility staff worked with a local engineering company to resize the drive unit. This change has significantly improved the uptime of the pumps and eliminated many of the "unusual" failure modes. Staff at the project are now working their way through the "items of interest" and implementing many of the group's recommendations.

A second benefit of the RCM review was a success directly related to the high temperature oil system. During the RCM Review the staff identified Failure Mode #61:

61	***Primary Oil***	Oil heater High Pressure alarm: Staff needs to confirm that there is indeed a High Pressure alarm on the oil heater when the PRV fails closed. This is a safety issue that can be mitigated with existing instrumentation

In following up on this item the plant staff confirmed that there was indeed a high pressure alarm; however, in checking the alarm, they discovered that the Pressure Relief Valves (PRV)'s were not the correct for this service. The Original Equipment Manufacturer (OEM) had supplied air service PRV's instead of liquid service. The process was shut down and immediately new PRVs were ordered and installed on the Heaters. The RCM review, allowed the staff to step through the features and failure modes of the equipment and identify and prevent a failure before it ever happened.

3.2 Critical Spare parts

There are many maintenance management practices that can help to properly manage risk. One area of maintenance management that must be included in a sound asset management program is the selection/identification of "critical spares." To achieve the desired availability of equipment, it is important to give proper consideration to having some spares parts/components available on-site. This is often implicitly considered by maintenance teams as an easy way to mitigate down time risks, thus leading to unnecessary amount of spares. One must be careful when evaluating critical spare parts. It is important that each spare part be properly vetted and justified through a proper risk based method.

Over the past 10 years equipment manufacturers have changed the way that they support their products. The days of having spares readily available for purchase have ended, in favor of the "just in time" manufacturing philosophy. In addition to less available stock, manufacturers have set some limitations on the timeframe that equipment is supported (planned obsolescence). These changes have created a need for operations and maintenance staff to better understand the challenges involved in obtaining needed parts.

Indeed, to be truly cost-effective the proper selection of critical spares has to be developed based on a methodical process. This process must take into consideration a cost-benefit/risk analysis that is aligned with the facility's goals and objectives. This analysis needs to include the following:

- Failure mode identification
- Consequences (safety, environmental, and economic) of not having the spare
- Lead time to procure
- Availability of spare (i.e. obsolete)
- Probability of failure
- Predictability of the failure
- Shelf life
- Storage considerations (space, environmental conditions, etc.)
- Cash flow

Each of these factors will now be considered in more detail.

3.3 Failure Mode Identification

A "failure mode" is an event that is reasonably likely to occur and that would cause the loss of an asset's functionality [1]. When developing a potential list of spare parts for an asset, an understanding of the equipment at the "failure mode" level is necessary.

The Chicago project identified 1784 failure modes in total. During the RCM review, staff decided which parts would be considered for spares. It is important that experienced operations and maintenance staff participate in the development of the failure modes. Special attention should be given to failure modes that are wear-related.

3.4 Consequences

Not all failure modes have the same impacts to the operations at a facility. Great care should be taken in evaluating the consequence of failure, utilizing a criticality review process (Figure 2). One should make an effort to monetize the risk of lacking a spare part when needed. For Chicago a project-specific matrix was developed and used to quantify lack of equipment availability into money. This allowed the staff to factor in items like cost of downtime, contract issues and safety into economic terms. This provided the framework upon which a proper spare part evaluation can be performed.

3.5 Lead Time

It is critically important that we understand the supply chain when it comes to spare parts. The availability and lead time for obtaining spare parts varies significantly based on the manufacturer/supplier and geographical location of the facility. When conducting a critical spares evaluation, one must always fully understand how long it will take to procure an item. Oftentimes excessive lead times may be reason enough to have a spare available at the facility. Initially a decision may be made not to stock a spare because a redundant asset is available. This decision may have to be reconsidered if one has to rely on solely the redundant system for a long period (while one waits for the spare to arrive). Many of the assets at the Chicago plant were custom built and manufactured overseas. Staff made contact with the vendors and determined the lead times on each of the items identified during the review.

3.6 Availability

Oftentimes equipment vendors discontinue product lines and support for products. Some of these changes are the result of market-driven forces, such as mergers and acquisitions within the manufacturing industry itself, while others are purely technologically-driven. Special consideration should be given to electronic components such as Variable Frequency Driver's (VFD) and I/O cards. In the Chicago review we discovered that some of the electronic components were no longer supported by manufacturers.

3.7 Probability of Failure

A proper critical spares evaluation needs to factor the probability of occurrence into the equation. This is usually done by experienced staff familiar with the common equipment in a similar operating context. There are published failure statistics that can also be used to benchmark this type of probability. Veolia has an enterprise Computerized Maintenance Management System (CMMS) that it uses to quantify the Mean Time Between Failures (MTBF). Understanding the failure rates will help in making a sound business decision as it relates to procuring spares.

3.8 Predictability of Failure

Some equipment/components will fail over time due to wear or age/fatigue. It may be possible to properly manage this type of failure by conducting periodic inspections to measure the overall condition of the asset. We identified components like chains, belts, sprockets, liners, and screws can be evaluated for wear and new parts can be ordered prior to loss of functionality.

Properly set up and monitored, condition assessment programs can minimize the amount of spares needed at a facility. Items like fuses, shear pins and electronics fail with little warning and should always be stocked if the asset is critical.

3.9 Shelf Life

Some components deteriorate with time (like probes, sensors, ball bearings, batteries and gaskets). These timing considerations should be factored into the decision-making process.

3.10 Storage Considerations

Prior to procuring any spare parts one must evaluate his/her location to determine if the facility has adequate storage facilities on site.

Consideration should be given to environmental conditions (humidity, temperature, sulfides, etc.) and the available space. Staff also needs to ensure that it has the proper systems in place to track the spares and to perform PM maintenance tasks on the inventory if needed (i.e., a spare pump may require that the shaft be rotated periodically to distribute lubrication and prevent shaft deflection). In Chicago we decided to store some of the spares on site and some in off site storage facilities. We had very good success in working with vendors in some cases to store items for us.

3.11 Critical Spares Methodology

Veolia Water utilized its internally developed critical spares methodology during the RCM review. The following is a small section of the spares analysis that was based on some contract terms and downtime estimates.

The contract is production-based and includes liquidated damages for downtime and loss of availability. What follows is a general description as to how the process was used to determine the need to stock a spare. For presentation purposes, only three scenarios are included.

3.11.1 Step One: "Monetize the Risk"

There are 4 process trains at this facility and three are needed to meet full capacity. The group determined that if all 4 trains were unavailable there would be a $1,250/hour cost of downtime. The group then evaluated loss of capacity and assigned a lower dollar number for some unit operations ($450/hr). After debating the liquidated damages clause of the contract and its impact on the business, the group settled on $225/hr as being the cost of loss of redundancy. This process was extremely rigorous and debated by the review team. The assumptions were sent in a memorandum to the senior management team.

3.11.2 Step Two: Failure Mode Evaluation

	Failure Mode	Local	System	Plant
1	Bearing seizes due to lack of lubrication	Bearing wears, gets hot and eventually fails	Bearing wears, gets hot and eventually fails. Standby pump starts and carries load	Loss of redundancy
2	Temperature transmitter fails	Transmitter fails and thermal oil pumps stop pumping	Oil system shutdown and OLC initiates a system shutdown	Loss of one train
3	Heat trace thermostat fails	Heat trace fails, water line freezes, resulting in loss of flow	Biosolids pumps trip on loss of seal water.	Complete plant shutdown

3.11.3 Step Three: Match Critical Failure Mode to Spare

Failure Mode	Spare
Bearing seizes due to lack of lubrication	Bearing
Temperature transmitter fails	Transmitter
Heat trace thermostat fails	Thermostat

3.11.4 Step Four: Determine Spare Cost and Lead Time and Calculated Downtime Cost

Spare	Lead time (in operating hours)	Downtime cost/hour	Cost of downtime due to lack of spare
Bearing	48	$225	$10,800/event
Transmitter	24	$225	$5,400/event
Thermostat	24	$1250	$30,000/event

3.11.5 Step Five: Adjust Downtime Cost for Failure Probability

Spare	Downtime Cost (DC) due to lack of spare	Failure Probability (FP)**	Downtime Cost adjusted (DC * FP)	Cost of spare and storage
Bearing	$10,800/event	.2	$2,160/year	$1,200
Transmitter	$5,400/event	.1	$540/year	$250
Thermostat	$30,000	.1	$3,000/year	$600

** The Failure Probability (FP) is based on the number of failures anticipated in a ten-year period (e.g. a .1 signifies a 10% chance of at least one failure in a 10-year period). This can be based on past failure data or available MTBF data from other sources. One must consider contractual terms and residual contract years when making spare part decisions. For example, if the contract end date is within a two-year period and the risk of not carrying a spare is tolerable, one should not stock the item.

3.11.6 Step Six: Step Six: Evaluate Cost of Spare versus the Downtime Costs

Spare	Cost of spare and storage	Downtime cost adjusted	Spare required (yes or no)
Bearing	$1,200	$2,100	Yes
Transmitter	$250	$540	Yes
Thermostat	$600	$1250	Yes

As with any type of analysis it is important that the review group uses a bit of common sense when making spare part decisions. One must factor into the decision-making process contractual obligations, contract term, and the practicality of storing the item. Oftentimes the decision may come down to a very small risk (<$1,500 can be mitigated by spending about the same amount of money). One needs to use sound judgment before money is spent on spares.

In total, we analyzed 1784 failure modes for critical spare purposes. In total we decided to stock 425 spares parts at a cost of $525,000. These spares mitigated over $33 million dollars worth of potential risk.

4. CONCLUSIONS

Veolia Water has introduced a number of innovations to the standard framework for delivering RCM that have improved the business case for delivery. These innovations include techniques for prioritizing and simplifying the analysis, as well as more reliable means for estimating maintenance frequencies and estimating the true cost-benefit of critical spares. The use of RCM in the water/wastewater industry has been adapted from the original methodology used by United Airlines. The changes to the process allow for speedier delivery and execution of RCM. Implementation of RCM on 100% of the assets is not practical, nor cost-effective in our industry.

Veolia Water has been able to demonstrate a link between doing RCM well and delivering excellence in asset management. Overall, the benefit to the business and our customers that result from implementing the Veolia Water RCM process is as follows:

- Better identification of risks. Many new, and often significant, operational risks are identified with recommendations to mitigate them.

- Improved maintenance plan. New maintenance plans are developed to specifically manage the risks of asset failure. Frequencies and resource levels are optimized.

- Gains in operational efficiency. Opportunities for improvement are proposed for operating scenarios that offer risk and efficiency benefits.

- List of critical spares. Spares needs are determined, based on cost-benefit assessments, including financial estimates of risk.

- Prioritized capital improvements. Investigations are recommended for a number of potential capital works items designed to enhance contract performance and minimize risk.

- Effective knowledge capture. The RCM captures both documented information as well as the tacit knowledge held by staff about how the systems work.

- Tool for operations training. Participating staff are trained in the methodology and learn in detail how their systems work.

The project staff has been busy implementing the recommendations that came about from the RCM review. The identification of the wrong PRV's on the high temperature oil system was a very important safety related finding, but the benefits of the program continue to add up. The project has already seen drastic improvements in equipment availability and uptime. Each maintenance task, spare part, and equipment check has been developed to manage risk and improve uptime. The staff at the plant continues to learn from equipment failure and has validated that the time spent on the analysis and spares has tremendously improved the overall operation. The following, Figure 3, is a trend of uptime of one train since the RCM reviews were completed.

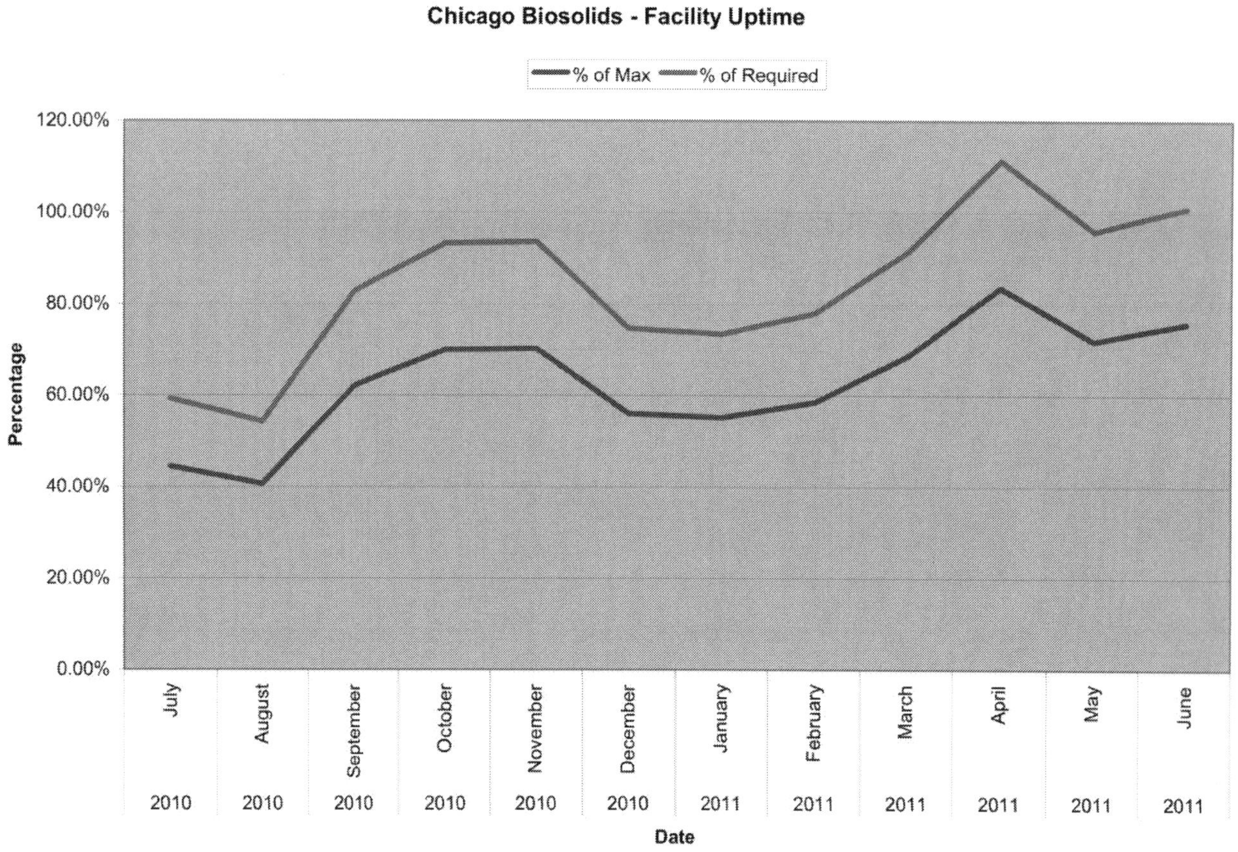

Figure 3, Uptime Performance Through June 2011

5. REFERENCES

[1] Moubray, John, *Reliability-centered Maintenance*, 2nd Edition, Industrial Press Inc., New York, New York, 1997.

[2] Nowlan, F. S.; Heap, H. F., *Reliability-centered Maintenance*, United Airlines and Dolby Access Press, sponsored and published by the office of Assistant Secretary of Defense, San Francisco, California, 1978.

[3] Smith, A.M.; Hinchcliffe, G.R., *RCM Gateway to World Class Maintenance*, Elsevier Butterworth Heinemann., Burlington, Massachusetts, 2004.

(44x) Characterization of the Emissive Power of Hybrid Methane-Air-Dust Flames

Brian Elias and Ali S. Rangwala, Fire Protection Engineering, Worcester Polytechnic Institute, Worcester, MA

Combustible dusts are commonplace in a wide variety of industrial settings ranging from coal mines to manufacturing facilities. The ignition of a dust cloud formed by the mixing of these combustible particles with the surrounding air may result in a deflagration or flash fire. The purpose of this study is to quantify the emissive power of the flame in order to better understand the combustion dynamics and the risk of burn injury associated with a dust flash fire.

Experimental tests are conducted by feeding combustible dust samples into a premixed dust-air burner apparatus to create a stabilized flame for analysis. The two dust types investigated in this study are Pittsburgh seam coal, and iron dust – both with sizes ranging from 70 to 90 um. In addition to the composition of the dust, other controlling parameters including dust concentration and turbulent intensity are also varied and their respective effect on the heat flux is reported. The total emissive power of the flame is calculated from the heat flux measurements in combination with flame surface area and an appropriate view factor.

9TH GLOBAL CONGRESS ON PROCESS SAFETY

Are You in Control of Process Safety?

Basis of Safety Assurance Can Provide the Answer

Graeme Ellis, Principal Lead Consultant
ABB Consulting
Daresbury Park, Warrington, WA4 4BT, UK
graeme.ellis@gb.abb.com

(Acknowledgement: Based on paper written by Phil Eames, ABB Consulting)

Prepared for Presentation at
American Institute of Chemical Engineers
2013 Spring Meeting
9th Global Congress on Process Safety
San Antonio, Texas
April 28 – May 1, 2013

UNPUBLISHED

Are You in Control of Process Safety?

Basis of Safety Assurance Can Provide the Answer

Graeme Ellis, Principal Lead Consultant
ABB Consulting
Daresbury Park, Warrington, WA4 4BT, UK
graeme.ellis@gb.abb.com

(Acknowledgement: Based on paper written by Phil Eames, ABB Consulting)

Keywords: Basis of Safety, Assurance, layers of protection, drill-down, vital signs, audit, safety critical elements

Abstract

Serious process safety accidents continue at an unacceptable frequency in the process industry, and senior managers are right to be concerned about whether all reasonable steps are being taken to control risks. Traditional methods for revalidation of Process Hazards Analysis and auditing of Process Safety Management systems can involve considerable time and resources. This paper will describe a 'Basis of Safety Assurance' approach that has been developed to provide a rapid view on the 'vital signs' related to the state of health of measures to prevent and mitigate major accident hazards. A representative set of scenarios for the site are selected and evidence gathered from documentation and site visits to confirm the robustness of the key layers of protection. Deficiencies related to plant, processes or people are identified resulting in a risk targeted set of improvement recommendations.

1. Introduction

Effective process safety management requires a thorough understanding of process risks so that measures to prevent and control them can be implemented and maintained. Structured techniques such as hazard identification (HAZID) and hazard and operability study (HAZOP) are used to identify major accident scenarios and decide how they will be prevented, controlled and mitigated. The description of major accident scenarios and the agreed layers of protection (or barriers) to reduce the risks to an acceptable level, is often termed the basis of safety.

The purpose of a process safety management system (PSMS) is to provide a framework of high level procedures (or risk control systems), in order to maintain protective equipment and procedures in a healthy state. Therefore many organizations, when they seek to assure themselves that process safety is being managed appropriately, base this assurance on detailed auditing of the PSMS.

However, process safety incidents are usually caused by failures of multiple risk control systems at a detailed level. The potential interactions between risk control systems are often not visible at the level of the PSMS, which is constructed of distinct generic elements (mechanical integrity, incident investigation, management of change etc.). Auditing of distinct elements may not identify the potential for an incident due to such interactions. Moreover, modern PSMS are tending to become more integrated with occupational safety management systems, environmental management systems and quality systems, making their role in the prevention of process safety incidents potentially less visible at system level.

Given that the prime concern in process safety management is to prevent process incidents from occurring, it makes sense to devote some assurance effort directly to the scenarios themselves and to their specific layers of protection. This is the essence of Basis of Safety Assurance. Whereas auditing addresses the component elements of the management system, Basis of Safety Assurance seeks to challenge the effectiveness of the process hazards analysis and robustness of the layers of protection for specific scenarios, thereby providing direct assurance that the risk reduction measures are in place and working.

This is not to suggest that auditing of the PSMS is not a necessary or effective tool. But focusing on an organization's key process incident scenarios can be a very direct and cost-effective way to provide assurance that process safety is being managed appropriately, complementing traditional audit-based assurance processes.

This paper describes the Basis of Safety Assurance technique and the experience of its use across a number of sectors in the process industry. It explains how the technique can be used to enhance an organization's process safety management audit and review program.

2. Limitations of Audit Based Assurance

Traditional process safety assurance or governance approaches are based on audits of the process safety management system (PSMS). However, in many organizations the components of the PSMS form part of a larger safety, health, environmental and possible quality management system, and may not be distinguished from components designed to control personal safety and health risks. This means that focus on process safety may diluted by or obscured by the larger management system.

Even if the PSMS is a stand-alone system, or is visible within a larger system, auditing focuses on the system as a whole and not on the control of specific risks. Auditing is often undertaken on an element-by-element basis, which is likely to miss the potential interactions between elements that characterize complex process safety events, typically represented by the "Swiss cheese" model. Moreover, even within a single element significant audit time is normally spent on understanding how the element is managed rather than on detailed field verification. Field verification on audits typically aims to confirm that the system is in compliance with requirements, for example that trip testing is being carried out to plan. It generally does not test that the system is effective in protecting against the risk of a specific scenario, for example the trip is identified as a layer of protection in HAZOP, there is a written test method, the trip test method is an "end to end", faults identified in previous tests have been corrected, etc.

3. Layers of protection and process safety "vital signs"

In his book on the management of complex process risks, James Reason [1] wrote of the need for proactive checking of the process or plant, or the 'system' as he describes it:

"*In the absence of bad outcomes, the best way - perhaps the only way - to sustain a state of intelligent and respectful wariness is to gather the right kinds of data. This means creating a safety information system that collects, analyses and disseminates information from incidents and near misses as well as from regular proactive checks on the system's vital signs. All of these activities can be said to make up an informed culture - one in which those who manage and operate the system have current knowledge about the human, technical, organisational and environmental factors that determine the safety of the system as a whole. In most important respects, an informed culture is a safety culture*".

The state of health of safety critical layers of protection is very much vital signs for the process in relation to major accident scenarios. If they are specified and designed correctly, maintained and tested, then the process is safe. The proactive checking that layers of protection are in place and healthy provides exactly the current knowledge of the state of the plant and the organization that Reason holds to be so important.

4. The 'Basis of Safety' Event

The testing of the state of health of layers of protection for specific major accident scenarios is the focus for a Basis of Safety Assurance event, a generic agenda for which is shown in Table 1.

Table 1. Basis of Safety Assurance Agenda

Phase	Day	Activity
Preparation	0	Arrange for immediate access to key site personnel and information during time on site
Introduction	1	Introduction to Process Safety on site. Review of major accident hazards Site tour Examination of Process Hazard Analyses Selection of 'Drill Down' scenarios
Drill Downs	2-N	Detailed drill downs Approx 4 scenarios per day including field visits
Feedback	Final	Preparation and delivery of feedback

The initial part of the event aims to understand the major accident hazards associated with the process by discussion, field visits and review of the process hazards analysis (HAZID and HAZOP documentation). This would typically take one day, during which time an assessment of the quality of the site's process hazards analysis and documentation of the basis of safety is also made. It culminates in the selection of specific major accident scenarios to be subjected to detailed assessment, sometimes termed a "drill-down" or "deep dive".

It is useful to select a range of different types of scenario where possible (fire, explosion, toxic release, environmental, etc.), so that as many different types of safeguard are subject to assessment. These then include for example; pressure relief systems, safety instrumented systems, passive and active fire protection systems, secondary and tertiary containment systems, etc.

The main part of the Basis of Safety event comprises the scenario assessments. These typically take around 2 hours per scenario, depending on the quality of the basis of safety documentation, the knowledge of the team and the speed of data retrieval, for example how fast plant records can be accessed to verify that maintenance and testing is taking place.

Findings from the scenario assessments provide the required assurance that layers of protection are healthy or lead to the generation of recommendations for improvement. Common findings across a number of scenario assessments can provide a useful indication of systemic issues, for example weaknesses in proof test procedures or compliance with critical equipment testing regimes. This part of the event complements conventional PSMS auditing.

5. The Scenario Assessment Technique

Most value can be gained from the scenario assessment technique if it is conducted as a team exercise. Experience has shown that the best results can be obtained by deploying two process safety specialists as assessors; one with a PSMS and operations background and the other with a plant engineering and asset integrity management background. This approach also helps from a leading and scribing standpoint since the event is recorded in a similar way to a process hazards analysis. In addition to the assessors, the involvement of process engineers, operating managers and maintenance engineers from the site is very important. This provides an understanding of the scenario and the location of information to verify that the identified layers of protection are in place and being maintained.

Since scenarios are normally selected from the site's process hazards analysis or basis of safety documentation, a scenario assessment normally starts with this documentation. The scenario is described and the relevant records are referenced, shown in the upper part of the recording form in Table 2.

It is important to note that the existing documentation may not fully describe the scenario, since the quality of HAZID and HAZOP documentation is heavily influenced by team composition, knowledge available, leader competence, availability of time and recording policy.

Table 2. Scenario Assessment Template (example)

Major Accident Hazard Drill Down Record
Scenario: Storage Tank V7706 (150m3) overfilled during (continuous) run-down and a large volume of solvent released into bund from the breather/emergency vent, subsequent ignition resulting in a flash fire/pool fire and potential for several fatalities.

PHA Reference: No specific scenario was identified for this material as incidents involving acetone were taken as the representative set. The acetone scenario for overfill was estimated to result in two fatalities.			
Safeguard	**Verification**	**Findings**	**Conclusions**
Monitoring of level transmitter by operators and dependent alarm function.	1. Calibration and maintenance of level transmitter and alarm LIA-7706. 2. SOP requiring level monitoring.	1. No regular maintenance is now performed on the level transmitter following the installation of the independent high level switches. Corrective maintenance is performed based on notifications raised by the operators. 2. There is no specific SOP relating to the operation of this solvent storage tank.	Consideration implementing routine level transmitter calibration on solvent storage tanks.
Independent high level switch with trip of run-down line.	1. Proof testing of high level trip LZA-7706H.	1. The trip is proof tested annually 2. The last proof test of this instrument was carried out on the 27/4/10. The proof test is an end- to-end test to a pre-defined test method including checklist. 3. No SIL assessment has been conducted.	Controls appear to be effective. However, conduct a SIL assessment for this scenario to verify that the proof test interval is adequate to provide the required level of protection.

It is therefore important that the assessors work with the team to identify all the relevant safeguards, which are recorded in the left hand column of the template shown in Table 2. Note that the term "safeguard" is used rather than "layer of protection" because the aim is to capture safeguards such as procedures and emergency plans as well as physical independent protection systems that are the focus of formalized Layers of Protection Analysis (LOPA) techniques.

Once safeguards have been listed the assessors ask - for each safeguard in turn - what can we check is in place? This is recorded in the "verification" column. Examples include procedure number and latest version, alarm test methods, maintenance routines and pressure safety valve test records.

Evidence for each verification is then sought and the findings recorded in the "findings" column of the template. It is important that documentary evidence is collected to ensure that the assessment is suitably probing and detailed.

Conclusions are then drawn for each safeguard. This may be a simple "Okay" if everything expected is in place, but may be a recommendation for improvement, a corrective action or a recommendation to consider whether a finding may be evidence of a systemic weakness. In viewing the robustness of the safeguards as a whole the team may determine that additional risk reduction is appropriate and make a recommendation to identify additional risk reduction measures. This would be required in a UK legislative context to demonstrate that risks have been reduced to "as low as reasonably practicable" or ALARP.

It is important to include a field visit for each scenario. This has two purposes, firstly to physically check that safeguards are installed in line with the design intent, for example that relief devices are in place and that passive fire protection and secondary containment are in good condition. Photographs of deficiencies provide high-impact evidence to site management. Secondly to talk to operators and maintenance technicians about their understanding of the process and of the major accident hazard that is being assessed. This gives an insight into the quality and experience of staff and the quality of process-specific training that they have received.

6. Experience with the Technique

There is now a significant body of experience of Basis of Safety Assurance across a number of process industry sectors. Companies have found it effective regardless of sector or scale of process hazard. Major hazards facilities are more familiar with the concepts but it is readily applicable to lower level process safety hazards, including those in the pharmaceutical and power generation sectors.

Some important advantages of using the scenario assessment "drill-down" technique are as follows:

- Drilling down into specific scenarios can identify significant risk exposures and weaknesses at a very detailed level, for example:
    - Trips tested only at commissioning;
    - Absence of detailed test methods and reliance on technicians' experience;
    - Trip testing not from sensor to final element (end-to-end);
    - Reliance on inexperienced 3rd parties for proof testing;
    - Demands on trips more frequent than the design basis.
- Examining multiple scenarios can provide evidence of whether there are systemic weaknesses.
- The technique is useful to identify weaknesses in process hazards analysis and in keeping records and basis of safety documentation up-to-date.
- The technique is a useful vehicle for providing training materials for specific process hazards.

7. Conclusion

Because it focuses on specific risks, Basis of Safety Assurance complements conventional PSM system audit activities. It helps to maintain a focus on major accident hazards and improve the understanding of site personnel and standard of documentation. It aligns with line management responsibilities for specific processes and assets, and it directly exposes risks of major accident hazards, thus providing real time assurance to senior management.

8. References

[1] Reason, J.T (1997), Managing the Risks of Organizational Accidents, Ashgate Publishing, Aldershot, UK.

9TH GLOBAL CONGRESS ON PROCESS SAFETY

PROCESS HAZARDS ANALYSIS, LAYERS OF PROTECTION ANALYSIS AND THE HUMAN FACTOR

Dr. Paul Baybutt
Primatech Inc.
50 Northwoods Blvd., Columbus, Ohio, USA
paulb@primatech.com

Prepared for Presentation at
American Institute of Chemical Engineers
2013 Spring Meeting
9th Global Congress on Process Safety
San Antonio, Texas
April 28 – May 1, 2013

UNPUBLISHED

PROCESS HAZARDS ANALYSIS, LAYERS OF PROTECTION ANALYSIS AND THE HUMAN FACTOR

Dr. Paul Baybutt
Primatech Inc.
50 Northwoods Blvd., Columbus, Ohio, USA
paulb@primatech.com

Keywords: Human factors, human failures, process hazards analysis, layers of protection analysis.

Abstract

Process hazards analysis (PHA) and Layers of Protection Analysis (LOPA) studies address human failures in operating and maintaining processes and the human factors that influence them. People perform PHA and LOPA studies and, therefore, such studies themselves are subject to various possible human failures. Much less attention has been paid to the human factors that influence the performance of PHA and LOPA than human factors that influence process hazards. Human failures in the performance of PHA and LOPA studies should be of significant concern to practitioners as such studies are difficult and time-consuming activities that place significant demands on participants which increases the chance that errors will be made. Human factors such as willingness to rely on the unsubstantiated opinions of others, groupthink, underestimation of the frequencies of low-probability, high-consequence events, and allowing a false sense of accomplishment to distract from implementing study results must be recognized and addressed.

This paper identifies and discusses various human factor issues that can influence the quality of PHA and LOPA studies covering preparing for, conducting, recording, documenting, and following-up on studies. Guidelines are provided to help minimize the extent to which these human factor issues may impair study quality.

1. Introduction

Process Hazards Analysis (PHA) [1, 2] and Layers of Protection Analysis (LOPA) [3, 4] address failures in processes that can result in hazard scenarios with adverse impacts on such receptors as people, property and the environment. PHA is used to identify hazard scenarios and LOPA is used to evaluate their risk. Often, human failures are causes of, or contributors to, hazard scenarios, and various human factors influence the rates of failure [5, 6]. Process safety regulations require that such human failures and human factors be addressed in PHA [7], and by implication in LOPA. Other failure types such as equipment failures and external events must also be addressed [2].

People perform PHA and LOPA studies and, therefore, such studies are subject to various possible human failures influenced by various human factors. Much less attention has been paid to the human factors that influence the performance of PHA and LOPA studies than human factors that influence process risks. Human failures in the performance of PHA and LOPA studies should be of significant concern to practitioners as they can have a significant adverse impact on study results. PHA studies involve considerable subjective judgment by team members. Indeed, LOPA was developed to provide a more rational and objective approach to making decisions on the tolerability of risk from hazard scenarios. However, subjective judgment still enters into the performance of LOPA studies.

This paper identifies and discusses various human factors that can influence the quality of PHA and LOPA studies covering preparing for, conducting (i.e. carrying out), recording, documenting, and following-up on studies. These human factors are not difficult to understand but, unfortunately, they are often ignored in the performance of studies. Guidelines are provided to help minimize the extent to which these human factor issues may impair study quality.

2. Framework for Human Factors

In attempting to identify human failures and the factors that influence them it is useful to employ a model [8]. Commonly, human failures are categorized as shown in Table 1 [2]. Typical human factors that may influence these failures include:

- Insufficient skill or knowledge
- Memory lapse
- Distraction
- Excessive workload
- Fatigue
- Time pressure
- Lack of or insufficient information
- Lack of, unclear or defective identification
- Incomplete, inaccurate or ambiguous procedure
- Roles / responsibilities not clear
- Split responsibilities
- Poor communication
- Handover problems
- Poor access
- Inadequate environment, e.g. lighting
- Need for simultaneous actions

Each stage, step and action in the performance of PHA and LOPA studies can be examined and possible human failures identified using Table 1. The impact of human factors such as those listed above can then be considered. Of course, PHA and LOPA studies are performed within a company's management system, culture, and physical environment which may all play a role. This paper does not seek to identify every human factors issue that may adversely impact the performance of PHA and LOPA studies. Rather, key issues that have been found to be of importance are addressed.

3. Human Factors in Preparing for PHA and LOPA Studies

Various human factor issues may impact this stage of PHA and LOPA studies.

3.1 Guidance on Study Performance

Without guidance, people proceed in whatever way they believe is appropriate or expedient. Some companies, study leaders, and team members may favor such an approach as it provides flexibility that is viewed as desirable. However, it is sure to lead to inconsistent studies as each team follows its own path. Such inconsistencies could lead to citations by regulators and legal difficulties in the event of litigation following accidents.

There are many ways of performing PHA and LOPA. Both regulatory and industry process safety standards are performance-based thus allowing flexibility in the way studies are performed. Consequently, if companies wish to perform consistent studies from one process to another, it is essential that the details of how the company wants studies to be performed be defined. Indeed, without such guidelines, a single PHA or LOPA study may contain internal inconsistencies as the team may make different interpretations and assumptions in different parts of the study. Formal, written guidelines are needed that are periodically reviewed and updated, as required. Study leaders need to be sensitive to questions from team members during studies regarding how the study should be conducted. If the company's guidelines do not address the issue and it is significant, the guidelines must be updated.

Guidance on the details of how to perform studies is available in the literature [2, 4] and through training courses [9, 10].

3.2 Team Leader Qualifications

Team leaders must be competent. Typical qualifications include:

- Technical expertise in the study method.
- Leadership / facilitation skills.
- Technical and leadership training.
- Understand processes and their operation.
- Comprehend engineering reference documents and drawings.
- Motivational / interpersonal skills.
- Communications skills with the team and management.
- Project management skills.
- Experience as a team member.
- No day-to-day responsibilities for the process being studied.
- Not an expert on the process under study.

A number of these qualifications relate to people skills and human factor issues. Unfortunately, some companies do not realize their importance. It is not sufficient for the team leader to be a technical expert in the study method. They must also be capable of effectively and efficiently facilitating a team of people in conducting the study. Arguably, particularly for PHA, this is more challenging than addressing the technical details of the study, particularly for the typical individuals designated as team leaders who tend to be engineers without the need for strong people skills in their regular work assignments. OSHA has noted that, "the team leader needs to be able to manage the team and the PHA study" [11]. Team leaders should be screened against these qualification criteria with emphasis on their aptitude for facilitation and communication. Promising candidates should be provided with suitable further training in facilitation and communication.

Desirable attributes for team leaders include:

- Patient
- Possess stamina
- Organized
- Diplomatic
- Imaginative
- Quick thinking
- Consensus-seeking
- Friendly and cooperative
- Able to read people
- Able to focus on multiple items simultaneously
- Gently authoritative
- Respected by team
- Able to help team members see matters in a new light
- Strong personality to control the team and drive the study but not overbearing

Few team leaders are likely to possess all of these attributes but prospective team leaders should be screened against these criteria to help ensure suitable people are assigned as team leaders.

It is not uncommon for companies to assign the process engineer, or perhaps the design engineer, as the team leader for a process being studied, in the belief that their knowledge of the process will help in the performance of the study. Unfortunately, this can create difficulties. Other team members may be unwilling to criticize the process as the leader will be perceived as owning it and team members may not want to risk offending the team leader. Consequently, issues with the process may not be identified as thoroughly as needed. This problem is compounded if the leader is defensive or unwilling to consider suggestions by team members. Such behavior will further discourage team members from contributing to the study. Furthermore, process or design engineers likely will have mindsets about their process. Mindsets are assumptions held by an individual which are so established that the individual does not recognize they are being made. Usually, they arise after an individual has worked in a process for some time. Current methods of working and existing levels of safety become viewed as acceptable and are no longer questioned. Therefore, it can be difficult to see the process in a fresh light. Consequently, team leaders should be chosen who do not have any day-to-day responsibilities for the process being studied. Indeed, OSHA has stated that team leaders should be "impartial in the evaluation" [11].

Similarly, other individuals who may have extensive knowledge of the process technology should not act as team leader. While they may not have any ownership of the particular process being studied, they may certainly have mindsets about the process. This may result in them not asking questions that another leader would ask, possibly resulting in missed scenarios in PHA or omissions in LOPA. Furthermore, team leaders who are viewed by the team as experts in the process or its technology may not be willing to ask certain questions in case they reveal a lack of expertise on their part causing loss of face, again resulting in incomplete studies.

3.3 Team Member Qualifications

Team members should collectively possess the knowledge and skill to identify hazard scenarios in PHA and analyze them in LOPA. They should also have a sense of ownership and responsibility for the process to ensure their commitment and motivation to identify and eliminate or reduce risks in the process.

Study teams are composed of groups of people with different personalities. People interact with each other in different ways dictated largely by their personalities and this impacts the chemistry of study teams. Usually, team members are selected for their technical skills and knowledge, as well as their availability, rather than their personality. However, team dynamics are very important for an effective and efficient study. Team chemistry can have a marked impact on the quality of a study, particularly in PHA, so attention should be paid to the selection of team members based not only on their technical qualifications but also their personal characteristics. OSHA has stated, in the context of PHA, that, "the selected team members need to be compatible" and "the team needs to be able to work together while benefitting from the expertise of others on the team or outside the team to resolve issues and to forge a consensus on the findings of the study and recommendations" [11]. Therefore, both PHA and LOPA team members should be screened against desirable attributes, including:

- Creativity
- Not afraid to express their opinion
- Not intimidated by working in group made up of different disciplines
- Willingness to participate
- Logical
- Alert
- Ability to concentrate
- Pays attention
- Focused
- Able to express themselves clearly
- Able to communicate technical issues in their area(s) of expertise to other team members

Few team members will display all of these attributes but, to the extent team members do not meet them, a conscious decision should be made as to whether the individual should be replaced with a more suitable person. In the event no one else is available, the team leader will need to plan for and be ready to deal with the consequent issues that may arise during the study.

It must also be recognized that different people assigned the same job responsibilities for a process may accomplish them in different ways. Such situations can arise, for example, when process operators have different levels of experience or display different attitudes towards work. It is not unusual to find that processes are operated in different ways by different shifts. Thus, there may be a need to reflect such variations by having more than one person from the same discipline as a team member.

Study teams should not consist entirely of people who know the process as groupthink can be a problem. Groupthink is a phenomenon in which a group of people think and make decisions in the same way thus discouraging creativity. Usually, it occurs when people in the group have worked together for a period of time and evolved a consensus view based on shared assumptions. Thus, it is beneficial to have an independent senior engineer as a team member, sometimes referred to as "cold eyes" or "fresh eyes", who does not have any prior experience with the particular process being studied so they can challenge assumptions made by other team members and contribute knowledge that may not be possessed by the team. This role can be played by an independent experienced team leader if care is exercised to avoid direct confrontation with team members.

Study teams that include supervisors and the supervised generally should be avoided as the supervised may feel inhibited from contributing freely for fear of retribution or making a poor impression. Similarly, while management participation is a good idea at the opening of the first session of a study as it may provide managers with a better understanding of the study, demonstrates management commitment, and provides an opportunity for a pep talk to be given by management to the team, it is not recommended for actual study sessions as it may intimidate team members and inhibit their contributions.

3.4 Team Size

In PHA studies, the team must be large enough both to provide the knowledge and skills needed for the study and to enable effective brainstorming. However, the team should not be so large that brainstorming is hindered by too many people attempting to contribute simultaneously. Typical effective team size is 5 – 9 people. Fewer than 3 or larger than 10 can create problems. Factors that influence team size are the complexity of the process and the expertise of individual team members. Team members are needed who together can provide the information needed to define the design intent for the process sufficiently. Team members who meet this criterion and whose number falls in the optimum range should be chosen.

LOPA studies do not involve any significant brainstorming so long as the supporting PHA studies provide the scenario details needed so staffing decisions are mostly driven by the need to have enough team members with sufficient technical expertise to perform the analysis.

3.5 Team Leader Authority

The team leader must be provided with the authority to take actions to safeguard the quality of the study, including postponing a study session if key team members are absent; needed information is not available, incomplete, or inaccurate; or appropriate facilities are not available. Similarly, the team leader must be able to terminate a study session early if the team is fatigued or there are other issues that may adversely impact study quality.

3.6 Reference Information

A considerable amount of information is needed to perform PHA and LOPA studies. Study teams must be able to rely on its accuracy and completeness. If issues with the quality of the information are discovered during the performance of a study, the team will be distracted by the need to identify, discuss, and correct the problems. If a significant number of inaccuracies are discovered in a reference document during the performance of the study, the document may need to be updated before the study can proceed, otherwise the team's performance will be impacted adversely owing to the difficulty of interpreting the document when it contains many marked corrections. Of course, updating documents will delay the study, which is undesirable once a study is underway as the delay likely will result in team members needing to revisit the part of the study already performed in order to refresh their memories before proceeding with the continuation of the study. Consequently, companies should ensure that study reference information is available, accurate, complete, and clear before a study commences.

LOPA studies require information on hazard scenarios which most often is taken from PHA studies. Unfortunately, many PHA studies provide insufficient detail to perform LOPA studies properly. Common issues are insufficiently clear initiating events, lack of scenario details, and lack of information on enablers. In such cases, the LOPA team must develop any needed information missing from the PHA. Not only does this require additional time and effort but it is also a significant distraction from performing LOPA. Furthermore, the LOPA results may be inaccurate as the LOPA team may not develop hazard scenarios correctly. PHA teams should follow guidelines to ensure appropriate information is available for LOPA [12].

Team leaders must not rely on the unsubstantiated opinions of others when assessing the quality of reference information. Critical information should be validated [13]. Some of the information needed may require a significant amount of time to develop, update or assemble. Therefore, compilation should begin well in advance of the first study session. It is useful to employ a checklist to keep track of information, its status, and its location.

While key documents, such as process drawings, should be provided as paper copies to each team member for convenience of reference, other documents are best provided electronically to avoid overwhelming the team with a large volume of paper.

3.7 Meeting Facilities

The arrangement of suitable meeting facilities may be seen as a mundane issue not worthy of much attention but this would be a serious mistake. The environment in which people work has a marked impact on their performance and every effort must be made to ensure an optimal environment for PHA and LOPA teams.

The meeting room must provide sufficient space for people to move around, adequate artificial and natural lighting, low noise levels, and suitable temperature, humidity, ventilation, and air quality. The location of the meeting room must minimize distractions and disturbances. It is a good idea to place a sign, "Do Not Disturb. PHA (or LOPA) in Progress." on the door. Managers must agree not to summon their people from the study, or otherwise interrupt the study sessions, except for true emergencies.

Computer software should be used to record study information. The study worksheet should be projected on a wall screen so team members can see entries as they are made. The font size of entries must be optimized, appropriate colors should be used, and a suitably large image should be projected for easy viewing. Computer projection avoids the need for subsequent editing of worksheets which is time-consuming and error-prone. It promotes greater involvement of team members in the conduct of the study and it facilitates use of guidelines, checklists and reference information which can be stored and retrieved electronically.

Good quality equipment must be provided. Failure of equipment during studies results in delays and wasted team time. Use of outdated equipment creates an impression that the study does not matter to management which may de-motivate the team.

Wall charts and paper hand-outs may assist team members during study sessions. A flip chart or white board, preferably the electronic type that allows recording, should be provided for impromptu use. Team member needs for beverages such as coffee and juices; snacks such as donuts, cookies and fruit; and food for lunch must be met, otherwise team members may suffer from inattention and low energy levels.

The meeting room should be set up conference style to encourage a sense of belonging to a team. The computer should be placed towards the head of the table with the projection screen at the opposite end, if the leader is scribing so the leader can look easily at both the team members and the screen being viewed by the team. If a separate scribe is used, the leader should stand close to the scribe near the bottom of the table and close to the projection screen to reinforce their control over the meeting and make it easy to point at the projected worksheet.

Video / tele / web conferencing is not recommended. The efficiencies and synergistic benefits of in-person meetings are lost and study facilitation is harder. Personal interactions are very important. However, exceptions may be necessary or acceptable in some cases, for example, when team members are separated geographically, when team members know each other well, or when consulting with a subject matter expert.

Audio or video recording of sessions is not recommended. Not only might it result legal problems but also it can inhibit participation in the study by team members who may be intimidated and unwilling to have their comments recorded.

4. Human Factors in Conducting PHA and LOPA Studies - Team Issues

A variety of human factors issues relating to the study team can affect the conduct of PHA and LOPA studies.

4.1 Study Foundation

The team leader must lay a foundation for a constructive study. By their nature, PHA and LOPA studies result in recommendations for process improvements. Usually, team members have ownership in the process and they may react negatively when problems are identified because they may see them as an implied criticism of their work. Leaders must explain that it is normal for studies to find needed improvements, even when processes are designed and operated by the most highly regarded people. The goal is to ensure that no one on the team feels threatened by critique of the process design, operation, etc.

Participants must help to ensure studies are performed on the process as it is actually constructed, operated and maintained. There is no point in looking at the design intent if it is ignored, reviewing drawings if the hardware is different, referring to written procedures if they are not followed, or relying on instruments if they have not been maintained or have been disconnected.

The operating practices actually followed may differ from the written operating procedures and process drawings, and other reference information may not be accurate. If such deviations are not identified, an otherwise high quality study may be performed on a process that does not actually exist. Team members usually are aware of such deviations but may be reluctant to identify some of them for fear of punishment. Consequently, management must agree to a "no punishment" policy with regard to the identification of process issues and team members must be convinced of management's sincerity and the validity of the policy.

4.2 Role of the Team Leader

The PHA team leader's role is to motivate the team, encourage brainstorming of hazard scenarios, and facilitate the study. The LOPA team leader's role is to guide the analysis of individual scenarios, focusing on the qualification of safeguards as Independent Protection Layers (IPLs).

Normally, PHA team leaders do not contribute entries for the worksheet, other than wording them for the scribe. The leader only contributes entries if the team needs to be stimulated or does not have sufficient technical expertise. Any such input should be provided *after* the team has had the opportunity to provide input, otherwise the team will rely on the leader to complete much of the study. If the team leader is a technical expert, they may act as a resource in providing information needed by the PHA team.

PHA team leaders may be tempted to join in the brainstorming, particularly if they are engineers. The temptation must be resisted, otherwise no one will be leading the study, time will be wasted, and study quality may suffer.

Team leaders ask many questions of the team during a study. They may fall into the trap of answering their own questions when team members are not immediately forthcoming in order to fill the silence, or because of time pressures to complete the study. The team may then rely on the team leader to complete much of the study.

Team leaders must be alert to these pitfalls. They should not lead studies where they have a vested interest in the outcome or are not impartial, or where they are a technical expert on the process, or have day-to-day responsibilities for it, as the chance of encountering these pitfalls increases significantly. The study scribe or another team member should be asked to alert the team leader if they notice these problems occurring.

4.3 Team Leader Responsibilities

Team leader responsibilities during the conduct of a study include:

- Ensuring the study is completed within the allotted time
- Keeping team members involved and energized
- Motivating team members
- Ensuring the team works together effectively
- Finding consensus for worksheet entries

4.3.1 Ensuring the study is completed within the allotted time

The team leader must maintain the study pace and keep the study focused. The leader must keep pushing the team while ensuring all key issues have been addressed. Judgment is critical to ensure optimum pressure on the team. How much pressure should be applied will vary from time to time during the study. The leader may also need to allow the pace to vary according to the needs of the team. The leader must keep the study moving while allowing sufficient discussion to satisfy the team and ensure sufficient opportunity to identify hazard scenarios in PHA and analyze them fully in LOPA. Ways to maintain the study pace include:

- Setting a schedule and keeping track of progress.
- Generally, avoiding team silences.
- Keeping the team thinking and moving.
- Learning to recognize when discussions are not fruitful.
- Being able to steer the team away from going off on a tangent.
- Avoiding repetition and "wheelspinning", i.e. going over the same ground repeatedly.

Study teams can digress, get bogged down, and lose focus, particularly in PHA.

Ways to keep a study focused include:

- Setting goals and objectives for each session.
- Reviewing progress periodically.
- Summarizing often.
- Posing questions on relevant issues.
- Directing attention to process drawings or other reference documents or information.
- Using checklists.
- Referring to the study purpose, scope and objectives statement.
- Pre-entering reference information into the worksheet.
- Ensuring study reference documentation is accurate, complete and the same version for all team members.
- Making documents easy to use, e.g. using color-coded P&IDs.
- Managing distractions.

Team leaders must understand common causes of distractions and develop plans to control them. Management agreement not to interrupt the team should be obtained. Unnecessary use of cell phones, pagers, radios, public address systems, computers, etc. should be avoided. A meeting room away from the facility may be beneficial to make it harder for the team to be interrupted and for team members to stop by their workplace on breaks, perhaps being delayed or forgetting to return to the study session on time.

Team leaders undoubtedly will feel pressure to keep studies on schedule. Teams are composed of people who have other important responsibilities for the process and the company will not want them to be away from their regular assignments for longer than necessary. There is an inherent conflict between meeting a schedule and allowing sufficient discussion of issues by the team. However, the quality of the study should prevail over schedule, and management must ensure that team leaders understand that they may extend the schedule, if needed.

4.3.2 Keeping team members involved and energized

Sometimes team members may stop participating in a study and just let other people contribute. They may have lost interest or their mind may have wandered but they are no longer an actively contributing team member and study quality may suffer as a consequence. Team leaders need to recognize such situations and act on them to bring the team member back into the study. Ways to keep team members involved and energized include:

- Watching for team members tiring.
- Calling breaks as needed.
- Using exercise in place, i.e. team members stand and follow the team leader for a minute or two in stretching, etc.
- Providing food / coffee etc. to maintain energy levels and attentiveness.
- Providing encouragement, as appropriate.
- Projecting the computer screen image of the worksheet on a wall screen so team members can see worksheet entries as they are made.
- Breaking up large processes into smaller ones so individual studies do not last as long.
- Encouraging spontaneity and building on ideas.
- Providing positive feedback by complimenting contributions.
- Asking other team members to comment on contributions.
- Stimulating participation.
- Using open-ended questions (not yes / no answers).
- Rewarding contributions.
- Exploring all suggestions made. Team members should be confident they will not be ignored or overruled in a perfunctory manner.
- Encouraging active debate.
- Resolving debates equitably so that team members agree with the way the resolution was handled and can accept it even if it is not exactly what they want.
- Ensuring everyone can hear all contributions.
- Ensuring all team members understand the discussions.
- Focusing on important and relevant issues.
- Managing the repetitive nature of studies.

PHA teams should focus on important and relevant issues by avoiding wasting time brainstorming routine hazard scenarios. Checklists and other aids can be used to focus on substantive items. Time should not be wasted on trivial matters. Leaders should remember the Pareto principle, or 80-20 rule, wherein roughly 80% of the significant consequences may be expected to come from 20% of the causes.

Unnecessary repetition should be avoided in PHA as it also wastes time. However, distinct hazard scenarios may often have similarities, e.g. while causes may be different, consequences and safeguards may be the same. The scenarios must still be documented as the risk rankings and recommendations may vary. Unfortunately, PHA teams may become frustrated by the apparent repetition. Ways to manage such repetition include:

- First, ensuring the team understands the need for repetition.
- Moving expeditiously in making such entries.
- Using PHA recording software features to speed up data entry.

- Not recording the same scenario in different nodes of a study. Using a reference, as appropriate.
- Using PHAs on similar processes as templates.
- Borrowing from common global nodes in other PHA studies.

The PHA team must think carefully about such repetitive entries. Treating them cavalierly may result in errors. Differences must be identified and captured.

4.3.3 Motivating team members

The team leader must ensure that team members are appropriately motivated throughout the study. Ways to motivate team members include showing graphic photographs of accidents, describing incidents that have occurred at the company, ensuring team members fully understand the consequences of catastrophic accidents including possible employee impacts not limited to health and safety but also job security, and providing a pep talk by management to instill enthusiasm and determination in team members at the outset of the study.

Setting goals for milestones to be reached in each study session, and requiring sign-off by team members on the study report can also be valuable. For longer studies, possible motivators include a weekly social gathering such as a pizza party, the provision of team hats or shirts, periodic visits by management to the study room or attendance at team lunches to provide encouragement, and recognition in the plant newsletter.

4.3.4 Ensuring the team works together effectively

At every opportunity, team leaders should model the behaviors they expect of team members, such as punctuality and the avoidance of side discussions, to set an example and avoid condoning undesirable behaviors. Since people interact in different ways according to their personalities, team leaders should also strive to understand personalities in order to be able to manage teams effectively. It is valuable for team leaders to be able to type personalities based on their observations of an individual's behavior to gain insights into how best to manage team members and their interactions. Various useful references are available [14, 15, 16].

Guidelines to help team leaders manage teams are provided in Table 2. Challenges faced by team leaders that are posed by team members and external factors are shown in Tables 3 and 4, respectively. Team leaders pose some of their own challenges, including staying aloof from brainstorming in PHA, particularly when the leader is an engineer and wants to solve problems, and the adequacy of their technical, people and management skills.

4.3.5 Finding Consensus for Worksheet Entries

Leaders must find a team consensus for each worksheet entry. Most entries are not controversial but in some cases there may be disagreement between team members which must be addressed, particularly in PHA. Consensus does not mean complete agreement, but that everyone can "live with it". Dissenting opinions are not recorded. Team leaders can seek consensus by:

- Explaining the importance of consensus at the first session.
- Not encouraging disagreements, although challenging assumptions is permissible.
- Not taking votes.
- Allowing discussion. The elapse of time may result in agreement.
- Promoting compromise.
- Phrasing PHA worksheet entries appropriately to reflect a middle position.
- Splitting PHA worksheet entries, as appropriate, to accommodate different points of view and accounting for their relative likelihood in the likelihood rating, e.g. injury versus fatality.

If consensus agreement cannot be reached, the issue can be recorded as an information need for further investigation.

4.4 Team Leader Challenges

Team leaders have many responsibilities and can easily become overloaded which may impair study quality. Leaders should delegate responsibilities to other team members wherever appropriate, such as study recording, quality control, and management of checklists and reference documents. It is particularly important to designate a team member as quality control manager so that independent checks are made throughout the study to confirm that study procedures are being followed.

Rarely will team leaders encounter ideal teams. The prevailing view is that in selecting team members their technical skills and knowledge outweigh issues with their personal characteristics, unless an issue is particularly serious. Consequently, the team leader should try and anticipate potential problems with team members and prepare for them, particularly in PHA studies. Team leaders should also try and anticipate other problems such as the unavailability of team members and logistical issues.

A key role for PHA team leaders is to encourage brainstorming and, thus, divergent thinking. However, the leader must be able to recognize when an issue has been aired sufficiently and further progress is not being made, so the team can be brought back to a focus by using convergent thinking, otherwise valuable team time will be wasted.

Team leaders necessarily must rely on considerable subjective judgment from team members in PHA, and to a lesser extent in LOPA. However, neither the leader nor other team members should rely solely on the unsubstantiated opinions of a single team member. Sometimes a team member with seniority and experience may offer a strong opinion, or a less experienced team member may try to advance an opinion using the strength of their personality. While the individual may be correct, it is important that other team members challenge the individual to substantiate their opinion and not just accept it as authoritative. In the absence of hard data and information, studies should be based on good engineering judgment. Indeed, OSHA has stated, "PHA is dependent on good judgement and the assumptions made during the study need to be documented and understood by the team and reviewer and kept for a future PHA" [11].

Team leader technical and facilitation skills may atrophy if leaders do not conduct studies regularly. Also, new developments occur regularly so refresher training is important.

4.5 Team Member Challenges

Various human factor issues may influence the behavior of team members which can impact studies adversely.

Sometimes team members are expected to perform most of their regular duties while participating in a study. Such pressure of work undoubtedly will affect their performance adversely, both in the study and their regular assignment. Team members must be relieved of their regular responsibilities sufficiently to allow participation in studies.

Team members may be confused by the technical details of studies, particularly if they are first-time participants. Team leaders need to guide team members through the process of conducting studies but, in order to avoid unnecessary confusion, the analysis should be performed consistently by doing things the same way throughout the study unless there is a good reason not to do so. In such cases, the reason for the departure from normal practice should be explained to the team.

Technical personnel may not like admitting to others that they do not know or understand something. This may produce unreliable contributions from team members who feel this way. Team leaders must stay alert and try to detect such situations. Sometimes, they can be detected when the individual involved becomes combative.

During longer studies, team members may disengage from participating in the study while staying physically present. This can occur for various reasons, including fatigue, being distracted by company or personal issues, alienation from other team members, and becoming angered by criticism from other team members. Some of these issues are within the control of the team leader; others are not. However, the leader must be alert to the development of such situations and take action to bring the individual back into the study. This can be accomplished, for example, by posing questions to the individual, asking them to comment on the contributions of others, or requesting that they take responsibility for some part of the study that requires active involvement, e.g. scribing.

Team members must not pre-judge matters. They should keep an open mind and avoid pre-conceived ideas. This is best accomplished by careful selection of team members supplemented by the team leader asking open-ended questions during the study and the presence of an independent senior team member who can challenge the assumptions of other team members.

Team members may have mindsets about the process. In such cases, team leaders must encourage the team to think divergently by asking open-ended questions or challenging team members to justify their views. Similarly, individual team members may be biased in various ways. The team leader should try to ensure that the bias does not enter into the study. Bias can be managed by utilizing other team members to comment on the biased view to help the individual recognize and move away from it.

Sometimes team members are unwilling to consider the unanticipated, such as some types of external events, e.g. objects dropped from lifting devices, or multiple failures, in PHA, even when they meet credibility criteria. In such cases, team leaders may need to re-explain the study guidelines and stress that conservative assumptions are made in PHA to be on the safe side. Also, it can be valuable to describe incidents of the type that is being doubted that are known to have occurred, particularly any from within the company.

Peer pressure can play a role in studies. Team members may be influenced by the views of others or they may be inhibited from expressing their real views by the presence of colleagues. Team leaders must help team members to overcome peer pressure. The importance of doing so should be emphasized at the outset of the study its negative effects explained. The team shares the common goal of helping to ensure the process is adequately safe and this goal can only be accomplished if team members contribute to the study freely. Leaders should be careful not to put someone on the spot when they offer dissenting views. Rather they should be encouraged to speak freely and independently. Sometimes, leaders may recognize through the body language of an individual that they disagree with what is being said by someone else but are reluctant to express their own views. In such cases, the leader should try to draw out the individual's views by first asking if anyone has any comments and, in the absence of a response, asking the individual if there is anything they would like to share.

Teams sometimes digress from discussion of pertinent matters as a consequence of the brainstorming process that is part of PHA studies. Some digressions may be relevant to the study, for example, description of the history of the process, but irrelevant digressions waste time and must be avoided wherever possible. A statement of the study purpose, scope and objectives can be used to help keep the team on track, ensure appropriate study content, ensure the study is complete, and avoid team members raising issues that are not relevant [2].

Unfortunately, some team members may be used to applying a philosophy in their work or personal life that x% performance is good enough where x may be as low as 75 and possibly no higher than 90. People may not have the opportunity to perform their regular work tasks as well as they would like to, or are capable of, as captured by statements such as "we need a Chevrolet not a Cadillac", and "we don't need to goldplate it". However, while this philosophy may be applicable to some activities, it should not be adopted in PHA or LOPA studies. Many lives may be a stake, and even a company's survival could be on the line, so the highest standards of performance are needed. Team leaders should seek to motivate team members to these levels of performance using the approaches described in this paper.

People have a tendency to take shortcuts, particularly when they become tired, bored, or believe they are falling behind schedule. Many studies are conducted under time pressures so this issue can be encountered with some frequency. For example, suggestions by the team to combine PHA nodes in order to speed up a study generally should be resisted unless there is a technical justification. Team leaders must be alert to shortcuts the team tries to take and forestall them by explaining to the team the consequences that would arise.

Studies are intellectually demanding and can exhaust team members [17]. Therefore, lengthy sessions should be avoided to prevent team member burnout. However, large gaps between sessions should be avoided in order to maintain study momentum.

4.6 Behavior of Team Members

There are both desirable and undesirable behaviors for team members during studies. No matter how well team members are screened for their suitability, there is always the possibility they will exhibit undesirable behaviors. Consequently, guidelines should be established for expected behaviors by team members, and rules developed to govern how studies will be conducted. Guidelines and rules should be tailored to the circumstances of each study and explained to team members at the outset. Typical guidelines for behavior are:

- Be punctual and attend all sessions.
- Do not criticize other team members.
- Be open-minded.
- Participate actively.
- One person speaks at a time.
- Avoid digressions.
- Don't reinvent the wheel, i.e. don't keep going over the same issue.
- Avoid brainstorming solutions to problems.
- Do not try to redesign the process during the study.
- Ignore minor typos by the scribe.
- Do not use mobile phones, pagers, radios or computers during study sessions.

Typical rules to govern how the study will be conducted are:

- No questions are stupid.
- No ideas are bad.
- Consensus is sought for worksheet entries.
- All participants are considered peers.
- Everyone contributes.
- Start sessions on time and end on time.
- Regular breaks will be taken.

Some companies have established behavioral guidelines that must be met by their employees, e.g. respect for other employees. Any such guidelines should be reviewed at the first study session and subsequently, as needed.

Even teams that usually work well together may have "off" days when they experience difficulty in making progress. They may have been influenced by the weather, events occurring in the workplace such as layoffs, or there may be no apparent reason. Team leaders need to be ready to deal with such situations, possibly by adjourning the session until the next day if there is no apparent underlying issue that can be addressed.

5. Human Factors in Conducting PHA and LOPA Studies - Technical Issues

Various human factor issues can impact different technical aspects of how PHA and LOPA studies are conducted.

5.1 Nodes in PHA

Node size can influence team performance. Large nodes will cause the team to become mired and make the identification of hazard scenarios difficult. Also, the analysis is more confusing. Very small nodes will cause the study to become repetitive and makes the big picture harder to see. Guidelines for optimally defining nodes should be followed [9].

Nodes should be selected so that their design intent can be defined adequately and is clear so that unambiguous deviations can be developed. For example, combining process lines with different intents into one node results in problems when deviations are developed as they will vary according to the different intents, e.g. different operating temperatures or line contents. Nodes must be defined to facilitate not impair the ability of teams to identify scenarios.

5.2 Design Intention in PHA

Fundamentally, PHA studies look for ways a process may deviate from the design intention. Deciding which aspects of design intention and which deviations may result in hazard scenarios within the study scope and objectives are critical steps in a study. However, the task is akin to looking into a crystal ball to predict the future. Commonly, PHA teams choose parameters that represent aspects of design intention as each node is considered. Eventually, the team reaches a point where they decide to move on based on the feeling that they have done enough work or spent enough time on the node or deviation, or because they are falling behind schedule, rather than that they have addressed all relevant aspects of node intention.

Alternatively, some team leaders identify parameters for consideration in each node in advance of the team sessions. In such cases, it is unlikely the team will want to consider the possibility of additional parameters whose analysis would prolong the study. Neither situation is satisfactory and can lead to the incomplete identification of hazard scenarios. The problem can be avoided by having the team define relevant aspects of design intention as each node is considered and then exploring deviations for them. It is more likely that important deviations will be identified and analyzed using this approach.

5.3 Causes in PHA

Causes, singly or in combination, are the initiating events of hazard scenarios. They are identified by team brainstorming. Often, initial brainstorming focuses mostly on equipment failures as most team members are engineers used to dealing with plant hardware. However, it is essential that human failures and external failures also be addressed so that scenarios with these other types of causes are not omitted. Brainstorming of causes is another area where the willingness of team members to be thorough may impact study quality. Consequently, once the initial brainstorming is complete, it is good practice to review equipment in the node under consideration, the roles of people in the node, and lists of utilities and natural and human-induced external events customized for the facility to ensure that all credible failures have been considered. This is a vital quality control check. If causes are missed, the identification of hazard scenarios will be incomplete.

The key issue for causes is deciding on their credibility. Unfortunately, credibility may mean different things to different people. The dictionary definition is "able to be believed" or "believable". For some people, it may mean the frequency of occurrence of an event is above their personal *de minimis* criterion. However, such criteria can vary widely depending on the person's experience. Guidelines should be established for determining the credibility of causes such as:

- Rely on past experience, both local, company and industry wide.
- Review reports of previous incidents or near misses.
- Inquire whether any team member has seen the cause before. If so, clearly it is credible.
- If failure data exist for the cause, it must be viewed as credible since it has actually occurred.
- Do not eliminate causes because there are existing safeguards. Safeguards can fail. The existence of multiple safeguards is an indication that the process designers thought the scenario credible.
- Do not assume that something cannot happen simply because it has not happened before.
- Consider using numerical criteria for screening when there is disagreement, e.g. a frequency of occurrence of one in a million / year. Such criteria should be related to the severity of the scenario and the tolerability of risk.
- Err on the side of conservatism to account for uncertainties and subjective judgment

Sometimes team members have difficulty relating to cause frequencies expressed as "once in N years", particularly when N exceeds 100 years, the nominal human lifetime. In such cases, it may be easier to re-frame the issue, for example, by considering N items in the field and asking how many failures could be expected in one year, or asking how many items it would take in the field to have one failure per year.

It is good practice to list causes first in the PHA worksheet and then determine if they are credible. Discussion of the credibility of each cause as it is identified should be avoided as it interferes with the brainstorming of other causes. The credibility of each cause should be addressed immediately after they have all been identified, or as entries in the next column of the PHA worksheet are considered.

Similarly, teams should identify all causes of a particular deviation before completing other worksheet columns. Moving back and forth across columns would adversely impact the brainstorming of causes as it would interfere with the thought processes of team members identifying causes.

Sometimes a cause is identified for a scenario that should be recorded in a later part of the PHA worksheet. The team should jump ahead and record it when first identified; otherwise it may be forgotten owing to the mental intensity of brainstorming causes.

5.4 Safeguards in PHA

Often, teams are inclined to provide unwarranted credit for human safeguards, such as operator response to alarms and process indications, and monitoring and surveillance of processes. People have a tendency to believe they and, by implication, other people, will always function at their highest level of performance when needed and be able to recover if they make a mistake. However, it is unlikely that this level of performance can be achieved consistently for a variety of reasons. People's attitudes and moods vary. Variable factors such as sleep deprivation, life distractions, and health influence performance. Something as common as a cold virus can have a significant adverse impact on an individual's performance. Also, such factors may impact different people to a different extent. Moreover, human failure rates are high [3], people do not think and act the same way every time they face the same set of circumstances, and events in a process may develop too quickly for people to respond. Furthermore, the stress of facing potentially catastrophic consequences may affect people's judgment. They may not be able to think clearly and may panic and make things worse. Also, in a short period of time, a very senior, highly-capable workforce can become a very junior, less-capable workforce through layoffs and retirements.

Additionally, training that supports human safeguards may not be effective; the experience of people may be limited or not applicable to the circumstances encountered; procedures governing human responses may not be well-written, may not be understood, or may not be used; and monitoring and surveillance may not be performed. Indeed, inadequate training, experience, or procedures, may be contributors to hazard scenarios. This may invalidate claims that they support human safeguards.

Moreover, companies may not want people to act as safeguards as it may place them close to hazards and expose them to injury as a scenario develops.

Companies must decide which, if any, types of human safeguards may be credited in PHA studies and not leave the decision in the hands of the team. Furthermore, people should not be the only safeguards for key, safety-related scenarios.

5.5 Risk Ranking in PHA

Processes are engineered so that the likelihood of accidents decreases with increasing severity of the consequences so that higher consequence accidents are less frequent. Of particular concern are accident scenarios with high severities and low likelihoods, i.e. rare and catastrophic events. Catastrophic scenarios are not expected statistically to occur within the lifetime of a process. Of course, that does not mean the process will not experience such an event. The statistical chance of observing them increases with plant population and operating time. Such scenarios are likely to involve multiple failures owing to the multiplicative property of the probabilities of the events that constitute such failures. Estimating the likelihood of rare events is challenging. The numbers involved are very small and the events are usually not observed. Measuring something that does not occur very often is difficult. PHA teams are likely to encounter difficulties in estimating the risk of such scenarios.

Risk ranking is performed by qualitatively estimating the severity, S, and likelihood, L, of hazard scenarios using subjective engineering judgment and combining them into risk estimates in a risk matrix where S is the severity of the consequence for the hazard scenario and L is the likelihood of all the events in the hazard scenario occurring together, including the cause, intermediate events, safeguard successes / failures, etc. that result in the scenario consequence. Estimates of S and L may not be consistent from team to team across the company, or even with the same study, unless considerable care is exercised to manage the subjectivity involved.

The estimation of consequences is easier than for likelihoods as most team members will have some appreciation of the possible spectrum of consequence severities that could occur in the process, from the benign to the catastrophic, as a result of their personal experiences or awareness of industry events. Estimating likelihoods up to once in 100 years is not too difficult for most people but becomes increasingly difficult for less frequent events. People have a reference point for frequencies of occurrence within the nominal human lifetime but have difficulty relating to lower frequencies and events which may not be within their realm of experience.

As for cause frequencies, when team members have difficulty relating to "once in N years", particularly when N exceeds 100 years, the nominal human lifetime, it may be easier to re-frame the issue, for example, by considering N processes and asking how many identical scenarios could be expected in one year, or considering how many identical processes it would take to have one scenario occur per year.

Human perception influences estimates of likelihood. A person's experience is reflected in their estimates. Usually, people underestimate the probability of an event they have not experienced and overestimate the probability of an event they have experienced. An attempt can be made to reduce subjectivity in likelihood estimates by requiring the team to start with the highest (most frequent) likelihood level and justify successive reductions to lower levels. However, if there is any doubt in assigning values, they should be assigned conservatively. The possibility of underestimating scenario likelihoods owing to the role of subjective judgment and human perception is one reason why some companies use LOPA to calculate the frequencies of scenarios that have high consequence severities, regardless of their estimated likelihoods.

The ability of the PHA team to perform risk ranking consistently is a function of the particular risk ranking scheme employed, including the number of levels used for severity and likelihood and how they are defined. However, these issues are beyond the scope of this paper.

Consistency in risk rankings is easier to accomplish when the levels of severity and likelihood are defined in terms that are understood by team members. It is valuable to calibrate risk ranking schemes by discussing with the team the range of scenarios that may be identified and how severities and likelihoods would be assigned for example cases that are representative of the range, particularly for the worst severity which is the limiting case for a catastrophic accident. Some companies provide guidance on assigning severities and likelihoods by specifying values to be used for different types of scenarios. Of course, the guidance must be tailored for each process and company.

Risk ranking is best performed as the PHA worksheet is completed when the details of the scenarios are fresh in the minds of the team members.

Team leaders may be tempted to suggest S and L values to the team when team members are uncertain or reluctant to provide values. Unfortunately, this practice may influence the values eventually suggested by the team and it should be avoided. The team may also simply agree with the leader's suggestions without giving serious consideration to the matter. Similarly, the scribe should not type in values until there is consensus or until directed to do so by the leader.

Team leaders should be watchful for team members trying to influence the risk ranking by the selection of particular severity and/or likelihood values owing to hidden agendas. For example, they may want to avoid the need for action items that may be assigned to them.

5.6 Recommendations in PHA

A significant problem in PHA can be the inclination of team members to spend significant time brainstorming recommendations, despite instructions to the contrary from the team leader. The primary objective of PHA is to identify problems, i.e. situations needing improved or additional risk reduction measures. Brainstorming recommendations wastes time and detracts from the primary objective. Usually, only obvious recommendations are recorded during PHA studies. Even then, the team should not try and identify every recommendation they can think of.

However, team leaders must recognize that PHA teams may have a need to brainstorm to a certain extent, particularly when a very serious scenario is identified, but, if significant brainstorming is required, the development of a solution usually should be deferred and a recommendation made for a follow-up investigation of possible solutions. It is useful to employ some rules so that team members understand when brainstorming solutions will be deferred, for example:

- More than 5 – 10 minutes of discussion is needed.
- Reference documents outside the PHA room must be consulted.
- People or expertise not present are needed.
- Calculations are needed.

The agreement of team members with these rules must be sought at the outset of a study.

Sometimes, the PHA team is charged by management with producing solutions for all problems identified. Usually, this occurs when the PHA team members are the same people who would have to develop problem solutions after the PHA is completed. However, it is a good idea to separate problem solution from problem identification as it not only reduces the time required for PHA but also, more importantly, avoids team members switching between the different thought processes required for the two different activities. This switching can detract from both activities and result in adverse impacts on them.

Another inclination of some PHA teams is to brainstorm worksheet entries for which they do not have all the information needed. The team leader must not let the team waste time in this way. The leader must identify the information needed, assign responsibility to a team member to obtain it, move on quickly, and only return to complete the worksheet entry when the information becomes available.

Team members may be unwilling to make recommendations, for example, if they believe recommendations may be assigned to team members or will need to be addressed before startup. In such cases, it is best to defer developing recommendations until the PHA has been finished. Assignment of responsibility for recommendations during the PHA may adversely affect participation by team members who may be reluctant to develop recommendations for which they will be assigned responsibility. Also, team members may debate assignments, claiming that someone on the team other than themselves should be assigned. This promotes discord within the team and conflict with the team leader and should be avoided. Responsibilities can be assigned once the study is finished.

5.7 Use of Checklists in PHA

PHA is intended to be a brainstorming exercise. Detailed checklists of technical issues can become a substitute for the creative thinking and analysis that are essential characteristics of a good PHA study. Team leaders should not let checklists constrain team members to consider only items on the checklists. Team members must think more broadly and "outside the box". Also, checklists can consume such large amounts of time and become so repetitive that any value from their use is negated. Therefore, they should be used judiciously.

5.8 PHA Revalidation

There are some additional human factor issues that are specific to PHA revalidation studies.

It is useful if the revalidation team leader is the same as for the previous PHA to help provide continuity. However, if the composition of the revalidation team is the same as for the previous PHA study, the team may not be willing to entertain modifications and may have mindsets based on the previous PHA. Moreover, previous PHA team members may not be available to participate in the revalidation study. A modified team is preferred. Participation by some of the previous team members can provide insights into the previous PHA while new team members can provide a fresh perspective. Previous team members may be inclined to make only minimal changes to the PHA in the belief that nothing has changed that much while new team members will be more willing to make modifications. New team members may identify hazard scenarios that the previous team overlooked while previous team members may have gained valuable experience in the interim which may benefit the revalidation study.

If there are open recommendations from the previous PHA, team leaders should be careful to preserve the confidence of the team in the importance of PHA to the company. If the team believes actions were not taken as a result of the previous PHA, they may be reluctant to take the revalidation PHA seriously. Management will need to convince the team that the company is indeed serious about PHA.

There can be a tendency for teams to try and avoid making edits to the worksheet in order to finish the study quickly. Team leaders need to avoid "breezing" through the worksheets without giving adequate consideration to possible modifications. A deliberate and systematic approach is needed. The team needs to stay focused. Team leaders should try to avoid jumping around in the worksheets, which is distracting to team members, and ensure team members understand what is being done.

5.9 Failure Rate Data Needed for LOPA

LOPA studies use data on initiating event failure rates, enabler multipliers, and IPL probabilities of failure on demand (PFDs) [3,4]. These data should reflect experience with the actual operating environment and regime for the process. Unfortunately, few companies currently have collected plant-specific data. Generic industry data are often used, supplemented by engineering judgment. Data taken from generic industry databases may not apply to specific processes and introduce a source of considerable subjective judgment into studies. Wherever possible, generic data should be adjusted to reflect local conditions and situations, e.g. the range of operating parameters, specific chemicals processed, internal and external operating environments, operating regime, testing and inspection frequencies used, maintenance regime, operator and mechanic training, and equipment design, quality and specifications. Conservative estimates are employed to account for uncertainties.

Human error rates may be underestimated by LOPA analysts. As in PHA, LOPA team members tend optimistically to assume people will perform at their highest level and save the day. While this is certainly possible, error rates at the other end of distribution should be used for conservatism.

Some failure data may combine events together, for example, initiating event frequencies with enabler multipliers or conditional modifier probabilities. Such confounded or conflated data result in unreasonably low scenario frequencies. Multiple counting must be avoided. Teams must be aware of this possibility and alert to what failures are included in the data used.

Some companies establish a standardized set of failure data for their plants and processes. Not only does this make it easier for teams to perform LOPA, but also it reduces one of the largest sources of subjective judgment by taking decisions on data values largely out of the hands of team members. Users of such data must give careful consideration to their true applicability given the scenario assumptions, otherwise, unrealistic scenario risk estimates may result that produce poor decisions. LOPA teams should be allowed to use more conservative values if they believe them to be appropriate but they should not be allowed to use less conservative values without providing written justification. Guidelines for initiating event frequencies and IPL PFDs with specified criteria for proof test method, proof test frequency, and some special considerations have been published [18].

5.10 Risk Values and Risk Tolerance Criteria - LOPA

When LOPA studies reveal significant gaps between calculated and tolerable risk, and significant process modifications or additions may be needed to close the gap, possibly at significant cost, analysts, perhaps under pressure from management, may be tempted to adjust failure rate data downwards to meet risk tolerance criteria and claim that the adjustments reduce conservatism in the data. This is poor practice and must be resisted. Humans have shown time and again that they can readily talk themselves into whatever may be expedient. The use of standardized data is a bulwark against this practice.

Various enablers may combine together to reduce the risk of a hazard scenario substantially. When trying to close a risk gap, teams must resist the temptation to search for additional enablers that can be credited or lower enabler multipliers. Some companies choose not to consider enablers that reduce the scenario frequency to avoid the potential for underestimation of scenario risk. However, rules can be used to address this issue. For example, no more than three such enablers can be claimed for any scenario, and the maximum risk reduction from all such enablers for any scenario is limited to 1×10^{-2}.

LOPA analysts must recognize that when very low scenario frequencies are calculated (say 1×10^{-6} / year or below), before any additional reduction measures have been considered, it is possible the analysis is in error. The initiating event frequency may be too low, too much credit may have been taken for enablers, dependencies between enablers may not have been recognized, too many IPLs may have been credited (common cause failures become more likely when there are more than three or four IPLs), IPL PFDs may be too low, or dependent failures may not have been addressed. People sometimes allow formal analysis to interfere with the application of common sense. Therefore, reality checks should be performed throughout LOPA. Input data, assumptions and results must all be checked to ensure they are reasonable.

5.11 Qualifying Safeguards as IPLs - LOPA

Safeguards must be qualified as IPLs if they are to be credited in LOPA. Common qualification criteria are effectiveness, independence, and auditability [1], although additional criteria are sometimes used [19]. For some types of IPLs, specific criteria are used, for example, for human action IPLs [3].

Application of qualification criteria can involve subjective judgment. In particular, LOPA teams should be thoroughly briefed on common cause failures and other dependencies that may invalidate independence of IPLs or other scenario elements. Overlooking dependencies will result in underestimates of risk. Many causes of dependencies are subtle and difficult for typical teams to recognize. Checklists of common sources and specific examples can help team members meet this challenge.

6. Human Factors in Recording and Documenting PHA and LOPA Studies

Lack of detail in recording PHA studies sometimes occurs owing to poor study guidelines, or laziness in not following them, making it difficult for others to comprehend the PHA after it is completed, even for members of the team that conducted the study. Inadequate recording complicates the performance of PHA revalidations.

Some team leaders utilize a scribe to record studies. Guidelines should be provided for scribes to ensure they do not diminish the quality of studies, e.g.

- Listen to the team leader and the team discussions.
- Follow the team leader's instructions.

- Respond only to directions from the team leader.
- Do not make worksheet entries or edits unless directed to do so by the team leader.
- Do not try to be the leader.
- Attend PHA and LOPA training, if possible.
- Try to anticipate needed entries.
- Know and understand guidelines for worksheet entries.
- Capture key parts of the discussions.
- Learn to use recording software.
- Do not play with the software.
- Do not move around the worksheet, project, or software unnecessarily.

Recording is the first responsibility of a scribe. Participation as a team member is a secondary responsibility. Individuals whose undivided attention is needed in the study should not be assigned the role of scribe.

Team leaders may consider pre-completing PHA and LOPA worksheets to save meeting time. While entering reference information such as node names in PHA and scenario information in LOPA is useful, team leaders should not complete any of the analysis as team members would be tempted to endorse what has been done and not think critically about the analysis. Furthermore, regulatory expectations may be violated [20].

Study worksheets alone are inadequate documentation. A comprehensive report is also needed to provide a permanent record of the study [2, 4].

7. Human Factors in Following-up on PHA and LOPA Studies

Proper management and follow-up of study recommendations is needed to ensure that study findings and recommendations are not neglected. Management must resolve recommendations resulting from studies, i.e. decide which recommendations to implement when they become *action items*. Sometimes the effort involved in completing a study can result in a false sense of accomplishment that distracts from resolving recommendations and implementing action items despite regulatory requirements to establish a system to handle PHA recommendations [7] (see Table 5) and similar expectations for LOPA recommendations. Completion of studies is only the first step. Their results must be acted upon in a timely fashion. Some keys to successful management of recommendations are:

- Assign responsibilities to specific individuals.
- Set deadlines.
- Provide management oversight and follow-up on assignments.
- Provide any needed resources.
- Get commitment from affected employees.
- Use a management system.

8. Conclusions

Usually, PHA and LOPA studies are conducted by teams of people. Study personnel must be adequately trained, competent, and provided with suitable study guidelines. Study quality can be adversely impacted by a variety of human factor issues at each stage and step. Consequently, actions should be taken to manage these issues. Famously, "errare humanum est" or "to err is human" [21], and it is an impossible challenge to ensure that none of the issues described in this paper will adversely affect a study. The goal should be to accomplish the best result possible under the circumstances faced for each study.

The first step in addressing these human factor issues is to recognize and understand the issues. Key issues have been identified and explained in this paper. The next step is to manage the issues, proactively where possible. Approaches to addressing key issues have been recommended. PHA and LOPA studies are not just technical analyses. Continuous effort is required during all their stages to help ensure that human factor issues do not undermine study quality.

9. References

[1] "Guidelines for Hazard Evaluation Procedures," 3rd Edition, *Center for Chemical Process Safety / American Institute of Chemical Engineers*, New York, NY, 2008.

[2] Baybutt, P., "Analytical Methods in Process Safety Management and System Safety Engineering – Process Hazards Analysis," in *Handbook of Loss Prevention Engineering*, J. M. Haight (ed), Wiley-VCH, 2013.

[3] "Layer of Protection Analysis," *Center for Chemical Process Safety / American Institute of Chemical Engineers*, New York, NY, 2001.

[4] Baybutt, P., "Analytical Methods in Process Safety Management and System Safety Engineering – Layers of Protection Analysis," in *Handbook of Loss Prevention Engineering*, J. M. Haight (ed), Wiley-VCH, 2013.

[5] Baybutt, P., Chapter 26, "Qualitative Hazard Analysis," *in Human Factors Methods for Improving Performance in the Process Industries*, D. Crowl (ed), CCPS / AIChE, 2007.

[6] Baybutt. P., "Layers of Protection Analysis for Human Factors (LOPA-HF)," *Process Safety Progress*, Vol. 21, No. 2, pps. 119 - 129, June, 2002.

[7] Federal Register, Volume 57, Number 36, pages 6355 to 6417, February 24, 1992, Final Rule, Process Safety Management (PSM) of Highly Hazardous Chemicals, 29 CFR 1910.119.

[8] Baybutt, P., "Human Factors in Process Safety and Risk Management: Needs for Models, Tools and Techniques," *International Workshop on Human Factors in Offshore Operations*, New Orleans (December, 1996).

[9] "Process Hazard Analysis (PHA) for Team Leaders," www.primatech.com./training/process-hazard-analysis-pha-for-team-leaders.

[10] "Layers of Protection Analysis (LOPA) for Process Safety Management (PSM)," www.primatech.com/training/layers-of-protection-analysis-lopa-for-process-safety-management-psm.

[11] OSHA Publication 3133, "Process Safety Management Guidelines for Compliance," 1994.

[12] Baybutt, P., "Conducting process hazard analysis to facilitate layers of protection analysis," *Process Safety Progress*, Vol. 31, No. 3, pps 282–286, 2012.

[13] Baybutt, P., "On the Validation of Safeguards for Process Hazard Analysis," *Process Safety Progress*, to be published.

[14] Kroeger, O., Thuesen, J. M., and Rutledge, H., "Type Talk, Type Talk at Work: How the 16 Personality Types Determine Your Success on the Job," *Delta*, 2002.

[15] Keirsey, D., "Please Understand Me II: Temperament, Character, Intelligence," *Prometheus Nemesis Book Company*, 1998.

[16] Dimitrius, J., and Mazzarella, M. C., "Reading People: How to Understand People and Predict Their Behavior," *Ballantine Books*, New York, 2008.

[17] Baybutt, P., "On the Ability of Process Hazard Analysis to Identify Accidents," *Process Safety Progress*, Vol. 22, No. 3, pps. 191 - 194, September, 2003.

[18] "Guidelines for Initiating Events and Independent Protection Layers for LOPA," *Center for Chemical Process Safety / American Institute of Chemical Engineers*, New York, NY, 2010.

[19] "Guidelines for Safe and Reliable Instrumented Protective Systems," *Center for Chemical Process Safety / American Institute of Chemical Engineers*, New York, NY, 2007.

[20] Baybutt, P., "Prework and Precompletion of Worksheets for Process Hazard Analysis," *Process Safety Progress*, Vol. 31, No. 3, pps 275–278, 2012.

[21] Pope, A., "An Essay on Criticism," *HardPress Publishing* 2010.

Table 1. Types of Human Failures

Type	Meaning
Omission error	Action is not performed
Commission error	Action is performed incorrectly
Extraneous act	Non-required action is performed instead of or in addition to required action
Violation (deliberate act)	Action that is prohibited, or different from that prescribed

Table 2. Guidelines to Help Team Leaders Manage Teams

Accept all inputs and make no value judgments. Everyone should feel free to contribute

Speak in measured and friendly tones

Read body language / control your own

Bring your sense-of-humor

Be vigilant to avoid inappropriate brainstorming of problem solutions

Try to hear all offered input

Use silence, if needed, to prompt input from the team

Keep control of the team
- Standing at the projection screen may help, if using a scribe
- Avoid loss of control to scribe, if using one
- Controlling the worksheet image on the screen via the computer and data projector provides a powerful but subtle way of controlling the discussion

Avoid development of side discussions
- Do not engage in them yourself
- Firmly insist that only one person speaks at a time

Park contentious or extraneous issues in the "idea garage" or "parking lot", i.e. on a flip chart for later consideration

Try to anticipate problems or unusual situations and prepare for them, e.g.
- Personnel
- Logistics
- Plant problems

Don't:
- Facilitate by "going around the table"
- Let the team word the worksheet entries. Listen to the offered input, word the entries yourself, and make sure the team is comfortable with the wording
- Allow the team to spend time making trivial changes to worksheet entries
- Go against the team's wishes
- Argue with team members
- Compete with team members
- Do or say anything that may be perceived as negative by the team members

Table 3. Challenges Posed By Team Members

Quiet person

Talkative person

Over-enthusiastic person

Argumentative person

Sarcastic person

Dominant individual

Arrogant individual

Skeptic, i.e. does not believe in the value of the study

Person with their own specific agenda

Disputes between team members

Personality conflicts

Peer pressure, e.g. operators may not want to admit they do not follow procedures if other operators are team members, or a team member may be unwilling to offer views different from the majority

"Dog-with-a-bone", i.e. even though an issue has been addressed, the team wants to open it up for further unfruitful discussion when it arises again

Poor attendance / punctuality

Bored / fall asleep

Not motivated (don't want to be there)

Control freak, i.e. want to control the team and/or the study

Long-winded person

Greater seniority / plant experience than team leader who may feel intimidated

People speak at the same time

Inadequate language skills

Use of offensive language

Unqualified people

Team members bring other work to sessions

Lack of accident experience influences team member perceptions

Desire to include very low probability events

Distrust procedures, e.g. worksheet wording

Want to go beyond defined scope and objectives

Groupthink, i.e. team members share common but possibly false beliefs

Team members have different cultures / languages

Table 4. Challenges Posed By External Factors

Interruptions by managers and others

Use of cell phones, pagers, radios, public address systems, computers

Poor facilities

Previous studies followed different procedures

Previous team leader had different style

Refusal to work or approve overtime

Union problems

Rotating team members

Substitute team members

Replacement team members

Major distractions, e.g. plant accident

Management pressure to limit recommendations

Conflict between stated and actual company policies, e.g. extent to which recommendations should be identified

Lengthy sessions scheduled

Table 5. OSHA Requirements for a System to Address PHA Findings and Recommendations.

Address promptly
Resolve in a timely manner
Document resolutions
Document actions to be taken
Develop written schedule for completion of actions
Complete actions *as soon as possible*
Communicate actions to persons whose work assignments are in the process and who may be affected

Benchmarking Industry Practices for the Use of Alarms as Safeguards and Layers of Protection

Todd Stauffer, PE
exida Consulting
64 N. Main Street, Sellersville, PA
tstauffer@exida.com

Dr. Peter Clarke, CFSE
exida Asia Pacific Pte Ltd
51 Goldhill Plaza, #21-08/09, Singapore
peter.clarke@exida.com

Prepared for Presentation at
American Institute of Chemical Engineers
2013 Spring Meeting
9th Global Congress on Process Safety
San Antonio, Texas
April 28 – May 1, 2013

UNPUBLISHED

Benchmarking Industry Practices for the Use of Alarms as Safeguards and Layers of Protection

Todd Stauffer, PE
exida Consulting
64 N. Main Street, Sellersville, PA
tstauffer@exida.com

Dr. Peter Clarke, CFSE
exida Asia Pacific Pte Ltd

Keywords: Alarm Management, ISA-18.2, Independent Protection Layers, Alarm Rationalization, Safety IPL Alarms, Operator PFD, Operator response to alarms, Safeguards, PHA, LOPA

Abstract

Operator response to alarms is a common risk reduction mechanism considered during layer of protection analysis (LOPA). Industry practices on how to treat alarms as independent protection layers can vary greatly. For example, some companies do not allow any credit to be taken for alarms during a LOPA (zero risk reduction), while others allow up to two orders of magnitude (risk reduction factor of 100, SIL 2) to be taken.

This paper discusses current industry practices around the use of alarms as safeguards and layers of protection as established by a recent benchmark survey of over 200 safety practitioners from around the world. Areas explored in the survey include: typical and maximum claimed risk reduction, considerations used to determine whether an alarm can be credited with risk reduction, how often IPL alarms are determined to be invalid or ineffective in operation, and practices for display and annunciation through a Human-Machine Interface (HMI). Key results and conclusions are presented as well as recommendations on where industry should focus on improvement.

1. Introduction

Alarms and operator response to them are one of the first layers of protection in preventing a plant upset from escalating into a hazardous event. When alarms fail as a layer of protection, catastrophic accidents, such as Milford Haven (UK), Texas City (USA), and Buncefield (UK) can be the result. At the Buncefield Oil Depot, a failure of a tank level sensor prevented its associated high level alarm from being annunciated to the operator. As the level in the tank reached its 'ultimate' high level, a second protection layer, an independent safety switch, failed to trigger an alarm to notify the operator and failed to initiate a trip which would have automatically shut off the incoming flow. The tank overflow and ensuing fire resulted in a £1 billion (1.6 billion USD) loss [1].

Treatment of alarms used as safeguards and protection layers has become an increasingly important topic for companies and regulatory agencies alike. For example, OSHA's Refinery

National Emphasis Program includes provision for citing a refinery if they claim an ineffective alarm as a safeguard or if the alarm design and implementation does not comply with RAGAGEP (Recommended and Generally Accepted Good Engineering Practice) [2].

The standard ANSI/ISA-18.2, *"Management of Alarm Systems for the Process Industries"* (ISA-18.2) provides guidance on how to design, engineer, implement and maintain an alarm system [3]. It is considered RAGAGEP by OSHA, so following its requirements and recommendations is critical for safety practitioners that want to use alarms as a layer of protection.

This paper documents the results from a survey that was conducted to benchmark the current practices used in industry for the management of safety-critical alarms (those that are used as safeguards and/or independent protection layers). The purpose of the paper is to allow companies to compare their own practices against industry benchmarks and best practices, as well as to highlight areas where companies can improve.

2. Survey Demographics

The survey was conducted over the period September 24th – October 5th, 2012. A total of 225 respondents participated in the survey, which consisted of a series of 26 questions. Relevant results are analyzed and presented for the three largest demographic groups described below in order to highlight differences based on region or industry.

Table 1. Survey Demographics

#	Region	% of Respondents	Industry	% of Respondents
1	North America	30%	Oil & Gas	55%
2	Europe	25%	Chemical	23%
3	Asia Pacific	18%	Engineering & Consulting	10%

3. Process Hazard Analysis (PHA)

Process Hazard Analysis (PHA) is a required activity of the IEC 61511 standard on functional safety and the OSHA Process Safety Management (PSM) regulation [4, 5]. There are numerous different techniques that can be used to perform hazard analysis, including What-If, Checklist, Hazard and Operability Study (HAZOP), and Failure Modes & Effects Analysis (FMEA). The HAZOP technique is one of the most commonly used in the process industry [6]. Some of the survey questions are specific to the use of the HAZOP method while others are generic in nature.

3.1 Alarms Identified as Safeguards

Survey respondents answered the following question:

Estimate the number of different alarms in your system that are typically identified as a Safeguard or Recommendation during the Process Hazards Analysis (PHA) process?

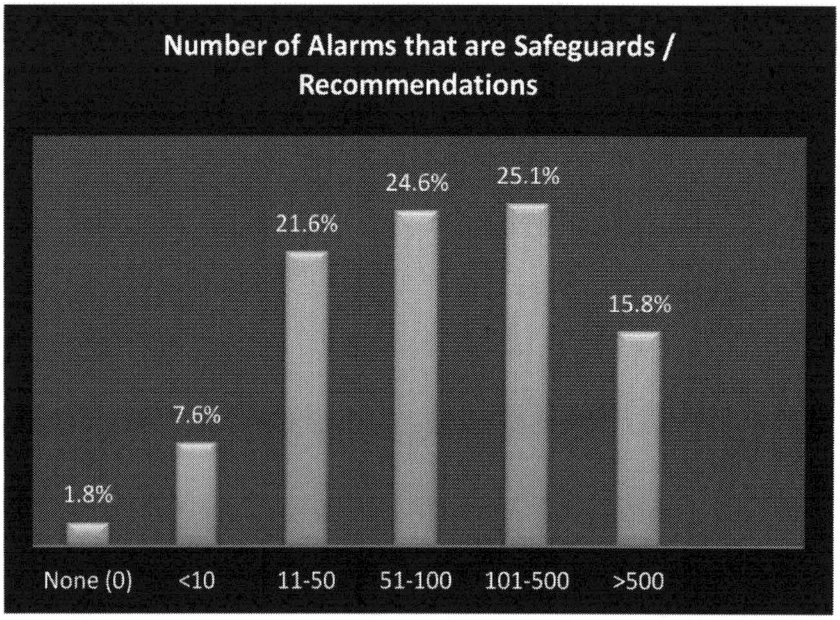

Figure 1. Number of Alarms that are Safeguards or Recommendations

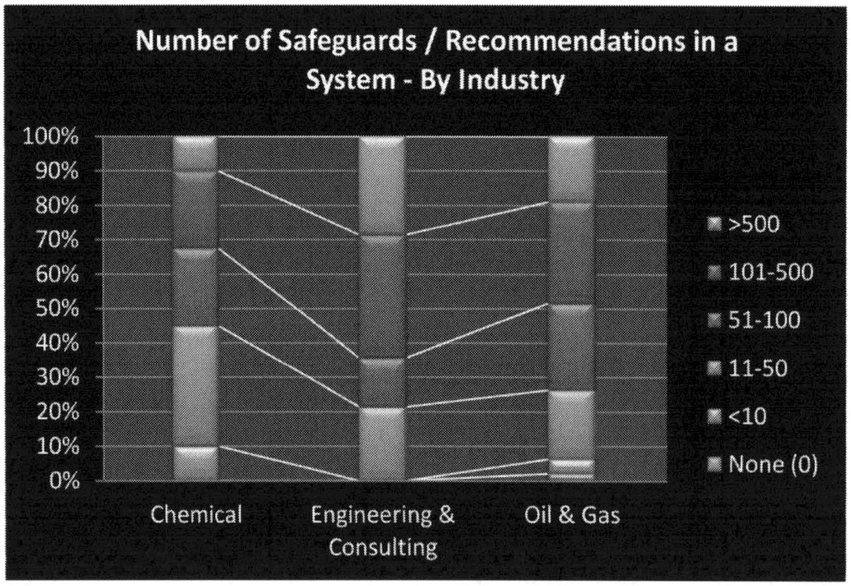

Figure 2. Alarms as Safeguards by Industry

Figure 1 shows that the majority of respondents (>65%) indicated that they have more than 50 alarms in their system that were identified as safeguards / recommendations during a PHA.

Figure 2 shows that the number of alarms identified as safeguards varies considerably by industry. In oil & gas, 73% of the respondents identify more than 50 alarms as safeguards in their system, whereas only 55% for chemical. This can be partly attributed to the size of the respective systems; the most common system size for respondents in the chemical industry was 2,000-5,000 I/O, whereas it was 5,000-10,000 I/O for those in oil & gas.

3.2 Analysis of HAZOP Cause / Consequence Pairs

Survey respondents answered the following question:

Estimate what percentage of cause / consequence pairs (in a Hazard and Operability Study) call for the use of an alarm as safeguard or recommendation?

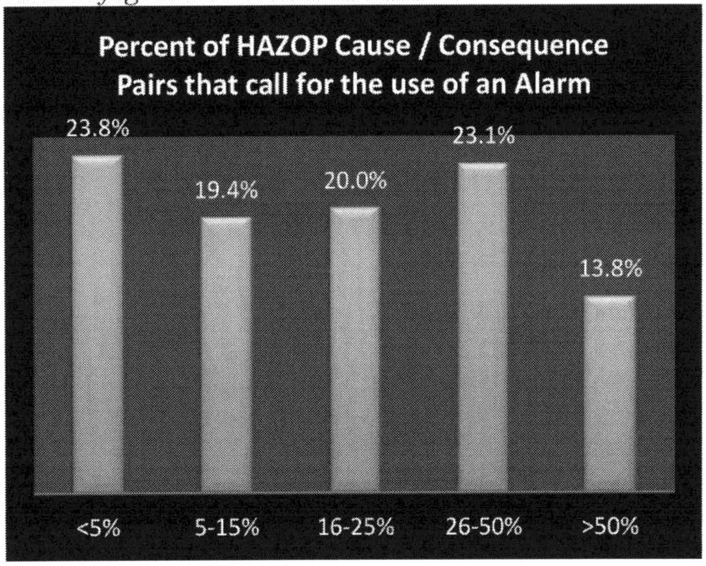

Figure 3. Percent of Cause / Consequence Pairs that call for the Use of An Alarm

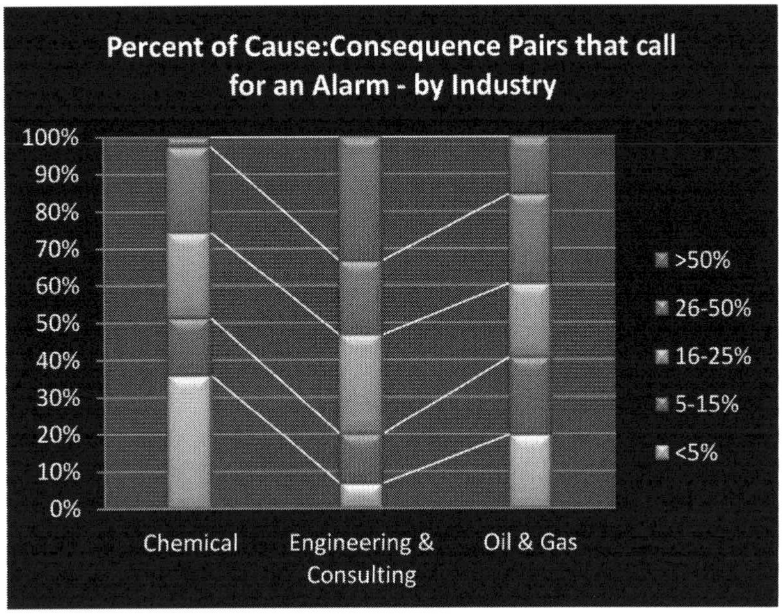

Figure 4. Percent of Cause / Consequence Pairs – By Industry

Figure 3 shows that the responses were relatively evenly distributed between the five choices. This indicates that there is significant variation in the percentage of HAZOP cause / consequence pairs that call for the use of an alarm. The same number of respondents answered "< 5%" as did "26-50%". One likely explanation for these results is variation in how the PHA / HAZOP process is carried out from company to company and the rigor with which all potential safeguards are documented.

Figure 4 and Table 2 shows that there is also significant variation by industry. On one end of the spectrum, 36% of the respondents in the chemical industry answered that <5% (a small minority) of cause consequence pairs call for an alarm, compared to only 7% in engineering and consulting. At the opposite end of the spectrum 33% of engineering and consulting respondents indicated that >50% (a majority) of cause consequence pairs call for an alarm, versus only 3% for chemical.

Table 2. Disparity in Alarms as Percentage of Cause / Consequence Pairs

Industry	% of Cause / Consequence Pairs that call for an alarm	
	< 5% (small minority)	> 50% (majority)
Chemical	36%	3%
Engineering & Consulting	7%	33%
Oil & Gas	20%	15%

3.3 Steps to Ensure Alarms Identified in a PHA are Valid and Effective

Survey respondents answered the following question:

When an alarm is identified as a safeguard or recommendation during a PHA, what steps are typically taken to ensure that it is a valid and effective alarm? Check all that apply.

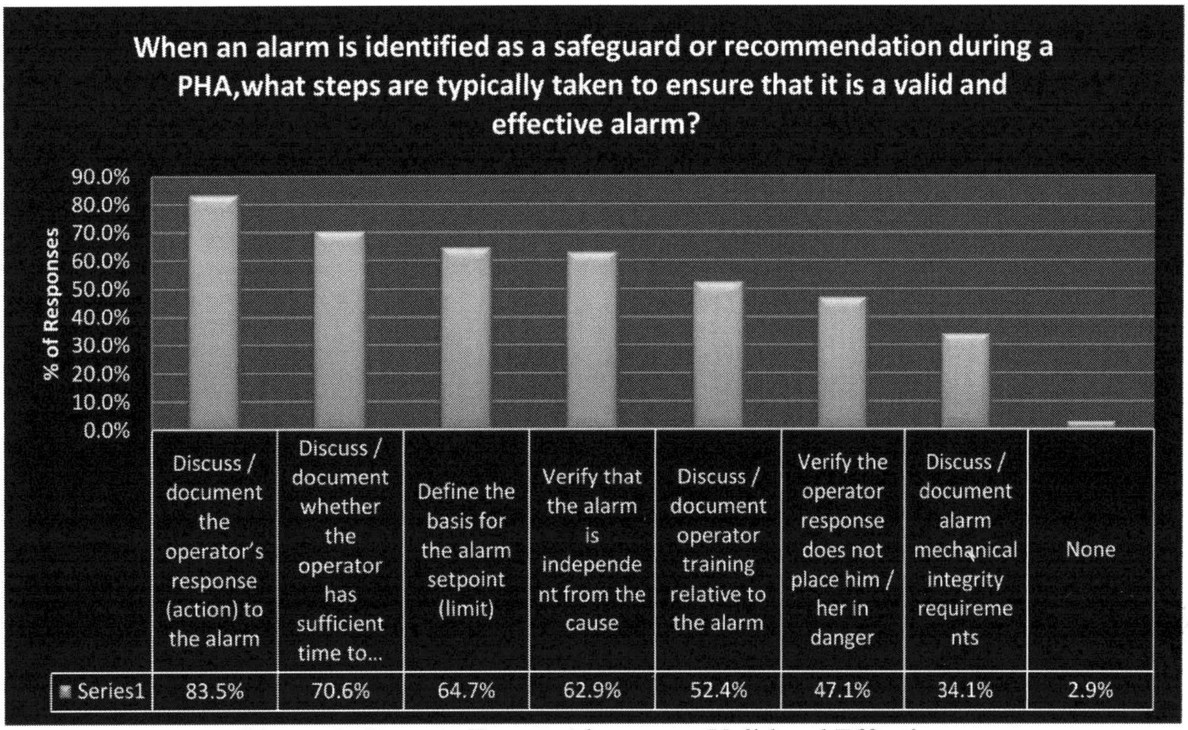

Figure 5. Steps to Ensure Alarms are Valid and Effective

Figure 5 documents the steps that are taken to ensure that an alarm which is identified as a safeguard or recommendation is valid and effective. Best practices such as those documented in ISA-18.2, by exida, and by the Center for Chemical Process Safety (CCPS) would suggest that the following activities at a minimum should be performed for an alarm that is used as a safeguard:

- **Discuss / document the operator's response (action) to the alarm** – According to ISA-18.2, if the alarm does not require an operator action, then it should not be considered a valid alarm. During the rationalization process, each alarm is subjected to this review [3].
- **Discuss / document whether the operator has sufficient time to respond** – This is another criterion which is reviewed during the rationalization process. If an operator does not have sufficient time to respond to prevent the consequences, then the alarm will not be effective and should not be considered a safeguard [6, 7, 8].
- **Verify that the alarm is independent from the cause** – This must be TRUE for the alarm to be considered a valid Independent Protection Layer, so it would make sense that it should also be applied to a safeguard when appropriate [6, 7, 8].
- **Verify the operator response does not place him / her in danger** – If the operator's response to the alarm places them in danger, then the alarm should not be considered a safeguard. The survey indicated that over half the respondents (52.9%) do not apply this criterion [8].

If the four criteria described above are accepted as best practice, then 100% of the respondents should have indicated that these steps are taken. Instead only 83.5%, 70.6%, 62.9% and 47.1% respectively indicated that they follow these best practices. Thus there is a gap between the actual practices used in industry versus those that are recommended and accepted as best practices. By not applying alarm management best practices upfront during the PHA / HAZOP process it is more likely that some of the alarms identified as safeguards will be proven to be invalid / ineffective during alarm rationalization or operation.

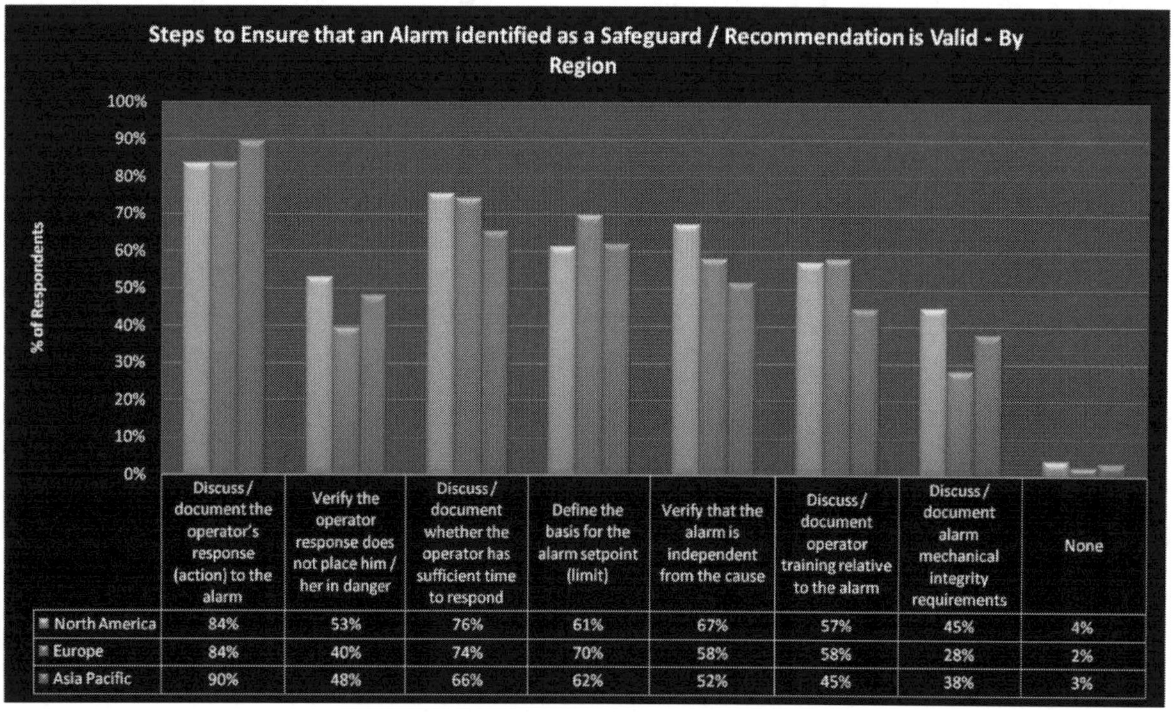

	Discuss / document the operator's response (action) to the alarm	Verify the operator response does not place him / her in danger	Discuss / document whether the operator has sufficient time to respond	Define the basis for the alarm setpoint (limit)	Verify that the alarm is independent from the cause	Discuss / document operator training relative to the alarm	Discuss / document alarm mechanical integrity requirements	None
North America	84%	53%	76%	61%	67%	57%	45%	4%
Europe	84%	40%	74%	70%	58%	58%	28%	2%
Asia Pacific	90%	48%	66%	62%	52%	45%	38%	3%

Figure 6. Steps to Ensure Alarms are Valid and Effective – By Region

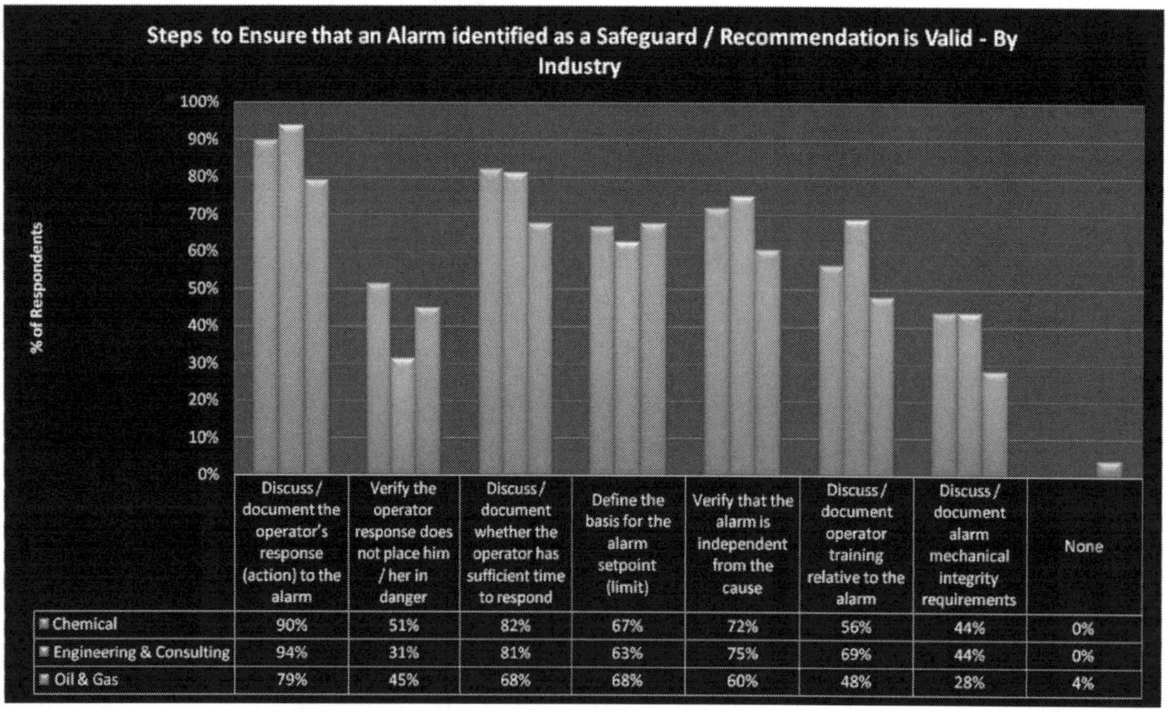

	Discuss / document the operator's response (action) to the alarm	Verify the operator response does not place him / her in danger	Discuss / document whether the operator has sufficient time to respond	Define the basis for the alarm setpoint (limit)	Verify that the alarm is independent from the cause	Discuss / document operator training relative to the alarm	Discuss / document alarm mechanical integrity requirements	None
Chemical	90%	51%	82%	67%	72%	56%	44%	0%
Engineering & Consulting	94%	31%	81%	63%	75%	69%	44%	0%
Oil & Gas	79%	45%	68%	68%	60%	48%	28%	4%

Figure 7. Steps to Ensure Alarms are Valid and Effective – By Industry

Figures 6 and 7 present the results based on region and industry. In Figure 6 the percent of North American respondents which indicated that they discussed mechanical integrity (MI) requirements (45%) was significantly higher than Europe (28%) and Asia Pacific (38%). This is likely from the strength of OSHA in the US in driving compliance to their Process Safety Management (PSM) regulation 1910.119 which includes requirements for the creation of a mechanical integrity program (a management system assuring equipment is inspected, maintained, tested and operated in a safe manner)[5, 7].

Of interest in Figure 7 is that engineering & consulting and chemical had higher response scores (greater compliance to best practices) than oil & gas for all categories except for one. This indicates that the understanding, acceptance and adoption of best practices may be higher here than in oil & gas.

3.4 Treatment of PHA Results

Survey respondents answered the following question:

After the PHA or HAZOP has been completed, what is done with the requirements for alarms identified as safeguards or recommendations? Check all that apply.

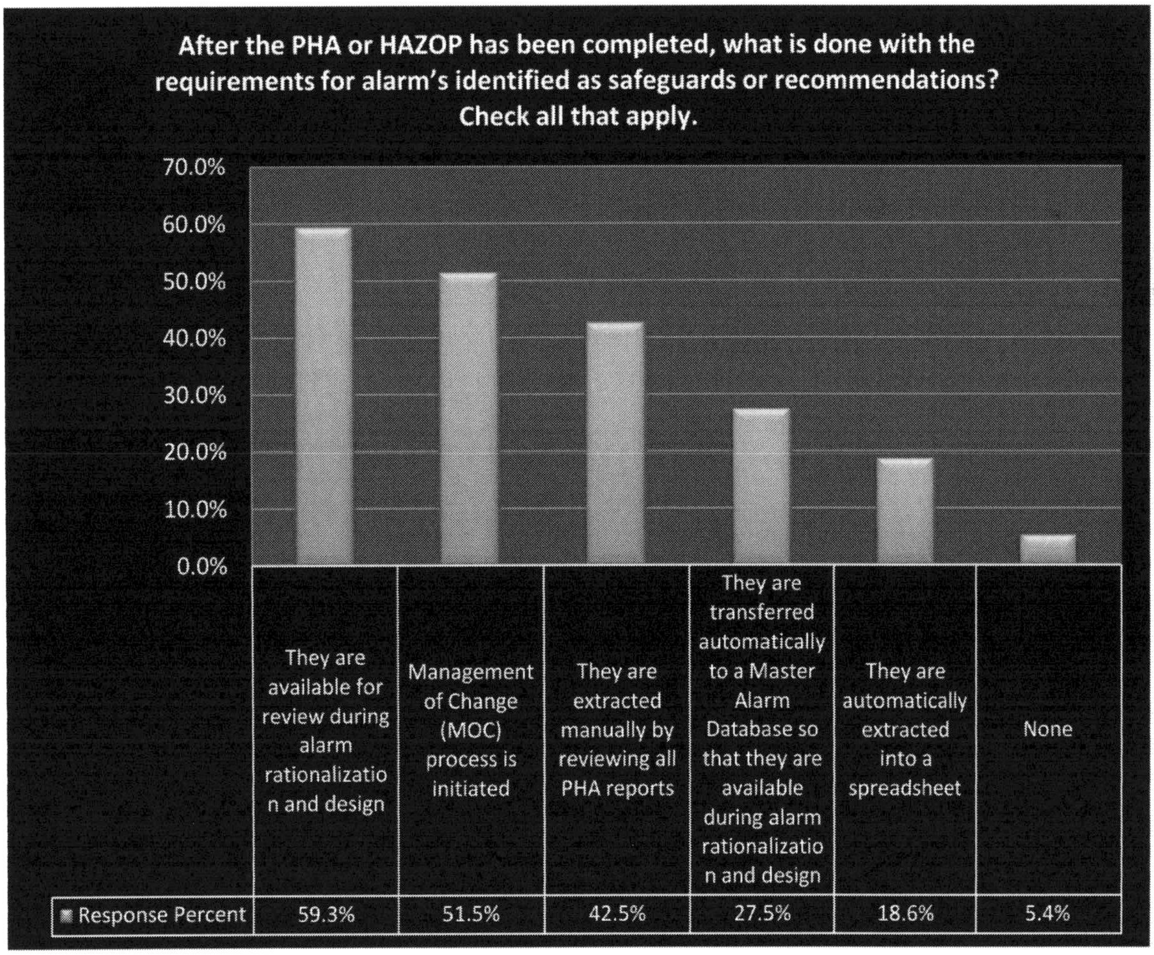

Figure 8. Treatment of PHA Results

The survey indicates that only 59.3% of respondents make the PHA results available during alarm rationalization. This figure should be 100% as alarm rationalization will revisit some of the topics that are covered during the PHA (establishing likely causes and consequences). This will improve the efficiency of the alarm rationalization and help ensure consistency between the alarm design and the PHA.

4. Layer of Protection Analysis (LOPA)

Layer of Protection Analysis is one of the most commonly used techniques for risk analysis. It is a method of analyzing the likelihood (frequency) of a harmful outcome event based on initiating event frequency and on the probability of failure of a series of independent protection layers capable of preventing the harmful outcome [6].

The primary goal of a LOPA is to determine if there are adequate protective devices or features in the process to produce a tolerable risk level. These protective devices or features are called Protection Layers or Independent Protection Layers (IPLs). Examples of potential protection layers include the mechanical integrity of a vessel, control loops and trips within the basic process control system (BPCS), operator intervention, a safety instrumented function, and physical relief devices.

It is important to note the difference between a safeguard and a layer of protection. A safeguard is any device, system or action that would likely interrupt the chain of events following an in initiating event. The benefit of some safeguards may not be able to be easily quantified because of lack of data, or uncertainty of whether it meets specific criteria such as independence, effectiveness, and auditability. An independent protection layer is a safeguard whose effectiveness can be quantified and which meets well-defined criteria. All IPLs are safeguards, but not all safeguards are IPLs [8].

4.1 Origin of Alarms Identified in a LOPA

Survey respondents answered the following question:

What percentage of the alarms that are considered during a Layer of Protection Analysis (LOPA) were identified during a PHA?

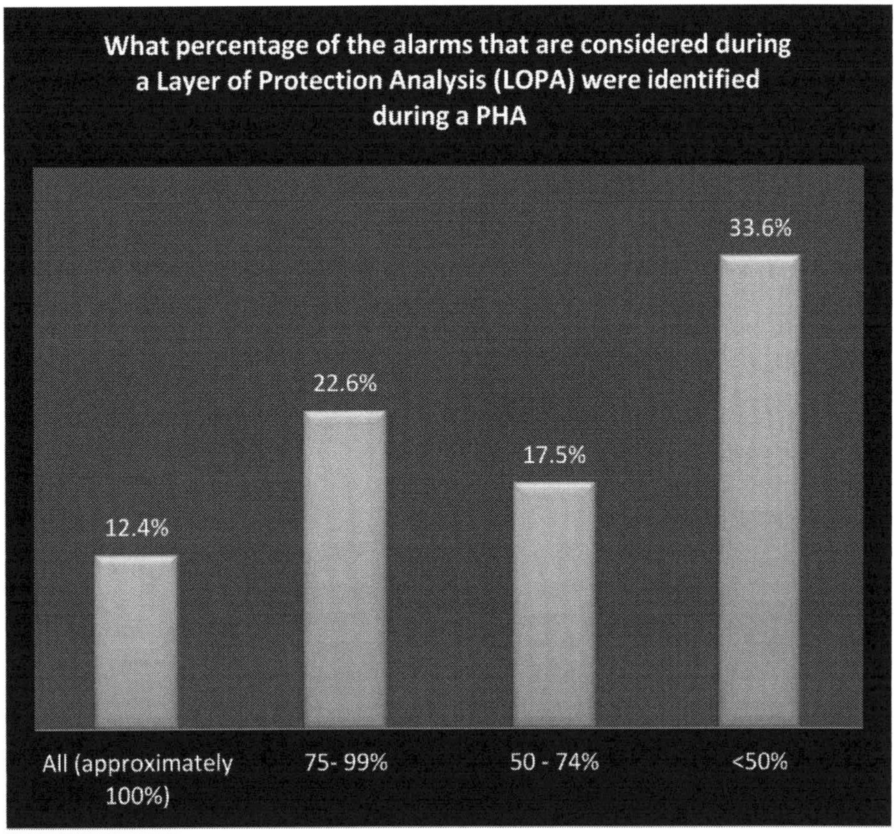

Figure 9. Percentage of LOPA Alarms Originating during a PHA

After the process hazards analysis has been completed, the results and recommendations are reviewed to determine which scenarios require further analysis to determine if there are adequate layers of protection, or if safety instrumented functions (SIF) will be needed to properly manage the risk. One industry reference defines a safeguard as "a potential protection layer that has yet to be evaluated in a LOPA to determine effectiveness and independence" [6]. Thus it would be expected that ideally 100% of the alarms that are considered in a layer of protection analysis would have first been identified as a safeguard / recommendation in the PHA. Figure 10 shows that this is far from the case in practice. Only 12.4% of the respondents indicated that all (100%) of the alarms in the LOPA had come from the PHA. Furthermore, 33.6% indicated that less than 50% of the time was the LOPA alarm identified during the PHA.

It is certainly possible that the LOPA may legitimately identify some alarms that were not considered during PHA. Allowing for this, one could consider the 75-99% response as acceptable. This leaves 51% of the respondents which appear to frequently identify new alarms during LOPA that were "missed" during the PHA. This would seem to indicate poor PHA practices are being used. Failing to identify alarms during a PHA could signal various issues, such as a lack of thoroughness, lack of documenting all safeguards in order to save time, or a lack of understanding of the process.

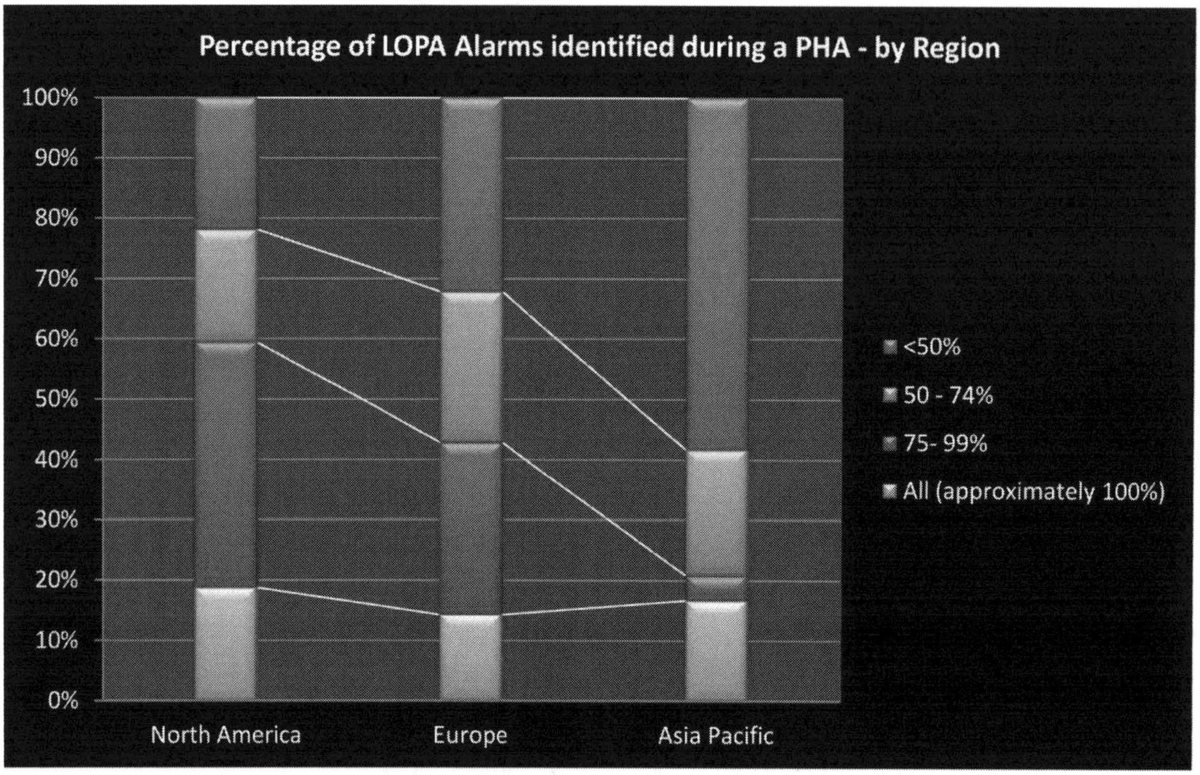

Figure 10. Percentage of LOPA Alarms Originating during a PHA – By Region

The results in Figure 10 demonstrate a significant disparity in the quality of the PHAs being conducted based on region. The percentage of "poor" PHA applications ranged from 41% in North America to 79% in Asia Pacific (where "poor" is defined as those that answered <50% or 50-74%). Note that the variation was much less significant when analyzed by industry.

4.2 Typical Risk Reduction for a Safety IPL Alarm

Survey respondents answered the following question:

What level of risk reduction (RRF) do you typically take for a Safety IPL alarm?

The effectiveness of an independent protection layer is typically characterized by assigning a probability of failure on demand (PFD), which is defined as the probability that it will fail to perform a specified function when called upon [8]. The risk reduction factor (RRF), which is a measure of how much a protective function reduces the frequency of the hazardous event, is the inverse of PFD [7].

$$\mathbf{RRF = 1 / PFD} \quad \text{[Eq. 1]}$$

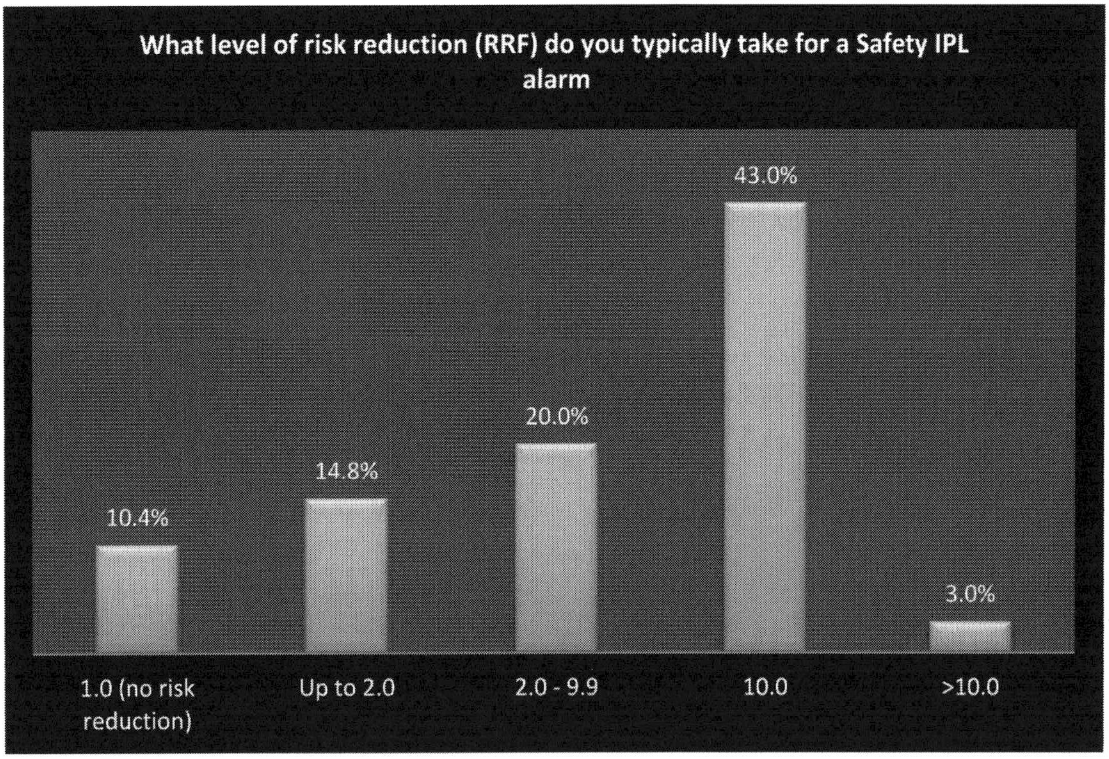

Figure 11. Typical Level of Risk Reduction for a Safety IPL Alarm

The level of risk reduction that can be taken for a Safety IPL Alarm (an alarm used as an independent protection layer) is an area of debate in the safety community. The debate originates because of the significant disparity that exists from plant to plant, unit to unit and person to person in the ability of an operator to prevent a hazardous situation form developing into an accident.

Figure 11 shows that the most common risk reduction factor taken for a Safety IPL Alarm is 10.0 (with 43% of the respondents). This corresponds to the risk reduction that is most commonly cited in the literature [4, 8]. Table 3 shows the correspondence between RRF, PFD, and Safety Integrity Level (SIL). It should also be noted that 10 % of the respondents claim no risk reduction for a Safety IPL alarm, while 20% claim a RRF between 2.0 and 9.9.

Table 3. Correspondence between RRF, PFD and SIL [4]

Risk Reduction Factor (RRF)	Probability of Failure on Demand (PFDavg)	Safety Integrity Level (SIL) potentially achievable
≤ 10	$\geq 10^{-1}$ to $\leq 10^{0}$	SIL 0
> 10 to ≤ 100	$\geq 10^{-2}$ to $< 10^{-1}$	SIL 1
> 100 to $\leq 1{,}000$	$\geq 10^{-3}$ to $< 10^{-2}$	SIL 2
$> 1{,}000$ to $\leq 10{,}000$	$\geq 10^{-4}$ to $< 10^{-3}$	SIL 3

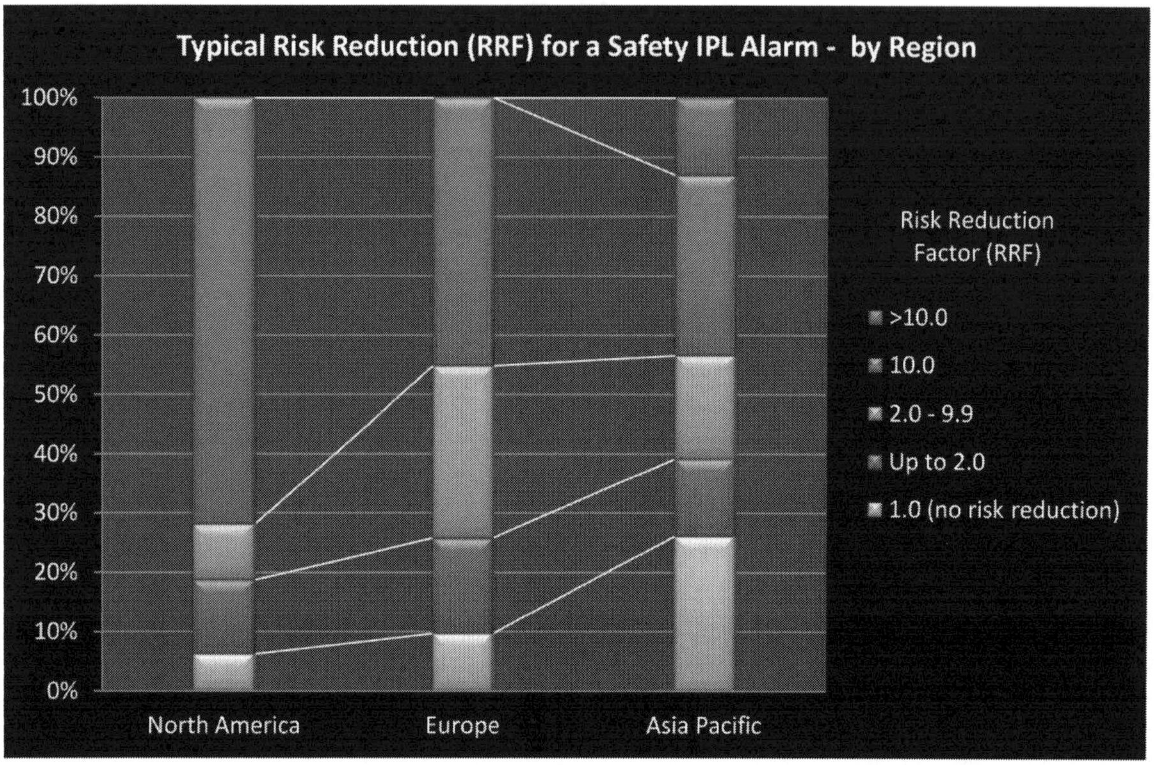

Figure 12. Typical Level of Risk Reduction for a Safety IPL Alarm– By Region

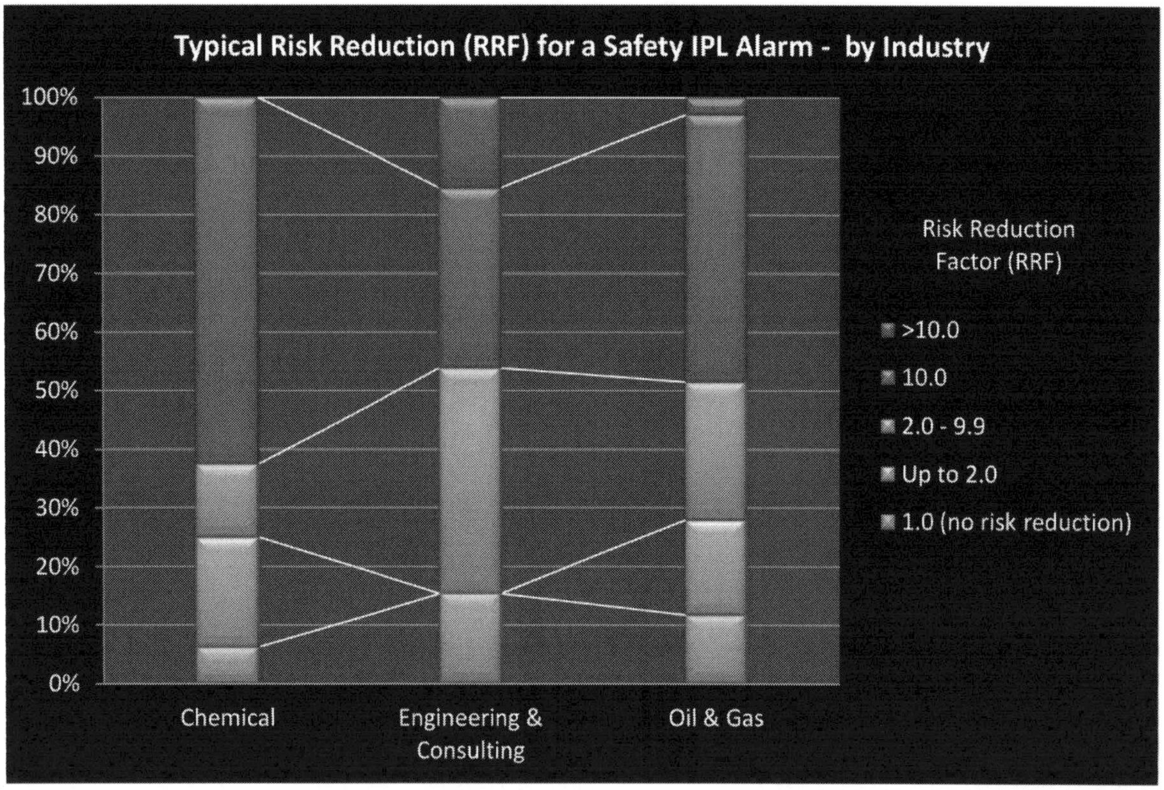

Figure 13. Typical Level of Risk Reduction for a Safety IPL Alarm– By Industry

Figures 12 and 13 show that the level of risk reduction varies considerably by region and by industry. For example in North America a clear majority (72%) use an RRF of 10, while in Asia Pacific only 30% use an RRF of 10. In Asia Pacific, a large percentage of the respondents are either very conservative (26% claim no risk reduction) or very aggressive (13% claim a risk reduction greater than 10).

It is also interesting to note that numerous respondents in the Engineering & Consultancy sector claimed to use RRFs that are not powers of 10 (i.e. in the 2.0~9.9 range). This suggests that quantitative LOPA techniques—which can make use of such RRF values—may be used frequently within this sector.

Risk reduction factors greater than 10.0 (PFD < 0.1) should be used sparingly if ever for Safety IPL alarms. As shown in Table 4, there are very few situations when it would be appropriate to use such a value. When it is believed to be appropriate, it is necessary to document a sound technical basis for that conclusion.

Table 4 – Simplified Technique for Estimating Operator Response [6]

Category	Description	Probability that Operator responds successfully	PFD	RRF
1	**Normal Operator Response** – In order for an operator to respond normally to a dangerous situation, the following criteria should be true: • Ample indications exist that there is a condition requiring a shutdown • Operator has been trained in proper response • Operator has ample time (> 20 minutes) to perform the shutdown • Operator is ALWAYS monitoring the process (relieved for breaks)	90%	0.1	10
2	**Drilled Response** – All of the conditions for a normal operator intervention are satisfied and a "drilled response" program is in place at the facility. • Drilled response exists when written procedures, which are strictly followed, are drilled or repeatedly trained by the operations staff. • The drilled set of shutdowns forms a small fraction of all alarms where response is so highly practiced that its implementation is automatic • This condition is RARELY achieved in most process plants	99%	0.01	100
3	**Response Unlikely / Unreliable** – NOT ALL of the conditions for a normal operator intervention probability have been satisfied	0%	1.0	1

Some alarm management practitioners have proposed that even a risk reduction factor of 10 should not be applied blindly without ensuring that specific alarm management requirements are / will be met, such as the following:

- The alarm system must be rationalized.
- Alarm system performance must be measured and proven to be adequate (based on industry-accepted KPIs) [9].

4.3 Maximum Risk Reduction for a Safety IPL Alarm

Survey respondents answered the following question:

In your experience, what is the maximum level of risk reduction (RRF) that has been taken for a Safety IPL alarm?

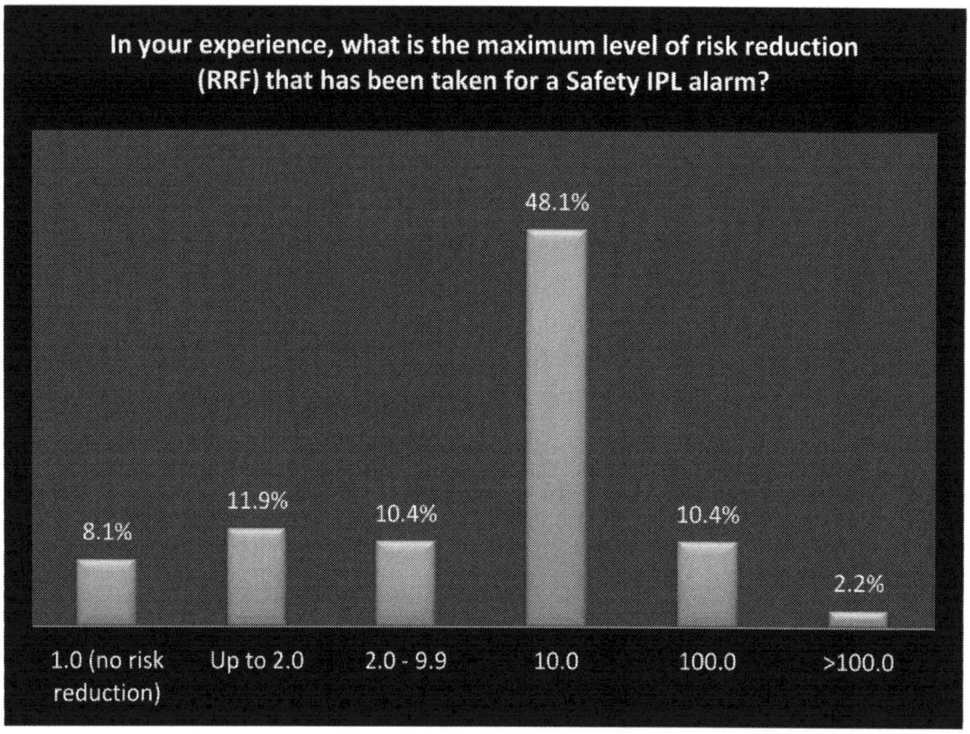

Figure 14. Maximum Risk Reduction for a Safety IPL Alarm

Figure 14 shows that a risk reduction factor of 10 was again the most popular response (48%). It is interesting to note that the percent of respondents that indicated 10.0 was the maximum risk reduction taken (48%) was slightly greater than the amount that indicated 10.0 was the typical value taken (43%) from the previous question. It also of interest to note that 12.6% of the respondents indicated a maximum RRF of 100.0 or greater.

4.4 *Considerations for Determining When an Alarm Can be Credited with Risk Reduction*

Survey respondents answered the following question:

What considerations are used to determine whether an alarm can be credited with risk reduction? Check all that apply.

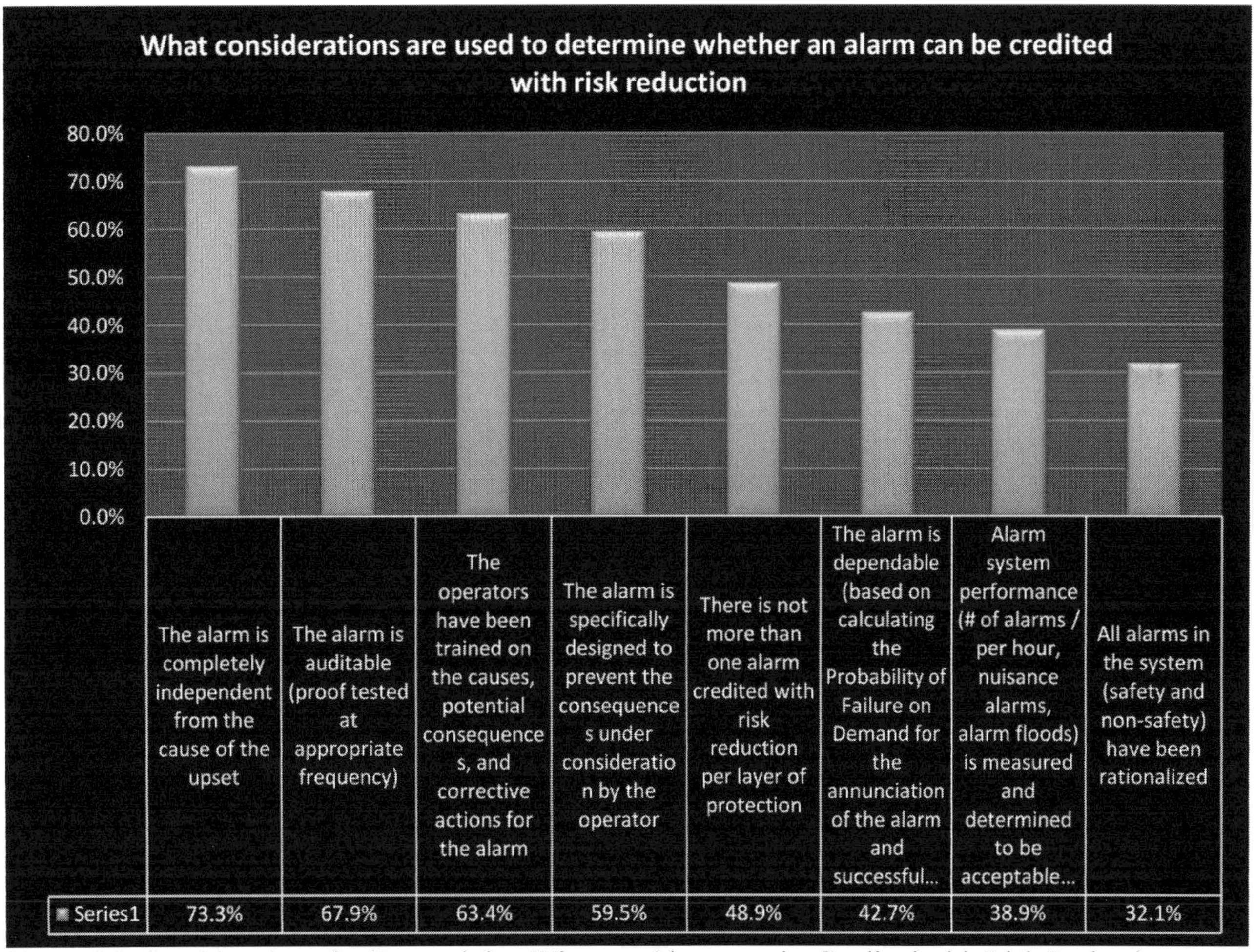

Figure 15. Considerations for Determining When an Alarm can be Credited with Risk Reduction

The general criteria for determining when a safeguard can be considered an IPL are well established in the literature and include the following:

Table 5 – Survey of General Criteria Used to Determine when a Safeguard can be used as an IPL

Layer of Protection Analysis [8]	**Practical SIL Target Selection [6]**	**Guidelines for Safe & Reliable Instrumented Protective Systems [7]**
Independent	Independent	Independence
Auditable	Auditable	Auditability
Effective	Dependable	Reliability, Integrity
	Specific	Functionality
		Access Security
		Management of Change

The generic criteria above have been used to create specific considerations that should be taken into account to ensure that an alarm can be credited with risk reduction (represented in Figure 15). A more detailed discussion about criteria applied to alarms can be found in Appendix A.

The presence of "nuisance" alarms – which are alarms that annunciate excessively, unnecessarily, or do not return to normal after the correct response is taken – can interfere with the operator's ability to detect and respond to safety IPL alarms. Standing alarms (lasting > 24 hours) and chattering alarms (points that go needlessly in and out of alarm on a frequent basis) are nuisance alarms that clutter the operator's display making it more difficult to detect a new alarm and increasing the chances that they might miss a critical alarm.

Alarm rationalization, which is the process of reviewing potential or existing alarms to justify that they meet the criteria for being an alarm, is a technique for ensuring the integrity of the alarm system and eliminating problems such as nuisance alarms, alarm overload and alarm floods . It includes defining and documenting the design attributes (such as priority, limit, type and classification) as well the cause, consequence, time to respond, and recommended operator response.

Since all of the criteria shown in Figure 15 have been cited as recommended best practices in the literature, it can be concluded that a large portion of safety practitioners are NOT following industry recommended practices (else the scores would be close to 100% for each consideration).

4.5 Invalid & Ineffective Safety IPL Alarms

Survey respondents answered the following question:

How often do you find that an alarm identified as an IPL is not valid, or is ineffective (does not provide the level of risk reduction expected)?

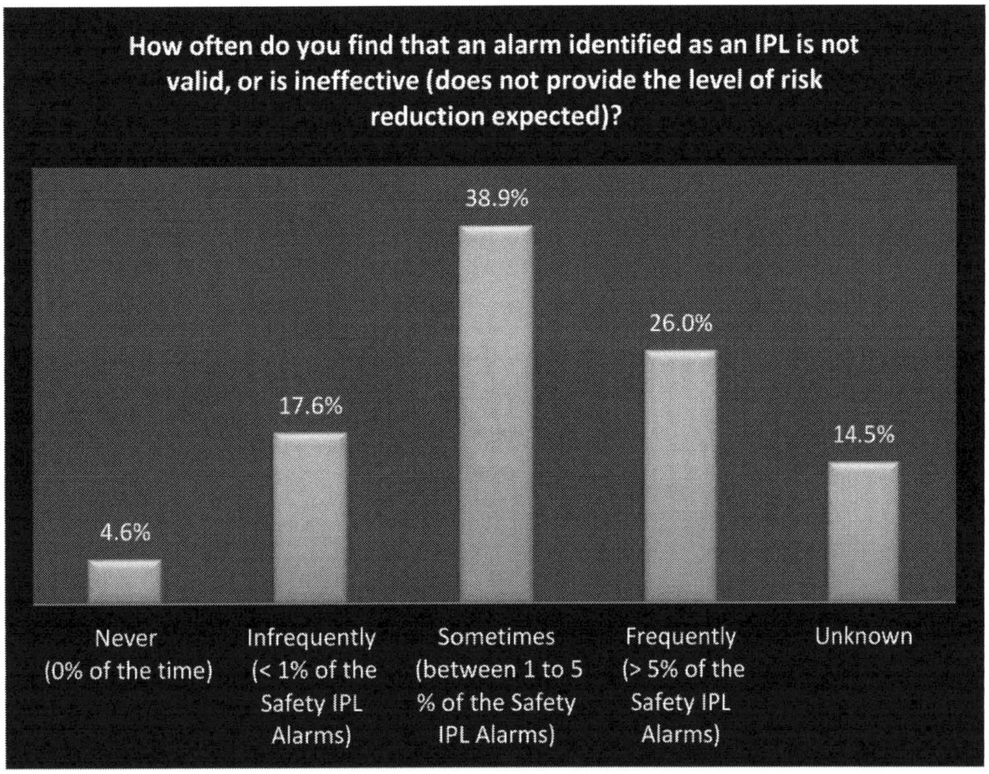

Figure 16. Frequency of Ineffective Safety IPL Alarms

Figure 16 shows how often a Safety IPL Alarm is found to be ineffective at providing the expected level of risk reduction. 65% of the respondents indicated that sometimes / frequently they find that an alarm is an ineffective IPL. This could create a situation where the actual risk reduction no longer meets or exceeds the company-defined tolerable risk level.

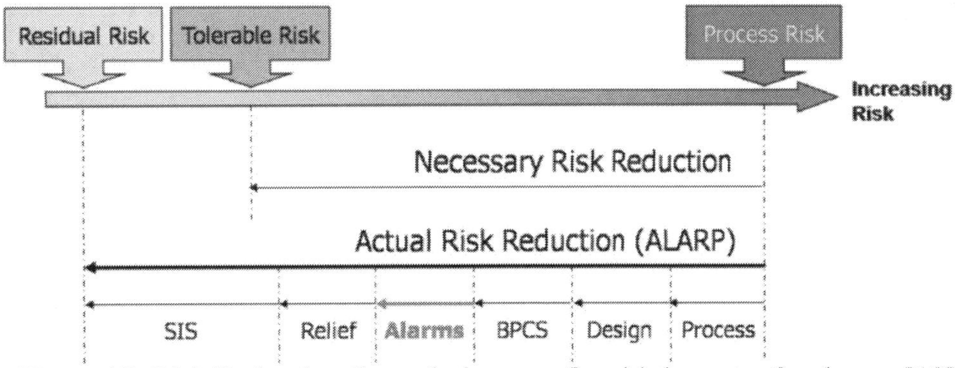

Figure 17. Risk Reduction through the use of multiple protection layers [10]

Figure 17 illustrates how the loss of risk reduction from an ineffective IPL alarm can have a ripple effect on the requirements for other layers of protection such as a safety instrumented function in an SIS. The higher the SIL, the more complicated and expensive is the Safety Instrumented System (SIS). A higher SIL may also require more frequent proof testing, which adds cost and can be burdensome in many plants [11].

One could surmise that this finding is partly caused by the gap in following best practices that exists as illustrated by Figure 15. A detailed discussion of failure modes of Safety IPL alarms is the subject of another paper [9].

4.6 Prioritizing Safety IPL Alarms

Survey respondents answered the following question:

What statement best describes how the priority of Safety IPL alarms are assigned?

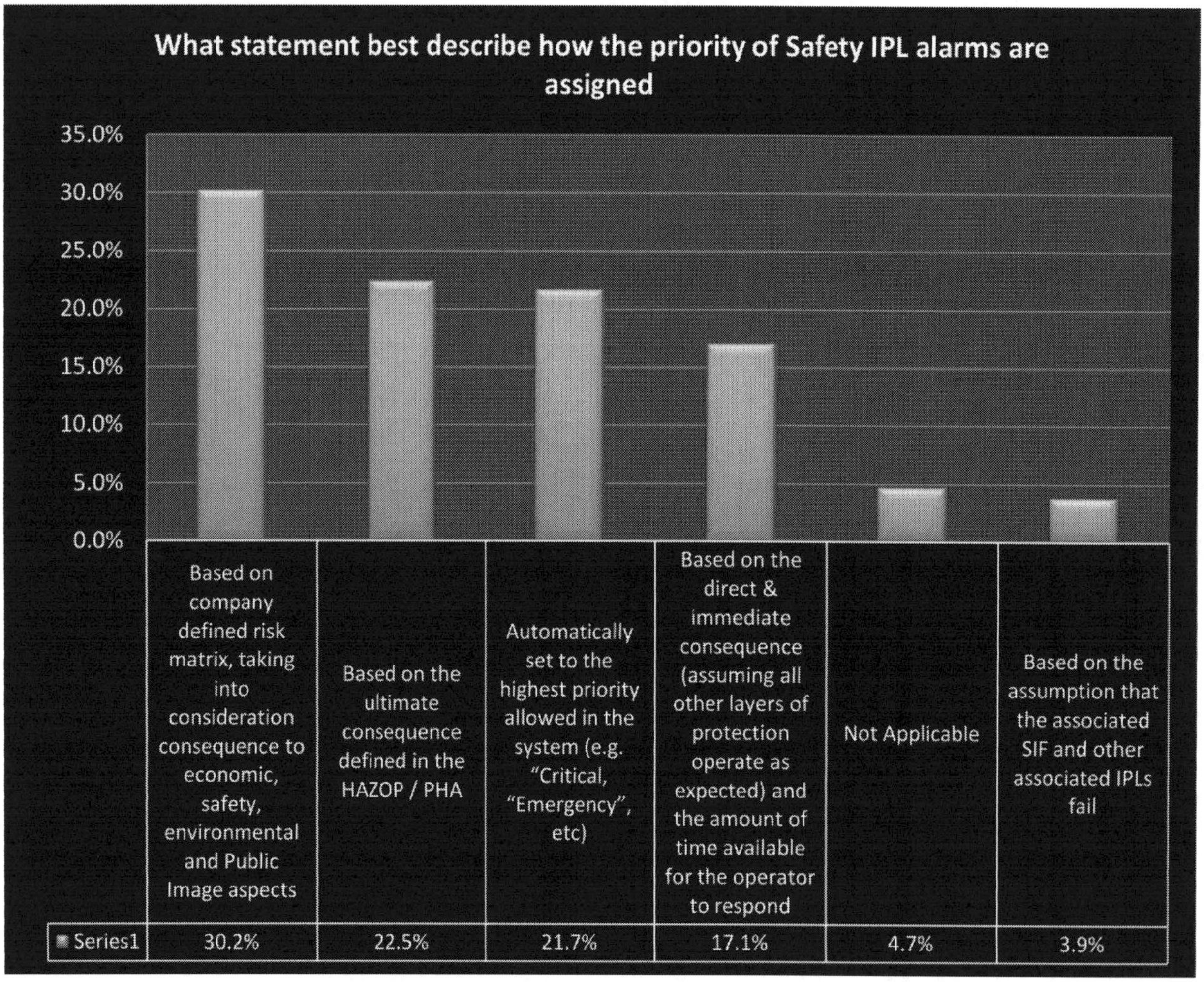

Figure 18. Methodology for Prioritizing Safety IPL Alarms

Alarm priority represents the importance assigned to an alarm within the alarm system to indicate the urgency of response. It helps the operator to know to which alarm to respond to first. Alarm priority is typically determined based on the severity of the potential consequences (in areas such as personnel safety, equipment damage, environmental, economic loss) and the time available to respond as shown in Table 6. Analysis of the severity of consequences is an activity that is common within the safety lifecycle. For a safety IPL alarm it is important to work with the direct (proximate) consequences and not the ultimate consequences which could occur after a series of failures [12, 13].

Consequences of Inaction				
Impact Areas	Consequence Category 1 (None)	Consequence Category 2 (Minor)	Consequence Category 3 (Major)	Consequence Category 4 (Severe)
Personnel	None	Minor or no injury, no lost time.	One or more severe injury(s).	Fatality or permanently disabling injury.
Environmental	None	Minor	Release which results in agency notification, permit violation or fine.	Significant release with serious offsite impact.
Financial	None	Impact to equipment or production <$500K.	Impact to equipment or production $500K to $5M	Impact to equipment or production > $5M
Operator Urgency (Time Available to Respond)				
Not Urgent (> 30 mins)	No Alarm	Re-engineer the alarm for urgency		
Prompt (15 to 30 mins)	No Alarm	Low	Low	Medium
Rapid (5 to 15 mins)	No Alarm	Low	Medium	High
Immediate (< 5 mins)	No Alarm	Medium	High	High*

Table 6. Example Alarm Priority Matrix

Figure 18 provides a view into how Safety IPL alarms are prioritized. As shown by Table 7, 48% of the respondents indicated that they use prioritization criteria which do not follow alarm management best practices.

Prioritization Criteria	**%**	**Compliance with Best Practices**
Based on company defined risk matrix, taking into consideration consequence to economic, safety, environmental and public image aspects	30.2%	***YES***
Based on the ultimate consequence defined in the HAZOP / PHA	22.5%	***NO***
Automatically set to the highest priority allowed in the system (e.g. "Critical, "Emergency", etc)	21.7%	***NO***
Based on the direct & immediate consequence (assuming all other layers of protection operate as expected) and the amount of time available for the operator to respond	17.1%	***YES***
Based on the assumption that the associated SIF and other associated IPLs fail	3.9%	***NO***

Table 7. Alarm Prioritization Results and Compliance with Best Practices

5. Human Machine Interface (HMI) Practices for Safety IPL Alarms

Safety IPL alarms are communicated to the operator through the Human Machine Interface (HMI). Once the alarm is annunciated, a series of steps must be performed by the operator to prevent escalation of the hazardous scenario and bring the process back to the normal operating range (reference) as shown in Figure 19.

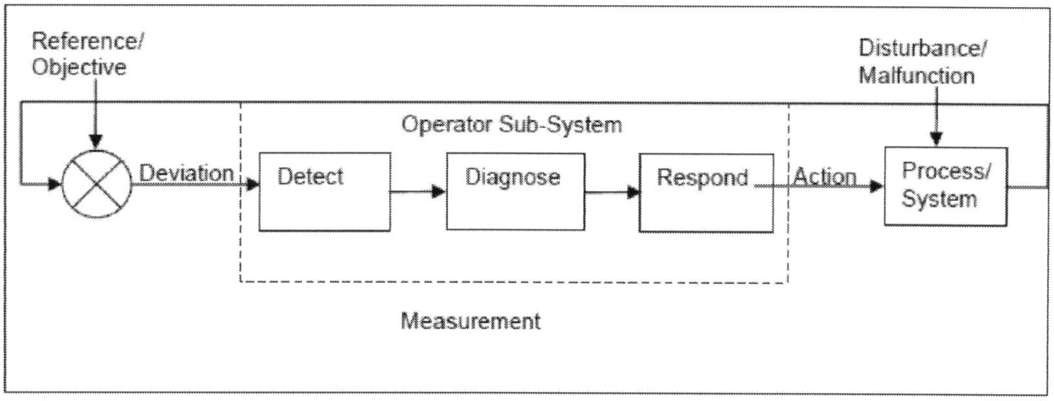

Figure 19 – Feedback Model of Operator Process Interaction [3]

For a successful outcome, the operator must proceed quickly through three stages of activity:

a) the deviation from desired normal operation is detected,
b) the situation is diagnosed and the corrective action determined,
c) the action is implemented to compensate for the disturbance.

The operator also continues to monitor the measurement as it returns to normal.

A well designed HMI should support situation awareness and ensure that the operator is able to quickly and repeatably ***detect***, ***diagnose,*** and ***respond*** within the operator response time. Operator response time represents the time from the activation of the alarm until the last moment the operator action will prevent the consequence (i.e., time available) [3]. Poor graphics, including alarm depiction deficiencies, have been identified as contributing factors to several major industrial accidents (such as Buncefield).

5.1 Display of Safety IPL Alarms

Survey respondents answered the following question:

What statement(s) best describes your current practice for display of Safety IPL alarms? Check all that apply.

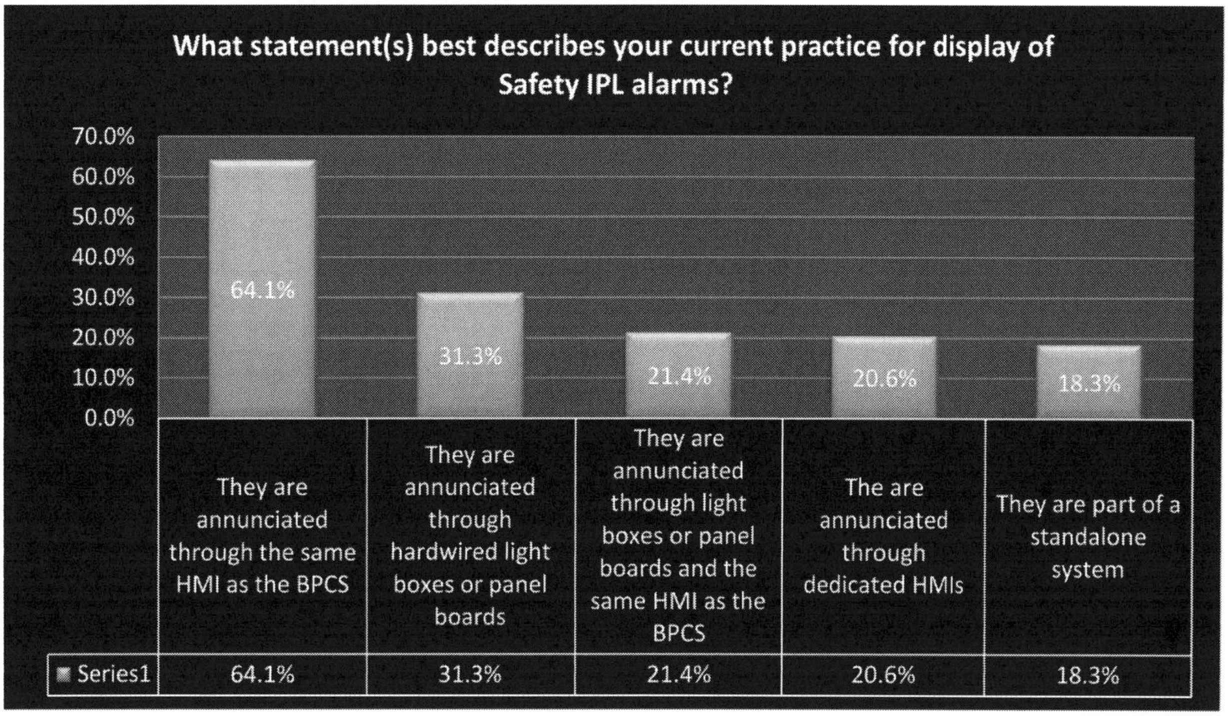

Figure 20. Display of Safety IPL Alarms

Safety IPL alarms and information can be presented to the operator in a number of different ways, including:

- Graphic displays on the basic process control system (BPCS) operator interface,
- Dedicated graphic displays on stand-alone video display units,
- Panel mounted graphic displays, and
- Panel mounted annunciators.

Figure 20 illustrates that a variety of architectures are used for the display of safety IPL alarms, with the most popular (64.1%) being annunciation through the same HMI as the BPCS.

Selected recommended best practices for display design include the following:

- Lightbox alarms, which provide an independent alarm display that can be typically seen by multiple operators within the control room, should be replicated in the BPCS interface for acknowledgement and logging purposes.
- Lightbox annunciators should be located close to the operator's work station or work areas so that it is visible from all locations where its information would be considered important [7].
- Graphic displays should be designed to maximize operator situation awareness and "pattern recognition" to aid in operator response.
- Graphic displays should be designed so the visibility of information is related to its operational importance; background information should be given low visibility, normal

plant measurements a medium visibility and abnormal conditions (values and states) should have the highest visibility.

- It is important that alarm state indications represent the presence of an alarm using not only color, but also symbols, patterns and/or text (8-12% of the male population is color blind).
- Alarm colors should be reserved for alarms only and not used for other functions within the HMI (such as process piping or equipment status). Alarm color coding should reflect the priority of the alarm.

5.2 Alarm Response Procedures

Survey respondents answered the following question:

Do you provide Alarm Response Procedures to the operator for safety IPL alarms?
If Yes, please indicate the format:

Alarm Response Procedures for Safety IPL alarms	**%**	**Format**	**%**
(Yes) - Provided	74%	Paper manuals	54%
		On screen display called up in context within the HMI	28%
		Call up files or displays on a dedicated computer (other than the HMI)	18%
(No) - Not Provided	26%		

Table 8. Use of Alarm Response Procedures for Safety IPL Alarms

Alarm response procedures typically include the following information:

- Likely cause(s) of the alarm
- Potential consequences of inaction
- Corrective action that is required by the operator to prevent the consequence
- Time available to respond
- Confirmation / Verification of the alarm condition [9, 11].

As shown in Table 8, 26% of the respondents do not provide alarm response procedures to the operator to help them respond to Safety IPL Alarms. This is inconsistent with the practices that should be followed to ensure that the operator response is effective and reliable / dependable [6, 7, 8].

For those that do provide alarm response procedures, 54% of the respondents indicated that they are provided in paper format. The use of printed (paper) manuals can be ineffective if they are not within immediate reach of the operator, are not kept up-to-date or require significant time for the operator to locate the relevant procedure. The ability to display the alarm response procedures in context within the HMI, which was selected by 28% of the respondents, is the most effective format and should be considered a best practice [9, 11]. Best practices assert that, "Operator response integrity can be improved by displaying operator action on request" [7].

6. Conclusion

Operator response to alarms can be used to reduce risk as a safeguard or as an independent protection layer. Survey results indicate that there is significant variation in the practices employed within industry for the management of safety-critical alarms. In some cases these variations are more significant when analyzing based on industry or region.

Analysis of survey results also revealed that there is significant room for improvement when it comes to the adoption of, and compliance with, industry best practices. In particular the following areas were identified:

- Improving the rigor and thoroughness of PHAs so that, for example, all alarm safeguards are identified and documented
- Verifying that an alarm identified as a safeguard or recommendation is likely to be valid and effective
- Ensuring that alarms credited with risk reduction meet the criteria established for them to be independent protection layers as cited in industry best practices [6, 7, 8]
- Understanding the implications and guidelines for assigning a risk reduction factor or probability of failure on demand to a Safety IPL alarm
- Prioritizing safety IPL alarms based on the ISA-18.2 standard and alarm management best practices
- Consider providing operators with alarm response procedures for Safety IPL alarms in context and within the HMI

Safety practitioners are encouraged to compare their own practices against the benchmark survey results and the best practices cited in this paper. This should highlight areas of improvement that can help improve the safety of the people and the processes they work with. It is also recommended that safety practitioners increase their knowledge of alarm management best practices such as those in ISA-18.2.

7. References

[1] "The Buncefield Incident; The final report of the Major Incident Investigation Board", Volume 2, Crown publishing, United Kingdom, (2008).

[2] Occupational Health and Safety Administration (OSHA), "Petroleum Refinery Process Safety Management National Emphasis Program", Directive CPL-03-00-010, Washington, DC, (2009).

[3] ANSI/ISA 18.00.02-2009 "Management of Alarm Systems for the Process Industries".

[4] ANSI/ISA-84.00.01-2004 Part 1 (IEC 61511-1 Mod) "Functional Safety: Safety Instrumented Systems for the Process Industry Sector".

[5] OSHA, "Process safety management of highly hazardous chemicals", 29 CFR 1910.119, Washington, DC, (1992).

[6] Hartmann, H., Scharpf, E., and Thomas, H., *Practical SIL Target Selection: Risk Analysis per the IEC 61511 Safety Lifecycle*, exida, Sellersville, PA, (2012).

[7] CCPS. *Guidelines for Safe and Reliable Instrumented Protective Systems.* Center for Chemical Process Safety. New York, NY. (2007).

[8] CCPS. *Layer of Protection Analysis: Simplified Process Risk Assessment.* Center for Chemical Process Safety. New York, NY. (2001).

[9] Stauffer, T. and Clarke, P., "Using Alarms as a Layer of Protection", AIChE 8th Global Congress on Process Safety, Houston, TX (2012).

[10] Hatch, D, and Stauffer, T., "Operators on Alert: Operator response, alarm standards, protection layers keys to safe plants", Intech, (September 2009).

[11] Stauffer, T. "Making the Most of Alarms as a Layer of Protection", Safety Control Systems Conference – IDC Technologies (May 2010)

[12] Stauffer, T., Sands, N., and Dunn, D., "Get a Life(cycle)! Connecting Alarm Management and Safety Instrumented Systems", ISA Safety & Security Symposium (2010).

[13] Hollifield, B., and Habibi, E., *Alarm Management A Comprehensive Guide (2nd Edition)*, ISA, Research Triangle Park, NC, (2011).

Additional references not cited:

[14] EEMUA 191, "Alarm Systems: A Guide to Design, Management and Procurement Edition 2". The Engineering Equipment and Materials Users Association (2007).

[15] Nimmo, I., "The Operator as IPL," Hydrocarbon Engineering, September 2005.

[16] Stauffer, T., Sands, N., and Dunn, D., "Alarm Management and ISA-18 – A Journey, Not a Destination", Texas A&M Instrumentation Symposium (2010).

[17] Suttinger, L. and Sossman, C., "Operator Action within a Safety Instrumented Function", WSRC-MS-2002-00091 (2002).

[18] "The Explosion and Fires at the Texaco Refinery, Milford Haven, 24 July 1994", HSE Books, Sudbury, U.K. (1995).

[19] "BP America Refinery Explosion" U.S. CHEMICAL SAFETY BOARD www.chemsafety.gov/investigations (2009).

Appendix A. Survey of Criteria for using Alarms as Layers of Protection

A.1 Guidelines for Safe and Reliable Instrumented Protective Systems [7]

Protection layers are known as IPLs are designed and managed to meet the following seven core attributes:

- **Independence** – the performance of a protection layer is not affected by the initiating cause of a hazardous event or by the failure of other protection layers;
- **Functionality** – the required operation of the protection layer in response to a hazardous event;
- **Integrity** – related to the risk reduction that can reasonably be expected given the protection layer's design and management;
- **Reliability** – the probability that a protection layer will operate as intended under stated conditions for a specified time period;
- **Auditability** – ability to inspect information, documents and procedures, which demonstrate the adequacy of and adherence to the design, inspection, maintenance, testing and operation practices used to achieve the other core attributes;
- **Access Security** – use of administrative controls and physical means to reduce the potential for unintentional or unauthorized changes; and
- **Management of Change** – formal process used to review, document, and approve modifications to equipment, procedures, raw materials, processing conditions, etc., other than "replacement in kind," prior to implementation.

Applying the seven core attributes to alarms, allows definition of specific recommendations and best practices. Alarms should only be used when the operator is expected to take a specified action, which is covered by operating procedure. Operators should be trained on how to respond to the alarm according to a written procedure. For most hazardous events, only one protective function can be claimed in the supervisory layer, irrespective of the number of indications or alarms. For an alarm to be classified as an IPL, it must meet the following three criteria:

- The alarm is independent of the initiating cause and other protective layers addressing the identified hazardous event.
- The alarm function, including inputs and outputs, is designed to provide the allocated risk reduction.
- There is sufficient time for the operator to detect a problem exists, to determine what to do and to take appropriate action necessary to return the process to normal operating limits.

The total operator response time should be less than one-half of the available process safety time. For a protective alarm, the process safety time is the time between the alarm occurrence and the hazardous event occurrence.

A.2 Practical SIL Target Selection: Risk Analysis per the IEC 61511 Safety Lifecycle [6]

- The sensor and logic solver used to activate the alarm must be at least 90 percent reliable and independent of the initiating event and other IPLs (independent)

- The alarm must be part of a well-rationalized alarm annunciation system such that the operator is not overwhelmed with too many alarms

- The alarm setpoint must be within the operating range of the sensor and may not be changed without permission and a change management procedure (dependable and auditable)

- The alarm must not be capable of being bypassed or inhibited and it must be annunciated in a control room that is continually manned when the process is operating (dependable)

- The operator must have adequate time to respond to the alarm. This response time includes the time it takes him to detect the alarm, diagnose what should be done, physically move to the final elements to be manipulated and execute the manipulation (dependable). For example, a high level alarm on a compressor suction drum will require the control room operator to acknowledge the alarm, determine the need to drain the drum, call the field operator, and request the action. Then the field operator must stop their current activity and physically go to the compressor, locate the correct drain valve, and then open the valve. This response time must also include the time it takes the operator to recover from making an incorrect decision or process manipulation – or come back into the control room to get a wrench to move a stuck valve!

- An alarm response procedure detailing the actions required by each type of operator (control room and field) must exist and be available to the operators. All operators must be trained, drilled and periodically audited on the procedure and its required actions (auditable)

- All operators must be capable, and willing, to make the correct intervention actions at least 90% of the time (dependable)

- The operators must have a final element to manipulate that is independent of the initiating event and other IPLs, including any SIFs (independent)

- The alarm must reveal the dangerous condition under all circumstances (specific)

- The proper functionality of the alarm must be periodically verified and documented (auditable)

- Alarm system performance must be measured and proven to be adequate (dependable). To ensure performance is acceptable it must be measured and compared to key performance metrics (targets) such as those defined in the ISA-18.2 standard.

Alarm Performance Metrics **Based upon at least 30 days of data**		
Metric	**Target Value**	
Annunciated Alarms per Time:	Target Value: Very Likely to be Acceptable	Target Value: Maximum Manageable
Annunciated Alarms Per Day per Operating Position	~150 alarms per day	~300 alarms per day
Annunciated Alarms Per Hour per Operating Position	~6 (average)	~12 (average)
Annunciated Alarms Per 10 Minutes per Operating Position	~1 (average)	~2 (average)
Metric	**Target Value**	
Percentage of hours containing more than 30 alarms	~<1%	
Percentage of 10-minute periods containing more than 10 alarms	~<1%	
Maximum number of alarms in a 10 minute period	≤10	
Percentage of time the alarm system is in a flood condition	~<1%	
Percentage contribution of the top 10 most frequent alarms to the overall alarm load	~<1% to 5% maximum, with action plans to address deficiencies.	
Quantity of chattering and fleeting alarms	Zero, action plans to correct any that occur.	
Stale Alarms	Less than 5 present on any day, with action plans to address	
Annunciated Priority Distribution	3 priorities: ~80% Low, ~15% Medium, ~5% High or 4 priorities: ~80% Low, ~15% Medium, ~5% High, ~<1% "highest" Other special-purpose priorities excluded from the calculation	
Unauthorized Alarm Suppression	Zero alarms suppressed outside of controlled or approved methodologies	
Unauthorized Alarm Attribute Changes	Zero alarm attribute changes outside of approved methodologies or MOC	

Table 10. ISA-18.2 Alarm Performance Metrics [3]

A.3 Layer of Protection Analysis: Simplified Process Risk Assessment [8]

- The indication for action required by the operator must be detectable. The indication must always be:
    - Available for the operator,
    - Clear to the operator even under emergency conditions,
    - Simple and straightforward to understand.

- The time available to take action must be adequate. This includes the time necessary to decide that the action is required and the time necessary to take the action. The longer the time available for the action, the lower the PFD given for human action as an IPL. The decision making for the operator should require:

- No calculations or complicated diagnostics,
- No balancing of production interruption costs versus safety

- The operator should not be expected to perform other tasks at the same time as the action required by the IPL, and the normal operator workload must allow the operator to be available to act as an IPL.

- The operator is capable of taking the action required under all conditions expected to be reasonably present. As an example, consider a proposed IPL where an operator is required to climb a platform to open a valve. If a fire (as the initiating event) could prevent this action, it would not be appropriate to consider the operator action as an IPL.

- Training for the required action is performed regularly and is documented. This involves drills in accordance with the written operating instructions and regular audits to demonstrate that all operators assigned to the unit cab perform the required tasks when alerted by the specified alarm.

- The indication and action should normally be independent of any alarm, instrument, SIF or other system already credited as part of another IPL or initiating event sequence.

The Impact of Human Factors on LOPA (and other risk assessment methods)

William G. Bridges, President
Process Improvement Institute, Inc. (PII)
1321 Waterside Lane
Knoxville, TN 37922
Phone: (865) 675-3458
Fax: (865) 671-4944
e-mail: wbridges@piii.com

Prepared for Presentation at
9th Global Congress on Process Safety
San Antonio, TX
April 28-May 1, 2013

UNPUBLISHED

The Impact of Human Factors on LOPA (and other risk assessment methods)

William G. Bridges, President
Process Improvement Institute, Inc. (PII)
1321 Waterside Lane
Knoxville, TN 37922
Phone: (865) 675-3458
Fax: (865) 671-4944
e-mail: wbridges@piii.com

Keywords: Systematic Error, human error, human factors, layer of protection analysis, LOPA, independent protection layer, IPL, initiating event

Abstract

All initiating events (IEs) and independent protection layers (IPLs) are inherently tied to human error. Human factors limit the risk reduction valves and must be taken into account to ensure true independence of IPLs and to ensure the probability of failure on demand (PFD) of the IPL is achievable in actual implementation. The paper explains the relationship between human factors and the resultant IE frequency and PFD and provides an overview of how to validate theses risk reduction values at a site. The paper also covers the more involved topic of dependent human errors in IPLs, such as high integrity SIS and other high reliability IPLs such as relief systems. Actual examples are provided to illustrate key learnings.

Humans can be the cause of an accident scenario (the Initiating Event [IE]) or humans can serve or participate as an independent protection layer (IPL). Humans can also contribute indirectly to raise the frequency of non-human IEs and to raise the PFD of non-human IPLs. Although textbooks[1,2] address human factors issues, this paper updates the current understanding of human factors, particularly from the perspective of IPLs and IEs.

Identifying and sustaining independent protection layers (IPLs) is the heart of LOPA. Each IPL must be:

- Independent of the initiating event (IE) and the other IPLs (including consideration of the complete system boundary for that subsystem, such as upstream and downstream block valves).

- Capable (big enough, fast enough, strong enough, etc., including consideration of the complete system boundary for that subsystem, such as upstream and downstream block valves).
- Maintained or kept in practice/service.
- Validated/proven to provide the probability of failure on demand (PFD) chosen.

And all of the above must be documented and audited periodically to ensure compliance with these definitions.

1. Human Error Fundamentals

Human errors are sometimes mistakenly called procedural errors. This is not any truer than saying all equipment errors are due to design errors. Throughout the past five decades of research and observation in the workplace on human error, we have come to know that human error probability depends on many factors. These factors are described in more detail elsewhere[3], and they include:

- *Procedure accuracy and procedure clarity* (the most-cited root cause of accidents):
  - ➢ A procedure typically needs to be 95% or more accurate to help reduce human error; humans tend to compensate for the remaining 5% of errors in a written procedure.
  - ➢ A procedure must clearly convey the information, and the procedure must be convenient to use.
  - ➢ Checklist features — Checklists should be used and enforced either in the procedure or in a supplemental document.
  - ➢ For human response IPLs, a checklist is not always practical; in fact, a troubleshooting guide (rather than a prescriptive, linear procedure) is usually required, since we do not always know what condition a process will be in before a serious problem begins that requires humans to respond (as a human-based IPL). Note that best practice is to have troubleshooting guides indexed by parameter or alarm; the most effective implementation has been to make these easily available from the BPCS screen (such as by selecting an option to display the troubleshooting guide when an alarm is triggered).
- *Training, knowledge, and skills*
  - ➢ Employees must be selected with the necessary skills before being hired or assigned to a department.
  - ➢ Initial training — There must be effective, demonstration based training for each proactive task and each reactive (e.g., response to alarm) task.
  - ➢ Ongoing validation of human action is required and must be repeated. For human IPLs, the action must be demonstrated to be "fast enough" as well. For the Site-Specific Data approach, validation can be done in actual performance (measuring errors in responses to actual alarms) or in drills/practice. Alternatively, the human performance can be validated by expert judgment, generic data, or by prediction. *Examples of methods for validating human IPLs (and estimating human IEFs) are provided later in this paper.*

   - Documentation — The actual performance of the humans or alternative validation approaches must be documented and retained to demonstrate the error rates chosen are valid.
- *Fitness for duty* — Includes control of many sub-factors such as fatigue, stress, illness and medications, and substance abuse.
- *Workload management* — Too little workload and the mind becomes bored and looks for distraction; too many tasks per hour can increase human error as well.
- *Communication* — Miscommunication of an instruction or set of instructions or of the status of a process is one of the most common causes of human error in the workplace. There are proven management systems for controlling communication errors.
- *Work environment* — Factors to optimize include lighting, noise, temperature, humidity, ventilation, and distractions.
- *Human system interface* — Factors to control include layout of equipment, displays, controls and their integration to displays, alarm nature and control of alarm overload, labeling, color-coding, fool-proofing measures, etc.
- *Task complexity* — Complexity of a task or job is proportional to the (1) number of choices available for making a wrong selection of similar items (such as number of similar switches, number of similar valves, number of similar size and shaped cans), (2) number of parallel tasks that may distract the worker from the task at hand (leading to either an initiating event or failure of a protection layer), (3) number of individuals involved in the task, and (4) judgment or calculation/interpolation, if required. For most chemical process environments, the complexity of the task is relatively low (one action per step), but for response actions (human IPLs) there are almost always other tasks underway when the out-of-bounds reading occurs or the alarm is activated, and as mentioned earlier, we cannot predict what state the rest of the process will be in when the alarm comes on.

In addition to the human factors listed, other considerations for use of a human as an IPL include (1) time available to perform the action and (2) physical capability to perform the action.

When considering human IEs and IPLs, the site must ensure that the factors above are consistently controlled over the long-term and that they are controlled to the same degree during the mode of operation that the LOPA covers. For instance, if the workers are fatigued following many extra hours of work in a two-week period (say 14 days straight days of 12 hours of work each day) leading up to restart of a process, then the human error rates can increase by 10 (or more) times during the subsequent startup[4,5,6,7].

Revealed versus Unrevealed Errors for Human. As with equipment failures, human errors can lead to a revealed fault in the system (e.g., the flow does not start) or to an unrevealed fault, such as the block valve downstream of a control valve is left closed, but the failure is not revealed until the control valve is used. If the error is revealed, then the error can be corrected or compensated for. If the restoration/correction time is sufficiently short, then the probability of being in the failed state is much less for a revealed failure than for an unrevealed failure, which is only discovered upon testing or inspection.

2. General Relationship of Human Factors to LOPA

Every risk assessment must consider the likelihood and effect of human factors. For LOPA, poor human factors can lead to higher human error rates that increase IE frequencies and that increase the PFD of a human IPL (and of other IPLs indirectly). Table 1 (adapted from an earlier paper[8]) on the next several pages summarizes the human factor issues that relate directly to LOPA. The table contrasts the impact of good and poor human factors on initiating event frequency (IEF) and on the PFD of human-based independent protection layers (IPLs).

If all of the factors were optimized, one would expect about 0.01 per task as an IEF and about 0.05 PFD (which would normally be rounded up to 0.1) for response to alarms or call for actions (human IPLs), but such an analysis must be made on a case-by-case basis, especially for the PFD of IPLs.

TABLE 1 Considerations for Getting Low Human Error Rates with Respect to IEs and IPLs

Issue	Requirement for claiming a low error rate for humans causing an accident (usually skipping a step or doing one wrong)	Requirement for using a low probability of failure on demand for a human as a protection layer against an accident (usually a response to alarm)	Comments
Time to perform action	NA	The total time to: detect deviation, diagnose the problem, decide on proper action, take action, and leave the vulnerable area must be less than the time required to reach the consequence of interest or to reach MART. The term "maximum allowable response time" (MART) is used throughout this guideline.	The reliability of the indication (sample & analysis in lab, field reading, etc.) or annunciation (alarm of the deviation) may limit the value for the human action IPL. Operator must be reliably available to respond quickly.
Capable	NA	It is physically possible to perform the control action required to forestall the consequence of interest, and the capability has been demonstrated and is not the limiting factor in the human action (e.g., the operator is strong enough to do the required task).	If the final control requires closing a manual block valve, the action and strength required must be verified in the field (some valves take several operators more than 1 hour to close, even with a valve wrench or cheater). This usually requires preventive maintenance or equipment use as part of a "critical action."

TABLE 1 Considerations for Getting Low Human Error Rates with Respect to IEs and IPLs

Issue	Requirement for claiming a low error rate for humans causing an accident (usually skipping a step or doing one wrong)	Requirement for using a low probability of failure on demand for a human as a protection layer against an accident (usually a response to alarm)	Comments
Procedure	Step must be in procedure. Not completing the step as stated and in that sequence (most aspects of errors of omission and commission) is the IE. The step is a critical step and is indicated as critical in the procedure or with a warning or caution. The procedure must follow "best practices" for control of human error; best practices relate to content accuracy, page format, and step presentation. Follow the current rules for procedure development and writing (Tew, 2010).	Step should be in a troubleshooting guide or similar contingency procedure (including emergency shutdown procedure), that is in paper form or reliably available on demand electronically (including via linkage to troubleshooting guides within a distributed control system [DCS] for an alarm point) that describes how to respond to a process deviation. The response procedure follows best practices as well; follow the current rules for procedure development and writing (Tew, 2010).	Note that we assume there is a 100% chance of skipping the required step or doing it substantially wrong if the step is not in the written procedure or if the step written is wrong or in the wrong sequence.
Checklist	Checklist in place and its use is enforced.	NA	Without a checklist, the error rate goes up by a factor of 3 to 5 times.
Initial Training	There is initial training focused on this step as a critical step.	There are initial training and drills on the response to a deviation; and the deviation is annunciated.	Note that we assume there is a 100% chance of skipping the required step or doing it substantially wrong if the step is not emphasized in initial training.

TABLE 1 Considerations for Getting Low Human Error Rates with Respect to IEs and IPLs

Issue	Requirement for claiming a low error rate for humans causing an accident (usually skipping a step or doing one wrong)	Requirement for using a low probability of failure on demand for a human as a protection layer against an accident (usually a response to alarm)	Comments
Ongoing validation (or practice) of human action	Steps must be practiced, but for seldom-used procedures, this practice (normally in the form of a talk-through for shutdown or startup of a continuous unit) can be just prior to use. Practice rate increases reliability per action (due to increase in skill), but Actual Use Rate also increases the number of opportunities for failure. These factors can offset each other completely. See limitations on error rates that account for "practice" versus "error rate per task demand" that are incorporated within the human IEs in Chapter 4.	Steps must be practiced routinely enough to ensure the reliability of the action under the increased stress caused by the alarm/indication. The time it takes to complete the action must be validated, to ensure response time limits are not exceeded. For the Site-Specific Data approach to validation (see Appendix C for details), each person who is expected to implement the action "on demand" must perform the practice/drill per the assigned frequency of about once per year. Per the Predicted Data approach to validation (see Appendix B for details), the PFD for the human IPL is estimated by calculations using factors based on the quality of the underlying management systems, procedures, and training; and based on the level of stress involved with the response task. Other methods of validation may also be used, as described in Chapter 3. It is normally possible to group similar IPLs along the classification of similar types of actions and similar available response times. When grouping is possible, then validating one action will in effect be validating all similar actions. Appendix C provides an example of grouping, if validating using Site-Specific drills of response actions.	Note that we assume there is a 100% chance of skipping the required step or doing it substantially wrong if the step is not practiced often enough (to be determined by the site).

TABLE 1 Considerations for Getting Low Human Error Rates with Respect to IEs and IPLs

Issue	Requirement for claiming a low error rate for humans causing an accident (usually skipping a step or doing one wrong)	Requirement for using a low probability of failure on demand for a human as a protection layer against an accident (usually a response to alarm)	Comments
Documentation of validation of human action	Documentation methods depend on the validation method chosen. Validation based on Expert Judgment or Predicted Data approaches would require the site to maintain sufficient documentation, including calculations and description of methods, for the individuals IEs or groupings of IEs. The Site-Specific Data approach to validation requires more effort since the site would need to document each practice (i.e., each startup and each shutdown) and document failures rates for "critical steps." For this same validation method, the site would keep the file of the validation associated with the task and also with each worker. Though more time consuming on documentation, this method may produce more side benefits that outweigh this cost.	Documentation methods depend on the validation method chosen. Validation based on Expert Judgment or Predicted Data approaches would require the site to maintain sufficient documentation, including calculations and description of methods, for the individuals alarm response actions. For the Site-Specific Data approach to validation, the site would need to document each practice (i.e., each response) and document failures rates for "critical response action." The site would also keep the file of the validation associated with the 'action' and also with each worker. If a sample plan method is used, the same data as above is still required, and further, the data should be checked each quarter to ensure against inadvertently using the same IPL and same operator too often.	Note that we assume there is a 100% chance of skipping the required step or doing it substantially wrong if the step is not practiced as often as necessary (to be determined by the site) or if there is no documentation of the completed refresher training or just-in-time training for an IE, or of drills for an IPL (if drills are part of the validation method).

TABLE 1 Considerations for Getting Low Human Error Rates with Respect to IEs and IPLs

Issue	Requirement for claiming a low error rate for humans causing an accident (usually skipping a step or doing one wrong)	Requirement for using a low probability of failure on demand for a human as a protection layer against an accident (usually a response to alarm)	Requirement for using a low probability of failure on demand for a human as a protection layer against an accident (usually a response to alarm)
Control of related physiological and psychological stressors (see Gertman, 2005, and Bridges, 2012b)	• Fatigue—Company policy and enforcement to limit fatigue, including limit of hours worked per day and per week, and including restrictions on hours worked outside of work. Can affect error rates by factor up to 20X. • Other fitness of duty issues—Company policy and enforcement to limit effect of alcohol and drug abuse, illness, prescription drug effects, and personal life stress effects. Can affect error rates by factor up to 50X. • Workload—Workload needs to be managed to optimize worker performance. Enough stress (time/task load-based), but not overstress. (Workload is also related to the complexity of the work. Complexity increases mental workload, even if the number of tasks per hour is not affected.) Workload evaluation may be necessary. Can affect error rates by a factor up to 10X. • Communication—The second most common cause of human error, the site must guard against miscommunication by having a policy and rules and follow-through for controlling verbal and visual communication. Rules include repeat-back on instructions received and use of common jargon. Poor control of communication rules can increase human error rates by 10X. • Work environment—Temperature, lighting, noise, distractions, ventilation, etc., has been optimized to improvement worker performance. Can affect error rates by a factor up to 5X. • Human-System Interface—Includes control-display-information integration, labeling, error-proofing designs, color-coding, alarm management, etc. Poor human-system interface control can increase human error rates by 10X or higher (especially if "norms" are violated, where a negative effect of 20-50X is assumed).		All of the listed human factors (beyond the procedure and training and skills listed above) have a large, independent influence on human error rates. A large negative swing in any one of the factors can increase human error rate by a factor of 3 to 20 times.

TABLE 1 Considerations for Getting Low Human Error Rates with Respect to IEs and IPLs

Issue	Requirement for claiming a low error rate for humans causing an accident (usually skipping a step or doing one wrong)	Requirement for using a low probability of failure on demand for a human as a protection layer against an accident (usually a response to alarm)	Requirement for using a low probability of failure on demand for a human as a protection layer against an accident (usually a response to alarm)
Continued—Control of related physiological and psychological stressors (see Gertman, 2005, and Bridges, 2012b)	• Complexity—Complexity of a task or job is proportional to the (1) number of choices available for making a wrong selection of similar items (such as number of similar switches, number of similar valves, number of similar size and shaped cans), (2) number of parallel tasks that may distract the worker from the task at hand (leading to either an initiating event or failure of a protection layer, (3) number of individuals involved, and (4) judgment or calculation/interpolation, if required. For most chemical process environments the complexity of the task is relatively low (one action per step), but for response actions there are almost always other tasks underway when the out-of-bounds reading or the alarm is activated. Complexity is difficult to predict (since it is not known when a human action will be needed), but higher complexity can increase error rates by 2X to 10X.		
Qualitative hazard evaluation	There has been a hazard evaluation of the scenario for skipping the step and doing it wrong and a hazard evaluation team has evaluated the consequences and safeguards.	The human action in this case is a safeguard that is called into play once a sequence of events is initiated. The human action (following the troubleshooting guide or the emergency shutdown procedure) is to make a change of state in the system to prevent propagation of the event to the stated consequence. The hazard evaluation has evaluated the human action (and any annunciation device unavailability) in judging that this "critical responsive action" is capable, reliable, and audited.	It is essential that a company find accident scenarios that can arise during all modes of operation, including startup, shutdown, emergency shutdown, and online maintenance.

TABLE 1 Considerations for Getting Low Human Error Rates with Respect to IEs and IPLs

Issue	Requirement for claiming a low error rate for humans causing an accident (usually skipping a step or doing one wrong)	Requirement for using a low probability of failure on demand for a human as a protection layer against an accident (usually a response to alarm)	Requirement for using a low probability of failure on demand for a human as a protection layer against an accident (usually a response to alarm)
Use of multiple successful human actions within a scenario	Making multiple mistakes (including shortcutting multiple steps) is the same probability as making a single mistake, if the "safeguard" steps amount to ensuring the same state in the system. The human who makes the errors sees this as performing the task by method B instead of method A. Method B could have different steps or less steps than method A.	If one person in a workgroup is assumed to make the mistake that is the initiating event, then it is assumed that no one in the workgroup can be part of the protection layers. Further, supervisors are also part of the workgroup, due to the trust/relationships that build quickly. Seemingly unrelated workgroups, such as maintenance and operations, usually build enough trust to rule out counting as separate groups. This quick development of trust among coworkers is part of inherent human nature and cannot be eliminated. For a group/individual to be counted as a safeguard against the errors by another human error, the detection and response group must not share staff, must have an incentive to find mistakes and report/respond, and near-miss reporting and auditing both indicate that independent measurement and action is occurring. An exception that is normally allowed is to re-use a workgroup or individual if the second action/response is separated by distance (location) and/or separated by enough time, and/or there is a reminder alarm for the same alarm (the alarm that was missed earlier) that recurs often enough in the allowed time to still avoid the consequence.	It is difficult to use a human response safeguard if a cause of the predicted accident sequence is a human error. A human reliability analysis (HRA) is typically done to verify the human protection layer probability in this situation.

3. Types of Human Errors of Importance in LOPA

Human error comes in many forms, but the primary ones of interest in LOPA are:

Human errors that directly relate to human IEs and human IPLs

- Errors in following proactive procedures, such as startup of a unit, that results in an initiating event of a LOPA accident scenario
- Errors in responding to a call for action; if performed correctly such actions would interrupt the scenario and prevent the consequence

Human errors that indirectly relate to LOPA component-based IEs and IPLs

- Errors when following management systems that lead to higher failure rates of components; although indirect, such errors can have impacts on risk that are orders of magnitude

The examples below illustrate both types of human error that directly affect LOPA. At the end of this paper are more details on human errors that indirectly related to IEs and IPLs, along with a Case Study

3.1 IE Caused by Human Error

Human errors that result in IEs of major process incidents are usually occur during startup; one definitive study (Rasmussen, 1989)[9] and a survey of the major accidents in the past 20 years[10] show that roughly 65% of **major** process accidents in chemical-related process plants occur during startup and on-line maintenance. Further, errors during normal shutdown or emergency shutdown together represent about 10% of the IEs caused by human errors. (Other studies show that most **minor releases**, such as packing leaks, occur during normal operation.) Human errors during normal operation result in less than 25% of the and include such issues as performing a step wrong during operation in manual mode, such as when a process controller is bypassed or by external impact (one major accident during normal operation occurred when a fork-truck hit and broke off a drain valve).

The two primary ways to find accident scenarios caused by human error (before the accident occurs) are:

- Prediction of the accident scenario in PHA of all modes of operation, especially startup and shut down procedures. Most PHA/HAZOPs world-wide are weak on finding accident scenarios during non-routine modes of operation. Overcoming this weakness is critical and related methods are explained in Chapter 9.1 of the *Guidelines for Hazard Evaluation Procedures, 3rd Edition* [11] and in an updated paper on the same topic[10].
- Getting a lot of near misses reported and investigated; usually having at least 15 times the number of near misses reported as accidents occurring is needed to have a 90% chance or higher of preventing the major accidents[12,13]. The ratio of near misses report to accidents occurring is relatively low at most companies world-wide, so this is low hanging fruit for many companies.

3.2 Human Errors as Related to Human Response IPLs

There are many types of IPLs, and some are more trustworthy than others, hence the difference in the PFD of IPLs. One possible type of IPL is a human IPL. These include preventive steps that may stop a scenario from progressing once it is initiated, but more typically the human IPLs are responses to alerts or alarms or troubling readings and sample results.

A 10^{-1} PFD value for a human response IPL indicates that the correct response occurs at least 9 out of 10 times (or no more than 1 wrong response for every 10 attempts). Most organizations will have identified many human responses involving a number of personnel as part of their LOPA studies. Some organizations believe that if they have "a procedure and a training program" in place, they can claim the PFD value of 0.1 for a human IPL. This is no truer for a human IPL than it is for an active component IPL. So, having a procedure and training are not the only requirements for a human IPL. As required for all IPLs, a human IPL must be ***validated***. The preferred approach to validation is Site-specific data where the site directly measures or tests the human response (under controlled conditions or drills). But other methods of validation include Expert Judgment, using data from other comparable settings (Generic Data approach), and the Predicted Data approach where estimation of the PFD of human IPLs is by a human error mathematical modeling, described later in this paper.

4. Approaches for Validating an IEF or PFD for a Human IE or Human IPL

For LOPA, the ways to estimate and validate human IEFs and validate PFDs of human IPLs include expert opinion, using industry values, calculation from human reliability factors, or direct measurement at the site.

4.1 Validating Human IEFs

To help minimize human errors that could be IEs, many companies will review a normal shutdown or normal startup just before its next use. This has proven one effective way to reduce human error in chemical-related industries. Nuclear power plants also perform such "just-in-time (JIT)" refresher training on normal shutdowns and normal startup procedures. If a site records the errors that are made during such tasks, then the company can validate the average IEF for that task and so the site will have a better estimate of the error probability. Further, if the measured rate is too high, they can take proactive steps to lower the error rate. Errors can be measured either in actual practice in the field or via simulators. They can also be roughly estimated by analysis of near misses reported (if the number of near misses reported is at least 15 times higher than the number of loss events or accident[12, 13]).

Below is an example of data from one site that was collected to estimate the IEF for mistakes made during normal startup:

> ***EXAMPLE - Initiating event frequency data collection and use:*** In one company, the operating group was able to provide the data on the following page for errors made during startup of various units. Based on this data, the average probability of human

error was 0.0071 and the number of errors per year that could lead to a known accident sequence was 0.17 (but note that if all errors are counted, the average error rate would be 0.29/yr). For basic LOPA, this value is rounded up to an IEF of 1/yr, for initiating events arising from these and similar procedures at the site. See Table 2 below.[14]

Table 2: Example of Determination of Human IEF from Site-Specific Data

Task for which a Potential Initiating Event exists	Number of steps in the procedure	Number of Uses of the procedure	Number of Human Errors Noted	Number of Years of Data	Number of Errors That Initiated a Potential Accident Sequence	Human Error Probability per Step (and one mistake will initiate a LOPA scenario)	Error Rate (probability of error per year), per scenario
1. Startup of Process A	46	4	2	6	1	0.0109	0.0073
2. Startup of Process B	71	5	3	6	1	0.0085	0.0071
3. Startup of Process C	55	4	2	6	2	0.0091	0.0061
4. Startup of Process D	56	4	0	6	0	0.0000	0.0000
					Average:	0.0071	0.0052
						IEF for LOPA for site (applies for any "human error initiated" LOPA scenario):	0.01/yr.

Chemical-related process plants should collect such data to understand and validate the initiating event frequency caused by human error. This data would then be used to validate the values for IEF used in LOPA and other risk assessments.

4.2 Validating the PFD of Human-Based IPLs

The probability of a human response being done wrong, not being done, or taking too long can be estimated different ways, as described in Figure 1 on the next page. Using site-specific data is preferred, as this is usually easier to defend. Note that CPQRA in Figure 1 refers to *Guidelines for Chemical Process Quantitative Risk Analysis.*[15]

All the human factors below were assumed to be nominal and were omitted in this simplified approach (including being set to a value of 1 for the predicted data portion of this example). IF for a given human IPL (or site) these are NOT nominal, THEN a full HRA is required, and this simplified estimation method cannot be used:

1. *Interface Factors:* The example approach assumes a suitable human-system interface exists such that information is presented in an easily assimilated form, alarms are prioritized and the number of simultaneous alarms is controlled to prevent channel capacity overload and to allow the human to easily perceive the alarm.
2. *Task Execution Factors:* The approach assumes that a procedure or troubleshooting guide exists for the alarm action that meets all of the requirements for such procedures or guidelines. The approach also assumes that the alarm will be considered a high priority demand by the human and will receive attention as quickly as can be practically achieved.
3. *Task Complexity Factors*: Task complexity is partly based on the number of steps and the level of cognitive processing required, and partly based on the number of different people involved and number of different interfaces involved to accomplish the task. Task complexity can also increase if the person is required to multitask, but this issue is addressed in item 5 below, Workload Factors. For this example, it is assumed that the people performing the task have the required knowledge, skill, and physical capability to complete it successfully, and that no more than two people have to communicate, and that the conditions of items 1 and 5 in this list of assumptions are satisfied.
4. *Training, Experience, and Familiarity Factors:* The approach assumes people performing the task are not allowed to work unsupervised until they have met the level, training, and experience requirement for the activity. Consequently, these training and experience levels are assumed within the HRA calculation. The approach also assumes the person performing the task is aware of any potential hazards, since such awareness should have been imparted during initial and follow-on training for the activity.
5. *Workload Factors:* The approach assumes the workload will not impact the ability of the human to respond adequately to the alarm since the prioritization of the alarm should override any other work in progress.

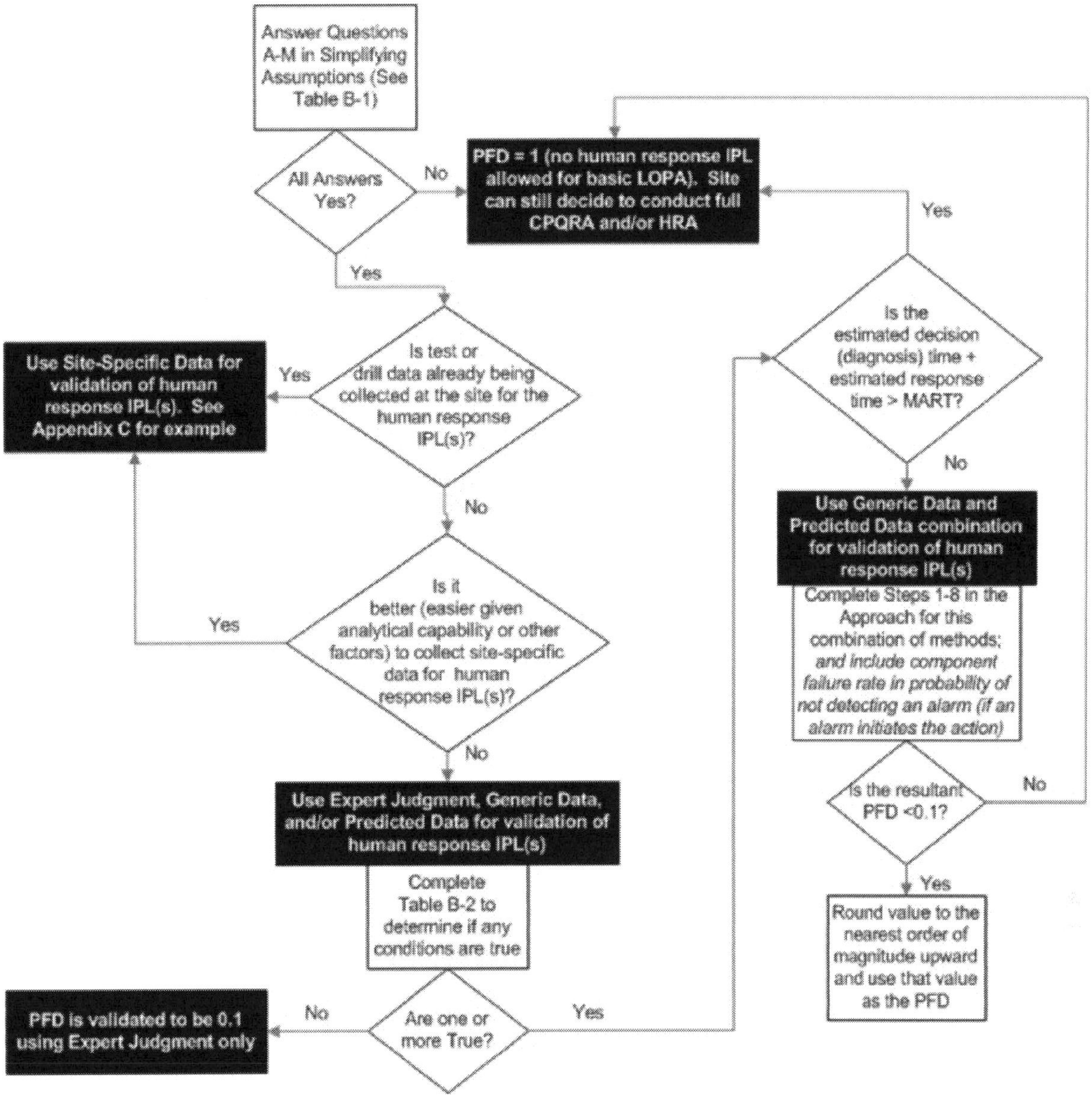

Figure 1. Example Decision Path for Validating Human Response IPL

To assure the simplifying assumptions are valid, a positive response to ALL of the following screening questions is needed to qualify use of this approach for validation of a human IPL:

TABLE 3 Determining the Calculation Approach to Be Used

Item	Criteria	Yes	No
A	Is this a human response to an alarm or call for action?		
B	Is the response action clearly documented in the related procedure or troubleshooting guide? (And does the documentation include enough guidance for making appropriate decisions?)		
C	Does the response procedure or troubleshooting guide follow best industry practice for content development and presentation? (See checklist of procedure writing rules in Tew, 2010, and other sources such as Bridges, 2007.)		
D	Is the procedure or troubleshooting guide kept up-to-date and accurate?		
E	Does the operator process and skills training include initial and refresher training on this specific response action?		
F	Is the alarm annunciated audibly and identified as an IPL?		
G	Can this procedure be done by ≤ 2 operators?		
H	Does the plant align to overtime practices and startup staffing expectations?		
I	Are supervisors required to screen workers for fitness-for-duty each day (looking for fatigue, mental stress, substance issues, and physical impairments)?		
J	Is the human response IPL completely independent of the initiating event and any other IPLs in the scenarios in which the human response IPL is credited?		
K	Is the annunciation system always active and constantly monitored (98%+ of time) by operator(s) trained in the response procedure and guideline? Is it confirmed that annunciation systems are Not turned off from time to time?		
L	Is the annunciation clear and distinct?		
M	Are the alarm and any instrumentation used in the response tested routinely enough?		
		If ALL are Yes, then Predicted Data method can be used for validation of the PFD	If ANY are No, use other methods for validation of the PFD of the human IPL

Again: IF the answers to A-M are not All Yes for a given human response IPL, THEN potentially a full HRA or actual drills are required to validate a human response IPL, and the simplified estimation approach for validation described in this appendix cannot be used.

Determine if Validation by "Expert Judgment Only" Is Allowed

The questions in Table 4 are used in the initial screening process. If ALL are False, *then a PFD of 0.1 is valid without further calculation.* If ANY of the statements in Table 4 are true, then the validation calculation approach that follows must be applied to verify that the required reliability is achieved.

TABLE 4 Determine *(Using Expert Judgment)* if Validation by Predicted Data Is Required

#	Criteria	True	False
1	Based on consensus of expert opinion, the operator has less than 15 minutes to successfully detect, diagnose, and perform the required action		
2	Operator response is required without explicit criteria and response instructions		
3	Critical or emergency responses involve multiple people		
4	Response actions provide no feedback that they are effective		
		If any are true, go through Predicted Data method for validation of the PFD	If all are false, use PFD = 0.1 (stop: no further calculation needed; Expert Judgment is sufficient for validation)

If Expert Judgment is not allowed as the sole method for validation, then the validation by simplified calculation (estimation) is required, as described in the next section; the example approach in the next section uses a combination of Generic Data and Predicted Data methods for validation. If the simplified calculation approach shown in the next section proves invalid or if a PFD greater than 0.1 is obtained, then other methods such as full HRA or direct error measurement at the site (Site-Specific Data approach to validation, as described later in this paper) may be used to validate the PFD of the human IPL.

Determine if the Expert Judgment Estimate of the Baseline Human IPL Time Is Less than the Maximum Allowable Response Time (MART)

The time available is critical to the reliability of any response activity. A shortage of time leads to hurrying and increased stress. In addition, under these conditions, any errors that occur may not be correctable. Ultimately, the action must be accomplished within the MART.

For a human response IPL, MART is the time from when the sensed parameter reaches the set point (and then perhaps a few moments later the alarm sounds, if alarmed) to the point of no return where the corrective action can no longer prevent the incident. This value is determined from process dynamics independent of any hoped-for human response. It includes any time delay in alarm activation and any time for automated actions (initiated by the operator) to occur.

This example of validation by predicted data requires three different time elements versus MART (maximum time available to stop the event):

A. *Detection time.* Time from when the parameter of interest exceeds the "safety" limit until the deviation is noticed by the human.
   - Detection and annunciation could be via a sensor and alarm, followed by sensory perception of the annunciation by the operator.
   - Detection could be by the operator taking a sample to the lab and then subsequent analysis and reporting of results by the lab technician. The detection time in this case includes time between sampling (at least one cycle) plus the time to take the sample plus the time to wait for analysis and perform analysis, plus the time to report the results to the appropriate operating staff.
   - Detection could be the operator noticing a problem during routine operator rounds, in which case the time since the previous rounds is the major portion of the time consideration. So, for rounds every four hours, the detection time is greater than or equal to four hours; but note that it is best to rotate operators every round to enhance vigilance.
   - Use Expert Judgment for to estimate detection time.

B. *Decision time* (time to decide what action to take; also called *diagnosis* time in HRA). The decision time was identified as a source of variability when people assessed the reliability of these activities. Consequently, the decision time is fixed within this validation method based upon the activity type. For purposes of alarms that a site would allow for LOPA, the decision time is normally less than one minute. But some HRA data developed for diagnosis time in control rooms (Swain, 1983) suggests that there is 90% chance the diagnosis will be correct if the worker in a nuclear power plant control room has at least 10 minutes to diagnose, and a 99% chance of correct diagnosis if they have 40 minutes. Because of these traditional values, the decision time is typically set at 10 minutes. However, for actions that require no or very little diagnosis or in simple process units, this value can reasonably be set to five minutes. Use Table 5.

TABLE 5 The Decision-Time Factor Assigned to the Different Activity Types

Activity Type	Decision Time (minutes)
Unambiguous cue in a continuously staffed control room or similar staffing near an alarm annunciation location, with simple process and little or no diagnosis (with a decision tool, such as a troubleshooting guide).	5
Unambiguous cue in a continuously staffed control room or similar staffing near an alarm annunciation location, with complex process unit that requires diagnosis to deduce the failure cause and the proper action to take (with a decision tool, such as a troubleshooting guide).	10
Requires diagnosis of a novel fault situation (cannot be used for IPL in LOPA).	Beyond LOPA

C. *Response time* (time to complete all the alarm response activities). This is the time required to complete the tasks that will prevent the undesired event, as detailed in the alarm response procedure (e.g., after the procedure has been chosen as the correct course of action). Use Expert Judgment for this estimate. (For comparison, this is the time that is measured directly by testing/drills in validation using Site-specific data, as discussed later in this paper)
Estimate the task response time: Using solicitation of expert opinion (including at least two senior operators from the unit and one senior process engineer or equivalent), develop an Expert Judgment estimate of the time to complete the response activities, given that the diagnosis is performed correctly.

IF: Detection Time (including any delays in a related instrument system) + *Decision time + Task response time > MART*

THEN: The human response IPL is not valid.

If the total human IPL time is too great, then the site may:

- Decide to use other methods such as full CPQRA or full HRA to estimate the response time and also to validate the human response IPL.
- Decide to redesign the human response IPL so that it can be done in less than the MART.
- Decide to redesign the system to eliminate or reduce the risk.
- Decide to install or upgrade other types of IPLs (such as IPS, which are faster to respond) in lieu of the human response IPL not being available (because it is currently invalid).

IF: Detection Time + Decision time + Task response time < MART

THEN: Proceed to the steps in the next section.

Validating Human Response IPLs by Prediction Data method

Another papers[16] have provided examples of how to validate the PFD of an IPL by Predicted Data using human reliability factors. Please refer to the referenced paper or to relative training courses (such as the LOPA course by Process Improvement Institute, Inc. [PII]) for further information on how this is done. Table 6 is an example of such a calculation using the approach taught within PII. The approach is not provided here.

Table 6 Example Calculation of Human Response IPL for High Temperature Alarm in Monomer Tank

Step #	Validation Step Description	Value	PSF (multiplier)	Probability	Perception, Diagnosis, and Response Activities
1	Activities Involved (type of action required) - **this sets the baseline probability**			0.018	Board operator acknowledges alarm & board operator activates fire monitor from control room without the need of field operator action. (Baseline human error + probability of alarm failing = 0.008+.01)
2	Number of steps	8	1		8 steps in the actual procedure, including immediate actions, diagnosis, and feedback checks
3	Number of activities or alarms per year	1	1.86		Estimate of 1, since operator is trained or tested annually on what to do (and since we expect much less than one actual event per year). (12/(1+0))^0.25
4	External Stressor Factors		6		Field operator does not have to go to monomer tank (hazardous zone) to verify the temperature; there is remote display from mutliple indicators in the control room; independent of alarm. Field operator does not have to go near the monomer tank. Control operator still under threat of taking wrong actions.
5	Communication Factor		1		Since the board operator is the one who receives the alarm and also the one who takes the primary action, miscommunication on the primary action is not possible.
6	Feedback Factor		0.1		Board operator would observe tank temperature after water spray application by field operator, and field operator is still required to observe water spray from a safe distance, so feedback is good.
			Interrim Result	0.020	<< Calculated PFD for this Human IPL
			FINAL RESULT	**0.10**	**<< PFD to use for this Human IPL (must be equal to or less than 0.1 to allow this Human IPL)**

Validating Human Response IPLs from Site-Specific Data

An approach for gather site-specific data from drills of responses to triggers of action has been described in previous papers[8, 14]. They incorporate collection of raw data on the response to triggers of a human IPL during drills. Data is collected on drill cards such as those shown below in Figure 2:

Human IPL Validation Test/Drill			
Response Task:	Max. Allowable Resp. Time (MART)		Response Time:
LAH for Tank 105	*15 minutes*		5:20 minutes
Date of Test:	Time/Shift:		Employee Number:
1/23/10	07:35/A		23122
	Pass/Fail:	Pass	

Figure 2 Example of card used to administer validation of a single human IPL

The results of such data collection is provided in Table 7 on the next page.

The previous papers[8,14] (presented in past GCPS) illustrate how this approach works and provides details on how to implement this approach and how to streamline the approach using a sampling of the alarms (grouped by similar response activities).

Table 7 Site-Specific Validation of Human Response IPLs

IPL No.	Response Task	Number of Test Performed	Average Response Time (minutes)	Maximum Average Response Time (minutes)	Number Failures	PFD (average)	LOPA PFD
Company A (USA), Site 1							
IPL 1	ABP: High Temp in Generator	6	2.3	10	0	0	0.1
IPL 2	ABP: Loss of Acid Flow to Generator	12	2.2	10	0	0	0.1
Company B (Canada), Site 1							
IPL 1	Low Seal Gas Pressure to Turbo-Exchanger Bearings	5	5.7	15	0	0	0.1
IPL 2	High Lube Oil Temperature – XX Compressor	5	6.3	30	0	0	0.1
IPL 3	Low Level Emergency Alarm – Steam Drum	5	4.9	15	0	0	0.1
IPL 4	High-High Lube Oil Temperature – XX Compressor	5	6.1	30	0	0	0.1
Company C (Malaysia), Site 1							
IPL 1	High Level on Ammonia Absorber Column	10	5.1	15	0	0	0.1
IPL 2	Low Temperature Alarm on Ammonia Compressor Discharge	10	4.9	15	0	0	0.1
IPL 3	Low Level in CO_2 Compressor Interstage KO drum	10	7.0	15	0	0	0.1
IPL 4	High Level on High Pressure Carbamate Heat Exchanger	10	6.1	15	0	0	0.1
IPL 5	High Pressure High Pressure Carbamate Condenser	10	5.0	15	0	0	0.1
IPL 6	High Pressure on Steam Controller to Rectifying Column Recirculation Heater	10	5.4	15	0	0	0.1

[a]Data provided by 3 companies (in the USA, Canada, and Malaysia)

5. Human Errors as Related to Impact on Non-Human IEs and IPLs (such as SIS and Relief Systems)

In addition to having a direct impact as an IE or as failure of a human IPL, human error can, of course, impact all other IEs and IPLs as well. This is because ALL process equipment are ultimately controlled and maintained by humans. Typically, a baseline human error rate of 1 error in 50 steps is reasonable, or for excellent control of human factors, a baseline error rate of 1 error in 100 steps may be achieved. This baseline error rate then can increase if the control of human factors slips. Fortunately, not all errors lead to negative or serious consequences. Below are some examples of how human factors control the reliability of all IPLs and IEs:

Example related to IE — ***Wrong materials of construction received:*** If the humans fail to control or detect errors during selection and use of materials of construction, then perhaps the wrong grade of steel or other material will be used in a highly corrosive environment and then the IEF will be many times higher than expected. Industry data from the 1990s indicated that the materials delivered are different than what we specify about 3% to 7% of the time (internal audit data from Process Improvement Institute, Inc. [PII] and from Swiss Reinsurance). Positive material identification (PMI) can detect and correct such failures to a large extent. Recent improvements in the quality assurance throughout the supply chain of alloys (especially with the advent of less expensive, handheld PMI devices) coupled with 100% PMI by the facility owners at receiving (and sometimes coupled with 100% PMI on as-welded in the plant) appears to lower the composite error rate to 0.1% to 0.5% (from PII data from several processes installed since 2005). This error reduction still leaves a large number of components that may be of the wrong materials and it is likely not possible for human-based detection to find the remaining deviations in materials of construction.

Example related to IPL — ***High integrity SIFs****:* Safety Instrumented Functions (SIFs) can be designed to provide risk reductions of 1, 2, or 3 orders of magnitude. However, even the best designed SIF can be negated if (1) the root valves for the sensors are left closed or (2) a bypass valve around an emergency isolation valve is left open (other errors are also possible, of course). In addition, the humans may be capable of revaluing trip points, etc., in the safety-rated PLC, given enough time to learn these shortcuts and given enough spurious trips to make them want to use such a shortcut. Although an SIL 4 SIF is allowed by IEC 61511, with a PFD of $<10^{-4}$ to $\geq 10^{-5}$, it is very unlikely for a process industry site to control human errors low enough to install or maintain an SIL 4 SIF. Similarly, SIFs with SIL of 2 and 3 require more rigorous design and maintenance practices to offset the chance of human intervention that can otherwise increase the composite PFD. A case study related to this example is provided at the end of this section.

Example related to IPL — ***Pressure relief systems****:* Many relief system IPLs can have a PFD of 0.01 or better, assuming the rest of the process does not interfere with or restrict the flow. However, closing a block valve upstream or downstream of a relief device will negate the entire value of the relief system. Excellent sustained control of human factors is necessary to maintain a PFD of 0.01 or better for a relief system because of the relatively high probability that a human will leave a block valve closed upstream or downstream of a relief valve. Common industry data for human error rates indicate that the error of leaving such a block valve closed will likely be the dominating failure mode for the relief system, so including a block valve upstream or downstream of a relief device (though allowed by ASME with administrative controls) will need to be evaluated by a site before assigning a PFD for the combination system (comprised of the relief system and block valves). The conservative PFD of 0.1 is therefore used as the default value for relief devices that have blocks valves upstream or downstream of relief devices, unless a human error analysis for the plant proves otherwise.

5.1 CASE STUDY: Human Error and the PFD of an SIF (based on a previous detailed paper[17])

The SIS standards of the mid-1990 through today recognized that systematic human errors have a deleterious impact on the PFD of an SIF. This effect can be either errors that exist at Time Zero or systematic errors while operating. The IEC standards qualitatively covered at length the need to control such errors. In ISA-TR84.00.02-2002[19], Equation 1a was provided which includes a system dangerous (D) failure (F)/error term. The equation is shown below:

$$PFD_{SIF} = \sum PFD_{Si} + \sum PFD_{Ai} + \sum PFD_{Li} + \sum PFD_{PSi} + \left[\lambda_F^D x \frac{T_i}{2} \right] \quad \text{[Eq. 1a]}$$

where PFD_{SIF} is the average PFD for a SIF. The first term in the equation is the contribution of the sensors, the second term is the contribution of the final elements, the third term is the contribution of the logic solvers, the fourth term is the contribution of the power supply, and the last term is the contribution of the dangerous system failures. But as stated in ISA-TR84.00.02[19], for this equation:

> "...the systematic failure is modeled as an error that occurred during the specification, design, implementation, commissioning, or maintenance that resulted in the SIF component being susceptible to a random failure. Some systematic failures do not manifest themselves randomly, but exist at time 0 and remain failed throughout the mission time of the SIF. For example, if the valve actuator is specified improperly, leading to the inability to close the valve under the process pressure that occurs during the hazardous event, then the average value as shown in the above equation is not applicable. In this event, the systematic failure would be modeled using $\lambda * Ti$. When modeling systematic failures, the reader must determine which model is more appropriate for the type of failure being assessed." [19]

This statement is accurate, but does not provide any practical guidance as to what systematic errors are most significant. In practice, most of the systematic error term results from human errors. These can include:

- Manufacturer contribution for certified equipment (Believed to be negligible relative to end user systematic errors)
- End user systematic errors:
    - Design and installation errors

  - Probability of being failed following proof test
  - Bypass during operation

Of these end user systematic errors, the dominating contribution is generally human errors that leave the protection failed at Time 0. These errors can occur during re-commissioning of an SIF following routine maintenance interventions such as:

- Leaving a root valve on an instrument closed
- Leaving a SIF in bypass, i.e.:
  - Bypassing the function due to a spurious trip and failing to remove bypass.
  - Bypassing the function for startup because the system dynamics require this, however, the designers missed this need during startup mode of the process resulting in an operational bypass that requires human intervention to remove the bypass rather than an automated design that removes the bypass.
  - Bypassing the final element and failing to remove bypass when the test or repair is complete

Therefore, a simple equation including the systematic error terms can replace Eq. No. 1a from ISA-TR84.00.02. The resulting improved equation is:

$$PFD_{SIF} \approx \sum PFD_{Si} + \sum PFD_{Ai} + \sum PFD_{Li} + \sum PFD_{PSi} + \sum PFD_{SYS\text{-}PROCi} + \sum P_{SYS\text{-}HUMi} \qquad \text{[Eq. A]}$$

Further, the overall systematic human error term can be expanded and written as:

$$P_{SYS\text{-}HUM} = P_{design\ error} + P_{installation} + P_{proof\ test\ error} + P_{bypassed} \qquad \text{[Eq. B]}$$

Of the four terms in this equation, the first two can be detected and corrected during initial commissioning steps for the SIF. Experience has shown that the last two terms, $P_{proof\ test\ error}$ and $P_{bypassed}$ are likely to dominate the $P_{SYS\text{-}HUM}$, though more industry data is needed to support this observation. Making the assumption that $P_{SYS\text{-}HUM}$ is dominated by $P_{proof\ test\ error}$ and $P_{bypassed}$, Equation D can be further simplified to:

$$P_{SYS\text{-}HUM} \approx P_{proof\ test\ error} + P_{bypassed} \qquad \text{[Eq. C]}$$

Experienced gained from many accident investigations and also from calculations, support the contention that for high SIL designs, the human errors during interventions $P_{proof\ test\ error} + P_{bypassed}$, dominate the calculated PFD_{SIF}. Unfortunately, most of the SIL verification calculations today use the truncated Eq. No. 1 (instead of 1a) from ISA-TR84.00.02: [19]

$$PFD_{SIF} = \sum PFD_{Si} + \sum PFD_{Ai} + \sum PFD_{Li} + \sum PFD_{PSi} \qquad \text{[EQ. 1]}$$

As a result, most SIL Verification calculations today ignore systematic errors when quantifying their risk reduction capability. This is equivalent to saying the system boundary for an SIF only includes the instrumented components (a subsystem), or Boundary A in Figure 3, instead of including the entire independent protection layer (IPL) system, shown as Boundary B in Figure 3. In LOPA and other quantitative risk assessments, the entire IPL system must be considered.

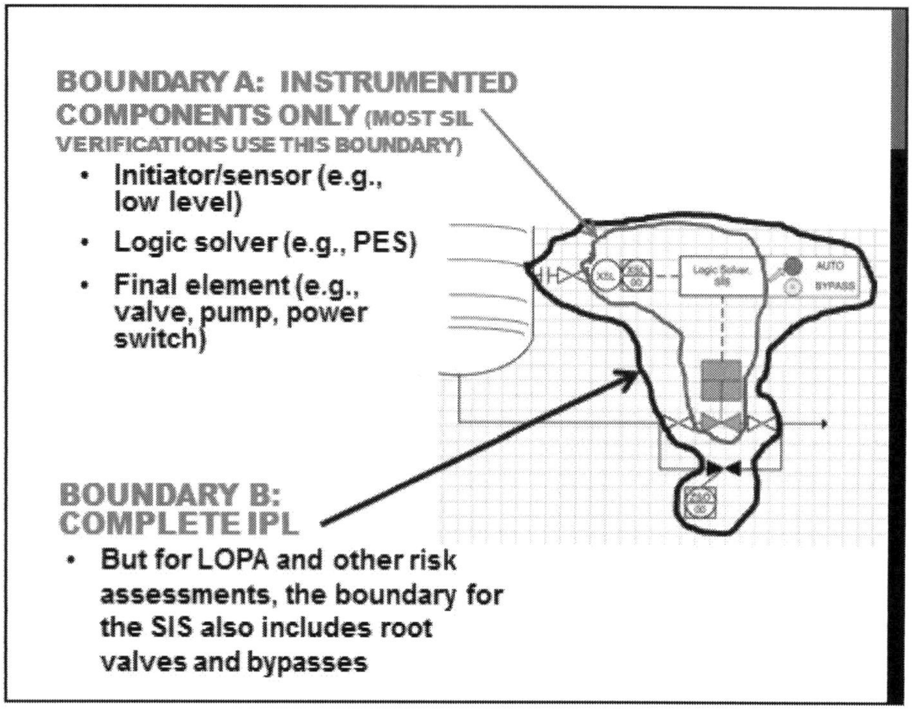

Figure 3: Boundary for SIF

For example, as mentioned earlier in this paper, if the IPL is a PSV, then the IPL system must include upstream and downstream features, such as isolation valves. Therefore, the probability of leaving an isolation valve closed should be included as a contribution to the overall PFD of the PSV IPL system.

Analyst performing LOPA should understand how to view and analyze the appropriate boundary for an IPL, including a SIF.

For the remainder of this Case Study, we will concentrate on the effect of including the systematic human errors for calculation of PFD_{SYS}, and the effect this has on the resulting SIL. For convenience, this paper arbitrarily sets $\sum PFD_{SYS\text{-}PROC} \approx$ zero. This is being done so the reader can better focus on the human error aspect of systematic errors. $PFD_{SYS\text{-}PROC}$ is worthy of its own paper as its significance is generally dependent on the process fluid properties and/or ambient conditions.

The next two sections of this paper provide a basis for (1) the baseline error rate for human error during interventions and (2) the error rates given coupling of activities, such as occur with redundant systems. Following that, simple examples are provided to help show the relative impact of including systematic human error terms in the calculation of PFD_{SYS}.

Human Error Probability for a Single Execution of a Rule-Based Task

To calculate $P_{SYS\text{-}HUMi}$, the type of tasks must be defined and the baseline error rate for such a task needs to be established. Note that with excellent control of all of the human factors, a company can begin to approach the lower limits that have been observed for human error. Excellent control of all human factors means a robust design and implementation of management systems for each human factor are achieved with a high level of operational discipline. The first well-researched publication detailing potential lower limits of human error probability was by Alan Swain and H Guttmann (NUREG-1278, 1983)[18] and by others. However, many times, the limits they referenced get used out of context. The lower limits in the NUREG-1278 assume excellent human factors, but such excellent control is rarely, if ever achieved. Additionally, some human errors listed by Swain and others were for a single error under highly controlled conditions, or on a "best day" instead of average error probability or rate over an average year of tasks. In general, Process Improvement Institute (PII) has found it best to use the average error probabilities as discussed in the following section.

Error Probability for Rule-Based Actions that are Not Time Dependent:

Actions that do not have to be accomplished in a specific time frame to be effective are not time dependent. It should be obvious then that these do not include response to alarms, or similar actions with time limits. Values listed below represent the lower limits for human error rates, assuming excellent control of human factors; these are expressed as the probability of making a mistake on any step:

- 1/100 - process industry; routine tasks performed 1/week to 1/day. *This rate assumes excellent control of all human factors. Most places PII visits, the workers and managers and engineers believe this is achievable, but not yet achieved.*
- 1/200 - pilots in the airline industry; routine tasks performed multiple times a day with excellent control of human factors. *This average has been measured by a few clients in the airline industry, but for obvious reasons they do not like to report this statistic.*
- 1/1000 - for a reflex (hard-wired) action, such as either proactive or minor corrective actions while driving a car, or very selective actions each day where your job depends on getting it right each time and where there are error recovery paths (such as clear visual cues) to correct the mistake. *This is about the rate of running a stop sign or stop light, given no one is in front of you at the intersection; the trouble is measuring this error rate, since you would have to recognize (after the fact) that you made the mistake.*

See earlier papers[3,20] for more details on this topic.

Adjusting the lower limit rates to estimate a baseline rate at a site

As mentioned earlier, the lower limit rates assume excellent control of human factors in the industry mentioned. Note that airline pilots have a lower error rate than what PII has measured in the process industry. This is due, in part, to the much tighter control by the airlines and regulators on factors such as fitness-for-duty (control of fatigue, control of substance abuse, etc.). Excellent control of human factors is not achieved in many organizations; therefore the human error rates will be higher than the lower limit, perhaps much as much as 20 times higher. Table 8 provides adjustment factors for each human factor. These factors can be used to adjust the lower limit of error rate upward or downward as applicable, but the factors should not be applied independently.

Table 8. SUMMARY TABLE of 10 HUMAN FACTOR CATEGORIES

Based in part on Gertman, D.; et. al., *The SPAR-H Human Reliability Analysis Method*, NUREG/CR-6883, U.S. Nuclear Regulatory Commission, Office of Nuclear Regulatory Research, Washington, DC, August 2005.[6] PII has modified the list slightly to account for general industry data and terminology and to incorporate PII internal data.

Human Factor Category	Human Factor Issue/Level	Multiplier for Cognitive & Diagnosis Errors
Available Time	Inadequate time	P(failure)=100%
(includes staffing Issues) – *for responses only*	Barely adequate time (≈2/3 x nominal)	10
	Nominal time (1x what is expected)	1
	Extra time (at least 2x nominal and >20 min)	0.1
	Expansive time (> 4 x nominal and > 20 min)	0.01
Stress/Stressors	Extreme (threat stress)	5
(includes staffing issues)	High (time pressures such as during a maintenance outage; issues at home, etc.)	2
	Nominal	1
Complexity & Task Design	Highly complex	5
	Moderately complex (requires more than one staff)	2
	Nominal	1
	Obvious diagnosis	0.2
Experience/Training	Low	10
	Nominal	1
	High	0.5
Procedures	Not available in the field as a reference, but should be	20
	Incomplete; missing this task or these steps	8
	Available and >90% accurate, but does not follow format rules ***(normal value for process industry)***	3
	Good, 95% accurate, follows >90% of format rules	1
	Diagnostic/symptom oriented	1
Human-Machine Interface (includes tools)	Missing/Misleading (violates populational stereotype; including round valve handle is facing away from worker)	20
	Poor or hard to find the right device; in the head calc	10
	Some unclear labels or displays	2
	Good	1
Fitness for Duty	Unfit (high fatigue level (>80 hrs/wk or >20 hr/day, no day off in 7-day period; or illness, etc.)	20
	Highly degraded fitness (high fatigue such as >15 hr/day, illness, injury, etc.)	10
	Degraded Fitness (>12 hr day and >72 hr/wk)	5
	Slight fatigue (>8 hr per day; ***normal value for process industry***)	2
	Nominal	1
Work Processes & Supervision	Poor	2
	Nominal	1
	Good	0.8
Work Environment	Extreme	5
	Good	1
Communcation	No communication or system interference/damage	10
	No standard for verbal communication rules ***(normal value for process industry)***	3
	Well implemented and practiced standard	1

For instance, even in the worst situations, we have not seen an error rate for an initiating event or initial maintenance error higher than 1/5, although subsequent steps, given an initial error can have an error rate approaching 1 due to coupling or dependency.

- 1/5 - highest error rates with poor control of human factors; this high rate is typically due to high fatigue or some other physiological or psychological stress (or combination). This is the upper limit of error rates observed with poor human factors and within the process industry. *The error rates in the Isomerization Unit the day of the accident at BP Texas City Refinery (CSB, 2007)[21] were about this rate. The operators, maintenance staff and supervisors had been working about 30 days straight (no day off) of 12 hour shifts.*

For the examples provided later in this paper ***will use a baseline error rate of 0.02 errors per step***, which is about average at the sites PII visited in the past 10 years. This could be justified based on the fact that most chemical process sites do not control overtime during turnarounds and/or do not have a system for controlling verbal communication of radios and phones. In addition, for critical steps such as re-opening and car-sealing the block valves under a relief valve after the relief valve is returned from maintenance is about 0.01 to 0.04[22]; plus, the average probability of making an unsafe error during maintenance of a relief is 0.02[23,24,25]. Both of these tasks have multiple checks and have rigorously enforced procedures (similar to what is done when servicing a SIF and when using bypasses for an SIF) and yet the human error probability remains about 0.02.

Human Error Probability for Multiple Executions of a Rule-Based Task

Coupled (dependent) Error Rates: Coupling represents the probability of repeating an error (or repeating success) on a second identical task, given that an error was made on the first task. The increased probability of failure on subsequent tasks given that an error has already been made is known as dependence. The list below provides some starting point guidance on values to use:

- 1/20 to1/90 - if the same tasks are separated in time and if visual cues are not present to re-enforce the mistake path. *This error rate assumes a baseline error rate of 1/100 with excellent human factors. If the baseline error is higher, then this rate will increase as well.*
- 1/2 - if same two tasks performed back-to-back and strong visual cue is present, and if a mistake is made on the first step of the two. *This error rate assumes a baseline error of 1/100 with excellent human factors. If there the baseline error is higher, then this rate will increase as well.*
- 8/10 - if same three tasks performed back-to-back and strong visual cue is present, and if a mistake is made on the first two steps of the three.
- Two or more people become the same as one person (with respect to counting of errors from the group), if people are working together for more than three days; this is due to the trust that can rapidly build.

These factors are based on the relationships provided in NUREG-1278[18] and the related definitions of weak and strong coupling provided in the training course by Swain (1993)[26] on the same topic. The following relationship is for errors of omission, such as failing to reopen a root valve or failing to return an SIF to operation, after bypassing the SIF. The values in Table 9 are based SPAR-H, 2005[6].

Table 9 Guideline for Assessing Dependence for a within-SIF Set of Identical Tasks (based partially on SPAR-H, 2005[6], and partially on field observations by PII)

Level of Dependence	Same Person	Actions Close in Time	Same Visual Frame of Reference (can see end point of prior task)	Worker Required to Write Something for Each Component
Zero (ZD)	No; the similar tasks are performed by different person/group	Either yes or no	Either yes or no	Either yes or no
Zero (ZD)	Yes	No; separated by several days	Either yes or no	Either yes or no
Low (LD)	Yes	Low; the similar tasks are performed on sequential days	No	Yes
Moderate (MD)	Yes	Moderate; the similar tasks are performed more than 4 hours apart	No	No
High (HD)	Yes	Yes; the similar tasks are performed within 2 hours	No	No
Complete (CD)	Yes	Yes; the similar tasks are performed within 2 hours	Yes	Either yes or no

One can readily conclude that staggering of maintenance tasks for different channels of the same SIF or for related SIFs will greatly reduce the level of dependent errors. Unfortunately, most sites PII visits do not stagger the inspection, test, or calibration of redundant channels of the same SIF or of similar SIF; the reason they cite is the cost of staggering the staff. While there is a perceived short-term higher cost, the answer may be different when lifecycle costs are analyzed.

Once the level of dependence is known, the probability of either repeat success or repeating errors on identical tasks can be estimated. For these probabilities, we use Table 10, which is a re-typing of Table 20-17 from NUREG-1278[18] and the similar table in SPAR-H[6].

Table 10 Equations for Conditional Probabilities of Human Success or Failure on Task N, given probability of Success (x) or Failure (X) on Task N-1, for Different Levels of Dependence

Level of Dependence	Repeating Success Equations (but shown as error probability)	Repeating Failure Equations
Zero (ZD)	$P_{Succes@N} = x$	$P_{Failure@N} = X$
Low (LD)	$P_{Succes@N} = (1+19x)/20$	$P_{Failure@N} = (1+19X)/20$
Moderate (MD)	$P_{Succes@N} = (1+6x)/7$	$P_{Failure@N} = (1+6X)/7$
High (HD)	$P_{Succes@N} = (1+x)/2$	$P_{Failure@N} = (1+X)/2$
Complete (CD)	$P_{Succes@N} = 1.0$	$P_{Failure@N} = 1.0$

To illustrate the impact (sensitivity) on PFD_{SIF}, we will look at a simple case and will not provide the details on the calculation of the component aspects of PFD_{SIF}, but instead will provide the results of PFD_{COMP} to be the value obtained by using Equation A, but without the systematic error terms (the same as using Eq. No. 1 from ISA-TR84.00.02). Then we will show a simple way to estimate the system human error term ($PFD_{SYS\text{-}HUM}$) and show the resulting impact on PFD_{SIF}. Figure 4 shows a candidate SIL 2 SIF.

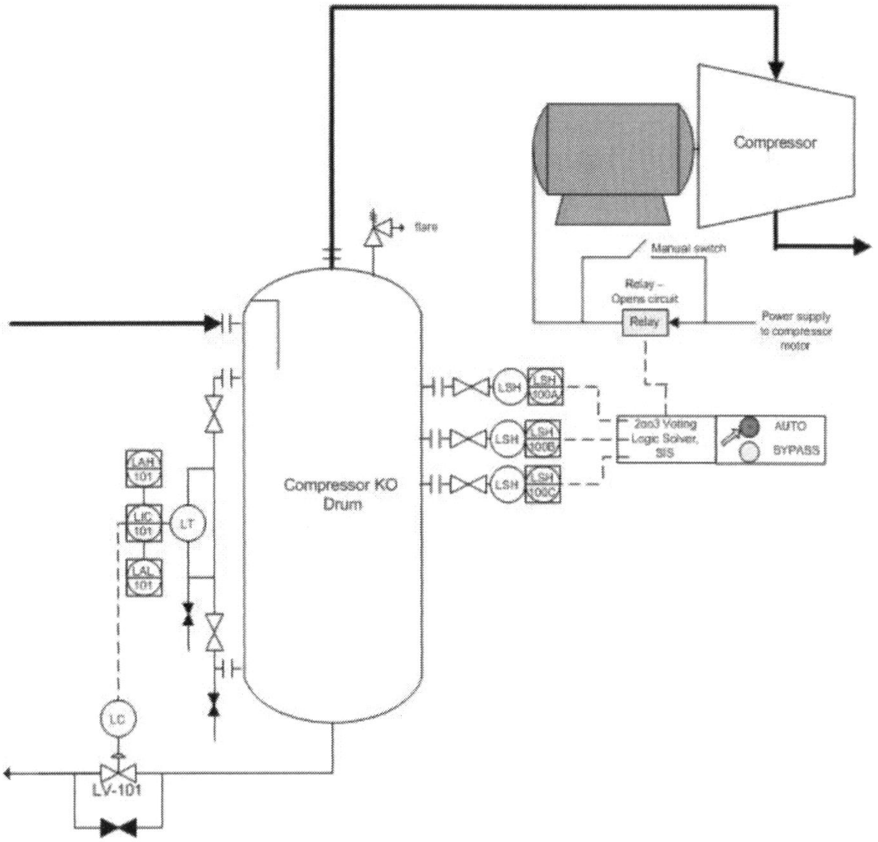

Figure 4. Example of SIL 2 SIF (high level trip of compressor motor)

For the SIL 2 SIF described in Figure 4, the component PFDs were estimated using standard, simplified equations for each, and using data available for component. For the case where the sensors are voted 2oo3, the PFD of the SIF without consideration of discrete systematic error yielded PFD_{COMP} = 0.008 (of which the 2oo3 voted sensor portion is 0.0025 and the 2oo3 voted logic solver is 0.003).

For this example, the term $\sum P_{SYS\text{-}HUM}$ is next estimated by summing the:

- Probability of leaving the level sensor/transmitters 2oo3 root valves closed, causing an unsafe failure. (This calculation is shown later.
- Probability of miscalibration of the level transmitter/switch. This calculation is shown later, but for this to be significant probability, all two of the three sensors/transmitters must be miscalibrated, unless there is comparison checking, then it would require miscalibration of all three transmitters.
- Probability of leaving the entire SIF in BYPASS after maintenance or after some other human intervention such as an inadvertent error or a necessity during startup; as before, we will use the base error probability of 0.02 as a starting point.
- Probability of leaving the relay bypass closed. As before, we will use the base error probability of 0.02 as a starting point.

To aid in the calculation of the probability of leaving 2oo3 root valves closed, we use an event tree to show the conditional probabilities for leaving Valve B closed, given Valve A is open or closed, and similarly, the conditional probability of leaving Valve C closed, given Valve A or B are closed or both Valve A and B are closed. Figure 5 shows the results of this calculation. For the branch probabilities, the equations for high dependency of the human actions were used (See Table 10); this reflects the more prevalent case of the maintenance of redundant channels being maintained on the same day, by the person, and that level valves are within the visual field of the worker. From Figure 5 the result for the probability of human error of leaving 2oo3 or 3oo3 of the root valves closed is 0.0129. But, the comparison checking between sensors/transmitters will alert the workers that a root valve is closed, so the only valid path is the 3oo3 path; the 3oo3 error case is the bottom row of the event tree in Figure 5. The probability of leaving all three root valves closed is 0.0077.

From the same figure, we can also extract the conditional probability of leaving 3oo3 sensors/transmitters bypassed; assuming comparison checking is in place to note deviations and correct the problem, only the case of 3oo3 errors is credible. This represents a strong recovery path for the previous errors. The 3oo3 error case is the bottom row of the event tree in Figure 5. The probability of miscalibrating all three sensors/transmitters is 0.0077.

Figure 5. Calculation of Conditional Probability of Opening Root Valves; with the Last Column Showing the Probability of Leaving Two or Three Valves Closed (using High Dependence Equations)

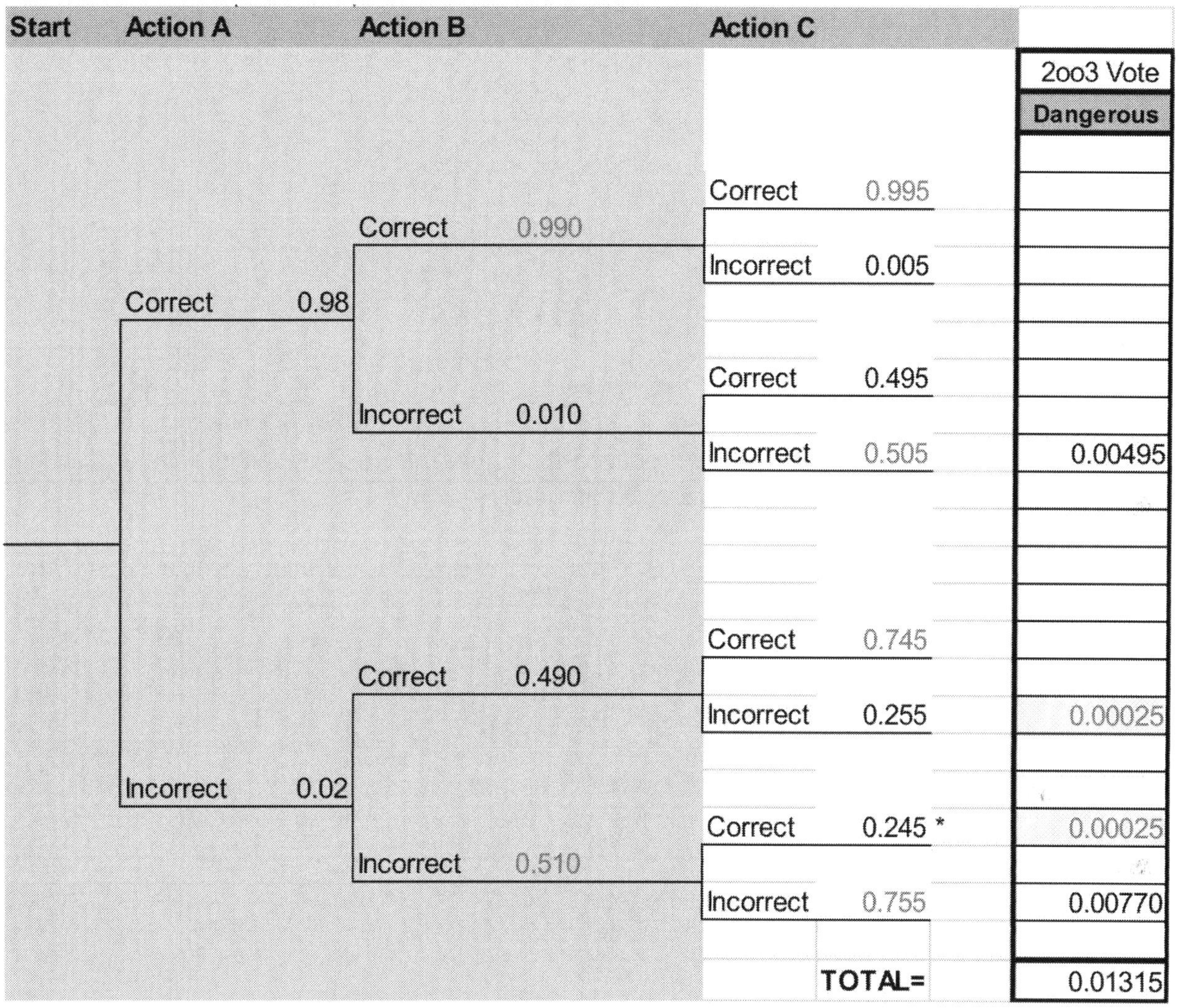

$$\sum P_{SYS\text{-}HUM} = 0.0077 + 0.0077 + 0.02 + 0.02 = 0.055$$

This would then give an overall failure probability for the SIF of:

$$\mathbf{PFD_{SIF} = PFD_{COMP} + P_{SYS\text{-}HUM}} = 0.008 + 0.055 = 0.063$$

Since the PFD is greater than 0.01, the instrumented system for high level protection in this example does not qualify as a SIL 2 SIF when accounting for human error probabilities related to interventions with the SIF.

Other variations are provided in an earlier paper, and are summarized below.[17]

One means to improve the reliability and independence of the instrumented system is to use a smart sensor/transmitter for the LSH which will detect null movement of the sensor reading, indicating the valve is closed on the tap is plugged. Another possibility is to implement a limit switch (or captive key system) on the root valve. There is a probability that these safeguards against human error will also fail or be bypassed by the staff, but assuming the probability of that failure is the same as other human errors for this example, 0.02, then the systemic human error drops to about zero as the probability of leaving the root valve closed is now ANDed with the probability of smart sensor/transmitter or limit switch failing,

$$\sum P_{SYS\text{-}HUM} = 0.0000 + 0.0077 + 0.02 + 0.02 = 0.048$$

In this case the revised PFD of the instrument system becomes:

$$\mathbf{PFD_{SIF} = PFD_{COMP} + P_{SYS\text{-}HUM}} = 0.008 + 0.048 = 0.056$$

Since the PFD is still greater than 0.01, the instrumented system for high level protection still does not qualify as a SIL 2 SIF when accounting for human error probabilities related to interventions with the SIF. But, we have reduced the errors related to dependent failures during checking of the sensors/transmitters as much as possible.

As another alternative (instead of using smart sensors/transmitters or instead of installing limit switches on the root valves) we can reduce potential dependent human error by staggering maintenance activities across different shifts. This would drop the dependence to Low. The result for the probability of human error of leaving 3oo3 of the root valves closed in 0.00016 (assuming that comparison of sensor readings alerts the workers that one root valve is closed).[17]

We can also calculate the conditional probability of leaving 3oo3 sensors/transmitters. As before, only the case of 3oo3 errors is considered credible, since it was assumed that sensor comparison checking was implemented where any transmitter not miscalibrated will provide the workers an opportunity to note the deviation and take corrective action to fix the problem; this represents a strong recovery path for the previous errors. The probability of miscalibrating all three sensors/transmitters is 0.00016.[17]

$$\sum P_{SYS\text{-}HUM} = 0.00016 + 0.00016 + 0.02 + 0.02 = 0.040$$

This would then give an overall failure probability for the SIF of:

$$\mathbf{PFD_{SIF} = PFD_{COMP} + P_{SYS\text{-}HUM}} = 0.008 + 0.042 = 0.050$$

Since the PFD is ***still*** greater than 0.01, the instrumented system for high level protection still does not qualify as a SIL 2 SIF when accounting for human error probabilities related to interventions with the SIF. The weak link in this design is again the human error probability of leaving either the relay bypass closed or the probability of leaving the entire SIF bypassed. This is a common concern on all SIF that have system bypasses. The most effective way to drop these error rates is to eliminate the capability for bypassing the relay and to eliminate the capability for bypassing the entire SIF. Or; we can install a parallel relay with a selector switch so that one relay (and only one) is aligned in the circuit to the motor of the compressor. This will likely drop the relay systemic human error probability from 0.02 down to 0.0004 or lower. The toughest bypass to eliminate is the one for the entire SIF. This is usually only feasible on batch systems or on continuous operations that can be shut down completely for each test interval. [17]

Sensitivity to Baseline Human Error Rate: Obviously, if the baseline human error probability increases to 0.04 due to extra fatigue or extra stress due to schedule constraints, the PFD of the systematic human error will increase substantially and the SIL 2 target becomes even less attainable. Likewise, if suitable operational discipline is adopted to reduce the baseline human error with independent performance measurement to validate the results, the human error rate will be reduced (though it is likely not possible to reduce the baseline human error probability enough to achieve a SIL 2 target, if a SIF bypass is present).

6. Acronyms Used

1oo2	One out of two voting architecture
1oo3	One out of three voting architecture
2oo3	Two out of three voting architecture
3oo3	Three out of three voting architecture
λ	Failure Rate
A	Final element
CD	Complete Dependence
COMP	Random hardware failure contributions to overall PFD
D	Dangerous
F	Failure/error term
HD	High Dependence
HRA	Human Reliability Analysis
IE	Initiating Event
IEF	Initiating Event Frequency
IPL	Independent Protection Layer
LD	Low Dependence
L	Logic Solver
LOPA	Layer of Protection Analysis
MOC	Management of Change
P	Probability
PES	Programmable Electronic System
PFD	Probability of Failure (dangerous) on Demand
PII	Process Improvement Institute, Inc.

PS	Power supply
PSV	Pressure Safety Valve
S	Sensor
SIF	Safety Instrumented Function
SIL	Safety Integrity Level
SIS	Safety Instrumented System
SYS	Systematic failure contributions to overall PFD
SYS-HUM	Systematic errors and failures generated by human error
SYS-PROC	Systematic errors and failures generated randomly by the process
TI	Proof Test Interval
ZD	Zero Dependence

7. Conclusion

As can be seen from the quantitative examples, systematic errors have the potential to significantly impact a SIF in a negative manner. In addition, SIL verifications performed today often do not account for this contribution to probability of failure. In such cases, it becomes increasingly likely that the risk reduction assumed by analysts (who rely upon a SIL 2 to have a PFD of 0.01 or lower) is not sufficient to satisfy corporate risk criteria when the actual risk reduction estimated for the IPL is being counted on, such as an SIF replacing a relief valve, as opposed to analyses that are simply performed on a comparative basis where consistency is more important than the actual numbers.

The paper points to the need for companies to begin:

- Accounting for systematic (and especially human systematic error probability) in SIL Verifications; otherwise the risk reduction factor from this IPL will be unrealistically optimistic.
- Taking a more in-depth look at the management systems and work process in place for operations and maintenance and their associated training and revalidation of performance.

Utilizing the mathematics presented, companies can gain insight as to the relative effectiveness of their practices and find areas where improvements can be made without adding any real cost. Just as improved human factors improve safety, this is one of those cases where improved performance if done properly with true operational discipline, should also improve reliability and plant availability.

8. References

1. *Guidelines for Preventing Human Error in Process Safety*, CCPS/AIChE, 1994.
2. *Human Factor Methods for Improving Performance in the Process Industries*, CCPS Concept Book, John Wiley, 2007.
3. Tew, R. and Bridges, W., *Human Factors Missing from PSM,* LPS/GCPS (AIChE), 2010.

4. University of South Australia (UniSA), Adelaide Centre for Sleep Research, SDA *Fact Sheet - AT03*, 2006.
5. U.S. NRC, *Fitness for Duty Programs, Preamble*, 10 CFR 26, Revised 2005.
6. Gertman, D.; Blackman, H.; Marble, J.; Byers, J. and Smith, C., *The SPAR-H Human Reliability Analysis Method,* NUREG/CR-6883, U.S. Nuclear Regulatory Commission, Office of Nuclear Regulatory Research, 2005.
7. Della Rocco, P.S., et al., *The Effects of Napping on Night Shift Performance,* Federal Aviation Administration, U.S. Department of Transportation, 2000.
8. Bridges, W., *LOPA and Human Reliability,* PPSS/GCPS (AIChE), 2010.
9. Rasmussen, J., *Chemical Process Hazards Identification*, Reliability Engineering and Safety Systems 24, Elsevier Science Publishers Ltd., UK, 1989.
10. Bridges, W., and Clark, T., *How to Efficiently Perform the Hazard Evaluation (PHA) Required for Non-Routine Modes of Operation (Startup, Shutdown, Online Maintenance)*, AIChE, 2011.
11. *Guidelines for Hazard Evaluation Procedures*, 3rd Edition, 2008, CCPS/AIChE.
12. *Guidelines for Investigating Chemical Process Incidents, 2nd Edition,* CCPS/AICHE, 2003.
13. Bridges, W., *Gains in Getting Near Misses Reported (Updated),* 8th GCPS, AIChE, 2012.
14. Bridges, W., and Clark, T., *LOPA and Human Reliability – Human Errors and Human IPLs (Updated),* 7th GCPS, Chicago, AIChE, 2011.
15. *Guidelines for Chemical Process Quantitative Risk Analysis (CPQRA), 2nd Ed*, CCPS/AIChE, 2000.
16. Stack, R., and Delanoy, P., *Evaluating Human Response to An Alarm for LOPA or Safety Studies,* LPS/GCPS, 2010.
17. Bridges, W., and Thomas, H., *Accounting for Human Error Probability in SIL Verification Calculations*, 8th GCPS, AIChE, 2012.
18. Swain, A., and Guttmann, H., *Handbook of Human Reliability Analysis with Emphasis on Nuclear Power Plant Applications,* Sandia National Laboratories, 1983 [this document became NUREG/CR-1278– *The Human Reliability Handbook*, guidelines from the US NRC on Human Reliability Analysis].
19. ANSI/ISA 84.00.01-2004 (IEC61511-1 Mod), *Functional Safety: Safety Instrumented Systems for the Process Industry Sector - Part 1: Framework, Definitions, System, Hardware and Software Requirements,* 2004.
20. Bridges, W., and Collazo-Ramos, G., *Human Factors and their Optimization*, 8th GCPS, AIChE, 2012.
21. U.S. Chemical Safety and Hazard Investigation Board, *Investigation Report, Refinery Explosion and Fire,* Report No. 2005-04-I-TX, March 2007.
22. *Guidelines for Initiating Events and Independent Protection Layers*, 2013 (pending), CCPS/AIChE.

23. Bukowski, Julia V. and Goble, William M., Villanova University, *Analysis of Pressure Relief Valve Proof Test Data, Process Safety Progress, AICHE, March 2009.*
24. Bukowski, Julia V. and Goble, William M., Villanova University, *Analysis of Pressure Relief Valve Proof Test Data: Findings and Implications, 10th Plant Process Safety Symposium, AICHE, 2008.*
25. Bukowski, Julia V., *Results of Statistical Analysis of Pressure Relief Valve Proof Test Data Designed to Validate a Mechanical Parts Failure Database,* Technical Report, September, exida, Sellersville, PA, 2007.
26. Swain, A., *Human Reliability Analysis*, Training Course, ABS Consulting (formerly JBF Associates), 1993.

Take the Guess Work Out of Testing Your Safety Instrumented System

Charles M. Fialkowski, CFSE
Siemens Industry Inc.
Spring House, PA
Charles.Fialkowski@siemens.com

Prepared for Presentation at
American Institute of Chemical Engineers
2013 Spring Meeting
9th Global Congress on Process Safety
San Antonio, Texas
April 28 – May 1, 2013

UNPUBLISHED

Take the Guess Work Out of Testing Your Safety Instrumented System

Charles M. Fialkowski CFSE
Siemens Industry, Inc.
Spring House, PA
Charles.Fialkowski@siemens.com

Keywords: Process Safety, safety instrumented systems, IEC 61508, ANSI/ISA 84, IEC 61511, Safety PLC, architectural constraints, proof tests, Safety Integrity Level, SIL, Probability of Failure on Demand, PFD

Abstract

A Safety Instrumented System (SIS) is used to implement one or more safety instrumented functions (SIFs). It is comprised of three main components; sensor(s), logic solver(s), and final control elements(s) with the purpose of monitoring potentially hazardous process operations and taking action if a dangerous condition is present.

While that may seem simple enough, it's often difficult to understand since these systems are typically dormant and may not be called to perform for months if not years (just how confident are you that the airbag in your car would still work after 10 years?). Due to this, the only way to evaluate its performance is to utilize statistics to calculate the probability of successful operation over time. This concept is the focus of performance-based standards such as IEC 61508, IEC 61511 and ANSI/ISA 84, which define how to calculate performance throughout the system's life cycle.

One commonly used method to improve the SIS performance is to shorten the functional 'proof' test interval which will have a quantitative impact on the calculated system performance. On paper this appears good; however, confusion exists on how to actually test a complex, microprocessor-based logic solver (i.e. safety PLC) and how effective that test really is (would you trust any mechanic to test the airbag in your car, and if so, how confident are you that they covered every possible failure?).

This paper questions the validity of periodic functional proof testing of microprocessor-based logic solvers and evaluates the role of diagnostics and redundancy. It will shed light on the fact it's virtually impossible to test a modern, state-of-art, safety PLC with any measure of success and/or performance improvement.

1. Introduction

It has long been considered (and practiced) that each safety instrumented function (SIF) be periodically proof tested to ensure the calculated safety performance is validated. The responsibility of conducting these tests has always been on the owner/operator with little to no guidance offered by the equipment manufacturer. This issue has become even more challenging when dealing with the complexity of today's programmable electronic system (PES), often referred to as a safety PLC (programmable logic controller).

In order to properly conduct a test, one must understand the following:

- How often is a test required (monthly, yearly, every 5 years, etc.)?
- How do you perform the test?
- How much coverage can you claim with the test (70%, 90% or 100%)?

Periodic proof testing of standard PLCs (used for safety functions) has always been extremely important due to their lack of diagnostics, but today's safety PLC manufacturers are claiming better than 99% diagnostics coverage, so what are the users really testing for? In order to better understand this, we must go back to first principles and understand why testing is needed and what value it provides.

2. Definition of a SIS

Clause 3.2.72 Part 1 from the ANSI/ISA 84 (modified IEC 61511) defines safety instrumented system (SIS) as: instrumented system used to implement one or more safety instrumented functions (SIF). An SIS is composed of any combination of sensor(s), logic solver(s), and final elements(s). The mission of this system is to take the plant to a safe condition when specified operating boundaries are violated.

Simply stated, it's a system made out of special components (inputs, control, and outputs) that monitor specific process control parameters (e.g. temperature, flow, pressure, level, etc.) for potentially dangerous conditions. If for any reason those parameters are exceeded, the system is designed to take the process to a pre-defined safe state. This implies that if a potentially dangerous condition is 'never' reached, the SIS will 'never' act. Thus, determining how well the system will perform when or if required becomes a very confusing issue.

For example, if you have a car that is ten years old and has never been in an accident, how confident would you be that the air bag would still work in the event of a collision? Or, if you were a five-year veteran fighter pilot and were issued one parachute your first day of service, and nobody had ever tested it in those five years, how confident would you be that it would still deploy if you needed it one day? How often would you have your parachute tested, who would test it, and how? These are all very significant questions when it comes to dormant systems.

3. Measuring the performance of a dormant system

When we say "dormant system" we do not imply an idle system, but quite the opposite. A simple analogy is like the guards at Buckingham Palace. They don't move or talk with the public; in fact, to many they don't even look real. However, we're pretty certain that if anyone were to unlawfully penetrate the palace, these once motionless guards would sense the disturbance, and stop it. An SIS stands dormant like the palace guards, executing only when a potentially condition is present. Deciding what process parameters are to be monitored, what the safety actions should be, and how well the system needs to operate are challenges for the process engineer who must reflect this in their functional and integrity specifications.

Functional safety performance standards were developed because it was apparent that prescriptive normalization was not practical for two obvious reasons:

a) all processes are not the same
b) prescriptive standardization is labor intensive as technology changes rapidly

As a result, performance-based standards were developed to help evaluate and calculate the required performance needed for dormant systems.

The ANSI/ISA 84 standard was initially released in 1996 and defined four (4) discrete Safety Integrity Levels (SIL 1 thru 4), where each SIL describes an order of magnitude improvement based on the average probability of a function failing to perform. This term is called the Probability of Failure on Demand (PFD).

Each Safety Instrumented Function (SIF) has the probability of functioning correctly, often referred to as availability (Figure 1). This same safety function also has a probability of failing and is divided up into two different types of failures, probability of failing safely (PFS) or probability of failing dangerously (PFD). The PFS would result in a spurious or nuisance trip of the system, even if it is a safe one.[1] The PFD meanwhile simply means that you've lost your safety protection, and may go undetected for months if not years, and may only be discovered during the periodic proof test.

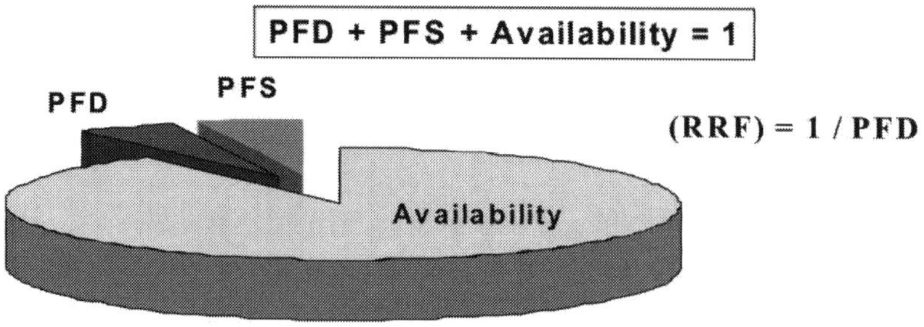

Availability: Probability of being successful
PFD: Probability of failing dangerously or "on demand"
PFS: Probability of failing to a safe state. (Spurious trips)

Table 1: Failure modes of a system

[1] Unnecessary shut downs may themselves be a source of hazard in certain applications.

PFD is not a constant value; rather it is a direct function of time. That is, if we consider a constant dangerous failure rate (λD) for PFD smaller than 0.1, the Probability of Failure On-Demand of a given system, can be simplified as:

$$PFD = \lambda_D * Ti \qquad [Eq. 1]$$

Where: Ti is the time between inspections.

And: λ_D is the dangerous failure rate (assumed constant)

Over time the PFD increases (worsens) as shown in Figure 2. In order to evaluate the performance of a SIF during its entire lifecycle, it is necessary to integrate and define a constant reference parameter. This is known as the PFD average (PFD_{AVG}).

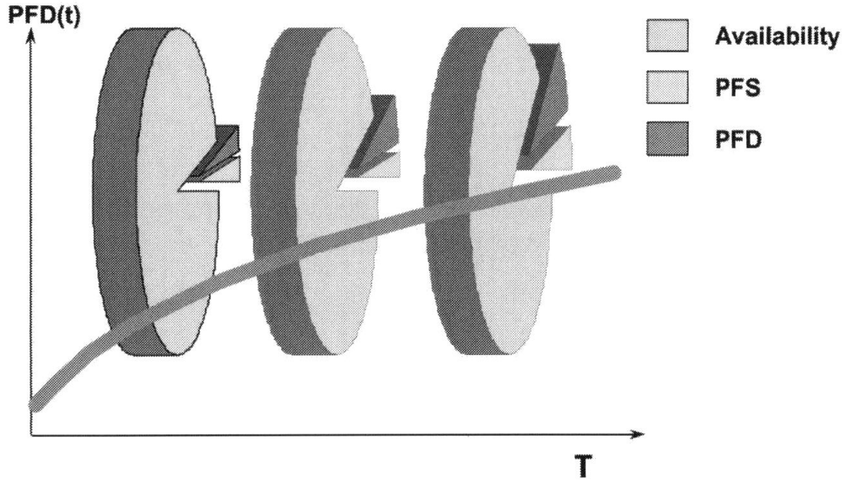

Figure 2: PFD changes as function of time.

Therefore; the PFD_{AVG} is defined as:

$$PFD_{AVG} = 1/T \int_0^{Ti} PFD\,(t)\,d(t) \qquad [Eq. 2]$$

This means that for one instrument in a SIF – in a simplex (non redundant) architecture often referred to as a One out of One (1oo1) -- the PFD_{AVG} will be given by:

$$PFD_{AVG} = ½\,Ti\,[\,\lambda_D \times Ti]^2 = ½\,\lambda_D\,Ti \qquad [Eq. 3]$$

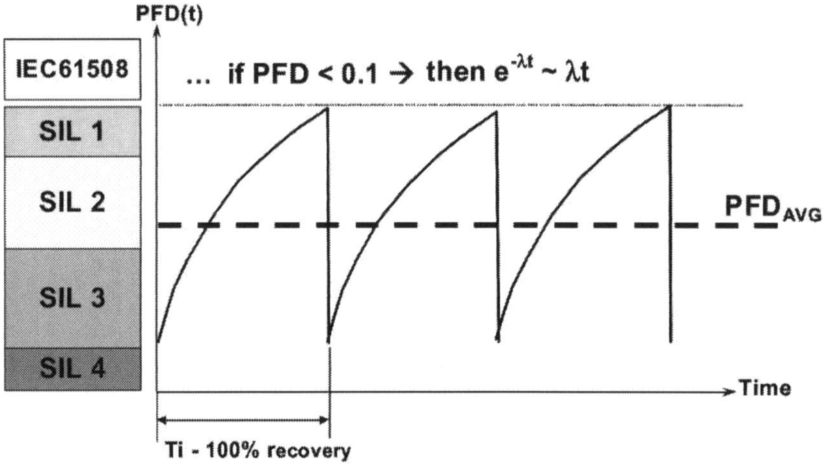

Figure 3: PFD curve for 1oo1 with perfect testing.

Figure 3 shows the curve for the PFD of a simplex (1oo1) component, assuming that the system is inspected, tested and recovered 100% (like new). The average value for this single component meets the safety performance equal to SIL2.

Table 1 describes the performance requirements for "on demand mode operation" equipment, (Ti >> Demand rate). This is consistent with dormant operation.

SIL	Probability of Failing on Demand (PFD_{AVG}) (Demand Mode Operation)	Risk Reduction Factor (FRR)
SIL 4	0.0001 - 0.00001	10000 a 100000
SIL 3	0.001 - 0.0001	1000 a 10000
SIL 2	0.01 - 0.001	100 a 1000
SIL 1	0.1 - 0.01	10 a 100

Table 1: Performance requirement for SIF "Low Demand" Mode (IEC 61508)

Unfortunately, 100% recovery is not realistic as this would assume perfect testing was conducted. Therefore, we need to breakdown the dangerous failure modes to further understand the impact of periodic proof testing. Dangerous failures (λD) can be broken down to include: dangerous detected (λDD), dangerous undetected (λDU), and dangerous never detected (λDN) failures, as shown in Figure 4.

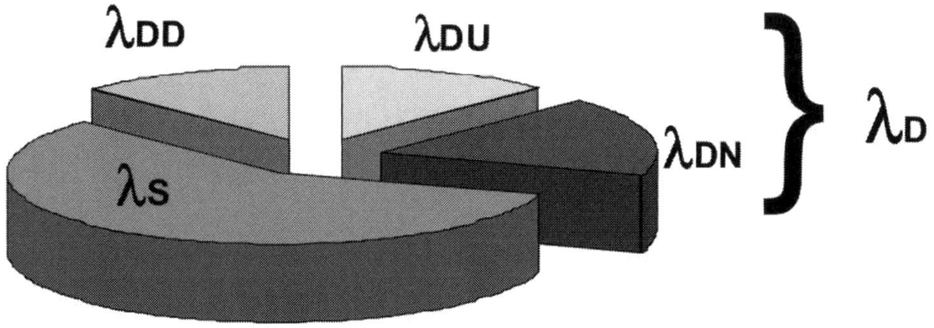

Figure 4: Failure Modes and associated failure rates.

Dangerous failures will ultimately determine the performance of your system (i.e. PFD_{AVG}).

1. λDD = Dangerous failures which are detected by automatic system diagnostics, and converted to safe failures will not negatively impact your PFD_{AVG} calculation. These will only affect the system during the time between automatic testing (typically very short).
2. λDU = Dangerous failures which are not detected by the system diagnostics, will affect the PFD_{AVG} until an alternative way of detecting such failures is used (a thorough manual test procedure).
3. λDN = Dangerous failures that are never be detected. These will affect the PFD_{AVG} throughout the entire life of the device.

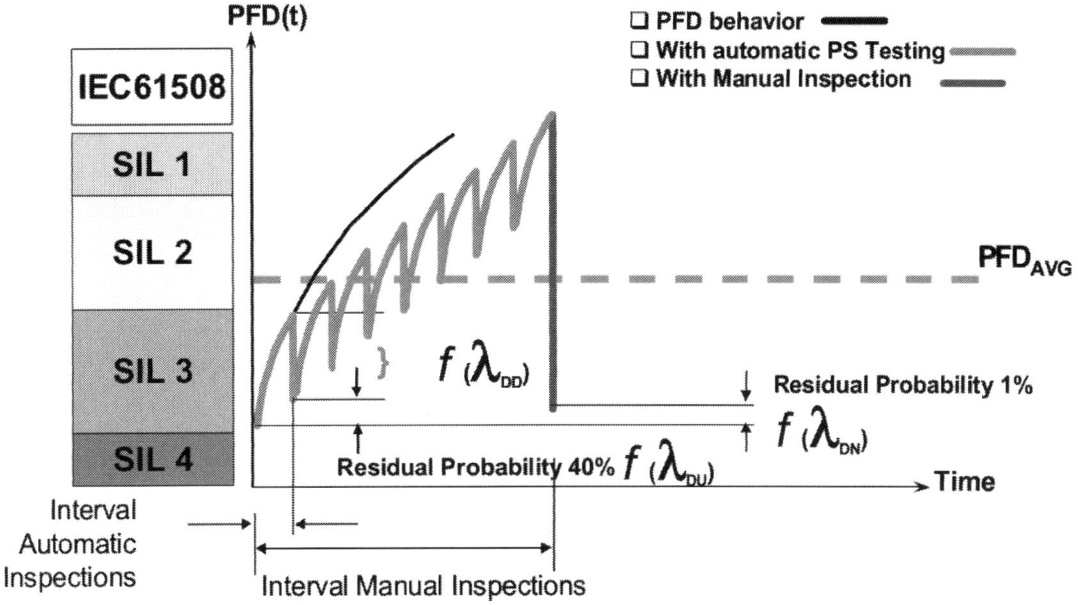

Figure 5: Performance impact with imperfect testing.

4. Calculating PFD_{AVG} with imperfect testing

Everything discussed so far applies to systems where all failure modes are well defined; their failure rates are well known, and their behavior under a failure is perfectly understood - these are identified as TYPE A devices. Devices such as on/off valves, electrical mechanical relays, mechanical switches, etc. are generally considered as TYPE A.

When the failure modes of a system are not well understood because of its complexity or its failure rates cannot be determined accurately or when its behaviour under fault is not completely understood, the system is classified as TYPE B.

Such cases have been recognized by today's safety standards where additional requirements have been specified to meet specific SIL. This is because even if a very thorough testing of the design is carried out to determine the PFD_{AVG} of the system, not all failures will be analyzed; thereby making dangerous undetected failures (λ_{DU}) equal to dangerous never detected failures (λ_{DN}).

That is, the dangerous undetected failures which are not detected during the systems automatic diagnostic testing will in fact NEVER be DETECTED, and therefore, manual proof testing becomes meaningless.

Table 2 shows how this concept translates within the IEC 61508, table of architectural limitation.

Type A vs. Type B Architectural Constrains

Safe Failure Fraction (SFF)	Hardware Fault Tolerance					
	0		1		2	
	Type A	Type B	Type A	Type B	Type A	Type B
< 60 %	SIL 1	Not Allowed	SIL 2	SIL 1	SIL 3	SIL 2
60 % - 90 %	SIL 2	SIL 1	SIL 3	SIL 2	SIL 4	SIL 3
90 % - 99 %	SIL 3	SIL 2	SIL 4	SIL 3	SIL 4	SIL 4
> 99 %	SIL 3	SIL 3	SIL 4	SIL 4	SIL 4	SIL 4

Table 2: IEC 61508 performance limit table

The Safe Failure Fraction (SFF), Hardware Fault Tolerance (HFT) and TYPE of equipment (A vs. B) must all be considered together for each sub-system when claiming a SIL level. This is done to prevent over realistic SIL levels, particularly when non-redundant architectures are being used.

The SFF measures the natural tendency of an instrument to fail safe or automatically detect dangerous failures. It is usually calculated using methods such as an FMEDA (failure modes and effect diagnostic analysis) FMEDA is a bottoms up approach designed to identify design flaws and unexpected results when components within the system fail.

For example, in a single component in a 1oo1 architecture (0 hardware fault tolerance) a Type A device has one level improvement over that of a Type B device unless both have extremely high diagnostic coverage (expressed as more that 99% SFF).

Claiming 99% diagnostic coverage for a Type B device is typically not achievable by "manual testing" as there would still have some failures which are not known or well defined (the fraction is so small) that they would be accepted as negligible.

5. Impact of Common Cause Failures

ANSI/ISA 84 defines a common cause failure as "failure, which is the result of one or more events, causing failures of two or more separate channels in a multiple channel system, leading to system failure"

Thus, common cause is a key factor that affects redundant systems. Its contribution is expressed as a percentage called the beta value (β), and since the dangerous undetected failure rate of a PLC becomes "never detected", it affects the system throughout its entire life.

As failure modes are not well understood and/or defined in PLCs, common cause becomes even more important.

For example, consider the PFD_{AVG} of an arrangement of two pressure switches (Type A equipment) for high safety performance.

In this case:

$$^{1oo2}\mathbf{PFD_{AVG}} = [2 \times (\lambda_{DD})^2 \times (MTTR + TIA/2)^2] + [((\lambda_{DU})^2 \times (TIM)^2) / 3] + [((\lambda_{DN})^2 \times Life^2) / 3] + [2 \times TD \times \lambda_{DU} \times (((TIM/2) + MTTR) / TIM)] + [\lambda_{DU} \times \beta \times TIM/2]$$

But for a PLC, (Type B equipment) where realistically it is recognized that:

1. Those dangerous undetected failures by automatic diagnostics cannot be detected by manual testing, (So no testing on line is performed)

2. $\lambda_{DU} = \lambda_{DN}$

Then,

$$^{1oo2}\mathbf{PFD_{AVG}} = [2\ (\lambda_{DD})^2 \times (MTTR + T_{IA})^2] + ([\lambda_{DN} \times Life]^2)/3 + \lambda_{DN} \times \beta \times Life$$

And, common-cause becomes a dominant contributor as it affects are present the entire life of the system.

This difference is around one order of magnitude, from the fact that manual testing is usually done yearly while PLCs usually last a couple of decades.

6. Conclusion

It is virtually impossible to manually test a complex system (such as a modern, IEC 61508 certified, safety-rated PLC) that its automatic diagnostics don't already test for. For years, end-users have struggled to understand the impact that periodic testing would actually have on their overall system performance. There are several reasons:

1. Type B components' failure modes are not well defined and are too complex to test.
2. Manual periodic testing is a good recommended practice for all Type A (or very simple Type B) devices.
3. Since manual periodic testing could only account for a fraction of failures that had not already been diagnosed by automatic diagnostics, PLCs should be sent back to the manufacturer for proper re-validation. Systems should be guaranteed to perform during that time (example 10 to 20 years) without depending on manual testing.
4. If the user is going to continue to periodically test their PLC and claim diagnostic credit, they must have clear direction from the manufacture on the following:
   - ✓ How often to test?
   - ✓ How to perform the test?
   - ✓ How much diagnostic credit can be claimed for this test?

Of course, one may still argue that functionally testing each SIF is good to determine if programming bugs are present, however the authors conclude that by following the good engineering practice of the safety lifecycle, both prior and post commissioning programming bugs would have be revealed during validation tests.

References

[1] Gruhn, Paul, Cheddie, Harry., "Safety Instrumented Systems: Design, Analysis and Justification" 2nd ed., 2006 ISA – The Instrumentation, Systems and Automation Society, Research Triangle Park, NC, USA

[2] Goble, William M, Cheddie, Harry., "Safety Instrumented Systems Verification, practical probabilistic calculations" 2005 ISA – The Instrumentation, Systems and Automation Society, Research Triangle Park, NC, USA

[3] ANSI/ISA 84.01-2004 (IEC 61511 Modified), Functional safety: Safety Instrumented Systems for the process industry sector, 2004, Instrumentation, Systems, and Automation Society, Research Triangle Park, NC, USA

Tips for the Creation and Application of Effective Operating Procedures

Colin R. Scholtz
Risk Management Professionals, Inc.
300 Goddard, Suite 200, Irvine, CA 92618
Colin.Scholtz@RMPCorp.com

Steven T. Maher, PE CSP
Risk Management Professionals, Inc.
300 Goddard, Suite 200, Irvine, CA 92618
Steve.Maher@RMPCorp.com

Prepared for Presentation at
American Institute of Chemical Engineers
2013 Spring Meeting
9th Global Congress on Process Safety
San Antonio, Texas
April 28 – May 1, 2013

UNPUBLISHED

Tips for the Creation and Application of Effective Operating Procedures

Colin R. Scholtz
Steven T. Maher, PE CSP
Risk Management Professionals, Inc.
300 Goddard Suite 200, Irvine, CA 92618
Colin.Scholtz@RMPCorp.com

Keywords: Operating Procedures, Process Safety Management (PSM), Risk Management Plan (RMP)

Abstract

Specific requirements exist in all Safety Management Systems Guidelines/Requirements (e.g., Process Safety Management, Risk Management Program, Safety and Environmental Management Systems) for the creation, content, quality review, and periodic update/certification of Operating Procedures. However, the creation and actual implementation of Operating Procedures is fraught with challenges that often result in safety deficiencies, regulatory citations, and in some cases, unfortunate tragedies.

Although Operating Procedures concepts involve the straightforward documentation of specific steps and precautions to take for safe and effective operation, many process facilities struggle with:

- securing the focus and specific content from operations personnel for the creation of quality procedures
- securing feedback from operations personnel if procedural steps do not coincide with actual practices
- ensuring the steps outlined in procedures avoid introducing additional process hazards
- creating procedures that are in a user-friendly format and written to facilitate training and understanding
- identifying the most effective level of information and depth to include in the procedure
- determining sensible levels of detail for consequences of deviation and the appropriate corrective actions
- implementing meaningful competency testing
- assigning the correct "skill set" to the development and update of Operating Procedures
- addressing all modes of operations, including defining appropriate responsibilities

Beyond the requirements, good-quality Operating Procedures are critical for encapsulating operational best practices for all of the different operating modes that the process may have to address and also provide a basis for ensuring consistent quality assurance. The objective of this paper is to convey an understanding of the challenges that must be considered with the development of Operating Procedures and provide tips and specific examples that will facilitate the creation and ongoing application of this important element of Safety Management Systems.

1. Introduction - Operating Procedures Defined

1.1 What are Operating Procedures?

As one element of Process Safety Management (PSM), Operating Procedures are a part of a management program designed to manage safety for facilities with hazardous materials or processes. In addition to risk reduction, Operating Procedures also provide a documented way to consistently perform a specific task while minimizing deviations. Generally, Operating Procedures, sometimes referred to as "Standard Operating Procedures" (SOPs) or "General Operating Procedures" (GOPs), are a collection of instructions on how to safely and consistently operate a process. Different facilities may have different interpretations of what constitutes an operating procedure and how it may differ from other forms of procedures or manuals. The following definition, taken from the Center for Chemical Process Safety's (CCPS) "Guidelines for Writing Effective Operating and Maintenance Procedures" [1], provides the concept that will be focused on in this paper.

> Operating Procedures: Written step-by-step instructions and associated information (cautions, notes, warnings, etc.) for safely performing a task within operating limits.

There are several key concepts in this definition that will be further highlighted and expanded upon in this paper, but it is important to note that Operating Procedures aim to address both the safety concerns of a process and keeping it within its operating limits or design intent. This intent is often captured in an operating manual. Though this term is sometimes used interchangeably with Operating Procedures, for the purposes of this paper the operating manual may contain a set of individual operating procedures as well as a collection of operating and process safety information which includes the operating intent [2].

A final distinction should be made between operating and maintenance procedures. While Operating Procedures focus on the performance of a process, maintenance procedures provide the instructions to perform general maintenance to ensure equipment operability, and typically fall under the Mechanical Integrity element of Process Safety Management.

1.2 Why are Operating Procedures Important?

Written operating procedures are an integral part of an effective Safety Management System (SMS) program that provide a number of benefits to the facility, its operators, and the general public. The key driving factor in the requirement for Operating Procedures is to avoid or minimize the occurrence of potentially disastrous incidents. As part of this,

Operating Procedures are required for both infrequent, high-risk tasks and more frequent, low-risk tasks that maintain the facility within its operating intent. As past incidents and near misses have shown, failures of Operating Procedures, often in conjunction with failures of other PSM elements, have led to hazardous events that have resulted in injuries and loss of life.

While reducing the risk of process safety incidents through minimizing human error is the main goal of Operating Procedures, a good program also provides operational benefits. These benefits include ensuring consistency of operations, which is important for facilities that have multiple operators operating the same system. Different personnel may operate the same system using different methods and with different limits in mind. A comprehensive set of Operating Procedures should capture those deviations obtained through general operating experience and should document the most effective method. Through an iterative process, Operating Procedures create a documented way to safely and efficiently operate the process.

2. Regulatory Requirements (RMP/PSM)

In order to establish a basis for the development of the Operating Procedures, this section will provide an overview of the Risk Management Plan (RMP) and PSM regulatory requirements for Operating Procedures. As shown in Figure 2.1, Operating Procedures (OP) make up a significant portion of regulatory citations. The additional elements listed are Process Safety Information (PSI), Process Hazard Analysis (PHA), Mechanical Integrity Inspection and Testing (MI – Insp. & Test), Recommendations (Recom.), Mechanical Integrity Operations (MI – Oper.), Management of Change (MOC)).

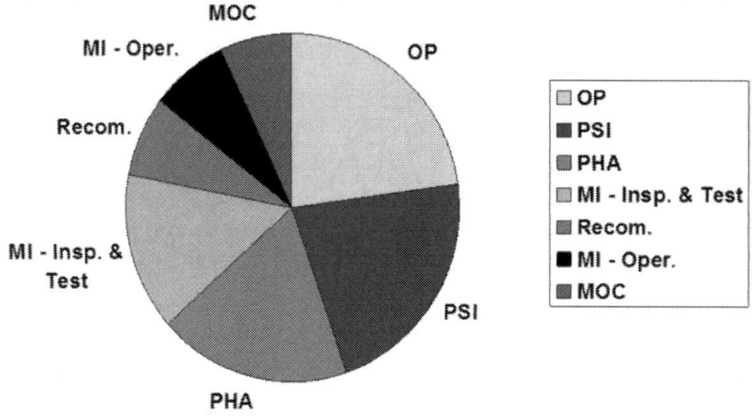

Figure 2.1 – Select OSHA Citations. [3]

The EPA's RMP Rule lays out specific requirements for Operating Procedures dependent upon program level. Program 2 (40 CFR §68.52) [4], the less stringent, outlines the basic requirements, while Program 3 (40 CFR §68.69), which is analogous to OSHA's PSM standard (29 CFR §1910.119(f)) [5], requires a more all-encompassing Operating Procedures program.

The following discusses the requirements for RMP Program 2 facilities. First, the facility must ensure that operating procedures address all of the modes of process operation, listed below.

- Initial startup
- Normal operations
- Temporary operations
- Emergency shutdown (ESD) and operations
- Normal shutdown
- Startup following a normal or emergency shutdown or a major change that requires a hazard review

In addition, the facility's procedures are required to address consequences of deviation and steps to correct or avoid deviations (collectively referred to as "consequences of deviation"). Procedures must also address inspections of process equipment. The final requirement is that Operating Procedures must be updated to be accurate whenever a change in the process occurs.

RMP Program 3 and OSHA PSM requirements build upon the Program 2 requirements to develop a more stringent Operating Procedures program. Expanding on the different operating modes, Program 3 and PSM require that procedures for emergency shutdown include the conditions under which an emergency shutdown is required and the assignment of shutdown responsibility. In addition to consequences of deviation, the procedures must also address the safety and health considerations of the chemicals in the process and any safety systems and their functions. The facility must also implement procedures that cover safe work practices, such as lockout/tagout, confined space entry, opening process equipment or piping, and control over entry into a stationary source. It is also important to note that the safe work practices also apply to any contractors working on the covered process. The final requirement for RMP Program 3 and PSM is that Operating Procedures must be certified annually to be current and accurate. The table below provides a comparison chart between Program 2 and Program 3/PSM.

Table 2.1 – RMP Program Comparisons

Procedure Requirements	RMP Program 2	RMP Program 3 / OSHA PSM
Written Operating Procedures	X	X
Address all operating modes	X	X
Emergency shutdown conditions		X
Operating limits		X
Consequences of deviation	X	X
Equipment inspections	X	
Safety and health considerations		X
Safety systems and their functions		X
Readily accessible		X
Current and accurate	X	X
Annual certification		X
Safe work practices		X

3. Development of Operating Procedures

3.1 Resources

Writing effective Operating Procedures, like most projects, requires identifying goals, establishing a schedule, and gathering and assigning resources. Gathering the proper resources provides a good foundation for developing an outline of the required procedures. Resources include technical information specific to the process and are typically comprised of the following:

- Operating manual and control narrative
- Piping and Instrumentation Diagrams (P&IDs)
- Process Hazard Analyses (PHAs)
- Material Safety Data Sheets (MSDS)
- Vendor Installation, Operation, and Maintenance manuals (IOMs)
- Relevant existing operating procedures

It is important to ensure this information is up-to-date prior to use. Older facilities may have undergone several upgrades or additions that may not have been captured in the technical information. Having the most accurate information generally means that fewer revisions will be required later on. Corporate standards can also serve as a basis for information to contain in the procedures. These can include company-wide safe work practices or Job Safety Analysis (JSA) exercises and may be incorporated by reference. For facilities already in operation, experienced operators and engineers are also a valuable resource that may be able to provide additional information gained from hands-on operation of the process. Compiling these resources helps get to the end goal of encapsulating the relevant knowledge for safe and effective operations.

Aside from facility technical information, the facility must also dedicate personnel resources to the procedure development team. The team usually consists of personnel with relevant expertise to the process and should also reflect the goals and expectations of the stakeholders of the procedures. The figure below provides an overview of a basic team.

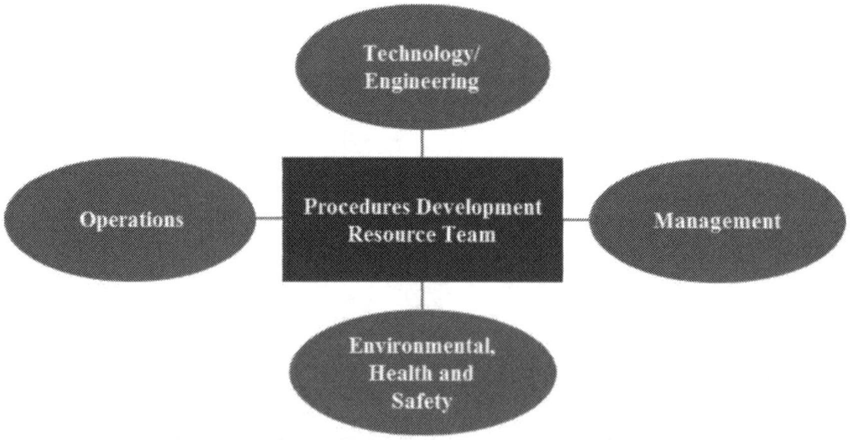

Figure 3.1 – Example Resource Team

As shown in the figure, the resource team is comprised of individuals with experience in the process and other stakeholders in the Operating Procedure program. This includes operations personnel (the end-user of the procedures), engineering, environmental, health, and safety (EHS), and management. Maintaining a balanced team with each group involved ensures that the end product will satisfy the goals of each of the stakeholders. In addition, other individuals may be involved through reviewing and clerical support to provide a resource for consistency.

3.2 Tips for Effective Procedure Development

It should be noted that there is no single, correct way to develop procedures and the needs and requirements will vary from facility to facility; however, this paper will provide a few general tips that may be applicable. Once all relevant information has been gathered, it must be put together in both a useful and useable way. It is important to keep in mind that the end user of operating procedures is always the operator. A format that is user-friendly to the operators will ensure that the procedures are a document they want to use. The following are some tips to consider when beginning the development:

- <u>Template:</u> The template should be approved by all stakeholders (operations, engineering, EHS, management, etc.). This ensures that all parties involved in the project, from development to continued use and updates, are able to get their desired information into the procedure without overloading it with superfluous information.
- <u>Format:</u> Easy readability and user-friendliness are goals to strive for when developing the procedure's format. In addition, consistent formatting across the facility's operating procedures will minimize possible deviations stemming from operators switching between differently formatted procedures.
- <u>Referencing:</u> More often than not, processes are intertwined, and often, operations in one unit will affect operations in another unit. To avoid creating long, cumbersome procedures, it is recommended to break down processes into separate units. Maintaining a template that is easily referenced by other procedures (and other policies in place at the facility) will help streamline jumping between operations and reduce clutter.
- <u>Depth of Information:</u> The most effective level of information should also be established, to streamline detail in procedure and prevent a lack of consistency. If a procedure presents too little information then it becomes more of a guide and less of a step-by-step procedure, which undermines its ability to control specific hazards. If too much detail is present, then the procedures may become cluttered with irrelevant and useless information that may be covered in operator training. Once the desired depth has been established, the training program should be adjusted to match the Operating Procedures.

To provide an outline of the procedure and make it easier to read, some procedures make use of the "T-Bar" format. In this layout, the page is divided into two columns (creating a "T" with the table header) with outline steps in the left column and specific details on the right. Table 3.3 provides an example of the T-Bar format.

Table 3.1 – Example T-Bar Format

Procedure		
Step		Details
1.	Fill the Tank.	Open the 2" valve. Close the valve when the level reaches 6'.
2.	Start the Pump.	Open the suction and discharge valves on the Pump. Start the Pump by turning the switch to the "On" position.
3.	Monitor conditions.	Monitor the level in the Tank. Ensure the Pump is functioning properly. Stop the Pump as described in Step 4 when the level reaches 1'.
4.	Stop the Pump.	Stop the Pump by turning the switch to the "Off" position. Close the suction and discharge valves on the Pump.

With this format, an operator can easily see the general actions to operate the process without getting bogged down by too many details. In scenarios where a few steps need to be quickly verified, an operator can skim down the left column to the desired action and verify the specific details. This format is also applicable to some procedures where multiple operators are required to perform the procedure. The left column could be substituted to contain the roles while the right column contains the assigned actions. This ensures that each role is clearly designated. In these cases, flowcharts may be helpful as a visual representation of how the task is to be completed [1].

After the format and template are established, then draft development may begin. First, each procedure must have a clearly defined purpose. The purpose serves as a focal point for the procedure and all information in the procedure should help the operator complete the defined task. Information contained in the procedure that does not help accomplish the purpose is typically extraneous material and can cause confusion.

Writing good Operating Procedures is in iterative process – drafts are developed, reviewed, and revised. As previously mentioned, the end goal is to get a straightforward, useable document that provides instructions to safely operate the process. As part of this, it is important to maintain clarity in each step of the procedure. A common shortfall of procedures occurs when a single "step" lists multiple actions to be performed. This increases the general clutter of the procedure and devalues the step. Consider the following example.

Table 3.2 – Example Step Comparison

Procedure A		Procedure B	
#	Step	#	Step
1.	Fill the Tank by opening the 2" valve. When the level reaches 6', close the valve.	1.	Open the 2" valve to fill the Tank.
2.	Open the suction and discharge valves of the Pump and then start the pump by turning the switch to the "On" position.	2.	Close the 2" valve when the level reaches 6'.

Procedure A		Procedure B	
#	Step	#	Step
3.	Monitor the level in the Tank.	3.	Open the suction and discharge valves on the Pump.
4.	When the level in the Tank reaches 1', stop the Pump and close the suction and discharge valves.	4.	Start the Pump by turning the switch to the "On" position.

While it may be clear in both examples how to perform the operation, Procedure B is more concise and in a quick reference is easier to follow. It should also be noted that each step in a procedure dictates an action, and these actions have a measurable outcome (e.g., it is measurable that the valve has been opened). It should also be noted that the capitalized "Tank" and "Pump" refer to the names of a specific tank or pump. Capitalizing and referring to the exact name ensures no confusion between different tanks or pumps.

To provide information that may help the operator in the next step, provide some operational background, or warn of potential hazards, notes, cautions, and warning statements are often used to designate the information as separate from a step. Table 3.2 provides information on the use of notes, cautions, and warning statements.

Table 3.3 – Notes, Cautions, and Warning Statements

	NOTE:	Notes are information only and are typically limited to: • information needed to perform the action safely • prepare the user for the unexpected • explain the reason for the step • reference other materials • help the user know when the action begins or ends • give other pertinent information regarding the action step.
	CAUTION:	Cautions are statements that are not part of the actual step but are used when an incorrect action is likely to lead to adverse consequences to unit operation, poor product quality, equipment damage, or harm to the environment.
	WARNING:	Warnings are statements that are also not part of the action step but used when an incorrect action is likely to lead to personnel injury or harm to the surrounding community.

4. Implementation and Ongoing Integrity

4.1 Successful Implementation

Operating Procedures, like the other elements of RMP/PSM, are a living program. Once a program has been put in place and procedures have been developed, continued involvement is necessary to maintain an effective program. The following are some management tools of effective Operating Procedures.

- Ownership/Control: Having a document owner ensures that procedures are taken care of and updated properly, which includes annual certification for RMP Program 3 and PSM facilities. RMP Program 2 facilities can benefit from document control by ensuring procedure accuracy and consistency.
- Accessibility: The RMP/PSM regulations require that operating procedures are accessible to those operating the process. Typically, controlled copies of the procedures are kept on the company's intranet, with printed, uncontrolled copies at key workstations. Maintaining electronic copies allows all employees to access the procedures for reference. In addition, accessibility control ensures that if changes are to be made to procedures, they are done so properly.
- Integration with Operator Training: Incorporating Operating Procedures into the operator training program helps familiarize operators with the process and procedures. Training also serves as a good avenue for securing feedback from operators for continuous improvement.
- Reviews and Audits: The RMP/PSM Compliance Audit is required every three years, but it is recommended that smaller, Operating Procedure-focused audits be incorporated into the annual certification program.

These tools will help ensure that the Operating Procedure program stays up to date and in-use at the facility. However, as the last bullet pointed out, continuous reviews are a reliable way to manage the content of procedures. Aside from audits, reviews may be initiated by other PSM elements such as Management of Change (MOC) or Pre Startup Safety Review (PSSR), Incident Investigations, or Process Hazard Analyses. Securing feedback during these reviews will also alert management to the need to retire obsolete procedure or the need of any additional procedures to be developed.

Over time, operators gain experience and operational knowledge of the process. Using management tools, this information could be extracted to help form typical operating ranges and troubleshooting guides, which help satisfy the operating limits and consequences of deviation required in procedures. Table 4.1 below provides an example.

Table 4.1 – Example Operating Limits

Operating Limits			
Parameter	**Limits**	**Consequences of Deviation**	**Steps Required to Correct or Avoid Deviation**
V-100 Level	30 – 70%	**Low** Potential gas blowby to Vessel V-200. **High** Potential liquid carryover to Compressor C-100.	Monitor level indicator LI-100 for low alarm LAH-100 or high alarm LAL-100. Adjust set point of level controller LIC-100 as necessary. Check level gauge LG-100 during rounds.

During initial procedure development, this information may be difficult to acquire. However, as operators become familiar with the process, this information could be tailored to the specific upsets that typically occur at the facility.

The following section provides a look at some common deficiencies in Operating Procedures as well as suggestions for improvement.

4.2 Common Deficiencies

As previously mentioned, many facilities are presented with challenges in their Operating Procedures. It is the objective of this section to point out some of the commonly deficient issues found in operating procedures. Listed below is a summary of these deficiencies, which are further elaborated in this section.

- Lack of annual certification
- Written procedures not synchronized with operator's actions
- All operating modes not addressed (especially temporary operations)
- ESD roles and responsibilities not clear
- Operating Procedures not current for all normal operations
- Procedures not in the language of the end user.
- Normal operating limits not documented
- Contractors not included

One of the most common difficulties for Program 3 and PSM facilities, especially large facilities with hundreds of procedures, is the annual certification of their procedures. It is often found that the procedures are not only lacking in annual certification but are outdated as well. The annual certification is required to ensure that the facility is aware that their procedures are up-to-date. A thorough review and update may not necessarily be needed if nothing has changed, however the facility must certify the procedures to show that they are being monitored. Table 4.2 provides an example of a sample procedure header to help track the annual certification.

Table 4.2 – Example Procedure Header

Operating Procedure for XXX				
Procedure: SOP-100	**Revision:** 01	**Effective Date:** **Next Review Date:**	May 2013 May 2014	**Owner:** Operations Manager
Objective: To complete Task XXX safely and efficiently.				

Another common deficiency is that Operating Procedures are not accurate for all modes of operation. Operations may constantly change throughout the life of a facility without being captured in the procedures. Though it may not be practical (or even possible) to address all the operating deviations that may arise, the procedures should match the general operating intent of the facility.

Many procedures also fail to address each mode of operation, as discussed under the regulatory requirements. For some facilities it is difficult to assign the operations performed to one of the modes of operations outlined in the regulations, particularly temporary operations. The objective of assigning operations to different operating modes is to ensure that all routine and non-routine operations are properly documented to reduce hazards. A facility may not have an operating procedure for a particular task, but it must be captured through a program (Operating Procedures, Job Safety Analysis, Management of Change, etc.) that will address all the potential safety and hazard concerns.

Synchronizing Operating Procedures with the operators' actions is also a challenge for facilities and could be reflective of several failures in the PSM system at the facility. In the Operating Procedures program, this could mean that the procedures have not addressed all oeprations or captured all relevant knowledge of the process. This could also reflect a deficiency in the Operator Training program, such as competency testing. Operators could be performing actions that deviate from the procedures because they have not been properly trained on why those specific steps outlined in the procedure are important. This is particularly important for hazardous processes where any deviations could potentially cause releases or exposure to hazardous environments.

When synchronizing procedures with operator actions (or vice versa) it is important to ensure that the procedures are written in the language of the end user. Procedures are typically written from an engineering standpoint which may not be useful to operations. Use of equipment tag numbers instead of the common name may lead to uncertainty, as shown in Table 4.3.

Table 4.3 – Example Equipment Clarification

Procedure A		Procedure B	
#	Step	#	Step
1.	Place PIC-237A in AUTO at 100 psig.	1.	Place PIC-237A Pressure Controller on Depropanizer Reflux Accumulator (V-237) in AUTO at 100 psig.

During a quick review of Procedure A, the operator may not be able to quickly recall what vessel PIC-237A controls. Adding the descriptor as shown in Procedure B eliminates all chances for confusion be specifying exactly which controller should be adjusted.

5. Conclusion

Although there are a number of challenges associated with the development and implementation of Operating Procedures, it is clear that they are an important element of RMP/PSM Programs and provide a number of benefits to the facility and its personnel. Without the proper management program in place, Operating Procedures will suffer from a variety of shortfalls leading to operational inconsistencies, regulatory citations, and possible process safety hazards. However, successful implementation will streamline operator training, reduce the risk of incidents, and ensure quality assurance.

6. References

[1] "Guidelines for Writing Effective Operating and Maintenance Procedures", Center for Chemical Process Safety, American Institute of Chemical Engineers, 1996.

[2] Sutton, I., "Writing Operating Procedures for Process Plants – A Practical Guide," Southwestern Books, 1995.

[3] Steinway, Seitz, Perry, and Siegel, "Before OSHA Comes Knocking...," Chemical Engineering Progress, March 2009.

[4] "Risk Management Programs for Chemical Accidental Release Prevention". Environmental Protection Agency, 40 CFR Part 68.

[5] "Process Safety Management of Highly Hazardous Chemicals". US Department of Labor, Occupational Safety and Health Administration, 29 CFR Part 1910.119.

Additional references not cited:

[6] American Institute of Chemical Engineers (AIChE) www.aiche.org/

[7] "Process Safety Beacon", CCPS, AIChE

9TH GLOBAL CONGRESS ON PROCESS SAFETY

Management Discipline: Defining a Process Safety Strategy

Jerry J. Forest
Celanese
1601 W. LBJ Freeway, Dallas, TX 75234
Jerry.Forest@Celanese.com

Prepared for Presentation at
American Institute of Chemical Engineers
2013 Spring Meeting
9th Global Congress on Process Safety
San Antonio, Texas
April 28 – May 1, 2013

UNPUBLISHED

Management Discipline: Defining a Process Safety Strategy

Jerry J. Forest
Celanese
1601 W. LBJ Freeway, Dallas, TX 75234
Jerry.Forest@Celanese.com

Keywords: Conduct of Operations, COO, Operating Discipline, OD, Engineering Discipline ED, Management Discipline, MD, strategy, strategic planning

Abstract

We often associate operational discipline with conduct of operations and define discipline as the tools to achieve repeatable results. While discipline in operations is primarily focused on the operator, engineers and management contribute to the success in achieving consistent results. A Plan-Do-Check-Act (PDCA) model, or Deming Cycle, can be used to represent the relationship among operational, engineering, and management disciplines. The management discipline part of the cycle involves making decisions (act) based on the output of the operations, and planning for success.

This paper will describe a modified strategic business analysis planning tool that managers can apply to process safety in order to achieve a desired vision and objective. Too often we are reactive to incidents and create activity lists from year to year in order to achieve objectives and goals. The process safety strategic business analysis tool is a structured approach of data analysis. It helps us to understand the company's internal and external environment in order to set strategic alternatives that will bring a company from its present state to the desired future state. By planning for and following a strategy, consistency in operations is achieved, and repeatable results can be obtained.

1. Introduction

In order to be successful in reducing process safety incidents, the causes of those incidents must be understood and eliminated. Since our manufacturing facilities are designed and operated by people, many of the causes are due to human error. Conduct of operations (COO) concerns the cultural aspects of an organization that are in place to define structural operational tasks designed to ensure that tasks are done correctly each time they are performed [1]. Management defines the process safety culture by setting expectations, providing accountability for performance, and allocating resources to achieve that performance [2]. Management must plan for success and set the strategic direction. This includes the COO tasks that are designed to eliminate human error.

This paper discusses a modified strategic business analysis process and tools to help the process safety manager define strategic alternatives for continuous improvement.

2. Management Discipline Model

COO is most often associated with the tasks that operators do when operating a processing or manufacturing facility. These tasks are actually elements of operating discipline (OD). They include making shift relief, writing shift notes, evaluating equipment, proper communication, and giving and receiving operating instructions. Haesle *et al.*, and the Center for Chemical Process Safety (CCPS) concept book Conduct of Operations and Operational Discipline present good overviews of OD [3][1].

But COO is more than OD. It also includes an element that consists of technical evaluation of process data, including data collected by operators in evaluation rounds and shift notes. These COO activities are often called engineering discipline (ED). Engineers or technical people bring predictability to OD by checking the human performance and acting on that data [4].

Managers often make the decisions to act on the data collected in OD and ED. In addition to acting on the data, managers should be planning for continuous improvement based on that data. Management discipline (MD) describes the activities that managers do to act on data and plan for continuous improvement in COO [4].

COO is therefore a combination of OD, ED, and MD; the combination of which describe a Plan-Do-Check-Act (PDCA) cycle as shown in Figure 1, first published in *Process Safety Progress* in 2012 [4]. When each element is in place, human error causes of incidents can be reduced.

3. Process Safety Strategic Planning

Planning for success can take many forms. It seems that often we are reactive to incidents and create long lists of items to correct. By being busy we think that we must be improving something. Another complaint often heard is that it is difficult to justify capital or expense spending for risk reduction. One cause might be that as technical people, we tend to communicate in technical terms rather than in business terms most often used by the decision makers who control budgets.

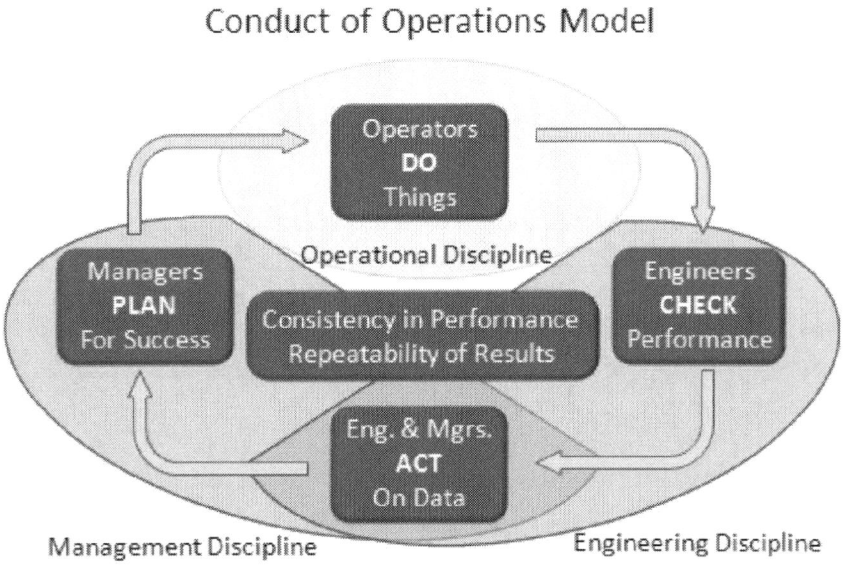

Figure 1: Conduct of Operations Model.

A way for mangers to plan for continuous improvement, break from the reactive cycle of activity lists, and provide a tool that is understood by business managers is by using a strategic business analysis (SBA) applied to process safety planning.

3.1 Strategic Business Analysis (SBA)

Strategic planning is having an understanding where you are as an organization; setting a vision for where you want to be; and developing strategic alternatives on how to achieve that vision. There are many tools that business managers use to understand the internal and external business environment, the competition, and other forces that might affect the vision or strategic alternatives. In this example the term SBA is used to collectively refer to the tools used to set the strategic plan, and these come from different sources and experience [5][6]. Analogous to Lean Six Sigma, no one tool is better than another and different tools might be used in different situations. For purposes of this paper, an 8-step process safety SBA is described. Table 1 illustrates the steps. The inputs, outputs, and process methodology for each are described and a case study is given at the end.

Table 1. The Process Safety SBA Process

Step	Description	Purpose
1	Beliefs	Align organization on fundamentals
2	Future State	Answers "where we are going"
3	Data Analysis	Answers "where we are at"
4	Risks Analysis	Pressure test; what could go wrong
5	Strengths, Weakness, Opportunities, Threats (SWOT)	Defines what we think of ourselves
6	Strategies	Preferred path and alternative paths
7	Action Plans	How we get there
8	Monitor & Improve	Check that plans are working

3.2 Beliefs & Future State

In order to make a strategic plan, an understanding of the existing organization is obtained to align the planning team on the fundamentals of where the organization is and how much work is required to achieve some future state. Therefore, define and write down what the organizational beliefs are.

For example, does your company really believe that all process safety incidents are preventable? If not, define exactly what the belief about incident prevention is so that you will align actions consistent with that belief. Other topics to examine in beliefs include: management systems, people and training, organizational structure, and process safety leadership. Use a simple brainstorming process. The output of this step provides a foundation for future steps, and helps define the organization's leveraging beliefs.

All strategic planning processes should include some definition of the desired future state, or vision. As Yogi Berra is often quoted, "If you don't know where you are going, you might wind up someplace else." The initial inputs to this process are a definition of company constraints (capital and expense, resources, etc.) and alignment with the overall company vision. The process can be a brainstorm session, or simply the given direction of senior management. The output is more than the vision, however. It should also include a consensus belief of where the organization believes it is going relative to peers, key performance indicator(s) (KPI's) to measure progress, and possible future scenarios that affect achieving the vision.

3.3 Data, Risk, and SWOT Analysis

3.3.1 Data Analysis

Data can come from several sources including process safety audits, incident investigation root cause analysis, and other leading indicators. It is not uncommon to complete an audit, develop a gap closure plan, and complete it. Similarly, after an investigation of a process safety incident is complete, we develop solutions that address causes and complete those actions. By stepping back and looking at this data as a whole, we can quickly identify common failure modes and management system weaknesses. Analysis of these lagging indicators should quickly surface the low hanging fruit from which to develop a strategy for improvement.

A methodology to group and objectively analyze and identify systemic management system failures was published in *Process Safety Progress* in 2011 [7]. It consists of structuring audit reports in a database type structure such that the subjective findings can be queried into logic groups. These groups can be used to describe the common systemic management system failures.

A simpler and more common type of data analysis is of the lagging indicators. Most companies document the results of incident investigation with various causal factor analyses. Various descriptive statistics coupled with graphical analysis can be performed on groups of data that will point to areas for improvement [8].

3.3.2 Risk Analysis
Understanding hazards and risks should be a part of any process safety strategic analysis. The inputs include the key beliefs developed earlier, industry data, and major risks from process hazard analysis. Typically, this tool involves a detailed look at data in order to expose scenarios that must be avoided, the critical process safety success factors, and what the competition is doing to be successful.

3.3.3 SWOT Analysis
A SWOT analysis is used in strategic planning to get a better understanding of both the internal and external business environment. Examination of a company's strengths and weaknesses results from analysis of existing internal beliefs. Similarly, analysis of the external business environment exposes external threats and opportunities.

This same type of analysis can be performed on process safety systems in order to surface existing beliefs in the company. In fact, by identifying beliefs, we are getting an idea of the process safety culture since culture is defined as a commonly held set of values, norms and beliefs [1]. Methodology to complete this type of survey includes use of a 5 point Likert scale and has been described in *Process Safety Progress* [9].

3.4 Develop Strategic Alternatives, Action Plans & Performance Monitoring

The results of the combined data, risk and SWOT analyses should show obvious areas for improvement. At this point, the strategic development team should analyze what strategies have worked in the past, and which have failed. Also note that the analysis will surface several areas to improve and therefore the team should develop strategic alternatives. Future conditions and assumptions may change. The organization needs to be in a position to change the strategy to match current conditions.

Note that strategic alternatives describe the overall improvement direction. Action plans define the activities aligned with the strategy, key milestones and KPI's to track success. The final step in execution of the strategy is monitoring performance so that changes in strategy and action plans can be made to optimize performance.

3.6 Process Safety SBA Summary

Table 2 summarizes the key inputs, processes, and outputs for each SBA step described.

Table 2. SBA Inputs, Processes, & Outputs

Step	Inputs	Process	Outputs
Beliefs	Experience, trends, fundamentals	Brainstorm	Common foundation for future work; Leveraging beliefs
Future State	Definition of company constraints, existing company vision	Brainstorm	Consensus beliefs of future scenarios, KPI's, overall vision
Data Analysis	Process Safety audit data, incident investigation root cause analysis, leading indicator data	Descriptive statistics, Pareto and other graphical analysis	Common failure modes, systemic weaknesses
Risks	Key beliefs, industry data, major	Sub-team analytical	Scenarios to avoid, critical process

Analysis	risks from PHA	work	safety factors, successful competitors
Strengths, Weakness, Opportunities, Threats	Results of the data and risk analysis	Survey	Key opportunities and strengths, competitive assessment, leveraging areas for improvement
Strategies	Results of data, risk and SWOT analysis	Assess past failures and successes, develop strategic alternatives, team meeting format	Defined process safety strategy and strategic alternatives
Action Plans	Strategic Alternatives	Team meeting format	Milestones, KPI's, action plans, communication plan
Monitor & Improve	Strategy, action plan	Existing performance monitoring processes	Update plans or strategy or choose new strategic alternative based on results

4. COO Strategic Planning Case Study

This SBA strategic planning process was used at Celanese to determine the process safety elements that needed most attention for the largest improvement. Discussed here are a few of the outputs from the tools used and one year results.

Most interesting were the results of the data analysis combined with the SWOT analysis. Descriptive statistics were applied to three years of loss of primary containment (LOPC) root cause studies. In summary, about half of all causes were human error, COO related. Of particular concern are the incidents involving operator line-up, operator evaluation, and failure to respond to alarm; all direct OD topics. This analysis also surfaced two additional areas for improvement: Mechanical integrity human error failures, and the need for improvement in incident investigation and assignment of causes. Audit finding analysis confirmed management system weaknesses in some areas of our COO implementation. Figure 2 shows a high level Pareto chart of the results of this analysis.

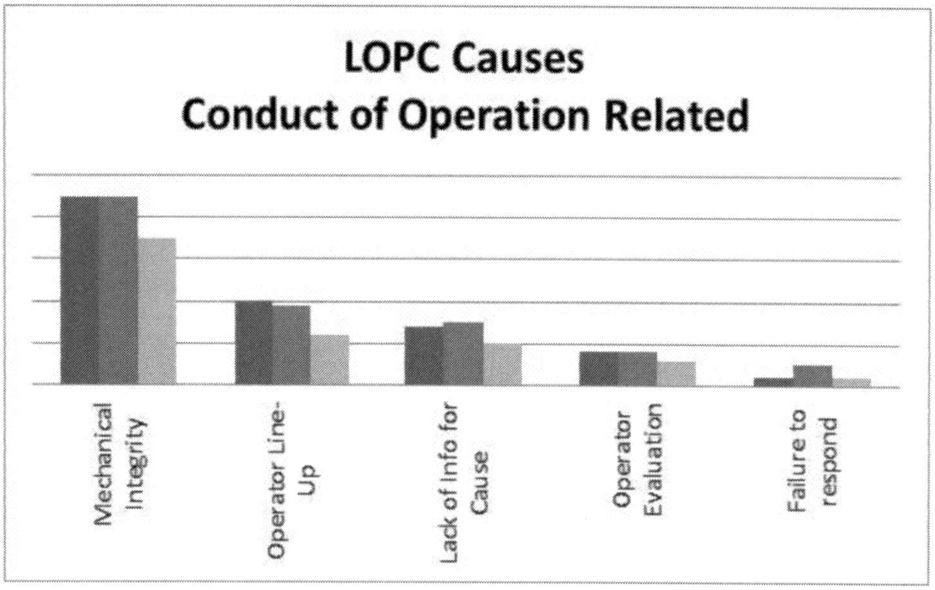

Figure 2: LOPC Causes Conduct of Operation Related

Analysis of lagging indicators also showed that for the time period studied, 80% of all incidents occurred at 5 sites.

The SWOT analysis surfaced a surprising result that the company, across the globe, indicated that there was an opportunity for improvement through internal networking with other sites and improving the process safety competency at the sites. The SWOT also showed a weakness in the area of process safety training and the need for improvement in how corporate procedures are executed at the site level. Once the strategy was set, the SWOT analysis helped define the action plan.

The data analysis gave rise to several strategic alternatives for process safety improvement. Table 3 illustrates four strategic alternatives for improvement.

Table 3. Strategic Alternatives

Objective	Strategy	Action Plan
Reduce # incidents	Human error reduction	1. COO workshops 2. COO Networks 3. Lessons Learned Project 4. Focused Site Support
Improve process safety competency	Management system improvement	1. Reformat procedures 2. Improve management process
Improve process safety competency	Improve leadership competency	1. Targeted annual training 2. New leader process safety certification process
Improve process safety competency	Improve operator competency	1. Improve and rollout process safety training for front line supervisors, operators and crafts.

Of course, each action is detailed with specific assignments, milestones, and KPI's.

Of particular interest for COO is the large number of incidents that are directly attributable to human error and the development of a targeted action plan to reduce human error. The overall improvement plan involved collection of COO best practices from across the company and development of a 2 day workshop to share those practices. Participants were asked to complete a gap analysis of their own site practices to the best practices and develop a closure plan specific for their site. We then followed up the action plans with network sharing of the practices, successes, and failures.

The result of implementation of the human error reduction strategy and action plans is reduction of COO related causes for process safety incidents from about half of the incidents to about one third in a one year time period. Most notable were improvements in operator line-up causes with various targeted programs on "walk-the-line" and pre-job focus on line-ups. While these results are encouraging, we recognize the need for continuous monitoring to show continued improvement, and make minor changes to the action plan and path forward based on the current conditions.

5. Conclusion

Conduct of operations is more than the operational discipline topics. In order to be most effective, it should include a management discipline element as well. The MD element acts on OD data and plans for continuous improvement.

A modified strategic business analysis can be used for process safety strategic planning. By analyzing incident and audit data, the common management system failures and human error related failures can be surfaced so that management can target specific COO areas for improvement. Tools such as risk and SWOT analysis will surface effective ways to implement action plans based on organizational belief's and what has worked before.

Since the second law of thermodynamics is still in place, it requires energy to monitor, maintain, and make changes to the strategic plan in order to achieve continuous improvement in process safety.

6. References

[1] Center for Chemical Process Safety, Conduct of Operations and Operational Discipline, American Institute of Chemical Engineers, Newark, NJ. 2011.

[2] OECD Environment, Health and Safety Chemical Accidents Programme, Corporate Governance for Process Safety, June 2012

[3] J. Haesle, C. Devlin, and J. McCavit. Improving process safety by addressing the human element, Process Safety Prog 28 (2009), 325-330.

[4] J. Forest. Management Discipline. Process Safety Prog 31 (2012), 334-336.

[5] Krajewski & Ritzman, *Operations Management Strategy and Analysis*, 4th ed., Addison-Wesley Publishing Company, Inc., Massachusetts, 1996

[6] Hutt & Speh, *Business Marketing Management A Strategic View of Industrial and Organizational Markets*, 5th ed., The Dryden Press, 1981

[7] J. Forest. Objective Analysis of Process Safety Audit Data with Microsoft Access. Process Safety Prog 30 (2011), 221-231.

[8] J. Forest & K. Kessler. Correlating Process Safety Leading Indicators with Performance. Process Safety Prog (2013), on-line edition.

[9] J. Forest. How to Evaluate Process Safety Culture. Process Safety Prog 31 (2012). 195-197.

9TH GLOBAL CONGRESS ON PROCESS SAFETY

Using Assessments to Improve Process Safety Culture

Russell A. Ogle,
Delmar "Trey" Morrison, III, and Sean J. Dee
Exponent, Inc.
4580 Weaver Parkway, Warrenville, IL 60555
rogle@exponent.com, tmorrison@exponent.com, and sdee@exponent.com

Prepared for Presentation at
American Institute of Chemical Engineers
2013 Spring Meeting
9th Global Congress on Process Safety
San Antonio, Texas
April 28 – May 1, 2013

UNPUBLISHED

Using Assessments to Improve Process Safety Culture

Russell A. Ogle,
Delmar "Trey" Morrison, III, and Sean J. Dee
Exponent, Inc.
4580 Weaver Parkway, Warrenville, IL 60555

Keywords: operational discipline, safety culture, game theory

Abstract

Deficiencies in conduct of operations and operational discipline (COO/OD) can lead to a decline in the quality of an organization's process safety culture and an increase in the likelihood of process safety incidents. A process safety assessment is one tool especially suited for identifying and correcting organizational safety culture deficiencies. Both AIChE/CCPS and API offer recent guidance documents on how to conduct effective assessments of process safety management systems.

This paper first introduces some basic concepts from game theory and applies them to the methodology of process safety assessments. A conceptual framework for the implementation of the process safety assessment process with balanced positive and corrective actions will be described. Assessment guidelines typically focus on the negative outcomes - correcting deficiencies. This paper argues that the assessment process must also use the positive outcomes - the compliance successes - to provide positive reinforcement to the process safety culture. Implementing this strategy is consistent with pragmatic management practices and is supported by empirical studies in behavioral science. Finally, this paper discusses how these ideas fit within the framework of AIChE/CCPS and API guidance documents.

1. Introduction

People are at the heart of any successful process safety management system. The term operational discipline refers to the commitment of each member of the organization to carrying out each task the right way, every time [1]. Deficiencies in operational discipline may become the precursors to a major catastrophic loss. Process safety assessments are an essential tool for evaluating the quality of the implementation of the process safety management system based on both lagging and leading metrics [2, 3]. Our focus in this paper is on the very specific use of the process safety assessment as a tool for identifying deficiencies in operational discipline. Here we use the term "assessment" to indicate a management system tool as opposed to the term "audit" which is a tool typically used for gauging regulatory compliance.

The theme of our paper is this: amongst their other important functions, process safety assessments should serve two important purposes: 1) to identify deficiencies in operational discipline as opportunities for improvement, and 2) to highlight those aspects of operational discipline that are implemented particularly well. The outcome of this approach is twofold: to correct undesirable behavior and to reinforce desirable behavior. In an assessment, it is easy to become focused on the deficiencies. However, if an organization wishes to create and sustain a high level of operational discipline in process safety, it is imperative that the organization creates a mechanism for recognizing and rewarding desirable behavior. The assessment is an excellent tool for that purpose.

The Center for Chemical Process Safety (CCPS), a directorate of the American Institute of Chemical Engineers (AIChE), has been very active in defining and promoting the use of process safety indicators [4, 5]. The CCPS references advocate the use of leading indicators as an early warning system to detect deficiencies in the process safety management system. The American Petroleum Institute (API) has published a guidance document API 754, which suggests several leading indicators that measure the quality of implementation of operational discipline and management system performance [6]:

1. Process Hazard Evaluations Completion
2. Process Safety Action Item Closure
3. Training Completed on Schedule
4. Procedures Current and Accurate
5. Work Permit Compliance
6. Safety Critical Equipment Inspection
7. Safety Critical Equipment Deficiency Management
8. Management of Change (MOC) and Pre Start-up Safety Review (PSSR) Compliance
9. Completion of Emergency Response Drills
10. Fatigue Risk Management

As indicated in API 754, these indicators are only suggestions and may need to be modified depending on the site-specific requirements of any given process plant. The specific methods for measuring these indicators are indicated in both the CCPS and API references. The key point here is that leading indicators of process safety are useful metrics for monitoring the state of the process safety culture. The realization of operational discipline, as measured by leading indicators, creates a strong process safety culture.

2. Why Game Theory?

Game theory is the scientific study of interactive decision-making [7]. It considers the decisions to be made by two or more parties in light of their motivations (expected payoffs), the information they have available to them, and with consideration for the underlying constraints that may be placed on their actions. Game theory is especially useful for studying the relationship between conflict and cooperation. Although it is an extremely powerful method for the mathematical modeling of interactive decisions, one does not have to employ the mathematics in order to reap the benefits of game theory. The effort of structuring an interactive decision situation into a game-theoretical context can yield important insights for developing practical solutions.

With regards to operational discipline, organizational requirements are imposed on individual employees. Compliance with these requirements imposes burdens of labor, attention, or thought on the employee. These burdens are absent if the employee chooses to not comply with the requirements. Thus, the employee's immediate interests are in conflict with the organization's interests. The optimal way to achieve cooperation is to structure a system of rewards and penalties to induce the desired (cooperative) behavior. The problem of how to structure a system of rewards and penalties to induce desirable behavior has often been the subject of the safety management literature. Many of these proposals have been based on psychology [8, 9]. They are typically focused on tactical level (routine, daily) behaviors. But this problem has also received much study in the realm of game theory, in both the context of economics and social science. Game theory takes a more strategic view of the problem, and is perhaps better suited to illuminating the role of upper management in promoting operational discipline.

Maintaining operational discipline at the individual employee level takes hard work, attention, and time. One would like to think that an organization's training programs, safety culture, and the individual employee's buy-in to those programs will be sufficient to overcome this burden. However, shirking one's responsibility is easier in the sense that it takes less effort. If the other employees in the organization invest the extra effort required to achieve good operational discipline, then any individual employee may be tempted to shirk their responsibility with the hope of not being discovered. Given the large number of opportunities to shirk responsibility, an employee doesn't have to continuously avoid responsible actions to achieve the reward of effort savings. If the behavior is continually rewarded, then it may increase in frequency and may spread in the organization. Getting the employee to cooperate and invest the necessary effort to achieve operational discipline means identifying the shirking employee and compelling them to invest this effort.

In the parlance of game theory, this problem is a social dilemma called the free rider problem. The free rider problem arises when someone takes advantage of a communal resource (like a safe work environment) without contributing to its upkeep. It can be modeled as a two-person game called the prisoner's dilemma. The shirking employee is one person and everyone else (the active contributors to operational discipline) are modeled as the second person [10]. The consequence of choosing to cooperate (invest effort in operational discipline) or shirk (do not invest effort in operational discipline) is called the payoff. Likewise for the organization, there is a positive payoff (benefit) if the employee decides to invest in operational discipline and a negative payoff (cost) if the employee shirks. From the employee's point of view, the payoffs can be summarized in the two-by-two matrix below.

In the matrix, there are four outcomes for the individual employee: Reward, Temptation, Sucker, and Penalty. Reward corresponds to both the individual employee and enough other employees cooperating to maintain operational discipline. In the absence of incidents, this reward is less tangible and may seem smaller. Temptation provides the highest payout for the individual employee because in addition to reaping the safety benefits of operational discipline, there are no costs to the shirking employee associated with time or effort as others pick up the slack. But, if an insufficient number of other employees carry out their organizational safety responsibilities, then the individual employee does not receive the benefits associated with good operational discipline. Instead, there is a small payout (penalty) associated with shirking that represents the employee not investing time or effort into organizational safety. This assumes that an incident does not affect the individual. Additionally, if an employee cooperates and fulfills their responsibilities but few others cooperate, then the individual receives the lowest payout and can be termed as a Sucker.

Everyone else in organization

Individual

	Enough employees cooperate	Not enough employees cooperate
Cooperate	Reward = 3	Sucker = 0
Shirk	Temptation = 5	Penalty = 1

The payoff values shown in the matrix are for illustration only. The main point of the scoring of the payoffs is to establish the ranking: Temptation > Reward > Penalty > Sucker.[1] The evaluation of payoffs in terms of expected utility is an important subject, but it is quite involved. We refer the reader to decision analysis texts for a discussion of practical strategies for the evaluation of utilities [11]. For our purposes, the ranking of the payoffs is sufficient. The following example applies the free rider problem to a common type of situation that may be experienced at a chemical plant.

Example: Operator Rounds

Charlie is an outside operator in an oxidation unit at a chemical plant. His shift duties are typically burdened by requirements to take hourly logs of analog temperature gauges on the reactor sidestream coolant system. During his two years in the process area, he hasn't seen a loss of control on the exothermic oxidation reactor that couldn't be quickly identified and addressed by the control room operator. The second-tier safety instrumented system is not called-upon either due to the quick reaction of the control room.

As a result, Charlie has started "penciling in" temperature values for the analog gauge during many stretches of hourly readings. After all, there is very little change in the readings throughout the shift. This is an obvious break in the operational discipline; nevertheless, it is a common example in industry. There are two other layers to keep the reactor running safely, and he hasn't seen any loss of control in the past. He sees his time as being more valuable in other areas of the work environment. However, with the time constant and fouling nature of the reactor, a slow drift in temperature at the analog gauge will indicate an impending loss of control in the reactor. If the layers of protection for the reactor do not react, then a thermal runaway reaction is possible. The outcome could be loss of containment of reactive or flammable materials that could lead to outside operator (i.e., Charlie) injury or fatality.

This situation can be evaluated in terms of the traditional free-rider problem. Charlie is one member of the overall process safety management system

[1] For technical reasons, it is typically assumed that the payoffs are structured such that Reward > ½ (Sucker + Temptation). This relation ensures that continued cooperation results in a higher total payout than alternation of cooperation and shirking. Thus the dilemma cannot be beaten by players taking turns receiving the temptation payout. See Appendix B in Axelrod (1984).

responsible for ensuring that the reactor operates safely. Other aspects of the overall safety system include the following:

- Control Room Operator, responsible for direct process state oversight
- Maintenance Engineer, responsible for scheduling temperature monitoring systems maintenance
- Instrumentation Technician, responsible for calibrating and verifying instrument performance
- Controls Engineer, responsible for programming temperature controls, alarms, and operator interfaces

Given the type of process, layers of protection, and the presence of outside operators, this situation can be graphically evaluated through a matrix approach. There are many potential ways to evaluate this problem. Here, we use the typical failure probabilities for basic process control system and operator error adapted from the CCPS Layer of Protection Analysis (LOPA) concepts [12]. Charlie's operational discipline is aggregated with the other generic layers identified above. There are two levels for Charlie's actions. For the rest of the Process Safety Management system, three levels are defined. "OD Intact" represents all other layers of protection in place and staff performing their duties. "OD - 1" represents failure of one of those layers, which may be consistent with a Control Room Operator error such as failing to respond to a silenced alarm. "OD - 2" represents two of those layers failing.

		OD Intact	OD - 1	OD - 2
Charlie	Hourly Readings	$0.1\times0.1\times0.1 = \mathbf{10^{-3}}$	$0.1\times0.1 = \mathbf{10^{-2}}$	$0.1 = \mathbf{10^{-1}}$
	Skips	$0.1\times0.1 = \mathbf{10^{-2}}$	$0.1 = \mathbf{10^{-1}}$	1

The matrix is a simplification to illustrate the potential relationships. The complexity of actual relationships between the layers of protection may be much more complicated for an actual process unit. From the matrix, four levels of probability for an incident are apparent. Given the process description, if you correlate a severity of outcome to each value, it may go something like this:

- $P = 10^{-3}$, process upset that is easily handled
- $P = 10^{-2}$, process upset / incident near miss
- $P = 10^{-1}$, loss of containment with outside operator injury
- $P = 1$, loss of containment with high likelihood of serious injury or fatality

In a repeated prisoner's dilemma, the promise of the future reward and the threat of a future penalty can provide enough incentive to promote cooperation provided the promise of repeated interactions is credible. If the game is played a finite number of times, the employee can benefit from shirking when the promise of future interactions has dwindled. If, however, the employee believes that the game will be played repeatedly without end, then the reward of continued cooperation minus the threat of future penalties outweighs the benefit of shirking. It makes more sense for the employee to cooperate. The essential idea is to have the reward for continued cooperation outweigh the benefit associated with using any combination of cooperation and shirking. . This is not only true mathematically, but it has also been demonstrated to be true in numerous empirical social science studies in both the field and in the laboratory [13]. The use of rewards and penalties to motivate individuals to actively cooperate and act in the best interests of the organization is also discussed at length in Schneier's book, where it is emphasized that actions speak louder than words [14]. In other words, it is not enough that upper management proclaims the importance of process safety. Upper management must reinforce the message consistently and continuously through active rewards for good performance (e.g., high implementation of leading indicators for process safety), and impose penalties for poor performance (poor implementation of leading indicators).

3. Using Assessments to Reinforce the Good and to Correct the Bad

The process safety assessment can and should serve many purposes. Our emphasis here is limited to the use of the assessment as a means of evaluating the extent of operational discipline with the intent of reinforcing desirable behavior and correcting undesirable behavior. The metrics for evaluating operational discipline are the leading indicators for process safety. The assessment data are classified into compliance and noncompliance (defect) categories. Then the assessment data are analyzed for identifying winners in each category, i.e., the individuals and groups who have the best track record for implementation of any given leading indicator. This information is then presented to upper management. Upper management can then reward these winners using any number of incentive programs, but the most important reward is recognition throughout the organization. Everyone in the organization must see that proper implementation of the leading indicators for process safety leads to rewards and recognition within the organization. Properly implemented, the rewards and recognition of those who excel in operational discipline can enhance their reputation in the organization. These rewards and recognition should be extended to the supervisors of the winners as well, so as to encourage all supervisors to emphasize the importance of operational discipline within the organization.

Similarly, the assessment will be useful in identifying those employees who have shirked their responsibility towards operational discipline. Deficiencies in operational discipline must be corrected through remedial education and appropriate penalties. Again, the penalties should be extended to some degree to the employee's supervisor. This can act as an incentive for supervisors to promote operational discipline within their group. The identities of those who fail to implement operational discipline should not be publicly revealed. However, it may be appropriate to reveal the departments with the highest and lowest degree of compliance. This can foster a constructive sense of competition to excel within the organization.

The status of operational discipline within the organization, and its improvement over time, can be monitored by tracking the degree of compliance with successful implementation of leading indicators for process safety. The lesson of the repeated prisoner's dilemma is that the successful path to cooperation is through repeated interaction, resulting in rewards or penalties as appropriate, in an effort to continually improve operation discipline. One final point to be made here is that the lagging indicators of process safety—number of incidents, number of leaks, or number of other failures of containment—should not be used for the evaluation of individual employee performance. This is because it is too easy to cheat by these metrics. Lagging indicators are a valuable measure of past and current performance, but they are difficult to employ for active control of the management system. Leading indicators are the superior management tool for monitoring operational discipline.

4. Conclusions

Operational discipline is a key component in the process safety culture in an organization. While process safety assessments serve many functions, the evaluation of leading indicators can reveal the state of operational discipline and become an important management tool for fostering commitment to process safety. Operational discipline is hard. Asking employees to demonstrate operational excellence is asking them to work even harder. Game theory offers a simple strategy for seeking cooperation from employees when shirking operational discipline would be so much easier. The game-theoretical solution is to provide a balance of rewards for cooperation and penalties for defection. Upper management can demonstrate its commitment to process safety by publicly rewarding and recognizing the best implementers of operational discipline, both the employees and their supervisors. To be successful, the recognition must be performed in a way to enhance the reputation of these cooperators. For those who shirk their responsibilities towards operational discipline, their inadequate performance should be addressed privately. It is essential that the shirkers feel the sting of penalties while also recognizing that rewards and recognition throughout the organization await those who strive for operational discipline

5. References

1. J.A. Klein and B.K. Vaughen. "Implement an Operational Discipline Program to Improve Process Plant Safety," Chemical Engineering Progress, pp. 49-52, June 2011.
2. Center for Chemical Process Safety, Guidelines for Auditing Process Safety Management Systems, Second Edition. American Institute of Chemical Engineers, New York, 2011.
3. Process Safety Site Assessments. American Petroleum Institute, Washington DC, December 2011.
4. Center for Chemical Process Safety, Guidelines for Process Safety Metrics. American Institute of Chemical Engineers, New York, 2009.
5. Center for Chemical Process Safety, Process Safety Leading and Lagging Metrics. American Institute of Chemical Engineers, New York, 2008.
6. API RP 754 Process Safety Performance Indicators for the Refining and Petrochemical Industries. American Petroleum Institute, Washington DC, April 2010.
7. J. Watson. Strategy: An Introduction to Game Theory, Second Edition. W.W. Norton & company, New York, p. 1, 2008.
8. E.S. Geller. The Psychology of Safety. CRC Press, Boca Raton Florida, 1998.
9. T.E. McSween. The Values-Based Safety Process. Wiley, New York 2003.
10. L. Fisher. Rock, Paper, Scissors: Game Theory in Everyday Life. Basic Books, New York, pp. 61-68, 2008.
11. R.T. Clemen. Making Hard Decisions: An Introduction to Decision Analysis, Second Edition. Duxbury Press, Belmont California, Chapters 13 and 14, 1996.
12. Layer of Protection Analysis, Simplified Process Risk Assessment. New York: American Institute of Chemical Engineers – Center for Chemical Process Safety; 2001.

13. R. Axelrod. The Evolution of Cooperation, Revised Edition. Basic Books, New York, 2006.

14. B. Schneier. Liars and Outliers: Enabling the Trust that Society Needs to Thrive. John Wiley & Sons, New York, pp. 1-14, 2012.

9TH GLOBAL CONGRESS ON PROCESS SAFETY

Using New Technology to Ensure Compliance and Improved Risk Management

Mike Neill
Petrotechnics USA Inc., President
Petrotechnics USA Inc. 10497 Town & Country Way
Suite 700, Houston, Texas 77024
Mike.neill@petrotechnics.com

Scott Lehmann,
Petrotechnics Ltd. Vice President Product Management & Product Marketing
Petrotechnics Ltd, Craigshaw Road
Aberdeen, AB12 3QH, UK
Scott.lehmann@petrotechnics.com

Prepared for Presentation at
American Institute of Chemical Engineers
2013 Spring Meeting
9th Global Congress on Process Safety
San Antonio, Texas
April 28 – May 1, 2013

UNPUBLISHED

Using New Technology to Ensure Compliance and Improved Risk Management

Mike Neill
Petrotechnics, US President
Petrotechnics 10497 Town & Country Way
Suite 700, Houston, Texas 77024
Mike.neill@petrotechnics.com

Keywords: Compliance, HSE, Operational Performance, Risk Management, New Technology

Abstract

With rising stakeholder expectations on performance, hazardous industries face substantial pressure to deliver sustainable operational improvements, and drive safety beyond mere compliance. Post Macondo, many organizations are reviewing and strengthening corporate policies and management systems to properly handle risk in their operations. The ability for these organizations to drive such initiatives, though, is a complex and time consuming challenge that goes far beyond rewriting procedures or improving training.

Better tools are needed to ensure the actions of people at the frontline align with corporate best practices and that these people have the necessary real time support for the daily decisions demanded in the dynamic world of operations. Given the complexity of global organizations in our industry today, many of which have grown through acquisition, aligning to standard practices is a recognized challenge. The traditional Operational Discipline approaches rely on developing a culture of conformance coupled with a stronger management system. Whilst this approach has significant merit it has lacked suitable supporting tools to help align practices whilst providing data on performance and generally improved surveillance. This paper will show how new technology platforms can transform the implementation and enforcement of a culture of conformance as well as establishing new methods of tracking performance in real time.

In this paper we will focus on the area where risk management decisions are made daily - front line work execution. Intervention is recognized as the biggest operational risk. It is the leading cause of HiPo's and second leading cause of fatalities. Since it involves people, equipment, tools, procedures and typically live plant, the need for Operational Discipline is paramount to drive safety primarily, but also the efficient use of resources. We will demonstrate how to convert the millions of man hours per year derived from thousands of people every day routinely intervening in plant operations to do repairs, maintenance, and interfering with standard operating procedure, into useful management data.

Using practical examples, we'll will explore some of the historical models for managing work at the front line and look at how technology has changed them over the past 10 years. In addition, we will look at some of the gaps which still exist in these business processes and highlight how new technology solutions can bridge these gaps and generate the necessary data to provide real time surveillance and feedback for all levels of the organization. Finally we will review powerful new performance measures which will dramatically change our ability to implement operational risk management in this space and compare them to conventional Process Safety Management and Personal Safety indicators.

1. Introduction

Organizations in hazardous industries have one overriding objective; deliver performance in a responsible and sustainable manner. Stakeholder expectations of operational performance have never been higher as the consequences of getting it wrong are significant. From reputational damage to substantial fines and penalties to serious threats to the financial viability of the organization, it's understandable that balancing performance against operational risk management is now a board level challenge.

This paper examines some of the challenges in managing frontline Operational Risk in Hazardous Plants. The source of the highest risk in these plants, however, is often overlooked in discussions about Process Safety Management (PSM). Process Safety conferences understandably focus on major accident hazards and the scenarios of which typically involve acute loss of primary containment resulting in a significant release, explosion, fire or toxic gas cloud. The consequences include loss of life, serious injury, major equipment damage, legal liabilities, and damaged reputation in the community. The strategies to prevent such events focus on proper design of primary containment barriers and maintaining those barriers through the life of the plant.

Implementing a robust PSM program significantly reduces the likelihood of a major event happening and can mitigate some of the consequences where these hazards will always remain the most significant.

As a result, most focus on proper design and traditional PSM approaches. In addition they make best efforts to improve operational performance and strengthen Operational Discipline initiatives supported by new policies, procedures, better rulebooks and further training. The expectation is that all safety related near misses and incidents will be substantially reduced and an assumption that behavioural compliance will follow suit, reducing the likelihood of unwanted outcomes.

Yet despite these efforts, which are often time-consuming and complex, improvement is often illusory or not sustainable. Take for example, the occurrence of a significant incident prompting major changes that lead to a complete re-write of standards and procedures. As a result, our HSE management standards and practices, the very systems developed to support safe and sustainable operations typically become so complicated as to restrict their utility to people in the Frontline- the people who need them the most.

While PSM initiatives and traditional approaches serve their purpose very well, this paper will address operational risk where we need to look beyond the major accident hazards and examine those hazards associated with the day-to-day working of the plant, as well as better tools that are needed to balance workload against risk according to policy.

We will focus on the area of "Front Line Work Management" which encompasses all aspects of inspection, maintenance, repair, construction and reinstatement of equipment on our plants.

These are daily activities carried out in and around live plants involving hundreds of thousands of man-hours on a typical facility. Not only does this type of work expose people directly to the surrounding process hazards but the work they do and the equipment being used can introduce new hazards to those doing the work, to other workers and to the plant itself. Furthermore, this type of work can often require primary safety barriers to be removed or disabled in order for the work to be carried out or to gain access to equipment. Whilst such removals and disablements are carried out under controlled conditions they are also concurrent with other ongoing work, with live plant operations and with other prevailing risk factors such as environmental and human factors. The cumulative effect on the overall operational risk is clearly quite complex, yet to date there have been few tools to help support the judgment of frontline supervisors.

The way such work has been managed on plants has not changed much in the last 30 years relying typically on paper forms to control permitted work, to perform Job Safety Analyses' (JSA's) or task risk assessment and to manage process/energy isolations. Such approaches generate very little data so our ability to proactively measure and control what is happening in this area has been limited. It is also an area which tends to lie "beneath the radar" for most organizational leaders who typically get involved only when or after something goes wrong. This situation is, however, changing. Technology is becoming ubiquitous in our daily lives both at work and at home and is driving a demand for real time communications, up to the minute awareness of situations and increasing volumes of data to measure performance. In our industry, recent events such as the Macondo disaster in the Gulf of Mexico have demonstrated the enormous cost of mismanaging operational risk. This has resulted in a stronger drive for corporate governance and assurance processes to ensure that people at the front line managing risk are making the right decisions.

Fortunately, new technology platforms are emerging which support business processes in these areas, generating high volumes of data and helping organizations better manage operational risk at the front line and provide the necessary assurance within the organization. Specifically, we will examine a new Operational Performance and Predictive Risk platform, "Proscient" that provides the capabilities to proactively manage workload against risk according to policy. To provide a better appreciation, understanding and management of performance and operational risk, it is necessary to begin with exploring some of the concepts in an operational context.

2. The Operational Context & Single Biggest Area of Operational Risk

Carrying out work at the frontline in hazardous industries involves three primary elements; the Plant, the People and the Interaction between the People and the Plant. The Interaction between the People and the Plant represents millions of man hours per year for the typical organization and thousands of people every day routinely intervening in plant operations to do repair and maintenance work. This work often results in the introduction of new hazards, inhibiting safety systems, isolating parts of the plant and interfering with standard operating procedures. The Interaction between People and the Plant, also known as Front Line Work Management, equals operating risk.

This Interaction includes every aspect of planning, controlling and executing all work tasks being carried out on hazardous plants. Some companies refer to Work Execution (WE) using different terms such as Control of Work, Work Management or Work Control. The majority of this work will be maintenance, repair or construction but it can also include seemingly benign routine tasks necessary to support the operation. The term encompasses all aspects of work management including permit control, isolations and reinstatements or Lock-Out Tag-Out (LOTO), Task Risk Assessment or Job Safety Analysis, Temporary Defeats/ Inhibits/ Override etc. In general, it includes all work which needs to be controlled due to its hazardous nature, the lack of familiarity of the people involved, or the need to coordinate multiple ongoing tasks and simultaneous operations.

Fig 1. The Interaction of People with the Plant = Operating Risk

Front Line Work Management is the single biggest cause of High Potentials (HiPo's) and is the primary source of fatalities (after road traffic accidents) according to a study carried out by a major oil company (2005). In this study three factors relating to 'Work Control' were identified: Planning of Work; Work Rules, Policies, Standards and Procedures; Management, Supervision and Leadership. A review of fatal industrial accidents over a 5 year period found that ninety percent (90%) had one of these elements as a contributory factor, seventy one percent (71%) had two of them and thirty percent (30%) had all three.[1] Similar statistics from other companies about fatalities and major incidents indicate the same contributory factors. Clearly getting WE right is critical to the safety of our plant workers.

If we look at some of the well-publicised major accidents such as Piper Alpha, Texas City, Buncefield and Phillips Pasadena, we see that they all involve varying degrees of failing in WE. Michael Ferrow (2011), the General Manager of Engineering and Technical Assurance at ConocoPhillips (UK), in his SPE paper entitled, *'An Operating Management System to Deliver*

Safe Operations' [2]furthers this by stating that "the other conclusion in looking at all of these accidents (Oppau, Germany; Feyzin, France; Flixborough, UK; Piper Alpha, UKCS; Texas City, USA; Buncefield, and UK; Deepwater Horizon, USA) was that in almost every case there were three clear contributors:

- Absence of adequate risk assessment and understanding
- Management of change
- Inadequate control of work at the workface" [2]

Process Safety Management approaches recommend creating multiple barriers to prevent unplanned releases but when we examine typical WE operations, we see that often these barriers meant to keep the plant safe are deliberately removed, safety devices inhibited and new hazards are introduced to the workplace in the course of carrying out work. For example, to access, replace or repair equipment it is often necessary to:

- Break containment of hydrocarbon bearing systems,
- Inhibit or take out of service, safety related devices such as alarms and shutdown systems in order to perform work without shutting down the plant,
- Access equipment which often involves working at height or having to perform heavy lifts over live plant

Such scenarios are daily occurrences on most plants.

More often than not, this work is taking place inside a live plant where production has to be maintained while the intervention work is carried out. By necessity, this work brings people into the plant area, thereby putting them at risk. The majority of frontline workers will be third party contractors less familiar with the hazardous plant on which they are working and perhaps less equipped to understand and address the process hazards surrounding them. Typically, multiple jobs of work need to be managed simultaneously on the same plant with the potential for work clashes and compounded risks, posing a further complication.

The types of hazards being managed in WE are wide ranging but examples of categories would include:

- Process Hazards, e.g. flammable/ explosive hydrocarbons under pressure
- Equipment hazards such as rotating equipment
- Occupational hazards such as working at height
- Dropped objects and environment hazards such as confined spaces, temperature, noise, etc

These examples are not comprehensive and the list of hazards is endless, as is the manner in which they can combine and escalate to a major incident. It is important to note the wide range of hazards and that this is not merely about 'slips, trips and falls' or those hazards with just the potential to cause minimum harm to individuals. Managing the risk posed by process hazards is paramount, particularly for those working in close proximity to and with less familiarity of such hazards.

Another important aspect of hazards introduced through WE is that they are dynamic and constantly changing from day-to-day and hour-to-hour depending on ongoing activities. Process hazards on the other hand, are relatively constant for a well-designed and well maintained plant operating steadily. Even with process upsets, the design of the plant and its control should be able to manage these without risk of loss of containment. Plant control systems are designed to operate within design limits and shutdown systems automatically take over if operators exceed these limits.

With each day typically presenting different hazards and risks and with inadequate tools to support the decision making process, the quantity and frequency of WE related incidents is not surprising. Neither are the operational inefficiencies associated with unplanned shutdowns, a priority work backlog, less than optimum plan accuracy, poor plan attainment; contractor wait time and little to no consistent understanding of the relationship between delivering performance and operational risk. The industry's traditional response has been to deem this a workforce culture or people problem; however, there are deeper underlying issues.

Most global corporations have a strong need to standardize their practices in this area. The next few sections will examine why this standardization is important and why it is very difficult to implement.

2.1 Why is Standardization so important within an organization? Why is it so hard to achieve?

Shareholders of corporations demand that their executives are controlling and directing activities of the company in accordance with regulatory laws and with their best interests in mind. Typically, this is achieved through the creation of policies, standards and sometimes detailed practices which must be adhered to at all levels of the organization. In the context of managing frontline hazardous work, this translates into common policies, standards and practices for managing specific hazards- hazards which should be the same in all places with similar activities. As Scotty in the TV series Star Trek often used to say, "You cannot change the laws of physics!" Similarly, in our plants around the world the laws of physics remain the same. Hence, the fundamental hazards of working at heights or breaking containment of hydrocarbon systems or heavy lifting should be the same the world over no matter where you are or what language you are speaking. If the hazards are the same, then it would make sense for a corporation to designate one way- the best way to properly manage that hazard. It should therefore be a relatively straightforward exercise for a multinational corporation to implement universal rules for all hazardous tasks, but in practice many, if not all, corporations struggle in this area.

A great many obstacles make it very difficult for large organizations to successfully align behaviors at the front line with global corporate policy and often the result is a simplification of the requirements to such general terms that their effectiveness is often reduced. Some of these barriers are created by the organization management structure itself. Autonomous Business Units are not designed to readily accept dictates from the corporate center. National Regulations may not align with corporate policy requiring you to do more or less than policy dictates. If a region has a long established industry with normalized and deeply embedded practices, it will be difficult to change. Sometimes, if it's a cultural practice, 'the way we do things around here', and local culture was to ignore a procedure, leadership not only ignores procedural violations, but may reward violations that led to improved efficiencies. Teams may even acquiesce because of the complexities involved with trying to change or evolve.

Local businesses may be small and informally run compared to other much larger operations, so strict compliance with prescriptive policy may seem excessive. There is also a wide recognition that culture and language can act as a significant barrier to adopting a single set of standardized policies. In an international operating environment, this can present significant challenges. Specifically, workers that must comply to policies not in their native or primary language or they are at greater risk of having an accident because of not having a full grasp of safety standards. The cost to comply itself may be a barrier. We know of one large organization who, a few years ago, wanted to mandate seatbelt use in all company vehicles globally until it was pointed out that in one country where the fleet of vehicles numbered in the thousands seatbelts were not fitted in their road tankers since that Country's vehicle regulations did not require it.

Standardization is also a critical approach when managing the integration of two organizations following a merger or acquisition. The new larger organization has to agree on single set of policies and standards if the organization is to work in harmony and with common goals. It is an impossible task to manage Operational Risk and Operational Discipline across a larger corporation with two separate sets of policies and practices. The benefits, however, of declaring a single standard or policy and aligning the organization around it is immense.

Some organizations use standardization in Operations as a deliberate strategy to leverage a competitive advantage. The philosophy is simple. Figure out the best way to manage your Operations effectively (i.e. safe and cost efficient) then ruthlessly standardize around these best practices. Most process industries are highly competitive and if you can figure out a way to repeatedly reduce your lifting costs or your processing costs per unit of production in a safe and sustainable manner then you are ahead of your competitor.

In the area of Front Line Work Management, the standard practices and policies may have been christened different names from the different companies but they typically have many common elements. Some of these names include "Golden Rules", "Life Saving Actions", "Cardinal Rules", etc.

The box on the right demonstrates some of the common elements of this type of approach. The practices under these titles may vary but the topics are the same. A common theme is to keep the "rules" simple, easy to communicate and closely aligned with corporate policy. Many exist as statements of that policy. The practices which go with these statements are more detailed and, as such, more difficult to achieve as corporate standards.

Common elements of "Golden Rules" approach to standardized practices:

- **Hot work**
- **Energy isolation / LOTO**
- **Working at height**
- **Heavy lifting**
- **Confined space entry**
- **Ground disturbance**

Furthermore, as organizations endeavor to unify its standards and harmonize processes across all its assets or in conjunction with its JV agreement, the difficulty is that good intention does not equate to actual compliance. For instance, a well written policy or procedure statement may be distributed in the form of email communication or otherwise, yet actual compliance or adoption of the policy is difficult to enforce or evaluate.

Andrew Hopkins, a well-renowned Associate Member of the National Research Centre for OHS Regulation, emphasizes the importance of this in one of his recent OHS papers, Risk Management and Rule Compliance Decision Making in Hazardous Industries (2011) "Clearly, rulemaking by itself is of little value in achieving safe operation. What is required is compliance with those rules. This in turn depends on adequate enforcement or accountability mechanisms. In other words, the rule-compliance approach requires both rule-making and rule-enforcement."[3]

2.2 Three Critical Gaps in Frontline Operations

In the pursuit of understanding these challenges and after 12 years of experience providing related solutions which has included some 12 million operational hours in over 350 installations, we have uncovered three critical operational gaps that all companies face in frontline operations: 1) Work Management Business Process; 2) Policy into Practice and 3) Proactively Managing Workload against Risk. The diagram below represents these gaps, two of which we are addressing and focusing on in this paper, Policy into Practice and Risk Management.

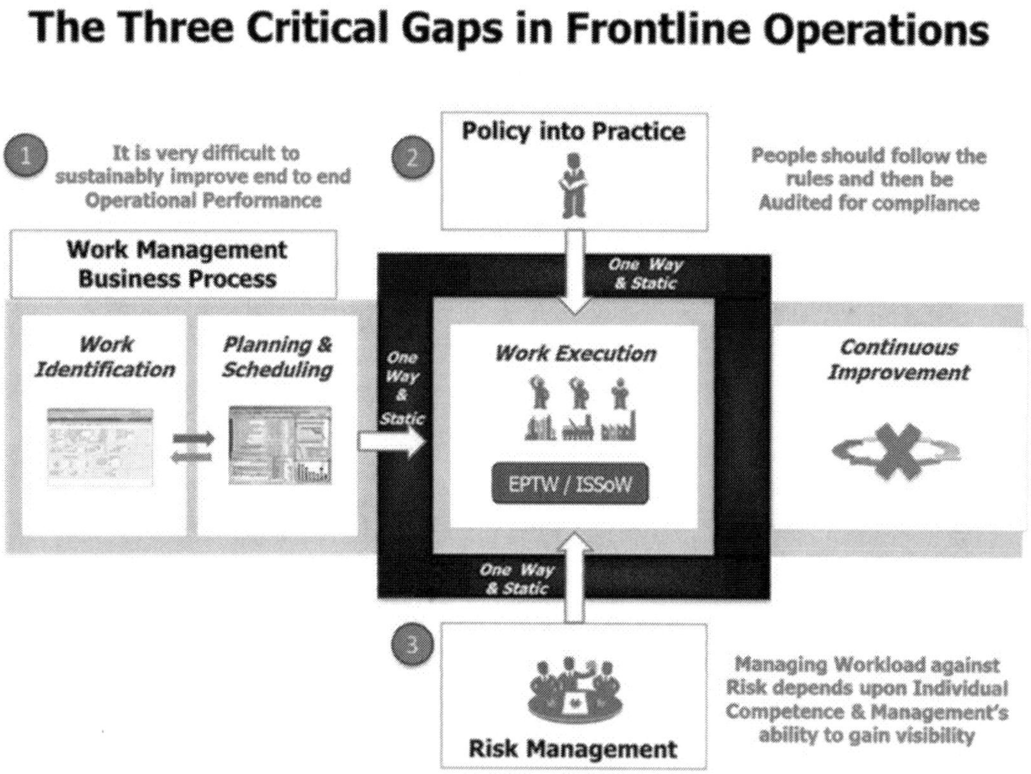

Fig 2 The Three Critical Gaps in Frontline Operations

To provide a more comprehensive understanding and impact of the last two gaps, it is necessary to provide insight into the first gap. The work management business process gap is related to the lack of integration and correlation between planning and execution.

2.2.1 Work Management Business Process Gap

The work management business process starts with the identification of work such as scheduled maintenance work that typically is held within maintenance management or ERP systems. This work is then combined with other work by planners to create a 14 day frozen schedule accounting for the all of the resource and logistics constraints. The plan is sent to operations who then build out the "safe work" plan to account for all of the additional tasks, sub-tasks, permits, isolations and other required safety constraints. It is only in the operations time and space that interdependencies between jobs and tasks can be identified and managed (i.e. common isolations, LOTO). Further complicating matters are Contractor Job Safety Analyses (JSA's) that contain specific expertise but require local knowledge of the plant and hazards to produce the appropriate task based risk assessment.

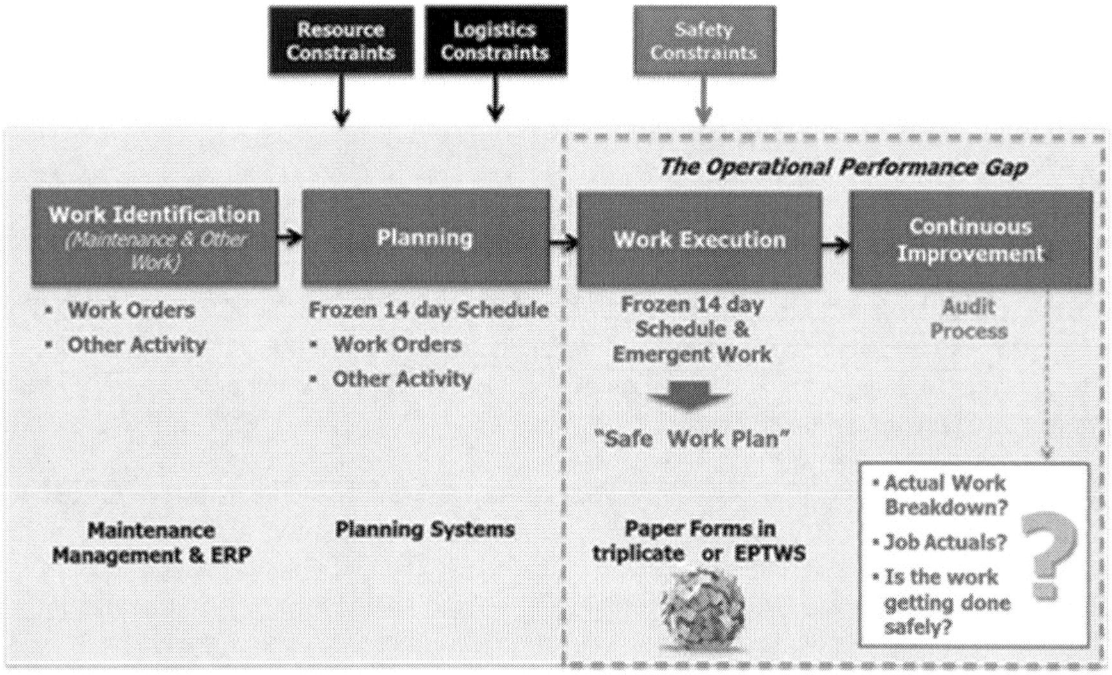

Fig 3. Work Management Business Process Gap

With the identification and planning stages being managed by MMS or ERP systems and the execution phase being managed by legacy paper forms in triplicate or basic standalone electronic permit to work (EPTW) systems at best, there is an operational performance gap as there is no end to end integration, no feedback data and no ability for sustained improvement. As valuable details on job breakdown and the actual time it took to complete the entire job, including all of its safety dependencies is lost; this gap ensures inconsistency in managing workload against operational risk.

2.2.2 Policy to Practice Gap

Following the plan, ensuring the execution of work is carried out accordingly to policy and consistently across the plant or organization is critical to the safety and success of operations.

Typically organizations have assurance over the PSM approaches and plant design that include policies, standards, practices, systems, data, KPI's/metrics and feedback. In this way, they have confidence and feedback that their policies are being followed and they have the means through the use of the right technology, the collection of data, to measure status and progress of their performance.

However, as Work Execution takes place and the PSM barriers meant to keep the plant operationally safe are temporarily disabled to carry out the critical maintenance work, there is ambiguity around the compliance of work being carried out according to policy.

Although there are some policies and standards to guide the frontline workers, they are limited often to paper based systems; there is no data, no metrics and as a result, no feedback loop of any consequence.

Rulebooks and guiding principles in these limited systems provided in a one way and static format absolutely ensure there will be divergence in the way they are delivered in operational practice even within the same plant.

As the industry traditionally relies on paper based system as its form of control in this area, there is no data. Paper based forms provide neither real data nor a window into on-going work leaving senior management with little to no oversight over the riskiest part of their operations. Without systems we do not have data and without data or feedback we cannot measure performance thus severely limiting management's effectiveness in driving improvements in operational performance and operational discipline.

Furthermore, the audit process is typically seen as adversarial with the auditors tasked with finding discrepancies or policy breach and the workforce being audited just trying to pass the audit so they can get on with their normal work activity. In the end what organizations have are audits whose value at best will be inconsistent and at worst questionable and an organization that likely knows it can do better but wonders why after so much effort and spend on trying to improve still doesn't move the dial.

From the difficulties described in achieving overall standardization, to the complexities of the hazardous environment such as the employee-contractor mix, the assumption of contractor competence on the part of the operator, the workforce culture; to the lack of automation, data and feedback; it is no surprise that even with the well-intended efforts and well-written policies compliance, adoption and even effectiveness of policies is questionable.

What is more is that with the increasing expectation of operational performance and work efficiency, it is possible workers can cut corners sacrificing safety and not follow the rules- something that is difficult to detect with a paper-based or limited digitised system.

2.2.3 Operational Risk Management Gap

This third gap lies with an organization's ability to ensure that it is consistently managing workload against risk according to policy and management's ability to gain visibility on this risk. When you strip away everything, the industry relies on individual competence and interpretation and application of the company's policies whether they are embodied in lifesaving rules, guiding principles or policy to get this balance right.

Effective Operational Risk Management is an important and highly specialised skill that demands careful balance between risk appetites, risk assessment, risk response decisions and risk awareness whenever operational work is executed. All too often it is a daunting task to provide the safest integrated response to the cumulative risks arising across multiple tasks in multiple locations as well as addressing conflicting production demands. It is inevitable that the necessary skills for effective risk management are constrained by entrenched local pressures and cultures, as well as the misleading bias that arises from an absence of incidents.

With the additional pressure of getting more work done, we often leave it up to individual competency to interpret how operational risk is identified and managed, leading to poor quality risk assessments being approved.

Despite the best efforts of training programs and policy documents to extend risk management competency frontline operations staff are still missing hazards, and a supervisor's ability to identify, understand and manage operational risk varies widely.

With disjointed and sometimes incomplete task and process hazard awareness & identification, there can be inadequate risk assessment leading to a higher likelihood for major error.

Frontline supervisors and workers lack robust tools to guide them to make the appropriate decisions. In other words, decisions are being made without the full information, decisions which could lead to catastrophic failure. As is always the case when there is no systematized process, the variance in how workload is managed against risk doesn't just vary from facility to facility but from frontline supervisor to frontline supervisor.

Virtually all of the industry's attempts to develop leading indicators related to operational risk are based on lagging metrics and or the wrong metrics. As a result, the ability to proactively manage workload against operational risk in frontline operations is at best very difficult.

3 How New Technology can Help

In the past 10-15 years we have seen some technology entering into this space. These started as simple electronic permit-to-work systems which typically digitized the paper processes with little integration. Later on, Integrated Safe Systems of Work (ISSOW) integrated Permit To Work (PTW) with Isolations / LOTO and Task Risk Assessment (TRA) / Job Safety Analysis (JSA). Until now, such systems have been deployed as point solutions in plants and while they have provided some great improvements in work efficiency and have helped standardize work flows, they have been limited in their capability to address some of the main gaps in frontline operations.

The new Operational Performance and Predictive Risk platform, Proscient, enables closure of these gaps as is shown below.

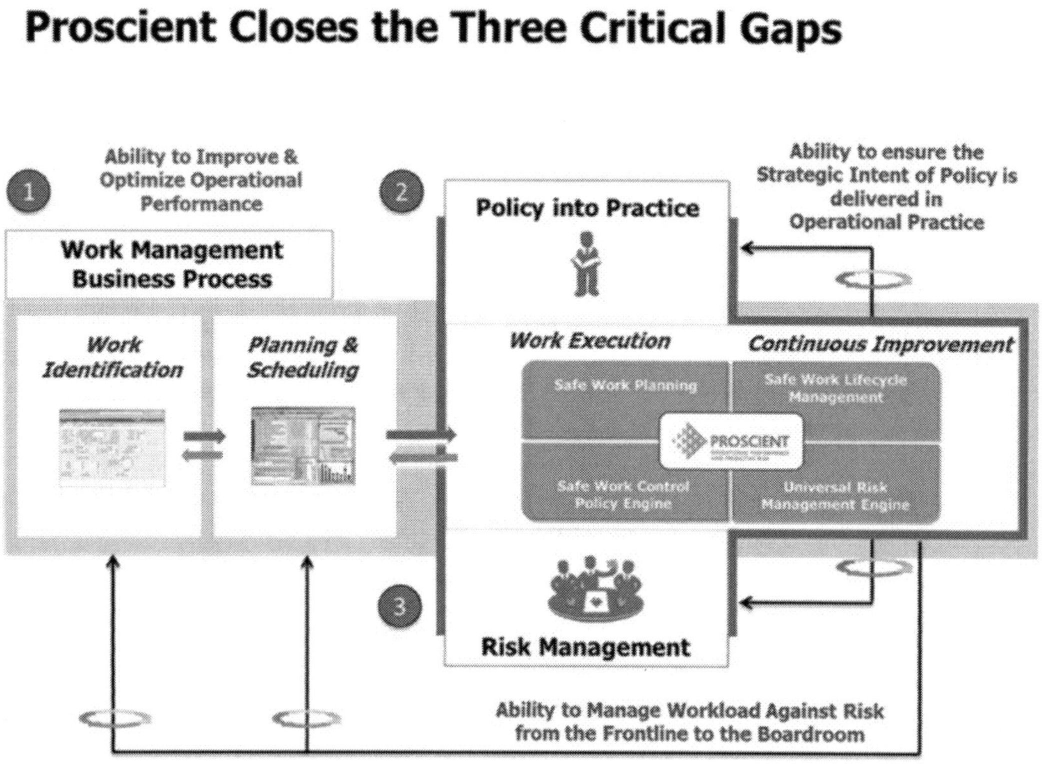

Fig 4. Proscient Closes the Three Critical Gaps

Proscient provides organizations the ability to improve and optimize operational performance. By systematizing the work execution business process, it drives efficiencies across the organization. When integrated with corporate IT, EAM , MMS and Planning systems it can provide the detailed work breakdown structure of all of the tasks, sub-tasks and safety dependencies it took to actually complete the job as well as the actual times for each of these elements.

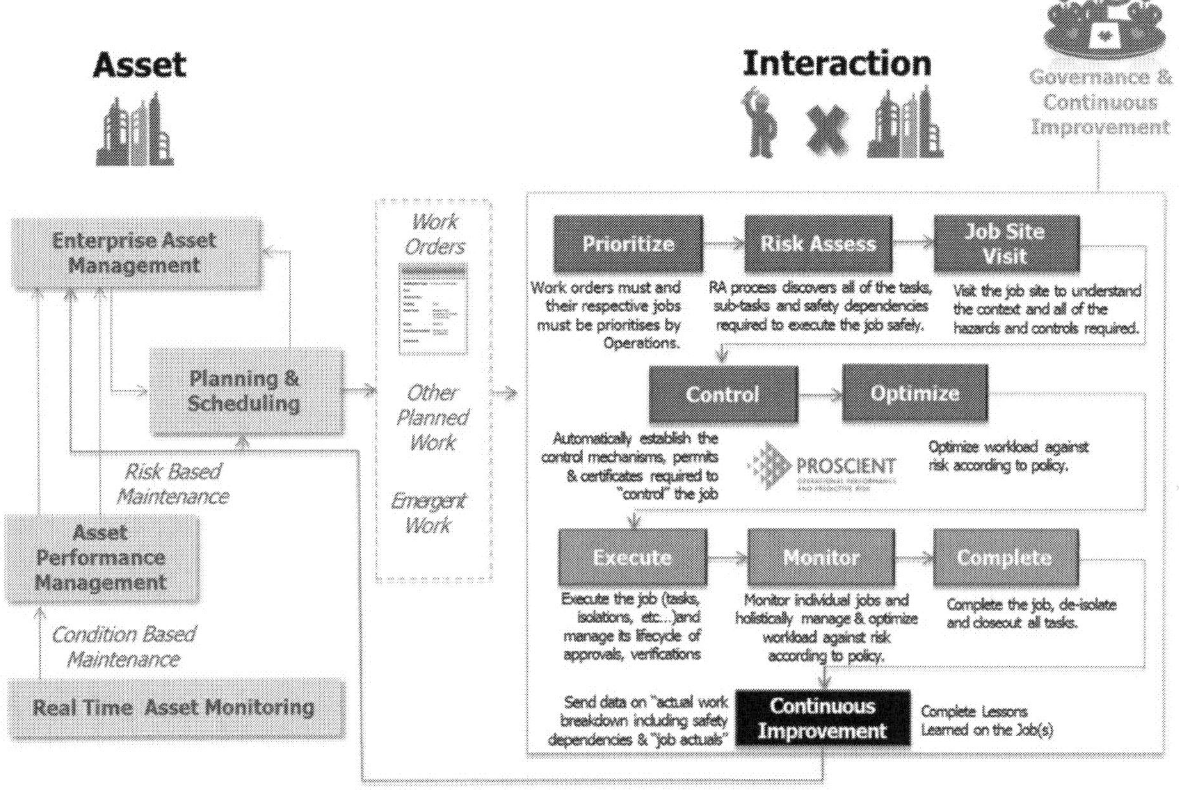

Fig 5. How Work is Executed with Proscient

These "results" of the actual execution of the work are fed back into the other systems thus providing a significantly better set of inputs which both Maintenance and Planners can use to substantially improve their subsequent planning. In more advanced customer environments, daily or real time updates to the progress versus the 14 day (or other defined time period) frozen plan can be provided to planners so they can proactively optimize the plan.

Proscient closes the Policy to Practice gap providing organizations the ability to ensure the strategic intent of their policies is delivered consistently in operational practice across the organization. Standard policy baselines can be embedded into a systematized process to ensure frontline workers and supervisors are supported and guided to the appropriate decisions. Proscient's flexibility allows organizations to be as prescriptive or high level as they need to be in terms of the baseline standards they deploy accounting for business model variances, regional regulatory requirements, JV requirements or the degree of local flexibility they are willing to allow.

The transparency and visibility the system provides in terms of how work was carried out provides insight both into the effectiveness of specific policy (e.g. how breaking containment work is performed) and the opportunity to drive continuous improvement. On a single plant basis as well as a comparative basis of similar plants, organizations will have the tools to understand how well work was executed against policy, areas where further investigation is warranted and insight on how to improve policy.

Proscient closes the Frontline Operational Risk Management gap as it gives organizations the ability to proactively manage workload against risk. By automatically building risk profiles based on planned work within the system it captures and making visible in a standard way a common understanding of the risk level of the work on a specific shift in a plant. It allows frontline supervisors to visualize what they already do in their heads every day when they managed workload against risk. They then have the tools to optimize workload against risk according to the organization's standards and rules. These risk profiles can then be cascaded up the organization to provide regional and senior management the visibility and the tools to proactively manage the organization to a more optimized and less risky performance.

3.1 Ensuring Compliance and Strategic Intent is delivered into Operational Practice

By capitalizing on the new technology and in particular, to strengthen and support Operational Discipline, the advantages of closing the Policy to Practice gap are unparalleled. In the example we will look at here, you can see how the new technology platforms can offer simple and flexible solutions to this dilemma. The diagram below shows how an enterprise deployment of the platform would allow rules to be set in a cascading hierarchy such that those rules set at a corporate level would be common for the whole hierarchy. Those set at a regional level would contain all the common corporate rules but allow each region to add their own specific rules for flexibility, maximum efficiency and effectiveness. At the plant level, rules can be set which are specific to the plant but include all corporate and regional rules. This hierarchy can even continue down to separate parts of the plant.

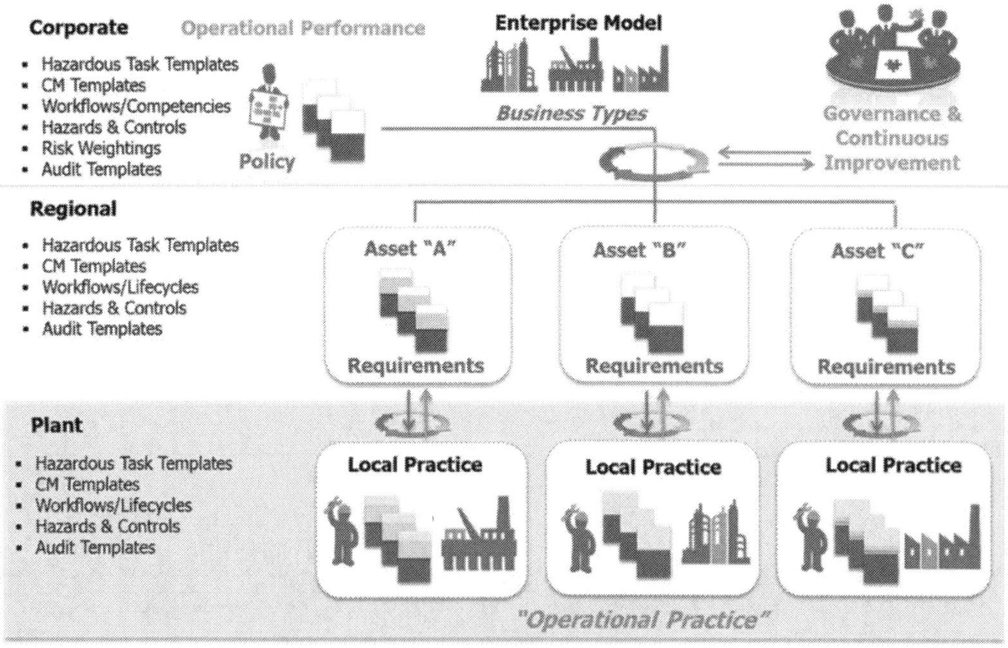

Fig 6. Embedding Policy into Practice

3.1.1 What Processes can be controlled in this hierarchical fashion?

Hazardous Task Templates- The way in which specific hazardous tasks are managed can be controlled from the corporate center using Hazardous Task Templates. These templates can be specified for particular hazardous tasks (e.g. working at height, heavy lifting over live plant, confined space entry, breaking containment, etc). These are common tasks carried out fairly frequently and each with a predictable set of hazards which will always be present though by no means represent all of the hazards. Within the template these hazards can be identified from the library of hazards and controls specified. Once a Template is created at the corporate level it is locked down and cannot be changed so for example, if the template is for working in a confined space one corporate control could be that positive isolations must be in place for any man-entry to a confined space. This rule would be enforced for all confined space work in every region and at every plant. At the regional level, these corporate templates can be modified to create regional templates but only additions can be made. None of the existing corporate hazards and controls can be eliminated. An example might be for a desert region where radiant heat is a problem most of the year. The corporate template for confined space work could be modified for the region adding the need for temperature monitoring and cooling. At the plant level there may be a local Fire Service which provides emergency rescue so the confined space entry template could be modified at that level for the local fire service to be on standby for rescue during any confined space entry event.

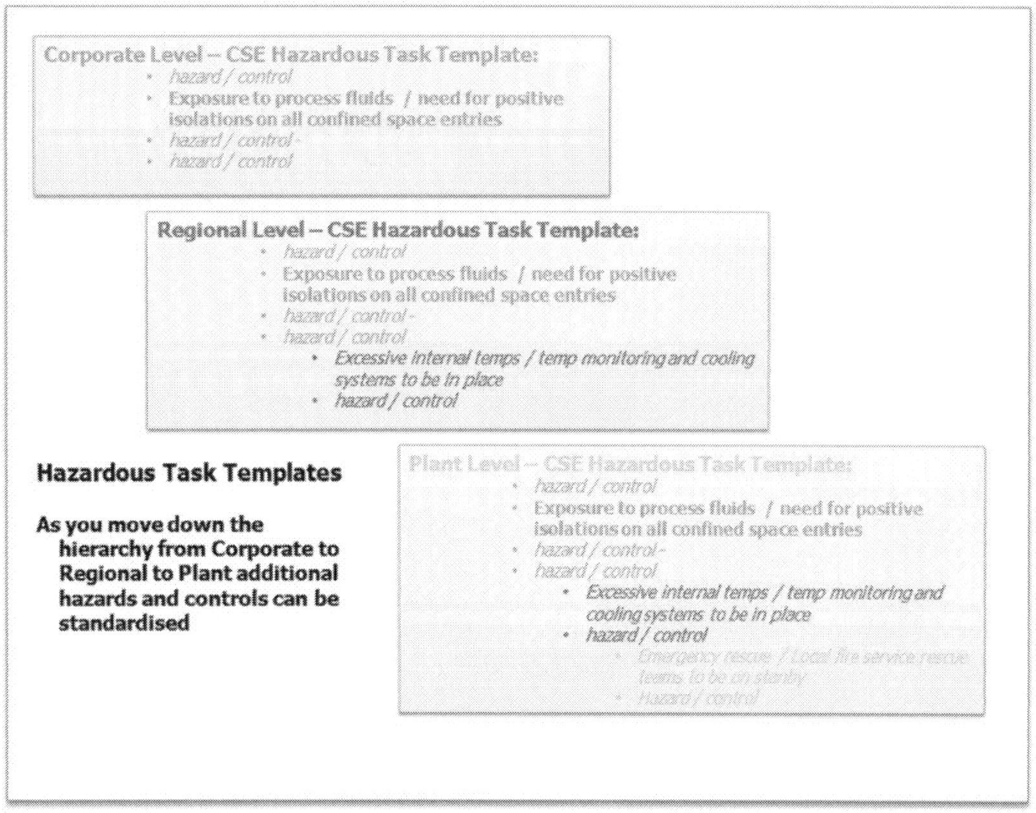

Fig. 7 Corporate, Regional and Local Hazardous Task Templates

It should be noted that these templates represent the minimum requirements for the task. They will be used to guide the task activities ensuring that the standards at all levels are applied. The expectation is that the person planning a task will also identify additional hazards and specify controls as appropriate for the specifics of the task, specific equipment, local conditions, tools being used, etc. In the CCPS book, Conduct of Operations and Operational Discipline, this is referred to as "thoughtful compliance" [4]. This eradicates the 'checklist' mentality ensuring that while the fundamentals are there, the individual can still interact with the system and really think about the hazards and control mechanisms required for the job.

Hazard / Control Libraries – These are the "libraries" in the system which are used to identify hazards and appropriate controls when carrying out a Risk Assessment. Common libraries can be driven from the corporate center in the same manner as Hazardous Task Templates. Similarly Regions and Plants can supplement these lists with regional hazards / controls. Where hazards may not be present in all regions or plants, the libraries can be filtered to only show those which are relevant. As such, permafrost hazards are not going to show as options for a plant in the desert and driving hazards shown in offshore facilities.

Work flows / Control Mechanisms – The new technology platform manages workflows for specific processes around Front Line Work Management. These workflows are driven through approval gates by various role authorities. Working in this way, procedures can be built into the system which ensure the proper routes through the process are followed and the correct permissions granted. As with the above, these workflows can be set at the different hierarchical levels. Regional or local rules can be added into the workflows without compromising the corporately defined workflows.

Examples of workflows managed by hierarchical Templates:
- **Permit to Work**
- **Lock out / tag out**
- **Permission to Deviate**
- **Temporary Defeating**
- **Routine working**

In this manner, the system ensures that rules are intelligently embedded within operational workflows and as Hopkins pointed out, the rule-making itself isn't sufficient; it must be enforced in pragmatic ways to facilitate successful compliance and discipline.

Organizations and senior management can be confident and assured that its strategic intent is translated efficiently and effectively into operational practice.

3.2 Driving Standardization

With policies, rules and regulations embedded into the operational workflows, all the task risk assessments take into account process hazards relevant to the location and equipment being worked on and are communicated to work crews. Frontline workers can be prompted by the system to be aware of process and hazard safety considerations and control mechanisms relevant to the area of work.

Through its scalable and enterprise nature, the system can be deployed across a plant, assets and even the enterprise, the standardized and harmonized approach cultivates a realistic and pragmatic approach - a tool that continually raises risk management capability and competencies, across the wide spectrum of employees.

From the central corporate centre, they can start to expel the wide variances and different ways of working that can foster subconscious bias toward inappropriate behaviors leading to unpredictable work execution, compliance failure notices, inaccurate planning as well as high levels of shutdown and incidents.

By automating and streamlining work processes and standardizing ways of working across the plant, assets or enterprise and yet being flexible enough to be pragmatic and appreciative of regional, local and cultural contexts, frontline workers flourish in the guidance supported by the tool. Over time this generates and sustains improved culture, discipline and behavior for operational excellence and enhanced safety

4 The Data

Not only does new technology provide organizations with the tools to close these three critical gaps, but they can establish effective feedback loops based on the automatic generation of data that can drive continuous improvement. Feedback loops are how we learn as they provide us with a sense of where we are at any given point versus where we want to be. Operational Performance and Predictive risk platforms allow an organization to actually know where it is in terms of its ability to manage workload against risk according to policy versus where it would like to be

As we have embedded and automated our frontline work processes in a consistent manner across our organizations and as work is being carried out, contextual data and role specific metrics can be extracted to provide insights into how work is actually carried out and the degree to which the frontline is executing work processes in line with the organization's policies, procedures and improvement initiatives.

From a planning perspective and in the Work Management Business Process Gap, frontline workers now, without having to do any additional work, except carry out their jobs as normal, now have the ability to feedback valuable data to Planners. The data inherently captured as part of routine work processes, establishes the complete job breakdown of all of the tasks, safety dependencies, isolations and the permits required to do each job and the actual time it took to complete "all of the job" during the Work Execution Phase. With tangible feedback on the actual job breakdown of what it took to complete the work order and the actual time it tool to complete the job fed back to the maintenance management and planning systems, there is actual high quality feedback to significantly improve future or in progress plans and maintenance schedules.

In the Policy to Practice Gap, data is also collected as a part of routine work processes and includes the type of work, the nature of work, the risk assessment types and details (hazards and controls selected), environmental conditions, human factors etc. The data provides rich contextual information of the work being done and its operating influences and environment. From this, it can tell us how well policy is being delivered in operational practice. Best practices can be consistently applied whether these are rules, regulations, processes or lessons learned.

With the visibility of how well policy is being delivered into practice, management has the ability to better understand, target and drive continuous improvement of policy.

By planning and scheduling the work accordingly, the system also generates role specific metrics that allow workers to see real-time contextual information driving better decision making to proactively manage workload against risk in time, space and risk, which we will explore later in the paper.

Predictive and Real time risk profiles based on cumulative risk are generated and can be viewed as they change throughout the shift/day/defined time period on individual plants, regions and for the entire enterprise. This visibility provides the opportunity to scrutinize and understand what is driving the risk.

Over time, the data collected as a part of these routine work activities can be used to provide visibility and accountability, as recommended by the International Association of Oil & Gas Producers. In their report entitled *Process Safety-Recommended Practice on Key Performance Indicators*, it states that "as such incidents [resulting from multiple, simultaneous barrier failures] do not occur very frequently, it can take a very long time to gather statistically relevant data on major incidents alone. Therefore, systems need to be implemented for consistent collection and analysis of data and related information on more than just major incidents." [5]

For the first time, management is able to access not just relevant historical data, but also new leading indicators and role specific of both operational risk and work efficiency.

4.1 Cumulative Risk Data

As we focus on operational risk management gap in this paper, it is important to explore the concept of cumulative risk. Even without data we may have some idea at the frontline leadership level of the risks we are carrying on a per-job basis but we don't have line of sight on the cumulative risk of all of the work we are performing at any given time. By better understanding what defines frontline cumulative risk and capturing the appropriate data, we have the key to developing valuable leading indicators that can determine risk trends and encourage better decision making. In frontline WE, managing cumulative risk applies not only to the sum of the parts of one job, including the hazards, controls, external factors, but to all of the daily tasks and procedures being carried out on an installation and their interaction and impact upon each other.

It is not unusual for 20 or more independent jobs to be in progress simultaneously within a defined area. While these individual tasks may have been independently risk assessed and authorised, there are dependencies and interactions between the jobs that need to be taken into consideration in order to reduce the potential overall risk of the combined jobs. Typical EPTW (Electronic Permit to Work) and even electronic PTW systems fall short in this area as they do not collect data in a consistent and efficient manner, nor do they allow users to manage work holistically across the plant, as they are focused on independently managing the lifecycle of each controlled job.

Let's consider a typical section of a plant, for example, the Alkylation unit of a refinery. For simplification, we will assume that the major process hazards which must be considered are:

- Large inventory of explosive and flammable hydrocarbons
- Volatile acid catalyst which is corrosive and toxic

On any given day the plant is live with a high throughput and these fluids are being contained in the pipes and vessels which have been designed for such service. Contingent systems are in place to detect loss of containment and to take action to mitigate any consequences such as: system shutdown and de-inventory; fire detection and water deluge; acid detection and water spray systems to knock down any release cloud. Emergency response plans will deal with evacuation, shelter and containment action to be taken by the Operations Team.

With steady conditions and a well maintained plant, the process hazards are relatively predictable. The containment and contingent measures significantly reduce the risk to a predictable, constant and acceptable level but do not eliminate them altogether. They are, however, not the only hazards to be considered. The additional hazards tend to be less constant as they vary from day to day, reflecting the reality of plant operations. Before we look at WE activities, there are some Human Factors aspects which could be considered 'hazardous' created by the Plant Operator team on shift at any time.

In the example shown below, there are typical Operations duties: sampling, chemical loading, preparing equipment for maintenance (LOTO), daily readings, permit issuing, gas testing, transfers, etc. Another typical example could be that the Shift Supervisor is on vacation and a Senior Operator is stepping up. Similarly, the Senior Operators role is being filled by a step up from the ranks. While not being abnormal, it does create a reduction in the overall experience and competence of the shift and an obvious increase in operational risk.

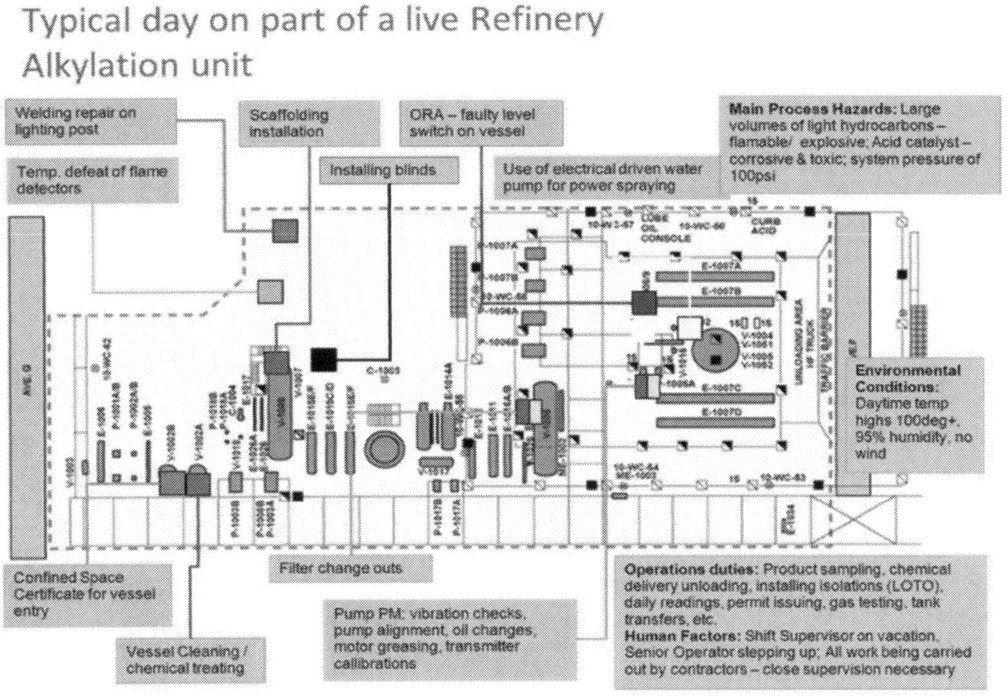

Fig 8. Example of Cumulative Risk on a Refinery Alkylation Unit

Human Factors and environmental conditions could also include contractors less familiar with the plant and its process hazards as well as prevailing working conditions – primarily climate related but also worker fatigue, travel times prior to arrival at the plant or distractions such as industrial action, wage/contract negotiations or personal crises at home. While these human and environmental factors can be difficult to measure they do have an impact and are often cited as contributing factors during incident investigations and should be recognised as impacting operational risk.

As a side bar, it is worth noting that all too often human factors are not considered as potential "hazards" during pre task risk assessments. Following an incident, however, a root cause frequently cited is Operator Error. Nancy Leveson, presenting in 2011 at the Massachusetts Institute of Technology2011 Conference, Reducing Accidents in the Oil and Gas industry, proposes that the traditional view is that human error is the main cause of incident and accidents, but system views find them, not solely the root cause of such, but rather the error can be the symptom of poorly defined processes and inadequate systems. Safety and performance is a control problem and this may be physical design, processes or social, not human error. [6]

Work activities, which vary from day to day, add a level of operational risk that is higher than for simply operating the plant. The residual risks posed by the process hazards are still present but there are now additional hazards to personnel for the Operations Team to manage. Each job is managed independently to minimize the residual risk posed by each hazard. The cumulative effect of each hazard's residual risk is seldom assessed formally but is tacitly acknowledged as being acceptable when a permit is issued for the work to commence.

The cumulative effect of multiple jobs coupled with the process hazards and any human factor or environmental hazards are seldom assessed in a formal manner on a day to day basis. Work schedules are assessed for clashes where two or more jobs may interfere with each another but we do not equip our plant supervisors with the right tools and techniques to properly assess these types of risks. In other words, decisions are being made without full information; decisions which sometimes can lead to disastrous consequences.

5 Improving Decision Making

With real data on what has occurred and more importantly with what is to come, the ability to move from reactive to proactive management is no longer a desired outcome but a distinct possibility. The consistent generation of data provides the capability to generate role specific metrics that are informative, actionable and improve decision making for each level of management from the frontline to the boardroom. An Operational Performance and Predictive Risk system empowers all levels of management to proactively manage towards operational excellence. The diagram below (Fig 9) demonstrates a small range of the possibilities.

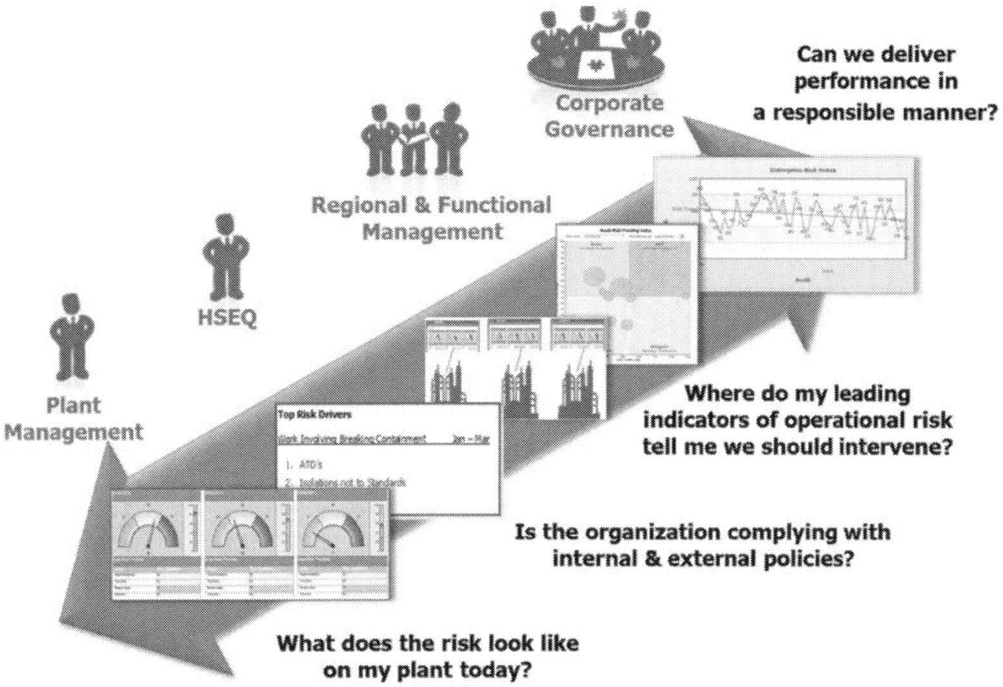

Fig.9 Empowering proactive management from the frontline to the boardroom

Frontline management

Frontline operations are where work must get done to ensure production targets can be met. It's where decisions are made every day to balance delivering that workload in service of production against the operational risk it entails. Traditionally we have placed significant pressure on the shoulders of those who issue, approve and manage work, based on limited paper systems to deliver performance and get day to day operational risk right.

With unparalleled visibility an Operational Performance and Predictive Risk system can provide frontline supervisors the ability to proactively manage and optimize work in time, space and risk according to corporately defined and embedded best practices. The visualisation of the 'actual' plan, reflecting the frontline realities, provides our people who issue, review and approve work further assistance to both structure and improve their decision making. It provides visible conflict detection capabilities to ensure that certain combinations of work are flagged to supervisors as well as hard rules to prevent specific combinations of work from occurring at the same time (e.g. Hot Work and Breaking Containment).

The diagram below (Fig 10) represents the dynamic visualisation of the safe work plan. While most organizations have clear standards for SimOps in terms of what multi task combinations can be done at the same time, we as an industry have essentially left it to our frontline supervisors to 'juggle' risk rather than provide them with the tools to properly 'manage' risk. The visualisation of the safe work plan provides an intuitive view into all of the jobs and their specific safety dependencies as well as the interdependencies between jobs. When work is moved in time frontline supervisors can see both the implications of their schedule optimizations in terms of "conflicts" as well as "cumulative risk" for the shift.

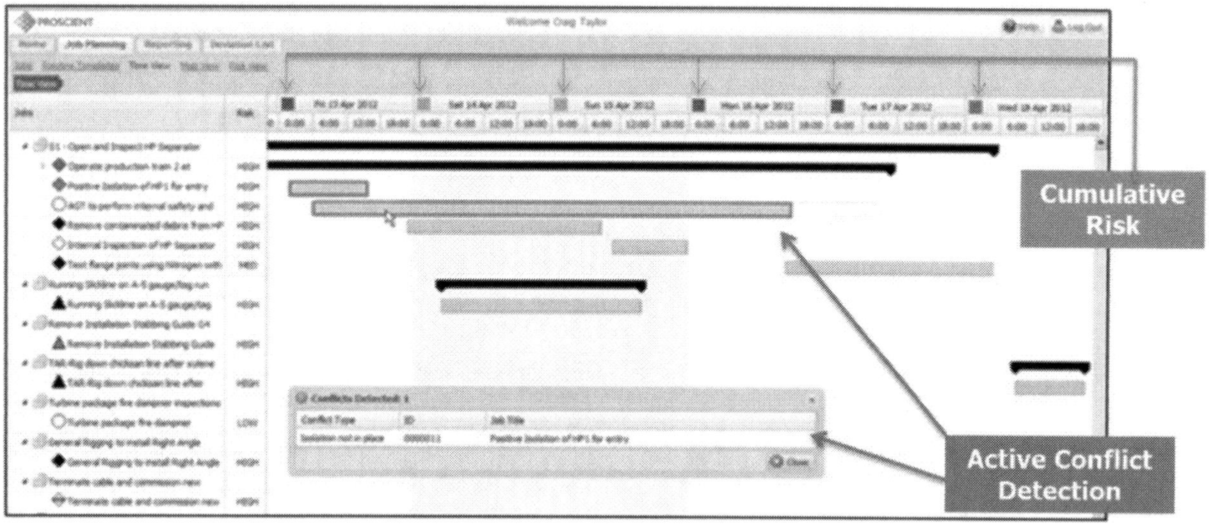

Fig 10. Optimize work in time with active conflict detection

The ability to visualise work in space (Fig 11) provides another dimension on how to improve SimOps and optimize work. We can see where work is occurring or planned during a defined period across our plant providing us with a holistic view. We can also see the process hazards in the area as well as any deviation from standards – operational risk (safety system inhibition or running equipment at levels outside normal conditions).

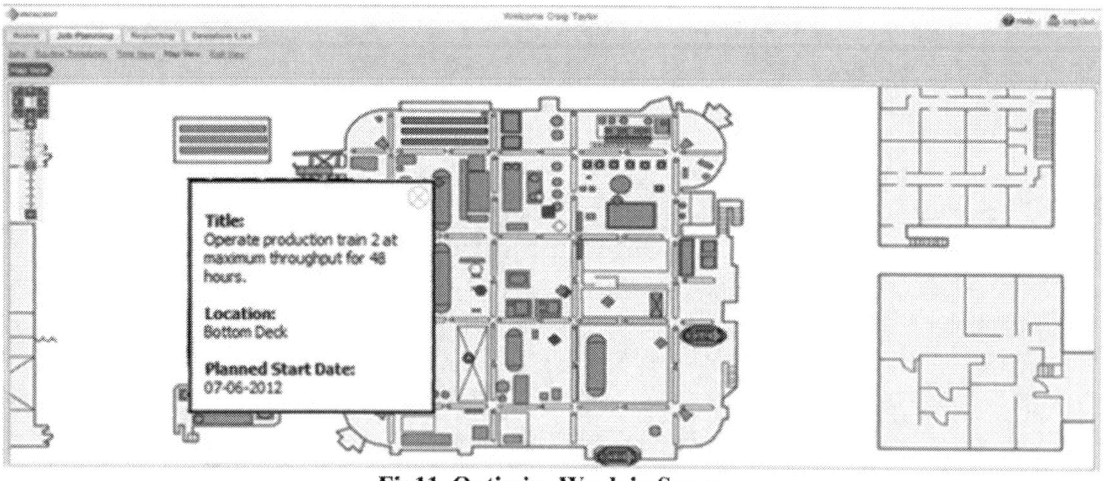

Fig11. Optimize Work in Space

The rich data inherent in building and completing work, risk assessments and interventions allow frontline supervisors to optimize work by predicting and proactively managing risk profiles. These leading indicators enable the adjustment of planned activities to minimise operational risk, while also optimising efficiency and providing visibility over real-time risk profiles across the plant (Fig 12). We can see the risk profile for the 'Plant' in the dial, the 'Distribution' of that risk in the LED bar across high medium and low risk jobs to ensure that a high risk job is not hidden by the average and finally we can look at the risk profile in each "area" of the plant and see the highest job risk profile for that area.

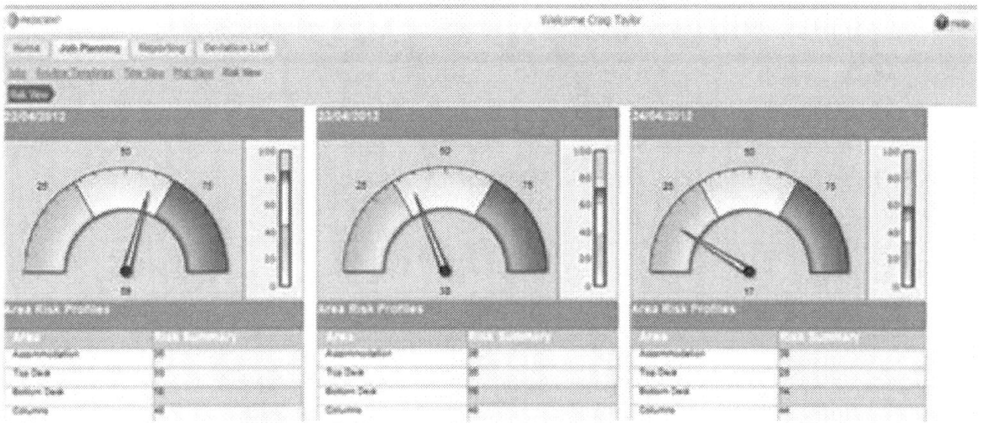

Fig 12. Optimize Work in Risk

These visualisation, optimization and risk management tools provide organizations the ability to:

- Understand the true nature of cumulative risk and avoid combining certain tasks
- Find alternative ways of performing high risk activities to reduce risk and enshrine these in best practice by incorporating them in the Enterprise major hazardous task templates and associated control mechanisms.
- Avoid high levels of risk during critical times like shift change over

At the Plant Level

Plant managers now have the ability to see what is coming in terms of planned work, its interdependencies and the risk profiles for each area of the plant and for the plant as a whole. It is a huge leap forward in terms of consequences management for plant managers. With more contextually relevant information in their hands, such as the risk profiles for each of the areas of the plant, the nature and location of the high risk jobs and the ability to see what is driving the risk, decision-making can be substantially improved. The prioritization of priority work and critical maintenance required to ensure production targets, can now be met and can be done in the context of an indicative level of operational risk.

Plant managers can also see which areas of the plant have the most risk for the day, based on planned and live work and these tools can be utilized to better optimize work while reducing operational risk. They can see the contribution of risk from high, medium and low risk jobs.

Fig 13 Plant and Area Risk Profiles

Such systems will allow plant management to identify cumulative risk issues during the near time planning of jobs. It highlight periods of excessive risk and either allow their Operations Team to simply move planned work packages in time in order to reduce the predicted risk to more acceptable levels or provide them with the right cues to dig deeper to find out why a high level of risk is being carried and what factors are specifically driving the risk on that job. They can also have a better view of the plant in terms of the barrier impairments impacting the frontline work as they can see those systems operating outside of normal parameters. Even with all of this information there may be situations where the decisions are made to carry a higher level of risk than usual albeit with the proper controls and contingencies in place, but at least decisions can be made with the accurate information and in the full context.

At the Corporate Level

At the corporate level in hazardous industries, organizations can move beyond rear view looking audits, anecdotes and lagging indicators. An Operational Performance & Predictive Risk system can dynamically move an organization to its front foot in terms of proactive management. Line of sight on frontline operations provides senior management with the insight and the ability to significantly improve corporate governance in the riskiest part of their operations.

Decision makers across organization can be provided with the benefits of the practical and meaningful data generated as a part of routine work processes. For the first time they can start to understand the relationship between performance and risk and proactively manage the organization to less risky and more optimized performance. With this level of visibility senior management can not only demonstrate performance and move beyond compliance but build the foundation for continuous operational improvement.

Comparisons between plants can be relatively straightforward, providing the ability to compare and contrast plant operational performance and operational risk. Senior management can now see over a period of a weeks and months, if one of their plants is consistently carrying too much cumulative risk, how that level of cumulative risk compares to the current organizational trends and whether that level of risk is outside the norm for that plant. The 'Executive Dashboard' (Fig. 14) provides a meaningful view of the data.

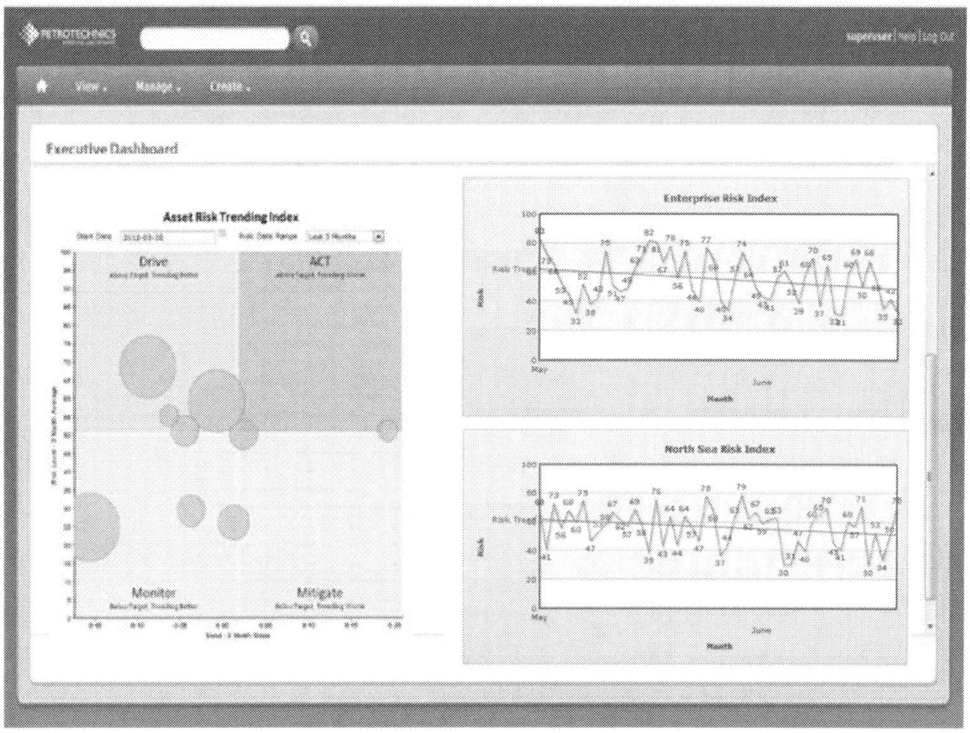

Fig .14 Executive Dashboard

- The Enterprise Risk Index (top right side of Fig 14) plots the daily risk scores for the organization and then trends the risk profile (red line) over the time period to provide management with an indicative view of the three month risk trend of the organization.

- Asset or Plant Risk Index (bottom right of Fig 14) provides local, regional and senior management can have an indicative understanding where each plant or asset sits and the direction it is heading.

- Asset Risk Trending Index shows (top left of Fig 14) the average risk level per asset over a time period versus its trending direction. The y-axis represents the level of risk and the x-axis represents the slope of for each Asset Risk Index trend line. So for example in this chart, any of the bubbles to the right of the y-axis have risk indexes that are trending upward to a more risky position and those to the left of the y-axis are improving. By seeing individual assets against their average risk score and trending direction, operational leaders understand where they need to act, mitigate, drive and monitor.

Senior management can use these enhanced indicators and improved organizational knowledge from the riskiest part of their operations, to enable their organizations to move from 'find and fix' to a culture of 'predict and prevent'.

5.1 How can such technology be used in response to an incident?

With an Operational Performance and Predictive risk system, proactive management can take on an entirely new dimension than heretofore even considered. Take for example the scenario where an incident has occurred within an organization. Traditionally, at best an email with preliminary findings and/or a warning might be sent out in the short term and then some time later a full investigation, its summary and key learnings make their way into the incident-accident system. A training program or initiative might be triggered months later to "address" the root causes of the incident even though at that time, the actual issues at the frontline may be something entirely different.

Organizational awareness and learning are paramount after an incident. Repetition of the same type of incident is all too common in hazardous industries and an incident often occurs too many times before its realised something needs to be done about it.

The importance of this cannot be emphasized enough. Professor Nancy Leveson in her US Senate Testimony sitting before the United States Senate Committee on Energy and Natural Resources (2011), summarized some of the common factors in major incidents, including the need for continual safety improvement and learning "Sometimes important causes of accidents are identified or problems detected during performance audits, but the information is never effectively used to redesign the social and physical components of the system. Why was the safety control structure ineffective in preventing the loss events? How can it be strengthened? A program of continual safety improvement needs to be created." *[7]*

With an Operational Performance & Predictive Risk system in place some of the prior actions would still continue, however the range of possibilities to become proactive empowers organizations. In the diagram below we can see some of the possible responses technology can offer to raise awareness of an incident across the organization as quickly as the organization's decision making process permits. Users doing a similar type of work to the work being done when the incident occurred can be presented with an advisory "at the time" they are about to do a similar job, wherever they are in the world. They could see a link to a preliminary incident report or advisory. This provides the real time contextual information and opportunity for senior management to positively influence behaviour where it matters most.

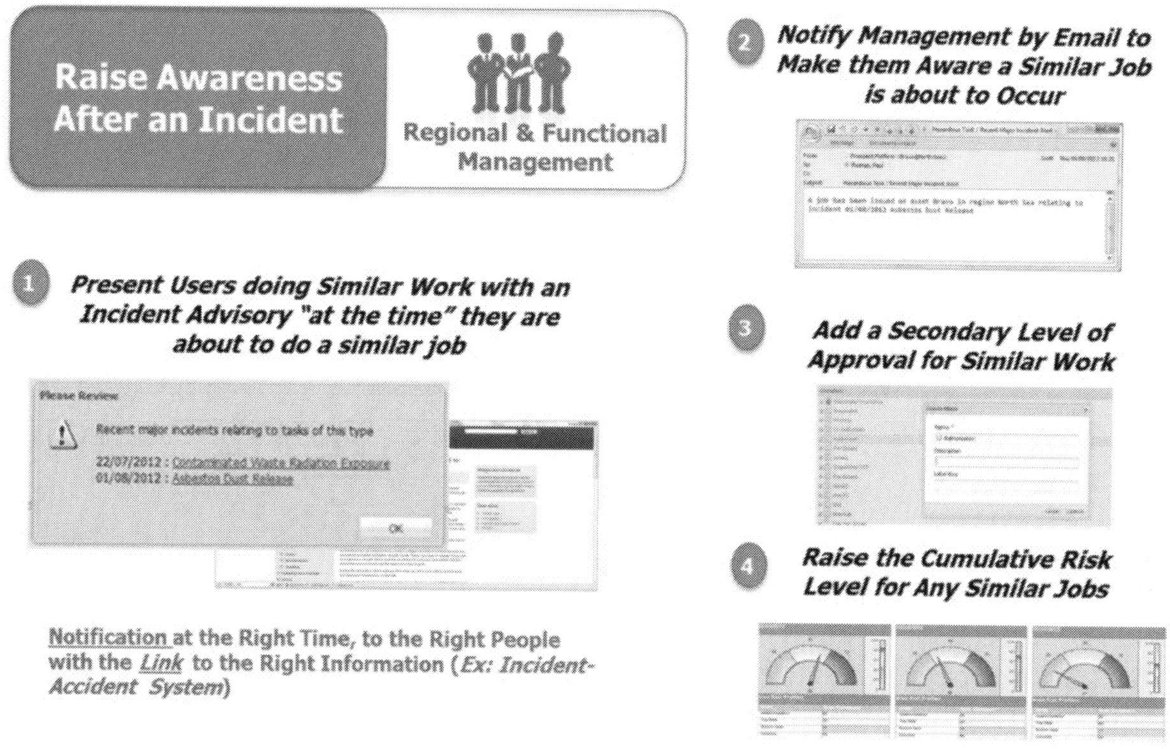

Fig 15. Organizational Awareness & Learning

Management can be alerted by email when any similar work is planned or about to occur throughout the organization providing advance notice. They can also add a secondary level of approval to all similar work such as requiring the plant manager to sign off on all similar work to ensure an extra level of oversight. And finally, similar type work can now carry a higher level of risk to drive risk profiles higher in order to influence frontline supervisor's optimization of work in time, space and risk.

Let's look at an example where a fatality has occurred to a worker working on a task which involved Confined Space Entry (CSE). This incident has occurred in the Middle East at a plant operated by a global company headquartered in Houston, Texas. Traditionally, the response to such an incident would be to put an investigation team together. Look at root causes for the incident, determine lessons which have been learned and come up with recommendations for changes in procedures, policies, etc. Communicating outcomes is usually through safety alerts or bulletins dispatched to all Operational Areas of the corporation with some recommendations. This process takes time and usually several weeks before some level of communication is made with the larger organization. Longer term responses could include rewriting policy or procedures covering confined space work and the results from these typically take months to implement

Often immediately following such an incident, questions are asked by senior management that are relevant but difficult to answer. How many of these types of activities do we have ongoing currently? What should we be saying to Plant Managers where CSE tasks are about to be performed in their plants? Do we know for certain that corporate policy on CSE is being implemented at all our plants? What is behind these questions is straightforward – what is our risk of this happening again and what steps can we take immediately to prevent it happening again?

With an operational performance and predictive risk system in place, Management have some powerful tools which can allow direct action within hours of such an incident taking place. Table 1 below shows some of the immediate action options available to Management through an Operational Performance and Predictive Risk platform

Table 1. Implementing Organizational Learning Immediately

Options for immediate Management Intervention following an incident	
Item	**Action**
Information gathering - Details of work activities	The system tells immediately the work task which was being undertaken. The risk assessment carried out and the specified controls and barriers for the job. Details of the isolation scheme in place and details of how the confined space had been certified. The people who authorised progression of the work at all stages is visible and traceable Details of other ongoing work on the same plant can also be reviewed.
Similar types of jobs	The system can be queried to find out how many other CSE tasks are being carried out at present. This will be across the whole enterprise – so all regions and all plants will be counted Queries can also be run to see future planned CSE tasks for the coming days
Immediate Intervention – workflow template	The corporate workflow template for CSE work can be altered so that all future CSE tasks would need to proceed only under the direction of the Plant Manager until further notice. A flag can be generated in the system to inform users of this temporary change in process and why such precaution is being taken. From that point onwards, all CSE approval workflows across global operations would follow this rule.
Immediate Intervention – Hazardous Task Template	If early indications have identified a hazard which has not been considered or a control which was not effective, the corporate template can be modified to make amends for this. From this point onwards when this template is used the modified hazards / controls will be in force.
Communication	Information flags can be set in the system to alert users to the changed workflow or template and explaining why this action was taken. Links to investigation systems can also be included.
Changed Process	Following the incident investigation, more permanent changes to the workflow, task template, Hazards / controls library can be made at the corporate level. All changes made will be reflected in the hierarchical layers below. For example, if the outcome of the investigation showed inadequate atmosphere

	monitoring, the changed procedure could mandate a specific type of monitor to be used on all CSE jobs. A change at the corporate level template to specify this would immediately be in effect across the whole enterprise.
Modified Risk Index	The algorithm which calculates the Risk Index involves a large number of variables, each with their own risk weighting. As a result of an incident, these risk weightings can be adjusted based on the investigation findings. If CSE issues have been seen in a number of incidents the risk weighting of this type of task can be increased to emphasise its importance in the overall risk make up.

From this example, it is easy to see how corporate insight and rules can be intelligently embedded into the system, ensuring there is benefit across the organization from frontline to boardroom. The possibility of integrating organizational learning in this quick, efficient, proactive manner, without having a long-drawn out implementation process provides the basis for operational excellence and much needed continuous improvement.

6 Driving Continuous Improvements

By providing the necessary supporting tools for workers and establishing an effective work environment via pragmatic processes that are intelligently embedded within workflows, the basis for proactively capturing key information and therefore monitoring, measuring, acting and improving all comes at ease.

Without the streamlined and automated data and feedback loops for each gap in a contextual manner, the only mechanism that remains to drive continuous improvement is auditing, which we've already discovered as adequate only. The issues is that it breeds a 'find & fix' culture and never cultivates a 'predict & prevent'

Audits identify the symptoms of problems, but infrequently address the causes. For example conventional inspections may show up a locked fire escape, but they don't say anything about why plant managers felt that this was an acceptable practice. Equally, audits may reveal shortcomings in record-keeping, but won't suggest the systemic improvements needed to upgrade processes and therefore drive improvements. If we are interested in continuous improvement, a tick-box mentality and approach can only take us so far.

With Operational Performance and Predictive Risk Systems, channeling feedback in real-time on each of the gaps described serves a greater purpose and allows anyone from the frontline to the boardroom to proactively intervene and change something before it goes wrong (rather than after) and to optimize the work and effectively balance workload against risk. For enhanced operational discipline and compliance, understanding policy adoption rates in real-time allows senior managers to act quickly if changes or improvements are to be made and embedded in an efficient and effective manner.

Overall, the instant feedback on these critical areas goes a long way and moves the organization to its front foot in terms of proactive management.

7 Summary & Conclusions

This paper has examined the often overlooked area of frontline work management which, data suggests, poses the greatest Operational Risk in our industry since it is the first or second major cause of fatalities and high potential incidents. We looked at traditional methods of risk management and assurance such as standardized practices and the challenges they face plus we examined three critical gaps existing in frontline Operations: The Work Management Business Process; putting Policy into Practice and Managing Workload against Risk. There have been few developments in these three areas in recent years and decades old practices which constrain our ability to realise performance improvements.

A new category of technology, the Operational Performance and Predictive Risk Platform which is being pioneered by a product called Proscient is changing this position. Deployed at the enterprise level, this platform integrates with traditional Maintenance Management / Asset Management systems to greatly improve work identification, planning and scheduling at the daily/hourly level. The hierarchical structure enables corporate policies to be configured as practice templates ensuring rigid conformance at the regional and local levels ensuring hazardous work is managed in a consistent manner at all times.

Introducing a fully systematized solution into this space which has historically been managed by paper forms results in vast quantities of data being automatically generated that can be instantly analysed and presented in ways that have never been seen before. Included in this is a risk index which is calculated in real time and can be aggregated across a whole site and even across many sites to show the overall Operational Risk across all global operations. This index can also be used in a predictive mode based on work which is planned 24hrs / 48hrs in advance. Ongoing and future work can now be viewed not just as a list of tasks but can be shown geographically, in relation to plant equipment, on a time axis as in a typical gaant chart and also from a risk perspective – the cumulative effect of all residual risks from planned tasks, prevailing process hazards and from environmental and human factors. Using this predictive view, optimization decisions on the immediate work schedule can now be carried out more effectively resulting in safer and more efficient operations. We also saw how such a system can provide more proactive and immediate response options in the event of an incident not just the traditional safety alert communications.

The ability to provide a real-time view of frontline activities across an organization's global operations is no longer a pipe dream. Working with high level dashboards summarising data from across the enterprise a new perspective on real-time operations is now possible which render the old concepts of leading and lagging indicators somewhat redundant. Real time and predictive indicators now become the standard performance measures and organizations can start to move from a "find and fix" mode of operating to one of "predict and prevent". With line of sight provided by new technologies, organizations now have the ability to see and understand the relationships between workload and risk and can manage the organization to a more optimized and less risky performance

References

[1] BP Control of Work Safety Booklet B, 2005

[2] FERROW, MICHAEL An Operating Management System to Deliver Safe Operations SPE, 2011, p.5.

[3] HOPKINS,ANDREW Risk Management and Rule Compliance Decision Making in Hazardous Industries, Working Paper 72 OHS Regulation 2010, p.3

[4] CCPS/ WILEY, Conduct Of Operations and Operational Discipline For Improving Process Safety In Industry – 2011

[5] PROCESS SAFETY, Recommended Practice on Key Performance Indicators, November 2011, p.10.

[6] LEVESON, NANCY A New Approach to Ensuring Safety in Software and Human Intensive Systems, 2009, p.23

[7] LEVESON, NANCY, US Senate Testimony on Energy and Natural Resources, 2011, P.3

Conduct of Operations Best Practice Networks

Marco Antonio Vela
Celanese
KM 12.3 Carretera Federal, Coatzacoalcos, Veracruz Mexico, 96400
Marco.Vela@Celanese.com.mx

Jerry J. Forest
Celanese
1601 West LBJ Freeway, Dallas TX, 75234
Jerry.Forest@Celanese.com

Prepared for Presentation at
American Institute of Chemical Engineers
2013 Spring Meeting
9th Global Congress on Process Safety
San Antonio, Texas
April 28 – May 1, 2013

UNPUBLISHED

Conduct of Operations Best Practice Networks

Marco Antonio Vela
Celanese
KM 12.3 Carretera Federal, Coatzacoalcos, Veracruz Mexico, 96400
Marco.Vela@Celanese.com.mx

Jerry J. Forest
Celanese
1601 West LBJ Freeway, Dallas TX, 75234
Jerry.Forest@Celanese.com

Keywords: Conduct of Operations, COO, OD, ED, MD, Networks, Best Practices

Abstract

Conduct of Operations is about reducing human error by consistent operation and repeatability of results. Celanese favors the use of checklist and forms for various conduct of operation activities such shift instructions, shift change, operator notes, and operator evaluation. These forms are an example of good practices that can be shared across our global operation to get consistent results.

Analysis of Loss of Primary Containment Incidents (LOPC) data indicates that about half of causes can be attributed to failure of conduct of operation systems. That statistic, combined with data obtained from a strength, weakness, opportunity, threat (SWOT) analysis indicates that that there is an opportunity for reduction in incidents by addressing the conduct of operations causes by sharing best practices through internal networks.

This paper will describe how to share conduct of operations best practices by the formation of a global best practice network. This network includes: periodic corporate conference calls, implementation of programs and milestones, a place to share information on a common intranet portal, and continued communication along all levels the organization.

1. Introduction

Conduct of Operations (COO) is about reducing human error by consistent operation and repeatability of results. The challenge for chemical producers is not just the implementation of a COO process at the manufacturing sites, but also how to sustain and provide an environment of continuous improvement.

COO is one of the first lines of defense in incident prevention. At Celanese, analysis of Loss of Primary Containment Incidents (LOPC) data indicates that about half of the causes can be attributed to human error created by failure of COO systems. From a corporate standpoint, correction of these causes becomes challenging due to the technological and geographic diversity within the company.

Celanese has established a communication process among sites around the globe to share the COO best practices. We believe that this global COO network is partially responsible for a 1 year reduction in human error COO causes by about one third. This paper will describe a process for implementing a global COO network.

2. Data Analysis and the Need for Networks

A strategic planning process that consisted of data analysis and internal and external environmental analysis was first used to develop strategic alternatives for process safety improvement. A more detailed analysis of the strategic planning process is presented at the CCPS 9th Global Congress on Process Safety. A high level summary is presented here to highlight the need for internal networking [1]. Some of the tools used in this process are shown in Table 1.

Table 1. Strategic Planning Tools

Tool	Inputs	Outputs	Process	Purpose
Data Analysis	Audit data, Leading and Lagging indicators	Common failure causes, areas for improvement	Descriptive statistics, Pareto analysis	To find and correct systemic incident causes
SWOT Analysis	Existing beliefs about process safety performance	Key opportunities and strengths, weakness and threats to process safety	Survey	To evaluate internal beliefs about the organization
Process Safety Cultural Analysis	Existing process safety systems	Common beliefs about process safety performance	Survey	To evaluate internal beliefs about the organization
External Environmental Analysis	Participation in trade groups, interaction with peers, monitoring the regulatory environment	Industrial trends in process safety	Various	To ensure we are working on the right things relative to industry

This paper will discuss lagging indicator analysis and the strengths, weakness, opportunity, threats (SWOT) tools for developing a strategy.

2.1 Lagging Indicators: Loss of Primary Containment (LOPC) Data Analysis

Analysis of LOPC common causes across a company can indicate if systemic weaknesses exist. Elimination of those causes can therefore prevent whole classes of incidents. Three years of LOPC causes were analyzed using standard descriptive statistics and charting. In summary, it was found that about half of all causes were human error, COO related. The breakdown of failures into the individual COO elements is shown in Figure 1.

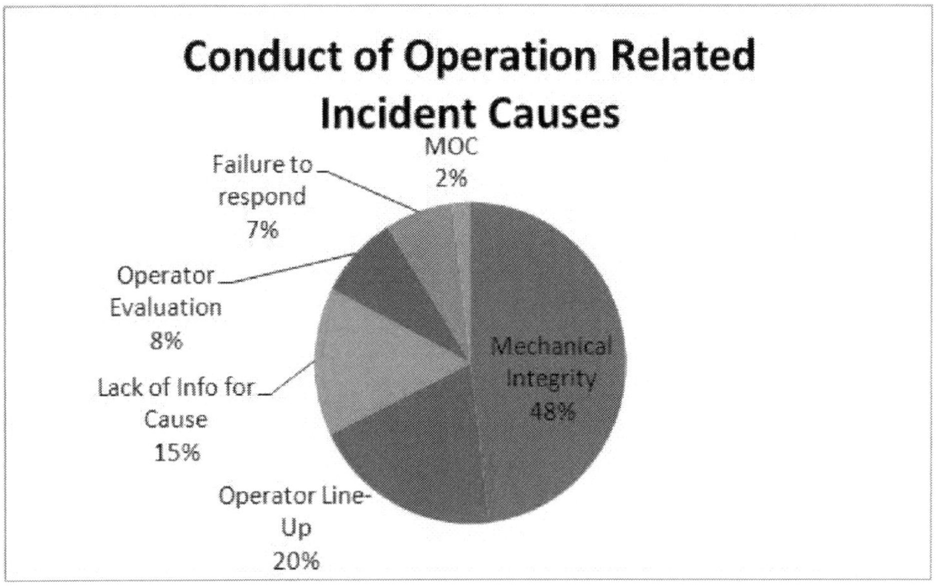

Figure 1: Conduct of Operations Related Incident Cuases

Of particular concern are the incidents involving operator line-up, operator evaluation, and failure to respond to alarm and conditions; all direct operational discipline topics. This analysis also surfaced two additional areas for improvement: Mechanical integrity human error failures, and the need for improvement in incident investigation and assignment of causes.

2.2 SWOT Analysis

A SWOT analysis is one tool used to assess both the internal and external environment of a company. The strength and weaknesses usually refer to what a company thinks about its own internal strengths and weaknesses. The opportunities and threats surface what a company thinks about external pressures that define opportunities and threats. Therefore, a SWOT can provide one indication of a company's culture by surfacing commonly held sets of believes. While usually considered a business tool, we have used SWOT analysis to determine the common beliefs across various business lines, technologies, and geography on process safety.

Among other things, the SWOT analysis showed an internal belief that we have an opportunity for improvement by improving the way we internally communicate best practices; we can best improve by networking.

2.3 Defining the Strategy

Analysis of lagging indicators shows that a common and correctable cause of process safety incidents is directly related to COO. Furthermore, internal believes about performance surfaced in a SWOT analysis of process safety indicated that an opportunity for improvement exists within the company by improving our own internal networking and sharing of best practices. This therefore defines one strategic alternative for improvement.

3. Planning for Global Networking

The corporate process safety steering committee chartered a team to implement a global COO network based on the strategic planning process. Objectives included providing a communication platform that could allow us to share COO topics related that would help each site representative in improving their understanding and application of each COO element. A second objective is to achieve flexible but consistent application of the practices globally.

3.1 Regional COO Workshops

The planning process used to initiate the best practice networks included "seeding" the organization with known practices that directly address the lagging indicator causes. A two day seminar was developed and delivered by region to introduce practices that could help sites with immediate improvement.

Topics for this workshop included elements of operational discipline (OD), engineering discipline (ED), and management discipline (MD) [2], with a focus on OD [3]. Among other topics, participants were given tools to help with shift relief, shift instructions, shift logs/notes, equipment evaluation, and abnormal events notification. Celanese favors the use of checklists and forms that detail what we wish operators to report. This helps to provide a consistent communication of what is important and facilitates evaluation by the technical staff that these OD elements are being performed correctly. The workshops were conducted on a regional basis: Asia, European Union (EU), and Americas.

Preparation by the participants for the COO workshop included a self-assessment of site performance to the internal COO corporate procedure. As a follow-up to the workshop, each site then had the responsibility to conduct a self-assessment against the best practices introduced during the workshop. An improvement plan was then tailored according to each site's needs.

The actions plans from the self-assessments then became a starting point for discussion in on-going regional networks.

3.2 Networking Communication System

It made best sense for our company to divide the networks into geographical regions that corresponded to the COO workshops: Asia, EU, and Americas. The main advantage of geographic regions is the time zone difference. In addition, it allowed for a cross section of technologies to be represented in sharing. The networks began with quarterly teleconferences so as to not burden manufacturing sites with too many meetings.

3.3 Develop a Share-Point Site

In order to upload, store and share the information that is generated during the network meetings, a share-point site was developed and the appropriate people given access. This common storage area ensures that this information will be always available for the employees that need access. Information posted and shared in this portal include: COO corporate procedures, best practices, COO network meeting information, sites action items, and training materials.

3.4 Site Representatives

Celanese has identified process safety champions at each site, these champions are invited to attend. In addition, we require a responsibility and accountability plan for each site. Those identified as having responsibility for COO should also participate.

3.5 Generic Meeting Schedule

The COO network charter indicated that the network should promote a proactive participation of each of the site representatives, as well as to build a meeting work agenda based on the company strategy and sites needs. To achieve this objective, a generic meeting agenda was created containing generic topics to be presented in each of the quarterly meetings:

- General COO topics to be shared with each sites (Led by a corporate process safety representative)
- Sites best practices sharing (Led by site COO contact)
- COO issues for feedback and recommendation from others (Promoted by site COO contact)
- Action plan & follow-up (All)

Each quarterly conference call is normally scheduled to be performed within a 60 minute timeslot.

4. Implementing the Network

For each network meeting, an agenda is developed that includes participants and presenters allowing time for the exchange of ideas and experiences among participants.

Each participant is responsible to present a specific topic and to prepare and develop a presentation. This presentation is then integrated into a general presentation consisting of all the topics that will be presented. Also the presenter has to promote an active participation of the meeting audience to generate ideas interchange and get adequate feedback to support and improve the sites COO system.

As indicated above, the promotion of an active participation in the COO network calls is important for establishing an open and trusting environment. This is necessary so that each of the participants can be comfortable with sharing their site best practices. It is during free sharing of practices and open exchange of ideas that provides the largest benefit.

Although site representatives are presenting topics relative to their own site need, the corporate representative will also include COO procedural topics for discussion. In this way, we know that best practices for all requirements of our standard are being covered.

Once the quarterly meetings have been established, is also important to promote regular communication between the site COO representatives and corporate process safety in order to provide support for the implementation of their action plans.

Table 2 illustrates some of the various COO topics discussed at our regional network meetings, and highlights the diversity of topics and areas for focus.

Table 2. Network Topics by Region

Region	Topic
Americas	Labeling
	Safe operating limits
	Fatigue management
	Safe Minimum staffing
	Tool management
EU	Shift handover
	Verification after maintenance
	Closed loop communication
	Housekeeping
	Process safety officer
	Lighting
	Shift meetings
Asia	Shift handover
	Safe operating limits
	Evaluation rounds
	Inhibiting/ disabling safety systems
	Alarm management

Note that while there is some overlap of regional topics, all topics and materials covered are available as resource material among the regional networks.

The regional topics do show some trends. For example, the Americas and EU region are focused on improving human factors; While Asia tends to focus on safety system improvement. These trends are an indication of the maturity of their COO system. The Americas and EU have had COO in place for years, while this process safety element is relatively new for our Asian sites.

Therefore, from a global perspective, it is important to consider an adequate balance between COO topics and the maturity of the participating sites and their region. This ensures that networking that will be beneficial for all sites based on individual and regional needs.

5. Check Performance

We began the networking effort in order to address a class of common causes; human error related to COO. The networking effort was begun in early 2012, and implemented throughout the year. COO related causes fell from a total of about half of the total causes of incidents from previous years to about one third of the total causes in 2012. Figure 2 shows the reduction in COO causes from 2012 from two prior years.

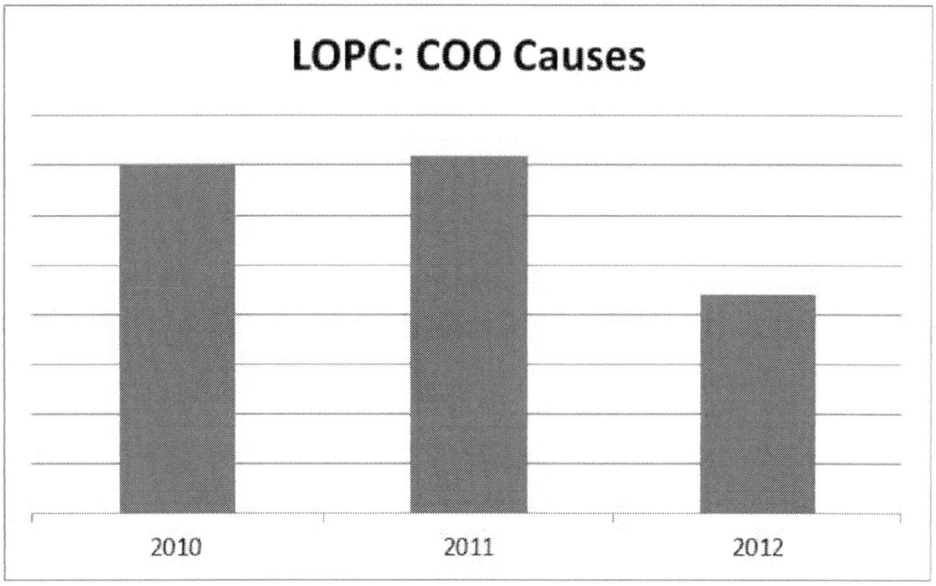

Figure 2: COO Cause Reduction after Network Implementation

5.1 Sustain Performance

The main incentive to participate in this program is to reduce the incidents with human error related causes. An interesting trend was seen in the data analysis that indicates a similar proportion of human error causes exist for both occupational safety and process safety incidents. Therefore, eliminating causes in one area may eliminate causes in both. Furthermore, we have also seen that sites around the globe are eager to share those practices that seem to improve performance, and are eager to adopt new ideas that have worked well at other sites.

Although we are pleased with the initial results of our efforts to address COO causes, we are concerned with maintaining the rate of improvement and eliminating this type of cause. In the spirit of the network, we solicited feedback from the participants to determine their view of effectiveness and how we might approve.

A simple survey was used to solicit feedback. A 5-point Likert scale was used.
The questions are:

- Do you consider that COO networking is a good tool to improve some of the aspects in your site?
- Has your site taken any benefit from the COO networking?
- Would you consider positive to keep the COO networking for the next year?
- Have you attended the COO meetings on a regular basis?
- Do you regularly use the COO portal?
- Are the quality and value of the presentations given adequate?

The results of the survey are shown in Figure 3.

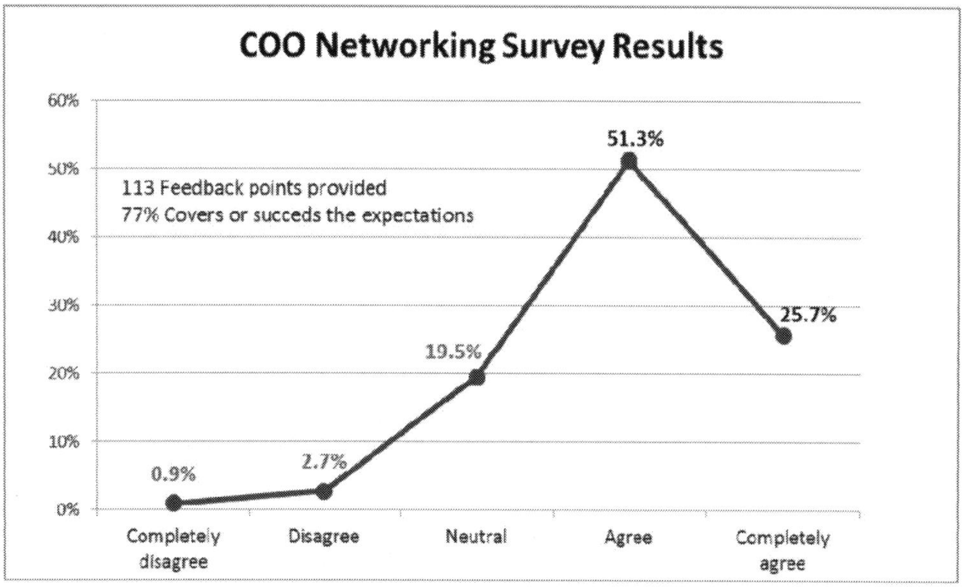

Figure 3: COO Networking Survey Results

This overall look at the process indicates that the network participants view the networks favorably and should be continued in their present form.

7. Conclusion

The first step in reducing process safety incidents is taking an honest look at areas that need improvement. Many tools exist to aid us in this identification; basic analysis of lagging indicators coupled with people's opinions can identify obvious areas to address. In our case, human errors related to conduct of operations surfaced as a common cause failure. Our process safety professionals told us that we have the largest opportunity for improvement by sharing of best practices and networking.

A structured approach in network implementation has led to an initial improvement of performance. Close monitoring of the networks will be needed to keep the material fresh, and relevant so that continuous improvement is realized.

Process safety incidents can happen at your site, but they can be prevented by following the process safety management systems.

8. References

[1] J. Forest. Management Discipline: Defining a Process Safety Strategy. Presented at the CCPS 9th Global Congress for Process Safety

[2] J. Forest. Management Discipline. Process Safety Prog 31 (2012), 334-336.

[3] J. Haesle, C. Devlin, and J. McCavit. Improving process safety by addressing the human element, Process Safety Prog 28 (2009), 325-330.

(94a) Management of Change Indicators

Clarissa Sossmeier da Silva, Safety, Health and Environment, BRASKEM - Petrochemical Basics Units, Triunfo, Brazil and **Americo Diniz**, Braskem, Camacari, Brazil

Management of Change Indicators

Management of change is an elementary pillar of o robust Process Safety Management Program once it aims to assure that all changes in installations, technology, work process and organization are performed in a controlled and structured way so that the risk to people, environment and assets are kept inside the company risk tolerance criteria. Basic Petrochemical Unit of South Brazil site have implemented the Braskem MOC Corporate Procedure in 2011, focusing in the extension of the existing process to new dimensions (such as technology and people changes) and in the use of electronic MOC process. Since then our site faced another important aspect that come together with the MOC implementation itself: the importance of having a reliable way to evaluate MOC effectiveness

In this paper it will be presented the work involved in the selection of MOC indicators that support to track process safety performance, considering best practices, insurance recommendations and our own experience. Besides that it will be shown our approach to guarantee that all changes are managed by a central group or a Guardian of the MOC process. It will also be discussed the main difficulties and our strategy related to the implementation of People Management of Change.

See more of this Session: Management of Change
See more of this Group/Topical: Global Congress on Process Safety

Organizational Change Management

John Wincek, CSP
Croda, Inc.
8 Croda Way
Mill Hall, PA 17751
john.wincek@croda.com

Molly R. Myers, P.E.
ioMosaic Corp.
93 Stiles Road
Salem, NH 03079
myers.m.nh@iomosaic.com

Henry Ozog
ioMosaic Corp.
93 Stiles Road
Salem, NH 03079
ozog.h.nh@iomosaic.com

Prepared for Presentation at
American Institute of Chemical Engineers
2013 Spring Meeting
9th Global Congress on Process Safety
San Antonio, Texas
April 28 – May 1, 2013

UNPUBLISHED

Organizational Change Management

John Wincek, CSP
Croda, Inc.
8 Croda Way
Mill Hall, PA 17751
john.wincek@croda.com

Molly R. Myers, P.E.
ioMosaic Corp.

Henry Ozog
ioMosaic Corp.

Keywords: organizational change, management of change

Abstract

It has long been acknowledged that when not properly evaluated and controlled, changes to physical equipment in a facility can lead to serious incidents with potentially severe consequences. Management of change (MOC) systems, replete with a variety of electronic systems, flow charts, and checklists, have been developed by a number of reliable organizations throughout the world to manage these physical changes. It is less commonly recognized that other types of changes such as changes in job responsibilities, loss of key personnel, or even changes in shift hours can have an adverse impact on process safety. These and other non-physical changes, collectively referred to as Organizational Changes, can lead to serious incidents with potentially severe consequences. Due to their focus on managing physical changes, most MOC systems have overlooked or only superficially address organizational change management (OCM) and the impact of organizational changes that affect process safety.

Organizational change is an unavoidable aspect of doing business. There is a large variety of changes which fit under this umbrella of organizational change. Any of these types of changes could result in catastrophic consequences if the changes are not successfully administered. Effective OCM procedures must include a system for managing potential modifications to a variety of organizational aspects.

The Center for Chemical Process Safety (CCPS) has published a new guideline book covering organizational change management. This presentation will highlight some of the key concerns related to organizational change management which are covered in this new publication and which should be included in a successful OCM program.

1. Introduction

OCM should encompass a variety of types of changes within an organization. Some of the types of changes to consider include the following:

- Modification of Working Conditions
  - Changes to the terms and conditions of employment such as hours, shifts, and allowable overtime (including temporary deviations)
  - Modifying the location, methods of communication, or time allocation for people
  - Staffing during extreme weather events, facility-wide emergencies, or turnarounds
- Personnel Changes
  - Change of plant management (e.g., plant manager, HSE manager)
  - Project changes (e.g. new project manager, replacement of a project engineer)
  - Replacing a subject matter expert (SME)/core competency (e.g., relief sizing, risk assessment, mechanical integrity, incident investigation, PHA facilitation, etc.)
  - Replacing the incumbent in a position that affects process safety (including corporate staff)
  - Strikes, work stoppages, downsizing/layoffs, slowdowns, etc.
  - Emergency response team staffing for facility-wide emergencies
- Task Allocation Changes
  - Job competency requirement changes
  - Relinquishing an individual's responsibilities for tasks without those tasks being reallocated
  - Requiring individuals to take on new responsibilities demanding skills and competencies unconnected with those previously required
  - Temporarily not filling a position (e.g., hiring freeze)
  - Temporary backfill for leave of absence, vacation, or temporary assignments
- Organizational Hierarchy Changes
  - Centralization or decentralization of job functions (engineering, maintenance, SHE, etc.)
  - Reorganization and de-layering the hierarchy
  - Changes to span of control
  - Linear vs. matrix organization
  - Acquisitions, mergers, and joint ventures
  - Changing service providers
- Organizational Policy Change
  - New and revised corporate process safety related polices/procedures
  - In-sourcing/out-sourcing of key departmental functions such as engineering design or maintenance
  - Major changes to policy or budgets for maintenance or operations
  - Staffing level policy changes (start-up, shut down, turnaround)
  - Changes to mission or vision statements

2. Organizational Change Management Program

Management of Change is an Element of one of the Management System Accident Prevention Pillars, identified in the CCPS publication *Guidelines for Risk Based Process Safety*. Most companies that handle hazardous materials have a Management of Change (MOC) program that addresses changes to equipment, facilities, and procedures. Some have included organizational changes in their existing MOC procedures or in a separate procedure. However, organizational changes, whether large or small can have less obvious but equally serious impacts on an organization's process safety performance. In many instances, it may take months or even years to realize the impact on process safety resulting from an organizational change. In some cases the impact has only been identified after a serious incident. Therefore it is important that these changes be reviewed to assess their impact on process safety.

Any management system should start with a clear policy from the highest level of the organization. This policy should state the principles, commitment and accountability of the organization and commit to proportionate consideration of all organizational changes, large and small, as even those with no apparent connection to safety need to be given due consideration to confirm whether or not they have impacts on safety.

The OCM procedure should have a clear definition of what constitutes an organizational change. In traditional MOC programs, a simple definition of a change is anything that is not "replacement in kind". For OCM the key issue is whether the change impacts the organization's ability to implement process safety critical tasks or to control process hazards Organizational Change is defined as any change in position or responsibility in an organization or any change to an organizational policy or procedure that affects process safety.

In order to effectively manage organizational change, a company needs to start with a procedure. This can either be a standalone procedure or modifications to an existing MOC procedure to incorporate organizational changes. The procedure should identify the types of changes covered by the procedure and should include a thorough risk assessment of the change to ensure that process safety does not suffer due to the change. Various types of organizational change can take place safely as long as they are properly managed.

An important aspect of OCM is communication and participation. The individuals affected by the change should be notified as early as possible in the OCM process to ensure their active participation and understanding of the need for the change. Additionally, internal customers should be consulted to ensure that all aspects of process safety management are considered and that no individual tasks fall through the cracks with the organizational change.

Any organizational change can have a significant impact on an individual's morale and focus. The sooner that affected individuals can provide feedback on the process, the more time management will have to address these concerns before the change is implemented. Downsizing or divestiture is one exception where communication of the change may not occur until after the change has been authorized for implementation. For these types of organizational changes, the communication part of the OCM process becomes even more important in order to ensure an orderly and safe transition.

The main component of an OCM procedure is the risk assessment (RA). The OCM risk assessment needs to identify potential risks resulting from implementation of the change as well as any risks associated with the process of implementing the change. The OCM risk assessment needs to consider the potential process safety impacts under all foreseeable conditions and scenarios, including:

- All activities required to operate and maintain the facility in a safe condition, such as operations, maintenance, and contractor activities
- All activities required for process safety management, such as process hazard analysis, mechanical integrity, incident and near miss investigations, and management of change
- Effective emergency response

As is true for a process hazard analysis (PHA) or hazard identification, the risk assessment is only as good as the team members (i.e. knowledgeable, skilled, and experienced). The risk assessment team should be led by someone trained in the methods and tools that will be used to identify and assess the hazards. At least one team member should be experienced and/or knowledgeable in the main positions or roles being impacted. At least one team member should be knowledgeable in the process safety management systems most impacted by the change(s). Due to the nature of the changes being evaluated, the tools used for evaluating an organizational change may be different from those used for evaluating a process change. Therefore, additional knowledge may be required on the OCM RA team, including human factors, organizational management, and labor relations.

An important set of baseline tools and information to assess how the change will impact the organization and specific job duties are detailed job descriptions (roles, responsibilities, and expectations (RREs)), and competencies (knowledge and skills). Typical duties/responsibilities for affected positions or individuals, if not formally documented, should be developed. One approach is to maintain a register of individuals and their tasks, roles and responsibilities related to mitigating major hazards, including contractors who may have a role in process safety. This will eliminate or reduce the level of effort to conduct this exercise as part of the OCM implementation process.

Once the risk assessment is completed, the findings should be summarized and an action plan developed. This action plan to address risk assessment findings should be part of an overall implementation plan developed for the change. The action plan should clearly identify the critical path for action items including which action items need to be completed before implementation of the change and which can be addressed after the change is implemented. Confirmation that these pre-implementation action items have been addressed should be part of the process to authorize implementation of the change. The implementation or transition plan should detail how the change from the old to the new organization will occur.

It is important to continue to monitor the impact of the change on the organization after the change is implemented. This can be accomplished by monitoring various safety and health indicators that are specific to the change and the potential risks that have been identified. Benchmarking these indicators should occur prior to implementation of the change in order to obtain a relevant comparison before and after the change.

3. Modification of Working Conditions

This category of changes includes planned actions such as modifications to working hours and shift schedules, the opportunity or requirement to work overtime, and relocation of personnel. It also includes changes to staffing arrangements in relation to start-ups, shutdowns, turnarounds, upsets, emergency conditions, and severe weather events, during which inadequate manpower can have devastating consequences.

When employees from a variety of disciplines work in close proximity, they develop relationships independent of their prescribed job duties. Casual interaction among maintenance staff, engineers, management, and control room employees fosters a sense of camaraderie and a feeling of accountability for each other. With this familiarity comes a more relaxed attitude toward communication. The separation of staff disciplines should only be considered after a thorough analysis of their interactions is performed. OCM procedures should assess potential impacts with regard to the efficiency of each to perform its duties.

It is common for shift employees to develop a cohesive team mentality and approach to their duties. In an efficient unit, the members of a shift operate as one entity, anticipating each other's moves and unconsciously cooperating on tasks. If changes are made to the structure of a shift without a thorough understanding of how that shift operates as a team, this solidarity can be disrupted. Coworkers may no longer have an inherent understanding of their responsibilities within the group as they relate to each other. If modifications are only made to some of the shift employees' schedules, this disturbance can escalate. Overlapping of shifts can mean that, in the event of an emergency, it is unclear which shift or which employees are responsible for certain duties. This can cause delays in response time, duplication of actions, and a general confusion at a time when efforts need to be streamlined. Making any of these types of changes without first having a clear picture of how employees interact with and rely on one another can be detrimental to both efficiency and safety.

Regardless of the location of a facility, there are bound to be occasional weather issues. Blizzards, hurricanes, severe thunderstorms, and tornadoes are some examples of natural phenomena that can have unpredictable impacts on operations. Emergencies within the facility, such as a fire or explosion, can also create extreme confusion and necessitate swift actions to prevent escalation of the emergency. When a unit or an entire facility only maintains the minimum number of employees required to complete job duties at any given time, unexpected upsets, such as erratic weather conditions or plant emergencies, can wreak havoc on a workforce that is already barely adequate. Completing a robust OCM evaluation each time changes are made to personnel numbers is imperative to ensure that the availability of adequate manpower during emergency situations is not neglected.

Changes to working conditions can take a variety of forms. Communication and training are key concerns associated with these types of changes. When teams of people are disrupted by changes to hours or location, communication pathways can change resulting in disruptive consequences. Both formal and informal communication between people must be encouraged and managed during such changes. Training may also need to be reviewed to ensure that these types of changes do not leave someone without adequate understanding and resources in an emergency.

4. Personnel Changes

One of the most common types of personnel change involves downsizing and layoffs. However, there are many other types of changes that should also be considered in this category, including retirements, replacement of key plant personnel such as the plant manager or a subject matter expert (SME), replacement of capital project personnel, and strike coverage.

When an experienced employee leaves the company, it is crucial to understand all of the roles he or she plays and ensure that these are adequately covered by remaining employees or his or her replacement. When considering the roles and responsibilities of an individual, be sure to include any unique skills, knowledge, and relationships he or she may have but which are not explicitly included in the basic job description.

When considering a downsizing, especially as it pertains to the actual operating staff, a comprehensive evaluation should be made to ensure there is no critical reduction in process knowledge or response times during emergency situations. The temptation to simply demand fixed percentage cuts from all departments must be avoided. The modern control room is a complex socio-technical system in which a number of factors influence the effectiveness of the operation, and thereby the safety of the facility. A baseline study should be done as well as a review of the proposed change to ensure that the new organizational structure does not reduce emergency capabilities below tolerable limits or introduce significant new hazards.

When the number of personnel is reduced due to downsizing or inability to fill a position, be on the lookout for issues related to overloading and fatigue. During any organizational change, but especially for downsizing, it is important to not only consider the implications of the revised organization, but to also think about the potential safety impacts during the transition. Whenever possible, phase in the changes to prevent a loss of control through over-complexity and to minimize peaks in workload. In addition to training, it is important to ensure ample support and/or supervision by competent people is available for anyone with new safety-sensitive work. Do not hurry through such a change before all necessary training is complete and any new safety measures are in place and functioning. There should be clear criteria available with regard to competency levels which should be verified before the organizational change takes effect.

Contingency planning for work force actions should carefully consider process safety implications. These types of changes are temporary changes which often do not have specified end dates. When approving a temporary change like this, a reasonable time frame should initially be selected for the change. The situation should be periodically re-evaluated and modifications made as necessary. Depending on the extent and duration of a change like this, it might be necessary to authorize multiple temporary changes until the situation is fully resolved. Do not overlook the fact that an additional organizational management of change may be necessary to return to the normal workforce if they have been off the job for a substantial period of time.

Personnel changes can encompass simple activities such as hiring a new engineer or complex changes such as operating a process plant during a union work stoppage. Each of these types of changes needs to be reviewed for its potential impact on process safety, if any. Simple changes may be handled with a streamlined procedure in which the manager documents the change, updates necessary documents, and ensures that essential training and communication takes place. However, more complex changes should be handled using the OCM procedure, including a risk assessment, to ensure that all necessary actions are taken to prevent process safety problems as a result of the change.

5. Task Allocation Changes

This type of change deals with reallocation of tasks to different individuals or the elimination of tasks that may have an impact on process safety. This can occur for a variety of reasons. The change can result in either underperformance or non performance of a task. The risk of this event obviously depends on the criticality of the task to process safety.

Position advancement occurs at all levels in the organization. Some are a result of attrition (e.g., retirement or departure) and some are by design (fast tracking of managers). The new skill requirements can be both task specific (becoming a maintenance coordinator) or more managerial such as leadership and people skills.

One of the major causes of acquiring more responsibility is the recent trend in industry to shrink the size of organizations to achieve the goal of ‘do more with less’. With fewer people in the organization, key tasks need to be redistributed to those remaining. This situation may result in some individuals getting assignments by default (somebody has to do it), after the logical (based on skills and experience) reassignments have been made. The impact of downsizing is often an increase in workload and/or span of work, or both, for individuals.

When it comes to accidents, organizational factors are conditions rather than causes. In risk assessment terms, they are enabling events. The impact of organizational change is often subtle and not immediately apparent. The effects can be degraded performance following increased workload or span of work. The change may also have disrupted unrecorded or informal activities or communications that contributed to safety. This is more likely when baseline competencies for assigned tasks are not documented.

The key aim of a risk assessment is to ensure that following the change, the organization will have the resources (human, time, information, etc.), competence, and motivation to ensure safety without unrealistic expectations of people. Task mapping can be a very useful tool in these situations.

Task allocation changes can occur as part of a larger organizational change or they can be considered a standalone change. Any time this type of change occurs it is essential to consider what happens to all of the safety-related tasks and who is responsible for each of them after the change. Task mapping is a good way to understand the complete impact of such a change. A risk assessment of the change may identify key actions which should be taken to ensure that all tasks are handled by competent personnel.

6. Organizational Hierarchy Changes

Reorganization or organizational redesign can come from a number of directions but the starting point is nearly always the need or opportunity to improve a company's financial performance. Indeed the company's financial survival may depend on how effectively and quickly this change can be executed.

For those primarily concerned with the financial costs and opportunities from the planned change it is good to note the severe financial consequences of process safety incidents, such as the Deepwater Horizon incident. According to a CCPS study, in the US, major industrial accidents cost an average of $80 million each and business interruption costs can amount to four times the cost of the property damage from an incident.

Given these warnings but also recognizing that change must and will occur, an OCM program should be utilized to manage such changes. Because there is generally greater vulnerability during organizational transition, special consideration should be given to process safety during and immediately after the transition occurs. Keep in mind that not only is a business being changed but people's lives are as well.

When planning reorganization, a company should ensure that no gaps are left in management and application of process safety elements. There are many elements to a strong process safety management program. They are all necessary and they all need at least one person responsible for ensuring they are being done effectively. These people are often referred to as Subject Matter Experts (SMEs). Each SME should be qualified and clearly charged with the task and responsibility.

The term, "Corporate memory loss," refers to the loss of knowledge during reorganization. Often downsizing is done using attractive retirement packages, thereby resulting in a disproportionate loss of experienced personnel. Such persons may include among others, shift supervisors, subject matter experts, process engineers, design engineers, and maintenance personnel. What these people know should be passed on before they leave. Nowhere is this more important than process safety. Ideally this is an on-going process since often a person's interest in passing along this knowledge diminishes quickly once they decide to leave or they are informed that they are leaving.

Task mapping is a good tool for ensuring that the new organization will have "all the bases covered". Checklists can also be helpful.

A topic closely related to hierarchy change is a change to the span of control. Sometimes this change occurs as a result of a significant reorganization, but it can also happen with simpler organizational changes. In this type of change, the scope of someone's job responsibilities is changed. If the span of control is increased, the types of responsibilities are the same, but the person may be expected to provide oversight, guidance, and expertise to more people or departments within the organization. If not handled correctly, this type of change can result in someone being "spread too thin". It is important to understand the demands of a job on a person's time before increasing the span of control. If the change to the scope of a job means that the person will be less available to the people who rely on them, then it may encourage people to find alternative options, some of which may not provide the same level of safety.

During a change in ownership, companies need to ensure that all aspects of process safety are functional and that everyone understands their roles, responsibilities, and resources available to get their jobs done completely and correctly. If there are gaps in systems due to the change in ownership, these need to be identified and adequate temporary systems put in place until the organizational change is complete.

Sometimes companies need to outsource various aspects of process safety management, such as maintenance. The initial decision to contract out such activities as well as the choice to change service providers should be considered an organizational change. Contracting out a process safety function can introduce many complications in terms of reporting structure, authorizations, and responsibilities. Even if you have these issues ironed out with one contractor, when you change service providers the same issues should be revisited with the new contactor.

7. Organizational Policy Changes

Although these changes may not be associated with immediate organizational changes, they can serve as enabling events which could adversely impact process safety within the organization. Because these types of changes do not involve actual changes to the people or tasks within the organization, they can easily be overlooked by an OCM procedure, but they should be included within the scope of an OCM process.

Many companies have high level policies and procedures which describe how they will address the many facets of process safety management. These are typically driven from a centralized corporate function and are applicable to all operating facilities within the company. On the surface it may seem like these are just typical policy changes which do not require an OCM. However, the specifics of the policy directives that are being made should be reviewed for potential impacts on the downstream organization. In some cases these policy changes could end up requiring changes to someone's span of control or adding additional tasks to their job function which they are not currently equipped to handle. If these new or revised policies require additional activities or documentation beyond what is currently expected, there may be significant resource burdens imposed on plant sites to comply with the new policy.

The risks associated with changes to policies or blanket changes to budgets is that the changes eliminate too many resources (people and/or money) resulting in an inability to properly carry out the necessary functions of a good process safety program. Although managers like to convey the message that everyone will just have to "do more with less" or "work smarter, not harder", there is a minimum resource level below which a facility cannot safely operate. Unless a company has a clear understanding of what this minimum resource level is, they will not be able to adequately argue against budget cuts that go too deep.

The associated activities of shutdowns, turnarounds, and startups present special challenges for process safety. Companies typically have special provisions or policies for additional staffing for these critical time periods. Changes to these special staffing policies should be considered as an OCM activity.

In many cases these types of policy changes are done by people situated very high in the organization who typically have many objectives to consider. In some cases these decision makers may not be familiar with process safety and its implications for the business. In this case, it may be necessary to educate these high level executives on what process safety is and how it may affect the business metrics that they may be more familiar with such as profitability and corporate image.

Policy changes should be carried out with process safety implications clearly understood. Once understood, those making the change may need to adjust the policy so that instead of being a detriment to process safety it is a benefit.

8. Acknowledgements

This presentation is based on the Center for Chemical Process Safety (CCPS) book, *Guidelines for Managing Process Safety Risks During Organizational Change*. We would like to acknowledge the direction and input provided by the Organizational Change Management Subcommittee members in developing this book.

Increasing the Agility of Process Safety Management Systems

Brian D. Rains
DuPont Sustainable Solutions
974 Centre Road; CRP-730
Brian.D.Rains@DuPont.com

Prepared for Presentation at
American Institute of Chemical Engineers
2013 Spring Meeting
9th Global Congress on Process Safety
San Antonio, Texas
April 28 – May 1, 2013

UNPUBLISHED

Increasing the Agility of Process Safety Management Systems

Brian D. Rains
DuPont Sustainable Solutions
974 Centre Road; CRP-730
Brian.D.Rains@DuPont.com

Keywords: PSM, Management of Change, MOC

Abstract

The Process Industry is constantly adapting to change. High prices of raw materials, new materials for construction, and advances in technology permit manufacturers to test new boundaries and explore new avenues. With this direction, however, operational risk is increased. In order to cope with this increased risk, it is vital that companies have a Process Safety Management (PSM) system that fully integrates Management of Change methodologies.

Within its operations, DuPont has developed a PSM system that integrates and manages technology, facilities and personnel change. While these must be considered singularly, it is also important for companies to build a holistic system that accounts for the interdependent nature of operational systems, and thus address process safety from every possible angle – training and skills development to hazard analysis and maintenance. This paper will detail a model that companies can use to effectively manage change and as needed, develop customized solutions for specific conditions and contexts.

1. Introduction

Deepwater drilling, hydraulic fracturing for gas and oil, proliferating oil sands production – new and more hazardous operations are on the increase in the oil and gas industry. Equally so in chemicals and mining operations, ever more challenging process conditions are being employed to optimize process and financial performance. Industry constantly has to adapt to change and this includes evolving technology, as well as changing market demand and pricing. With these changes come more difficult and operational risks. When coupled with the trend to larger, consolidated and costlier assets, these increased risks can lead to higher incident and accident rates and impact environmental protection, as well as challenge business if a disaster occurs.

But the process industry's efforts to minimize risk have steadily improved and major process safety incidents are becoming more rare events. When they do occur, however, they become the subject of intense media scrutiny, impact assessments and investigation into the root causes of the incident.

While the loss of lives is the most devastating, such catastrophes may also have serious environmental and financial consequences. The pollution caused by the April 2010 Macondo well blowout had a severe economic effect on several countries in the Gulf of Mexico. It led to the temporary suspension of wider drilling activities with the resultant loss of jobs and forced BP to suspend dividend payments for a year. The total financial cost of the catastrophe is estimated to be in excess of US$ 37.2 billion. Both independent and BP enquiries have come to the conclusion that the disaster was caused by critical technical and organizational failings, some of which are considered to apply across the industry.

The Macondo incident featured signals of process abnormality. Although senior management had visited the rig just hours before the disaster when several indicators of insufficient well control had failed to be corrected by the rig crew, the senior management team was more focused on occupational safety than hazard-critical process safety operations at the time. A series of well design changes had been made, but were considered to be within the range of acceptable operational changes. After the disaster, both the National Commission Inquiry and BP found that overall Management of Change (MOC) and related communication was severely insufficient and contributed to the catastrophe.

The HSE enquiry into the 2005 Buncefield oil storage depot explosion in the UK arrived at a similar conclusion. The incident investigation found that the operators' management systems at Buncefield were inadequate in several respects. Among those listed were Management of Change procedures that were not adequately applied to bund projects. Changes during design and construction were also not reviewed in terms of impact on the ability of the bund to retain liquids during an incident.

Clearly, inadequate management of change was not solely responsible for any of these incidents, but it was a contributing factor. Events such as the Macondo blowout and the Buncefield explosion are relatively uncommon, but their severity highlights how important it is to learn from them. How can management of change failures be eliminated to prevent similar events in the future?

Despite good process and behavioral safety management in the oil and gas industry, these rare accidents still occur. The industry needs to be flexible enough to adapt its PSM programs to ever-changing safety needs, putting in place best controls that exceed legal requirements and, through the use of new ideas, go beyond the current norm.

2. Why is PSM so Challenging?

Even a moderate facility performs thousands of PSM related tasks and manages a vast amount of data round the clock every day of the year. Aging facilities require ongoing maintenance and renewal, which, of necessity, entail the updating of equipment and introduction of new materials. All such changes to processes – from new products to plant improvements – must be evaluated. This requires a tremendous attention to detail and operational discipline.

At DuPont, our PSM Managing System is fully integrated into our Operational Strategy. It is not considered as a separate safety priority, but as an integral approach to safe and efficient operations. Successful PSM implementation requires both a strong safety culture and strong PSM technical execution.

We are well aware of the leading operational risk factors. Among the oft cited primary challenge of "insufficient time and inadequate resources", the list of key risk factors also includes the perceived competing pressures of production, quality, cost and safety; the lack of strong visible leadership commitment to safety; an incorrect identification and/or evaluation of operational risks; the failure to manage process safety risks differently to workplace safety hazards; ineffective audits; and a failure to comply with procedure.

One risk factor that is also always on our radar is the specter of inadequate Management of Change. Management of Change is clearly part of the operational risks that must be controlled and eliminated to run a safe facility. This paper focuses on ways of increasing the agility of process safety management systems by fully integrating change management.

3. Change Management is Critical to Process Safety

Processes throughout all industries are constantly undergoing change to increase yield, improve quality, reduce cost, enhance safety, etc. The safe, effective and thorough management of these changes is vital to the overall success of any manufacturing operation to ensure changes take place without process-related incidents.

While all aspects of change management across multiple facilities must be considered singularly, it is important for companies to build a holistic system that accounts for the interdependent nature of operational systems, and thus addresses process safety from every possible angle – from training and skill development to hazard analysis and maintenance.

At DuPont, we introduced an annual review of our process safety management system in the late 1960s to ensure that process additions or changes did not create new hazards. We therefore consider change management as an integral part of the cyclical activities of process safety management practices that interlock with behavioral safety management. All the 14 PSM technical elements on our PSM wheel are highly interdependent and cannot be successfully managed as stove pipes or independent activities. See figure below.

DuPont Process Safety Management Model

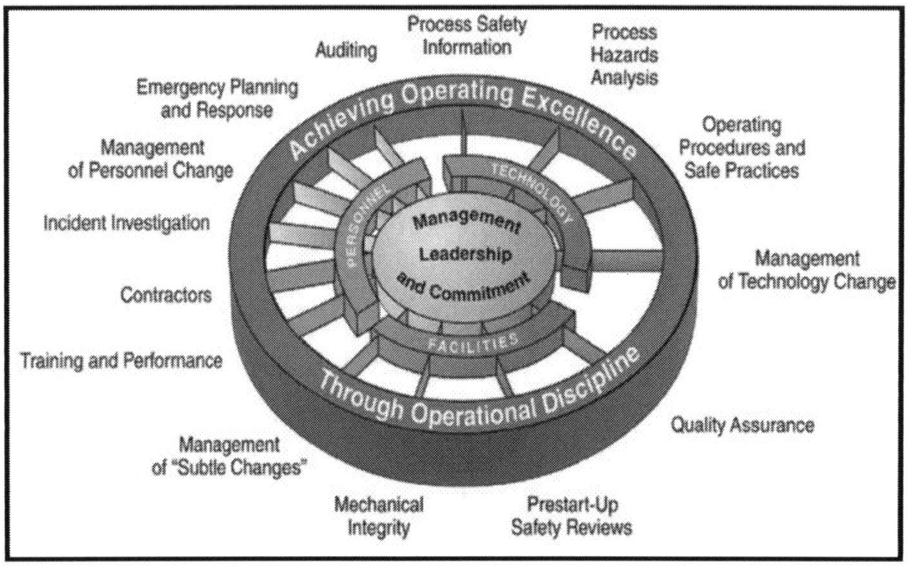

Our process safety management system forms a continuous "to-do" loop that refers operators back to the fundamental safety questions tackled at start-up. All aspects of PSM, and that includes Management of Change, have multiple knock-on effects.

As process safety focuses on putting in place management systems and controls to identify and control the hazards of any given manufacturing process in three key areas – technology, facilities, and personnel – Management of Change protocols must also address these three distinct elements. If any of them have undergone change, whether to a small or large degree, they need to be re-evaluated to assess whether safety has been compromised. Although many of the principles are the same and can often be managed by shared resources, they are maintained as separate elements to emphasize the importance of change management, but also because approval processes, requirements and the individuals involved are often different. Lastly, documentation systems and/or ownership for all three types of change management are frequently separate and distinct.

Management of Change can be understood as a systematic procedure for the analytical review of changes (including subtle changes) to documented process technology, facilities and personnel, taking into consideration potential new hazards that may be introduced to the process, system or operation, and their elimination or control.

While many companies will have Management of Change procedures in place, errors occur when it comes to identifying what constitutes a change. Even a small change can have a catastrophic effect. We constantly need to ask the question: "Is what we are doing really a change?" For example, is increase in capacity throughput rate a change or not? Can this increase be managed without an MOC process?

4. Management of Change - Technology

Most serious incidents take place within six months of a significant change being introduced, but sometimes the effect can take years to manifest itself. Several years ago, an internal inspection of a tank at DuPont identified a damaged pipe. The engineers who were aware of the situation believed that the facility could be restarted and the damaged pipe repaired during the next turnaround window. This decision was taken without following a strict MOC review, as the engineers did not realize that operating the plant with the damaged pipe was indeed a process technology change. The damaged pipe was eventually a contributing cause to an incident.

Pertinent examples of potential technology changes in the oil & gas industry include, but are not limited to:

- Changes in the composition of the crude oil beyond previously approved/authorized limits.
- Changes to raw material specifications including the introduction of new materials.
- Changes in the known hazards associated with a material (e.g. if new carcinogenic information becomes available).
- Operations of the unit at higher temperatures, flows or pressures than previously reviewed/approved.
- New data on possible reactions from the inadvertent mixing of chemicals in the process.
- Re-design of the process flows and/or its associated instrumentation.
- Changes to safety alarm or interlock settings.
- Re-rating of process alarms or interlocks.

Changes to documented process technology potentially invalidate prior operations risk assessments and create new hazards. Accordingly, all changes to the documented process risk information must be subjected to the same rigorous review that is applied to new processes. This is how we proceed at DuPont. Because much of the technology that DuPont operates in its facilities has been developed internally, there is often a high degree of process optimization (technology and facility changes) that takes place once a new facility becomes operational.

Before any change is made, we review, seek authorization and compile documentation about the planned change to existing process technology. Special emphasis is placed on the safety review portions of a change by including the appropriate technical and operations resources familiar with the type of proposed change.

Best practice management of technology changes provides the foundation for identifying and understanding potential new hazards. Sites need to establish and implement written procedures to manage all temporary changes such as tests and trials, emergency changes, as well as permanent process technology changes.

At DuPont, our documentation considers and addresses the purpose of the change and takes into account the scope and desired results. It highlights the technical basis for the change and includes information on which element of process technology is being changed and how. This information is based on lab results, engineering evaluations and vendor or supplier data, so that it is clear whether the change is feasible and can be implemented safely. Other information included in the documentation covers the safety, health and environmental impact of the change.

5. Repeating the Process Hazard Analysis

At this point, we also look at whether a complete or partial new Process Hazard Analysis (PHA) is required. When some level of PHA activity is specified, the emphasis is on the identification of new hazards, hazardous events and the routes to and protections against those events. If appropriate, changes impacting existing layers of protection (e.g., fire safety system design) and facility siting (e.g., introduction of new buildings, removal of buildings, and changes to building occupancy or use) are also addressed. We therefore recommend conducting a PHA for those changes that may exacerbate existing hazards or introduce new or additional process hazards.

Apart from considering the need for a renewed PHA, documentation of an authorized technology change also includes a description of the change, how this will affect operating procedures and how personnel will need to be trained to cope with this change. At DuPont, it will also look at how the change must be communicated, set time and quantity limits and detail who has approval and authorization responsibility.

The change should then be trialed. We use test authorizations (TAs) to safely manage pre-planned temporary operations of our processes that fall outside our procedures and limits. These test authorizations will be used for the start-up of new equipment, but also for other changes such as new or altered process equipment, or if there is a change in a critical raw material supplier.

After the trial, results and recommendations are documented. We note how process technology is being modified and how it will impact existing safety systems and lines of defense, how operating procedures will change and consider what recommendations applicable Process Hazard Analyses make. We also put in place a tracking system to ensure the management of the change is fully completed.

We also carry out a change of design (COD) process to safely manage changes to our facilities, particularly if the facility is not a direct replacement in kind. This covers addition of new processing equipment from pumps to tanks, but also a change like the re-routing of piping because it affects the process due to changes in pressure, for example.

Once we have planned a change of technology, trialed it and prepared the necessary documentation, we develop a written procedure for implementing the change. This includes a definition of the change, sets out who is involved in making the change and outlines the authorization and review responsibility levels. In order for the procedure to be effective, it is vital to obtain the "buy in" of everyone involved, so that the proposed safeguards do not just remain on paper.

As a next step, we update related existing procedures and process safety information and then set about training and informing all personnel concerned in the change. Finally, we monitor and audit operations to check the Management of Change procedure is being implemented correctly.

It is important that all process changes are reviewed by experienced personnel with relevant knowledge of the change and that are they are authorized by responsible management. All personnel must understand what the change is and where to seek information, review and authorization. When communicating change to personnel, it is important to consider all applicable groups, from operations to mechanical and emergency response teams, as well as people from other operating units.

We also amend our training program to ensure it is up-to-date with any changes. In doing so, we will consider the need to train for non-routine instruction. Again, our documentation will describe the training required and who is responsible.

By putting in place adequate reviews, authorization and documentation, repeating the PHA if necessary, and, importantly, communicating the change to all personnel concerned, we ensure that PSM processes are up-to-date with the latest change, so reducing the risk of surprise incidents.

6. Management of Change - Facilities

While the alteration in the design of a process technology may be an obvious change to spot, there are large, as well as subtle or minor changes to process equipment and facilities in the field, which can lead to undesirable conditions and events. Facilities changes can be defined as all modifications to a facility, which may or may not involve large revisions or capital expenditure. They relate to any change in the equipment design basis of the facility. This goes from changes in PSM Critical Equipment classification of process equipment, changes to the material of construction for PSM Critical Equipment, substitution of different spare parts with different specifications for PSM Critical Equipment to changes in the performance specifications or design measurement tolerances of PSM Critical Equipment, but also covers new facilities or a change of supplier, if it is not manufactured to the same specifications as the originally reviewed and approved equipment.

Fundamentally, all changes to facilities or equipment documented within Process Technology which are not "replacement-in-kind," as well as temporary "in service" repairs, that are actual changes, form part of the category of changes to facilities. All these changes should receive appropriate review and authorization to ensure they are carried out safely.

When it comes to clarifying what constitutes a facilities change however subtle, written procedures are helpful. They allow personnel to understand what a facilities change is and to recognize the difference between a technology change and a subtle change. Written procedures can also provide information on the systems that are in place to manage and implement each type of change.

Some examples of facilities change include, but are not limited to, the following:

- Materials of construction.
- New trim size of a valve.
- Rerouting a piping system during field modifications or replacement.
- Temporary piping supports, jumpers, or connections.
- Replacement of sections of piping insulation with an alternate material.
- Installation of an automatic valve to replace a manual valve with the same function.
- Removal or modifications to fireproofing insulation.

Before any changes, however minor, are made, they should be assessed for their impact on safety, health and the environment, so that the associated process safety management requirements are understood and established. If required, operating procedures should also be adapted, and affected personnel offered adequate training. A follow-up system should be put into place to ensure all required PSM elements are completed before implementation. Sites should therefore conduct periodic audits of process equipment and facilities to ensure that all changes receive adequate review and approval.

It is important that requirements for implementing the change, including process safety considerations, be documented up front when the proposal is conceived and expenditures authorized. Implementation of the modification is then contingent upon completion of the PSM requirements.
The sequence for implementing facilities changes without adversely affecting process safety should be structured along the following lines. Once a change to facilities is conceived, a management of facilities change form should be prepared and authorized. If necessary, procedures should be adapted and personnel trained. Only then should the change be carried out and completed. The final step in the process is to prepare a closing document and obtain authorization. We have a similar process in place to govern changes which affect that most influential element of safety – people.

7. Management of Change - Personnel

Imagine an experienced employee has just resigned. He has critical PSM knowledge, skills and capabilities. You have a successor in place, but despite his abilities, he lacks the requisite PSM knowledge that is specific to your organization. You would not put him on the job without prior training. At DuPont, our recommendation is to provide long-term training and development, assess the newly assigned member of staff for competency over a 90 day period and review performance annually. That way, we ensure we maintain PSM knowledge and functional capacity.

MOC – Personnel is related to changes in personnel in roles that are deemed PSM critical roles. DuPont is one of the pioneers of management of personnel change. Any promotion, retirement, long-term disability or illness, new hire, or an intra- or inter-site transfer, leads to a loss of or change in knowledge. Yet a facility relies on a minimum level of collective experience and knowledge among its operating, maintenance and technical team to function well and safely.

Significant changes to site organizational or functional structures that may occur as a result of restructuring, centralization or decentralization of key functions, changing service providers or outsourcing, new operations initiatives that require reassigned resources, or business realignment can also affect PSM program implementation in terms of accountabilities, program management, and personnel in vital PSM roles and functions. These changes should be anticipated, analyzed, and transitioned effectively to maintain responsibilities and sustain or reassign knowledge and functional capacity within the site line organization.

At DuPont we recognize that people are the essential ingredient that links all elements of process safety management. It is vital to maintain a minimum level of specific direct process experience, as well as knowledge and skill in managing process safety within the site line and support organizations. We therefore believe it to be indispensable that sites understand the required skills, knowledge, and capabilities for personnel in key PSM roles.

Indicators of weak management of personnel change to watch out for include poor audit findings, incidents, chronic performance problems, surprise deviations from existing policies, cyclic or highly variable results, missing documentation or overdue PSM actions due to personnel changes, as well as management systems with a high degree of people dependency.

DuPont works with Management of Personnel Change guidelines to ensure that a minimum level of PSM knowledge is maintained. We train newly assigned personnel in the principles and features of process safety and review their proficiency after training is complete. We also provide additional support if knowledge and experience that are beyond local control are lost.

Our overarching philosophy is to develop PSM expertise within sites for "routine" needs and activities. At DuPont, site and higher levels of management are responsible for developing and implementing criteria and guidelines regarding Management of Personnel Change. These guidelines define the minimum proficiency required in process safety management and establish individual skill levels required for the operation of hazardous processes. They also establish a plan for the emergency replacement of individuals in critical positions.

Since changes may be introduced into a process or facility by people in many different roles (e.g. engineers, operators, mechanics, supervisors), a periodic MOC training program is provided to educate employees on the definition of change and the site or area modification procedures and approvals. This training may involve a review of site-specific MOC policies, incident case histories, as well as roles and responsibilities. We check knowledge and proficiency through a written test, in field demonstrations and/or through an oral panel review by the technology guardian and experienced peers.

Overall, we follow a life-cycle pattern of managing personnel changes to ensure we have well-trained, experienced personnel in operating and maintenance positions and we extend this to contractor personnel.

Although we have discussed the three different elements of change management – technology, facilities and personnel – separately, they do of course overlap. Take for instance the "technology guardian" concept, which DuPont uses. This concept serves to identify key, long-term technical personnel and ensures that they are maintained in place for many years. They become critical in the evaluation of all technology and facility changes and also provide stability when PSM critical personnel changes are under consideration.

8. Customizing MOC Practices to Specific Conditions and Contexts

However good PSM and Management of Change systems may be, they cannot be universally applied under a "one size fits all" approach. Naturally, decision-making levels should be appropriate to the extent of the change and the degree of related risk, so that large changes should involve more senior management levels than small changes, which have less impact on safety.

Several examples of potential areas of customization are listed below:

- Changes in technology will require different reviewers and approvers than changes in facilities or personnel.
- Significant changes, particularly to technology and facilities, such as ones that are investment intensive, or require large-scale instrumentation design changes, introduce operation condition amendments or lead to re-rating of plant capacity, will frequently require a more thorough review than more common, smaller changes. The review for such large changes may include corporate technology and/or facility leaders and potentially also sister plants which share common technology or equipment types.
- Depending on the complexity of the change, the risk analysis tool used in the review step will differ. This can range from a full "HAZOP" review for a significant design change to a fault tree analysis if the change is prompted by an actual incident investigation and recommendation. A simple change may only need a "What If" risk review.
- Certain operational risk situations will require modification of existing procedures. Take for example emergency work versus routine work. Emergency changes could trigger a modified approval process for speed, which in turn can require different documentation. It must do so, however without compromising the risk review process or the risk mitigation process.
- The complexity of the Pre-Startup Safety Review (PSSR) Checklist may be

customized to suit the type and degree of change that is being made.

9. The Positive Impact of MOC

As previously mentioned, the Management of Change cannot be separated from PSM. Even small changes can disrupt a well-functioning process safety management system. Management of Change must therefore be treated as an integral part of process safety. This makes it difficult to evaluate MOC's contribution to overall safety in isolation. Effective MOC can however be said to support the smooth running of the continuous cycle of PSM procedures.

A well run process safety system will significantly reduce the risk of incidents, accidents and catastrophes that harm people, facilities, the business, and the environment. Associated with such incidents are costs of property damage, business interruption, litigation, incident investigation, regulatory penalties and loss of market share. Aside from averting the negative consequences, efficient PSM that integrates good MOC practices has a positive effect on shareholder value. As we all know, good process safety helps companies demonstrate corporate responsibility. Investors perceive companies that display such corporate responsibility as a lower risk. In addition, corporate responsibility has a definite influence on corporate image and branding, while also increasing credibility and company reputation. Local communities are likely to be less concerned about such companies operating in their area.

Furthermore, consistent process safety produces good quality products, on time, and at a lower cost. It leads to reliable equipment, which in turn ensures better productivity, as there is less downtime. Employee morale is bolstered, loyalty and retention strengthened. Effective PSM, which is only possible through well-executed MOC, increases credibility in the investment world, with the workforce and in the local community and so provides companies with the sustained "right to operate".

10. Conclusion

Over the years we have all learned from major incidents. One of these learnings is that the challenge for managing change well is to be flexible and to identify the change correctly. Following these guidelines will allow oil and gas companies to adapt MOC practices to any given condition and ensure safe process operation.

The bottom line of using a Management of Change system is that all modifications:

1. Receive prudent review, including process safety considerations;
2. Are appropriately documented; and
3. Are authorized by appropriate levels of management.

By developing a MOC model that takes account of all these points and integrates it into a holistic process safety management system, it is possible for oil and gas companies to minimize risk and improve their overall operational safety. This does require continuous commitment by senior management and regular re-evaluation and modification of the existing process safety management program to ensure all changes that can affect safety are addressed. Good process safety management is a never-ending cycle, not a system that can be put in place and be expected to serve, untouched, under the ever-changing circumstances the oil and gas industry faces.

Technical Risk – It's Not a One-Size-Fits-All Proposition

Kumar (Chris) Israni, P.Eng, M.S
ERM
15810 Park Ten Place, Houston TX 77084
chris.israni@erm.com

Joshtin Cooper
ERM
15810 Park Ten Place, Houston TX 77084
joshtin.cooper@erm.com

Keywords: Risk Assessment, HAZOP, What-if, Bow-tie, HAZID, ALARP, FMEA

Abstract

Critical aspects of plant design and operations (new capital projects or existing facilities) are identifying, assessing, and mitigating technical risk. A proactive approach is best—an approach that strives to protect vulnerable risk receptors such as human life, assets, and environment. The nature of these receptors varies among regions as do the types of plant operation and associated parameters. Therefore, the risk solution for a specific set of conditions may not be a quick and easy fit for similar facilities or operations.

The process of identifying and assessing technical risk uses various qualitative and quantitative techniques such as HAZOPs, HAZIDs, bow-tie diagrams, What-if analysis, FTAs, ETAs, QRAs, etc. Outcomes of each technique are increased insights to effective mitigation that may currently exist and proposed mitigation measures to add, where required. The biggest misconceptions in the process industry are that the techniques are interchangeable and that a one-time application of a technique is adequate to manage a facility's risk for a lifetime. Each technique has specific applications and limitations.

In fact, new facilities and those that have been operating for several years always have the potential to harm human life, assets, and environment in and around the facility. Hence, the success of implementing a successful technical risk program rests with the choice of the appropriate technique or combination of techniques based on the project or operational scope. When implemented, the techniques must have the right group synergy by way of team preparation and participation. Equally important is addressing the outcomes of the techniques in a timely manner, providing closure and the path forward, and bearing in mind that a proactive (rather than reactive) approach serves the prime purpose of protection. This paper provides a refresher on the relevant techniques for technical risk identification, assessment, and mitigation, along with guidance on the validity and applicability of those techniques to various plant situations and specific to the scope.

1. Introduction

Technical risk assessments (TRAs) are performed frequently for new and startup projects and periodically for existing facilities. The main objective of these assessments is to identify how potential hazards may impact the lives of the employees, neighboring residents, the environment in which they operate, plant production, and the business assets.

One of the largest drivers behind these TRAs is to ensure compliance with government regulations such as Occupational Health and Safety Administration (OSHA) 1910.119 [1] and Environmental Protection Agency 40 CFR Part 64 [2]. Companies are constantly working to stay in accordance with the law by implementing most current policies, regulations, and best practices. In addition, facilities incorporate lessons learned from incidents that have occurred at their location, within the corporation, or in other industries.

To assist with the objectives of protection, this paper briefly describes some of the TRA techniques. While applying these techniques, some people hold a common misconception that managers can pick any of the techniques available. Another misconception is that performing certain types of TRAs one time at a facility can take care of the facility's risk needs for a lifetime.

However, although many different techniques can be used to perform a risk assessments, each of them have its own advantages on certain applications as compared to the others. The techniques that will work for one company are not necessarily ideal for another. A "cookie cutter" approach cannot be used to perform a successful TRA.

2. Techniques

The various risk assessment techniques can be classified into two broad categories – qualitative and quantitative. While applying these techniques, the key to success is ensuring that they are used proactively with adequate preparation and planning. These studies once completed, may result in action items that will then need to be addressed. The subsequent subsections herein briefly describe these techniques, most of which are widely known for their applications [3-10]. These sections include views from literature and industrial experience.

2.1 Qualitative

Qualitative risk assessment techniques focus on brainstorming in a workshop format with participation from individuals of various disciplines. This participation assists in identifying potential hazardous scenarios that could impact human lives, environment, and assets. The output from these studies provides information about prevailing risks and suggestions which if implemented, will reduce the risks. These studies are based on the relative comparison of risks by indicators such as low, medium, and high; hence they are termed as 'qualitative risk assessments." To improve this understanding, facilities may use risk matrices that use simple numbers as risk indicators. No detailed numerical calculations are performed, and qualitative techniques focus on single point failures—they do not take into account multiple jeopardies.

2.1.1 HAZID Study

Hazard identification (HAZID) study is a high-level, team-based technique to identify and qualitatively assess hazards and risks. HAZID is the first step in risk evaluation and management. Being a high-level study, this study requires minimum inputs, which implies that it can be used in the early stages of design.

A HAZID study helps to determine:

- What are the hazards and associated accident events?
- What are the potential consequences associated with accident events?
- What are the preventive and mitigating safeguards in place to prevent and recover from accident events?
- Are these preventive and mitigating safeguards adequate?

The objectives of a HAZID study are to:

- Identify the hazards that have the potential to impact the health and safety of personnel or result in damage or impacts to the environment
- Identify the safeguards in place to manage the hazards
- Risk rank the likelihood and consequences associated with the realization of the hazardous events
- Capture recommendations that help manage the risks associated with the hazardous events

The technique is very popular in the offshore industry. A similar method, preliminary hazard analysis, is used in the early stages of design in the downstream and chemical industry. This technique also can be used to compare risks among various processes at a site to determine priorities for detailed risk assessments.

2.1.2 Bow-tie Diagrams

The bow-tie method is a diagrammatic representation of major accident events (MAEs) that provides a simple and effective way to view the escalation of a hazard and the safeguards in place that help to prevent, detect, control, or mitigate the hazard.

A bow-tie analysis serves to provide:

- Awareness of improvements needed to procedures, training, or equipment
- An effective, living risk-management tool—including the diagrams in an emergency procedures manual
- An aid in risk-management training and explaining the roles of staff members
- A tool to demonstrate risk-management procedures to shareholders or regulatory auditors

The objectives are to:

- Link safety-critical failure events directly to consequences
- Prescribe measures to prevent and recover from system events or failures

Figure 1 shows the general layout of a bow-tie diagram.

Figure 1: Layout of a Bow-tie Diagram

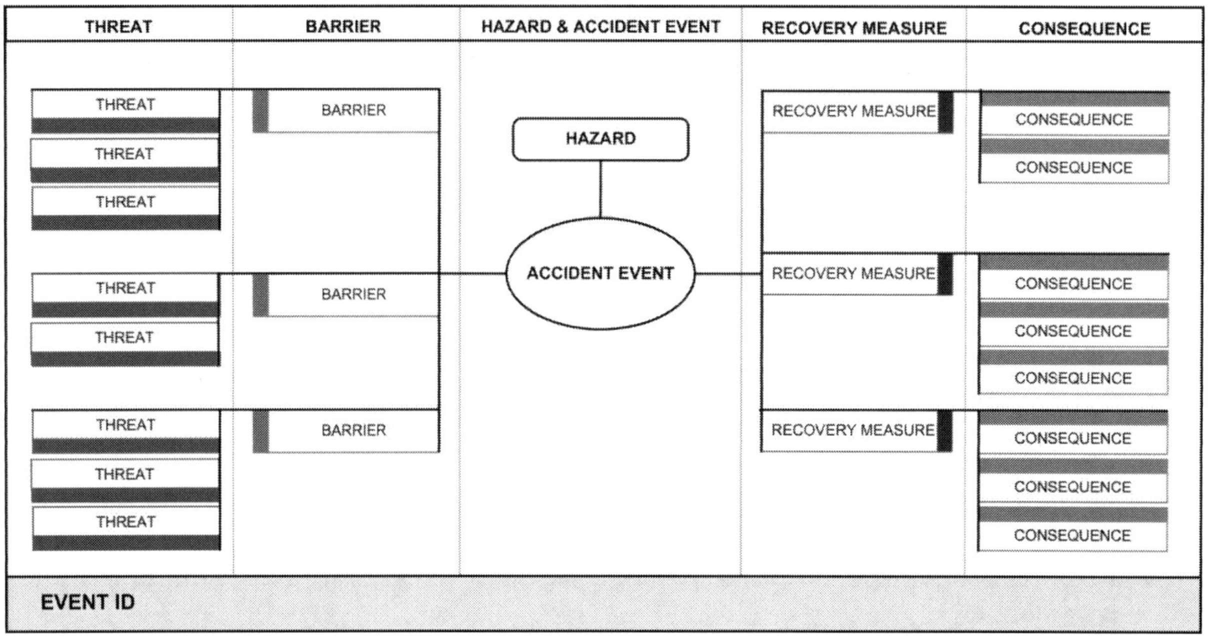

2.1.3 ALARP Workshop

As low as reasonably practicable (ALARP) workshops are brainstorming sessions attended by relevant project design and operations personnel to identify risk-reduction measures (RRMs). The objectives of an ALARP workshop are to demonstrate that residual risk is as low as reasonable practicable (assessed against defined acceptability criteria as illustrated by the example in Figure 2) and to verify whether the cost of reducing a risk is grossly disproportionate to the benefit gained.

Figure 2: Example of an ALARP Triangle

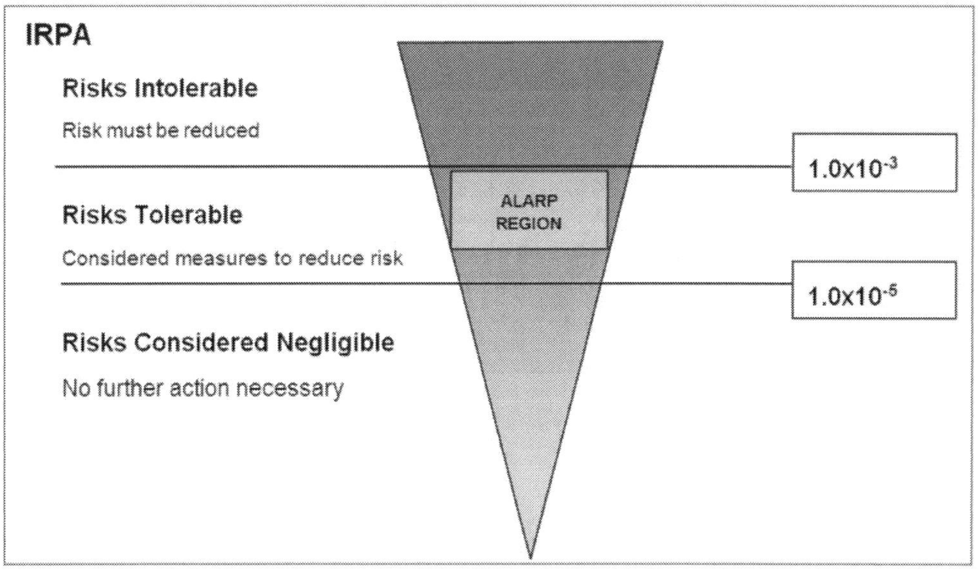

This concept is very prevalent in large capital projects in the offshore industry.

2.1.3 HAZOP Study

A hazards and operability (HAZOP) study is one of the most widely used techniques. It focuses on identifying hazards and operability issues caused by deviations from the design intent. A HAZOP study is carried out in a workshop setting, similar to a HAZID, by applying deviations to process conditions and operating parameters to identify causes of such situations which could lead to potential hazardous scenarios.

A hazards and operability (HAZOP) study is one of the most widely used techniques. Similar to a HAZID, a HAZOP study is carried out in a workshop setting with a multidisciplinary team. However, a HAZOP focuses on process hazards and operability issues caused by irregularities or deviations from the design intent that could lead to potential hazardous scenarios.

HAZOP studies are known for the level of detail that workshop teams tend to get into and hence their long durations. The technique is well respected in oil and gas and chemical facilities. A successful HAZOP study is achieved when accurate drawings such as process flow diagrams (PFDs) and piping and instrumentation diagrams (P&IDs) are available. Hence for design projects, HAZOP studies are highly recommended in the later stages of the project – when the design is clearly defined and during detailed engineering. For existing facilities, HAZOP studies can be conducted any time, but they are recommended during the periodic process hazard analysis (PHA) revalidation required by regulatory agencies.

2.1.4 What-If Analysis

What-if analysis also is a team-based technique used to identify and assess hazards and risks. The approach differs from HAZOP in that it does not focus specifically on operational parameters. What-if analysis offers flexibility to the team by allowing brainstorming in a manner that is not very sequential—where team members randomly consider questions about irregularities in the process. This makes the technique less rigid than the HAZOP method, but also it is relatively quicker to complete and achieve similar results. Due to less rigidity in the brainstorming, there exists a possibility of details being overlooked, which renders the study less comprehensive than a HAZOP study.

It is a widely used technique when estimating risks on less complicated processes, when minor changes to the process are required, or even when estimating risks by reviewing nonprocess situations such as plant layouts.

The success of a What-if analysis is very dependent on team participation and team experience. Its optimum use is when standard checklists are used in conjunction with the What-if questions.

2.1.5 FMEA

Failure modes and effects analyses (FMEAs) are used to determine how components may fail and the resulting effects of those failures. The FMEA method requires the identification of the items or components in the system to be analyzed, followed by clarification and description of the function of each component. Then the possible modes of failure of the components are identified along with the possible consequences of those failures.

The objectives of an FMEA are to:
- Identify possible failures modes caused by high or increased failure rates and their consequences to assets and deferred production
- Assess the risk associated with erosional velocity and other critical variables

The approach is similar to the techniques discussed above but is recommended to determine failure scenarios for equipment and components. This technique is widely used in the automotive, aerospace, medical devices industries and equipment manufacturing industries.

2.1.6 ENVID Study

The environmental impact identification (ENVID) technique is very similar to the HAZID, but its focus is limited to identifying environmental risks from facilities. It gives more detailed results than HAZID studies.

2.2 Quantitative

These techniques focus on the numerical aspects of risk, by assigning numbers with significant digits to the potential impacts, based on scenarios identified. They do account for multiple jeopardies.

2.2.1 QRA

Quantitative risk assessment (QRA) is the systematic development of numerical estimates of the expected frequency and/or consequence of potential accidents associated with a facility or operation based on engineering evaluation and mathematical techniques.

The objectives of a QRA are to:

- Identify hazards and risks associated with operations
- Perform consequence modeling to evaluate frequency and consequences of hazardous events
- Assess the individual and societal risks—risks for site workers and adjacent land use
- Compare risks against locally and internationally recognized criteria
- Recommend risk remedy measures to reduce risks to a tolerable (ALARP) level

The QRA technique is very sensitive to the finalized scope of the study, which determines the duration of the study. Hazardous scenarios from previously conducted qualitative techniques form the input to QRAs. These studies require detailed numerical calculations, using various release modeling tools to estimate the consequences. Available controls also are reviewed for their reliability, including failure rate data, thereby calculating frequency of the undesired events. This combination of the consequence and frequency figures provides the level of risk associated with the incident.

2.2.2 FTA and ETA

Fault tree analyses (FTAs) are deductive techniques that focus on failure of systems. This technique is used in scenarios where multiple causes are considered to contribute to failure of systems. Data from FMEAs can complement FTAs.

Event tree analyses (ETAs) focus on the possibilities of various incidents occurring in complex processes. This analysis is performed by looking at the possible outcomes due to the success (availability) and failure (non-availability of protective controls.

These two techniques are used for reliability studies. Reliability, availability, and maintainability (RAM) analysis is used to quantify the availability of the system and to provide a basis to optimize the facility. Using reliability data, probabilities of system failure are developed that incorporate maintainability parameters (e.g., repair time and maintenance crew deployment) to determine overall availability of the system.

2.2.3 FRA

Fire risk assessments (FRAs) use the results from modeling the potential consequences of different types of fires—jet fires, pool fires, etc.—and combine the results with the release frequency calculations to determine the risk to key targets and personnel. Then this information is used to identify risk drivers by ranking the events in order of risk to determine the characteristics and likelihood of fire events and their associated potential impact on people and facilities, such as manned spaces and large flammable inventories.

The objectives of an FRA are to

- Determine the characteristics and likelihood of fire events and their associated potential impact on people and plant facilities, such as manned spaces and large flammable inventories
- Provide consequence and frequency information for subsequent risk studies, such as the explosion risk analysis, QRA, and SIMOPs (simultaneous operations) QRA
- Verify compliance with legislation and regulation

2.2.4 ERA

Similar in concept to FRAs, explosion risk assessments (ERAs) are essential modeling studies carried out to evaluate the effects of potential explosions at facilities, due to ignition of gas leaks.

The objectives of an ERA are to:

- Identify targets of interest
- Define explosion scenarios
- Model overpressures at targets of interest

3. Applications and Misconceptions

Section 2 lists only a few of the popular qualitative and quantitative studies techniques, and each technique described has its specific application and advantage over the others.

The basic concept in the various risk studies follows the principles of

- Identifying potential hazards
- Causes leading to these hazards
- Resultant consequences that may harm people, environment, assets, and/or reputation
- Existing safeguards
- Actions or recommendations to minimize the consequences or eliminate the causes, if required.

The successful risk identification in a technical risk study depends on the amount of preparation by the facilitator and team contribution to the study. A crucial part of the preparation depends on the type of risk technique that is selected.
Important in choosing the type of risk technique to be use is that the decision process involve such questions as

- What is the intended outcome or objective of the risk study?
- What type of documentation is available?
- For projects, in what stage is the design?
- For existing facilities, is the study a re-do or revalidation?
- Is the chemical process a batch operation or continuous?
- Is the new process in question based on a new/unique formulation or is it similar to existing processes?
- The type of technology used – is that new or proven?

- Does the regulation specify the use of certain techniques?

To assist in the decision making on the choice of risk techniques, figure 3 provides an illustration from the offshore industry.

Figure 3: ERM's Approach to Risk – Offshore Industry

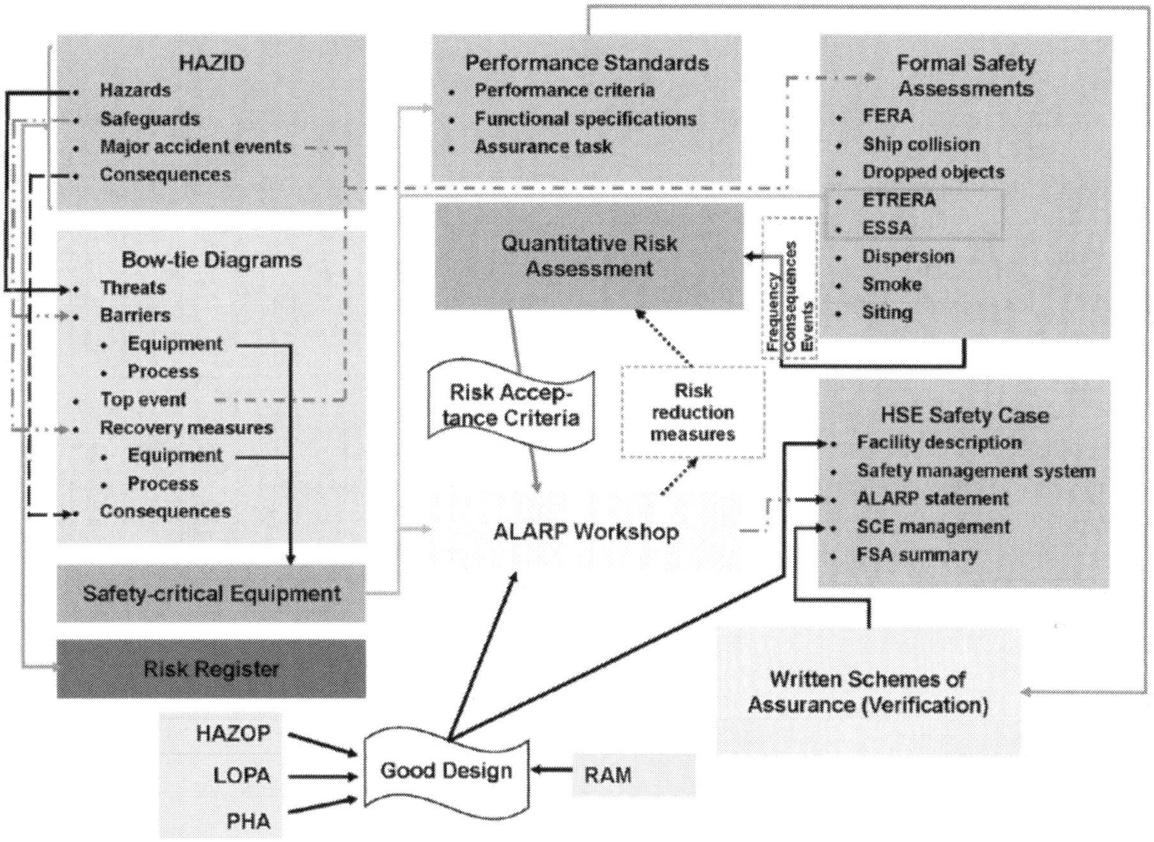

A common misconception though, is that either a one-time application takes care of all the risk needs for a facility or site or that a "take-your-pick" route works fine. The latter is a decision broadly based on schedule rather than what is appropriate for that stage of the project or process. As Figure 3 shows, one risk technique feeds into another. Hence the notion of "one size fits all"—implying that using the most detailed method and the facility does not need any other risk applications—is incorrect. This misconception may be due to lack of awareness at the facility on the advantages and disadvantages of the various techniques or the rush to get things done on a new design project or a maintenance activity at an existing facility.

The subsections below illustrate some real-time examples based upon the authors' experiences in conducting TRAs

3.1 Examples

3.1.1 When to Choose a HAZOP Over a What-if?

The conventional HAZOP focuses on changes to design and operational parameters. The technique is a fairly detailed approach, yielding favorable results. Regular participants in HAZOPs even consider it to be exhaustive. On the other hand, the What-If technique provides users flexibility by allowing them to choose the concerns to be assessed. If these concerns focus on operational parameters, then the study can be considered to simulate a HAZOP with limitations.

There was a need to conduct a risk study at a new constructed facility that manufactured additives for the oil and gas industry. This location had detailed P&IDs, process descriptions, and several other forms of process safety information. The facility's management decision was to conduct a What-if and checklist analysis, but the design team was relatively inexperienced, During day 1 of the What-if analysis, any questions raised by some of the team members led to a defensive response on the part of the design group and the study faced a very slow start. Modifying the course to a HAZOP study made it easier for the team to brainstorm on deviations such as high pressure, low pressure, high temperature, etc.

The team even focused on additional deviations such as startup and shutdown conditions, maintenance issues, and a human factors checklist.

This led to a very comprehensive discussion for a new facility with an inexperience team.

3.1.2 When to Choose a What-if Over a HAZOP?

At an existing chemical manufacturing, due to several corrosion issues on piping, a decision was made to replace piping to a higher specification. All process parameters were to remain unchanged. The facility was subject to OSHA 1910.119 and so it had records of all HAZOP studies conducted to date.

For a situation such as change in pipe specification, there was not need to go through the entire HAZOP rigor. This was a management of change (MOC) situation and a risk evaluation was to be completed. With the right expertise as part of the team, the risk requirement was completed using a What-If checklist, where relevant questions posed to the team focused on the change in pipe specification.

3.1.3 How to HAZOP an Operating Procedure

By following the HAZOP method, a set of operating instructions also can be assessed for potential issues leading to human error. Every step or sequence or instruction is to be considered as a single node, and deviations need to be generated for this node using guidewords such as

- SOONER
- LATER
- AS WELL AS

- LESS
- MORE
- NO
- OTHER THAN

This guideword and node combination essentially forms the deviations, which are then subjected to the standard HAZOP approach. Only the deviations that make sense are analyzed. The resulting analysis yields flaws in the instruction, which once corrected can minimize the potential for human error.

3.1.4 How to Conduct an Environmental Study Using a Bow-tie Diagrams [9]

Typically bow-tie diagrams are conducted to provide a pictorial representation to the hazard analysis process. If translated to a column format, the bow-tie diagrams resemble a HAZID worksheet. In this case, modifying the bow-tie objective brought the focus to primarily environmental releases and their effect on the surrounding environment for an upcoming oil and gas facility.

The results are show in below in Table 1. This change in approach facilitated an easy understanding of the ENVID concept to the workshop team and served as a great training tool for operations.

Table 1: ENVID Table to Fit the Bow-tie Approach (based on a recent study; risk ranking not shown) [9]

Cause	Prevention and Detection Barriers	Environmental Event	Controls/ Mitigation	Consequence	Recommendations
1. Diesel engine exhaust	1. Routine maintenance and inspection	1. Air Emissions	1. Monitoring for black smoke. Shut down for repairs if excessive.	1. Release of pollutants to surrounding environment (Particulates, SO_x, NO_x, CO_2)	2. Review helicopter exhaust parameters in later stages
2. Third-party equipment	2. Engineering				3. Review supply boat exhaust properties in later stages
3. Specific equipment	3. Equipment selection to code				8. Verify that drilling contractor equipment will not exceed emissions limits.

4 Conclusions

As can be seen in the above examples, the different types of risk assessment have their own advantages. Hence the success of any TRA rests on the choice of the most appropriate technique based on the risk and safety professional's experience, scope of the study, and the desired objective.

Teams must bear in mind that, in addition to choosing the most appropriate technique, team participation is essential in qualitative studies. A comprehensive qualitative study that identifies relevant scenarios can be directly integrated into the quantitative studies.

In addition, it is critical to understand that due to time and scheduling reasons facilities may opt for a single detailed study, eachof which relies on quality documentation and results from preceding studies (if applicable). Hence it is not a case of performing just one study identifying all risks – emphasizing the assertion that TRAs are not a one-size-fits-all solution. The nature of the studies depend on documentation available, type of industry and its associated regulations, the required scope and objective based on the stage of the project or facility. In turn, success of a study depends on choosing the optimum technique, having the right amount of preparation by an experienced facilitator or risk and safety professional and, of course, team participation.

5. References

[1] OSHA CFR 1910.119 Process Safety Management of Highly Hazardous Chemicals, 1992

[2] EPA 40 CFR Part 68 Risk Management Program (RMP) Rule

[3] Center for Chemical Process Safety (CCPS), "Guidelines for Hazard Evaluation Procedures" 3rd Edition, Hoboken, New Jersey: John Wiley and Sons, 2008

[4] American Institute of Chemical Engineers, Center for Chemical Process Safety, "Plant Guidelines for Technical Management of Chemical Process Safety (Revised Edition)", New York: Center for Chemical Process Safety of the American Institute of Chemical Engineers, 1992

[5] Center for Chemical Process Safety (CCPS), "Guidelines for Risk Based Process Safety", Hoboken, New Jersey: John Wiley & Sons, Inc. 2007

[6] Center for Chemical Process Safety (CCPS), "Guidelines for Chemical Process Quantitative Risk Analysis" 2nd Edition, Hoboken, New Jersey: John Wiley and Sons, 1999

[7] N. Hyatt, Dyadem, "Guidelines for Process Hazards Analysis, Hazards Identification & Risk Analysis". Boca Raton, Florida, USA: CRC Press LLC.

[8] Saud, Yaneira, Israni, Chris, Goddard, Jeremy "Application of cause-consequence diagrams in operational risk assessment", paper presented at American Society of Safety Engineers Conference, Denver June 2012

[9] Jones, Fred and Israni, Chris, "Environmental Impact Assessment Using Bowtie Methodology", paper presented at Society of Petroleum Engineers Conference, Perth, Australia, September 2012

[10] Saud, Yaneira and Israni, Chris, "Bowtie diagrams in downstream hazard identification and risk assessment", manuscript under review for publication, Process Safety Progress Journal, American Institute of Chemical Engineers, 2013.

Additional references not cited:

[1] Hudson, Patrick, Leiden University of the Netherlands & Delft University of Technology, The Netherlands "Integrating Organization Culture into Incident Analyses: Extending the Bow-tie Model". SPE International Conference on Health Safety and (2010), Vol. 4, 2662-2674

[2] OSHA 3132 Process Safety Management Informational Booklet, 2000

[3] American Petroleum Institute, API Recommended Practice 75, Recommended Practice for Development of a Safety and Environment Management Program for Offshore Operations and Facilities, 3rd edition, Washington, DC: API, May 2004

[4] Introduction to Bow-tie Diagrams. Avoid 3 common mistakes in major accident event prevention & recovery (seminar brochure). IRC Risk and Safety, 2010

[5] N. Gobeau and X.X. Zhou. Evaluation of CFD to predict smoke movement in complex enclosed spaces. Application to three real scenarios: an underground station, an offshore accommodation module, and building under construction. Research Report 255. Prepared for Health and Safety Laboratory for the [U.K.] Health and Safety Executive, 2004

[6] D. Kuzmin. Introduction to Computational Fluid Dynamics. (n.d.) Institute of Applied Mathematics, University of Dortmund. J.E. Vinnem. Offshore Risk Assessment. Principles, Modelling and Applications of QRA Studies. 2nd edition. London: Springer Verlag London Limited, 2007

[7] Hazard Consequence Modeling & Visualization. Pacific Disaster Center, (n.d.).

[8] OGP Risk Assessment Data Directory, Report No. 434-8. Mechanicallifting failures. International Association of Oil and Gas Producers, March 2010

[9] Nomenclature for Hazard and Risk Assessment in the Process Industries. Prepared by a Working Party of the Engineering Practice Committee of the Institution of Chemical Engineers, (n.d.)

[10] American Institute of Chemical Engineers. Guidelines for Hazard Evaluation Procedures. Second edition with Worked Examples. Hoboken, New Jersey: John Wiley & Sons, Inc., 1992. Preparing safety reports: Control of Major Accident Hazards

[11] Regulations. United Kingdom Health and Safety Executive, 1999.A Guide to Measuring Health & Safety Performance. [U.K.] Health & Safety Executive, December 2001

[12] Verification that safety critical elements are "suitable" at the commencement of a verification scheme. SPC/Enforcement/174. U.K. Health and Safety Executive, (n.d.)

[13] International Standard ISO 17776 Petroleum and Gas Natural Industries – Offshore production installations- Guidelines on tools and techniques for hazard identification and risk assessment, 15 October 2000

[14] 30 CFR Part 250 Subpart S, Safety and Environmental Management Systems in Federal Register 63610, Vol. 75, No 199, Washington, DC: U.S. Department of the Interior, 15 October 2010

Kevin D. Watson

- Team Lead – Major Capital Project Support Group
- Energy Technology Company – Chevron, U.S.A.
- B.S. Chemical Engineer – Missouri University of Science & Technology, Registered Professional Engineer, Process Engineering, Process Safety and Process Risk Engineering
- Personal Notes:
  - Wife, Susan, an Oklahoma University mechanical engineer; daughter, Grace Ann, in her junior year in art school (School of the Art Institute of Chicago), go figure…
  - I like bike riding (Trek Navigator) and smoking an occasional cigar, a bit ironic and to be clear, I do not do both at the same time…
  - There was a time when I could really play the saxophone. Now, just have a fine appreciation for Jazz & The Blues

Robert Dayton

- Staff Process Risk Engineer– Major Capital Project Support Group
- ETC / HES / Safety Tech Unit / MCP Support Group
- B.S. Engineering Technology (mechanical specialty) 1981, M.S. Safety Engineering, 1986, Texas A&M University.
- Personal Notes:
  - Married 29+ years to wife Julie who has supported me in the ups and downs of the oil industry.
  - Two kids: Son 25 years old who is in 4th year of Medical School at UTMB. Daughter that is in her Junior year at Concordia University Texas in Austin.
  - Hobbies include travelling and World War II history.
  - Worked for Chevron for 6 years.

Preventing the Process Safety Problems of Tomorrow, Today!

Kevin Watson
Team Lead – Major Capital Project Support Group
Energy Technology Company – Chevron

Robert Dayton
Senior Process Risk Engineer – Major Capital Project Support Group
Energy Technology Company – Chevron

Preventing the Process Safety Problems of Tomorrow, Today!

The most <u>cost effective</u> time for implementing process safety design elements is during the <u>planning stages</u> of the project management process.

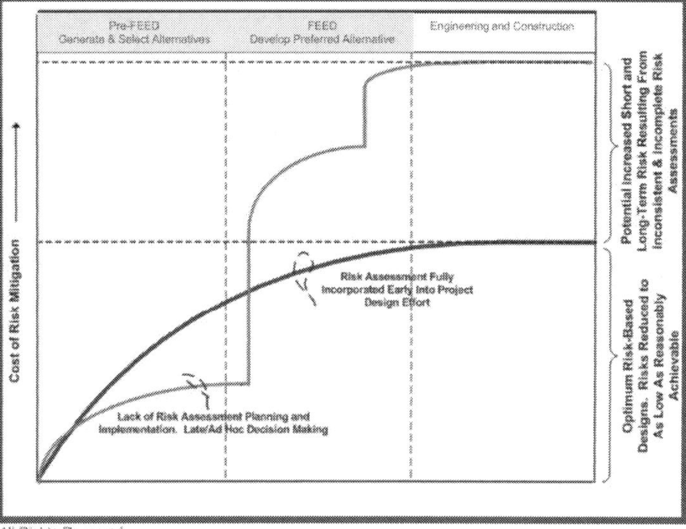

Preventing the Process Safety Problems of Tomorrow, Today!

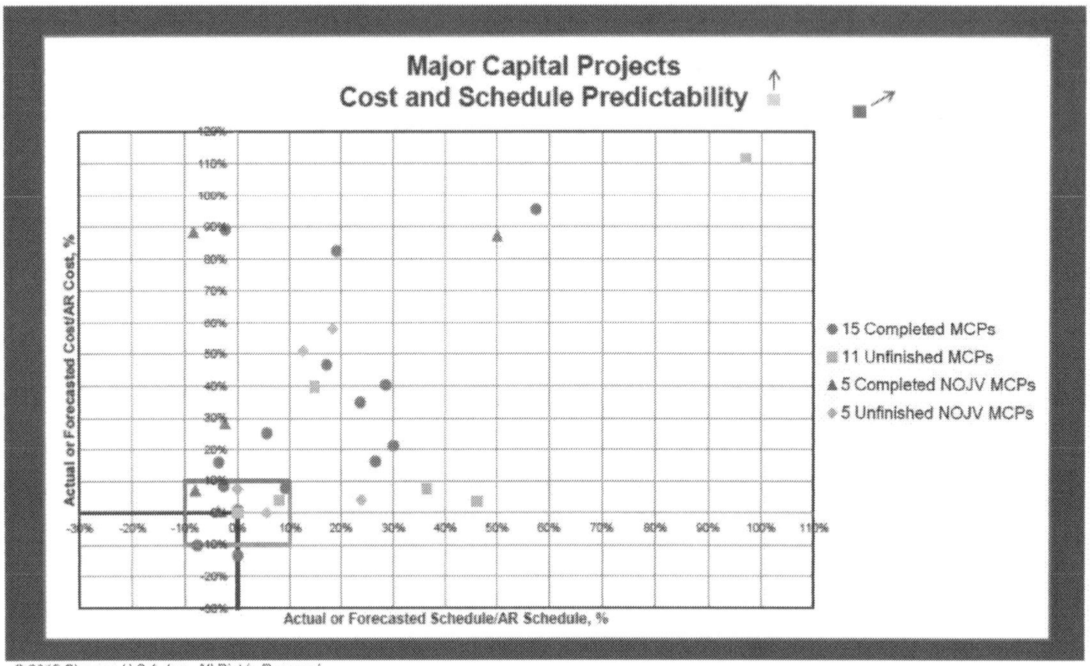

Preventing the Process Safety Problems of Tomorrow, Today!

- Delay executing a proper risk management strategy and a premium is paid to fix elevated risk gaps:
  - Discovered very late in the design. Still a chance to recover but again, premium paid to fix it
  - Discovered during the execution of what was thought to be a final design. At least it hasn't been built yet; tough choices to make
  - Worse yet, process safety design issues discovered after the facility has been constructed. Startup delayed, production deferred and fingers are pointed
  - After start-up . The penultimate premium to pay.
  - Discovered as a root cause during an incident investigation...
- So what's a reasonable timing strategy for success? First, keep the end in mind...

Preventing the Process Safety Problems of Tomorrow, Today!

Vision of Success for Process Safety / Risk Management

- A proper risk assessment strategy is implemented, ranging from assisting with determining a preferred alternative to the delivery of a "ready for construction" design
- The design is risk assessed,
- Elevated risks are uncovered and gaps are closed following a preferred risk control hierarchy, and
- The risks are considered managed as per the risk management process

7

Preventing the Process Safety Problems of Tomorrow, Today!

Knowing what the end looks like, again, what's a reasonable timing strategy?

- Example – The timing strategy for the Process Hazards Analysis (PHA) Methodology
- The strategy is a reasonable analog for other process safety deliverables
- This includes ensuring that the results from the plan deliverables are incorporated into the design (risks uncovered; gaps are closed)
- So what does the timing strategy for the PHA methodology look like?

8

Preventing the Process Safety Problems of Tomorrow, Today!

- The primary purpose of the PHA is to have a multifunctional team of experts confirm that a final process design and its anticipated integration into Operations, properly address the potential for process safety incidents. At Chevron, this purpose aligns with a documented expectation (keeping the end in mind and the vision of success):

  *"A process is in place to comprehensively assess and evaluate safety, health, and environmental, asset integrity and reliability risks… **Assessments conducted early in project phases shall be re-evaluated during final detailed design to determine whether mitigation plans have been implemented**."*

- The company expectation sets the planning and timing strategy, that is, if one uses the tools early, be prepared to use them again when final design information is obtained
- So what does Final mean???

9

Preventing the Process Safety Problems of Tomorrow, Today!

- Timing of the final process design risk assessment is based on the availability of the information needed to perform the PHA. This includes but is not limited to:
  - Process Flow Diagrams (PFD) with Heat and Material Balances
  - Piping and Instrument Diagrams (P&ID), including specification breaks
  - Piping and Instrument Diagrams for vendor-supplied packages
  - Equipment Data Sheets (Vessels, Pumps, etc.)
  - Relief Device Sizing Calculations and Relief System Basis and Sizing
  - Etc.
- As a circular thought, a final risk assessment of a final design cannot occur until the design is final. Is this the end of Front End Engineering & Design (FEED)? Early Detailed Design / Engineering?
- **It is when the above information is available, anything less than what is needed is a compromise to the strategy!**
- So how does a project team find design issues / elevated risks early enough so when the final risk assessment is conducted, there aren't any process safety show stoppers or there is enough time to close the gap??

10

Preventing the Process Safety Problems of Tomorrow, Today!

- Preliminary PHA Studies as Design Tools
  - With incomplete design information and with documented assumptions, Design Checklists can be used to identify hazard scenarios that need further definition
    - API RP-14C
    - API RP-14J
    - Equipment specific checklist
    - Etc.
  - Checklists take less resources to execute and are not necessarily facilitated sessions
  - Risk rankings are not needed to look for, as an example, lower pressure rated block valves downstream of a pressure specification break
  - These tools are ideal for a more focused, early issue P&ID review

11

Preventing the Process Safety Problems of Tomorrow, Today!

- The Preliminary PHA / Level of Protection Analysis (LOPA) as a Design Tool
  - The purpose of the PHA / LOPA process is to identify the numbers of Independent Protection Layers (IPLs) and their corresponding reliability required to manage identified risks
  - It is performed, most often, in conjunction with the Hazard and Operability (HAZOP) methodology in FEED, as the Preliminary PHA technique
  - Ideally, the Preliminary PHA / LOPA Study should be performed after utilizing the Design Checklist tools and having those results reflected into the P&IDs used for this preliminary risk assessment

12

Preventing the Process Safety Problems of Tomorrow, Today!

- Understand that the Preliminary PHA / LOPA Study is conducted using the same requirements as if the design were to be considered final
  - As an example if a relief valve size is not available during the study, no credit is given to it as a protection device and the scenario will be risk ranked accordingly
  - Most likely, a recommendation will be written to ensure that the relief valve is designed for the appropriate scenario (remember the Design Checklist reviews?)
  - One should expect a Preliminary PHA / LOPA Study, performed with less than the required design data, to have a relatively high risk profile. This should find the show stoppers
  - The risk profile should move to a more manageable level when the appropriate design is completed and the data is made available during the final risk assessment

13

Preventing the Process Safety Problems of Tomorrow, Today!

- Other Strategies of Interest
  - Emphasize design discipline when the final design risk assessment has been performed. This usually means implementing an MOC process with the main purpose to limit design changes. Remember, the final design should be, in essence "ready for construction" so design changes should be made difficult to approve
  - Don't confuse Change Management with HES Management of Change
  - To repeat, ensure that the results from the plan deliverables are incorporated into the design (risks uncovered; gaps are closed) when moving from preliminary assessments to final assessments.
  - The project team needs to work with the receiving business unit to determine and agree to what is considered a final design risk assessment and any associated documentation to support that assessment.

14

Preventing the Process Safety Problems of Tomorrow, Today!

- Summary / Conclusions
  - Delay executing a proper risk management strategy and a premium is paid to fix elevated risk gaps
  - The Company expectation sets the planning strategy, that is, if one uses the tools early, be prepared to use them again when final design information is obtained
  - For example, use Design Checklists for preliminary assessments, they are less resource intensive and risk ranking isn't required. Have the results from these studies reflected into the P&IDs used for the Preliminary PHA / LOPA study
  - It's OK that preliminary risk assessments have a relatively high risk profile. Having design details completed and incorporating the results from preliminary studies, the final design assessed will usually demonstrate the risk is adequately managed
  - Include cost and schedule estimates to account for resources needed to engineer solutions to the risk gaps uncovered during the various planning stages. This includes the final risk assessment of a final design. Elevated risks will be uncovered, even at this stage, and gaps will need to be closed to ensure a well risk managed, "ready for construction" design

15

Preventing the Process Safety Problems of Tomorrow, Today!

Questions?

16

Critical P&ID Reviews Using HAZOP/LOPA Methodology – Overkill or Underrated?

Mr. John T. Perez, P.E.
Cognascents Consulting Group, Inc.
11777-A Katy Fwy Ste 438
Houston, TX 77079
john.perez@cognascents.com

Mr. Andrew Madewell
Cognascents Consulting Group, Inc.
11777-A Katy Fwy Ste 438
Houston, TX 77079
andrew.madewell@cognascents.com

Prepared for Presentation at
American Institute of Chemical Engineers
2013 Spring Meeting
9th Global Congress on Process Safety
San Antonio, Texas
April 28 – May 1, 2013

UNPUBLISHED

Critical P&ID Reviews Using HAZOP/LOPA Methodology – Overkill or Underrated?

Mr. John T. Perez, P.E.
Cognascents Consulting Group, Inc.
11777-A Katy Fwy Ste 438
Houston, TX 77079
john.perez@cognascents.com

Mr. Andrew Madewell
Cognascents Consulting Group, Inc.

Keywords: Process Safety Management (PSM), HAZOP, LOPA, Process and Instrumentation Drawing (P&ID), Process Hazard Analysis (PHA), Inherently Safer Design (ISD).

Abstract

Prior to moving from define phase to execution phase on a given project, the typical project protocol is to perform a critical process and instrumentation drawing (P&ID) review as a transition from "issued for information or preliminary design" to "issued for design or HAZOP".

The objectives of a critical P&ID review are to (1) identify commercially-positive design changes early in the project lifecycle to reduce downstream implementation costs, (2) identify operational concerns to ensure optimal resolution and long-term operational effectiveness, and (3) identify safety vulnerabilities for early application of inherently safer design (ISD) principles where possible and early identification of safety integrity levels (SIL) for required safety-instrumented systems (SIS).

Companies perform critical P&ID reviews using various evaluation methods, including P&ID symbology/element checklists, process hazard analysis (e.g. Checklist, HAZID, and HAZOP), and "cold-eyes" review. Each of these approaches provides structure to a critical P&ID review; however, each also provides opportunities for commercial, operability, and safety vulnerabilities to remain in the design when applied independently.

The authors posit use of a hybrid HAZOP/LOPA process hazard analysis methodology to conduct critical P&ID reviews (1) results in a more rigorous evaluation and (2) reduces the likelihood for commercial, operability, and safety vulnerabilities to persist into the design/execution phase of projects. In this paper, the authors present the advantages and disadvantages of using a hybrid HAZOP/LOPA approach for critical P&ID reviews. They also provide recommendations to enhance the effectiveness of the HAZOP/LOPA methodology when applied to critical P&ID reviews. Specifically, the authors provide examples of commercial benefits realized, enhanced operational insight, ISD successes, and application pitfalls when applying a hybrid HAZOP/LOPA PHA methodology to a critical P&ID review.

The target audience for this paper includes project managers, project engineers, EH&S managers, PSM coordinators, and operators; however, anyone involved with small or large capital projects may also benefit from this paper's content.

1. Background and Purpose

A Process and Instrumentation Drawing (P&ID) is arguably the most important document for any given process across the oil, gas, and chemical industries. P&IDs provide piping, instrumentation, process flow, process control, process chemistry, equipment, relief/effluent system, and safety system design details. Critical operational and maintenance decisions are made every day at offshore platforms, gas terminals, refineries, and chemical plants with the aid of P&IDs. Before P&IDs are used for operational and maintenance purposes, they must be developed as effective representations of the design proposed for construction and operation.

P&IDs are typically developed during the Define phase of the project life cycle (see Figure 1.1 below). More specifically, P&IDs are usually one of the primary deliverables of the Front-End Engineering and Design (FEED) stage of a capital project. The activity used by a project team to issue a P&ID for design is called a critical P&ID review. The objectives of a critical P&ID review are to (1) identify commercially-positive design changes early in the process life cycle to reduce downstream implementation costs, (2) identify operational concerns to ensure optimal resolution and long-term operational effectiveness, and (3) identify safety vulnerabilities for early application of inherently safer design (ISD) and risk management principles, which may afford early identification of safety integrity levels (SIL) for required safety-instrumented systems (SIS).

Figure 1: Typical Project Life Cycle

Companies perform critical P&ID reviews using various traditional evaluation methods; including P&ID symbology/element checklists, process hazard analysis (PHA), and “cold-eyes” review. Each of these traditional approaches provides structure to a critical P&ID review; however, each also provides opportunities for commercial, operability, and safety vulnerabilities to remain in the design when applied independently.

When PHAs are used to facilitate a critical P&ID review, the traditional methods employed are Checklist, Hazard Identification study (HAZID), or Hazard and Operability study (HAZOP). Recently, the authors facilitated critical P&ID reviews using an emerging hybrid PHA methodology. The hybrid PHA approach combines the HAZOP methodology with the Layer of Protection Analysis (LOPA).

This paper provides a comparison between traditional approaches to conducting critical P&ID reviews and the emerging hybrid HAZOP/LOPA PHA methodology. The authors provide information from recently-facilitated critical P&ID reviews to help illustrate the (1) differences between traditional approaches and the hybrid HAZOP/LOPA PHA methodology and (2) advantages and disadvantages of various approaches. Specific to employing a hybrid HAZOP/LOPA PHA methodology to conduct critical P&ID reviews, the authors provide examples of realized commercial benefits, enhanced operational insight, ISD successes, and application pitfalls. The authors argue the use of a hybrid HAZOP/LOPA PHA results in a more rigorous critical P&ID review; and reduces the likelihood for commercial, operability, and safety vulnerabilities to persist into the Execute phase of capital projects. The authors also provide recommendations to enhance the effectiveness of the hybrid HAZOP/LOPA PHA methodology when applied to critical P&ID reviews.

2. Traditional Approaches to Critical P&ID Reviews

2.1 Traditional Non-PHA Approaches

2.1.1 Symbology/Element Checklist

Project teams often use symbology/element checklists to conduct critical P&ID reviews. Most checklists detail the P&ID symbology and elements required by a governing standard or guidance document. These standards and guidance documents may be controlled internal corporate documents. Some companies simply adopt standards and guidance documents published by professional organizations, which have achieved Recognized and Generally Accepted Good Engineering Practice (RAGAGEP) status by the company's industry (oil, gas, refining, or chemical).

Checklists vary by company and oftentimes by business units within a company. Some checklists are simple lists of required elements, such as valve types, line numbers, continuation arrows, title blocks, piping specifications, equipment blocks, instrumentation designations, and control and alarm information. Other checklists include questions for the reviewer to answer and verify, such as the following:

1. Are all valve types and sizes accurately depicted?
2. Are effective mechanisms for isolation, venting, bleeding, and de-energization readily available and accessible?
3. Are the governing design cases for pressure relief devices indicated on the drawing?
4. Are safe limits for critical process parameters indicated on the drawings?

For example, the checklist used by project teams in the upstream business unit may be different than the checklist used by the refining or chemical business units. Furthermore, the checklist used by North American operations within the upstream business unit of Company X may be different than the checklist used by European and African operations within the same upstream business unit of Company X.

Checklists may be applied by a single individual or by a team with a representative from each discipline (e.g. process, mechanical, electrical, instrumentation, controls, flow assurance, maintenance, and operations). As the project approaches the end of the Define phase or the end of the FEED stage, each person responsible for reviewing the checklist or a portion of the checklist against the P&IDs can do so independent of other persons' schedules. All changes identified by application of the checklists are made to the P&IDs before issuing them for design.

2.1.2 Cold-Eyes Review

Another traditional non-PHA approach is the cold-eyes review. This approach may incorporate a checklist (as described above) to guide the reviewer, but the primary goal of a cold-eyes review is to obtain an objective perspective and evaluation from a person or persons independent of the project. Time constraints are usually imposed on the reviewers to ensure closure of the activity and provide adequate time to update P&IDs to reflect changes identified by the cold-eyes review.

2.1.3 3-D Model Review

With the advent of effective 3-D modeling of processing units, most capital projects require a "virtual" walkthrough during the FEED stage. 3-D models show design elements unavailable on a 2-D drawing. For example, the 3-D model identifies:

1. Proposed piping routes and lengths;
2. Potential interferences between piping, equipment, and instrumentation elements;
3. Accuracy of fitting connections; and
4. Accessibility and availability of critical field elements, such as safety showers/eyewash stations, self-contained breathing apparati, and fire monitors.

Some project teams have combined the required 3-D model review with the checklist-driven critical P&ID review to ensure timely delivery of P&IDs out of the FEED stage of a capital project.

2.2 Traditional PHA Approaches

2.2.1 Checklist

The Checklist PHA methodology is similar to the symbology/element checklist approach as it also uses a prescribed element set to evaluate the P&IDs against. Checklist PHAs are normally performed by a multi-disciplined team towards the end of the FEED stage of a capital project. The team as a rule consists of personnel

The primary difference between the two checklist approaches is the prescribed element set. Checklist PHAs use prescribed element sets focused on process safety and risk management. A Checklist PHA is more focused on identifying, eliminating, or mitigating process hazards. Checklist PHAs usually employ questions similar to the following:

1. Are effective mechanisms for isolation, venting, bleeding, and de-energization readily available and accessible?
2. Are the governing design cases for pressure relief devices indicated on the drawing?
3. Are safe limits for critical process parameters indicated on the drawings?

In contrast, symbology/element checklists typically prescribe P&ID requirements with respect to symbology, drawings elements, and installation/design conventions (which may include process safety and risk management considerations).

2.2.2 HAZID

The HAZID PHA methodology is a structured brainstorming exercise designed to identify hazard scenarios early in the process life cycle. A HAZID can use guidewords and parameters along with nodes (system sections or procedural tasks) to focus the team's discussion and scrutiny. A HAZID does not incorporate risk assessment as it is usually performed before the completion of technical risk characterization calculations and analyses. Most actions from HAZID reviews involve further risk assessment to better understand identified hazard scenarios. Similar to other PHA efforts, HAZIDs are performed by a multi-disciplined team.

2.2.3 HAZOP

The HAZOP PHA methodology is another structured brainstorming exercise performed by a multi-disciplined team. Unlike the HAZID PHA methodology, a HAZOP uses risk assessments to detail out the cause and consequence of identified hazard scenarios. Risk acceptance criteria is applied to determine Likelihood and Consequence rankings within a given risk matrix. Safeguards are not accounted for in assigning the Consequence value; however, safeguards are taken into account when determining the Likelihood value. Where the qualitative risk-ranking recognizes a process safety or risk management vulnerability, the team crafts a recommendation to resolve the gaps.

Due to the information required to perform a HAZOP, it is often deemed impractical to employ this PHA methodology for critical P&ID reviews. It is uncommon for all of the risk assessment information required to fully characterize the hazard scenario in a HAZOP to be complete in time to execute the critical P&ID review during Define or FEED.

3. Emerging HAZOP/LOPA PHA Approach to Critical P&ID Reviews

The authors recently performed critical P&ID reviews using a hybrid HAZOP/LOPA PHA methodology. From conversations with (1) other consultants serving the oil, gas, and chemical industries and (2) colleagues employed by a company within the oil, gas, or chemical industry, the authors have determined a hybrid HAZOP/LOPA PHA approach to critical P&ID reviews is not common. Nonetheless, the use of the hybrid PHA approach is emerging as a comparable, perhaps enhanced, alternative to the already-mentioned traditional approaches to critical P&ID reviews.

By adding the LOPA methodology, the hybrid PHA approach also evaluates proposed Independent Protection Layers (IPLs) for adequacy to eliminate and/or mitigate the identified hazard scenario. HAZOPs and LOPAs are typically not used for design purposes; however, they can be valuable tools to assess the adequacy of proposed IPLs by quantifying the required risk reduction factors (RRFs) of proposed safety-instrumented functions (SIFs).

3.1 Example 1 of Critical P&ID Review Facilitated By HAZOP/LOPA PHA

One of the critical P&ID reviews facilitated by the authors using the hybrid HAZOP/LOPA PHA methodology involved an upstream subsea asset. There were approximately 22 unique nodes and the effort was completed in approximately seven days. The average number of persons in attendance each day was approximately 12. The approximate cost of the effort to the project was estimated by the authors at $150,000, which incorporates time to prepare, attend, and document the critical P&ID review.

Approximately 60 actions were generated using the HAZOP/LOPA PHA methodology. Several significant design changes were identified as being necessary to eliminate and/or safeguard against high-risk hazard scenarios. One action specified adding a completely redundant hydraulic line with separate pumps. The cost to implement this one design change was minimal as the critical P&ID review was performed towards the end of Define. It is unlikely traditional approaches would have uncovered the required change. The cost to implement the design change had it been uncovered during the formal HAZOP/LOPA PHA of the Execute phase would have eclipsed the total cost of the critical P&ID review estimated at $150,000.

3.2 Example 2 of Critical P&ID Review Facilitated By HAZOP/LOPA PHA

A second critical P&ID review facilitated by the authors using the hybrid HAZOP/LOPA PHA methodology involved another upstream asset. The scope included both the subsea and topsides designs. There were approximately 65 unique nodes and the effort was completed in approximately two months. The average number of persons in attendance each day was approximately 15. The approximate cost of the effort to the project was estimated by the authors at $850,000, which incorporates time to prepare, attend, and document the critical P&ID review.

Approximately 400 actions were generated using the HAZOP/LOPA PHA methodology. Once again, several significant design changes were identified as being necessary to eliminate and/or safeguard against high-risk hazard scenarios. Significant ISD opportunities were identified to safeguard against gas blow-by scenarios, which require pipe design modifications. Additional ISD opportunities were identified regarding chemical injection pressures and vendor skid specification requirements. Once again, the cost to implement these design changes was minimal as the critical P&ID review was performed towards the end of Define. It is unlikely traditional approaches would have uncovered the required ISD changes as the effort required complex, in-depth analysis of multi-system interactions. As before, the cost to implement the design changes had they been uncovered during the formal HAZOP/LOPA PHA of the Execute phase would have eclipsed the total cost of the critical P&ID review.

4. Comparison of Traditional and Emerging HAZOP/LOPA Approaches

The authors present the observed intrinsic advantages and disadvantages of the various approaches in the following tables. The approach comparisons have been grouped by the following three criteria:

- Implementation – factors evaluated include speed to implementation, ease of implementation, and information required for implementation;
- Quality – factors evaluated include expertise demanded by approach, diversity of personnel demanded by approach, breadth and depth of scope demanded by approach, and potential for early identification of inherently safer design opportunities;
- Financial Impact – factors evaluated include potential for early identification of significant cost-impacting (either positive or negative) changes and overall cost to complete the critical P&ID review.

Table 1. Implementation Advantages and Disadvantages

Approach	Advantages	Disadvantages
Symbology/Element Checklist	• Ability to implement quickly as implementation is not dependent on coordinating multiple schedules • Prescriptive nature affords execution by lesser experienced personnel	• Requires controlled checklist to already be developed against all applicable standards and guidance documents (i.e. all applicable RAGAGEP)
Cold-Eyes Review	• Utilizing resource pool independent of project; hence, project resources can continue to work on project deliverables	• Requires controlled checklist/procedure to already be developed against all applicable standards and guidance documents (i.e. all applicable RAGAGEP) • Time to complete is contingent on availability of resources not assigned to project
3-D Model Review	• 3-D perspective can accelerate visualization and understanding of piping, equipment, and instrumentation design and layout	• Requires group meeting rather than individuals working at their workspace with flexibility to work around other commitments • Review pace dependent on efficiency of virtual walkthrough
Checklist PHA	• Faster than other PHA methodologies • Prescriptive nature affords execution by lesser experienced personnel	• Requires group meeting rather than individuals working at their workspace with flexibility to work around other commitments • Requires controlled checklist to already be developed against all applicable standards and guidance documents (i.e. all applicable RAGAGEP)
HAZID PHA	• Faster than other HAZOP PHA methodologies	• Requires group meeting rather than individuals working at their workspace with flexibility to work around other commitments • Brainstorming nature not as prescriptive or efficient as checklist-driven approaches

HAZOP PHA	• Accelerates focus and delivery of critical process safety information, which is sometimes not generated until Execute	• Requires group meeting rather than individuals working at their workspace with flexibility to work around other commitments • Brainstorming nature not as prescriptive or efficient as checklist-driven approaches • Effective implementation dependent on availability and technical accuracy of process safety information, including thorough risk assessments
HAZOP/LOPA PHA	• Accelerates focus and delivery of critical process safety information and SIL determination for critical IPLs, which is sometimes not generated until Execute	• Requires group meeting rather than individuals working at their workspace with flexibility to work around other commitments • Brainstorming nature not as prescriptive or efficient as checklist-driven approaches making it the slowest of PHA methodologies

Table 2. Quality Advantages and Disadvantages

Approach	Advantages	Disadvantages
Symbology/Element Checklist	• Checklists can be tailored to accommodate varying degree of depth and breadth • Prescriptive nature drives consistent quality	• Not specifically focused on hazard identification and risk assessment • Prescriptive review only as comprehensive as standards and guidance documents (i.e. not a brainstorming exercise) • Prescriptive nature affords lesser experienced personnel to perform review • Does not provide insight to SIL requirements of critical IPLs • Requires further risk analysis to fully define risk envelope and subsequent safeguarding/mitigation measures
Cold-Eyes Review	• More people involved; hence, more eyes reviewing design • Objective evaluation from person(s) independent of project and design • Potentially greater depth of expertise reviewing the design	• Generally not focused on hazard identification and risk assessment • Personnel not familiar with project history and design considerations • Does not provide insight to SIL requirements of critical IPLs • Requires further risk analysis to fully define risk envelope and subsequent safeguarding/mitigation measures

3-D Model Review	• More in-depth with respect to mechanical, structural, ergonomic, and layout aspects of process design • More people involved; hence, more eyes reviewing design • Greater depth of expertise and team diversity affords more in-depth review of process design	• Generally not focused on hazard identification and risk assessment • Process tends to narrow focus of review on specific elements rather than systems and interactions between equipment • Does not provide insight to SIL requirements of critical IPLs • Requires further risk analysis to fully define risk envelope and subsequent safeguarding/mitigation measures
Checklist PHA	• Checklists can be tailored to accommodate varying degree of depth and breadth • Prescriptive nature drives consistent quality • Focused on hazard identification • More people involved; hence, more eyes reviewing design • Greater depth of expertise and team diversity affords more in-depth review of process design	• Generally not focused on risk assessment • Prescriptive review only as comprehensive as standards and guidance documents (i.e. not a brainstorming exercise) • Does not provide insight to SIL requirements of critical IPLs • Requires further risk analysis to fully define risk envelope and subsequent safeguarding/mitigation measures
HAZID PHA	• Structured brainstorming focuses on identification of hazard scenario for further risk analysis • Brainstorming nature affords more creative and in-depth review of process design • More people involved; hence, more eyes reviewing design • Greater depth of expertise and team diversity affords more in-depth review of process design	• Not as in-depth a review as HAZOP PHA approach • Requires further risk analysis to fully define risk envelope and subsequent safeguarding/mitigation measures • Does not provide insight to SIL requirements of critical IPLs

HAZOP PHA	• Structured brainstorming focuses on identification of hazard scenario along with comprehensive characterization of hazard scenarios via input from completed risk assessments • Brainstorming nature affords more creative and in-depth review of process design • HAZOP methodology affords systems-based review along with facility siting and human factors checklist and 3-D model reviews • More people involved; hence, more eyes reviewing design • Greater depth of expertise and team diversity affords more in-depth review of process design	• Does not provide insight to SIL requirements of critical IPLs • Quality of analysis is dependent on quality of information inputs
HAZOP/LOPA PHA	• Structured brainstorming focuses on identification of hazard scenario along with comprehensive characterization of hazard scenarios via input from completed risk assessments • Brainstorming nature affords more creative and in-depth review of process design • HAZOP methodology affords systems-based review along with facility siting and human factors checklist and 3-D model reviews • More people involved; hence, more eyes reviewing design • HAZOP/LOPA methodologies look at facility siting, human factors, and SILs of IPLs • Greater depth of expertise and team diversity affords more in-depth review of process design	• Quality of analysis is dependent on quality of information inputs

Table 3. Financial Impact Advantages and Disadvantages

Approach	Advantages	Disadvantages
Symbology/Element Checklist	• Not as labor and time intensive as other approaches making it most cost-effective approach	• Does not afford early identification of process safety and risk vulnerabilities
Cold-Eyes Review	• Not as labor and time intensive as PHA approaches making it more cost-effective	• Does not afford early identification of process safety and risk vulnerabilities

3-D Model Review	• Affords early identification of layout modifications, which can be costly if not identified until later project stage	• Labor and time intensive making it less cost-effective than non-PHA checklist and cold eyes review
Checklist PHA	• Affords early identification of non-compliant process safety elements of the design, which are usually not identified until final HAZOP/LOPA of P&IDs at the end of Execute phase	• Labor and time intensive making it less cost-effective than non-PHA approaches, but more cost-effective than other PHA approaches
HAZID PHA	• Affords early identification of hazard scenarios, which are usually not identified until final HAZOP/LOPA of P&IDs at the end of Execute phase	• Labor and time intensive making it less cost-effective than non-PHA and checklist PHA approaches, but more cost-effective than HAZOP and hybrid HAZOP/LOPA PHA approaches
HAZOP PHA	• Affords early identification of process safety and risk vulnerabilities as well as safeguarding requirements, which are usually not identified until final HAZOP/LOPA of P&IDs at the end of Execute phase	• Labor and time intensive making it less cost-effective than non-PHA and other PHA approaches, but more cost-effective than hybrid HAZOP/LOPA PHA approach
HAZOP/LOPA PHA	• Provides greatest opportunity to (1) identify commercially-advantageous design changes early in the project lifecycle to reduce downstream implementation costs, (2) identify operational concerns to ensure optimal resolution and long-term operational effectiveness, and (3) identify safety vulnerabilities for early application of inherently safer design (ISD) principles where possible and early identification of safety integrity levels (SIL) for required safety-instrumented systems (SIS).	• Labor and time intensive making it most costly approach

5. Recommendations for More Effective Critical P&ID Reviews

As presented, each approach has its advantages and disadvantages. Based on experience and observation, the authors recommend using multiple approaches to conduct critical P&ID reviews on all projects, not just large capital projects.

Prior to reviewing the P&IDs using a hybrid HAZOP/LOPA PHA methodology, the project team should subject the P&IDs to a checklist-driven review to ensure compliance with corporate and regulatory standards and guidance documents. This review should verify whether the proposed design meets the requirements of specific mechanical, metallurgical, structural, electrical, instrumentation, process control, and process safety standards and guidance documents. Actions to ensure compliance with all applicable standards and guidance documents should be resolved prior to the next phase of the critical P&ID review – the HAZOP/LOPA PHA.

Once the checklist review has been performed and all compliance and risk assessment actions have been resolved, then a HAZOP/LOPA PHA should be performed. A 3-D model review should be incorporated into the Facility Siting and Human Factors nodes required by the HAZOP methodology. Special care should be taken to ensure evaluation of the 3-D model against the electrical area classification drawings.

The phased approach described in this section requires special information considerations. More specifically, the proposed fusion of critical P&ID review approaches requires greater technical accuracy and scope of information. Significant ***preliminary*** process design, operational insight, and risk analyses are required to properly employ the HAZOP and LOPA methodologies. The following ***preliminary*** documents and models are a sampling of the envisioned inputs to the proposed phased approach to critical P&ID reviews:

1. Material safety data sheets;
2. Fire and explosion models;
3. Dispersion models of potential release scenarios;
4. Chemical/Material reactivity matrices;
5. Mass and energy balances;
6. Safe upper and lower limits for critical process parameters;
7. Operating philosophies;
8. Materials of construction;
9. Electrical area classification drawings;
10. Relief and effluent system design bases;
11. Ventilation system design basis;
12. List of design codes and standards applicable to the process;
13. Cause and effect matrices for safety controls, alarms, and interlocks (SCAI);
14. Equipment specifications;
15. Sparing philosophies (primarily used in upstream subsea asset development);
16. Electrical one-line diagrams;
17. Plot plans;
18. Safety shower/eyewash location drawings; and
19. Vendor-provided skid package information.

The authors acknowledge and appreciate the resources and time required to produce the information required to facilitate the proposed phased approach. The authors recognize the challenge associated with balancing (1) business excellence against (2) operational risk management; hence, the authors propose the allowance of practical flexibility when defining the scope and level of detail for all inputs to the proposed phased critical P&ID review. The design is not supposed to be final coming out of the Define or FEED stage of a project; however, ***preliminary*** calculations and models can provide valuable insight at this phase of the project.

6. Conclusion

Hazard identification and risk analysis (HIRA) are not activities exclusive to formal tasks. Such as vapor cloud explosion modeling, facility siting studies, human factors assessments, HAZOPs, and LOPAs. HIRA must be inherent to each activity and deliverable of each phase of a project's life cycle. Although more costly and more difficult to implement on the front-end, multiple approaches should be utilized to conduct critical P&ID reviews on all projects – not just large capital projects. Using multiple approaches provides greater structure, detail, and certainty to a critical P&ID review; thereby affording identification of more commercial, operability, and safety opportunities earlier in the project life cycle. Using multiple approaches amplifies the ISD opportunities afforded by simple, surface checklist-driven reviews and complex, in-depth PHA-driven reviews.

More specifically, the authors recommend (1) subjecting the P&IDs to a checklist-driven review to ensure compliance with corporate and regulatory standards and guidance documents; (2) reviewing the P&IDs using a hybrid HAZOP/LOPA PHA methodology; (3) incorporating a 3-D model review into the Facility Siting and Human Factors nodes required by the HAZOP methodology. By adding the LOPA methodology, the hybrid PHA approach also evaluates proposed Independent Protection Layers (IPLs) for adequacy to eliminate and/or mitigate identified hazard scenarios. The authors recognize HAZOPs and LOPAs are typically not used for design purposes; however, these methodologies can be valuable tools to assess the adequacy of proposed IPLs by quantifying the required risk reduction factors (RRFs) of proposed safety-instrumented functions (SIFs).

7. References

[1] Center for Chemical Process Safety, *Guidelines for Hazard Evaluation Procedures*, Second Edition, Copyright © 1992, American Institute of Chemical Engineers, 345 East 47th Street, New York, NY 10017.

[2] Center for Chemical Process Safety, *Guidelines for Risk Based Process Safety*, Copyright © 2007, American Institute of Chemical Engineers, Published by John Wiley & Sons, Inc., Hoboken, 111 River Street, New Jersey 07030.

[3] Spencer, P.E., Michael, "P&ID reviews promote productivity, safety and quality," *Processing*, September 2012, www.ProcessingMagazine.com, accessed 17 January 2013.

9TH GLOBAL CONGRESS ON PROCESS SAFETY

Using Near Misses to Improve Risk Management Decisions

Sean J. Dee,
Brenton L. Cox, and Russell A. Ogle
Exponent, Inc.
4580 Weaver Parkway, Warrenville, IL 60555
sdee@exponent.com, bcox@exponent.com, and rogle@exponent.com

Prepared for Presentation at
American Institute of Chemical Engineers
2013 Spring Meeting
9th Global Congress on Process Safety
San Antonio, Texas
April 28 – May 1, 2013

UNPUBLISHED

Using Near Misses to Improve Risk Management Decisions

Sean J. Dee,
Brenton L. Cox, and Russell A. Ogle
Exponent, Inc.
4580 Weaver Parkway, Warrenville, IL 60555

Keywords: enter a few keywords here to aid searching

Abstract

In the process safety literature it is often claimed that the analysis of near-miss incidents can improve process safety performance. However, empirical research has demonstrated that near-miss events are interpreted as marginally successful, leading to riskier behavior due to lower perceived risk In other words, experiencing a near-miss incident can result in the false conclusion that the original risk was over-estimated. To be effective, the analysis of near-misses must be grounded in an objective evaluation of the event, the outcome, and the severity of the alternative outcome's consequences.

In this paper several examples are presented where a near-miss incident preceded a serious incident with significant human injury or property damage. If the near-miss had been investigated, properly evaluated, and appropriate corrective actions been implemented, the more severe incident would not have occurred. A recurring theme in these examples is the underestimation of consequence severity, i.e., the misinterpretation of the near-miss as a high-probability, low-consequence severity scenario rather than a narrowly averted low-probability, high-consequence severity scenario. A deliberate consideration of the lower probability, more severe consequence scenario would have facilitated a more thoughtful evaluation of corrective actions.

A methodology based on game theory is presented for evaluating the impact of near-misses on risk perception. A game-theoretical model is used to demonstrate the need for shifting payoffs away from lagging process safety indicators and towards leading indicators. An equally important outcome of this analysis is the effective communication of the risk at operational and managerial levels of the organization.

1. Introduction

The evaluation and management of near miss events is a critical component of successful process safety management programs.[1],[2] A near miss is categorized as an abnormal event that has little

[1]D. R. Morrison, M. Fecke, J. D. Martens, "Migrating an incident reporting system to a CCPS process safety metrics model," *Journal of Loss Prevention in the Process Industries*, Volume 24, NO. 6, 2011; pp. 819–826.

immediate impact on a process or system, but has the potential to cause much more serious consequences.[3] Near misses can be further classified as vulnerable (an event that results in perceived vulnerability of the system) or resilient (an event that results in the perceived resilience of the system).[4] The near miss classification feedback loop, and its effect on the perceived risk is illustrated in Figure 1. A near miss event can be evaluated from two perspectives. In one case, a near miss can be viewed as showing the vulnerability of the system, which can lead to a higher perceived risk when faced with the realization that a system almost experienced a high consequence critical failure. However, it is also possible that experiencing a near miss can reinforce the resilience of the system, when it is realized that despite all the errors and mishaps that led to the near miss, the system was still protected against the hazard. It has been argued that resilient near misses not only lower an individual's perceived risk, but also can result in riskier behavior when faced with futures decision that require risk evaluation of the same hazard. Conversely, vulnerable near misses have been proposed to prompt risk mitigation in future behavior, suggesting that near miss classification can have a dramatic effect on a process safety management program. To foster a strong process safety culture, it is essential that the perception of resilient near misses must be overcome. As Adrian Sepeda has said, the process safety culture must be reinforced by "constantly maintaining a sense of vulnerability and avoiding complacency."[5]

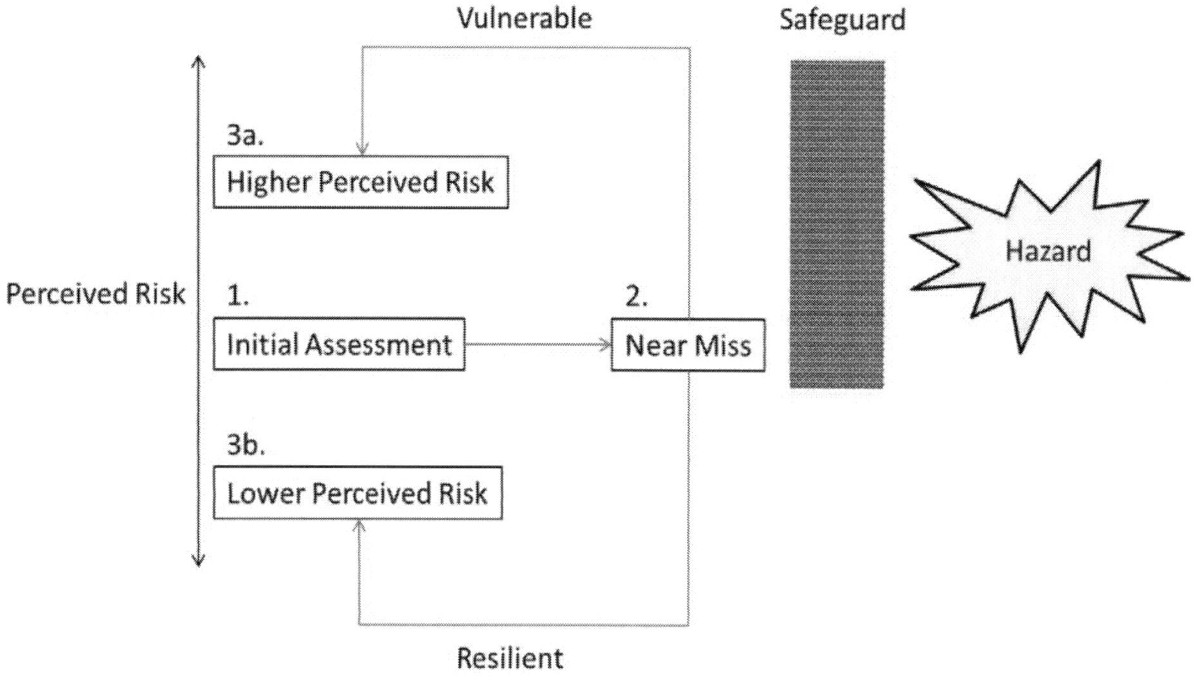

Figure 1. Near miss classification feedback loop.

[2]S. D. Lakhiani, "Near miss reporting behavior and the safety climate—A scenario based study," American Institute of Chemical Engineers, 2011 Spring National Meeting, 7th Global Congress on Process Safety, March 13–16, 2011.
[3] J.R. Phimister, U. Oktem, P.R. Kleindorfer, and H.Kunreuther, "Near-Miss Incident Management in the Chemical Process Industry," *Risk Analysis*, Volume 23, NO. 3, 2003, pp. 445-459.
[4] C.H. Tinsley, R. L. Dillon, M.A. Cronin, "How Near-Miss Events Amplify or Attenuate Risky Decision Making, *Management Science*, Volume 58, NO. 9, 2012, pp. 1596-1613.
[5] A. Sepeda. "Understanding Process Safety Management," *Chemical Engineering Progress*, Volume 106, NO. 8, 2010, pp. 26-33.

One valuable tool that can be used to understand how near miss event interpretation and valuation affect future behavior is game theory. Game theory uses a mathematical approach to understand the interactions of players, given the potential payoffs and penalties associated with their decisions. Game theory has been applied to problems in many areas including process safety management.[6] This work aims to use game theory to understand the impact of near miss classification and evaluation on the behavior of operations in a process safety management program. First, the decisions, and payouts of an operator in a game against the system are determined. Then the impact of changing the game payout structures given different near miss classifications and player attitudes on the game outcome is examined. Three case studies are reviewed to highlight specific game scenarios. Finally, solutions and improvements to process management programs are proposed.

2. Game Construction

2.1 Basic Format: Dominated Strategies

The day-to-day operations of a plant, process, or system were viewed as a two player dual decision game. The players were operations and the system, and their decisions were to either perform their tasks correctly (denoted as C for the operator and c for the system) or deviate to produce an upset (denoted as D for the operator and d for the system). These player classifications are meant to represent the internal (operations) and external (system) components of a process. Payouts for both players were varied from -1 to 1. Payouts, in this sense, correspond to benefits and punishments for each player in relation to the performance of the overall system. A payout of 1 indicates positive results. For example, an operator may get a bonus, promotion, or raise for the plant running successfully. Negative payouts represent discipline, such as an operator losing their job due to poor performance, or the system failing and shutting down. A payout of 0 indicates no net positive or negative result.

The first format for the construction of the game theory matrix set the payouts for each player directly through their individual decisions. The resulting game and expected payouts from this approach are shown in Figure 2. Since the expected payouts of performing tasks correctly (E[C]) is always higher than the expected payout from deviation (E[D]) the operations strategy of C strictly dominates D. Regardless of the probability of system upset (P(d)), there is never a scenario where operations would benefit from deviating. Cell (C, c) is also the only Nash Equilibrium in this scenario, since it represents a best response for both players

In the matrix of Figure 2, cell (D, c) corresponds to a near miss event, where an operational upset (human error, equipment failure, malfunctioning safe guards, etc.) occurs, but the resulting system payout is not affected. If an operator does not identify the near miss, then it is likely that operations would wrongly assume there had been no system upset, and the Nash Equilibrium in (C, c) is the result. From the operational perspective, the near miss event only results in a negative payout if the near miss is correctly identified as a deviation from strategy C,

[6] A. Meel, W.D. Seider, U. Oktem, "Analysis of management actions, human behavior, and process reliability in chemical plants. II. Near-miss management system selection" Process Safety Progress Volume 27, NO. 2, 2008, pp. 139-144.

highlighting the importance of identifying when a near miss occurs at the operational level. Another key concept for evaluating near misses is to understand the driving force to recognize and prevent near misses from occurring. This driving force can be defined as the payout difference between (C, c) and (D, c). In Figure 2, if operations identifies the near miss, then it has the maximum driving force to rectify the issue (1 – -1 = 2). In the resulting environment, near miss identification, investigation, and corrective action are only executed when the near miss is recognized as a negative payoff by operations.

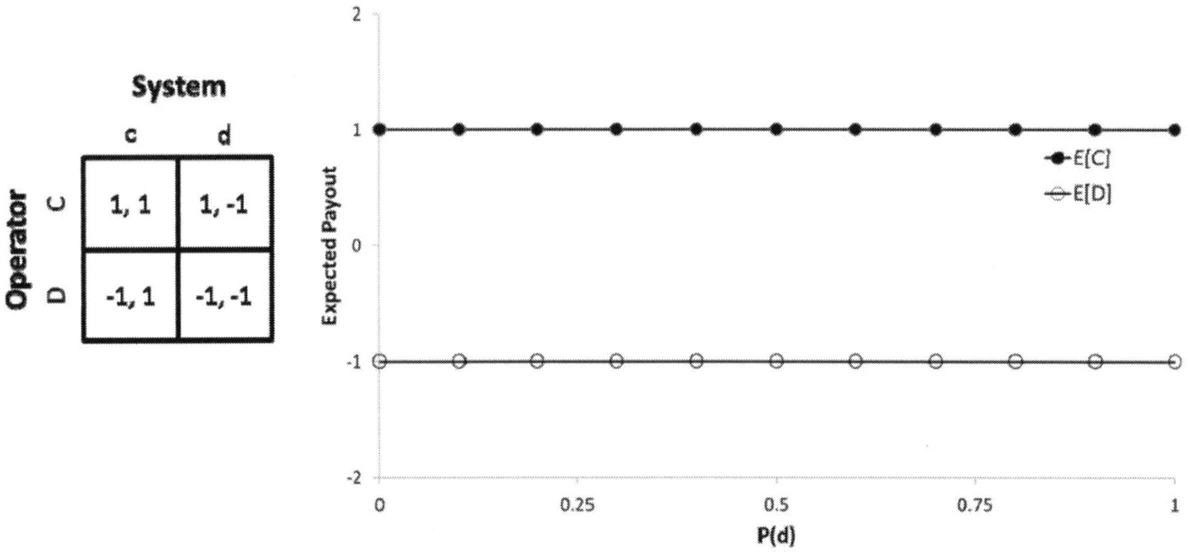

Figure 2. Basic format for game theory application to plant performance.

2.2 Game theory behind elimination of operational near miss evaluation

The first key component to a near miss management system is the identification of near miss events when they occur. From a game theory perspective, one way near misses identification can be compromised is by changing the payoffs such that the payout from an operational perspective is based solely on system performance. This theory can be applied to situations where operations assume their payoffs are directly tied to system performance, and feel their actions have no impact on their payouts from the game. For example, plants that evaluate operational performance on the sole basis of lagging process safety indicators such as lost work days, process downtime, or recordable safety incidents are at risk to produce this payoff structure. In this scenario, there are two Nash Equilibria in cells (C, c) and (D, c) as shown in Figure 3. Since the near miss cell (D, c) is a Nash Equilibria, there is no driving force to move out of that cell when a near miss occurs because both players have played a best response to each other. As shown in the expected payout for operations, C no longer strictly dominates D, and the expected payouts for strategies C and D are equivalent. The slope of the line represents the influence of the probability of the system response of d on the payout for a given operational strategy. In this scenario, there is no driving force for correction of near miss events because they are viewed exactly the same as (C, c), and operations decisions have no impact on the output of the game. One strategy for shifting these payoffs towards encouraging desirable behavior is to focus on leading indicators of process safety rather than lagging indicators.

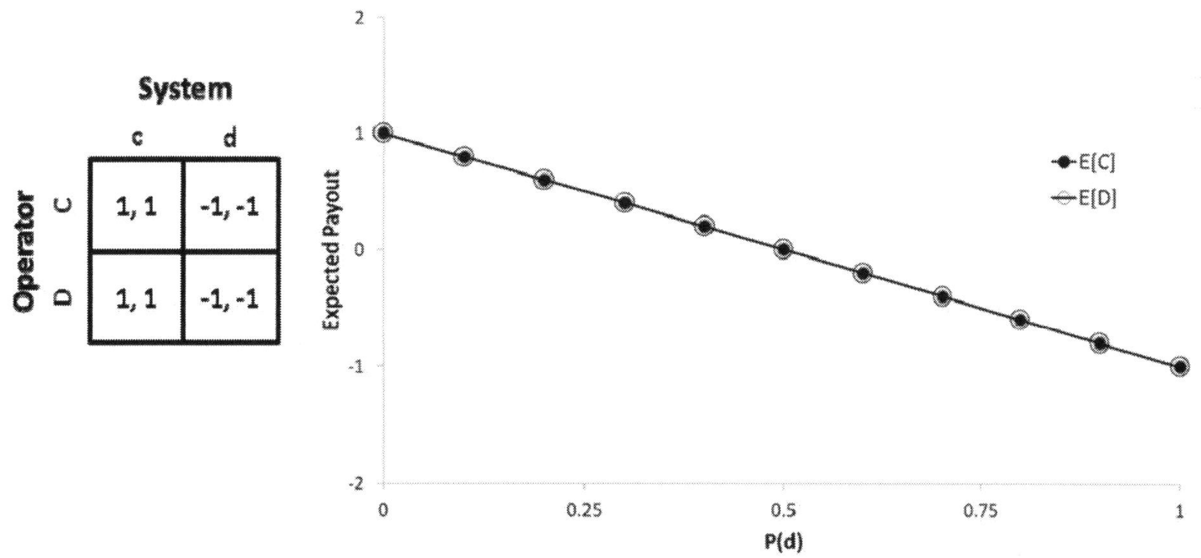

Figure 3. Game results when operational payouts are based only on system performance.

2.3 Implications of near miss payoff evaluation

As shown in Figure 3, changing the payoff structure of cell (D, c) can have a dramatic impact on the driving force for evaluation and correction of near misses. The classification of near misses as events that show the vulnerability versus the resilience of a system can have a drastic effect on how individuals perceive the risk associated with future decisions regarding the same hazard.[4] In an extensive study on resilient and vulnerable near misses, it was concluded that resilient near misses lead to positive feelings regarding lower probability of risk leading to complacency, while vulnerable near misses evoke negative associations that lead to risk mitigation behavior.[4] In other words, experiencing a resilient near miss induces a positive payout, because the avoidance of the potential loss and decrease in perceived risk is viewed as an agreeable outcome. Thus, both probability of system failure (P(d)), and the valuation of the near miss outcome (E[C]) play a role in near miss characterization. These two criteria for a near miss event can be represented and understood individually using a game theory approach.

A generalization of the game theory model discussed in Figures 1 and 2 is shown in Figure 4. The operational payout associated with the near miss event is represented as x in cell (D, c), which can be any number in the range [D, C]. The resulting expected payoffs for different x values are then described by the series of lines corresponding to E[D, x] over the range of probabilities for a negative system response (p(d)). The driving force for investigation and evaluation of a near miss is the difference in payout for (C, c) and (D, c). Research has suggested that individuals who experience near miss events may recalculate the initial hazard probability based on intuitive application of Bayes Theorem.[7] Thus, as near miss events are experienced repeatedly, regardless of whether they are vulnerable or resilient, it is possible that an individual will adjust the probability of experiencing the more serious negative consequence (P(d)) downward. In Figure 4, as P(d) decreases, the expected payout increases, thereby reducing the

[7] R. L. Dillon, C.H. Tinsley, "How near-misses influence decision making under risk: A missed opportunity for learning," *Management Science*, Volume 54, NO. 4, 2008, pp. 1425-1440.

driving force for evaluating a near miss. Additionally, if x = 1 the second Nash Equilibria is introduced into the game, which eliminates the driving force for evaluation of near miss events as previously discussed. Therefore, the two critical factors for a near miss management system are proper assignment for probability of critical failure (P(d)), and operations payout associated with system recovery (x).

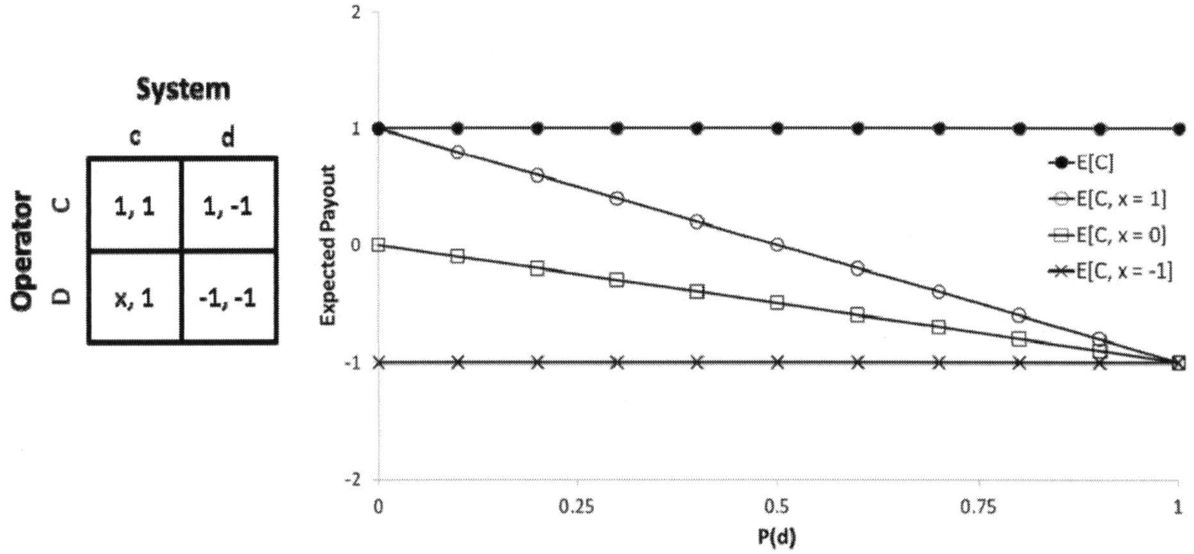

Figure 4. Generalized case for game theory approach to near misses

Properly maintaining the probability for critical failure of a system or process can be difficult. Process safety management includes several layers of safeguards which prevent catastrophic failures. While these safeguards are important to the safety and reliability of the process, they can lead to an underestimation of P(d) for a near miss in the context of the resilience of the entire safety management system. Take the simple example of a crane that drops a load that could potentially hit a worker below. In this near miss, two safe guards were in place. First an automatic stop function was installed on the crane, which has a failure rate of 1%. There is also a restricted area policy under the crane, which is 97% effective. Thus, if the crane drops 1 out of every 100 loads, the system P(d) is 0.0003%, which is several orders of magnitude smaller than the individual failure rates of the components involved in the near miss. The system P(d) shrinks even further if it is internally adjusted using Bayesian analysis as loads are lifted without incident. Therefore, evaluation of near miss events requires that the failure probability be evaluated on an individual basis, and past history be negated in risk assessment.

The assessment of an adjusted payout for a near miss event helps distinguish the differences between resilient and vulnerable near misses. When a resilient near miss is experienced, it is uncommon to experience any form of negative payoff. By definition, reinforcement of the resiliency of the system is likely to confirm bias in the ability of the system to respond positively to minor upsets, when in actuality the minor upsets are signs of the susceptibility of the system. In contrast, a vulnerable near miss is much more likely to induce negative feelings which correspond to an adjusted payoff that is lower than the payoff of normal. This adjustment downward increases the driving force for investigation of near miss events. Therefore, while experiencing a near miss event may change the perceived probability of the hazard, a vulnerable near miss that creates an adjusted negative payout can still lead to mitigation behavior. While it is important to always investigate incidents from an objective point of view, the outcome of the investigation must never become the basis for rationalizing away the significance of the near miss event. To maintain a strong process safety culture, near miss events should be regarded as an opportunity to reinforce a sense of vulnerability within all employees.

3. Case Studies

A game theory approach is a useful viewpoint of a near miss management system for two primary reasons. First, it is necessary to understand from an operation performance viewpoint, what constitutes a near miss event that shows the necessity to report the event for investigation. Understanding how valuation and probability of failure are analyzed internally can lend itself to education programs or best practices to ensure proper reporting of near miss events. More importantly, once a near miss is reported, there needs to be a concrete understanding of the severity of the event and the role it could play in a system failure.

The U.S. Chemical Safety and Hazard Investigation Board (CSB) is a useful resource for relevant case studies regarding accidents in the chemical processing industry. It is important to note that the approach, analysis, and interpretation of the following three case studies does not constitute agreement or disagreement with the findings, conclusions, or recommendations of the CSB reports, but are only meant to provide real world examples where conscious consideration of game theory principles could have led to more effective near miss management. The following three case studies highlight the importance of relating operations and system payouts, and the principles or proper valuation, and probability assessments.

3.1 Vinyl Chloride Monomer Explosion

The CSB published a full report detailing a fire and explosion at the polyvinyl chloride (PVC) facility that resulted in multiple fatalities and injuries.[8] The facility manufactured PVC through polymerization of highly flammable vinyl chloride monomer (VCM). The CSB determined the root cause of the incident was an operator who mistakenly opened the drain valve of a full, heated, and pressurized PVC reactor by bypassing the valve interlock. It was proposed that the operator intended to empty a nearby reactor that was undergoing cleaning, but when the bottom valve did not open the operator assumed the valve was not functioning properly, and bypassed the interlock to open the valve manually. The released contents of the reactor subsequently ignited, resulting in the explosion and fire at the processing plant.

The CSB investigation revealed two previous process hazard analyses (PHA's) had been conducted within the 15 years preceding the fatal incident. The risks associated with bypassing safety interlocks without supervisor approval was noted in the initial PHA, and when the PHA was later revisited the inadvertent transfer of contents from a reactor was specified as a high consequence event that had already occurred once at the plant. The CSB report highlighted a Risk Matrix for employee safety that designated decreasing risk tolerance as the consequence and frequency an event increased. This is the same concept discussed in the game theory

[8] Vinyl Chloride Monomer Explosion, U.S. Chemical Safety and Hazard Investigation Board. Report NO. 2004-10-I-IL, March 2007.

approach which suggests that near misses must be evaluated with reduced payouts (higher consequences) and higher system failure probability (higher frequency). This case highlights the need for operations to understand the payout and probability associated with their work. Among the contributing causes to the event, the CSB highlighted a lack of written guidelines for matching safeguards and risk, lack of written standards for managing interlocks, and the unpreparedness of the employees in the event of a major VCM release.

3.2 Electric Arc Furnace (EAF) Explosion

While cases similar to the VCM explosion highlight a need to educate operations of the risks associated with a process, proper valuation of process payoffs can also be problematic when near misses are properly identified. The CSB recently released a report detailing the EAF explosion at a calcium carbide plant.[9] The facility manufactured calcium carbide (CaC_2) from coke and lime (CaO) in an electric arc furnace. On the day of the incident, the EAF over pressurized and released debris, hot gases, and molten calcium carbide onto the second floor of the furnace building. The explosion knocked out a window in a nearby control room, severely burning two operators who died within 24 hours from burn injuries. While the CSB was unable to determine an exact cause, it found several likely scenarios that would introduce cooling water into the furnace, which would lead to over pressure events called "blows". These "blows" were considered normal low consequence incidents, and happened several times over the 20 years leading up to the incident. However, only two reports were filed relating to "blows". The first incident detailed a large overpressure event that blew in the window of the control room. As a result, the facility replaced the window with reinforced glass, which was damaged in a subsequent overpressure event. The later incident prompted the facility to replace the single pane glass with double pane reinforced glass, which was blown in during the fatal event detailed in the report. The CSB noted the attitude surrounding "blows" constituted a normalization of deviance, which occurs with similar major catastrophic events that are the subject of CSB reports. The EAF explosion report noted, "In spite of past warnings from alarms, near-misses, or lower severity incidents, each of these companies failed to control a hazard, and a high severity incident occurred later."

The EAF case illustrates the importance of properly evaluating the possible negative payoffs associated with a near miss that transitions into a catastrophic failure. In the above example, the payoff was valuated from the standpoint of replacing a window, rather than a major over pressure incident. As such, window replacement and improvement was used as the corrective actions, rather than determining the cause of "blows", preventing over pressure events, or relocating the control room out of a potential blast radius.

3.3 Aluminum Dust Explosion

Another incident which the CSB attributed to normalized deviance was an aluminum dust explosion at a facility that manufactured aluminum alloy wheels.[10] Scrap metal from the primary

[9] Electric Arc Furnace Explosion, U.S. Chemical Safety and Hazard Investigation Board, Report NO 2011-05-I-KY, February 2013.
[10] Aluminum Dust Explosion, U.S. Chemical Safety and Hazard Investigation Board. Report NO. 2004-01-I-IN, September 2005.

manufacturing facility was recovered through a process in which it was milled into chips and melted in a furnace. Aluminum dust generated in the process was collected through a dust collection system. Smoke from the melt furnace was drawn out of the building by a fume hood, through a steel duct, ultimately reaching a fume separator.

The chip feed system to the furnace was being restarted after a fire had occurred in the fume duct. After about 10 minutes of operation, aluminum chips were noticed falling out of the spark box in the dust collection system. As employees attempted to stop the chip feed again, a fireball erupted from beneath the furnace fume hood. There was one fatality and two burn injuries, one of which was critical. The CSB stated that the explosion likely originated from the dust collector, but the specific ignition source could not be identified. The report indicated that a secondary deflagration blew the roof off of the building and a structure fire led to significant damage.

The CSB reported that a root cause of the incident was a failure to address why the chip system was releasing excess dust. Duct fires were a "frequent event" and dust fires were common whenever the chip feed was started. Previously, a smoldering fire had occurred in the fume hood which was attributed to accumulated dust and chips on top of the hood, and resulted in a deflagration during an attempt to put it out. However, it was not clear whether management was made aware of these occurrences, and there was no record of any investigation of the events. Had an effective incident recognition and investigation management system been employed, such as an analysis of near misses discussed above, the fatal incident might have been avoided. Instead of viewing fires as isolated incidents, the common contributing cause of the fires could have been identified and addressed.

4. Solutions and Improvements for Near Miss Management

The game theory approach described herein has illustrated that both probability assessments and payout structures can change the driving force for implementation of corrective actions related to near miss evaluations. Solutions to these types of issues should, therefore, focus on preventing improper probabilistic assessments of system failure, or adjust payouts downward to stimulate follow through on corrective actions. These ideas can be enforced both cognitively and physically. Prevention of risk probability adjustment can be accomplished by regularly auditing processes internally, to reinforce risk probability even with exceptional safety track records. The probability of failure can also be maintained through training, or rotation of the teams on specific jobs. Investigation of near miss events as learning opportunities, which have no net positive or negative payout, can create an environment where payouts from near misses are not over or under valued. Beginning assessments by outlining a worst case scenario can ensure that all near misses are evaluated from a reduced payout standpoint, regardless of the vulnerable or resilient nature of the event. Tangible forms of payout adjustment can be dangerous, because if the penalty associated with a near miss event is too large, or permanent, then near misses may go unreported. One solution could be inclusion of near misses as recordable incidents, which are transitioned from a recordable to a learning event when investigation and corrective action implementation are finished.

After identification and evaluation of a near miss incident, a valuable near miss management system still requires proper analysis and implementation. Casual factor charting analysis to help illustrate the individual components of a system, and their relationship to a near miss event that produces either a low or high consequence scenario are still necessary as part of the investigation. Additionally, a hazard-barrier analysis approach can reveal safeguards that lead to safer processes. However, in both of these approaches, it is necessary to make a conscious, systematic decision to consider the near miss event in a broader context and the role it can play in a complete system failure. Otherwise, investigation and analysis run the risk of further biasing the evaluation of the event and process as being resilient rather than vulnerable.

5. Conclusions

Near miss events were evaluated from a game theory perspective to understand the factors that drive their identification, reporting, and implementation of their corrective action. It was determined using game theory, that the driving force for correction of near misses is reduced when near miss events are viewed as having payouts similar to normal operation, or when the probability of the system failing catastrophically is low. Two case studies were presented that illustrated that a proper near miss management system must include a driving force for proper hazard identification and corrective action implementation in order to be effective. Lastly, solutions to prevent underestimation of risk probability and over valuation of near miss events were discussed.

6. References

[1] D. R. Morrison, M. Fecke, J. D. Martens, "Migrating an incident reporting system to a CCPS process safety metrics model," *Journal of Loss Prevention in the Process Industries*, Volume 24, NO. 6, 2011, pp. 819–826.

[2] S. D. Lakhiani, "Near miss reporting behavior and the safety climate—A scenario based study," American Institute of Chemical Engineers, 2011 Spring National Meeting, 7th Global Congress on Process Safety, March 13–16, 2011.

[3] J.R. Phimister, U. Oktem, P.R. Kleindorfer, and H.Kunreuther, "Near-Miss Incident Management in the Chemical Process Industry," *Risk Analysis*, Volume 23, NO. 3, 2003, pp. 445-459.

[4] C.H. Tinsley, R. L. Dillon, M.A. Cronin, "How Near-Miss Events Amplify or Attenuate Risky Decision Making," *Management Science*, Volume 58, NO. 9, 2012, pp. 1596-1613.

[5] A. Sepeda. "Understanding Process Safety Management," *Chemical Engineering Progress*, Volume 106, NO. 8, 2010, pp. 26-33.

[6] A. Meel, W.D. Seider, U. Oktem, "Analysis of management actions, human behavior, and process reliability in chemical plants. II. Near-miss management system selection," *Process Safety Progress*, Volume 27, NO. 2, 2008, pp. 139-144.

[7] R. L. Dillon, C.H. Tinsley, "How near-misses influence decision making under risk: A missed opportunity for learning," *Management Science*, Volume 54, NO. 4, 2008, pp. 1425-1440.

[8] Vinyl Chloride Monomer Explosion, U.S. Chemical Safety and Hazard Investigation Board. Report NO. 2004-10-I-IL, March 2007.

[9] Electric Arc Furnace Explosion, U.S. Chemical Safety and Hazard Investigation Board, Report NO 2011-05-I-KY, February 2013.

[10] Aluminum Dust Explosion, U.S. Chemical Safety and Hazard Investigation Board. Report NO. 2004-01-I-IN, September 2005.

9TH GLOBAL CONGRESS ON PROCESS SAFETY

Keeping the Memory Alive, Preventing Memory Loss That Contributes to Process Safety Events

Barry Throness
ConocoPhillips
Calgary, Alberta
Barry.Throness@conocophillips.com

Summary of thesis prepared for Heriot Watt University M Sc submission

Prepared for Presentation at
American Institute of Chemical Engineers
2013 Spring Meeting
9th Global Congress on Process Safety
San Antonio, Texas
April 28 – May 1, 2013

UNPUBLISHED

Keeping the Memory Alive, Preventing Memory Loss That Contributes to Process Safety Events

Barry Throness
ConocoPhillips
Calgary, Alberta
Barry.Throness@conocophillips.com

Keywords: process safety events, memory loss, taxonomy, domain, learning activity, overview drawing

Acronym listing
HSE – Health, Safety and Environment
LOPC – loss of primary containment
MOC – management of change
PJHA – pre job hazard assessment
PSE – process safety event
RCA – root cause analysis

Abstract

Recurring process safety events (PSEs) are a real concern to the energy industry. Contributing causes to these events are quite often very similar. It appears as though the learnings from past events are not retained in the memories of the workforce, setting the stage for accidents to repeat. Even with best practices available to prevent such recurring accidents, these events continue to happen again and again. It seems as if something is missing, in order to effectively use the knowledge gained from so many past disasters and near misses, to prevent further PSEs.

It was desired to develop a tool to aid ConocoPhillips Canada (CPC) in preventing memory loss that is contributing to PSEs. Some of the world's worst process safety accidents were reviewed to gather common learnings, and investigation reports of CPC past PSEs were analyzed to determine how prevalent the issue of memory loss is within the company. Best practices to prevent such memory loss were researched and found to be readily available, and yet for some reason, memory loss issues are very widespread. The cognitive sciences were looked to for an answer on how memories are developed and effectively retained. The field of education was researched, to determine how leading educators effectively teach learning to achieve high levels of memory retention. Through this the taxonomy table, a tool that has been used by educators to enhance teaching and learning for many years, was discovered. Then, effective safety communication methods that target memory retention were explored. All researched information was finally tied together, into a learning curriculum, consisting of various activities. These activities were constructed to advance the learning process towards an objective that had been carefully developed using the taxonomy table guidelines. This objective was 'for the workers to integrate past process safety learnings to prevent future process safety events.'

1. Introduction

Far too many PSEs occur in the energy industry that should be easily preventable. According to Gerstein (Ref 6, px), organizations continually fail to learn from past errors and disasters, allowing accidents to repeat across industry and even with the same companies. Renowned process safety expert Trevor Kletz refers to many process safety accidents that he investigated, where he could have simply pulled a list of contributing causes from the files of previous accidents already investigated but forgotten by many. (Ref 8, p1) And Mannan (Ref 12, chap27, p18) states that the memory within an organization that should help to prevent process safety accidents decays thus allowing accidents to repeat.

To determine if the same concerns hold true for CPC, a review was undertaken of past PSEs. The company has a rigorous incident investigation program in place and yet still encounters recurring PSEs. A detailed review of these past investigations was undertaken, to gain an understanding of the type of events that have been occurring and if memory loss has played a role. It was then determined, through an interview process involving several key company individuals, that the memory of past PSEs was waning, especially after approximately three years. At three years, the events seemed to be all but forgotten unless there had been a personal involvement with the event.

Industry has been learning from PSEs for many years, and many best practices are available to help remember lessons learned. There are many safety experts working hard to implement these best practices, and yet accidents repeat, with very similar contributing causes, for both industry and CPC. Even with best practices in place, memory loss continues to contribute to PSEs.

In an attempt to determine why memories of these past events seem to diminish so quickly, the cognitive sciences were explored to better understand how memories are developed and retained. This led to an understanding of techniques that help to establish memories with better retention, such as original awareness, and developing triggers within the memory for future recall. Research on memory retention led to various methods of communication that can enhance the ability for a memory to be retained. It was learned that, in general, safety messages are not communicated in an effective manner to enable memory retention.

Now understanding the problem at hand, a solution was sought, that would help to teach lessons of the past in a manner that allows for effective memory retention, and through this the ability to improve process safety. The field of education was researched for best practices that enable learning. A taxonomy for learning, teaching, and assessing has been developed by experts in the field of education, and employed for many years to help advance learning needs. This taxonomy was explored further to determine how it could be used to help CPC enhance memory retention and process safety.

Learnings from industry, CPC, best practices, the cognitive sciences, and effective communication were incorporated into a learning curriculum within this taxonomy tool. The resulting activities are designed to develop required knowledge through specific cognitive actions. Key components of these activities are:

- One page drawing overview summaries of past process safety events
- Combine learnings of past events from industry and CPC into process safety awareness training

In summary, not adequately learning from lessons of past PSEs has proven to be a critical issue that hampers organizations within the energy industry. The problem is not that we don't know what causes these accidents, but that we do not effectively teach the lessons learned in a manner that allows the memory to be retained within the organization. The objective of these activities is to do just this – to teach lessons learned so that the knowledge is stored effectively in the memory and is able to be retrieved when it will aid in preventing PSEs.

2. Project Methodology

To determine tools or a technique to enable an energy company to "keep the memory alive" and not forget the lessons learned from past PSEs, the first step is to understand the problem at hand. Then one can move forward with developing a method to solve the problem. This method has to be carefully developed, looking at how others have attempted to solve the problem, and what may be missing from the 'formula' that others have utilized. The new method of solution then needs to be implemented, with opportunities to assess its effectiveness and adjust technique as required to continually improve the process. This section describes the methodology used to gain an understanding of the problem and then develop the solution.

A simplified overview of the methodology is shown below.

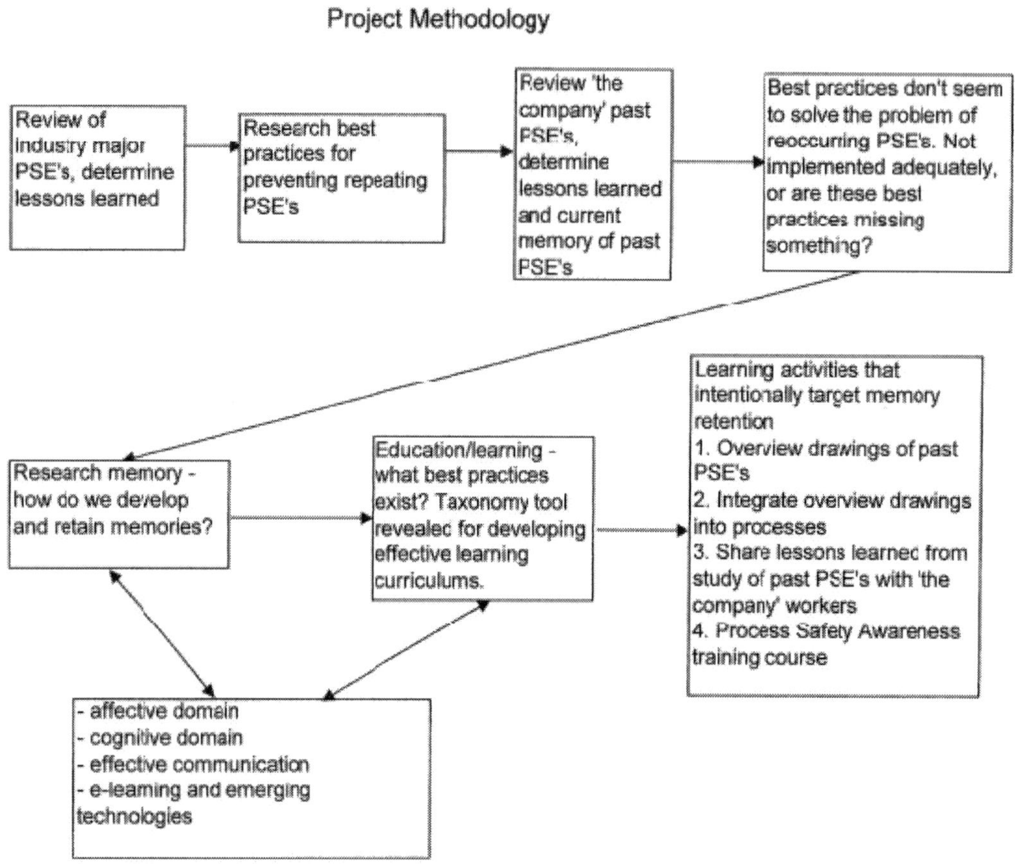

It was first desired to gain an understanding of the problem of memory loss as it relates to the energy industry. Research looked at various significant PSEs that occurred throughout the industry in all areas of the world. Particular importance was paid to large disasters that had numerous fatalities or significant impact to the surrounding community. Many of these disasters led to changing regulations in the industry. A high-level look at common contributing causes and accidents with recurring contributing causes was undertaken. Lessons learned were gathered for comparison between these industry disasters.

Best practices, as portrayed by leading process safety experts and organizations, were then researched. Before attempting to fix a problem, it is advantageous to understand how other experts have attempted to solve the problem in the past. Materials from renowned process safety expert Trevor Kletz, *Perry's Chemical Engineers' Handbook*, and various information available from the Center for Chemical Process Safety were researched, specifically for best practices to enable an organization to retain memory of past PSEs.

CPC was then assessed to determine how memory loss may be contributing to PSEs. It was determined to analyze past PSEs, consisting of both accidents and near-misses, as these near-misses only needed circumstances to change ever so slightly to alter the course from near-miss to PSE. This information would then be used in an attempt to assess the level of memory, within CPC, of past PSEs. This information indicated there was potentially a memory problem contributing to PSEs, but more information was still required to confirm this. It was determined that an interview process would be needed to further assess this, and several key individuals were a part of this past PSE review. Analyzing results of interviews showed there was a definite lack of memory relating to past PSEs, similar to what research indicates throughout the industry. Known best practices seemed to tell us what we ought to do to improve memory, and yet, past lessons are forgotten resulting in recurring process safety events. It seemed as if something was missing; how to effectively remember and learn from the past.

It was not known how to solve this dilemma, and so the cognitive sciences were explored for an answer. Research into how the brain functions, how memory works and how humans learn was carried out, hoping that knowledge in this field would lead to a solution that would in essence help to improve memory of the past and thus reduce process safety events. This research did lead to some best practices in memory retention, and also to the study of effective teaching and learning. The field of education has been researching methods for effective teaching and learning for many years and are considered leaders in this area. It was desired to learn what some best practices are for teaching, learning and assessing, and how to build learning activities into the development of a tool or technique to improve how CPC can best learn to keep the memory of the past alive and in everyday use. Through discussion with leaders in the field of learning, the taxonomy tool, an aid used by educators to develop learning curriculums since 1956, was researched.

Methods of communication to aid in effectively delivering the material to be taught and remembered were also researched, that would complement available knowledge on how the memory works and how humans learn. This led to determining methods for effectively displaying written and pictorial communication that can vastly improve memory retention. Improvements that can be gained through the use of various forms of e-learning were researched, and the outcome is very promising. There is almost unlimited potential in this field but, at this time, data is lacking to substantiate the potential benefits. It needs to be pursued, albeit cautiously, because the usage is increasing across the world so rapidly. Best practices for developing and structuring learning courses to enable memory retention were researched. All this information was utilized, with the taxonomy tool, to develop a learning curriculum for 'the company'. An objective was first created, that was "Students should learn to integrate past process safety learnings to prevent future PSEs". The learning curriculum, comprised of activities that were designed to achieve this objective, was then developed. The lessons learned from past industry accidents, industry best practices, learnings from CPC RCAs and interviews, and knowledge gained from memory and communication research were all taken into account to build activities that would achieve the objective.

3. Results

3.1 Energy Industry Review of Past PSEs

The energy industry has long been struggling to prevent the recurrence of process safety accidents. An overview look at some of the past significant accidents shows some alarming trends in common contributing factors, and an inability to learn from the past. Incidents reviewed included Piper Alpha, Longford, Texas City, Flixborough, Bhopal, Mexico City, Pasadena, and others. Results of this work aligned with a report by Derek Park, of *Oil and Gas IQ*, entitled Never Say Never Again. (Ref 14, p1) Park refers to the relevance of a famous warning quoted by

George Santyana that "Those who cannot remember the past are condemned to repeat it." The report details how this warning has been unheeded by the energy industry, exampled by various past accidents.

So the question is asked, "why can't industry learn from significant accidents in different organizations?" Why do we have to see recurring accidents, so similar and with the same contributing causes? Trevor Kletz, discusses this in his book, '*Lessons From Disaster, How Organizations Have no Memory and Accidents Recur.*' But he also discusses how serious accidents repeat in the same organization, given time, as memory of an accident fades.

3.2 Industry Best Practices to Help Remember the Past

The energy industry has learned a lot from accidents of the past. In fact, Kletz states "it is not the lack of knowledge that prevents our safety record being better that it is, but a failure to use the knowledge that is available, much of it acquired as the result of past accidents." (Ref 8, p96) And yet, the industry continues to have accidents with common contributing causes. So what is the knowledge that is available from the experts in the process safety field – what are best practices that, if followed, should help to prevent these recurring accidents?

Perry's Chemical Engineer's Handbook, a world-famous reference for the process safety industry, reiterates what Kletz states, saying that we know how to prevent most accidents, but they occur because we do not utilize the information that is available to us. Over time, the recommendations made following an accident are forgotten, procedures are allowed to lapse, changes are made to equipment and the accident is just waiting to happen again. (Ref 15, p26-7)

Looking specifically at a very significant process safety accident that occurred at Longford, Victoria, Australia in 1998, where an explosion at the facility killed two workers and the city of Melbourne lost its gas supply for a two week period. One of the key findings of the investigation was that Esso's various failures were due to 'an absence of mindfulness.' Study of high reliability organizations, such as nuclear, aerospace and airline, show that organizations within these industries are always mindful of the possibility of disaster. This culture allows them to never 'put their guard down' but always to be on the lookout for the precursors to a disaster. (Ref 7, p7) Because of this lack of mindfulness at Esso, the organization did not actively search within for warning signs of pending accidents, or look outside of itself for lessons that could be learned from other organizations in the industry. (Ref 7, p9) This thinking led the organization to believe that, prior to the Longford explosion, they were competent and skillful in the area of process safety simply because they had not suffered disasters in the past. This then led to justification to reduce effort and redundancy in this area. This type of culture within an organization leads to the result that 'current success makes future success less probable.' (Ref 7, p141) It is very important for a company to have a healthy sense of vulnerability, to believe that something that hasn't yet caused a problem can still be a dire incident in the making. (Ref 3, p5,8) Organizations need to be open to risk and the possibility that failure can occur to them. Being risk-blind is at the root of most tragedies. (Ref 6, p64)

3.3 Review of CPC Past PSEs

The investigation reports for eighty-four PSEs were reviewed.

The data reviewed gave a strong indication that memory loss is playing a role in recurring contributing causes and process safety events. It does not, however, show specifically where this is contributing to PSEs. How can the company have such a repeat of contributing factors leading to recurring PSEs, if the organization were truly remembering the past PSEs? Something seemed amiss and the answer was thought to perhaps lie within the workers. As Kletz has stated, "Organisations do not learn from the past or, rather, individuals learn but they leave the organisation, taking their knowledge with them, and the organisation as a whole forgets." (Ref 8, p1) Plans were then made to attempt to determine the level of memory of workers. Interviews were carried out to help determine how prevalent memory loss is with regards to process safety events that have occurred at CPC.

Interviewees were well aware of all events within the last three years. Beyond three years, interviewees were unaware of the event unless one of three reasons were existing:

- They had been in their current role and thus personally involved in the event and the recommendations that ensued.
- The event was of such significance that it resulted in substantial changes within the business.
- A recent event, either with CPC or with a peer company, had created discussion and triggered someone's memory of the past event

It is concerning that these three reasons that enable memory do not in themselves represent methods that enable memory retention very far into the future. The first – personal involvement, will pass when the individual retires or moves into a different position. The second – event significance to the company, will wane as time progresses, and several interviewees expressed this exact concern. And the third reason – event recently remembered due to a similar failure, is exactly what this study is attempting to solve; events repeating because the past has been forgotten.

3.4 So What's Missing?

Thus far we have review of industry literature showing that there are various best practices available to prevent process safety events, and yet significant accidents still occur, often with contributing causes that are repeated, or not learned, from other significant process safety accidents. The review of CPC RCAs shows a similar trend, and the subsequent interviews with various key personnel has proven that there is a problem with memory of past process safety events. It is not, however, based on the fact that nothing has been done in an attempt to remember the past events. In many instances, good investigations have been done, bulletins issued, training implemented and recommendations carried out. It seems then, that what has been done to remember lessons learned has been ineffective in transferring this knowledge and keeping the memory alive.

Best practices from various references highlight many different methods that will aid in remembering lessons learned from previous accidents. One such practice is simply stated - to not forget about the old accidents, and review them from time to time. This does not, however, address the importance of how the communication of these events benefits with the use of memory retention techniques. Without an intentional method to effectively deliver the lessons learned in a manner that yields optimal retention, individuals will quickly forget and the learnings will not be built into improvements within the organization. This only leads to repeating the contributing causes that will continue to lead to further similar process safety events. This is one very important aspect of learning from past events that seems to have been missed by industry and CPC.

3.5 Memory

Memory is defined as "the faculty of recalling or recognizing previous experiences" or "the length of time over which remembrance reaches." (Ref 16) So the questions are asked, "what makes a memory?" "What makes a memory stick in an individuals mind?" "What makes a memory stick within an organization?" Memory expert Harry Lorayne, in *The Memory Book*, explains that people often claim to forget something, when in fact they never actually remembered it in the first place. The problem is that people do not become "originally aware" of the information, or create an association in their mind that allows the information to be effectively remembered. (Ref 11, p6) Information needs to make its way into long-term memory to be effectively stored for retrieval at a later date. (Ref 4) Long-term memory is one of three types of memory that people are equipped with, the others being sensory memory and working memory. (Ref 5, p24) Understanding how the brain functions through utilization of these different memory types may be an important piece to implementing effective memory retention and recall of process safety learnings, and was thus explored further.

A simplistic depiction of these three memory types is shown below:

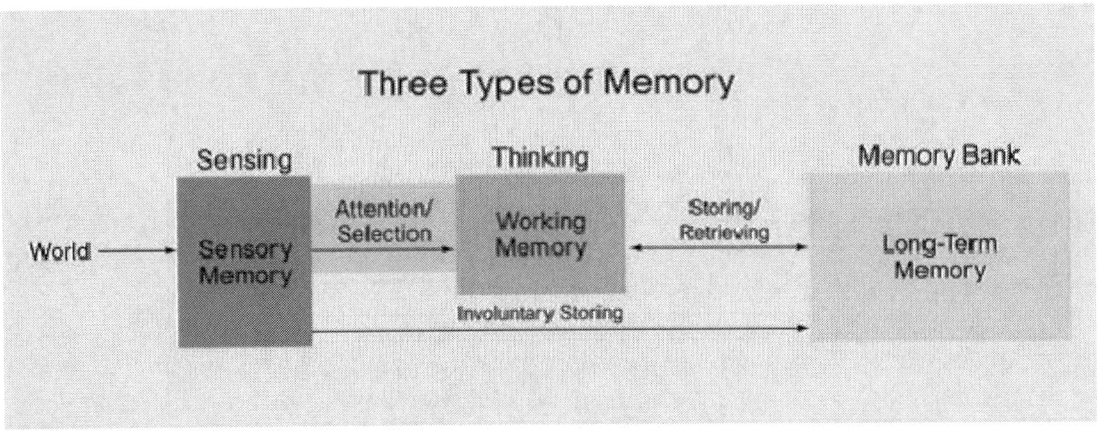

Based on Mayer (2003); Moreno & Mayer (2007); Marois (2005); and Miyake, et al (1999).

Sensory memory is enacted when triggers from the outside world are captured through the human senses. They are stored in sensory memory involuntarily and can transfer from here into either working memory, if they are thought about, or into long-term memory, if not thought about. Working memory, as indicated by its name, is where the brain does the thinking, although it is limited by capacity. Memories are only in the working memory while being thought about, and then proceed into long-term memory. Long-term memory has essentially unlimited storage capacity, and works in parallel with the working memory. The working memory retrieves memories from the long-term memory as needed and available. (Ref 5, p9)

Further detail of the thinking processes within the brain:

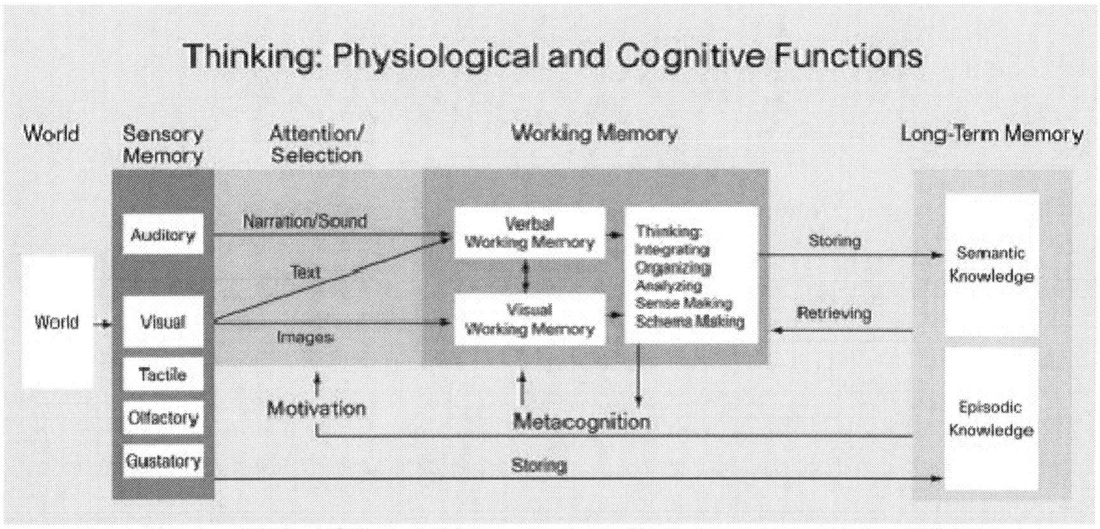

Based on Mayer (2003); Moreno & Mayer (2007); Marois (2005); and Miyake, et al (1999)

The working memory has two different components to it, the verbal working memory, and the visual working memory. The verbal working memory includes inputs such as noises, conversations, and writings. The visual working memory includes inputs such as drawings, pictures, and physical movements. The working memory buffers and processes these two different input sources separately, and these two information streams work together to build understanding. The working memory is very much limited by its capacity, as researchers believe that the verbal working memory has a capacity of seven objects, while at the same time the visual working memory is limited to approximately four objects. According to Sweller, cited in *Cisco*, (Ref 5, p9) cognitive overload can result from either of these buffers being overfilled. However, if verbal and visual working memories are successfully utilized together in creating a memory, there are more triggers in place to recall this memory, and any of these triggers will help to recall this entire memory. (Ref 5, p10)

The sensory memory takes in any aspect of the world captured through the human senses. If the mind pays attention to this sensory memory, it then transfers into working memory. If no special attention is paid to the sensory memory, it then moves into long-term memory as episodic knowledge. (Ref 5, p10) Episodic knowledge tends to degrade quickly. The other type, known as semantic knowledge, arises from knowledge that passes through the working memory into long-term memory. This knowledge has much better retention that the episodic knowledge, due to the links or triggers created in working memory that allow it to be pulled back from long-term memory when needed. (Ref 5, p10)

3.6 Learning

According to Bransford, cited in *Cisco*, research into the cognitive sciences of how people best learn reveals three important methods that enhance learning: (Ref 5, p11)
1. Learning is improved when existing knowledge is built upon to create new concepts.
2. Learning is improved when the learning topic has a direct personal meaning to them.
3. Learning is improved when students are forced to use existing knowledge to learn from failures and attempt to predict outcomes.

The study of cognitive sciences then led to research in the field of education or more specifically, the discipline of teaching and the subsequent learning that is desired as a result of teaching efforts. Bloom's taxonomy is a tool used by educators throughout the world since its inception in 1956, and referred to as one of the leading methods to influence education, both in America and the world. (Ref 1, pxx1) A revision to the 1956 B.S. Bloom's (editor) *The Taxonomy of Educational Objectives, The Classification of Educational Goals, Handbook I: Cognitive Domain* has been recently published. It is *Abridged Edition, A Taxonomy for Learning, Teaching, and Assessing, A Revision of Bloom's Taxonomy of Educational Objectives (2001)* and incorporates new knowledge that has influenced the education process since the writing of the Bloom's original text. (Ref 1, pxxII)

The taxonomy is an educational aid that follows a defined process. It utilizes the taxonomy table, (Table 1) that lays out the interrelationships between the cognitive and knowledge dimensions. This process helps teachers set objectives and then model or plan their training in a manner that facilitates achievement of the objective. Instruction should then be intentional, set out to achieve the objective. (Ref 1, p3)

The objective is first developed that contains a specific verb and noun. The verb, which describes the intended cognitive process, is chosen using the Cognitive Process Dimension table. (Table 2) A noun, that describes the knowledge that is expected to be acquired by the students, is chosen and fit to the Knowledge Dimension Table. (Table 3) At this point the objective statement is created. (Ref 1, p4) The objective is then placed on the taxonomy table, at the intersecting point of the cognitive and knowledge dimensions.

Table 1: (Ref 1, p28. Reprinted by permission of Pearson Education, Inc., Upper Saddle River, New Jersey)

The Knowledge Dimension	The Cognitive Process Dimension					
	Remember	Understand	Apply	Analyze	Evaluate	Create
Factual Knowledge						
Conceptual Knowledge						
Procedural Knowledge						
Meta-cognitive Knowledge						

Table 1: The Revised Taxonomy Table

The objective statement is paramount as it is the foundation for the process, which then builds around the desired outcome. It describes the intended result that is wanted to be achieved. It does not describe the means to attain the end result. (Ref 1, p17) This is instead attained through learning activities. These activities are developed, that will help achieve the objective, and are added to the taxonomy table, at the appropriate location that depicts the knowledge and cognitive attributes of each activity.

It is very important that the 'affective domain' is targeted as a component of the learning activity. The affective domain is the category of learning where one develops attitude and motivation. As this is developed, the student better receives the knowledge and becomes motivated to respond to it. This can even lead to creation of a personal value to the learning and, in turn, enhanced commitment. As this affective domain advances, learning in the cognitive domain improves as the student desires to learn more. Increased learning in the cognitive domain then further develops the affective domain. In this manner, the two domains compliment each other, leading to enhanced learning and retention. (Ref 5, p1,4)

According to Anderson and Krathwohl, effective use of the taxonomy table can help educators in at least three ways:

1. Focused attention on objectives leads to a deeper understanding of the learning question.
2. Taking this deeper understanding of objectives and then applying the thought process to determine how the objective fits to the taxonomy table, calculating both the knowledge and cognitive dimension of the objective, leads to better decisions on instruction and assessment of the students attainment of the objective.
3. Help determine how well the objectives, instructional activities, and assessment can compliment each other. (Ref 1, p95)

When the process is performed effectively, it should lead to memory retention and then allow for the student to use what was learned to solve new problems – effectively linking past learnings to future improvements. (Ref 1, p63)

Learning begins with thoughtful preparation of the message to be taught, and then effectively communicating this message. Dr. TJ Larkin, of Larkin Communication Consulting, is an expert on communication and further specializes in safety communication. According to Larkin, industry has done a poor job of effectively communicating safety messages to the workforce. Some of the key errors that have been made include:

- Providing too much information. According to O'Reilly, cited in Larkin (Ref 10, p3), there is a point where too much information leads to poorer performance.
- Information is presented at too high a grade level for the audience to comprehend. According to Aldridge and Brownson, cited in Larkin (Ref 10, p5), the average reading level of U.S. adults is 7th grade. And Larkin (Ref 10, p5), citing DuBay notes that safety communication at oil refineries nears a 16th grade level, while a grade level of 5 to 7 should be targeted to achieve what is considered effective communication.
- Words and pictures are poorly organized, yielding a safety message that is inadequately presented. Pictures should be simple, even hand-drawn, to capture the main points that need to be communicated. Utilizing a simple hand-drawn drawing to represent a complex event can increase comprehension by as much as 600 percent. (Ref 9, p8)

Effective safety communication should have three layers – simple picture, simple text, and technical information that is separate but available for those who desire or need to delve deeper into the subject. (Ref 9, p8) Diagram 1 shows an example drawing that is designed with simple picture and simple text.

Diagram 1 Overview Drawing of Incident (Ref 9, p8 with permission)

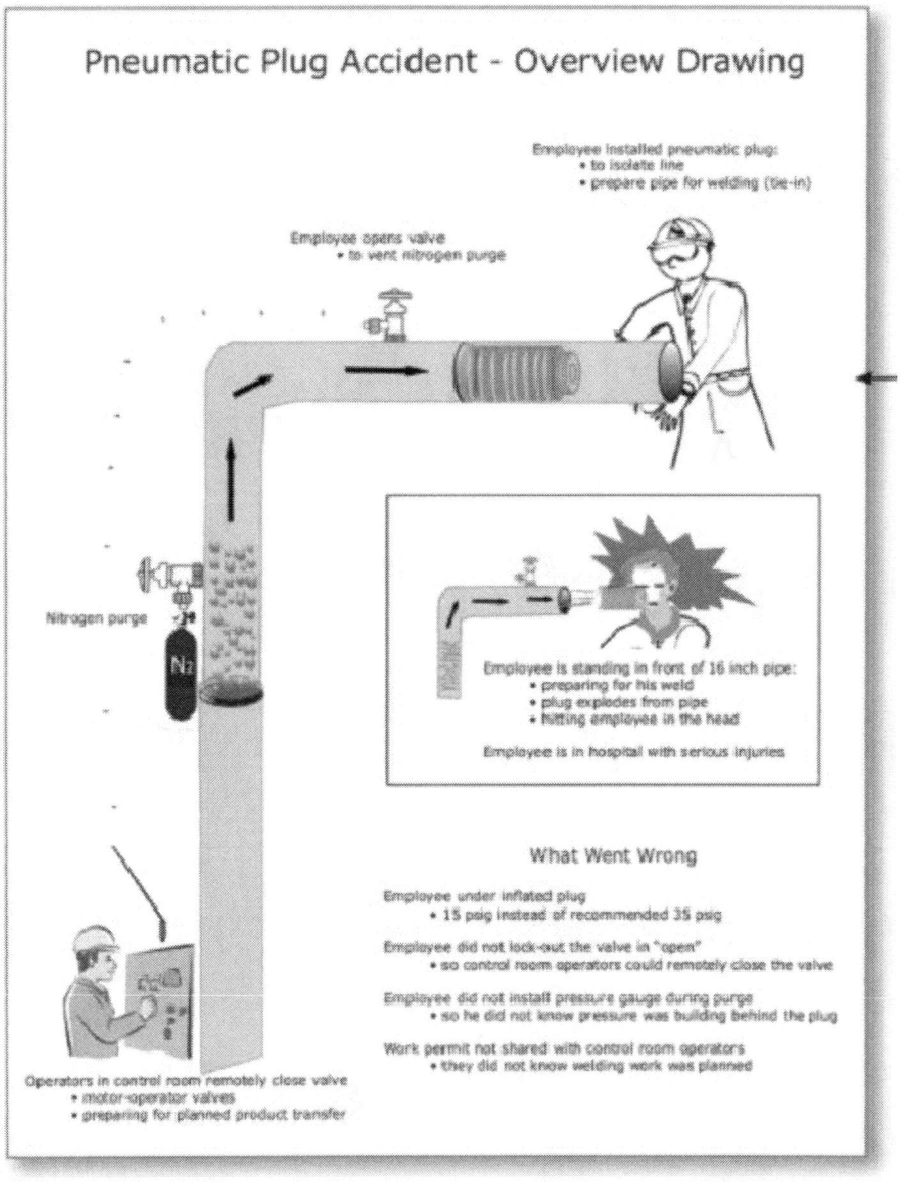

Another communication method that can assist in making a memory stick is to utilize speculation in the review of an incident. According to Morris, cited in Larkin (Ref 9, p9), speculating about "what could have happened" as opposed to a "facts only" approach can enhance learning by 70 percent. Diagram 2 shows a sample drawing that includes speculation.

Diagram 2 Overview Drawing Showing Speculation (Ref 9, p9 with permission)

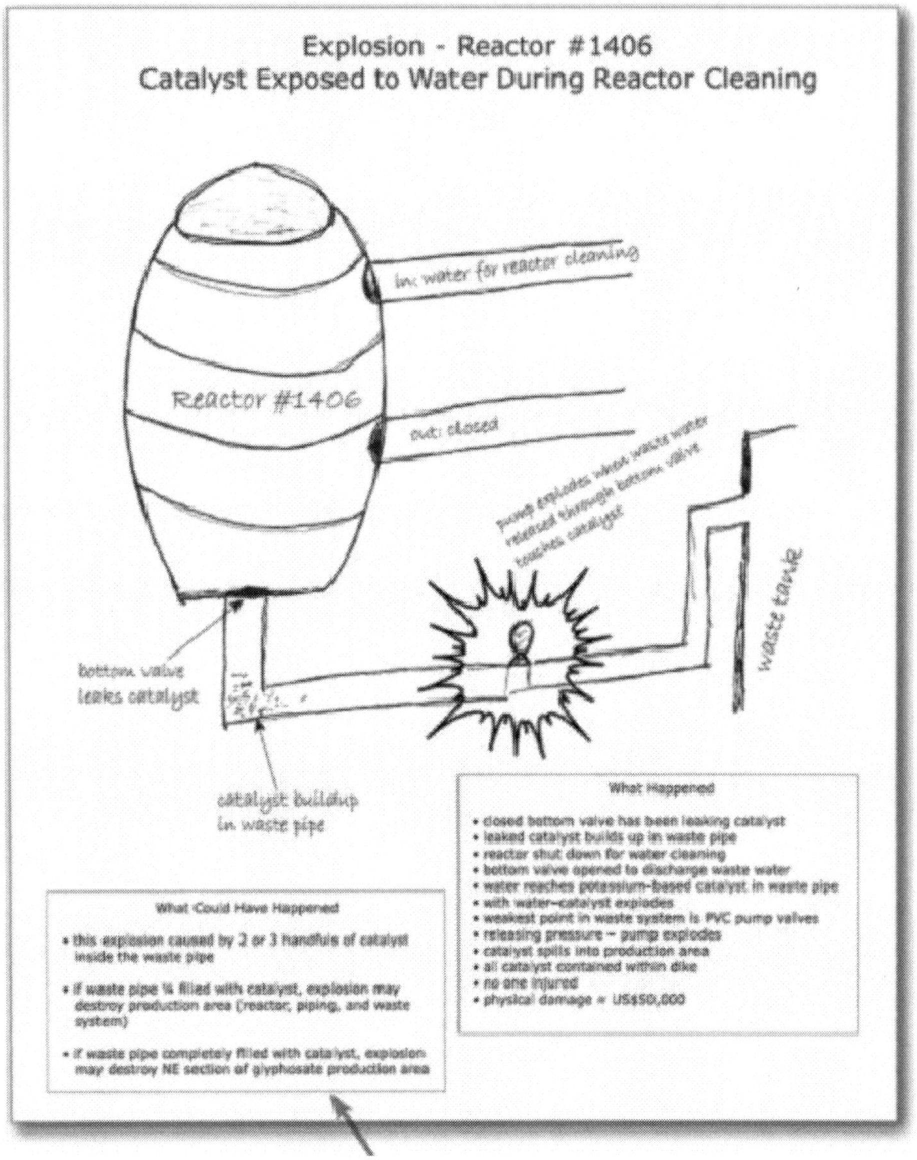

Chunking is also an effective learning strategy. Chunking is a term that was formulated by Harvard psychologist George A. Miller in 1956. It revolves around the belief that smaller bits of information taken in by the working memory allow it to process information in an optimal manner, enabling an easier integration into the long-term memory. The Research Institute of America has found that memory retention increases significantly when the chunking concept is applied. (Ref 2). Larkin's simple drawings build around the chunking concept in the manner in which incident information is presented. This is seen in Diagrams 1 and 2, with information grouped into small bits around the page.

3.7 Applying The Taxonomy Table to CPC

Effective use of the taxonomy table should allow for an intentional learning curriculum to be built for CPC to help prevent memory loss that contributes to PSEs. The goal is to build learning activities within the taxonomy table, utilizing the knowledge gained thus far from:

- Research into major industry accidents.
- Learnings from CPC RCAs.
- Industry best practices.
- Memory retention techniques.
- Emerging e-learning technologies.
- Effective communication methods that target affective and cognitive domains.

The learning activities are designed to achieve an objective that is carefully constructed and placed within the taxonomy table. The objective statement is comprised of a verb and a noun. First, the noun was determined to be 'past process safety learnings'. This refers to remembering past PSEs from CPC, their potential impact on workers, the contributing causes relating to the various type of RCAs the company has encountered, and the lessons learned pertaining to the underlying root, or latent causes. It is also important to utilize the learnings from past industry accidents, so those same lessons do not have to be learned 'the hard way'.

Then a verb is chosen, from the Cognitive Process Dimension Table (Table 2). This table lists and describes nineteen specific cognitive processes. From this table, under the category of 'Analyze', the verb 'integrating' has a strong fit with the desired intent. The definition for integrating is "Determining how elements fit or function within a structure" and the example given is "structure evidence in a historical description into evidence for and against a particular historical explanation." These descriptors fit well with the intent of this study – "determining how elements fit or function within a structure" aligns with determining how previous RCAs fit into a workers role and how various process safety hazards fit into the larger operation of the company. The example "structure evidence in a historical description into evidence for and against a particular historical explanation," fits into the process safety events and how they can have an impact on process safety in the future. More simply stated, it is the desire to have workers able to analyze events that have occurred in the past to make improvements in the future.

Table 2 (Ref 1, p. 67, 68. Reprinted by permission of Pearson Education, Inc., Upper Saddle River, New Jersey)

The Cognitive Process Dimension		
CATEGORIES &COGNITIVE PROCESSES	**ALTERNATIVE NAMES**	**DEFINITIONS AND EXAMPLES**
1. REMEMBER—Retrieve relevant knowledge from long-term memory		
1.1 RECOGNIZING	Identifying	Locating knowledge in long-term memory that is consistent with presented material (e.g., Recognize the dates of important events in United States history)
1.2 RECALLING	Retrieving	Retrieving relevant knowledge from long-term memory (e.g., Recall the dates of important events in United States history)
2.UNDERSTAND—Construct meaning from instructional messages, including oral, written, and graphic communication		
2.1 INTERPRETING	Clarifying, paraphrasing, representing, translating	Changing from one form of representation (e.g., numerical) to another (e.g., verbal) (e.g., Paraphrase important speeches and documents)
2.2 EXEMPLIFYING	Illustrating, instantiating	Finding a specific example or illustration of a concept or principle (e.g., Give examples of various artistic painting styles)
2.3 CLASSIFYING	Categorizing, subsuming	Determining that something belongs to a category (e.g., Classify observed or described cases of mental disorders)
2.4 SUMMARIZING	Abstracting, generalizing	Abstracting a general theme or major point(s) (e.g., Write a short summary of events portrayed on a videotape)
2.5 INFERRING	Concluding, extrapolating, interpolating, predicting	Drawing a logical conclusion from presented information (e.g., In learning a foreign language, infer grammatical principles from examples)
2.6 COMPARING	Contrasting, mapping, matching	Detecting correspondences between two ideas, objects, and the like (e.g., Compare historical events to contemporary situations)
2.7 EXPLAINING	Constructing models	Constructing a cause-and-effect model of a system (e.g., Explain the causes of important 18th Century events in France)
3. APPLY—Carry out or use a procedure in a given situation		
3.1 EXECUTING	Carrying out	Applying a procedure to a familiar task (e.g., Divide one whole number by another whole number, both with multiple digits)
3.2 IMPLEMENTING	Using	Applying a procedure to an unfamiliar task (e.g., Use Newton's Second Law in situations in which it is appropriate)

4. ANALYZE—Break material into its constituent parts and determine how the parts relate to one another and to an overall structure or purpose		
4.1 DIFFERENTIATING	Discriminating, distinguishing, focusing, selecting	Distinguishing relevant from irrelevant parts or important from unimportant parts of presented material (e.g., Distinguish between relevant and irrelevant numbers in a mathematical word problem)
4.2 ORGANIZING	Finding coherence, integrating, outlining, parsing, structuring	Determining how elements fit or function within a structure (e.g., Structure evidence in a historical description into evidence for and against a particular historical explanation)
4.3 ATTRIBUTING	Deconstructing	Determine a point of view, bias, values, or intent underlying presented material (e.g., Determine the point of view of the author of an essay in terms of his or her political perspective)
5. EVALUATE—Make judgments based on criteria and standards		
5.1 CHECKING	Coordinating, detecting, monitoring, testing	Detecting inconsistencies or fallacies within a process or product; determining whether a process or product has internal consistency; detecting the effectiveness of a procedure as it is being implemented (e.g., Determine if a scientist's conclusions follow from observed data)
5.2 CRITIQUING	Judging	Detecting inconsistencies between a product and external criteria, determining whether a product has external consistency; detecting the appropriateness of a procedure for a given problem (e.g., Judge which of two methods is the best way to solve a given problem)
6. CREATE - Put elements together to form a coherent or functional whole; reorganize elements into a new pattern or structure		
6.1 GENERATING	Hypothesizing	Coming up with alternative hypotheses based on criteria (e.g., Generate hypotheses to account for an observed phenomenon)
6.2 PLANNING	Designing	Devising a procedure for accomplishing some task (e.g., Plan a research paper on a given historical topic)
6.3 PRODUCING	Constructing	Inventing a product (e.g., Build habitats for a specific purpose)

The objective is to have workers take past process safety learnings and integrate this knowledge into the situations encountered in the workplace to prevent future PSEs. So an objective is created; "Students should learn to integrate past process safety learnings to prevent future PSEs." And it is now fitted against the Knowledge Dimension (Table 3) to see where it should be placed on the Taxonomy table.

Table 3 (Ref 1, p.46. Reprinted by permission of Pearson Education, Inc., Upper Saddle River, New Jersey)

The Knowledge Dimension

MAJOR TYPES AND SUBTYPES	EXAMPLES
A. FACTUAL KNOWLEDGE—The basic elements students must know to be acquainted with a discipline or solve problems in it	
AA. Knowledge of terminology	Technical vocabulary, musical symbols
AB. Knowledge of specific details and elements	Major natural resources, reliable sources of information
B. CONCEPTUAL KNOWLEDGE—The interrelationships among the basic elements within a larger structure that enable them to function together	
BA. Knowledge of classifications and categories	Periods of geological time, forms of business ownership
BB. Knowledge of principles and generalizations	Pythagorean theorem, law of supply and demand
BC. Knowledge of theories, models, and structures	Theory of evolution, structure of Congress
C. PROCEDURAL KNOWLEDGE—How to do something, methods and inquiry, and criteria for using skills, algorithms, techniques, and methods	
CA. Knowledge of subject-specific skills and algorithms	Skills used in painting with watercolors, whole-number division algorithm
CB. Knowledge of subject-specific techniques and methods	Interviewing techniques, scientific method
CC. Knowledge of criteria for determining when to use appropriate procedures	Criteria used to determine when to apply a procedure involving Newton's second law, criteria used to judge the feasibility of using a particular method to estimate business costs
D. METACOGNITIVE KNOWLEDGE—Knowledge of cognition in general as well as awareness and knowledge of one's own cognition	
DA. Strategic knowledge	Knowledge of outlining as a means of capturing the structure of a unit of subject matter in a textbook, knowledge of the use of heuristics
DB. Knowledge about cognitive tasks, including appropriate contextual and conditional	Knowledge of the types of tests particular teachers administer,

knowledge	knowledge of the cognitive demands of different tasks
DC. Self-knowledge	Knowledge that critiquing essays is a personal strength, whereas writing essays is a personal weakness; awareness of one's own knowledge level

First comparing the descriptors for Conceptual and Factual knowledge, it was determined that the objective had a stronger link to conceptual knowledge, as it is important for various workers to understand the underlying causes of the PSEs that have occurred, and not just remember the facts of the events. This aligns with the general statement for conceptual knowledge – "the interrelationships among the basic elements within a larger structure that enable them to function together" (from Table 3) The subtype from this table that aligns with the objective knowledge is "knowledge of classifications and categories." There is great value if knowledge attained by the workers enables them to understand the different type of PSEs, process safety hazards and contributing causes, and are able to classify and categorize these, then integrate this knowledge into various tasks and situations they encounter. In some instances, the desired knowledge falls within the Procedural Knowledge type. This knowledge type was assessed and compared to Conceptual Knowledge, but it was determined that in most cases the worker needs to understand the situation and potential hazard, and use this knowledge to determine how to best prevent the process safety hazard from becoming a PSE, rather than attain procedural knowledge to apply to the situation at hand.

In summary then, the objective is "Students should learn to integrate past process safety learnings to prevent future PSEs," and this is placed on the taxonomy table intersecting 'Analyze' and 'Conceptual' knowledge.

The end result now is preventing future PSEs which is really the true desire. The process of determining an objective and how it fits within the taxonomy framework yields an increased understanding of what needs to be attained (Ref 1, p5), this statement proven by the careful analysis and development of objective statement within this report.

Now that an objective is created, it is imperative to develop a learning curriculum that is appropriately designed to achieve the objective. The stated objective can be seen to be in cell 4B of the CPC taxonomy table. (Table 4)

It was observed through the interview process, and discussions with various operators, that there is a lack of knowledge, or memory, of past RCA investigations and events that have occurred. This lack of factual knowledge reflects the current state of the organization, and spotting this current state on the taxonomy table positions this current state at 1A – deficient in memory of factual knowledge. This is the initial state, the starting point for the learning activities, which now need to be developed to progress the company from here to the objective in cell 4B.

Knowledge needs to advance from factual to conceptual, while cognitive skills will need to be developed to remember, followed by understand, then apply and finally allow the student to analyze conceptual knowledge.

Table 4 CPC Learning Activities Within the Taxonomy Table

	The Cognitive Process Dimension					
The Knowledge Dimension	1. Remember	2. Understand	3. Apply	4. Analyze	5. Evaluate	6. Create
A.Factual Knowledge	Activity 1: Intended to help students recognize past PSEs and store in long term memory for effective retrieval	Activity 2: Intended to help students understand how past PSEs specific to their job function have developed	Activity 3: Intended to help students implement their understanding of contributing factors.			
B.Conceptual Knowledge		Activity 4: Intended to help students understand the underlying causes of PSEs.		Objective: Students should learn to integrate past process safety learnings to prevent future PSEs		
C.Procedural Knowledge						
D.Meta-Cognitive Knowledge						

3.8 Learning Activities

Learning activities need to be successful to attain the objective. They need to be designed to enhance memory retention, but the success of learning activities is also strongly linked to how effectively the affective domain is tapped into. The affective domain is known to consist of attitude and motivation. According to Edwards and Porter, cited in Miller (Ref 13, p1), "a student's attitude toward a given course or subject area can be a contributing factor to his achievement in it." The ensuing learning activities are designed to grow attitude in the student, increasing the motivation to want to learn and subsequently increase memory retention. Miller, citing Simonson and Maushak (Ref 13, p6) highlights six guidelines that support increasing attitude through instruction. These are quoted here, with comments specific to how these apply to the learning activities developed for CPC, in parentheses:

- "Make the instruction realistic, relevant, and technically stimulating." (Have learning activities comprised of real scenarios, graphically stimulating, relevant to their job, intended to pique interest)
- "Present new information." (Even though utilizing past PSEs, these are new to most. Present the information in a non-typical manner that catches the students attention)
- "Present persuasive messages in a credible manner." (Showing the impact of what has or can really happen to them, in a short, to the point persuasive manner)
- "Elicit purposeful emotional involvement." (Stressing what bad things can really happen to them, appealing to the student's sense of emotion, and what can be done to prevent it)
- "Involve the learner in planning, production or delivery of the message." (The interview process was used to query what was felt the effectiveness of Larkin's overview drawings would be – all queried felt they would be very effective. These drawings are being further enhanced to target the affective domain. Workers that were personally involved in past PSEs were enlisted to ensure the portrayal is accurate and meaningful to others)
- "Provide post-instruction discussion or critique opportunities." (This will be done post-learning activities as part of review and adjust process)

The students should have an inherent motivation to learn as much as possible about past PSEs, because of the personal significance these events can have on them. Still, the affective domain needs to be intentionally targeted with the learning curriculum.

Activity 1 - The first learning activity needs to address the initial concern, that past RCA investigations seem to be unknown to the workforce and, if known at the leadership level, are only retained in memory for a short time unless personally involved at the time of the event. This activity should draw on the knowledge gained from research of effective communication, memory retention and best practices within the industry. The aim at cell 1A is to remember factual knowledge – in this case it can be stated that that the intent is to recognize past PSEs and store in long-term memory for effective retrieval. According to Anderson, (from Table 2), recognizing involves being able to pull relevant information from long-term memory to compare it with current information. Industry best practices highlight the importance of reviewing past PSEs, keeping the memory of these alive and sharing with new employees. Larkin suggests a one page overview drawing that summarizes an incident. (Diagram 1) This is good in that it is organized in chunks, is written simplistically, and utilizes both verbal and visual memories in a manner that forces information presented to pass through the working memory, creating more triggers to allow for recall at a later date when needed. It focuses attention on to the specifics of the incident in a manner that creates an original awareness, and should help to elucidate the hazards associated with the event.

Now, specifically targeting the affective domain can even further enhance memory retention. It is desired to motivate the student, to appeal to his attitude and therefore develop an inherent 'want' to learn the lessons that are available to be learned from the past PSEs. To achieve this, an education company specializing in elearning was enlisted, to reach out to the affective domain of students. Diagram 3 and Diagram 4 are simplified portrayals of past PSEs at CPC. The three-dimensional enhancement and inclusion of speculating "what else could have happened?" should appeal to the affective domain. Then, as the affective domain becomes more engaged, the cognitive domain improves the memory retention.

Drawings such as these are now to be developed for many past process safety events. Various methods of delivery will be utilized to improve usage and retention. Review at Health, Safety and Environment (HSE) meetings, email messaging to mobile phones, as well as postings to HSE web page for increased accessibility of the workforce will all be utilized. The key is to bring back review of old incidents, in a manner that allows the memory to stick in the minds of the workers and to be able to be recalled for comparison to current information workers may be faced with (in working memory) as process safety hazards are encountered. In summary, Activity 1 teaches what has happened, in a manner set out to be remembered.

Diagram 3 - Tank Lid Incident - Final Overview Drawing

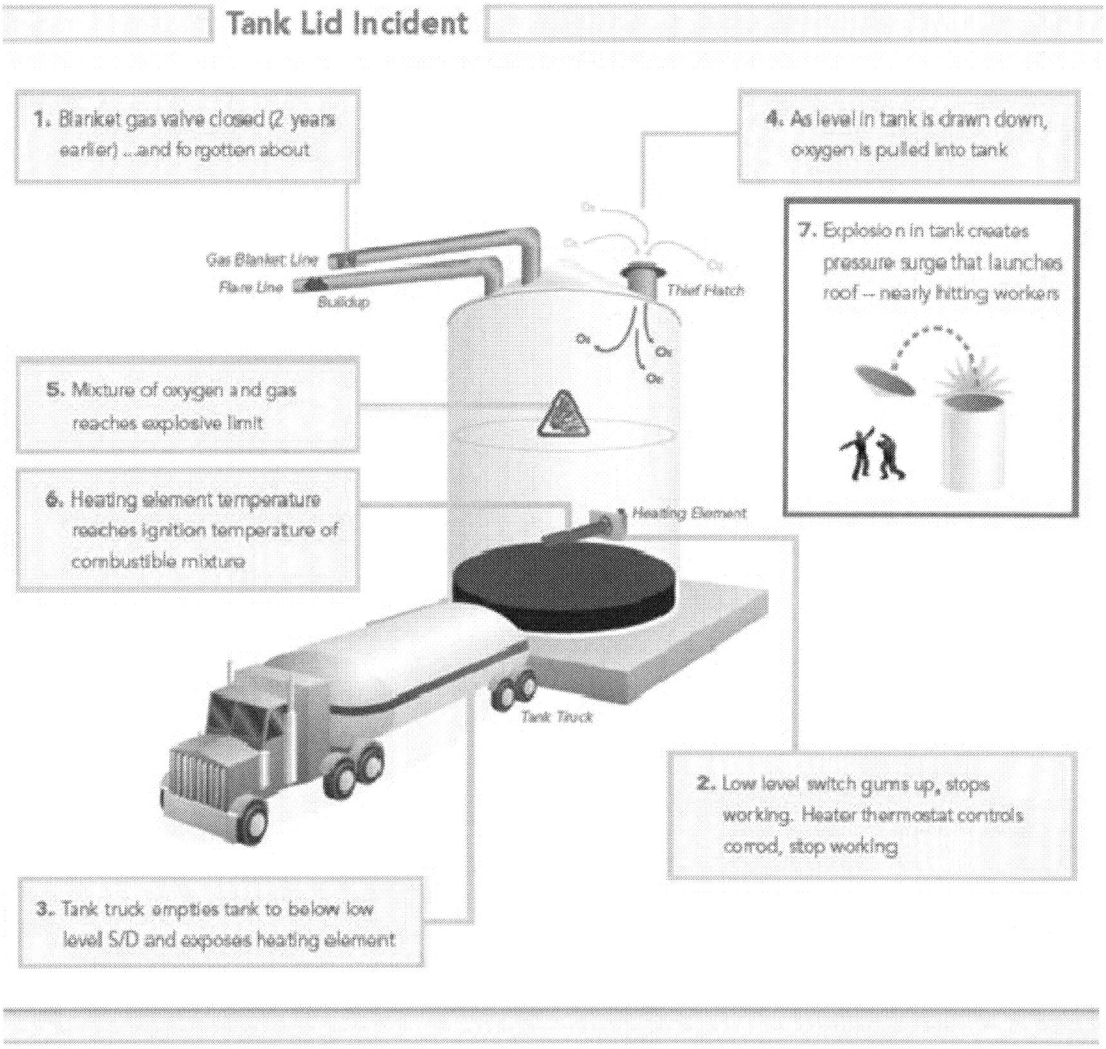

! What Happened

Blanket gas valve closed:

- Wasn't re-opened at end of initial job
- Site personnel never became aware of closed valve

Maintenance practices did not detect:

- Gummed up low level switch
- Corroded heater thermostat controls

Communication with trucking company did not address:

- Site alarms for low tank level
- Hazards of loading from tank with heater

? What Else Could Have Happened

Over time, the presence of oxygen led to a buildup of crud and blockage of the flare line. This could have led to:

- A similar explosion event under different operational conditions releasing large amounts of sour gas
- Thief hatch relieving sour gas leading to personnel knocked down

Diagram 4 - Gas Release During Welding Incident - Final Overview Drawing

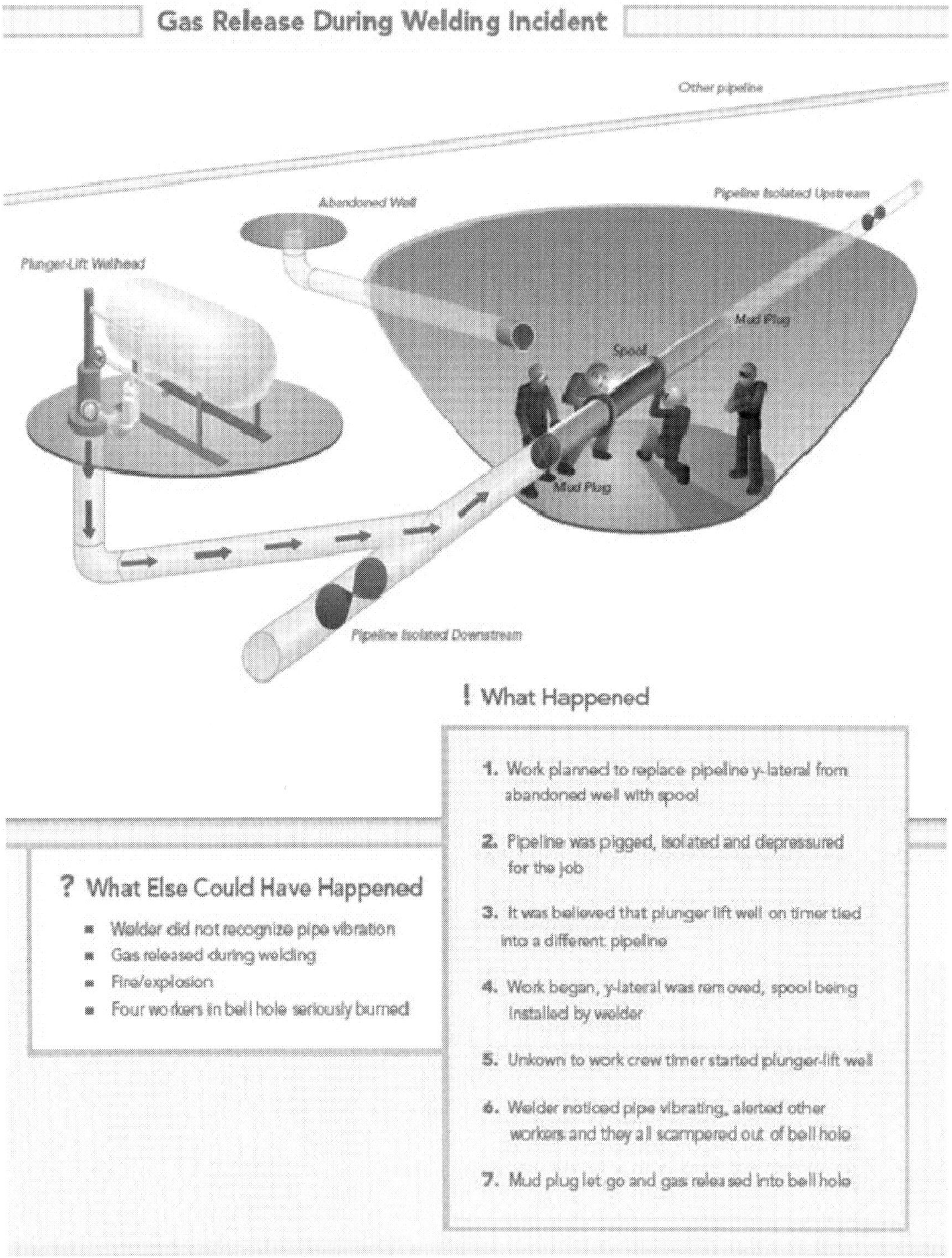

Activity 2 – Activity 1 is the starting point to get the facts into long term memory for future retrieval when needed. The cognitive ability gained from Activity 1 in itself is not sufficient to drive significant improvement in reducing recurring PSEs. To advance knowledge towards the objective, "Students should learn to integrate past process safety learnings to prevent future PSEs", requires the memory to advance from remembering information to having the ability to make sense of this information to solve new problems. The learning needs to advance to meaningful learning, where the knowledge gained is used to solve problems. (Ref 1, p64) This learning activity falls within cell 2A, understanding factual knowledge. It is intended to help students understand how past PSEs specific to their job function have developed. It has been shown that students often solve problems by comparing a present situation to a past situation, and applying the solution of the past to the current state. (Ref 1, p65) Activity 2 drives towards this, by integrating the overview drawings into various processes within CPC. Some examples of how this may be accomplished are now described.

The Construction Inspector group has an on-boarding process for new inspectors, where all new inspectors go through an in-house training program. Through discussions with the supervisor for this group they are planning on utilizing the overview drawings that are applicable to their work within this on-boarding process. In this way, the past PSE is reviewed, and discussion regarding the contributing factors and prevention methods should allow the student to understand the factual knowledge of the event. Diagram 4 is an example that can be implemented. The Inspector will then be well-aware of a PSE that has occurred in the past, and, in the event he is working on a pipeline abandonment project, he can apply the solution of the past to the current state. Also the Abandonment and Operations groups can create a link to this drawing within their Management of Change process, ensuring this PSE is reviewed prior to similar work being done.

Another method of integrating the overview drawings into a process is to link an overview drawing specific to a maintenance activity into an electronic maintenance generation system. Then, when the personnel responsible for the work are preparing for the job, they will review the PSE that has occurred in the past under similar circumstances, talk through the hazards, and determine what they may need to do to prevent a similar event from occurring.

One other example would be to link an overview drawing to a safe operating procedure, and in this way include review of a past PSE related to similar work as part of the procedure review.

Activity 2 should lead the students to understand factual knowledge as the PSEs are integrated into various processes. To summarize Activity 2, it teaches what to prevent.

In essence, Activities 1 and 2 utilize the same tool, the overview drawings, but Activity 2 implements them within a process. Activity 1 is introducing the students to the various PSEs through the summary drawing reports, while Activity 2 involves a deeper understanding as the students are now developing classifications and understanding the different contributing causes.

These activities build on the three important methods that enhance learning, and were referenced in section 3.6. (Ref 5, p11)

1. Learning is improved when existing knowledge is built upon to create new concepts. (The activities 1 and 2 are building on the workers existing knowledge of the process.)

2. Learning is improved when the learning topic has a direct personal meaning to them. (The activities 1 and 2 show the workers how past events have impacted others just like them, in situations they have or will encounter. They have appealed to the affective domain.)
3. Learning is improved when students are forced to use existing knowledge to learn from failures and attempt to predict outcomes. (The activities 1 and 2 build on prior knowledge and past failures, speculating forces the workers to predict outcomes.)

Activity 3 – This activity should now lead to applying the factual knowledge gained. It involves helping the students understand that they have the tools available to prevent PSEs, and implement their understanding of the contributing factors as part of this prevention. A summary course outlining the learnings from study of the company and industry PSEs, as well as information gathered from the interviews is to be developed. This will focus on the top process safety hazards that have contributed to PSEs at CPC, and the key prevention techniques available to the workforce.

The course will be built around the following information:
A. Top process safety hazards that have contributed to CPC PSEs (in no particular order) and lessons learned:
1. Pipeline Strikes.
2. Liquids that can freeze and expand in your pressure equipment.
3. Atmospheric tanks and overpressure events.
4. Pressure Relief Valve and blowdown piping subject to reactionary forces at relief.
5. Pressured fittings and attachments.
6. Preparation of equipment for maintenance.
7. Operation of Pig traps.
8. Plunger lifts.
9. Hydrates.
10. Water Hammer.

B. Key prevention techniques
1. Procedures.
2. Hazard Recognition.
3. Management of Change and Pre-Startup Safety Review
4. Personal Safety Involvement.
5. Learning from the past.
6. Sense of vulnerability.

Activity 3 will also utilize the key learnings to promote memory retention, including the concept of chunking, writing to a lower grade level, using both verbal and visual memory cues, etc. The training will be designed to appeal to the affective domain and to enhance memory capture. In summary, Activity 3 teaches ways to prevent PSEs. Activity 3 also begins to transcend the student's knowledge from factual towards conceptual.

Activity 4 – The previous three activities have been dealing primarily with factual knowledge. The intent of activity 4 is to develop an understanding of conceptual knowledge, or how to think about the operation in a manner that avoids the initiation of the contributing factors in the first place. To best learn from past process safety learnings and prevent the recurrence of PSEs, a solid understanding of process safety and a culture that strives to manage the hazards is paramount. Activity 4 is designed to help the students understand what process safety is, the underlying causes that contribute to PSEs, and what each worker needs to do to help prevent incidents. Gaining an understanding of conceptual knowledge, and learning how to apply this wisdom is the aim of this activity. A process safety awareness course, strategically designed to utilize the knowledge gained regarding effective communication and memory retention will achieve this.

This activity touches on many cells within the taxonomy table, but lies primarily in cell 2B - understand conceptual knowledge. However, it should also have a means of applying conceptual knowledge. Activity 4 teaches how to avoid. This activity will be developed with assistance from a training company that can help to integrate all the knowledge gained throughout this study.

When the learning activities are added to the taxonomy table, it can be seen that there are no empty rows or columns as the process advances from initial state to the objective. This is good, as an entire row or column that is vacant may indicate an alert and other learning activities may need to be added. (Ref 1, p7)

As these four learning activities are completed it should lead to attainment of the objective, which falls in cell 4B for analyze conceptual knowledge. The objective was "Students should learn to integrate past process safety learnings to prevent future PSEs".

As the taxonomy table was utilized to analyze the learning activities, it can be seen that the objective cannot be met with the development or utilization of only one tool, but will require several tools. This is in the form of a learning curriculum made up of different learning activities, that each help to increase knowledge and cognitive processes to the point where the objective can be attained.

X Conclusion

Not adequately learning from lessons of past process safety events has proven to be a critical issue that hampers organizations within the energy industry. This has led to recurring process safety events that often arise from very similar contributing causes. These contributing causes are typically not new learnings to investigators but rather a regurgitation of what was previously learned through past events.

Best practices to improve memory of process safety learnings are available, and yet the problem still exists. The question is therefore asked; is something missing within these suggestions for improving memory within industry, and within CPC?

It was determined the problem is not that we don't know what causes these accidents, but instead that the lessons learned are not taught in an effective manner that allows the memory to be retained within the organization. Lessons learned need to be taught in a way that intentionally targets memory retention. To achieve this, knowledge of how memory works needs to be combined with effective, proven communication techniques. This then should be applied in concert with proven learning methods to achieve a learning that sticks in memories of the workforce. The field of education is a leader in learning methods, and one tool that has proven as an effective aid is the taxonomy table. This and the methodology behind it were used to construct learning activities that built on the knowledge gained from the research of the following:

- Industry PSEs.
- CPC PSEs.
- Industry best practices regarding process safety lessons learned.
- Memory development and retention.
- Effective safety communication.

The primary learning activities developed through this process were:
1. A one page overview drawings of past process safety events.
2. Process safety awareness training, developed through the combined learnings of past events from industry and CPC.

It is imperative that these learning activities are constructed to develop both the affective and cognitive domains. Specifically, they need to target memory retention, using the knowledge gathered from research of memory and effective communication. Then, an organization must build these learning methods into processes that allow the memories to remain fresh throughout the future and effectively taught to new workers. With time, an organizational culture will be developed that is more aware of process safety and vigilant to what can be done to avert process safety events.

X+1. References (examples given to show format)

1. Anderson, L., Krathwohl, D., Airasian, P., Cruikshank, K., Mayer, R., Pintrich, P., Raths, J. and Wittrock, M. (2001) A Taxonomy for Learning, Teaching, and Assessing, a Revision of Bloom's Taxonomy of Educational Objectives, Abridged Edition, New York: Addison Wesley Longman Inc.

2. Brink, Julie (2010) Improve E-Learning with Chunks and Bits, trainingmag.com. Available from http://trainingmag.com/article/improve-e-learning-chunks-and-bits (Accessed 29 May 2012)

3. Center for Chemical Process Safety (2012) Recognizing Catastrophic Incident Warning Signs in the Process Industries, New Jersey, John Wiley & Sons, Inc.

4. Cherry, Kendra. (2012) 'Explanations for Forgetting', About.com Psychology (weblog). Available from http://psychology.about.com/od/cognitivepsychology/tp/explanations-for-forgetting.htm (Accessed 24 April 2012)

5. Cisco (2008) Multimodal Learning Through Media: What the Research Says (online). San Jose,CA:Cisco. Available from http://www.cisco.com/web/strategy/docs/education/Multimodal-Learning-Through-Media.pdf (Accessed 15 April 2012)

6. Gerstein, Marc and Ellsberg, Michael (2008) Flirting With Disaster, New York: Sterling Publishing Co., Inc

7. Hopkins, Andrew (2000) Lessons From Longford, Sydney, Australia: CCH Australia Limited

8. Kletz, Trevor (2003) Lessons From Disaster, How organisations have no memory and accidents recur, Rugby, Warwickshire, UK: Institution of Chemical Engineers

9. Larkin, Dr TJ and Larkin, Sandar (2012) Communicating Safety, The Silver Bullet, 2nd ed., Larkin Communication Consulting. Available from www.Larkin.Biz (Accessed May 3, 2012)

10. Larkin, Dr TJ and Larkin, Sandar (2007) You Know Safety But Admit It.....You Don't Know Communication, Larkin Communication Consulting. Available from www.Larkin.biz (Accessed May 3, 2012)

11. Lorayne, Harry and Lucas, Jerry (1974) The Memory Book, New York: Random House Publishing Group

12. Mannan, Dr., PE, CSP Sam, (2005). Lees' Loss Prevention in the Process Industries, Hazard Identification, Assessment and Control (Third Edition) Elsevier Inc.

13. Miller, M. (2005) Teaching and Learning in Affective Domain. In M.Orey (Ed.), Emerging perspectives on learning, teaching, and technology. Available from http://projects.coe.uga.edu/epltt/ (Accessed 27 July, 2012)

14. Park, Derek (2012) Oil & Gas IQ, Never Say Never Again (online). Available from http://www.oilandgasiq.com/integrity-hse-maintenance/white-papers/special-report-macondo-never-say-never-again/ (Accessed May 14, 2012)

15. Perry, Robert H. and Green, Don W. (1997) Perry's Chemical Engineers' Handbook, 7th ed., New York: McGraw-Hill

16. Webster's Desk Dictionary of the English Language, (1990) New York: Portland House.

9TH GLOBAL CONGRESS ON PROCESS SAFETY

Black Swans, White Swans and Fifty Shades of Grey: Remembering the Lessons Learned from Catastrophic Process Safety Incidents

John F. Murphy, PE
Process Safety Services
2304 Kenya Lane
Punta Gorda, FL 33983
hamjfm@embarqmail.com

James Conner
Eiger Consulting, LLC
15906 El Dorado Oaks Dr.
Houston, TX 77059
jim.conner8@att.net

Prepared for Presentation at
American Institute of Chemical Engineers
2013 Spring Meeting
9th Global Congress on Process Safety
San Antonio, Texas
April 28 – May 1, 2013

UNPUBLISHED

Black Swans, White Swans and Fifty Shades of Grey: Remembering the Lessons Learned from Catastrophic Process Safety Incidents

John F. Murphy, PE
Process Safety Services
2304 Kenya Lane
Punta Gorda, FL 33983
hamjfm@embarqmail.com

James Conner
Eiger Consulting, LLC
15906 El Dorado Oaks Dr.
jim.conner8@att.net

Keywords: Black Swan, Gray Swan, White Swan, lessons-learned, corporate memory,

Abstract

Last year at the 8th Global Congress for Process Safety we presented a paper entitled, "Beware of the Black Swan: The Limitations of Risk Analysis for Predicting the Extreme Impact of Rare Process Safety Incidents." The paper discussed the difficulties of predicting black swan events -- rare but catastrophic occurrences which continue to happen in the chemical processing industries despite the focus that has been brought to process safety over the past 30-40 years. Once black swan events occur, after investigations and development of lessons learned, they become white swans. By white swans we mean these type of events become more predictable and as we do hazard identification and risk analysis we become cognizant of the similar potential scenarios in new plants and plant modifications.--at least for a while. This paper discusses how white swans slowly become greyer as time goes on if the lessons learned from black swan events are not kept fresh. Examples of black swans that become white swans and then become grey as time goes on will be discussed. Ideas of how to keep the white swans from becoming grey with time, including the key role of Management, will also be discussed.

1. Introduction

As we discussed in last year's paper, "Beware of the Black Swan: The Limitations of Risk Analysis for Predicting the Extreme Impact of Process Safety Incidents," the concept of the black swan is not new. The Greek philosopher Aristotle first used the term white swan as an example of necessary relations and the term black swan as the expression of the improbable.[1] In his recent book entitled "*The Black Swan, The Impact of the Highly Improbable,* Nassim Taleb discusses the extreme impact of rare and unpredictable events (black swans) and the human tendency to find simplistic explanations of these events after they occur. According to Taleb the black swan event has three attributes: "First is an outlier, as it lies outside the realm of regular expectations. Second, it carries an extreme impact (unlike the bird). Third, in spite of its outlier status, human nature leads us to concoct explanations after the fact, making the previously unimaginable event to be both explainable and predictable." [2] The paper goes on to illustrate examples of black swan events that have occurred in the chemical processing industries including their catastrophic impacts. The major points of the paper were (1) that although black swan events are unpredictable, we must continue to search for them, and (2) we must learn from black swan events that occur so they do not occur again. Some ideas on preventing black swans were offered including more imaginative process hazard analysis, installing safeguards to prevent low perceived probability, high consequence incidents, learning from near miss events, improving process safety management systems, implementing inherently safer technology when feasible, and maintaining a sense of vulnerability. [3]
This paper discusses how we often do not learn from black swan events, and even where we do learn from those events, we often lack effective systems for keeping the lessons-learned fresh and alive over the long term.

2. Black Swans, White Swans, and Fifty Shade of Grey

All too often we in the Chemical and Refining Industries do not retain the lessons learned from Black Swan events. A Black Swan event was not foreseen, and in many cases believed to not be possible. But in the aftermath of a Black Swan event, root cause analysis will uncover the previously unforeseen pathway to the event. Once the pathway is known, the Black Swan becomes a White Swan. Going further, we can now use the previously unforeseen pathway to help us recognize the potential for similar events as we conduct hazard identification and risk analysis on new processes and process modifications. The Black Swan pathway plus the recognition of similar pathways constitute some of the most valuable lessons-learned from a Black Swan event. Unfortunately, these valuable lessons-learned are often forgotten over time and the white swan becomes greyer and greyer unless efforts are made to keep the lessons-learned fresh.

A classic example illustrating this concept begins with the 1974 vapor cloud explosion that occurred in Flixborough, UK and the lessons learned about the siting and blast-resistance of control rooms, laboratories, offices, and other occupied buildings located within or adjacent to high-hazard chemical plants/refineries.

On the afternoon of Saturday, June 1, 1974 a catastrophic vapor cloud explosion occurred at the Nypro (UK) Ltd. chemical plant in Flixborough, UK. Twenty-eight employees were killed and 36 injured, and the plant was totally destroyed. There were an additional 56 injuries in the local community coupled with significant damage to homes and businesses. This was a black swan event because up to this time the potential for such a devastating impact of a vapor cloud explosion had not been realized. After a thorough investigation by the government, the technical cause of the incident was determined to be the failure of an inadequately engineered bellows expansion joint and the release of approximately 30 tons of cyclohexene, a flammable gas that subsequently ignited resulting in a vapor cloud explosion[4]. As with most major incidents there were a number of significant root causes (failure of several management systems including hazard evaluation, management of change, and asset integrity) and many lessons learned. For the illustration of the grey swan concept we will focus on two of the lessons learned.

Of the 28 people that were killed, 18 were in the control room. The lesson s-learned were that control rooms and other occupied buildings must be kept away from major hazards (such as large quantities of flammable materials) and where this is not feasible, such buildings must be constructed such that if an explosion does occur, the building will withstand the blast and not put people at risk who are inside the building. At the time of this incident, John worked at Dow and Dow's response to Flixborough was to develop engineering guidelines (called Loss Prevention Principles), that offered calculation methods to estimate the consequences of vapor cloud explosions and criteria for designing control rooms and other occupied buildings that could be affected by such explosions. Separation distances were also recommended from process hazards to control rooms and other occupied buildings. Other major chemical companies took similar action. The chemical process industry did recognize and learn the lesson of occupied buildings from Flixborough, turning the Black Swan of the hazards of occupied buildings into a White Swan. But over time, for a variety of reasons, the white swan turned grey for many companies in the chemical process industries.

On the afternoon of October 23, 1989, a massive explosion and a major fire occurred at the Phillips Chemical Company at Pasadena, Texas. The consequences of this event included 23 fatalities, 130 injuries, and major destruction of the involved plant. After thorough investigation, the immediate technical cause of the incident was determined to be the inadvertent release of approximately 40 tons of flammable material during a maintenance activity. The flammable cloud ignited, resulting in a catastrophic explosion and fire. The control room was totally destroyed and workers inside the control room were killed. The root causes of the incident were failures in management systems including hazard assessment, plant layout and separation distances, location and construction of control rooms, and many others.[4] Why had the lessons learned from the 1974 Flixborough incident of the importance of locating control rooms away from major inventories of flammable materials and of constructing occupied buildings to protect the occupants not been applied? In their investigation report on the Phillips incident, OSHA made the following observation: "Overall, the US reaction to the Flixborough event was minimal".[5] The white swan of Flixborough lessons-learned had turned dark grey. But now that we had experienced both Flixborough and Phillips, surely the white swan would remain white and not again go back to grey.

On March 23, 2005 an explosion and major fire occurred during a start up after a turnaround in the ISOM Unit of BP's Texas City, Texas refinery. The investigation determined that the immediate cause was a process upset releasing approximately 13,000 pounds of flashing liquid hydrocarbon out of an elevated release stack in a manner described by witnesses as a "geyser". The liquid fell to grade and created a large vapor cloud that was ignited by a nearby pickup truck which had been left with its engine idling. The fire and explosion resulted in 15 contract employee fatalities, 170 injuries and major property damage. All of the fatalities (and many of the injured) occurred inside a group of temporary wooden trailers occupied by contract employees working on the turnaround of a completely unrelated unit.[4] These occupied temporary buildings (the trailers) were located too close to the process hazard and not constructed to protect the occupants. The white swan had become dark grey again.

---

3. Keeping the White Swan White

How can we keep the white swan white? To answer this, we must address those issues which are known to be reasons why organizations fail to capture learnings from incidents, or when they do capture them, why they forget them over time.

Drive incident investigations to latent root cause

Over the years the chemical and refining industries have learned that determining the immediate technical causes of an incident is essential but is not sufficient to capture the learnings needed to prevent recurrence. Our experience with incident investigations in Dow, in Celanese, and with the Chemical Safety Board has convinced us that root causes of all major process safety incidents are management system deficiencies. To eliminate repeat incidents we must understand and correct both the technical causes as well as the deficiencies in our Process Safety management systems. Deficiencies in management systems are often referred to as latent root causes. In the examples discussed above, the technical issues of separation distances between occupied buildings and process hazards as well as the appropriate construction of occupied buildings located in high-hazard areas were not addressed because of inadequate management systems to identify process hazards and implement safeguards to manage the risk. These management system failures were present in Flixborough, but the Flixborough incident is most commonly remembered as a failure of Management of Change. Without good investigation to uncover all root causes, and sharing of those lessons-learned, the Black Swan never turns white, but will remain grey.

Keep the lessons-learned alive through story-telling

Psychologists have determined that people learn best by listening to stories rather than trying to remember isolated facts. Once a black swan event happens, it can be remembered as a white swan through story-telling. Many major chemical companies have had black swan events that have been told and retold in an effort to remember the lessons-learned from these events. John was a member of the Reactive Chemical Committee at Dow's Michigan Division and the Chair of the Reactive Chemicals

Committee at the Dow Freeport Texas site. The purpose of the Reactive Chemical Committees at Dow was to review all new process chemistry and modifications to existing process chemistry to ensure that proper safeguards are in place to prevent future reactive chemical incidents. All members of the Reactive Chemicals Committee and process safety organization are told the story of how the Dow Reactive Chemical Program came to be.

On December 12, 1966 at the Dow plant in Freeport, Texas, propargyl bromide and chloropicrin were being blended with a mixture of dichloropropane and dichloropropene in a tank car when one of the tank cars exploded, causing the other tank car to explode. The combined explosions left a large crater, 87 ft. long by 60 ft. wide by 11 ft. deep. The blast caused serious damage to a wooden wall building and a concrete building. Two tank cars, one tank truck and a length of railroad were destroyed. There were three fatalities; two operators and a truck driver.

The energy release of the explosion was estimated as the equivalent of between 6.6 and 14 tons of TNT. Much of the evidence was destroyed, but it is thought that a valve closed causing a gear pump to run dry igniting a mixture of chloropicrin and propargyl bromide. The immediate cause of the accident was the mixing of chemicals in equipment not designed for this process and probably a pump running hot and dry. A contributing cause of the incident was a lack of understanding of the chemistry involved by the people supervising and doing the work.[6]

The story has been told again and again at Dow in an attempt to keep the white swan white. Dow's Board of Directors at the time established the Reactive Chemical Program to prevent future incidents. The program is part of the Dow Chemical's process safety program and continues to be supported by top management illustrating Dow's commitment to a strong positive process safety culture.

At the 2004 Loss Prevention Symposium, Pete Lodal presented another story that has been told and retold at the Eastman Chemical Company. On October 4, 1960 Tennessee Eastman's aniline manufacturing plant exploded resulting in 16 fatalities, 48 hospitalized, and more than 250 people requiring medical treatment. The October 4 explosion occurred in the nitrator - the base of the nitrobenzene column - and the associated thermosyphon reboiler and sludge evaporator. Nitrobenzene concentration in the system had increased to the point that it penetrated the single-phase, shock sensitive boundary. An unknown disturbance such as a valve closing or pump starting caused the mixture to detonate resulting in a fire ball of 150 ft. with a TNT equivalency estimated as six tons.[7]

Eastman instituted its Process Safety Review Committee structure immediately after this event which continues to this day. Each site has a committee chaired by the site's highest ranking official that meets monthly to review process safety and reactive chemistry issues at the site.[6] This organizational structure has kept many white swans white with an order of magnitude reduction in the rate of serious incidents that has been sustained for decades. Pete has told this story many times in an effort to keep the swan white and maintain a sense of vulnerability in the company.

There are a number of excellent resources available to companies where past incident "stories" are captured.

- The textbooks by Trevor Kletz (*What Went Wrong* and *Still Going Wrong*) are excellent reading for new engineers and in control rooms.[8,9]

- The Chemical Safety Board has done an excellent job with storytelling with its animated videos describing the incidents and lessons learned from their investigations. The videos are available for download at www.csb.gov.

- CCPS has also done an excellent job telling stories with the publication of The Beacon each month in *Chemical Engineering Progress.* The Beacon is a one page incident description and lessons-learned story utilizing pictures and diagrams targeted at the operations level. It is meant to be posted in control rooms to keep the story alive and keep the swan white. You can also subscribe to The Beacon by contacting CCPS.

- CCPS has also published a book entitled *Incidents that Define Process Safety.* This resource is excellent for all process safety and manufacturing people.

- The Case Histories session of the Global Congress on Process Safety has attempted to keep the white swans white for many years. The Eastman story discussed above was presented by Pete Lodal at the Case Histories session of the 2004 Loss Prevention Symposium.

Implement practices and systems to surface PSM incidents which need to be investigated ….. particularly near-misses

Companies need to establish clear definitions for Process Safety incidents and implement effective systems for investigation. It is critical that these definitions clearly spell out near-misses, and that investigations are done thoroughly on near-misses. Many companies have excellent systems for reporting and tracking of personnel injuries and environmental excursions. It is common to find companies that limit their capture of process safety events to fires and explosions, missing the tremendous learning opportunity that exists in their near-misses for process safety events.

In the 10 years that preceded the 2005 BP Texas City ISOM vent stack release and explosion, records show that there were 10 release incidents involving the vent stack. Eight of these ten incidents were serious enough to result in ground-level vapor clouds or flames at the point of venting. But only two of these incidents had been investigated as a safety concern, and the BP organization was unaware of this pattern of incidents on the vent stack until after the explosion, when CSB investigators pieced it together from a variety of sources.[10]

In his book regarding the BP Texas City event entitled "Failure to Learn", author Andrew Hopkins commented "Had this historical record been assembled and publicized, it would have been clear that it was only a matter of time before one of these incidents culminated in a major accident". He further went on to state "Near misses are often described as free lessons, but it takes organizational commitment to ensure that such lessons are remembered and embedded into operations".[10]

To help companies establish definitions for meaningful PSM near-miss events, guidelines for PSM reporting metrics, which include definitions for near miss events, have been published by the Center for Chemical Process Safety (*Guidelines for Process Safety Metrics*) and by the American Petroleum Institute (*Recommended Practice 754: Process Safety Indicators for the Refining and Petrochemical Industries*).[11,12] Companies which adopt these or similar comprehensive systems for identifying PSM incidents, and then investigating those incidents to capture lessons-learned will have a much better chance of capturing white swan lessons-learned, and then keeping those lessons-learned from later turning grey.

Implement effective systems for broad sharing of lessons-learned

Companies must have effective systems to share lessons-learned to all parts of the organization where they might be applicable. Many companies can show effective sharing of an incident within the fence of the location where the incident occurred, but fewer can show effective sharing of the same incident to other company operating units. And fewer still can show effective systems for capture and sharing of lessons learned from applicable incidents in other companies. Let's look at a tragic and recent example.

> In April 2010, the Macondo well in the Gulf of Mexico suffered a blowout. At the time of the blowout, the crew was replacing mud in the well riser with seawater, in preparation for moving the rig to a new location. The well blowout was attributed by the investigation to an inadequate cement job coupled with a decision to not perform a cement-evaluation test after cementing was completed. The driller for the Macondo well was Transocean, Ltd., the owner of the well was BP, and Halliburton was contracted to assist with cementing of the well. The blowout of the Macondo well was preceded by three remarkably similar incidents where black swans were turned to white, but then not effectively shared - missed opportunities to capture learnings from other companies, as well as within the same company:[13]
>
> 1. Four months prior to the Macondo blowout, a well operated by Shell in the North Sea suffered a near blowout while replacing the riser mud with seawater in preparation for moving the rig. The drilling company for the well was none other than Transocean. The seriousness of the incident is reflected in the fact that the rig was ordered abandoned during the

emergency. Fortunately in this incident, the blowout preventer (BOP) on the ocean floor successfully activated and prevented a disaster (the BOP failed to activate on the Macondo blowout). Shortly after the event, Transocean developed a PowerPoint presentation on the lessons-learned and issued an operations advisory within the North Sea region. But these lessons-learned and advisories did not reach the Transocean crew on Macondo. An opportunity that might have prevented Macondo was lost.[14]

2. Eight months prior to the Macondo blowout, the worst offshore platform spill in Australia history occurred when the Montara well suffered a blowout. The operator for the well was PTTEP Australasia. As with Macondo, Halliburton had been contracted to assist in the cementing of the well. The investigation determined that the cementing of the well had not been done properly, and a critical test of the cement job had been bypassed – a sequence of events strikingly similar to what happened on Macondo. But while this incident occurred months prior to Macondo, the official report and conclusions as to what caused the Montara blowout were not available to the Macondo team as they were not issued until after the Macondo blowout (i.e. this was not a failure by BP in any way, but rather a missed opportunity by the entire industry, impacted by the manner in which the involved parties shared information during the investigation process. The official investigation report strongly criticized PTTEP for withholding of information).[15] Once again, an opportunity that might have prevented Macondo was lost.

3. Eighteen months prior to the Macondo blowout, BP suffered a blowout on a well in the Caspian Sea. The blowout was attributed to a failure of the cement job, but the lessons-learned were "not assimilated by BP engineers … in the Gulf of Mexico". Why they did not is the subject of much speculation, but for whatever reason, another opportunity that might have prevented Macondo was lost.[16]

Three major incidents prior to Macondo with root causes strongly similar to Macondo, and all of which involved one of the three companies involved in Macondo. But the lessons-learned (capturing the white swan) did not make it to Macondo, the swan stayed gray, and one of the world's worst offshore incidents happened when it could have been prevented.

Implement practices and systems to minimize "corporate memory loss" due to changes in personnel

Trevor Kletz wrote that "… organizations have no memory. Only people have memories and after a few years they move on, taking their memories with them".[17]

Change in personnel is inevitable, whether it be through retirements, or transfers within or outside the company. This constancy of change, with its impact on process safety expertise and particularly on corporate PS memory at the point of need, presents a challenge for all companies. In 2001, member companies of the Center for Chemical Process Safety participated in a series of roundtable discussions and surveys which were designed to gauge the level of concern with loss of corporate memory in the Process Safety arena, and the extent to which companies were having success in addressing it. Among the conclusions in the final report[18], were:

- 15% of survey respondents said they have seen an internal investigation that pointed to loss of company PSM knowledge as a causal factor or root cause in the incident.
- The problem is getting worse and not better.
- Many companies have tried various approaches to stem the loss of process safety knowledge and expertise.
- Very few companies have formal, documented management systems to deal with aspects of prevention of corporate process safety memory loss.

While the focus of this effort was PS knowledge in general, clearly one of the key elements in PS knowledge is learnings from past incidents – learnings which are key to keeping swans white and not allowing them to turn grey.

Trevor Kletz – supplementing his quote above – added that "Ultimately, accident prevention depends on educating people at all levels of an organization about previous accidents and how they occurred".[19] There are many ways to approach this. We will share the example of three companies which are addressing this issue of educating key leadership in strikingly similar fashion.

- DuPont has a system by which line supervision newly assigned to chemical processes involving hazardous materials must develop and demonstrate a minimum degree of proficiency within a period not to exceed 3 months.[18]
- Dow implemented its Reactive Chemical Program in 1967. One of the key elements in the program is the requirement that all newly assigned leaders in manufacturing demonstrate understanding and competency in reactive chemistry in their area of responsibility within 90 days of their appointment. This demonstration of competency is made via an oral review conducted by a selected panel of process experts, peer leaders, and higher level managers. Included in the review is a demonstration of knowledge of learnings from past incidents.[20]
- Celanese began its New Leader Training program in 2010. Similar to Dow's program, it requires competency demonstration within 90 days of appointment by going before a review board of process experts, peer leaders, and higher level managers. Scope includes certification of unit leaders (like Dow) but also includes new Site Managers/Directors and Operations Vice-Presidents. Content for the certification is, for the person's scope of responsibility, hazards of the process, layers of protection for the hazards, and management controls to insure integrity of those layers of protection. Past incident learnings are a key part of the review.

Note that all three of these companies have gone beyond simply teaching leaders – they have incorporated a requirement of demonstration of competence (i.e. certification).

Certification of leaders is not common in our industry. But with the growing evidence that severe process safety incidents are strongly rooted in management system failures, these companies would appear to be leading industry on this issue. Clearly, the programs described insure that leaders are sensitized to the white swan learnings from the past before they are in a position to make operating unit decisions ----- an important step in helping white swans not move to grey.

Incorporate lessons-learned into guidelines, standards, and compliance requirements at the company, industry, and national levels

Company and industry policies and standards are essential elements in capturing key learnings from incidents and disseminating those learnings which will keep white swans from turning gray. One reason why the lessons learned about control room and occupied process building location and construction were forgotten was that there were not adequate industry standards and guidelines developed to capture these lessons. CCPS published its first edition of *Guidelines for Evaluating Process Plant Buildings for External Explosions and Fires* in 1996, before the BP Texas City incident describe earlier, but apparently these guidelines were not thought to apply to temporary buildings and trailers. A few companies developed appropriate internal standards and guidelines for occupied buildings, but most did not. The recently published CCPS *Guidelines for Evaluating Process Plant Buildings for External Explosions, Fires, and Toxic Releases, 2nd Edition* and API's Recommended Practices RP752 (*Management of Hazards Associated with Location of Process Plant Permanent Buildings*) and RP753 *(Management of Hazards Associated with Location of Process Plant Portable Buildings*) are now available and will play a large role in preventing the occupied building white swan from again turning gray.[21,22,23]

And in a similar role, national standards exist to set a minimum level of requirements for operation of petrochemical and refining plants. These standards (e.g. the Process Safety Management Standard, CFR 1910.119 in the United States, and the Seveso Directives in the European Union) were developed based on a perceived need to capture key learnings from major petrochemical events in a manner that would require adherence to practices judged to be capable of preventing recurrence – another system for keeping white swans from turning grey.

4. *Conclusions:*

Black swan events become white swans as lessons are learned from root cause investigation and, for a while, we recognize the potential for similar events as we conduct hazard identification and risk analysis. But unless efforts are made to insure all root cause lessons are captured, and that these lessons-learned are kept fresh, the white swan becomes grey and the potential for similar incidents increases. We can keep the white swans from turning gray through a combination of techniques:

- ***After the black swan event, drive incident investigations to capture all root causes***

- ***Keep the lessons-learned alive through story telling***
- ***Implement practices and systems to surface PSM incidents which need to be investigated ….. especially near-misses***
- ***Implement effective systems for sharing of lessons-learned***
- ***Implement practices and systems to minimize "corporate memory loss" due to changes in personnel***
- ***Incorporate lessons-learned into guidelines, standards, RAGAGEPs and compliance requirements at the company, industry, and national levels***

These suggestions are not a complete answer to making sure white swans are captured and kept white, and it is likely that your organization has found other means to address this issue that work well and can be included in this list. In the end, it is not important how we keep White Swans from becoming Gray Swans --- only that we do it, and do it effectively and continuously. As seen through our examples, this is a significant issue facing our industry. Our hope is that we have given you some ideas which might help in your situation. And we welcome any ideas for improving on this work.

References

[1] "The Black Swan Theory," Wikipedia, 2011

[2] N. N. Taleb, *The Black Swan: The Impact of the Highly Improbable,* Random House Digital, New York, 2010.

[3] John F. Murphy, Jim Conner, *Beware of the Black Swan: The Limitations of Risk Analysis for Predicting the Extreme Impact of Rare Process Safety Incidents,* **Process Safety Progress,** Volume 31, No. 4, December 2012.

[4] CCPS, *Incidents that Define Process Safety,* [2008], AIChE, New York.

[5] Elizabeth Dole, Report to the President "*Phillips 66 Company Houston Chemical Complex Explosion and Fire*", U.S. Dept. of Labor, Occupational Safety and Health Administration, April 1990

[6] John F. Murphy, *Dow's Reactive Chemicals Evaluation Process*", Process Plant Safety Symposium, October 26-27, 1998, Houston, Texas.

[7] Peter N. Lodal, *Distant Replay: What Can Re-Investigation of a 40-year-old Incident Tell You?,* **Process Safety Progress,** Volume 23, No. 3, September 2004.

[8] Trevor Kletz, *What Went Wrong?: Case Histories of Process Plant Disasters and How They Could Have Been Avoided,* 5th ed (2009) Butterworth-Heinemann/IChemE

[9] Trevor Kletz, *Still Going Wrong! Case Histories of Process Plant Disasters and How They Could Have Been Avoided,* (2003) Gulf Professional Publishing

[10] Andrew Hopkins, *Failure to Learn: the BP Texas City Refinery disaster,* (2009), CCH Australia, Ltd.

[11] CCPS, *Guidelines for Process Safety Metrics [2009], AIChE, New York*

[12] API Recommended Practice, *AP 754: Process Safety Indicators for the Refining and Petrochemical Industries,* [2010], American Petroleum Institute, Washington D.C.

[13] Andrew Hopkins, *Disastrous Decisions: The Human and Organizational Causes of the Gulf of Mexico Blowout,* (2012), CCH Australia Ltd.

[14] Robbie Brown, *After Another Close Call, Transocean Changed Rules,* **The New York Times,** 16 August, 2010
[15] David Borthwick, *Report of the Montara Commission of Inquiry,* (June 2010), Commonwealth of Australia
[16] Tim Webb, *BP suffered blowout on Azerbaijan gas platform*, **The Guardian**, 16 December 2010
[17] Trevor Kletz, *Still Going Wrong! Case Histories of Process Plant Disaster and How They Could Have Been Avoided*, (2003) Gulf Professional Publishing
[18] Arendt, J.S. "*Retaining Corporate Process Safety Memory*", unpublished paper, May 2001 Technical Steering Committee Workshop-Denver, Center for Chemical Process Safety, March 2003.
[19] Andrew Hopkins, *Failure to Learn: the BP Texas City Refinery disaster*, (2009), CCH Australia, Ltd., page 8.
[20] Kenan Stevick, "*Reactive Chemical / Process Hazard Analysis: New Leader Review Process*", 5th International Forum for Work Safety, September 02, 2010, Beijing, Peoples Republic of China
[21] CCPS, *Guidelines for Evaluating Process Plant Buildings for External Explosions, Fires, and Toxic Releases, 2nd Edition* [2012], AIChE, New York.
[22] API Recommended Practice, *RP 752 Management of Hazards Associated with Location of Process Plant Permanent Buildings* [2009], American Petroleum Institute, Washington D.C.
[23] API Recommended Practice, *RP 753 Management of Hazards Associated with Location of Process Plant Portable Buildings*, [2007], American Petroleum Institute, Washington D.C.

Lessons Learned from Fire in Air Supply Ducts

Al Waller
FMC Corporation
Philadelphia, PA
alvin.waller@fmc.com

Michelle Brown, CSP
FMC Corporation
Philadelphia, PA
michelle.brown@fmc.com

Prepared for Presentation at
American Institute of Chemical Engineers
2013 Spring Meeting
9th Global Congress on Process Safety
San Antonio, Texas
April 28 – May 1, 2013

UNPUBLISHED

Lessons Learned from Fire in Air Supply Ducts

Al Waller
FMC Corporation
Philadelphia, PA
alvin.waller@fmc.com

Michelle Brown, CSP
FMC Corporation
Philadelphia, PA
michelle.brown@fmc.com

Keywords: Lessons learned, fire, HEPA filter, air filter, thermal inertia

Abstract

In 2012, loss of air flow caused activation of interlocks to shut off steam and electrical heat inputs to an air heating system that was part of a drying/milling operation. Shortly thereafter, upstream filters and air ducts ignited causing more than US$500K in damage and business interruption. Investigations concluded that the interlocks functioned as designed on loss-of-airflow but were not sufficient to prevent overheating the inlet air filters (including HEPA filters) by the residual heat in the air heaters. The two main lessons learned from this incident apply to a broad range of operations in the CPI:

1) Mundane unit operations such as air filters that have combustible components must be evaluated as thoroughly as the rest of the chemical process.
2) Designers and PHA leaders must recognize the potential of thermal inertia of heaters (and coolers) to heat (or cool) a stagnant volume of fluid when flow is interrupted. Gas streams are particularly susceptible due to the low heat capacity relative to the heat capacity of a metallic heat exchanger. Interlocking the supplies (to steam and electrical heaters in this incident) is often necessary but may not be sufficient.

Other lessons "re-learned" include the importance of learning from a previous near miss, the importance of limiting escalation aspects of this fire, and the importance of preventative maintenance.

1. Introduction

A drying and milling operation experienced a fire inside its air supply ducts in 2012. Total losses exceeded US$500k, including replacement of air duct components and business interruption. An internal investigation was conducted with assistance from a consulting company by first focusing on the source of the fuel and then on the ignition mechanism.

Figure 1 below is a simplified diagram of a portion of the system; instrumentation and controls are omitted for clarity. Ambient air is pulled into the system by negative draft where it is filtered through pre-filters then HEPA filters. This filtered air is then heated via a steam heat exchanger then an electrical resistance heater to approximately 200C. This hot air then contacts damp combustible powder to convey it through a drying zone and into a cyclone. The distance between the HEPA filters and the steam heater is less than 2 meters while the distance from the electrical heater to the damp solids entry point is approximately 25 meters. Section 3.2 will more completely describe the air duct system and the process mills.

FIGURE 1 – Sketch of Air Duct for Dryer System

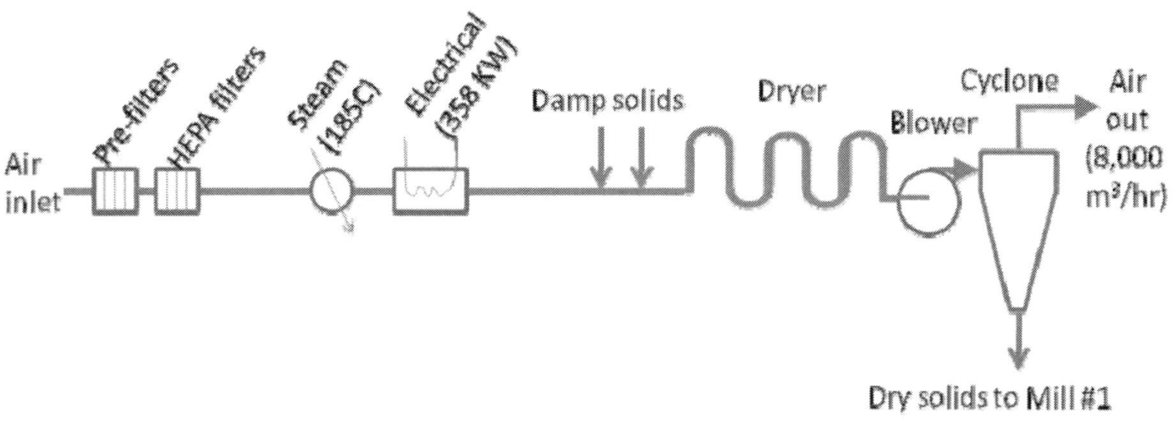

FIGURE 2 – Fire In Progress

The blower drive belts failed in September 2012 resulting in rapid loss of airflow so interlocks shut off steam and power to the heaters. Approximately 17 minutes later a portion of the air duct internally ignited.

2. "New" Lessons Learned

Forensic evidence pointed to the air filters as the source of fuel for the fire. Damage was limited near the air filters in Figure 1 plus there were no process solids or process ash upstream of the entry points for damp solids. The steam heater and electric heater were confirmed as tripped off. Calculations indicate that the residual heat (thermal inertia) was sufficient to ignite the air filters.

2.1 Filters As Fuel

The specific pre-filters and HEPA filters used at the incident site were rated for maximum continuous operation at 70C and 80C, respectively. Tables 1 and 2 list the key aspects of combustion testing of new pre-filters and HEPA filters. All components were susceptible to decomposition at 200C or less with the glass fiber and polyester component of the HEPA filter supporting a self-propagating flame in the oven test. Estimated energy release from combustion of the air filters was consistent with the damage observed from the incident.

Table 1. Stability of Air Filters

Unit	Component	Stability in 200C Oven Test	Stability in 500C Oven Test
Pre-filter	Glass fiber & polyester	Exothermic self-heating at 200C (to 600C). 47% weight loss	Exothermic self-heating (to 600C), 56% weight loss
HEPA Filter	Glass fiber & polyester	Exothermic self-heating at 160C (to 600C), self-propagating flame. 23% weight loss	Exothermic self-heating (to 600C), self-propagating flame. 39% weight loss
	Polyurethane sealant	38% weight loss	97% weight loss

Table 2. Estimated Air Filters Combustion Contributions

Unit	Component	Mass (kg)	Combustible Mass, from 500C Oven Test (kg)	Available Energy Release	
				(MJ)	(% of Total)
Pre-filter	Glass fiber & polyester	6.0	3.3	88	5%
HEPA Filter	Glass fiber & polyester	52.8	20.5	539	31%
	Polyurethane sealant	52.8	51.5	1089	63%

2.2 Residual Heat As Ignition Source

Recorded temperatures from the incident were not helpful in quantifying the residual heat remaining in the heaters after heating sources were tripped off. Recorded temperatures from a prior normal shutdown with the blower operating at 8000 cubic meters per hour were useful to calculate residual heats in the heaters as summarized in Figure 3 and Figure 4. The steam heater retains approximately 48 MJ while the electrical heater retains approximately 15 MJ.

FIGURE 3 - Steam Heater Residual Heat

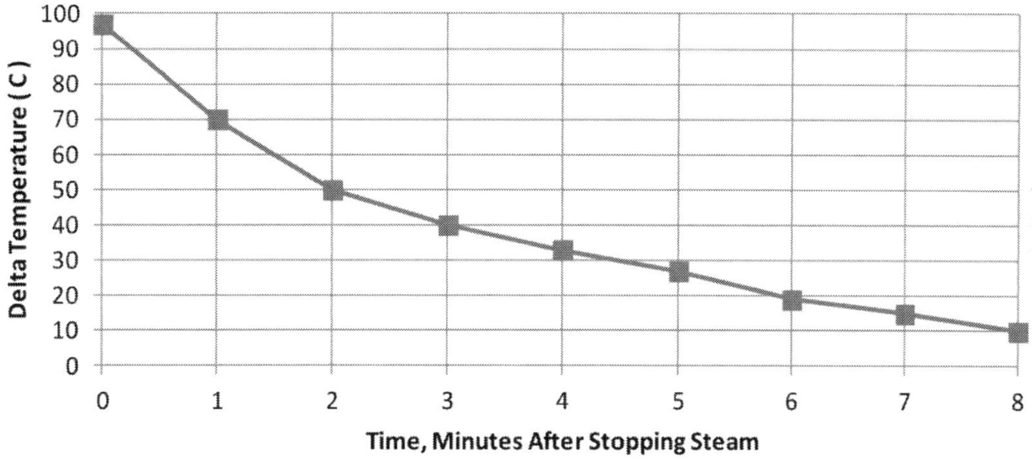

FIGURE 4 - Electrical Heater Residual Heat

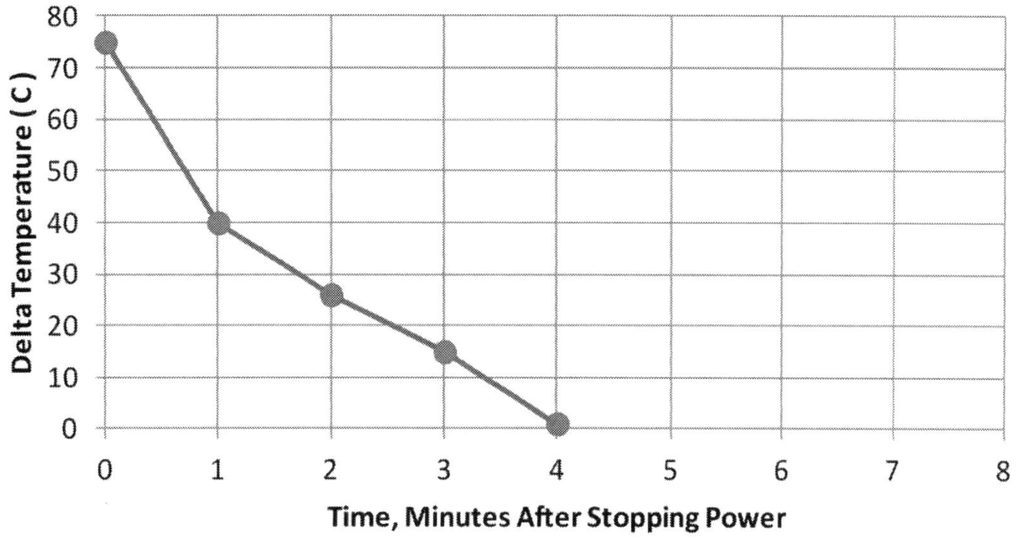

For the incident initiated by blower failure, this combined residual heat of 63 MJ can increase the temperature of stagnant air in the heater portion of the duct to well over 200C resulting in expansion driving the hot air backwards into the air filters. Alternatively, for this incident this combined residual heat could establish a convection current of dense cool air flowing forward at the bottom of the air filter duct, rising in the hot heaters zone, and hot air flowing backwards at the top of the air filter duct to the air inlet. Either incident mechanism results in excessively hot air in air filters that are rated for 70C or 80C maximum temperatures. The Standard Operating Procedures (SOP) for startup recognized the need to start the blower before heating and the SOP for normal shutdown recognized the need to keep the blower running until the air cooled down. Interlock designs recognized the necessity of tripping heating sources on loss-of-airflow. However tripping heating sources on loss-of-airflow is not sufficient to prevent the hazard of hot air reverse flow; therefore, an additional safeguard is needed. For this system, a new vertical vent duct with actuated valve was added between the heaters to open on loss-of-airflow as the primary additional safeguard to provide chimney-effect cooling of the heaters.

3. "Old" Lessons Re-Learned

In addition, lessons from this air duct fire have been refreshed multiple times and are widely known in the CPI.

3.1 Need to Investigate Near Misses and Minor Incidents

Prior to the September 2012 incident, a separate minor incident occurred during a normal shutdown. Some charring of the HEPA filters was attributed to a leaking steam heater isolation valve. The response was to repair the leaking steam valve; thus, the opportunity to investigate this precursor incident was missed.

3.2 Need to Limit Potential Escalation

The fire damage from the September 2012 incident extended beyond the Dryer air duct system (shown in Figure 1 above). The damage was much more severe to the other air filters (pre-filters and HEPA filters), other heat exchangers, and other ductwork interior surfaces as shown in Figure 5. The 3 other blowers in Figure 5 stayed on-line following the Dryer airflow failure and trip of the Dryer air heaters. Hot fumes of pyrolysis/combustion products from the Dryer air filters were pulled into the connected ductwork by these other blowers resulting in direct escalation of the fire or perhaps the Mill #2 electrical heater ignited fumes which backflashed to the air filters. While the Dryer air filters likely experienced restricted oxygen availability, the other air filters had ample oxygen to support combustion. The common air intake duct reduced the initial capital cost by a very small amount compared to separate intake ducts yet added unnecessary complications that led to major escalation. The system was rebuilt with separate air intakes.

FIGURE 5 – Sketch of Air Ducts for Complete System

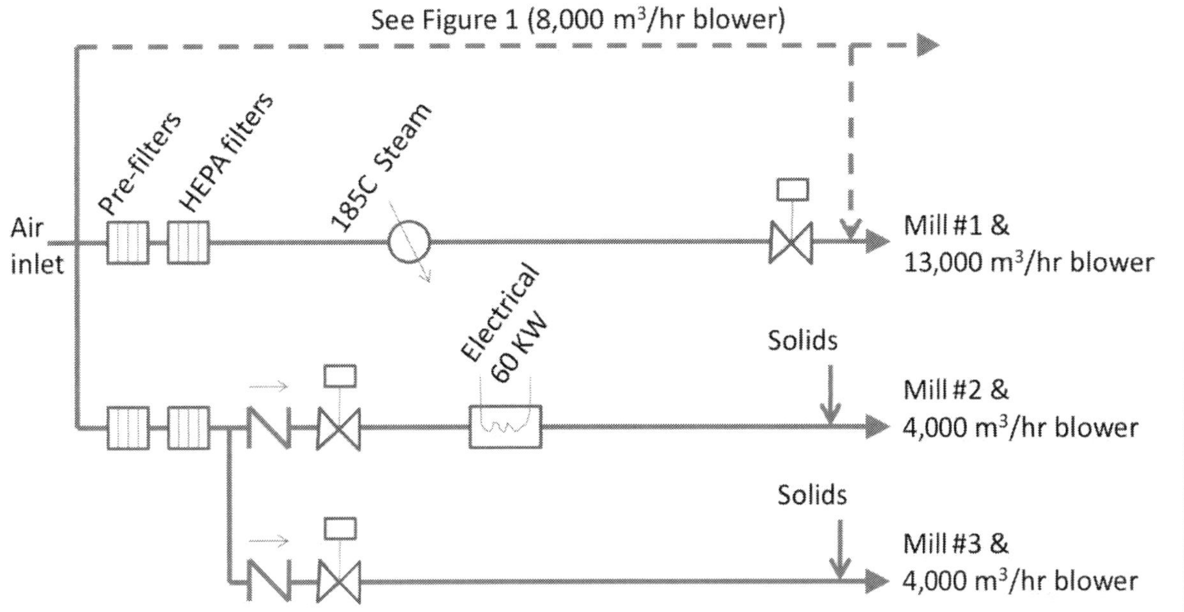

Modest external damage was also visible from smoldering combustible dust on the surfaces of some ducts due to accumulation. Housekeeping practices need to limit combustible dust accumulations below 3 mm (1/8-inch).

3.3 Need to Maintain Equipment Properly

Finally, the blowers did not have a specific installation procedure to ensure belt alignment plus they were not on a preventative maintenance program.

4. Conclusion

Those who forget the past are doomed to repeat. The main lessons learned from this incident apply to a broad range of operations in the CPI:

1) Mundane unit operations such as air filters that have combustible components must be evaluated as thoroughly as the rest of the chemical process.
2) Designers and HAZOP leaders must recognize the potential of thermal inertia of heaters (and coolers) to heat (or cool) a stagnant volume of fluid when flow is interrupted. Gas streams are particularly susceptible due to the low heat capacity relative to the heat capacity of a metallic heat exchanger. Interlocking the supplies (to steam and electrical heaters in this incident) is often necessary but may not be sufficient.
3) Near misses and minor incidents should be thoroughly investigated; an opportunity was missed at this site to prevent the major incident.

4) Keep it simple and keep areas clean. Significant escalation occurred due to a combined air duct that added unnecessary interactions between the systems. Minor escalation occurred from accumulated combustible dusts on exterior surfaces.
5) Maintain equipment properly. Rigorous installation procedures (such as belt alignment) and preventative maintenance improve both safety performance and on-stream reliability.

5. References

Additional references not cited:

[1] Kletz, Trevor. "What went Wrong? - Case Histories of Process Plant Disasters and How They Could Have Been Avoided" (5th Edition, 2009), Elsevier.

[2] DOE (U.S. Department of Energy), "Nuclear Air Cleaning Handbook" Chapter 10 – Fire Protection, DOE-HDBK-1169-2003 (2003).

Hydrogen Sulfide Exposure from Molten Sulfur – A Forgotten Hazard?

Mark J. Viz,
Sean J. Dee, Russell A. Ogle, and Brenton L. Cox
Exponent, Inc.
4580 Weaver Parkway, Warrenville, Illinois 60555
mviz@exponent.com, sdee@exponent.com, rogle@exponent.com,
and bcox@exponent.com

Prepared for Presentation at
American Institute of Chemical Engineers
2013 Spring Meeting
9th Global Congress on Process Safety
San Antonio, Texas
April 28 – May 1, 2013

UNPUBLISHED

Hydrogen Sulfide Exposure from Molten Sulfur – A Forgotten Hazard?

Mark J. Viz,
Sean J. Dee, Russell A. Ogle, and Brenton L. Cox
Exponent, Inc.
4580 Weaver Parkway, Warrenville, Illinois 60555

Keywords: molten sulfur; hydrogen sulfide; railcar; tank car; tank trailer; process safety management; PSM; hazardous materials; hazmat; PHMSA; OSHA; NA2448

Abstract

Molten sulfur is one of the most frequently transported hazardous materials (hazmat) in North America. Molten sulfur is typically transported at an elevated temperature by rail tank cars or over-the-road tank trailers as a Class 9 "Miscellaneous Hazardous Material" under the identification number NA2448. Perhaps the most prominent hazard of molten sulfur is that it is shipped at an elevated temperature, which can cause thermal burn injuries. A less obvious hazard is the potential release of hydrogen sulfide (H_2S), a highly toxic gas.

In this paper we first summarize the key features of the H_2S-molten sulfur equilibrium that are relevant to this hazard. Then we address the relative risk of H_2S exposure from an analysis of molten sulfur incident release statistics. Next, we discuss a few case studies in which vented H_2S gas from molten sulfur led to unexpected consequences for handling personnel. We close our discussion with suggestions for how to possibly prevent H_2S exposure through a combination of engineering controls, operating procedures for both loading and unloading operations, air monitoring, and personal protection equipment.

1. Introduction

Molten sulfur is one of the most widely used global feed stocks. In 2010, the global demand for sulfur reached 78.3 million tons, with elemental sulfur accounting for 66% of global production.[1] With growing demand in the Middle East and Asia, continued use of sulfuric acid, and a growing need for nickel, sulfur production is projected to grow steadily over the next decade.[2] In 2011, elemental sulfur was a $1.6 billion dollar industry in the United States, with 8.1 million tons of production.[3] Additionally, over the last thirty years there has been a significant shift in the production of molten sulfur, moving from mining of geologic sulfur deposits (Frasch process) to environmental byproduct sulfur (Claus process). This technology shift has increased the H_2S

[1] Messick, D. L., World Sulfur Outlook, The Sulfur Institute, 2010.

[2] Kitto, M., Sulfur Supply Demand Balance: The Outlook to 2015, British Sulfur Consultants, 2005.

[3] Apodaca, L. E., Sulfur, U.S. Geological Survey, Mineral Commodity Summaries, January 2012.

hazard in molten sulfur. Given the livelihood of the global sulfur market and the change in the source of the material, the need to communicate and control the hazards associated with the handling of molten sulfur is more critical than ever.

The hazard of dissolved H_2S in molten sulfur became widely recognized in 1963 with the disappearance of the molten sulfur tanker, the Marine Sulfur Queen.[4] This tanker disappeared without a distress signal off the coast of Florida with all 39 crew members. The U.S. Coast Guard searched for the vessel but was only able to recover a few life preservers and some floating debris. Reports came to light of frequent fires on the sulfur tanker and the frequent odor of rotten eggs.[5,6] Investigators from various countries performed headspace measurements in molten sulfur transportation vessels and found H_2S concentrations approaching the minimum flammable level. Recommendations were published that the dissolved H_2S content of molten sulfur should be maintained at 10 parts per million (ppm).[7]

As early as 1970, the U.S. Department of Transportation (USDOT) issued a rule under docket HM-70, that even though it was intended to address requirements for rail tank cars and tank trailers used to ship H_2S as its own commodity, clearly emphasized the hazards associated with exposure to H_2S. The comments to the federal rule state that H_2S is a "severe irritant and is highly toxic"; that exposure leads to "olfactory fatigue"; that H_2S is "heavier than air" and "tends to 'pool' rather than to dissipate into the atmosphere."[8]

Later, in the mid-1980s, the USDOT issued a Final Rule under docket HM-198 that clearly identified two significant hazards associated with the transportation of molten sulfur: the presence of unknown but potentially significant levels of H_2S and that the elevated temperature of molten sulfur could cause burns.[9] As reported in the Final Rule: "Several incidents involving the transport of molten sulfur, including three in the state of California since 1985, have emphasized the need for regulation of this material. Accidents involving spills near Barstow County and Culver City in 1986, and another in Benecia in 1985, demonstrated a lack of hazard communication. In each of these incidents, lack of swift and accurate identification of the material led to delays in handling the incidents and, in at least two instances, contributed to injuries sustained by persons on the scene."

A review of available Material Safety Data Sheets (MSDSs) for molten sulfur suggests that the range of H_2S concentrations and corresponding safe loading, unloading, and handling procedures for molten sulfur vary considerably among producers and shippers. As a consequence, the handling processes may be performed in a manner ignorant of the possible high concentrations of H_2S. According to several MSDSs, the primary hazards associated with molten sulfur are

[4] King, F.W., "Taking H_2S from liquid sulfur-why and how," Oil and Gas Journal, pp. 164-165, June 1974.

[5] Anon., "Investigations: The Queen with the Weak Back," TIME, March 8, 1963.

[6] U.S. Coast Guard, Marine Board of Investigations, "Disapearance of the SS Marine Sulphur Queen at sea on or about 4 February 1963 with the presumed loss of all persons on board," March, 1964.

[7] Lagas, J.A., "Stop emissions from liquid sulfur," Hydrocarbon Processing, pp. 85-89, October 1982.

[8] 35 FR 18919 et seq., December 12, 1970, docket no. HM-70, "Hydrogen Sulfide Gas in Cargo Tank Trucks and Tank Cars."

[9] 53 FR 17158 et seq., October 21, 1986, docket no. HM-198, "Molten Sulfur and Molten Materials Generally."

thermal burns, sulfur-fueled fires, dust explosions, exposure to H_2S, and, in the event of a sulfur fire, inhalation of sulfur dioxide.[10,11,12,13,14] The hazard of H_2S is not prominently communicated in many of these current MSDSs.

Elemental sulfur is typically transported in a molten state in railcars or tank trailers. Commonly, the H_2S hazard is reported as a potential risk, because the concentration of H_2S within the vapor space of the railcar, tank trailer or stationary storage or process vessel is specified as a range or threshold value. The actual concentration of H_2S present in the vessel headspace will likely change due to temperature variations and agitation (as occurs in railcars and tank trailers during movement), and the potential exposure will be dependent on the producer, shipper or consignee's handling procedures and best practices. The authors' own field experiences have also shown that particular to the unloading of molten sulfur railcars, unloading personnel can have a false sense of safety when opening the top fittings or manway cover to prepare a tank car for steaming. Unloading personnel occasionally believe that the solidified "crust" of molten sulfur that is usually present at the surface, especially in cold weather, will somehow prevent any significant release of H_2S. Our own investigatory experience has shown that this is not always true.

Occupational exposure to H_2S during handling can generally be categorized as an example of a low probability, high consequence risk. This is supported by a review of the Pipeline and Hazardous Materials Safety Administration (PHMSA) and Occupational Safety and Health Administration (OSHA) incident report databases, which confirms that over the last 20 years, there have been six OSHA-reported fatalities from H_2S exposure associated with molten sulfur containers and over the last 24 years, there have been no PHMSA-reported fatalities from molten sulfur releases (while the commodity is still under the jurisdiction of the federal hazmat transportation regulations).[15,16] Our collective experience based on our investigations of numerous hazmat releases is that the tendency is for successful management of low probability, high consequence hazards to lead to complacency and the occasional elimination of seemingly superfluous safeguards.

[10] Hovensa, Sulfur MSDS No. 1794, 2006.

[11] Tesoro, Safety Data Sheet: Sulfur, Molten, 2012.

[12] Keyera, Sulfur, Molten MSDS, 2011.

[13] Devon, Material Safety Data Sheet: Sulfur (Liquid), 2010.

[14] Marathon, Material Safety Data Sheet: Marathon Sulfur, 2010.

[15] Note that there can be overlap between the release incident reporting required by OSHA and PHMSA based on where the incident occurred and what function was being performed when the release occurred. So, for example, an accidental release of molten sulfur involving an injury or fatality that occurred while a tank car was connected to plant apparatus during unloading might be reported to OSHA and not to PHMSA even though releases during unloading can be covered by PHMSA regulations as well. Efforts to harmonize the regulations and reporting requirements promulgated by a variety of federal agencies has been the subject of federal rulemaking in dockets such as HM-223, "Applicability of the Hazardous Materials Regulations to Loading, Unloading and Storage," Docket No. PHMSA-98-4952 (HM-223), under which PHMSA issued a Final Rule on June 1, 2005.

[16] Note that at times, exposure to H_2S can result in a prolonged hospitalization that results in fatality at a date after the original paperwork related to regulatory agencies is submitted.

According to the USDOT PHMSA "Incident Reports Database," there have been 428 reported releases of molten sulfur in transportation between the years 1989 to 2012, inclusive. For all 428 reported releases, no fatalities and 19 injuries were reported. A portion of these injuries were related to exposure to fumes and odors that resembled "rotten eggs", a description commonly used for H_2S. Additionally, incident reports indicated that H_2S gas was released at various times while the commodity was in transit, being loaded or unloaded, or while being stored in transit. Although there may be some overlap between exposure incidents reported in PHMSA's Incident Reports Database and the analogous Accident Investigation Summaries from OSHA (OSHA Form 170), which are maintained in OHSA's Integrated Management Information System (IMIS), the OSHA database includes six reported fatalities from H_2S exposure associated with the handling of molten sulfur over the last twenty years.

Our review of H_2S exposure incidents is intended to act as a reminder of the importance of employing a hierarchy of safeguards when managing high consequence hazards like H_2S.

2. Technology Background

A major technological shift in the production of sulfur has resulted in an increased potential for exposure to H_2S. For most of the 20th century, the Frasch process was used to recover sulfur from underground, elemental deposits.[17] In the Frasch process, superheated water is used to melt solid deposits of sulfur which are driven to the surface with compressed air for recovery. However, with the growth of the refining industry, the supply of elemental sulfur from the desulfurizing of sour crude oil and natural gas overtook production from the Frasch process in 1970. The Claus process, which can produce elemental sulfur from feed stocks containing up to 20% H_2S, is now the preeminent production process for producing sulfur.

To convert H_2S to elemental sulfur, the Claus process uses a two-step process, first utilizing a high temperature thermal step, followed by a lower temperature catalytic step. In the thermal section, air is added to H_2S at temperatures in excess of 1800°F, to covert one-third of the H_2S to sulfur dioxide (SO_2) as shown in Equation 1.

$$H_2S + 3/2\ O_2 \rightarrow SO_2 + H_2O \qquad \text{[Eq. 1]}$$

The resulting SO_2 then reacts with the remaining H_2S in an endothermic equilibrium reaction to form elemental sulfur vapor as shown in Equation 2.

$$2H_2S + SO_2 \leftrightarrow 3/2\ S_2 + 2\ H_2O \qquad \text{[Eq. 2]}$$

To drive conversion in the endothermic equilibrium reaction shown in Equation 2, S_2 vapor is removed through condensation. As the vapor cools, the S_2 equilibrates with the S_6, and S_8 forms through multiple equilibrium reactions.

If one-third of the sulfur is consumed in Reaction 1, then the remaining two-thirds is consumed in Reaction 2 resulting in stoichiometric consumption of H_2S. However, several side reactions usually lower the conversion to SO_2 in Reaction 1, subsequently lowering the overall H_2S

[17] Greenwood, N. N. and Earnshaw, A., Chemistry of the Elements, 2nd ed., 1997.

conversion. This highlights both the importance of the air to H_2S feed ratio as a process control variable, and the need for a secondary catalytic reactor to further convert the H_2S that is not consumed in the thermal section.

The catalytic section of the Claus process is used to convert H_2S as well as side products including COS and CS_2 that are formed from side reaction in the thermal section. The catalytic section runs at significantly lower temperatures than the thermal section (inlet 400°F - 450°F, outlet 600°F - 630°F), but uses a Clause catalyst (usually aluminum (III) or titanium (IV) oxide) to covert H_2S to elemental sulfur as shown in Equations 3 and 4.[18,19]

$$2\ H_2S + SO_2 \rightarrow 3/6\ S_6 + 2\ H_2O \qquad [Eq.\ 3]$$

$$2\ H_2S + SO_2 \rightarrow 3/8\ S_8 + 2\ H_2O \qquad [Eq.\ 4]$$

The effluent of the catalytic section can be cycled through additional catalytic beds, or condensed to a final product stream at temperatures between 265°F - 350°F.

The one common feature to all of these chemical reactions is the constituent H_2S. While the Claus process can achieve 96% - 97.5% conversion with one thermal reactor, and three catalytic reactors in series, the solubility of H_2S in molten sulfur adds an additional processing constraint. The solubility of H_2S in molten sulfur actually increases with temperature up to approximately 425°F. Solubility remains approximately constant from 425°F until 630°F, where it begins to drop with further increases in temperature.[20] Thus, in the catalytic reactors and condensers, there is still an appreciable amount of H_2S dissolved in the molten sulfur. If the hot sulfur is then stored in a tank, as the tank cools, the dissolved H_2S will release into the headspace, increasing the concentration to potentially dangerous levels. If the sulfur continues to cool, then a crust will form as the surface of the product reaches temperatures below 240°F. This solid crust can serve as a physical barrier protecting personnel from potential H_2S exposure. In cases where a crust is observed, special consideration should be noted to avoid agitation that could disrupt the solid interface, resulting in a potential H_2S vapor exposure. However, it should be noted, that formation of a protective crust should not be regarded as a reliable safety indicator, since crust formation is not correlated to the concentration of H_2S in the vapor space.

The headspace concentration of H_2S that can be achieved with molten sulfur in a transportation or storage container is influenced by four primary factors: the dissolved H_2S concentration, the liquid temperature, agitation of the liquid, and the presence of any organic contaminants. To maintain an H_2S concentration below its lower flammable limit (LFL), typical specifications for the dissolved H_2S concentration range from 5 to 15 ppm. At the temperature range at which molten sulfur is typically handled, the LFL corresponds to a headspace concentration of 3.4%, or

[18] Goar, B. G. and Fenderson, S., Fundamentals of Sulfur Recovery by the Claus Process, presented at The Laurence Reid Gas Conditioning Conference, Norman, Oklahoma, March 3, 1996.

[19] Sulfur and Hydrogen Sulfide Recovery, Kirk-Othmer Encyclopedia of Chemical Technology, John Wiley & Sons.

[20] Fanelli, R., Solubility of Hydrogen Sulfide in Sulfur, Industrial and Engineering Chemistry, pp. 2031-2033, September 1949.

34,000 ppm (at room temperature the LFL for H_2S is 4.3%).[21] From the viewpoint of respiratory exposure, this is a potentially instantaneous lethal dose.

3. Hazard Communication

The MSDSs for molten sulfur identify several hazards. Common features were noted in the 19 MSDSs reviewed for this study. First, nearly all of the surveyed MSDSs noted the potential dangers associated with H_2S and listed H_2S as a material present in molten sulfur. The amount of H_2S present was usually noted as a small amount; descriptions included "trace", "< 0.1%", "< 1%", "< 500 ppm", and "< 1% (variable)". It was also common for MSDSs to include additional information about the potential hazards of H_2S exposure, the acceptable exposure limits (which even today are still a subject of debate), and to identify the need for proper ventilation during storage and handling.

In the MSDS review, there is significant confusion and room for improvement regarding the description of PPE and unloading procedures. In addition to H_2S hazards, respiratory protection also included consideration for dust hazards since molten sulfur can solidify to a yellow solid when cooled. In most MSDSs, it is noted that respiratory protection should be used, but only if the concentrations of H_2S are unknown/above regulated levels, or other engineering controls are not sufficient to control the exposure. One MSDS stated that respiratory protection was not commonly required when handling molten sulfur. While several MSDSs noted that unloading/opening of tanks should be done in well ventilated areas, it was not noted if proper ventilation is a sufficient engineering control to protect workers from potential exposure to H_2S. One MSDS included a picture of a dust mask in the recommended PPE, which is appropriate for the potential dust hazard, but not for H_2S exposure.

In addition to not specifying when a respirator should be worn, very few MSDSs had detailed recommendations for unloading procedures related to the H_2S hazard in molten sulfur. As previously discussed, most of the MSDSs recognized several key components to safe unloading: proper ventilation, respiratory protection if the H_2S concentration was unknown or above acceptable limits, and confined space warnings. Air monitoring was not commonly listed as a recommended engineering control in the MSDSs. Additionally, recommendations regarding venting of tank headspace were not usually recommended before sampling or unloading. This omission, however, could be the result of the inclusion of opening the tank car or tank trailer before steaming as part of the unloading procedure. Details such as these regarding unloading might not typically be specifically addressed in the MSDS. Other MSDSs specified venting, only if the product was excessively heated or exposed to elevated temperatures. The potential evolution of H_2S into the vapor space can be driven both by temperature increases in the presence of water or by temperature decreases which drive dissolved H_2S out of the molten sulfur. Including specific directions for when respiratory protection is required and directions for proper unloading safety controls before handling/unloading could improve the hazard communication for H_2S in molten sulfur.

[21] NFPA 655, Standard for Prevention of Sulfur Fires and Explosions, National Fire Protection Association, Quincy, MA, Section A.5.1, p. 17, 2007.

4. Review of PHMSA and OSHA Incident Data

Two meaningful sources of incident data regarding H_2S exposure from molten sulfur handling are the PHMSA Incident Reports Database and the Accident Investigation Summaries from OHSA Form 170. Although there may be some overlap in incident reporting between these two databases, in general the PHMSA database contains mandatory reports that involve the release of a regulated hazardous commodity while in transit, while being loaded or unloaded, or while in storage during transit. The OSHA database contains mandatory reports that involve incidents or accidents that have resulted in the injury or fatality of an employee. The differences between the type of incidents reported in each database and the corresponding limitations in the reports should be understood. For example, personnel may have been briefly exposed to H_2S shortly after a tank car has been opened for steaming or unloading, but if there was no release of molten sulfur, the incident would not necessarily need to be reported in the PHMSA database. In a similar manner, if a release of molten sulfur occurred after a transport vessel had been connected to plant apparatus but it did not involve the unloading equipment, this incident might only be reported in the OSHA database, and then only if a reportable injury or fatality occurred as a result. In this manner, then, the review and subsequent use of the information provided in these reports should be considered with care and properly qualified when attempting to estimate occurrence probabilities and severities.

Bulk shipments of molten sulfur in North America are typically identified by the number "NA2448." Using the PHMSA Incident Reports Database, we performed a search of reportable incidents involving NA2448 by all modes of transportation (air, water, highway and rail) over the entire range of dates for which electronic records are available. From 1989 to 2012, the database contained 428 reports of unintended releases of molten sulfur, all of which occurred either by rail or highway. In these 428 reports, no fatalities were reported and 19 injuries were reported. An interesting finding is that of the 428 reports, 43 of them were described (based on the responder's description provided in Part VI of the Form DOT F 5800.1) as being initially discovered based on the detection of a strong odor. In addition, the 43 reports that explicitly identified "odor" in the description all involved rail tank cars. Of these 43, three of them further described the odor as being that of "rotten eggs," which is the typical way that people refer to the unique smell of H_2S.

It is compelling that roughly 10% of the molten sulfur releases reported from 1989 to 2012 indicate that they were initially detected by the "odor" of sulfur, and in a few cases, the more telling odor of "rotten eggs." Many of these instances occurred when a rail tank car was technically "in transit" but was not "in movement," such as when a railcar is in a classification yard. The irony involved in the discovery of these situations is that while the inhalation hazard of the H_2S is a high consequence hazard, had it not been for personnel noticing the odor coming from these tank cars, detection of the lading loss might not have occurred.

For releases that occurred in rail transit, the most common causes were a corroded or otherwise breached rupture disk, an improperly secured manway cover (usually involving a number of loose or missing bolts) or an improperly seated gasket. These causes create concern for railcar handling personnel because unless they can detect a potential exposure by smell or see the physical indicators of lost lading (such as crusted sulfur on the outside of the tank car), they would likely approach the railcar assuming that it was completely sealed. This might lead to personnel being placed in danger if they approach an unsealed railcar that is venting high concentrations of H_2S that are not detected quickly enough for personnel to detect and identify the hazard and approach the railcar using more appropriate PPE.

The majority of releases from tank trailers involved human error on the part of the driver who failed to properly monitor the loading operation and allowed the tank trailer to be overfilled. Reports involving overfilling do not refer to sulfur or H_2S odors, likely because the loss of lading is immediately apparent by visual observation.

A review of the Accident Investigation Summaries from OSHA revealed five cases that noted an H_2S release from molten sulfur tanks during unloading.[22] A total of 16 employees were involved in the five incidents, with six total fatalities. There are several common features related to each of these incidents. First, all five incidents occurred in the fall or winter season in locations where the ambient temperatures are low. Second, none of the five incident reports makes mention of the exposed employees wearing proper respiratory protection, and two investigations noted that employees were not wearing respirators. In two of the incidents, the exposure to H_2S resulted in loss of consciousness which caused a fall for the employees who were not wearing fall protection.

The most recent OSHA report of an accident related to H_2S exposure from molten sulfur occurred in 1995. An employee was taking samples from a molten sulfur tank truck. After opening the manway cover, he reached approximately 15 inches into the tank to collect the sample. A respirator was not provided, and even though he was on a nearly nine-foot tall platform, he was not wearing fall protection. The employee was found unconscious on the gravel covered ground approximately 10 minutes after taking the sample. An autopsy listed blunt force injuries from the fall as the cause of death, but testing revealed H_2S levels up to 800 ppm using similar unloading procedures on site. OSHA maintains that H_2S was a causal factor related to the fall.

In addition to potential fatalities due to falls, H_2S can also cause poisoning or asphyxia. In 1993, three employees were unloading a molten sulfur railcar. Two employees were on a platform on top of the railcar, while the third employee was on the ground hooking up steam lines. The third employee found his two coworkers unconscious on the platform shortly after they had opened the manway cover on the railcar. The employee who opened the manway cover was found dead on arrival, the other employee on the platform was hospitalized for several days, while the third employee was examined and released.

5. Safeguards to Prevent or Control H_2S Exposures

H_2S exposure during the handling of molten sulfur can be generally categorized as a low probability, high consequence event based on an assessment of the data available from PHMSA and OSHA. However, these incidents were of high consequence due to the fatalities that have

[22] Incident reports from www.ohsa.gov. A total of 172 reports were reviewed with "hydrogen sulfide" as a key word. Five reports were related to molten sulfur tanks. Those incident reports are numbers 109097600, 109713388, 17713199, 106387822, and 1383249.

been reported. Managing this or any low probability, high consequence risk may benefit from the use of multiple safeguards for a variety of reasons.

Suggested safeguards to consider, most of which can be implemented independently or together, are:

- Closed-loop venting and treatment of H_2S vapor (e.g., adsorption systems, flares);
- Administrative controls (e.g., standard procedures and training for unloading, buddy system);
- External ventilation of unloading areas;
- Air monitoring systems with alarms;
- Respiratory protection (PPE) sufficient to the hazard level or that ensures the safety of unloading personnel when the hazard level is unknown.

In fact, some states have mandated safety procedures associated with the handling of H_2S, including specifications for personal protective equipment, venting requirements, and equipment specifications.[23] As always, the best system of safeguards will be site-specific and tailored to the capabilities of site personnel. It is also important to consider the consequences of any changes prior to their implementation, even the addition of safeguards, lest any new hazards be introduced.

6. Recommendations

The safe handling of molten sulfur requires understanding and planning regarding the potential release of H_2S. Herein, we have discussed the technical reasons for H_2S hazards in molten sulfur, reviewed accident statistics, and relevant safety documentation. The following recommendations might be considered when handling molten sulfur.

- Sulfur supply contracts, which typically include requirements for a dry weight analysis of the sulfur based on minimum or maximum amounts of ash, carbon, arsenic, selenium and possibly other elements, should also address maximum levels of H_2S. In addition, these requirements should address when and how the product sample should be taken and then analyzed, especially if it is determined that the H_2S content is strongly dependent on sample temperature and storage time.
- Documentation regarding the shipping, receipt, and unloading of molten sulfur should be maintained per regulatory and "best practices" requirements. These documents, as well as sulfur supply contracts, might include warnings regarding the potential increase in H_2S concentration during shipment.

[23] For example, California Title 8, Section 4414 and Washington Administrative Code 296-79-250.

- Development of site-specific loading and unloading procedures should be written and followed given due consideration to the assessment and management of the hazards associated with H_2S in molten sulfur. These procedures might include:
    - Forced exhaust ventilation requirements;
    - Proper temperature, humidity, and air monitoring regarding H_2S concentrations at tank inlets, outlets, and bulk concentrations within the unloading area;
    - Proper PPE for burn protection and H_2S exposure, possibly including a requirement for a SCBA respirator and an escape bottle;
    - Controlled tank headspace venting before sampling or unloading;
    - Consider fall protection requirements knowing that non-lethal exposure to H_2S can induce a loss of consciousness and potential injuries related to falls.

Finally, it is emphasized that there is typically no fixed set of safeguards that will work at every facility. A well-designed system of safeguards to protect workers from H_2S exposure should likely consider the personnel, equipment, and safety management system at that particular facility.

Calcium Stearate Deflagrations – A series of unfortunate events

Albert Ness
ABS Consulting
10301 Technology Drive
Knoxville, TN 37932
aness@absconsulting.com

Konanur Manjunath
Dow Chemical Co.
Freeport, Tx
KManjunath@Dow.com

Prepared for Presentation at
American Institute of Chemical Engineers
2013 Spring Meeting
9th Global Congress on Process Safety
San Antonio, Texas
April 28 – May 1, 2013

UNPUBLISHED

Calcium Stearate Deflagrations – A series of unfortunate events

Albert Ness
ABS Consulting
10301 Technology Drive
Knoxville, TN 37932
aness@absconsulting.com

Konanur Manjunath
Dow Chemical Co.
Freeport, Tx
KManjunath@Dow.com

Keywords: Calcium Stearate Solids, Combustible Dust, Bulking Brush Discharge, Particle Size Distribution, Resistivity, Bonding, FIBC, Inertion

Abstract

A series of deflagrations occurred in a plant unloading Calcium Stearate from bulk bags. Each one had a different root cause and illustrates the importance of management systems for ignition control and the need to evaluate a system for many types of ignition sources. The first event was caused by static discharge from an ungrounded tote. When upgrading the system Type C bags were specified for the product and a second deflagration occurred due to static discharge from and ungrounded bag. In the third deflagrations cause was most likely a conical pile discharge. The presence of fines and coarse material plus the use of a larger cubic vessel probably enhanced the likelihood of the brush discharge.

.

1. Introduction

This paper describes three deflagrations that occurred in a bulk bag unloading system that had previously run for over six years without incident.

1.1 Product data for Calcium stearate

Table 1 lists data for the Calcium Stearate that was being used in the facility.

Table 1 Calcium Stearate Properties

Parameter	Value
D_{50}	500 microns
Minimum Ignition Energy (MIE) (< 75 μ fraction)	8 mJ
Volumetric resistivity	$14 \quad 10^{14}$ Ω-m

2. Unloading station description

The unloading station consisted of a bulk bag of Calcium Stearate (CaSt) hung u above a feed hopper. An Iris valve was used to control the flow rate of the CaSt into the hopper. A flexible feeding chute was attached from the hopper outlet to the lid of a 2500 L tote. Local exhaust ventilation was provided at the feed hopper and tote.

Figure 1: Unloading station for calcium stearate bulk bags

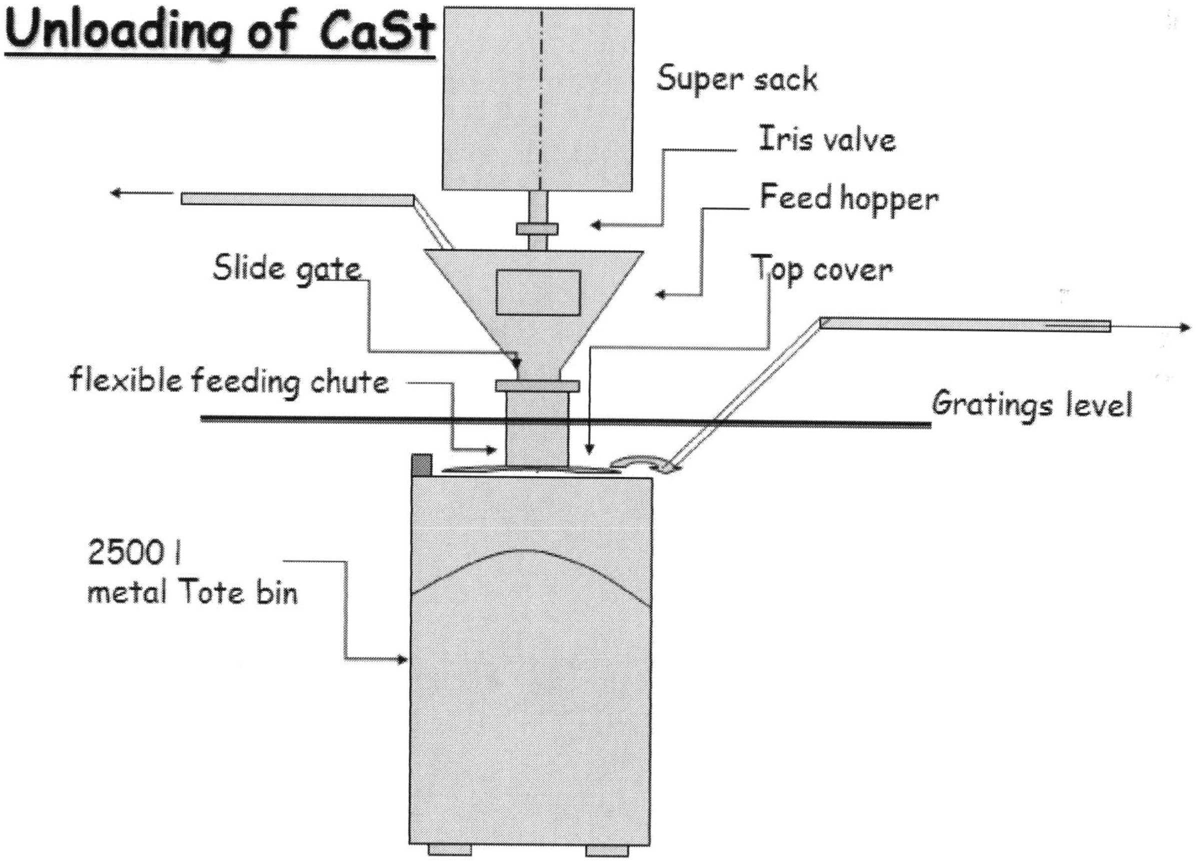

2.1 Event #1 while unloading

At the end of the loading cycle a deflagration occurred in the head space of the tote. The fire was quickly extinguished and little actual damage occurred. Investigation showed that the Feed hopper and the lid of the tote were not grounded and that the flexible chute was not static dissipative. Additionally, it was discovered that the tote vacuum system was not used. Figure 2 shows where the damage occurred.

Figure2: Unloading station at the time of event 1

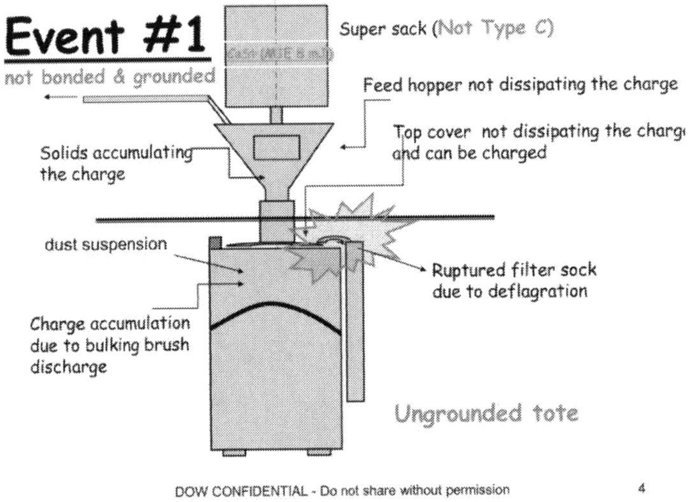

With the vacuum system inoperative, fines were able to stay suspended in the head space of the tote. A calculation of the capacitance of the lid of the lid showed that it had a capacitance of 180 pico Farads. Figure 3 shows that at that level, a static discharge energy of about 30-40 mJ for a charge of 20 EV, well above the 8 mJ MIE of the fraction under 75 microns, was possible. Therefore it was concluded that the most likely cause of the deflagration was a spark between the lid and tote ignition suspended fines in the tote.

Figure 3: Calculated electrostatic build up on process equipment

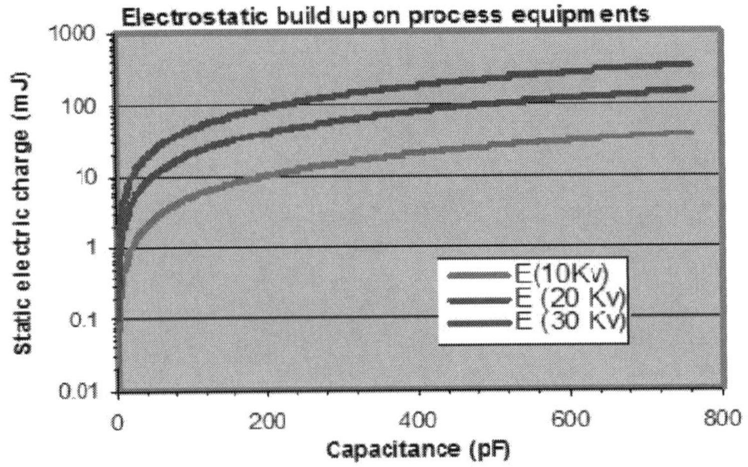

2.2 Event #2

After the first event, the metal equipment was grounded and the tote vacuum system was put back into operation. Also, it was decided to specify use of Type C bags for the bulk bags. Previously, the bulks bags were type B, that is, an all plastic bag that cannot be grounded, but has a breakdown voltage of 4kV. This allows for discharge of energy to the air. A Type C Bag has continuous conductive threads are sown into these bags, hence continuity is provided if the bag is grounded.

After starting up the improved system, a second deflagration occurred. Upon investigation, it was found that; 1) the bulk bag had not been grounded and 2) the Iris valve was not grounded due to its Teflon coating and the powder buildup on it. The failure to ground the new bulk bag is a reminder of the expression "No good deed goes unpunished". It is also a reminder of the need for a good Management of Change System. In the case of this change, the operator's training should include instruction on the hazard of NOT grounding the Type C bag.

2.3 Event #3 while unloading

Prior to restarting the unit, a ground verification system was installed to ensure the bulk bag was grounded before charging. Figure 4 shows where the damage occurred.

Figure 4: Unloading station at the time of event 3

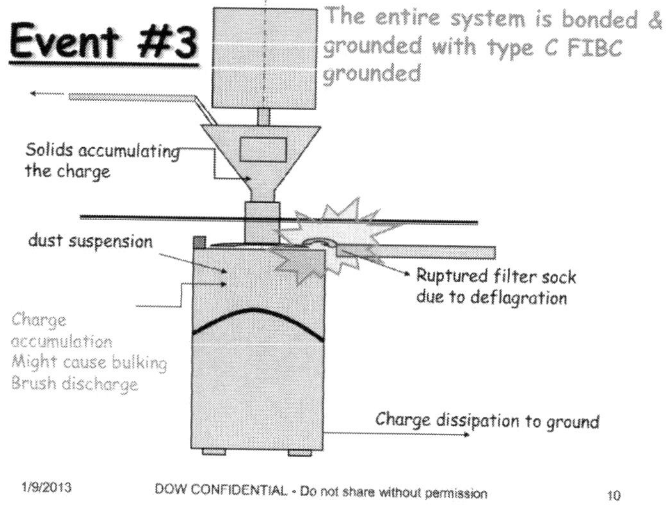

At this point, there was no obvious source of ignition. Several experts were called in and a more thorough investigation was done. Findings were:

1. The FIBC was examined for continuity of grounding. Except for the area of the bag that was damaged by the fire, the continuity was acceptable. Testing of FIBCs at the supplier found all the bags were conductive.
2. The CaSt in the FIBC from the last event was somewhat smaller than in the initial events. Table 2 shows the difference in size between the CaSt particles from the first and third events. The material from the third event was not out of spec, it was considered

acceptable material by the plant. The key point here is that normal variation in the raw material can affect the degree of hazard.

Table 2: Size Comparison

Cumulative Vol. %	Event 1	Event 3
90	800	800
50	500	425
10	200	150
% < 75 μ	2-3	~5

3. Fines collected from the top of a different tote had a D50 of approximately 10 μ and were found to have an MIE of 3 mJ.
4. The dust extraction system was again found to have been off.

2.4 Bulking or Conical Brush Discharges

Whenever solids are filled into containers, bulk bags and so on, bulking brush discharges are developed. It is the type of discharge observed on the conical pile of a heap of powder; thus it is often called conical pile discharge. Bulking brush discharges have an apparent maximum effective energy of 10-25 mJ (with respect to dust ignition), and are believed to be responsible for dust explosions in grounded silos [5, 17, 18].

Britton [17] states that bulking brush discharges cannot always be practically eliminated in equipment above about 1 meter diameter when non-conductive powders are transferred.

The following scenarios can lead to bulking brush discharges:

1. Fines in the suspension that coat the conductive wall rendering it less conductive or nonconductive. Now we have a situation of a system that is not grounded in practice. As an example the fines may be all fraction < 10 microns and the MIE for such a fraction will be much lower than the tested sample (< 74 microns). Even though this is a hypothetical situation, one needs to consider if the attrition is a factor in product processing.
2. Fines are in suspension in the free head room of the vessel and the filling continues with the mixture containing coarser particles. As the particle size increases, the equivalent energy increases. See Figure 3. This combination with fines to super fines and coarser particles is a concern as this can result in an incident.

Scenario 2 perfectly describes the situation that occurred in the third event.

2.4 Causes

The root cause of the third events is the high fines content of the CaSt in the FIBCs. The lack of dust extraction may have led to a suspension of the fines in the head space of the metal tote bin. Ignition could have been caused by either a defective Type C FIBC, or a conical brush discharge.

Testing of the existing bag was inconclusive due to the damage it sustained. Experts in static electricity and combustible solids believed that the conical pile discharge could not be ruled out.

Potential mitigation for conical brush discharges can include:

- Flow rate reduction
- Use of active neutralizers for relatively small silos.
- To minimize the probability of ignition, the first consideration
- Minimize the sub-200 mesh fines concentration to increase the MIE to greater than 20 mJ. Inertion, especially if process design or powder properties make deflagrating venting and suppression undesirable alternatives. Inerting is often considered where the powder MIE falls below the 3-10 mJ range.

In this application, inerting was chosen as the fire and explosion prevention technique

3. *Concluding remarks*

This series of events illustrates many points.

Not fully understanding the source of hazards can lead to jumping to obvious, but not comprehensive, conclusions in an investigation. It pays to consult the appropriate subject matter experts even for seemingly simple events.

No good deed goes unpunished. Changes made to improve the safety of a system can introduce new hazards. In this case the switch to Type C bags introduced a new potential ignition source. Training of operating personnel should include not only what the change is and how the procedures may change, but also on new hazards introduced.

Even on a simple system, it pays to consult with subject matter experts on potential hazards. The effect of variations of particle size on MIE and the concept of conical pile discharges, may very well not be known to process designers.

8. REFERENCES

1. Bartknecht, W., **Dust Explosions**, Berlin: Springer-Verlag, 1989.
2. Glor, M., **Electrostatic Hazards in Powder Handling**, Letchworth UK: Research Studies Press Ltd., 1988.
3. "**Recommended Practice on Static Electricity**", NFPA 77, 2003.
4. "**Recommended Practice for the Classification of Flammable Liquids, Gases, or Vapors and of Hazardous (Classified) Locations for Electrical Installations in Chemical Process Areas**" NFPA 497, 2004
5. "Recommended Practice for the Classification of Combustible Dusts and of Hazardous (Classified) Locations for Electrical Installations in Chemical Process Areas" NFPA 499, 2004
6. "Standard for the Prevention of Fire and Dust Explosions from the Manufacturing, Processing, and Handling of Combustible Particulate Solids, NFPA 654, 2006

7. Eckhoff, R.K., **Dust Explosions in the Process Industries**, London UK: Butterworth-Heinemann, 2003.
8. Britton, L.G., **Avoiding Static Ignitions Chemical Operations**, AIChE, CCPS Concept Book, 1999, Publication G-67
9. Jaeger, Norbert and Siwek, Richard; "**Prevent Explosions of Combustible Dusts**", Chemical Engineering Progress, June 1999
10. T.H. Pratt **Electrostatics ignitions of fires and explosions**. Center for Chemical Process safety. AIchE Industry technology Alliance, USA, 1997.
11. L. Britton Private communications, November , 2003
12. M. Glor **Ignition hazard due to static electricity in particulate processes**, Powder Technology 135–136 (2003) 223– 233

Effectiveness of improvised gas absorption techniques for emergency responders at releases of toxic gases

Hannes Kern
Montanuniversität Leoben
Franz-Josef Strasse 18, 8700 Leoben
hannes.kern@unileoben.ac.at

Matthias Taferner, Harald Raupenstrauch
Montanuniversität Leoben
Franz-Josef Strasse 18, 8700 Leoben
matthias.taferner@stud.unileoben.ac.at, harald.raupenstrauch@unileoben.ac.at

Prepared for Presentation at
American Institute of Chemical Engineers
2013 Spring Meeting
9th Global Congress on Process Safety
San Antonio, Texas
April 28 – May 1, 2013

UNPUBLISHED

Effectiveness of improvised gas absorption techniques for emergency responders at releases of toxic gases

Hannes Kern
Montanuniversität Leoben
Franz-Josef Strasse 18, 8700 Leoben
hannes.kern@unileoben.ac.at

Matthias Taferner, Harald Raupenstrauch
Montanuniversität Leoben
Franz-Josef Strasse 18, 8700 Leoben
matthias.taferner@stud.unileoben.ac.at, harald.raupenstrauch@unileoben.ac.at

Keywords: toxic gas release, emergency response, absorption techniques

Abstract

In September 2011 a spill of nitric acid at a galvanic company in Trieben, Austria caused the release of large amounts of nitrous fumes. An IBC tank with 1000 liter of nitric acid (53%) was flipped over by a forklift and cracked open in the top area. The content was fully released and lead to serious corrosion effects on storage racks combined with the production of critical amounts of nitrous fumes. 27 workers had to undergo medical treatment and a storage and chemical handling area of several hundred square meters was filled by nitrous fumes with concentrations of up to 100 ppm.
Emergency responders used an improvised gas absorption system to clear the area of toxic fumes. Outside the sealed area concentrations of nitrous fumes well below the AEGL-2 (4h) value for nitrous dioxide where measured. In the course of the analyses of the response the effectiveness of improvised gas absorption techniques was investigated.
For the investigations chlorine and ammonia in combination with a mobile exhauster, (10.000 m^3/h) commonly used by emergency responders in Austria were used. Different methods of gas absorption where investigated and tested upon their practical relevance.

1. Introduction

Toxic gases are commonly used in industrial and public areas, both as base products for chemical production and as technical gases for various purposes. In Europe gases like chlorine or ammonia are often used as disinfecting agents or coolant in cold storage warehouses. In February 2012 a release of several hundred kilograms of ammonia from a cold storage warehouse in Vienna (Austria) injured 62 Persons, 32 firefighters and 4 emergency medical technicians right in the busy city center of Vienna. Firefighters were able to contain the released ammonia within the warehouse and tried to absorb and disperse it with water by using water curtains and water shields. Half a year earlier, on September 5th 2011 the spillage of 1000 liter nitric acid 53% in a metallurgical production site Trieben (Austria) caused the occurrence of significant amounts of nitrous fumes in a large storage area. The nitrous fumes were produced due to the reaction of the acid with the material of storage racks and metal materials in the surrounding area. The nitrous fumes were contained in the storage area, but ventilation procedures were very critical due to the close surrounding neighborhood. At the incident in Trieben 27 workers were injured. Due to a lack of knowledge regarding improvised gas absorption techniques experiments with ammonia, chlorine and different absorption agents were carried out. The experiments were supported by the Hazmat Team of the Graz Fire Brigade.

2. Application and testing of absorption techniques

2.1 General overview on techniques for gas defense

In the field of gas defense and control different techniques are applied by emergency responders. Depending on the strategy followed by the responding crews and the situational demands there are three main categories of technical and strategic measures that can be used during a response situation dealing with toxic or hazardous gases. These methods are described in the following sections.

2.1.1 Active methods

As "active methods" all response techniques can be considered that aim to directly fight and disperse gas clouds or change their direction of spread. Most of the active response methods are based on technical devices that are able to generate large volume flows of water or/and air that are directed against the spreading gas cloud. Depending on the size of the gas release water nozzles, water monitors or different sized ventilators are used for the active response approach. Appliances like the LUF 60 (firefighting support vehicle) combine a high volume rate ventilator (90.000 m^3/h) with a circular water dispersion system and provide therefore an aerosol cloud that shows better kinetic effects than the sole air stream. For water soluble gases also absorption effects can be achieved by this method.

2.1.1 Passive methods

"Passive methods" aim to contain the released gas cloud or try to protect areas from gas by a restriction of spread. An example for such systems are, water shields normally used to protect from radiative heat during firefighting operations. Also known are water curtains mainly consisting of fire hoses with different numbers of upward directed spray nozzles. Water curtains are more flexible in application. Water shields and water curtains are available with adjustable flow rates and for different operating pressures. The effectiveness largely depends on the spilled gas type and the nozzle and flow configuration of the chosen appliance.

2.1.1 Absorption techniques

Absorption techniques for emergency responders are in general not well established. For different purposes the effects of water on water soluble gases is known and well understood. The application of knowledge from industrial processes and their applicability for emergency responders is scientifically not well reviewed. Commercial portable gas absorbers are established in some industrial emergency response crews but mostly have very low volume rates. For water soluble gases some absorption techniques concerning the use of water aerosols or spray streams are known but often improvised. For chlorine spills the effects of sodium thiosulfate were investigated and have proved to be very effective [1,2], Other chemicals for absorption purposes regarding emergency response have not been studied very well. Effluents from absorption processes could cause hazardous situations too, but absorption in general offers the chance to reduce the three dimensional problem of a hazardous gas release to a two dimensional problem of a hazardous liquid.

2.2 *Response at Trieben (Austria)*

On September 5th 2011 a forklift at a metallurgical production site in Trieben (Austria) by accident dropped an IBC with 1000 liter 53% of nitric acid in the chemical storage area. The local fire brigade was called at about 11:00 am and arrived about five minutes later. The incident commander noticed during size up operations orange/brown fumes inside the cellar of the building. Most of the cellar was affected, about 200m² were directly contaminated with nitric acid. It is assumed that before the fire brigade was called, workers tried to dilute the nitric acid with up to 3000 liter of water, which caused severe a deterioration of the situation. About 30 minutes later the hazmat team arrived and started measurements with detector tubes. Values of around 100 ppm for nitrous fumes were measured in all areas of the cellar. The nitrous fumes occurred from the reaction of nitric acid with the material of the storage racks in the storage area. The lower parts of the storage racks were principally protected by a coating material but at some points the coating was damaged. Emergency responders over nearly 24 hours tried to contain the spill with chemical binding agents and chemical vacuum cleaners. Due to the contamination with nitrous fumes gastight suits were used which lead to huge logistical demands. The normal working time for the cleaning teams was around 45minutes to an hour, they were supplied by air hoses from a mobile compressor for respiratory air. Due to the close vicinity of the surrounding neighborhood it was not possible to vent the nitrous fumes directly in the atmosphere. A mobile exhauster was placed a spray nozzle was fixed near the exhaust opening of the ventilator. The flow rate of the water stream was around 200 l/min at 6 bar. The exhauster had an output of around 20.000 m³/h. During the exhaust process the concentration of nitrous fumes was constantly monitored and well below 8,2 ppm (AEGL 2, 4h for NO_2) at the direct vicinity of the exhauster. The effect might have not only occurred through absorption effects but also because of a dilution effect of excess air at the input side of the exhauster.

2.3 *Absorption tests with the MSA Auer COMVEX Ventilator*

To investigate the potential of improvised gas absorption techniques the applicability of the MSA Auer COMVEX Ventilator was chosen as an example. The COMVEX Ventilator is used by many Austrian fire departments and offers the possibility to mount a water spray system on the ventilator that is normally used for the production of high expansion foam.

2.3.1 MSA Auer COMVEX [3]

The MSA Auer COMVEX Ventilator mainly consists of an air fan and tubing for suction and exhaust purposes. The fan has an air consumption of 10.000 m^3/h and a power of 2,6 kW. Additional to the fan, a top piece for the production of high expansion foam is available (Pic.1). The foam generating kit consists of short tube with five spraying nozzles and a mixer to be able to ad foaming agents. At an operating pressure of 4-8 bar the water consumption of the foam generating kit is around 100 l/min. Mixing rates of 1-3% are possible.

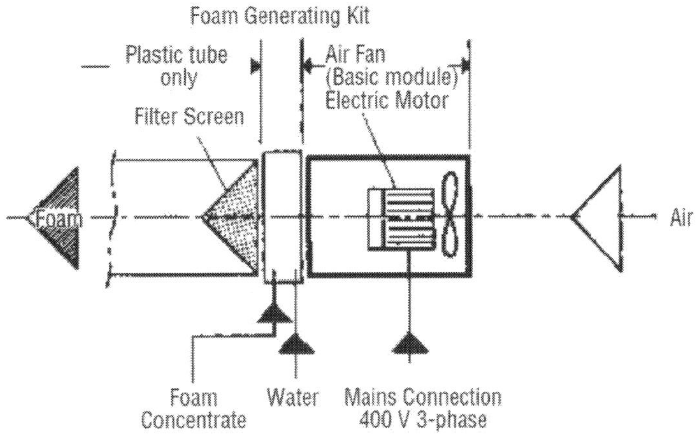

Pic. 1: Principal layout of the COMVEX Ventilator system [3]

2.3.2 Sodium thiosulfate as absorption material

Many absorption materials for technical purposes are well known. For emergency response purposes it is crucial to have absorption materials that present a low hazardous potential and have a good availability in large quantities. Logistically it is not possible to store large amounts of different absorption chemicals for specific tasks. For high risk chemicals (concerning toxicity and used quantities) it may be reasonable to foresee special measures. In case of chlorine that has a high toxic potential and is commonly used for chemical production processes and disinfection purposes, sodium thiosulfate gives good results as absorption agent in combination with water [2].
Sodium thiosulfate in an aqueous solution reduces elementary chlorine to chlorine anions. The reaction is shown in Equation 1:

$$2Na^+ + S_2O_3^{2-} + 4Cl_2 + 13H_2O \rightarrow 2Na^+ + 2HSO_4^- + 8\,H_3O^+ + 8Cl^- \text{ [Eq. 1]}$$

One mol of sodium thiosulfate is able to reduce 4 mol of elementary chlorine and has therefore a high potential to be used as absorption agent. Sodium thiosulfate pentahydrate has a solubility of around 680 g/l at 20 °C which equals a concentration of about 2.7 mol/l.

2.3.2 The MSA Auer COMVEX as improvised absorber

The focus of the investigations was the absorption of ammonia and chlorine at concentrations around 200-300 ppm. The ventilator was equipped with a suction tube (10 m) and an equally built exhaust tube. Tests with ammonia where carried out with inlet concentrations of 250 and 290 ppm. The ammonia feed stream was provided by a bottle of pressurized ammonia, the concentration was monitored by detector tubes. At the start of the experiment a constant ammonia stream was adjusted and measured at the exhaust tube while the ventilator was operating. When constant concentration conditions were reached, the injection of water started and the gas concentration was measured again. At the first test the ammonia concentration dropped from 250 ppm to 100 ppm (33.3%). The second run lead to a concentration of 150 ppm at a starting concentration of 290 ppm (48.3%).

Pic. 2: Improvised absorption system during experiments with chlorine

Due to the good solubility of ammonia in water, the tests were only carried out with water. The second series of experiments was carried out with chlorine in the same way. The first test with chlorine was started at a concentration of 490 ppm, the injection of water caused no measurable reduction of the concentration. In the second experiment with chlorine, a 2,7 mol/l solution of sodium thiosulfate was added to the water stream by the mixer of the foam generating kit (Pic.2). A mixing ratio of 3% was chosen. The use of sodium thiosulfate resulted in a reduction of the chlorine concentration from 260 ppm to 100 ppm (61.5%).

3. Conclusion

The effectiveness of gas absorption techniques is not very well known by emergency responders in Austria. Examples from real incidents show that the availability of gas absorption techniques may improve the effectiveness of emergency response at toxic gas releases. It is clearly not possible to have ready gas absorption appliances and agents for a wide range of specific problems. Scientific research on available improvised techniques and the development of technical improvements may enhance their effectiveness. Experiments with sodium thiosulfate in cases of chlorine spills show an effectiveness of over 60% at low chlorine concentrations. Keeping in mind that the investigated ventilator/nozzle system was not designed for gas absorption in general, the results may still leave a lot of room for improvements. Future research will focus on the improvement of the effectiveness of different methods for the control of toxic gas releases.

4. References

[1] Engelhardt, F., "Chlorabsorption an fallenden Tropfen und Übertragung auf Wasservorhänge," *Bergische Universität –Gesamthochschule Wuppertal*, Dissertation Juni 2002

[2] Bauch, M., Peine, T., "Versuche zur Erhöhung der Effizienz beim Niederschlagen von Chlor," *Brandschutz - Deutsche Feuerwehr Zeitung*, NO. 1, 2010

[3] "MSA AUER COMVEX Portable Ventilator and High Expansion Foam Generator - Instruction Manual," MSA AUER

9TH GLOBAL CONGRESS ON PROCESS SAFETY

Case Studies in Process Safety: Lessons Learned from Software-Related Accidents

Terry L. Hardy
Director, Safety & Risk Management
Great Circle Analytics, LLC
1238 Race Street
Denver, CO, 80206
thardy@gcirc.com

Prepared for Presentation at
American Institute of Chemical Engineers
2013 Spring Meeting
9th Global Congress on Process Safety
San Antonio, Texas
April 28 – May 1, 2013

Case Studies in Process Safety: Lessons Learned from Software-Related Accidents

Terry L. Hardy
Director, Safety & Risk Management
Great Circle Analytics, LLC
1238 Race Street
Denver, CO, 80206
thardy@gcirc.com

Keywords: Safety Analysis, Software Safety, Computing Systems, Process Safety Management

Abstract

Software and automation can play a significant role with respect to safety in the chemical process and energy production industries. Because computing systems are increasingly used to control critical functions, software may directly contribute to an accident. On the other hand, software can also be used as part of the hazard control strategy to reduce risks, and computing systems can provide valuable information to help make safety decisions. The importance of including software as part of an organization's efforts to analyze and manage hazards and risks seems clear, but for many organizations software is not effectively incorporated into process safety efforts. This paper reviews lessons learned from accidents and incidents to illustrate the potential for a software-related accident even when process safety management tools and techniques are used. This discussion is intended to provide insights to help improve process safety and software safety efforts.

1. Introduction

Process Safety Management generally refers to the application of management principles and systems to the identification, understanding, and control of process hazards to protect employees, facility assets, and the environment. Process Safety Management is focused on prevention of, preparedness for, mitigation of, response to, and restoration from catastrophic releases of chemicals or energy from a process associated with a facility. In Process Safety Management, a process is defined as any use, storage, manufacturing, handling, or the on-site movement of highly hazardous chemicals, or combination of these activities. Process Safety Management is regulated by OSHA through the Process Safety Management (PSM) Standard.

Software and computing systems are becoming essential to the operation of most industries, often through the automation of tasks, and as such can have a major impact on process safety. Software and computing systems offer tremendous advantages that encourage the use of automation. Software can allow complex operations to be performed, and can perform complicated calculations. Computing systems are reliable in the sense that there are a small number of components to perform a large number of actions. Computing systems are also flexible and can be modified to fit the application. From a safety perspective, automation can reduce the risks to operators by separating them from dangerous environments such as those in hazardous chemical processing facilities. In addition, automation can help by performing tasks that might pose a threat to humans, such as lifting large objects or operating high-speed cutting tools.

Software, computing systems, and automation have disadvantages as well. A major disadvantage to computing systems is that they often introduce increased complexity to systems, making them harder to understand, operate, maintain, and test. This complexity can also introduce new failure modes and interactions. Complex systems have a greater number of requirements, more interfaces, more subsystems, and more logical operations, increasing the potential for error. Computing systems can create a situation where no single operator can immediately foresee the consequences of a given action in the system. Computing systems can also increase coupling. Tight coupling occurs when processes are highly integrated and intrinsically time-dependent; for example, once a process has been set in motion, it must be completed within a certain period of time. Tightly coupled systems allow little time for operators to take action to prevent adverse effects. In addition, automation may require that operators think about their tasks in new ways, and may change the fundamental role of the operator from one who physically performed tasks to someone who monitors and communicates information. Finally, software will only do what a programmer has told it to do. Therefore, unlike humans, software cannot react to an abnormal situation unless specifically told to.

In spite of the fact that software is such an important part of complex systems, the analysis of hazards and risks from software has been inconsistent in the process industries. Safety analyses have historically been hardware-focused. Therefore, many analysts may not understand how to incorporate software into their system hazard analyses, and evaluators of those analyses may not understand what should be included and assessed. Organizations may be focused on compliance to regulations, which often do not address software, and as a result organizations may not properly assess or mitigate software risks. Organizations need to increase the attention given to addressing and analyzing the potential for hazards related to software and computing systems.

Software includes computer programs, procedures, scripts, rules, and associated documentation and data pertaining to the development and operation of a computer. Software can be developed by the organization implementing the system, by an outside software developer, or may be purchased as Commercial Off-The-Shelf (COTS) software. Software safety encompasses not just the software but also the computing system. A computing system includes the software and supporting hardware, sensors, effectors, humans who interact with the system, and data necessary for successful operation. Examples of computing systems include Programmable Logic Controllers (PLC) and Supervisory Control and Data Acquisition (SCADA) systems.

Through the review of hundreds of software-related accidents and incidents, and the author's personal experience in this discipline, common themes and lessons have been found.[1] This paper will use case studies to illustrate those themes and lessons learned. The discussion will center on software safety as part of a broader process safety effort and will include recommendations that can be used to improve the safety of software-driven systems.

2. Lessons Learned: Software and Computing System Related Accidents

This section discusses a number of lessons learned related to accidents and incidents where software and computing systems were contributors. These accidents are taken from detailed reports and investigative summaries from multiple industries and organizations, and are included to broadly illustrate hazards and risks in software and computing systems. Note that in discussing these accidents, this paper does not intend to oversimplify the events and conditions that led to the accidents, or place blame on any individuals or organizations. There is rarely a single identifiable cause leading to an accident. Accidents are usually the result of complex factors that include hardware, software, human interactions, environments, and procedures. The descriptions are meant to provide examples of where the analysis process failed in some way. Readers are encouraged to review the complete accident and mishap investigation reports referenced here to understand the often complex conditions and chain of events that led to each accident discussed here.

2.1 Decisions made in the acquisition and planning phases of development can profoundly affect safety

Planning typically involves trade-offs between many different facets of the program, including cost, schedule, performance, and safety. Poor planning can lead to unexpected safety consequences, and many safety decisions are actually made in the planning and acquisition phase. For example, inadequate resources may be allocated to the software safety effort. This can result in a failure to perform hazard analyses and identify safety requirements early in the program when these activities have the most impact. Therefore, safety personnel should be included in early phases of a program, especially when developing contractual requirements, and software safety considerations should be included in schedule and resource discussions.

On August 11, 1985, an Institute, West Virginia chemical production facility leaked methylene chloride and aldicarb oxime, toxic chemicals used to manufacture the pesticide Temik. Six workers were injured, and more than a hundred residents living near the facility were sent to the hospital. The toxic chemical leak occurred when a tank overheated after a steam valve failed. Due to the failure, steam leaked into the heating jacket of a storage tank containing a mixture of methylene chloride and aldicarb oxime. The methylene chloride vaporized from the heating, leaving concentrated aldicarb oxime to settle on the bottom of the tank. Operators later pumped the material from the storage tank for processing. They thought they had emptied the tank, but concentrated aldicarb oxime remained in the storage tank because a level gauge was broken and the operators performed no additional checks. The concentrated aldicarb oxime continued to be heated over several days by the steam in the heating jacket, ultimately resulting in a runaway chemical reaction days later. The high pressure from the reaction led to release of aldicarb oxime through a flare, and large amounts of the chemical were also released to the atmosphere when a burst disc failed.

The root causes of the leak were found to be the use of a tank that was not designed to hold aldicarb oxime, faulty level indicators, defective safety valves, mistaken transmission of steam to a tank, and failure of control room operators to notice critical pressure and temperature changes. In addition, a computerized monitoring system contributed to the severity of the accident. Earlier in the year, the company had installed a new air-monitoring system that could automatically identify and detect releases of hazardous gases. That computerized system could also predict whether the gases would be contained within the release area or would migrate and potentially harm the public. This information was important because it could be used to mitigate risks through evacuation or other measures if a vapor cloud appeared to be spreading. Company officials initially stated that the computer system had failed and therefore they were not informed of the lack of containment of the aldicarb oxime gas cloud. However, they later admitted that they had not purchased a version of the software programmed to detect aldicarb oxime. Had the company invested in the more expensive version of the software, they could have predicted that a cloud of aldicarb oxime would not have been contained in the production area. However, the company had only ordered the basic software model that did not contain aldicarb oxime properties.[2, 3]

2.2 Hazard causes may not include software and computing systems

Many organizations are improving their safety analyses with respect to software. However, software safety still lacks sufficient numbers of qualified practitioners, and the methods for performing comprehensive analyses are not universally known. In many cases, software causes may not be included in hazard analyses. Or if they are, the software causes may be stated in general terms in hazard analyses, such as including a generic hazard cause called "software error," instead of defining specifically what software functionality can lead to an undesirable outcome. In addition, the software hazard analyses may not pay enough attention to those cases where the software works exactly as intended, but the implemented functionality was unsafe.

On October 18, 2005, an explosion occurred at a natural gas decompression/recompression facility near Empress, Alberta. Extensive damage occurred to the facility, but no injuries were reported. The facility decompressed and conditioned high-pressure gas from a pipeline to achieve pressures and temperatures consistent with what was needed in the Natural Gas Liquids (NGL) stripping facilities. At the time of the accident the facility's "A" PLC lost communication with the NGL facility control room. Transportation Safety Board of Canada (TSB) investigators found that a processor card had failed, shutting down the PLC. When the PLC failed its outputs were de-energized. However, some of the microprocessor functions required an energized signal to initiate physical shutdown of some functions. These functions included main and auxiliary lube oil pumps, purge/damper valves, and suction and discharge valves. As a result of this condition, the pumps did not shut down and valves did not close in a safe position when the PLC failed. Operators in the NGL control room received an alarm indicating that the communications were lost. They investigated the problem and believed that the failure in communication had triggered an emergency shutdown of the facility. The operators believed that the emergency shutdown meant that all components were in a safe state. However, because the recompressor motor did not shut down, a surge event occurred in the pipe. This surge caused a pipe to break, leading to a natural gas leak. The leaking natural gas ignited, resulting in an explosion. The TSB noted that the loss of a PLC had not been identified as a safety concern in the hazard analyses prior to the accident.[4]

2.3 Risks may be underestimated or optimistically evaluated

The risk assessment process helps decision makers in their risk decisions (e.g., accept the risk without change, reduce the risk to an acceptable level through hazard elimination and mitigation, transfer the risk, or forego the activity), assists in justifying the acceptance of residual risk, and allows communication of risks throughout the organization. A risk assessment can either be qualitative or quantitative, although the emphasis in process safety is often on qualitative risk assessment. Evaluating software and computing system risk can be difficult, especially if an extensive history does not exist with the automated equipment. Risk must be carefully evaluated based on known technical factors, such as design complexity and maturity, degree of system testing, use of unproven technologies, potential for unexpected human intervention, and so on. Accidents have illustrated unfounded optimism in the use of software to assure safety.

On April 7, 2000, a pipeline failure occurred at the Chalk Point Generating Station in Maryland. The pipeline released approximately 140,400 gallons of fuel oil into surrounding wetlands, Swanson Creek, and the Patuxent River. The cost of the environmental response and cleanup exceeded $71 million. The U.S. National Transportation Safety Board (NTSB) determined that the probable cause of the accident was a fracture in a buckle in the pipe. The buckle went undiscovered because data from an in-line inspection tool were interpreted incorrectly. The NTSB stated that a contributor to the severity of the accident was inadequate operating procedures and practices for monitoring the magnitude of flow through the pipeline. When the leak occurred the 12-inch-diameter pipeline was being cleaned. An automated pipeline monitoring system existed, but this automated system could not adequately monitor pipeline conditions during this operation because the locations of the meters, pressure-sensing points, and temperature-sensing points had not been configured to be in the direct liquid flow path. Therefore, the automated system did not provide alarms or other information to operations personnel. Field personnel manually recorded tank level measurements at the upstream and downstream points, but did not use this information to evaluate whether any fuel oil had been lost. The manual monitoring procedures and lack of alarms from the automated system led to a 7-hour delay in responding to the leak. The field personnel finally realized there was a problem when the fuel pump began to cavitate. Using hand calculations the crew then realized that they had not received over 3000 barrels of fuel oil and they shut the pipeline down. Following the accident the company installed a SCADA system with software-based leak detection and radar tank gauges. In this case the operators underestimated the risks associated with a leak and overestimated their ability to detect a failure.[5]

2.4 Hazard controls may rely only on good software processes and testing

Hazard controls are devices and approaches to mitigate risk presented by a hazard by reducing either the severity of the hazard or the probability of its occurrence. Strong hazard control design follows the design order of precedence, where the first approach is to try to design out the hazard or minimize the risks through design selection. Many hazard reports will focus on the implementation of good software processes, or extensive unit testing, to prove that the design is safe. However, the software processes and testing will not help if the software design is flawed with respect to safe system operations (including hardware, software, processes, humans, and environments). In these cases the software may operate exactly as designed, but the operation may be unsafe.

On February 11, 2003, an employee of a manufacturer in Gonzales, Texas was fatally injured while performing maintenance on a reaction tank. The U.S. Mine Safety and Health Administration (MSHA) determined that the cause of the accident was a failure to close and secure a manual gate valve for a steam line and a failure to place the batch PLC in the stop mode. The company was a surface clay mill that purchased clay and blended, refined, milled and processed the material into products used in paints, inks, and grease. On the day of the accident the employee had been informed that there had been a product change in one of the batch processing systems. The employee was assigned to perform cleanup duties on a reactor tank. Two valves controlled steam entry into the tank: a manual gate valve and a butterfly valve with an automatic pneumatic actuator. The PLC controlled the functioning of the batch system based on sensors that monitored material flow. At the time of the accident the PLC was in "slurry hold" mode. In this mode the system was programmed to actuate the steam valve when the clay slurry level reached 5.5 feet. An aluminum extension ladder used by the employee caused the level sensor to falsely sense that slurry was in the reactor, which resulted in the PLC sending a command to open the steam valve. Because the manual valve had been left open, steam at 350° F then entered the tank, fatally burning the employee.[6]

2.5 Organizations may fail to consider the possibility of hazard controls not working

Hazard controls are never 100 percent reliable. Therefore, it is important for the analyst to consider what could happen if the controls fail. Consideration should be given to unexpected hardware, software, and human interactions, and a "defense in depth" approach should be used where multiple, independent methods are used to prevent a hazard from becoming an accident. Controls should be carefully evaluated to identify what could happen if they do not work as expected, especially if those controls rely heavily on software and personnel actions.

On February 18, 2009, an employee was fatally injured at a coal preparation plant in the Hunter Valley region of New South Wales, Australia. The accident occurred when 10 tons of waste rock were inadvertently released from the reject bin and fell onto the cabin of the employee's truck. At the plant, raw coal was extracted from the mine and usable coal was separated from waste rock. The waste rock was transferred approximately 2 kilometers on conveyers to the reject bin. The waste rock was then loaded from the reject bin onto trucks and hauled away. The process of loading the trucks with waste rock was controlled by a PLC system. The PLC system included truck detection sensors, traffic lights, bin capacity sensing, and remote control, hand-held transmitters used by the truck drivers. On the day of the accident the truck driver drove his truck under the reject bin delivery chute. A signal was sent from the handheld remote control to command the chute to open. The accident report stated that it was not clear whether the signal was sent inadvertently or intentionally. Opening the chute required that two of three lines of truck detection sensors be blocked in addition to a command from the remote control to assure that the truck was in the correct location. Each sensor line contained three sensors, and all three sensors had to be blocked for the entire line to be considered as blocked. At the time of the accident the truck was obscuring one line of sensors, and a second line of sensors was obscured by dirt on the lenses and therefore was not working correctly. Because two of the sensor lines were blocked and the remote control signal had been sent, the PLC automatically opened the reject bin chute door and dropped 10 tons of material on the truck cab before the driver had safely cleared the chute, resulting in the fatal injury.[7]

2.6 *Testing may not provide information on subsystem and component interactions or hazard control operations*

The safety verification and validation process is intended to determine that the design solution has met all the safety requirements (verification) and that the correct system is built (validation). The verification and validation process, if performed correctly, will provide evidence that risk has been reduced. Testing is an important part of the verification and validation process, and comprehensive software testing should be conducted throughout the development cycle. Testing should include not only nominal conditions based on requirements but also abnormal conditions such as improper inputs, inputs outside expected conditions, inputs exactly at and near the boundary conditions, and inputs stuck at some value. Safety-critical software must include full system integration testing of end-to-end events. That testing must include stressing of the software, and should include interactions of the software with hardware, humans, and environments. A number of accidents have occurred when no component failed in the conventional sense, but the interaction of components caused a system failure. Testing must include verifications of hazard controls, and it should assure that redundancy works when needed.

On October 24, 2002, a grinder exploded at a limestone processing plant in Foreman, Arkansas. An operator was killed when flammable waste fuel covered him and ignited. The operator had started the pump for solid waste fuel processing when the accident occurred. The MSHA stated in its report that the cause of the accident was that the safety monitoring system designed to shut off the waste fuel system pump had not been maintained so that it functioned properly. Kilns were used in processing activities at the plant, and these kilns were heated by burning coal, natural gas, and liquid waste fuel. The liquid waste fuel was delivered by truck or railcar and pumped into large storage tanks. From the storage area it was pumped through a grinder to reduce the particle size of the solids in the fuel. Two independent systems monitored and controlled the waste fuel delivery. A Foxboro Intelligent Automation Distribution Control System monitored and recorded normal operating parameters. The Foxboro also issued audible and visual alarms that were available at the plant control room. A PLC provided basic startup and shutdown of the system and responded to commands from the Foxboro. On the day of the accident the Foxboro sensed that the fuel delivery pressure was low, apparently due to blockage in the line. As designed, the Foxboro sent a command to the PLC to shut down the pumps. However, the PLC failed to respond and the pumps kept running. Three months prior to the accident this PLC had been installed; this was supposed to be a simple replacement of an older PLC of similar capability. However, the Foxboro had not been connected to the newer PLC, and the connections remained to the older non-functioning PLC. The complete system had never been tested with the new PLC. A system test had been scheduled three days prior to the accident but had been aborted when a pump failed during the test; the test had never been rescheduled. The accident report stated that the blockage may have broken free just prior to the accident. With the pumps running, the pressure elevated significantly and a "water hammer" effect caused overpressurization in the system at the grinder. The grinder was torn loose from its base, spraying fuel and pulling loose a 480-volt cable that ultimately served as an ignition source.[8]

2.7 Software change management and hazard analyses processes may not be integrated

Engineering by its very nature is an activity that requires change. Changes can come about for a variety of reasons, including the discovery of problems during development, changes in requirements, or routine upgrades of software or hardware. As the development cycle proceeds and as an organization gains more operational experience, new hazards are often uncovered and some hazards may no longer be relevant. If the hazard analysis is not updated to reflect these changes then resources may be expended on previously identified hazards that may no longer be relevant, and new hazards may not be discovered as the design matures. In addition, routine changes and modifications to software and computing systems must be analyzed for potential hazards as part of the broader hazard analysis process.

On June 10, 1999, a 16-inch-diameter pipe carrying gasoline ruptured in Bellingham, Washington. The ruptured pipeline released approximately 237,000 gallons of gasoline into a nearby creek, according to the NTSB. That gasoline then ignited, burning approximately 1½ miles along the creek. Three people died in the fire. One home and the Bellingham water treatment plant were also damaged in the accident.

The NTSB investigated the accident and determined that the probable cause of the rupture was damage to the pipe during a modification project performed in 1994. This damage weakened the pipeline, making it susceptible to rupture under increased pressure in the pipe. The NTSB also stated that inspections of the pipeline during the project were inadequate and the company did not identify and repair damage. The report noted that in-line pipeline inspection data should have prompted the company to excavate and examine that section of pipeline, but the company failed to perform such work after reported anomalies. The SCADA system computers also played a role in the accident. The SCADA system was used for operation of the pipeline, for example to open and close valves remotely as required or to operate pumps as needed. Just prior to the accident the operator was preparing to initiate delivery of gasoline to a terminal in Seattle, diverting delivery from another facility. During the process of switching delivery destinations, the pressure in the pipeline began to increase, which is a normal condition but one that required the operator to start a pump to reduce pressure. However, when the operator tried to start that pump, the SCADA system failed to execute the start command issued by the operator. The operator soon found that the SCADA system was unresponsive to any commands, something that had never happened before. The report stated that, "Had the controller been able to start the pump at Woodinville, it is probable that the pressure backup would have been alleviated and the pipeline operated routinely for the balance of the fuel delivery." Instead, the pressure in the pipe increased, and the increased pressure likely caused the damaged pipe to rupture.

The cause of computer system failure was likely a change made to the system database just prior to the accident. The NTSB accident report stated that the SCADA system administrator entered new records into the live database at the time of the accident. The system administrator however did not check the records or test the system software to see if those changes introduced any problems. The computing system problem could not be replicated after the accident and therefore the cause of the anomaly could not be definitively identified. The report stated, "The Safety Board concludes that, had the SCADA database revisions that were performed shortly before the accident been performed and thoroughly tested on an off-line system instead of the primary on-line SCADA system, errors resulting from those revisions may have been identified and repaired before they could affect the operation of the pipeline."[9]

2.8 Human-software interactions have significant safety implications that are often underestimated

Humans interact with hardware and software in a variety of ways. A number of accidents have occurred where these interactions have been a significant contributing cause. Of special concern is when a new computing system is implemented that changes what is expected of the operator in normal and emergency situations.

On August 28, 2008, an explosion occurred at a pesticide manufacturing plant in Institute, West Virginia. Two workers were killed in the explosion and eight others were injured. The U.S. Chemical Safety and Hazard Investigation Board (CSB) found that the explosion was the result of a runaway chemical reaction. The company was starting up a methomyl unit for the first time after several months of down time for maintenance to install a new computer control system and reactor vessel. Normal operations called for dissolved methomyl and waste chemicals to be fed into a preheated residue treater vessel partially filled with solvent. The treater allowed the methomyl to decompose safely, after which it was mixed with other waste chemicals and used to fuel facility boilers. On the day of the accident a methomyl-solvent mixture was added prematurely to the residue treater vessel before solvent had been added to the tank. This event occurred when operators were troubleshooting equipment problems during the startup. This mixture was supposed to be added to the vessel only after that vessel was filled with clean solvent and heated to a minimum safe temperature. An interlock system existed as part of the automatic feed control system to prevent inadvertent introduction of methomyl. This interlock was password-protected to prevent inadvertent override, but operators intentionally overrode the interlock. Bypassing the interlocks had apparently been a practice condoned by management in the past as a workaround for operational problems. Once the methomyl decomposition reaction started it could not be stopped, and the pressure rapidly rose in the vessel due to gas from the reaction, leading to the explosion. Uncrystallized methomyl existed in the tank due to equipment problems, and this material greatly increased the methomyl concentration in the residue treater, contributing to the runaway reaction.

The CSB stated that the company initiated this process startup before the company had completed critical checks, including valve lineups, equipment checkouts, and computer calibrations. The CSB said that the operating procedure for the startup was complex, but this procedure had not been reviewed or approved. In addition, the company had not performed training on the new computer system put in place as part of the maintenance. This new computer system offered significant improvements to help automatically control the operation. For example, the control system included graphical display screens that simulated the process flow. But the system was also complex and the modifications changed the way operators interacted with the system. The operators thought the screens were difficult to navigate, and responding to troubleshooting alarms was difficult. The accident report stated that, had the operators received adequate training on the new computer system, they may have been able to recognize problems in operation before the explosion.

The report faulted the company for not performing an adequate pre-startup safety review. System operators did not participate in the safety review, and review checklists showed items as completed when they were not. The CSB stated that the company had also failed to perform a thorough Process Hazard Analysis. The report stated that the Process Hazard Analysis was performed quickly because management had not allotted sufficient time for analysis. In addition, the CSB stated that the Process Hazard Analysis included invalid assumptions, and said that the team did not apply the tools properly, resulting in unmitigated accident scenarios.[10]

2.9 Support software, including models and simulations, may be as critical to safety as control software

While the focus of software safety efforts is usually on software directly controlling an application, support software may contribute to an accident or to the effectiveness of the emergency response. Examples of support software include databases used for maintenance activities, software to estimate load stability, models that provide design calculations or assurance information, and so on. A failure to analyze the hazards associated with this support software can result in unforeseen system failures.

On November 12, 2008, a 2 million gallon liquid fertilizer tank at in Chesapeake, Virginia collapsed. Two workers performing welding operations at the site were seriously injured and an adjacent neighborhood was partially flooded as a result of the accident. The CSB found that the company had not assured that welds to replace vertical joints met accepted industry standards, and the CSB faulted the company for its failure to perform inspections of the welds. The company was also faulted for not having proper procedures in place for filling the tanks following major facility modifications. In its report, the CSB also noted that the contractor hired by the company to calculate the maximum fill height had used some faulty assumptions. The maximum liquid level was supposed to be calculated in part based on the minimum measured shell thicknesses and the extent of the weld inspection (full, spot, or no radiography). The contractor used the maximum (not minimum) measured thickness, and assumed full inspection of the welds. These assumptions led to an overestimation of the allowable liquid level. The tank failed at a fill level of 26.74 feet, below the calculated maximum of 27.01 feet. The CSB also noted a number of previous overfilling accidents. The CSB found 16 other tank failures at nine facilities in other states between 1995 and 2008. These 16 failures resulted in one death, four hospitalizations, one community evacuation, and two releases to waterways. Eleven occurred due to defective welding.[11]

3. Recommendations for the Software System Safety Process

There is a great deal of information in the literature on how to improve software development processes. However, the above themes and lessons learned indicate that improving software and computing system safety is more than just providing additional training or improved software development techniques or tools. Improving software safety requires a change in the way process safety practitioners think. Software safety must be fully integrated into the process safety analysis process. And process safety practitioners must begin to ask critical questions about their software safety efforts. Some examples of these questions include the following:

- Do plans reflect how business is really done? Are plans reviewed? Do plans have unrealistic schedules or resource allocations? Are software and computing systems considering in the planning and acquisition processes? Poor or unrealistic plans may reflect an organization that does not truly place a priority on safety activities.
- Is there a convincing story that the safety analysis is complete? Is there a sufficient description of the system to support an analysis, including the software and computing system? Are computing systems treated as "black boxes"? Have software causes been thoroughly evaluated using a systematic approach, or are the software causes listed as "software error" with no further explanation? Do the hazard analyses include support software, such as design models or databases? Failure to show that the problem is being looked at systematically could indicate that there are holes in the analysis, with potentially significant problems overlooked.
- Are the hazard reports detailed enough? Are causes descriptive? Are the recommendations clear and sufficiently detailed? Does the logic make sense and is it complete? Do controls match up with the causes, showing a one-to-one or many-to-one relation? Lack of detail could be an indication of insufficient knowledge of the system, or lack of information on the system.
- Do the risk assessments have a logical basis? Have the risk assessments considered software complexity, maturity, testing, use of unproven technologies, etc.? Are assessments overly optimistic based on what is known about the computing system? A failure to provide a basis for your risk assessments may result in unrealistic assessments.
- Are the hazard controls primarily procedural rather than design changes, safety features, or devices? Is there an overreliance on alarms and signage? Do software controls only rely on good software processes? Is there an overreliance on humans and software to "save the day"? Overreliance on operational controls may indicate a weak safety design.
- Are hazard control recommendations being implemented? Can the selected hazard control strategy actually be implemented and verified? Is the control strategy so complex that it will be impossible to determine whether it will work when needed? Complex controls, overlapping control strategies, or inadequate implementation may be an indication of a weak safety design.
- Has the risk assessment truly considered the worst case? What is the basis for the likelihood levels? Is the risk analyzed only for steady-state operation, or have startups and shutdowns also been considered? Failure to provide good answers to these questions indicates a potential misunderstanding of the risk.
- Are problems found in test and design included in the hazard reports and factored into the design? Have problems and incidents been fully investigated? Does a process exist for tracking action items, including whether recommendations have been implemented? Are changes to software and computing systems properly factored into the hazard analysis? Failure to incorporate problems, changes, and corrective actions is an indication of the potential to miss serious design flaws.

These questions help to identify whether the hazard analysis process is robust. However, we must also ask questions specifically related to the use of software and computing systems in complex systems. The best questions come from real-world examples of accidents where software has been a contributor. Some examples of questions are as follows, and others can be found in the literature.[12]

- Have safety-critical software, commands, and data been identified?
- Do hazard controls for software-related causes combine good practices and specific safeguards?
- Is software and system testing adequate, and do tests include sufficient off-nominal conditions?
- Is the computing system design overly complex?
- Is the design based on unproven technologies?
- What happens if the software locks up?
- Are the sensors used for software decisions fault tolerant?
- Has software mode transition been considered?
- Has consideration been given to the order of commands and potential out of sequence inputs?
- Will the software and system start up and shut down in a known, safe state?
- Are checks performed before initiating hazardous operations?
- Will the software properly handle spurious signals and power outages?

These are by no means all the questions a decision maker should ask, and positive answers to these questions provide no assurance that an accident will be prevented. These questions should encourage critical thinking about safety and generate additional questions to provide further insight on system risk. A failure to ask these questions could mean that the potential for an accident is higher than we had assumed.

It is up to all stakeholders to look for those conditions that could lead to an accident and to recognize that the worst can happen. This means we should all express concerns about software safety management and engineering when necessary based on our knowledge, experience, and judgment, including lessons learned from accidents. We must ask questions to understand the potential for harm, to understand the steps taken to assure that the risks have been reduced, and to assure that there is proof that hazard controls are effective. We do not have to be software experts to ask questions, and in fact a lack of expertise could be an advantage when trying to understand how a system works. Most importantly, when thinking of software and computing system safety, we must think of the system and its interactions between software, hardware, humans, processes, and environments.

4. Conclusion

Process Safety Management can provide immense benefits in reducing operational risks in the chemical process and energy production industries. By proactively identifying hazards, assessing and characterizing risks, and taking actions to reduce those risks, organizations can prevent accidents and reduce the potential for death, injury, property damage, and environmental impacts. Given the importance of software and computing systems in the design and operation of many systems, software must be included as part of the broader process safety effort. However, the accidents provided in this paper, along with many others, illustrate that software is often not systematically considered as part of process safety activities. For example, software may not be identified as a hazard cause, risks may be underestimated, software-related hazard controls may be oversimplified, testing may not include sufficient safety scenarios, and system changes may not be factored into the design or hazard analysis. Process safety practitioners must take advantage of available software and computing system lessons learned, going beyond examples presented here, to improve their own safety efforts and, most importantly, to prevent accidents.

5. References

[1] Hardy, T.L., *Software and System Safety: Accidents, Incidents, and Lessons Learned,* AuthorHouse, 2012.

[2] Pool, B., "System Wrongfully Blamed in Union Carbide Leak: SAFER Fights Off Dark Cloud of Bad Publicity," *Los Angeles Times,* August 20, 1985.

[3] Schlager, N., *Breakdown: Deadly Technological Disasters,* Visible Ink Press, 1995.

[4] Transportation Safety Board of Canada, "Programmable Logic Controller Failure, Foothills Pipe Lines Ltd. Decompression/Recompression Facility, BP Canada Energy Company Empress Natural Gas Liquids Facility Near Empress, Alberta 18 October 2005," Report Number P05H0061, July 12, 2006.

[5] U.S. National Transportation Safety Board, "Rupture of Piney Point Oil Pipeline and Release of Fuel Oil Near Chalk Point, Maryland, April 7, 2000," Pipeline Accident Report NTSB/PAR-02/01, July 23, 2002.

[6] U.S. Mine Safety and Health Administration, "Report of Investigation: Fatal Other Accident (Steam Burns), February 11, 2003, Southern Clay Plants & Pits Southern Clay Prod. Inc., Gonzales, Gonzales County, Texas," Mine I.D. No. 41-00298, 2003.

[7] State of New South Wales (Australia), Department of Industry and Investment, "Fatality involving David Hurst Oldknow Ravensworth Underground Mine Coal Preparation Plant Reject bin 802 18 February 2009," May 2010.

[8] U.S. Mine Safety and Health Administration, "Report of Accident : Exploding Vessels Under Pressure Accident, October 24, 2002, Foreman Quarry and Plant, Ash Grove Cement Company, Foreman, Little River County, Arkansas," Mine I.D. No. 03-00256, 2003.

[9] U.S. National Transportation Safety Board, "Pipeline Rupture and Subsequent Fire in Bellingham, Washington, June 10, 1999," Pipeline Accident Report NTSB/PAR-02/02, October 8, 2002.

[10] U.S. Chemical Safety and Hazard Investigation Board, "Pesticide Chemical Runaway Reaction Pressure Vessel Explosion, Bayer CropScience, Institute, West Virginia, August 28, 2008," Report No. 2008-08-I-WV, January, 2011.

[11] U.S. Chemical Safety and Hazard Investigation Board, "Investigation Report: Allied Terminals, Inc. – Catastrophic Tank Collapse, Allied Terminals, Inc., Chesapeake, Virginia, November 12, 2008," Report No. 2009-03-I-VA, May 2009.

[12] Hardy, T.L., *Essential Questions in System Safety: A Guide for Safety Decision Makers,* AuthorHouse, 2011.

9TH GLOBAL CONGRESS ON PROCESS SAFETY

Technical Report on Chevron 2012 Pipe Rupture and Fire

Steve Cutchen, Lauren Grim
United States Chemical Safety and Hazard Investigation Board
1 Denver Federal Center Way
Building 20, Room A1419
Denver, CO 80225

Roger Evans, Don Holmstrom, Amanda Johnson, Dan Tillema, Kelly Wilson, Mark Wingard
United States Chemical Safety and Hazard Investigation Board

The CSB is an independent federal agency charged with investigating serious chemical incidents. The agency's board members are appointed by the President and confirmed by the Senate. CSB investigations look into all aspects of chemical incidents, including physical causes such as equipment failure as well as inadequacies in regulations, industry standards, and safety management systems. Please visit the CSB's website at www.csb.gov. For further information contact Communications Director Hillary Cohen at (202) 446-8094 or Sandy Gilmour at (202) 251-5496

Prepared for Presentation at
American Institute of Chemical Engineers
2013 Spring Meeting
9th Global Congress on Process Safety
San Antonio, Texas
April 28 – May 1, 2013

UNPUBLISHED

Technical Report on Chevron 2012 Pipe Rupture and Fire

Steve Cutchen, Lauren Grim
United States Chemical Safety and Hazard Investigation Board
1 Denver Federal Center Way
Building 20, Room A1419
Denver, CO 80225

Roger Evans, Don Holmstrom, Amanda Johnson, Dan Tillema, Kelly Wilson, Mark Wingard
United States Chemical Safety and Hazard Investigation Board

Keywords: corrosion, sulfidation corrosion

Abstract

The U.S. Chemical Safety and Hazard Investigation Board (CSB) and the California Division of Occupational Safety and Health (Cal/OSHA) released a technical evaluation report on piping samples taken from the Chevron Refinery in Richmond, California, where a hydrocarbon release and massive fire occurred on August 6, 2012. The report, prepared by Anamet, Inc., concluded the piping ruptured due to severe sulfidation corrosion, and that tested pipe samples showed a very low concentration of corrosion-inhibiting silicon.

1. Background

On August 6, 2012, a hydrocarbon release and resulting fire occurred at the Chevron U.S.A. Inc. Richmond Refinery in Richmond, California The incident occurred when a combustible hydrocarbon liquid known as "gas oil" leaked from an 8-inch pipe connected to an atmospheric crude oil distillation column in the refinery's crude unit. Workers initially noted the leak and were in the process of attempting to diagnose the source of the leak in the still-operating crude unit when the pipe ruptured catastrophically. Due to the high temperature, in excess of 600 degrees Fahrenheit, and physical properties of the material in the equipment, the gas-oil immediately formed a large hydrocarbon vapor cloud.

The CSB determined that nineteen Chevron employees were engulfed in a vapor cloud formed by the hydrocarbon release. Eighteen employees escaped before the fire started and one employee escaped without injury after the fire ensued. The fire burned for approximately 5 hours, and water was applied to the rupture site for 23 days. The incident resulted in six minor injuries. Production at the crude unit was suspended for over six months after the incident. More than 15,000 residents in the surrounding area sought treatment at area medical facilities as a result of the release and fire. Post incident on-site visual examination by Chevron, Cal/OSHA, and the CSB revealed extensive thinning of the pipe wall at the location of the rupture.

2. Sample Testing and Report

The evaluation, performed by Anamet, Inc. of Hayward, California was designed and overseen by the CSB, the California Division of Occupational Safety and Health Administration (Cal/OSHA), the United Steelworkers union, and Chevron. The evaluation was funded by the CSB. The report concludes that the 8-inch steel pipe, from a section designated as 4-sidecut which was installed in 1976, ruptured due to severe sulfidation corrosion, and that tested pipe samples showed a very low concentration of corrosion-inhibiting silicon.

The report cites wall thinning due to sulfidation corrosion as the cause of the piping failure. In crude oil distillation, the report notes, naturally occurring sulfur and sulfur compounds are available to react with steel components, particularly plain carbon steels. Corrosion rates vary according to the sulfur content of the oil being processed, temperature, and other factors, including silicon content and other materials in steel pipe.

The report concludes, "The reported service conditions of the #4 Crude Unit 4-sidecut piping, the composition of the carbon steel, presence of thick sulfide scale on the inside surfaces of the pipe, and generally uniform wall thinning indicate that sulfidation corrosion was the cause of the 8-inch 4-sidecut rupture."

The report noted that the failed pipe section was subjected to a higher corrosion rate than neighboring components due to low silicon content in the pipe. The report states that "Experience has shown that silicon in carbon steel is known to inhibit sulfidation" when the concentration is above a threshold value. Chemical analysis of the Chevron pipe showed the silicon concentration of the ruptured section to be far below this value, resulting in a significantly higher rate of sulfidation corrosion.

The report, which contains numerous photographs of the corrosion and pipe rupture, also notes the presence of an inward deformation toward the inside surface of the pipe. The CSB has determined this deformation was likely caused by a fire pike used during the plant's emergency response. A photograph depicts a tight fit between the fire pike tip and this inward deformation of the failed pipe.

3. Report Findings

The Anamet report provides the following conclusions:

1. Rupture of the 8-inch 4-sidecut resulted from wall thinning caused by sulfidation corrosion.
2. Post rupture corrosion destroyed fracture morphology that could have indicated the rupture initiation site. However, rupture likely originated in the thinnest region and initially followed a longitudinal path driven by the hoop stress. Consequently, an area of likely rupture initiation was identified.

3. The wall thickness of the ruptured section was less than all the other sections of the 8-inch 4-sidecut that were evaluated.
4. Chemical analysis showed the silicon concentration of the ruptured section was 0.01-wt%.
5. Six of twelve specimens (50%) from unique sections of the 8-inch 4-sidecut, 12-inch 4-sidecut, and 2011 12-inch samples were found by chemical analysis to have a silicon concentration of less than 0.1-wt%.
6. Deformation of a small region of the 8-inch 4-sidecut rupture edge was consistent with mechanical force applied from the outside surface of the pipe, possibly from a fire pike, such as samples E-099 and E-082-2, or other pointed object. This deformed region was located within the area of likely rupture initiation.

4. CSB Future Plans

The CSB investigation to determine the root causes of the incident is ongoing. The CSB will release a report detailing its findings and recommendations to key stakeholders later this year. The CSB, Cal/OSHA, the USW, and Chevron are cooperating under an agreement to test 4-sidecut carbon steel piping from Chevron's refinery in El Segundo, California, south of Los Angeles.

AUTHOR INDEX

Aguiar, Leonardo Salim Damian....329
Aguilar-Otero Sr., Jose R....249
Aguirre, J. Luis....600
Allawi, Ahmed J....1
Allford, Lee....54
Arnaldos, Josep....460, 587
Bahiense, Ana....321
Barrera, Carlos A....634
Baybutt, Paul....688
Bearrow, Mike....307
Bennett, Raymond H....603
Bernechea, Esteban....460, 587
Bishop, Nikki....344
Bradshaw, William....96
Bridges, William....470, 752
Brigman, James....358
Burban, Peter M....452
Burgess, John....637
Cadena, Jaime E....124,
Cai, Tianxing....292
Caputo, Maria....321
Carreto, Victor H....12
Clarke, Peter....722
Conner, Jim....980
Cooper, Joshtin....900
Cox, Brent L....1002
Cox, Brenton L....937
Cravens, Robert....176
Cristea, Nicholas N....645
Cummings, David E....80, 484
Cutchen, Steve....53, 1044
Da Silva, Clarissa Sossmeier....875
Davis, Scott G....613
Dayton, Robert....913
De Oliveira, Thiago Souza....372
Dee, Sean J....824, 937, 1002
Deivasagayam, Karthick....356, 409
Dennehy, Martin A....67
Denton, Jerad....53
Dewees, David J....617
Díaz, Juan C....495
Diniz, Americo....875
Duran, Luis....509
Ebrahim, Fareed....254
Edel, Matthew....203
Eisenberg, Jonathan....391
Elias, Brian....679
Evans, Roger....1044
Farrell, Thomas....562
Fegley, Kenneth....67
Fialkowski, Charles....792
Filho, Salvador Ávila....601
Fink, William....614
Forest, Jerry....815, 865
Frawley, Thomas....357, 437
Freeman, Raymond....410
Gómez, Jorge Mario....124
Gomez, Manuel R....253
Graeme, Ellis....680
Grim, Lauren....1044
Grubbe, Deborah L....522
Guld, Jeppe....96
Gutierrez, Eliecer....599
Hardy, Terry....1029
Herrera, Carolina....614
Holmstrom, Donald....53, 1044
Horn, Brad....203
Hoyle, Bill....53
Huang, Szu-Ying....459
Israni, Kumar....900
Iv, Peter J. Ralbovsky....615
Jain, Prerna....176
Jiang, Chunming....656
Johnson, Amanda....1044
Johnson, Ryan....664
Johnson, Thomas....357, 437
Jones, Jerry L....251
Jung, Seungho....138
Kadri, Shakeel....33, 67
Kane, Kara....53
Karthikeyan, Muthukumarasamy....401
Kemal, Abid....634
Kern, Hannes....382, 523, 1022
Khajehnajafi, Shahryar....157
Khan, Farheen....175
Khan, Mohammad Naved....358
Klippel, Alexander....230
Knijff, Piet....54
Krywonos, Mike....509
Kumar, Ajay....405
Kusche, Christian....230
Lakhiani, Sunil D....175
Laskar, Sandipan....274
Le, Hai....392
Leblanc, John A....263
Lehmann, Scott....833
Li, Xinrui....459
Lione, Diego Da Silva....329
Liu, Yi....392, 635
Ludwig, Keith....218
Luebeck, Hope....264
Luna Sr., German....190, 254, 345
Lutostansky, Elizabeth....138
Lyon, Chris....53
Mackenzie, Cheryl....53
Madewell, Andrew....921
Maher, Steven T....202, 802
Manjunath, Konanur....1013
Mannan, M. Sam....12, 392, 459, 584, 635
Marriott, James....636

AUTHOR INDEX

Massaguer, Cristian 587
Mateo, Alvin 67
Meel, Anjana 157
Melhem, Georges 533
Mittal, Pawan 405
Morgan, Christina 253
Morrison, Delmar 175, 824
Mulcahy, Marybeth 53
Munoz, Felipe 124, 495, 599
Murphy, John F. 980
Myers, Molly 876
Neill, Mike 833
Ness, Albert I. 1013
Nguyen, Ha 584
Norton, Kristin D. 202
Nour, Michael A. 202
Nourai, Farshad 612
Ogle, Russell A. 824, 937, 1002
Osterberg, Paul 357, 437
Ozog, Henry 876
Pathak, Swapnil 358
Perez, John T. 921
Peters, Glen 67
Pineda-Solano, Alba 12
Polcyn, Michael A. 603
Powell, Joseph B. 28
Prasad, Hv 401
Pridy, Cindie 176
Prueter, Phillip E. 617
Rains, Brian D. 112, 887
Ramaswamy, Ravi 176
Ramesh, V. M. 111
Ramos, Manuel A. 124
Rangwala, Ali S. 679
Raupenstrauch, Harald 382, 523, 1022
Richardson, Joshua 584
Rodríguez, Sergio E. 124
Sakthivel, R. 111
Saud, Yaneira 612
Scheid, Marc 230
Schmelzer, Peter 54
Scholtz, Colin 802
Schork, Joan M. 138, 218
Schröder, Volkmar 230
Seguí, Xavier 460
Shah, Ghanshyam 284
Sheppard, Dave 344
Sherburne, Carol 437
Sierra, Lina M. 124
Sliva, Dan 33
Smith, Dustin J. 147, 637, 645
Smith, Gregory 470
Smith, Robert C. 560, 561
Snoeys, Jef 562
Soczek, Charles A. 242, 586
Souza Jr., Maurício Bezerra 329, 372
Spijker, Christoph 523
Stauffer, Todd 722
Taferner, Matthias 1022
Throness, Barry 949
Tillema, Dan 1044
Tschirschwitz, Rico 230
Vaez, Najmeh 612
Valdes, Olga Juliana Reyes 12
Van Wingerden, Kees 613
Vanommeren, James 67
Vaz Jr., Carlos André 329, 372
Vela, Marco Antonio 865
Velasco, Nubia 599
Victoria, Jorge 599
Villaça, João Márcio Almeida 329
Vingerhoets, Jim 562
Viz, Mark J. 1002
Vyas, Renu 401
Waller, Alvin 994
Wang, Zheng 656
Waters, Steven 664
Watson, Kevin 913
Wilkins, John 147
Wilson, Kelly 53, 1044
Wimmer, Bruno 382
Wincek, John 876
Wingard, Mark 1044
Xu, Qiang 292
Yarlott, Marc W. 664
Yelisetty, Suresh 612
Yuan, Jiwu 656
Zhang, Jiaqi 635
Zhang, Yu 218
Zhao, Xiangdi 656
Zirker, Stefan 230